Physics and Chemistry
of Small Clusters

NATO ASI Series

Advanced Science Institutes Series

A series presenting the results of activities sponsored by the NATO Science Committee, which aims at the dissemination of advanced scientific and technological knowledge, with a view to strengthening links between scientific communities.

The series is published by an international board of publishers in conjunction with the NATO Scientific Affairs Division

A	**Life Sciences**	Plenum Publishing Corporation
B	**Physics**	New York and London
C	**Mathematical and Physical Sciences**	D. Reidel Publishing Company Dordrecht, Boston, and Lancaster
D	**Behavioral and Social Sciences**	Martinus Nijhoff Publishers
E	**Engineering and Materials Sciences**	The Hague, Boston, Dordrecht, and Lancaster
F	**Computer and Systems Sciences**	Springer-Verlag
G	**Ecological Sciences**	Berlin, Heidelberg, New York. London,
H	**Cell Biology**	Paris, and Tokyo

Recent Volumes in this Series

Series B: Physics

Physics and Chemistry of Small Clusters

Edited by

P. Jena
B. K. Rao and
S. N. Khanna

Virginia Commonwealth University
Richmond, Virginia

Plenum Press
New York and London
Published in cooperation with NATO Scientific Affairs Division

Proceedings of a NATO Advanced Research Workshop and International Symposium on
the Physics and Chemistry of Small Clusters,
held October 28–November 1, 1986,
in Richmond, Virginia

Library of Congress Cataloging in Publication Data

International Symposium on the Physics and Chemistry of Small Clusters
(1986: Richmond, Va.)
Physics and chemistry of small clusters.

(NATO advanced science institutes series. Series B, Physics; v. 158)
"Published in cooperation with NATO Scientific Affairs Division."
Bibliography: p.
Includes index.
1. Cluster theory (Nuclear physics)—Congresses. 2. Atomic structure—Congresses. 3. Electronic structure—Congresses. 4. Metal crystals—Congresses. 4. Metal crystal—Congresses. I. Jena, P. II. Rao, B. K. III. Khanna, S. N. IV. Title. V. Series.
QC793.3.S8I587   1986   539′.6   87-14199

ISBN 978-1-4757-0359-7   ISBN 978-1-4757-0357-3 (eBook)
DOI 10.1007/978-1-4757-0357-3

© 1987 Plenum Press, New York

Softcover reprint of the hardcover 1st edition 1987

A Division of Plenum Publishing Corporation
233 Spring Street, New York, N.Y. 10013

PREFACE

Recent advances in experimental techniques now enable researchers
to produce in a laboratory clusters of atoms of desired composition from
any of the elements of the periodic table. This has created a new area
of research into novel materials since clusters cannot be regarded either
as a "large" molecule or as a fragment of the bulk. Both experimental
and theoretical studies are revealing unusual properties that are not observed in solid state environments. The structures of micro-clusters are
found to be significantly distorted from the most symmetric arrangement,
some even exhibiting pentagonal symmetry commonly found in icosahedric
structures. The unusual stability of certain clusters, now described as
"magic number species", shows striking similarities with the nuclear shell
structure. The relative stabilities of clusters depend not only on the
composition of the clusters but also on their charged states. The studies
on spontaneous fragmentation of multiply charged clusters, commonly referred
to as Coulomb explosion, illustrate the role of electronic bonding mechanisms
on stability of clusters. The effect of foreign atoms on geometry and
stability of clusters and the interaction of gas atoms with clusters are
showing promise for an indepth understanding of chemisorption and catalysis.
The magnetic and optical properties are dependent not only on cluster size
but also on its geometry. These findings have the potential for aiding
industry in the area of micro-electronics and catalysis.

The purpose of the "International Symposium on the Physics and Chemistry
of Small Clusters", held in Richmond, Virginia from October 28 - November
1, 1986, was to bring together researchers from different disciplines such
as Physics, Chemistry, Mathematics, and Engineering to discuss the various
challenging problems of both fundamental and technological importance.
The symposium consisted of eight plenary sessions and two poster sessions.
There were thirty-one invited papers and 126 contributed poster presentations
presented by nearly 220 scientists from 26 countries. The proceedings
of this symposium constitute this book.

The discussions and presentations included a number of topics:

.Preparation and Characterization: beams, matrices,
zintl ions and other methods.
.Atomic Structure and Simulation: geometry, magic
numbers, kinetics.
.Thermodynamic Properties: melting, specific heat,
vibrations.
.Electronic Structure and Stability: binding energy,
ionization potentials, density of states, charge distribution, fragmentation.
.Electrical and Magnetic Properties: superconductivity,
hyperfine interactions, magnetism, transport properties.

.Optical Properties: UV, IR, and Raman spectroscopy.
.Modeling of Defects by Clusters: energetics, electronic
 and structural properties.
.Applications: adsorption, catalysis, cluster-support
 interaction, astrophysics.

The studies on atomic structure, using x-ray absorption spectroscopy
and electron microscopes not only revealed the existence of five-fold symmetry but also the changes in inter-atomic distance as clusters grow.
A video presentation graphically illustrated the dynamics of formation
and growth of clusters. Theoretical calculations based upon ab initio
quantum chemical and density functional methods as well as semi-empirical
techniques were presented to illustrate the evolution of the geometries
and interatomic spacings of clusters as a function of size. These properties were also studied using the molecular dynamics method. The
photofragmentation of neutral and charged clusters studied both experimentally and theoretically illustrated the preferred channel of the fission
as well as the predominance of magic number clusters in the fission product.
The electronic properties such as electron spin densities, binding energies,
hyperfine interaction, and magnetic behavior were described experimentally
through ESR, NMR, and Mössbauer techniques and theoretically through electronic structure calculations. The cluster spectroscopy was probed by
infrared, ultra-violet, and Raman spectroscopy. Theoretical explanation
was provided through methods based upon classical, quantum mechanical,
and fractal formulations. The interaction of adsorbates with cluster surfaces generated considerable excitement since the reactions of clusters
to a gas molecule exhibited strong dependence on cluster size. Implications of these studies on catalysis were discussed. Multi-disciplinary
aspects of the cluster field was stressed throughout the symposium. The
study of clusters as a way of understanding the age-old problem of evolution of solid state electronic and structural properties was highlighted.

The symposium was in the planning stage for nearly two years. We
owe our gratitude to a large number of colleagues and institutions who
have assisted us in the organization of the meeting. We wish to thank
the members of the International Advisory Board for their help in selecting the topics and speakers for the symposium. We are grateful to the
members of the Local Organizing Committee for spending countless numbers
of hours during the planning stage. We wish to thank the conferees for
the high quality of their participation in the invited, and poster presentations, and exchange of ideas. Our special thanks are to Ms. Barbara
Martin, who was responsible for every arrangement that went right.

This symposium was made possible by a generous grant from the North
Atlantic Treaty Organization. We acknowledge, with gratitude, financial
assistance from Virginia Commonwealth University, which acted as the symposium's host, and Philip Morris, U.S.A. The symposium was also supported
in part by grants from the National Science Foundation, Department of
Energy, National Aeronautics and Space Administration, and the Air Force
Office for Scientific Research.

P. Jena
B. K. Rao
S. N. Khanna

Richmond, Virginia
   January, 1987

CONTENTS

ATOMIC STRUCTURE

DYNAMICS

STABILITY AND FRAGMENTATION

ELECTRONIC STRUCTURE AND PROPERTIES

OPTICAL PROPERTIES

VAN DER WAALS CLUSTERS

INTERACTION WITH ADSORBATES

ASTROCHEMISTRY

PHOTOELECTRON SPECTROSCOPY OF METAL CLUSTER ION BEAMS

O. Cheshnovsky[a], P. J. Brucat[b], S. Yang[c], C. L. Pettiette[c], M. J. Craycraft[c], and R. E. Smalley

Rice Quantum Institute and Department of Chemistry
Rice University
Houston, Texas 77251

INTRODUCTION

Metal clusters in the size range of a few to several hundreds of atoms are one of the most interesting frontiers of molecular physics and chemistry.   One of the reasons for this interest is that they may serve usefully as molecular models of bulk surfaces[1].   Somewhere within this 2-1000 size range it is expected that metallic behavior will set in: the sparse set of well-defined single quantum levels of the atom becoming a more complicated set of molecular orbitals in the 2 and 3 atom clusters, and then evolving (slowly?) toward the band structure of the bulk metallic phase. Somewhere along the way, the catalytic properties of these clusters will begin to resemble that of surfaces of the bulk metal.   Detailed study of the physical and chemical nature of these small objects may therefore play a key role in the development of a fundamental and predictive understanding of metallic surfaces and heterogeneous catalysis.

---

[a] Department of Chemistry, University of Tel Aviv, Tel Aviv, Israel.
[b] Department of Chemistry, University of Florida, Gainesville, Fla. 32611.
[c] Robert A. Welch Predoctoral Fellow.

In spite of such high interest, and in spite of the remarkable progress made over the past 10-15 years in the production and study of isolated metal clusters both in matrices and in molecular beams, it is still true that hardly anything is known with certainty about the details of the electronic structure of small metal clusters. At this writing, the only clear exceptions to this rule of general ignorance are a variety of metal dimers (such as Na_2, Cu_2, Cr_2, Ni_2, etc.) and only a few metal trimers (particularly Na_3, and Cu_3)[2,3].

Within the past 5 years one of the techniques that showed most promise for reducing this level of ignorance in the electronic structure of metal clusters was the use of multi-color laser resonant photo-ionization of the clusters in a supersonic molecular beam[4]. In some cases this technique (also called resonant two photon ionization, or simply "R2PI") proved to be sensationally effective: not only giving a detailed high resolution electronic spectrum of a particular cold cluster, but simultaneously giving a full isotope analysis as well. In the right hands, other techniques such as laser induced fluorescence, LIF, have also produced clear, firm insights into the electronic structure of small clusters both in beams and in matrices[2-3,5].

However, these LIF and R2PI techniques share a common weakness: they both require the presence of an excited electronic state with sufficiently long lifetime that it can serve as the basis for detection (either by simply emitting a spontaneous photon in the case of LIF, or by triggering an efficient ionization of the cluster as required in R2PI). While in diatomics such metastable excited electronic states often exist -- particularly below the first dissociation threshold, in larger metal clusters they unfortunately prove to be quite rare. To put it simply: metals don't fluoresce. At least in this one respect, metal clusters begin to behave like metals far more rapidly than we would wish.

In the case of the metal cluster ions, one can design spectrometers which avoid this requirement of a metastable excited level simply by using photofragmentation as a detector of the spectral absorption event. Some initial work along these lines was reported earlier from this group for the case of a few diatomic transition metal positive ions, and there is certainly promise here for a rather rich future of spectral studies[6].

But even here where techniques may be developed to routinely record the electronic absorption spectrum of small metal cluster ions, it is likely that only the smallest clusters will ever display a spectrum that is richly informative. High resolution molecular spectroscopy is powerful only when the observed spectrum can be interpreted as a detailed consequence of a simple quantum mechanical model with a few parameters. In the case of even quite small transition metal clusters, however, it is quite possible that hundreds of excited electronic states can contribute to the spectrum. The mutual vibronic interactions of this dense manifold can easily produce such a chaotic jumble of levels that no sane spectroscopist would even attempt its detailed analysis. For medium sized clusters one could still avoid this problem by working in the infrared, but here the way is blocked fairly effectively by the extremely low absorption cross-sections expected for such states.

What is needed is a spectroscopy that is just as applicable for a nearly macroscopic object as it is for a single atom. Such a spectroscopy would not focus on the properties of individual quantum levels, but rather on bands of levels, their average spatial form, and their density as a function of energy. Most readers will readily recognize this as a description of ultraviolet photoelectron spectroscopy, UPS, a spectroscopy that is almost unique its richness of application both in the microscopic world of atoms and molecules, and the macroscopic world of bulk surfaces.

The purpose of this article is to review recent progress in the development of this critical technique of photoelectron spectroscopy of metal clusters, with emphasis on the very recent work of our group at Rice University using the general technique of pulsed metal cluster beams produced by laser vaporization.

EXPERIMENTAL

Photoelectron spectroscopy had been recognized since the very earliest days of the supersonic beam studies as a vitally important potential application of the new laser-vaporization method of preparing cluster beams. Given sufficiently intense metal cluster beams in a high vacuum apparatus, one can readily imagine coupling to a conventional high resolution photoelectron spectrometer. Of course there would be the problem of the low duty cycle of the pulsed beam apparatus, but at least in principle this could be solved by using a high repetition rate laser and modifying the gas flow to produce a quasi-continuous metal cluster beam[7].

The biggest problem with photoelectron spectroscopy of neutral cluster beams, however, lay in the fact that clusters of many different sizes are present in such a beam at the same time and place. Either some method would have to be found to discriminate between the photoelectrons coming from the various clusters, or one would have to be content with only a composite photoelectron spectrum, averaged over all cluster sizes and compositions in the beam. One such method would be to detect the photoelectron spectrum in coincidence with a time-of-flight mass analysis of the corresponding photo-ions. With the present generation of synchrotron light sources, such an experiment is quite feasible; but it would easily rank as the most elaborate and difficult experiment ever performed on a supersonic beam.

As described below, a much more simple solution is to obtain the photoelectron spectrum by photodetachment of the negative cluster ion. Since the target cluster is now an ion, it may be selected by mass prior to the photoelectron experiment, thereby eliminating all confusion caused by other clusters in the original beam. In addition, as we shall see below, the photoelectron spectrum of the negative clusters has the additional advantage of providing a direct measure of the "band gap" between the highest occupied orbital (HOMO) and lowest unoccupied orbital (LUMO) in the corresponding neutral metal cluster.

A. The Cold Metal Cluster Ion Source

As has been described in considerable detail in a number of recent publications from this group[7,8] and others[9], the pulsed laser-vaporization cluster beam source may be configured to produce not only the neutral clusters, but quite intense cluster ion beams as well. With the intense laser conditions typically used in these sources metals are generally expected to vaporize initially into a highly ionized plasma which then begins to recombine rapidly as it is cooled through collisions with the (ca. 1 atm) helium carrier gas travelling over the target surface at near sonic velocity. A key feature of these sources is that the carrier gas density above the metal target at the moment of laser vaporization is sufficiently high that high densities (ca. 10^{10} to 10^{14} cm^{-3}) of small metal clusters are generated within the first few microseconds. The presence of these small neutral clusters allows some of the original free electrons of the laser-vaporization plasma to survive as negative charge carriers by attaching to become negative cluster ions. Because of their far higher mass, these negative cluster ions have a much slower recombination rate with the remaining positive ions than the original free electrons. As a consequence, the original laser-induced plasma is largely recombined in the dense helium carrier gas flow, <u>except</u> for a small fraction of charge carriers which persist as relatively massive negative and positive cluster ions.

This model suggests that the more readily a metal clusters in the early, hot conditions immediately following laser vaporization, the more intense will be the cluster ions found in the resultant supersonic beam. In fact this is just what is observed. Niobium is a highly refractory metal with a very strongly bound dimer which promotes the clustering process, and niobium cluster ion beams are found to be quite readily made using the vaporization laser alone. On the other hand, copper clusters far less readily in such a laser-vaporization source, and one is hard pressed to find any copper cluster ions in the resultant beam. This variation in the behavior of various elements in the cluster source is even more pronounced in the case of main-group elements such as carbon and silicon, both of which cluster far more readily than even niobium, and produce correspondingly far more intense cluster ion beams.

Even though elements such as copper are more difficult to use in generating cluster ion beams using the vaporization laser alone, one can compensate somewhat by using an additional laser directed into the throat of the supersonic nozzle. As originally reported by this group[8], an ArF excimer laser beam of 2-10 mj cm^{-2} will generate a shower of low energy electrons when directed at the front of the supersonic nozzle just as the metal clusters emerge. Most of these low energy electrons arise from photoionization of the metal clusters themselves. These electrons can then attach efficiently to the metal clusters in the beam to produce a new inventory of negative clusters. Together with the positive cluster ions, these negative charge carriers constitute a neutral plasma which is then free to cool through collisions with the expanding helium carrier gas. At densities of $> 10^{10}$ ions cm^{-3}, the Debye screening length of such a plasma is so short that this cold cluster ion plasma is unaffected by stray electrostatic and magnetic fields as the cluster beam travels through the various vacuum chambers of the apparatus. Such a technique therefore provides a very general beam source of cold cluster ions suitable for a wide variety of experiments.

B. Cluster Ion Extraction & Time-of-Flight Mass Filtering

Given the pulsed nature of the laser-vaporization and cluster ion generation technique, it is natural to turn to pulsed extraction techniques to separate the cluster ions from the neutral beam, and to use time-of-flight to accomplish the requisite selection of a particular cluster ion for study. In our apparatus[10], the negative cluster ions are extracted at a right angle to the molecular beam by a 1000 V pulse applied as the clusters pass into a two-stage acceleration grid arrangement of the type suggested by Wiley and McLaren[11]. After suitable electrostatic aiming and focusing optics, this negative cluster ion beam is directed down a 2 meter flight tube at the end of which is mounted the photoelectron spectrometer. Mass selection is accomplished by a pulsed 3-grid mass gate located at the end of this flight tube, just prior to the detachment region of the spectrometer. All timing of the various lasers, nozzles, and pulsed grids is completely controlled by computer-driven, nanosecond-precision pulse generators.

C. The Magnetic TOF Photoelectron Spectrometer Design

Unlike photoelectron spectrometers designed for study of neutral species, this negative cluster UPS device cannot afford the luxury of high target densities. Even if there were no limits on the number density of negative clusters of a particular mass we could generate, space charge in the detachment region of the photoelectron spectrometer provides a stringent limit on the resolution attainable. For 10 meV resolution, one must not exceed a negative cluster ion density of 10^5 cm^{-3}. Since the laser-vaporization cluster source runs at a rather low repetition rate (10 pps), any reasonable photoelectron spectrometer design must collect a substantial fraction of all the detached photoelectrons, and it must be non-dispersive, (ie. the entire photoelectron spectrum must be collected for every laser shot).

Luckily, there are a number of reasonable designs available, some using electrostatic fields to channel the electrons to the detector[12], some using magnetic fields[13,14], and some entirely field-free[15-17]. All use time-of-flight measurement of the photoelectron energies. Although we have used a number of these, only the most promising design is discussed in

detail here.   It is one which uses a combination of solenoidal magnetic
fields to parallelize the photoelectron trajectories, and transmit them
along the flight tube to the detector much as a fiber optic array performs
the same function for light.   A schematic diagram of this spectrometer
design is shown in Figure 1.

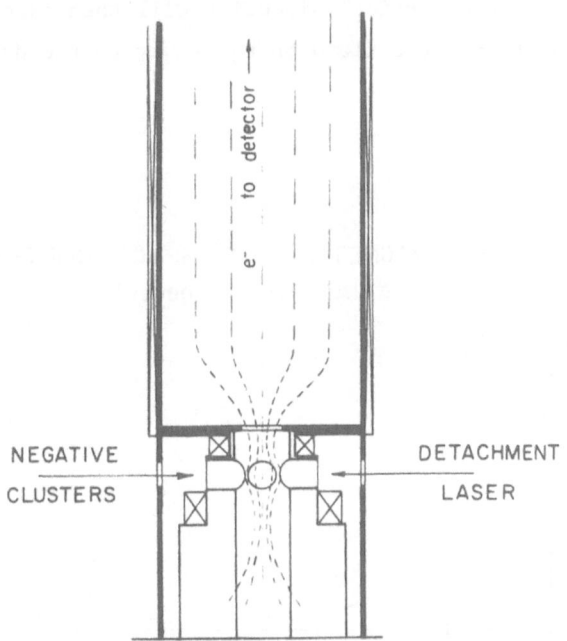

FIGURE 1.   Schematic of one design for a time-of-flight
photoelectron   spectrometer which uses axially symmetric
magnetic fields to confine and focus the photoelectrons.   The
negative cluster ion beam is   pulsed in from the left.   The
small solenoids mounted immediately above and below the detachment region produce an 850 gauss field which increases to
nearly 1000 gauss as the detached electrons pass through the
center of the top solenoid.   The long time-of-flight tube is
also wound with a solenoidal coil throughout its length so as
to maintain a 10 gauss guiding field.   The 100 fold decrease in
magnetic   field   serves   to   parallelize   the   photoelectron
trajectories.

Many of the key concepts for such a magnetic design of a photoelectron spectrometer can largely be found in a paper by Kruit and Read published several years ago[14], although their spectrometer design is far too restrictive in size to be useful for study of negative cluster ions. The main idea is to detach the photoelectron in a region of high magnetic field, B_i, and to configure the apparatus such that these photoelectrons rapidly but smoothly transit into a long flight tube which has a low magnetic field, B_f. For example, in the design shown in Fig. 1, the photoelectrons are detached by laser irradiation of the negative clusters in between two small solenoids of copper wire. The axial magnetic field generated by these solenoids by a pulsed current is shown graphed in Figure 2. The cross-hatch on Fig. 2 indicates the center of the photoelectron detachment zone, which can be seen to be at roughly 850 gauss. The long flight tube leading to the detector is held at 10 gauss. Photoelectrons initially traveling in the vertical direction will then experience a 100 fold decrease in axial magnetic field on their way to the detector.

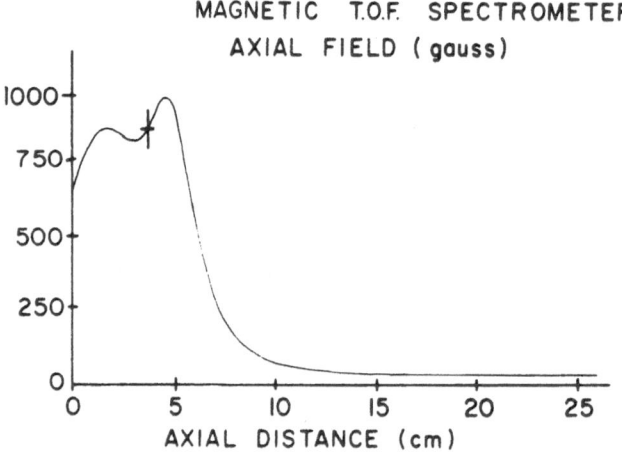

FIGURE 2. Plot of the axial magnetic field magnitude generated by the solenodial magnets of Fig. 1. The center of the photodetachment zone is marked by the cross-hatch at an axial distance of 3.5 cm.

As Kruit and Read discuss[14], under conditions of a smooth transition from the high to the low magnetic field, the electron motion is approximately adiabatic. Such motion conserves two quantities, the total energy, E, and the classical angular momentum, L:

$$E = E_c + E_b \tag{1}$$

$$L = (2m/e) \; E_c/B \tag{2}$$

where m = the electron mass,

  e = the electron charge,

  Ec = the energy in the cyclotron motion,

  Eb = the energy of motion along the magnetic field,

and B = the magnitude of the local magnetic field.

As the electron moves down the flight tube, B decreases from the high initial value of B_i to the final, flight tube value of B_f. Equation (2) then requires that the energy in the cyclotron motion, Ec, must also decrease. Since the total energy is always conserved for motion in a magnetic field, energy must therefore flow from the cyclotron degrees of freedom into motion along the flight tube axis. In other words, the effect of moving out in a diverging magnetic field is to parallelize the electron trajectories.

The result is that suitably configured solenoidal magnetic fields can produce a very simple photoelectron spectrometer design with collection efficiences approaching 50% at a fractional energy resolution approaching the ratio B_i/B_f in the limit of very long flight tube lengths.

The design shown in Fig. 1 uses a split solenoid with an asymmetric field so that one may control the effective acceptance angle of the spectro-meter. By lowering the field from the rear solenoid, one can make the spectrometer increasingly selective to only a narrow initial acceptance cone, thereby permitting studies of the polarization dependence of the detachment process.

RESULTS and DISCUSSION

Using this magnetic design, along with a variety of others over the past few months, we have recorded extensive surveys of the photoelectron spectra of negative metal clusters both of copper and niobium. Copper clusters were selected as the major topic of study since our previous measures of their ionization potentials[18], IP and electron affinities[10]

EA, gave clear predictions as to how the photoelectron spectra should behave as a function of cluster size. These IP and EA measurements showed a pronounced even/odd alternation as a function of cluster size, with the even cluster showing the higher IP's and the correspondingly lower EA's.

As discussed in previous publications from this group[10,18], this even/odd alternation suggests that the even copper clusters are ground state singlets with a rather broad HOMO--LUMO "band" gap. The neutral odd clusters, on the other hand, act as though they are ground-state doublets, with a less tightly bound HOMO. The negative ions for the even clusters should then be ground state doublets with the extra electron occupying the LUMO at the top of the "band" gap. Photodetachment of these even numbered Cu_n^- clusters should then reveal the full band gap, while the photoelectron spectrum of the odd clusters should be much different, with very little indication of a gap in the band structure at all.

Figure 3 shows that these expectations are born out in the actual photoelectron spectrum of copper clusters in the 10-12 atom range. Here the onset of the photoelectron spectrum is shown for the three clusters for data obtained with a magnetic TOF spectrometer design similar to Fig.1, using the 3rd harmonic of the Nd:YAG laser (3.492 eV) as the detachment photon. Note that both Cu_{10}^- and Cu_{12}^- show a strong initial peak,

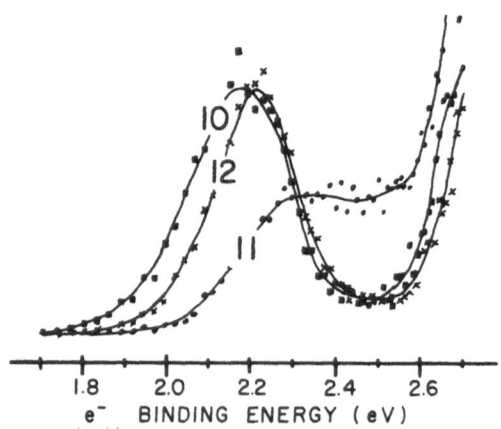

FIGURE 3. Photoelectron spectrum for Cu_{10}^-, Cu_{11}^-, and Cu_{12}^- taken in a spectrometer of the type shown in Fig. 1. The detachment laser photon energy was 3.492 eV.

followed by a gap of about 0.5 eV. The odd cluster, Cu_{11} , however, gives
a photoelectron spectrum onset shifted to higher energies, and shows little
evidence of a gap. The onsets of these UPS profiles agree well with the
previously published electron affinity results for these clusters[10].

Figure 4 shows this same even/odd alternation continues to be seen in
the photoelectron spectrum of larger copper clusters. Here more of the
photoelectron spectrum is seen to higher energies, and the even/odd
alternation of pattern is quite apparent. This data was again taken with

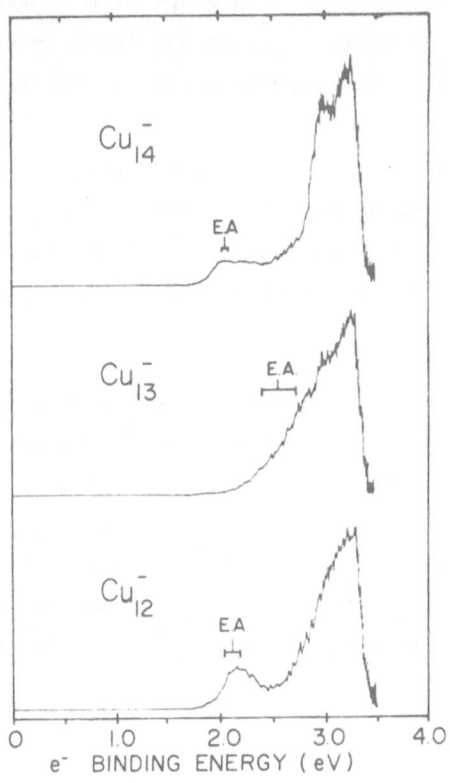

FIGURE 4. Photoelectron spectra for larger copper clusters
taken in a photoelectron spectrometer which used a central
guiding wire at 0.25 volt potential with respect to the
flight tube. Note the prominent even/odd alternation in the
spectral pattern. The brackets marked "E.A." correspond to
the best estimates of the electron affinity from previous
photodetachment studies of this laboratory. The sharp drop
in photoelectron signal seen here for binding energies within
0.2 eV of the 3.492 eV detachment laser may be an artifact
due to small stray electric fields in the spectrometer.

the 3.492 eV detachment laser, but used an electrostatic alternative to the magnetic design shown in Figure 1. This alternative design featured a thin central wire along the axis of the flight tube. When operated at a small positive voltage (0.25 V) with respect to the flight tube walls, such a design is rather equivalent to a graded-index optic fiber[12]. We used it in these initial photoelectron studies as an independent check that the photoelectron spectra were free of artifacts.

Photoelectron spectral results similar to those of Fig.4 were obtained for all clusters in the 10-23 atom size range. Although for some clusters (particularly Cu_{16}^-) the effect is less distinct, the even/odd alternation in the photoelectron pattern is seen throughout, with only a slow narrowing of the apparent band gap. The Cu_{20}^- cluster was found to have a particularly large gap in the UPS spectrum, suggesting that the 20 (4s) electron neutral cluster, Cu_{20}, is particularly stable. Such a special stablility of a 20 electron structure is predicted (along with 2,8, and 40) as a spherical shell closing in the simple jellium model of metal cluster electronic structure[19]. It is interesting to note in this regard that Cu_{21}^+ was found to be a specially stable positive cluster ion in the early work from this group on the laser photoionization of copper clusters[18]. Neglecting the 3d electrons, Cu_{21}^+ is, of course, also a 20 valence electron system.

While this work was in progress, we were made aware of the beautiful recent results of Leopold, Ho, and Lineberger[20] who have recorded high resolution photoelectron spectral of Cu_1^- through Cu_{10}^- using a new flowing afterglow, continuous cluster ion source connected to a conventional photoelectron spectrometer with a hemispherical energy analyzer. In all cases we have checked (Cu_5^-, Cu_6^-, Cu_8^-, and Cu_{10}^-), our data is in good qualitative and quantitative agreement with their spectra.

CONCLUSIONS

Although the new results reported here are quite striking, the most significant aspect is not so much the new information on copper clusters as is the simple fact that such experiments can be done at all. What once seemed to be a nearly impossible experiment has now been done twice (at Rice and at the University of Colorado) in the space of a few months. For the low-lying states of small metal clusters it now appears clear that high resolution photoelectron spectral will soon be available for a wide range -

of elements. The pulsed photoelectron spectrometer discussed in this paper has the disadvantage (at least in its current incarnation) of lower resolution, but it has several key advantages by the very fact that it is pulsed. Chief amongst these is that pulsed detachment lasers can be used, and it will be quite straightforward to generate the vacuum ultraviolet photon flux necessary to truly do UPS experiments on a wide range of clusters -- experiments that can directly probe the band structure of these clusters for the first time. Using the pulsed supersonic cluster ion source, chemisorbed clusters can also be studied. Their UPS spectra should be particularly revealing when compared to the vast literature on the subject for bulk chemisorbed surfaces.

ACKNOWLEDGEMENT

We wish to thank W. C. Lineberger and D. G. Leopold for communication of their photoelectron spectral results on copper clusters prior to publication. Our research on the properties of bare metal clusters is supported by the U.S. Department of Energy, Division of Chemical Sciences, the Robert A. Welch Foundation, and the Exxon Educational Fund. The apparatus used here is primarily supported by the National Science Foundation for the study of chemisorbed species on metal cluster ions.

REFERENCES

1. R. E. Smalley, in Comparison of Ab Initio Quantum Chemistry with Experiment: State of the Art, edited by R. J. Bartlett (Reidell, New York, 1985).

2. D. R. Salahub, Adv. Chem. Phys., vol. 69, in press.

3. M. D. Morse, Chem. Rev., in press.

4. a) T. G. Dietz, M. A. Duncan, M. G. Liverman, and R. E. Smalley, J. Chem. Phys. 73, 4816 (1980).
   b) R. E. Smalley, Laser Chem. 2, 167 (1983).
   c) M. D. Morse and R. E. Smalley, Ber. Bunsenges. Phys. Chem. 88,228,1984.

5. W. Weltner, Jr., and R. J. Van Zee, Ann. Rev. Phys. Chem. 35, 291 (1984).

6. P. J. Brucat, L-S. Zheng, C. L. Pettiette, S. Yang, and R. E. Smalley, J. Chem. Phys. 84, 3078 (1986).

7. S. K. Loh, D. A. Hales, and P. J. Armentrout, Chem. Phys. Lett. 129, 527 (1986).

8. L-S. Zheng, P. J. Brucat, C. L. Pettiette, S. Yang, and
   R. E. Smalley, J. Chem. Phys. 83, 4273 (1985).

9. L. A. Bloomfield, M. E. Geusic, R. R. Freeman, and W. L. Brown,
   Chem. Phys. Lett. 121, 33 (1985).

10. L-S. Zheng, C. M. Karner, P. J. Brucat, S. Yang,
    C. L. Pettiette, M. J. Craycraft, and R. E. Smalley,
    J. Chem. Phys. 85, 1681 (1986).

11. W. C. Wiley and I. H. McLaren,
    Rev. Sci. Instrum. 26, 115 (1955).

12. N. S. Oakey and R. D. MacFarlane,
    Nuclear Instrum. and Methods 49, 220 (1967).

13. K. E. Norell, P. Baltzer, B. Wannberg, and K. Siegbahn
    Nuclear Instrum. and Methods 227, 499 (1984).

14. P. Kruit, and F. H. Read, J. Phys. E.16, 313 (1983).

15. S. R. Long, J. T. Meek, and J. P. Reilly
    J. Chem. Phys. 79, 3206, 1983.

16. K. Kimura, Adv. Chem. Phys, 60, 161 (1985),
    and reference cited therein.

17. L. A. Posey, M. J. DeLuca, and M. A. Johnson,
    Chem. Phys. Lett. in press.

18. D. E. Powers, S. G. Hansen, M. E. Geusic, D. L. Michalopoulos,
    and R. E. Smalley, J. Chem. Phys. 78, 2866 (1983).

19. a). W. D. Knight, K. Clemenger, W. A. de Heer, W. A. Saunders,
        M. Chou, and M. L. Cohen, Phys. Rev. Lett. 52, 2141 (1984).,
    b). W. D. Knight, W. A. de Heer, and W. A. Saunders,
        Z. Phys.D 3, 109 (1986), and references cited therein.

20. D. G. Leopold, J. Ho, and W. C. Lineberger, J. Chem. Phys.
    in press.

ATOMIC STRUCTURE OF SMALL CLUSTERS : THE WHY AND HOW OF THE FIVE-FOLD

SYMMETRY

J. Farges, M.F. de Feraudy, B. Raoult and G. Torchet

Laboratoire de Physique des Solides, LA 02, Université de

Paris-Sud, 91405 Orsay, France

ABSTRACT

Particles with pentagonal symmetry which have been observed by microscopy in metal deposits possess the normal fcc structure, as revealed by electron diffraction. However, at the first stage of their formation, a nucleus with anomalous structure was existing, this structure being related with the nucleus shape. The structure characteristic of the icosahedral shape has been effectively observed by electron diffraction in Ar clusters comprising a few hundreds of atoms. For a given number of atoms this structure proves to be more stable than that of small fcc microcrystals since at the expense of some elastic distortions and twin formation, it allows a better surface energy. However, this structure cannot be grown up to macroscopic sizes due to the compression of the central region by external layers.

INTRODUCTION

In 1966, Ino was surprised, while he was examining micrographs from gold thin films evaporated onto NaCl crystals[1], to discover a great number of five-fold clusters, a few tens of Å in diameter, showing decahedral or icosahedral external shapes, i.e. five-fold axis which are well-known to be incompatible with a crystalline structure. Later on, such anomalous shapes have been observed frequently for several fcc metals such as Au, Ag, Ni, Pt or Pd, prepared by means of various experimental techniques : condensation onto a substrate or in a rare gas vapor, precipitation from a chemical solution or electrodeposition.

Hence a lot of questions have been asked concerning these particles with five-fold symmetry. This paper mainly deals with what it is possible to say about the origin and the structure of these particles.

THE FORMATION OF PENTAGONAL RINGS

When packing N spheres in order to get the maximum coordinance, it is easy to note that the solution is unique up to N = 5. Considering the center of the spheres, one finds successively the dimer (N = 2), the equilateral triangle (N = 3), the regular tetrahedron (N = 4) and the bitetrahedron (N = 5), as shown in Figure 1. In the case N = 6, there are 2

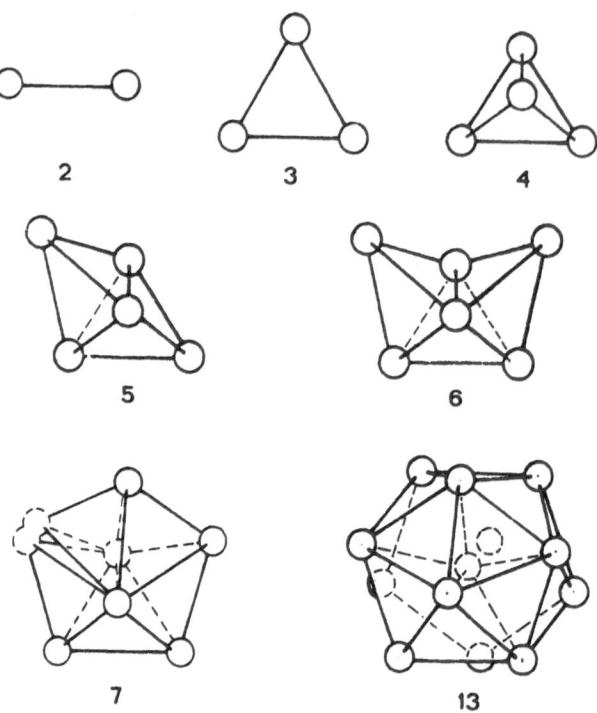

Fig. 1. Growth sequence of small monoatomic arrangements with N atoms. The construction of the decahedron N = 7 and that of the icosahedron N = 13 are explained in the text.

isomers with the same coordinance : the octahedron and the packing of 4 adjacent tetrahedra (Fig. 1). When one more atom has to be placed, the latter isomer is far more favourable since its reentrant angle offers a good location. If the tetrahedra were made of hard spheres, they would be regular and thus would let an angular gap of 7.35°, as suggested in Fig. 1 (N = 7). It is obvious that in the case of actual atoms, a more stable configuration can be obtained by slightly distorting the tetrahedra so as to form a regular pentagonal bipyramid (decahedron). In this way, a higher coordinance is obtained at the expense of a small elastic distortion. Consequently, this bipyramid has one more bond than any of the possible arrangements of seven atoms and is clearly the most stable one.

Going on with the addition of new atoms, the number of possible isomers is found to increase very rapidly with N. As an example, Hoare and Mac Innes[2] have shown that for N = 13, the use of a Lennard-Jones potential leads to 988 stable isomers. A particularly interesting arrangement is obtained precisely for N = 13, by adding one atom on each adjacent face surrounding one vertex of the N = 7, thus forming a pentagonal ring on the axis of which an additional atom is placed. Again at the expense of a slight elastic distortion of the tetrahedra, the 12 surface atoms are located at the vertices of a regular icosahedron so that this model has more bonds than any other arrangement of 13 atoms. For example, each of the 12 surface atoms has five surface nearest neighbors, instead of four in the case of a hcp or fcc model. It is worthwhile to note that the bipyramid and the icosahedron show respectively one and six five-fold axis and that in both cases, this symmetry allows the arrangement to increase its binding energy at the expense of some elastic strain.

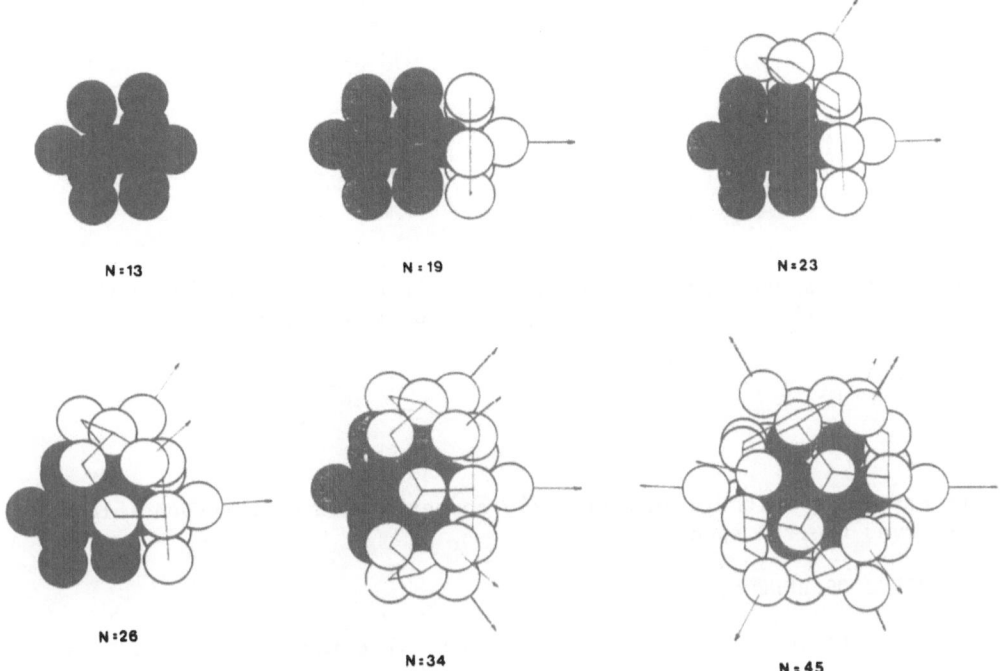

Fig. 2. Several models in the DIC sequence from the primitive icosahedron (N = 13) to the last model (N = 45). Pentagonal rings indicated by a pentagon are successively constructed on the primitive icosahedron with one additional (white) atom on the five-fold axis.

Following the preceding method, it is possible to continue the growing sequence above N = 13 by adding pentagonal rings around each vertex of the icosahedron. The procedure does not warrant the obtention of the isomers with greatest stability. However, it will be amply justified later in that it leads to the first part of the "magic" number sequence found in mass spectroscopy experiments. Furthermore, constructing models with this method ensures that the amorphous local order, which is that of the smallest Ar clusters observed in a cluster beam as reported previously[3], will be preserved. The arrangement N = 19 is thus obtained by adding one pentagonal ring around a five-fold axis and one additional atom on it. This double icosahedron (DIC) shown in Fig. 2 is formed by two interpenetrating icosahedra sharing two central atoms and one pentagonal ring. A second ring can be placed around another axis of the icosahedron so that it shares a maximum number of atoms with the first added ring. As seen in Fig. 2., three more atoms are sufficient to form a second ring, which leads, with an additional atom on the axis, to the N = 23 arrangement. In the same way, two supplementary atoms are sufficient to form a third ring and then to give, with the additional atom, the N = 26 arrangement. Going on with this construction, further arrangements are found : N = 29, N = 32 and N = 34. The latter model N = 34, shown in Fig. 2, has a five-fold symmetry axis and is made of 7 interpenetrating icosahedra forming 16 DIC. Once all possible pentagonal rings have been formed, the primitive icosahedron is completely covered. The resulting arrangement, N = 45, is shown in Fig. 2. The first well-defined N values which have been found, in this DIC sequence, namely 13, 19, 23, 26, 29, 32, 34, account perfectly for the first maxima in the mass spectrum of Ar clusters[4] reproduced in Fig. 3. But these experiments do not reveal particular maxima at the larger N values found in this sequence. This means that above N = 34, another arran-

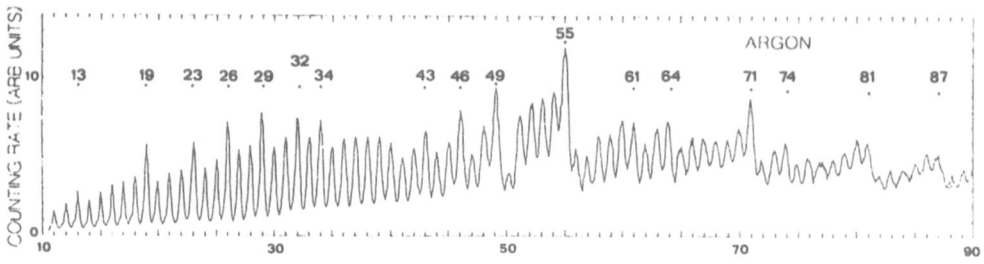

Fig. 3. Mass spectrum taken from I.A. Harris et al (1984) giving the
intensity vs the number of atoms in charged Ar clusters.

gement may exist which is energetically competitive and is responsible for
the observed blurred spectrum.

MULTILAYER ICOSAHEDRON

This different arrangement has been described for the first time by
Mackay [5] in 1962. It has the external shape of a regular icosahedron,
with its 12 vertices and its 20 faces, but this time, its size can be
extended to infinity. Despite the perfect five-fold symmetry of this arrangement, its internal local order is nearly that of a crystal. The internal
structure of this new arrangement, which will be called multilayer icosahedral (MIC), can be described as the packing of 20 identical tetrahedra
possessing a common vertex and connected with each other through adjacent
faces, each of which forms twinning planes. The constituting tetrahedra
are not perfectly regular because the three radial edges are shorter by
approximately 5% than the three radial edges. In each tetrahedron, atoms
are located in planes parallel to the surface, just as in the fcc structure, but the radial interatomic distances are here 5% shorter than the
tangential ones. An icosahedron with n complete layers possesses n+1 atoms
on each edge just as the constituting tetrahedra (Fig. 4), the smallest
in the sequence being the 13-atom icosahedron (n = 1) with two atoms on
each edge. It is easy to calculate the total number of atoms N corresponding to a MIC model with complete outer layer, knowing that the layer
of index n contains $(10 \, n^2 + 2)$ atoms. Thus a second sequence of well-
known "magic" numbers is found, namely 13 - 55 - 147 - 309 - 561 - 923...
As an example, the maximum at N = 55 is clearly visible in Fig. 3.

From a crystallographic point of view, the structure of the constituting tetrahedra can be described in two different ways : it is considered
either as the rhombohedral structure [6] suggested in Fig. 5, or as a
distorted fcc structure. Although these descriptions seem to be a priori
equivalent, they lead to different conclusions when increasing the number of layers. The first one implies that due to the crystalline structure, there are no strains inside the icosahedron and thus the arrangement may be grown up to macroscopic size while keeping its structure.
On the contrary, the second one implies that due to the tensile stresses
existing in each atomic plane parallel to the external face, planes belonging to the same layer exert important strains on each other which
result in radial strains pointing towards the center of the icosahedron,
a region which is thus highly compressed. In order to test this compression, several MIC models made of atoms interacting through a Lennard-
Jones potential have been allowed to relax freely [7] to the point where

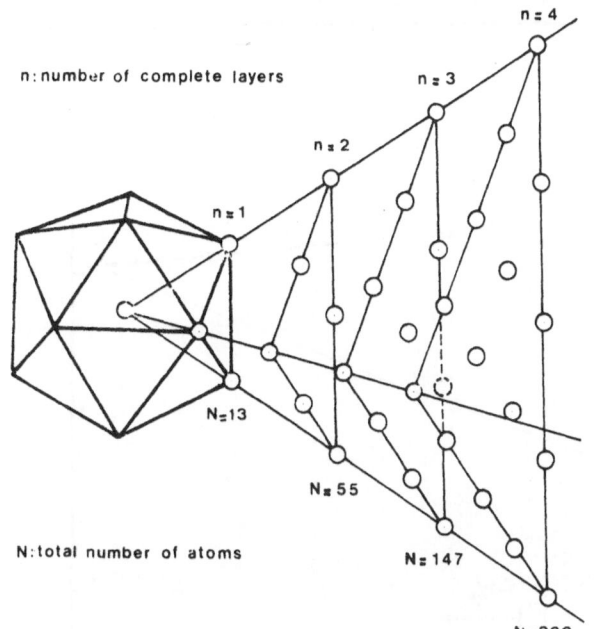

n: number of complete layers

n = 1

N = 13

N = 55

N: total number of atoms

N = 147

N = 309

n = 2

n = 3

n = 4

Fig. 4. Successive MIC models are constructed by growing each of the 20 tetrahedra constituting the primitive icosahedron.

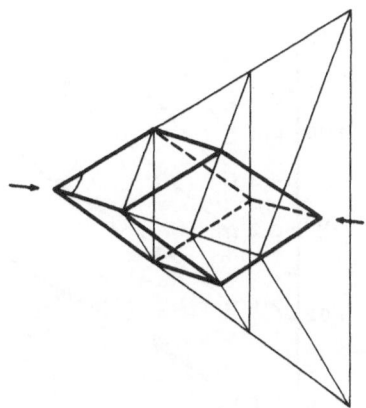

Fig. 5. Rhombohedral unit cell extracted from a multilayer tetrahedron. The deformation indicated by arrows increases the indicated angle from 60° (fcc structure) to 63.26° (MIC structure)

the force exerted on each atom was reduced to about 2×10^{-14}N. Results concerning the radial relaxation are shown in Fig. 6. Distances d_n between the vertices of layers $n - 1$ and n are given as a function of $N^{-1/3}$, a quantity which varies inversely as the mean model diameter, i.e. as the surface to volume ratio. In a given model d_n decreases as one approaches the center thus indicating a radial compression. It can be noted that once one moves away from the surface, the distances d_n become less than 1.0917 r.u., this being the value between nearest-neighbor atoms in the bulk Lennard-Jones crystal. As a consequence of this and contrary to the conclusion drawn from the crystalline description, the strain relaxation certainly prevents a distorted fcc structure from growing up to macroscopic size while keeping its internal structure. Furthermore, the possibility for a monoatomic compound, giving macroscopic fcc crystals, to adopt the MIC structure when it forms small clusters even appears questionable. However, it can be understood that if the MIC structure were to be adopted, it would be because its gain in surface energy would surpass its loss in elastic and twin energy. In addition to this, it can be noticed that multilayer pentagonal bipyramids have been introduced by Bagley [8]. They can be described either as a crystalline structure, this time orthorhombic, or as distorted fcc tetrahedra and they could be discussed in the same way as precedingly for the multilayer icosahedra.

ARGON CLUSTERS

At this point, the question to be asked should be the following : have clusters with MIC structure ever been observed ? The answer is yes they have, but in the unique case of rare gas clusters produced in a free

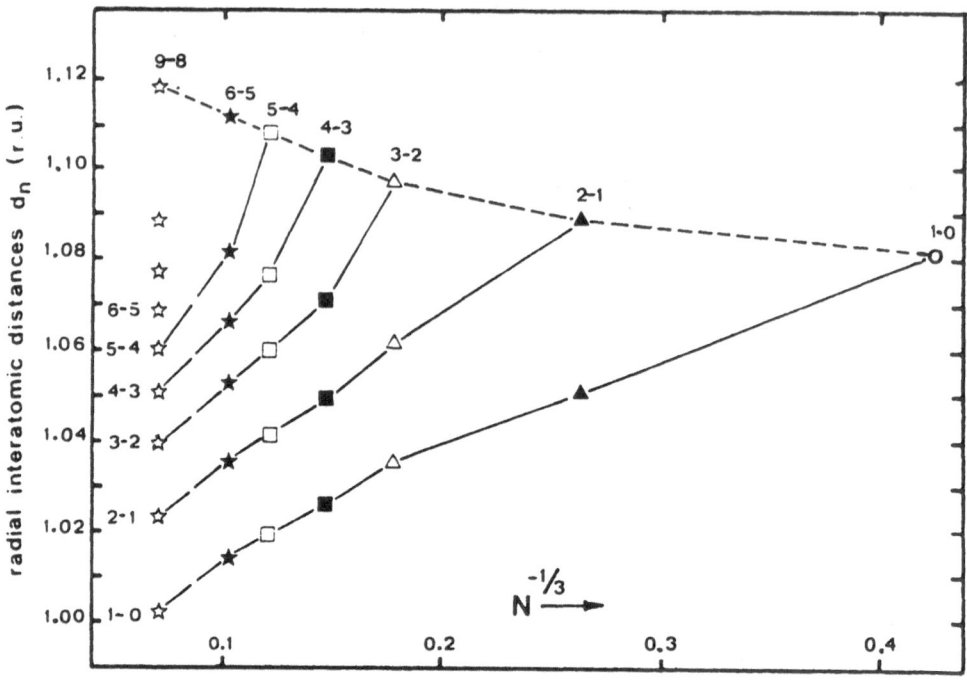

Fig. 6. Distances d_n between atoms situated on a five-fold axis,
belonging to layers n and n-1 (n = 0 refers to the central atom),
calculated for seven relaxed MIC models with complete layers. From
the right to the left, similar symbols refer to a given model, i.e.
successively N = 13, 147, 309, 561, 923 and 2869. The quantity $N^{-1/3}$
varies inversely as the mean model diameter. 1 r.u. = 3.405 Å.

jet expansion and observed while they travel into vacuum. Small clusters of
several metals are likely to adopt this structure, as will be commented
later, but they have not been observed until now.

A free jet is obtained by expanding a gas with pressure p_o through a
small hole into a chamber at low pressure. Clusters are formed by homogeneous nucleation during the expansion and grow by monoatomic addition.
Their mean size is easily increased by increasing p_o. One or two diaphragms
are located on the jet axis in order to produce a cluster beam downstream.
Clusters travel into vacuum and the beam is crossed perpendicularly by a
50 kV electron beam. Scattered electrons are recorded onto a photographic
plate in the form of a powder pattern characteristic of the cluster
structure.

When clusters are small, patterns show that they have an amorphous
structure[3]. When the mean cluster size is increased, patterns are
different[9] and those recorded for p_o = 3.3, 6 and 9 bar are shown in
Fig. 7. For higher inlet pressures (p_o > 12 bar), patterns show crystalline lines which can be indexed according to the fcc structure. Such lines
are absent from the patterns p_o = 3.3 and 6 bar which must then be attributed to a noncrystalline structure. In front of these experimental
patterns, diffraction functions have been reproduced in Fig. 7 calculated
for a MIC model with 147 and 420 atoms, respectively. The almost perfect
agreement between experimental and calculated curves proves that for the

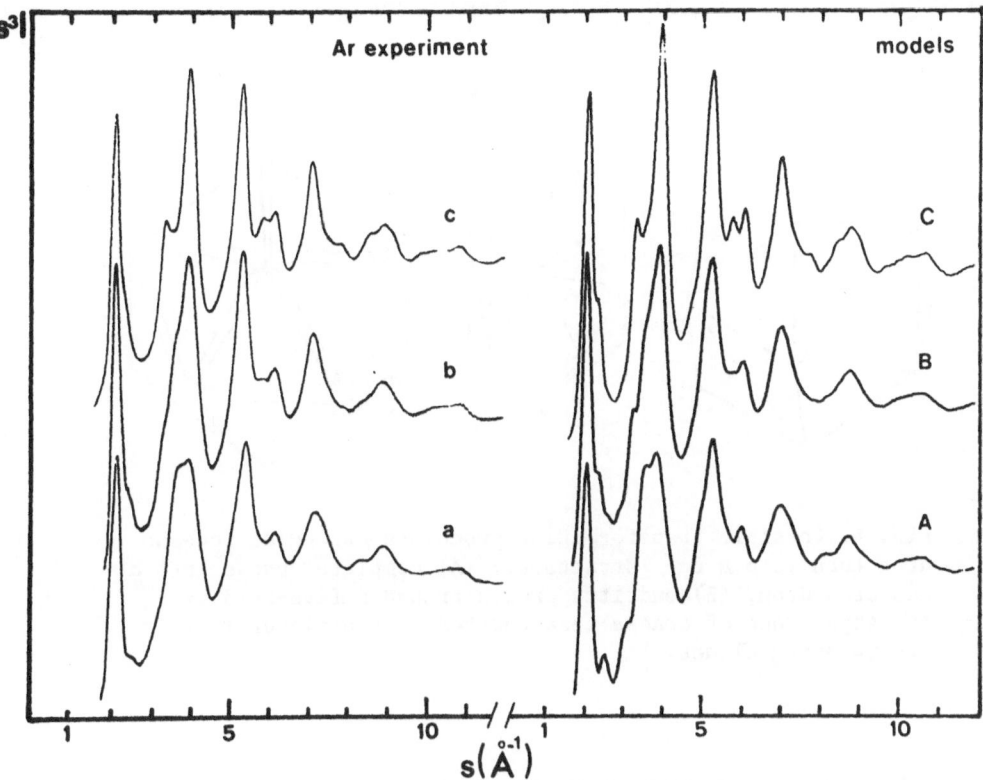

Fig. 7. Left side : Ar diffraction patterns recorded for several
inlet pressures p_o = 3.3 (a), 6 (b) and 9 bar (c).
Right side : diffraction function calculated for several MIC models
N = 147 (A), 420 (B). Curve C is the superposition of two weighted
functions of MIC (0.70) and fcc (0.30) models.

considered number of atoms, Ar clusters have the relaxed MIC structure.
Two more informations can be deduced from the high resolution of the details visible on the patterns :
1) the cluster size distribution must be sharp and 2) all clusters possess
the same structure true to the model without any defect. It is fascinating
to imagine that when p_o = 3.3 bar, all clusters in the beam are almost identical to the emblem of this Symposium !

The upper pattern p_o = 9 bar in Fig. 7 corresponds to clusters with
an estimated mean size $\overline{N} \simeq$ 1000, as deduced from the extrapolation of preceding results. It cannot be accounted for by any of the MIC models, whatever the considered size. On the contrary, the superposition of both MIC
and f c c structures provides a very good agreement as shown in Fig.7. A
more detailed analysis of the patterns p_o > 9 bar reveals that the appearing f c c structure is highly twinned. The limit of stability of Ar MIC
clusters is then estimated to be about $\overline{N}$ = 1000.

In order to know what is occuring inside the clusters when they reach
and cross the limit size, it would be necessary to take a picture of the
beam at p_o = 9 bar. Obviously, this is impossible, and one has then to
guess what is happening. Fig. 8 gives three schematic pictures among all

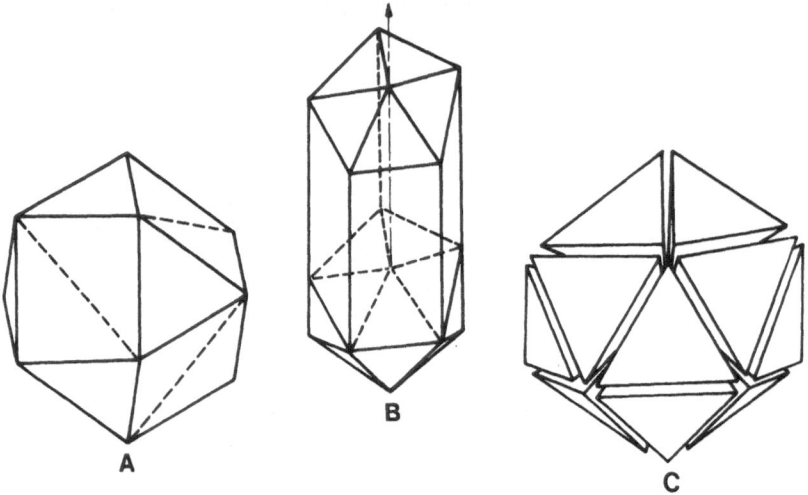

Fig. 8. Possible transformation processes making an icosahedron turn into a fcc structure : (A) complete change into a cuboctahedron, (B)dendritic growth around a five-fold axis, (C) appearance of cracks (exaggerated on the picture) along the twinning planes.

the possible transformations which would lead to a fcc pattern,
In the first one (A), the icosahedron turns directly into a cuboctahedron[5]
Twelve faces, joined two by two, are transformed into 6 square faces, as indicated by dashed lines in Fig. 8A, while the remaining eight keep their equilateral shapes. Such a transformation can occur without any diffusion process, each atom keeping the same nearest neighbors. Despite its geometrical interest, this process in unlikely to occur in Ar clusters since it does not allow any twin formation.

The second picture (Fig. 8B) shows a dendritic growth around an icosahedral axis. The dentritic faces are then parallel to the considered axis and consist in [100] planes. A similar process could lead to the formation of a pentagonal bipyramid grown up around the same axis, the external faces being then [111] planes, just as in the icosahedron case. Due to their pentagonal symmetry, these structures involve strains, however much weaker than those existing in the icosahedron.

The third picture (Fig. 8C) suggests that cracks are produced inside the icosahedron. In the picture, they are represented as produced along each twinning plane, thus allowing the tetrahedra to be regular. The angular gap is then 2.87°. These cracks could as well concern only a few twinning planes or even be produced inside a single tetrahedron. In these cases, the angular gap can reach 7 or 8°, which makes it possible to insert a dislocation plane when the icosahedron has more than seven layers, a size corresponding precisely to about 1000 atoms.

METALLIC CLUSTERS

The description of icosahedral models by Mackay in 1962 has been hardly noticed at the time and the present revival of interest for these models is not only due to aesthetic or geometrical reasons but mainly to the

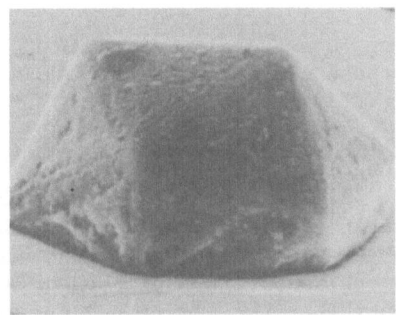

Fig. 9. Micrograph taken from ref. 13 showing a Ni cluster – 250 Å in diameter – with decahedral shape, presented in the direction of its five-fold symmetry axis. Equal thickness fringes are visible around the axis. This cluster has been produced by chemical reduction in liquid phase.

Fig. 10. Ag microcrystal – 10 µm in diameter – produced by electrodeposition. Only half an icosahedron has grown suggesting a primitive icosahedral nucleus located on the support. Micrograph obtained by C. Digard and G. Maurin, Université Paris VII.

fact that icoahedral – or decahedral – shapes have been observed in electron micrographs of several metals known as fcc metals. Many striking – and beautiful – micrographs have been published and Figs. 9 and 10 show two of them.

Many questions have been asked about such pentagonal clusters and three particular aspects will be briefly discussed in the following paragraphs.

Stability

The most significant test for cluster stability is their behaviour when they are brought to equilibrium by a suitable annealing. In a collection of gold clusters 40 to 80 Å in diameter, an annealing process makes the proportion of clusters with five-fold symmetry, often called MTP (Multiplied Twin Particles), increase from some 70% to almost 100%[10]. A similar result is expected for Ag. On the contrary, for Ni, Pd and Pt particles, the proportion is decreased from some 10% to zero. A straightforward conclusion can thus be drawn : in the considered size range – a few thousands of atoms – Au (and probably Ag) icosahedra are stable meanwhile Ni, Pd and Pt ones are metastable. These results have to be related with the particularly low twinning energy of Au and Ag. On the other hand, a high twinning energy prevents Al from forming clusters with five-fold symmetry.

Structure

Whereas microscopy observations have allowed to discover the existence of metallic icosahedral clusters, only diffraction techniques may provide informations about their structure. Unfortunately the obten-

tion of diffraction patterns from small metallic clusters, either supported or produced in a beam[11], involves considerable experimental difficulties. To our knowledge, the best studies concerning Au have been achieved
on powder patterns from clusters of 120 Å diameter[12] and by microdiffraction
from an icosahedral cluster of 250 Å[13]. The somewhat surprising result is
that these clusters show the nondistorted fcc structure.

Formation process

Two processes for the formation of pentagonal metallic clusters
have been previously suggested. One consists in the successive twinning of
tetrahedra and the other in the addition of individual atoms[14]. Since
pentagonal clusters have been observed in a wide range of sizes, there is
no doubt that the second process is much more frequent. Furthermore, keeping in mind that clusters in an Ar beam are thermodynamically stable and
that their structure has been elucidated, one may infer that any pentagonal
cluster, whatever its size, originates in a stable nucleus with the relaxed
MIC structure. An additional argument to support this idea can be found in
Fig. 9 where it is visible that only half an icosahedron could grow since
the substrate has prevented the other half from growing. This particular
feature is found in all clusters produced by electrodeposition.

Above some critical size depending on the metal, the primitive
icosahedron is deformed as a consequence of the compression experienced
by its central region : strains are then relaxed through the formation of
cracks and insertion of dislocation planes. Such dislocations are likely to
have been observed recently[15]. The structure becomes fcc and the defects
even favour the cluster growth. The important point is that the cluster is
able to grow up while keeping its icosahedral shape since when a new atomic
layer nucleated onto a face has reached one edge, it can develop over the
adjacent face without necessity of any new nucleation[14].

REFERENCES

1. S. Ino, J. Phys. Soc. Jap., 21 : 346 (1966).
2. M.R. Hoare and J.A. Mc Innes, Adv. Phys.,32 : 791 (1983).
3. J. Farges, M.F. de Feraudy, B. Raoult and G. Torchet, J. Chem. Phys.,
   78 : 5067 (1983).
4. I.A. Harris, R.S. Kidwell and J.A. Northby, Phys. Rev. Letters, 53 :
   2390 (1984).
5. A. Mackay, Acta Crystallogr., 15 : 916 (1962).
6. C.Y. Yang, J. Cryst. Growth, 47 : 274 (1979).
7. J. Farges, M.F. de Feraudy, B. Raoult and G. Torchet, Acta Crystallogr.,
   A38 : 656 (1982).
8. B.G. Bagley, Nature, 225 : 1040 (1970).
9. J. Farges, M.F. de Feraudy, B. Raoult and G. Torchet, J. Chem Phys.,
   84 : 3491 (1986)
10. C. Solliard, Ph. Buffat and F. Faes, J. Cryst. Growth 32 : 123 (1976)
11. B.G. De Baer and G.D. Stein, Surf. Sci., 106 : 84 (1981).
12. C. Solliard, thèse n° 497, Lausanne (Switzerland) (1983).
13. M. Brieu, thèse n° 1260, Toulouse (France) (1986).
14. M. Gillet, J. Cryst. Growth,36 : 239 (1976).
15. L.D. Marks and D.J. Smith, J. Cryst. Growth, 54 : 425 (1981).

STUDY OF PRODUCTION TECHNIQUE FOR METALLIC

ULTRA FINE PARTICLES USING ARC ENERGY

Takeshi Araya, Yoshiro Ibaraki, Susumu Hioki,
Ryoji Okada and Masatoshi Kanamaru

Mechanical Engineering Research Laboratory
Hitachi Ltd.
502 Kandatsu-cho, Tsuchiura-shi, 300 Japan

1. INTRODUCTION

Ultra fine particles (referred to as UFP here) have drawn considerable attention in recent years as possible new function materials. As such, they have been the subject of intensive research in various fields.[1] The authors have studied the UFP production technique using arc energy looking closely at the electromagnetic force resulting from arc current. Through this study a method for producing UFP with higher efficiency was developed. The results of that investigation are covered in this report.

2. APPARATUS AND EXPERIMENTAL METHOD

The UFP production apparatus used for this experiment consists of a UFP generating chamber, arc power source, collection portion, etc. as illustrated in Fig. 1, with the external appearance shown in Fig. 2. To produce

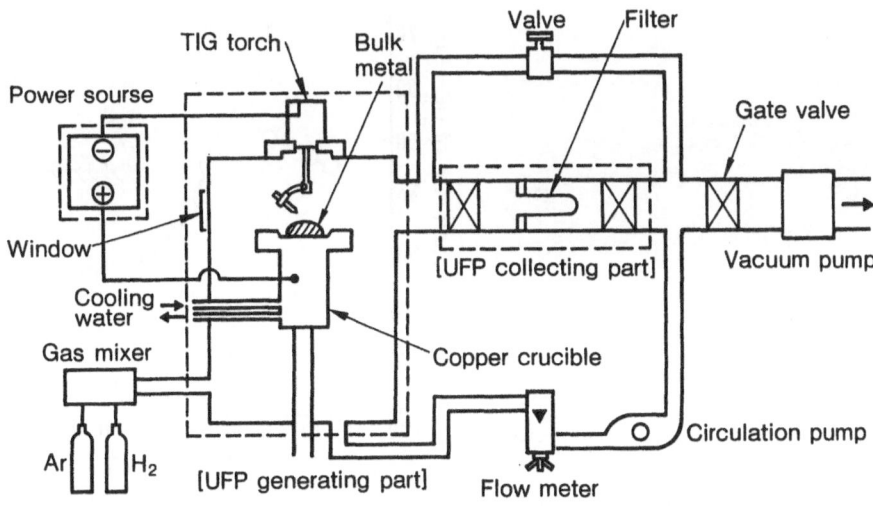

Fig. 1 Construction of UFP production apparatus

UFP, argon/hydrogen mixed gas should be sealed up to the atmospheric pressure after vacuum-discharging the interior of the unit, then the gas inside the unit should be circulated by the circulating pump. The arc should be made between the cathode tungsten and the anode bulk metal. The generated UFP moves with the circulated gas and should be collected in a filter. The tungsten electrode, hydrogen concentration of atmospheric gas, and arc current were systematically changed during the experiment. The bulk metal weight decrease was assumed to be generated UFP volume.

3. PRODUCTION RATES

Production rates of UFP are affected by several parameters with the strongest influence coming from electrode angle, hydrogen gas concentration and arc current. This will be discussed in sections 3.1 and 3.2 below. In section 3.3, the production volume of various materials measured with the most suitable conditions for producing UFP will be described.

3.1 Influence of Electrode Angle

The generated volume of UFP was measured by changing the angle θ between the mother material and the electrode in an atmosphere containing 50% hydrogen. The UFP was generated at arc current $I_a=150A$ by using the mother material, Ni, and maintaining a gap between the mother material and the electrode tip at L=6 to 8 mm. As a result, generated volume for the UFP was found to reach its maximum value in the vicinity of electrode angle $\theta=40°\sim45°$ as shown in Fig. 3.

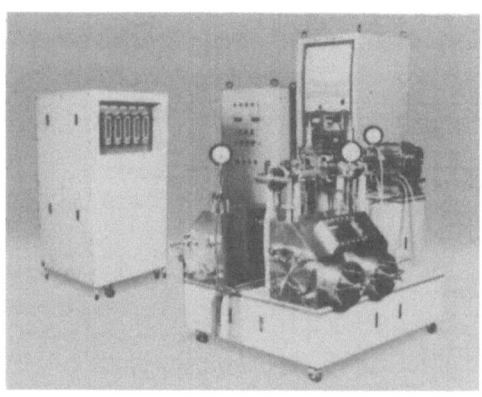

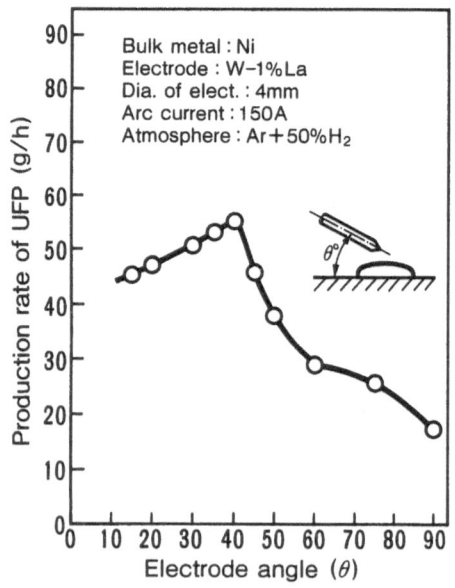

Fig. 2   External appearance of UFP   Fig. 3   Relation between electrode angle
          production apparatus   and UFP production rates

3.2 Influence of Arc Current

The UFP production rate when changing the arc current and the hydrogen gas concentration while keeping the electrode inclination at a constant angle (θ = 45° and 75°) is shown in Fig. 4. Production rate increased following an increase in the hydrogen gas concentration and the arc current. When the electrode angle was nearly perpendicular to the bulk metal (θ=75°), indicated by the dotted line, the production rate decreased at the transient current if the arc current increased. However, the transient phenomenon did not occur by tilting the electrode. Consequently, the production rate can be increased by enlarging the arc current and hydrogen concentration at a proper electrode angle.

3.3 Production Rates of Various Metals

Table 1 shows the measured results of production rates for various metals by the conventional process (θ=75°) and by the new process when the electrode is tilted (θ=45°). Here, arc current Ia was 200A, and the atmospheric gas was Ar + 50%H$_2$.

The calculated ratio of these two processes shows that the new process is at least 1.5 times higher and at most 26 times higher than the conventional process. This can be clearly seen in the table.

A typical example of electron microscopic photograph Ni is shown in Fig. 5. Average diameter of UFP is 20 ~ 30 nm.

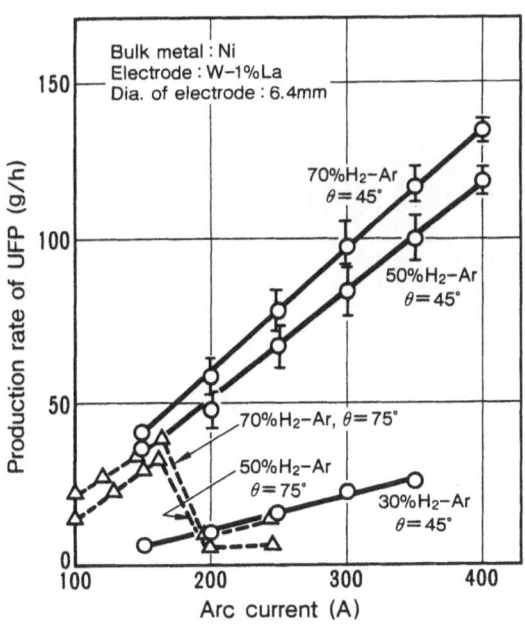

Fig. 4 Influence of arc current, atmosphere and
electrode angle on UFP production rates

Table 1. Production rates of conventional process and
new process with various metals

| Metals | UFP production rate (g/h) I=200A | | Ratio: New process / Conventional process |
	Conventional process (θ=75°)	New process (θ=45°)	
V	0.98	25.3	25.8
Cr	23.5	159.2	6.8
Fe	9.1	129.9	14.3
Co	25.7	67.1	2.6
Ni	4.9	47.3	9.6
Cu	6.4	9.9	1.5
Nb	1.8	3.6	2.0
Mo	0.55	4.93	9.0
Ta	1.98	7.50	3.0
W	0.28	1.24	4.4

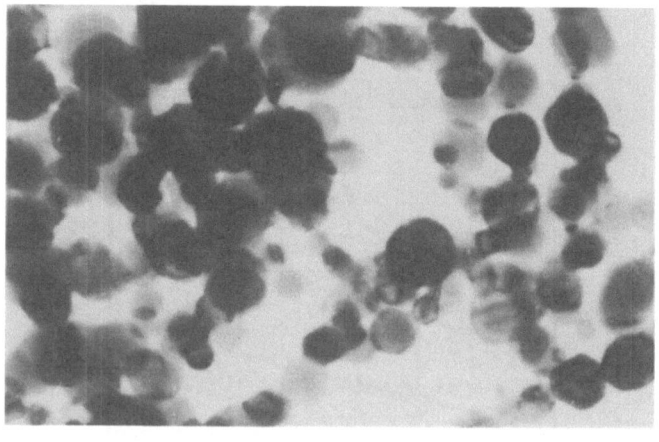

├─────────────────────┤
100nm

Fig. 5. Electron microscopic photograph of Ni UFP

4. MECHANISM OF PRODUCTION

The formation phenomenon of UFP is generally explained by the hydrogen dissolution/release phenomenon according to the research done by Uda.[1] The arc profile is bell shaped as shown in Fig. 6(a). Using this conventional process, production rate is very low when the arc current exceeds 170A, as shown by $\theta=75°$ in Fig. 4.

By decreasing the arc current, the transient current value can be seen where the production rate increases suddenly, as observed in the same figure. The arc profile shows a concentrated anode arc on the metal surface as shown in Fig. 6(b). It was assumed that the metal surface vaporized due to the concentrated arc energy, and that transfer of metallic vapor on the plasma stream to the electrode caused the production increase.

The tilted electrode angle caused a change in the arc profile as shown in Fig. 6(c), and the production rate increased with an increase in the arc current, shown by $\theta=45°$, $Ar + 50 \sim 70\%H_2$ in Fig. 4.

The UFP production phenomena illustrated in Fig. 7, may be explained as follows:

(1) the arc on the metal surface is concentrated by the thermal pinch effect of hydrogen gas, and increases the energy density;

(2) surface temperature rises due to the reaction heat ΔHf generated by the thermal combination of hydrogen;

(3) the arc is blown by the electromagnetic force caused by current electrification channel change;

(4) the blown arc generates the plasma stream.

Consequently, the bulk metal surface was vigorously vaporized and the vapored metal was transferred by the plasma stream. On this occasion, the metal vapour pressure decrease on the surface accelerated metal vapouring. Accordingly, it was concluded that this phenomena greatly increased UFP production rates.

 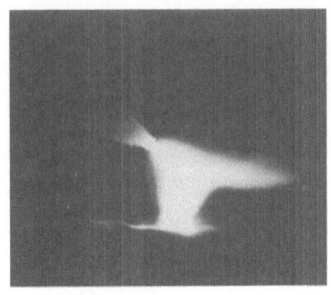

(a) conventional arc   (b) concentrated arc   (c) brown arc

Fig. 6. Typical examples of arc profile

5. CONCLUSION

As a result of studying the UFP production technique using a hydrogen/argon gas atmosphere, it was found that (1) UFP can be produced with high efficiency by tilting the electrode to an adequate angle, and (2) the produced volume can be increased considerably by enlarging the arc current and hydrogen concentration.

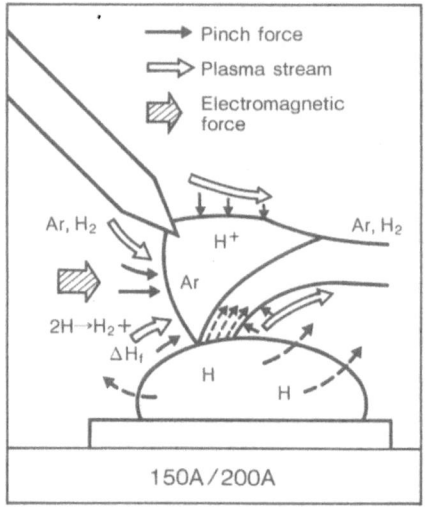

Fig. 7. Illustrated model of production phenomena

6. REFERENCE

1. M. Uda, Journal of the Japan Welding Society, Vol. 54 No. 6, 1985, p. 16

ELECTRON DIFFRACTION STUDY OF SUPERSONICALLY GENERATED CLUSTERS

Yael Z. Barshad* and Lawrence S. Bartell

Department of Chemistry
University of Michigan
Ann Arbor, MI 48109

ABSTRACT

Clusters of cyclopropane, n-butane, octafluorocyclobutane, neopentane, and hexafluorobenzene were produced in flows with neon carrier gas through a Laval nozzle. All gave liquid-like diffraction patterns except for C_4F_8 which gave Debye-Scherrer rings corresponding to 90 Å crystals. The algorithm RISM accounted qualitatively but not quantitatively for the structures of the liquid-like clusters, perhaps because the clusters are too cold and dense for the approximations in RISM to be satisfied.

INTRODUCTION

Diffraction studies of clusters produced by the condensation of vapors in supersonic flow provide results of potential value in several areas of research. For one thing, they can help answer long-standing questions about homogeneous nucleation and the growth of critical nuclei. For another, they offer an opportunity to test whether current theories of molecular fluids adequately account for cold, dense, supercooled liquids. The present research touches upon the first question insofar as it augments recent studies[1] showing that condensation to liquid-like clusters is neither unnatural nor rare. Prior diffraction investigations of simpler molecular systems had all found solid microclusters. Having identified systems of simple, symmetrical molecules that yield microdrops appearing to be liquid, the present research addressed the second question. One of the most attractive of the current theories of molecular liquids is the Reference-Interaction-Site-Model (RISM)[2,3] based on the Percus-Yevick approximation that makes it tractable and feasible to apply to systems of some complexity. While it is known to yield quite good results for fluids that are not too dense or cool,[4] it is less clear that it applies accurately at contrary conditions such as those of condensates generated in nozzle flow.[5] How well RISM accounts for cold molecular clusters is examined in the following.

*Present address: Isotope Department, The Weizmann Institute, Rehovot 76100, Israel

EXPERIMENTAL

Vapors of the subject compounds were seeded into neon carrier gas and expanded through a miniature Laval nozzle (glass, no. 6).[6] Clusters transmitted through a skimmer were probed by a 70 keV electron beam. Electron diffraction patterns were measured and converted into s-weighted cluster structure functions

$$sM(s) = s \ I_{cl}/I_{at}{}^{el}$$

where I_{cl} and $I_{at}{}^{el}$ represent, respectively, the <u>intermolecular</u> interference intensity contributions and the smooth background of elastically scattered atomic intensity. Initial conditions were varied over small ranges, characteristically 2 to 3 atm or 3 to 4 atm for stagnation pressures and mole fractions of 0.125 or 0.25 for hydrocarbons, 0.04-0.06 for fluorocarbons.[7]

RISM COMPUTATIONS

Pair correlation functions from which electron diffraction intensities could be calculated were derived from the RISM algorithm of Johnson and Hazuomé[3] incorporating continuous, rather than hard-sphere potential functions. In all cases computations were carried out using pairwise additive atom-atom potential functions very similar to those of Williams[8,9] which were derived from crystal structures and energies. In the case of cyclopropane, $CH_2 \cdots CH_2$ group interaction functions close to those of Jorgensen[10] were also tried, requiring that zero-potential sites for hydrogen be included to make diffraction calculations possible.

Difficulties were often encountered in obtaining convergence to a stable solution of the RISM equations when the Williams potential parameters and physically reasonable densities and temperatures were fed into the computations. It was assumed that the Yaw's representation[11] of density as a function of temperature gives a plausible extrapolation of liquid densities to the cluster temperatures. Some adjustment of potential parameters and/or densities had to be made. These will be documented elsewhere.[12] Commonly, densities had to be adjusted downwards toward room temperature values instead of being left at values corresponding to the 120-200 K temperatures of the clusters.

RESULTS

Four of the five systems studied, namely cyclopropane, n-butane, neopentane, and hexafluorobenzene, gave clusters whose structure functions corresponded to those of a liquid. These structure functions are shown in Figs. 1-3, excluding the butane results. n-Butane is complicated by rotational isomerism and will be discussed elsewhere.[12] Octafluorocyclobutane, on the other hand, yielded microcrystals whose Debye-Scherrer diffraction pattern is illustrated in Fig. 4. In subsequent research[13] carried out by Harsanyi and Valente under a wider range of conditions, solid clusters of neopentane could be obtained reproducibly instead of liquids when the mole fraction was reduced somewhat, and two additional solid forms of C_4F_8 besides that manifested in Fig. 4 were seen.

Also plotted in Figs. 1-3 are the results of the RISM computations. In addition to the runs whose results are plotted were a great many more carried out using a variety of temperatures, densities, and potential constants. This effort to improve the agreement between experiment and theory led to no results significantly better than those in Figs. 1-3.

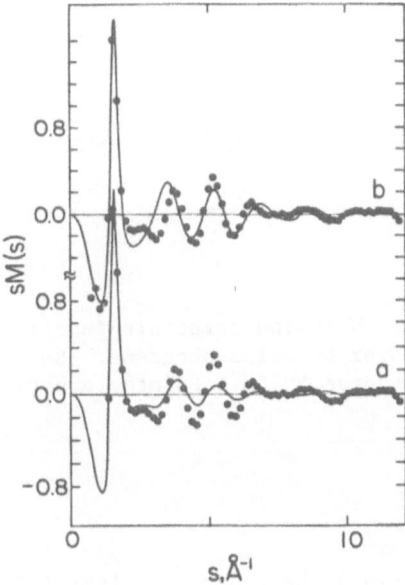

Fig. 1. s-Weighted electron diffraction
         structure functions for cyclo-
         propane. Solid curves, RISM
         calculations (a) via atom-atom
         potentials, (b) via group inter-
         actions. Points, experiment.

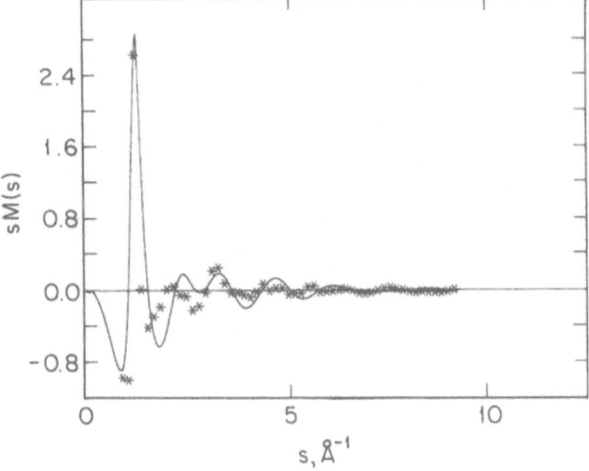

Fig. 2.   s-Weighted structure functions
          for neopentane. Solid curve,
          RISM. Points, experiment.

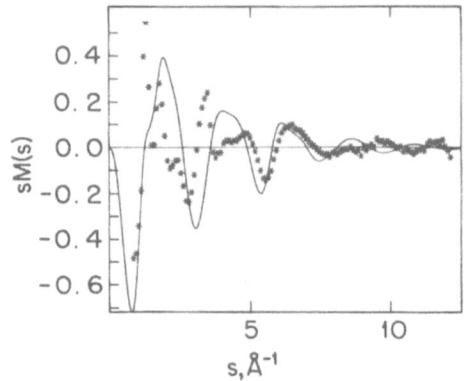

Fig. 3. s-Weighted structure functions
for hexafluorobenzene. Solid
curve, RISM. Points, experiment.

DISCUSSION

The single example among the present systems that gave crystalline
clusters, C_4F_8, is known to have a complex solid state behavior. Heat ca-
pacities[14] and NMR spectra[15] reveal five different solid state phases at
1 atm as temperature is lowered. No information on crystal structures is
available. Although the electron diffraction rings recorded at high resolution (the lower curve of Fig. 4) can be indexed as bcc reflections
implying a cell constant of 10.24 Å and six molecules per cell, arguments

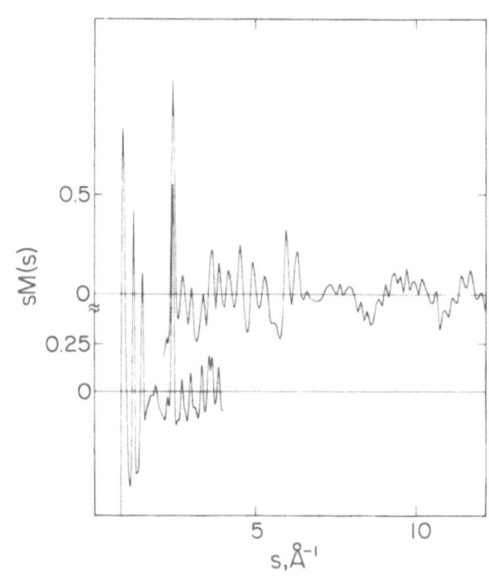

Fig. 4. s-Weighted structure functions
for octafluorocyclobutane. Lower curve recorded at high resolution, upper curve at low
resolution.

can be advanced[12] that suggest this interpretation is spurious or indicative of crystalline disorder. The reflections in the upper curve of Fig. 4 were recorded at too low resolution to be helpful in analyses except to verify that the ordering in the microcrystals is enormously higher than in the hydrocarbon analog C_4H_8 whose crystals yield only two diffuse Bragg reflections.[16]

According to an empirical relation discussed elsewhere,[1,17,18] clusters of n-butane are close to the melting point, neopentane and perfluorobenzene are roughly 100° below the melting point, and cyclopropane is intermediate in degree of supercooling. Because slight changes in conditions give crystalline clusters of neopentane,[13] it is difficult to imagine that neopentane, or any of the above liquid-like cases, are rigid, frozen, glassy solids instead of true liquid microdrops. It is known that great supercooling is considerably easier to achieve and maintain in minute droplets than in the bulk.[19] It is not unreasonable to inquire, then, whether statistical mechanics, as expressed in the algorithm RISM, can account for the electron diffraction patterns. Figures 1-3 show the results.

Structure functions calculated via RISM display elements of similarity with the observed cluster diffraction patterns. Nevertheless, in none of the examples is a satisfactory agreement found. What appears to be a characteristic aspect of the disagreement in these cases, and in an earlier study[20] of benzene clusters as well, is manifested most clearly in the radial distribution function computed by Fourier inversion of the diffraction data.[12] Although RISM generates features coinciding in internuclear distance with the peaks in the experimental curves, the RISM features are considerably more washed out. It is difficult to reconcile the disagreement on the basis of assuming that the clusters are submicroscopic (ca. 10^2 Å) instead of bulk, that they are amorphous solids instead of true liquids, or that the Williams-type potential functions are inadequate. We conclude, tentatively, that even though RISM is known to reproduce fairly well the diffraction data for some liquids at higher temperatures and lower densities, it is less satisfactory for cold, dense liquids. Computations are under way to find whether the classical theory of molecular liquids exemplified in Monte Carlo calculations is any more successful than RISM in accounting for the structures of clusters nucleated in supersonic expansions.[21]

ACKNOWLEDGEMENT

This research was supported by a grant from the National Science Foundation. We thank A. H. Narten and E. Johnson for providing their version of the RISM program. We express our appreciation to E. J. Valente for much help and advice and to R. Ramani for assistance in recording diffraction patterns of n-butane. We gratefully acknowledge a generous allocation of computing time from the Michigan Computing Center.

REFERENCES

1. For a recent review, see L. S. Bartell, Chem. Rev. 86:491 (1986).
2. L. J. Lowden and D. Chandler, J. Chem. Phys. 59:6587 (1973); ibid. 61:5228 (1974).
3. E. Johnson and R. P. Hazoumé, J. Chem. Phys. 70:1599 (1979).
4. A. H. Narten, E. Johnson, and A. Habenschuss, J. Chem. Phys. 73:1248 (1980); A. Habenschuss, E. Johnson, and A. H. Narten, J. Chem. Phys. 74:5234 (1981).
5. C. G. Gray and K. E. Gubbins, "Theory of Molecular Fluids," Clarendon, Oxford (1984), Vol. 1.

6. E. J. Valente and L. S. Bartell, J. Chem. Phys. 79:2683 (1983).
7. Y. Z. Barshad, Ph.D. Thesis, University of Michigan (1985).
8. D. E. Williams, Acta Cryst. A30:71 (1974).
9. D. E. Williams and D. J. Houpt, Acta Cryst. B42:286 (1986).
10. W. L. Jorgensen, J. D. Madura, and C. J. Swenson, J. Am. Chem. Soc. 106:6638 (1984).
11. C. L. Yaws, "Physical Properties, a Guide," McGraw-Hill, New York (1977).
12. Y. Z. Barshad and L. S. Bartell, to be published.
13. L. S. Bartell, L. Harsanyi, and E. J. Valente, Proceedings of this symposium.
14. G. T. Furakawa, R. E. McCoskey, and M. L. Reilly, J. Res. Nat. Bur. Stds. 52:11 (1954).
15. E. Szczésniak and J. R. Brookman, Mol. Phys. 48:1221 (1983).
16. C. F. Carter and D. H. Templeton, Acta Cryst. 6:805 (1953).
17. J. Farges, M. F. de Feraudy, B. Raoult, and G. Torchet, Surf. Sci. 106:95 (1981).
18. C. E. Klots, Phys. Rev. Lett. (submitted).
19. J. P. Dumas, C. R. Acad. Sci. Paris C. 284:25C (1977).
20. E. J. Valente and L. S. Bartell, J. Chem. Phys. 80:1451 (1984).
21. L. S. Bartell, L. Sharkey, and X. Shi, to be published.

CONTROL OVER THE STRUCTURE OF CLUSTERS GENERATED IN SUPERSONIC FLOW

Lawrence S. Bartell, Laszlo Harsanyi, and Edward J. Valente

Department of Chemistry
University of Michigan
Ann Arbor, MI 48109

ABSTRACT

Clusters of a variety of volatile substances have been generated in flow with carrier gases through a miniature Laval nozzle. Cluster structures were investigated by electron diffraction and found to be liquid-like in some cases, crystalline in others. Solid clusters tended to occur in experiments with substances exhibiting plastic crystalline phases. In a number of examples the cluster structure could be controlled by the flow conditions. Factors governing the type of cluster formed are discussed.

INTRODUCTION

The aim of this research has been to shed light on the structural chemistry of clusters generated from the vapor phase by homogeneous nucleation during supersonic expansion. Recent work[1-5] shows that this technique can lead to a diversity of cluster types. Although the clusters produced depend critically upon the physical properties of the molecular systems examined, we find that the experimentalist can often exercise a significant degree of control over the way the molecules organize into clusters. In view of the fundamental importance of nucleation in many scientific fields, it seemed worthwhile to investigate the structural characteristics of clusters formed from a wide range of substances in an attempt to uncover the governing principles. To this end we have carried out electron diffraction studies on a variety of compounds, approximately doubling the number of cluster systems that have been subjected to this method of analyses. The virtues and limitations of diffraction analyses were reviewed recently.[5] In the following we examine the results obtained to date, including those of prior studies at Michigan and elsewhere, and identify several considerations relating cluster structures to properties of materials and to nozzle expansion conditions.

EXPERIMENTAL

Condensation in Nozzle Flow

Factors governing nucleation conditions have been reviewed elsewhere.[5,6] Molecular weight of carrier gas, subject gas concentration, initial pressure, initial temperature, and nozzle design are all important. Most research in the first laboratory to undertake structural studies of clusters (Orsay) has involved free jet expansions. Alternatively, almost all of the studies conducted in the other two contributing laboratories (Northwestern and Michigan) have made use of miniature Laval nozzles. Laval nozzles greatly slow down expansions and, as a rule, induce nucleation at appreciably lower pressures and higher initial temperatures. In flows with carrier gas, simple computations[7,8] suggest that critical supersaturation is achieved earlier in the flow, and at warmer temperatures, the higher is the initial total pressure and/or mole fraction of the seeded subject gas up to a certain point. Large mole fractions tend to inhibit nucleation. Furthermore, several arguments[7-9] suggest that increasing the molecular weight of the carrier gas enhances clustering except, possibly, in the case of very light subject gases. Applications of models of one-dimensional flow indicate[10] that cluster temperatures and velocities should lag very little behind those of the accelerating, cooling, carrier gas once cluster growth is complete. It often turns out in practice, however, that cluster temperatures lag far behind the carrier temperature, and that the type of phase nucleated and grown correlates only roughly with the computed nucleation temperature. Moreover, some inversions of efficiency of carrier as a function of molecular weight have been seen in the present research. Clearly, the science of cluster formation is at an early stage of its development. Nevertheless, application of the above simple considerations can provide broad, rough guidelines. In any event, it has been found for a significant fraction of the cases investigated to date that small, easily controlled changes in flow conditions can alter the type of cluster formed. This topic will be discussed subsequently.

Approximately 80% of the clusters tabulated in the next section were studied solely, or in overlapping research, by the Michigan laboratory. In all examples from this laboratory, a small Laval nozzle (glass, no. 6)[11] at room temperature was used. Pressures ranged from 2 to 7 atm, and subject mole fractions, from 0.01 to 0.25. Exact conditions will be tabulated elsewhere in the cases not previously documented. Neon was the most common carrier gas but helium, neon, and argon were sometimes all applied. Systems were subjected to conditions believed to condense most of the seeded vapor into comparatively large clusters ($\sim10^2$Å). Clusters were probed by electrons within 10^{-4}s of nucleation.

Electron Diffraction Analyses

Characteristics of the various diffraction apparatuses for cluster research have been described elsewhere.[5] For the present purposes the analyses of diffraction records were restricted to an identification of the type of phase encountered, namely: were the clusters liquid-like or solid and, if solid, could more than one type of molecular packing be recognized.

RESULTS

Clusters analyzed by electron diffraction to date[1-5,11-17] are listed in Table 1 with an identification of cluster type, or types, observed for a given substance. Only a few previously described clusters, all metallic, have been excluded from the table, partly because their mode of formation was quite different from those of the compounds listed. To the extent that they could be interpreted, however, the structural characteristics of the metal clusters seemed to be not greatly different from those of the rare gases. Also tabulated are certain physical properties[18-21] related to the phases nucleated. One index is whether or not the higher temperature solid phases are known to be plastic crystalline. Lack of an entry

Table 1. Clusters formed by homogeneous nucleation,[a] and related physical properties.[b]

Liquid-like Clusters

Subst	$\Delta S_m/R$	ODIC[c] ?	$(T_B-T_m)/T_B$	$(T_m-T_{cl})/T_m$
CS_2	3.28	---	0.49	0.01
CHF_3	4.31	---	0.41	0.15
$CHCl_3$	5.05	yes	0.37	0.19
CH_2Cl_2	4.05	---	0.43	0.10
CH_2Br_2	4.66	---	0.40	0.06
CH_3OH	2.18	---	0.48	-0.16
CF_3Cl	---	---	0.52	-0.06
CF_4	0.94	yes	0.39	0.15
C_2H_5OH	3.81	---	0.55	-0.40
C_3H_6[d]	4.50	---	0.39	0.20
C_3F_8	4.70	---	0.62	-0.28
C_3H_6O[e]	3.39	---	0.46	-0.04
nC_4H_{10}	4.05	---	0.51	0.01
C_5H_{10}[f]	0.41	yes	0.44	0.11
C_6H_6	4.24	---	0.21	0.36
C_6F_6	5.01	---	0.21	0.30
C_6H_{14}[g]	0.40	yes	0.46	-0.02
C_6H_{14}[h]	0.66	---	0.56	-0.22
$SiCl_4$	4.57	no	0.38	0.14
CCl_4[m]	1.19	yes	0.29	0.30
$C(CH_3)_4$[m]	1.53	yes	0.09	0.47

Solid Clusters

Subst	$\Delta S_m/R$	ODIC[c] ?	$(T_B-T_m)/T_B$	$(T_m-T_{cl})/T_m$
Ne	1.64	yes	0.10	0.58
Ar[i]	1.69	yes	0.04	0.56
Kr[i]	1.69	yes	0.04	0.54
Xe[i]	1.71	yes	0.03	0.54
N_2[i]	1.37	yes	0.18	0.48
CH_4[i]	1.25	yes	0.16	0.48
C_2F_6	1.86	yes	0.11	0.44
C_4F_8[j]	1.43	yes	0.13	0.39
C_6H_{12}[k]	1.12	yes	0.21	0.37
SF_6[j]	2.72	yes	-0.08	0.48
SeF_6[j]	3.59	yes	-0.05	0.50
TeF_6[j]	3.63	yes	-0.00	0.53
CO_2	4.63	---	-0.17	0.55
H_2O[l]	2.64	---	0.27	0.32
NH_3	3.47	---	0.18	0.39
SiF_4	4.64	yes	0.04	0.35
CCl_4[m]	1.19	yes	0.29	0.30
$C(CH_3)_4$[m]	1.53	yes	0.09	0.47

[a] All monatomic, diatomic, and triatomic solid clusters and CH_4 are reported in references 1, 2, 13-17. The remainder of the substances as well as Ar, Kr, and Xe in reinvestigation represent Michigan work, references 5, 12, and this research.
[b] References 18-21 plus a few estimated quantities.
[c] Orientationally disordered (i.e., plastic) crystals.
[d] Cyclopropane.
[e] Acetone.
[f] Cyclopentane.
[g] Neohexane.
[h] 2,3-Dimethylbutane.
[i] Icosahedral packing for small clusters, fcc for large.
[j] Two or more different crystal packings observed, depending on conditions.
[k] Cyclohexane.
[l] Amorphous network for small clusters, hexagonal microcrystals for large.
[m] Seen both as liquid clusters and microcrystals, depending on conditions.

in the table signifies ignorance of the present authors, not a negative result. The entropy of melting ΔS_m and the fractional range of liquid existence at one atmosphere, $(T_B - T_m)/T_B$, where T_B is the normal boiling point and T_m the melting point, are straightforward. On the other hand, cluster temperature, T_{cl} in the entry $(T_m - T_{cl})/T_m$, is supposed to represent the (possibly hypothetical) temperature of clusters in their liquid form. This temperature does not represent a measured quantity but is, rather, a value calculated according to

$$T_{cl} = 0.04 \; \Delta E_{vap}/R, \tag{1}$$

where the energy of vaporization is computed, via the representation of Yaws,[18] at the temperature T_{cl}. Equation (1) is a rough rule of thumb of the form suggested by several authors.[5,13,22]

DISCUSSION

Table 1 is quite consistent with the intuitive expectation that compounds with a low entropy of melting and a stable plastic crystalline phase, a small or null range of liquid existence, and a cluster temperature well below the melting point, will tend to yield solid clusters. The tabulated physical characteristics are not unrelated to each other, of course (although a glance at the table shows that there is a substantial independence of the different entries). Materials with particularly low entropies of melting must, of course, be heated to comparatively high temperatures before their standard free energies of fusion vanish and allow melting. Entropy of melting of clusters of polyatomic molecules can only be low if the low temperature, brittle crystalline phase has undergone one or more solid state transitions to acquire some of the disorder normally associated with melting. Indeed, in the cases of SF_6, SeF_6, TeF_6, and C_4F_8, two or more different crystalline packings have been seen in the clusters, depending upon carrier, concentration, and pressure. Low mole fractions and heavier carrier gas favored the low temperature phases of the hexafluorides. In fact, the high temperature bcc phase of TeF_6, a volatile phase with a small range of thermal stability, was never seen. One or another of two lower temperature phases were observed, depending reproducibly upon flow conditions.

Why the plastic crystalline property is correlated with the nucleation of solid clusters deserves a few comments. As rationalized above, this property is naturally associated with low entropies of fusion and a relatively high melting point and, hence, the formation of crystalline aggregates. From another perspective, molecules in plastic crystals tend to be spherical or quasispherical. They are unusually free in their rotational and, in some cases, translational motions.[20,21] Therefore, if condensed into a cold, chaotic globule, they are freer to reorganize into a crystalline mass than molecules of more irregular shape.

Octafluorocyclobutane is an unusual case with five different solid state phases,[23,24] one of which is stable only over a two degree temperature range! As yet no studies have been made of the crystal structures of any of the phases. In supersonic expansions, depending upon flow parameters, three distinctly different diffraction patterns were seen. These patterns (Debye-Scherrer powder patterns) were rather diffuse because of the submicroscopic sizes of the crystals, and have not yet been analyzed in terms of lattice constants. One curious anomaly is that the cluster sizes increased as carrier molecular weights decreased, possibly because of specific interactions between the microcrystals and the heavier carriers. One crystal phase is suspected of having large cavities.

Two solid forms each for the rare gases and for N_2 and CH_4 have been seen and studied extensively by the group at Orsay.[1,2,13,15] The clusters of those substances are all very similar. When clusters are small, composed of fewer than, perhaps, 800 molecules, they are characterized by an underlying icosahedral motif. Such a structure is a particularly efficient packing arrangement for small aggregates of soft spheres but it cannot be propagated into a true crystal. Larger clusters, readily generated with higher initial pressures, are cubic-closest packed. Small clusters of water tend to be cold, amorphous networks, and larger ones, hexagonal crystals.[17] Crystal order is improved if the clusters are annealed by passage through a shockwave.[25]

In two cases, so far, namely CCl_4[3] and $C(CH_3)_4$, it has been possible to nucleate either liquid-like clusters or, if the pressure is raised modestly, microcrystals. As the bulk solids are well-known examples of plastic crystals[21] with especially low entropies of fusion, and the liquid existence ranges are appreciable but not large, this behavior is quite natural. It should be noted that heat of condensation can warm cluster temperatures by well over 50° after the onset of nucleation if the mole fraction is substantial. No doubt other simple systems exhibiting similar cluster properties will soon be found.

When liquid existence range is large, and cluster temperature is not greatly below the melting point, liquid clusters can be expected and are found, as can be seen in Table 1. Intermediate cases in which the various tabulated indicators do not all point in the same direction are more difficult to predict. Even though a single indicator is inadequate as a predictor of cluster phase, a three-dimensional plot taking the three numerical indicators of Table 1 as coordinates shows two quite distinct regions. Points representing liquid and solid clusters tend to be on opposite sides of the inclined surface

$$0.1 \; \Delta S_m/R + 0.85(T_B-T_m)/T_B - 0.52(T_m-T_{cl})/T_m = K$$

where $K \approx 0.34$ for common conditions. Under more extreme conditions leading to colder clusters, K might be increased somewhat.

Although volume change on melting[20] conceivably might be correlated with a predisposition to form one or another type of cluster, no systematic correlation of this type was found. Activation energies for molecular reorientation and migration undoubtedly play a crucial role, and one that is manifested in the physical properties listed in Table 1 only to a limited extent.

Further studies are in progress to test a wider variety of substances, to model the process of nucleation and cluster growth, and to carry out Monte Carlo simulations of some of the more interesting cluster systems.[26] In the course of this research some relationships have already emerged that elucidate, in a modest way, what types of clusters are likely to be formed. Moreover, rough guidelines have been found that suggest how to generate alternative forms by changing nucleation conditions. Much needs to be learned, however, before a thorough understanding is at hand.

ACKNOWLEDGMENT

This research was supported by a grant from the National Science Foundation. We thank Kathleen Nolta for her able assistance in carrying out the research.

REFERENCES

1. J. Farges, M. F. de Feraudy, B. Raoult, and G. Torchet, J. Chem. Phys. 84:3491 (1986).
2. J. Farges, M. F. de Feraudy, B. Raoult, and G. Torchet, J. Chem. Phys. 78:5067 (1983).
3. E. J. Valente and L. S. Bartell, J. Chem. Phys. 80:1458 (1984).
4. L. S. Bartell, E. J. Valente, and J. C. Caillat, J. Chem. Phys. (in press).
5. L. S. Bartell, Chem. Rev. 86:491 (1986).
6. O. F. Hagena, Z. Phys. D (in press).
7. T. Chmielewski and P. M. Sherman, AIAAJ 4:68 (1966).
8. E. J. Valente and L. S. Bartell, J. Chem. Phys. 80:1451 (1984).
9. O. F. Hagena and H. v. Wedel, in "Rarefied Gas·Dynamics," M. Becker and M. Fiebig, ed., DFVLR, Ponz-Wahn (1974), Vol. II, F. 10-1.
10. P. P. Wegener, in "Nonequilibrium Flows," P. P. Wegener, ed., Marcel Dekker, New York (1969).
11. E. J. Valente and L. S. Bartell, J. Chem. Phys. 79:2683 (1983).
12. L. S. Bartell and Y. Barshad, J. Chem. Phys. (submitted).
13. J. Farges, M. F. de Feraudy, B. Raoult, and G. Torchet, Surf. Sci. 106:95 (1981).
14. S. S. Kim and G. D. Stein, J. Colloid Interface Sci. 87:180 (1982).
15. J. Farges, M. F. de Feraudy, B. Raoult, and G. Torchet, Ber. Bunsen-Ges. Phys. Chem. 88:211 (1984).
16. G. Torchet, H. Bouchier, J. Farges, M. F. de Feraudy, and B. Raoult, J. Chem. Phys. 81:2137 (1984).
17. G. Torchet, J. G. Schwartz, M. F. de Feraudy, and B. Raoult, J. Chem. Phys. 79:6196 (1983).
18. C. L. Yaws, "Physical Properties, a Guide," McGraw-Hill, New York (1977).
19. D. R. Stull, E. F. Westrum, and G. C. Sinke, "The Chemical Thermodynamics of Organic Compounds," Wiley, New York (1969).
20. A. R. Ubbelohde, "The Molten State of Matter," Wiley-Interscience, Chichester (1978).
21. N. G. Parsonage and L. A. K. Staveley, "Disorder in Crystals," Clarendon, Oxford (1978).
22. C. E. Klots, Phys. Ref. Lett. (submitted).
23. G. T. Furakawa, R. E. McCoskey, and M. L. Reilly, J. Res. Nat. Bur. Stds. 52:11 (1954).
24. E. Szcześniak and J. R. Brookman, Mol. Phys. 48:1221 (1983).
25. G. Torchet, J. Farges, M. F. de Feraudy, and B. Raoult, "Rarefied Gas Dynamics," R. Campargue, ed., CEA, Paris (1979), Vol. II.
26. L. S. Bartell, L. Sharkey, and X. Shi, to be published.

ANNEALING OF FINE POWDERS: INITIAL SHAPES AND GRAIN BOUNDARY MOTION

J. Bernholc
Department of Physics, North Carolina State University
Box 8202, Raleigh, North Carolina 27695

Peter Salamon
Department of Mathematical Sciences, San Diego State
University, San Diego, California 92182

R. Stephen Berry
Department of Chemistry and the James Frank Institute
University of Chicago, Chicago, Illinois 60637

ABSTRACT

This paper describes the evolution of shapes of powder particles during the first stage of sintering, assuming initial spherical shapes. Although this model has been proposed in 1949, it is solved exactly for the first time in the present paper. Several paradigmatic cases are considered, e.g. pure surface transport and pure grain boundary and volume transport, as well as the extensions of the original model to mixed transport. The influence of grain boundary tension forces on the shape of the sinter is also investigated. The new results differ substantially from earlier approximate ones for both small and large neck sizes. Based on these results we propose a new, experimentally testable mechanism for grain boundary annealing in second and third stage sinters.

I. INTRODUCTION

Many important metallic and ceramic materials are formed by annealing of fine powders (sintering) /1/. One usually divides the sintering process into three stages. In the first stage, small non-overlapping necks are formed between the powder particles. In the second stage, the necks start to overlap, but the pores are still connected. Finally, in the third stage, the pore connectivity is lost and the material attains its final density or porosity. Much of the utility of powder processing comes from the ability to tailor the density of the sinter, and thereby its weight/strength ratio or the size of its pores, to a specific application. A fascinating aspect of powder processing is the fact that the sintered matter largely preserves its shape, which can be rather intricate, during the second and third stage processing, where volume shrinkage of 30% is common. This aspect of powder processing remains unexplained to date.

Sintering occurs under the influence of surface tension forces. A simple model of first stage sinters was developed by Kuczynski in 1949 /2/. His model and its simple extensions form the basis for the analysis of sintering today. Kuczynski has also presented approximate solutions to the model, which are still widely used. Several workers attempted to improve the accuracy of these solutions, either by more elaborated analytical approximations /3/ or by brute force numerical solutions /4/. The accuracy and the range of validity of the original solutions has also been debated.

In the present paper, we give a preliminary account of the results of an exact solution of Kuczynski's model and compare the results with the original approximate solutions. We also study the extension of the model which takes into account the effects of grain boundaries at the interfaces between the particles. These grain boundaries anneal out in the second and third stage of the sintering process. We propose a new experimentally testable mechanism for their disappearance.

II. KUCZYNSKI'S MODEL AND ITS VARIATIONAL SOLUTION

Kuczynski has modeled the neck formation in first stage sinters in a simple two-sphere model (see fig 1). He assumed that for sufficiently small neck sizes the cross section of the neck can be approximated by a circle of radius rho. Two limiting cases were studied: 1) the matter filling the neck coming exclusively from the surface of the spheres, thereby decreasing the sphere radius but leading to no shrinkage for a space filling arrangement of spheres; and 2) assuming shrinkage but no surface transport, i.e. the sphere radii remained constant, but the center to center distance of the originally touching spheres diminished. In the second case the matter for the neck formation came exclusively from the overlapping parts of the spheres. The radius rho is equal to x2/2 and x2/4, respectively, for the two cases. (We set the radii of the spheres at unity, since extensions to other radii follow simply by scaling.)

Improvements to Kuczynski's solutions were proposed by several authors /3-4/. The analytical solutions centered on an improved approximation to the shape of the neck, and fully numerical solutions were also carried out. However, the analytical work still required approximations, the accuracy of which was difficult to estimate. The numerical methods required solutions of two-dimensional differential equations under a numerically imposed constraint of volume conservation, and significant numerical problems were encountered. For these reasons, a considerable uncertainty still exists in the literature concerning the true nature of the solutions even for the highly idealized two-sphere model.

The problem at hand is the minimization of the surface area of the neck for a given distance between the spheres under the constraint of preservation of volume. The Lagrangian for this problem is

$$L = f(1 + f'^2)^{1/2} + \lambda f^2 \qquad (1)$$

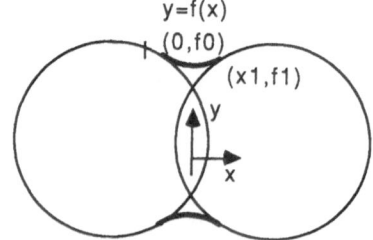

Fig. 1 Cross section of a two-sphere model of first stage sinters.

where $f(x)$ describes the shape of the surface of the neck (see fig. 1). Using the Euler-Lagrange equations we obtain

$$f^1 = [f^2/(H - \lambda f^2)^2 -]^{1/2} \qquad (2)$$

where H and λ are constants, determined from the equation

$$H = f_1^2(1 + \lambda) = f_0(1 + \lambda f_0) \qquad (3)$$

(see fig 1 for notation). The variationally determined surface obtained by revolving the graph of the function $f(x)$ about the axis connecting the centers of the spheres has constant mean curvature, as it should, since the interfacial pressure must be the same across the full neck surface. The value of the mean curvature is equal to the Lagrange multiplier λ. It should be pointed out that the approximations to the function $f(x)$ employed by Kuczynski and others do not have constant mean curvature.

The one-dimensional differential equation (2) is easily solved numerically after the constants H and lambda have been determined from equation (3). The distinction between the surface, the grain boundary and volume and the combined transport, as well as the requirements of an dihedral angle constraint enter through the boundary conditions applied to eq (2). (Time-independent calculations cannot distinguish between matter transported via grain boundary or volume diffusion.) A full description of the present solution will be published elsewhere /5/.

III. RESULTS AND DISCUSSION

We have studied several paradigmatic cases of neck formation, both for the two-sphere geometry as well as for a close-packed array of spheres. In this paper we will limit the discussion to the two-sphere problem and consider the cases of 1) pure grain boundary/volume transport, 2) pure surface transport with and without grain boundary term, and 3) a mixed case where the both surface and grain boundary/volume transport are allowed and the system can choose the thermodynamic path which will minimize the surface free energy. In all the cases presented here, the variational formulation (1)-(3) allowed us to solve the problem exactly. In this section we consider only spheres with unit radii, since the results can be trivially scaled to other sphere sizes.

We begin with the case of pure grain boundary/volume transport. In this case, the sphere radii remain fixed and the neck volume is equal to the volume of the overlapping part of the spheres as the sphere centers approach. In figure 2A we compare the true neck volume to the one calculated by Kuczynski as a function of neck thickness. It is clear from the figure that the Kuczynski approximation is qualitatively correct but quantitatively poor for this case, with relative errors ranging from 200% for the smallest neck sizes (beyond the resolution of the figure) to approximately 50% for neck sizes of the order of unity. For the case of pure surface transport Kuczynski's approximations work much better, with the relative error ranging from 50% to 2% (see figure 2B). Again, the largest relative errors are at the smallest neck sizes. This manifests itself more clearly in the analysis of the mean curvature, e.g. fig 2C shows the true mean curvature and Kuczynski's mean curvature at the center of the neck for the case of surface transport. Since the differences in the mean curvature provide the driving force for sintering (a surface with a constant mean curvature is at equilibrium), the kinetic equations derived with help of Kuczynski's solutions, which are often used to predict sintering rates in the first stage, are likely to be quantitatively inaccurate even at the very beginning of the sintering process. It is not clear, however, if the qualitative conclusions derived from the previous kinetic analyses are affected as well. We also studied the influence of grain boundary tension on the neck thickness by comparing the true neck volumes as a function of neck radius with and without a dihedral angle constraint. In figure 2D, the results with the 160 degrees dihedral angle constraint are compared to the unconstrained ones. It is clear that the presence of the grain boundary significantly affects the neck size only at the larger neck sizes.

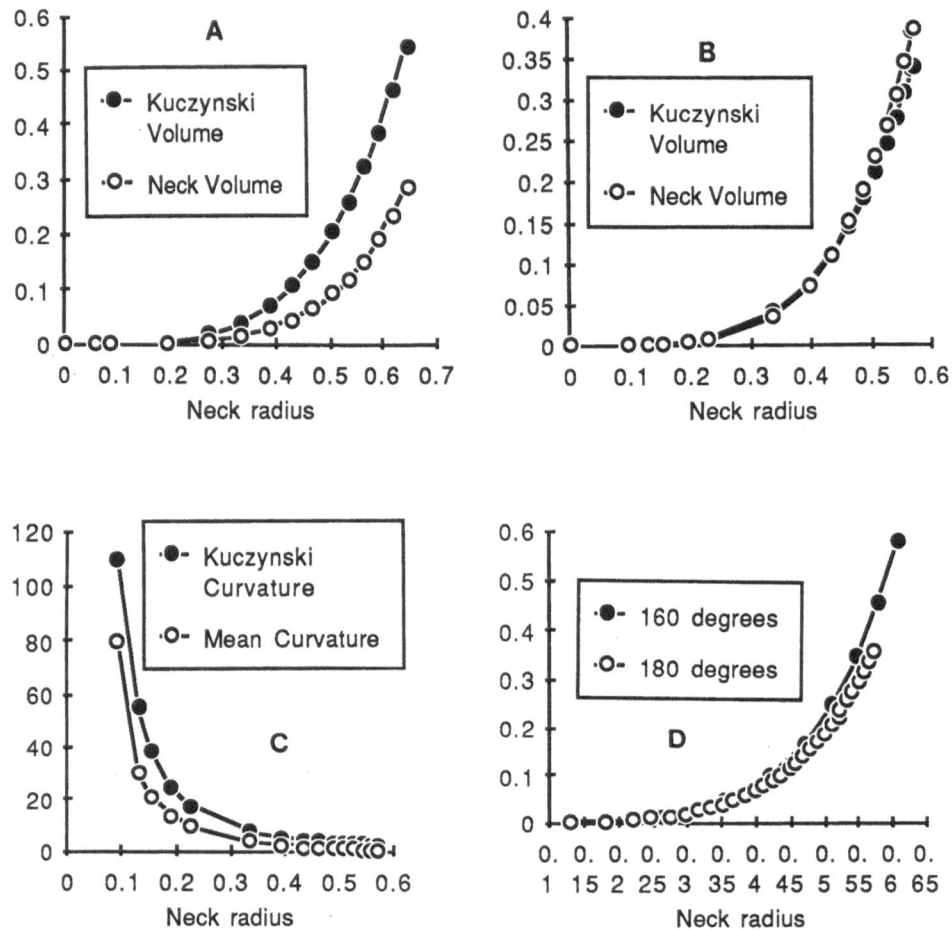

Fig. 2 A-B) Comparison of the true neck volume to Kuczynski's volume for the case of: A) pure grain boundary and volume transport, B) pure surface transport. C) Comparison of the true mean curvature with Kuczynski's mean curvature at the center of the neck. D) Comparison of the true neck volumes for the cases of 180 and 160 degrees dihedral angles.

Kuczynski's model has up to now been considered a very good approximation to the true solution for small neck sizes. This regime has been particularly difficult to test numerically, because of the smallness of the required mesh sizes in the two-dimensional integration schemes used by previous workers. Also, the preservation of volume constraint is very difficult to maintain for very small necks in the previous schemes. Only because this constraint is explicitly incorporated in our solution, and because we have reduced the numerical work to solving a one-dimensional differential equation, have we been able to obtain accurate results in this experimentally very important regime.

In order to study the more realistic case of a mixed transport, we allowed the system to follow an optimum path for minimizing the surface area (proportional to surface free energy) by either grain boundary/volume and/or surface transport. Somewhat to our surprise, the best way to reduce the surface energy for necks smaller than the diameter of the spheres was to increase the overlap of the spheres, i.e. grain boundary and volume transport is energetically most favorable. One should note, however, that the volume diffusion coefficients are smaller than the grain boundary diffusion coefficients, which in turn are smaller than the surface diffusion coefficients. The process of minimization of the surface free energy should thus be strongly dominated by kinetics. Microscopic effects of surface and interface morphology may also be important. We are planning to extend this work to these processes in the near future.

IV. A MECHANISM FOR THE DISAPPEARANCE OF GRAIN BOUNDARIES IN SINTERS

As the particles are sintered, the material solidifies, the pore space shrinks and the grain boundaries between the particles anneal out. We propose a new mechanism for the disappearance of the grain boundaries during the second and the third stage of sintering. We assume that the initial powder consists of particles of a variety of sizes and that annealing produces a broad distribution of particle sizes during the first and second stages of sintering. We then consider the interface between a large and a small particle (fig 3A). Because of the existence of the grain boundary at the interface, the free energy of the system contains a term $A\Delta f$ where A is the area of the grain boundary and Δf is the excess free energy of the grain boundary per unit area. During the initial stage of the annealing, the grain boundary is pinned between the two particles (fig 3A), since movement in any direction would increase the grain boundary area and therefore its free energy. When the neck is filled, however, the grain boundary unpins and moves towards the surface of the combined particle (fig 3B). The driving force for this motion is $dA/dx*\Delta f$, the rate of decrease of the grain boundary free energy. Depending on the magnitude of this driving force, we expect that the grain boundary motion is either frictional or diffusive in character. In the frictional case, the velocity rather than the acceleration will be proportional to the driving force, while in the diffusive case the grain boundary motion will be governed by the one-dimensional diffusion equation with a force term. We are currently in the process of calculating neck shapes and the rate of change of the surface area as a function of both the distance to the surface of the combined particles and the relative particle sizes.

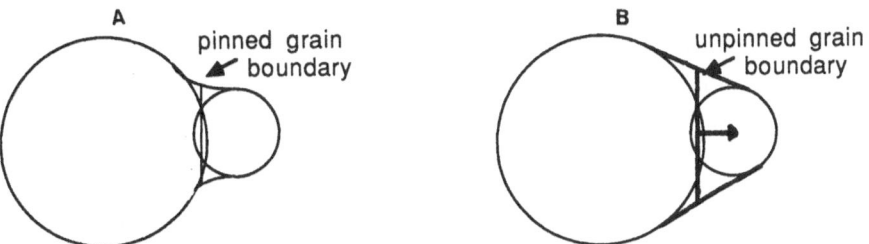

Fig. 3   Grain boundary annealing: A) The grain boundary is pinned between the two particles while the neck is being filled; B) The neck is filled and the grain boundary can anneal out.

V. SUMMARY

We have obtained a general solution for neck growth between spherical particles. The present solutions are the first ones to fully incorporate the preservation of volume constraint. The calculated neck surfaces have constant mean curvature as required for a time-independent solution. We have investigated several paradigmatic cases of first stage sintering. In particular, the solutions to Kuczynski's model show that the previous approximate solutions, which are frequently used for calculation of both neck shapes and neck formation kinetics, are only qualitatively correct for both small and large neck sizes. The relative errors are actually largest for small neck sizes, where the previous solutions were thought to be most accurate. We have also proposed a new mechanism for the disappearance of grain boundaries in sinters. Assuming that the first and second stage sintering produces a broad distribution of particle sizes, we have considered a grain boundary between a large and a small particle and shown that the grain boundary moves out to the surface under the influence of a grain boundary tension force. A testable one-parameter theory is developed from which one can calculate the rate of disappearance of grain boundaries during sintering.

REFERENCES

1. For a review see H. E. Exner and E. Arzt, p. 1886, in Physical Metallurgy, 3rd Ed., edited by R. W. Cahn and Haasen (Elsevier, Amsterdam, 1983).

2. G. C. Kuczynski, Trans. AIME 185, 169 (1949); see also R. L. Coble, J. Am. Cer. Soc. 41, 55 (1958).

3. R. M. German and Z. A. Munir, Metallurg. Trans. 6B, 289 (1975), ibid, 6A, 2229 (1975); F. B. Swinkels and M. F. Ashy, Powder Metallurg. 23, 1 (1980).

4. F. A. Nichols and W. W. Mullins, J. Appl. Phys. 36, 1826 (1965); F. A. Nichols, ibid, 37, 2805 (1966); R. M. German and J. F. Lathrop, J. Mater. Sci. 13, 921 (1978).

5. P. Salamon, J. Bernholc, R. S. Berry, M. Carerra, J. Jellinek and F. Amar, to be published.

CONTINUOUS MASS SELECTED CLUSTER ION PRODUCTION USING A LIQUID METAL ION SOURCE

S. D. Berry

AT&T Bell Laboratories
600 Mountain Avenue
Murray Hill, New Jersey 07946

ABSTRACT

A continuous source of mass-selected cluster ions is highly desirable for investigations requiring large numbers of clusters. An apparatus capable of producing a variable energy DC beam of mass-selected cluster ions and landing them on a solid substrate has been developed based on a liquid metal ion source. Clusters emitted from such a source into a cone of approximately eight degrees half-angle are focused by a multiple-element electrostatic lens into a nearly parallel beam and mass separated by a 60 degree magnetic sector to resolve individual cluster species. A second electrostatic lens/decelerator focuses the mass-resolved beam through a 0.125 inch diameter gridded aperture. The ion energy at this point is variable and can range from 0 to 2000 eV, and collection of clusters on a solid substrate is possible with landing energies ~20 eV or less. Characterization of the transport of clusters through the system is described. In addition, preliminary results have been obtained for collection of cluster ions on silicon surfaces.

INTRODUCTION

The electronic and physical structure of metal clusters has been the subject of increasing interest in recent years. The small (<200 atoms) metal cluster lies intermediate between individual atoms and molecules on one hand, and solid crystalline material on the other, and the study of electronic and physical structure as a function of increasing cluster size should be highly useful in elucidating the transition to a solid. Many of the techniques of surface analysis such as photoelectron spectroscopy or electron microscopy could be used to study small metal clusters if either a large enough volume density of clusters could be assembled in an ion trap or a high enough surface density of clusters could be collected on a substrate for observation. Recently work in this area has been reported, for example, using either partially mass-resolved clusters[1] or very small clusters[2].

The experimental study of metal clusters has been greatly enhanced by the development of the laser vaporization type of source[3] which has made possible efficient generation of clusters of a wide variety of elements and compounds. However, while it produces large numbers of clusters per pulse, it has a much lower time-averaged value. For this reason an alternative arrangement for generation and collection of clusters involving continuous production of metal clusters is desirable.

One such alternative is provided by a liquid-metal ion source (LMIS). The LMIS, chiefly known for its ability to produce finely focused high-brightness ion beams[4], also produces cluster ions at higher emission currents[5]. While the DC current of cluster ions from a liquid metal ion source can be relatively high ($\sim$nA) in a single sized cluster, the emission of cluster ions is quite broad in both energy (typical widths $\sim 100-200$ eV) and angle (emitting into a cone of about ten degrees half-angle[5]). By combining electrostatic lenses designed to handle wide beam angles with magnetic separation of individual cluster species, the latter disadvantage has been largely overcome, and relatively high DC currents of cluster ions at low final energies have been achieved.

EXPERIMENTAL APPARATUS

Figure 1 contains the schematic layout of the cluster ion apparatus. The LMIS[6], shown in the top inset, is mounted on a five-way manipulator (three translation directions and two rotations) 10 mm before a 0.125 inch aperture. The liquid tip of the source is held at a potential which determines the beam energy. The tip faces an extraction aperture which applies the electric field necessary to pull the liquid into a cone against surface tension forces and initiate the ion emission process. The operating potential of the extraction aperture is typically 3-6 kV negative with respect to the source needle tip. The grounded aperture plate which follows the extraction aperture in Figure 1 is used to measure the ion current emitted outside of the acceptance cone of the lens. The extraction potential is adjusted through feedback to keep the current collected on the aperture constant.

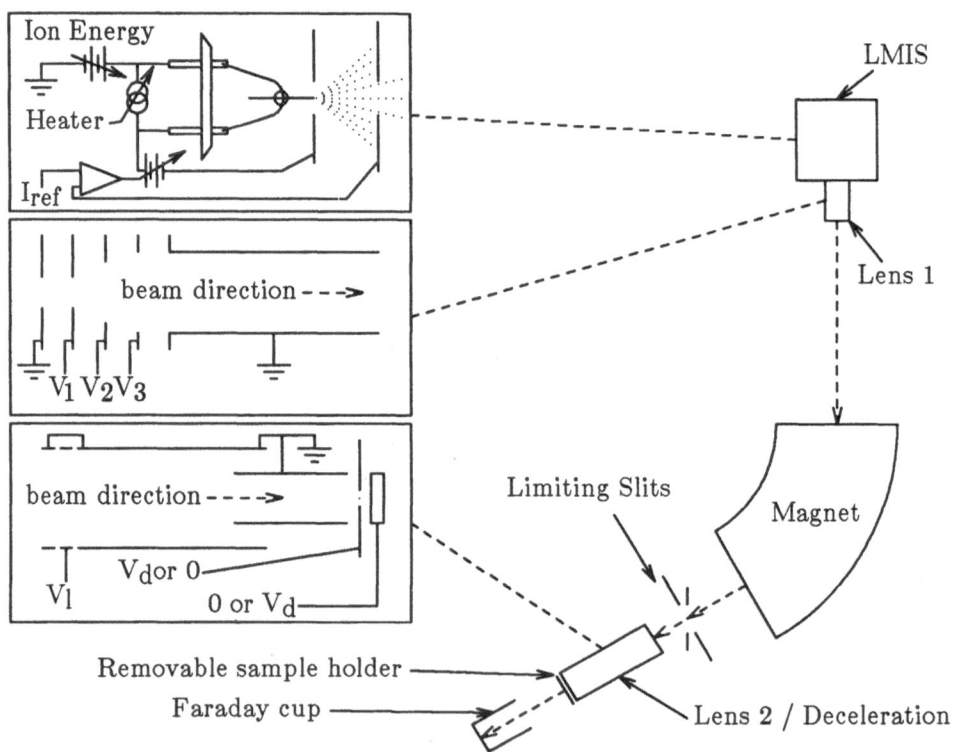

Figure 1. Schematic diagram of the mass selecting ionic cluster apparatus.

Immediately following the current aperture is a five element aperture/tube lens shown schematically in the middle inset of Figure 1. The outer aperture and final tube are held at ground, while the three inner elements are maintained at voltages of about 60-80% of the beam potential. Due to the wide angle of collection and the aberrations of electrostatic lenses at these large angles[7], this lens produces only an approximately parallel beam of cluster ions.

Following a drift region of 60 cm the ions enter a 60° bending magnet capable of field strengths to about 12 kG, with a bending radius of 45 cm. With a typical beam energy of 4.0 keV, the magnet can bend singly-ionized Ga_n^+ clusters with n as large as 50. Located at the focal point of the magnet (assuming a source at infinity) is a set of four movable collimator jaws which limits the beam size and energy spread of the selected clusters.

The second lens and deceleration system (lower inset) is located 10 cm beyond the slit system. This lens is a simple tube einzel which focuses the mass resolved cluster ions through a 0.125 inch gridded aperture. This aperture is used in two modes: in the first, for tuning purposes, a deceleration potential is placed on the aperture and the beam is reaccelerated into a Faraday cup capable of measuring currents as low as 2.0 fA. In the other mode the aperture is grounded but followed by a sample 2 mm beyond, which is held at the deceleration potential. This is the method used to collect low energy deposition samples.

RESULTS

Initial experiments were carried out to characterize the electrostatic lens elements omitting the magnetic separation. The source and first lens were as described before, with the second lens/deceleration unit placed approximately 40 cm distant in a straight line with a 0.5 inch limiting aperture located midway between the two lenses. Following the second lens unit was the Faraday cup.

Even though lens design parameters indicate that the best way to operate an electrostatic lens is in the accel/decel mode (where the charged particle is first accelerated, then decelerated in focusing), it was found that the more reliable way to operate the first lens was in the inverse mode (decel/accel). This was due to two effects - the lens elements could be operated at much lower potentials, and thereby avoid breakdown problems, and the cluster beam energy could be raised, giving a greater throughput of ions. This mode of operation was not found necessary for the second lens, however, so the lens was operated in the accel/decel mode to allow a larger entering beam size. Best values of ion throughput were found for lens voltages of about 60-80% of the beam potential, with the first active element at a somewhat higher potential than the other two.

In this test set-up, it was found that about 25-35% of the ion beam passing through the current monitoring aperture could be transported through the second lens. This was somewhat greater than initial expectations based on preliminary computer simulations of the focal properties of these lenses.

Figure 2 shows typical deceleration curves taken using the complete system incorporating magnetic deflection. The final current reaccelerated into the Faraday cup is shown versus the increasing deceleration voltage applied to the final gridded aperture. The shape of the measured curve is described very well as the convolution of a gradual defocusing effect due to beam blowup before the deceleration aperture and the energy distribution of the ions emitted by the source. For cluster ions, the results are essentially the same as for the atomic ions, taking into account the increased energy spread of the emitted cluster ions[5].

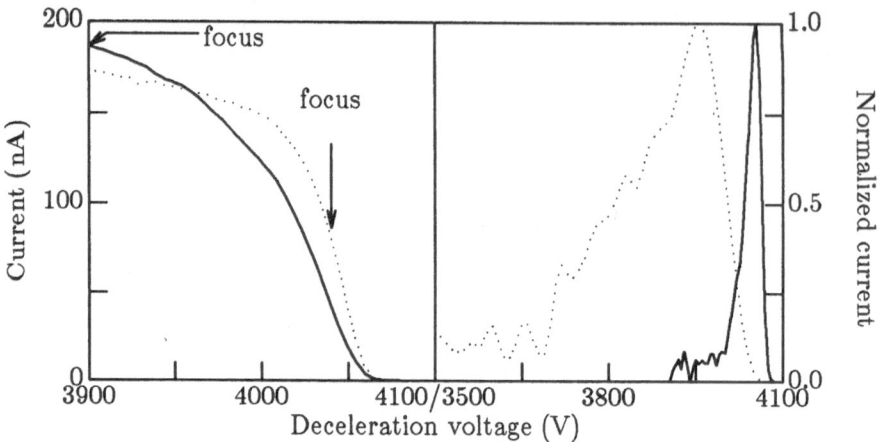

Figure 2. (a) Current versus deceleration voltage for 4.0 keV Ga^+ ions focused at deceleration voltages of 3900 V and 4040 V respectively. (b) Differential lineshape for Ga^+ (solid line) and Ga_4^+ plotted versus deceleration voltage. Curves were calculated from 4040 V curve in (a) for Ga^+ and from an equivalent curve for Ga_4^+ plotted on the same energy scale to allow comparison of the differing energy distributions. The Ga_4^+ curve used for the calculation was smoothed using a 3 point moving average to suppress source current glitches.

Collection efficiency during deceleration and landing was tested using Rutherford backscattering (RBS) to measure the surface density of the collected material. The gridded aperture was held at the final deceleration voltage to initially focus the system, and then grounded while the same deceleration voltage was applied to a silicon sample held 2 mm downstream from the aperture, collecting low energy ions for several hours. The sample was then analyzed for gallium surface coverage as a function of the position on the sample. The results of this procedure are displayed in Figure 3. The projection of the aperture approximately follows the 30% contour level.

Table I presents the results of the available current of selected cluster ions assuming either no deceleration, or a final energy of ~ 3 eV per atom. Because the present method of deceleration rejects cluster ions with energies lower than the deceleration potential, only those ions lying in the high-energy tail of the energy distribution for each species are measured. By incorporating an energy filter which can select a narrow band of energies located at the peak of the distribution, an increase of 1-2 orders of magnitude in available current should be possible.

PRELIMINARY DEPOSITION RESULTS

Using the RBS technique described above, samples were collected for both Ga_4^+ and $Ga_{\sim 28}^+$ clusters deposited at energies $\lesssim 3$ eV per atom. (The number of atoms in the larger cluster ion size is known only approximately due to the inability to completely resolve single cluster ion species for heavier clusters). Typical atomic densities of 6.9×10^{13} and 6.8×10^{12} atoms/cm^2 were found respectively for n=4 and ~ 28. This is equivalent to 1.7×10^{13} and 2.4×10^{11} clusters/cm^2, and represents cluster currents of 13 particle pA and 100 particle fA passing through the final gridded aperture. Results are still incomplete concerning the variation of surface density across the collection area for these two cluster cases.

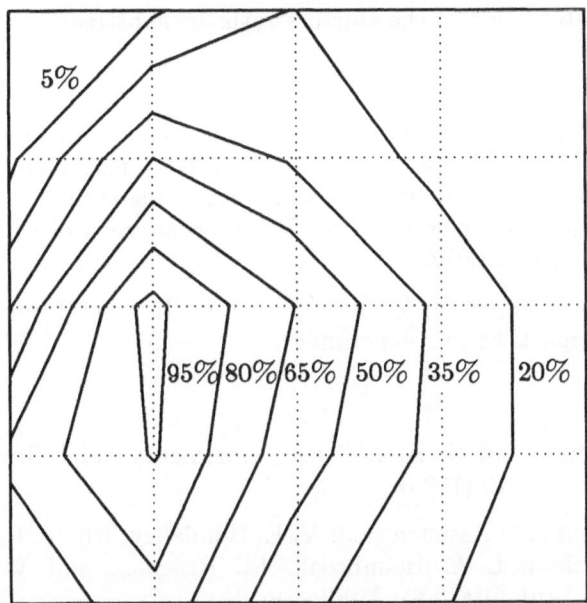

Figure 3. Contour level representation of the 5×5 array of surface density measurements obtained using RBS for 4.0 keV Ga$^+$ ions decelerated to a final energy < 20 eV. Grid line intersections mark the position of each measurement (1.0 mm spacing both horizontally and vertically) which determined the average Ga surface density over a circular area 1.0 mm in diameter.

Table 1. Available cluster ion currents.

Atoms in cluster	No deceleration voltage		Final energy ~3 eV/atom	
	Source current (μA)	Ion current (nA)	Source current (μA)	Ion current (pA)
1	20.0	218		
	80.0	304		
2	20.0	1.7	20.0	20.0
	80.0	4.0		
3	20.0	0.10	40.0	5.0
	80.0	0.77		
4	40.0	0.10	60.0	8.2
8	40.0	0.0016		
~16	20.0	0.0037	30.0	0.02
~28	60.0	0.0003	60.0	0.03

The RBS measurements do not give any information pertaining to the integrity of the collected clusters. Further analysis of similar samples by other techniques will be necessary to determine the cluster survival probability after the impact of landing and the interaction of the clusters with the substrate.

SUMMARY

In summary, an apparatus has been designed and built capable of producing and collecting low energy beams of mass resolved metal cluster ions on solid substrates. Preliminary indications are that reasonable atomic surface densities can be achieved in collection times ≤ 1 day. However, survival of clusters after landing has not yet been demonstrated.

I would like to thank W. L. Brown and D. L. Barr for fruitful discussions pertaining to all aspects of this experiment.

REFERENCES

1. G. K. Wertheim, and S. B. DiCenzo, Comments Solid State Phys. (GB) v. 11, no. 5, p. 203–19 (1985).

2. L. A. Heimbrook, M. Rasenen, and V. E. Bondebey, Chem. Phys. Lett. **120** (3), 233-38 (1985); L. A. Heimbrook, M. Rasenen, and V. E. Bondebey, J. Phys. Chem., submitted for publication.

3. P. Dhez, P. Jaegle, S. Leach, and M. Velghe, J. Appl. Phys. **40**, 2545 (1969); D. E. Powers, S. G. Hansen, M. E. Geusic, A. C. Puiu, J. B. Hopkins, T. G. Dietz, M. A. Duncan, P. R. R. Langridge-Smith, and R. E. Smalley, J. Phys. Chem. **86**, 2556 (1982); J. B. Hopkins, P. R. R. Langridge-Smith, M. D. Morse and R. E. Smalley, J. Chem. Phys. **78**, 1627 (1983).

4. R. L. Seliger, J. W. Ward, V. Wang, and R. L. Kubena, Appl. Phys. Lett. **34**, 310 (1979); W. L. Brown, T. Venkatesan, and A. Wagner, Solid State Technol. **24**, 60 (1981).

5. A. R. Waugh, J. Phys. D **13**, L203-8 (1980); S. P. Thompson and A. von Engel, J. Phys. D **15**, 925-931 (1982); D. L. Barr, to appear in Proceedings of the 30th Intl. Symp. on Electron, Ion and Photon Beams, 1986.

6. R. Clampitt, K. L. Aitken, and D. K. Jeffries, J. Vac. Sci Technol. **12**, 1208 (1975); G. L. R. Mair and A. von Engel, J. Appl. Phys. **50**, 5592 (1979).

7. The reader is referred to any elementary text on electrostatic optics.

DANGLING BONDS RECONSTRUCTION AT THE CORE OF A 90° PARTIAL DISLOCATION IN SILICON: A THEORETICAL STUDY

Aldo Amore Bonapasta, Andrea Lapiccirella
and Norberto Tomassini

ITSE - IMAI, A.d.R. National Research Council
Via Salaria Km. 29.500, CP 10 00016 Monterotondo St., Italy

Simon L. Altmann and Kenneth W. Lodge

Department of Metallurgy and Science of Materials
University of Oxford
Parks Road, OX1 3PH Oxford, England

ABSTRACT

The energetics of bond reconstruction at the core of a 90° partial dislocation in silicon has been studied by means of HFR-MO-LCAO-SCF computations on Si5H10, Si9H18 and Si10H18 model molecular clusters. These studies show the reconstructed geometry to be the most favourable one.

INTRODUCTION

In silicon and germanium, 60° dislocations, dissociated into 90° and 30° partials separated by a stacking fault, are known to play an important role in determining the structural and electronic properties of the deformed material (1-3).

In two previous papers (4,5) the structure and the associated bands of a 90° partial dislocation had been studied, however definitive conclusions had not been achieved. In the first paper the investigation was based on an empirical Valence Force Field (VFF) model and the Extended Huckel Theory - Large Unit Cell Approach (EHT-LUCA) method (4,6,7). Two different core geometries were considered in that work: one derived assuming dangling bond reconstruction, fig. 1a, and the other built on the opposite hypothesis (i.e. unreconstructed dangling bonds), fig. 1b. The energies derived for the two core geometries from the VFF treatment favour the reconstructed geometry by 2.66 eV per lattice vector, the bond-formation energy being the most important term in minimizing the total energy (4). It is desirable, however, to pay further attention to the analysis of dangling bond reconstruction within the core of the dislocation in order to confirm or otherwise the trends given by the VFF method. To this end, the methods of ab-initio molecular quantum chemistry can be considered very valuable when

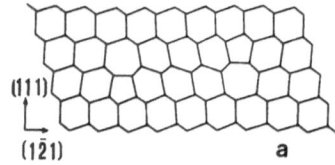 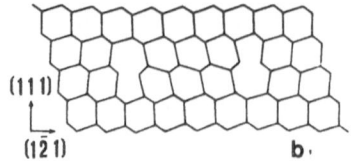

Fig. 1 (1,0,-1) projection of the 500 silicon atom clusters generated by the VFF method (see ref. 4). a) reconstructed bonding topology. b) unreconstructed topology.

dealing with structural and electronic properties of atoms localized within the dislocation region, if small cluster representing the defect (8-16) can be extracted from the lattice.

In the present paper the choise of the cluster shapes will be considered first. The reconstructed and unreconstructed cluster total energies, derived by ab-initio methods, will then be discussed.

METHODS OF CALCULATION

All the ab-initio Hartree Fock Roothan - Molecular Orbital - Linear Combination of Atomic Orbitals - Self Consistent Field (HFR-MO-LCAO-SCF) calculations were carried out by means of GAUSSIAN 80 (17). The STO-3g basis set (18) was used extensively, the more rich 4-31g one (19-20) whenever possible (see ref. 16, 21 and 22 for a discussion about the accuracy of these basis sets). The Pople and Nesbet (23) open shell treatment was employed where unpaired electrons were found.

RESULTS

Choice of the Cluster Geometry

The model geometries have been extracted from the VFF calculations performed on two different 500-atoms clusters containing two 90° partial dislocations with their Burger vectors one opposite to the other (4). Two different minimum-energy geometries, shown in figs. 1a and 1b, had been then generated on the basis of the reconstructed and unreconstructed topologies, respectively. The dangling-bond reconstruction takes place on each of the (111) planes shown in figs. 1a and 1b, which contain the pairs of silicon atoms carrying the dangling bonds. It is therefore possible to limit the search for a significant cluster shape to one of the above mentioned planes, neglecting interplanar interactions as a first approximation. The lattice strain energy has to be considered as the other important term in determining the dangling bonds reconstruction, in addition to the bond formation energy. As a consequence, three different clusters shape have been chosen (pointing attention to the two dislocation cores appearing in fig. 1) so that both energy terms appear in them. Hydrogen atoms were used as "saturators" for all the selected cluster models (see ref. 24 for a comprehensive discussion of this point). The hydrogen atoms were generated with bond and torsional angles equal to those of the parent silicon atoms in the 500-atoms clusters, whereas the Si-H distance was taken equal to that one in gaseous disilane (1.492 Å). The most important shapes will be shown selecting different sets of numbered atoms from the biggest model cluster. Figs. 2a and 2b correspond to the reconstructed and the unreconstructed structures of this cluster respectively. The shape of the first model, Si5H10, is extracted from the planes in figs. 1a and 1b by taking the sil-

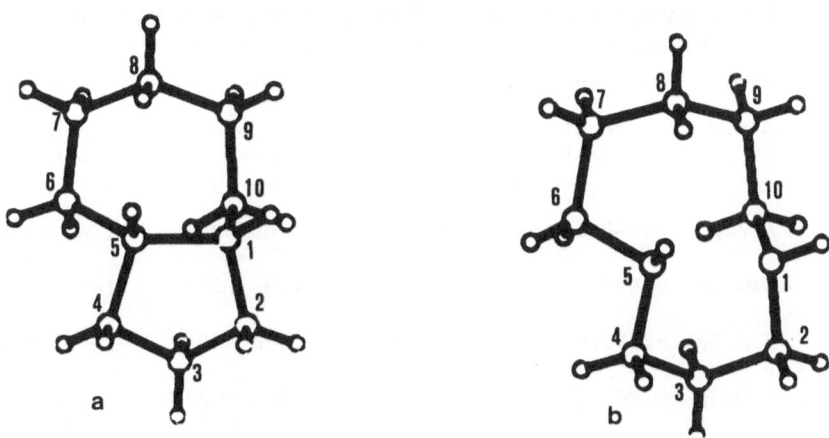

Fig. 2   Ball and stick model of the Si10H18 cluster.
          a) reconstructed model b) unreconstructed model.

icon atoms around a five-member ring. Atoms numbered 1-5 in fig. 2 represent
this model. The second model is based on the same five-member ring shape in
which the four hydrogen atoms bonded to the silicon atoms 1 and 5, (see
figs. 2a and 2b), have been substituted by four silyl fragments producing
a Si9H18 cluster. The silyl geometries have been taken, as usual, from the
500-atoms clusters. The third model, Si10H18, is based on a ten-member ring
shape extracted from the planes shown in figs. 1a and 1b by taking the silicon atoms of two adjoining five- and seven-member rings. This model, represented by all atoms in figs. 2a and 2b, includes within itself those atoms
more severely affected by deformation.

<u>Energetics of the Reconstruction Process</u>

     It can be said, on intuitive chemical terms, that the energetics of the
dangling bond reconstruction at the dislocation core is mainly determined by
the interplay of two major competing energy contributions, the elastic and
the bond-formation energies, acting on the two different reconstructed and
unreconstructed model geometries. The SCF calculations on the chosen clusters
give total energies which contain within themselves a very good description
of the bond formation term and an estimate of the strain whose accuracy is
directly related to the dimensions of the chosen models. Therefore, a good
basis for the discussion of the reconstruction problem is given by the analysis of the total-energy differences of the proposed reconstructed and unreconstructed models.

     The treatment of a bond formation problem implies the possibility that
the reacting species undergoes an electronic transition. Because of this
reason, in the present case, the total energies for the reconstructed and
unreconstructed models were computed assuming two different electronic configurations: a singlet, closed shell, configuration and a triplet, open
shell, one. Table I presents all total energy results.

     The first cluster to be discussed is the Si5H10 one, for which the
most complete set of calculations is available. In this case, the absolute
minimum energy is always coupled with the reconstructed models. Furthermore,
the singlet electronic state is the most stable one for the reconstructed
model whereas the triplet state is the favoured one in the unreconstructed
geometry. An estimate of the electron correlation energy, a factor which is

Table I. Total energies (in a.u.) of the selected model clusters for all the computational procedures carried out.

Model Cluster	Basis Set	Reconstr. Singlet	Reconstr. Triplet	Unreconstr. Singlet	Unreconstr. Triplet
Si5H10	4-31g	-1448.5691	-1448.4012	-1448.4847	-1448.4916
Si5H10	STO-3g	-1433.8448	-1433.6516	-1433.7237	-1433.7514
Si9H18	STO-3g	-2580.9433	-2580.7845	-2580.8565	-2580.8846
Si10H18	STO-3g	-2866.6009	---	-2866.5027	---

important in a bond formation problem, is in progress. Preliminary results substantially agree with the present ones (24). The actual ΔE, (E rec.- E unrec.), evaluated by taking into account the minimum energy electronic configurations, are -2.11 eV and -2.46 eV for the 4-31g and the STO-3g calculations respectively. These quantitative ΔE estimates agree among themselves and they clearly indicate that the reconstructed geometry is the favoured one. The STO-3g basis is able to reproduce the same trends generated by the more accurate 4-31g one; which means that it can be used with some confidence whenever the 4-31g calculations are unavailable.

The second batch of computations were performed on the Si9H18 cluster, fig. 4. This model has one main additional feature with respect to the Si5H10 one: the silicon atoms carrying the danglings are totally surrounded by Si-Si back bonds. The results parallel those obtained for the Si5H10 cluster: the absolute minimum energy corresponds to the reconstructed model in the singlet electronic configuration and ΔE equals -1.6 eV.

The last set of calculations have been performed on Si10H18 clusters. In this case the most strained silicon atoms, correctly contributing to the energy per (111) plane, are included in the model. Unfortunately only the STO-3g singlet computation was possible, this led to ΔE equal to -2.67 eV. It is however possible to estimate a final ΔE of -1.91 eV, thus supporting the dangling bonds reconstruction, by adding to the above mentioned value a singlet-triplet transition energy of 0.76 eV estimated by averaging the results of Table I for the unreconstructed models.

All the ΔE's computed up to now clearly indicate the reconstructed model as the energetically favourable one, but several problems must be solved before the reconstruction problem could be considered completely solved. These are mainly connected with the small size of the proposed model clusters as compared with that of the deformed solid. It is, however, possible to apply a correction for the finiteness of the proposed clusters. A reasonable approach is to obtain a rough estimate of cluster size effects by merging the present results with those coming from the VFF work (4). If one takes the 4-31g ΔE, -2.11 eV, computed for the Si5H10 cluster as a good estimate of the bond formation energy and adds to it the VFF value, 0.85 eV (4), for the difference in strain energy per lattice vector between the reconstructed and unreconstructed 500 atom clusters, one arrives to a final -1.26 eV total energy difference per lattice vector, a value which still indicates reconstruction. Manipulating in the same way the STO-3g results, one arrives at the same conclusion (total STO-3g energy per lattice vector equal to -1.6 eV). It must be noted in the above argument that the strain energy contribution of the silicon atoms within the five-members ring is accounted for twice, clearly affecting quantitatively the previous energy estimate, but if this bias were taken into account, it would modify the estimated values above by making the reconstructed model even more favourable, because the value of ΔE, -2.11 eV, used above for the band formation energy should be even more negative.

Clearly, the above-mentioned assumption, on whose basis this estimate is made, is far from accurate (the attempt to partition the total energy derived from ab initio quantum mechanical calculations into different energy contributions based on intuitive chemical concepts is always questionable), notwithstanding plausible trends can be deduced from its use. Then, it is possible to conclude that the resulting energetics confirm the hypothesis of reconstruction put forward by previous VFF calculations (4) and even the semi-quantitative agreement is satisfactory.

REFERENCES

1. P.B. Hirsch: Jour. de Physique, Paris 40, C6, 27, (1980).
2. P.B. Hirsch: Jour. of Microsc. 118, 3, (1980).
3. R. Jones: Jour. de Physique, Paris 40, C6, 33, (1979).
4. K.W. Lodge, S.L. Altmann, A. Lapiccirella and N. Tomassini: Phil. Mag. B 49, 41, (1984).
5. J.R. Chelikowsky and J.C.H. Spence: Phys. Rev. B 30, 694, (1984).
6. S.L. Altmann, A. Lapiccirella and K.W. Lodge: Int. J. Quant. Chem. 23, 1057, (1983).
7. S.L. Altmann, A. Lapicirella, K.W. Lodge and N. Tomassini: J. Phys. C 15, 5581, (1982).
8. A. Redondo, W.A. Goddard III, T.C. Mc Gill and G.T. Surratt: Solid State Commun. 20, 733, (1976).
9. A. Redondo, W.A. Goddard III, T.C. Mc Gill and G.T. Surratt: Solid State Commun. 21, 991, (1977).
10. G.T. Surratt and W.A. Goddard III: Phys. Rev. B 18, 2831, (1978).
11. K. Hermann and P.S. Bagus: Phys. Rev. B 20, 1603, (1979).
12. A.C. Kenton and M.W. Ribarsky: Phys. Rev. B 23, 2897, (1981).
13. L.C. Snyder and Z. Wasserman: Surf. Science 71, 407, (1978).
14. L.C. Snyder and Z. Wasserman: Surf. Science 77, 52, (1978).
15. L.C. Snyder, Z. Wasserman and J.W. Moskowitz: Jour. Vac. Sci. Technol. 16, 1266, (1979).
16. A. Amore Bonapasta, C. Battistoni, A. Lapiccirella, E. Semprini, F. Stefani and N. Tomassini: Il Nuovo Cimento Sez. D 6, 51, (1985).
17. J.S. Binkley, R.A. Whiteside, R. Krisnan, R. Seeger, D.J. De Frees, H.B. Schlegel, S. Topiol, L.R. Kahan and J.A. Pople: GAUSSIAN 80, QCPE Program No.406,437 Indiana University (1980).
18. W.J. Hehre, R.F. Stewart and J.A. Pople: J. Chem. Phys. 51, 2657, (1969).
19. R. Ditchfield, W.J. Hehre and J.A. Pople: J. Chem. Phys. 54, 724, (1971).
20. L.C. Snyder and Z. Wasserman: Surf. Science 71, 407, (1978).
21. C. Battistoni, A. Lapiccirella, E. Semprini, F. Stefani and N. Tomassini: Il Nuovo Cimento Sez. D 3, 663, (1984).
22. J.A. Pople: in Application of Electronic Structure Theory edited by H.F. Schefer III pag. 1, (1977), Plenum Press New York and London.
23. J. A. Pople and R.K. Nesbet: J. Chem. Phys. 22, 571, (1954).
24. To be published.

ISOLATION OF MONODISPERSED GOLD CLUSTERS OF CONTROLLED SIZE

Eugine Choi and Ronald P. Andres

School of Chemical Engineering
Purdue University
Lafayette, IN 47907

ABSTRACT

Transmission electron micrographs of 1.9 nm diameter Au clusters deposited on an amorphous carbon surface yield direct verification of the narrow size distribution of clusters generated using a Multiple Expansion Cluster Source (MECS).

INTRODUCTION

The catalytic activity and selectivity of ultra-small supported metal clusters, containing only a few atoms, are expected to be sensitive both to the number of atoms in the cluster and to the degree of interaction between the clusters and their support. The ideal way of studying these two effects would be to produce model catalysts for which: 1) cluster size is controlled and uniform, 2) clusters are supported on a well characterized surface, and 3) surface coverage is controlled to prevent cluster-cluster interaction.

The Multiple Expansion Cluster Source (MECS) being developed at Purdue University provides a flexible method for producing an aerosol of ultra-small metal clusters with controlled size and narrow size distribution[1,2]. This source can be adapted for production of model catalysts by expanding a sample of the aerosol into a vacuum chamber to produce a cluster beam and by directing this beam onto a flat substrate. Previous attempts to make model catalysts by this technique, however, have resulted in formation of raft-like cluster aggregates on the support[2,3]. The present study explores the conditions necessary for isolation of individual MECS clusters on an amorphous carbon surface.

APPARATUS

The MECS is composed of three sections (see Figure 1): 1) a oven for evaporating the metal, 2) a quench zone for rapid cooling of the metal vapor and 3) a condensation reactor for growing the clusters.

The metal is evaporated in a resistively heated carbon tube, which is pressurized with He. The evaporated metal atoms and diatoms diffuse through a stagnant region in the oven, are entrained in the He carrier and expand sonicly into the quench zone. The rate of metal evaporation is controlled by controlling the temperature and pressure in the oven. Oven temperature determines the equilibrium vapor pressure above the metal pool, while oven pressure controls the rate of diffusion through the stagnant region between the metal pool and the oven orifice.

In the quench zone, the hot flow from the oven is mixed rapidly with room temperature He or Ar to cool the mixture. Ar quench gas was used in making Au clusters for the present study. The cool mixture once again expands sonicly into the condensation reactor where cluster growth occurs.

In the condensation reactor, metal clusters are formed by condensation of metal atoms onto the diatoms present in the mixture flowing from the quench zone. The extent of cluster growth in this fast flow aerosol reactor is controlled by throttling the mechanical pump (7.2 liter/sec) that evacuates the system, thereby, varying the pressure and the residence time in the reactor. Changing the reactor pressure does not affect the conditions in the quench zone or the oven because of the sonic expansions that separate the three regions. Thus, metal evaporation, metal cooling, and cluster growth can be controlled independently for optimizing cluster production and for controlling cluster size.

At the exit of the condensation reactor, a small portion of the total flow is sampled through a capillary leak into a vacuum chamber maintained at 10^{-4} to 10^{-6} torr. This final expansion stops cluster growth and also forms a uniform velocity cluster beam. The inert gas flow through the capillary is handled by a large oil diffusion pump (70,000 liter/sec).

The cluster beam is intercepted by a film thickness monitor (Inficon XTM), which records the mass arrival rate at its surface. A deceleration cell filled with He gas surrounds the film thickness monitor. This deceleration cell serves two important functions: 1. rejection of uncondensed monomer from the cluster beam and 2. characterization of mean cluster size. By increasing cell pressure successively larger clusters can be prevented from reaching the detector surface.

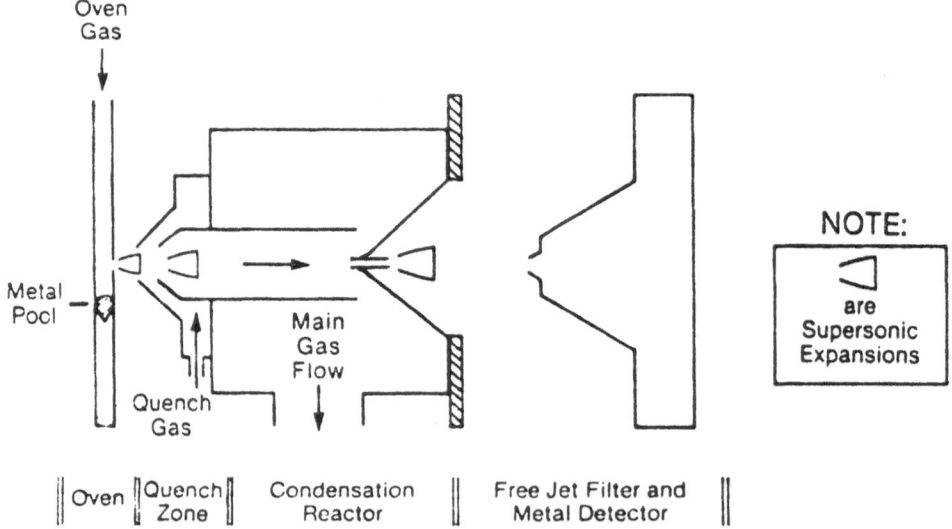

Figure 1. Schematic Diagram of MECS

Sample deceleration data are shown in Figure 2. The low pressure
portion of the curve, where detector signal drops rapidly with pressure,
is due to the deceleration and rejection of uncondensed metal atoms. This
is followed by a portion of the curve where detector signal remains
constant. In this pressure region the uncondensed monomer has been
eliminated from the cluster beam, but all the clusters still reach the
detector. The last part of the curve corresponds to cluster rejection.
Cluster deceleration has been modeled by Nguyen and Andres[4]. Their model
provides a quantative relationship between the mean stopping pressure and
the mean cluster diameter.

Supported samples are prepared by moving the film thckness monitor
out of the beam path and exposing a clean substrate surface to the cluster
beam. The deceleration cell pressure is always adjusted to reject
uncondensed monomer before supported samples are made.

EXPERIMENT

Au clusters were deposited on amorphous carbon for study by TEM. The
clusters were made with the following operating conditions: 1) oven
pressure = 100 torr, 2) oven temperature = 2100 °K, 3) initial reactor
pressure = 2 torr, and 4) throttled reactor pressure = 18 torr. Ar quench
gas and He oven gas at a 4 to 1 molar ratio were used. The substrates
were thin amorphous carbon films (approximately 3 to 5 nm thick). These
films were prepared by evaporating carbon onto a freshly cleaved NaCl
crystal. The thin films were transferred to nickel TEM grids which had
been covered with a thick carbon film (approximately 20-40 nm thick)
containing numerous holes (approximately .5 um in diameter)[5]. The TEM
micrographs were made using a JEM 200 at 210,000 times magnification.

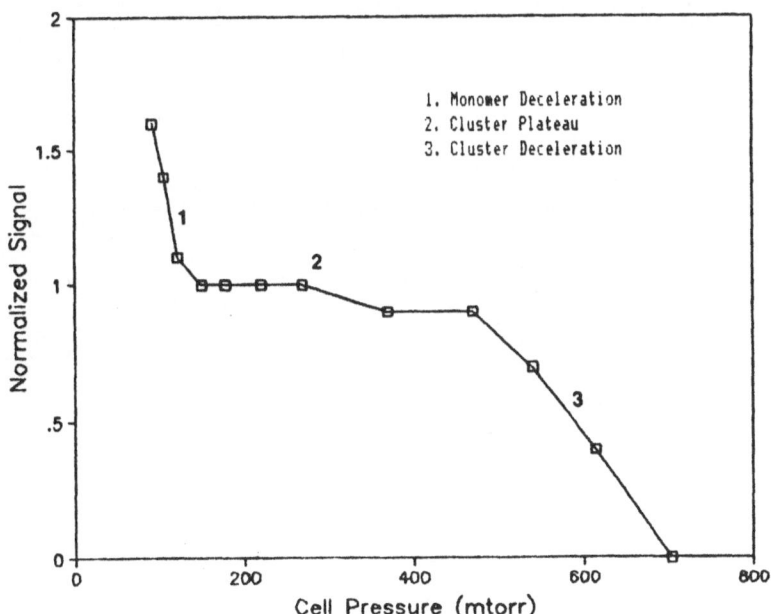

Figure 2. Deceleration Curve

RESULTS AND DISCUSSIONS

At an atomic coverage of 4.9×10^{13} atoms/cm^2, 2.4×10^{11} clusters/cm^2 were observed on the carbon surface. A photomicrograph (Figure 3) shows that the clusters are all approximately the same size. The mean cluster diameter obtained from this micrograph is 1.9 nm. This agrees with the mean cluster diameter found by measurement of the mean deceleration pressure (1.9 nm) and that found by assuming that the clusters, which on the average contained 210 atoms/cluster, are spheres having the same density as bulk Au (1.9 nm).

A cluster size distribution obtained from the photomicrograph in Figure 3 is presented in Figure 4. This experimental distribution is well modeled by a normal distribution with a mean of 1.9 nm and a variance of 0.37 nm. The width of this distribution is due to two effects: 1) the underlying distribution in cluster size and 2) the uncertainty in determining cluster diameter from a photomicrograph image. The latter is believed to contribute a variance of as much as 0.25 nm to the experimental distribution.

The 1.9 nm Au clusters remain isolated up to a coverage of about 5×10^{11} clusters/cm^2 (1.0×10^{14} atoms/cm^2). Larger cluster agglomerates begin to form when the coverage is higher. The growth of the agglomerate particles of both Ag and Au has been studied on amorphous carbon[5]. Two tentative conclusions can be drawn from these studies: 1) Ag clusters are substantially more mobile than are Au clusters and 2) the larger the cluster the greater its mobility in the size range of 5 to 250 atoms/cluster.

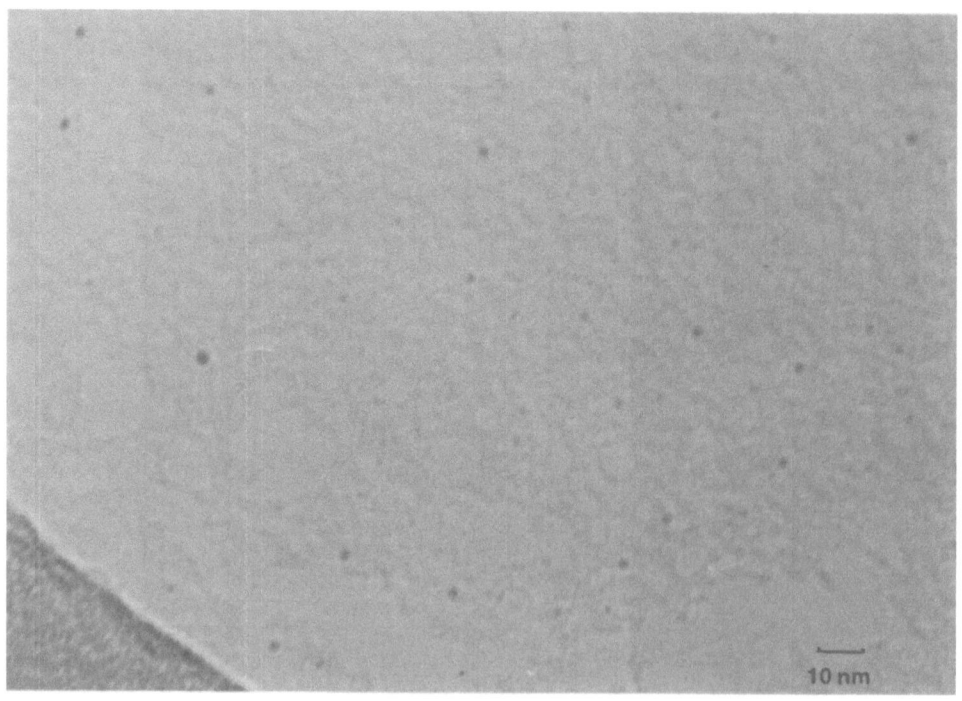

10 nm

Figure 3. Electron Micrograph of 1.9 nm Au Cluster Deposition on Amorphous Carbon (Exp. #60327)

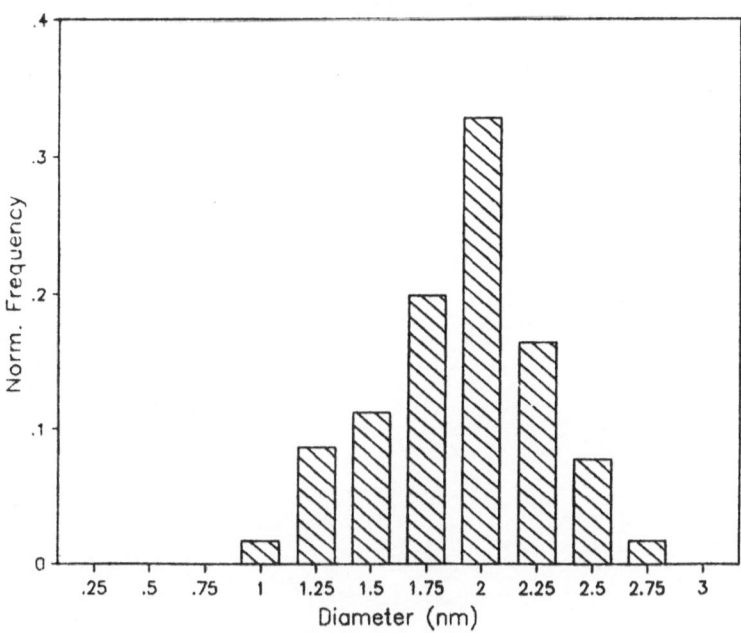

Figure 4.   Au Cluster Size Distribution (Exp. #60327)

CONCLUSIONS

Transmission electron micrographs of 1.9 nm diameter Au clusters produced by a MECS and deposited on an amorphous carbon surface show that isolated single clusters can be obtained by cluster beam deposition when the coverage is below 5×10^{11} clusters/cm^2. The size distribution of the supported clusters is narrow as expected. At higher coverage, raft-like aggregate particles are formed on the surface.

ACKNOWLEDGMENTS

This research was supported by the National Science Foundation thru grant DMR 87-16988 of the MRL program. The electron micrographs were made with the assistance of Professor N. Otsuka and Dr. B. P. Gu.

REFERENCES:

1. R. S. Bowles, J. J. Kolstad, J. M. Calo, and R. P. Andres, Generation of Molecular Clusters of Controlled Size, Surface Science, 106:117 (1981).
2. R. S. Bowles, S. B. Park, N. Otsuka, and R. P. Andres, Generation of Supported Metal Clusters of Controlled Size, J. Molecular Catalysis, 20:279 (1983).
3. M. Listvan and R. P. Andres, Production and STEM Examination of Controlled-size Silver Cluster, in:"Proceedings of 41st Annual Meeting of the Electron Microscopy Society of America," G. W. Bailey, ed., San Francisco Press, Inc., San Francisco, 338-339 (1983).
4. T. K. Nguyen and R. P. Andres, Fokker-Planck Description of the Freejet Deceleration Flow, in:"Rarefied Gas Dynamics, Progress in Astronautics and Aeronautics 74, Part 1," S. S. Fisher, ed., American Institute of Aeronautics and Astronautics, New York, 627 (1981).
5. E. Choi, M. S. Thesis, School of Chemical Engineering, Purdue University, (1986).

STRUCTURES OF C_5 AND C_6

David W. Ewing and Gary V. Pfeiffer

Department of Chemistry, John Carroll University
Cleveland, Ohio 44118
Department of Chemistry, Ohio University
Athens, Ohio 45701

INTRODUCTION

Recent experiments involving laser vaporization of graphite, followed by supersonic expansion,[1-4] have rekindled interest in the structures of small carbon clusters. While theoretical calculations have firmly established the structure(s) of C_4,[5-8] relatively little theoretical work has thusfar been reported for C_5 and C_6, most such studies being small model calculations on diamond and graphite. There has been, however, a MINDO/2 geometry search on small carbon clusters which included C_5 and C_6[9]. In that work C_5 was found to be trigonal bipyramidal, and C_6 was found to be a distorted octahedron of C_{2v} symmetry. The available experimental data, from matrix isolation studies, indicate that C_5 and C_6 are linear.[10-11]

METHODS OF CALCULATION

Calculations were performed using the Gaussian 82 system of programs.[12] RHF calculations were done for singlet states; UHF calculations were done for higher multiplicities. The D95 and D95* carbon basis sets, which are stored internally in Gaussian 82, were utilized. The D95 basis set is Dunning's (9, 5)→[4,2] double-zeta basis set.[13] The D95* basis set augments D95 with a set of single primitive d orbitals having exponents of 0.75. Six component d- functions were employed.

Geometries were first optimized at the Hartree-Fock level of theory using the D95 basis set (HF/D95), then refined at HF/D95*. Berny optimizations were employed throughout. For the four lowest energy isomers of C_5, electron correlation was included via full(core orbitals included) second order Moller-Pleset perturbation theory, using the D95* basis set at the D95* optimized geometry. This level of theory is referred to as MP2/D95*. For many of the clusters studied here, ground state electron configurations had previously been determined.[14,15] Various states of all clusters were investigated at the HF/D95 level to determine the ground state.

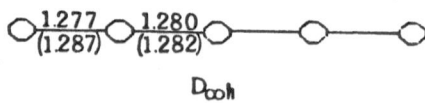

$D_{\infty h}$

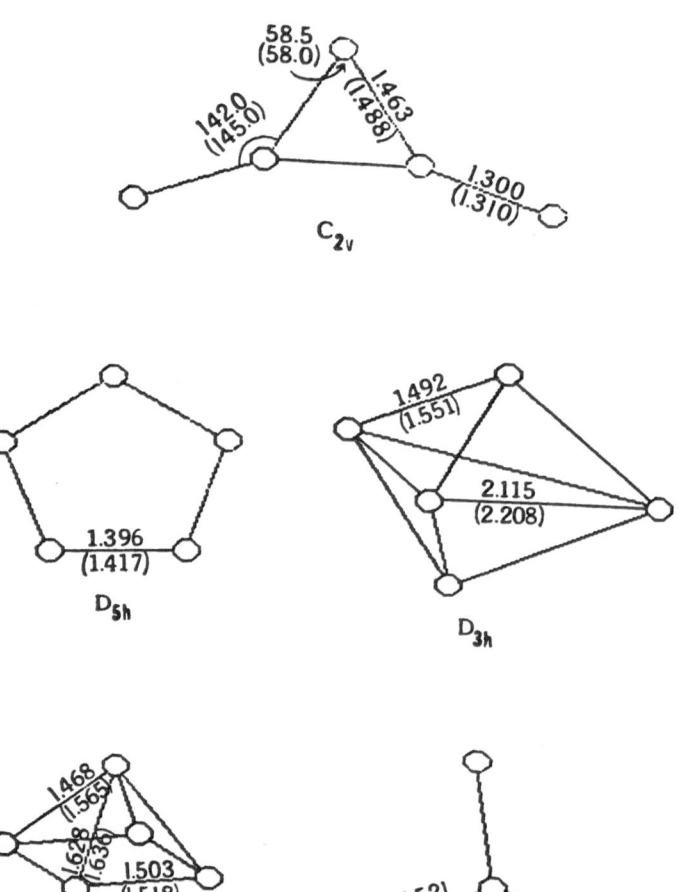

Figure 1. Structures of C_5 showing D95* and D95 (in parentheses) optimized geometries. Distances are in Å. Angles are in degrees.

Table 1. Total electronic energies (a.u.) for optimized C_5 clusters.

Cluster[a]	HF/D95[b]	HF/D95*[b]	HF/D95*[c]	MP2/D95*[c,d]
$D_{\infty h}$	−188.96351	−189.03440	−189.03468	−189.69468
C_{2v}	−188.84503	−188.93774	−188.93886	−189.56357
D_{5h}	−188.85348	−188.96077	−188.96287	−189.54882
D_{3h}	−188.68863	−188.85604	−188.86702	−189.54976
C_{2v}'	−188.68557	−188.82354	−188.83860	−−
T_d	−188.63882	−188.68418	−188.68440	−−

[a]See Figure 1.
[b]HF/D95 geometry.
[c]HF/D95* geometry.
[d]Core orbitals included in the correlation calculation.

Table 2. Electronic configurations and spin states for the ground states of C_5 clusters.

Cluster[a]	Configuration	Spin State
$D_{\infty h}$	$\ldots(1\pi_u)^4(6\sigma_g)^2(5\sigma_u)^2(1\pi_g)^4$	$S = 0$
C_{2v}	$\ldots(1b_1)^2(7a_1)^2(5b_2)^2(8a_1)^2(1a_2)^2$	$S = 0$
D_{5h}	$\ldots(3a_1')^2(1a_2')^2(3e_1')^2(3e_2')^2(1e_1'')^2$	$S = 3$
D_{3h}	$\ldots(2a_2'')^2(1e'')^4(4a_1')^2(3e')^4(5a_1')^2$	$S = 0$
C_{2v}'	$\ldots(1a_2)^2(7a_1)^2(3b_1)^2(3b_2)^2(8a_1)^1(4b_2)^1$	$S = 1$
T_d	$\ldots(4a_1)^1(3t_2)^3(1e)^2(4t_2)^3(1t_1)^3$	$S = 6$

[a]See Figure 1.

C_5 CLUSTERS

Figure 1 depicts the various isomers of C_5 investigated. These are labeled according to molecular point groups. Since there are two clusters which have C_{2v} symmetry, these are labeled C_{2v} and C_{2v}'. Shown on the figure are the D95* and D95 (in parentheses) optimized geometries. As expected, including polarization functions in the basis set shortened all bond lengths. For the three most stable isomers, $D_{\infty h}$, C_{2v}, and D_{5h}, all geometric parameters were allowed to relax.

Table 1 gives total electronic energies for the optimized clusters shown in Figure 1. Energies for D95* calculations done at D95 geometries are included in Table 1 to show that nearly all of the increased stability in the D95* optimized clusters comes from the inclusion of polarization functions, rather than from the reoptimization of the clusters. MP2 calculations were not performed for the two lowest energy isomers.

It is concluded from Table 1 that C_5 is linear. The linear form is found to be the most stable isomer at every level of theory employed here. At the MP2/D95* level it is lower in energy than any other isomer by more than 60 Kcal/mol. This result has been confirmed by very recent <u>ab initio</u> calculations by Raghavachari.[16] In contrast to C_5, C_4 has two stable forms, linear and rhombic.[5-8]

Table 2 gives the ground state electronic configurations and spin states of the C_5 clusters depicted in Figure 1. The energies reported in table 1 were calculated with these electronic configurations. The method of determining these ground states has been discussed elsewhere.[14].

The only other geometry search on C_5 at the time of this writing is a MINDO/2 study.[9] While linear C_5 was not predicted to be the most stable isomer, the bond lengths calculated in that study, 1.25 Å (outer C=C) and 1.29 Å (inner C=C), are in good agreement with the present work (Figure 1). As has been discussed elsewhere, linear carbon molecules all have nearly equal bond lengths, consistent with the structure for C_5 of :C=C=C=C=C:.[15]

C_6 CLUSTERS

Figure 2 depicts the various isomers of C_6 investigated. The style of Figure 2 is like that of Figure 1. Table 3 gives total electronic energies for the optimized clusters shown in Figure 2. MP2 calculations were not performed on any of the C_6 clusters.

The data in Table 3 indicate that C_6 is linear. Distortions of these perfect geometries have not been explored, however, as was done for C_5. There may also be more geometries of C_6 to try. More work on C_6 is needed. Indeed Raghavachari has very recently found that a singlet state of cyclic C_6 is lower in energy than linear C_6.[16]

Table 4 gives the ground state electronic configurations and spin states of the C_6 clusters depicted in Figure 2. The energies reported in Table 3 were calculated with these electronic configurations.

CONCLUSIONS

It has been shown here that C_5 is linear. This is in agreement with the interpretation of recent matrix isolation data.[11]

ACKNOWLEDGMENTS

The authors wish to thank Mr. Robert Kozel and the staff of the John Carroll University Computer Center for their technical assistance. Mrs. Elaine DiCillo is gratefully acknowledged for her help with this presentation.

Table 3. Total electronic energies (a.u.) for optimized C_6 clusters.

Cluster[a]	HF/D95[b]	HF/D95*[b]	HF/D95*[c]
$D_{\infty h}$	−226.78377	−226.87954	−226.87988
D_{6h}	−226.66201	−226.76592	−226.76682
D_{4h}	−226.33880	−226.57558	−226.60171
D_{3h}	−226.25274	−226.37634	–

[a] See Figure 2.
[b] HF/D95 geometry.
[c] HF/D95* geometry.

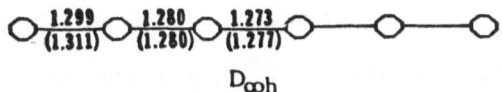

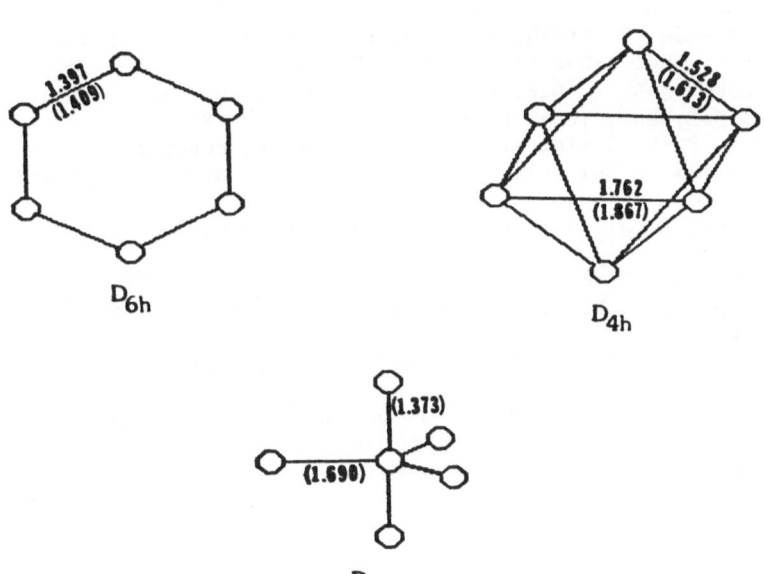

Figure 2. Structures of C_6 showing D95* and D95 (in parentheses) optimized geometries. Distances are in Å.

Table 4. Electronic configurations and spin states for the ground states of C_6 clusters.

Cluster[a]	Configuration	Spin State
$D_{\infty h}$	$\ldots(6\sigma_u)^2(7\sigma_g)^2(1\pi_g)^4(2\pi_u)^2$	$S = 1$
D_{6h}	$\ldots(3a_{1g})^1(1b_{2u})^2(1a_{2u})^2(3e_{1u})^2$ $(2b_{1u})^1(3e_{2g})^2(1e_{1g})^4$	$S = 3$
D_{4h}	$\ldots(2a_{2u})^2(2b_{2g})^2(4a_{1g})^2(1e_g)^4$ $(1b_{1g})^2 (5a_{1g})^2(3e_u)^4$	$S = 0$
D_{3h}	$\ldots(6a_1')^2(3e')^4(3a_2'')^2(1e'')^4(4e')^2$	$S = 1$

[a]See Figure 2.

REFERENCES

1. E. A. Rohlfing, D. M. Cox and A. Kaldor, J. Chem. Phys. _81_, 3322 (1984).
2. L. A. Bloomfield, R. R. Freeman and W. L. Brown, Phys. Rev. Lett. _54_, 2246 (1985).
3. L. A. Bloomfield, M. E. Geusic, R. R. Freeman and W. L. Brown, Chem. Phys. Lett. _121_, 33 (1985).
4. M. E. Geusic, T. J. McIlrath, M. F. Jarrold, L. A. Bloomfield, R. R. Freeman and W. L. Brown, J. Chem. Phys. _84_, 2421 (1986).
5. R. A. Whiteside, R. Krishnan, D. J. Defrees, J. A. Pople and P. von R. Schleyer, Chem. Phys. Lett. _78_, 538 (1981).
6. A. V. Nemukhin, A. I. Demeat'ev, A. I. Kolesnikov, N. F. Stepanov and V. I. Polyakov, Teor. Eksp. Khim. _19_, 715 (1983).
7. Z. Z. Wang, R. N. Diffenderfer and I. Shavitt, paper presented at 39th Symposium on Molecular Spectroscopy (Ohio State University, (1984).
8. D. H. Magers, R. J. Harrison, and R. J. Bartlett, J. Chem. Phys. _84_, 3284 (1986).
9. Z. Slanina and R. Zahradnik, J. Phys. Chem. _81_, 2252 (1977).
10. K. R. Thompson, R. L. DeKock, and W. Weltner, Jr., J. Am. Chem. Soc. _93_, 4688 (1971).
11. W. Kratschmer, N. Sorg and D. R. Huffman, Surf. Sci. _156_, 814 (1985).
12. J. S. Binkley, M. J. Frisch, D. J. DeFrees, K. Raghavachari, R. A. Whiteside, H. B. Schelgel, E. M. Fluder and J. A. Pople, Department of Chemistry, Carnegie-Mellon University.
13. T. H. Dunning, Jr., J. Chem. Phys. _53_, 2823 (1970).
14. D. W. Ewing, Ph.D. Thesis, Ohio University (1985).
15. D. W. Ewing and G. V. Pfeiffer, Chem. Phys. Lett. _86_, 365 (1982).
16. K. Raghavachari, J. Chem. Phys. (in press).

THE STRUCTURE OF SMALL GOLD CRYSTALS

Peiyu Gao and H. Gleiter*

Fritz-Haber-Institut der Max-Planck-Gesellschaft
Faradayweg 4-6, D-1000 Berlin 33
*Universität des Saarlandes, D-6000 Saarbrücken

ABSTRACT

The structure of small gold crystals with diameters between 2 and
6 mm generated by the inert gas evaporation technique (GET) has been investigated by high resolution electron microscopy. The structures observed range from a single crystal (about 56%) to singly or doubly twinned
particles and multiply twinned particles containing partial dislocations.
The tendency to form perfect single crystals increased with decreasing
size. The size-dependence of the microstructure observed may be caused
by particle collision and controlled by the energy of the grain boundaries
formed between colliding crystals.

INTRODUCTION

The structure of small particles is significant for the understanding
of the nucleation of thin films and of catalysts as well as for explanation of the particles' properties. Since Allpress and Sanders [1] and
Mihama and Yasuda [2] observed abnormal 111 spots in electron diffraction
patterns of epitaxial Au films, many results concerning the structure of
such small metal particles have been published in the last two decades
[3-11]. The results obtained may be summarized as follows. Small (about
20 nm) metal particles may be singly or multiply twinned [3,4,5] involving
parallel or radial twins. The generation of these twinned structures may
be visualized in terms of the interaction of small nuclei formed before
crystal coalescence occurred. For very small (about 3 nm) particles of
noble metals, the icosahedral structure seems to be of lower energy than
the fcc structure [12-14]. The misfit strains associated with the growth
of icosahedral particles may result [8,9] in a uniformly distorted crystal
lattice structure (rhombohedral or orthorhombic instead of fcc), or in the
incorporation of dislocations [9], stacking faults [11] or gaps [10].

One of the problems associated with growing small particles epitaxially
on single-crystal substrates is the uncertainty about the substrate effects
on the initial structure and morphology. Furthermore, the majority of particles investigated so far were larger than 5 nm, giving limited information about the early stages of growth. The purpose of the present study
was to overcome these limitations by investigating the structure and morphology of gold particles with a diameter of 2-6 nm generated by the inert
gas evaporation technique, GET.

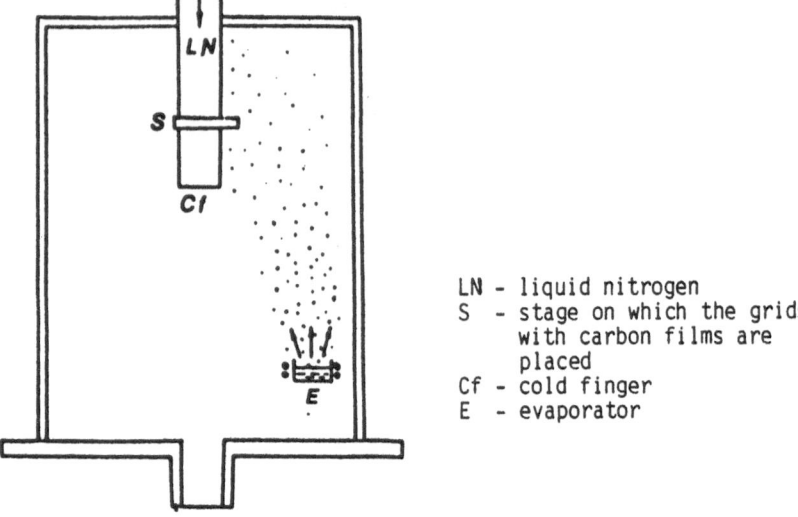

LN - liquid nitrogen
S - stage on which the grids
     with carbon films are
     placed
Cf - cold finger
E - evaporator

Fig. 1   Schematic drawing of arrangement used to obtain
          the fine gold particles

To obtain information about the fine structure of very small particles
directly from electron diffraction patterns, several efforts had been
made [15,16], but there are still some difficulties. The phase grating
approximation for the crystal specimen and electron optical image formation
theory have been applied to study the relation between the electron diffraction pattern and the optical diffraction pattern obtained from the
electron micrograph of the crystal specimen. For very thin objects
Cowley [17] proved that the electron wave may be considered to travel
straight through the object undergoing only a phase change proportional
to the electrostatic potential in the object. Fortunately, in the last
15 years, the development of electron optics in both theory and practice
provided a useful method for identifying the fine structure of ultrafine
small particles [18-27]. It has proved that the optical diffraction patterns and Fast Fourier Transform (FFT) patterns of an atomic resolution
electron micrograph taken under optimum conditions are equivalent to the
selected area electron diffraction patterns. Even when the specimen is
not very thin and dynamic scattering cannot be disregarded, the diffraction spots formed by the FFT represent the spacing of the lattice plane
[27].

EXPERIMENT

The scheme of the device for producing particles with a diameter of
2-6 nm by GET is shown in Fig. 1. Gold (99.99%) was evaporated from an
electrically heated tungsten boat into a 99.9996% helium atmoshpere at
low pressure (0.5 KPa). The particles formed in the helium atmosphere
were collected on a carbon-coated grid (carbon film thickness 4 nm) and
were stored at 77 K.

The gold particles were investigated in the DEEKO-100 electron microscope (100 kV) of the Fritz-Haber-Institut of the Max-Planck-Gesellschaft,
West Berlin. The DEEKO-100 electron microscope has a shperical aberration
of 0.5 mm and a point-to-point (line) resolution of better than 0.2 nm
(0.14 nm) respectively. The micrographs were recorded as through-focal
series on the TV monitor of an image intensifier. Imaging processing
was done by digitizing the micrographs on a VAX computer system. The

resolution limit in the computer system is 0.068 nm for one pixel. All digitized electron micrographs shown are taken at underfocus of 3 units of Scherzer focus, e.g. 129 nm.

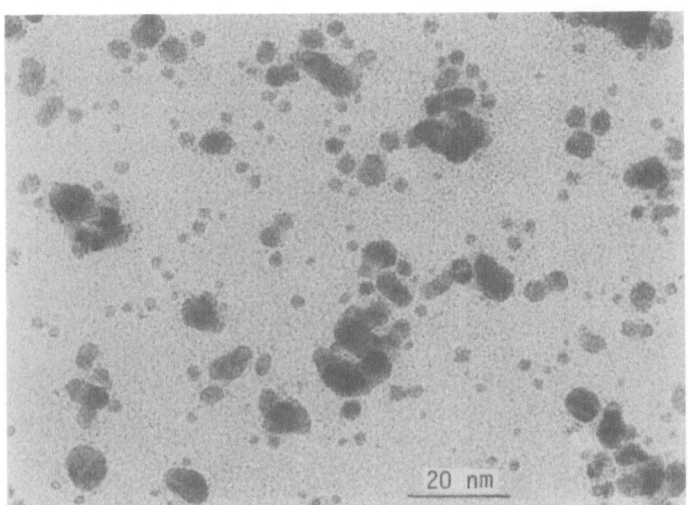

Fig. 2   TEM micrograph of gold particles produced by GET

Fig. 3   High resolution electron micrograph of small gold
           particles, size less than 3 nm, exhibiting single
           crystals with a micro-twin-band indicated by arrow

RESULTS

Electron diffraction experiments gave little evidence for any pre-ferred orientation of the Au particles with respect to the carbon film. The size of the particles varied from 1.5 nm to about 20 nm (Fig. 2).

High resolution electron micrographs of particles with diameters of 2 nm to 6 nm were recorded and subsequently processed in the VAX computer system. The position of the atomic column in the HREM can be recognized as a white spot. Among the 263 particles, the microstructure of which was studied, about 56% had a single crystal morphology, about 42% were singly or doubly twinned; only a few (about 1%) of them were of the multiply

a   b

c   d

Fig. 4   High resolution electron micrograph of small gold
particles, approximate size less than 4 nm, and their
diffraction patterns by FFT indicating single and
double twin

twinned type. A large proportion (about 80%) of smaller particles (size
about 2 nm) exhibited a single crystal structure. For the large particles
(size about 20 nm) no single crystal structure has been observed. In Fig.3
a particle containing a single layer micro-twin-band may be seen both in
the image and in the corresponding diffraction pattern. Several examples
of singly and doubly twinned particles are shown in Fig. 4. In Fig. 4a
a few layers of twinned atoms touched on a main part of the particle. The
multiply twinned particles have mostly been observed in the particles
larger than 5 nm in diameter. One of the multiply twinned particles is
shown in Fig. 5. This particle contained a partial dislocation (arrow).
The angle between the two spots of 111 plane nearby on the corresponding
diffraction pattern in Fig. 5 is calculated to be 7°34' which is consistent
with model calculations [3,10].

DISCUSSION

The experimentally observed relationship between the size of the gold
particles and their microstructure may be understood as follows. Previous
experiments [28-30] have shown that particle formation during inert gas
evaporation occurs by nucleation and growth in a nearby spherical region
surrounding concentrically the vapour source. Once a small particle is
nucleated in the helium atmoshphere, it may increase its size by the following two processes: by incorporation of additional gold atoms at surface
sites from the surrounding vapour or by particle collision between neighbouring crystals, e.g. due to Brownian movement. As two colliding particles
have different crystallographic orientations, a grain boundary is formed
between them. The energy of their boundary can be minimized by a relative

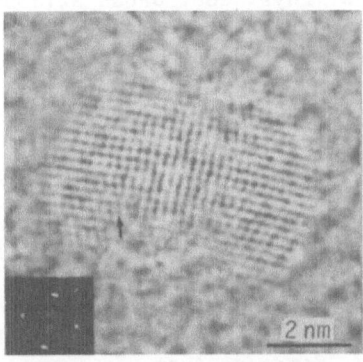

Fig. 5   High resolution electron micrograph and its diffraction
           pattern of MTP's. Two spots nearby in the diffraction
           pattern are separated by $7^\circ 34'$

rotation of the two particles with respect to one another so that a low
energy boundary results. In the case of noble metals, this particle rotation has been demonstrated to occur at temperatures far below the melting point (if the particle diameter is a few nanometers) and was shown
to lead preferentially to the formation of twin boundaries, unless the
particles can reorient so that s single crystal results. For large (e.g.
20 nm) particles - formed by a multi-collision process or a sequence of
several subsequent collisions - multiple twinning is expected, as was observed. On the other hand, small (e.g. 2 nm) particles involving no collisions during growth are likely to exhibit no twins, which agrees with the
observations. Simultaneous to the formation of twin boundaries by particle/
particle collisions, there occurs also a boundary annihilation process by
boundary migration which prevents all twin boundaries formed during growth
from bein retained in the final microstructure. Due to the free energy of
a twin boundary there exists a force on this interface, which has been
demonstrated [31,32] to migrate the interface out of the particles so that
a single crystal results. In fact, the motion of the twin out of the particles had been observed in situ in the electron microscope. Hence the
final twin boundary density may be controlled by the rates of boundary
formation and boundary annihilation.

ACKNOWLEDGEMENTS

    The authors acknowledge Dr. W. Kunath's support in image processing;
Dr. B. Tesche kindly provided the thin carbon films. We are grateful to
Mr. K. Weiss for operating the DEEKO-100 TEM and to Dr. G. Lehmpfuhl for
helpful discussion. Prof. Dr. E. Zeitler has continuously supported this
study by helpful discussions. The financial support of MPG and the ALCOA
Foundation is gratefully acknowledged.

REFERENCES

[1] J.G. Allpress and J.V. Sanders, Phil. Mag. 10:645 (1964)
[2] K. Mihama and Y. Yasuda, Phys. Soc. Japan 21: 1166 (1966)
[3] S. Ino, J. Phys. Soc. Japan 21: 346 (1966)

[4] S. Ino and S. Ogawa, J. Phys. Soc. Japan 22:1365 (1967)

[5] N. Uyeda, M. Nishino and E. Suito, J. Coll. Interface Sci. 43:276(1973)

[6] K. Yagi, K. Takayanagi, K. Kobayashi and G. Honjo, J. Cryst. Growth 28:117 (1975)

[7] K. Heinemann, M.J. Yacaman, C.Y. Yang and H. Poppa, J. Cryst. Growth 47:177 (1979)

[8] M.J. Yacaman, K. Heinemann, C.Y. Yang and H. Poppa, J. Cryst. Growth 47:187 (1979)

[9] L.D. Marks, A. Howie and D.J. Smith, in: Electron Microscopy and Analysis, T. Mulvey, ed., The Institute of Physics, Bristol (1980), p. 397

[10] M. Gillet, Surface Sci. 67:139 (1977)

[11] L.D. Marks and D.J. Smith, J. Microscopy 130:249 (1983)

[12] J.G. Allpress and J.V. Sanders, Aust. J. Phys. 23:23 (1970)

[13] Y. Fukano and C.M. Wayman, J. Appl. Phys. 40:1656 (1969)

[14] S. Ino, J. Phys. Soc. Japan 27:941 (1969)

[15] W. Bryan Monosmith and J.M. Cowley, Ultramicroscopy 12:177 (1984)

[16] R.H. Geiss, Appl. Phys. Lett. 27:174 (1975)

[17] J.M. Cowley, Acta Cryst. 12:367 (1959)

[18] J.G. Allpress, J.V. Sanders and A.D. Wadsley, Phys. Status Solidi, 25:541 (1968)

[19] S.Iijima, J. Appl. Phys. 42:5891 (1971)

[20] S. Iijima, Acta Cryst. A29:18 (1973)

[21] N. Uyeda, T. Kobayashi, E. Suito, Y. Harada and M. Watanabe, J. Appl. Phys. 46:1581 (1972)

[22] H. Hahimoto, T. Tanji, A. Ono and A. Kumao, J. Electron Microsc. 24:212 (1975)

[23] R. Clarke and G. Thomas, Proc. Electron Microsc. Soc. Am. (1976) 492

[24] R. Gronsky, R. Sinclair and G. Thomas, Proc. Electron Micros. Soc. Am. (1976) 494

[25] T. Tanji and H. Hashimoto, Acta Cryst. A34:453 (1978)

[26] Y. Yokota, M. Tomita, H. Hashimoto and H. Endoh, Ultramicroscopy 6:313 (1981)

[27] M. Tomita, H. Hashimoto, T. Ikuta, H. Endoh and Y. Yokata, Ultramicroscopy 16:9 (1985)

[28] C. Fritsche, F. Wolf and A. Schaber, Z. Naturforsch. 16a:31 (1961)

[29] C. Grangvist, R. Buhrman, J. Appl. Phys. 47:2200 (1976)

[30] A. Tholen, Acta Met. 27:1765 (1979)

[31] R. Sun, C. Bauer, Acta Met. 18:639 (1970

[32] B. Rath, H. Hu, Trans. AIME 245:1577 (1969)

GRAPH THEORY DERIVED METHODS FOR THE STUDY OF METAL CLUSTER BONDING TOPOLOGY: APPLICATIONS TO POST-TRANSITION METAL CLUSTERS

R.B. King

Department of Chemistry
University of Georgia
Athens, Georgia 30602, U.S.A.

ABSTRACT

Graph-theory derived methods have been used to study the chemical bonding topology of clusters of post-transition metals such as Sn, Pb, Sb, and Bi thereby suggesting that similar structure and bonding ideas apply to both condensed phase and gas phase clusters of these metals.

INTRODUCTION

Several years ago we developed a graph-theoretical approach for the study of the bonding topology in polyhedral boranes, carboranes, and metal clusters.[1,2] Subsequent work has shown this approach to be very effective in relating electron count to cluster shape for diverse metal clusters using a minimum of computation. Examples of metal clusters treated effectively by this approach include condensed phase post-transition metal clusters[3] as well as osmium carbonyl clusters,[4] gold clusters,[5,6] platinum carbonyl clusters,[5,7] and rhodium carbonyl clusters having fused polyhedra.[8,9]

Recent experimental work by Duncan and collaborators[10,11,12] indicates that laser vaporization of post-transition metals such as tin, lead, antimony, bismuth, and their alloys in a supersonic molecular beam apparatus[13] leads to gas phase metal clusters having unusual local maxima in their cluster size distributions. This paper summarizes an initial attempt to relate these "magic numbers" to the graph-theory derived and related metal cluster bonding models thereby demonstrating apparent connections between the structure and bonding of metal clusters in the condensed and gas phases.

BACKGROUND

The topology of chemical bonding can be represented by a graph in which the vertices correspond to atoms or orbitals participating in the bonding and the edges correspond to bonding relationships. The eigenvalues x of the adjacency matrix of such a graph are related to the Hückel theory molecular orbital energies E and the Hückel parameters α and β by the following equation[1,2,3,9,14,15]:

$$E = \frac{\alpha + x\beta}{1 + xS} \tag{1}$$

Positive and negative eigenvalues x thus correspond to bonding and antibonding orbitals, respectively.

The two extreme types of chemical bonding in metal clusters may be called edge-localized and globally delocalized.[1,2] An edge-localized polyhedron has two-electron two-center bonds along each edge of the polyhedron and is favored when the numbers of internal orbitals of the vertex atoms match the vertex degrees. (The number of edges meeting at a vertex is its degree.) Since vertex atoms in metal clusters normally use three internal orbitals for cluster bonding, cluster polyhedra exhibiting pure edge-localized bonding normally have only degree three vertices (e.g., the tetrahedron, cube, or various prisms). A globally delocalized polyhedron has a multicenter core bond in the center of the polyhedron and is favored when the numbers of internal orbitals do not match the vertex degrees.

One of the major triumphs of the graph-theory derived approach to the bonding topology in globally delocalized systems is the demonstration of the close analogy between the bonding in two-dimensional planar polygonal aromatic systems such as benzene and that in three-dimensional boranes and carboranes based on deltahedra without tetrahedral chambers.[1,2,9] In both cases the three internal orbitals from each vertex are partitioned into two twin internal orbitals and a single unique internal orbital. In the two-dimensional planar polygonal systems the twin internal orbitals overlap pairwise to form the so-called σ-bonding network around the circumference of the polygon and the unique internal orbitals overlap cyclically (C_n graph) to form the so-called π-bonding network. In the three-dimensional deltahedral systems the twin internal orbitals overlap pairwise in the surface of the polyhedron and the unique internal orbitals form a multicenter core bond (K_n graph) at the center of the polyhedron. Both of these types of bonding are found in post-transition metal clusters.

The globally delocalized deltahedra with n vertices have $2n + 2$ skeletal electrons with $2n$ of these electrons arising from the surface bonding and the remaining 2 electrons occupying the single molecular orbital arising from the multicenter core bond.[1,2] Electron-rich polyhedra having more than $2n + 2$ apparent skeletal electrons have one or more non-triangular faces whereas electron-poor deltahedra having less than $2n + 2$ apparent skeletal electrons have one or more tetrahedral chambers. Several examples of so-called nido electron-rich polyhedra having precisely $2n + 4$ skeletal electrons are encountered in post-transition metal clusters. All but one of the faces of nido polyhedra are triangles. If a deltahedron having $2n + 2$ skeletal electrons is regarded as topologically homeomorphic to a sphere, then a nido polyhedron having $2n + 4$ skeletal electrons is topologically homeomorphic to a sphere with a hole in it corresponding to the single non-triangular face.

The general approach for considering metal cluster bonding models involves calculating the number of available skeletal electrons for comparison with the numbers of skeletal electrons required to fill the bonding molecular orbitals for various cluster shapes and bonding topologies. If vertex atoms furnishing the normal 3 internal orbitals are considered, then Sn and Pb vertices like BH, $Fe(CO)_3$, and C_5H_5Co vertices are donors of 2 skeletal electrons and Sb and Bi vertices like CH, $Co(CO)_3$, and C_5H_5Ni vertices are donors of 3 skeletal electrons.[3] Such relationships provide isoelectronic and isolobal analogies of bare post-transition metal clusters to planar aromatic hydrocarbons, polyhedral boranes, and/or transition metal clusters having external carbonyl, hydrocarbon, and/or phosphine ligands.

Condensed Phases

The condensed-phase bare post-transition metal cluster ions reviewed by Corbett[16] and discussed partially in an earlier paper[3] have been found by X-ray crystallography to have the following structures for which the indicated chemical bonding topologies may be proposed:

Square. Bi_4^{2-}, Se_4^{2+} and Te_4^{2+} isoelectronic and isolobal with the cyclobutadiene dianion with 14 skeletal electrons, 8 for the 4 σ bonds and 6 for the π-bonding.

Tetrahedron. $Pb_2Sb_2^{2-}$ with 12 skeletal electrons for localized bonds along the 6 edges of the tetrahedron analogous to organic tetrahedrane derivatives, R_4C_4.

Trigonal bipyramid. Sn_5^{2-}, Pb_5^{2-}, and Bi_5^{3+} with 12 skeletal electrons analogous to the $C_2B_3H_5$ carborane.

Capped square antiprism. Ge_9^{4-}, Sn_9^{4-}, and Pb_9^{4-} with the 2n + 4 = 22 skeletal electrons required for an n = 9 vertex C_{4v} nido polyhedron having 12 triangular faces and one square face.

Tricapped trigonal prism. Ge_9^{2-}, and $TlSn_8^{3-}$ with the 2n + 2 = 20 skeletal electrons required for an n = 9 vertex globally delocalized D_{3h} deltahedron analogous to $B_9H_9^{2-}$ and Bi_9^{5+} anomalously having 22 rather than the expected 20 skeletal electrons suggesting[3] incomplete overlap of the unique internal orbitals directed towards the core of the deltahedron.

Bicapped square antiprism. $TlSn_9^{3-}$ with the 2n + 2 = 22 skeletal electrons required for an n = 10 vertex globally delocalized D_{4d} deltahedron analogous to that found in the $B_{10}H_{10}^{2-}$ anion.

Gas Phase

The following gas phase post-transition metal clusters have been observed in the greatest abundances in the laser vaporization experiments using a supersonic molecular beam apparatus[10,11,12]:

Neutral bismuth clusters. Bi_4 with 12 skeletal electrons postulated to have a tetrahedral structure like P_4.

Cationic bismuth clusters. Bi_3^+ isoelectronic with the cyclopropenyl cation $C_3H_3^+$ and thus postulated to have an equilateral triangular structure; Bi_5^+ with 14 skeletal electrons isoelectronic with B_5H_9 and thus postulated to have a nido square pyramidal structure; Bi_7^+ with the requisite number of skeletal electrons for an edge-localized 7-vertex polyhedron having 11 edges, 6 faces, 6 vertices of degree 3, and one vertex of degree 4.

Anionic bismuth clusters. Bi_2^- and Bi_5^- isoelectronic with the paramagnetic NO and the cyclopentadienide anion, C_5H_5, respectively.

Tin and lead clusters. The neutral 2n apparent skeletal electron systems E_7 and E_{10} (E = Sn, Pb) postulated to have capped octahedron and 3,4,4,4-tetracapped trigonal prism structures, respectively, each having C_{3v} symmetry and one tetrahedral chamber.

<u>Mixed post-transition metal clusters</u> Pb_5Sb_4 and Sn_5Bi_4 isoelectronic with Sn_9^{4-} and Bi_9^{5+} observed in condensed phases[16] and postulated to have a nido C_{4v} capped square antiprism structure analogous to Sn_9^{4-}; Pb_3Sb_2, Pb_4Sb_2, and Pb_5Sb_2 having $2n + 2$ skeletal electrons and isoelectronic with the deltahedral carboranes $C_2B_3H_5$, $C_2B_4H_6$, and $C_2B_5H_7$ known[17] to have trigonal bipyramid, octahedron, and pentagonal bipyramid structures, respectively.

These examples suggest that principles of structure and bonding which are applicable to condensed phase metal clusters may also be applied to gas phase post-transition metal clusters arising from molecular beam experiments. However, further very difficult experimental work is needed on the gas phase post-transition metal clusters in order to obtain evidence confirming or disproving the structures postulated above.

ACKNOWLEDGMENT

I am indebted to the U.S. Office of Naval Research for partial support of this work and to Prof. Michael Duncan for helpful discussions and providing information on his experimental results in advance of publication.

REFERENCES

1. R.B. King and D.H. Rouvray, <u>J. Am. Chem. Soc.</u> 99:7834 (1977).
2. R.B. King <u>in</u>: "Chemical Applications of Topology and Graph Theory," R.B. King, ed., Elsevier, Amsterdam (1983).
3. R.B. King, <u>Inorg. Chim. Acta</u> 57:79 (1982).
4. R.B. King, <u>Inorg. Chim. Acta</u> 116:99 (1986).
5. R.B. King <u>in</u>: "Mathematics and Computational Concepts in Chemistry," N. Trinajstić, ed., Ellis Horwood, Chichester (1986).
6. R.B. King, <u>Inorg. Chim. Acta</u> 116:109 (1986).
7. R.B. King, <u>Inorg. Chim. Acta</u> 116:119 (1986).
8. R.B. King, <u>Inorg. Chim. Acta</u> 116:125 (1986).
9. R.B. King, <u>Int. J. Quantum Chem.</u> in press.
10. R.G. Wheeler, K. LaiHing, W.L. Wilson, and M.A. Duncan, <u>Chem. Phys. Lett.</u>, in press.
11. R.G. Wheeler, K. LaiHing, W.L. Wilson, J.D. Allen, R.B. King, and M.A. Duncan, submitted for publication.
12. M.E. Geusic, R.R. Freeman, and M.A. Duncan, submitted for publication.
13. T.G. Dietz, M.A. Duncan, D.E. Powers, and R.E. Smalley, <u>J. Chem. Phys.</u> 74:6511 (1981).
14. K. Ruedenberg, <u>J. Chem. Phys.</u> 22:1878 (1954).
15. H.H. Schmidtke, <u>J. Chem. Phys.</u> 45:3920 (1966).
16. J.D. Corbett, <u>Chem. Rev.</u> 85:383 (1985).
17. R.N. Grimes, "Carboranes," Academic Press, New York, N.Y. (1970).

LASER VAPORIZATION AND PHOTOIONIZATION STUDIES OF TIN AND LEAD CLUSTERS

K. LaiHing, R. G. Wheeler, W. L. Wilson, and M. A. Duncan

Department of Chemistry
School of Chemical Sciences
University of Georgia
Athens, GA   30602

ABSTRACT

Molecular beams of bare metal clusters containing up to 30-40 atoms of either tin or lead are prepared by laser vaporization in a pulsed nozzle source. Laser photoionization mass spectra of these clusters are characterized by magic number patterns. Laser wavelength and power dependence studies, and comparisons to previous electron impact data, probe stability in both neutral and cation clusters.

INTRODUCTION

Mass spectrometric studies of metal clusters in molecular beam environments are often characterized by "magic numbers," or local maxima and minima in size distributions.[1-4] However, the origin of these patterns is not well understood and has been the subject of some controversy.[4-5] Abundances may be influenced by a variety of factors, including the mechanism of metal vaporization, the kinetics of cluster growth, the stability of packing arrangements and bonding in the clusters formed, and fragmentation patterns in electronic impact or photoionization detection. Any conclusions drawn from mass spectral data, therefore, must include a consideration of all these factors.

In this report we describe a study of magic numbers in the laser photoionization mass spectra of lead and tin clusters produced by laser vaporization. These systems have been studied previously using inert gas condensation in an oven/beam source for cluster production and electron impact ionization for detection.[2,3] By a correct choice of laser wavelength and power, photoionization reproduces the magic number patterns observed with electron bombardment. However, not all features are observed under the same laser conditions. The comparison of lead and tin clusters under various photoionization conditions provides new insight into magic numbers in these systems, and suggests contributions from both neutral cluster and cation fragment stabilities.

EXPERIMENTAL

Clusters for these experiments are produced by excimer laser vaporization (193, 248 or 308 nm) in a pulsed nozzle source like those described previously.[6] Growth in these sources occurs under collision-dominated conditions near room temperature, so that thermodynamic stability should have a strong influence on cluster abundances. At the end of the growth region, clusters are cooled supersonically in an expansion of the helium buffer gas. Downstream from the source, clusters are probed with excimer laser photoionization (248, 193 or 157 nm) in a time-of-flight mass spectrometer.

As observed for other cluster species, multiphoton absorption and fragmentation are efficient in tin and lead clusters, and mass spectral intensities are extremely sensitive to ionization laser conditions. Extensive power dependence studies were therefore conducted at each laser wavelength used for photoionization. In the limit of low laser power, it is generally possible to eliminate, or at least limit, multiphoton absorption. Under these conditions, only clusters ionized by single photon absorption will be detected and relative abundances will not change with further reduction in power. Wavelength dependent studies in the low power limit provide a probe of cluster ionization potentials. Near threshold photoionization without multiphoton absorption will also limit fragmentation (parent ion production dominates), and these conditions provide a probe of neutral cluster abundances. In the other extreme, moderate to high laser powers, especially at longer wavelengths, are expected to cause extensive multiphoton absorption and fragmentation. Under these conditions, only stable fragment ions survive further fragmentation, and abundances are expected to favor stable cluster cations. Using these guidelines and spectra at various wavelengths, it is usually possible to develop a consistent picture of neutral clusters and fragmentation.

RESULTS AND DISCUSSION

The mass spectra obtained for lead and tin under neutral sensitive conditions are shown in Figure 1. Photoionization was accomplished with a fluorine excimer laser (157 nm; 7.9 eV) at a fluence of 0.1 mj/cm^2. Laser power and wavelength studies show that small tin clusters (N=2-10) have ionization potentials (IPs) between 7.9 and 6.4 eV, while larger tin clusters and all lead clusters have IPs less than 6.4 eV. Mass spectral focusing in this data emphasizes the center mass region (around N=10) in both spectra. However, in addition to this broad focusing effect, local maxima at N=7 and N=10 are clearly observed in both spectra. A slight, but reproducible, maxima is also observed for lead at N=13. Local minima occur for Sn_{13}, Sn_{14} and Sn_{17} as well as for Pb_{14}, Pb_{18} and Pb_{20}. Mass spectra under cation favoring conditions are shown in Figure 2. Photoionization for these spectra was accomplished with an ArF excimer laser (193 nm; 6.4 eV) at a fluence of 10 mj/cm^2. Local minima for tin are observed at N=14, 17 and 20, while lead clusters exhibit a maxima at N=13 and minima at 14, 18 and 20.

The most significant difference between Figures 1 and 2 is the presence and absence of maxima at N=7 and 10. Therefore, these magic numbers are attributed to stable neutral clusters but not to cluster cations. For comparison, ArF radiation in the low power limit can also be used for single photon ionization of lead clusters (all IPs are < 6.4 eV), and the same maxima are observed. In contrast to this neutral cluster result, tin minima at 14 and 17, as well as the lead maxima at 13 and minima at 14, 18 and 20, are observed under both neutral and cation conditions. The pattern for lead is similar to that observed in rare gas atom van der Waals clusters and has been explained by a simple atom packing model.[2b] It is therefore understandable that this pattern is observed for both neutrals and cations. Some of the features associated with this model are observed for tin (i.e. minima at N=14).

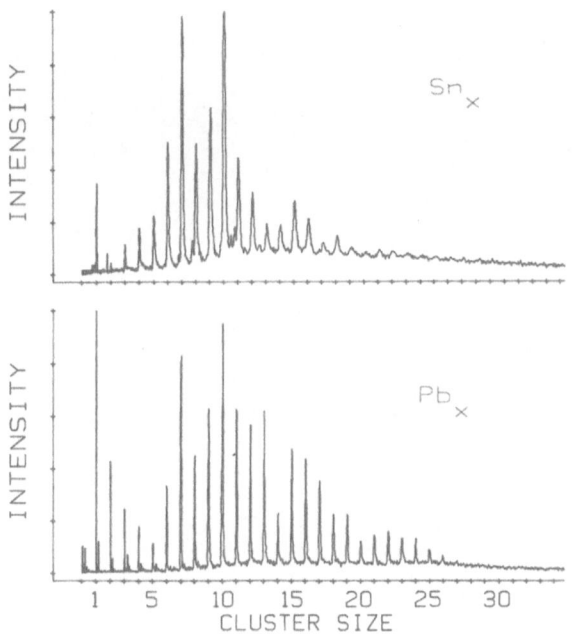

Fig. 1. Photoionization mass spectra under neutral favoring conditions (157 nm; 7.9 eV; 0.05 mjoule/cm^2).

The comparison of these data to previous lead and tin experiments provides useful insight into the significance of magic numbers in these systems. In the case of lead clusters, where best comparisons are possible, an oven/beam source and electron impact (35 eV) detection produce maxima at 7, 10 and 13 and minima at 14, 18 and 20.[2b] This data, therefore, illustrates features observed in both neutral and cation-sensitive photoionization experiments. Tin data is only available at higher electron impact energies (70 eV) for clusters up to N=16,[3] but these conditions also produce the minima at N=14 observed under our cation conditions. It therefore seems that high energy electron impact ionization tends to favor cation production (just like high powered photoionization), while lower energy electron impact (< 10 eV) could be used as a probe of neutral cluster abundances (just like low power photoionization). These same ideas have been suggested by Recknagel and coworkers.[7] Allowing for detection differences, this comparison also suggests that the two cluster production techniques (oven with inert gas condensation and laser vaporization) produce the same abundance pattern for lead clusters. This fact strongly favors cluster stability, rather than growth kinetics, as the determining factor for magic numbers in this system (and for tin also, by comparison).

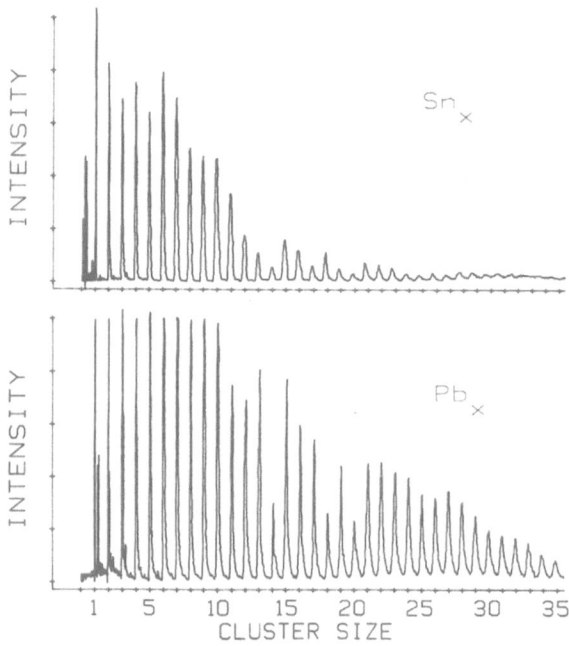

Fig. 2. Photoionization mass spectra under cation fragment conditions (193 nm; 6.4 eV; 10 mjoule/cm^2).

It is also useful to compare tin and lead systems to clusters of other group 14 materials (C, Si, Ge). Because of high IPs relative to available lasers, C and Si clusters have not been photoionized under neutral sensitive conditions. Cation conditions, however, find maxima for C_{11} and C_{15},[8,9] for Si_6, Si_7, Si_{10}, Si_{11}[9] and for Ge_6, Ge_7, Ge_{10}, as well as minima for Ge_{13}, Ge_{17} and Ge_{20}.[3,10] Neutral photoionization conditions produce maxima at Ge_6 and Ge_{10}.[10] Therefore, cation or neutral maxima near 6 or 7, and especially neutral maxima at N=10, seem to be general characteristics of group 14 clusters. Both

Sn and Ge systems have cation minima at 13, 17 and 20, which have switched over to a maxima at 13 and minima at 14, 18 and 20 for lead. Thus, cluster stability seem to be influenced by both bonding patterns and atom packing considerations for these systems, with atom packing considerations becoming most important for lead. More detailed comments about these magic number trends and their causes will be possible when full scale quantum mechanical calculations become available for these larger systems. In the meantime, electron counting theories used for clusters of these same elements in condensed phases[11] may provide useful insight into electronic structure and bonding in these systems.

ACKNOWLEDGMENTS

This research was supported by the U.S. Army Research Office. Additional support was received through the Office of Naval Research.

REFERENCES

1. W. D. Knight, K. Clemenger, W. A. DeHeer, W. A. Saunders, M. Y. Chow and M. L. Cohen, Phys. Rev. Lett. $\underline{52}$, 2141 (1984).
2. a) K. Sattler, J. Muhlbach and E. Recknagel, Phys. Rev. Lett. $\underline{45}$, 821 (1980); b) K. Sattler, J. Muhlbach, P. Pfau and E. Recknagel, Phys. Lett. $\underline{87A}$, 415, 418 (1982).
3. a) T. P. Martin and H. Schaber, J. Chem. Phys. $\underline{83}$, 855 (1985); b) T. P. Martin, J. Chem. Phys. $\underline{83}$, 78 (1985).
4. J. C. Phillips, Chem. Rev. $\underline{86}$, 619 (1986).
5. E. Schumaker, M. Kappes, K. Marti, P. Radi, M. Schar and B. Schmidhalter, Ber. Bunsenges. Phys. Chem. $\underline{88}$, 220 (1984).
6. J. B. Hopkins, P.R.R. Langride-Smith, M. D. Morse and R. E. Smalley, J. Chem. Phys. $\underline{78}$, 1627 (1983).
7. D. Schild, R. Pfaum, K. Sattler and E. Recknagel, to be published.
8. E. A. Rohlfing, D. M. Cox and A. Kaldor, J. Chem. Phys. $\underline{81}$, 3322 (1984).
9. L. A. Bloomfield, M. E. Geusic, R. R. Freeman and W. L. Brown, Chem. Phys. Lett. $\underline{121}$, 33 (1985).
10. J. R. Heath, Y. Liu, S. C. O'Brien, Q. Zhang, R. F. Curl, F. K. Tittel and R. E. Smalley, J. Chem. Phys. $\underline{83}$, 5520 (1985).
11. R. B. King, Inorg. Chim. Acta $\underline{57}$, 79 (1982).

LASER VAPORIZATION AND PHOTOIONIZATION OF GROUP IV AND V INTERMETALLIC CLUSTERS

R. G. Wheeler, K. LaiHing, W. L. Wilson, and M. A. Duncan

Department of Chemistry
School of Chemical Sciences
University of Georgia
Athens, GA 30602

ABSTRACT

Sn/Bi and Pb/Sb mixed-metal clusters have been produced in the gas phase with laser vaporization and studied with laser photoionization mass spectroscopy. Magic numbers are observed for certain neutral cluster species which are isoelectronic to stable post transition metal cluster ions studied previously in condensed phases. This correspondence between gas phase and condensed phase clusters has not been observed previously, and may have significant implications for the stability and structure of gas phase clusters.

INTRODUCTION

The recent development of laser vaporization/molecular beam technology[1] has resulted in numerous studies of gas phase metal cluster molecules.[2,3] Various experiments have examined properties such as cluster structure,[3,4] ionization potentials,[5] chemical reactivity,[2,6] and ion fragmentation.[7] Prior to the development of these rather exotic techniques, however, clusters had already been studied for many years in condensed phases. For example, borane,[8,9] carborane,[10] transition metal carbonyl,[11] and post transition metal ionic cluster systems[12,13] are well characterized. In general, condensed phase clusters have limited volatility and are coordinately saturated with ligands so that detailed comparisons with bare metal gas phase species are not possible. However, ionic clusters of the post transition metal elements (Sn_5^{2-}, Pb_9^{4-}, Bi_9^{5+}, etc.) are ligand-free, consisting of charged metal polyhedral networks accompanied by counterions. In this report, we describe neutral gas phase counterparts to these condensed phase cluster ions. These results establish one of the few existing links between gas-phase and condensed-phase cluster chemistry.

The structure and bonding patterns of post transition metal ion clusters show a strong similarity to those of carborane clusters.[12,13] Consequently, although detailed molecular orbital studies have been conducted, many of the salient features of bonding in these systems are described by simple valence electron counting schemes.[12] In condensed phases, stable electron configurations are achieved with multiply charged cluster species. Gas phase experiments, however, produce primarily neutrals and cluster elements must be specifically chosen to produce isoelectronic analogs of ionic species for comparisons. In condensed phase cluster ions, group IV and V elements respectively behave analogously to (i.e. are "isolobal" to)[14] the BH and CH building blocks of carboranes. Therefore, alloy clusters combining these elements provide the logical test cases for stable neutral clusters in the gas phase.

EXPERIMENTAL

Mixed metal clusters for these experiments are produced by excimer laser (248 nm) vaporization of 1:1 alloys in a pulsed nozzle source. Laser photoionization was accomplished at several wavelengths (157, 193, 248 nm) over a range of laser powers to investigate ionization potentials and fragmentation processes. As described in our accompanying paper, high energy radiation in the low power limit provides a probe of neutral cluster abundances while limiting fragmentation. Low energy radiation at moderate-to-high power enhances fragmentation and probes stable ion fragments. Following photoionization cluster ions are mass analyzed with a computer controlled time-of-flight spectrometer system.

RESULTS AND DISCUSSION

Figure 1 shows the mass spectra of Sn/Bi and Pb/Sb clusters under neutral sensitive photoionization conditions (157 nm; 0.05 mjoule/pulse). Intensities in these spectra are determined by the density of each size cluster formed, by relative ionization efficiencies, and by the slowly varying mass spectral focusing function. Cluster growth in our source occurs under collision-dominated conditions near room temperature, so that thermodynamic stability should have a strong influence on cluster abundances. Detection efficiency should be roughly size independent if ionization potentials (IPs) are lower than the photon energy used (7.9 eV). Pure metal clusters of tin, lead, Sb, or Bi all have IP's less than 7.9 eV for clusters larger than 3 atoms. Therefore, although we cannot completely rule out the presence of other cluster species not ionized under these conditions, we believe the abundances shown represent the primary stable neutral clusters formed.

Figure 2 illustrates the corresponding mass spectra under cation-sensitive conditions (193 nm; 10 mjoules/cm^2). As shown, these spectra are characterized by enhanced abundances at mass features very different from those in Figure 1. Relative abundances in Figure 1 do not change with reduction in laser power. Reduction in power under cation conditions, however, causes an increase in relative abundances of higher mass features, clearly indicating the effects of fragmentation. Therefore, we believe the maxima in Figure 2 reflect stable cluster cations formed by fragmentation of larger species. This method favors smaller clusters since larger ones are highly fragmented.

In contrast to mass spectra of transition metal cluster alloys,[15] the spectra presented here under either set of conditions do not represent statistical combinations of the component elements. In Figure 1 for example, nearly all of the more abundant species are even-electron neutral molecules containing two or four atoms of antimony or bismuth. In Figure 2 the more abundant species contain three or

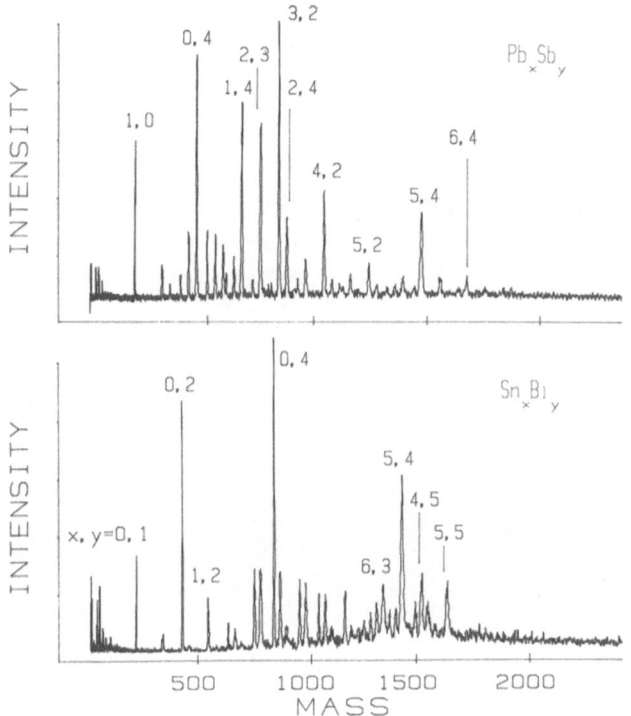

Fig. 1. Photoionization mass spectra under neutral favoring conditions (157 nm; 7.9 eV; 0.05 mjoule/cm^2.

five atoms of antimony or bismuth and are even-electron species as cations. A close comparison of the spectra under neutral and cation conditions shows that these terms are only approximate; some clusters appear under both conditions (e.g. Pb_2Sb_3). However, power dependences for these species confirm their assignments to parent neutral molecules or cation fragments. In every case, the general rule of even-electron species seems to be observed. Of the stable neutral species indicated by these data, the tetramers of antimony and bismuth are well known species and dominate their respective pure component cluster distributions.[16] Several of the remaining abundant neutrals and cations represent molecules related to condensed phase cluster ions and to known carborane clusters. Pb_3Sb_2, $Pb_2Sb_3^+$ and $Sn_2Bi_3^+$ are all isoelectronic to the 5-atom 12-skeletal electron species Sn_5^{2-} and Pb_5^{2-} (trigonal bipyrimid structures).[13] Pb_4Sb_2 and Pb_5Sb_2 are isoelectronic to the carboranes $C_2B_4H_6$ and $C_2B_5H_7$ (octahedral and pentagonal bipyrimid structures, respectively).[10] Their isoelectronic cations $Sn_3Bi_3^+$ and $Sn_4Bi_3^+$ are also observed. More importantly, by far the largest neutral features at higher mass for both alloys are the 9-atom 22-skeletal electron species Pb_5Sb_4 and Sn_5Bi_4, which

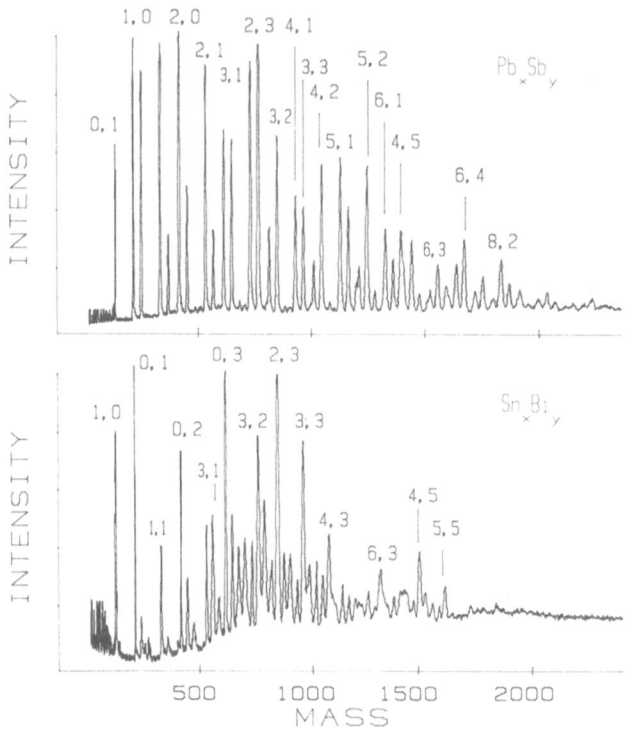

Fig. 2. Photoionization mass spectra under cation fragment conditions (193 nm; 6.4 eV; 10 mjoule/cm^2.

are isoelectronic to the well known ions Ge_9^{4-}, Sn_9^{4-} and Pb_9^{4-} (C_{4V} unicapped square antiprism structures).[13] The isoelectronic cations $Pb_4Sb_5^+$ and $Sn_4Bi_5^+$ are also observed.

The relationships between condensed phase cluster ions and gas phase neutral clusters suggested by these data is significant. Structures of the condensed phase ions have been measured with X-ray crystallography and with NMR spectroscopy.[13] In essentially every case, cluster ions with the same number of atoms and skeletal electrons have the same structures. Unfortunately, gas phase cluster spectroscopy is much more difficult and there are no detailed structural data except for metal dimers and trimers. Although there are no guarantees that gas phase clusters will have the same structure as their isoelectronic analogs in condensed phases, the condensed phase data should certainly provide initial estimates for gas phase structures. Additionally, the concepts of metal cluster bonding and stability that have been developed over the years, including the use of electron counting theory,[13,14] may also prove useful for treatments of gas phase post transition metal clusters. We are continuing to investigate these ideas with mass spectroscopy of other group IV/V alloys (e.g. Sn/As) and resonant two-photon electronic spectroscopy of these alloy species.

ACKNOWLEDGEMENT

This research was sponsored by the U.S. Army Research Office. Additional support was received through the Office of Naval Research.

REFERENCES

1. a) T. G. Dietz, M. A. Duncan, D. E. Powers and R. E. Smalley, J. Chem. Phys. 74, 6511 (1981); b) J. B. Hopkins, P.R.R. Langridge-Smith, M. D. Morse and R. E. Smalley, J. Chem. Phys. 78, 1627 (1983).
2. M. D. Morse, Chem. Rev., to be published.
3. J. C. Phillips, Chem. Rev. 86, 619 (1986).
4. W. Weltner and R. J. van Zee, Ann. Rev. Phys. Chem. 35, 291 (1984).
5. a) E. A. Rohlfing, D. M. Cox, A. Kaldor and K. Johnson, J. Chem. Phys. 81, 3846 (1984); b) R. L. Whetten, M. R. Zakin, D. M. Cox, D. J. Trevor and A. Kaldor, J. Chem. Phys. 85, 1697 (1986).
6. R. L. Whetten, D. M. Cox, D. J. Trevor and A. Kaldor, Surf. Sci. 156, 8 (1985).
7. P. J. Brucat, L. S. Zheng, C. L. Pettiette, S. Yang and R. E. Smalley, J. Chem. Phys. 84, 3078 (1986).
8. a) E. L. Muetterties and W. H. Knoth, "Polyhedral Boranes," Marcel Dekker, New York, 1968; b) R. W. Rudolph, Acc. Chem. Res. 9, 446 (1976).
9. K. Wade, Adv. Inorg. Chem. and Radiochem. 18, 1 (1976).
10. R. N. Grimes, "Carboranes," Academic Press, New York, 1970.
11. a) E. L. Meutterties, T. N. Rhodin, E. Bond, C. F. Bruckner and W. R. Pretzer, Chem. Rev. 79, 91 (1979); b) F. A. Cotton and R. A. Walton, "Multiple Bonds Between Metal Atoms," J. Wiley and Sons, New York, 1982; c) G. Schmid, Structure and Bonding 62, 51 (1985).
12. R. B. King, Inorg. Chim. Acta 57, 79 (1982).
13. J. D. Corbett, Chem. Rev. 85, 383 (1985).
14. R. Hoffman, Angew. Chem. Int. Ed. 21, 711 (1982).
15. E. A. Rohlfing, D. M. Cox, R. Petkovic-Luton and A. Kaldor, J. Phys. Chem. 88, 6227 (1984).
16. R. G. Wheeler, K. LaiHing, W. L. Wilson and M. A. Duncan, Chem. Phys. Lett, in press.

A THEORETICAL STUDY OF CARBON CLUSTERS: EQUILIBRIUM GEOMETRIES AND ELECTRONIC STRUCTURES OF C_n

L. S. Ott and A. K. Ray

Department of Physics, University of Texas at Arlington
Arlington, Texas 76019

INTRODUCTION

In recent years, there has been widespread interest in the theoretical investigation of clusters of atoms (Schaefer 1975, Echt et al. 1981, Muhlback et al. 1982, Messmer 1982, Knight et al. 1984, Ray et al. 1985, Ray 1986). The understanding of the chemistry and physics of clusters is of signal importance in many diverse areas, such as catalysis and combustion. Such clusters are also often used to model bulk solid-state phenomena and their surfaces. Newly developed experimental techniques also make it possible to study the transition from cluster to bulk behavior. Much of the earlier work has been on metal clusters, specifically alkali metal clusters (Knight et al. 1984, Clemenger 1985), where "magic numbers" have been identified in the relative populations of aggregates of atoms. In a cluster experiment, the existence of these magic numbers are identified as local maxima in the cluster abundance spectrum at certain "magic" cluster sizes. For these metals, the spherical jellium model (Knight et al. 1984, Chou et al. 1984) provide a good understanding of the existence of magic numbers, predicted to occur at 2, 8, 20, 40, etc. Similar work has been reported on the formation of "magic clusters" for molecules (Beuhler 1982) and rare gases (Echt et al. 1981). In the rare gases, the existence of magic numbers can be explained by Mackay icosahedral packing (Phillips to be published) and the magic sizes of these clusters are predicted to be at 13, 19, 55, 147, etc. Recently, however, there has been significant theoretical and experimental work on semicounductor systems such as small Si, Ge and C clusters (Tsong 1984; Bloomfield et al., 1985, Martin et al. 1985, Heath et al. 1985 Raghavachari 1985, Tomanek et al. 1986, Bernholc and Phillips to be published). The theoretical work on Si clusters has involved both accurate ab-initio molecular orbital calculations and pseudopotential calculations. As has been pointed out by Bernholc and Phillips (to be published), for covalently bonded materials, the directional nature of the bonding places severe constraints on the atomic arrangement in the cluster. These constraints are very different for carbon compared to silicon or germanium. In this paper we report on detailed ab initio quantum chemical investigation of carbon clusters. Apart from physical and chemical applications, the study of carbon clusters is also of astrophysical interest since such clusters have been identified in carbon stars and in comet tails. For this purpose, the currently available theoretical and experimental results on carbon clusters are reviewed first.

On the theoretical side, only the physical and electronic structure of C_2, C_3 and, to a lesser extent, of C_4 are fairly well established. It is well-known that C_3 is most stable in the linear structure and ab-initio Hartree-Fock configuration interaction calculations predict that $^1\Sigma_g^+$ is the ground state for both C_2 and C_3 (Ransil 1960, Clementi et al. 1962; Fougere et al. 1966, Liskow et al. 1972, Peric-Radic et al. 1977, Romelt et al. 1978, Whiteside et al. 1981). For C_4, Clementi (1961), using a single-zeta STO basis, has calculated the ground state of linear C_4 to be $^3\Sigma_g^-$ while Whiteside et al. (1981) have considered linear, cyclic and bicyclic conformations as possible ground-state structures for C_4. They found that at the highest level of theory, bicyclic structure with two π electrons is the most stable structure. For larger clusters, there exist several theoretical calculations at various levels of approximation. Pitzer and Clementi (1959) have performed Huckel MO calculations on linear C_n (n = 2, 3, ..., 17) molecules and predicted that C_n clusters with n-odd would be singlets and those with n-even triplets. Hoffman (1966) has shown, using extended Huckel theory, that for C_n molecules (n < 9) linear structures are more stable than closed structures. These extended Huckel calculations also predict increased stability for monocyclic rings of regular polygons over linear chains above n = 9. Ewing and Pfeiffer (1982) have performed ab-inito Hartree-Fock calculations for linear C_n molecules (n = 2, 3, ..., 6) using both double-zeta and double-zeta plus polarization basis sets. Their calculations indicate that very little bond length alternation occurs in these molecules and that linear C_n molecules with n-odd have singlet ground states and those n-even have triplet ground states. On the experimental side, Rohlfing et al. (1984) have produced carbon clusters ranging in size from 2 to nearly 200 atoms by laser vaporization of a solid graphite rod within the throat of a high pressure pulsed nozzle. Their results are in good agreement with the recent electron impact appearance potential measurements for C_2 and C_3 (Gupta et al. 1979) and with the theoretical calculations of Ewing and Pfeiffer (1982). Their data also indicate that the ionization potentials for clusters as large as C_{12} remain high (9.98 - 12.84 eV). It is also to be noted that three different sets of experiments (Rohlfing et al. 1984, Furstenau et al. 1981, Bloomfield et al. 1985) show several magic numbers in the C_2 - C_3 range. Recently (Rao et al. 1986) there also has been a systematic study of the ground state energies, equilibrium geometries, electronic and magnetic properties and relative stabilities of neutral carbon clusters consisting of up to four atoms. The work reported here concentrates on C_2 - C_4 clusters and extensions are then made to larger clusters.

GEOMETRY OF CLUSTERS

Clusters investigated in this work are taken from different studies available in the literature. Triangular, equidistant and general C_3 clusters are investigated. Structures considered for C_4 are indicated in Figure 1 and include, for example, tetrahedral (tet), triangular (tri), square (sq), linear equidistant (eq), linear consisting of two diatomic molecules (mol), general linear (lin), oblong formed by two parallel diatomic molecules (obl), T shaped (T), rhombic formed by two diatomic molecules (rho) and parallelopiped (para). For the higher clusters, only linear and cyclic ring structures are considered.

COMPUTATIONAL METHOD AND RESULTS

Our calculations are based on accurate all-electron ab initio molecular-orbital techniques, specifically Hartree-Fock theory. One of the primary considerations involved in these calculations is determination of the type of basis set to be used. Gaussian-type basis sets used in ab initio mole-

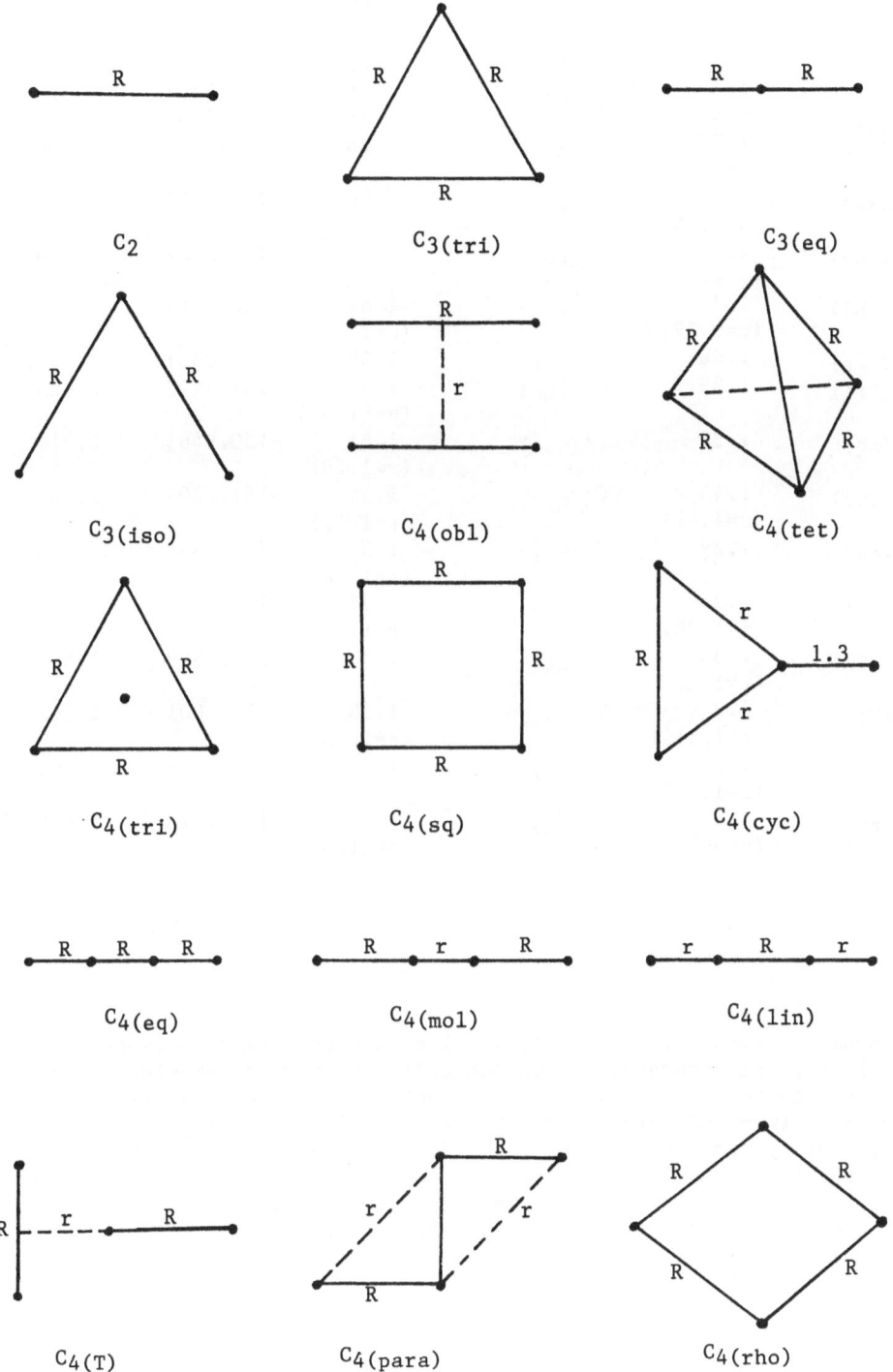

Figure 1. Geometry of two, three and four atom Carbon Clusters.

Table 1. Binding Energy Calculations for $C_2 - C_4$ Clusters

Cluster	Initial R (A)	Final Geometry	Final R	E_{TOT} (a.u.)	E_b (eV)
C_2	1.35	C_2	1.25	−75.3792	0.2377
C_3(tri)	1.4	C_3(eq)	1.28	−113.3552	1.8785
C_3(eq)	1.4	C_3(eq)	1.28	−113.3552	1.8785
C_3(iso)	1.35 (α=53.5°)	C_3(eq)	1.28	−113.3552	1.8785
C_3(iso)	1.5 (α=67.6°)	C_3(eq)	1.28	−113.3552	1.8784
C_3(iso)	1.5 (α=90°)	C_3(eq)	1.28	−113.3552	1.8785
C_4(obl)	1.3 (r=1.27)	C_4(obl)	1.61 (r=1.29)	−150.9463	1.516
C_4(tet)	1.46	C_4(tet)	1.59	−150.8337	0.7497
C_4(tri)	1.89	C_4(rho)	1.43 (α=61.6°)	−151.1478	2.885
C_4(sq)	1.38	C_4(obl)	1.61 (r=1.29)	−150.9463	1.516
C_4(cyc)	1.43 (r=1.47)	C_4(cyc)	1.38 (r=1.45)	−151.1001	2.561
C_4(eq)	1.29	C_4(mol)	1.3 (r=1.28)	−151.1359	1.8488
C_4(mol)	1.3 (r=1.28)	C_4(lin)	1.33 (r=1.2)	−150.9021	1.215
C_4(lin)	1.33 (r=1.3)	C_4(eq)	1.3	−151.1359	1.8488
C_4(T)	1.3 (r=1.314)	C_4(cyc)	1.38 (r=1.45)	−151.1001	2.561
C_4(para)	1.3 (r=1.804)	C_4(eq)	1.3	−151.1359	2.804
C_4(rho)	1.5 (α=66°)	C_4(rho)	1.426 (α=61.59°)	−151.1478	2.885

cular orbital computations usually involve some compromise between computational cost and accuracy. A considerable increase in computational efficiency can be achieved if the exponenets of the Gaussian primitives are shared between different basis functions. At the split valence level, this is usually exploited by sharing primitive exponents between s and p functions for the valence funcitons. Accordingly, for carbon, we have used here the so-called 6-31G* basis and 3-21G for larger clusters. These computations were made by the Program GAMESS written by DuPuis et al. (1979). The results of all our calculations are in Table 1-3. The binding energy is calculated from

$$E_b = (NE_c - E_{cN})/N \qquad\qquad (1)$$

and the first ionization potential is obtained from Koopmans' theorem

$$IP = -E_{homo} \qquad\qquad (2)$$

For C_2 our results were in good agreement with those of Ewing and Pfeiffer (1982). The results of Rao (et al. 1986), using UHF-CI, and a 6-31G basis

Table II. Ionization Potentials for C_2 - C_4 Clusters

Cluster	First Ionization Potential	Second Ionization Potential	Dipole Moment
C_2	0.4485	0.4485	0.0
$C_{3(eq)}$	0.4649	0.4649	0.008
$C_{4(obl)}$	0.3595	0.4065	0.0
$C_{4(tet)}$	0.4280	0.4280	0.0
$C_{4(rho)}$	0.4146	0.4502	0.0
$C_{4(cyc)}$	0.4014	0.4403	0.6596
$C_{4(mol)}$	0.3804	0.4904	7.705
$C_{4(lin)}$	0.3571	0.3571	0.0
$C_{4(eq)}$	0.3804	0.4904	0.0006

Table III. Total Energy Calculations for Linear and Cyclic C_n Clusters

n	Total Energy (Linear) (a.u.)	Total Energy (Rings) (a.u.)
5	-187.962757	-187.732672
6	-225.579712	-225.458217
7	-263.202913	-263.096317
8	-300.831631	-300.730224
9	-338.444812	-338.325007
10	-	-376.095979
11	-	-413.626959
12	-	-451.261608
13	-	-488.907341
14	-	-526.610000
15	-	-564.210285
16	-	-601.790804

set give a bond length of 1.35 Angstroms and a higher binding energy.

Our results show conclusively that the equilibrium geometry for C_3 is a linear structire. This is consistent with all the literature (Rao et al. 1986, Clementi et al. 1962, Romelt et al. 1978, Ewing et al 1982, and Whiteside et al. 1981).

We conducted an extensive study of C_4 structures, wherein many possible geometries were considered. Rao et al. (1986) found that the equilibrium structure for C_4 was a rhombus. Our calculations confirm this. Whiteside et al. (1981) also considered a rhomibic structure with similar results. However, their calculations resulted in higher binding energies for linear structures, which contradicts our findings, although our study does conclude that a linear C_4 structure is possible. The results of the dipole moment calculations indicated in Table II give also some indication of the relative stabilities of these clusters.

The total energies for the larger C_n clusters are reported in Table 3. The claculations are done at optimized geometries for C_4, C_5 and C_6 in the sense that bond lengths were varied until no further lowering of the total energy of the molecule was obtained. For larger clusters, no further optimizations were performed since the optimized bond lengths for C_6 was found extremely close to those of C_5. The results are only reported up to n = 9, since ring structures were found to be energetically more favorable for larger clusters. Based on our calculations, C_4, C_6 and C_8 clusters have triplet ground states at the STO-3G and the 3-21G level. C_5, C_7 and C_9 clusters have singlet ground states at the 3-21G level but C_7 and C_9 have triplet ground states at STO-3G level. It is claimed that the prediction of the wrong gtound states for C_7 and C_9 at the STO-3G level is merely a manifestation of the inadequacy of the basis set. In general, linear C_n molecules have single ground states for n odd and triplet ground states for n even. As we add atoms, we introduce a new σ_g and a new σ_u orbital to the set of occupied orbitals. No existence of magic number is indicated and the cohesive energy, or binding energy per atom, appears to converge to an approximate value of 78 kcal/mole, which is significantly less than the cohesive energy of graphite, which is 170 kcal/mole (Zavitsanos et al. 1973). Our ab initio all electron calculations indicate that cyclic rings are preferred over linear chains for C_n clusters, n > 9. This confirms the predicitons of the semi-empirical calculations of Pitzer and Clementi (1959) in that a transition to neutral π bonded rights should occur around C_{10}. We also find the existence of magic number at n = 10, 14, i.e. C_{10}, C_{14}.

ACKNOWLEDGEMENT

    We thank the Robert A. Welch Foundation (Grant No. Y707) for partial financial support of this work.

REFERENCES

Bernhold J. and Phillips J. C., to be published.
Bernhold J. and Phillips J. C., to be published.
Beuhler R. J. and Friedman L. 1982 J. Chem. Phys. 77:2549.
Bloomfield L. A. Freeman, R. R. and Brown W. L. 1985 Phys. Rev. Lett.
        54:2246.
Bloomfield L. A., Geusic M. E., Freeman R. R. And Brown W. L. 1985 Chem.
        Phys. Lett. 121:33
Chou M. Y., Cleland A. and Cohen M. L. 1984 Solid St. Comm. 52:645.
Clemenger K. 1985 Phys. Rev. B. 32:1359.
Clementi E. 1961 J. Am. Chem. Soc. 83:4501.
Clementi E. and McLean A. D. 1962 J. Chem. Phys. 36:45.
Dupuis M., Spangler D. and Wendoloski 1979 NRCC Report.
Echt O., Sattler K and Recknagel E. 1981 Phys. Rev. Lett. 47:1121.
Ewing D. W. and Pfeiffer G. V. 1982 Chem. Phys. Lett 86:365.
Fougere P. F. and Nesbet R. K. 1966 J. Chem. Phys. 44:285.
Furstenau N. and Hillenkamp F. 1981 Int. J. Mass. Spect. 37:155.
Gupta S. K. and Gringerich K. A. 1979 J. Chem. Phys. 71:3072.
Heath J. R., Liu Y., O'Grien S. C. Zhang Q. L. Curl R. F., Smalley R. E.,
        and Tittel F. K. 1985 J. Chem. Phys. 83:5520.
Hoffmann R. 1966 Tetrahedron 22:521.
Knight W. D., Clemenger K. de Heer W. A., Saunders W. A., Chou M. Y. and
        Cohen M. L. 1984 Phys. Rev. Lett 52:2141.
Koopmans T. 1934 Physica 1:104.
Liskow D. H. Bender C. F. and Schaefer H. F. III 1972 J. Chem. Phys 56:5075.
Martin T. P. and Schaber 1985 J. Chem. Phys. 83:855.
Messmer R. P. 1982, Springer Series in Chemical Physics 20:315 (eds.
        Vanselow R. and Howe R., published by Springer-Verlag, New York,
        1982).
Muhlback J., Sattler K. and Recknagel E. 1982 Phys. Lett. 87A:418.

Peric-Radic J., Romelt J., Peyerimhoff S. D. and Buenker R. J. 1977 Chem. Phys. Lett. 50:344.

Philips J. C. Chem. Rev. to be published.

Pitzer K. S. and Clementi E. 1959 J. Am. Chem. Soc. 81:4477.

Raghavachari K. and Logovinsky V. 1985 Phys. Rev. Lett. 55:2583.

Ransil B. J. 1960 Rev. Mod. Phys. 32:245.

Rao B. K., Khanna S. N. and Jena P. Solid St. Comm. 58:53.

Ray A. K., Fry J. L. and Myles C. W. 1985 J. Phys. B. 18:381.

Ray A. K. 1986 J. Phys. B. 19:1253.

Ray A. K. 1986 Bull. Am. Phys. Soc. 31:682.

Rohlfing E. A. Cox D. M. and Kaldor A. 1984 J. Chem. Phys. 81:3322.

Romelt J., Peyerimhoff S. D. and Buenker 1978 Chem. Phys. Lett. 58:1.

Schaefer H. F. 1975 J. Chem. Phys. 62:4815.

Tsong T. 1984 Appl. Phys. Lett. 45:1149 and 1984 Phys. Rev. B 30:4946.

Tomanek D. and Schluter M. A. 1986 Phys. Rev. Lett. 55:2853.

Whiteside R. A., Krishnan R., Frisch M. J., Pople J. A. and Schleyer von R. P. 1981 Chem. Phys. Lett. 30:547.

Whiteside R. A., Krishnan R., Defrees D. J., Pople J. A. and Schleyer von R. P. 1981 Chem. Phys. Lett. 78:538.

Zavitsanos P. D. and Carlson G. A. 1973 J. Chem. Phys. 59:2966.

MULTIPHOTON STIMULATED DESORPTION:

THE GEOMETRICAL STRUCTURE OF NEUTRAL SODIUM CHLORIDE CLUSTERS

R. Pflaum, K. Sattler and E. Recknagel

Fakultät für Physik, Universität Konstanz

D-7750 Konstanz, West Germany

ABSTRACT

It is shown that multiphoton induced fragmentation can serve as a tool for structural investigation of clusters. Sodium chloride clusters are generated by condensation in warm or cold helium gas. In mass spectra of warm clusters taken by multiphoton ionization magic numbers show up which are different from the anomalies reported so far, and which are independent from the photon energy. They are explained by electronic excitation of sodium halide admolecules desorbing before ionization and leaving rectangular neutral cluster cores. In contrast, cold clusters show the well-known magic numbers of ion cuboids if they dissociate after ionization.

INTRODUCTION

The geometrical configurations of alkali halide clusters have been discussed extensively, but experimental information was restricted to cluster ions: Anomalies arising in cluster mass spectra due to dissociation of clusters after ionization yield information about the stability and structure of alkali halide cluster ions. For sodium chloride, the magic numbers produced by delayed dissociation of cluster ions sputtered from solid surfaces /1/ are the same as those produced by electron induced dissociation of cluster ions which occurs when neutral clusters are ionized with electrons of sufficient energy /2/. They correspond to cluster ions having the shape of rectangular parallelepipeds, which will be denoted as cuboids below. Information on neutral clusters' structure could not be obtained, because the size distribution of clusters ionized close to the ionization threshold is smooth /2/. This could be attributed to a deficiency of any preferred configurations or it can be an indication that the influence of desorption during the condensation process is negligible. It will be shown that the latter is correct and that neutral sodium chloride clusters can be described as cuboid cluster cores with or without admolecules on their surface.

EXPERIMENT

Clusters are generated by quenching the sodium halide vapor (typ. 0.2 Torr at 1030 K) in helium gas of typ. 1 Torr at 70 K ("cold") or 300 K ("warm"). They are extracted into vacuum and analyzed in a time-of-flight mass spectrometer. Ionization is performed by photons (multiphoton ionization =: MPI) and electrons (for reference spectra) alternately with 50 laser pulses and 1000 electron pulses per second. For each type of ionization a spectrum is accumulated within the memory of a multichannel analyzer /3/. The laser pulse (duration 10 nanoseconds FWHM) is generated by an excimer pumped dye laser producing some 4 mJ per pulse, or 0.1 mJ after frequency doubling (UV). For a focal diameter of 0.1 mm this corresponds to a peak power of some 5 GWcm^{-2} (visible) or 150 MWcm^{-2} (UV).

Unlike electron ionization, which is performed at currents too low for a significant probability that one cluster may be hit by two electrons, MPI is possible only for photon currents that provide two photons within several femtoseconds. With the laser pulse parameters given above and a nearest-neighbour-distance of 2.7 Å, a cluster of size n will be illuminated by $n^{2/3} \cdot 125,000$ (visible) or $n^{2/3} \cdot 1,500$ (UV) photons during the laser pulse. Besides the ionization, electronic excitation plays an important role, and this is when alkali halides behave quite different from more covalent salts: The first electronically excited state of an alkali halide molecule is a weakly bound van-der-Waals state with the electron being transferred to the Na (fig. 1) /4/. Molecules dissociate after photon absorption, as the shallow minimum in the upper state cannot be reached by vertical transitions. Higher electronically excited states are reported to be completely repulsive /5/.

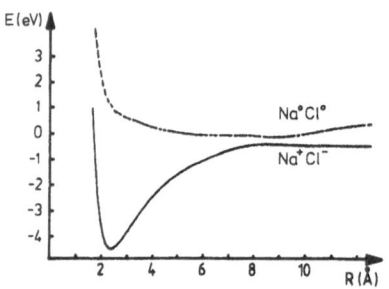

Fig. 1.

Ground state and electronically excited state of NaCl, determined by <u>ab initio</u> calculations /4/.

Therefore a molecule which is excited electronically for instance on the surface of a cluster can be assumed to result in two neutral atoms on the cluster surface. These are bound to the cluster by the polarization of their electron shells caused by the cluster's electrostatic field (ion induced dipole interaction). This bond is much weaker than the ionic bond before the excitation. The bond energies are some 40 meV for Cl and 400 meV for Na atoms /6/. The energy required for electronic excitation, from this reason, is approximately given by the ionic bond energy (Madelung energy) of the ion pair under consideration. We note that a molecule, within an ionic cluster, is an arbitrary pair of nearest-neighbour-ions. The Madelung energy was calculated for ion pairs on and within several configurations for a nearest-neighbour-distance of 2.7Å. As an example, fig. 2 displays the results for (NaCl)$_{24}$.

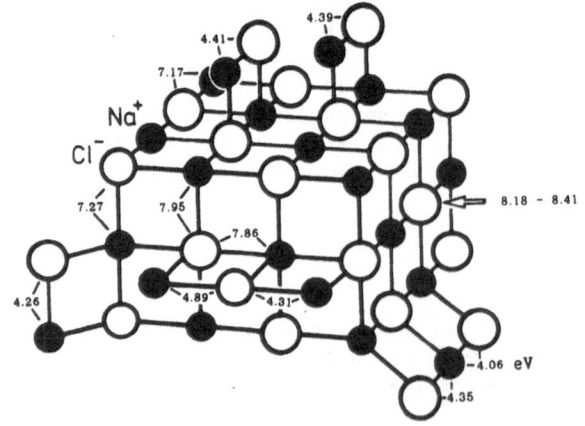

Fig. 2.

Energy neccessary to excite an ion pair into an atom pair, calculated for a nearest-neighbour-distance of 2.7 Å by evaluation of the cluster lattice sum and comparison with bulk Madelung constant and lattice energy (7.97 eV).

From these considerations three main energy regions are associated with the excitation of ion pairs, given by: core- or partially-core-molecules (6.9-8.4 eV), admolecule pairs (6-6.3 eV) and single admolecules (4-4.9 eV). This means the clusters are completely transparent for photons below 4 eV, and especially the cluster cores are transparent below 7 eV, which is still lower than the bulk energy gap given by 8.97 eV /9/. Within the laser pulse, we thus have to consider the three processes listed below, according to their approximate threshold energies:

1.) Electronic excitation of admolecules on cluster cores ($\geq$ 4 eV)

2.) Excitation of core molecules ($\geq$ 7 eV)

3.) Ionization (i.e. neutralization of one chlorine, $\geq$10 eV /10/)

At a given photon energy of 3 eV, for instance, 1.) will be performed by a two-photon-process, while 2.) will require three and 3.) four photons. In contrast to electron ionization, which can be described by one scattering process, laser ionization provides the facility for all three processes to occur within one cluster during the laser pulse, eventually more than one time, as 10 nanoseconds pulse duration is a long time compared with lattice vibrations. This results in a considerable probability for electronic excitation of admolecules during the rise time of the laser pulse, before the laser power reaches the level neccessary for ionization. Whether or not the resulting neutral adatoms desorb from a cluster will depend on the cluster temperature.

RESULTS & DISCUSSION

MPI of warm clusters, as shown in fig. 3a, produces deep intensity minima which are not observed in the reference spectrum taken by electron ionization (fig. 3b). The <u>maxima</u> observed at n = 18, 20, 24, 32, 40, 48, 50, 56 and 60

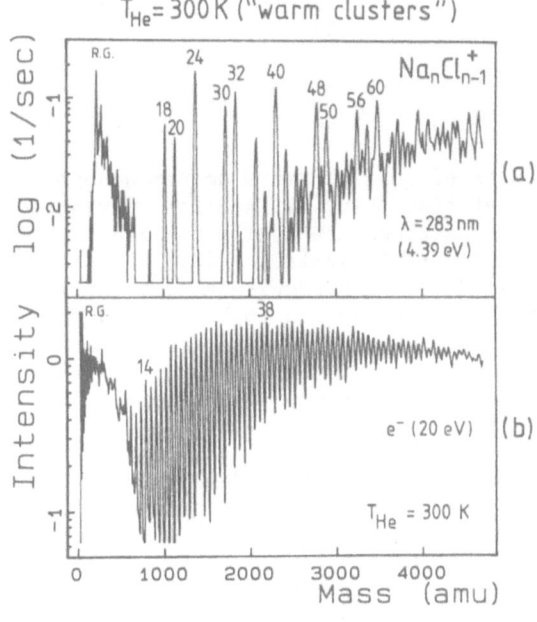

Fig. 3.

MPI mass spectrum (a) and electron reference spectrum (b) of "warm" NaCl clusters (condensed in water-cooled helium). Warm clusters show distinct magic numbers, n = 18, 20, 24, 30, 32, 40, 48, 50, 56, 60 and 72 corresponding to <u>neutral</u> $(NaCl)_n$ cuboids of 3x3x4, 2x4x5, 3x4x4, 3x4x5, 4x4x4, 4x4x5, 4x4x6, 4x5x5, 4x4x7, 4x5x6 and 4x6x6 single ions.

correspond to <u>neutral</u> cuboids consisting of 3x3x4, 2x4x5, 3x4x4, 3x4x5, 4x4x4, 4x4x5, 4x4x6, 4x5x5, 4x4x7 and 4x5x6 single ions (Na$^+$ or Cl$^-$) /11/. They are essentially the same for photon energies beween 2.19 and 4.4 eV /12/ and include all compact neutral cuboids in this size range.

We attribute the appearence of these magic numbers to multiphoton stimulated desorption of molecules from <u>neutral</u> cuboid clusters, <u>before ionization</u> which is possible as a result of electronic excitation of admolecules as described above. The important role of cluster temperature on the desorption of Cl (neutral) was experimentally confirmed by the stoichiometry of ionized clusters: Electron ionization of cold clusters generates $(NaCl)_n^+$, whereas ionization of warm clusters generates ions $(Na_nCl_{n-1})^+$ only /2/. In the case of MPI, the time scale of desorption is crucial: If the cluster is sufficiently warm, neutral adatoms can desorb from the cluster within a time much shorter than the duration of the laser pulse. Thus, <u>warm</u> clusters are primarily stripped off their admolecules, leaving compact cluster cores for ionization which, from the measured amount of constituents, are identified as cuboids.

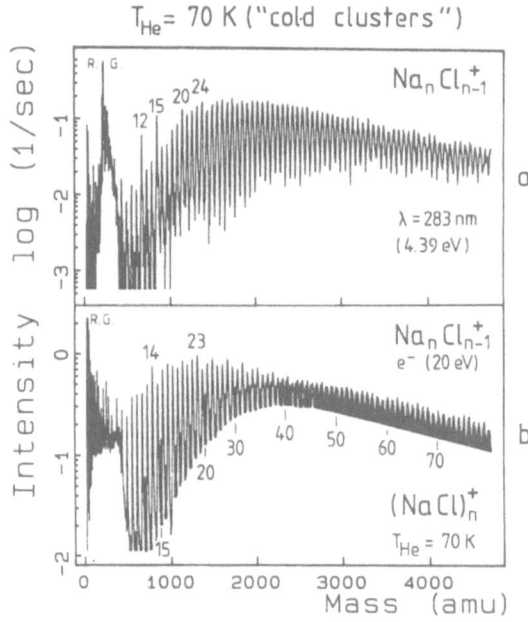

T$_{He}$= 70 K ("cold clusters")

Na$_n$Cl$_{n-1}^+$

λ = 283 nm (4.39 eV)

a

Na$_n$Cl$_{n-1}^+$
e$^-$ (20 eV)

$(NaCl)_n^+$

T$_{He}$ = 70 K

b

Intensity log (1/sec)

Mass (amu)

Fig. 4.

MPI mass spectrum (a) and electron reference spectrum (b) of "cold" NaCl clusters (condensed in LN$_2$-cooled helium). The maxima in the small size region, n = 12, 15, 20 and 24, correspond to neutral $(NaCl)_n$ cuboids of 2x3x4, 2x3x5, 3x3x4 and 2x4x5 single ions. In the electron reference spectrum (b) the lower cluster temperature is reflected by the appearence of stoichiometric ions $(NaCl)_n^+$.

In contrast, the excited admolecules of <u>cold</u> clusters do not desorb at once. Therefore all cluster sizes are detected (fig. 4). Another consequence is that cold clusters with neutralized Na adatoms are easily ionized (IP(Na) = 5.1 eV in comparison with 10 to 12 eV for NaCl clusters /10/). Therefore ionization and excitation of additional admolecules require almost the same energy, and we are not forced to assume all admolecules to be excited before ionization, as we have to do for warm clusters. Nevertheless, maxima at n = 12, 15, 20 and 24 corresponding to neutral 2x3x4-, 2x3x5-, 3x3x4- and 2x4x5-cuboids show a remaining influence of desorption <u>before ionization</u> within the small clusters.

With lowered photon energy, i.e. higher photon fluxes necessary for ionization (fig. 5), different maxima emerge which correspond to the magic numbers of cluster <u>ions</u>. At 2.19 eV for instance, the 3x3x3 $(Na_{14}Cl_{13})^+$ cube ion is the only line in the low mass range. This is attributed to <u>dissociation after ionization</u> resulting from a slow heating of the cluster ions which were initially cold: If the excited admolecules, i.e. adatom pairs, do not desorb at once, they can relax to their ground state by electron transfer. In this case, they transfer their excitation energy of some 5 eV to the cluster lattice. Under the assumption that these 5 eV are distributed equally on the 6n-6 vibrational degrees of freedom of, for instance, an n = 14 cluster, the temperature rises from 70 to more than 750 K. This leads to immediate desorption of all other excited adatoms and finally can cause dissociation of the cluster ions before mass analysis, but after the end of the laser pulse.

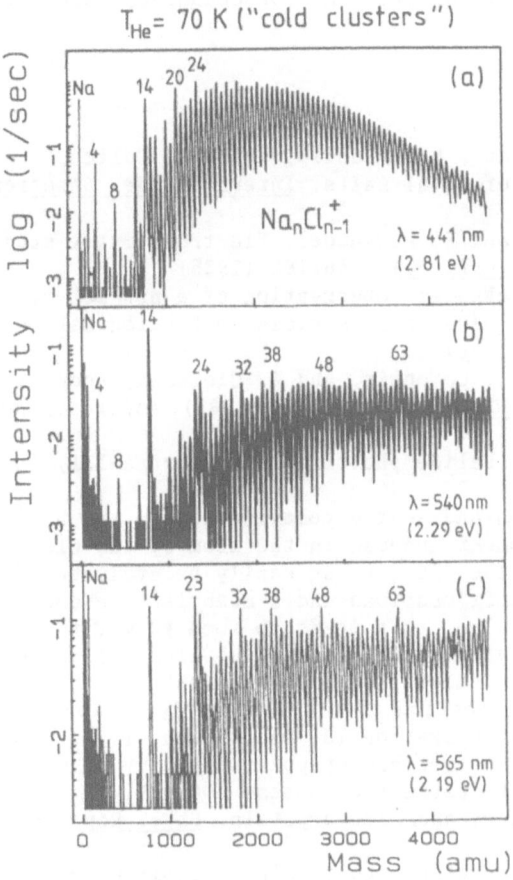

Fig. 5.

MPI mass spectra of "cold" NaCl clusters, taken at 441 (a), 540 (b) and 565 nm (c). With lowered photon energy, the magic numbers of cluster ions $(Na_nCl_{n-1})^+$ n = 14, 23, 38 and 63 increase in intensity.

We would like to point out that resonance enhancement does not create the magic numbers in our study. For evidence see figs. 3 and 4. One would expect more pronounced resonance enhancement for cold clusters, because in this case intensity anomalies are due to spectroscopic differences between subsequent cluster sizes, and because these differences can take larger values when the clusters are rovibronically less excited. Instead, the anomalies observed are much stronger for warm clusters. Evidence against a strong influence of resonance enhancement also comes from the similarity of MPI mass spectra from different sodium halides, NaCl, NaBr and NaI, taken at the same photon energy. The anomalies within the MPI mass spectra are essentially the same /12/ although the molecular ionization potentials of NaCl, NaBr and NaI differ by more than 0.5 eV.

SUMMARY

MPI of sodium halide clusters has shown the existence of <u>exceptionally stable neutral clusters</u> which can be enriched by multiphoton stimulated dissociation of warm clusters during the laser pulse. The magic numbers obtained in this way confirm rectangular configurations. The dissociation which is found to depend on the cluster's initial temperature is explained by multiphoton excitation of admolecules resulting in neutral adatoms bound to the cluster by ion induced dipole interaction.

ACKNOWLEDGEMENT

This work was supported in part by the Deutsche Forschungsgemeinschaft

REFERENCES

1. T.M Barlak, J.E.Campana, J.R. Wyatt, B.I. Dunlap, and R.J. Colton, Secondary ion mass spectrometry of metal salts, <u>Intern J. Mass Spectrom. Ion Phys.</u> 46:523 (1983)
2. R. Pflaum, P. Pfau, K. Sattler, and E. Recknagel, Electron impact studies on sodium halide microclusters, <u>Surf. Sci.</u> 156:165 (1985)
3. R. Pflaum, K. Sattler, and E. Recknagel, Observation of a new set of anomalies in mass spectra of CsI cluster ions close to the ionization threshold, <u>Phys. Rev. B</u>33:1522 (1986)
4. P.K. Swaminathan, A. Laaksonen, G. Corongiu, and E. Clementi, Accurate theoretical modeling of NaCl, <u>J. Chem. Phys.</u> 84:867 (1986), and references therein
5. T.-M. R. Su, S. J. Riley, Alkali halide photofragment spectra III, <u>J. Chem. Phys.</u> 72:6632 (1980)
6. For a given cluster configuration and adatom polarizability, the bond energy provided by the dipole moment induced in the atom by the electrostatic field of the ions within the cluster can easily be evaluated /7/. Using rectangular equidistant configurations and a mean lattice constant between the molecular and bulk value, this leads to some 40 meV for a chlorine atom located on a cluster surface. This value lies well between chlorine bond energies which were calculated for different rectangular configurations, including the distortion of cluster configuration and chlorine bond after chlorine neutralization (J. Diefenbach, T. P. Martin, unpublished). Due to the ten times higher polarizability of neutral sodium /8/, its bond energy can be estimated to some 400 meV.
7. T.P. Martin, Alkali halide clusters and microcrystals, <u>Phys. Rep.</u> 95:167 (1983)
8. R.R. Teachout and R.T. Pack, The static dipole polarizabilities of all the neutral atoms in their ground states, <u>Atomic Data</u> 3:195 (1971)
9. D. M. Roessler, W. C. Walker, Electronic spectra of crystalline NaCl and KCl, <u>Phys. Rev.</u> 166:599 (1968)
10. T.P. Martin, Electronic, vibrational and structural properties of alkali halide clusters, <u>Physica B</u> 127:214 (1984)
11. Whereas neutral sodium chloride clusters are composed of equal numbers of Na^+ and Cl^- ions, and therefore contain an even number of atoms, cluster ions $(Na_nCl_{n-1})^+$ always consist of an odd number of atoms. Therefore with neutral cluster cuboids at least one edge length is given by an even number, but with cluster <u>ion</u> cuboids all edge lengths have to be given by odd numbers.
12. R. Pflaum, K. Sattler and E. Recknagel, Multiphoton stimulated desorption, subm. for publication

MASS AND TEMPERATURE MEASUREMENT IN PURE VAPOR EXPANSION OF METALS AND SEMI-METALS

J.G. Pruett, H. Windischmann, M.L. Nicholas and P.S. Lampard

The Standard Oil Company

4440 Warrensville Center Rd., Cleveland, Ohio 44128

INTRODUCTION

The formation and characterization of clusters has recently become a subject of considerable interest not only from a fundamental but also a practical point of view.[1] The early experimental studies have concentrated on determining the size dependence of the structural, electronic and magnetic properties of clusters to bridge the gap between atomic and bulk properties. The unique properties of clusters have led to application in areas of fusion refueling[2], catalysis[3] and thin film deposition.[4]

Two common methods for producing large clusters are by vaporizing a solid in the presence of a cold buffer gas[5,6] and by direct supersonic expansion of a material already a gas at room temperature.[7] It has been claimed that large metal clusters can be formed in a neat expansion at a few torr vapor pressure from a heated crucible without the presence of a buffer gas.[8] Using electrostatic energy analysis and electron microscroscopy[9] it has been estimated that silver clusters as large as 500-2000 atoms are formed for a source stagnation pressure of 10 torr and 1 mm nozzle diameter. Films deposited from this type of source have unique properties[10], relative to conventional evaporation, which have been ascribed to the presence of large clusters.

The purpose of this work is to determine the cluster size and distribution from a pure-vapor source using time-of-flight mass spectrometry and to determine the dominant cooling mode of the expanding vapor using fluorescence spectroscopy.

EXPERIMENTAL PROCEEDURE

Two series of experiments were performed in this study. In the first series, the mass spectrum of several materials was measured. In the second series, the final temperature of a tellurium vapor jet was measured.

Mass Analysis

The mass spectrum of material was recorded using a standard linear time of flight mass spectrometer shown in Figure 1. The crucible was a pyrolytic graphite closed cylinder with a screw on cap and a capacity of 10 ml.

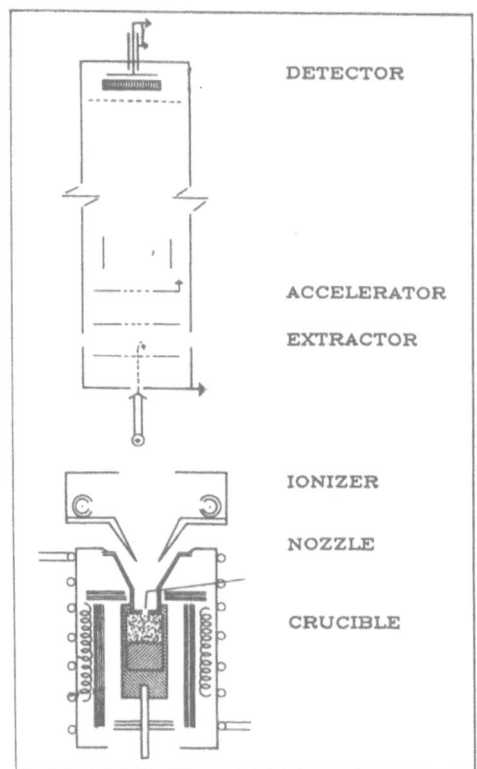

Figure 1. Schematic of pure vapor source and time of mass spectrometer.

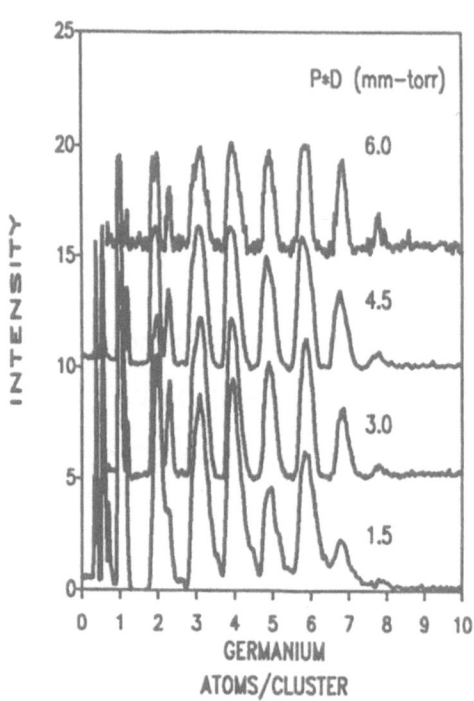

Figure 2. Ge mass spectrum taken at various P*D values. Low masses not to scale. Baselines offset for clarity.

The nozzle diameter, D, ranged from 0.5 to 2.0 mm. The crucible was radiantly heated to a maximum temperature of 2100 °C with a carbon heater (for refractory materials) or to a maximum of 1500 °C with a tungsten wire heater (for lower melting point materials).

The nozzle flow was directed into an electron bombardment ionizer which was biased at 2 kV, and whose electron energy was set to 100-150 V. The resulting ion stream was accelerated to the grounded opening of the time of flight spectrometer. As the vapor entered the mass spectrometer the ions were retarded with a 1.9 kV grid, and stopped with a 2.1 kV grid, preventing all ions from entering the accelleration region of the ion optics. A 300 volt step was added to the retarding grid with a rise time of 20 nsec, ejecting all ions in the draw-out region into the acceleration region. After drifting 125 cm, they were detected by an inline channelplate multiplier. The time of flight signal was then collected with a transient digitizing scope, averaged and mass converted. The mass resolution was only about 200, but this was adequate for the present cluster size studies. For one experiment, neutral material entering the draw-out region was photoionized using 193 nm light from an excimer laser, rather then electron beam ionized. For the lower melting point materials, the ions were formed in the draw-out region with a pulsed electron gun.

Temperature Measurement

To obtain the tellurium dimer excitation spectrum, the mass spectrometer was replaced by a light baffle and photomultiplier placed at a right angle to the vapor stream. The light from a pulsed dye laser operating on Coumarin 420 was directed into the vapor flow and tuned over the B-X band of the tellurium dimer. The fluorescence was filtered to remove remaining scattered laser light. The pulsed fluorescence decay signal was sampled by a boxcar integrator and digitized. Care was taken to avoid saturation of the fluorescence signal and of the photomultiplier gain. The resulting spectra were compared with simulated spectra with varying rotational contour temperatures until a match with the experimental spectra was obtained. This method was confirmed by obtaining an excitation spectrum of tellurium dimer in a heated bulb at a known temperature, and comparing it with a simulated spectrum at that temperature. Vibrational temperatures were then obtained by comparing fundamental band and hot-band intensities in the bulb spectrum and in the various jet expansion spectra.

RESULTS AND DISCUSSION

Mass Spectra

Figure 2 presents the mass spectrum of germanium obtained at a stagnation pressure, P, of 3 torr and various nozzle sizes ranging from a diameter, D, of 0.5 to 2.0 mm , with a fixed throat length of 1.0 mm. The largest germanium cluster observed is 8 atoms at a P*D value of 6. The distribution generally peaks at the monomer and decreases significantly for higher masses. The signal of the monomer to background noise was 10^4. The intensity of the monomer was typically 30-50X that of the dimer. The four and six atom cluster signals are stronger than those of their nearest neighbors, consistent with results obtained from a laser vaporization source in this laboratory and that of others. The three peaks located to the left of the monomer are the double, triple and quadruple ionized germanium monomers. High resolution spectrometry of these peaks revealed the five isotopes in proper relative proportion. Experiments

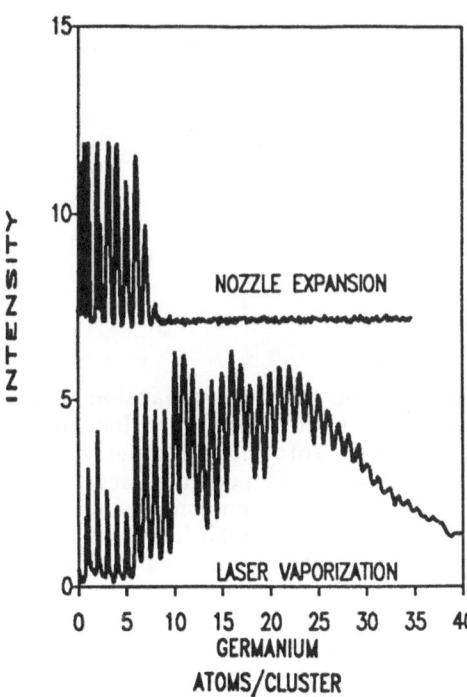

Figure 3. Comparison of Ge mass spectra for laser vaporization and pure gas expansion.

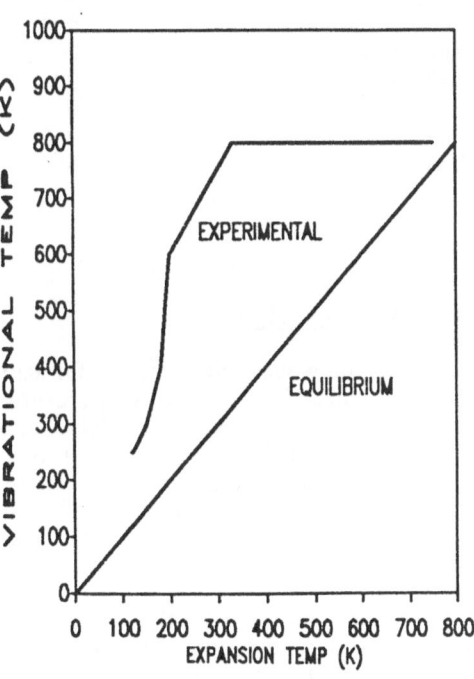

Figure 4. Vibrational temperatures of Te dimer at various expansion temperatures showing lack of equilibrium.

performed with photo rather than electron impact ionization produced similar cluster sizes.

We observed the same cluster size and distribution for a 1mm nozzle diameter source as for a crucible with a 13 mm opening, indicating that the mass spectrum observed for the smaller nozzles represents the equilibrium distribution of clusters for a germanium source operated at these temperatures.

Figure 3 presents a comparison of the germanium mass spectrum obtained for two types of sources. The top spectrum is from a pure vapor source operated at P*D= 6.0 mm-torr and ionizer conditions as described above, while the lower trace is from a laser vaporization gas aggregation source of the type described in reference 5. The results show that the laser-vaporization source readily produces clusters as large as 40 atoms. Both traces show similar intensity patterns up to 6 atoms, while the laser-vaporization source shows additional peaks for larger cluster sizes. Under optimum conditions we have seen slicon clusters as large as several hundred atoms.

Table I summarizes the maximum cluster size detected at the indicated P*D values for a variety of materials. In all cases there was no change of the mass distribution as P*D was varied.

Table I. Maximum Cluster Size Observed from a Pure-Vapor Source

	P*D (mm-torr)	Maximum Cluster Size
Germanium	6	8
Silver	400	9
Aluminum	3	2
Tellurium	90	5
Selenium	6	8
Zinc	240	2
Magnesium	60	2

Expansion Temperature

Since there was no evidence for cluster growth, the tellurium expansion was used to characterize the jet temperature. Rotational simulations revealed that spectral band shapes were sensitive enough to obtain rotational temperatures accurate to about 50 °K over the range 100-1000 °K. It is generally accepted that because of the small rotational energy spacing, the relaxation kinetics of the rotations of a jet cooled molecule is fast enough to reach equilibrium with the translational energy bath,[11] so we take our rotational temperatures to be a measure of the expansion temperature.

Analysis of the vibrational hot band intensities revealed that the vibrational degree of freedom did not relax at the same rate as did the rotational degrees of freedom, so that vibrations are not in equilibrium with rotation and translation. Similar effects have been the rule in jet cooled spectroscopies.[11] Figure 4 shows the effect by plotting the vibrational temperature observed versus the temperature of each expansion. Equilibrium between vibration and rotation would result in the data lying along the diagonal. It is clear that vibrational relaxation does not occur until the rotational temperature reaches 300 °K or below. A room temperature rotational bath was obtained by a source pressure of 10 torr with a nozzle diameter of .6 mm.

The lack of cluster growth is rationalized as follows. The beginning of cluster growth is promoted by an increase in the probability of having a critical sized nucleus, which is then able to spontaneously grow due to the lowering of its free energy of formation with increased size.[12] The probability of finding a critical nucleus in an equilibrium environment is given by its Boltzman probability based on its free energy of formation.

As thermodynamic conditions shift (such as T and P in the supersonic expansion), the density of critical nuclei will change only if enough time elapses at the new conditions that equilibrium can be approached. In the case of cluster size changes, chemical reactions are required, releasing significant energy as vibration, which must be cooled for cluster stability to occur. The cooling rate is limited by the tendency of the vibrational degrees of freedom to relax in their bath of cold translation and rotation. Thus, unless vibrational relaxation occurs, new critical sized nuclei cannot be formed. In that case, unless the critical sized nuclei exist as a significant fraction of the vapor in equilibrium with the liquid, no significant population of larger clusters can be formed.

CONCLUSION

Mass spectrometric measurement in combination with temperature measurement of a pure-vapor source indicate that the expanding gas is rotationally cooled to about 130°K but not substantially vibrationally cooled. The data indicate that while small clusters are present for several different materials, no large clusters were produced for a source operated under a wide variety of conditions. Large germanium and silicon clusters were observed in a laser vaporization gas aggregation source.

REFERENCES

1. a. International Meeting on Small Particles and Inorganic Clusters, Paris. Proceedings in J. Phys.,**38**,Colloq. C-2 (1977).
    b. Second International Meeting on Small Particles and Inorganic Clusters, Lussane. Proceedings in Surf. Sci,**106**,1-608,(1980).
    c. Discussion Meeting of the Deutsche Bunsen-Gesellschaft fuer Physikalische Chemie, Koenigstein, Taunus (1983). Proceedings in Ber. Bunsenges. Phys. Chem.,**88**(1984).
    d. International Symposium on Small Particles and Inorganic Clusters, Berlin. Proceeding in Surf. Sci.,**156**(1985).
    e. International Symposium on The Physics and Chemistry of Clusters, Richmond, Virginia, U.S.A., Oct.(1986). This Journal

2. E.W. Becker,H.D. Falter, O.F. Hagena, P.R.W. Henkes, R. Klingelhofer, H.O. Moser, W. Obert, and I. Poth, in: "Fusion Technology", Pergamon Press, 331(1979).

3. M.E. Geusic, M.D. Morse, S.C. O'Brien, and R.E. Smalley, Rev. Sci. Instr.,**56**,2123(1985).

4. T. Takagi, I. Yamada, M. Munori, and S. Kobiyama, Proc. 2nd Int. Conf. on Ion Sources, Wien (Oesterreichische Studiengesellschaft fur Atomenergie) 790,(1972).

5. T.G. Dietz, M.A. Duncan, D.E. Powers, and R.E. Smalley, J. Chem. Phys.,**74**,6511(1981).

6. K. Sattler,J. Muehlbach and E. Recknagel, Phys. Rev. Lett.,**45**,821(1980).

7. E.W. Becker, K. Bier, and W. Henkes, Z. Phys.,**146**,333(1956). For a Review see P.P. Wegerer and B.J.C. Wu in "Nucleation Phenomena", Advances in Colloid and Interface Science, vol. 7,ed. A.C. Zettelmeyer, Elsevier, Amsterdam, 325(1977).

8. I. Yamada and T. Takagi, Thin Solid Films,**80**,105(1981).

9. I. Yamada, H. Takaoka, H. Inokawa, H. Usui, S.C. Cheng, and T. Takagi, Thin Solid Films,**92**,137(1982).

10. For a comprehensive collection of papers dealing with films produced by a pure vapor source,referred to as Ionized Cluster Beam (ICB) deposition, see Proc. Int. Ion Engineering Congress, Ed. T. Takagi, Kyoto, Japan (1982,1983,1984,1985).

11. D.H. Levy, Annu. Rev. Phys. Chem. **31**,197(1980).

12. M. Volmer and A. Weber, Z. Phys. Chem., **119**, 277 (1925).

FINE STRUCTURE OF MASS SPECTRUM OF SODIUM CLUSTERS: GIANT-ATOM CONCEPT

Susumu Saito and Shuhei Ohnishi

NEC Fundamental Research Laboratories

Miyazaki, Miyamae-ku, Kawasaki 213, Japan

ABSTRACT

We propose a giant-atom cluster model for sodium clusters, which explains the fine structure of the mass spectra of sodium clusters, that is, the non-shell-closing magic numbers. The giant-atom concept is another aspect of the shell model for alkali-metal clusters. A typical example of the giant atom is Na_2, which can be regarded as a giant rare-gas atom because it has a shell-closing electronic structure, $(1s)^2$, in the shell model. An icosahedron of this giant rare-gas atom, $(Na_2)_{13}$, is expected to be stable and corresponds well to the non-shell-closing magic number, 26, in the mass spectrum of sodium clusters. Another example is a giant alkali-metal atom, Na_{19}, which has a spherical core and one s-electron, and the dimer $(Na_{19})_2$ explains the peak at 38. Some other examples are also enumerated and found to coincide to the mass spectrum of sodium clusters. This fact supports that the reaction of sodium clusters is analogous to that of ordinary atoms, since the cluster reaction at the adiabatic expansion of sodium gas rules the relative abundances of clusters.

I. INTRODUCTION

Recently, clusters attract great interests of physicists, since their physical properties are expected to be different from those of their bulk phases. Their geometorical structures, however, can hardly be observed directly, especially in the case of small clusters whose numbers of atoms are less than 10^2. Therefore, their structures have usually been conjectured by time-of-flight mass spectra. The number of atoms of the abundant cluster, that is, the number which gives the peak of mass spectra, is called "magic number." For example, 13 and 55 are clear magic numbers of xenon clusters[1] and coincide to the numbers of atoms of the icosahedron. Hence, xenon clusters are thought to have icosahedral packings. Theoretical approaches to the cluster geometry using the first-principle calculations have also been done.[2-6]

In the case of alkali-metal clusters, however, magic numbers have revealed the validity of the simple geometrical model, that is, a shell model.[7] In a bulk metal, electrons bearing metallic bonds are delocalized and behave as nearly free electrons. In a shell model for metal clusters, valence electrons are also delocalized but confined to a finite region. A recent experiment using Secondary-Ion-Mass-Spectrometry (SIMS) has reported that the shell model is also applicable to noble metal clusters.[8]

Magic numbers of sodium clusters, 2, 8, 20, 40, 58, and 92, have been explained by a spherical shell model,[7] in which valence electrons are moving independently. Magic-number sodium clusters are found to have shell-closing electronic configuration. Recent calculations using a spherical jellium-background model for alkali-metal clusters including an electron-electron interaction also support the above shell-closing magic numbers.[9-11] Mass spectra of sodium clusters (Figs. 1 and 2 in Ref. 7), however, show other, non-shell-closing magic numbers, 12, 14, 26, 30, 36, and 38. The relative abundances of these non-shell-closing magic numbers are usually less than the above shell-closing magic numbers. Peaks at 36 and 38, however, have nearly the same hight as the peak at 34, which is the shell-closing magic number. These fine structures are clearer in sodium clusters than in potassium clusters.[12] Here we report that these non-shell-closing magic numbers indicate the validity of another aspect of the spherical shell model for sodium clusters, that is, a "giant-atom" concept.[10]

It is well known that the sequence of electronic energy levels in an infinite spherical potential well is given by

$$1s, \quad 1p, \quad 1d, \quad 2s, \quad 1f, \quad 2p, \quad 1g, \quad 2d, \quad 1h, \quad 3s, \quad \ldots \qquad (1)$$

In the case of a simple effective potential, 1h and 3s are in the reverse order but lower levels are in the same order as (1). In a spherical jellium model,[13] the sequence is the same as (1). Symmetries of the lower levels, s, p, d, etc. also appear in the case of the Coulomb potential, that is, in the atomic enegy levels, although the level sequence of the Coulomb potential is quite different from (1).

Many physical properties of atoms are decided by the symmetries and occupation numbers of the highest occupied levels. This is the reason why the periodic law for elements exists. Since the electronic energy levels for some sodium clusters possess the common feature with atoms from the viewpoint of a shell model, it is expected that there are some analogous features between atoms and sodium clusters. Certain sodium cluster will behave as a "giant atom." This giant-atom analogy brings us important information about the reaction of sodium clusters and, in turn, about their mass spectrum. The sodium cluster beam in the time-of-flight experiment is usually produced by the pulsed laser vaporization of the bulk within a pulsed supersonic nozzle. Relative abundances of clusters will mainly be decided by the reactivity for fusion of clusters when the sodium gas is adiabatically expanded. Shell-closing clusters will have giant inert-gas atom properties and scarcely react to others and many of them will survive at the adiabatic expansion. Therefore, they are really observed abundantly in the time-of-flight mass spectra.

The giant-atom feature of sodium clusters is found to be certified by the fine structure of the mass spectra. Some non-shell-closing magic numbers can be interpreted by the giant-atom concept.

II. GIANT RARE-GAS ATOM

The first example is a giant rare-gas atom. Such sodium clusters as Na_2 and Na_8 have spherical shell-closing electronic structure, $(1s)^2$ and $(1s)^2(1p)^6$, respectively, in a shell model. So, these clusters are expected to have properties of rare-gas atoms.

Magic numbers of rare-gas, xenon clusters[1] are 13, 19, 25, 55 and so on. In the case of argon clusters, the observed magic numbers are almost the same.[14] Most of these magic numbers are explained by icosahedral packings. The observed magic numbers 13, 55, and 147 correspond to the icosahedron. The number 19 corresponds to the pentagonal-pyramid-capped icosahedron. Model calculations using the Lennard-Jones potential also support the icosahedral packings for rare-gas clusters.[15] The mass spectrum of xenon clusters also shows slight peaks at 6 and 7 before 13. The magic number 7 is thought to correspond to a pentagonal bipyramid.

Hence, when the Na_2 cluster is considered a giant rare-gas atom, $(Na_2)_6$, $(Na_2)_7$, $(Na_2)_{13}$, and $(Na_2)_{19}$ are expected to have relatively large abundances. These coincide well with observed non-shell-closing magic numbers of sodium clusters, 12, 14, 26, and 38. The clear and important peak corresponding to an icosahedron, 26, is also observed for potassium clusters.[12]

III. GIANT ALKALI-METAL ATOM

Next example is a giant alkali-metal atom. The Na_{19} cluster has one s-electron and spherical core, $(1s)^2(1p)^6(1d)^{10}$, in a shell model. Therefore, Na_{19} corresponds to a giant alkali-metal atom. From this point of view, the non-shell-closing magic number 38 is attributed to $(Na_{19})_2$, since 2 is the magic number of the alkali-metal cluster. The next magic number expected for the giant alkali-metal cluster is, however, too large, $(Na_{19})_8 = Na_{152}$.

It is interesting to point out that this giant alkali-metal concept has the scaling property of sodium clusters. Since the very shallow well also have the same level sequence as (1) from 1s to 2d,[16] $(Na_{19})_{19} = Na_{361}$ cluster is expected to have a spherical core and one s-electron if 2s electrons of giant alkali-metal atom, Na_{19}, are treated as valence electrons moving in a spherical well. So, the $(Na_{19})_{19}$ cluster turns out to be a giant alkali-metal atom. Na_{361} may be, however, too large for a simple shell model because in the larger cluster the crystal-field splitting easily affects the electronic shell structure.[11]

IV. GIANT TRANSITION-METAL ATOM

Further fascinating examples of giant atoms consisting of sodium atoms can be proposed. The Na_{18} cluster is the shell-closing cluster in a shell model. In the experiment, however, Na_{18} is not an abundant cluster. This discrepancy shows that its electronic configuration is not $(1s)^2(1p)^6(1d)^{10}$ but some amount of electrons are in the 2s level instead of 1d. The energy difference between 1d and 2s is small in ordinary simple spherical wells. They even become degenerate for the harmonic potential well.[17] Therefore, Na_{18}, if its electronic configuration is $(1s)^2(1p)^6(1d)^8(2s)^2$, may act as a giant-nickel atom, because the nickel atom has an electronic configuration $(1s)^2(2s)^2(2p)^6(3s)^2(3p)^6(3d)^8(4s)^2$. Since the Ni_3 cluster has been found to be an abundant cluster,[18] the small step observed at 54 in the mass spectrum of sodium clusters[7] may be attributed to the trimer of giant nickel atoms, $(Na_{18})_3$.

One of the most interesting features of giant atoms is the possibility of the magnetism, since such highly degenerate levels as d, f, g, and h easily appear in comparison with an ordinary atom. Although the effect of the non-spherical perturbation due to the geometrical structure will really remove their degeneracies in most cases,[10] some clusters can hold their high spin configurations. For example, an icosahedral cluster, Na_{13}, possesses the same level structure as the spherical shell model. 1s, 1p, and 1d states correspond respectively to A_{1g}, T_{1u}, and H_g under the icosahedral symmetry. Hence, Na_{13}, which cannot be classified to any ordinary group of elements, may have a high-spin configuration.

V. GIANT HYDROCARBON

Another interesting example is hydrocarbon. Na_4 cluster has $(1s)^2(1p)^2$ configuration and may be regarded as a giant carbon atom, since the electronic configuration of the carbon atom is $(1s)^2(2s)^2(2p)^2$. Hence, Na_4 cluster is now designated by C^*. Na atom can be considered a giant hydrogen atom H^* because of its electronic configuration, $(1s)^1$. Then, methane $C^*H^*_4$ is the first candidate for the stable cluster and is really observed as an abundant cluster, Na_8. Ethane $C^*_2H^*_6 = Na_{14}$, propane $C^*_3H^*_8 = Na_{20}$, butane $C^*_4H^*_{10} = Na_{26}$, benzene $C^*_6H^*_6 = Na_{30}$, and cyclohexane $C^*_6H^*_{12} = Na_{36}$ are also observed abundantly.

The correspondence between the giant-hydrocarbon series and mass spectra is different from other giant-atom examples at the point that not only the fine structure but also the main structure of mass spectra, that is, shell-closing magic numbers are included. Hence, the relation between the viewpoint from the simple spherical shell model and that from the giant hydrocarbon is an interesting problem. In the case of the giant methane, the deepest electronic energy level is expected to be A_1, a non-degenerate level, and the second is T_2, the three-fold degenerate level, because of the T_d symmetry. These correspond to the simple shell model for Na_8, which has the deepest 1s level and the three-fold degenerate 1p level above it. In other cases, the direct correspondence of energy levels between giant hydrocarbon and the shell model cannot be expected because geometries of hydrocarbon structures differ from the sphere. In considering the cluster reaction, however, the giant hydrocarbon is one possible viewpoint and mass spectra may show above correspondences.

VI. CONCLUDING REMARKS

From the viewpoint of giant atoms, some clusters can be seized in several different ways, one of which have not to be adopted but each of which has a complementary aspect of others. Same-size clusters detected in the time-of-flight experiment have not always undergone the same aggregation process. For example, $Na_{38} = (Na_{19})_2 = (Na_2)_{19}$. The former is the dimer of giant alkali-metal atoms and the latter is the pentagonal-pyramid-capped icosahedron of giant rare-gas atoms. This may make the non-shell-closing magic number 38 rather clear.[12]

Although the good correspondence between the giant-atom feature of sodium clusters and their mass spectra supports the analogy of the cluster reaction to the atom reaction, in some cases the static analogy between the real cluster (or molecule) and the giant-atom cluster (molecule) can also be considered, as a giant methane. In that case, non-spherical geometries of sodium clusters are taken into account to some extent. It has been reported that the ellipsoidal deformation from the sphere well explains the fine structure of mass spectra of sodium clusters.[19] Although such static approach may not be sufficient to interpret the mass spectra of various clusters,[20,21] the geometries of some giant-atom clusters resemble ellipsoidal shapes. For example, the giant rare-gas icosahedron, $(Na_2)_{13} = Na_{26}$, will resemble a prolate ellipsoid because the giant rare-gas atom, Na_2, really has two centers.

The level sequence of a giant atom is not the same as an ordinary atom and rather resembles that of the semiconductor-heterostructure superatom,[22] which has been proposed recently.[23,24] This new level sequence will give rise to not only ordinary properties of atoms but also new properties characteristic of giant atoms. The giant-atom feature propesed here for alkali-metal clusters may be valid for other metal clusters, and gives us a new aspect of cluster physics.

REFERENCES

1. O. Echt, K. Sattler, and E. Recknagel, Phys. Rev. Lett. 47, 1121 (1984).

2. J. S. Martins, J. Buttet, and R. Car, Phys. Rev. B 31, 1804 (1985).

3. B. K. Rao and P. Jena, Phys. Rev. B 32, 2058 (1985).

4. G. Pacchioni and J. Koutecky, J. Chem. Phys. 84, 3301 (1986).

5. K. Raghavachari, J. Chem. Phys. 84, 5672 (1986).

6. S. Saito, S. Ohnishi, C. Satoko, and S. Sugano, J. Phys. Soc. Japan, 55, 1791 (1986).

7. W. D. Knight, K. Clemenger, W. A. de Heer, W. A. Saunders, M. Y. Chou, and M. L. Cohen, Phys. Rev. Lett. 52, 2141 (1984).

8. I. Katakuse, T. Ichihara, Y. Fujita, T. Matsuo, T. Sakurai, and H. Matsuda, Int. J. Mass Spect. and Ion Processes 67, 229 (1985).

9. M. Y. Chou, A. Cleland, and M. L. Cohen, Solid State Commun. 52, 645 (1984).

10. Y. Ishii, S. Ohnishi, and S. Sugano, Phys. Rev. B 33, 5271 (1986).

11. F. R. Fedfern and R. C. Chany, Phys. Rev. B 32, 5023 (1985).

12. W. D. Knight, W. A. de Heer, K. Clemenger, and W. A. Saunders, Solid State Commun. 53, 445 (1985).

13. W. Ekardt, Phys. Rev. B 29, 1558 (1984).

14. I. A. Harris, R. S. Kidwell, and J. A. Northby, Phys. Rev. Lett. 53, 2390 (1984).

15. M. R. Hoare, Adv. Chem. Phys. 40, 49 (1979).

16. L. D. Landau and E. M. Lifshitz, "Quantum Mechanics," Pergamon Press, Oxford (1977).

17. M. G. Meyer and J. H. D. Jensen, "Elementary Theory of Nuclear Shell Structure," John Wiley and Sons (1955).

18. J. B. Hopkins, P. R. R. Langridge-Smith, M. D. Morse, and R. E. Smalley, J. Chem. Phys. 78, 1627 (1983).

19. K. Clemenger, Phys. Rev. B 32, 1359 (1985).

20. J. Bernholc and J. C. Phillips, Phys. Rev. B 33, 7395 (1986).

21. C. Satoko (private communication).

22. T. Inoshita, S. Ohnishi, and A. Oshiyama, to be published in Phys. Rev. Lett. 57.

23. H. Watanabe, "The Physics and Fabrication of Microstructures and Microdevices," ed. M. J. Kelley and C. Weisbuch, Springer-Verlag, Heidelberg (1986).

24. H. Watanabe and T. Inoshita, Optoelectronics: Devices and Technologies, 1, 33 (1986).

THE Xα(R) METHOD : A SEMI-EMPIRICAL METHOD WITH α VARYING WITH THE

INTERNUCLEAR DISTANCE

T.J. Tseng

Department of Physics
Chung Yuan Christian University
Chung-Li, Taiwan 320, Republic of China

Based on a model function, a new method for the molecular Xα(R) calculations, is proposed with parameter α varying with the internuclear separation. An application has been carried out for the N_2 molecule. Results show that the calculated equilibrium internuclear distance, the total and the dissociation energies are all better than those results obtained by the conventional Xα computations where α is a constant throughout the whole potential energy curve.

INTRODUCTION

Since the Xα method[1,2] introduced two decades ago, the method has been extensively applied to atoms and molecules. Review articles have been reported by several groups[3-6]. The essential feature of this method is the usage of the α scaled exchange correlation potential which replaces the exact one in the Hartree-Fock [HF] method. In the Xα molecular calculations, values of α were usually determined by the atomic energy criteria[7-9]. The value so determined is used through the whole internuclear separations. In other words, the parameter α does not vary with the internuclear distance.

Konowalow and co-workers[10-13] proposed a method for determining the α parameter to be used in the Xα method based on the molecular rather than atomic considerations. The α value was obtained by requiring the calculated internuclear distance of the molecular potential curve at the minimum to be equal to the experimental value. The value of α so determined was then used throughout all internuclear separations. In other words, α is still a constant over the entire potential energy curve. Moreover, the total energy of molecules so calculated may be too low by using this method, molecules such as N_2 is an example[14].

The present work is to propose a method for determining α where α varies with the internuclear separation. The computation is done on the nitrogen molecule, results are compared with those[15,16] by using the conventional Xα calculations.

METHOD

At present, we consider only diatomic molecules, the total energy of such systems under the Xα(R) scheme can be expressed as (energies in Rydberg, distances in a.u.)

$$<EX\alpha(R)> = \sum_i n_i \int u_i^*(1)\ f_1\ u_i(1)\ dv_1 + \int \frac{\rho(1)\rho(2)}{r_{12}}\ dv_1 dv_2$$

$$- \frac{9}{2}(\frac{3}{4\pi})^{\frac{1}{3}}\alpha(R)\int [\rho_\uparrow^{\frac{4}{3}}(1) + \rho_\downarrow^{\frac{4}{3}}(1)]\ dv_1 + \int \frac{2Z_A \cdot Z_B}{R}|\psi|^2 d\tau \quad \ldots\ldots(1)$$

where $f_1 = -\nabla_1^2 - 2Z/r_1$, $\rho(1) = \sum_i n_i u_i^*(1) \cdot u_i(1)$, $\rho_\uparrow(1)$ is the total charge density of up spin electrons. Other symbols have their usual meanings which can be found in Reference 5. It should be noted that the parameter α is now a function of the internuclear distance instead of a constant as it has been considered conventionally.

The proposed α(R) for a diatomic molecule AB should meet the following requirements :

(1) When $R\to\infty, \alpha(R)\to\alpha_s$, where α_s is defined as $\alpha_s = \dfrac{E_A\alpha_A + E_B\alpha_B}{E_A + E_B}$ $\ldots\ldots(2)$

E_A and E_B are total energies of atoms A and B, respectively. α_A and α_B are the α values of each atom. These values can be taken from, e.g., Schwarz[7], Lindgren and Schwarz[8] or Tseng, Hong and Whitehead[9].
(2) When $R\to0$, $\alpha(R)\to\alpha_u$; where α_u is the α value of the united atom.

In one of our previous studies[17], we have shown numerically that the internuclear separation at the maximum of the exchange correlation energy, i.e. the third term on the right hand side of e.q.(1), is greater than that at the minimum of the total energy. On the other hand, the calculated internuclear equilibrium distances by the Xα method were usually larger than the experimental ones[18]. Based on these facts, we then require that the α(R) value is maximized at Re, the equilibrium internuclear separation. The minimum of the calculated potential energy curves will then be expected to shift to the left. Namely, it should be smaller than the conventional Xα calculated Re value; or in other words, it should be closer to the experimental result.

A function similar to the Morse potential[19] but "inverted" (as shown in Fig. 1) is proposed

$$\alpha(R) = p - q\ (1-e^{-\frac{1}{Re}(R-Re)})^2 \quad \ldots\ldots\ldots\ldots(3)$$

parameter p and q can be determined by requirements (1) and (2) stated earlier. If one sets $\alpha=\alpha_{Re}$ at R=Re, it will be obvious, as from eq.(3), that $p=\alpha_{Re}$. From requirement(1), one can easily obtain $q=\alpha_{Re} - \alpha_s$, eq.(3) can then be expressed as

$$\alpha(R) = \alpha_{Re} - (\alpha_{Re} - \alpha_s)(1 - e^{-(\frac{R}{Re} -1)})^2 \quad \ldots\ldots\ldots(4)$$

α_{Re} can be expressed in terms of α_s and α_u by considering the requirement (2). Namely when $R\to0$ one has, from eq.(4),

$$\alpha_u = \alpha(R=0) = \alpha_{Re} - (\alpha_{Re} - \alpha_s)(1 - e)^2$$

or $\quad \alpha_{Re} = 0.51217\ (2.95249\ \alpha_s - \alpha_u) \ldots\ldots\ldots\ldots(5)$

It can be shown that, from eq.(4),

$$\frac{d\alpha}{dR}\Big|_{R=Re} = 0 \quad \ldots\ldots\ldots\ldots\ldots\ldots\ldots(6)$$

$$\frac{d^2\alpha}{dR^2}\Big|_{R=Re} = \frac{-2(\alpha_{Re} - \alpha_s)}{R_e^2} \quad \ldots\ldots\ldots\ldots\ldots(7)$$

The present model of α function leads to α_{Re} to be greater than α_s, therefore $\alpha(R)$ has its maximum value at R=Re. The value of Re can either be taken from the experimental result or be calculated theoretically.

APPLICATION TO N_2 MOLECULE

The multiple scattering $X\alpha(R)$ method with muffin-tin potential approximation[18] is applied to the nitrogen molecule. The molecular space is divided into three regions : region I contains atomic spheres that they are in touching contact with each other, and a single outer sphere is surrounding the atomic spheres. The intersphere space is region II, and space outside of the sphere is region III.

α_s and α_u are taken to be 0.75197 and 0.72751, respectively[7]. These values were obtained by setting the total atomic energies of the nitrogen and the silicon atoms to be equal to their HF energies. The equilibrium internuclear distance 2.068 a.u. of N_2 is taken from the HF molecular calculations[20]. From eq.(5), one may then obtain the value 0.76450 for α_{Re}. The α value at any internuclear distance R can be calculated by eq. (4). The α value so obtained is used for all three regions in the $X\alpha(R)$ method.

The ground state of $N_2(1\sigma_g^2\ 1\sigma_u^2\ 2\sigma_g^2\ 2\sigma_u^2\ 1\pi_u^4\ 3\sigma_g^2)$ molecule is calculated. Values of α determined from eq.(4), are given in Table 1 and Fig. 1. Total energies and exchange correlation energies are listed in Table 1. Total energies, dissociation energies and the equilibrium internuclear separations from various sources are given in Table 2 and Fig 2.

Table 1. Total and exchange correlation energies (in Ry.) of the ground state of N_2 in $X\alpha(R)$ scheme with α determined from eq.(4)

R(a.u.)	$\alpha(R)$	$-E(exc)$	$-E(tot)$
1.5	0.76325	27.8487	213.8422
2.0	0.76449	27.2291	216.1335
2.5	0.76405	26.8342	217.0954
3.0	0.76285	26.5394	217.4835
3.5	0.76137	26.3250	217.5979
3.6	0.76107	26.2897	217.6020
3.7	0.76077	26.2563	217.6026
3.8	0.76047	26.2250	217.5999
4.0	0.75988	26.1687	217.5872
5.0	0.75731	25.9883	217.4662
6.0	0.75543	25.9052	217.3549
7.0	0.75417	25.8709	217.2854
8.0	0.75335	25.8514	217.2429
9.0	0.75283	25.8403	217.2168
10.0	0.75250	25.8336	217.2041
∞	0.75197*		217.1850*

*References (7) and (21)

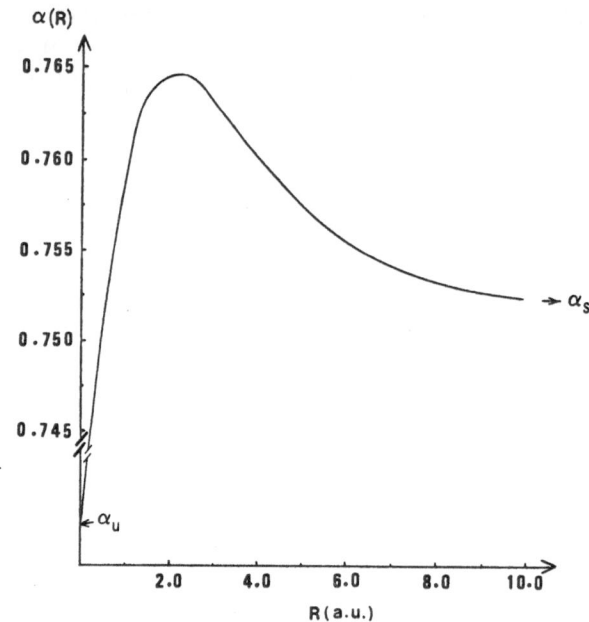

Fig.1. The empirical α values as a function of the internuclear separation determined from eq.(4) used in the Xα(R) N_2 calculations with $\alpha_{Re}=0.76450$, $\alpha_u=0.72751$ and $\alpha_s=0.75197$.

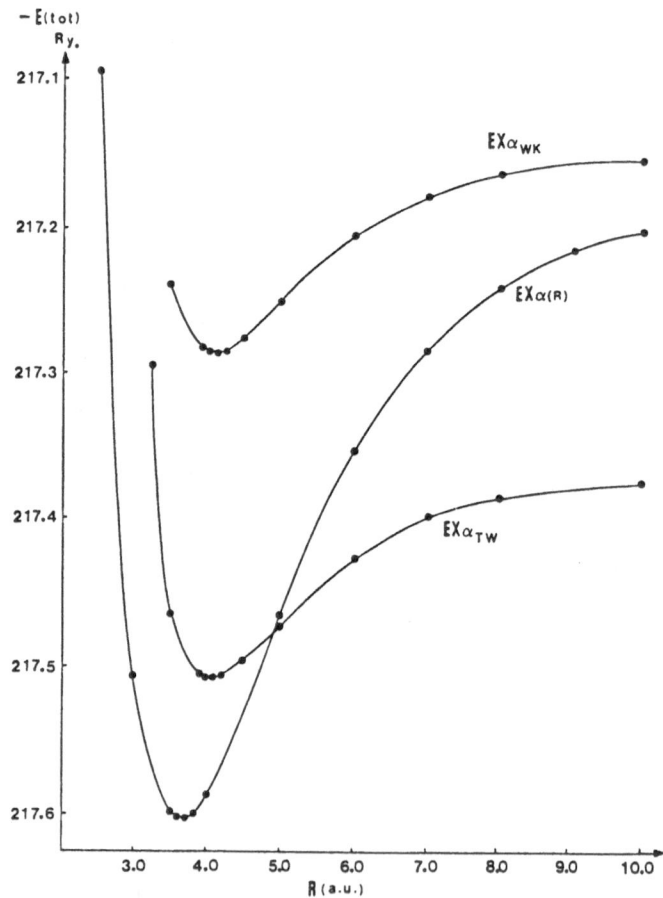

Fig.2. Total energies (in Ry.) of the ground state of N_2 in Xα(R), Xα_{WK}, and Xα_{TW} calculations.

From Fig. 1. it shows clearly that the α value varies with the internuclear distance as expected. The curve shows that α varies drastically with R when R is smaller than Re. It falls off gradually as R increases in the region of R>Re, and it approaches $α_s$ for the very large internuclear distance.

From Table 1, it shows that while the total energy, E(tot), has a minimum around 3.7 a.u., the exchange correlation energy, E(exc) does not have a maximum. This is different from the Xα calculations[17] where E(exc) has a slight hump at the internuclear distance which is greater than the equilibrium distance appearing in E(tot).

Table 2 or Fig. 2 shows that the dissociation energy of the N_2 molecule obtained from the present calculation is 5.68 eV, which is much better than those results obtained by using the conventional Xα method. It is also better than the HF result, 5.27 eV, calculated by Cade et al[20]. In other words, as comparing to the experimental value, the present result is improved by about 8% over the HF one.

As it is expected, the minimum of the N_2 potential energy curve by the Xα(R) calculation, shown in Fig. 2, does shift to the left comparing to the Xα calculations. The present result of Re is improved by about 10%, however it is still quite off from the experimental or HF value.

It is also worth mentioning that the total energy -217.6026 Ry. of the N_2 molecule calculated in the present work is also better than those obtained by using the conventional Xα method. It is simply because the value 0.76077 of α(R) at R=3.7 a.u. in the present work is greater than either 0.75118 used by Weinberger and Konowalow[15] or 0.75753 used by Tseng and Whitehead[16] in their Xα calculations.

Table 2. Total energies (in Ry.), equilibrium internuclear separations, and dissociation energies of the ground state of N_2

Data Source	Re(a.u.)	-E(tot)	De(eV)	Reference
Experiment	2.068	—	9.91	(22)
MC-SCF	—	—	8.02	(23)
HF	2.01	217.9856	5.27	(20)
$MSXα_{WK}$[a]	4.1	217.2882	1.79	(15)
$MSXα_{TW}$[b]	4.1	217.5072	1.79	(16)
$MSXα(R)$[c]	3.7	217.6026	5.68	present

a. α = 0.75118 for all R.

b. α = 0.75753 for all R.

c. α varies with R. Values of α are given in Table 1.

CONCLUSIONS

A new method, Xα(R), with α value varying with the internuclear distance determined semiempirically is proposed. For the diatomic molecules, this new method is able to give a smaller calculated equilibrium internuclear distance than the one obtained by the conventional Xα calculations where α is a constant for different internuclear separations. The dissociation energy calculated by this new method is much better than those results obtained from the usual $X\alpha$ calculations. The method can also be applied to the polyatomic molecules provided the proper dissociation limits are known.

ACKNOWLEDGMENTS

The author wishes to thank Miss S.M. Young, Miss S.C. Kao and the Computer centre of National Taipei Institute of Technology for their valuable help in computations. Thanks also go to the NSC of the Republic of China for the support under the contract NSC 74-0208-M033-08.

REFERENCES

1. J.C. Slater, J. Chem. Phys., $\underline{43}$, S228 (1965).
2. K.H. Johnson, J. Chem. Phys., $\underline{45}$, 3085 (1966).
3. K.H. Johnson, J.G. Norman Jr. and J.W.D. Connolly, "Computational Methods for Large Molecules and Localized States in Solids," Eds. F. Herman et al., Plenum Press, New York (1972).
4. J.C. Slater and K.H. Johnson, Phys. Rev., $\underline{85}$, 844 (1972).
5. J.C. Slater, "Quantum Theory of Molecules and Solids, Vol.4. The Self-Consistent Field for Molecules and Solids," McGraw-Hill, New York (1974).
6. K.H. Johnson, "Annual Review of Physical Chemistry," Eds. H. Eyring et.al., Annual Reviews Inc., Palo Alto, Calif., (1975).
7. K. Schwarz, Phys. Rev., $\underline{B5}$, 2466 (1972).
8. I. Lindgren, and K. Schwarz, Phys. Rev., $\underline{A5}$, 542 (1972).
9. T.J. Tseng, S.H. Hong and M.A. Whitehead, J. Comput. $\underline{1}$, 88 (1980).
10. D.D. Konowalow and M.E. Rosenkrantz, Chem. Phys. Lett., $\underline{44}$, 321 (1976).
11. D.D. Konowalow and M.E. Rosenkrantz, Chem. Phys. Lett., $\underline{49}$, 54 (1977).
12. M.E. Rosenkrantz and D.D. Konowalow, Int. J. Quantum Chem., $\underline{12}$, 707 (1977).
13. M.E. McAdon and D.D. Konowalow, J. Chem. Phys., $\underline{71}$, 3089 (1979).
14. In order that the potential curve of N_2 has a minimum around 2.1 a.u., we found that the α parameter was to be around 1.5 and the total energy was -237.9093 Ry.
15. P. Weinberger and D.D. Konowalow, Int. J. Quantum Chem. Quantum Symp., $\underline{7}$, 353 (1973).
16. T.J. Tseng and M.A. Whitehead, J. Comput. Chem., $\underline{2}$, 38 (1981).
17. T.J. Tseng, Chung Yuan J., $\underline{14}$, 53 (1985).
18. J.C. Slater, Int. J. Quantum Chem. Quantum Symp., $\underline{8}$, 81 (1974).
19. P. Morse, Phys. Rev., $\underline{34}$, 57 (1929).
20. P.E. Cade, K.D. Sales and A.C. Wahl, J. Chem. Phys., $\underline{44}$, 1973 (1966).
21. J.B. Mann, Los Alamos Sci. Lab. Rep., LA-3690, 1967.
22. D.E. Gray, Ed., "American Institute of Physics Handbook," (Mc Graw-Hill, New York 1972), sec. 7g.
23. H.F. Schaeffer, III, "The Electronic Structure of Atoms and Molecules," (Addison-Wesley, Reading, Mass., 1972).

ATOMIC RESOLUTION STUDY OF STRUCTURAL REARRANGEMENTS IN METAL CLUSTERS

Amanda K. Petford-Long, N.J. Long, David J. Smith[#], L.R. Wallenberg[*] and J.-O. Bovin[*]

Center for Solid State Science, Arizona State University
Tempe, AZ 85287, USA
[*]Inorganic Chemistry 2, Chemical Center, University of Lund
Box 124, S-221 00, Lund, Sweden

ABSTRACT

Dynamic events in small particles of Au, Pt, Rh and Ru have been observed inside the electron microscope. Depending on the particle size and the beam current density, the small clusters undergo structural rearrangements, with rapid changes in orientation and shape with respect to the beam direction. Various forms of surface activity have been recorded including atom hopping, twinning and atomic "clouds" above some surfaces. In a fraction of a second, a 2.5nm ruthenium crystal changed its internal structure from cubic-close-packing to hexagonal-close-packing. Twinning was comparatively common, even in particles with diameters of 2nm or less. From hot stage observations the speed of structural rearrangements was found to be temperature dependent.

INTRODUCTION

The technique of high-resolution electron microscopy (HREM) has recently been used to study dynamic events on the atomic scale in small metal particles of Au[1-5] and Pt.[6] It was established that the incident electron beam caused rapid structural rearrangements of the particles to take place, and surface atom hopping was also observed. In the present study, these results are compared with subsequent observations of small crystals of Rh and Ru nominally made at room temperature. Studies were also made using a 100kV lower-resolution electron microscope, with the entire specimen heated to several hundred °C, but with an electron beam of low current density so that heating effects due to the electron beam could be neglected.

EXPERIMENTAL PROCEDURE

The particles of Rh, Ru and Pt were fabricated as 55-metal-atom clusters with organic ligands (diameter ~0.9nm), as described previously for Au crystals.[7] For electron microscope observation, these clusters were usually collected on holey carbon support films resting on Mo or Cu grids.

The atomic resolution studies were executed using a JEM-4000EX high resolution electron microscope, and the particle motion and structural rearrangements were recorded in real time using video facilities attached to the microscope, as described elsewhere.[4] The current density of the

---

[#] Also at Department of Physics, Arizona State University.

incident electron beam used was typically in the range of 20 to 50Acm^{-2}. All the images shown in the following sections were recorded from single video-tape frames although frame-averaging was applied in some cases to the original tape.

In order to establish the extent to which the structural rearrangements and particle motion were temperature-dependent, further experiments were carried out in a 100kV Philips 400ST electron microscope which was equipped with a specimen-heating holder. A video recording and viewing system identical to that used on the JEM-4000EX allowed real-time recordings to be made of dynamic changes in the metal particles. During heating, the electron beam current density was kept as low as feasible (<2Acm^{-2}), so that distinctions could be made between heating of the sample by the electron beam and 'thermal' heating of the particles and substrate. Care was taken to examine only those particles resting on areas of support film in good thermal contact with one of the Mo grid bars, and thereby ensure that the particles really were being subjected to external heating during these observations.

RESULTS

Gold

Small Au particles with diameters of less than 8nm underwent rapid dynamic changes of their internal structure. The rapidity of motion depended on a number of factors including the beam current density and the extent of contact between the Au and the substrate film.[3] The rate of change was very similar for Au supported on either Si or C films. As shown in fig. 1, the same particle could change from single crystal to twinned to multiply-twinned within a few seconds. Both decahedral and icosahedral multiply-twinned particles (MTP) were observed. When the Au particles were heated in the 100kV microscope, no motion was apparent until the temperature exceeded 500°C.

Platinum

The behaviour of the platinum crystals under electron irradiation was very similar to that of the gold particles:
1. Small crystals (diameter <4nm) underwent internal structural rearrangements, as illustrated by the particle in fig. 2. This particle is initially a single crystal as seen in 2a). It then develops twins as shown in figs. 2b) and 2c). Structural rearrangements of this type only occurred in crystals which were over holes in the substrate film and in poor contact with the film surface. Pt crystals were never observed to form multiply-twinned structures.
2. Atomic columns at the crystal surface 'hopped' from one surface site to another, resulting in progressive rearrangements of the surface shape. This activity was observed even for particles with diameters of more than 8nm. An example is shown in fig. 3, in which a flat (100) surface develops steps due to the addition of extra atomic columns at the surface.
3. Atomic "clouds" which interacted with the crystal have been seen outside (100) and (110) surfaces. (A more detailed report on the observation and nature of atomic clouds outside the surfaces of Au particles can be found elsewhere.[2])

Rhodium

Rh proved to be more stable under the electron beam than either Pt or Au. Surface hopping was frequently observed, as shown in fig. 4, but small particles did not usually undergo internal structural rearrangements when they were in contact with a large 'sheet' of the substrate film. However, after a short period of exposure to the electron beam (typically 30 - 40 minutes), areas of the carbon film began to etch away, leaving Rh particles

attached to thin strands of partially graphitised carbon. Small crystals
(diameter < 3nm) were then observed to undergo very rapid changes in
structure and orientation with respect to the electron beam direction, as
well as structural alterations. Single crystal, twinned and multiply-
twinned structures (both icosahedral and decahedral) were observed.

The original 55-atom clusters had in general coalesced to form larger
crystals, although a few of the original 0.9nm clusters were observed.
Clouds were occasionally seen outside some surfaces of the Rh particles,
and an example of this is shown in fig. 5.

One Rh sample was studied using the heating holder. The Rh crystals
remained immobile until the specimen was heated to a nominal temperature of
~800°C. Observation of the electron diffraction pattern indicated the
temperature at which the particles began to change orientation, and observation of high magnification images indicated the temperature at which the
small particles appeared to change shape.

Ruthenium

Unlike the metals discussed above, Ru was frequently found as very
small (~0.9nm) crystals, suggesting that coalescence of particles did not
take place as easily for Ru as it did for Au, Pt and Rh. Structural
rearrangements also did not occur as readily for Ru as for Pt, and it was
often necessary for some of the substrate film to be etched away before any
dynamic effects were observed. However, once a Ru crystal started to
become mobile, it then usually underwent rapid fluctuations in structure,
as illustrated by the images shown in fig. 6. The same particle displays
single crystal, twinned and multiply-twinned icosahedral structures, and in
fig. 6b. the internal stacking of the particle has changed from cubic-
close-packed (ccp) to hexagonal close-packed (hcp). This type of close-
packing change has never been observed previously at atomic resolution,
although the present study indicates that it can also occur for Cd and Rh.
Note that bulk Ru and Cd are hcp, whilst bulk Rh is ccp.

When a Ru sample was heated in the 100kV microscope, the Ru particles
remained immobile until the temperature was greater than 1000°C. At this
high temperature, image drift was a serious problem and it was not possible
to obtain images of the small particles of sufficient quality to be certain
that structural changes were occurring. The changes in the electron
diffraction pattern clearly indicated particle movement.

DISCUSSION

From the observations of the dynamic events in the particles, it was
not always obvious whether the particles were changing structure or whether
they were just changing orientation with respect to the beam direction.
However, it is not physically possible to obtain images from the same
particle which correspond to a single crystal and to an icosahedral MTP
unless internal structural rearrangements have occurred.

All of the small metal particles examined underwent internal and
surface structural rearrangements, but the ease with which these
rearrangements occurred varied considerably, depending on the metal under
examination. All of the metals were prepared on both carbon and silicon
support films. There were no obvious differences in behaviour due to the
different substrate film material; the extent of contact between the particle and the substrate seemed far more important in influencing the
particle behaviour.
The heating experiments confirm that the particle mobility is
temperature-dependent, and indicate that the temperature at which mobility
is initially observed is related to the melting temperature (T_m) of the
metal. As previously determined [8,9], T_m for small metal particles is

reduced compared with the bulk melting temperature, to an extent which is dependent on the particle size. It was observed that Au (T_m bulk 1063°C) began to move at ~500°C, whereas Rh, with a melting temperature of 1966°C did not show any signs of mobility until heated until ~1000°C.

There are several possible electronic excitation mechanisms whereby small metal particles could acquire sufficient energy from the electron beam to become mobile and undergo the structural rearrangements observed here. The heating experiments suggest these are equivalent to an increase in temperature; i.e. one effect of the electron beam on the particles is to heat them. It is significant that we are able to observe such dynamic events on the atomic scale, since it may well be that real catalysts, often composed of supported metal particles, are undergoing similar changes. Thus HREM could represent an alternative method for replicating, and following, the behaviour of such materials.

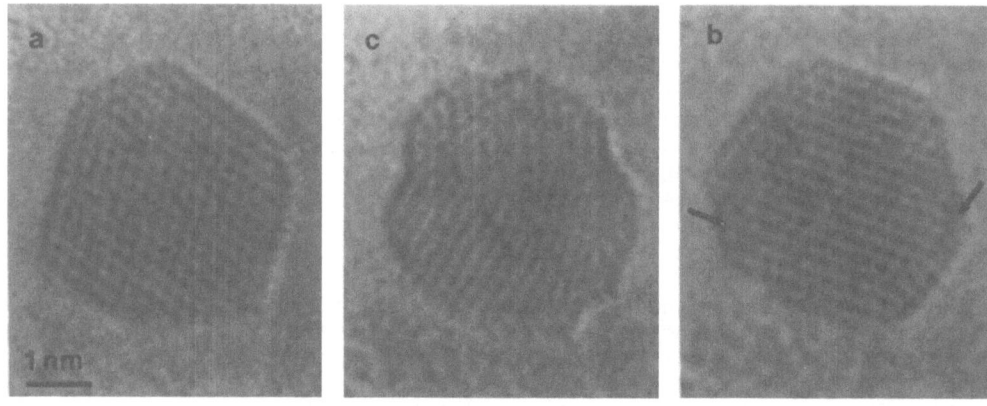

Fig. 1 Structural rearrangements of a small Au crystal supported on amorphous silicon as photographed directly from the TV monitor without frame-averaging. Note the twin planes arrowed in b) and the reentrant notches visible in c).

ACKNOWLEDGEMENTS

This research was supported in part by Arizona State University Faculty Grant-In-Aid and by the Facility for High Resolution Electron Microscopy, in the Center for Solid State Science at Arizona State University, established with support from the National Science Foundation (Grant DMR-8306501). Support was also received from the Swedish Natural Research Council (E-EG 3914-116 and K-KU 3914-120) and from the National Swedish Board for Technical Development (DNR 84-3515). Prof. G. Schmid, Essen, Germany is gratefully acknowledged for providing the cluster samples.

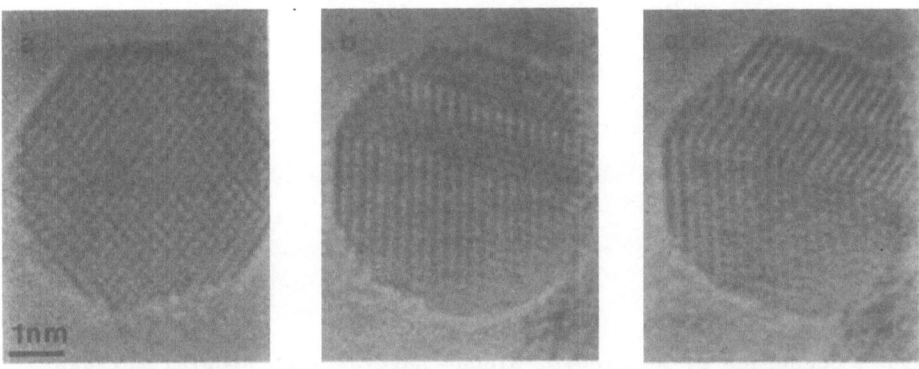

Fig. 2 Rearrangements of a small Pt crystal which occur during microscope observation. Note the twin boundaries visible in b) and c).

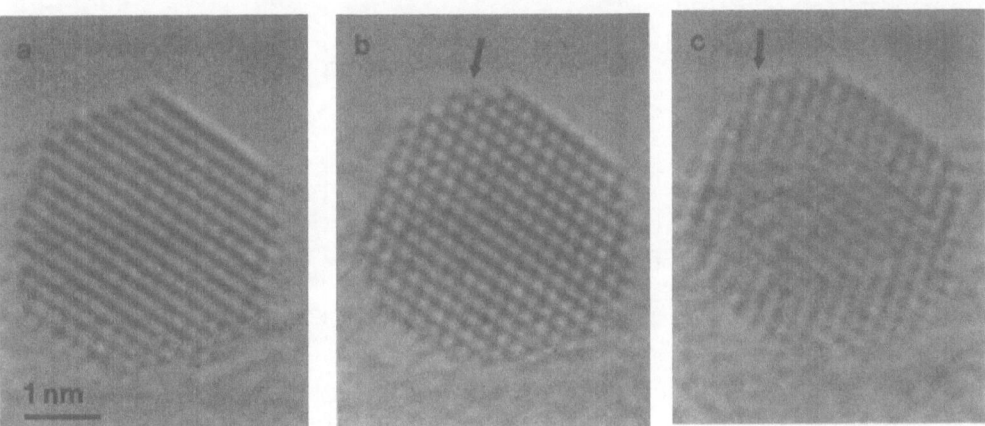

Fig. 3 Images of small Pt particle, photographed directly from the video monitor, showing the addition of extra atom columns (arrowed) at a (100) surface. Total time elapsed: less than 0.5s.

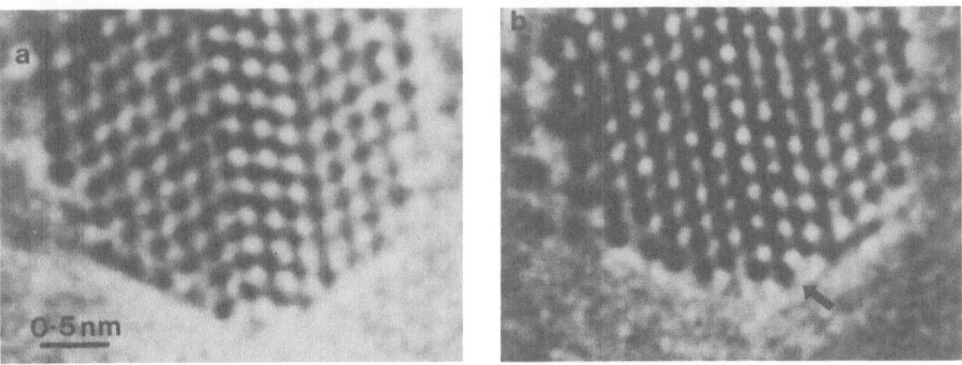

Fig. 4 Atomic 'hopping' on the surface of a small Rh crystal.

Fig. 5 Occurence of an atomic "cloud" outside a Rh crystal surface.

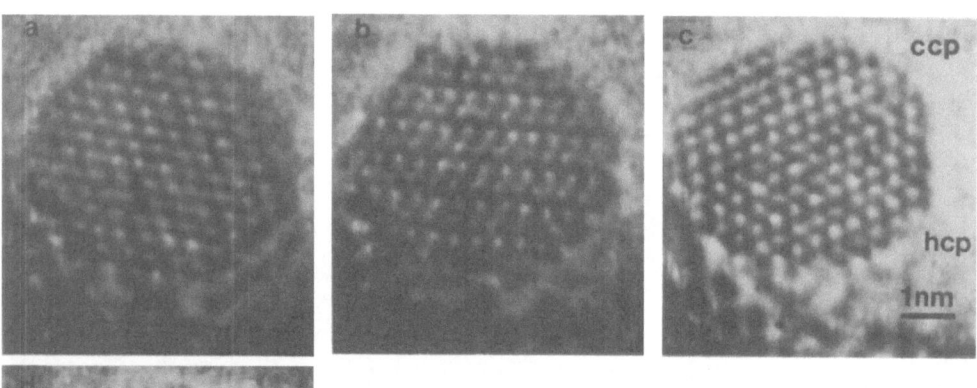

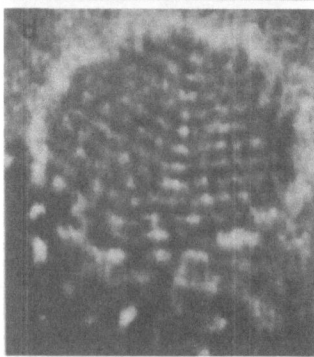

Fig. 6 Series of images of Ru crystal, recorded over a 2-second time interval, showing single crystal (a), twinned (c) and multiply-twinned-decahedral structures, as well as a change in internal packing from cubic-close-packed to hexagonal-close-packed (b).

REFERENCES

1. R. Wallenberg, J.-O. Bovin and D.J. Smith, Naturwiss., 72: 539 (1985).
2. J.-O. Bovin, R. Wallenberg and D.J. Smith, Nature, 317: 47 (1985).
3. D.J. Smith, A.K. Petford-Long, L.R. Wallenberg and J.-O. Bovin, Science, 233: 872 (1986).
4. S. Iijima and T. Ichihashi, Phys. Rev. Lett., 56: 616 (1986).
5. S. Iijima, J. Electron Microsc., 34: 249 (1986).
6. L.R. Wallenberg, J.-O. Bovin, A.K. Petford-Long and D.J. Smith, Ultramicroscopy, 1986 (in press).
7. R. Wallenberg, J.-O. Bovin and G. Schmid, Surf. Sci., 156: 256 (1985).
8. Ph. Buffat and J.-P. Borel, Phys. Rev. A, 13: 2287 (1976).
9. J. Ross and R.P. Andres, Surf. Sci., 106: 11 (1980).

A FRAGMENT CLUSTER STUDY OF POLYACETYLENE

J. A. Darsey, N. R. Kestner[#], and B. K. Rao[*]

Tarleton State University, Stephenville, TX 76402
[#]Louisiana State University, Baton Rouge, LA 70803
[*]Virginia Commonwealth University, Richmond, VA 23284

ABSTRACT

Using CH fragments to obtain an optimized cluster, a model calculation has been made for polyacetylene. The model reproduces the bond alternation as expected due to Peierls condition. After removing the errors due to a finite length of the chain, a complete rotational potential energy surface has been generated for the cis-transoidal isomer of polyacetylene. The potential depends upon rotations about two successive single bonds in the backbone. The potential surface indicates the possibility of formation of a super helix.

INTRODUCTION

Polyacetylene was first synthesized from acetylene by Natta[1] and coworkers in 1958. Using similar catalyst but different conditions, Shirakawa[2-4] and co-workers succeeded in preparing polyacetylene films with metallic luster. Polyacetylene, synthesized under varying polymerization conditions, can lead to molecular weights ranging from about 500 to 220,000 and having diverse morphologies and structures which seem to strongly influence the electronic properties of this material.[5]

Polyacetylene has attracted the attention of experimentalists and theoreticians because of its high dc conductivity which is unusual for polymers. This has led to many actual uses and numerous potential uses have also been proposed.[6] However, the origin of conductivity and related properties are still not completely understood. However, it is known[5] that structural defects and doping can enhance the conductivity in trans-polyacetylene.

Thus far, experimental and theoretical studies have focussed mostly on trans-polyacetylene.[7-9] Limited work on cis-polyacetylene has also been done.[10-13] It was found[13] that the most stable conformation for the cis-transoidal (CT) of polyacetylene was where the dihedral angle for rotation about a single bond is equal to 117°. For the trans-cisoidal (TC) conformation, this angle is 42°. A question remains whether a stable conformation exists for the simultaneous rotations about two consecutive single bonds in CT and TC conformations. This current study is aimed at generating an energy map for the (CT) conformation of polyacetylene in order to partially answer this question.

For the present work, the universally accepted bond alternating model (i.e., the Peierls instability condition[14] of polyacetylene $(CH)_x$ has been adopted. There are primarily three configurations that polyacetylene can be found in; the trans-isomer, the trans-cisoidal isomer (TC) and the cis-transoidal isomer (CT). This work is concerned primarily with the (CT) isomeric configuration as illustrated in Fig. 1.

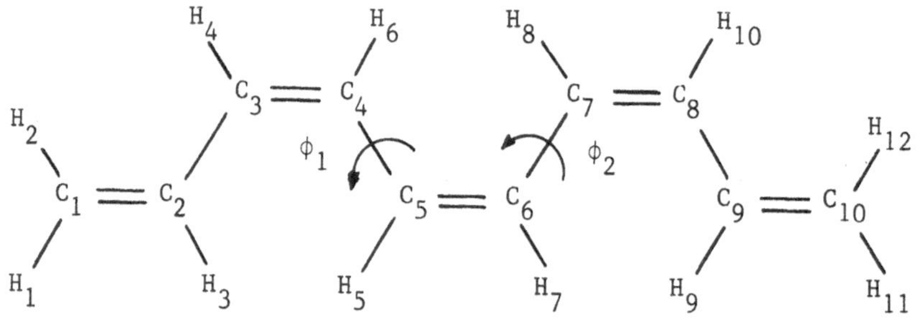

Fig. 1   Schematic representation for the model of cis-transoidal polyacetylene showing the bond rotations.

In previous works[11,13], we looked at the rotational potential energy function for the rotation ϕ_1 of cis-transoidal polyacetylene. It was found that a rotation of $\phi_1 = 117°$ produced the most stable structure of $(CH)_x$. It was also determined that the conformational isomer where $\phi_1 = 117°$ and $\phi_2 = 121°$ was stable. This structure, if extensively propagated throughout the polyacetylene chain, would produce a tightly wound 2*3/1 helix with a chain unit cell length[12] of 4.43A°. However, in order to ascertain how stable this conformation was, it was necessary to produce a complete potential energy surface for the simultaneous rotation of ϕ_1 and ϕ_2. This potential energy surface is shown in Fig. 2.

There are many calculational methods possible for generating potential energies. In fact, several workers have used different approaches in the study of polyacetylene.[12,15-18] We have chosen the ab initio self-consistent field linear combination of atomic orbitals-molecular orbital (SCF-LCAO-MO) method for this study. In this, the molecular orbitals and the energies are calculated in a self-consistent manner starting with an initial set of input orbitals. All the one and two electron integrals are explicitly calculated as required in the ab initio SCF procedure. The input atomic orbitals are represented by linear combination of several Gaussian functions, suitably chosen to represent the atoms. More detail on this calculational procedure can be found elsewhere.[19,20]

Modelling a polymer like polyacetylene requires that a suitable finite representation be chosen. For the present work, a fragment cluster approach was used. Using many CH units in a chain and varying the C-C bond lengths in the backbone a model for polyacetylene $(CH)_x$ was created. To avoid the errors due to the finite length of the chain (which leaves dangling bonds at the ends) extra hydrogen atoms were added at the ends. The model simulates the polyacetylene as a large molecule and, upon optimization of the C-C bond lengths, it produced the bond alternation as expected to be arising from Peierls instability[14]. It was noticed that upon increasing the value of x, the number of CH units in the chain, the structure stabilized at about x=8. Therefore, final calculations were performed for a 10 unit chain.

Since the computational procedure is based upon the SCF-LCAO-MO scheme, one must be careful about the choice of the input basis set. For the present calculation we have used the atomic functions as described before[13]. As usual, extra s and p functions were added to the relevant hydrogen atoms (shown as H_3, H_5, H_6, H_7, H_8 and H_9 in Fig. 1) along the chain. It is easy to notice that these hydrogen atoms have the greatest chance of interacting with each other and, therefore, need extended basis sets. With this representation, the rotational potential energy surface was produced for all values of simultaneous rotations ϕ_1, and ϕ_2 about the two consecutive single bonds as shown in Fig. 1.

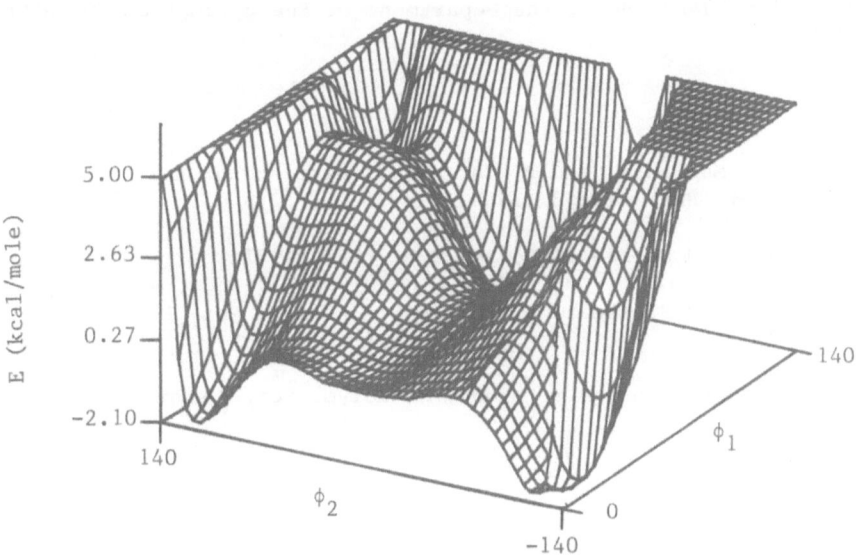

Fig. 2   Rotational potential surface for polyacetylene.

RESULTS AND DISCUSSION

Figure 2 presents the results of this ab initio study. The zero of the potential was chosen for the energy coresponding to the planar CT conformation. Therefore, all conformations with energies less than zero, for example, would be favored over the planar CT. The opposite would be true for conformations with energies greater than zero. As expected from before[13], minima were found at $\phi_1 = 0°$, $\phi_2 \cong 120°$ (represented as [0, 120]) and its symmetric analog [120,0]. Normally, without any bond-bond interctions and the consequent steric hinderances, one would expect the [120,120] to be an even deeper minimum. Such a situation would lead to the formation of a tight helical structure as mentioned before[13]. However, bond-bond interactions do exist in this case and, therefore, a simple additive picture will not be applicable.

A quick analysis shows this clearly. For the rotation about ϕ_1, when ϕ_1 approaches 120° the distance between H_3 and H_8 is optimum to give rise to the energy minimum. Upon increase of ϕ_1, H_3 and H_8 come closer and the steric hinderance increases the energy. With $\phi_1 = 120°$, if one increases ϕ_2 from zero, H_3 and H_8 do not stay at the optimum distance anymore. Therefore, the conformation moves out of the potential minimum. This is observed clearly from Fig. 2. At [117,121] and its complimentary symmetry points, minima do occur. But they are shallow (1.27 kcal/mol only) and they occur at a level of 1.98 kcal/mol above the [0,0] state instead of being below. Therefore, the probability of the formation of the "tight" helix[13] is very low. In comparison to this minimum, the [0,120] conformation provides an energy gain of 3.85 kcal/mol. Hence, this would be very prominent in the final structure of CT and would show up as a "super" helix. A Monte Carlo analysis using this potential energy surface is under way to study the probable structures of CT under various conditions.

ACKNOWLEDGMENT

This work was supported in part by the Center for Energy and Mineral Resources of Texas A&M University and the Organized Research Grant of Tarleton State University (JAD) and by the Department of Energy contract DE-AS05-77-ERO5399 (NRK).

REFERENCES

1. G. Natta, G. Mazzanti, and P. Carradini, Alti Accad. Naz. Lincei Rend. Sci. Fis. Mat. Nat. 25, 2 (1985).

2. H. Shirakawa and S. Ikeda, Polym. J. 2, 231 (1971).

3. H. Shirakawa and S. Ikeda, J. Polym. Sci. Polym. Sci. Polym. Chem. Ed. 12, 11 (1974).

4. T. Ito, H. Shirakawa and S. Ikeda, J. Polym. Sci. Polym. Chem. Ed. 12, 11 (1974).

5. James C. W. Chien, "Polyacetylene", (Academic Press, N. Y., 1984), pages 19 - 23.

6. A. G. MacDiarmid and A. J. Heeger, Synth. Met. 1, 101 (1979).

7. I. B. Goldberg, H. R. Crowe, P. R. Newman, A. J. Heeger, and A. G. MacDiarmid, J. Chem. Phys. 70, 1132 (1979).

8. B. R. Weinberger, J. Kaufer, A. Pron, A. J. Heeger, and A. G. MacDiarmid, Phys. Rev. B 20, 233 (1979).

9. S. Lefront, L. S. Lichtman, H. Temkin, D. B. Fitchen, D. C. Miller, G. E. Whitehall, and J. H. Burlitch, Solid State Commun. 29, 191 (1979).

10. Y. Tomkiewicz, T. D. Schultz, H. B. Brom, A. R. Taranko, T. C. Clarke, and G. B. Street, Phys. Rev. B, 24, 4348 (1981).

11. B. K. Rao, J. A. Darsey, and N. R. Kestner, J. Chem. Phys. 79 (3), 1377 (1983).

12. E. Cernia and L. D'Ilario, J. Polym. Sci. Polym. (Chem. Ed.) 21, 2163 (1983).

13. B. K. Rao, J. A. Darsey, and N. R. Kestner, Phys. Rev. B 31 (2), 1187 (1985).

14. R. E. Peierls, "Quantum Theory of Solids", (Clarendon Press, Oxford, 1955).

15. J. L. Bredas, R. R. Chance, R. Silbey, G. Nicolas, and Ph. Durand, J. Chem. Phys. 75, 255 (1981).

16. J. L. Bredas, R. R. Chance, R. H. Baughman and R. Silvey, J. Chem. Phys. 76, 3673 (1982).

17. W. K. Ford, C. B. Duke, and A. Paton, J. Chem. Phys. 77, 4564 (1982).

18. H. Shirakawa, T. Ito, and Ikeda, Makromol. Chem. 179, 1565 (1978).

19. J. A. Darsey and B. K. Rao, Macromolecules 14, 1575 (1981).

20. W. J. Hehre, L. Radom, P. v. R. Schleyer, and J. A. Pople, "Ab Initio Molecular Orbital Theory", (John Wiley, N.Y. 1986).

AN APPROXIMATE AB-INITIO THEORY OF METAL CLUSTERS

G. Das

Physical Chemistry Division
National Chemical Laboratory
Pune 411008, India

ABSTRACT

We propose an ab-initio approach to the study of metal clusters of both intermediate and large-size. The method differs from standard full-scale ab-initio calculations in the introduction of localized orbitals and in emphasizing the charge-transfer terms as the most dominant effect in metallic bonding.

I. INTRODUCTION

A satisfactory treatment of clusters by way of configuration interaction involves very large numbers of configurations, even for the smallest clusters. It is, therefore, essential that we look for simplifications that are at once drastic and simplified.

One of the sources of complications is the significance of the charge-transfer terms. They stand out as the most important terms for transition metal clusters as shown recently by Das and Jaffe [1]. Their calculations on the dimers V_2 and Cr_2 using "symmetrically" orthogonalized localized orbitals, indicate that the charge-transfer terms together can have a weight as large as that of the neutral terms. However, since the individual matrix elements are small it is still possible to retain a "second-order perturbation"-type approach. This is precisely what has been developed recently by Das [2] for large metal clusters. In this work a similar treatment based on localized orbitals is presented for intermediate-size clusters. Also an analysis of interaction in large metal clusters in terms of one-, two- and three-body effects is presented.

II. SMALL CLUSTERS vs. LARGE CLUSTERS

It is now a fairly well-established [3] assumption that bond-lengths increase sharply as one goes from dimers to large clusters, stabilizing quickly for $N > 4$. In clusters with smaller bond-lengths all the usual bonding forces - electrostatic, charge-transfer, exchange and dispersion - are important. For larger clusters, however, the bonding takes place principally through the electrostatic and the charge-transfer processes, the exchange and dispersion playing a secondary role. One can show by general arguments that given the above fact, viz. the secondary role of the exchange forces, the charge-transfer contributions calculated on the basis of a properly spin-symmetruzed wave-function are very nearly the same as that calculated with a wavefunction that represents an antiferromagnetically aligned spin-arrangement. While a rigorous proof does not seem to exist in the literature, one can use the following reasonability arguments. Consider an antiferromagnetically ordered arrangement such as

$$\uparrow_1 \quad \downarrow_2 \quad \uparrow_3 \quad \downarrow_4 \quad \uparrow_5$$
$$\downarrow_{10} \quad \uparrow_9 \quad \downarrow_8 \quad \uparrow_7 \quad \downarrow_6$$

i.e., a zeroth order wavefunction

$$\Psi_o = \phi_1 \alpha \, \phi_2 \beta \, \phi_3 \alpha \, \phi_4 \beta \, \phi_5 \alpha \, \phi_6 \beta \cdots \phi_{10} \beta \tag{1}$$

Using the nearest-neighbor (nnb) charge-transfer functions such as

$$\Psi_{12}^{+-} = (\phi_1 \alpha \, \phi_1 \beta) \, \phi_3 \alpha \, \phi_4 \beta \quad - - - - - \tag{2}$$
$$\Psi_{12}^{-+} = (\phi_2 \alpha \, \phi_2 \beta) \, \phi_3 \alpha \, \phi_4 \beta \quad - - - -$$

etc. and a first-order perturbation language, the total contributions from the charge-transfer terms can be written as

$$\Delta E_{cT} = V_{1,2} + V_{2,3} + \cdots + V_{9,10}$$
$$+ V_{1,10} + V_{2,9} + \cdots + V_{5,6} \tag{3}$$

Using, on the other hand, a zeroth-order function such as:

$$\Psi_o' = 2^{-1/2} \left[\phi_1 \alpha \phi_2 \beta \, \phi_3 \alpha \cdots \phi_{10} \beta - \phi_1 \beta \phi_2 \beta \, \phi_3 \alpha \cdots \phi_{10} \alpha \right] \tag{4}$$

for example, which maximizes the contribution from the pair of sites 1 and 10, one, however, gets the same total contribution as above since while the (1, 10) contribution is doubled, the (1, 2) and (9, 10) contributions are actually halved (of course, neglecting the exchange effects). Consider now a zeroth order wavefunction for a cluster of the size 2n. We shall consider an $M_c = 0$ state. The total number of terms is clearly $(2n)!/(n!)^2$. Using equal weight for all terms and signs that lead to the largest contributions for the nnb pairs, the total charge-transfer contribution is

$$\left[\frac{(n!)^2}{(2n)!} \right] \left[\frac{(2n-2)!}{\{(n-1)!\}^2} \right] \cdot f \cdot 4V \tag{5}$$

assuming that there exist f nnb pairs and all of them give identical contributions. This gives fV/(1-1/2n), while a straight-forward antiferromagnetic ordering gives fV.

III. ANSATZ FOR INTERMEDIATE-SIZE CLUSTERS

We shall concern ourselves here with those clusters for which the exchange effects, though small, are not negligible compared to the charge-transfer effects. For such cases, clearly, an antiferromagnetic zeroth order state is not adequate. It is necessary to consider some or all of the neutral terms belonging to different kinds of spin alignment. For clusters of size

$N \lesssim 10$, it is practicable and, indeed, desirable that all the spin-flip terms are included. However, for larger systems, it is imperative that we take recourse to some judicious selection of these terms. Since charge-transfer and not exchange will still be the dominant effect, it is natural to expect that the spin-flip terms in Ψ_o representing smaller and smaller number of adjacent pairs (i.e., pairs of adjacent sites) with opposite spin will also be the less important.

Let Ψ_o represent a set of 'neutral' terms, Ψ_1 a set having one charge-transfer, Ψ_2 one having two charge-transfers, etc. Then the total wavefunction can be written as:

$$\Psi_o' = a_o \Psi_o + a_1 \Psi_1 + a_2 \Psi_2 + \cdots \tag{6}$$

For large enough cluster-size we approximate this by a recurring expression

$$
\begin{aligned}
\Psi_o' &= a \Psi_o + b \Psi_1' \\
\Psi_1' &= a \Psi_1 + b \Psi_2' \\
\Psi_2' &= a \Psi_2 + b \Psi_3'
\end{aligned}
\tag{7}
$$

where

$$a^2 + b^2 = 1 \tag{8}$$

The corresponding energy is written in a Raleigh-Schroedinger (RS) like form:

$$\mathcal{E} = \mathcal{E}_o - \frac{\langle \Psi_o | \mathcal{H} | a \Psi_1 + b(a \Psi_2) \rangle^2}{\mathcal{E}_1 - \mathcal{E}_o} \tag{9}$$

where the denominator $\mathcal{E}_o' - \mathcal{E}_1$ is approximated as $\mathcal{E}_1 - \mathcal{E}_o$ with

$$
\left.
\begin{aligned}
\mathcal{E}_o &= \langle \Psi_o | \mathcal{H} | \Psi_o \rangle \\
\mathcal{E}_1 &= \langle \Psi_1 | \mathcal{H} | \Psi_1 \rangle
\end{aligned}
\right\}
\tag{10}
$$

Minimizing (4) with respect to a,b under the constraint (3) gives

$$\frac{b}{a^2 - b^2} = \frac{\langle \Psi_o | \mathcal{H} | \Psi_2 \rangle}{\langle \Psi_o | \mathcal{H} | \Psi_1 \rangle} \tag{11}$$

The energy is also minimised with respect to the expansion coefficient $\{a_{i,n}\}$ of the functions Ψ_i:

$$\Psi_i = \sum_n a_{i,n} \Phi_{i,n} \, , \quad i = 0, 1, 2 \tag{12}$$

where $\{\Phi_{i,n}\}$ represent charge-transfer of the i-th order involving various sites. However, orbitals making up the $\{\Phi_{i,n}\}$ are optimized by way of solving the corresponding Fock equations.

IV. LARGE CLUSTERS : AN ENERGY EXPRESSION CORRECT TO SECOND-ORDER IN DIFFERENTIAL OVERLAP (SODO)

Let Φ_i^o be the localized wavefunction of the atom at the i-th site of a metallic cluster. We assume (for simplicity) that there is only one valence electron per site and that the dominant part of the wavefunction consists of an 'antiferromagnetic' ordering of spins in the sense that each site is surrounded by its nnb's having opposite spin. Although such a state cannot be a pure spin-state, it can be shown that a spin-symmetrized state built from it leads to an energy that differs from the non-symmetrized value by $O(1/N)$, N being the size of the cluster.

Let the valence orbital on the site i be denoted as ϕ_i^o. We construct out of $\{\phi_i^o\}$ an orthonormal set $\{\phi_i\}$ by a symmetric procedure such as the Lowdin orthogonalization scheme. Let the corresponding orthogonalized 'site'-states be $\{\Phi_i\}$. We now assume the wavefunction of the cluster in the following recursive form:

$$\Psi_n = A\left\{\phi_n \Psi_{n-1,n} + \sum_{j \subset nnb(n)} (a_{nj} \Phi_n^+ \Phi_j^- + b_{nj} \Phi_n^- \Phi_j^+) \times \Psi_{n-2,nj}\right\} \quad (13)$$

The definition of $\Psi_{n-1,i}$ ($\Psi_{n-2,ij'}$) is that they are constructed in the same way as Ψ_n, with the same values of $\{a_{k\ell}\}$, $\{b_{k\ell}\}$, but with the site i (both the sites i and j) missing. The site functions Φ_i^+ and Φ_i^- (with 0 and 2 electrons occupying ϕ_i respectively) represent the positive and negative ions at the site i. Then, as shown in Ref. 1, one obtains an energy expression correct to SODO as follows, $\{n\}$ denoting the cluster :

$$\begin{aligned}
\mathcal{E}_n = &\sum_{i \subset \{n\}} \epsilon_i + \frac{1}{2}\sum_{i \neq j \subset \{n\}} \langle \rho_i \| \rho_j \rangle \\
&+ \sum_{i \neq j \subset \{n\}} \left\{ a_{ij}^2 \left[\epsilon_i^+ + \epsilon_j^- - \epsilon_i - \epsilon_j + \langle \rho_i^+ + \rho_j^- - \rho_i - \rho_j \| \sum_{k \subset \{n\} \neq i,j} \rho_k + \langle \rho_i^+ \| \rho_j^- \rangle \right] \right. \\
&\left. + 2\sum_{i \neq j \subset \{n\}} a_{ij} \langle \phi_i + \sum_{k \neq i, \subset \{n\}} J_k | \phi_j \rangle \right\}
\end{aligned} \quad (14)$$

where ϵ_i, ϵ_i^+ and ϵ_i^- are the site energies corresponding to Φ_i, Φ_i^+ and Φ_i^- respectively. $\langle \rho_i \| \rho_j \rangle$ is the electrostatic energy between the sites i and j. The operator J_k is given by

$$J_k = \frac{z_{eff,k}}{|\underset{\sim}{r} - R_k|} + \int \frac{\rho_k^{el}(\underset{\sim}{r}')\, d\underset{\sim}{r}'}{|\underset{\sim}{r} - r'|} \quad (15)$$

where $z_{eff,k}$, R_k define the effective charge and the geometry of the k-th site respectively and ρ_k^{el} is the electronic charge density. Minimizing (14) with respect to $\{a_{ij}\}$ we get the energy correct to SODO :

$$\mathcal{E}_n = \sum_{i \in \{n\}} \epsilon_i + \frac{1}{2} \sum_{i \neq j \subset \{n\}} \langle P_i \| P_j \rangle$$

$$- \sum_{i \neq j \subset \{n'\}} \frac{\langle \phi_i | h_i + \sum_{k \subset \{n'\}, \neq i} J_k | \phi \rangle^2}{[\epsilon_i^+ + \epsilon_j^- - \epsilon_i - \epsilon_j + \langle P_i^+ + P_j^- - P_i - P_j \| \sum' P_k \rangle + \langle P_i^+ \| P_j^- \rangle]^2} \quad \text{(16)}$$

Minimization of (16) with respect to the orbitals is an important step. In this context we shall first describe a procedure (a variant of the Lowdin's orthogonalization procedure) to obtain an orthonormal set of orbitals. We notice first that since exchange and charge-transfer effects are considered only between electrons on the same site or belonging to nnb's, orthogonalization should, therefore, be considered to the same approximation. We adopt the following symmetric orthogonalization process as the most reasonable choice. Denoting N_i (N_{ij}) as the set of nnb's of the site i (i and j simultaneously), we write $\{\phi_i\}$ in terms of $\{\phi_i^o\}$ as (η_i being normalization factor):

$$\phi_i = \eta_i \left(\phi_i^o + \sum_{k \subset N_i} \alpha_{ik} \phi_k^o \right) \quad \text{(17)}$$

and impose only the conditions

$$\langle \phi_i | \phi_j \rangle = 0, \quad j \subset N_i$$
$$\alpha_{ij} = \alpha_{ji} \quad \text{(18)}$$

This leads to

$$\delta_{ij} + 2 \alpha_{ij} + \sum_{k \subset N_{ij}} \left(\alpha_{ik} \delta_{jk} + \alpha_{jk} \delta_{ik} \right) \approx 0 \quad \text{(19)}$$

where

$$\delta_{ij} \equiv \langle \phi_i^o | \phi_j^o \rangle \quad \text{(20)}$$

With this scheme of orthogonalization Eq. (16) within the SODO approximation reduces to

$$\mathcal{E} = \sum_i \epsilon_i^o + \sum_{i>j} \left(\delta_{ij} + \frac{Z_{eff}^2}{|R_i - R_j|} \right) + \sum_{i \neq j} u_{ij}$$

$$+ \sum_{i,j \subset N_i} \alpha_{ij}^2 \left[u_{jj} + u_{ij} + 2\delta_{ii} \right]$$

$$- \sum_{i,j \subset N_i} \frac{\left[\langle \phi_i^o | J_j^o | \phi_i^o \rangle + \alpha_{ij} (\delta_{ij} + \delta_{jj} + 2u_{ij}) \right]^2}{\Delta + \frac{Z_{eff}^2}{|R_i - R_j|} + 2u_{ji} + \sum_{k \neq i,j} (\delta_{jk} + u_{jk} - \delta_{ik} - u_{ik})} \quad \text{(21)}$$

$$+ 3 \sum_{i,j \subset N_i} \alpha_{ij}^2 \sum_{k \neq j} \left(u_{ik} + \delta_{ik} \right)$$

where

$$g_{ij} = \iint \frac{\rho_i^{(0)\,el}(r) \; \rho_j^{(0)\,el}(r')}{|r - r'|} dr\,dr'$$

$$u_{ij} = -Z_{eff} \int \frac{\rho_i^{(0)\,el}(r)}{|r - R_j|}\,dr \Bigg\}\qquad(22)$$

$$\Delta \equiv \epsilon^+ + \epsilon^- - 2\epsilon = I.P. - E.A.$$

We must, however, recognize that since $g_{ij} + u_{ij}$ is small, the three-body terms occurring in (21) can be neglected in the same spirit as the SODO approximation. Important three-body terms will, however, reappear when the orbitals are optimized for the cluster allowing polarization. Assuming that the polarization or optimization contributions are of the same order of magnitude as the charge-transfer terms, we shall neglect the latter from the energy expression (21) in optimizing ϕ_i^c. The corresponding Fock operators are clearly of the form:

$$F_i = F_i^o + V_i^{ext} \qquad(23)$$

where F_i^o has the same from as the atomic Fock operator at the site except that the orbitals are distorted and V_i^{ext} is simply the external field due to the atoms at the other sites. We carry out the optimization by solving the Fock equations $F_i \phi_i = E_i \phi_i$ iteratively.

ACKNOWLEDGEMENT

This work was supported in part by grants from the Army Research Office (DAAG 29-85-K-0244).

REFERENCES

[1]   G.P. Das and R.L. Jaffe, Chem. Phys. Letters 109 (1984) 206

[2]   G.P. Das, Chem. Phys. Letters 114 (1985) 309

[3]   B.K. Rao, P. Jena and D.D. Shillady, Phys. Rev. B 30 (1984) 7293

ALKALI CLUSTERS: STRUCTURE, STABILITY, LARGE AMPLITUDE MOTION AND CHEMICAL PROPERTIES

Manfred M. Kappes, Martin Schär, Chahan Yeretzian,
Ulrich Heiz, Arthur Vayloyan and Ernst Schumacher

Institute of Inorganic and Physical Chemistry
University of Bern, CH-3012 Bern, Switzerland

ABSTRACT

Extensive measurements of ionisation potentials and relative thermodynamic stabilities for pure alkali and alkali/heteroatom species have provided much data pertinent to the electronic structure of simple metal clusters. Due to poorly characterised internal temperatures, an understanding of the relative magnitude of electronic versus geometric structure effects remains tenuous. Particle specific studies of chemical reactivity are ongoing.

INTRODUCTION

Since the development of high flux beam methods for their generation several years ago[1,2] it has proved possible to obtain extensive experimental information on the global electronic structure of neutral alkali clusters. Of these, the larger species (M_x, x>10) are produced with significant but as yet unquantified internal excitation (a function of particle size and formation history). As large alkalis have (1) many low lying excited electronic states (coexisting isomers - in thermodynamic equilibrium?) and (2) easily accessible freezing and melting points[3] (transition to large amplitude motion), temperature information is crucial in trying to relate an ensemble property as determined from experiment to intrinsic size-dependent electronic structure changes[4].

RESULTS AND DISCUSSION

In addition to polarisabilities[5], two other electronic structure related properties have been studied over comparatively large cluster size ranges: vertical ionisation potential and relative thermodynamic stability. The experimental determination of ionisation thresholds is straightforward[6,7]. Numbers resulting can be interpreted classically in terms of a spherical droplet model[6,7]. While significant deviations from this classical model occur for clusters with less than ten atoms, agreement for larger species is remarkably good, indicating that (a) the assumption of spherical symmetry is viable, (b) the molecular equivalent to the Fermi level (HOMO) does not shift dramatically with particle size and (c) the most important single effect determining the size dependence of ionisation potential is surface curvature.

In a supersonic expansion, cluster formation occurs by multiple aggregation/evaporation cycles, ensuing distributions being typically intractable convolutions of the associated kinetics and thermodynamics. Analysis of cluster distributions resulting from mixed lithium/sodium expansions shows that high collision number expansions have terminal cluster distributions in which thermodynamic stability effects become resolvable[8]. Consequently mass spectra obtained for pure alkali expansions at similar stagnation conditions should allow insight into system thermodynamics *if* measured ion abundances can be translated into precursor neutral cluster compositions. This is a multi parameter problem which depends on instrument response, cluster ionisation cross section and cluster ionisation induced fragmentation cross section. At this writing instrument response and fragmentation are well understood[9]. There are as yet no determinations of relative photoionisation cross sections for alkali clusters, but a set of related measurements supports the assertion that ionisation cross section changes monotonically with particle size (going roughly as the geometric cross section)[10]. The dramatically multimodal ion distributions observed in photoionisation mass spectra obtained for neutral alkali cluster beams (figure 1) can then be explained in terms of islands of enhanced stability among the precursor neutrals.

The jellium model rationalizes this series of dominant clusters purely in terms of the electronic structure peculiar to a finite spherical free electron metal[2]. Stability is predicated by the total number of valence electrons, with negligible geometric contributions to total energy. This approach works quite well in explaining the numerology of abundance maxima but runs into problems when it's predictions are compared to dynamic system response properties such as ionisation potentials or polarisabilties[7,5,11,12].

A finite system composed of atoms can strictly never have spherical symmetry[13]. Why then does this symmetry work so well at describing the series of observed maxima - at least at a qualitative level? In an attempt to gauge the influence of geometry on cluster stability we have performed a series of experiments in which individual alkali atoms were replaced by elements with differing atomic radius and more than one valence electron[14]. Then, according to the jellium model, "magic" numbers should be shifted by an amount appropriate to the heteroatom valence electron count. Heteroatoms studied so far include Mg, Ca, Ba, Sr, Eu, Yb and Zn. Table I presents a summary of preliminary results (subject to confirmation in several cases by ongoing ionisation potential determinations) while figure 1 provides an example of such a measurement: a photoionisation mass spectrum obtained for a coexpansion of sodium and ytterbium metal vapors. Many systems yield heterocluster (M_xN) "magic" numbers which are inconsistent with the model[14]. The data suggest that for the corresponding mixed clusters, geometric effects are large enough to induce level switching relative to homonuclear alkali species even - at internal temperatures for which large amplitude motion is expected to average out most structure.

The particle specific gas-phase characterization of neutral cluster chemistry is a challenging area. Some inroads have been made into the

TABLE I.

NEUTRAL ABUNDANCE MAXIMA IN ALKALI/HETEROATOM CLUSTER BEAMS[a]

Metals (M,N)	Maxima(M_nN)	Heteroatom (N) Configuration
Na/Li	9, 19, 39	$..2s^1$
K/Li	9, 19	"
Na/Mg[b]	6-8, 18	$3s^2$
Na/Ca	6, 18	$4s^2$
Na/Sr	6, 16-18, 38	$5s^2$
Na/Ba	6, 16	$6s^2$
Na/Zn	8, 18	$3d^{10}4s^2$
Na/Eu	6, 16	$4f^76s^2$
Na/Yb	6, 18, 38	$4f^{14}6s^2$
K/Mg	8, 18	
K/Zn	8, 18	
K/Hg	8, 19	$5d^{10}6s^2$

[a]Determinations of neutral abundance maxima are based on photoionisation mass spectra which show dramatically multimodal ion distributions (typically greater than >10x change between M_nN^+ and $M_{n+1}N^+$). Note that ionisation potentials and cross sections as well as ionisation induced fragmentation cross sections have as yet not been determined in each case. For homonuclear alkali and mixed alkali clusters, neutral abundance maxima are observed at M_8, M_{20}, M_{40}, M_{58}..
[b]Abundance maxima specified as a range indicate that there are no major differences in ion signal among the clusters listed (subsequent dramatic reduction).

reactions of transition metal clusters by way of injection pulse reactors (which have also been demonstrated with continuous reagent flows)[15]. Aside from Fe_x and Nb_x hydrogenation reactions in which reaction rate can be correlated with cluster ionisation potential[16], there have been few examples of dramatically size dependent variations in chemical reactivity which could be rationalized in terms of known geometric or electronic structure variations.

There is very little reactivity data available for alkalis. We have recently intiated a survey of gas-phase alkali cluster chemistry. In this experiment a sodium cluster beam (Na_x, $x<41$) is crossed at right angles with a reagent gas effusion (pick-up configuration), 2 cm downstream from the cluster source. Cluster abundances and neutral product species entrained on axis are analysed by photoionisation mass spectroscopy. As this experimental configuration is subject to kinematic constraints, we plan to compare results with collision cell measurements. At this writing interactions with Ar, N_2, CO_2, N_2O, O_2, SF_6 and CCl_4 have been studied. Apart from cluster size dependent scattering and collisional dissociation, no reactions are observed for Ar, N_2 and CO_2. Rapid reactions and dramatic size dependent changes in reaction cross section occur for N_2O, O_2, SF_6 and CCl_4. Product species observed (and partially characterized) include Na_xO, Na_xS, Na_xF and Na_xCl ($x<6$). A thorough understanding of size effects will require (a) unambiguous elucidation of reaction pathways (by

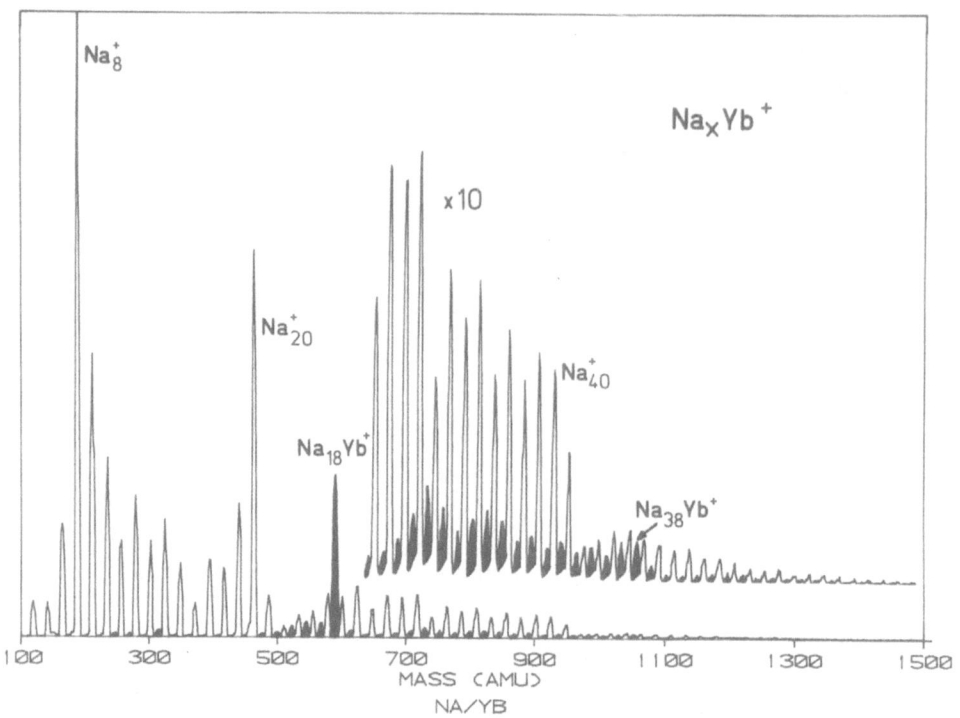

FIGURE 1. Photoionisation mass spectrum obtained upon irradiating a Na/Yb cluster beam with the full output of a 1 kW Xe/Hg arc lamp. Clusters were generated by coexpansion of sodium and ytterbium metal vapors from a high temperature oven at 800 °C. Cluster formation was enhanced by using a 0.6 mm throat diameter, 30 degree conical nozzle. The multimodal cluster ion abundances observed reflect abundance maxima occurring at Na_8, Na_{20} and Na_{40} among homonuclear neutral species and at Na_6Yb, $Na_{18}Yb$ and $Na_{38}Yb$ among heteronuclear clusters containing one ytterbium atom.

varying reagent cluster beam composition), (b) determination of relative ionisation cross sections for reactants and products (to obtain actual neutral abundances),(c) measurement of reactive and non-reactive scattering cross sections, and (d) an understanding of translational and internal energy dependence. These experiments are in progress.

ACKNOWLEDGEMENTS

This work was supported by the Swiss National Science Foundation under research grant number 2.431.84.

REFERENCES

1. M. Kappes, R. Kunz and E. Schumacher; Chem. Phys. Lett., 91, 413 (1982).
2. W. Knight, K. Clemenger, W. de Heer, W. Saunders, M. Chou and M. Cohen, Phys. Rev. Lett., 52, 2141 (1984).
3. G. Natanson, F. Amar and R. Berry, J. Chem. Phys., 78, 399 (1983); R. Berry, J. Jellinek and G. Natanson, Chem. Phys. Lett., 107, 227 (1984); J. Jellinek, T. Beck and R. Berry, J. Chem. Phys., 84, 2783 (1985); J. Borel, Surf. Sci., 106, 1 (1981).
4. M. Kappes and E. Schumacher, Proceedings of the 6th International Conference on Liquid and Amorphous Metals, Garmisch-Partenkirchen, 1986.
5. W. Knight, K. Clemenger, W. de Heer and W. Saunders; Phys. Rev.B, 31, 2539 (1985).
6. A. Herrmann, E. Schumacher and L. Wöste, J. Chem. Phys.,68, 2327 (1978); M. Kappes and E. Schumacher in "Electronic and Atomic Collisions", edited by J. Eichler, I. Hertel and N. Stolterfoht (Elsevier, Amsterdam, 1984).
7. M. Kappes, M. Schär, P. Radi and E. Schumacher, J. Chem. Phys., 84, 1863 (1986).
8. M. Kappes, P. Radi, M. Schär, C. Yeretzian and E. Schumacher, Z. Phys. D, 3, 115 (1986); M. Kappes, M. Schär and E. Schumacher, J. Phys. Chem., in press.
9. M. Kappes, M. Schär, E. Schumacher and A. Vayloyan, Z. Physik D, submitted for publication.
10. In this volume, C. Brechignac and P. Cahuzac report experiments which probe for stability islands in potassium cluster cations. Relative to measurements which nominally probe for neutral stability, maxima are now observed shifted to K_{n+1}^+, where n=8,20.
11. W. Saunders, K. Clemenger, W. de Heer and W. Knight, Phys. Rev. B 32, 1366 (1985).
12. W. Ekardt and Z. Penzar, Solid State Commun., 57, 661, 1986.
13. Various calculations at the unrestricted Hartree-Fock level have generated equilibrium structures for small alkalis which are far from spherical. See for exmaple: J. Koutecky and P. Fantucci, Chem. Rev. 86, 539 (1986); B. Rao and P. Jena, Phys. Rev. B 32, 2058 (1985); J. Martins, J. Buttet and R. Car, Phys. Rev. B 31, 1804 (1985).
14. M. Kappes, P. Radi, M. Schär and E. Schumacher, Chem. Phys. Lett., 119, 11 (1985); C. Yeretzian, Diploma Thesis, University of Bern, 1986.
15. A. Kaldor, D. Cox, D. Trevor and M. Zakin, Z. Physik D, 3, 195 (1986).
16. M. Geusic, M. Morse and R. Smalley, J. Chem. Phys., 82, 590 (1985); R. Whetten, D. Cox, D. Trevor and A. Kaldor, Phys. Rev. Lett., 54, 1494 (1985); S. Richtsmeier, E. Parks, K. Liu, L. Pobo and S. Riley, J. Chem. Phys., 82, 3659 (1985); R. Whetten, M. Zakin, D. Cox, D. Trevor and A. Kaldor, J. Chem. Phys., 85, 1697 (1986).

A UNIFIED EXPLANATION OF MAGIC NUMBERS IN SMALL CLUSTERS

G. S. Anagnostatos

Tandem Accelerator Laboratory
National Research Center of Natural Sciences "Demokritos"
Aghia Paraskevi-Attiki, 153 10 Greece

INTRODUCTION

Small clusters with specific numbers of atoms, called magic numbers, exhibit particular stability properties in comparison to other cluster sizes. The identification of magic numbers in the literature, however, is rather confusing, since different sets of such numbers have been reported for different kind of atoms. For example, for Xe clusters[1] the numbers 13, 19, 25, 55, 71, 87, 147, while for Ar and Kr the numbers[2] 14, 16, 19, 21, 23, 27, and 14, 16, 19, 22, 25, 27, 29, 39, 75, 87, respectively, and for Na the numbers[3] 2, 8, 20, 40, 58, 92 have been reported. The theoretical explanation of magic numbers in the literature is not unique as well. It follows two independent reasonings, one[3] for the last set of above numbers (Na clusters; magic numbers as a result of decoupled 3s electrons driven by a central potential) and another[1] for the other three sets (Xe, Ar, Kr clusters; magic numbers as a result of packing of spheres). Moreover, even for the same kind of clusters, theoretically expected magic numbers are not supported experimentally[1] and experimentally observed numbers are not covered by theory[1].

The purpose of the present communication is to contribute to the resolution of the rather unpleasant situation described above. First, for the magic numbers in the Na clusters we give a different explanation from the existing one[3], but one which is consistent with the explanation for the Xe, Ar and Kr clusters[1,2]. Second, we make theoretical predictions and experimental observations more consistent.

The key to our unified explanation of magic numbers is the softness of spheres standing for atoms. Specifically, we argue that when atoms can be sufficiently well presented by soft spheres, the packing-of-spheres assumption[4] leads to a set of magic numbers (i.e., 1, 13, 55, 147, 309, 561), while when atoms can be sufficiently well presented by hard spheres, the packing-of-shells assumption[5] leads to another set of magic numbers (i.e., 2, 8, 20, 40, 58, 90).

MAGIC NUMBERS FOR ATOMS PRESENTED BY SOFT SPHERES

In Fig. 1 we show the close-packing of spheres equilibrium geometries for soft spheres, as has been presented in Ref. 4, i.e., presented by nested icosahedra standing for closed shells. Close-packed spheres eventually form the edges, the faces, and thus the icosahedron-shell itself. Each cumulative

number of spheres of all such shells from the first (only one sphere) up
to a particular one corresponds to a magic number. Parts (a)-(e) in Fig. 1
posses 2-6 nested shells,respectively, snd the corresponding magic numbers
are 13, 55, 147, 309, and 561. While the surface spheres are in contact
with one another, all interior spheres overlap with one another and with
the surface spheres[4], a fact which is permissible for soft spheres.

   Numbers shown on each of the exterior shells in Fig. 1 (a)-(e) mark
spheres standing at the corners, at middles of edges, and the middles of
faces of the corresponding icosahedron. The corners of the first icosahedron (Fig. 1(a)) are the only spheres of that icosahedron-shell and thus
are not marked at all. The corners of the second icosahedron are marked
and its middles of edges as well. (Spheres at the middles of faces do not
exist.) The numbers 12 at the corners show the number of vertices of an icosahedron, while the numbers 6 mark those six middle points (out of the thirty
edges) forming a regular octahedron. During the procedure of filling up this
second icosahedron, when the number of spheres is just 6 or 12, these spheres arrange themselves at the vertices of the marked octahedron or icosahedron, respectively. Thus, the addition of these spheres give rise to symmetric equilibrium partial fillings which correspond to the semi-magic numbers
19(=13+6) and 25(=13+12), which are verified experimentally[2]. Spheres fill
all thirty middles of edges of the icosahedron after the semi-magic number
25 and thus fill up all remaining places of this shell (i.e., 55=25+30). In
similar way one could utilize the numbers of Fig. 1(c)-(e) to derive the semi-magic numbers 75, 87, etc.

   While the magic numbers reported in Ref. 2 for Ar and Kr do not coincide with those of Ref. 1 for Xe clusters, the series of magic numbers for all
these three kinds of clusters have many common members, particularly after
the previous discussion that N=19, 25, 75 and 87 are semi-magic numbers. The
absence of N=13 and 55 in Ar and Kr clusters is noticeable. However, the presence of N=19 and 25, and N=75, respectively, implies their hidden existence
as explained earlier. The existence of non-common members in the reported
series of magic numbers for Xe, Ar, Kr could be attributed to differences
among the experimental set-up involved and particularly to different behavior of neutral and ionized atoms. The latter constitute what we actually
measure in our experiments and the final ion spectrum would be representative of the relative stability of cluster ions.

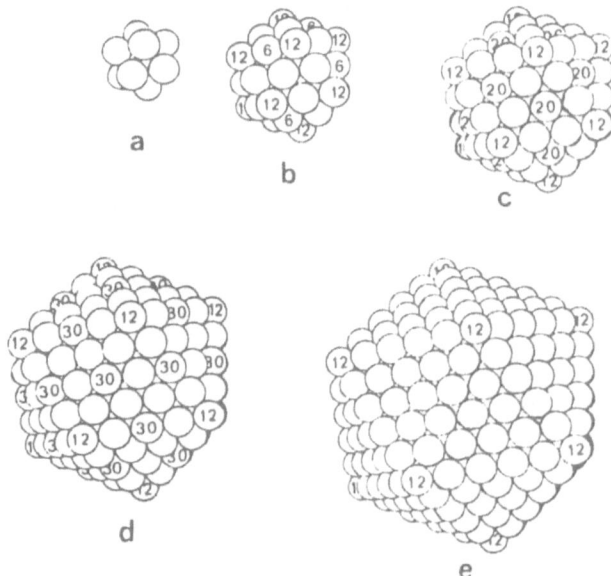

Fig.1. Close-packing-of-spheres equilibrium geometries:Nested icosahedra
       standing for closed shells of soft-sphere atoms.

MAGIC NUMBERS FOR ATOMS PRESENTED BY HARD SPHERES

In Fig. 2 we show the different polyhedra assumed as successive shells in contact for the structure of small clusters in the case where hard spheres stand for atoms. Each such polyhedron is an equilibrium[6] polyhedron which permits an equilibrium of particles at its vertices for any sort of force among particles and, in particular here, minimizes their mutual repulsion. The mutual separation of the polyhedral vertices depends on the number of particles in the shell which defines the kind of polyhedron employed. The relative orientation of polyhedra in different shells is such that their sets of particles are in equilibrium with each other[6]. (As a result their relevant polyhedral axes of symmetry coincide). That is, the equilibrium geometry of small clusters of hard sphere atoms is made up of equilibrium polyhedral shells in contact and in equilibrium with each other. Their order proceeds from the polyhedron with the smaller number of vertices to the polyhedron with the larger number of vertices and their relative orientation is consistent with reciprocity[7]. In general, both the order of polyhedra employed and reciprocity are consistent with the close-packing of hard spheres.

At the top of each block in Fig. 2 presenting a polyhedron, the name of the polyhedron is given, while at the bottom of the block two specific numbers are shown. The first, in parenthesis, presents the number of vertices for the particular polyhedron, while the second, in brackets, gives the cumulative number of vertices of all previous and this polyhedron. As one can see from the numbers in brackets, the cumulative numbers appearing in Fig. 2 (where successive shells are completed) are 2, 8, 20, 40, 58 and 90, that is, they precisely present the magic numbers of the second set except the last one. That is, 90 instead of 92 appears here as a magic number. From the experiments in Fig. 1 of Ref. 3, however, 92 might not be considered as a magic number according to the following reasoning. Each peak presenting a magic number in this figure follows after a smaller peak and is followed by a smaller peak as well. However, 92 does not follow the first, but only the second criterion. Further experimentation is needed to distinguish between 90 and 92.

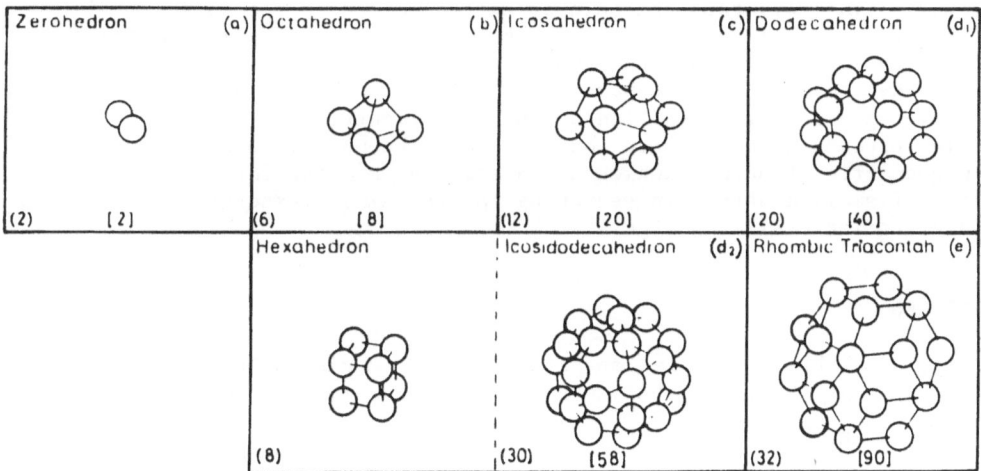

Fig. 2. Close-packing-of-<u>shells</u> equilibrium geometries:Nested equilibrium polyhedra standing for closed-shells of hard-sphere atoms. Shells in contact:Zerohadron with octahedron and hexahedron, octahedron with icosahedron and dodecahedron, hexahedron with icosidodecahedron, and icosidodecahedron with rhombic triacontahedron. Reciprocal pairs shown:Octahedron and hexahedron, icosahedron and dodecahedron, and icosidodecahedron and rhombic triacontahedron.

As becomes obvious drom Fig. 2 up to N=40, all previously formed poly-hedral shells coexist in size and form. After N=40, however, the last shell in N=40, i.e., the dodecahedron (which does not posses stable equilibrium[6]), does not coexist with the new coming shells. That is, when the icosidodeca-hedron is added out of the twenty vertices of the dodecahedron, only eight vertices forming a cube remain, which shrink in radial size and come in con-tact with the interior zerohedron. The formation of a cube is not favored earlier, since a hexahedron (cube) does not posses stable equilibrium by itself[6] and a suitable interior shell to lend stability does not exist. (Out of its eight vertices only four can be in contact with the interior zerohedron). When the icosidodecahedron has been included in the structure, however, the spheres of the hexahedron are all in contact with those of the icosidodecahedron which possesses a stable equilibrium by itself[6]. This pe-culiarity of the cluster structure, i.e., including the hexahedron or not, shows up in the peaks[3] of N=10, 12 (and perhaps 14) where the extra 2 and 4 atoms after the magic number 8 correspond to those vertices of the hexa-hedron which are in packing (contact) with the two spheres of the zerohe-dron.

The present interpretation of the magic numbers of the second set (Na clusters) is completely different from the one proposed in Ref. 3 based on the decoupled 3s electron, i.e., on electron properties. Peaks at N=18, 34, 68 and 70 proposed by that explanation are not supported by the experiments (see Ref. 3). Specifically, there is not at all a magic-like peak at N=18, while the magic effect at N=34 cannot be distinguished from N=26, 30, 36, and 38. The latter is also valid for N=68 and 70 in comparison with neigh-boring peaks. Also, a decoupled 3s electron explanation of the second set makes no connection with the valid close-packing of spheres explanation of the first set of magic numbers. In contrast, the present explanation by u-sing the close-packing of shells concept goes along very well with the ex-planation of the first set of magic numbers*. The only difference between the two explanations, is the softness of the sphere-like atoms.

CONCLUSIONS

Through the present paper a unification of the understanding of magic members in small clusters has been presented, based on the softness of spheres standing for atoms. The equilibrium geometry of magic numbers for soft sphere-like atoms results from close-packing of spheres, while for hard sphere-like atoms from close-packing of shells. Both packings lead to magic and semi-magic numbers justified by the experiments.

The present communication could prove to be useful in deriving equili-brium geometric structures which could serve as a guide for computer simu-lations of small clusters in searching for the minimum-energy configuration for a given large size cluster.

REFERENCES

1. O. Echt, K. Sattler, and E. Recknagel, Magic numbers for sphere packings : Experimental verification in free xenon clusters, Phys. Rev. Lett. 47:1121(1981).
2. A. Ding and J. Hesslich, The abundance of Ar and Kr micro-clusters generated by supersonic expansion, Chem. Phys. Lett. 94:54(1983).
3. W. D. Knight, K. Clemenger, W.A. de Heer, W.A. Saunders, M. Y. Chow, and M. L. Cohen, Electronic shell structure and abundances of sodium clusters, Phys. Rev. Lett. 52:2141(1984).

* The deviation of the related but different to the second set magic numbers reported in Ref. 8 (i.e., N=2, 7, 19 and 38) has been explained in Ref.3.

4. J. A. Barker, The geometries of soft-sphere packings, J. Phys. (Paris), Collog. 38:C2-37(1977).

5. G. S. Anagnostatos, Isomorphic shell model for closed-shell nuclei, Int. J. Theor. Phys. 24:579(1985).

6. J. Leech, Equilibrium of sets of particles on a sphere, Math. Gazelle 41:81(1957).

7. H. S. M. Coxeter, "Regular Polytopes", The MacMillan Co., New York (1973).

8. M. M. Kappes, R. W. Kunz, and E. Schumacker, Production of large sodium clusters (Na_x, $x \leq 65$) by seeded beam expansions, Chem. Phys. Lett. 91:413(1982).

A MOLECULAR DYNAMICS STUDY OF SILICON CLUSTERS

Estela Blaisten–Barojas

Instituto de Física, Universidad Nacional Autónoma de México, Apartado Postal 20–364, 01000 México D.F., México

D. Levesque

Laboratoire de Physique Téorique et Hautes Energies, Université de Paris–Sud, 91405 Orsay, France

ABSTRACT

The structural properties of neutral and charged silicon clusters of moderate size were obtained from a molecular dynamics simulation using Stillinger–Weber model potential [Phys.Rev.B 31,5262(1985)]. Cluster equilibrium configurations resulting from quenches initiated at finite temperatures and ended at low temperatures established different growth sequences for neutral and charged clusters. Cooling and heating experiments were carried out showing that the transition from liquid-like systems to solid-like structures is very smooth. A model of an amorphous five layer film is constructed using the 14-atom cluster as elemental building block.

PACS numbers:36.40.+d,61.50.-f

1. INTRODUCTION

Molecular dynamics simulations have been extremely successful in describing the process of melting and freezing of atomic systems.[1,2] Specifically, the microscopic mechanism underlying these processes is a consequence of the classical interatomic force field used to describe the system. To explore such situations, interatomic potentials based on small deviations about bulk equilibrium structures are inadequate. Moreover, for covalent materials pair potentials alone are insufficient to describe these equilibrium structures.[3] Very few simulations have been carried up to now using n-body terms in the potential model, both in 3D or lower dimensions, and both in the termodynamic limit or for finite systems. Also, except for the triple-dipole terms, almost no 3-body class of potentials have been analyzed in all of its details.[4]

The purpose of our work[5] is to provide a computational probe of the Stillinger–Weber (SW) potential function[3] in the description of annealing experiments on moderate-sized (up to 32 atoms) neutral and charged silicon clusters. The SW potential includes 2- and 3-body terms and its parameters were adjusted to fit the silicon properties of condensed phases: bond length, cohesive energy, crystallization into the diamond lattice and melting temperature. The model is also good to satisfy qualitatively the Lindemann melting criterion for solids and to reproduce the property of shrinking when silicon melts. Recently other intermolecular potential including 2- and 3-body terms or directional bonds[6] have been proposed for silicon, although no dynamical study has been provided for any of them.

A finite temperature description of cluster properties obtained from molecular dynamics calculations is always useful when there is not a great deal of experimental data available. Although solid-liquid like phase changes are expected to occur for finite systems at temperatures lower than in the bulk material, and the transition is expected to be "broad" due to the finite size, still, this has to be tested. Also, increasing the cluster size should predict the crossover to bulk-like structures. Silicon was chosen as a prototypical semiconductor for which the largest number of cluster experiments have been reported.[7,8]

In section 2 a brief description of the SW potential is given as well as a description of the model used to simulate neutral and charged clusters. The finite temperature behavior of the cluster total energy, coordination number and structure is discussed in section 3. Section 4 contains the results concerning the geometrical configurations that the clusters acquire at low temperatures as a result of the annealing experiments. The binding energy as a function of the cluster size is also reported. Finally, the paper is closed in section 5 with a discussion on the possibility to form larger aggregates using the very stable clusters generated in the simulation.

2. THE POTENTIAL MODEL

In the present work, the structural energy of the N-atom cluster is a sum of 1-, 2- and 3-body potentials,

$$V(\vec{r}_1, \vec{r}_2, \ldots, \vec{r}_N) = \sum_i^N v_1(\vec{r}_i) + \sum_{i<j}^N v_2(\vec{r}_i, \vec{r}_j) + \sum_{i<j<k}^N v_3(\vec{r}_i, \vec{r}_j, \vec{r}_k). \qquad (1)$$

We have investigated two cases. In the first case the one-body terms are zero to represent neutral clusters. In the second case the $v_1(\vec{r}_i)$ terms model approximately the interaction between an ionized atom with all other atoms in the cluster. The ionization takes place on one atom creating a positive charge q and inducing an energy

$$U^{(\text{ind})} = \sum_i^N v_1(\vec{r}_i) = -\frac{1}{2} \sum_{j=2}^N \frac{q^2 \alpha}{r_{1j}^4} \qquad (2)$$

where the atomic polarizability α is considered isotropic and the contributions of induced dipoles to the electric field were neglected. The values $q = -e$ and $\alpha = 7.2\text{Å}$ were adopted,[9] when not stated otherwise.

In both cases the 2- plus 3-body terms were modeled using the SW potential. This potential is a seven parameter function, where four parameters (A, B, p, q) belong to the pairwise interaction

$$v_2(r) = \begin{cases} A\left(Br^{-p} - r^{-q}\right) e^{1/(r-a)}, & \text{if } r < a, \\ \\ 0, & \text{otherwise,} \end{cases} \qquad (3)$$

and where the cutoff distance a is common to both 2- and 3-body terms. The 3-body contribution is represented as a function of two bond lengths l_1, l_2 and the included angle θ_{12}. The assumption on this term is to be separable, *i.e.*, the dependence on three interatomic distances is represented by products of functions $f(l)$ of each bond length:

$$f(l) = \begin{cases} \sqrt{\lambda}\, e^{\gamma/(l-a)}, & \text{if } l < a, \\ \\ 0, & \text{if } l > a, \end{cases} \tag{4}$$

such that

$$v_3(l_1, l_2, l_3) = \sum_{s<t=1}^{3} f(l_s)\, f(l_t)\, (\cos\theta_{st} + 1/3)^2. \tag{5}$$

For silicon, the 3-body potential was constrained to have a minimum at the tetrahedral angle and confined to within a 3.77Å cutoff. This short range makes it very attractive for molecular dynamics simulations. The assumption of separability gives a local classical bonding picture of the atomic interactions. The angular variation in Eq.(5) is a special case of an expansion of V_3 in Legendre polynomials.

The choice of the seven parameters in this work is the same as used in reference (3). This choice allows two minima of the SW for the $N = 3$ cluster: the equilateral triangle which is the more stable ($V = 1.4786$eV) and the isosceles triangle with one tetrahedral angle of 109.47 follows in stability ($V = 1.4447$eV). When the parameter is increased from its value of 21 to 54 only the isosceles geometry remains as minimum of the potential function.

3. ANNEALING OF CLUSTERS. FINITE TEMPERATURE RESULTS

· Molecular dynamics (MD) at constant energy was used throughout. The units of length and energy were chosen to be $\sigma = 2.0951$Å and $\epsilon = 2.167$eV/atom-pair. The Newtonian equations of motion were solved using the Verlet[10] method with a time step of 7.66×10^{-16}s at a reference number density $\rho_0 = 0.46\sigma^{-3}$.

The annealing and quenching experiments were carried out in the following way. First, an initial configuration was chosen and the clusters equilibrated at the highest possible temperature. Second, the clusters were cooled in a step-like procedure at rates ~ 100K/ps, such that each intermediate step in a quench was an equilibrated state of the clusters. The change to each new temperature was obtained by scaling the final velocities of the prior equilibrated state. Third, the clusters were led through a cycle of reheating followed by cooling to $T = 0$ to assure that the system was not trapped in metastable states at low temperatures.

The typical behavior of E, mean total energy per atom along a quench, is illustrated in Fig. 1 for the neutral 13- and 14-atom clusters (open symbols). Similar curves for other cluster sizes have been reported previously[5] indicating extremly smooth changes of E between the low and high temperature limits. Each point is an average over 1.5 ps, but at $T = 375$K longer runs (7 ps) were performed before cooling to almost zero temperature. Black dots show a path followed by the system with a 5 time faster quenching rate. Clusters with odd number of atoms had a tendency to end the quench in different conformations of quasi-degenerate mean potential energy. Specifically, for the 13-atom cluster, three geometries were detected as shown in Fig. 1, although more quenches ended in case (a).

All clusters changed shape with temperature because the atoms have a high movility during a quench, as shown for the 14-atom cluster in the top part of Fig. 2. Eventually at high temperatures the cluster "evaporated" by loosing one atom after certain "life time", *i.e.*, the neutral 14-atom cluster lost one atom after 4.8 ps at $T = 3168$K. The mean coordination

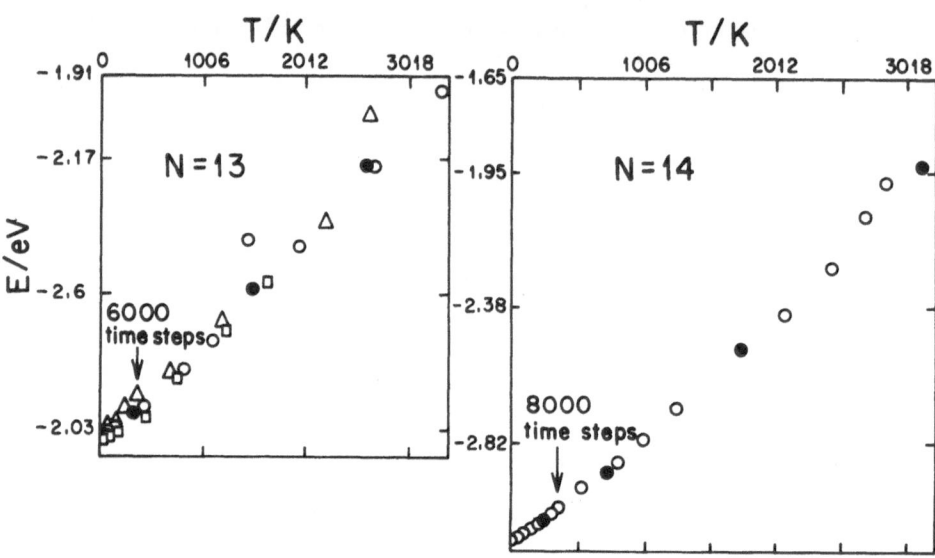

Fig. 1. Mean total energy per particle as a function of temperature. For the 13-atom cluster quenches ended in 3 energies: *(a)* open dots ($E = 3.044$eV at $T = 155$K); *(b)* open squares ($E = 3.0639$eV at $T = 157$K); *(c)* open triangles ($E = 3.0225$eV at $T = 161$K). Black dots stand for MD runs at faster quenching rate.

number n_c and the mean square distance of the atoms from the cluster center of mass $< R^2 >$ are illustrated in Fig. 2. The averaged quantity n_c, is calculated from the pair correlation function $g(r)$, namely

$$n_c(r_{bond}) = 4\pi\rho_0 \int_0^{r_{bond}} r^2 g(r)\, dr. \tag{6}$$

The pair correlation function is a time average of $n(r)$, the number of atom pair distances r averaged over all directions of $\vec{r}$:

$$g(r) = \frac{n(r)}{4\pi\rho_0 r^2}. \tag{7}$$

The bond distance r_{bond} is extracted from the position of $g(r)$ first minimum (2.94Å). As it is seen in Fig. 2, a region of temperatures where the cluster has a coordination number larger than 3 without increasing its size is apparent, possibly indicating a more liquid-like behavior. The cluster is shrinking. At higher temperatures the cluster increases its mean size, melts and finally temperature induced fragmentation occurs.

It is not clear from an analysis of Fig. 1 and 2 at what temperature the cluster melts, although the structural characteristics are certainly different at low than at high temperatures. A discussion on the definition of a "liquid cluster" has been given[5] in terms of the temperature changes of structural quantities such as the pair correlation function or the

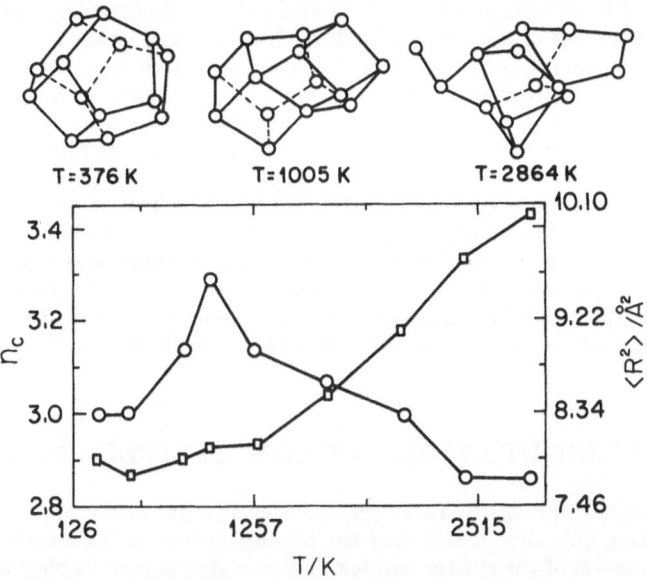

T=376 K   T=1005 K   T=2864 K

Fig. 2. Mean coordination number (dots) and mean square-radius from the cluster center of mass (squares) as a function of temperature for the 14-atom cluster. Instantaneous cluster configurations at three temperatures are shown.

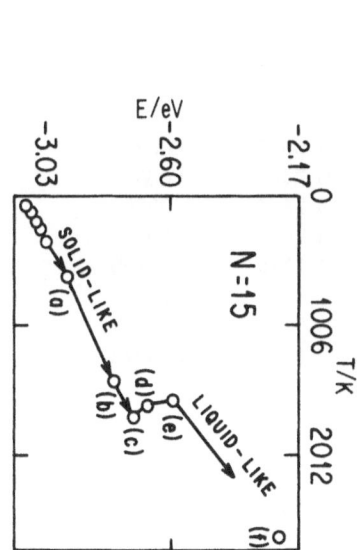

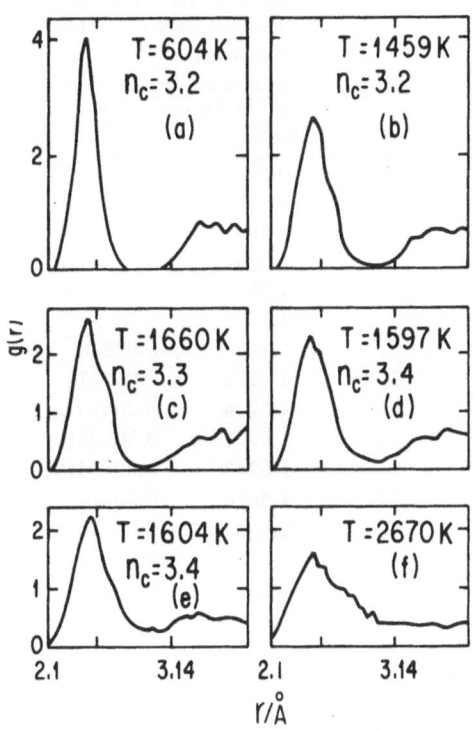

Fig. 3. Mean total energy per atom as a function of temperature while heating the neutral 15-atom cluster

Fig. 4. Pair correlation function for the temperature steps (a) to (f) in Fig. 3.

density distribution of bond angles. Let us state again the arguments as a matter of completeness for the specific case of the neutral 15-atom cluster. Fig. 3 shows a selected heating experiment for the neutral 15-atom cluster in which an abrupt change in the total energy is evident. In this diagram, the low temperature branch can be seen as a solid-like phase, the high temperature branch can be associated to a liquid-like phase and the region where the abrupt change takes place should correspond to a coexistence of phases. To see this effect more closely, in Figure 4 is given the evolution through the coexistence region of the pair correlation function for states labeled (a) to (f) in Fig. 3. The first peak grows broader and lower, when going from low to high temperatures. The second coordination shell looses progressively its structure from state (a) to (e) and the gap between first and second coordination shells tends to disappear with increasing temperature. From this analysis it can be said that in state (a) the cluster has structural characteristics of a solid phase, whereas in state (e) the $g(r)$ has changed to be more like that of a liquid phase.

4. CLUSTER CONFIGURATIONS AT LOW TEMPERATURE

The low temperature limit of the mean potential energy per atom, V_0, reached after a quench is an interesting quantity. It is in fact the binding energy of the cluster. The values of this energy as a function of the cluster number of atoms are given in Table 1 as well as the final temperatures reached in each simulation. In certain cases more than one final potential energy was obtained as listed in the table with labels (a) through (c), whereas in reference (5) an average was reported.

Table 1. Mean potential energy per atom at low temperature, V_0, as a function of the number of atoms in the cluster. Energies are in units of ϵ, $T^* = k_BT/\epsilon$, k_B is Boltzmann's constant.

N	NEUTRAL CLUSTERS $V_0(N)$	T^*	CHARGED CLUSTERS $V_0(N)$	T^*
3	−0.6828			
4	−0.9375		−1.5785	0.00129
5	−0.9998		−1.6842	0.00176
6	−1.0880	(a)	−1.6909	0.00187
	−1.0906	(b)		
7	−1.1786		−1.6965	0.00244
8	−1.3128	0.00640	−1.7457	0.00376
9	−1.3179	0.00627	−1.7544	0.00302
10	−1.3716	0.00530	−1.7627	0.00350
11	−1.3737	0.00608(a)	−1.7719	0.00160
	−1.3712	0.00504(b)		
12	−1.4075	0.00679	−1.7783	0.00412
13	−1.4047	0.00615(a)	−1.8105	0.00213
	−1.4139	0.00640(b)		
	−1.3948	0.00640(c)		
14	−1.4375	0.00524	−1.8074	0.00227
15	−1.4310	0.00539	−1.8150	0.00227
16	−1.4346	0.00593(a)	−1.8049	0.00407
	−1.4432	0.00346(b)		
17	−1.4409	0.00392		
32	−1.5469	0.00644		

The binding energies of the neutral series were completed for $N < 8$ with values of the minimum potential energy as obtained by the steepest-descent method. Nothing dramatic is observed at a first sight. However, in the neutral series there is a quantitative change in the slope at around $N = 10$ or 11 suggesting that a sudden change in the way of aggregation takes place for that cluster size. An analysis of the derivative as a function of N (Fig. 5) shows a definite even-odd alternation in the cluster stability: clusters with even number of atoms are easier to be formed. The neutral 14-atom has the lower derivative and stands as a very stable cluster. For charged clusters the aggregation takes place in a steplike fashion with a slightly higher stability when $N = 13$ and 15.

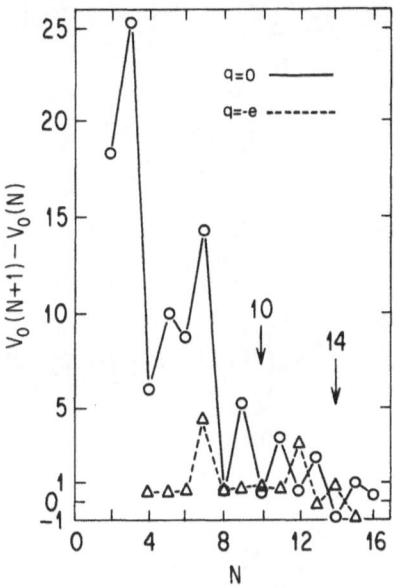

Fig. 5. Derivative of the binding energy per atom as a function of N.

The cluster geometries corresponding to the energies in Table 1 are given in Fig. 6 for the neutral sequence. These geometries are hollow clusters with all atoms on the surface. The small neutral clusters tend to have a coordination of three for all of its atoms, until at $N = 15$ an icipient 8-member unit connecting two four coordinated atoms builds up (Fig. 6).

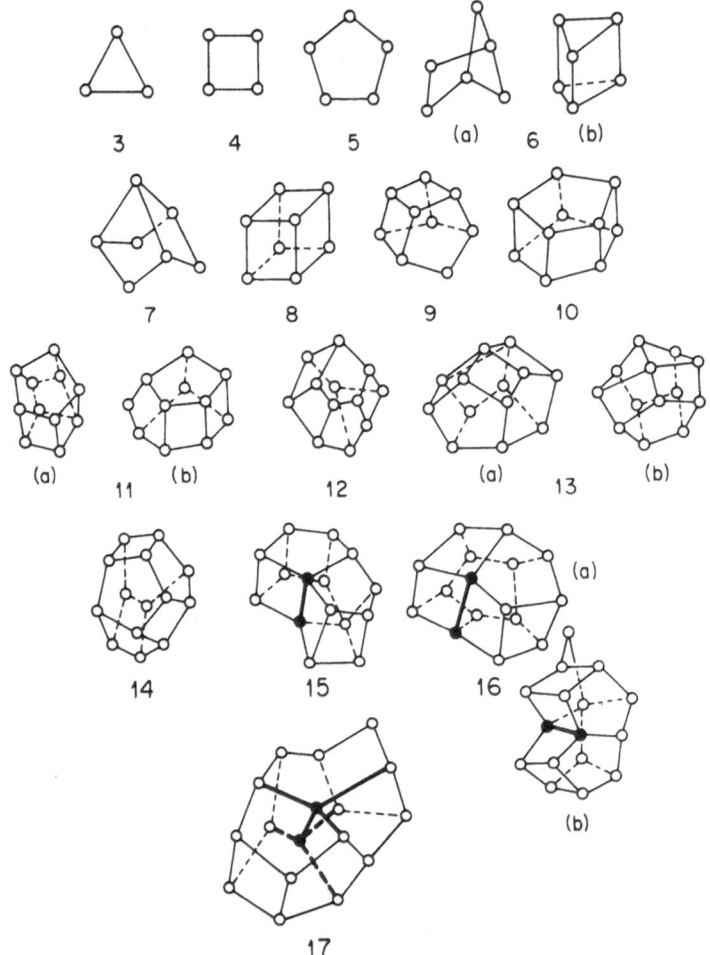

Fig. 6. Sequence of geometrical configurations of neutral clusters at low temperatures as obtained from the molecular dynamics simulation. Bonded atoms are at distances < 2.94Å.

When the cluster acquires a size of 17 atoms the tetrahedral seed is clearly present, although the dihedral angle is still distorted. In Fig. 7 it is shown the evolution of $g(r)$ as a function of N at very low temperatures. It is interesting to point out that the cluster $g(r)$ carries over to the bulk "inherent" pair correlation function of the supercooled silicon melt (reproduced from reference 3 by dotted lines in Fig. 7c). This inherent $g(r)$ is the result of mapping the geometrical configurations of a system with 216 atoms onto nearby potential energy minima. The seeds to build up those minimum-energy configurations seem to be already present in the small clusters analyzed in this study.

The charged cluster geometries corresponding to the energies in Table 1 are given in Figure 8. This sequence is qualitatively different from the neutral sequence since the charged atom tends to assume a coordination of 4 around itself. Therefore, a distorted 5-member tetrahedral unit is built even for the very small clusters. The first 8-member unit is formed when the cluster reaches a size of 14 atoms. The 16-atom cluster presents already a 6-atom ring on its surface.

It is interesting to show how an increase in the charge affects the final geometry obtained in the quench. This is seen in Fig. 9, where the geometries and energies of the 13-atom cluster are given as a function of increasing charge. This is one of the more stable clusters in the charged sequence. When the charge is small the most stable neutral geometry is picked up by the system. But as soon as the radial dependence in the potential is stronger, the cluster looses its spherical shape in behalf of the formation of one 5-member tetrahedral unit.

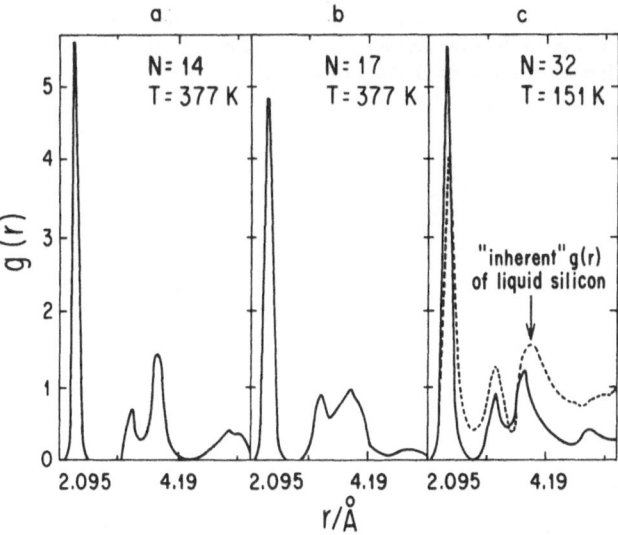

Fig. 7. Pair correlation function for the neutral 14-, 17-, and 32-atom clusters.

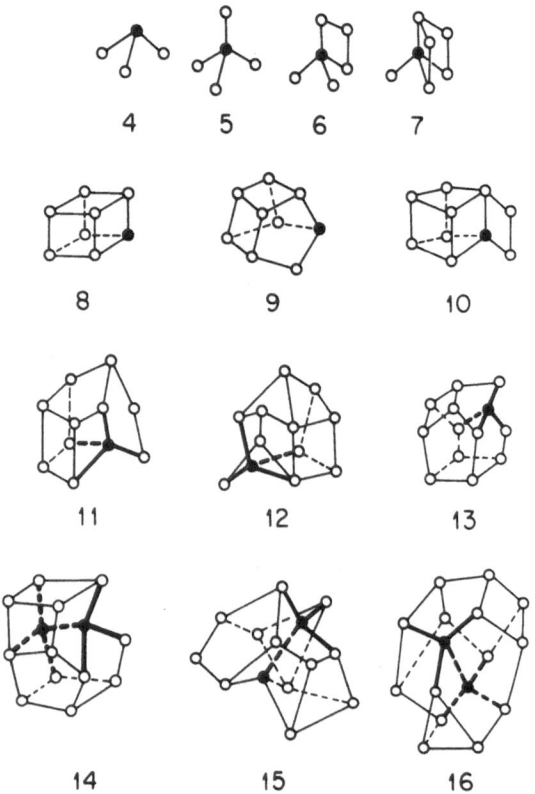

Fig. 8. Sequence of geometrical configurations of charged clusters at low temperatures as obtained from the molecular dynamics simulation. Black dots indicate the charged atoms. Bonded atoms are at distances < 2.94Å.

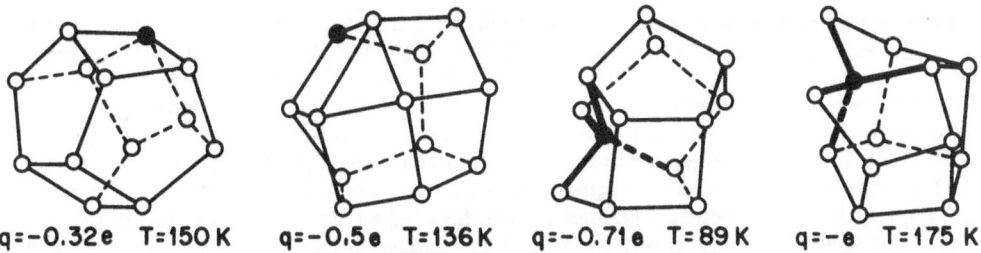

q=-0.32e T=150 K   q=-0.5e T=136K   q=-0.71e T=89 K   q=-e T=175 K

Fig. 9. Low temperature configurations of the charged 13-atom cluster as a function of the charge. Black dots indicate the charged atom.

5. DISCUSSION

The clusters generated in this work, specifically the more symmetric ones, are good candidates for the construction of larger aggregates made up of these elemental building blocks. Various noncrystalline models can be built. Lets consider the neutral 14-atom cluster in Fig. 2 at $T = 376$K. The cluster has several symmetry operations, namely: C_3, $3C_2$, $3\sigma_d$ and σ_h. It has 6 pentagonal and 3 square bent faces. A 2D hexagonal lattice can be built with this unit with the following prescriptions: *1)* since the three C_2 axes of every unit are on the same plane, the C_3 axes of all the units are parallel; *2)* every C_2 axis is common to two units such that neighbor clusters are connected through the atoms on the square faces. As a result, only pentagons are left on the surface and the cluster centers of mass form an hexagonal lattice, as shown in Fig. 10.

This network has included hexagonal holes, which in turn form a triangular lattice by themselves. This gives a model of an amorphous silicon film with five atomic layers and a surface made up only by pentagons. The stability of the film increases with a deformation of square onto rhomboidal faces.

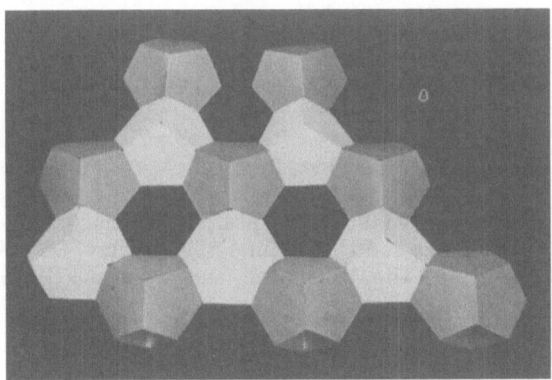

Fig. 10. Model of a 5 atom layer built with the most stable 14-atom cluster.

The conclusions from this work are the following. First, the clusters found in the early stages of growth do not have locally the tetrahedral symmetry of bulk silicon, the crossover starts to show at $N = 17$ for neutral clusters and $N = 14$ for charged clusters. For small clusters it seems that underlying stable structures more liquid-like dominate the process of growth. Second, using the same potential model neutral and charged clusters aggregate differently. The neutral sequence shows a slight increase in stability for the 10- and 14-atom clusters, whereas the charged sequence indicates the 13- and 15-atom clusters as more stable. Third, these stable cluster geometries can be used as elemental units to built larger aggregates of amorphous structures. Finally, since changes in the potential parameters can give rise to selected cluster geometries at zero temperature, *i.e.*, for $N = 3$ the equilateral triangle is a minimum depending on λ, it would be convenient to fit these parameters not only on bulk properties, but also on well established properties of small clusters. In this simulation we have tried to produce the clusters much in the same way as the bulk crystal is produced. However, this mechanism might not be the best to simulate the formation of amorphous clusters. Better simulations would be possible if more experimental knowledge of the parameters controlling a laboratory quench were available.

AGNOWLEDGMENTS

We thank Prof. T. P. Martin and Dr. A. Gonzalez for helpful discussions. This work was supported by Consejo Nacional de Ciencia y Tecnología, México, grant No. PCCBBNA-022643 and joint contract with Centre National de la Recherche Scientifique, France.

REFERENCES

1. J.P. Hansen and R.J. Mc Donald, "Theory of Simple Liquids" (Academic Press, New York 1976); F.F. Abraham, Rep. Prog. Phys. 45, 1113 (1982).

2. F.H. Stillinger and T.A. Weber, Kinam 3A, 159 (1981); Phys. Rev. A 25, 978 (1982); *ibid*. 28, 2408 (1983).

3. F.H. Stillinger and T.A. Weber, Phys. Rev. B 31, 5262 (1985).

4. E. Blaisten–Barojas and H.C. Andersen, Surface Sci. 156, 548 (1985); E. Blaisten-Barojas, Kinam 6A, 71 (1984).

5. E. Blaisten–Barojas and D. Levesque, Phys. Rev. B 34, 3910 (1986).

6. P.N. Keating, Phys. Rev. 145, 637 (1986); T. Takai, T. Halicioglu and W. Tiller, Scripta Metall. 19, 709 (1985); R. Biswas and D.R. Hamann, Phys. Rev. Lett. 55, 2001 (1985); J. Tersoff, *ibid*. 56, 632 (1986); S. Saito, S. Ohnishi, and S. Sugano, Phys. Rev. B 33, 7036 (1986).

7. T.P. Martin and H. Schaber, J. Chem. Phys. 83, 855 (1985).

8. L.A. Bloomfield, R.R. Freemen and W.L. Brown, Phys. Rev. Lett. 54, 2246 (1985).

9. R.R. Teachout and R.T. Pack, Atomic Data 3, 195 (1971).

10. L. Verlet, Phys. Rev. 159, 98 (1967).

ELECTRON LOCALIZATION AND EXCITATION

DYNAMICS IN SMALL CLUSTERS

Uzi Landman, R. N. Barnett and C. L. Cleveland

School of Physics
Georgia Institute of Technology
Atlanta, Georgia 30332

Dafna Scharf and Joshua Jortner

School of Chemistry
Tel Aviv University
69978 Tel Aviv, Israel

1. INTRODUCTION

Structural, electronic, dynamic and chemical characteristics of materials depend primarily on the state (phase) and degree (size) of aggregation. Small clusters often exhibit unique physical and chemical phenomena, of both fundamental and technological significance, and provide the opportunity for exploration of the transition from molecular to condensed matter systems. Particularly, investigations of the correlations between physical properties and degree of aggregation allow elucidation of the development of collective phenomena responsible for phase transformations[1,2] (such as nucleation, melting, and structural transitions), studies of the excitation dynamics and the kinetics of reactive processes[3,4] (such as fragmentation) and of the energetics and dynamics of electron attachment,[5,6] solvation phenomena[7] and physical processes induced by electron attachment.

In this paper we focus on the energetics and dynamics of electronically excited clusters and on electron attachment and solvation in clusters. Theoretical studies of clusters were hampered by the relatively large number of particles which renders the adaptation of molecular science techniques rather cumbersome, while the lack of translational symmetry prevents the employment of solid-state methodology. These problems are alleviated by computer simulations,[8,9] where the evolution of the system is simulated, with refined temporal and spatial resolution, via direct numerical solution of the equations of motion and are in a sense computer experiments which open new avenues in investigations of the microscopic origins of physical phenomena. We demonstrate the versatility and wealth of information obtained from computer simulations via classical molecular dynamics (MD) as well as quantum path-integral molecular dynamics (QUPID) calculations. The dynamics of electronically excited clusters and vibrational predissociation phenomena induced by excitation of inert-gas clusters[3] are discussed in Section 2. The QUPID method and it's applications to studies of electron attachment and localization in alkali-halide clusters[5] and of cluster isomerization induced by electron attachment are described in Section 3. Electron attachment to water clusters is discussed in Section 4.

2. DYNAMICS OF ELECTRONICALLY EXCITED CLUSTERS

The processes of energy acquisition, storage and disposal in clusters are of considerable interest for the elucidation of dynamic processes in finite systems, whose energy spectrum for electronic and nuclear excitations can be varied continuously by changing the cluster size.[10] In this context, an important issue involves the consequences of vibrational excitations of clusters. Such relaxation processes fall into two categories: (i) Nonreactive vibrational energy redistribution in the cluster, which does not result in dissociation, and (ii) reactive dissociation or vibrational predissociation. The mechanisms of vibrational energy acquisition by a cluster can involve collisional excitation, optical photoselective vibrational excitation or electronic excitation followed by the degradation of electronic energy into vibrational energy. In charged clusters, the vibrational excitation resulting in both nonreactive and reactive relaxation, can originate from ionization followed by hole trapping in inert-gas clusters[11] and from electron attachment to alkali halide clusters.[5,6] In neutral clusters, it was found, using MD simulations,[12] that the dissociation of Ar_n (n = 4-6) clusters can be accounted for in terms of the statistical theory of unimolecular reactions, which implies the occurrence of vibrational energy randomization in small clusters. The nonreactive and reactive processes induced by the degradation of electronic energy into vibrational energy in clusters have not yet been elucidated. An interesting problem in this category involves the dynamic consequences of exciton trapping in rare-gas clusters (RGXs), which is the focus of our study. Extensive information is currently available regarding exciton trapping in solid and liquid inert gases.[13] Exciton trapping in the heavy rare-gas solids, i.e., Ar, Kr, and Xe, exhibits two-centre localization, resulting in the formation of electronically excited, diatomic inert-gas excimer molecules. Electronic excitation of a RGC, R_n, is expected to result in an exciton state, which subsequently becomes trapped by self-localization. Although the details of the energetics and spatial charge distribution of excitons in finite RGCs have not yet been explored, some information can be drawn from the analogy with the lowest electronic excitations in solid and liquid rare gases.[13] The two lowest, dipole allowed, electronic excitations in RGCs can adequately be described in terms of tightly bound, Frenkel-type excitations with a parentage in the $^1S_1 \rightarrow {}^3P_1$ and $^1S_0 \rightarrow {}^1P_1$ atomic excitations which are modified by large nonorthogonality corrections.[13] The process of exciton trapping in the heavy RGCs of Ar, Kr and Xe involves the formation of the diatomic excimer molecule R_2^*, which is characterized by a substantial binding energy at a highly vibrational state. Energy exchange between the R_2^* excimer and the cluster in which it is embedded involves two processes.

(1) Short-range repulsive interactions between the expanded, Rydberg-type excited state of the excimer and the other cluster atoms result in a dilation of the local structure around the excimer, leading to energy flow into the cluster.

(2) Vibrational relaxation of the excimer induces vibrational energy flow into the cluster.

The vibrational energy released into the cluster by processes (1) and (2) may result in vibrational predissociation.

We have explored the dynamic implications of exciton trapping in RGCs by conducting classical molecular dynamics (MD) calculations[8,9] on electronically excited states of such clusters. As model systems we have chosen, Ar_n (with n = 13,55) and mixed $Xe_m Ar_{n-m}$ (m = 1,2 and n = 13,55) clusters. The ground states of the RGCs were described by Lennard-Jones pair potentials $V(r) = 4\epsilon[(\sigma/r)^{12} - (\sigma/r)^6]$ with the well-depth and distance parameters appropriate for Ar_o-Ar[14] (121K, 3.4Å), Ar_o-Xe[15] (177.6K, 3.65Å) and $Xe-Xe$[16] (222.3K, 4.10Å), respectively. In the electronically excited state the excimer potential is represented by a Morse curve, $V(r) = D_e\{exp(-2B(r/R_e - 1))]\} - 2 exp[-B(r/R_e - 1))]\}$ with the parameters D_e, R_e and B taken as 9125.3K, 2.319Å and 5.12 for $Ar_2{}^*$[17] and 11609K, 2.99Å and 4.228 for $Xe_2{}^*$.

An important consequence of the electronic excitation involves the drastic modification of the interaction between the excimer and the ground-state atoms. On the basis of the analysis of Xe^*-Ar interactions[15] (for which a description in terms of a 6-12 LJ potential with $\epsilon = 92.8K$ and $\sigma = 4.13Å$ is adequate) the Ar^*-Ar potential for each of the constituents of the excimer has been described in terms of a Lennard-Jones potential with the parameters ϵ^* and σ^*. We have taken for the energy $\epsilon^* = \epsilon$, while the distance scale ratio $\bar{\sigma} = \sigma^*/\sigma$ has been chosen in the range $\bar{\sigma} = 1.0-1.2$. The appropriate Ar^*-Ar interaction is characterized by[15] $\bar{\sigma} = 1.10-1.15$, reflecting the enhancement of short-range repulsive interactions in the electronically excited Rydberg-type state.

Following equilibration of the ground state system, electronic excitation was achieved (at the time t = 0) by the instantaneous switching on of the excimer potential between a pair of nearest neighbor atoms and of the potentials between the excimer and the ground-state atoms comprising the rest of the cluster (using a fifth-order predictor-corrector method, the integration time-step in the ground state was 1.6×10^{-14} sec and in the excited state 1.6×10^{-16} sec).

We consider first the dynamics of excited homonuclear clusters, $Ar_{13}{}^*$ and $Ar_{55}{}^*$. In figures 1a and 1b we show an overview of the dynamics of the nuclear motion following the electronic excitations in $Ar_{13}{}^*$ which is expressed in terms of the interatomic distances. The excimer exhibits a large amplitude motion in a highly excited vibrational state, while all the other interatomic distances increase, indicating the initiation of the escape of the ground-state cluster atoms. Insight into the energy flow from the excimer into the cluster is obtained from the time dependence of the kinetic energy (KE), the potential energy (PE) and the total energy of the excimer (fig. 2). The strong oscillations in the PE and KE clearly indicate the persistence of the vibrational excitation of the excimer over a long time scale. Further detailed information concerning the implications of this energy flow on the cluster dissociation was inferred by considering the composition and the energetics of the "main fragment", i.e., the fragment which consists of the excimer together with ground-state atoms, with all the nearest-neighbour separations being smaller than 3σ, beyond which all interatomic interactions are negligibly small. The total energy within the main fragment was partitioned into two separate contributions: (i) The energy of the "reaction centre", which consists of the excimer PE and KE together with half of the sum of the potential energy of the Ar^*-Ar interactions, and (ii) the energy of the "bath subsystem", which involves the KE of the ground-state Ar atoms, the potential energies of the Ar-Ar interactions and half of the sum of the potential energies of the Ar^*-Ar interactions. The time evolution of the various contributions to the total energy (fig. 3) portray the energy flow from the excitation centre into the "bath". However, the energy per atom does not equilibrate. Discontinuities (i.e., "steps") in the energy plots of fig. 3 mark the dissociation of the main fragment, with the decrease of the total energy corresponding to the KE of the ground state atoms dissociating from it. A

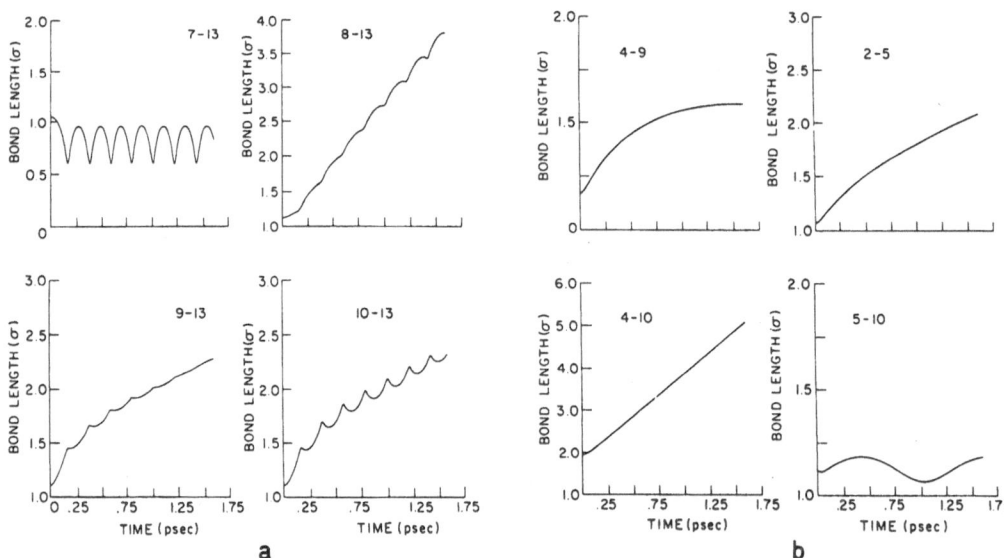

Fig. 1. Time dependence of interatomic distances in the electronically excited Ar_{13}^* cluster. The insert shows the ground-state equilibrium structure at 24K. The labelling of atoms is indicated. The dashed atoms 7 and 13 form the excimer. Distances are in units of σ. The Ar^*-Ar interaction is characterized by $\bar{\sigma} = 1.20$. (a) Interatomic distances within the excimer and between the excimer atoms and some ground state atoms. (b) Interatomic potentials between some ground-state atoms.

cursory examination of the time evolution of the composition of the main fragment (fig. 3) clearly indicates that the major fragmentation process involves the sequential stepwise dissociation, i.e., "evaporation", of single ground-state atoms from the main fragment. The escape of the excimer from the main fragment has not been encountered. The dissociation process is dominated by the magnitude of the excited-state potential scale parameter $\bar{\sigma}$. For realistic values[15] of $\bar{\sigma} = 1.10$–1.20, the threshold for cluster dissociation is exhibited on the time scale of $\sim$ 2-20 psec (fig. 4).

From these MD results the following picture concerning vibrational energy flow and reactive dynamics of the Ar_{13} cluster emerges. The temporal persistence of the vibrational excitation of the excimer (fig. 2) and of the "reaction centre" (fig. 3) corresponds to a "mode selective" excitation of the excimer, with vibrational energy redistribution within the cluster being precluded by two effects. First, the difference in the characteristic frequencies of the (high frequency) dimer motion and the (low frequency) motion of the cluster. Second, the local dilation of the cluster structure around the excimer, which is induced by the short-range excimer-cluster repulsive interactions. The vibrational energy flow from the dimer into the cluster (figs. 2 and 3) consists of two stages:

(A) Ultrafast vibrational energy transfer due to repulsion, which occurs on the time scale of $\sim$ 200 fsec (figs. 2 and 3). This energy transfer process is dominated by the magnitude of the scale parameter σ.

(B) "Slow" energy transfer on the time scale of tens of picoseconds (for $\overline{\sigma}$ = 1.2) and up to hundreds of psec (for $\overline{\sigma}$ = 1.0) due to vibrational relaxation of the excimer. The dynamics of the cluster induced by these energy transfer processes involves reactive vibrational predissociation, as is apparent from figs. 3 and 4. This state of affairs is, of course, drastically different from that encountered in infinite systems, where a nonreactive process prevails when the phonon modes of the system are excited. A cursory examination of the dissociative dynamics (fig. 4) of the Ar_{13} cluster following excimer formation indicated that two reactive processes prevail.

(i) A fast stepwise "evaporation" of Ar atoms is exhibited on the time scale of $\sim$10 psec. This proceess is induced by the energy transfer process (A). Subsequently, an additional reactive process appears (fig. 4), which involves,

(ii) Slower vibrational predissociation of Ar atoms on the time scale $\geq$ 10 psec. This dissociative process is induced both by energy transfer processes (A) and (B).

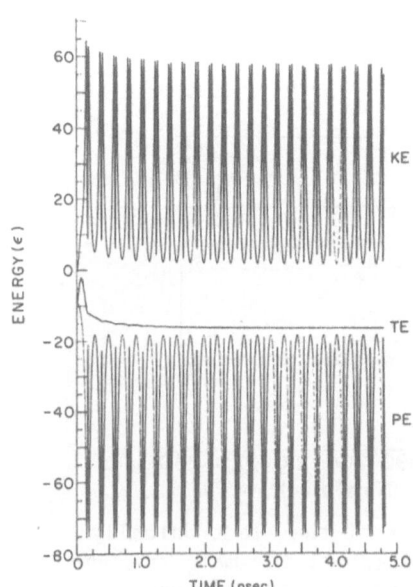

Fig. 2. Time dependence of the potential energy (PE), the kinetic energy (KE) and the total energy (TE) of the bare excimer in Ar_{13} ($\overline{\sigma}$ = 1.20). Energies are given in units of ϵ.

It is imperative to note that the short-time "explosion" of the electronically excited cluster is induced by energy transfer due to short-range repulsive interactions. When these interactions are switched off by taking σ = 1.0, only mechanism (B) is operative for vibrational energy flow into the cluster and the cluster dissociative process, which again occurs by stepwise "evaporation" occurring on the time scale of 100-1000 psec. The appropriate excited-state repulsive physical parameter characterizing excimer-cluster interactions in RGCs is $\overline{\sigma}$ = 1.1-1.2, and we expect the occurrence of energy flow predissociation induced by excited repulsive interactions to occur on the time scale $\sim$ 10 psec.

In order to investigate the dependence of the nature and rate of energy redistribution processes following excitation, on the degree of aggregation, we have performed similar situations for an Ar_{55} cluster, with the excimer located at either the center or at the outer shell of the cluster. These simulations revealed that while the ultrafast vibrational energy flow from the vibrationally-excited excimer to the rest of the cluster is operative (see Fig. 5), no fragmentations are observed, i.e., nonreactive vibrational energy redistribution, characteristic to infinite condensed matter systems, prevails. This of course is a direct consequence of the large number of densely spaced available "bath" modes. We note however that following the ultrafast energy flow which occurred on the 0.1-0.3 psec time scale, the subsequent vibrational relaxation rate is rather slow, due to the frequency mismatch and local dilation effects.

We conclude this section with a few observations on mixed clusters.

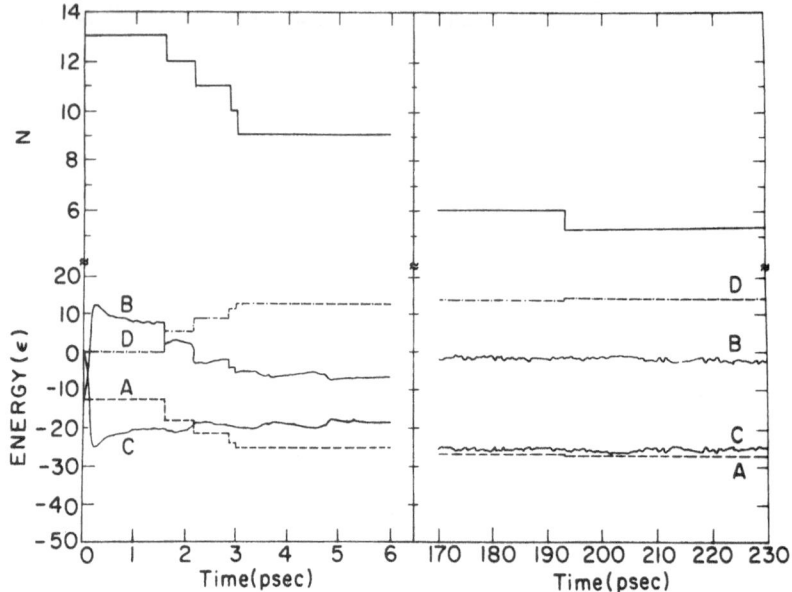

Fig. 3. Time evolution of fragmentation dynamics of the electronically excited $(Ar)_{13}$ cluster ($\bar{\sigma} = 1.15$). (A) Total energy of the main fragment. (B) Total energy of the "bath subsystem". (C) Total energy of the "reaction centre". (D) Kinetic energy of the dissociated atoms. (E) Number of atoms in the main fragment. The steps in curves (A), (D) and (E) mark the stepwise dissociation of individual Ar atoms from the main fragment.

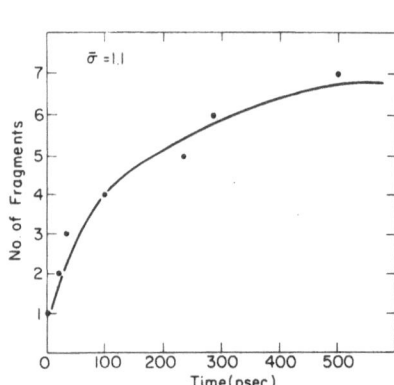

Fig. 4. The time evolution of the fragmentation of the electronically excited $(Ar)_{13}$ cluster ($\bar{\sigma} = 1.10$). Note the sequential "evaporation" of the single ground-state Ar atoms.

We have performed extensive studies of $Ar_n Xe_m^*$ cluster, with n = 13, 55 and m = 1, 2, with the Xe atoms located at either the center or on the outer shell. In the case of $Ar_{12}Xe^*$, with the Xe^* centrally located, the excited atom escaped from the cluster on a time scale of 50+20 psec. In view of the fast escape, we predict that the radiative decay will be characteristic of the free excited atom rather than from the cluster-shifted species. Curiously, we observed no such phenomenon for $Ar_{54}Xe^*$ or $Ar_{53}Xe_2^*$ (either centrally or peripherally locating the excitations). In these systems the effectiveness of the cluster mediated vibrational relaxation dominates, aided by characteristics of the interaction potentials and the heavy mass of the impurity (the defect modes are embedded within the vibrational manifold of the host cluster).

3. ELECTRON LOCALIZATION AND CLUSTER ISOMERIZATION INDUCED BY ELECTRON ATTACHMENT

The physical and chemical phenomena associated with the attachment of an excess electron to a molecular cluster are of considerable interest

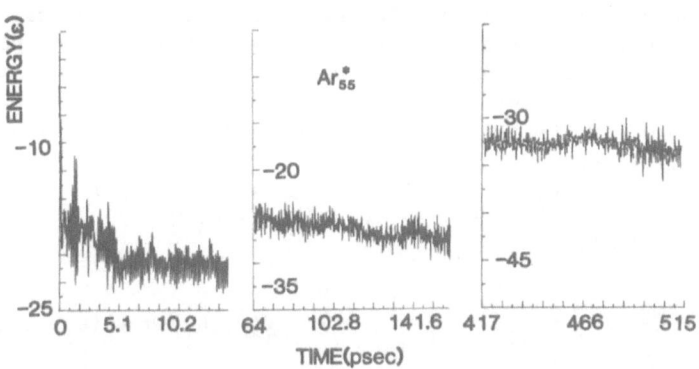

Fig. 5. Time dependence of the total internal energy of the bare excimer in Ar_{55} ($\bar{\sigma} = 1.1$).

because of two reasons. Firstly, quantum phenomena are expected to be pronounced in such finite systems, where the electron wavelength is comparable to the cluster size. Secondly, the excess electron can serve as a probe for the nuclear motion and dynamics of the cluster.

Electron attachment may occur via reactive or non-reactive channels where in the latter, electron attachment is not accompanied by the breaking of either intermolecular or intramolecular bonds, and can result in either bulk or surface states of the excess electron. From the point of view of general methodology, the formation of electron surface states on clusters is facilitated by the large surface to volume ratio of clusters. From the technical point of view, the formation of such surface states is dominated by the structure of the cluster and by the nature of the electron-cluster interaction. Internal localization of the electron on the other hand tends to occur in moderately large clusters which contain an anion vacancy. We demonstrate[5] these modes of localization through investigations of electron attachment to $[Na_{14}Cl_{13}]^{+}$ and $[Na_{14}Cl_{12}]^{++}$.

When a nonreactive attachment of an excess electron to a cluster results in a localized state, which has no parentage in the atomic (or molecular) states of the individual cluster constituents, the cluster nuclear structure may undergo substantial reorganization.[5] Investigations of the energetics of cluster isomerization induced by electron attachment are facilitated by our studies of $e^{-} + (NaCl)_4$ at a wide range of temperatures.[18] The alkali-halide clusters (AHCs) were chosen because of three reasons. Firstly, the nature of interionic interactions in these clusters is well understood.[19,20] Secondly, there exists an abundance of model calculations on both neutral and charged ACHs.[19] Thirdly, quite extensive information is available on electron-alkali cation[21,22] (M^+) and electron-halide anion (X^-) interaction.

3.1 The Quantum Path Integral Molecular Dynamics Method (QUPID)[5,22-25]

The structure and energetics of an e-AHC has been explored using the quantum path-integral molecular dynamics (QUPID) method. This approach, which rests on a discrete version of Feynman's path integral method, provides a powerful method for the study of these systems. In this scheme the quantum problem is isomorphic to an appropriate classical problem, with the excess electron being mapped onto a closed flexible polymer of P points. The isomorphism becomes exact as $P \to \infty$. The practical applicability of the computational method rests on the development of numerical algorithms, which achieve convergence with manageable values of P.

175

The Feynman path integral formulation of quantum statistical mechanics[26] allows a derivation of an approximate expressioon for the partition function, Z, for a system consisting of a quantum particle (mass m and coordinate $\vec{r}$), interacting with a set of N classical particles (whose phase space trajectories are generated by classical equations of motion) via a potential $V(\vec{r}, \vec{R}_1, \ldots, \vec{R}_N)$,

$$Z = \left(\frac{mP}{2\pi\hbar^2\beta}\right)^{3P/2} \int d\vec{r}_1 \ldots d\vec{r}_P \quad d\vec{R}_1 \ldots d\vec{R}_n \, e^{-\beta V_{eff}}, \qquad (1a)$$

where

$$V_{eff} \equiv \sum_{i=1}^{P} \frac{mP}{2\hbar^2\beta^2}(\vec{r}_i - \vec{r}_{i+1})^2 + \frac{1}{P}\sum_{j=1}^{N}\sum_{i=1}^{P} V(\vec{r}_i, \vec{R}_j) + V_c(\vec{R}_1, \ldots, \vec{R}_N). \quad (1b)$$

V_c is the interaction potential between the classical particles, and $\beta = (k_B T)^{-1}$.

Eqs. (1) establish an isomorphism[23] between the quantum problem and a classical one in which the quantum particle is represented by a flexible periodic chain (necklace) of P pseudoparticles (beads) with nearest-neighbor harmonic interactions with a temperature dependent spring constant, $Pm/\hbar^2\beta^2$. The above expression for the partition function is exact as $P \to \infty$. In practice, the finite value of P employed in a calculation is chosen to yield convergent results and depends upon the temperature and characteristics of the interaction potential, V.

The average energy of the system at equilibrium is given by

$$E = \frac{3N}{2\beta} + \langle V_c \rangle + K + \frac{1}{P}\langle \sum_{i=1}^{P} V(\vec{r}_i)\rangle, \qquad (2a)$$

where

$$K = \frac{3}{2\beta} + \frac{1}{2P}\sum_{i=1}^{P} \langle\frac{\partial V(\vec{r}_i)}{\partial \vec{r}_i} \cdot (\vec{r}_i - \vec{r}_P)\rangle . \qquad (2b)$$

and the angular brackets indicate statistical averages over the probability distribution as defined in Eq. (1). The first two terms in Eq. (2a) are the mean kinetic and potential energies of the classical component of the system. The quantum particle kinetic energy estimator,[25] K, consists of the free particle term ($K_f = 3/2\beta$) and a contribution due to the interaction (K_{int}). Finally, the last term in Eq. (2a) is the mean potential energy of interaction between the quantum particle and the classical field.

A measure of the quantum character of the system may be inferred from the relative contribution of K_{int} to the kinetic energy K. The magnitude of K_{int} depends on two factors, the gradients of the potential at the location of the beads and the bead spatial distribution. For a classical particle whose thermal wavelength, $\lambda_T = (\beta\hbar^2/m)^{1/2}$, approaches zero, the Gaussian factor in Eq. (1a) (see first term in Eq. (1b)) reduces to a delta function ($\delta(\vec{r}_i - \vec{r}_{i+1})$, for all i) and the necklace collapses to a single point, representing a classical particle. Under this circumstance $K_{int} = 0$. Generally, the degree of quantal character depends upon the interplay between the spatial extent of the particle (which in the above formulation is related to the spatial distribution of the pseudo-particles) and a characteristic length associated with the rate of change of the potential.

Another convenient measure of the quantum character of the system and the degree of localization of the quantum particle is provided by the complex time correlation function,[28]

$$R^2(t-t') = \langle|\vec{r}(t) - \vec{r}(t')|^2\rangle , \qquad (3)$$

where $|\vec{r}(t) - \vec{r}(t')|$ is the distance between two points (beads) on the electron path separated by a "time" $(t-t') \varepsilon (0,\beta\hbar)$. The value at $t-t' = \beta\hbar/2$, $R \equiv R(\beta\hbar/2)$, characterizes the breadth of the bead distribution, and yields the correlation length. Note that for a free particle, where the bead distribution obeys Gaussian statistics, $R_f = \sqrt{3} \lambda_T/2$. For a localized state, when fluctuations in $R(t-t')$ are inhibited, $R(t-t')$ is dominated by the ground state,[28] and as a function of $(t-t')$, $R^2(t-t')$ starts from zero at $t-t' = 0$ and rapidly achieves a constant value, with $R < R_f$ (i.e., it is independent of "time" except for $t-t'$ close to 0 and βh). The rise time, τ, is a (rough) measure of the excitation energy $\Delta E \equiv E_1 - E_0$, (where E_1 is a Boltzmann weighted average of the manifold of excited states), given by $\Delta E = \hbar/\tau$. A delocalized state, on the other hand, is characterized by a dependence of $R^2(t-t')$ on $(t-t')$ over the whole range $(0,\beta\hbar)$.

The formalism described above is converted into a numerical algorithm by noting the equivalence[22] between the equilibrium statistical averages over the probability distribution given in Eqs. (1), and sampling over phase space trajectories, generated by a classical Hamiltonian,

$$H = \sum_{i=1}^{P} \frac{m^* \dot{\vec{r}}_i^2}{2} + \sum_{I=1}^{N} \frac{M_I \dot{\vec{R}}_I^2}{2} + \sum_{i=1}^{P} [\frac{Pm}{2\hbar^2\beta^2} (\vec{r}_i - \vec{r}_{i+1})^2 + \frac{V(\vec{r}_i)}{P}]$$
$$+ V_c(\vec{R}_1,\ldots,\vec{R}_N) \quad, \tag{4}$$

where m* is an arbitrary mass chosen such that the internal frequency of the necklace, $\omega = [mP/(m^*\beta^2\hbar^2]^{1/2}$, will match the other frequencies in the system, and M_I is the mass of a classical particle. For a description of the application of the method to a wide variety of quantum many-body systems of chemical and physical interest the reader is referred to a recent review.[29]

3.2 Electron Localization in Charged Clusters

To investigate the modes of electron localization in cluster we have performed QUPID simulations of electron interaction with $[Na_{14}Cl_{13}]_{19}^+$ and $[Na_{14}Cl_{13}]^{++}$, at 300K. The structure of the bare clusters is cubic (with small distortions) and the doubly-charged one contains an internal anion vacancy. The ions were treated as classical particles interacting via coulomb attraction and Born-Mayer repulsion.[19,20]

The interaction of the electron consists of a sum of electron-ion potentials: pure Coulomb repulsion for the e-anion interaction[22] and a pseudo-potential $-e^2/R_c$ for $r \leq R_c$ and $-e^2/r$ for $r > R_c$ for the e-cation[22] (for Na^+, $R_c = 3.22a_0$). Based upon the examination of the stability of the variance of the kinetic energy contribution K_{int}, the number of "electron beads" was taken as $P = 399$. Employing an integration step of $\Delta t = 1.03 \times 10^{-15}$ sec, long equilibration runs were performed $(1-2\times10^4 \, \Delta t)$. The reported results were obtained via averaging over $8\times10^3 \, \Delta t$ following equilibration.

In Figs. 6a and 6b we present our results for the equilibrium electron charge distribution (small dots), and for the nuclear configuration of the clusters. The energetics of these systems may be summarized in terms of the electron affinity, E_A, of the cluster

$$E_A = E_B^e + E_c, \tag{5}$$

which is obtained by summing the electron binding energy

$$E_B^e = 3/2\beta + K_{int} + P^{-1} \sum_{i=1}^{P} \langle V_e(r_i) \rangle \tag{6}$$

and the cluster reorganization energy

 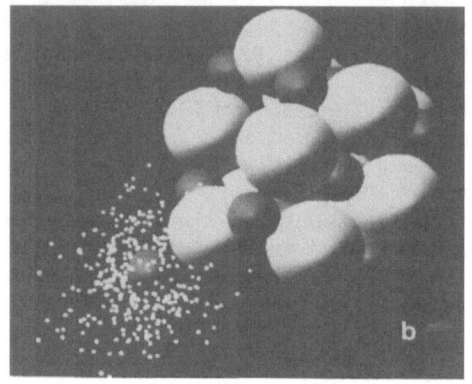

Fig. 6. Cluster configurations for (a) e-$[Na_{14}Cl_{12}]^{++}$ and (b) e-$[Na_{14}Cl_{13}]^{+}$, exhibiting internal and external electron localization, respectively. Large and small bells, Cl^- and Na^+, respectively. Small dots represent the electron distribution.

$$E_c = \langle V_{AHC} \rangle - \langle V_{AHC} \rangle_0 , \qquad (7)$$

where $\langle V_{AHC} \rangle_0$ is the potential energy of the "bare" AHC in the absence of the electron.

Table I

Energetics of Electron attachment to charged clusters (in Hartrees).

	$[Na_{14}Cl_{13}]^{+}$	$[Na_{14}Cl_{12}]^{++}$
E_B^e	−0.1594	−0.2490
E_c	0.1055	0.0799
E_A	−0.0539	−0.1691

As is apparent from Figs. 6a that the $[Na_{14}Cl_{12}]^{++}$ cluster stabilizes an internally localized electron state, with the ionic configuration similar to that found in a bulk F^- center defect, establishing the dominance of short-range interactions for internal electron localization.

A drastically different localization mode is obtained in the $[Na_{14}Cl_{13}]^{+}$ system (Fig. 6b), where a novel surface state is exhibited. We refer to this state as a cluster surface localized state.[5] Such surface states were considered for macroscopic alkali halide crystals by Tamm about fifty years ago, but were never experimentally documented. Finally, it is of interest to note that the total energy of e-$[Na_{14}Cl_{12}]^{++}$ is close to that of the $[Na_{14}Cl_{13}]^{+}$, whereupon the electron (internal) binding energy in the cluster is similar to that of a negative ion.

3.3 Cluster Isomerization Induced by Electron Attachment

While for charged clusters of intermediate size electron localization is not accompanied by major structural rearrangement, (see section 3.2) such effects may be found for smaller systems.[9] Moreover, we found that electron attachment may induce configurational transformations is small neutral ionic clusters at relatively low temperatures, and that the equilibrium configurations of the cluster with the attached electron, do not necessarily relate to those of the parent neutral cluster.

Our QUPID simulations of electron interaction with Na_4Cl_4 clusters at various temperatures, employed the same interaction potentials described in section 3.2. In these studies we have used constant-temperature simulations,[30] with a velocity form of the Verlet algorithm. The numbers of "electron beads", P, were taken as 1000, 520, 520 and 250 for T = 50, 575, 750 and 1000K, respectively. Prolonged runs were performed to assure proper sampling of the accessible phase-space and thus numerically reliable averaged quantities.

Equilibrium configurations at various temperatures are shown in Fig. 7. at low T (T $\leq$ 500K) the electron distribution is extended with an equal probability charge distribution around all the four Na^+ ions, and the ionic configuration is similar to that of the low-T neutral cluster. It is likely that in the low temperature domain, symmetry considerations dominate and determine the mode of localizations. At the intermediate temperature domain, 500K $\leq$ T $\leq$ 750K, a configurational change of the negative cluster is exhibited (see Fig. 7b). This is a distorted planar configuration and is similar to the stable structure of the related classical Na_4Cl_5 cluster, which we found in simulation performed in the range 50 <T < 1100K. At the high-T domain, 750 < T < 1200K, a coexistence of several isomers is found. At $\sim$ 750K a boat-like configuration (Fig. 7c) dominates (in this range the elongated chain is also found) and at T $\sim$ 1000K the

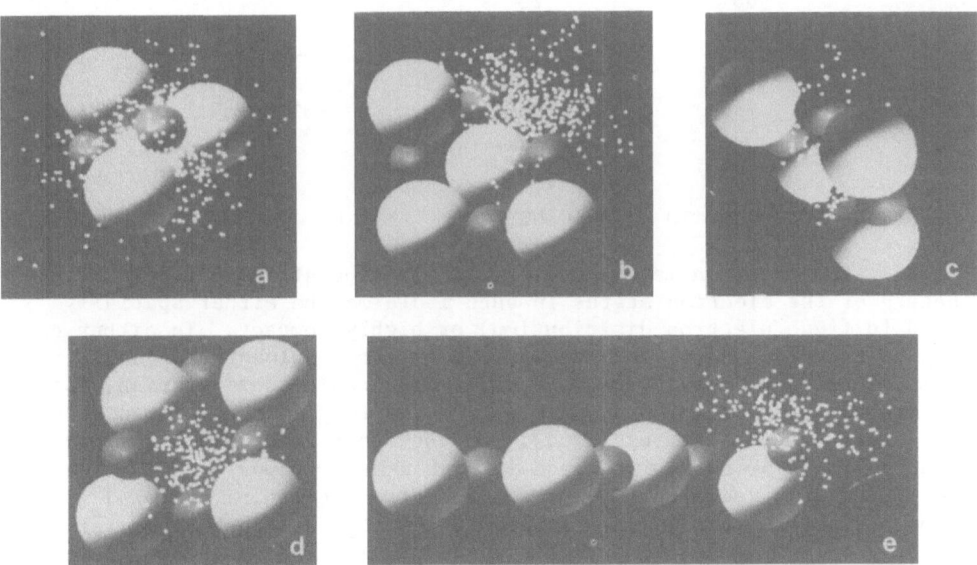

Fig. 7. Cluster configurations for e-Na_4Cl_4. (a) 50K, (b) 575K, (c) 750K, (d,e) 1000K. See caption to Fig. 6.

elongated chain in coexistence with a planar ring (7d and 7e) are found. We observe a tendency toward spatial localization of the electron upon increase in T. Furthermore, the presence of the excess electron induces two types of configurational modifications which are either quantitatively or qualitatively different from those in the neutral cluster:

(i) The localized excess electron can play the role of a pseudo-negative ion, with appreciable kinetic energy, which is overwhelmed by the electron potential energy, leading to new nuclear configurations which have no counterpart in the neutral cluster.

(ii) The partial neutralization of a single positively charged ion by the excess electron results in the appearance of the high-T configuration of the neutral parent cluster at substantially lower temperatures for the negatively charged cluster. The first effect is observed in the intermediate T range, while both effects are exhibited in the high-T domain.

The electron affinity, EA, of the $[Na_4Cl_4]$ cluster at low T, is calculated to be $0.029 \pm 8 \times 10^{-3}$ Hartrees, which is considerably lower than the EA of the intermediate size positive ionic clusters. Further energetics and other characteristics of the clusters are given in Table II.

Table II

Total energies of the neutral and negatively charged clusters. The calculated thermal wavelength, λ_T, localization parameter $R(\beta\hbar/2)/R_f$, and degree of quantum character[31] $Q \equiv (K - K_f)/K$ (Eq. 2b), are given. Energy and length in atomic units. Variances in parenthesis.

	50K	575K	750K	1000K
$E(Na_4Cl_4)$	-1.0244	-1.0010	-1.0049	-0.9978
$E(Na_4Cl_4^-)$	-1.053_3 (8×10^{-3})	-1.00_{-2} (2×10^{-2})	-0.99_{-2} (2×10^{-2})	-0.96_{-2} (3×10^{-2})
λ_T	68	23	20	17
R/R_f	0.09	0.33	0.39	0.42
Q	99%	95%	93%	90%

4. ELECTRON ATTACHMENT TO WATER CLUSTERS

A solvated electron is a bound excess electron state in a polar fluid. The nature of the electron states in such media may be either spatially extended (diffuse electron distribution) or highly compact. In either case binding of the electron may be accompanied by large solvent reorganization. A water molecule, which has a closed shell electronic configuration, isoelectronis to Ne, does not bind an electron (as long as unusually distorted nuclear configurations are excluded). On the other hand, electrons can be trapped in liquid or solid water and in high density vapor.[7,32-34] The nature of the solvated electron and the formation mechanism of $(H_2O)_n^-$ have been the subject of various theoretical treatments from continuum dielectric models to quantum-mechanical calculations.[4,35] While these studies have advanced our knowledge, the recent addition of quantum simulations (see section 3) to our theoretical arsenal opens new avenues in studies of bulk[36,37] and cluster[5,38] solvation.

A key issue in modeling the system is the choice of interaction potentials. Fortunately, for small water clusters, interaction potential functions which provide a satisfactory description for a range of properties are available.[39,40] We have used the RWK2-M model[40] for the intra and inter-water interactions. Less is known about the electron-water interaction.[37,38] We have constructed a pseudo-potential which consists of Coulomb interactions and a term which excludes the electron (represented by the beads, see sec. 3) from the region of the molecular water electronic charge density. The effect of the exclusion principle is modeled as a kinetic energy contribution,[41]

$$V(\vec{r};\vec{R}_3,\vec{R}_0,\vec{R}_{H1},\vec{R}_{H2}) = \frac{2qe}{|\vec{r}-\vec{R}_3|} - \sum_{j=1}^{2} \frac{qe}{|\vec{r}-\vec{R}_{Hj}|} + \frac{e^2 a_0}{2}(3\pi^2\rho)^{2/3},$$ (8a)

$$a_0{}^3 \rho(\vec{r},\vec{R}_0,\vec{R}_{H1},\vec{R}_{H2}) = 8e^{-3|\vec{r}-\vec{R}_0|/a_0} + \sum_{j=1}^{2} e^{-3|\vec{r}-\vec{R}_{Hj}|/a_0},$$ (8b)

where $q = 0.6$; $\vec{R}_3$ is the location of the negative charge in the RWK2-M model, and $\vec{R}_0$, $\vec{R}_{Hj}$ are the locations of the oxygen and hydrogen nuclei, respectively. We note that in the RWK2-M model the dipole and quadrapole of the water molecule are well described. Contours of the electron-H_2O potential (in the plane containing the three nuclei, the oxygen located at (0,0)) and of the electronic density (which fits well that obtained by Hartree-Fock calculations,[42] in the regions of interest),are shown in Fig. 8.

In our Quantum Path Integral Molecular Dynamics Simulations (QUPID, sec. 3), the electron "necklace" contained 1024 beads (P), and the atomic constituents were treated classically (there is no evidence for an isotope effect on solvation). The energetics and other characteristics of the systems, at 15.8K, are summarized in Table III, and certain cluster configurations are shown in Fig. 9. In order to test our newly proposed electron-water pseudo-potential we have performed QUPID calculations (T = 15.8°K) for a $(H_2O)_8^-$ cluster with the water molecules arranged, and held fixed, as in the SCF calculations of Rao and Kestner.[35] In this configuration the oxygen atoms of the four water molecules in the first coordination shell form a tetrahedron and the molecules are oriented with their dipoles pointing towards the center of the cavity. Our results compare very favourably

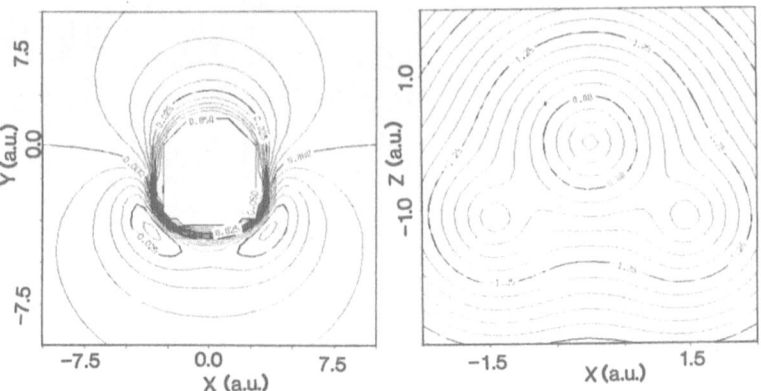

Fig. 8. Contours of electron-water interaction, left, and of the electronic density ($Log_{10}\rho(r)$), right, see Eqs. 8, in the plane containing the nuclei. The oxygen is located at the origin.

Table III

Energetics of $(H_2O)_n^-$, n = 2, 8 and 16, clusters. The values in square brackets refer to the neutral clusters, and in round brackets to standard deviation. The stars indicate metastable clusters, i.e., configurations which lasted for a relatively long "time" in the QUPID simulations without a discernable trend in the quoted properties. Energies and lengths in Hartrees and Bohr radii, respectively. EBE is the electron binding energy; and E_c the cluster reorganization energy.

	$(H_2O)_2^{-*}$	$(H_2O)_{16}^{-*}$
PE(inter)	−0.0097 (.0003)	−0.197 (.002)
[neutral]	[−0.0101 (.0002)]	[−0.270 (.002)]
PE(intra)	0.0006 (.0002)	0.018 (.001)
[neutral]	[0.0006 (.0002)]	[0.022 (.002)]
E_c	0.0004	0.069
PE (e-H_2O)	−0.0024 (.0007)	−0.084 (.002)
K_{int}	0.0018 (.0007)	0.037 (.001)
EBE (vertical)	−0.00055 (.0008)	−0.047 (.003)
EBE (adiabatic)	−0.00016	+0.022
R_g	21	5.2
R_T	242	175
$R(\beta\hbar/2)/R_f)$	0.25	0.06

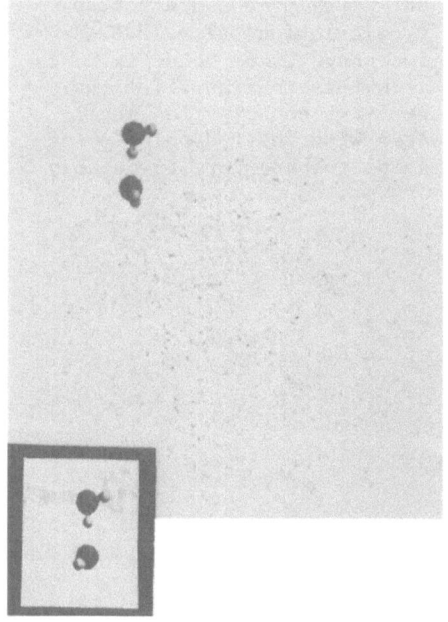 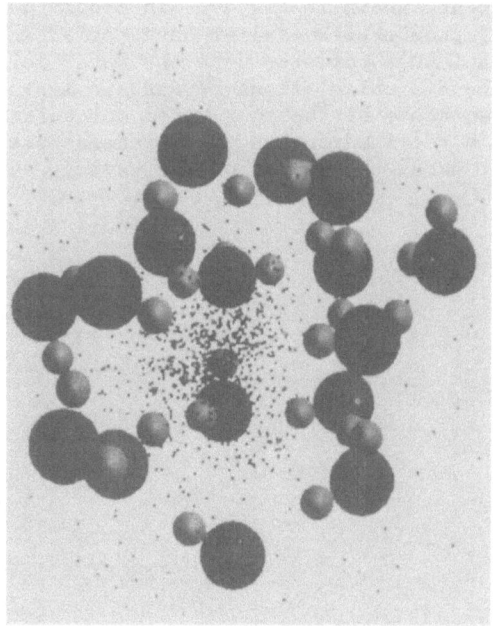

Fig. 9. Configurations of $(H_2O)_n^{-*}$, n = 2, 16, and $(H_2O)_2$ in the inset, at 15.8K. Note configurational change of the dimer upon e-attachment and the diffuseness of the electron distribution (small dots).

with the scaled SCF results [35] for the vertical electron binding energy of
-0.024 Hartrees as compared to our value of -0.018 Hartrees. This comparison provides evidence for the adequacy of the electron water pseudopotential given in Eq. 8.

Having established the merit of our model via comparative studies, as
described above, we turn to studies on $(H_2O)_n^-$ (n=2,16) with full dynamics
of the atomic as well as the necklace components. Our results for $(H_2O)_2^-$
suggest the existence of a metastable state of the negatively charged dimer
in which the water nuclear configuration fluctuates between that of the
neutral $(H_2O)_2$ cluster (see insert in Fig. 9) and the configuration shown
in Fig. 9 (left). The water dipole moment in this state fluctuates between
~ 1.2 and 1.63 a.u. In the metastable state the vertical EBE $\simeq -15$ meV, the
radius of gyration of the electron necklace $Rg \simeq 21$ a_o, and associated with
it is a large cluster reorganization energy $E_c \simeq 11$ meV. It is of interest
to note that this value of the vertical binding energy is in correspondence
with experimental estimates (see ref. 33, Haberland et al. (1984)).
Kestner and his associates, employing the high-dipole moment configuration
have performed SCF calculations obtaining a vertical EBE $\simeq -9$ meV. In the
fully equilibrated system the neutral dimer nuclear configuration is maintained, and the electron distribution is very diffuse, $Rg \approx 35$ a_o, with a
vertical EBE $= 4$ meV. The degree of electron localization is also reflected in the values of thermal wavelength,[5] R_T, and the localization parameter
$R(\beta\hbar/2)/R_f$ defined in Section 3.1. In our simulations of the larger cluster we started from a $(H_2O)_{16}^-$ cluster in which the molecular cage was formed by condensing the H_2O molecules around a point (classical) electron.
Subsequently the point electron was replaced by a 1024-bead necklace. The
system stabilized in the configuration shown in Fig. 9 (right) with the
properties given in Table III (under $(H_2O)_{16}^{-*}$). The vertical EBE in this
state of 1.3 eV is comparable to the experimental value.[34] After a prolonged QUPID run the molecular cage confining the electron opened up and a
more diffuse surface state, with an EBE (vertical) $\simeq 0.11$ eV and EBE (adiabatic) $\simeq 0.09$ eV developed. In further studies the dependence on cluster
size and temperature and possible improvements of the interaction potential
will be explored.

Our calculations employed realistic potentials i.e., inter and intramolecular as well as an electron-H_2O pseudopotential which incorporates the
essential physics of the interaction, and goes beyond oversimplified descriptions such as point-dipole or mere Coulomb interactions. We demonstrated that quantum simulations provide significant insight into the energetics and mechanisms of electron solvation in clusters. In further
studies the dependence on cluster size and temperature and possible
improvements of the interaction potential will be explored.

Acknowledgement. We gratefully acknowledge helpful conversations with N.
R. Kestner on water clusters, and the invaluable assistance of V. Mallette
in the preparation of the figures and A. Ralston in the preparation of the
manuscript. This work is supported by the U. S. Department of Energy under
grant No. FG05-86ER45234 (to UL), and by the US Army through its European
Research Office (to JJ).

References

1. J. Jellinek, T. L. Beck and R. S. Berry, J. Chem. Phys. 84, 2783
    (1983) and references therein.
2. J. Luo, U. Landman and J. Jortner, these proceedings.
3. D. Scharf, J. Jortner and U. Landman, Chem. Phys. Letts. 126, 495
    (1986).
4. J. Jortner, Ber. Bunsenges. Physik. Chem. 88, 188 (1984).
5. U. Landman, D. Scharf, and J. Jortner, Phys. Rev. Letts. 54, 1860
    (1985).

6. T. D. Mark and A. W. Castleman, Jr. in Adv. Atomic and Mol. Phys. (D. R. Bates and B. Bederson, Eds.), Vol. 20 (1984).
7. See papers in J. Phys. Chem. 88 (1984).
8. F. F. Abraham, J. Vac. Sci. Technol. B2, 534 (1984).
9. U. Landman et al., Mat. Res. Soc. Symp. Proc. Vol. 63, 273 (1985).
10. See papers in Ber. Bunsenges. Physik. Chem. 88, 188 (1984).
11. H. Haberland, Surf. Sci. 156, 305 (1985); J. J. Saenz, J. M. Soler and N. Garcia, ibid. 156, 121 (1985).
12. J. W. Brady and J. D. Doll, J. Chem. Phys. 73, 2767 (1980).
13. J. Jortner, E. E. Koch and N. Schwentner in: Photo-Physics and PhotoChemistry in the Vacuum UV, S. P. McGlynn et al., Eds., (Reidel, Dordrecht, 1985), p. 515.
14. W. D. Kristensen, E. J. Jensen and M. J. Cotterill, J. Chem. Phys. 60, 4161 (1974).
15. I. Messing, B. Raz and J. Jortner, J. Chem. Phys. 66, 2239 (1977).
16. A. E. Sherwood and J. M. Prasunitz, J. Chem. Phys. 41, 429 (1964).
17. K. T. Gillen, R. P. Saxon, D. C. Lorentz, G. E. Ice and R. E. Olson, J. Chem. Phys. 64, 1925 (1975).
18. D. Scharf, U. Landman and J. Jortner (to be published).
19. T. P. Martin, Phys. Rep. 95, 167 (1983).
20. F. G. Fumi and M. P. Tosi, J. Phys. Chem. Solids 25, 31, 45 (1964).
21. R. W. Shaw, Phys. Rev. 174, 769 (1968); see also QUPID Studies of e^-+Na$^+$ and (NaCl)$^-$ in D. Scharf, J. Jortner and U. Landman (Chem. Phys. Letts. 1986, in press).
22. M. Parrinello and A. Rahman, J. Chem. Phys. 80, 860 (1984).
23. D. Chandler and P. G. Wolynes, J. Chem. Phys. 79, 4078 (1981).
24. B. De Raedt, H. Sprik and H. L. Klein, J. Chem. Phys. 80, 5719 (1984).
25. D. Chandler, J. Phys. Chem. 88, 3400 (1984).
26. R. P. Feynman and A. R. Hibbs, Quantum Mechanics and Path Integrals (McGraw-Hill, New York, 1965).
27. M. F. Herman, E. J. Bruskin and B. J. Berne, J. Chem. Phys. 76, 5150 (1982).
28. A. L. Nichols, D. Chandler, V. Singh and D. M. Richardson, J. Chem. Phys. 81, 5109 (1984).
29. B. J. Berne and D. Thirumalai, in Ann. Rev. Phys. Chem. (to appear, 1986).
30. J. R. Fox and H. C. Anderson, J. Phys. Chem. 88, 4019 (1984).
31. U. Landman, R. N. Barnett, C. L. Cleveland and P. Norlander, in "Tunneling", Eds. J. Jortner and B. Pullman (Reidel, 1986).
32. J. Phys. Chem. 88, 3699-3914 (1984); Electrons in Fluids, J. Jortner and P. N. Kestner, Eds. (Springer, N. Y., 1973).
33. H. Haberland, H. G. Schindler and D. R. Worsnop, Ber. Bunsenges, Chem. 88, 270 (1984); J. Phys. Chem. 88, 3903 (1984); J. Chem. Phys. 81, 3742 (1984).
34. J. V. Coe, D. R. Worsnop and K. H. Bowen (J. Chem. Phys., 1986).
35. B. K. Rao and N. R. Kestner, J. Chem. Phys. 80, 1587 (1984) and references therein.
36. M. Sprik, R. W. Impey and M. L. Klein, J. Chem. Phys. 83, 5802 (1985).
37. C. D. Jonah, C. Romero and A. Rahman, Chem. Phys. Lett. 123, 209 (1986).
38. A. Wallquist, D. Thirumalai and B. J. Berne, J. Chem. Phys. 85, 1583 (1986).
39. J. R. Reimers, R. O. Watts and M. L. Klein, Chem. Phys. 64, 95 (1982).
40. J. R. Reimers and R. D. Watts, ibid. 85, 83 (1984). This paper as well as the description of the potential in ref. 39 (Eq. 13 and Table 1) contain several ambiguities and typographical errors. When corrected we reproduce their results.
41. J. N. Bardsley, case studies in Atomic Physics 4, 299 (1974); G. G. Kleiman and U. Landman, Phys. Rev. B8, 5484 (1973).
42. C. W. Kerr and M. Karplus in Water, F. Franks, ed., (Plenum, N. Y., 1972), p. 21; M. W. Ribarsky, W. D. Luedtke and U. Landman, Phys. Rev. B32, 1430 (1985).

MELTING AND FREEZING OF MICROCLUSTERS

R. Stephen Berry, Thomas L. Beck and Heidi L. Davis

Department of Chemistry
The University of Chicago
Chicago, Illinois 60637

Julius Jellinek

Chemistry Division
Argonne National Laboratory
Argonne, Illinois 60439

INTRODUCTION AND BACKGROUND

One of the most intriguing unsolved problems in the physical sciences is the elucidation at the atomic level of freezing and melting. Despite the great advances in the description of the critical region, the "simple" first-order phase transition remains as elusive a problem as ever. This topic has belonged, traditionally, to condensed matter science, so that it has been addressed with the tools we use to study bulk materials. This in turn means that virtually all attempts to understand freezing and melting have been carried out in schemes based on infinite numbers of atoms. Some simulations done in the 1970's, however, gave powerful signals that we could learn a great deal about freezing and melting from the study of small clusters. We shall review these briefly and then go on to describe how following those signals has indeed led to new insights into the nature of the equilibrium between solids and liquids. This journey has exposed new phenomena we should expect to see in finite clusters, phenomena in a sense richer and more varied than the simple first-order phase transition between bulk solids and bulk liquids.

Molecular dynamics (MD) simulations of argon clusters at constant energy were used by McGinty (1973) to calculate diffusion coefficients; because he saw no sharp break, he argued that the clusters show no sharp phase transition but seem to have solid-like and liquid-like forms at low and high energies. Cotterill and his collaborators argued also for solid-like and liquid-like forms but with a smooth transition between them (Cotterill et al., 1973; Damgaard Kristensen et al., 1974, 1975). Then Briant and Burton (1973, 1975a,b, 1976) inferred from their MD results at constant energy that the two forms, solid and liquid, not only have recognizable, separate existence for clusters of seven or more argon atoms; they found s-shaped regions in their curves of mean kinetic energy per particle $<K>$ or mean temperature $<T>$, as functions of total energy per particle, the caloric curves. They argued that these regions are analogous to the s-shaped region in the isotherms of the van der Waals fluid, and therefore indicate first-order phase transitions.

Lee et al., (1973), using Monte Carlo (MC) simulations of argon, also reported a smooth change as the cluster energies were raised from a range of solid-like behavior to one of liquid-like behavior. Isothermal Monte Carlo simulations of argon clusters showed no indications of s-shaped regions in the caloric curves (Briant and Burton, 1975a; Etters and Kaelberer, 1975, 1977; Etters et al., 1977a,b; Etters and Danilowicz, 1979; Kaelberer and Etters, 1975). However Etters and Kaelberer argued that a sharp transition could be distinguished in the form of a discontinuity in the slopes of the caloric curves. In constant-volume, isothermal MC simulations of clusters of 13, 19 and 55 Lennard-Jones particles, Nauchitel and Pertsin (1980) found a two-phase region for the 55-atom cluster which they interpreted as a solid 13-atom core with a 42-atom liquid shell, on the basis of the radial density distribution and the heat capacity; in the same region, they saw the cluster also take on both icosahedral and face-centered cubic forms. The smaller clusters, they reported, exhibit solid-solid transitions but not surface melting. Nauchitel and Pertsin take issue with Kaelberer and Etters in saying that the 55-atom cluster displays a coexistence region, rather than a sharp change from one form to the other in the sense that the caloric curve exhibits a portion in which the two-phase form is present, rather than changing from solid clusters to liquid clusters (but possibly through a region in which the two coexist). One important point they make is that two forms of a cluster of fixed size may have a coexistence region because the free energy difference between them, ΔF, is finite, so that they may coexist in equilibrium in a ratio given by $\exp(\Delta F/kT)$. They do not touch on the issue of conditions for the local thermodynamic stability of each phase. In a recent MC calculation, Quirke and Sheng (1984) have disputed the occurrence of any sharp "phase" change or any discontinuity the slope of the caloric curve for the 13-particle cluster, and report both a smooth caloric curve and a smooth, peaked heat capacity. They also studied the angular distribution function, which they report for 33 K as that of a hot, expanded solid but not a liquid, which the results of Etters and Kaelberer would imply.

Experimental evidence on phase behavior of clusters is still meager. A recent interpretation (Eichenauer and LeRoy, 1986) of a spectroscopic study of argon clusters containing a molecule of sulfur hexafluoride (Gough, et al., 1983) argued that the shift and broadening of a vibrational band gives evidence for coexistence of solid and liquid forms. Stace (1983) has examined the fragmentation of argon clusters containing a molecule of dimethyl ether; the dependence of the fragmentation pattern on the conditions of formation of the cluster seems to be evidence for the existence of solid-like and liquid-like forms, for clusters containing up to 22 argon atoms. Electron diffraction of clusters of argon (Farges, et al., 1973, 1983, 1986) and of several organic molecules (Heenan and Bartell, 1983a,b; Valente and Bartell, 1984a,b) gives evidence for liquid-like or at least amorphous forms, and in a few cases, of the possible coexistence of two forms. At this time, no experiment has been performed that was designed to examine the transitions between solid-like and liquid-like clusters, or the possibility of their coexistence.

By a combination of quantum statistical mechanics and numerical simulation, it has become possible to interpret the previous simulations, to get new insights into the nature of the equilibrium between liquid-like and solid-like forms of clusters and perhaps to begin to achieve a deeper understanding of melting and freezing. Experiments to test the ideas discussed below have not yet been performed but we can begin to realize what can be learned from them. The next section describes the quantum statistical theory, and the third section discusses recent MD and MC simulations related to that theory.

A QUANTUM STATISTICAL MODEL OF MELTING AND FREEZING OF CLUSTERS

The essential argument is this: Clusters are like molecules. If the component atoms or molecules of a cluster vibrate with small amplitudes around their equilibrium positions and the entire cluster rotates as a nearly rigid object, then the cluster is solid-like; this is the way ordinary molecules usually behave. If the atoms comprising the cluster move more freely, like those of a very nonrigid molecule, then the cluster is liquid-like. The argument then proceeds from general but not universal properties of the densities of states of the two forms and of how they are related. Unless a cluster is liquid-like at however low a temperature it is observed, as with Li_3, Na_3 or small helium clusters, i.e., so long as its zero-point amplitudes of oscillation do not carry the cluster from one potential minimum to another, the density of states for the solid form can be expected to be higher at low energies than for the liquid form. As we look higher on the energy scale, we find the densities of states of both forms increase, but that of the liquid grows much faster than that of the liquid form. As we look higher on the energy scale, we find the densities of states of both forms increase, but that of the liquid grows much faster than that of the solid, so that even the <u>cumulative</u> density of states of the liquid eventually passes that of the solid at sufficiently high energy.

The densities of states--the energy levels and their degeneracies--are precisely what is needed to compute the partition functions $Z(T)$ of the two forms as functions of temperature. From the partition functions, it is an elementary exercise to compute the thermodynamic functions, particularly the free energy. The more stable form is of course the one with the lower free energy. However we only now raise the question of whether either or both forms is stable; if a phase has a relative stability then it is meaningful to compare its stability to that of another phase which must also have a relative or local stability. The logic of the argument goes thus: if we assume that there can be two forms, like phases, each with local stability, then we can compute properties of those forms from their partition functions and compare the properties with those generated by simulations. The degree of elaboration necessary to bring the quantum statistical calculations into rough agreement with the simulations tells us how much specificity is required in our description of matter to account for the solid-liquid equilibrium behavior shown by the simulations. Then we can ask what conditions are necessary and sufficient for the local stability that had to be postulated to carry out the calculations, and explore the consequences of those conditions.

The first step was carried out by Natanson et al., (1983) for the clusters of N argon atoms, with N taking various values from 3 to 100. The densities of states for the solid-like form were taken from models of various levels of sophistication, from a simple, N-independent Einstein model to a detailed fitting of the quantum spectrum that has been extracted from classical simulations. For the liquid-like form the density of states was taken to be that of the (3N-3)-dimensional isotropic harmonic oscillator as developed by Gartenhaus and Schwartz (1957) for a liquid-like model of a nucleus. The energy levels for the limiting cases can be connected in a correlation diagram, if one assumes that parity, particle permutation symmetry and angular momentum are conserved (Kellman and Berry, 1976; Kellman et al., 1980; Amar et al., 1979; Ezra and Berry, 1982). The importance of the correlation diagram will become apparent at the next stage. The outcome of the calculations by Natanson et al., is that many of the results from the simulations of Briant and Burton, particularly their 1975 work, and from Etters and Kaelberer, can be rather well recovered with all but the very simplest of models for the two forms of argon clusters. Only the Einstein model with a single vibrational frequency for solid clusters of all sizes fails to reproduce the shape of the curve of "equilibrium temperatures", i.e., temperatures for which the free energy of the solid and liquid N-cluster are equal, as a function of N. The ranges ΔT for which one might observe the two forms present in equilibrium together are rather well reproduced, although the simulations show the fluctuations with N that reflect characteristics of clusters of particular sizes, such as stability or instability. The free energies, surface free energies and surface tensions are also moderately well replicated. In short, rather uncomplicated physics is capable of describing fairly well the solid-liquid equilibrium of small clusters.

The rationale for the stability of two forms of the cluster is the next problem. Local stability implies a minimum in the free energy, but with respect to what? Certainly we want to hold the temperature constant. The quantity we want is found in the correlation diagram; it is the parameter that quantifies the abscissa. Defining a parameter for that axis can be done in terms of the ratio of two excitation energies, (with a multiplier of 2) namely $\gamma \equiv 2\Delta E_r/\Delta E_v$, where ΔE_r is the lowest excitation energy to a state of the cluster which would be the first rotational excitation in the rigid limit, and ΔE_v is the lowest excitation energy to a rotationless, vibrationally excited state. This is analogous to the parameter introduced by Yamada and Winnewisser for linear triatomic and polyatomic molecules (1976). This definition is useful but it would be valuable to have a definition with a clearer microscopic significance, comparable to the internuclear distance in a correlation diagram of the energy levels of a diatomic molecule, or the crystal field strength in a correlation diagram of the energy levels of anion in a crystal.

Having the parameter, the free energy can be treated as a function of both that parameter γ and the thermodynamic variable of temperature (Berry et al., 1984a,b). We do not know how the density of states varies across the range of γ but we do know something about its general form. First, as defined, $0<\gamma<1$, with 0 the rigid limit and 1 the nonrigid limit. Second, every state appearing at $\gamma = 0$ must also appear somewhere along the energy scale for every other value of γ, including 1. This means that the low-energy levels have positive slope, on average across γ, and the high-lying levels have negative slope, on average. If T is so low that only the positive-slope levels are significantly populated, then the partition function is a monotonic, decreasing function of γ and the free energy is a monotonic, increasing function, and only the solid-like form of the cluster is stable at such a temperature. At higher temperatures, for values of γ near 1, the increasing density of levels with negative slope makes the partition function $Z(T;\gamma)$ assume a smaller and smaller slope, until a temperature T_f is reached at which $Z(T_f;\gamma)$ has a slope of zero. The free energy must show the same behavior, but with opposite slope of course. Thus below T_f only one form of the cluster is thermodynamically stable but above T_f two forms are locally stable. Thus above T_f the condition for coexistence is met. At still higher temperatures, the negatively-sloping levels dominate the partition function for all values of γ and the partition function and free energy are again monotonic functions of this parameter, with Z having a positive slope and F having a negative slope. At such high temperatures, only the liquid form is stable. Because both Z and F are smooth functions of T and γ, and they turn from having two minima to having only one at high T, there must be a temperature T_m at which $\partial F/\partial\gamma=0$, which is the upper limit of stable existence for the solid-like form. The temperatures T_f and T_m are called the freezing temperature and the melting temperature, respectively, for reasons now obvious. Between them must be the temperature T_{eq} at which the free energies of the individual forms are equal for clusters of a fixed size. Of course T_f, T_m and T_{eq} are all functions of N. We now can see that for finite systems there is no logical necessity for the freezing temperature to be the same as the melting temperature, although the two must come to essentially a single value for large N because the free energy difference ΔF between forms at any $T \neq T_{eq}$ must be so large that one form totally dominates over the other. Only at T_{eq} can we hope to see both forms of a large cluster. Our theory, at this stage, cannot yet tell us what "large" is. Calculations (Berry et al., 1984a,b) for a model something like Ar_5 show the expected behavior at T_f. No calculations from partition functions have yet been done of T_m.

Here then is a new phenomenon, a characteristic of "phase" equilibrium for finite systems that is different from a first order phase transition of bulk matter, or of a bulk transition of any order. The existence of two sharp temperatures with a coexistence range between them is simply unlike the phase equilibrium of bulk matter, although the convergence of T_f and T_m to a common value for large enough N not only ties this phenomenon to something familiar; it gives us new insight into the nature of the freezing/melting temperature of bulk matter.

To test the validity of this result requires not only laboratory experiments. Because true thermodynamic stability is the central point here, and any laboratory experiment is vulnerable to the charge of metastability, simulations, particularly Monte Carlo simulations must be a necessary part of the testing of this theory. One point more should be clarified before we discuss the simulations. It may seem at first sight that the picture just represented violates the phase rule. In fact it does not because the two forms of a cluster are more like isomers and therefore like components than like bulk phases.

SIMULATIONS OF ARGON CLUSTERS

To explore the theoretical predictions just discussed and probe more deeply into the nature of the rigid and nonrigid forms of clusters, we embarked on an extensive simulation program, using isoergic MD, isothermal MD and isothermal MC, all classical, to treat clusters of argon atoms. The isoergic MD has been used for clusters of a variety of sizes up to N of 33, with concentration on 7 and 13. The isothermal studies have been restricted to 13-atom clusters. All have been done with pairwise Lennard-Jones potentials.

The simulations of Ar_7 were done initially to test the hypothesis (Natanson et al., 1983) that melting can be usefully characterized by a breakdown in the separability of two time scales, a shorter one for vibration of the system about a collective potential minimum, and a longer one for passage among minima. This hypothesis seems well borne out now by simulations with larger clusters of several kinds (Stillinger and Weber, 1981, 1982, 1983a,b; Weber and Stillinger 1984; LaViolette and Stillinger, 1985) and now by complementary simulations of Ar_7 (Amar and Berry, 1986). Passage from the solid-like region of temperature to the liquid-like region is indeed associated with large increases in the rates of passage between potential minima. The calculations of Stillinger and Weber trace this by following the time history of the energy of the potential minimum above which the system moves, from instant to instant. Amar and Berry use a much smaller system in order to be able to follow every isomerism, even between geometrically equivalent structures. The isomerism rates, in ns^{-1}, for Ar_7 at 14.98, 15.83, 16.38, 17.71 and 19.37 K are, respectively, 0.20, 1.00, 1.01, 4.34 and 19.94, respectively. Furthermore this cluster shows the same two-phase behavior we see for the 13-particle cluster.

Argon forms tightly bound icosahedral clusters of 13 atoms. This is perhaps the most elegant and simplest demonstration of the phase equilibrium predicted by the quantum statistical theory, perhaps too elegant because many other clusters have somewhat more complex behavior. Molecular dynamics simulations at a series of constant energies (Jellinek et al., 1986) demonstrate a low-energy region in which the cluster is clearly solid-like in every respect. It has a well defined geometry and equilibrium structure, the mean square displacements of the atoms are only a small fraction of the interparticle distance, the velocity autocorrelation function shows a "bounce back" characteristic of elastic solids, and the Fourier transform of the velocity autocorrelation function has essentially no density of states around zero frequency. At high temperatures, the cluster is clearly liquid. All these characteristics tell us that: the mean square displacement grows linearly with time until it reaches the size of the cluster, and its slope yields a diffusion coefficient; the structure is unrecognizable, the "bounce" is nearly missing from the velocity autocorrelation function and its Fourier transform shows clearly the presence of soft, near-zero-frequency modes.

Between is a range of temperature in which the cluster passes back and forth at irregular intervals between the solid-like and liquid-like forms. This is a very important point: the 13-atom cluster, and some but not all others, spend long intervals in each form and require only brief intervals for the passage between "phases". This tells us that an ensemble of Ar_{13} clusters will look in an experiment like a collection of solids, or of liquids, or of a mixture of solid and liquid clusters, depending on the temperature. In the coexistence range the clusters spend long enough in each form to develop well-defined average properties appropriate to that "phase". The temperature or mean kinetic energy distribution in the coexistence range is bimodal, with two Gaussian peaks, corresponding to a cold liquid and a hot solid (recall that the system is isoergic, not isothermal) and the two parts of the distribution may be thrown together to yield long-time averages or binned separately. If the long-time average is taken, we find that the caloric curve of $<T>$ vs. E is smooth, without s-shaped portions, like those of Etters and Kaelberer and unlike those of Briant and Burton. The latter could be reproduced if we used short runs of only 10,000 steps instead of 500,000 or 1,000,000. If the two Gaussian subdistributions are taken separately, the caloric curve develops two smooth branches for a sharp and discrete portion of the temperature scale. That is, the equation of state must be double-valued in that region; the separate branches are straight-forward continuations of the corresponding branches in the normal regions.

Not all clusters display such bimodal curves or two-phase behavior. Clusters of 8, 14 and 17 argons (Beck et al.) pass back and forth between solid-like and liquid-like forms, when their temperatures lie in their coexistence ranges, too rapidly to exhibit the properties of well-marked phases. Instead, they seem to be "slush-balls". Others, such as the 15-atom and 19-atom clusters, are much like Ar_7 and Ar_{13}, with bimodal distributions of temperature and well-marked "phases".

Isothermal simulations of Ar_{13} (Davis et al.) bring the final confirmation insofar as simulations can, of the correctness of the picture of a sharply-bounded coexistence range. The mean potential energy of this cluster has a bimodal distribution in isothermal MC simulations, for temperatures in the coexistence range. The caloric curve developed from the MC simulations is superimposable on that from the isoergic MD simulations. The angular distribution function in the coexistence range is also composed of two components, one for a solid-like form and the other for a liquid-like form; at the temperature for which Quirke and Sheng report their angular distribution (Quirke and Sheng, 1984) their result is precisely the average of the two. Thus even the recent results of Quirke and Sheng, which were originally interpreted to imply that there is not a sharp solid-liquid distinction for this cluster, is in reality a demonstration of a very special sort of sharp bounding, but of a coexistence band, not of a transition temperature.

ACKNOWLEDGMENTS

This research was supported by a Grant from the National Science Foundation.

REFERENCES

Amar, F., Kellman, M.E. and Berry, R.S., 1979, J. Chem. Phys. 70:1973.
Amar, F. and Berry, R.S., 1986, J. Chem. Phys. (in press).
Beck, T.L., Jellinek, J. and Berry, R.S. (in preparation).
Berry, R.S., Jellinek, J. and Natanson, G., 1984a, Chem. Phys. Lett. 107:227.
Berry, R.S., Jellinek, J. and Natanson, G., 1984b, Phys. Rev. A 30:919.
Briant, C.L. and Burton, J.J., 1973, Nat. Phys. Sci. 243:100.
Briant, C.L. and Burton, J.J., 1975a, J. Chem. Phys. 63:2045.
Briant, C.L. and Burton, J.J., 1975b, J. Chem. Phys. 63:3327.
Briant, C.L. and Burton, J.J., 1976, J. Chem. Phys. 64:2888.
Cotterill, R.M., Damgaard Christensen, W., Martin, J.W., Petersen, L.B. and Jensen, E.J., 1973, Comput. Phys. Comm. 5:28.
Damgaard Christensen, W., Jensen, E.J. and Cotterill, R.M., 1974, J. Chem. Phys. 60:4161.
Davis. H.L., Jellinek, J. and Berry, R.S. (in preparation).
Eichenauer, D. and LeRoy, R.J., 1986 (submitted to Chem. Phys. Lett.).
Etters, R.D. and Kaelberer, J.B., 1975, Phys. Rev. A 11:1068.
Etters, R.D. and Kaelberer, J.B., 1977, J. Chem. Phys. 66:5112.
Etters, R.D., Danilowicz, R. and Dugan, J., 1977, J. Chem. Phys. 67:1570.
Etters, R.D., Danilowicz, R. and Kaelberer, J.B., 1977, J. Chem. Phys. 67:4145.
Etters, R.D. and Danilowicz, R., 1979, J. Chem. Phys. 71:4647.
Ezra, G.S. and Berry, R.S., 1982, J. Chem. Phys. 76:3679.
Farges, J., Raoult, B. and Torchet, G., 1973, J. Chem. Phys. 59:3454.
Farges, J., deFaraudy, M.F., Raoult, B. and Torchet, G., 1983, J. Chem. Phys. 78:5067.
Farges, J., deFaraudy, M.F., Raoult, B. and Torchet, G., 1986, J. Chem. Phys. 84:3491.
Gartenhaus, S. and Schwartz, C., 1957, Phys. Rev. 108:482.
Gough, T.E., Knight, D.G. and Scoles, G., 1983, Chem. Phys. Lett. 97:155.
Heenan, R.K. and Bartell, L.S., 1982a, J. Chem. Phys. 78:1265.
Heenan, R.K. and Bartell, L.S., 1982b, J. Chem. Phys. 78:1270.
Jellinek, J., Beck, T.L. and Berry, R. S., 1986, J. Chem. Phys. 84:2783.
Kaelberer, J.B. and Etters, R.D., 1977, J. Chem. Phys. 66:3233.
Kellman, M.E. and Berry, R.S., 1976, Chem. Phys. Lett. 42:327.
Kellman, M.E., Amar, F. and Berry, R.S., 1980, J. Chem. Phys. 73:2387.
LaViolette, R.A. and Stillinger, F.H., 1985, J. Chem. Phys. 83:4079.
Lee, J.K., Barker, J.A. and Abraham, F.F., 1973, J. Chem. Phys. 58:3166.
McGinty, D.J., 1973, J. Chem. Phys. 58:4733.
Natanson, G., Amar, F. and Berry, R.S., 1983, J. Chem. Phys. 78:399.
Nauchitel, V.V. and Pertsin, A.J., 1980, Mol. Phys. 40:1341.
Quirke, N. and Sheng, P., 1984, Chem. Phys. Lett., 110:63.
Stace, A.J., 1983, Chem. Phys. Lett. 99:470.
Stillinger, F.H. and Weber, T.A., 1981, Kinam 3A:159.
Stillinger, F.H. and Weber, T.A., 1982, Phys. Rev. A 25:978.
Stillinger, F.H. and Weber, T.A., 1983a, Phys. Rev. A 28:2408.
Stillinger, F.H. and Weber, T.A., 1983b, J. Phys. Chem. 87:2833.
Valente, E.J. and Bartell, L.S., 1983a, J. Chem. Phys. 80:1451.
Valente, E.J. and Bartell, L.S., 1983b, J. Chem. Phys. 80:1458.
Weber, T.A. and Stillinger, F.H., 1984, J. Chem. Phys. 81:5089, 5095.
Yamada, K. and Winnewisser, M., 1976, Z. Naturforsch. A31:134.

STRUCTURE AND MELTING OF ARGON CLUSTERS ON A SUBSTRATE

I.L. Garzón, M. Avalos and E. Blaisten-Barojas

Instituto de Física
Universidad Nacional Autónoma de México
A.P. 2681, 22830 Ensenada, Baja California, México

ABSTRACT

The structural and thermodynamic properties of 13-atom argon clusters adsorbed on a substrate are studied using Molecular Dynamics simulations. The cluster atoms interact among themselves via a Lennard Jones potential, whereas the cluster-substrate interaction is modeled by a continuous model for the adsorbent surface. Heating and cooling processes on the system show that the adsorbed clusters exhibit solid and liquid like phases as well as a coexistence region of both of them. The strength of the cluster-substrate interaction determines the mechanism of desorption in the region of high temperatures.
PACS Nos: 36.40.+d, 64.70.Dv, 68.40. +e, 68.45.-v.

To study both theoretically and experimentally the physical properties of small clusters is a problem of current scientific and technological interest.[1] Computer simulations provide a mean to perform "ideal experiments" on these finite size systems with the specific advantage to show the effect of well controlled parameters on the cluster production. Following this line of work, it has been shown that isolated argon clusters exhibit phase changes from solid-like to liquid-like structures as well as a temperature region where both phases coexist.[2] The production of such systems in molecular beam experiments together with photofragmentation spectroscopy offers a possibility to test the above results.[3] The theory of clusters adsorbed on substrates has received less attention, even though a priori this is a simpler experimental technique. In this communication we report the results from a Molecular Dynamics study of 13-atom argon clusters adsorbed on a substrate.. Our main emphasis is to give more insight on the clusters structural and thermodynamic properties as a function of the cluster-substrate interaction.

The system consists of 13 atoms interacting among themselves via Lennard-Jones potentials as to form a cluster, and this cluster is adsorbed on a substrate. The two parameters of the pairwise interaction are E and σ. The cluster-substrate model is given by

$$V_s = 4E_s \sum_{i=1}^{N} [(\sigma_s/y_{is})^{12} - (\sigma_s/y_{is})^3] \qquad (1)$$

where E_s and σ_s are parameters measuring the strength of the interaction and the hardness of the core between cluster and substrate. The y_{is} are the perpendicular distances from each cluster atom to the substrate. This is a continuous model for the cluster-substrate interaction.

Molecular Dynamics at constant energy was used to calculate structural properties and to simulate various melting experiments. The Newton equations of motion were solved using the Verlet algorithm[4] with a time step of 0.01τ ($\tau=(m\sigma^2/E)^{\frac{1}{2}}$, m is the atomic mass). Energy and distance units of E and $r_o = 2^{1/6}\sigma$ and $T^* = k_B T/E$ were used throughout.

To follow the effect of the substrate on the phase changes of the cluster we considered three stregths E_s/E of the cluster-substrate potential as to simulate weak (0.2), intermediate (0.5) and strong (0.8) interactions. The process of heating and cooling the clusters was always carried out in presence of a frozen substrate at T=0. The temperature of the system was changed by scaling the velocities of an equilibrated and previous calculation. Every heating (cooling) process was done in a steplike way, allowing in each step of the process for an equilibration time of 20τ and a further run of 50τ to average the calculated quantities. Clusters were heated slowly at a rate of approximately $\Delta T^*/\Delta t = 1.7 \times 10^{-4}\tau^{-1}$. In most cases the initial configuration of the cluster atoms was an icosahedron and the substrate was kept at a fixed initial distance of $-1.8\ r_o$.

In Figure 1 we show the mean kinetic energy as a function of the mean total energy for the free 13-atom cluster and the three values of E_s/E. For the free cluster, Figure 1.a, it is observed the solid, liquid and coexistence regions for low, high and mean kinetic energy in agreement with the results of Jellinek et. al. [2] Figures 1.b,c, and d show a similar behavior when the cluster is adsorbed on a surface with various intensities. This result shows that the cluster exhibits the solid, liquid and coexisting phases even near the adsorbent surface. In the case of strongly adsorbed clusters a structural phase change takes place in the solid-like region of the caloric curve, although the dispersion of points is large. This result is made evident when we calculate the mean coordination number as a function of temperature.[5] In Figure 2 we show this dependence for weakly and strongly adsorbed clusters. The difference among the two cases is now clear. For weak coupling, Figure 2.a, the mean coordination number remains constant up to temperatures around 0.25 and rather abruptly acquires a smaller value at higher temperatures. This indicates a phase change from a solid-like to a liquid-like regime, since for Lennard Jones systems the coordination number decreases when melting occurs. Figure 2.b, corresponding to strong cluster-substrate coupling is specially interesting since a plateau shows up at intermediate temperatures pointing to an additional phase change in the solid-like region.

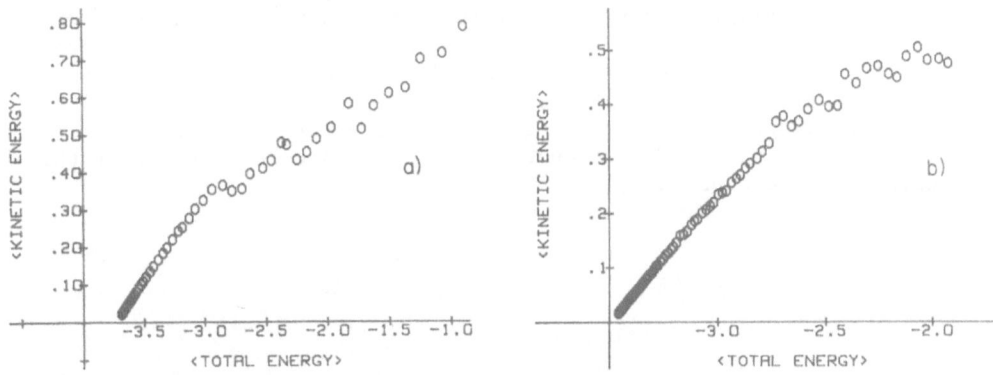

Fig. 1. Mean kinetic energy as a function of the mean total energy.
(a) Free cluster; (b) cluster adsorbed on a weak substrate,
Es/E = 0.2; (c) cluster adsorbed on an intermediate substrate, Es/E = 0.5; cluster adsorbed on a strong substrate,
Es/E = 0.8.

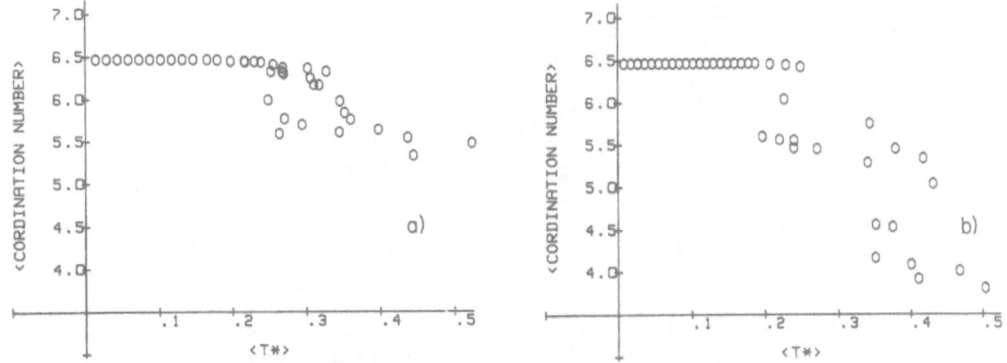

Fig. 2. Temperature dependence of the mean coordination number for
the adsorbed cluster on a (a) weak substrate, Es/E = 0.2;
(b) strong substrate Es/E = 0.8.

To make sure that a structural change is taking place, we draw in Fig. 3.a the cluster configurations for different temperatures. Indeed an ordered, two layer cluster is apparent at intermediate temperatures before melting. At high temperatures the completely disordered configurations characteristic of the liquid-like behaviour are obtained. Figure 3.b shows the cluster configuration changes for the weak cluster-substrate case in the same temperature region. The absence of the 2-D phase indicates that this is an effect of the substrate.

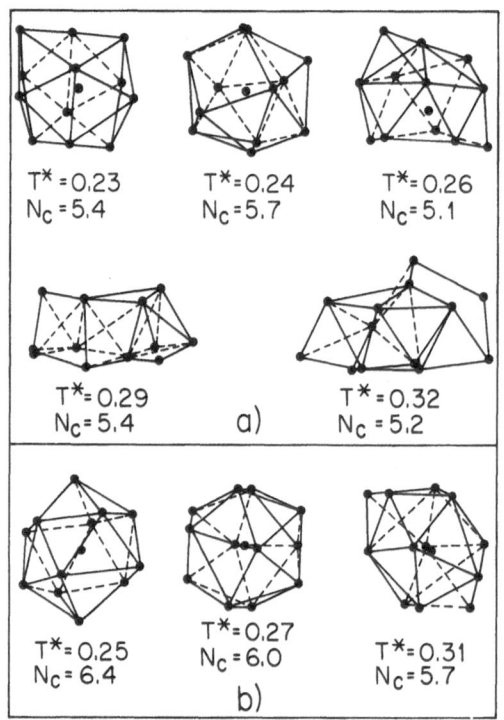

Fig. 3. Cluster configurations at different temperatures for the adsorbed cluster on a (a) strong substrate, $Es/E = 0.8$; (b) weak substrate, $Es/E = 0.2$. N_c is the coordination number calculated from the atom coordinates of the cluster configurations.

At high temperatures the cluster desorbs from the surface. This mechanism is also dependent on the cluster-substrate interation. For weak interaction the cluster desorbs as a whole, whereas for Es/E = 0.5 and 0.8 the cluster "evaporates" atom by atom.

Simulations of cooling processes show as well that the V_s potential affects the cluster desorption from the substrate.[6] In fact, when the system was cooled starting with a configuration corresponding to an isolated cluster in the liquid phase at T*=0.28 it was found that: 1) for the weak interaction case, Es/E = 0.2, at T*=0.25 the cluster desorbed as a whole. Further decrements of the temperature resulted in a redistribution of the energy in such a way as to increase the cluster center of mass velocity at the expenses of a lost of potential energy. The cluster used this energy in reconstructing the icosahedral structure; 2) for the stronger cluster-substrate interaction the desorption was not observed and the cluster changed into a solid-like phase while adsorbed on the surface.

In conclusion, these simulations show that phase changes in small clusters are also possible in the presence of a substrate. The details of structural changes depend on the strength of the cluster-substrate interaction. This interaction is also important in the desorption mechanism.

ACKNOWLEDGEMENT

This work was supported by Consejo Nacional de Ciencia y Tecnología, México, under grant PCCBBNA-022643.

REFERENCES

1. T.P. Martin, Phys. Rep. 95, 167 (1983); Angew. Chem. 98, 197 (1986).
2. C.L. Briant and J.J. Burton, J. Chem. Phys. 63, 2045 (1975);
   E. Blaisten-Barojas and H.C. Andersen, Surf. Sci. 156, 548 (1985);
   J. Jellinek, T.L. Beck, and R.S. Berry, J. Chem. Phys. 84, 2783 (1986).
3. J. Farges, M.F. de Feraudy, B. Raoult, and G. Torchet, J. Chem. Phys. 84, 3941 (1986).
4. L. Verlet, Phys. Rev. 159, 98 (1964).
5. E. Blaisten-Barojas, Kinam 6, 71 (1984).
6. I.L. Garzón, M. Avalos and E. Blaisten-Barojas, to be published.

DYNAMICS OF ATOMIC COLLISIONS ON HELIUM CLUSTERS

Jürgen Gspann and Roland Ries

Kernforschungszentrum und Universität Karlsruhe
7500 Karlsruhe, Federal Republic of Germany

ABSTRACT

Differential scattering of Cs atoms from ^{4}He clusters indicates strong interaction at all relative velocities between 140 and 350 m/s. Quantum fluid properties seem to be masked by the particular dynamics at the surface of the ^{4}He clusters.

INTRODUCTION

Free helium clusters formed in nozzle expansions[1] obtain very low internal temperatures by evaporation cooling. The estimated values are 0.4 K for ^{4}He clusters, and 0.15 K for ^{3}He clusters.[2] As bulk liquid ^{4}He becomes superfluid below the λ-temperature of 2.17 K, the ^{4}He clusters should be superfluid. Hence, impurity atoms can be expected to move frictionless inside ^{4}He clusters if their velocity is lower than about 60 m/s, i.e. below the velocity needed to create elementary excitations. In a crossed beam experiment, we have investigated the interaction of ^{4}He clusters with Cs atoms, varying the relative velocity by changing the angle of beam intersection. Cesium atoms are calculated to have a negative binding energy of about 260 K in bulk liquid ^{4}He.[3] To overcome this repulsion, Cs atoms have to have a velocity of impact of about 180 m/s. In the present investigation, we have looked for the angular distribution of the scattered Cs atoms.

EXPERIMENTAL

The experimental setup described earlier[4] has been equipped with an additional Langmuir-Taylor detector in order to measure the scattered Cs atoms. In the plane of the two beams, the detector can be turned upon the axis of the beam intersection. The helium cluster beam is pulsed by a rotating disk chopper to yield pulsed signals of scattered atoms. The signal amplitudes are of the order of 10^{-4} times the cesium beam intensity.

RESULTS AND DISCUSSION

Scattered Cs atoms are observed solely in the region of the ^{4}He cluster beam. With the given sensitivity, no scattered Cs is observed in the angular range between the two beams. Only a part of the atoms scattered out of the Cs beam is found in the helium cluster beam, however.

We assume the scattered Cs atoms to be dragged along with the helium clusters. The detection efficiency for these atoms buried in helium clusters is lower than that for naked atoms, presumably.

The mean relative velocities have been changed from about 140 to 350 m/s. No peculiarities show up in the range form 180 to 240 m/s. Hence, we have to conclude that Cs atoms able to penetrate into the clusters seen unable to escape again, under the given conditions, even if their motion should be frictionless inside the clusters. The particular surface dynamics either at the entrance or at the opposite cluster surface seem to mask the internal quantum fluid properties of the ^{4}He clusters.

REFERENCES

1. E.W. Becker, R. Klingelhöfer, and P. Lohse, Strahlen aus kondensiertem Helium im Hochvakuum, Z. Naturforschung 16a:1259 (1961).

2. J. Gspann, Electronic and Atomic Impacts on Large Clusters, in: "Physics of Electronic and Atomic Collisions", S. Datz ed., North-Holland Publ. (1982).

3. K.E. Kürten, M.K. Ristig, Atomic impurities in liquid helium, Phys. Rev. B27: 5479 (1983).

4. J. Gspann, R. Ries, ^{3}He and ^{4}He clusters colliding with Cs atoms: total cross sections for varying impact speed, Surf. Sci 156: 195 (1985)

ISOMERIZATION AND MELTING OF

SMALL ALKALI-HALIDE CLUSTERS

Jia Luo and Uzi Landman
School of Physics
Georgia Institute of Technology
Atlanta, Georgia   30332

Joshua Jortner
School of Chemistry
69978 Tel Aviv University, Israel

The nature of phase transformations in small alklai-halide clusters is investigated using molecular dynamics simulations. On the basis of systematic studies of clusters of variable sizes, it is concluded that the kinetics and dynamics of the transformations depend on system size. For small clusters, distinct, diffusionless, isomerization occurs. Intermediate size clusters exhibit hierarchical kinetics with isomerizations preceding the onset of diffusion and eventual melting. True sharp melting transition and solid-liquid coexistence is found only for relatively large clusters. This sequence of physical processes versus size can be modeled as a gradual opening of the accessible phase-space and the coalescence of time scales distinguishing intra-well and inter-well dynamics.

1. INTRODUCTION

The nature of phase transformations in finite systems is a subject of considerable conceptual and practical interest.[1-6] While the monotonic decrease in the melting temperature with decreasing particle size has been long observed,[5] satisfactory theoretical understanding of the melting transition, in bulk as well as small cluster systems, is lacking.[3] Since thermodynamics is founded on the premise of the thermodynamic limit, the applicability of a thermodynamic description to finite systems is in question. Indeed for small systems the magnitude of fluctuations precludes a sharp definition of a phase transition. The construction of theoretical models of melting of small clusters is further complicated by the intrinsic inhomogeneity of the system (exterior, surface, versus interior atoms), and by the spatial and system size dependencies of the spectrum of motional degrees of freedom.[6]

Molecular dynamics (MD) simulations, where the evolution of a physical system is simulated, with refined temporal and spatial resolution, via a direct numerical solution of the equations of motions, alleviate certain of the major difficulties which hamper other theoretical approaches, thus opening new avenues in investigations of the microscopic dynamics and kinetics of physical processes.[7,8]In this study we have investigated, using

MD simulations, the energetics, structure and dynamics of alkali-halide clusters of variable sizes [(NaCl)$_n$, for n = 4, 16 and 108] over a wide temperature range. Our simulations show that the very nature of the phase transformation and the underlying physical processes involved, depend on the size of the system. We conclude that at the lower end of the size spectrum, the phase space of the system is characterized by a small number of stable configurations (solid isomers) between which the system transforms in a diffusionless manner, with temperature dependent rates and branching rations. As the system size is increased the number of accessible conformers increases leading to a hierarchial kinetics of isomerization events which exhibits itself as a broadening of the transition region. For these small clusters coexistence is between solid isomers rather than inter-phase (solid-liquid) coexistence. The latter develops for clusters of sufficient size, characterized by a dense spectrum of accessible states, separated by thermally surmountable barriers.[8] Under these circumstances conventional melting is observed as a sharp transition, the separation of time-scales for inter-well and intra-well dynamics ceases,[2,3] and true solid-liquid coexistence is found.

2. RESULTS AND DISCUSSION

We have performed constant energy MD simulations for (NaCl) clusters, with interionic potentials consisting of a point Coulomb term and a Born-Mayer repulsion.[9] At each temperature the system was equilibrated for a prolonged period (typically 10^{-10} sec) after which data was collected (typically for $\sim 10^{-9}$ sec). The integration time step was chosen as 3 x 10^{-15} sec to assure energy conservation.

In Fig. 1 the caloric curves (total energy, U, versus kinetic energy, EK) for (NaCl)$_n$ with n = 4, 16 and 108 (a, b and c, respectively) are shown. (The kinetic temperature is defined as $k_B T = 2U/3(N-2)$. Throughout, energy is expressed in atomic units 1a.u. = 2 Rydbergs). For the n = 4 cluster the break in the curve corresponds to isomerization between a cube and ring conformations (see Fig. 2). The development of a hirerchy of structural transformations is demonstrated for n = 16, in Fig. 1b. The lower temperature break corresponds to diffusionless solid to solid isomerization while the higher temperature break is accompanied by the development of disorder and diffusion characteristics (see Figs. 3,10). The caloric curve shown in Fig. 1c for n = 108 cluster is markedly different from those of the smaller systems. We observe in it a "Van der Waals loop phenomenon" with the true melting point (solid-liquid coexistence, see Fig. 4) at 835K. While gradual increase in the total energy leads to the lower (solid) branch of the caloric curve, terminating ultimately and resulting in the liquid branch, the coexistence point (marked by x) is achieved by providing the system via scaling of the momenta, when on the solid branch at 835K, with sufficient energy to reach the horizontal line (melting of part of the system was observed to be initiated at the surface, progressing inwards). The structural "phase diagram" of the smaller clusters (n = 4 and 16) is seen to be characterized by branches corresponding to stable isomers, separated by regions of isomer-coexistence. Associated with the latter is a time scale dictated by the residence times at the different isomers. A sample of the time evolution of the kinetic energy of the n = 4 cluster, at T = 698 (coexistence region in Fig. 1a), exhibiting a transformation between the ring and cube isomers (see Fig. 2) is shown in Fig. 5a. This record was obtained by short time averaging[3] of KE over sets of 100 time-steps. The distribution of short-time averages of the kinetic energy (for that point on the caloric curve), shown in Fig. 5b, is trimodal, with each of the peaks corresponding to a distinct structure as noted (transition state (L), cube and ring). Note that the choice of the averaging time span is of importance in obtaining the resolution

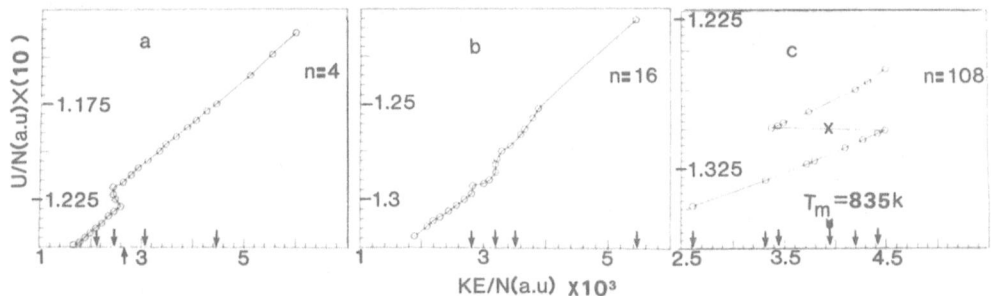

Fig. 1. Caloric curves for n = 4,16,108. Arrows denote temperatures in increasing order: (a) 589, 698, 740, 874, 1249K; (b) 636, 726, 806K, 1684K; (c) 555, 705, 730, 890, 938. Note change in character of transition upon increase in size.

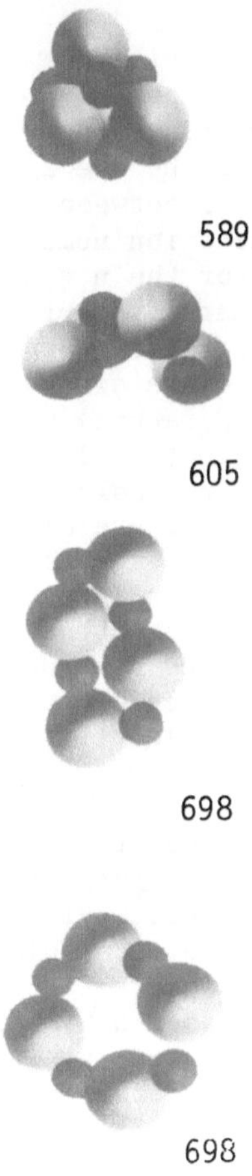

589

603

605

603

698

726

698

726

698

1684

555

811

892

Fig. 2 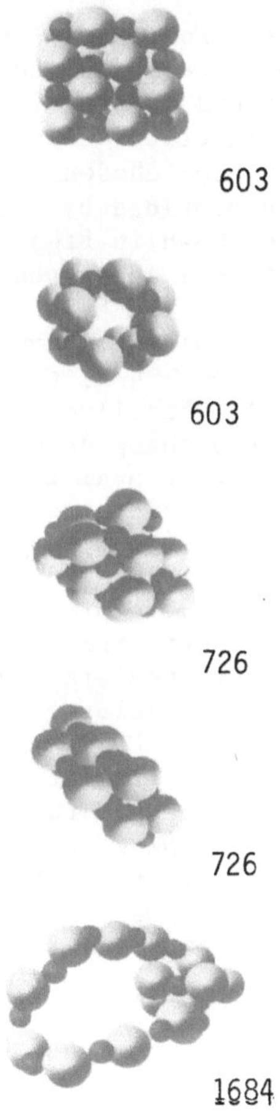   Fig. 3 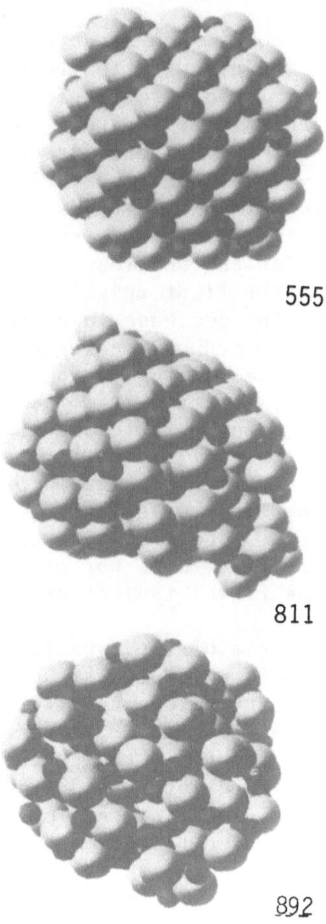   Fig. 4

Figs. 2-4. Cluster geometries for (NaCl), n = 4, 16 and 108 at the denoted temperatures. Large and small balls, Cl⁻ and Na⁺ ions, respectively.

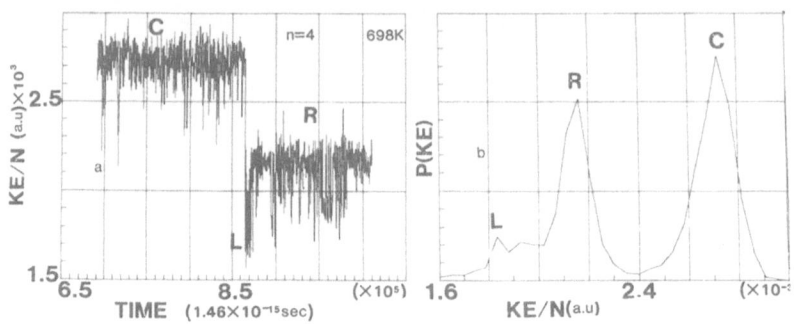

Fig. 5. (a) Sample of KE vs. time for (NaCl) showing isomerization from cube (C) to ring (R), through a ladder (L) transition state, conformations. (b) Short-time average, KE distribution, P(KE).

which affords such structural identification. Away from the coexistence region the distribution becomes, monotonously, unimodal. Further structural characterization of the systems is provided by the function $n(r)$, related to the pair correlation function, and defined as the average number of atoms whose distance from an atom chosen as origin is between r and $r + \Delta r$. Additional information is provided by the coordination number function $C(R) = \int_0^R n(r')dr'$. These are shown in Figs. (6-8) for the n = 4, 16 and 108 clusters, respectively. It is evident that the sharp isomerization for the smallest cluster is accompanied by a sudden decrease from 3 to 2 in the first neighbor coordination while for the intermediate size cluster the decrease in coordination (from ~ 4 at the low temperature) across the transition region is gradual. Finally for the large (n = 108) cluster the melting is characterized by a sharp drop in the first neighbor coordination (from ~ 4.5 to ~ 3). The structural transformations are also clearly seen by following the evolution of features in $n(r)$ as a function of temperature. In particular, except for the n = 4 cluster, upon increase in EK the first peak grows broader and decreases in magnitude and further coordination shells lose their structure (more pronounced for the n = 108 cluster). Note also that this characteristic trend, which corresponds to liquification, does not occur for the intermediate cluster (n = 16) till past the first, isomerization, break in the caloric curve.

A probe of the dynamical nature of the transition is provided by the density of states (power spectrum) calculated as the fourier transform of the velocity autocorrelation function.[10] These are given in Figs. (9-11) for n = 4, 16 and 108, respectively. First, we notice the trend of stabilization at higher temperature of the structural form (solid isomers for the smaller clusters and liquid for the large one) which possesses a higher density of low frequency modes. This can be understood in terms of the vibrational contribution to the free energy,[9] which the systems attempts to minimize. Secondly, for the n = 4 cluster we do not find any indication of soft mode or onset of diffusion (see value near $\omega = 0$ in $I(\omega)$ plots). For the intermediate size cluster, however we do observe what may be construed as soft mode development and onset of diffusion at temperatures <u>above the first, isomerization, break</u> in the caloric curve (see Fig. 1b), thus corroborating our distinction between isomerization and melting for intermediate size clusters. True diffusion character is observed for the n = 108 cluster both at coexistence and on the liquid branch.

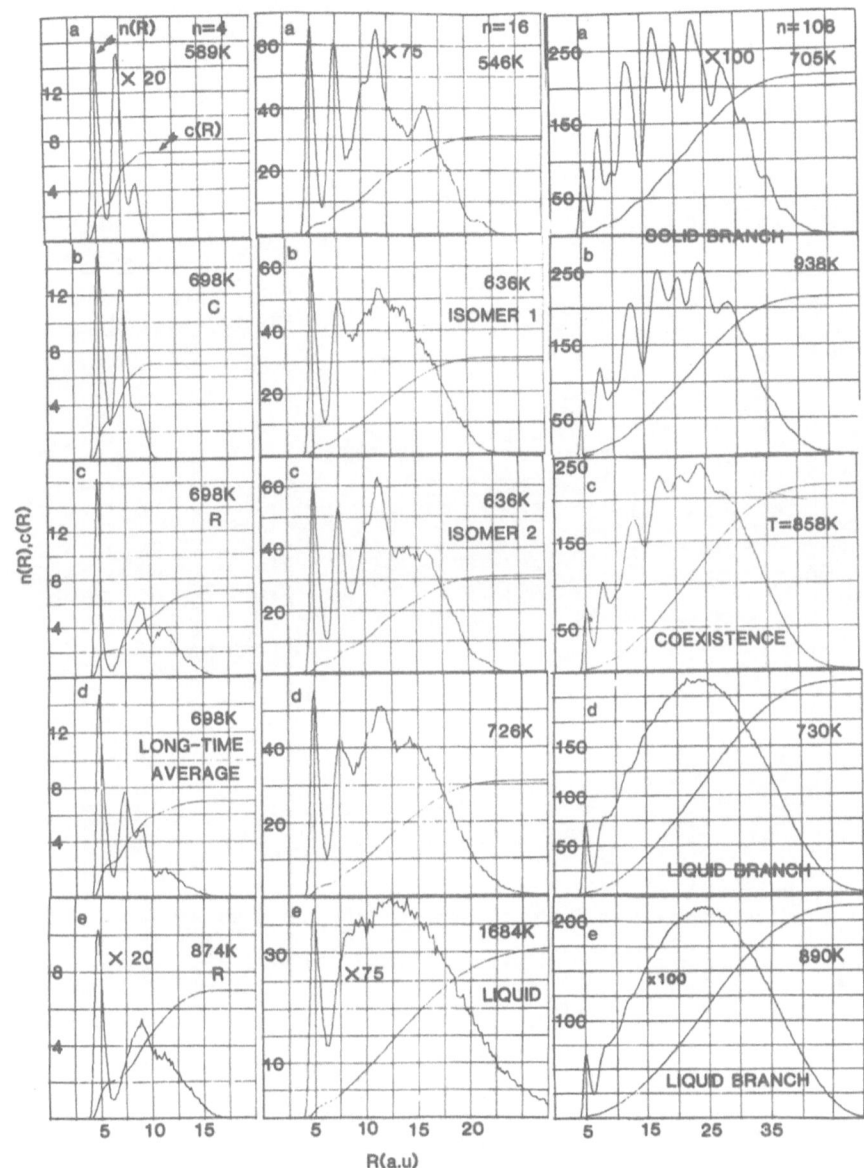

Figs. 6-8. Particle pair distribution, n(R), and coordination number C(R), functions for n = 4, 16, and 108 at the indicated temperatures. Note dependencies on size and on temperature.

REFERENCES

This work was supported by US DOE Grant DE-FG05-86ER45234.

1. T. L. Hill, Thermodynamics of Small Systems, (Benjamin, N.Y., 1963).
2. G. Natanson, F. Amar and R. S. Berry, J. Chem. Phys. 78, 399 (1983).
3. J. Jellinek, T. L. Beck and R. S. Berry, J. Chem. Phys. 84, 2783 (1986); see also references to earlier work here and in ref. 2.
4. R. W. Hockney and J. W. Eastwood, Computer Simulation Using Particles, (McGraw-Hill, N.Y., 1981), pp. 488-498.
5. P. R. Couchman and W. A. Jessor, Nature 269, 481 (1977).
6. P. R. Couchman and C. L. Ryan, Phil. Mag. A37, 369 (1978); R. Balian and C. Bloch, Ann. Phys. N.Y. 60, 401 (1970).

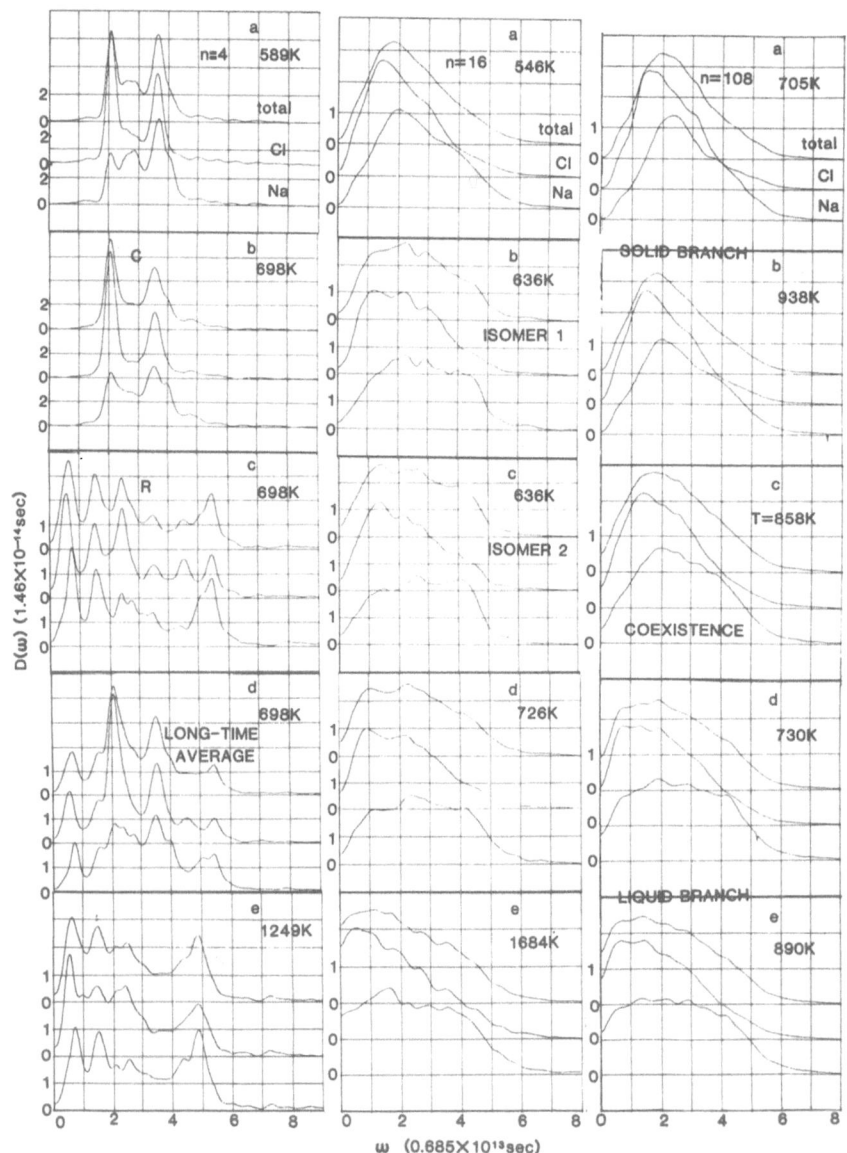

Figs. 9-11. Densities of states for n = 4, 16 and 108 at the denoted
temperatures. Note shift to lower frequencies with increase
in temperature, and onset of diffusion for n = 16 and 108 at
higher temperatures.

7. For recent reviews see: F. F. Abraham, J. Vac. Sci. Technol. B2, 534
    (1984); U. Landman et al. in Mat. Res. Soc. Proc. 63, 273 (1985).
8. This is reminiscent of the picture advanced by F. M. Stillinyer and T.
    A. Weber for a liquid, see: Kinam A5, 159 (1981); Phys. Rev. A25, 978
    (1972).
9. T. P. Martin, Phys. Rep. 95, 167 (1983), with the parameters given by
    F. G. Fumi and M. P. Tosi, J. Phys. Chem. Solids 25, 31, 45 (1964).
10. D. A. McQuarrie, Statistical Mechanics, (Harper and Row, N. Y., 1976),
    Chap. 21, 22.

MINIMUM ENERGY STRUCTURES OF $Br_2^- Ar_N$ CLUSTERS:

IMPLICATIONS FOR DYNAMICS

François G. Amar

Department of Chemistry
University of Maine
Orono, Maine   04469

I. INTRODUCTION

The dynamics of photexcited dihalide species such as I_2 has been extensively studied (1) in small van der Waals molecules with emphasis placed on the role of intramolecular energy transfer (2). In fluids, the geminate recombination dynamics of I_2 excited to several electronic states has been studied both experimentally by picosecond spectroscopy and theoretically by molecular dynamics (MD) and other techniques (3). In order to bridge the gap between van der Waals molecules and condensed phases it is necessary to study small clusters of between 5 and 100 molecules in molecular beams. Our previous MD simulation study of photoexcitation dynamics of $Br_2 Ar_N$ neutrals under beam condition demonstrated a gradual transition between the small molecule limit and the condensed phase limit as the numberof argon atoms was increased from 6 to 70 (4). For clusters of less than about 20 atoms, the Br_2 caging dynamics is extremely sensitive to cluster geometry and loss of argon atoms occurs by impulsive energy transfer from Br_2. For two shell cluster (70 atoms), loss of argon occurs by evaporation after the whole cluster is heated.

It is difficult of perform size-selective experiments of neutral clusters in this intermediate size range since laser induced fluorescence becomes insensitive to size and mass spectrometry introduces large uncertainties about the origin of the cluster size distribution due to fragmentation effects. In order to overcome these drawbacks, the Lineberger group (among others) has turned to supersonic beam studies of ionic clusters whcih dissociate following laser excitation. Cluster distributions observed in their tandem mass spectrometer reflect reaction dynamics (not ionization dynamics). Time-of-flight methods permit the observation of specific cluster sizes. Among the systems which have been investigated by the Lineberger group are $(CO_2)_N^+$ (5,6), $Br_2^- (CO_2)_N$ (7), and $Br_2^- Ar_N$ (8).

In this paper I present results of Monte Carlo (MC) and MD simulations of the photodissociation, caging and recombination dynamics of Br_2 in argon clusters containig fewer than 20 atoms. Minimum energy structures of various size clusters of $Br_2 Ar_N$ are identified--their geometries and binding energies are discussed below. The reaction dynamics of these clusters are then followed by MD after a simulated photoexcitation of the minimum energy configuration.

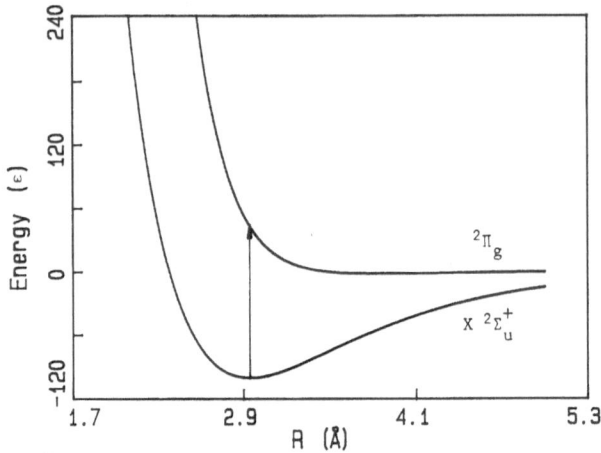

Fig. 1 Morse curves for the electronic states of Br_2^-. X state: D_e=12069 K, r_e=2.94 Å, α=1.237 Å^{-1}. $^2\Pi_g$ state: D_e=290 K, r_e=3.8 Å, α=1.952 Å^{-1}.

II. POTENTIAL MODEL

In a $Br_2^-Ar_N$ cluster, three types of contributions to the potential must be considered. First, for Br_2^- itself, the ground X $^2\Sigma_u^+$ and first excited $^2\Pi_g$ potential curves are required since the reaction of interest is a photodissociation. Figure 1 shows these two electronic states as Morse potentials fit to the relativistic effective core potential calculation of Wadt and Hay (9). Second, both the Ar-Ar and the Ar-Br contributions to the dispersion part of the potential are represented by pairwise sums of Lennard-Jones functions with ε_{Ar-Ar}= 119.8 K, σ_{Ar-Ar}= 3.405 Å and ε_{Ar-Br}= 143 K and σ_{Ar-Br}= 3.51 Å just as in the neutral species (4). The LJ parameters for argon, hereafter denoted ε and σ, are taken to be the units of energy and length in this discussion. Finally, an ion-induced dipole contribution to the potential is computed as follows: Each bromine atom is assigned a charge of 0.5 electron located at its center; then each argon is given its point polarizability of 1.64 Å^{-3}. The ion-induced dipole energy (and the corresponding contribution to the force) is then calculated self-consistently; that is, the induced dipole on an argon depends not only of the coulomb field of the ion but also on the field due to induced dipoles on neighboring argons. Thus the induced dipoles must be calculated iteratively (10).

III. MINIMUM ENERGY STRUCTURES

The minimum energy structures of $Br_2^-Ar_N$ with $2 \leq N \leq 13$ have been located using an MC procedure to sample configuration space and then to quench configurations chosen at intervals along the Markov chain to the nearest local minimum. The geometric structures of the lowest minima are pictured in Figure 2. At first, the argon atoms add around the waist of Br_2^-, forming a regular pentagon at N=5. Between N=6 and N=9, added atoms first perturb the central ring and then form a partial second ring of atoms around the Br_2^- axis. Then at N=10, the cap site along the Br_2^- axis becomes more favorable than the fifth ring site and

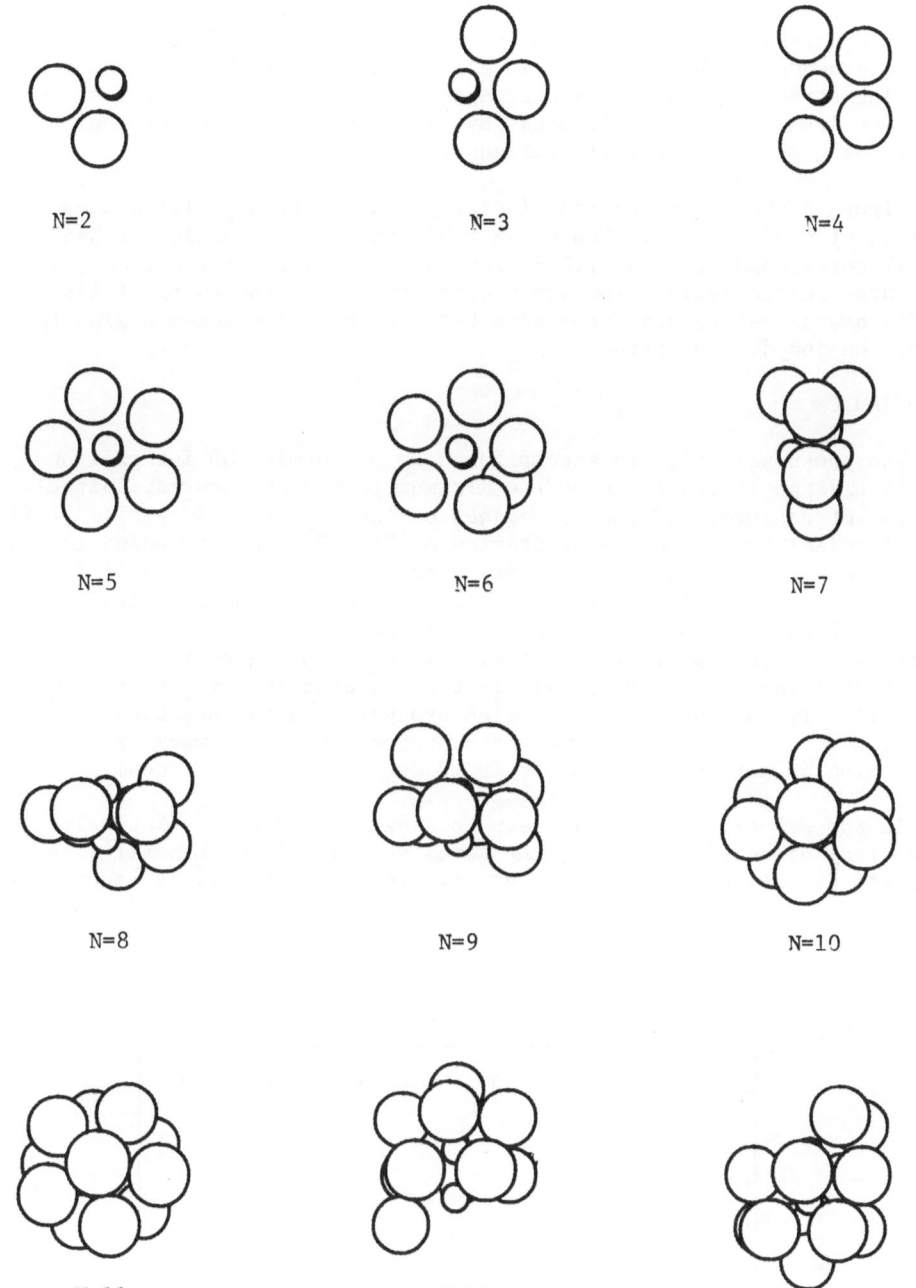

Fig. 2 Minimum energy structures of $Br_2^-Ar_N$, $2 < N < 13$. The Br atoms are drawn half size to distinguish them from argon. In N=10 and N=11 the bromine bond is perpendicular to the plane of the paper.

only in $Br_2^-Ar_{11}$ does one see two complete pentagonal rings plus the cap atom. Beyond N=11, the third pentagonal ring begins to appear until a complete doubly capped structure forms at $Br_2^-Ar_{17}$. The structures can be classified (4) as <u>uncapped</u> up to N=9, <u>half-capped</u> between N=10 and N=15, and <u>fully-capped</u> for N=16 and larger. There are major differences between the neutral and ionic geometries, due in large measure to the repulsive nature of the induced dipole--induced dipole interactions. One consequence of the repulsive interactions is that the uncapped ions form ring structures while the corresponding neutrals maximize the number of Ar-Ar van der Waals bonds by clumping on one side of the Br_2, forming several partially filled rings (4).

Figure 3 shows the reduced binding energy of $Br_2^-Ar_N$ clusters as a function of cluster size. The curve exhibits distinct maxima at N=5 and at N=11 corresponding to a filled ring and a complete half-capped structure, respectively. The ionic contribution to the energy falls rapidly beyond N=5 because repulsion between induced dipoles begins to cancel the ion-dipole attraction.

IV. DYNAMICS

The photoexcitation is accomplished by performing MD integration with initial positions given by the minimum energy structures, initial momenta set to zero, and the Br_2^- placed on the $^2\Pi_g$ repulsive wall. This is a classical Franck-Condon excitation of 144.32ϵ corresponding to 832 nm. photons (near infrared). The Br_2^- bond energy is initially 43.58 ϵ above the gas phase dissociation limit. Figure 4a-d shows several features of the Br_2^- bond as a function of time for N= 5,7,8,10 respectively. The upper trace (labeled i) of each figure is the Br_2^- bond length; the middle trace (ii) is the Br_2^- bond energy; and the lower trace (iii) is the sum of Ar-Br forces projected on the Br_2^- bond direction (the integral of this force projection is the work of compression or extension of the Br_2^- bond due to external forces).

In the N=5 cluster, Br_2^- is seen to dissociate with a relatively large final velocity. For N=7, the final velocity is much smaller as evidenced by the small slope of r vs. t. For N=8, the Br_2^- is caged, but

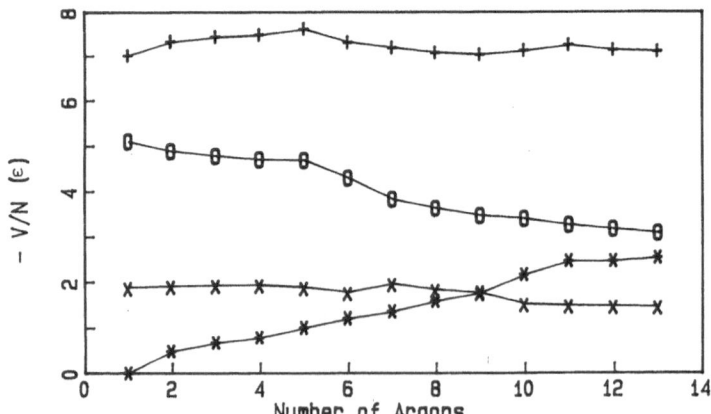

Fig. 3 $Br_2^-Ar_N$ binding energy per argon (exclusive of Br_2^- bond energy): total energy (+); sum of Ar-Ar LJ terms (*); sum of Ar-Br LJ terms (x); ion-induced dipole energy (0).

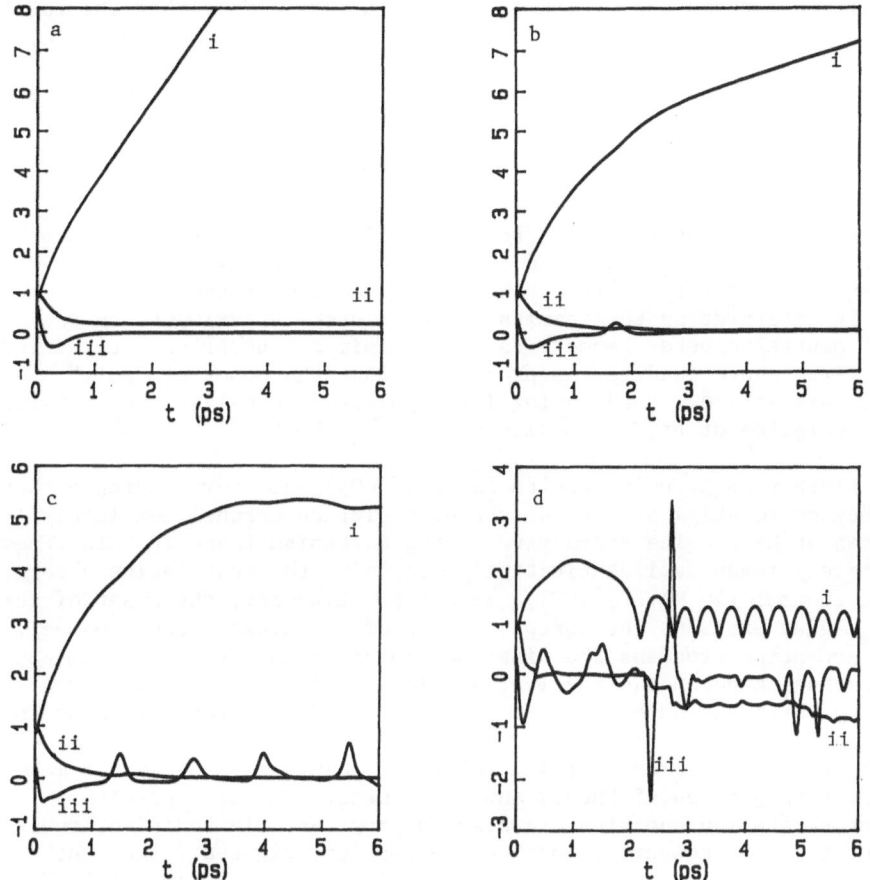

Fig. 4   Panels a–d correspond to clusters with 5,7,8,10 argons. In each
figure (i) is the Br$_2^-$ bond length in σ: (ii) is the bond energy
in units of 48ε; (iii) is the force projection in reduced force
units (48ε/σ = 1), see text.

the bond extends to a maximum of 5 σ. Finally, in the N=10 case, one of
the Br atoms collides with the cap atom and then recombines (crossover
back to the X state is forced when r = 2σ). The periodicity in the bond
length trace of Figure 4d corresponds to large amplitude oscillatons
near the dissociation limit of the X state. The negative spikes in the
force projection curves correspond to compression forces exerted by the
cluster on the Br$_2^-$ bond. The positive spikes are due to Ar atoms which
have intervened between the bromines, resisting the compression of the
Br$_2^-$ bond. It can be seen especially clearly in the N=10 plot (Figure 4d)
that abrupt decreases in the Br$_2^-$ bond energy coincide with sharp spikes
in the force projection curve. This is evidence for the impulsive
nature of the energy transfer to the solvent atoms in small clusters.

The N=10 case is an example of a classic kinetic cage effect caused
by collison with an external cage or cap atom. The N=8 case, on the
other hand, has no counterpart in the neutral Br$_2$Ar$_N$ clusters. Br$_2^-$Ar$_8$
is caged by attractive forces due to the strong ion-induced dipole
interaction. The combination of decreased excitation energy and
increased interatomic binding (with respect to the neutral) yields the

possibility of attractive caging in ionic clusters. Note that the total Ar-Br binding energy for N=7 is 40.766 ϵ while for N=8 it is 43.885 ϵ. These values bracket the excitation energy of 43.58 ϵ. For an N=8 cluster initially at zero temperature it is energetically impossible for Br_2^- (in this model) to dissociate at this excitation energy.

V. CONCLUSIONS

Ionic cluster geometries can be quite different from those of the corresponding neutral species and the photoexcitation dynamics can consequently be of very different character. One must use extreme caution in interpreting the results of ion cluster experiments on the basis of neutral cluster geometries. The minimum cluster structures for $Br_2^- Ar_N$ given above lead to the prediction that the onset of "soft" caging occurs at N=8. Hard caging (or repulsive caging) starts at N=10. Thus dissociation of Br_2^- effectively ceases at N=10.

The Lineberger group's results for $Br_2^- (CO_2)_N$ clusters indicate that caging begins (weakly) at N=6 and the dissociation channel completely shuts down at N=12. The attractive caging mechanism found in this study qualitatively (even semiquantitatively) explains the weak caging observed in small thermal (50 K) $Br_2^- (CO_2)_N$ clusters. Likewise, the onset of repulsive caging explains the abrupt closing of the dissociation channel. Thermal averaging broadens the distribution of sizes for which a given caging mechanism becomes effective; small variations in photon energy could also shift the onset of soft caging up or down by one argon or so.

Statistical averaging over initial phase points is underway to make more direct comparison of theory and experiment. Simulations with different excitation energies are also in progress. In addition, several refinements of the potential model are being investigated. The most important of these is a correct treatment of the dissociation of Br_2^- into $Br + Br^-$ rather than $Br^{1/2-} + Br^{1/2-}$; insofar as the neutral Br is less tightly bound at large r, significant differences in dissociation behavior may appear.

ACKNOWLEDGEMENT

This research was supported by a grant from Research Corporation.

REFERENCES

1. D. Levy, Advances in Chemical Physics (Wiley, New York, 1981) vol. 47, pp. 323-362.
2. J.A. Beswick and J. Jortner, Advances in Chemical Physics (Wiley, New York, 1981), vol. 47, p. 363ff and references contained therein.
3. J.T. Hynes, Ann. Rev. Phys. Chem. (Academic Press, New York, 1985), vol. 36, pp. 574-578 for a review of this topic and recent references.
4. F.G. Amar and B.J. Berne, J. Phys. Chem. __88__, 6720 (1984).
5. M.L. Alexander, M.A. Johnson, and W.C. Lineberger, J. Chem. Phys. __82__, 5288 (1985).
6. M.L. Alexander, M.A. Johnson, N.E. Levinger, and W.C. Lineberger, Phys. Rev Lett. __57__, 976 (1986).
7. W.C. Lineberger, M.L. Alexander, N.E. Levinger, and M.A. Johnson, presented at New York ACS Meeting--Symposium on Structure, Spectroscopy, and Dynamics of Atomic, Molecular and Ionic Clusters, abstract 16 (Division of Physical Chemistry), April 14-18, 1986.
8. W.C. Lineberger, private communication.
9. W.R. Wadt and P.J. Hay, J. Chem. Phys. __82__, 284 (1985).
10. J.G. Gay and B.J. Berne, Phys. Rev. Lett. __49__, 194 (1982).

STEEPEST-DESCENT QUENCHES AND THE MELTING OF MICROCLUSTERS

Thomas L. Beck and R. Stephen Berry

Department of Chemistry and The James Franck Institute
The University of Chicago
Chicago, Illinois 60637

ABSTRACT

Extensive molecular dynamics simulations have been performed on clusters of argon atoms to investigate the microscopic mechanism for the melting transition. The steepest-descent quench technique has been employed along the trajectories at energies in the transition region and in the liquid portion of the caloric curves. The potential minima most frequently accessed at the various energies are thus obtained. The coexistence and magic number phenomena are explained in terms of the geometrical structures accessed and the mechanisms for the motions between structures. Pentagonal structures dominate in the melting of these small clusters.

INTRODUCTION

Computer simulations carried out in the middle 1970's and subsequent simulations have revealed that clusters of as small as three argon atoms undergo a rather sudden transition from a rigid form to a nonrigid form as the energy or temperature of the clusters is increased.[1-7] The simulations were performed with both the constant energy molecular dynamics (MD) and the constant temperature Monte Carlo (MC) methods. In the MD method, static and dynamical quantities may be calculated along a trajectory at a given energy; the melting transition is indicated by the onset of a linearly increasing mean square displacement and a nonzero intensity of the power spectrum at $\omega=0$. Both of these quantities are indicative of the persistent translational motion occuring in the liquidlike, diffusive clusters. On the other hand, in the MC simulations phase space averaged quantities are calculated by a random sampling in the canonical ensemble at a constant temperature. The relative rms bond length fluctuations were calculated by Etters and Kaelberer as a function of increasing temperature and displayed a sharp fourfold increase over a small temperature range, implying the onset of frequent rearrangements in the clusters.[2-4] These early studies focused on the solid and liquid limit forms and did not examine the range of energies or temperatures over which the melting transition takes place in a detailed fashion.

In related research, Hoare and Pal studied the potential minima of various growth sequences of geometrical isomers in the rare gases.[8] Different growth sequences were started with pentagonal, tetrahedral, and the fcc seed structures, and the potential energies of the various isomers were calculated. Over the size range N=7 to N=22, the structures with pentagonal symmetry resulted in the lowest potential energy structures. The N=7 pentagonal bipyramid, the N=13 icosahedron, and the N=19 double icosahedron are the three "closed shell" structures of pentagonal symmetry in the growth sequence. Intermediate sized lowest energy structures correspond to decorations of the faces of these three structures. The Ar_{13} icosahedron is a structure of particularly high stability in the growth sequence, and has now been implicated in a large range of phenomena where local ordering is possible but long range translational ordering is not observed.[9]

MOTIVATION

Recently Jellinek, Beck, and Berry[10] and Beck, Jellinek, and Berry[11] carried out extensive MD simulations of the melting transition of clusters in the size range N=7 to N=33. This work is also briefly reported on in another chapter of this volume.[12] In summary, several clusters in the size range N=7-33 (7,9,11,13,15,19, and perhaps others) display a coexistence of distinct forms over a well-defined energy range. This coexistence range is evidenced by the onset and disappearance of a bimodal distribution in the short time averaged kinetic energies (T_s, which is the average kinetic energy over a 500 step interval). Below an energy E_f, only the solidlike form is observed. Above a different, higher energy E_m, only the liquidlike form is observed. In the intermediate energy range, the bimodal distribution is observed with the relative size of the low temperature peak increasing with increasing energy. Meaningful averages can be calculated over each of the two peaks in the bimodal distribution because a given form persists for many vibrational periods before jumping over to another form. The high temperature form in the bimodal distribution exhibits the properties of the solid limit form near melting while the low temperature form displays properties more like the liquid limit form. It was therefore concluded that the caloric curve has a two-valued form over the coexistence energy range.

A quantity ΔT was defined as the difference between the highest mean kinetic energy at which the solidlike form is observed, T_m, and the lowest mean kinetic energy at which the liquidlike form is observed, T_f. These two temperatures correspond to the locations of the peak maxima of the solid and liquid forms, respectively, in the bimodal distribution at the limit energies E_m and E_f. The N=7, N=13, and N=19 clusters display particularly large ΔT values relative to clusters of similar size, and the Ar_{13} and Ar_{19} clusters melt at mean temperatures (long time averaged kinetic energy, T_l) considerably higher than the neighboring cluster sizes. These clusters can thus be called magic numbers for the melting transition in that the original solidlike structures are particularly stable to the onset of rearrangements as the energy of the clusters is increased.

When ΔT is plotted versus N, however, several peculiar features are observed. First, while the N=7,13, and 19 clusters give large coexistence ranges, the N=8,14, and 20 clusters display no bimodal distribution for the T_s values as the clusters melt. The N=9 and N=15 clusters do result in a bimodal distribution, although the coexistence range is narrower than for the N=7 and N=13 cases. Second, no clusters above N=19 (the N=20,22,26, and 33 clusters were studied) resulted in a bimodal distribution. Third, for the clusters that do display the above described coexistence, the ΔT value is large for N=7 and decreases until the N=13 cluster is reached. At N=13 ΔT increases; then it decreases again until another increase at N=19. These data further indicate the high stability properties expected for the magic number clusters. Clearly, however, the coexistence phenomenon is not a smooth function of N, but depends in some way on the particular structure and dynamics of a given cluster size.

The current study is an initial report on an attempt to explain the observed melting behavior in simulations of small clusters by examining the underlying dynamics of the phenomenon. The technique of steepest-descent quenches has been coupled to the MD method to follow the motion of the clusters over the mutidimensional potential surface as the clusters begin to melt. Close connections will be drawn to the potential minimum searches of Hoare and Pal,[8] as the onset of melting is reflected by the ability of the clusters to begin to access other isomers (either permutational or geometrical), many of which were predicted in ref. 8. Several isomers with geometries and binding energies different from those predicted by Hoare and Pal have also been found in our simulations.

Amar and Berry have recently applied the steepest-descent quench technique to the study of the melting transition in Ar_7.[13] The four isomers reported by Hoare and Pal[8] for N=7 were observed in the Ar_7 cluster as it began to melt. The bimodal distribution corresponds to a coexistence of the solidlike pentagonal bipyramid with the other three less stable, more liquidlike isomers. Amar and Berry argued that, based on a simple model for the various isomers, the trajectories resulted in a nearly ergodic sampling at high (nonrigid limit) energies, but in the transition region, the various isomers were not sampled in a completely ergodic fashion. Barrier effects play a key role as the cluster begins to melt.

METHOD

The details of the basic constant energy MD technique employed in our simulations are presented elsewhere.[10,11] The initial conditions for the quench studies in the melting region were obtained from the careful heating curves in our simulations of the melting transition, and trajectories of approximately 50,000 steps ($\tau = 10^{-14}$sec) in length were run. This length of time corresponds to many hundreds of vibrational periods in the clusters, and allows the observation of many transitions between the forms most frequently accessed in the transition region. It is not long enough, however, to obtain converged average rate constants for the frequency of transitions between the forms.

In the steepest-descent quench technique, the equation of motion is:[14]

$$\dot{\mathbf{r}} = -\nabla\Phi$$

This equation corresponds to the motion of a completely damped oscillator. The configuration of the cluster at a given point along the trajectory is quenched to the nearest minimum by solving the above differential equation for the given set of coordinates. In our simulations, we quenched the clusters at 2000 step intervals along the trajectory, which is frequent enough to observe the transitions between forms and yet avoid a substantial number of multiple rearrangements. The simulations were performed at three energies for each cluster size; two runs were performed in the portion of the caloric curve where the fluctuations in the bond lengths quadruple in magnitude and one simulation was performed in the liquidlike limit phase. Of course, in the limit solid form, all quenches result in the equilibrium low energy structure. For the clusters that display the bimodal distributions in the kinetic energies, the two lower energy runs were performed at energies where the bimodal distribution occurs, while the third higher energy run was performed where the broad unimodal distribution of the liquid form is observed.

Three important pieces of information concerning the nature of the motion over the potential surface can be obtained from the quench studies. First, one obtains the exact potential energies of the nearest potential minima over which the system passes along a trajectory. Second, one can follow the mechanism of the transition from one form to another by constructing a time sequence of the cluster unquenched and quenched structures. Thereby, one obtains the reaction coordinate for the transition. Third, if very long runs are performed, the average rate constants for transition between the various forms can be calculated, thus obtaining estimates for the barrier heights between the various forms. In our current study, we have focused on two of these problems; we perform runs long enough to observe the important structures accessed as the clusters melt and then determine the mechanism for the transitions. As we mentioned above, 50,000 step trajectories were run. Runs of at least $1 \cdot 10^6$ steps will be necessary to obtain reliable rate constants, which are long enough to be approaching the limit of computational restraints. The problem of obtaining the barrier heights for the transitions will be addressed in a future study.

RESULTS

We will first present the results for the Ar_{13} quench studies, as this cluster provides a particularly clear example of several important aspects in the melting process. At energies below the onset of the bimodal distribution in the T_s values, all quenches lead to the low energy icosahedral structure. The potential energies of these structures correspond exactly to the values obtained by Hoare and Pal for the icosahedron.

In the portion of the caloric curve where the bimodal distribution is observed, the quenches resulted in two types of structures. The quenches which occurred when the T_s value fell in the high temperature portion of the bimodal distribution led to the icosahedral structure (Fig. 1c. Also, recall that since the total energy is conserved, the low potential energy structure will result in a higher kinetic energy). As long as the cluster remains in the high temperature form, the quenches lead to the same relative arrangement of atoms; no bonds are broken nor new bonds created.

Alternatively, quenches occurring while the cluster is in the low temperature portion of the bimodal distribution lead to one type of structure: the icosahedron with one atom removed and placed on one of the remaining triangular faces of the icosahedron (particle-hole pair structure, fig. 1d). The binding energy of this structure is less than that for the icosahedron, but greater than that for any of the alternate structures in the growth sequences of Hoare and Pal. This structure, then, is the closest potential minimum that the system can "find" as rearrangements begin to occur. Most often the relative arrangement of atoms in this particle-hole structure is also maintained prior to transition back to the icosahedron. However, periodically the cluster will pass from one low temperature structure to another before jumping back to the icosahedral form. This behavior is reflected in the power spectrum calculated for the two forms; the spectrum of the low temperature form shows a small diffusive motion ($I(\omega=0)>0$) and a higher density of low frequency modes than the spectrum of the high temperature form. We also mention here that the structures prior to quenching have the same general shape as the quenched structures, although they are slightly distorted due to the kinetic energy of the unquenched structure. Quenching does not drastically alter the shape of the cluster.

It would be tempting to predict that the reaction coordinate for the transition from the icosahedron to the particle-hole structure occurs by the removal of one atom from one of the vertices of the icosahedron and placement of this atom on the surface of the remaining atoms. In fact, the opposite occurs. One of the twelve "surface" atoms of the icosahedron is "squeezed out" in a highly collective motion, leaving behind a hole on the opposite side of the cluster. The barrier to rearrangement is high due to the high cost in energy as the hole is opened.

At a higher energy, in the liquid portion of the caloric curve, quenches result in several irregular isomers of lesser binding energy than the two isomers mentioned above, along with the icosahedron and the particle-hole structure(Fig. 1e). The cluster now has enough kinetic energy to carry it over the barriers to rearrangement between many different isomers on a timescale not much longer than the vibrational period.

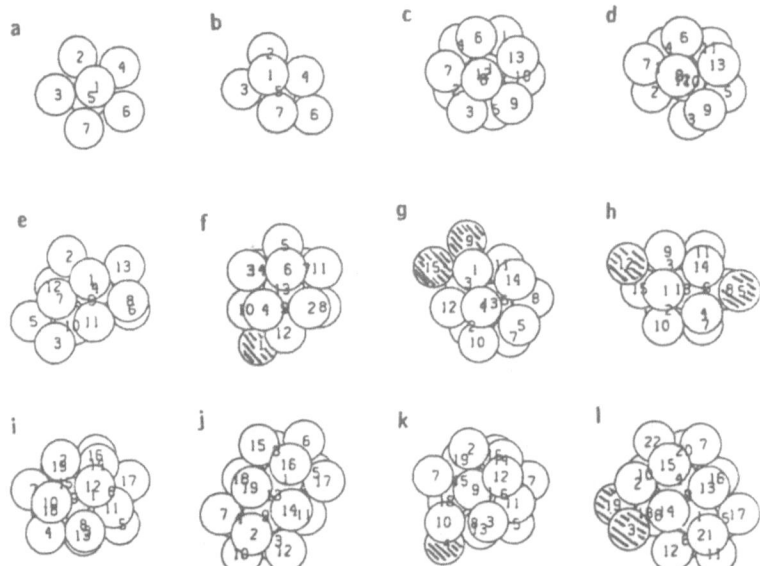

Figure 1. Representative quenched structures: a)N=7 pentagonal bipyramid in the coexistence region. b)N=7 coexistence structure based on the N=6 octahedron. c)N=13 icosahedron in the coexistence region. d)N=13 particle-hole pair structure in the coexistence region. e)N=13 irregular structure observed in the liquid phase. f)N=14 structure based on the icosahedron. g)N=15 coexistence structure based on the icosahedron. Here the two "surface" atoms are located on adjacent faces. h)N=15 coexistence structure with the two atoms on separated faces of the icosahedron. i)N=19 coexistence double icosahedron. j)Structure in i rotated by 90 degrees. k)N=19 coexistence structure with a hole in the double icosahedron. Atom 4 is on the surface of the structure. l)N=22 structure in the transition region. Atoms 3, 18, and 19 decorate the surface of the N=19 double icosahedron.

The higher frequency of passage in the liquid form raises an important issue in the melting transition, i.e. the separation of timescales for the various motions.[15] As the Ar_{13} cluster begins to melt, it spends long periods of time relative to the vibrational period in either the icosahedral or the particle-hole structure. The frequency of transition between the forms is much less than the vibrational frequency. This fact, then, allows one to calculate meaningful averages for each of the two forms separately. When this averaging is performed, the icosahedral structure displays the properties of the limit rigid form while the particle-hole structure behaves in a fashion more reminiscent of the liquid limit form. It was thus concluded in ref. 10 that the equation of state has a two-valued form in the region of the caloric curve where these two forms are observed.

At the higher energy in the liquid portion of the caloric curve, however, the rearrangements occur on a timescale on the order of one or several vibrational periods. The cluster now does not become trapped in one potential well for long periods of time but is constantly passing over the barriers to rearrangement. Large scale diffusive motion becomes apparent in the power spectrum as the frequency of motions between minima on the potential surface approaches a large fraction of the vibrational frequency (say 10%). The cluster nevertheless holds together for thousands of vibrational periods prior to any evaporation.

The Ar_{14} cluster has a melting temperature(22K) much lower than Ar_{13}(34K), even though the N=14 structure has a higher binding energy per atom. No bimodal distribution of the T_s values occurs as it melts. This behavior can be explained in terms of the mechanism for the rearrangements. When quenches are performed in the transition region (the portion of the caloric curve where the rms bond length fluctuations quadruple), only one structure is observed: the icosahedron with one atom on the surface(Fig. 1f). Two types of motion occur in the clusters; the surface atom can move about on the surface of the icosahedron, or it often inserts into the icosahedral core resulting in the removal of one of the twelve atoms on the vertices of the icosahedron. The key point here is that both forms of motion described above have low barriers separating the potential minima. The insertion of the surface atom into a hole created in the icosahedral core has a much lower barrier than the analogous process in Ar_{13} since the hole is being filled as another atom is squeezed out of the icosahedron. No bimodal distribution is observed since all the isomers have the same potential energy. The motion of one atom over the "surface" of the icosahedron is not truly surface melting since this atom frequently inserts into the icosahedral core.

The Ar_{15} cluster has a lower melting temperature (21K) than the Ar_{13} cluster, but does display a bimodal distribution in the T_s values. Again, this behavior is easily explained in terms of the mechanism of the reactions. In this case two types of quenched structures are observed in the transition region: the icosahedron with two atoms placed on adjacent faces and the icosahedron with two atoms placed on separated faces(Fig. 1g,h). The bimodal distribution results from a coexistence of these two types of structures with differing potential energies. Again, both motion over the surface of the icosahedron and insertion into the core are observed; no clear surface melting is apparent. Quenches in the liquid phase for the Ar_{14} and Ar_{15} clusters result in many different irregular structures, as the energy is now large enough for large amplitude motion over large portions of the potential surface.

Similar behavior is observed for the other magic number clusters, Ar_7 and Ar_{19}. Ar_7 displays a bimodal distribution in the T_s values and a large ΔT value, while the N=8 cluster results in no bimodal distribution. Quenches in the coexistence region for N=7 result in the pentagonal bipyramid for the high temperature form, and the other three structures predicted by Hoare and Pal for the low temperature form(Fig. 1a,b). This behavior is discussed in ref. 13. The N=8 cluster, however, quenches to one of two structures: the pentagonal bipyramid plus one atom and the fcc octahedron plus two atoms on adjacent faces. These structures have nearly equal binding energies and thus no bimodal distribution is observed. N=9 does exhibit a bimodal distribution and quenches most frequently reveal the pentagonal bipyramidal seed structure plus two atoms, either on adjacent faces or separated.

The Ar_{19} cluster displays a bimodal distribution. Quenches reveal that the high temperature form is the low potential energy double icosahedron while the low temperature form corresponds to several different isomers which are usually the double icosahedron with one atom removed and placed on the surface of the structure(Fig. 1i,j,k). No clusters studied above N=19 display a bimodal distribution, presumably due to the large number of isomers available with similar potential energies. However, one can often observe the N=19 double icosahedron and N=13 icosahedron as underlying structures in the larger clusters as they begin to melt. For example, the N=22 cluster maintains the double icosahedron plus three atom structure as it begins to melt(Fig. 11). Again, highly collective motions are observed, and no clear surface diffusion is apparent.

DISCUSSION

The melting of small clusters in the size range N=7 to N=33 is dominated, then, by underlying structures of pentagonal symmetry. The three closed shell structures, Ar_7, Ar_{13}, Ar_{19} are magic numbers for the melting transition in that all three display particularly large coexistence temperature ranges, and the Ar_{13} and Ar_{19} clusters melt at temperatures much higher than clusters of similar size. These effects arise from the high stability of the equilibrium low energy structures due to high barriers to rearrangement to the nearest available isomers. When single atoms are added to the N=13 and N=19 structures, the coexistenece behavior is lost, due to the alteration of the potential surface.

In all of the clusters of intermediate size between the magic numbers, the pentagonal seed structure is observed as the dominant underlying structure. Exchange can occur between atoms on the surface and atoms in the pentagonal core, but the potential surface which the cluster accesses during the melting transition generally includes one of the structures with pentagonal symmetry. No clear motion is observed on the surface of the icosahedral core, although frequently atoms will migrate on the surface prior to insertion.

These findings relate closely to the potential minimum searches of Hoare and Pal in that the structures accessed during melting correspond closely to the structures predicted in their work. However, new structures have been found which did not fit into the pentagonal or tetrahedral growth sequences. The coexistence behavior predicted by analytical models[16] and simulations results from a coexistence of these isomers. The solidlike form is the more stable, low potential energy structure, while the liquidlike form consists of one or several of the less stable, asymmetric structures. These less stable forms display properties which are like the liquid limit form, while the more stable, high temperature form behaves more like a hot solid structure.

The dominance of the structures of pentagonal symmetry in the melting of small clusters is further evidence of the importance of these structures in cases where local ordering is important. Further elucidation of the important structures accessed in the melting of small microclusters will be presented in a forthcoming study.[17]

ACKNOWLEDGMENTS

This research was supported by a grant from the National Science Foundation.

REFERENCES

[1] C.L. Briant and J.J. Burton, J. Chem. Phys. **63**, 2045 (1975).

[2] R.D. Etters and J.B. Kaelberer, Phys. Rev. A **11**, 1068 (1975).

[3] J.B. Kaelberer and R.D. Etters, J. Chem. Phys. **66**, 3233 (1977).

[4] R.D. Etters and J.B. Kaelberer, J. Chem. Phys. **66**, 5112 (1977).

[5] D.J. McGinty, J. Chem. Phys. **58**, 4733 (1973).

[6] W.D. Kristensen, E.J. Jensen, and R.M.J. Cotterill, J. Chem. Phys. **60**, 4161 (1974).

[7] N. Quirke and P. Sheng, Chem. Phys. Lett. **110**, 63 (1984).

[8] M.R. Hoare and P. Pal, Adv. Phys. **20**, 161 (1971).

[9] see, for example P.J. Steinhardt, D.R. Nelson, and M. Ronchetti, Phys. Rev. Lett. **18**, 1297 (1981).

[10] J. Jellinek, T.L. Beck, and R.S. Berry, J. Chem. Phys. **84**, 2783 (1986).

[11] T.L. Beck, J. Jellinek, and R.S. Berry, (in preparation).

[12] R.S. Berry, T.L. Beck, H.L. Davis, and J. Jellinek, in *The Physics and Chemistry of Small Clusters*, (Plenum, New York).

[13] F. Amar and R.S. Berry, 1986, J. Chem. Phys. (in press).

[14] F.H. Stillinger and T.A. Weber, Phys. Rev. A **25**, 978 (1982).

[15] G. Natanson, F. Amar, and R.S. Berry, J. Chem. Phys. **78**, 399 (1983).

[16] R.S. Berry, J. Jellinek, and G. Natanson, Phys. Rev. A. **30**, 919 (1984).

[17] T.L. Beck and R.S. Berry, (in preparation).

PHOTOFRAGMENTATION AND STABILITY IN

SEMICONDUCTOR MICROCLUSTER IONS

Louis A. Bloomfield

Department of Physics
University of Virginia
Charlottesville, VA 22901

Abstract

Details of current photofragmentation experiments and theory in the semiconductor materials (principally C, Si, Ge, and GaAs) are discussed. The observed dissociation behaviors of semiconductor cluster ions, particularly of carbon and silicon, are substantially in agreement with the current theories for their structures and binding energies. The principal experimental "magic" cluster sizes (C_3, Si_6^+, Si_{10}^+, and Si_{10}^-) have also been found to be relatively stable in a variety of theoretical studies.

1. Introduction

In the past three years, semiconductor clusters have become the focus of intense experimental investigations in several laboratories.[1-22] Unlike the more extensively studied metal and van der Waals clusters, the semiconductors form tightly bound clusters with properties which reflect the nature of their directional covalent bonds. Among these properties are the ease with which semiconductor atoms coalesce into clusters, their appearance in "magic" cluster sizes which depend on charge state (either neutral clusters or positive or negative ions), and, as will be the focus of this paper, their complex dissociation behavior when induced to fragment into smaller units.

The basic interest in semiconductor clusters can be traced to their role as an intermediate between atom and solid. Somewhere in the spectrum of clusters should be the origins of band structure and the transition from molecular behavior to that of a bulk solid. Nowhere should this transition be more evident than in the semiconductors, where band gaps, surface states, and fundamental excitations are matters of extreme interest in hundreds of solid state laboratories throughout the world. A clear understanding of the properties of semiconductors clusters will add considerably to the growing knowledge of the properties of semiconductors.

Semiconductor clusters are not a recent discovery. The production of silicon, germanium, and carbon clusters with thermal sources was first reported in the 1950s.[23] However, the advent of laser vaporization cluster sources[24] has permitted the general production of semiconductor clusters and cluster ions which are larger and have much lower internal temperatures than can be created with conventional thermal cells. Furthermore, it is now possible to generate clusters of binary semiconductors such as gallium arsenide (GaAs)[8,18,20] and indium phosphide (InP),[18] although the stoichiometric balance between the two elements is difficult to control.

Unfortunately, semiconductor clusters, like all clusters, are extremely difficult to characterize directly. They are produced in such small quantities and in such an inhomogeneous manner as to force virtually all measurements of their structure to be indirect and subject to interpretation.

They are unsuitable for most molecular diagnostic techniques which require molar quantities of uniform samples and are similarly unsuitable for most solid state techniques which require a supported, instrumentable sample. With a few notable exceptions such as reaction studies, ESR, and ultra-high resolution electron microscopy, only a handful of particle counting techniques have yielded data on clusters and many of the most informative of these have failed for clusters of more than 2 atoms.

One of the few experimental techniques which has yielded useful information about the larger semiconductor clusters is photofragmentation.[25] Photofragmentation is a particle counting technique which has been used with reasonable success to study the binding of simple molecules. In this scheme, well characterized sample particles are exposed to an intense, monochromatic beam of light (typically a laser) which prompts them to photodissociate. The photofragments are then characterized as precisely as possible and details of the breakup process are deduced. By varying the light wavelength and intensity while studying the fragment channels, binding properties and cross sections can be extracted. More sophisticated experiments have been done with small molecular clusters,[26-42] in which the kinetic energies of the fragment particles are measured and the dissociation mechanisms and unimolecular decomposition times can be established.

This paper will focus on the details of current photofragmentation experiments and theory in the semiconductor materials (principally C, Si, Ge, and GaAs). Such measurements have generally been performed on cold cluster ions which have been mass-selected into a uniform sample and are then photodissociated by a visible or ultraviolet laser beam. The photofragmentation cross sections and daughter branching ratios provide insight into the nature of the bonds between atoms. The experiments have so far served to supplement "magic" number data as an indication of those clusters that are most stable, as well as to give some bounds to the actual binding energies involved and barrier potentials which must be negotiated in the unimolecular decomposition process. Unfortunately, such binding energy measurements are subject to ambiguities due to multiphoton absorption events. Nonetheless, there is now considerable data accumulated to serve as a guide to theoretical calculations of cluster structure. Such calculations[43-60] have been undertaken for silicon, germanium, and carbon clusters to yield results which agree reasonably well with experiment. As a result, the physical geometries of several semiconductor clusters are thought to be known.

2. Experimental Techniques

2.1 Photodissociation

Photodissociation of molecules is a fundamental part of photochemistry, but has become particularly important in recent years in at least two principal forms: infrared multiphoton dissociation and visible/ultraviolet single photon dissociation.

Infrared multiphoton dissociation was discovered in 1971 by Isenor and Richardson,[61] who found that a focused CO_2 laser pulse could produce visible light in the midst of a gas cell filled with a resonantly absorbing medium such as NH_3. Part of this light could be attributed to a nearly instantaneous, unimolecular decomposition of the exposed gas followed by luminescence from the excited fragments. The energy needed to produce the decay came from the multiphoton absorption of infrared photons, through a vibrational-rotational ladder of states.[62]

The process of infrared multiphoton excitation and dissociation resembles a thermal one, in which energy is slowly poured into a molecule until it reaches the dissociation threshold. At low laser intensities, the molecule will tend to dissociate via the lowest energy decay mode. As the peak intensity of the radiation increases, the molecule can become more highly excited before it decays and new decay channels may "open up." Thus, by shortening the laser pulse without increasing its fluence, a different fragment spectrum can be produced.

The goal in visible/ultraviolet single photon dissociation is quite different. Only a single photon of a specific wavelength is absorbed and the fragment distribution and energies are recorded. Dissociation thresholds can be established by varying the light wavelength until a transition from single to multiple photon absorption is observed. Anisotropies in the decay directions can indicate the decay time, and final state kinetic energies can indicate the type of dissociation mechanism involved.[26] Ideally, the experiments would involve a conventional light source, with virtually no possibility of multiphoton absorptions, however all cluster studies have required lasers in order to obtain sufficient monochromatic light intensities.

To insure that only a single laser photon is involved, one must observe a dissociation probability which is truly linear in laser fluence. Unfortunately, the two regimes, infrared and visible, are not as different as one might hope. Infrared multiphoton excitation works because every photon is absorbed resonantly into a virtual continuum of excited states. The concept of cross section gives way to Poisson statistics and fluence dependence measurements yield no useful information. Huge numbers of resonant states are also present for visible photons, so that in laser based studies, it is often difficult to distinguish single photon events from multiphoton ones involving several very different cross sections. The smallest cross section is the rate limiting step and produces a deceptively linear fluence dependent transition probability.

To reduce the confusion of multiphoton absorption, one must work at very low radiation intensities and small transition probabilities. For sufficiently low levels, none of the cross sections will saturate and each will contribute to the fluence dependence of the transition probability. Multiphoton events will then appear non-linear as a function of laser fluence, although noise and counting statistics may become more significant sources of error.[16]

2.2 Semiconductor Cluster Ion Production

Because of the need to mass analyze both the parent cluster and its daughter fragments, virtually all cluster photodissociation experiments have been performed on cluster ions. A number of successful photodissociation experiments have been performed on gas[26-42] and metal cluster ions.[63-69] The most detailed of these studies have been performed on gas ions produced in supersonic expansions, and have yielded very useful information on the decay mechanisms, branching ratios, and decay times of the parent ions. Metal cluster ions, produced in ion sputtering and laser vaporization sources, have been studied using somewhat simpler experimental procedures, not sensitive to final state energies or spatial distributions. These experiments have also produced very interesting results on the relative stabilities and branching behaviors of the ions involved.

The semiconductor cluster ion photodissociation experiments discussed in this paper have been performed using laser vaporization cluster sources (LVCS). LVCS, initially used by Smalley, et al.[24] to produce transition metal clusters, will produce clusters of virtually any material. In this technique, the surface of a sample is vaporized by an intense, pulsed laser beam. Simultaneously, an inert carrier gas is directed over the sample to entrain the hot vapor, cool it, and permit the growth of clusters. The mixture travels through a short channel and then expands into a vacuum. The resultant supersonic expansion further cools the clusters, so that rotational temperatures of 10-100° K and vibrational temperatures of 20-200° K are quite possible.

The cluster ions used for the experiments have been obtained either by photoionizing the neutral clusters produced in an LVCS or directly in an LVCS, in a manner recently demonstrated by Bloomfield, et al.[5] The latter technique has the advantage that the cluster ions are cooled by the carrier gas and supersonic expansion.

2.3 Photofragmentation Techniques

Once the semiconductor cluster ions have been produced, they must be mass separated, photofragmented, and then reanalyzed for the appearance of daughter fragment ions. Schematic diagrams of the instruments used by the groups at AT&T Bell Laboratories (Freeman, et al.)[4,12,16] and Rice University (Smalley, et al.)[20] are shown in Fig. 1a and 1b, respectively. The apparatuses are quite similar, apart from small differences in the ion optics used.

The cluster ion beam is permitted to drift, field free, into the center of a pulsed acceleration/extraction plate system. A high voltage is then applied to the plates and the cluster ions within are accelerated to a uniform high kinetic energy (~2kV). The different cluster sizes reach different final velocities and begin to disperse in space as they travel down the bore of the time-of-flight spectrometer. After travelling approximately 1 m, the mass separation is sufficient

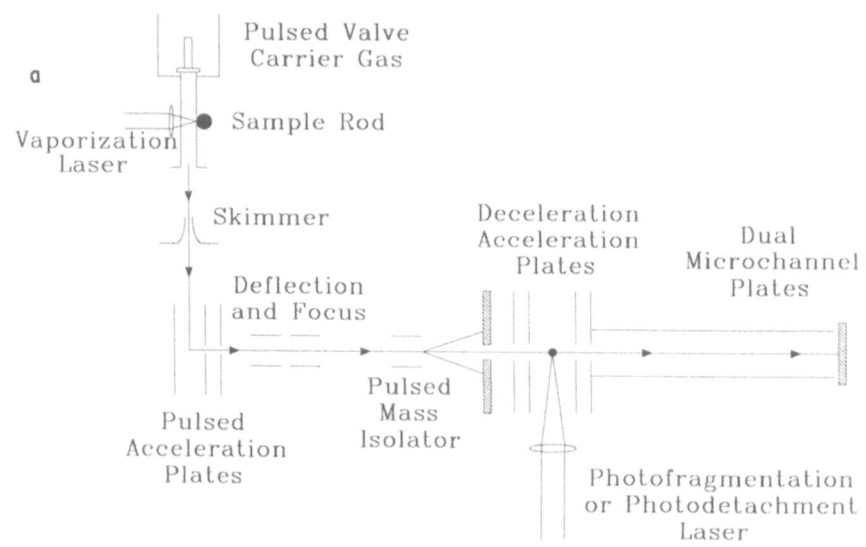

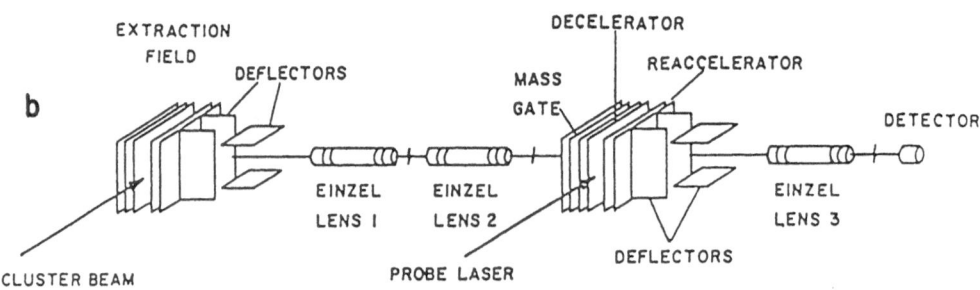

Fig. 1. Experimental Apparatuses used by: a) Freeman, et al.[4,12,16] and b) Smalley, et al.[20]

to isolate single cluster sizes, either by deflecting all undesired clusters out of the beam or by repelling them with a pulsed grid. The remaining, purified cluster size is then decelerated to roughly 1 kV and exposed to a photodissociating laser pulse. The resultant parent and daughter ions are reaccelerated to high potential (the neutral fragments are not detected) and are once more dispersed according to mass. In the case of negative ions, photodissociation is accompanied by photodetachment,[70,71] so that the detection of a bare electron is also important and will be discussed later.

2.4 Data Analysis

The observed signals consist of the parent cluster ions, preceded by any fragment ions created during exposure to the laser, in order from lightest to heaviest. An example of such a signal for C_{18}^+ is shown in Fig. 2.[16] The C_{18}^+ peak appears inverted due to a background subtraction technique and represents the depletion of the original C_{18}^+ beam as it is converted to fragment ions. Among the prominent fragment ions are C_{10}^+, C_{11}^+, C_{12}^+, and C_{15}^+.

The populations of the fragment ions are observed to vary dramatically with the fluence (time integrated intensity) of the photodissociating laser.[4,16] This effect is illustrated in Fig. 3, where the populations of the C_{15}^+ and C_{12}^+ fragments are plotted as functions of laser fluence.[16] In the case of C_{15}^+, the fragment population depends linearly on fluence for fluences of less than 4 mJ/cm². Saturation occurs above this level and is followed by a population decrease at fluences above 10 mJ/cm². In contrast, the population of C_{12}^+ appears quadratic with laser fluence. It must be pointed out that throughout this example, the uniform, 10 ns pulse from a 248nm excimer laser was used as the dissociation source.[16] As with infrared multiphoton

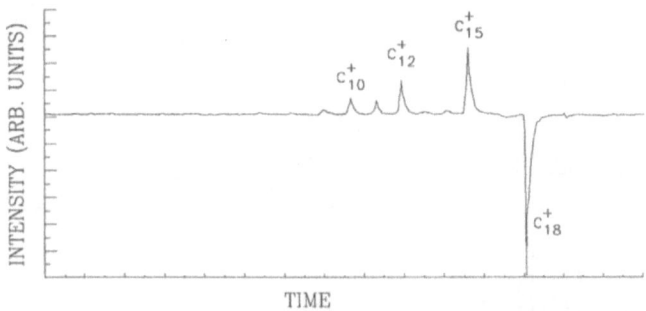

Fig. 2. Time-of-flight mass spectrum of the photofragments from C_{18}^+ photodissociation measured with 248nm light and a photodissociating laser fluence of approximately 12 mJ/cm^2.[16]

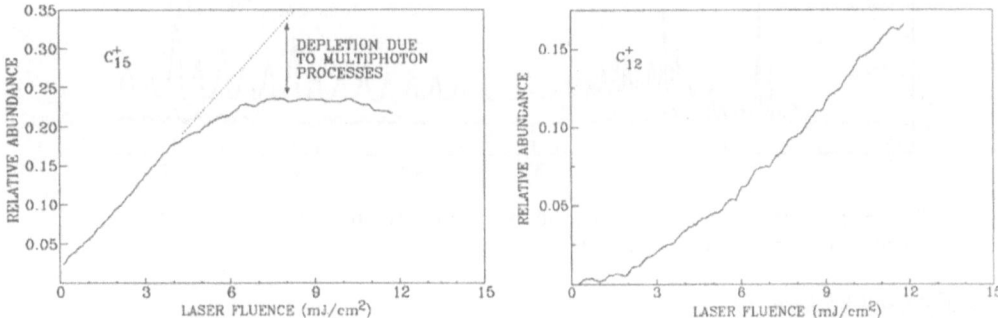

Fig. 3. Laser fluence dependence of the intensity of the C_{15}^+ and C_{12}^+ photofragments from C_{18}^+ photodissociation with 248nm light.[16]

dissociation, pulse duration and shape alterations can completely change the contribution of any multiphoton processes, even when the overall fluence is constant.

The standard interpretation of these two plots is that C_{15}^+ is formed in a single photon dissociation from C_{18}^+, while C_{12}^+ is the result of a sequential decay involving C_{15}^+ as an intermediate particle.[4,16] The saturation and ultimate decline in the C_{15}^+ population at high laser fluences represents the further conversion of the first generation C_{15}^+ fragments into second generation C_{12}^+ fragments. A less likely model, involving multiphoton absorption followed by extensive breakup will be discussed later.

It is clear that by studying the fluence dependence of the population of cluster fragments, it is possible to determine which of them has resulted from a single photon absorption. Relatively low laser fluence must be used throughout the measurements in order to avoid the deceiving multiphoton absorptions discussed in Section 2.1. Single photon branching ratios can thus be obtained for each initial cluster ion size and at any given wavelength. Such measurements, along with relative and absolute photodissociation cross sections associated with them, give considerable insight into the binding energies and structures of the cluster ions.

3. Experimental Results

There is a now substantial collection of data on the photodissociation behavior of carbon, silicon, germanium, and gallium-arsenide clusters ions, both positive and negative. There are also a number of interesting and related studies which will be included in the following discussions. Most apparent among the differences between gas/metal and semiconductors clusters is that the latter photodissociate by breaking into two large pieces, whereas the former tend to evaporate an atom or molecule at a time (with rare exception[69]). In all cases, however, the cross sections for photodissociation have been found to be large.[4,12,16,69] As a result, this process must be considered carefully during other optical studies, such as photoionization.

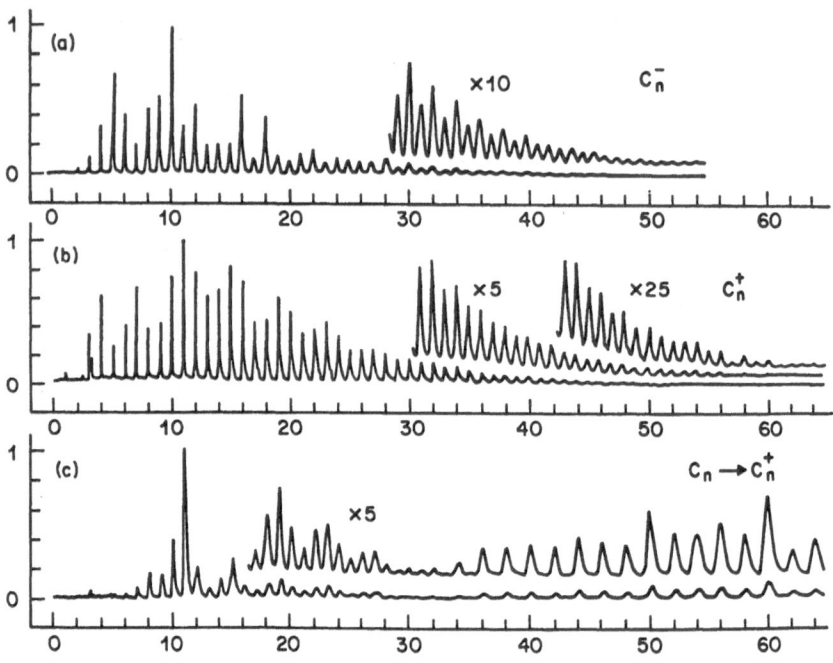

Fig. 4. Cluster spectrum of carbon: (a) C_n^- obtained directly from the source; (b) C_n^+ obtained directly from the source; (c) C_n^+ obtained by ionizing C_n with 193nm radiation.[5]

3.1 Carbon Cluster Ions

Carbon clusters have been the focus of unusual interest during the past year, in part because of hopes for resolving the long standing question of linear vs. cyclic geometry for the small clusters[72,73] and in part because of the possible existence of a spherical C_{60} molecule (Buckminsterfullerene).[7] Carbon clusters have proven to be the easiest semiconductor clusters to produce in an LVCS,[1,2,5] and may well be the most extensively studied semiconductor clusters during the next few years.

Carbon cluster and cluster ion production has been reported by at least five groups,[2,5,7,11,17] including two which have produced carbon cluster ions directly in FT-ICR (Fourier Transform Ion Cyclotron Resonance) cells for reaction and collision studies.[11,17] To date, photodissociation experiments have been performed only on the positive cluster ions and include measurements of branching ratios, photodissociation cross sections, and photodissociation energy thresholds.[12,16]

Spectra of carbon cluster ions produced directly in an LVCS and by photoionizing (at 193nm) the neutral clusters appear in Fig. 4.[5] The positive photoion spectrum closely resembles that first produced by Kaldor, et al.[2] and exhibits strong "magic" numbers and an absence of odd sized clusters with more than 30 atoms. The direct positive ions have less pronounced "magic" numbers and include large odd numbered clusters. In short, photoionization at 193nm appears to increase the contrast between relatively strong and weak peaks, probably as a result of photofragmentation during photoionization.[5] Relatively high photoionization thresholds for large odd sized clusters may well be responsible for their absence following photoionization. The negative clusters also have strong "magic" numbers, which are different from those of the positive ions.[5]

Photodissociation experiments have been performed by Freeman and coworkers on positive carbon cluster ions, using the techniques discussed in Sections 2.3 and 2.4.[12,16] An excimer laser, operating at 248nm and 351nm, served as the photodissociation source. The principal observation is that all clusters ions of more than 6 atoms decay via the loss of a neutral C_3 fragment. C_5^+ and C_4^+ also decay to C_3, but the high electron affinities of C and C_2 force much of the C_3 to appear as a positive ion (C_3^+). For example, C_{18}^+ decays to C_{15}^+ in a single step (see Fig. 2) and a second photon may reduce it to C_{12}^+. This tendency to produce C_3 during photofragmentation was first observed indirectly by Kaldor, et al. during photoionization studies of carbon clusters.[2]

Table 1. Photofragment branching ratios for photodissociation of carbon cluster ions with 248nm and 351nm light at a fluence of approximately 2 mJ/cm². The 351nm deta is in the brackets. Differences between the branching ratios at 248nm and 351nm of less than 5% of the total should not be considered significant.

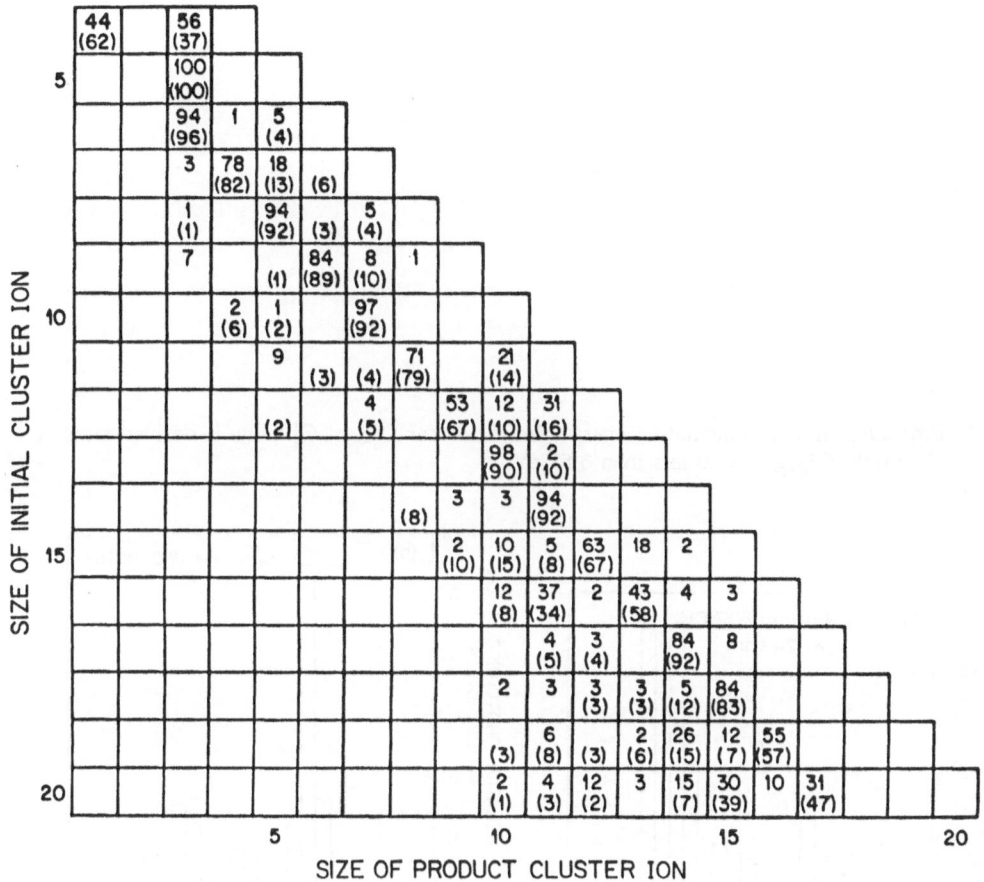

SIZE OF INITIAL CLUSTER ION (vertical axis, values 5, 10, 15, 20)

SIZE OF PRODUCT CLUSTER ION (horizontal axis, values 5, 10, 15, 20)

Branching ratios for the decays of C_4^+ through C_{20}^+ at 248nm and 351nm appear in Table 1.[16] The loss of C_3 appears as a strong diagonal in this table.

The dominance of the C_3 loss channel is believed to be due to the low energies involved in this decay mode. The C_3 species is known to be very stable. It appears that the photodissociation is not a direct transition to a repulsive state (specifically cutting free a C_3 "subunit" of the larger cluster), but is a transition to a bound excited state followed by a statistical unimolecular decomposition.[16] In this respect it resembles multiphoton infrared dissociation with only a single photon. The relative independence of the branching ratios to laser wavelength gives further support to the statistical decay model, although one would expect to see subtle changes in branching ratios as one exceeds thresholds for higher energy decay modes.

These experiments have also pláced limits on the photofragmentation thresholds for many of the carbon cluster ions by determining which clusters photofragment following the absorption of a single photon at either 248nm or 351nm.[16] Results of these studies appear in Fig. 5. In three cases, C_4^+ at 351nm and C_3^+ at both wavelengths, insufficient fragmentation was observed to establish the fluence dependence of the photodissociation (linear or higher-order), although they are believed to be multiphoton.

By measuring the photodissociation which resulted from a known low laser fluence, Freeman, et al. have obtained approximate values for relative photofragmentation cross sections.[16] These values appear in Fig. 6 and show considerable variation from cluster to cluster, particularly between C_9^+ and C_{10}^+, the range at which some theories have predicted a change from linear to cyclic geometry for the clusters.[72]

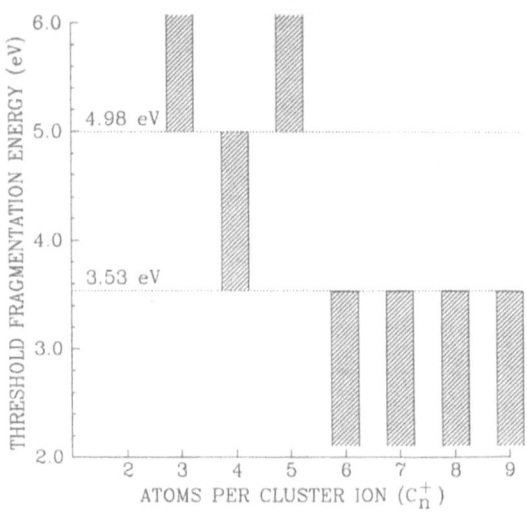

Fig. 5 Bracketed photofragmentation energy thresholds for C_{3-20}^+. C_3^+ value is derived from refs. [75,76].[16] C_{10-20}^+ are all less than 3.53 eV.

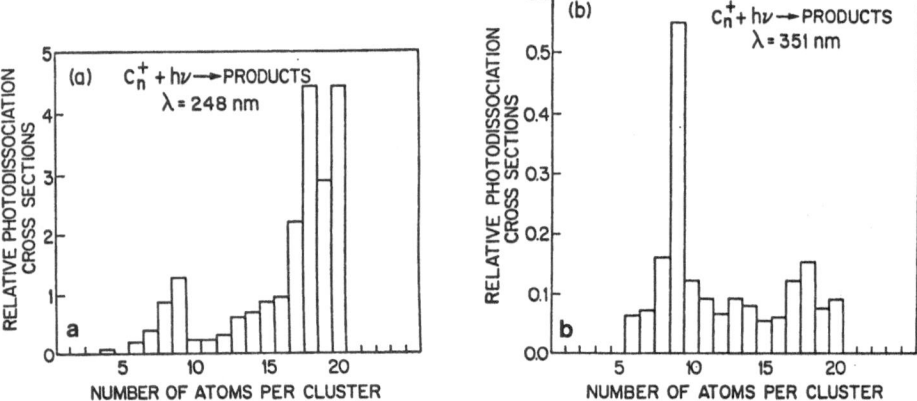

Fig. 6 Photodissociation cross sections for the photofragmentation of carbon cluster ions at a) 248nm and b) 351nm. Approximate absolute values for the cross sections can be derived from the relative values by multiplying the scale by 10^{-17} cm². [16]

Having obtained the cross sections for photodissociation of the cluster ions, it is possible to review the proposition, made in Section 2.4, that photodissociation proceeds sequentially: C_{18}^+ absorbs a photon and decays to C_{15}^+, which in turn absorbs a photon and decays to C_{12}^+. An alternative model is the absorption of two photons followed by decay directly from C_{18}^+ to C_{12}^+. However, by combining measurements of fluence dependence (e.g. Fig. 3) with the photodissociation cross sections, the role of C_{15}^+ as an intermediate seems fairly certain.[16] There are, however, some cases in which the cross section for photodissociation of the intermediate state seems to be several times too large.[16] These anomalies could arise because the daughter ions are produced internally excited with larger densities of states and elevated cross sections for further absorptions.[16]

3.2 Silicon

Several photodissociation measurements have been performed on silicon positive and negative cluster ions,[4,20] supplemented by at least two studies of reactivities using FT-ICR.[11,17,22] Like carbon, silicon cluster ions demonstrate a rich collection of "magic" numbers which reflect the complex molecular structures involved. Spectra of the cluster ions produced directly in an LVCS appear in Fig. 7, along with a spectrum of photoionized neutral clusters (193nm).[5] Again, the

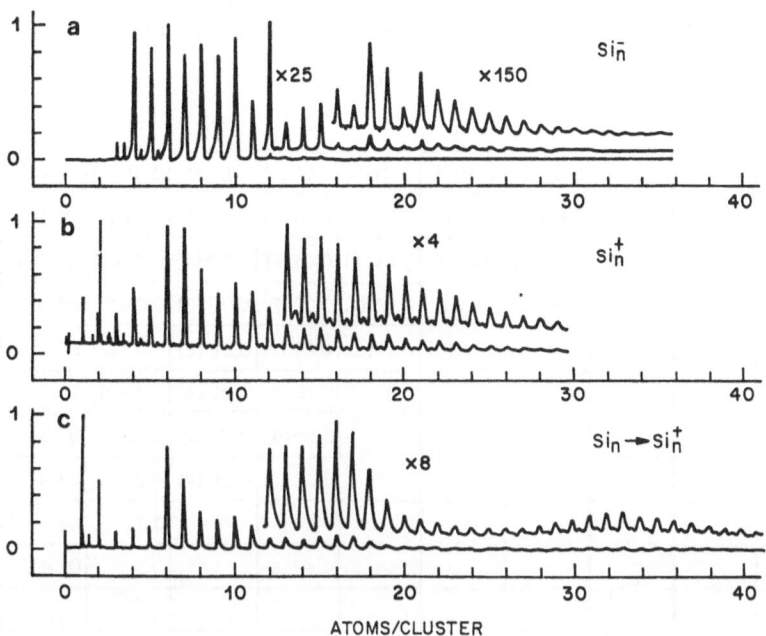

Fig. 7 Cluster spectrum of silicon: (a) Si_n^- obtained directly from the source; (b) Si_n^+ obtained directly from the source; (c) Si_n^+ obtained by ionizing Si_n with 193nm radiation.[5]

photoionization process seems to increase the contrast between strong peaks and weaker ones, probably as the result of photofragmentation.

Photofragment branching ratios have been obtained for positive and negative silicon cluster ions by Freeman, et al.[4] and Smalley, et al.[20] respectively and appear in Tables 2 and 3. The positive cluster ion study was done with cluster ions formed by photoionizing neutrals. As a result, the cluster ions where internally hotter than those derived directly from an LVCS. This higher internal temperature is not likely to affect the measurements because of the observed insensitivity of fragmentation branching ratios to laser wavelength. If cluster internal energy mattered significantly, laser wavelength should also matter.

For both positive and negative cluster ions, photofragmentation tends to produce very specific daughter ions. For positive clusters, Si_6^+ and Si_{10}^+ are favorite daughters or "magic" fragments. For negative clusters, Si_{10}^- is unusually common among all fragments. That Si_{10} is "magic" both as a positive and negative cluster and is also a "magic" number during formation in an LVCS (see Fig. 7) suggesting that it is an unusual geometric structure which is particularly stable against decay.[4,20] Theoretical studies which show this to be the case will be discussed in Section 4.2. Table 3 also indicates several negative cluster size/wavelength pairs at which fragmentation is single photon.[20] Some upper bounds for fragmentation energies are thus established.

Total photofragmentation cross sections at 532nm have been measured for the positive clusters and are given in Fig. 8.[4] These measurements are further evidence for the unusual nature of Si_{10}^+ as well as of Si_4^+ and Si_6^+, all of which are local minima in cross section.

In the case of negative cluster ions, photodissociation competes with the process of photodetachment. For wavelengths which are only slightly above the one photon threshold for photodissociation, photodissociation appears to be the primary decay mode.[20] However well above threshold, electron detachment appears to dominate. Values for the photodetachment thresholds have been reported by Smalley, et al.[20] This photodetachment may also be a useful technique for producing mass selected beams of neutral clusters.

An indirect observation of the fragmentation of silicon clusters has also been reported by Smalley and coworkers.[6] They observed photodissociation following photoionization of the

Table 2. Branching ratios for the fragmentation of Si_n^+ initial states into Si_m^+ final states.[4]

INITIAL CLUSTER

	Si_2^+	Si_3^+	Si_4^+	Si_5^+	Si_6^+	Si_7^+	Si_8^+	Si_9^+	Si_{10}^+	Si_{11}^+	Si_{12}^+
Si^+	1.00	0.25(3)	0.17(3)	0.05(1)	0.00(1)	0.01(1)	0.00(1)	0.00(1)	0.00(1)	0.00(1)	0.00(1)
Si_2^+		0.75(3)	0.18(3)	0.05(1)	0.04(1)	0.02(1)	0.01(1)	0.00(1)	0.00(1)	0.00(1)	0.02(1)
Si_3^+			0.65(5)	0.08(1)	0.05(1)	0.01(1)	0.01(1)	0.02(1)	0.01(1)	0.02(1)	0.00(1)
Si_4^+				0.82(2)	0.21(2)	0.03(1)	0.07(2)	0.06(2)	0.09(2)	0.09(2)	0.05(2)
Si_5^+					0.70(4)	0.11(2)	0.07(2)	0.08(2)	0.03(2)	0.13(3)	0.05(2)
Si_6^+						0.82(2)	0.28(4)	0.39(3)	0.64(4)	0.29(3)	0.05(2)
Si_7^+							0.55(4)	0.19(3)	0.12(3)	0.36(4)	0.08(2)
Si_8^+								0.26(4)	0.07(2)	0.04(2)	0.12(3)
Si_9^+									0.04(2)	0.03(2)	0.09(2)
Si_{10}^+										0.04(2)	0.42(5)
Si_{11}^+											0.13(3)

FINAL CLUSTER

Table 3. Fragmentation channels of Si_x^- and Ge_x^-. The daughters are listed in the order of decreasing intensity. **Bold** face type indicates that photofragmentation to this particular daughter is a one photon process with the bold face wavelength laser.[20]

Parent Size x	Daughter Si_y^+ of Si_x^+ Size y	Laser (nm)	Daughter Ge_y^+ of Ge_x^+ Size y	Laser (nm)
9	None	532,353,414	5	426
10	None	532,460,440,414,353	4,6,5	426,414
11	5,4,6	460,**440**,**414**,353	5,4,6	440,**414**
12	**6**,5	**440**,414,353	6	426
13	6,7	414,353	6,7	**426**,353
14	7,10,8	414,353	7	426
15	**9**,5	440,**414**,353	9,5	426,414
16	10	440,414,353	10	426
17	10	440,353	**10**	**426**,353
18	12,6,7,9,11	440,414,353	12,9,6,11,7,10,5	426
19	9,10,6	414,353	9,6,10	426
20	**10**	500,440,**414**,353	10	426
21	5,9,10,11,14,15	414,353	11,5,7,10	426
22	10,16,6,9,5	353	16,9,10	426

neutral clusters, with a strong tendency to produce clusters in the range of 6 to 11 atoms.

3.3 Germanium

The behavior of germanium clusters is remarkably similar to that of silicon. Fragmentation studies on the positive and negative cluster have been carried out by Freeman, et al.[74] and Smalley, et al.[20] respectively with results which differ only slightly from those in silicon. The same "magic" fragments were observed, Ge_6^+, Ge_{10}^+, and Ge_{10}^-. The branching channels for negative ion photodissociation appear in Table 3.[20] Some differences in channel intensities between silicon and germanium negative ions were reported.

3.4 Gallium Arsenide

The study of clusters of binary, III-V semiconductors such as GaAs and InP is relatively recent.[8,18,20] At present, only the photodissociation of negative GaAs clusters has been studied directly by Smalley, et al.[20] Unlike the elemental semiconductors discussed previously, GaAs

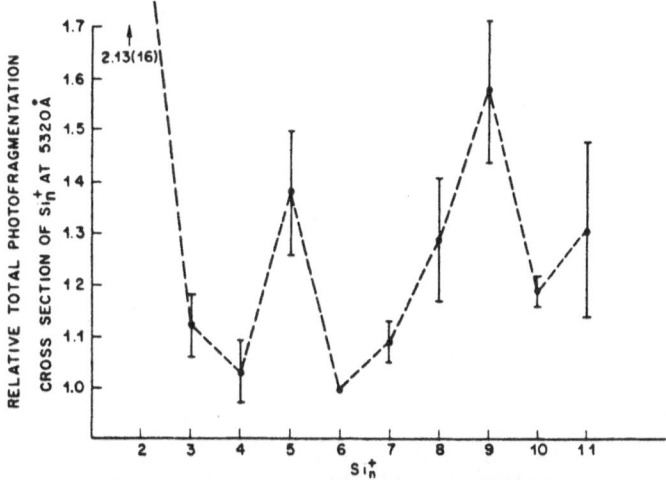

Fig. 8 Relative total photofragmentation cross sections of Si_n^+ at 532nm. Si_6^+ is used as a reference. The absolute total photofragmentation cross section for Si_6^+ is $7(3) \times 10^{-19}$ cm^2.[4]

ions fragment by a non-fission process: one, two, or more atoms may be eliminated. There is, however, a strong tendency to produce fragments with an odd number of atoms. The higher electron affinities of the odd numbered clusters make them "magic," since one would gain less energy by putting an extra electron on an even sized cluster. Unimolecular decomposition tends to produce the lowest energy, odd sized cluster ions.

The positive cluster ions have also been observed to evaporate during photoionization in such a way as to leave them with more gallium than arsenic atoms.[8] As the intensity of the photoionizing laser is increased, all of the gallium rich peaks increase in amplitude while the arsenic rich peaks diminish.

4. Discussion

4.1 Carbon

There have been a number of theoretical studies made on carbon clusters, particularly of C_{60} in the wake of experimental evidence that C_{60} may exist as a hollow spherical shell.[47,49,52-55,58,60,72,73] At present, the definitive theoretical work on the smaller carbon clusters (C^n for n=2-10) is that of Ragahavachari and Binkley (RB).[60] They report calculated structures and binding energies for neutral and ionic carbon clusters, along with proposed fragmentation energies for the ionic clusters. The qualitative agreement is very good, although the values for fragmentation energies disagree by several electron volts.

According to RB, the ground states of the odd sized neutral and ionic carbon clusters are linear (C_3, C_5, C_7, and C_9) while their more weakly bound even neighbors are monocyclic rings (C_4, C_6, C_8, and C_{10}). Previous calculations had predicted a linear structure for all neutral clusters up to approximately 10 atoms,[72] although a cyclic structure for C_4 had been predicted more recently.[73] RB are less certain of the structure for the ionic clusters C_6^+ and C_8+ and it is possible that they are linear. In addition to determining structure, RB have predicted ionization potentials for the neutral clusters, all of which fall in the range of 9.3-12.1 eV. It is interesting to note that C_3 has a slightly greater electron affinity than C, which is why C_4^+ produces a significant amount of C$^+$ when it photodissociates.

In Fig. 9, RB have predicted the energy required to photodissociate the positive carbon cluster ions through their most favorable channel, the loss of neutral C_3 (except C_3^+ for C_5^+ photodissociation). The values for C_3 through C_5 are in reasonable agreement with those obtained by Freeman, et al. (see Fig. 5).[16] For C_6 and above, the predicted values are several eV higher than those observed experimentally. It is also possible to derive experimental values for the thresholds by combining binding energy measurements[75] with photoionization studies.[76] These derived values for the thresholds are in good agreement with the photodissociation experiment, except for C_7^+, for which the derived value is approximately 6.8 eV (vs. <3.51 eV).

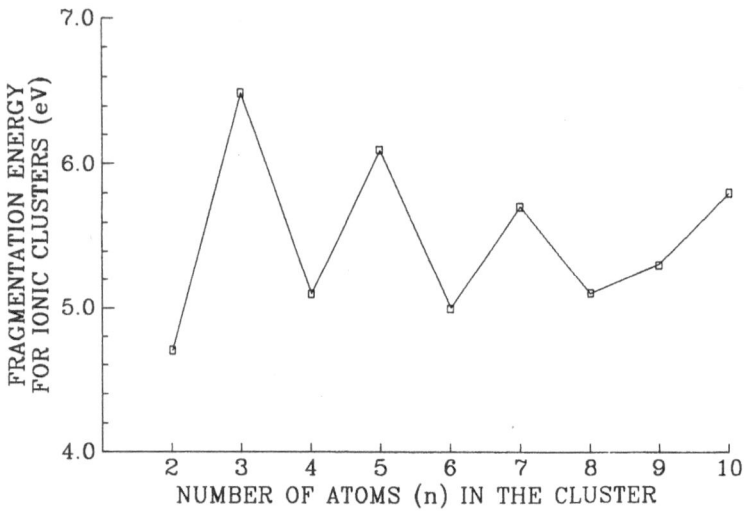

Fig. 9 Fragmentation energy for the most favorable channel for ionic carbon clusters C_n^+ vs. the number of atoms n.[60]

One explanation for such discrepancies between experiments and theory for the larger clusters has been suggested by recent FT-ICR experiments by O'Keefe and coworkers.[11,22] In measuring the reactivity of carbon cluster ions produced by laser vaporization, they have observed what appear to be two different isomers of C_7^+, one of which reacts easily to various reagents and one of which is essentially inert. The photodissociation threshold measurements of Freeman, et al. would be sensitive only to the higher energy, most fragile isomer of C_7^+ and could therefore yield a dissociation threshold value which is substantially too low for the more stable ground state.[16] An explanation of the other discrepancies will await further theoretical and experimental investigations.

It is also interesting to point out that in their FT-ICR studies, O'Keefe, et al. have also studied the collision induced dissociation of small carbon cluster ions.[22] As in photodissociation, the overwhelming decay mode is loss of a neutral C_3 fragment.

4.2 Silicon

As with carbon, there are now a number of theoretical works on the structure of small silicon clusters (n=2-10).[44-48,50,51,56,57,59] Among the results of these studies is the prediction that Si_4, Si_6, and Si_{10} should be "magic" numbers,[44,48,50,56,59] just as observed in the experiments.[4,20] Also, the calculated geometries of the clusters differ considerably from their microcrystalline equivalents: the bond lengths are shorter and the structures are more compact than would be allowed by the bulk diamond lattice.[44,48,50]

Raghavachari and Logovinsky have calculated the energy required to fragment silicon positive cluster ions for n=2-7 based on ground state energies (Fig. 10)[44] and have found local maxima for Si_4^+ and Si_6^+. These same cluster ions (Si_4^+ and Si_6^+) have small photodissociation cross sections (Fig. 8)[4] and were found to have low reactivities in FT-ICR studies by two groups: Mandich, et al.,[15] and Creasy and O'Keefe.[21]

A very recent molecular dynamics study has been carried out by Feuston, et al.[59] to study the high temperature behavior of silicon clusters. While their study does not use the sophisticated quantum-chemical approach employed in ground state calculations, it does include temperature. There principal results are the relative stabilities of Si_4, Si_6, and Si_{10} at high temperatures and the tendency of larger clusters to break into fragments of these sizes.

Perhaps the most remarkable silicon cluster is Si_{10}, the ground state of which has been calculated to be tetra-capped octahedron (Fig. 11).[48,50] This structure has the same symmetry as the bulk diamond lattice, although the nature of the bonding and the exact positioning of the atoms is substantially different. Its compact shape and high binding energy make it an obvious

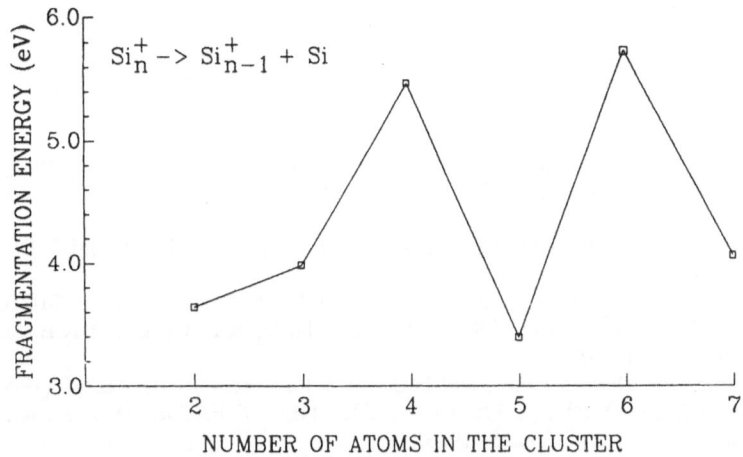

Fig. 10. Fragmentation energy (eV) for the reaction $Si_n^+ \rightarrow Si_{n-1}^+ + Si$ vs the number of atoms n for Si_2^+-Si_7^+.[44]

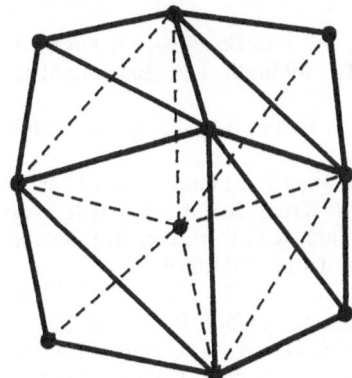

Fig. 11. Calculated ground state structure for Si_{10}.[48,50]

candidate for a "magic" number and "magic" fragment, as observed experimentally in both positive and negative cluster ions.

4.3 Germanium and Gallium Arsenide

Few calculations have been performed on clusters of germanium[43,57] and the author is unaware of any done on gallium arsenide. Work of germanium has found that it is quite similar to silicon, apart from the absolute scale of binding energies and bond lengths. Like silicon, the small germanium clusters are compact, non-bulk-like particles.

5. Conclusion

We have reviewed the techniques involved in photodissociation of small clusters and their application to clusters of semiconductors. The observed dissociation behaviors of semiconductor cluster ions, particularly of carbon and silicon, are substantially in agreement with the current theories for their structures and binding energies. Photodissociation appears to proceed via a statistical unimolecular decomposition following the absorption of a photon. Several values for the energy required to cause photodissociation have been obtained and are in fair agreement with theoretical results. The principal experimental "magic" cluster sizes (C_3, Si_6^+, Si_{10}^+, Si_{10}^-) have also been found to be relatively stable in a variety of theoretical studies. Finally, the cross sections for photodissociation have been found to be large and so this process must be taken into account whenever optical interactions with clusters are studied.

REFERENCES

1. N. Furstenau and F. Hillenkamp, Intern. J. Mass Spect. Ion Phys. 37:135 (1981).
2. E.A. Rohlfing, D.M. Cox, and A. Kaldor, J. Chem. Phys. 81:3322 (1984).
3. T.T. Tsong, Appl. Phys. Lett. 45:1149 (1984).
4. L.A. Bloomfield, R.R. Freeman, and W.L. Brown, Phys. Rev. Lett. 54:2246 (1985).
5. L.A. Bloomfield, M.E. Geusic, R.R. Freeman, and W.L. Brown, Chem. Phys. Lett. 121:33 (1985).
6. J.R. Heath, Y. Liu, S.C. O'Brien, Q.-L. Zhang, R.F. Curl, F.K. Tittel, and R.E. Smalley, J. Chem. Phys. 83:5520 (1985).
7. H.W. Kroto, J.R. Heath, S.C. O'Brien, R.F. Curl, and R.E. Smalley, Nature 318:162 (1985).
8. S.C. O'Brien, Y. Liu, Q. Zhang, J.R. Heath, F.K. Tittle, R.F. Curl, and R.E. Smalley, J. Chem Phys 84:4074 (1986).
9. W. Begemann, K.H. Meiwes-Broer, and H.O. Lutz, Phys. Rev. Lett. 56:2248 (1986).
10. Y. Liu, S.C. O'Brien, Q. Zhang, J.R. Heath, F.K. Tittel, R.F. Curl, H.W. Kroto, and R.E. Smalley, Chem. Phys. Lett. 126:215 (1986).
11. S.W. McElvany, W.R. Creasy, and A. O'Keefe, J. Chem. Phys. 85:632 (1986).
12. M.E. Geusic, T.J. McIlrath, M.F. Jarrold, L.A. Bloomfield, R.R. Freeman, and W.L. Brown, J. Chem. Phys. 84:2421 (1986).
13. I. Plesser, Z. Vager, and R. Naaman, Phys. Rev. Lett. 56:1559 (1986).
14. J.R. Heath, S.C. O'Brien, Q. Zhang, Y. Liu, R.F. Curl, H.W. Kroto, F.K. Tittel, and R.E. Smalley, To Be Published.
15. M.L. Mandich, W.D. Reents, Jr., V.E. Bondybey, J. Phys. Chem., In Press.
16. M.E. Geusic, M.F. Jarrold, T.J. McIlrath, L.A. Bloomfield, R.R. Freeman, and W.L. Brown, Z. Phys. D. In Press.
17. R.D. Knight, R.A. Walch, S.C. Foster, T.A. Miller, S.L. Mullen, and A.G. Marshall, Chem. Phys. Lett., To Be Published.
18. V.E. Bondybey, W.D. Reents, Jr., M.L. Mandich, J. Chem. Phys., To Be Published.
19. S.C. O'Brien, J.R. Heath, H.W. Kroto, R.F. Curl, and R.E. Smalley, To Be Published.
20. Y. Liu, Q.-L. Zhang, F.K. Tittel, R.F. Curl, and R.E. Smalley, To Be Published.
21. W.R. Creasy and A. O'Keefe, To Be Published.
22. S.W. McElvany, B.I. Dunlap, and A. O'Keefe, To Be Published.
23. R.E. Honig, J. Chem. Phys. 22, 126 (1954); R.E. Honig, J. Chem. Phys. 22:1610.
24. D.E. Powers, S.G. Hansen, M.E. Geusic, A.C. Puiu, J.B. Hopkins, T.G. Dietz, M.A. Duncan, P.R.R. Langridge-Smith, and R.E. Smalley, J. Phys. Chem. 86:2556 (1982).
25. See for example: R.C. Dunbar, in "Molecular Ions: Spectroscopy Structure and Chemistry", T.A. Miller and V.E. Bondybey, eds., North Holland, Amsterdam (1983).; J. Moseley and J. Durup, J. Chim. Physique 77:673 (1980).
26. M.F. Jarrold, A.J. Illies, and M.T. Bowers, J. Chem. Phys. 79:6086 (1983) and References Therein.
27. M.F. Jarrold, A.J. Illies, and M.T. Bowers, J. Chem. Phys. 81:222 (1984).
28. M.F. Jarrold, L. Misev, and M.T. Bowers, J. Chem. Phys. 81:4369 (1984).
29. M.A. Johnson, M.L. Alexander, and W.C. Lineberger, Chem. Phys. Lett 112:285 (1984).
30. M.F. Jarrold, A.J. Illies, and M.T. Bowers, J. Chem. Phys. 81:214 (1984).
31. A.J. Illies, M.F. Jarrold, W. Wagner-Redeker, and M.T. Bowers, J. Phys. Chem. 88:5204 (1984).
32. M.F. Jarrold, A.J. Illies, and M.T. Bowers, J. Chem. R.E. Smalley, To Be Published.
32. M.F. Jarrold, A.J. Illies, and M.T. Bowers, J. Chem. Phys. 82:1832 (1985).
33. M.F. Jarrold, A.J. Illies, and M.T. Bowers, J. of Am. Chem. Soc. 107:7339 (1985).
34. M.L. Alexander, M.A. Johnson, and W.C. Lineberger, J. Chem. Phys. 82:5288 (1985).
35. L. Misev, A.J. Illies, M.F. Jarrold, and M.T. Bowers, Chem. Phys. 95:469 (1985).
36. A.J. Illies, M.F. Jarrold, W. Wagner-Redeker, and M.T. Bowers, J. Am. Chem. Soc. 107:2842 (1985).
37. M.F. Jarrold, A.J. Illies, W. Wagner-Redeker, and M.T. Bowers, J. Phys. Chem. 89:3269 (1985).
38. H.-S. Kim, M.F. Jarrold, and M.T. Bowers, J. Chem. Phys. 84:4882 (1986).
39. P.C. Engelking, J. Chem. Phys. 85:3103 (1986).
40. M.L. Alexander, M.A. Johnson, N.E. Levinger, and W.C. Lineberger, Phys. Rev. Lett. 57:976 (1986).
41. H.-S. Kim, M.F. Jarrold, and M.T. Bowers, J. Phys. Chem. 90:3584 (1986).
42. H.-S. Kim and M.T. Bowers, To Be Published.

43. G. Pacchioni and J. Koutecky, Ber. Bunsenges. Phys. Chem. 88:242 (1984).
44. K. Raghavachari and V. Logovinsky, Phys. Rev. Lett. 55:2853 (1985).
45. K. Raghavachari, J. Chem. Phys. 83:3520 (1985).
46. J.C. Phillips, J. Chem. Phys. 83:3330 (1985).
47. S. Saito, S. Ohnishi, and S. Sugano, Tech. Rep. of ISSP A1608 (1985).
48. S. Saito, S. Ohnishi, C. Satoko, and S. Sugano, J. Phys. Soc. Japan 55:1791 (1986).
49. B.K. Rao, S.N. Khanna, and P. Jena, Sol. St. Comm. 58:53 (1986).
50. K. Raghavachari, J. Chem. Phys. 84:5672 (1986).
51. K. Balasubramanian, Chem. Phys. Lett. 125:400 (1986).
52. A.J. Stone and D.J. Wales, Chem. Phys. Lett. 128:501.
53. R.C. Haddon, L.E. Brus, and K. Raghavachari, Chem. Phys. Lett. 125:459 (1986).
54. R.L. Disch and J.M. Schulman, Chem. Phys. Lett. 125:465 (1986).
55. M. Ozaki and A. Takahashi, Chem Phys. Lett. 127:242 (1986).
56. D. Tomanek and M.A. Schluter, Phys. Rev. Lett. 56:1055 (1986).
57. G. Pacchioni and J. Koutecky, J. Chem. Phys. 84:3301 (1986).
58. K. Raghavachari, R.A. Whiteside, and J.A. Pople, J. Chem. Phys., In Press.
59. B.P. Feuston, R.K. Kalia, and P. Vashishta, To Be Published.
60. K. Raghavachari and J.S. Binkley, To Be Published.
61. N.R. Isenor and M.C. Richardson, Appl. Phys. Lett. 18:224 (1971).
62. Y.R. Shen, "The Principles of Nonlinear Optics," Wiley, New York (1984) pp. 437-465.
63. P. Fayet, L. Woste, Spect. Int. J. 3:91 (1984).
64. P. Fayet and L. Woste, Surf. Sci. 156:134 (1985).
65. P.J. Prucat, L.-S. Zheng, C.L. Pettiette, S. Yang, and R.E. Smalley, J. Chem. Phys. 84:3078 (1986).
66. M.R. Zakin, R.O. Brickman, D.M. Cox, K.C. Reichmann, D.J. Trevor, and A. Kaldor, J. Chem. Phys. 85:1198 (1986).
67. P.J. Brucat, L.S. Zheng, C.L. Pettiette, S. Yang, and R.E. Smalley, J. Chem. Phys. 84:3078 (1986).
68. M. Broyer, G. Delacretaz, P. Labastie, J.P. Wolf, and L. Woste, Phys. Rev. Lett. 57, 1851 (1986).
69. P. Fayet and L. Woste, Zeit. Phys., In Press.
70. L.-S. Zheng, P.J. Brucat, C.L. Pettiette, S. Yang, and R.E. Smalley, J. Chem. Phys. 83:4273 (1985).
71. L.A. Bloomfield, M.E. Geusic, R.R. Freeman, and W.L. Brown, in "Electronic and Atomic Collisions," D.C. Lorents, W.E. Meyerhof, J.R. Peterson, eds., Elsevier, Amsterdam (1986) p. 807.
72. K.S. Pitzer, E. Clementi, J. Amer. Chem. Soc. 81:4477 (1959); S.J. Strickler, K.S. Pitzer, in "Molecular Orbitals in Chemistry Physics and Biology," B. Pullman and P.O. Lowden, eds., Academic Press, New York (1964); R. Hoffman, Tetrahedron 22:521 (1966).
73. R.A. Whiteside, R. Krishnan, D.J. DeFrees, J.A. Pople, P. von R. Schleyer, Chem. Phys. Lett. 78:538 (1981).
74. L.A. Bloomfield, R.R. Freeman, W.L. Brown, Unpublished.
75. J. Drowart, R.P. Burns, G. DeMaria, M.G. Ingram, J. Chem. Phys. 31:1131 (1959).
76. H.M. Rosenstock, K. Draxl, B.W. Steiner, J.T. Herron, J. Phys. Chem. Ref. Data 6, Supplement No. 1 (1977).

ATOMIC AND ELECTRONIC STRUCTURES OF SEMICONDUCTOR CLUSTERS

S. Ohnishi and S. Saito

Fundamental Research Laboratories, NEC Corporation
Miyazaki, Miyamae-ku, Kawasaki 213, Japan

C. Satoko

Institute for Molecular Science
Myodaiji, Okazaki, Aichi 444, Japan

S. Sugano

Institute for Solid State Physics, University of Tokyo
Roppongi, Minato-ku, Tokyo 106, Japan

ABSTRACT

We discuss the magic number and its origin of clusters of group-IV atoms, semiconductor clusters, based on the non-spherical model potential study and electronic structure calculations. Stable structures determined by the model potential for the sp^3-hybridized atom are found to be classified into two groups of having six- and five-membered rings, the crystalline and amorphous series, respectively. Magic numbers are 6,10,14,18... for the crystalline series and 5,10,12,16,18,20... for the amorphous series. Detailed reconstruction mechanisms of the magic number clusters are discussed by the analysis of electronic structures of Si_6, Si_{10}, Si_{14}, and Si_{26} of the crystalline series and Si_5 and Si_{20} of the amorphous series. It is proposed that the driving force for the reconstruction of clusters of the crystalline series comes from the significant contraction of triangles formed by three atoms in the Si(111) surface induced by the interaction between dangling bonds.

1. Introduction

One of the most important concepts of the small cluster physics is the "magic number" for the size of a cluster, which has been intensively investigated by recent experiments of mass spectral analysis.[1] Different kinds of series for the magic number have been obtained depending on the nature of chemical bond of atomic elements:13,19,25,55... for rare-gas (Xe) atoms[2], and 8,18,20,34... for alkali atoms[3], for example. The magic numbers for group-IV clusters have also been reported to be 6,10,.....[4] It is of great interest to understand the physical mechanism of the formation of stable structures with the corresponding magic number based on the bonding nature of each atomic element. In this aspect, group-VIII, rare-gas clusters are rather easy to understand: magic numbers are interpreted to correspond to icosahedral packing by using model potential of Lennard-Jones type, which comes from the inactive nature of electronic structure. Magic numbers for group-I clusters have been investigated intensively on the basis of the shell model for s-electrons with a jellium background of a constant density[3,5]: the shell model for an aggregate of alkali atoms explains the origin of the magic number very well as associated with the shell-closing electron numbers.

Contrary to the clear picture of these two atomic elements, clusters of group-IV atoms, semiconductor clusters, have not been well understood theoretically because of their complicated non-spherical nature of interaction potential and their strong dependence of covalent bonding on geometrical structure. The first step to analize possible structures of semiconductor clusters will be to build the molecular models by sp^3-hybridized atoms which have "four hands" in tetrahedral co-ordination. Saito et al.[6] has succeeded in formulating the corresponding non-spherical model potential mathematically, by which two geometrical series of stable structures have been obtained: crystalline (six-membered-ring) and amorphous (five-membered-ring) series, of which magic numbers are 6,10,14,18... and 5,10,12,16,18,20... respectively. The magic numbers of crystalline series obtained theoretically agree very well with experimental ones.[4] The essential point of this model potential study is that the total energy is proportional to the number of bonds so that stable structures tend to have connected rings. This bond energy concept does explain the magic number of semiconductor clusters. Theoretical treatments are summarized in Sec.2. However, it should be noted that our model potential study does not take into account the interaction between dangling bonds, which will be shown to play an important role for the reconstruction of the clusters built by the sp^3-hybridized atoms. Other model potential studies concerning stable structures of semiconductor clusters have been reported[7,8], which are based on the molecular dynamical approach by use of two- and three-body interaction potentials of Stillinger-Weber type.[9] However, it is not so clear if the characteristic feature of the dangling bond states was fully taken into account.

The next step for characterizing semiconductor clusters seems for us to be the analysis of the effect of the dangling bond interaction on the stable structures which are determined by our model potential study. At first, we examine electronic energy levels of dangling bond states based on the simple model Hamiltonian of the bond-orbital-model by fixing geometrical structure. As is discussed in Sec.3-1, energy levels consisting of dangling bond states are concentrated near the Fermi energy, which implies the structural instability. Then, first principle calculations based on the linear-combination-of-atomic-orbital (LCAO) method with the local density functional scheme (Xα potential for the exchange-correlation term) have been carried out in order to find out the physical mechanism of the reconstruction for several Si clusters. For Si_6 and Si_{10} clusters, equilibrium structures have been reported by Saito et al.[10] where two kinds of distortion modes for each cluster were optimized within the restriction of keeping original symmetries. Elaborate calculations to determine geometrical structures for Si_n of n=2-7,10 have been also performed by Raghavachari and Logovinsky[11] on the basis of quantum chemical approach. And, also, model Hamiltonian study by Tomanek and Schluter[12] has discussed several geometries of Si clusters. The stable structures determined theoretically for Si_6 and Si_{10} in this way are essentially the same as long as a small distortion of symmetry-breaking modes for Si_6 is ignored. The main point we would like to clarify in this paper is how one can illustrate a physical picture of the role of dangling bonds played in the reconstruction of semiconductor clusters. For this purpose, the analysis of the force field on each atom of the cluster seems to be most suitable. Actually, Saito et al.[10] proposed that the concept of the "triangle contraction" due to the interaction of dangling bonds is very usefull in understanding the reconstractions of Si_6 and Si_{10} clusters. In Sec.3-2, further discussions are given by analysing the calculated results for Si_6, Si_{10}, Si_{14}, and Si_{26} for crystalline series and Si_5 and Si_{20} for amorphous series.

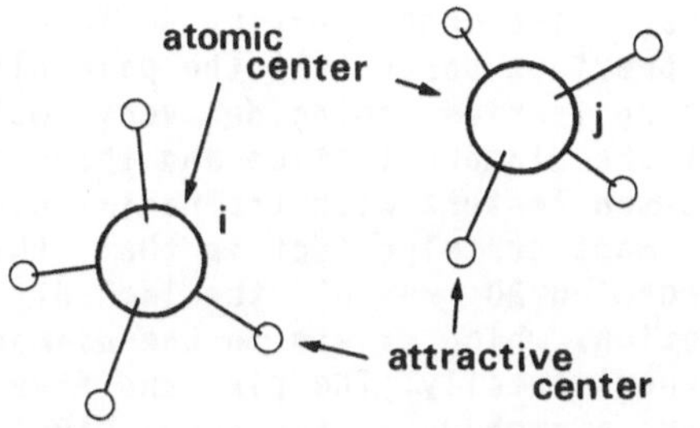

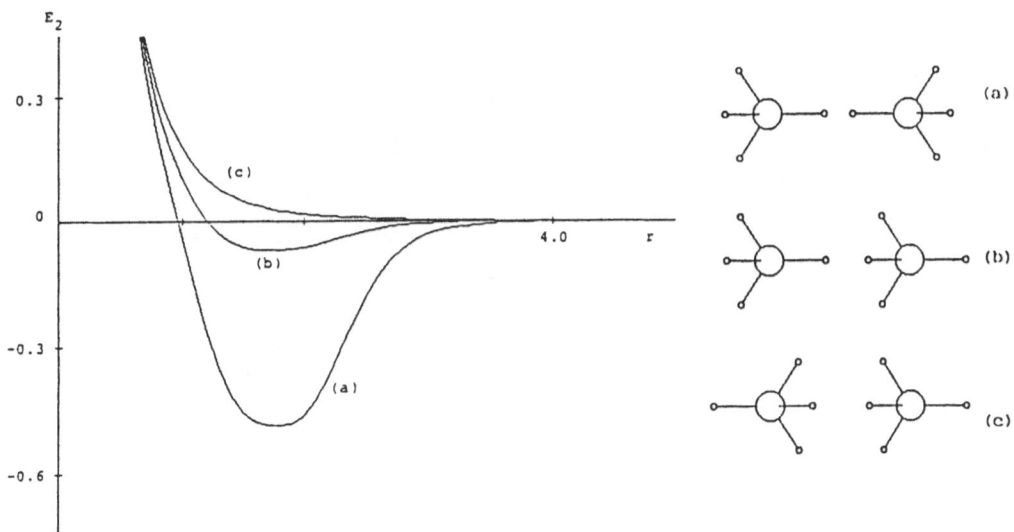

Fig.1.   Interaction potential is given by $U_{rep}(r_{ij}) + U_{attra}(r_{ij}, \alpha_i, \alpha_j)$. $r_{ij}=r_j-r_i$ is relative coordinate. α_i is Euler angle of atom i. U_{attra} depends on the relative distance between atomic center and attractive centers.

Fig.2.   Potential curves for three different cases: (a) each atom is directed to the other, (b) one atom directed but the other opposite the partner, and (c) each atom opposite to the other.

2. Atomic Structures

The new model potential for describing the interaction between sp^3-hybridized atoms[6] is schematically shown in Fig.1. Each atom has four attractive centers which interact with other atomic centers. A short-range repulsive force of Yukawa type is introduced so as to keep a certain distance between atomic centers. In case the i-th atom sits on one of the attractive centers of the j-th atom and at the same time the j-th atom on one of the those of the i-th atom, the i-th and j-th atoms are connected tightly. This corrsponds to the situation of sharing a covalent bond bewtween the i-th and the j-th atoms. Mathematical details are already described[6]. In Fig.2, the typical potential curves are shown for three cases of atomic configurations. In Fig.3-1 and 3-2, the stable structures of clusters with the atomic number n<21 which minimize the total energy are illustrated for the six-and five-membered-ring series, respectively. It has been pointed out in the previous paper that the pair distribution functions of the six-membered-ring series coincide very well with the radial distribution function of the diamond lattice and those of the five-membered-ring series have the common feature with the radial distribution functions of amorphous Ge: the most striking fact is that the pair distribution function of a dodecahedron, n=20, reveals the lack of the third peak of the radial distribution function, which exists in the diamond lattice but not in amorphous Ge observed experimentally. The six- and five-membered-ring series are called crystalline and amorphous series respectively. The dodecahedron of n=20 cluster may be an element of "amorphon" under suitable circumstances of saturating unoccupied dangling bonds in amorphous semiconductors.

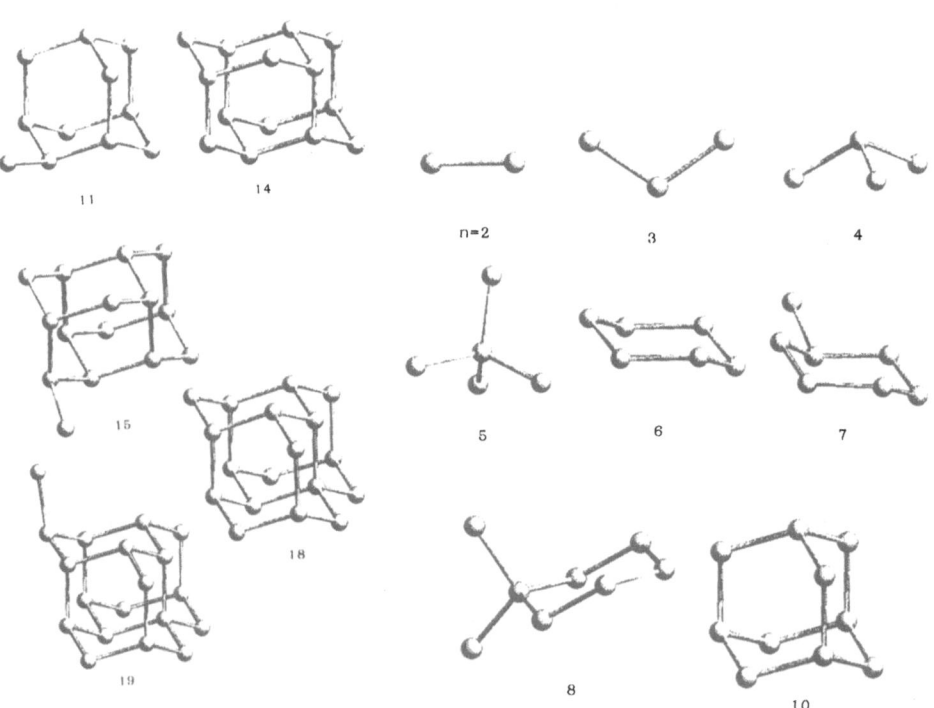

Fig.3-1. Stable structures for the crystalline series.

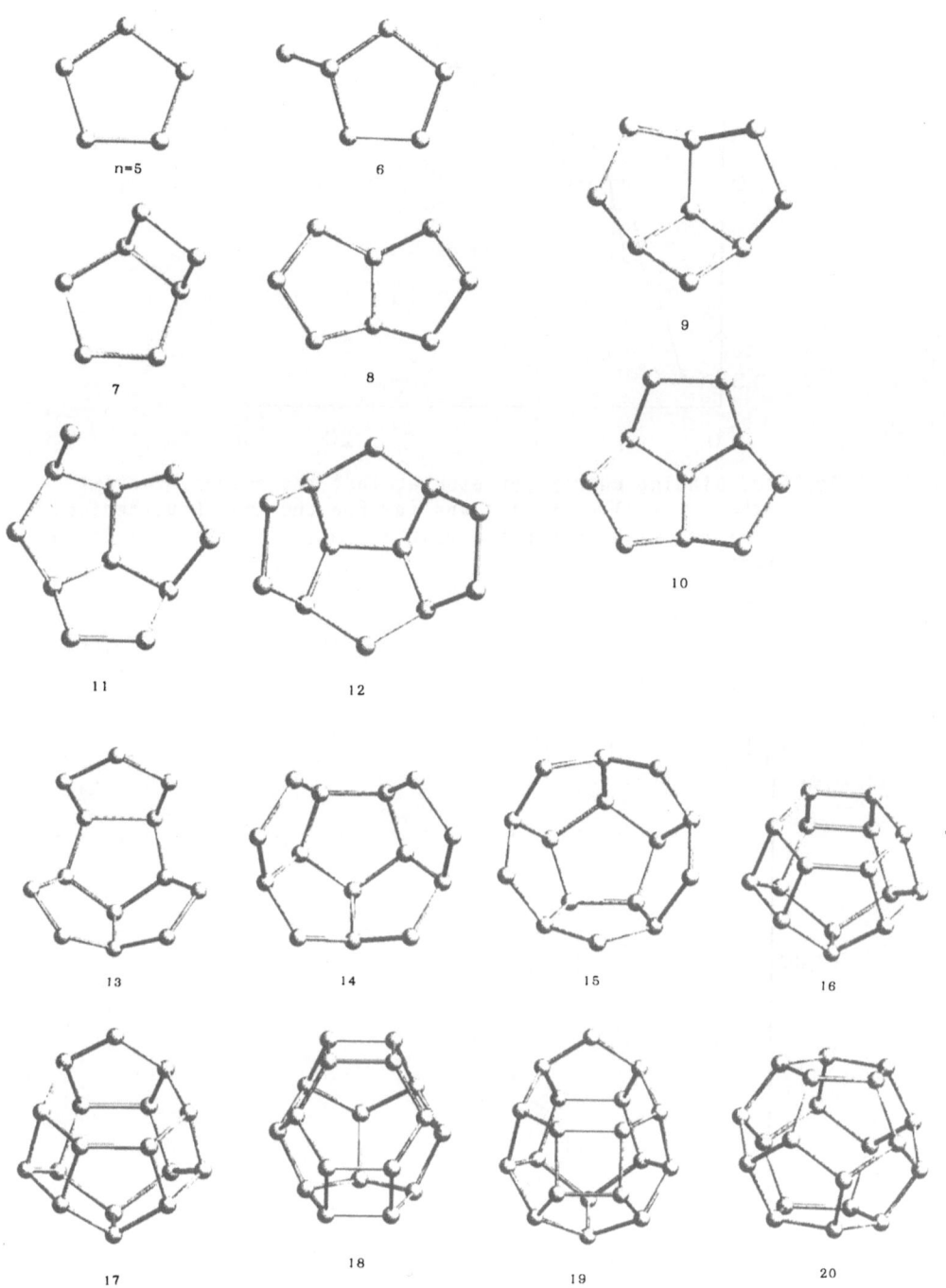

Fig.3-2. Stable structures for the amorphous series.

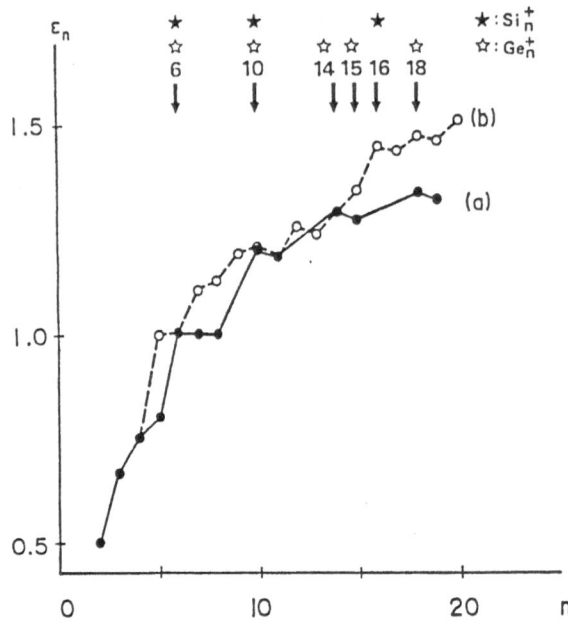

Fig.4.   Calculated binding energy per atom against the number of atoms
(a) for the crystalline series and (b) for the amorphous series.
Numbers written with arrows are observed magic numbers of Si and
Ge clusters.[4]

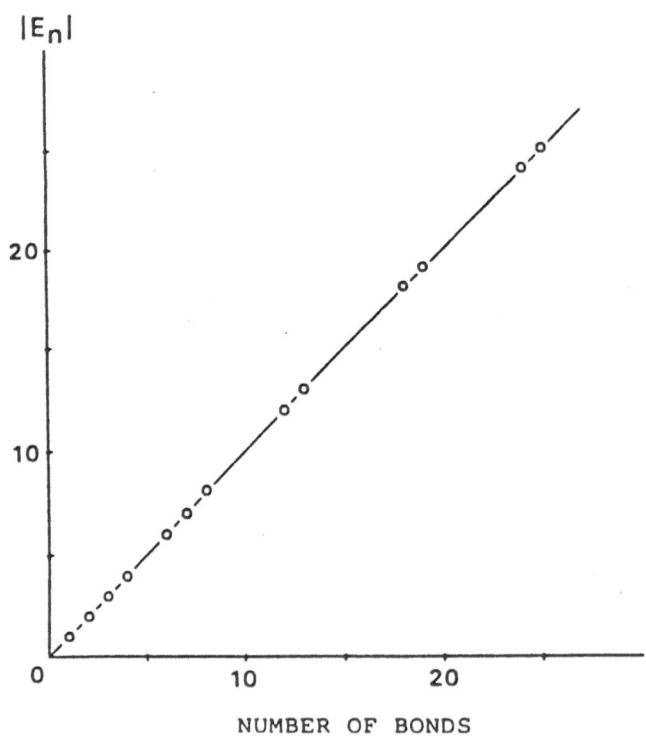

NUMBER OF BONDS

Fig.5.   Total energy against the number of bonds.

Binding energy per atom against n for both series is shown in Fig.4. Local peaks are considered to correspond to the magic numbers, which are 6,10,14,18 for the crystalline series and 5,10,12,16,18,20 for the amorphous series. The magic numbers of crystalline series agree very well with experiments. In Fig.5, total energies are plotted against the number of bonds. The linear curve indicates that the "bond energy" concept is valid and explains the magic number of semiconductor clusters.

3. Electronic Structures

3-1. Bond-Orbital-Model Study

The characteristic feature of semiconductor clusters found by model potential calculations is the fact that the cluster may be regarded as a multi-surface system with many dangling bonds sticking out from the surfaces. We now restrict our attention to the electronic states of these dangling bonds. In this place, we pick up a typical cluster of n=10 for the crystalline series, which is the magic number cluster, and that of n=20 for the amorphous series. We apply the bond-orbital-model where the interaction term of dangling bonds is introduced. Transfer parameters are schematically shown in Fig.6. Energy levels for n=10 and 20 clusters with and without dangling bond interactions are shown in Fig.7 and Fig.8, respectively. Two points should be noted. (1) Energy levels of dangling bond states concentrare near the Fermi level. High density of states near the Fermi level induces structural instability. (2) Energy level structures are sensitive to the transfer parameters of dangling bond interaction. However, it does not seem to be strong enough to reduce the high density of states at the Fermi level without changing geometrical structures.

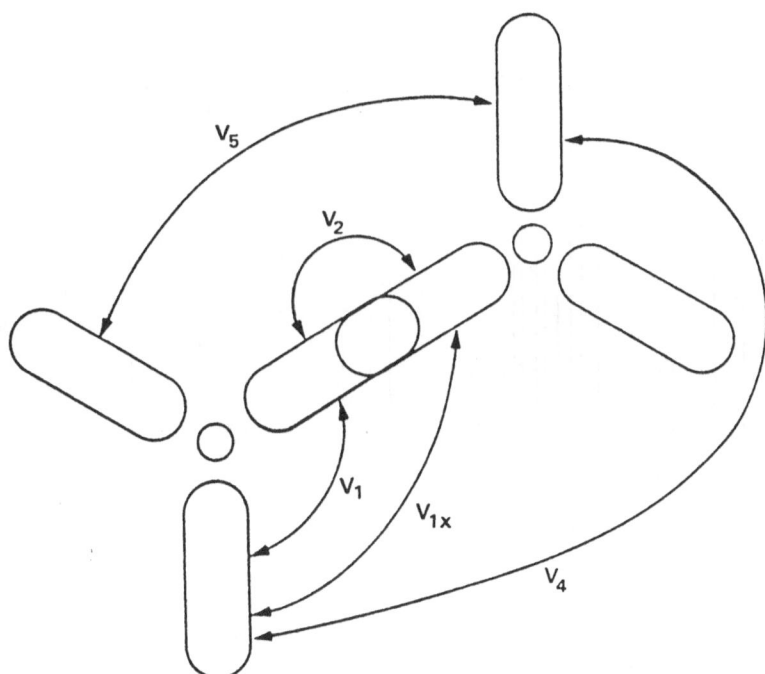

Fig.6. Transfer parameters of the bond-orbital model. As for the parameters between dangling bonds, only nearest-neighbour dangling-bond pair is taken into account.

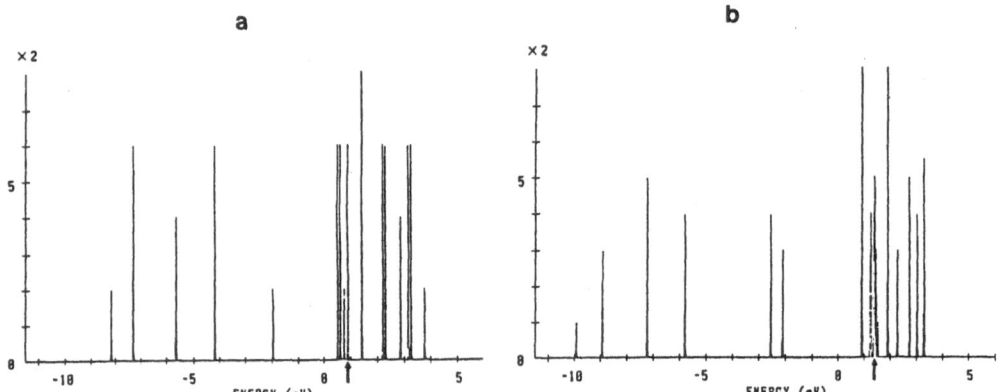

Fig.7. Density of states of (a) crystalline-series Si_{10} and (b) amorphous-series Si_{20}. Arrows indicate the Fermi levels.

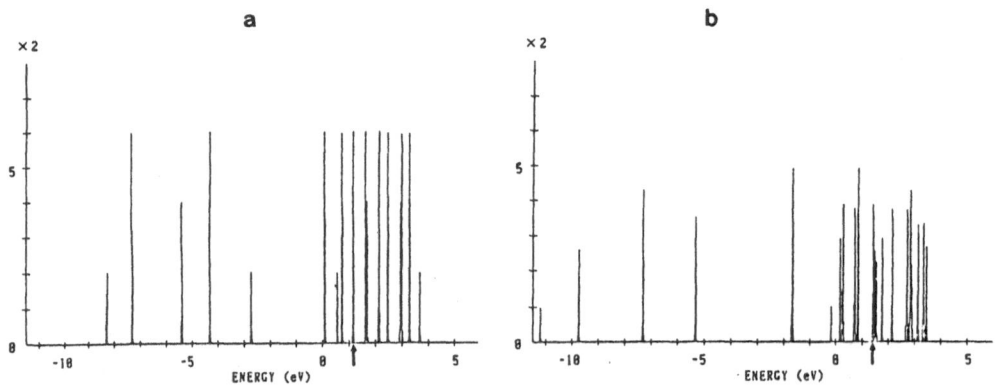

Fig.8. Density of states of (a) Si_{10} and (b) Si_{20} with additional transfer parameters between dangling bonds.

3-2. LCAO-Xα-Force Calculation

In order to clarify the physical mechanism of the reconstruction of semiconductor clusters, we have performed self-consistent local density functional calculations based on the LCAO-Xα-force method[13] for clusters of Si_6, Si_{10}, Si_{14}, and Si_{26} of the crystalline series and Si_5 and Si_{20} of the amorphous series. From the calculated force field acting on each atom, stable structures of the n=6,10, and n=5,20 cases are optimized by the condition of forces=0 by keeping the original symmetries of D_{3d}, T_d, D_{5h}, and I_h, respectively. The calculated geometrical structures of reconstracted Si_6 and Si_{10} agree very well with other calculations.[11,12] As for Si_6,

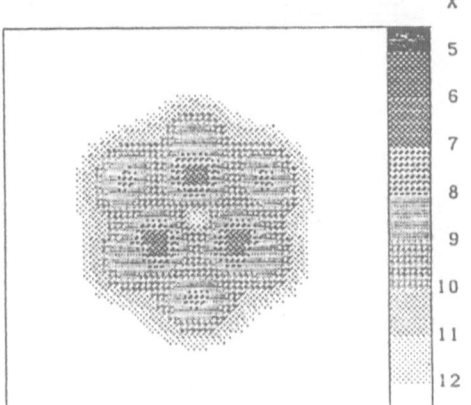

Fig.9-1 Highest-occupied-state electron density on the triangle of unreconstructed Si_{10} cluster.

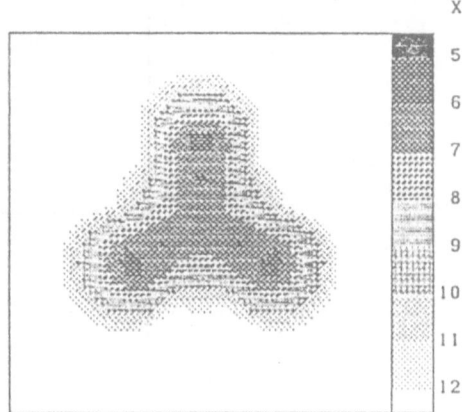

Fig.9-2 After reconstruction.

however, a small deviation from the D_{3d} symmetry has been reported.[11] The important conclusion obtained in this calculation is that the driving force for the reconstruction comes from the contraction of triangles formed by three atoms in the Si(111) surface, which is induced by the strong interaction between three dangling bonds. Charge densities on such triangle planes are shown in Fig.9-1 and 9-2 for the unreconstracted and reconstracted structures of Si_{10}, respectively. As shown in Fig.10, a large deformation induces a large energy gap between the highest occupied state and the lowest unoccupied state, which indicates the stability of the reconstracted structure of Si_{10} cluster.

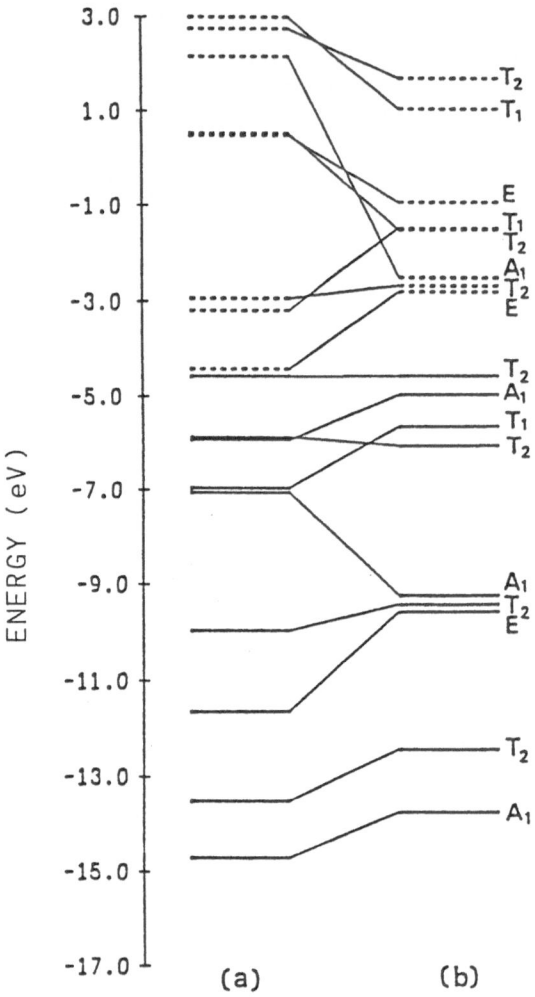

Fig.10. Energy levels of the electronic states of Si_{10} cluster of (a) unreconstructed and (b) reconstructed cases.

In Fig.11, the geometrical structures and the force field acting on the atoms in the cluster surfaces of n=10,14, and 26 unreconstructed clusters are illustrated. It should be noted that Si_{14} and Si_{26} clusters are regarded as the composite of Si_{10} adamantanes: there are two and four adamantanes for n=14 and 26 clusters, respectively. The calculated force fields for these two clusters are understood as the superposition of the force field for the adamantane of Si_{10} cluster. There should exist a similar contraction of the triangles in the (111) plane.

As for Si_8 and Si_{20} of the amorphous series, the equilibrium atomic distances do not change appreciably: only 1% contraction for Si_{20}. The electronic structures of Si_{20} is shown in Fig.12. Since the highest occupied state is degenerate because of the high symmetry and the unsaturated dangling bonds, the stable structure may be distorted.

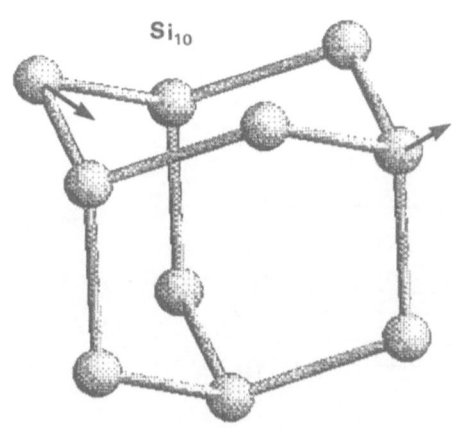

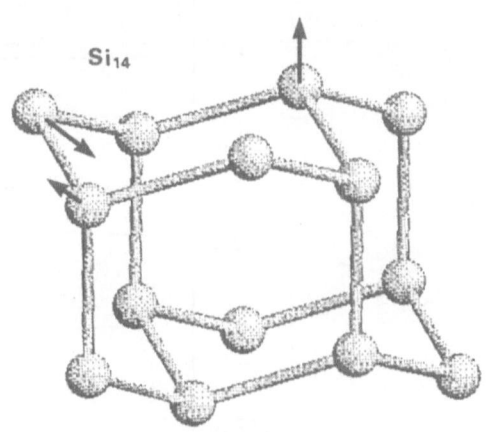

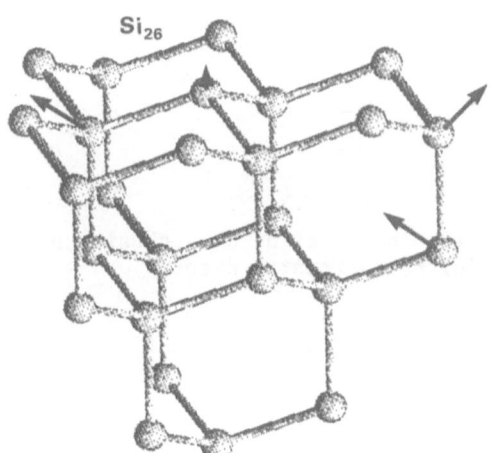

Fig.11. Calculated force fields for Si_{10}, Si_{14}, and Si_{26} clusters. Forces on atoms represented by arrows are shown for different atomic types, two for Si_{10}, three for Si_{14}, and four for Si_{26}.

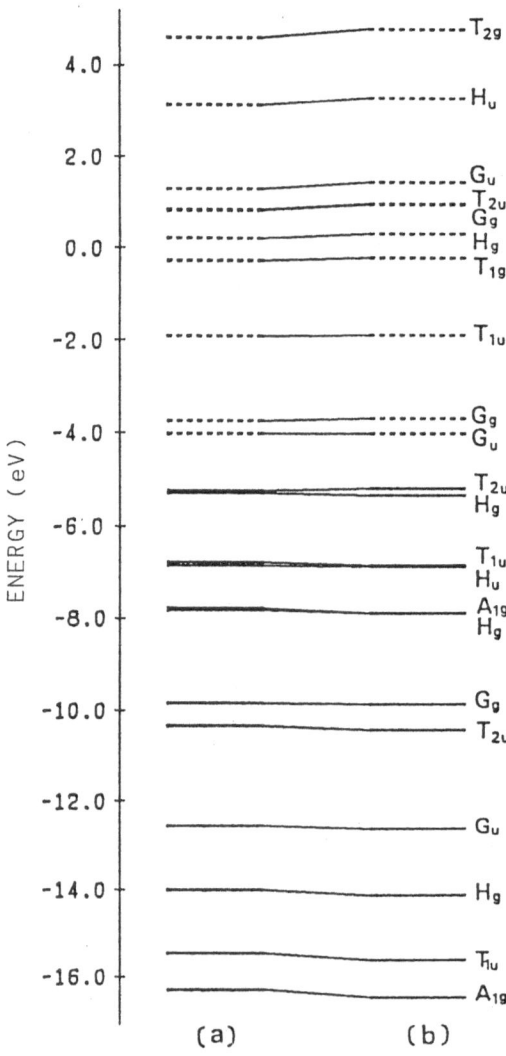

Fig.12. Energy levels of the electronic states of Si₂₀ cluster of (a) unreconstructed and (b) reconstructed cases.

4. Discussion

One of the most interesting results from the model potential study is the appearence of the amorphous series. Since the number of rings and bonds of the amorphous series is the same or more than that of the crystalline series of the same size n, the energy gain due to making bonds is larger than the energy loss due to the distortion: in fact, the bond angle of the dodecahedron is 108°, which is nearly the same as the one, sp^3 bond angle, for the crystalline series of 109°. The total energy of clusters of the amorphous series is generally lower than those of the crystalline series because of the excess number of bonds. The LCAO-X -force calculations, however, show that structural deviations from clusters of the crystalline series are large. It is considered that the energy gain due to the reconstruction makes the clusters of the crystalline series much stabler than those of the amorphous series.

It is of interest to understand the magic numbers and the stability of larger clusters of the crystalline series by superimposing force fields of the adamantanes. Calculated results by the LCAO-Xα-force method for Si_{14} and Si_{26} supported our superposition model. It should be noted that the interaction between dangling bonds strongly depends on the geometrical configuration of the triangles on the cluster surface. As shown in Si_{26}, no force acts on the atom sitting on three equivalent triangles. This fact may be related with the critical size of the cluster of having different properties from the bulk crystal.

REFERENCES

1. Proc. 2nd Int. Meet. Small Particles and Inorganic Clusters, Lausanne, 1980, Surf.Sci. 106 (1981).
   Proc. 3rd Int. Meet. Small Particles and Inorganic Clusters, Berlin, 1984, Surf.Sci. 156 (1985).
2. O. Echt, K. Sattler and E. Recknagel, Phys. Rev. Lett. 47:1121 (1981).
3. W. D. Knight, K. Clemenger, W. A. de Heer, W. A. Saunders, M. Y. Chou, and M. L. Cohen, Phys. Rev. Lett. 52:2141 (1984).
   M. Y. Chou, A. Cleland, and M. L. Cohen, Solid State Commun. 52:654 (1984).
   W. D. Knight, W. A. de Heer, K. Clemenger, and W. A. Saunders, Solid State Commun. 53:445 (1985).
4. L. A. Bloomfield, R. R. Freeman, and W. L. Brown, Phys. Rev. Lett. 54:2246 (1985).
   T. P. Martin and H. Schaber, J. Chem. Phys. 83:855 (1985).
5. Y. Ishii, S. Ohnishi, and S. Sugano, Phys. Rev. B 33:5271 (1986).
6. S. Saito, S. Ohnishi, and S. Sugano, Phys. Rev. B 33:7036 (1986).
7. E. Blaisten-Barojas and D. Levesque, Phys. Rev. B 34:3910 (1986).
8. B. P. Feuston, R. K. Kalia, and P. Vashisthta, private communication.
9. F. H. Stillinger and T. A. Weber, Phys. Rev. B 31:5262 (1985).
10. S. Saito, S. Ohnishi, C. Satoko, and S. Sugano, J. Phys. Soc. Japan 55:1791 (1986).
11. K. Raghavachari and V. Logovinsky, Phys. Rev. Lett. 55:2853 (1985), K. Raghavachari, J. Chem. Phys. 84:5672 (1986).
12. D. Tomanek and M. A. Schluter, Phys. Rev. Lett. 56:1066 (1986).
13. C. Satoko, Chem. Phys. Lett. 47:490 (1981), Phys. Rev. B 30:1754 (1984).

MAGIC NUMBERS OF ALKALI HALIDE CLUSTERS

J. C. Phillips

AT&T Bell Laboratories
Murray Hill, New Jersey 07974

ABSTRACT

Recent experiments have shown that the most stable sodium halide NaX neutral and ionic clusters are cube shaped, but that the most stable CsI clusters may be either cube-shaped (high ionization energies) or have some other structure (low ionization energies). This other structure is identified here from its observed set of magic numbers in the context of a physical model which distinguishes CsI from NaX.

INTRODUCTION

Alkali halide cluster ions $(M_nX_{n-1})^+$ produced and detected in secondary-ion mass spectrometry[1] are found with enhanced abundance at "magic numbers" $n = 14, 23, 38$ and 63. From these numbers compact cube-like structures ((3 x 3 x 3) for $n = 14$) are inferred. When different preparation procedures are used, rectangular neutral cluster magic numbers are observed[2] at $n = 18, 20$ (2 x 4 x 5), 24, 30, 32, ..., 72 for NaX clusters (X = Cl, Br, I). These differences are well understood and are explained by electronic excitation of admolecules desorbing before or after the clusters are ionized for detection. A different situation holds for magic numbers of CsI cluster ions.[3] Here the magic numbers are found to depend on photon energy and are those of cuboids only at high photon energies. At low photon energies a neutral cluster magic number sequence $n = 7, 10, 13, 17, 19$ and 25 emerges. Here I propose a model to explain the origin of this novel sequence.

The principal difference between NaI and CsI is that the polarizability of the Cs^+ ion is ten times that of the Na^+ ion.[4] As a result of this greater polarizability the structural role of the $M^+ - X^-$ ionic attractive energy is reduced compared to that of the covalent X-X framework energy. I therefore suggest that the X atoms form one of the frameworks that can be derived from sphere packing.[5] The M atoms are left to form a liquid-like structure in which various sites are occupied with fractional probabilities. An analogy can be made to crystalline solid electrolytes, such as α-AgI and $RbAg_4I_5$, where the cations form a liquid-like structure within an I lattice.[6] The formation of such structures is closely correlated with the very large lattice polarizabilities which are found in crystals (such as β-AgI) with soft or low-frequency optic vibrational modes.[7]

FRAMEWORK MAGIC NUMBERS

Most of the magic numbers in the observed series[3] n = 7, 10, 13, 17, 19 and 25 are easily obtained from a pentagonal growth sequence.[5] For sphere packing n = 7 is a pentagonal bipyramid which is conveniently described on a layer-by-layer basis as 1-5-1, each layer being normal to the pentagonal axis. The next stable structure is 1-5-1-5-1 (n = 13), which is a regular icosahedron. The structure 1-5-1-5-1-5-1 (n = 19) is the double icosahedron shown in Fig. 1.

Fig. 1. The double icosahedron, 1-5-1-5-1-5-1 (n = 19), is shown with the four axial atoms shaded.

Clearly n = 25 is also part of this sequence. We can obtain n = 17 from n = 19 either by removing the two end atoms, which makes the structure more spheroidal, or by removing the two interior axial atoms, which makes it more compact. Finally n = 10 is surely a double square pyramid, 1, 4, 4, 1, also described as an archimedean antiprism capped with two half octahedra.[5]

Where are the Cs^+ cations now placed? As a simple example consider Cs_7I_7. For this structure the Cs ions are inserted as alternate layers between the I ions (Cs numbers in parentheses): 1(3) 5(4) 1. Thus on the average the Cs^+ ions in these two layers form a triangle and a square. However, the triangle and square are not in angular registry with the I pentagon. This means that there will be many Cs^+ sites — in this case 5 in each layer — which are only fractionally occupied (0.6 and 0.8 in this case). An exactly parallel situation holds for solid electrolytes.[6] In this case for statistical purposes the 10 sites cannot actually be distinguished and each has an occupation probability of 0.7.

Next let us consider a more complex example, $Cs_{19}I_{19}$. Again the Cs ions are inserted on alternate pentagonal layer sites denoted by (5), thus 1(5) 5 (5') 1 (5'') 5 (5'') 1(5') 5 (5) 1. We notice that there are 30 sites for 19 ions, so that the average occupation probability is 0.6. Moreover, there are three inequivalent layers, which may have slightly different occupational probabilities. The average number of cations per layer is nearly three, and alternating cation layers are probably similar in their angular orientation about the cluster axis to staggered triangles. One of the six cation layers contains four cations, but the position of this layer is statistically distributed.

Alternating solid and "liquid" layers, with the "liquid" layer sites fractionally occupied, have a lower free energy at higher temperatures because of their large configurational entropy.[6] At present our knowledge of interatomic forces is insufficient to enable us to predict, even in bulk crystals, whether such phases will be stable compared to conventional crystal phases. However, the phenomenological trends in various solid electrolytes are clear.[7] Moreover the additional softening found in clusters because of their high surface/volume ratios favors the formation of such "electrolytic" phases in clusters for a material (such as CsI) which does not form such a phase in the bulk crystal.

CONCLUSION

A simple model has been proposed to explain the novel sequence n = 7, 10, 13, 17, 19, and 25 observed[3] for $(CsI)_n$ clusters. The same sequence is observed for Pb_n^+ clusters[8] and that series can be explained by the same structural model. An obvious implication of the present model is that AgI (and possibly

CuI) clusters should also exhibit a similar sequence corresponding to "liquid" cations.

References

1. J. E. Campana, T. M. Barlak, R. J. Colton, J. J. DeCorpo, J. R. Wyatt, and B. I. Dunlap, Phys. Rev. Lett. *47*, 1046 (1981).

2. R. Pflaum, K. Sattler and E. Recknagel (unpublished).

3. R. Pflaum, K. Sattler, and E. Recknagel, Phys. Rev. *B33*, 1522 (1986).

4. J. R. Tessman, A. H. Kahn and W. Shockley, Phys. Rev. *92*, 890 (1953).

5. M. R. Hoare and P. Pal, Adv. Physics *20*, 161 (1971). In describing the pentagonal growth sequence these authors omit the present $n = 17$ structure. This is because they consider only growth by atom addition, not both growth and evaporation, and because their model with short-range forces only contains no large energy associated with surface tension, such as would be important for CsI and Pb clusters.

6. W. Andreoni and J. C. Phillips, Phys. Rev. *B23*, 6456 (1981).

7. J. C. Phillips, J. Electrochem. Soc. *123*, 934 (1976).

8. J. Mühlbach, P. Pfau, K. Sattler, E. Recknagel, Z. Phys. *B47*, 233 (1982).

SPECTROSCOPY OF Na_3 AND EXPERIMENTS ON SIZE-SELECTED METAL CLUSTER IONS

M. Broyer[1], G. Delacrétaz[2]*, P. Fayet[2], P. Labastie[1], Ni Guoquan[2],
W.A. Saunders[2], R.L. Whetten[3], J.-P. Wolf[2] and L. Wöste[2]

[1]Laboratoire de Spectrométrie Ionique et Moléculaire,
 (associé au CNRS no. 171), Université Lyon I, Bât. 205,
 43, Bd du 11 Novembre 1918, 68622 Villeurbanne Cedex, France

[2]Institut de Physique Expérimentale
Ecole Polytechnique Fédérale de Lausanne
PHB-Ecublens, CH-1015 Lausanne, Switzerland

[3]Department of Chemistry and Biochemistry
Solid State Science Center
University of California, Los Angeles 90024

*Present address : Central Analytical Department
                   Ciba Geigy LTD
                   CH-4002 Basel, Switzerland

ABSTRACT

The spectroscopy of Na_3 has been systematically investigated with
three different techniques: two-photon ionization (TPI), depletion
spectroscopy (DS), and stimulated emission pumping (SEP). Four excited
electronic states have been found in the range from 850 nm to 330 nm. The
lifetime measurement of these states suggests that the highly excited ones
are partially or totally predissociated. For this reason depletion spectroscopy has been used, and reveals the complete structure of the predissociated C-state. The Na_3 ground state has been investigated using
stimulated emission pumping. This allows a precise comparison with the
calculations.
Photofragmentation patterns of Al_n^+ are measured in a size-selected
sputtered cluster ion beam. The results exhibit Al^+ as the most probable
fragment for the cluster sizes n = 3, 5 and 7. This shows that the monomer
has a lower ionization potential than the cluster, as explained by recent
calculations. Ion-molecule reactions of nickel clusters and carbon monoxide
allow the formation of carbonyl compounds. The stoichiometry of these products correlates extremely well with simple electron counting rules.

INTRODUCTION

Cluster science lies at the interface between many traditionally
separated disciplines, incorporating aspects of molecular spectroscopy,
analytic and synthetic chemistry, surface science, solid state physics,
nuclear physics, etc., in the explanation of the diverse properties of

aggregated matter the size of a few atomic diameters. To the benefit of the field, the richness of chemical and physical properties displayed by clusters and the great promise for useful application of cluster research has drawn specialists from many disciplines. Indeed, it is the wide variety of talent devoted to cluster research which has been responsible for its extremely rapid development.

In this paper we shall explore selected aspects of cluster research which, though diverse, are united in the exploration of the physical and chemical properties of metallic clusters. In the first section of the paper we discuss recent spectroscopic results on Na_3. Spectroscopy is perhaps the most precise method of determining electronic structure and thus provides a rigorous test of electronic structure calculations.

In the second section, we discuss results from recent experiments on size-selected cluster ions. Photofragmentation studies of Ag and Al cluster ions reveal information about their electronic structure, and thus relate to both their, chemical and physical properties. In comparison, gas-phase reactivity studies of size-selected nickel clusters inform directly about the chemical properties of the clusters, from which their structural and electronic properties can be determined.

SPECTROSCOPY OF Na_3

Despite intense interest in the spectroscopy of clusters, the only clusters larger than dimers for which gas phase spectra have been measured are Na_3[1] and Cu_3[2]. Here we shall briefly summarize some recent spectroscopic investigations of the neutral sodium trimer using two-photon ionization (TPI), depletion spectroscopy (DS), and stimulated emission pumping (SEP).

The experimental apparatus has been described in detail elsewhere[3]. Briefly, the clusters are produced in supersonic expansion of Ar (nominal

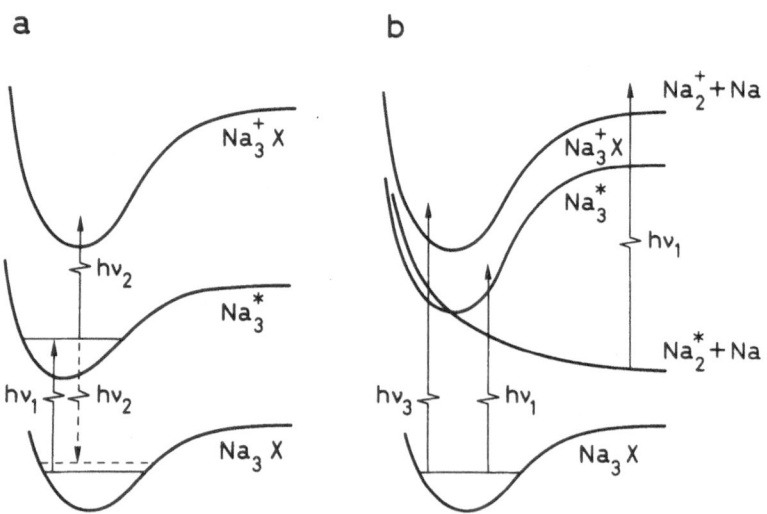

Fig. 1 : Scheme of the three techniques used for the spectroscopy of Na_3. (a) Two Photon Ionization (TPI) and Stimulated Emission Pumping (SEP) (dashed line). (b) Depletion spectroscopy (DS).

pressure 10 bar) seeded with 10 to 100 mbar of sodium through a 50 μm diameter nozzle. The cluster and laser beams intersect at right angles at the entrance to a quadrupole mass spectrometer. By setting the quadrupole to pass the appropriate mass, Na_3^+ (TPI, DS, and SEP), or Na_2^+ (as a photo-fragmentation product) signal can be recorded.

The principals of the experiments we have performed are illustrated in Figures 1a and 1b. The Na_3 molecules are electronically excited with a tunable laser $h\nu_1$. In the resonant TPI scheme (Figure 1a), the excited molecules are ionized with a second laser, $h\nu_2$, and the photoions are detected directly. If $h\nu_2$ is the proper spectral region, it can stimulate transitions from the intermediate state back to the ground-state. This is the principle of the SEP experiment. In the DS experiment (Figure 1b), which is used to probe the dissociative states of Na_3, and ultraviolet laser, $h\nu_3$ (308 nm) directly ionizes Na_3 and monitors the remaining population in the molecular beam. The photofragmentation products Na and Na_2 can also be detected by increasing the power of $h\nu_1$ sufficiently to ionize them by multiple step processes.

Results

Four electronic excited states have been observed by resonant TPI spectroscopy in the wavelength range 330 to 850 nm (Figure 2). They are labeled the A-, B-, B'- and C-states, in ascending order. Of these, only the C-state shows any evidence of being predissociative. The A-state shows only a short vibrational progression, thus implying it probably has a geometry similar to that of the ground-state. Lifetime measurements indicate it is not predissociated (Table I).

Table I : Na_3 lifetimes.

A state	60 ± 10 nS
B state	14 ± 5 nS
B' state	7 ± 3 nS
C state	7 ± 3 nS

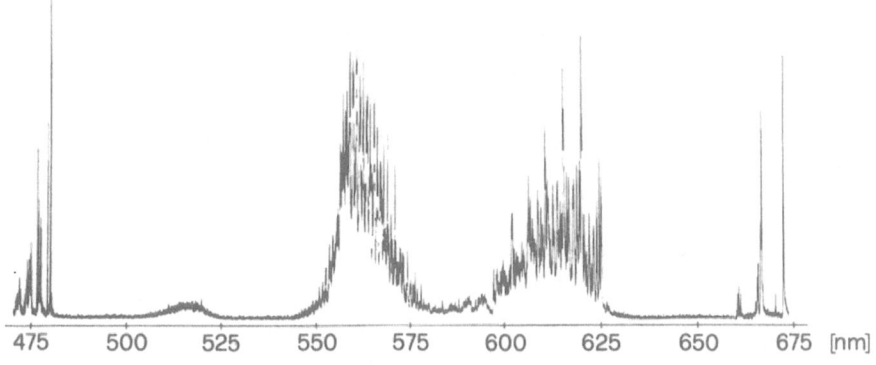

Fig. 2 : Excitation spectrum obtained by TPI.

Of the excited electronic states, the B-state is, so far, the best understood[4]. The salient features of the band can be classified as : (1) a long progression of nearly equally spaced bands (ω_0 = 128 cm^{-1}) which are split into doublets and (2) accompanying each member of the main progression there is a series of closely-spaced bands fanning out from the doublet.

It can be shown that this spectrum corresponds to the vibronic pseudo-rotation spectrum predicted by Longuet-Higgins in the case of large Jahn-Teller distortions. In that case, a simple pattern emerges given by

$$E(u,j) = (u + 1/2)\,\omega_0 + Aj^2$$

where u = 0, 1, 2, ... corresponds to the distortion amplitude and $|j|$ = 1/2, 3/2, ... to the internal pseudorotation. The quantum number j is half integral because the electronic wavefunction changes sign in a degenerate state (adiabatic sign-change theorem) upon each complete internal rotation.

The B' system is very complex and not yet well understood. The slow transition between the structure of the B system and the B' system suggests that the B' system corresponds to transitions toward a second Born-Oppenheimer surface of the same electronic excited state.

C-state is not fully observed by TPI because it is partially pre-dissociated (Figure 3). Lifetime measurements of the C-state (Table I) suggest it is predissociated. The complete structure of the state has been observed by DS and an interpretation of it as a weakly distorted state is presently under investigation.

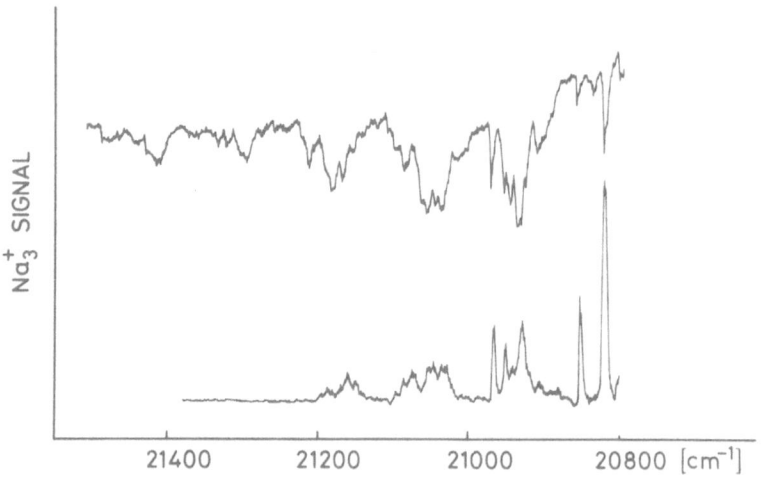

Fig. 3 : C-state spectrum obtained by TPI (lower) and DS (upper).

Another state, the dissociative D-state, has been observed by both DS and as structure on the Na$_2^+$ signal[5]. No two photon ionization signal is found for this state. Its structure is similar to that of the ionic ground state, implying the Rydberg levels of Na$_3$ may be mostly predissociated.

Theoretically, the most interesting state of the trimer is its ground-state. By altering the cluster source conditions we have been able to measure the ground-state vibrational frequencies by observing the hot-bands in the spectroscopy of the B-state [6]. We determine fundamental vibrational frequencies $\omega_1 = 139$ cm^{-1}, $\omega_2 = 49$ cm^{-1}, and $\omega_3 = 87$ cm^{-1}. These compare favorably with the values $\omega_1 = 142$ cm^{-1}, $\omega_2 = 58$ cm^{-1}, and $\omega_3 = 94$ cm^{-1} calculated from results of Martins et al. by Thompson [7]. However, hot band measurements are useful only for the lowest lying energy levels of the ground-state. To explore the higher lying levels, we have begun SEP experiments using the C-state as an intermediate state. The SEP spectrum (Figure 4) shows the rich structure of the ground state, especially the mixing of the fundamental frequencies and their relative anharmonicities. Precise interpretations of the SEP spectra are being explored.

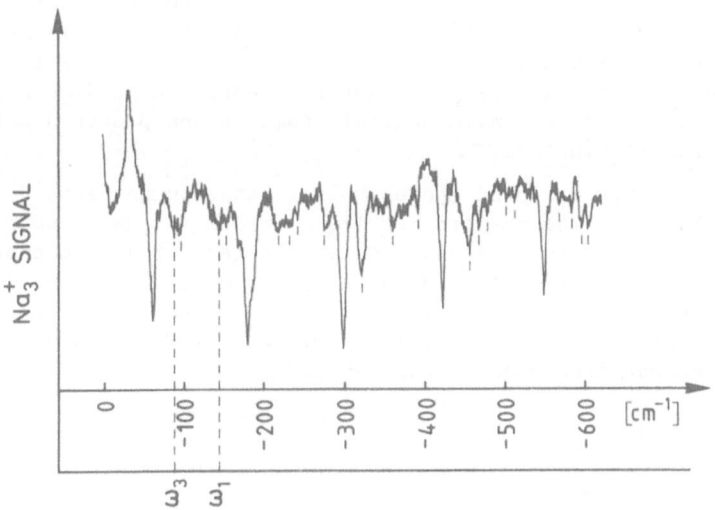

Fig. 4 : SEP spectrum using the 20'846 cm^{-1} band of the C-state as the intermediate level. The intense harmonic progression is a perturbation of the Na$_3$ signal resulting from space charge from simultaneous TPI of Na$_2$, whose B-band lies in this spectral region.

EXPERIMENTS ON SIZE-SELECTED METAL CLUSTER IONS

Molecular beam expansion techniques are well established as valuable tools for investigating and cluster properties. However, the experiments are handicapped in two regards, (i) the range of materials which can be studied with them is limited, and (ii) fragmentation phenomena are difficult to measure without mass selection prior to experimentation.

For these reasons, we have developed a SIMS cluster ion source coupled with a triple-quadrupole arrangement to study the properties of metal cluster ions. In this section, we shall discuss photofragmentation and chemical reactivity studies of size selected metal clusters. The photofragmentation data are shown to be consistent with recent calculations of cluster ionization energies [8]. The ionization energies, in turn, are known to reflect on the chemical properties of the clusters [9]. From the chemical reactivity studies, on the other hand, we are able to determine a correspondence with the physical structures of the clusters.

Apparatus

The apparatus has been described in detail elsewhere [10]. Briefly, metal clusters sputtered from an ambient temperature target (typically by 17 keV Ar^+ ions with an on-target current of 2 mA) are first energy filtered and then introduced to the first quadrupole, where they are mass filtered. The clusters interact with the reactant gas or a laser beam in the second quadrupole, which is operated in a non-discriminative ion-trapping mode. The third quadrupole is used to analyze the fragmentation or reaction products.

Photofragmentation of Ag and Al clusters

As shown above for Na_3, photofragmentation is a reliable spectroscopic tool for studying cluster properties. The relative distribution of photo-fragments also gives useful information about electronic structure. For instance, the results of Bloomfield et al.[11] show that Si_6^+ is an especially abundant fragmentation product, in agreement with recent calculations of Si cluster stabilities [12]. Photofragmentation studies have also shown that evaporation of single neutral atoms is the prefered fragmentation pathway for Fe clusters [13].

We have recently measured the photofragmentation patterns for size selected Al clusters. The results for Al_N^+, with N = 3, 5, 7 and 9 are shown in Figure 5. For these measurements, the laser beam was chopped and the ion signal was recorded phase-sensitively. Thus, the parent ion signals are negative, corresponding to their depletion by the laser beam, while fragment ion signals are positive. It is evident from the data that Al^+ is a prefered fragmentation product for N = 3, 5, and 7.

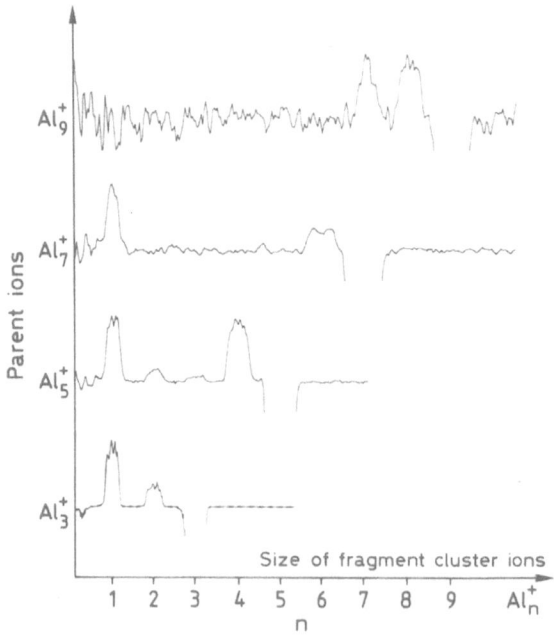

Fig. 5 : Fragmentation signals of aluminum cluster ions, Al_n^+, n = 3, 5, 7 and 9. The atomic ion is an abundant product for the smaller clusters.

For comparison, we display the fragmentation curves obtained for Ag_N^+, N = 3, 5, 7 and 9 in Figure 6 [10,14]. Ag clusters preferentially fragment into a charged product containing an even number of electrons, and except for Ag_3^+, the most abundant product is the next lowest odd cluster. These observations are in accord with the known odd-even alternations in the binding energies of noble metal clusters [15].

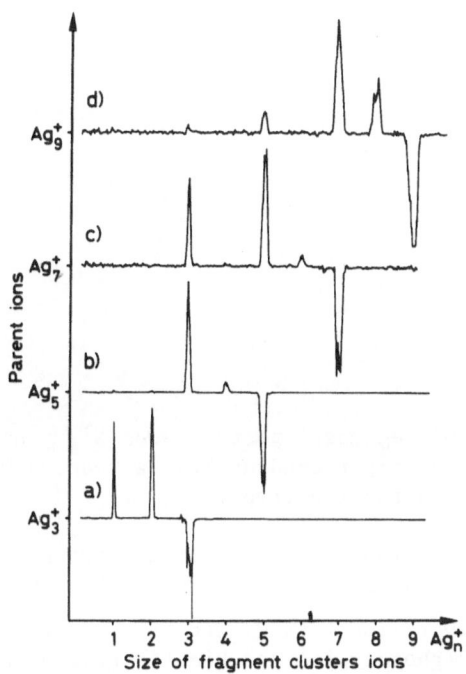

Fig. 6 : Fragmentation signals of silver cluster ions, Ag_n^+, n = 3, 5, 7 and 9. Fragments with even numbers of electrons are favored.

The striking difference between the results for the Ag and Al clusters is explained by recent calculations [8] which show that the Al atom has a lower ionization energy than any cluster up to Al_6. Thus, it is energetically favorable for an Al atom to carry away the charge from a fragmenting cluster, in contrast to the cases of Ag and Fe, where the ionization energies of the atoms are higher than those of the clusters [16-17].

Ni cluster carbonyls

The chemical reaction of size selected clusters also tell about their structural properties. In collaboration with M.J. McGlinchey [18], we have recently measured the reaction products of size-selected Ni cluster ions with CO. Shown in Figure 7 is the mass spectrum which results when Ni_4^+ is reacted with CO at a pressure of 1.3×10^{-3} mbar in the ion trap. Both carbonyls, $Ni_4(CO)_k^+$ and carbides, $Ni_4C(CO)_1^+$ are observed. The mass spectrum shows all the carbonyls (k = 1 to 10) with an especially enhanced ced abundance at k = 9. No carbonyl is observed above k = 10, indicating that 10 CO ligands saturate the cluster.

Similar mass spectra have been recorded for the Ni_x^+, with x = 2 through 13. With x = 6, the maximum k = 13, for x = 10, the maximum k = 18. The stoichiometry of the carbonyls can be understood using methods based on empirical electron counting rules. For example, a tetrahedral cluster will be maximally stabilized when the total number of valence electrons is 60. This total is made up from the metal valence electrons (10 per Ni atom) and those contributed by the ligand (2 per CO molecule). Thus, $Ni_4(CO)_{10}$ contains the appropriate number of electrons to be stabilized

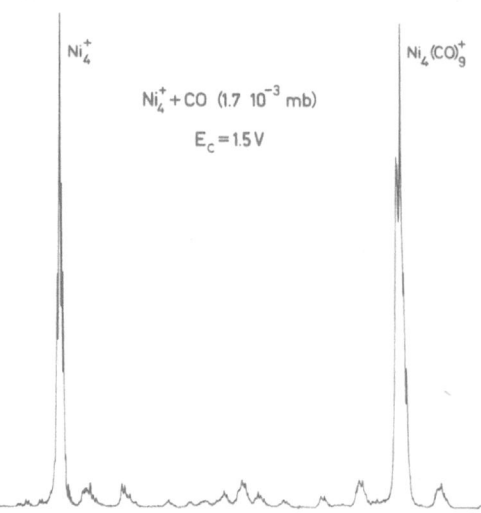

Ni_4^+

$Ni_4^+ + CO$ (1.7 10^{-3} mb)

$E_c = 1.5 V$

$Ni_4(CO)_9^+$

Fig. 7 : Resulting mass spectrum when Ni_4^+, selected
          by the first quadrupole, is reacted with CO
          gas in the ion trap.

in a tetrahedral configuration. The geometrical arrangements of all the Ni
clusters as determined by similar methods are to be published elsewhere [18].

Recently, in collaboration with C. Brechignac, we have begun studying
the photofragmentation (photolysis) and kinetic growth of the Ni cluster
carbonyls.

CONCLUSIONS

In this paper we have discussed some recent experimental results on a
wide range of systems. The richness of cluster properties demands a battery
of experimental and theoretical tools, not all of which can be applied to
any single cluster system with a guarantee of success. However, the boundary
between the physical and chemical properties of clusters is not rigid. Thus,
knowledge about the electronic states of Na_3 or from the photofragmentation
of cluster ions helps to understand their chemical properties. Conversely,
what we learn from the chemical properties of clusters helps to understand
their physical properties.

ACKNOWLEDGMENTS

The authors wish to thank the Swiss National Science Foundation for
supporting this project.

One of us (WAS) acknowledges support by NSF grant INT-8603428.

REFERENCES

1. A. Herrmann, S. Leutwyler, E. Schumacher and L. Wöste, Chem. Phys. Lett.
   62 : 216 (1979).
   G. Delacrétaz and L. Wöste, Surf. Sci. 156 : 770 (1985).
2. M.D. Morse, J.B. Hopkins, P.R.R. Langridge-Smith and R.E. Smalley,
   J. Chem. Phys. 79 : 5316 (1983).
3. G. Delacrétaz, Thesis 603, EPFL (1985).
4. G. Delacrétaz, E.R. Grant, R.L. Whetten, L. Wöste and J. Zwanziger,
   Phys. Rev. Lett. 56 : 2598 (1986).
5. M. Broyer, G. Delacrétaz, P. Labastie, J.P. Wolf and L. Wöste,
   Phys. Rev. Lett. 57 : 1851 (1986).

6. M. Broyer, G. Delacrétaz, P. Labastie, J.P. Wolf and L. Wöste, to be published in J. Phys. Chem..

7. J.L. Martins, R. Car and J. Buttet, J. Chem. Phys. 78 : 5646 (1983). T.C. Thompson, G. Izmirlian, S.J. Lemon, D.G. Truhlar and C.A. Mead, J. Chem. Phys. 82 : 5597 (1985).

8. T.H. Upton, Phys. Rev. Lett. 56 : 2168 (1986).

9. R.L. Whetten, D.M. Cox, D.J. Trevor and A. Kaldor, Phys. Rev. Lett. 54 : 1494 (1985).

10. P. Fayet and L. Wöste, Surf. Sci. 156 : 134 (1985).

11. L. Bloomfield, R. Freeman and W. Brown, Phys. Rev. Lett. 54 : 2246 (1985).

12. D. Tomanek and M.A. Schlüter, Phys. Rev. Lett. 56 : 1055 (1986).

13. P.J. Brucat, L.-S. Zheng, C.L. Pettiette, S. Yang and R.E. Smalley, J. Chem. Phys. 84 : 3078 (1986).

14. P. Fayet and L. Wöste, Spectros. Int. J. 3 : 91 (1984).

15. G. Blaise and G. Slodzian, C.R. Acad. Sci. Paris, Ser. B 266 : 1525 (1968), I. Katsuke and T. Ichihara, Int. J. of Mass. Spect. and Ion Proc. 67 : 229 (1985), and W. Begemann, K.H. Meiwes-Broer and H.O. Lutz, Phys. Rev. Lett. 56 : 2248 (1986).

16. J.L. Martins and W. Andreoni, Phys. Rev. A 28 : 3627 (1983).

17. E.A. Rohlfing, D.M. Cox, A. Kaldor and K.H. Johnson, J. Chem. Phys. 81 : 3846 (1984).

18. P. Fayet, M.J. McGlinchey and L. Wöste, J. Amer. Chem. Soc. (to be published).

DENSITY FUNCTIONAL CALCULATION OF THE FRAGMENTATION OF NEUTRAL, SINGLY AND DOUBLY-IONIZED SPHERICAL JELLIUM-LIKE METALLIC MICROPARTICLES

M.P. Iñiguez, L.C. Balbás and J.A. Alonso

Departamento de Física Teórica
Universidad de Valladolid, Valladolid, Spain

ABSTRACT:

The fragmentation of neutral and ionized metallic clusters is studied by using the density functional method and the spherical jellium model. The peculiarities of the fragmentation are controlled by the tendency for fragments with a "magic" number of electrons, mainly 2 or 8. We have also obtained that, provided that we consider parent clusters with the same number of electrons, the fragmentations of neutral and ionized clusters are very similar.

INTRODUCTION

In a previous paper |1| we have studied the fragmentation of doubly-ionized Sodium or Magnessium clusters into two singly-ionized fragments using the Density Functional formalism |2| and the spherical jellium model. In this model the valence electrons move in a common effective potential with spherical symmetry about the cluster center. The model has been successful in describing the relative abundance (magic numbers) of Na and K clusters produced in supersonic expansions of the metallic vapours |3, 4| . Assuming that the relative abundances are correlated to the cohesive energies, the result of the model is that a magic number corresponds to a number of atoms such that the number of electrons in the cluster is just enough to fill electronic shells. We must point out that the jellium model is more appropriate for studying hot clusters, like those presumably produced in Knight's experiments. Recent experiments by Schumacher |5| show that geometrical structure effects begin to appear in the mass distribution of Na clusters which are presumably produced in a colder state.

In this work we extend our previous calculations to study the fragmentation mode of $(Na_n)^{2+}$ clusters into Na_p and $(Na_{n-p})^{2+}$ fragments. In addition we explore the fragmentation of Na_n and $(Na_n)^{+}$, with $n \leqslant 90$ atoms. Fragmentation is a phenomenon that can occur during the process of production and detection of clusters, especially because the clusters have to be ionized prior to the detection in a mass spectrometer.

RESULTS

We have studied the following fragmentation reactions

$$Na_n \longrightarrow Na_p + Na_{n-p} \tag{1}$$

$$(Na_n)^+ \longrightarrow Na_p + (Na_{n-p})^+ \tag{2}$$

$$(Na_n)^{2+} \longrightarrow Na_p + (Na_{n-p})^{2+} \tag{3}$$

with $n \leqslant 90$. This complements our previous study of the reaction

$$(Na_n)^{2+} \longrightarrow (Na_p)^+ + (Na_{n-p})^+. \tag{4}$$

The heats of reaction $H^o(n;p)$, $H^1(n;p)$ and $H^2(n;p)$, corresponding to the reactions (1), (2) and (3) respectively, have been calculated by difference of total energies of parent and fragments. That is

$$H^o(n;p) = E\left[Na_p\right] + E\left[Na_{n-p}\right] - E\left[Na_n\right]. \tag{5}$$

The fragmentation is spontaneous if $H^o < 0$ and needs an activation energy if $H > 0$. Analogous expressions hold for H^1 and H^2.

If we consider, for instance, reaction (1), with fixed n and varying p, the smallest $H^o(n;p)$ gives the most favorable fragmentation channel. We have obtained that the most favorable channel is controlled by the tendency of the fragments to be magic, that is, to have filled electron shells. This confirms our earlier findings obtained from the study of reaction (4). In the size range considered here the closed-shell numbers are 2, 8, 18, 20, 34, 40, 58 and 68. Shell effects are more important for small clusters, and in fact it is generally the small fragment the one found to be magic, with 2 and 8 electrons as the most important cases. There are exceptions, however, in which the big fragment is the magic one, as in the cases n = 3, 9, 13, 14 and 15 for reaction (1). Obviously the smaller fragment can not be magic for Na_3, and for the other four exceptions the magic fragment has 8 electrons, which reveals the competition between the fragmentation channels p = 2 and p = 8. In a similar way the most probable channels for reactions (2) and (3) lead to a magic fragment which is the small one. This happens with few exceptions, these occurring for n = 15, 16, 20, 78 and n = 5, 6, 7, 8, 9, 10, 11, 17, 20 for reactions (2) and (3) respectively. Calculations by Ishii et al |6| including spin polarization and self-interaction corrections also indicated that fragments with 2 or 8 electrons appear preferentially in the fragmentation of Na_n clusters with $n \leqslant 25$.

An interesting observation about the most favorable path in reactions (2) and (3) is that the charged fragment is the big one and the neutral is the small one in practically all cases. This can be explained by a conducting sphere picture, the electrostatic energy of a charged cluster varying inversely with the sphere radius. Thus for the most favorable channel the small fragment is at the same time magic and neutral, and this occurs with very few exceptions.

In Figure 1 we represent H^o and H^1, for the most favorable channel, as a function of the number of electrons in the parent cluster. A similar representation for H^2 appears as the upper curve in Figure 2. It turns out that the values of H^o, H^1 and H^2 are nearly equal. This certainly happens because in the most favorable channel the small (and neu-

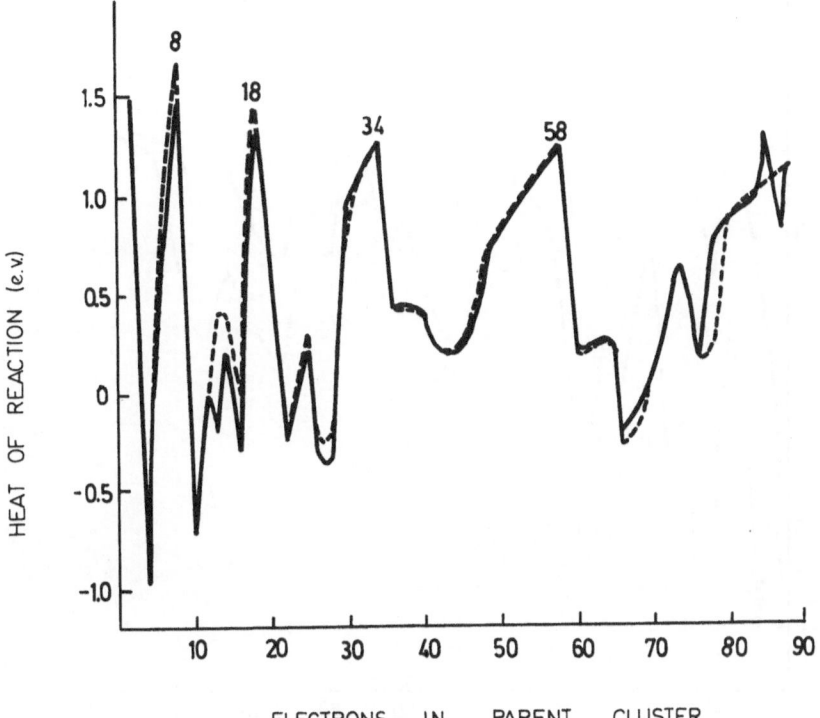

Fig. 1. Heat of reaction for the most favorable channel versus the number of electrons in the parent cluster. ——————— reaction
(1); — — — reaction (2)

tral) fragment generally is the same for these three reactions. Since
the horizontal axis in the figures is the number of electrons in the parent cluster, then the similarity observed occurs between the heats of
the following reactions

$$Na_N \longrightarrow Na_p + Na_{N-p} \tag{6}$$

$$(Na_{N+1})^+ \rightarrow Na_p + (Na_{N-p+1})^+ \tag{7}$$

$$(Na_{N+2})^{2+} \rightarrow Na_p + (Na_{N-p+2})^{2+} \tag{8}$$

for fixed N (and p = 2 or 8 generally). From the equality of heats of
reactions (6) and (7) it is easy to show the relation

$$I[Na_N] + E_S[Na_N] = I[Na_{N-p}] + E_S[Na_{N-p}], \tag{9}$$

where I is the first ionization potential and E_S is the sublimation
energy, $E_S[Na_N] = E[Na_N] - E[Na] - E[Na_{N-1}]$.

A comparison of the heats of the most favorable channels of reactions (3) and (4) is given in Figure 2. Reaction (4) is always preferred.
This agrees with one's intuition and reflects the conducting sphere
behaviour of the clusters. Nevertheless, the difference in energy

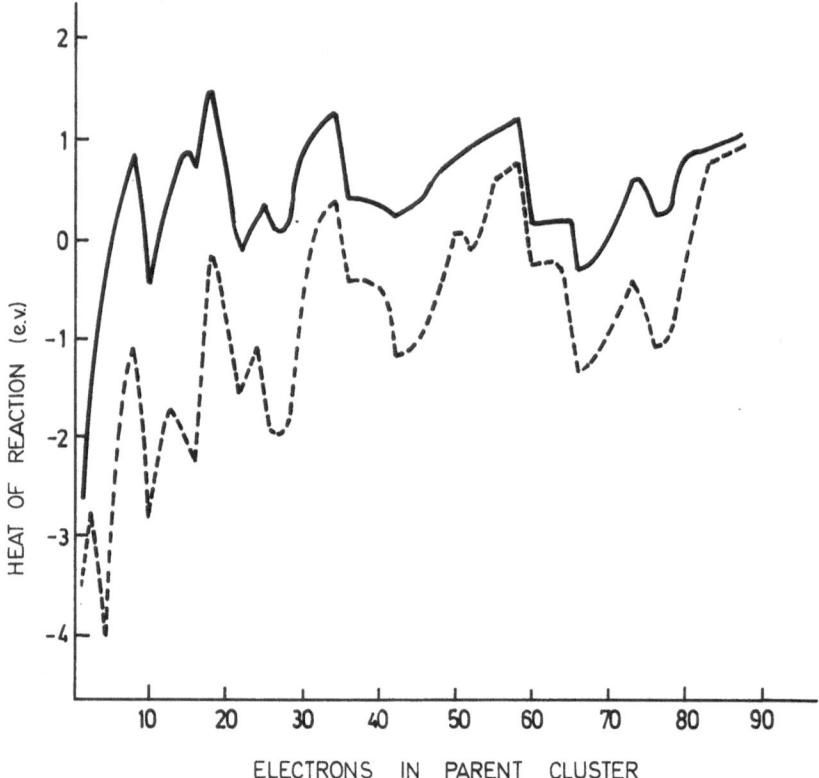

Fig. 2. Heat of reaction for the most favorable channel versus the number of electrons in the parent cluster. ————— reaction (3)
— — — — reaction (4).

between the two processes disminishes somehow as n grows.

CONCLUSIONS AND COMMENTS

The fragmentation of neutral and charged clusters according to reactions (1), (2) and (3) has been studied. We have obtained that formation
of magic fragments dominates these processes. Also, the modes and the
heats of these reactions are practically equal if we take parent clusters with the same number of electrons, which correspond to reactions
(6), (7) and (8).

If double ionization occurs during the process of cluster detection then the spontaneous fragmentation which occurs for most of the
$(Na_n)^{2+}$ clusters leads to an increase in the population of $(Na_3)^+$ and
$(Na_9)^+$ clusters. However the energy required to form $(Na_n)^{2+}$ is large
and this process can easily be avoided under the usual experimental conditions. The fragmentation of $(Na_n)^+$ clusters is spontaneous for only a
few values of n, as shown in Figure 1. Consequently reaction 2 will have
a rather small influence in modifying the original population of neutral
clusters. Induced fragmentation of Na_n and $(Na_n)^+$ clusters is possible
with a rather small activation energy. This process could modify the
original population, enhancing the presence of neutral Na_2 and Na_8 clus-

ters. However this process can be avoided to a large extent by carefully selecting the photon energies used for ionization.

ACKNOWLEDGEMENTS

This work was supported by CAICYT: Grant 3256-83.

REFERENCES

1   M.P. Iñiguez, J.A. Alonso, M.A. Aller and L.C. Balbás, Phys. Rev. B34:2152 (1986).
2   P. Hohenberg and W. Kohn, Phys. Rev. 136: B864 (1964).
3   W.D. Knight, K. Clemenger, W.A. de Heer, W.A. Saunders, M.Y. Chou and M.L. Cohen, Phys. Rev. Letters 52: 2141 (1984).
4   M.Y. Chou, A. Cleland and M.L. Cohen, Solid State Commun.52: 645 (1984).
5   E. Schumacher, Proc. 6th Internat. Conf. Liquid and Amorph. Metals (Garmisch, 1986). To be published in Z. Phys. Chem.
6   Y. Ishii, S. Ohnishi and S. Sugano, Phys. Rev. B33: 5271 (1976).

EXPERIMENTS ON SPUTTERED CLUSTERS AS PROBE OF METAL CLUSTER ION STABILITY

W. Begemann, S. Dreihöfer, K. H. Meiwes-Broer and H. O. Lutz

Fakultät für Physik, Universität Bielefeld
4800 Bielefeld 1, West Germany

ABSTRACT

A variety of metal cluster ions is produced by sputtering under UHV conditions. Mass selected beams of these species undergo collision induced fragmentation (CIF) in rare gas and oxygen gas targets under single collision conditions, the fragments being analyzed in a time-of-flight reflectron. The striking similarity between fragmentation rates, the abundances in mass spectra, and unimolecular decay rates suggest that such measurements directly reveal the cluster ion stabilities.

1. INTRODUCTION

Experiments on gas phase clusters have overcome the "preparative phase", i.e. in many cases the species can be produced with reliable and reproducible methods. Effort is now concentrating on physical and chemical properties of mass-selected, isolated clusters. Mass selection requires them to be charged; therefore, we prefer to work on charged clusters without having to fear that ionization processes might lead to a misinterpretation of experimental results. The investigated species are sputtered metal-cluster ions, the spontaneous behaviour of which revealed a strong relationship between cluster size specific unimolecular decay rates and ion abundances in the spectra /1/, and turned out to be a measure of the cluster's stability. A largely independent method to study cluster stabilities is their collision induced fragmentation (CIF) in thin gas targets, i.e. under single collision conditions. A similar method has first been applied to hydrogen cluster ions by van Lumig et al. /2/ and yielded fragmentation cross sections as function of collision energy and mass.

The experiments described here investigate CIF of 1.8 keV mass selected Al_n^+, Cu_n^+ and W_n^+ in low pressure rare gas and oxygen gas targets. The aim is to obtain information concerning the stability of these clusters.

2. EXPERIMENT

The apparatus is described in detail elsewhere /3/. In short (see Fig. 1), cluster ions are produced in a UHV machine by 20 keV Xe_2^+ ion beam bombardment of thick metal targets (spot diameter 2 mm, 100 $\mu A/cm^2$). The (positively) charged secundary ions are accelerated over a preselected distance, undergo velocity (mass) selection by means of a Wien filter and reach the scattering chamber after having traversed a drift tube. For the suppression of neutral clusters the drift tube behind the Wien filter is tilted by an angle of $\alpha = 2 \ldots 5^o$.

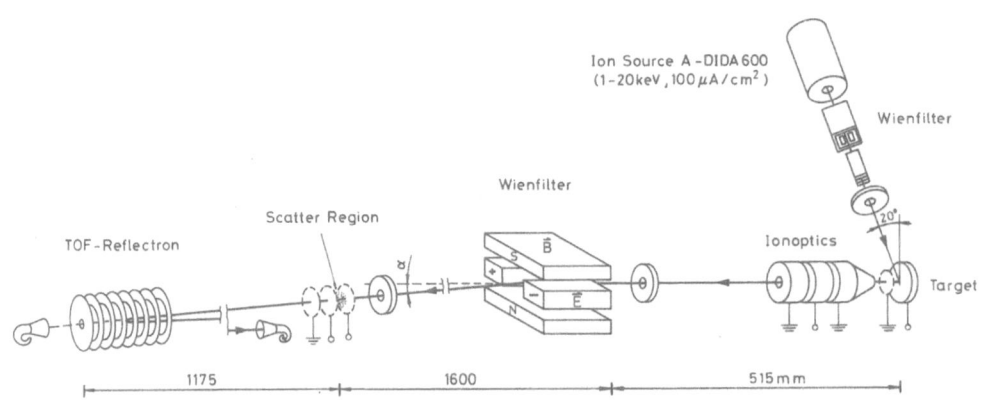

Fig. 1: Experimental setup

Mass identified cluster ions, pulsed by either a grid system inside or by a pair of chopper plates in front of the scattering chamber, impinge onto the $1 \ldots 5 \times 10^{-5}$ mbar gas target. Surviving parent ions as well as all charged fragments which have been scattered by an angle of not more than ~2° with respect to the parent ion direction will be reflected in the time-of-flight (TOF) reflectron and detected by a tandem channel plate. Besides offering several advantages, as improvement of the mass resolution and the possibility to study metastable decompositions /4/, the reflectron turnes out to be useful for investigations of CIF: After exitation in a collision event, a cluster ion might decompose and thus loose a considerable amount of kinetic energy whereas the velocity remains nearly unchanged (small angle scattering). As the penetration depth into the reflector is a function of energy, the fragments will be well separated from the parents and appear as narrow lines in the TOF spectrum.

Note that we employ <u>doubly</u> differential pumping between the scattering chamber and the target chamber when reactive collision gas is used. A contamination of the sputter target by e.g. a partial pressure of only ~1×10^{-7} mbar oxygen changes the cluster ion production conditions severely so that a considerable amount of metal-oxides is sputtered; even the bare metal cluster intensities will be affected.

3. RESULTS AND DISCUSSION

Before analyzing the CIF, the ion beam composition is studied by chopping the cluster beam with the Wien filter switched off.

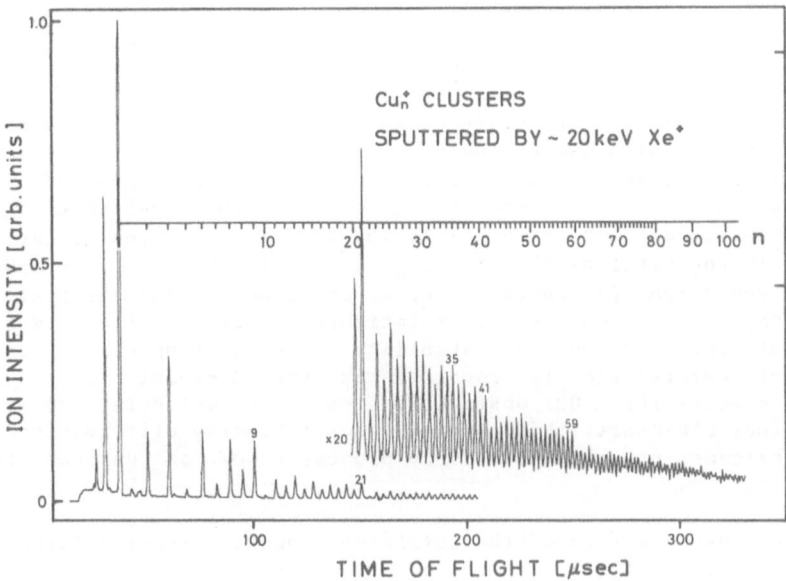

Fig. 2: TOF spectrum of sputtered Cu_n^+.

As typical examples Figs. 2 and 3 show the mass-resolved parts of TOF spectra of sputtered Cu_n^+ and Mo_n^+. It should be noted that the recorded spectra reveal the beam composition several μs after the production process, i.e. when the clusters have reached the chopper. We showed earlier that at this position (~1.9 m away from the target) the metastable decomposition of 1.8 keV sputtered metal cluster ions has largely ceased /1,3/. Recording the ion beam composition at much earlier times by e.g. chopping the primary Xe^+ ion beam results in different spectra.

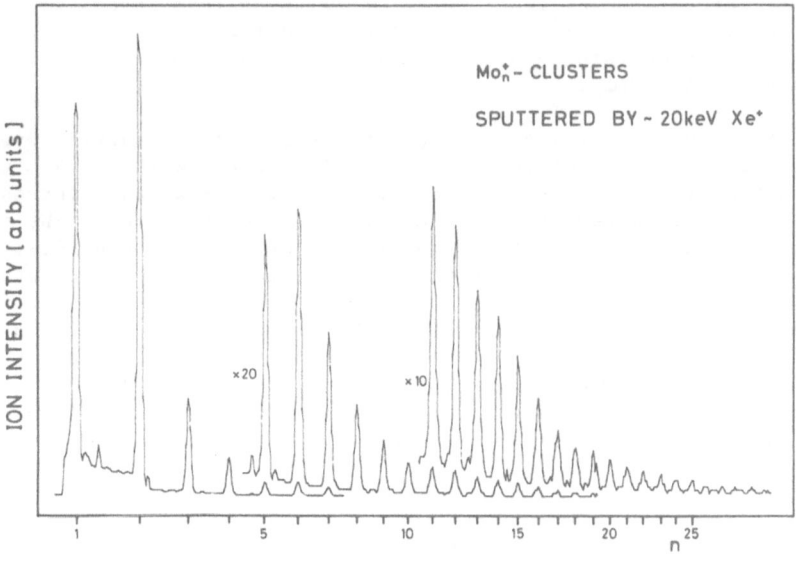

Fig. 3: TOF spectrum of sputtered Mo_n^+.

The Cu_n^+ intensities show discontinuities with increasing cluster size similar to those recorded by other methods /5,3/. We confirm the regions of high abundances at n = 9, 21, (35), 41 and 59. The even-odd intensity variations of sputtered Cu_n^+ have been attributed to the increased stability of clusters possessing spin-paired binding electrons. When recording <u>negative</u> ions we obtain a similar spectrum: The even-odd effect remains, but outstanding lines are shifted to n-2 for Cu_n^-.

In Mo_n^+ very little structure disturbs the exponential decrease of the ion intensity with increasing number of atoms n. The same holds true for W_n^+. There are some intermediate cases: E.g. aluminium and silicon cluster ions do not show a regular even-odd structure, but there are pronounced intensity drops after n = 7,14 and 23 for Al_n^+, and after n = 6 for Si_n^+. In the cases of Al_n^+ and Cu_n^+ all shell closings as proposed by Ekardt /6/ and Knight /7/ appear as prominent lines in the spectra. This is a strong indication for a relationship between the intensity of <u>sputtered</u> cluster ions and their stability. Such a connection has been pointed out before and is regarded as the dominant factor in the <u>production</u> process /8/. Our observations on the unimolecular decay /1/, however, indicate that the complete ion intensity distribution is the result of cascades of metastable decay processes of an unknown initial distribution.

For the investigation of the collision induced fragmentation (CIF) the Wien filter is utilized. Pure beams of mass-identified cluster ions, now pulsed, hit onto the low pressure (typically $1...5 \times 10^{-5}$ mbar) gas target. Parent clusters as well as collision-induced fragments appear as well resolved sharp lines in the TOF mass spectra almost irrespective of their original kinetic energy spread.

Much informations can be obtained by this method, including attenuation and fragmentation cross sections as function of n, collision energy and target gas as well as chemical reaction paths into defined product channels. Three different examples of the CIF of 1.8 keV metal cluster ions are given in Fig. 4.

The collision systems are quite different in their fragmentation behaviour. The system W_n^+-Kr (Fig. 4a) shows monotonous decay into W_i^+, i<n. Mainly large fragment ions are formed with uniform intensity distributions. A similar behaviour is e.g. observed for Mo_n^+ clusters; other metal cluster ions containing atoms with an even number of electrons should behave analogously.

The CIF of Cu_n^+ in Ar and O_2 (Fig. 4b) results in a marked even-odd oscillation of the fragment intensity for all parents with n > 3. Closer inspection shows that (for Ar) also the sums of the fragment intensities oscillate with the number of atoms in the parent cluster ions. We expect a similar behaviour of CIF in other monovalent atom metal clusters. The O_2 target enhances the CIF cross section into selected decay products.

For Al_n^+ (Fig. 4c) a smooth fragment distribution is observed with the exception of the outstanding Al^+ fragment intensity. The CIF cross section in O_2 is throughout larger than in Ar. The extraordinary fragmentation into Al^+ is consistent with recent measurements of low (1 eV center-of-mass) energy collisions of Al_n^+ in oxygen /9/: The 'reaction-induced' fragmentation results in strong production of Al^+ for all parents with $n<13$. The unimolecular decay measurements, on the other hand, do not give any hint that sputtered Al_n^+ decomposes by Al^+ emission /1/. We conclude that dissociation into Al_n^+ proceeds fast, i.e. less than 10...50 ns after the excitation (sputter process or collision).

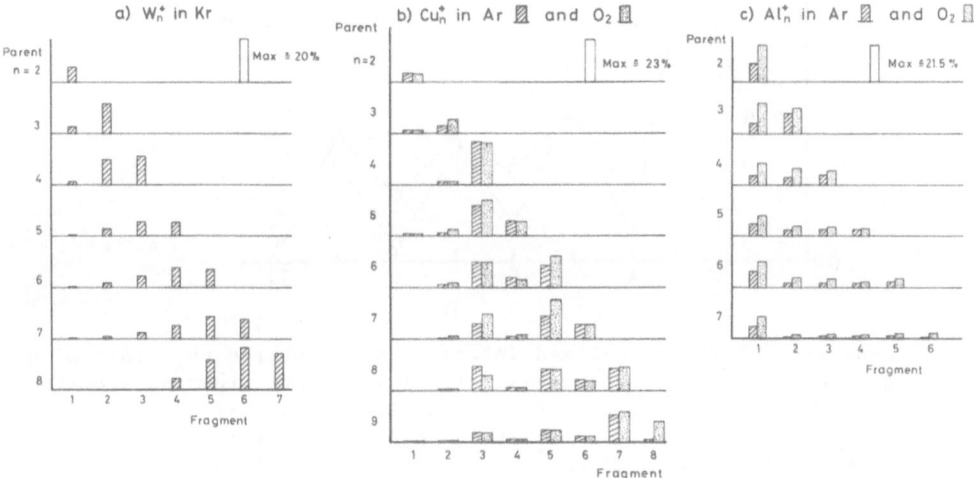

Fig. 4: Intensities of collision-induced decay products of mass identified parent cluster ions into defined fragments. Gas targets are 5×10^{-5} mbar rare gas (striped bars) or O_2 (dotted bars). The bar heights give relative intensities, i.e. the intensity of the fragment divided by the sum of the surviving parent and all fragment intensities.

As the detection angle of our apparatus is small ($\approx 10^{-2}$ rad) we investigate small angle collisions in which energy is transferred in a grazing process; the excited cluster ion decomposes mainly by metastable fragmentation in which the excitation energy relaxes during cascades of evaporations. Therefore the fragmentation patterns show broad mass distributions. At first glance this might seem to be in contrast to the observations of the metastable decompositions of sputtered metal cluster ions where mainly $(n-1)$ decay has been seen /1,3/. However, in those measurements the experimental time window opened >50 ns after the production ($\equiv$ excitation) so that really fast processes could not be seen. The method of fragmentation analysis by means of a reflectron integrates over all "stable" decay products originating in a time window which opens directly at the collision (exitation) event and closes when the ions reach the reflector.

CIF is a direct measure of the cluster's stability. Accordingly there should be a strong relationship between CIF rates and ion abundances in mass spectra of sputtered clusters. This is indeed the case as is shown in Fig. 5 for the example of Cu_n^+:

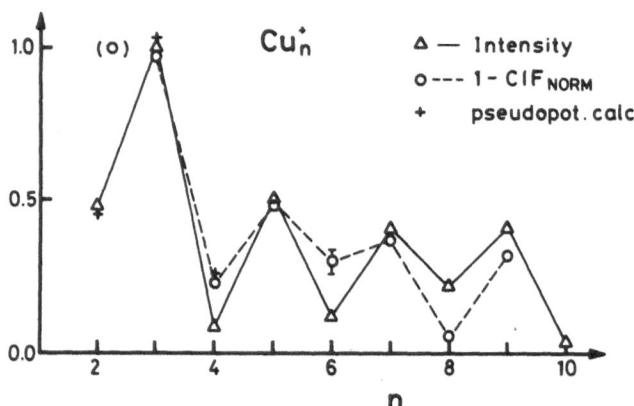

Fig. 5: Comparison of the normalized intensity of sputtered Cu_n^+ (Δ) with the complement of the sums of all fragments stemming from the respective parent cluster after collision in an Ar gas target, $1-CIF_{Norm}$. (X) denotes results of pseudopotential calculation of the binding energies of Cu_n^+, normalized to E_B of Cu_3^+ (= 3.15 eV) /10/.

The striking similarity between ion abundances and $1-CIF_{Norm}$ suggests that both, the CIF at small angle collisions and the beam composition of sputtered, <u>relaxed</u> Cu_n^+ are independent measures of their relative stability. The calculations of Cu_n^+ binding energies by Flad et al. /10/ fit into this pattern.

The unimolecular decay rates of Cu_n^+ /1/ cannot be compared directly to the data of Fig. 5 as their exponential dependence on the degrees of freedom generates a steep slope with n. Nevertheless there exists a qualitative accordance between the oscillations of the CIF cross sections and the decay rates.

It would be interesting to compare rare gas CIF patterns of metal cluster ions to the photofragmentation (PF) of the same species. Unfortunately there are no results available yet for the systems described above. Instead we compare the PF behaviour of Ag_n^+, the spectrum of which is quite similar to that of Cu_n^+, to the CIF of Cu_n^+ in a rare gas target. The PF of Ag_5^+ by 488 nm light /11/ results in a fragment pattern which is very similar to our CIF results of Cu_5^+, see Fig. 4. This agreement may be typical for cases in which fast ($\tau < 10^{-13}$ sec) photofragmentation, i.e. photodissociation, can be excluded. After randomization of the excess energy into vibration the cluster will have lost its memory of the exciting process, thus decomposing according to the rules of the metastable decay.

We expect that direct photodissiciation would result in different patterns. In addition, resonant multiphoton absorption might lead to increased PF cross sections of selected cluster ions.

We conclude that experiments on sputtered cluster ions, yielding a) mass spectra, b) unimolecular decay rates and c) collision induced fragmentation patterns, obtained under single collision conditions, are complementary independent methods for the investigation of the metal cluster ion stabilities.

REFERENCES

1. W. Begemann, K. H. Meiwes-Broer, and H. O. Lutz, Phys. Rev. Lett. 56:2248 (1986).
2. A. van Lumig and J. Reuss, Int. J. Mass Spectrom. Ion Phys. 25:137 (1977).
3. W. Begemann, S. Dreihöfer, K. H. Meiwes-Broer, and H. O. Lutz, Z. Phys. D 3:183. (1986).
4. U. Boesl, H.J. Neusser, R. Weinkauf, and E.W. Schlag, J. Phys. Chem. 86:4857 (1982)
5. G. Blaise, G. Slodzian, C. R. Acad. Sc. Paris, Ser. B 266:1525 (1968).
    I. Katakuse, T. Ichihara, Y. Fujita, T. Matsuo, T. Sakurai, and H. Matsuda, Int. J. Mass Spectr. Ion Proc., 67, 229 (1985)
6. W. Ekardt, Phys. Rev. B 29:1558 (1984).
7. W. D. Knight, K. Clemenger, W. A. de Heer, W. A. Saunders, M. Y. Chou, and M. L. Cohen, Phys. Rev. Lett. 52:2141 (1984).
8. P. Joyes, P. Sudrand, Surf. Sci. 156, 451 (1985)
9. M. F. Jarrold and J. E. Bower, to be published
10. J. Flad, G. Igel-Mann, H. Preuss, and H. Stoll, Chem. Phys. 90:257 (1984).
11. P. Fayet, L. Wöste, Spectros. Int. J.3:91 (1984)

VELOCITY MEASUREMENTS OF BISMUTH CLUSTERS :

VELOCITY SLIP AND FRAGMENTATION PROCESSES

M. Broyer[*], B. Cabaud[**], A. Hoareau[**], P. Melinon[**],
D. Rayane[*] and B. Tribollet

[*] Laboratoire de Spectrométrie Ionique et Moléculaire
(U.A. CNRS n°171)

[**] Département de Physique des Matériaux
(U.A. CNRS n°172)

Université Cl. Bernard - Lyon I, Campus de La Doua
69622 Villeurbanne Cedex, France

ABSTRACT

Bismuth clusters are produced by the inert gas condensation technique. The clusters velocity is studied systematically as a function of
the helium pressure. A large velocity slip is observed between He and the
clusters. This velocity slip depends on the mass cluster and allows us to
select in size the neutral clusters. Moreover this mass selectivity enables
us to study the fragmentation of bismuth clusters (as a function of the
ionizing energy).

*   *   *

The inert gas condensation technique[1,2] has been demonstrated to be
a very efficient method to produce small metallic clusters M_n in the range
$10 < n < 100$. While the clusters produced in seeded molecular beams or by
laser vaporisation in a supersonic beam are known from spectroscopic measurements to be very cold, nothing is known about the temperature of
clusters produced by the inert gas condensation method. In the present
paper, we report velocity measurements of small bismuth clusters. These
experiments allow us to deduce the translational temperature of the inert
gas. For the clusters, we observe a velocity slip phenomenon. This phenomenon precludes the cooling of the clusters in the expansion. This velocity
slip also allows a rough mass selection on neutral clusters and enables us
to study the fragmentation processes by electron impact ionization.

The cluster source is very similar to that developed by Sattler et
al.[1] The bismuth metal is heated in an opened crucible. The helium gas
vector, initially to the liquid nitrogen temperature is introduced
through a metallic ring located just above the crucible. The outer mantle
of the oven chamber is cooled to liquid nitrogen temperature. The condensed metal clusters are carried away by the gas stream through a 2 mm
nozzle into a chamber where a Roots pumping unit maintains a pressure of
about 10^{-1} Torr. After passing a 0.8 mm collimator diameter the clusters

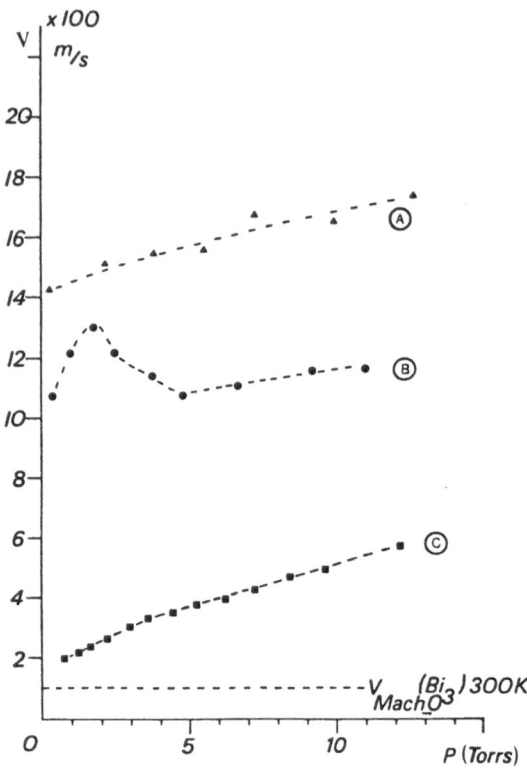

Fig. 1.   Velocity measurements as a function of the helium pressure.
(A) helium without bismuth in the crucible ;
(B) helium with clusters ;
(C) Bi_3.
The dashed curves are just interpolations between experimental
points.

enter an intermediate 10^{-4} chamber and finally are transferred to a high
vacuum (2.10^{-7} Torr) chamber as a well collimated beam.

In order to perform velocity measurements, the clusters beam is
chopped at 90 Hz and detected at a 70 cm distance by means of a quadrupole mass spectrometer axially directed in the beam. The quadrupole is
then tuned on the mass of a given cluster or of a helium atom and the ions
number is recorded as a function of the time. In all the experiments, the
temperature of the bismuth crucible is constant and the velocity measurements are performed as a function of the helium pressure.

The initial temperature of the helium-clusters mixing before the
expansion is very difficult to determine ; the helium gas is introduced
at 77 K (liquid nitrogen temperature) and the walls of the first chamber
are also cooled at 77 K (see Fig. 1). However this chamber also contains
the opened crucible at a temperature of 1100 K. The temperature of the
helium gas is probably intermediate between these two temperatures and
depends on the equilibrium in the chamber. In order to estimate the
initial helium temperature, we have performed velocity measurements with
the oven at 1100 K but without bismuth. The results are shown in Figs. 1A
and 2. They can be fitted by assuming an isentropic expansion[3] and an
initial helium temperature $(T_{He})_i$ = 300 K. The corresponding theoretical
curves are also shown in Fig. 2. The agreement with the experimental
results is good and a translational temperature of about 50 K is achieved
for the helium.

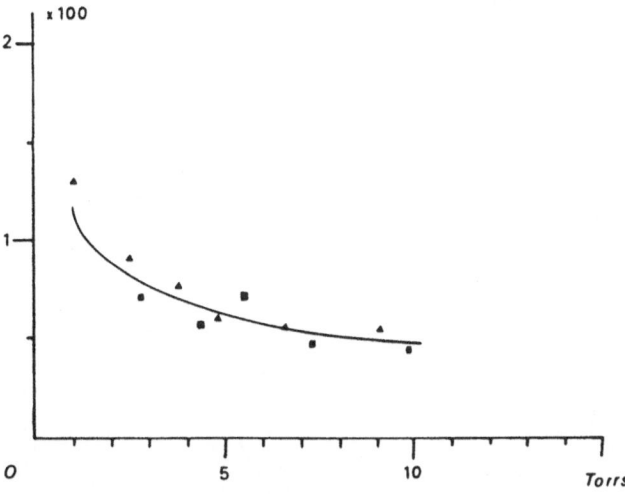

Fig. 2. Translational temperature of helium.
■ helium without clusters ;
▲ helium with clusters.
The solid curve corresponds to theoretical temperatures assuming
isentropic expansion.

Figure 1B shows the helium velocity in presence of the bismuth
clusters and illustrates the slowing down of the helium atoms by the
bismuth clusters. The helium translational temperature can also be measured from the velocity distribution and is very close to the temperature
obtained without clusters.

Figure 1C shows also the evolution of the Bi_3 velocity as a function
of the helium pressure. The observed increase of the velocity corresponds
to the driving of the bismuth clusters by the helium. The results can be
interpreted by a rough model taking into account the energy transfer
during each collision and assuming isotropic collisions in the mass center.
The collision cross-section between Bi_3 and helium is roughly the Bi_3
geometrical section. With this model the velocity increase of Bi_3 is
roughly linear as a function of the helium pressure because it is proportional to the number of collisions.

This velocity slip phenomenon precludes the cooling of the clusters
during the expansion. In our rough model, we obtain a slight heating of
the bismuth clusters as a function of the helium pressure. This is due to
the isotropic character of energy transfer in the mass center during the
collision. Similar results have been obtained by Anderson[4] in the expansion
of gas mixtures when the velocity slip is very large.

In principle the Bi_3 velocity distribution could lead to the measurement of the Bi_3 translational temperature. However the observed velocity
distributions at high helium pressure are very large and correspond to non
realistic translational temperatures. This is due to the fragmentation of
large clusters having slower velocity by electron impact ionization. The
fragmentation increases as a function of the electron energy and we indeed
observe a broadening of the velocity distribution at higher electron energy
(Fig. 3).

In the case of Bi_4, we clearly observe a second velocity distribution
at higher electron energy (Fig. 4). It corresponds to the time of flight
of bigger clusters fragmented by electron impact ionization. In our rough

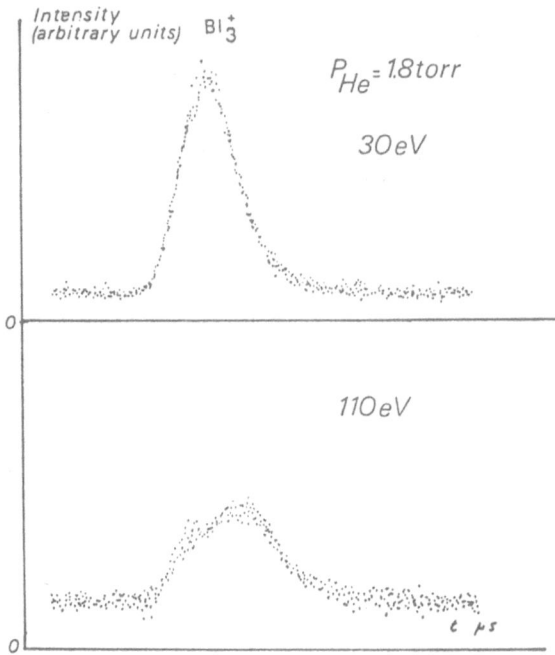

Fig. 3. T.O.F. of Bi_3^+ for two ionizing electron energies.

model of velocity slip, the mean size of the second distribution can be estimated to correspond to about 15-20 atoms. By a similar study we have shown that the Bi_5^+ clusters observed in the mass spectra are almost only due to fragmentation. Our results indicate that the cooling of large bismuth clusters (strongly heated by electron impact) occurs by fission (or evaporation of Bi_3^+, Bi_4^+ and Bi_5^+) rather than by evaporation of a neutral atom.

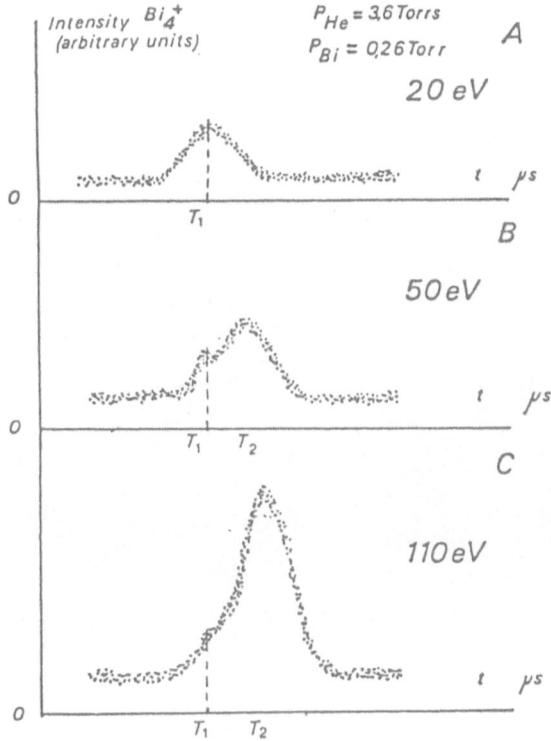

Fig. 4. T.O.F. of Bi_4^+ for three ionizing electron energies.

REFERENCES

1. K. Sattler, J. Mühlbach and E. Recknagel, Phys. Rev. Lett. 45:824 (1980).
2. H. Schaber and T.P. Martin, Surface Science 156:64 (1985).
3. A. Kantrowitz and J. Grey, Rev. Sci. Instr. 22:328 (1951) ;
   R. Campargue, PhD Thesis, Orsay (France).
4. J.B. Anderson, Entropie 18:33 (1967).

FRAGMENTATION OF SILICON MICROCLUSTERS

B.P. Feuston[*], R.K. Kalia and P. Vashishta

Argonne National Laboratory
Argonne, Illinois 60439

[*]University of Cincinnati
Cincinnati, Ohio 45221

A detailed study of Si_N (N=2-14) clusters has been performed through molecular dynamics (MD) simulation and the method of steepest descent quench (SDQ) with the Stillinger-Weber three-body potential. The approach is intrinsically a finite-temperature method which investigates all underlying mechanically stable structures visited by the system in the equilibrium state. Results indicate that the magic numbers, relatively stable clusters Si_N (N=4,6,10), are determined by the topology and energetics of high energy bound structures rather than the structure and ground state energies at zero temperature.

INTRODUCTION

In the past several years there has been growing interest in the properties of small clusters of atoms.[1-5] Inert gas clusters are known to form predominantly 13-, 55-, etc. atom clusters explained by stable icosahedral packing.[5] More recently there has been an effort to determine the geometrical arrangements and electronic configurations for Group IV elements, C, Si, Ge, and Sn.[6-9] Silicon and germanium have both been found to form unusually stable 6- and 10-atom clusters. Si_4 and Ge_{14} have also been found to be particularly stable.[6-7] Bloomfield, Freeman, and Brown[6] (BFB) performed a photofragmentation experiment on ionized silicon, measuring relative cross-sections and individual fragmentation channels for clusters containing up to 12 atoms. Fragmentation occurred upon exposing charged silicon clusters to an intense beam of laser radiation. The temperature of the cluster prior to dissociation reaches the order of the melting temperature of bulk silicon (T ~ 2000 K). Bloomfield et. al. report relatively low total photofragmentation cross sections for Si^+_4, Si^+_6, and Si^+_{10} in addition to the unusually common occurrence of Si^+_6 and Si^+_{10} from the fragmentation of Si^+_{7-11} and Si^+_{12}, respectively. Hence, N=4,6,10 are said to be the "magic numbers" for silicon. In an attempt to explain the existence of these magic numbers, we have chosen an intrinsically finite-temperature method, a molecular dynamics (MD) simulation approach. In conjunction with the steepest-descent quench (SDQ), one can determine global ground-state structures and study the energetics of small clusters. The results of our MD simulation for Si_{2-14}, using the Stillinger-Weber[10] 3-body potential, indicate that the presence of magic numbers is determined by the topology and the energetics of high-energy bound structures and not by the structure and ground-state energies at zero temperature.

METHOD OF STEEPEST DESCENTS AND METHOD OF CALCULATION

The trajectory of individual atoms in the equilibrium state are completely determined by the MD simulation. The instantaneous positions of N-atoms are represented by a 3N-dimensional vector in configuration space. As time evolves this vector traces out a trajectory with each point completely specifying the 3N coordinates of the system. By uniquely assigning each point on this trajectory to a potential minimum, corresponding to a particular geometry, one can determine the underlying structure of the system at each time step. The assignment of each point in configuration space to a local potential minimum is a mathematically well defined problem that can be solved through the method of steepest descents.[11] The steepest-descent quench (SDQ) assigns each point in configuration space to the first potential minimum encountered when descending from that point along the steepest available path. When SDQ is invoked the instantaneous positions are mapped onto a new point in configuration space where the total kinetic energy of the system is zero and the new positions correspond to the zero-force configuration of the local potential minimum.

An exhaustive search was carried out to (i) enumerate all accessible, mechanically stable configurations, particularly the global ground-state structure, and the relative probability of occupying any particular configuration at finite temperatures, (ii) find the highest energy (E_f) and temperature (T_f) that the system would remain bound, and (iii) determine the fragmentation spectra, for Si_{2-14} microclusters. We have defined the fragmentation energy (E_f) to be the highest energy that the cluster remains bound for 25,000 MD time steps before the onset of fragmentation. The fragmentation temperature (T_f) is determined by averaging the total kinetic energy over many time steps (typically 11,000 steps) during a MD run at the highest energy, E_f. The ground-state energy and structure were determined from the lowest of the approximately 1000 potential minima identified for each Si cluster. Of course, only a small subset of these potential minima were numerically different corresponding to distinctly different accessible structures. The relative probability of visiting any particular structure, at finite temperatures below T_f, was found through SDQ performed periodically on systems generated by monotonic heating from the ground-state. In this way one finds only those states accessible from the ground state. A schematic diagram of the MD simulation in conjunction with the SDQ is shown in Fig. 1.

GROUND STATE STRUCTURES

The ground-state configurations of Si_3, Si_4, and Si_5 have the following symmetrical and planer geometries: the equilateral triangle, the square, and the regular pentagon, respectively. It should be noted that the isosceles triangle, consisting of only two bonds forming a perfect tetrahedral angle, is only slightly higher in energy than the ground-state structure of Si_3. The loss in the 2-body binding energy of this metastable state with respect to the lowest-energy configuration due to the decrease in the number of bonds is offset by the 3-body interaction. It is not surprising the pentagon has the lowest energy for Si_5 since the angle between adjacent bonds ($108°$) is only $1.5°$ smaller than the perfect tetrahedral angle. The "squashed" trigonal bipyramid, Si_5, whose energy is higher than the ground state of the pentagon by less than 1%, was also commonly found during steepest-descent quenches from intermediate temperatures.

The ground-state structure of Si_6 is the first to have 3-dimensional geometry with all atoms 3-fold coordinated. The symmetrical stacking of two equilateral triangles, similar to a wedge (Fig. 2a) form the Si_6 structure while the Si_7 ground-state configuration can be thought of as capping each of the three edges on the base of trigonal pyramid. One can form the symmetric ground-state geometry of Si_8, a perfect cube, by

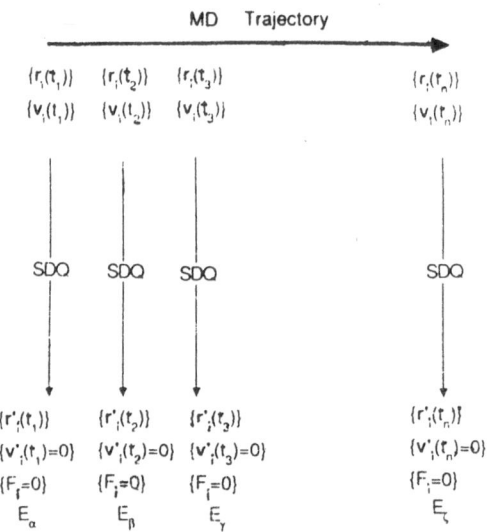

FIG. 1 Schematic representation of the steepest-descent quench (SDQ) performed in parallel with the molecular dynamics (MD) simulation. MD generates positions $\{r_i\}$ and velocities $\{v_i\}$ at each time step through integration of Newton's equation of motion. The SDQ maps the instantaneous positions to the local minimum E_α where the force $\{F_i\}$ on each particle vanishes and the velocities are to zero. This mapping does not in any way effect the continuation of the MD trajectory.

reconstructing a face-capped Si_7 structure. The addition of one atom to the edge of the Si_8 cube forms the lowest-energy structure for Si_9. Allowing this structure to relax breaks the edge-capped bond forming two identical, nearly perpendicular, non-planar pentagons. The ground-state geometry for Si_5 may be used to obtain Si_{10} in the same way the ground-state structures for Si_N (N=2,3,4) can be used to obtain those of Si_{2N} with the lowest-energy configuration for Si_{10} made of two symmetrically stacked pentagons.

Reconstruction of an atom capped to a pentagon face on the Si_{10} ground-state structure results in the lowest-energy configuration for Si_{11}. The ground-state structure for Si_{11} is the smallest to contain a 4-fold coordinated atom. The symmetrical ground-state figure for Si_{12} can perhaps be most easily described by the addition of a Si_2 dimer to a square face of the Si_{10} ground state. The ground-state geometry for Si_{12} has four identical pentagons bound in stacked pairs orientated perpendicular to one another and pointing in opposite directions. Similar to Si_{11}, the ground-state structure of Si_{13} contains one 4-fold coordinated atom. The configuration is best described by the edge-sharing of four identical pentagons to create a 4-sided cone with the small end forming a square and the open end capped by an atom connected to each of the four protruding points. The Si_{14} ground-state geometry similar to all even-numbered clusters, is very symmetrical consisting of six identical pentagons and three perfect squares with all atoms lying on the surface having only three bonds.

Several features appear common to all Si_N (N=3-14) microclusters interacting through the SW 3-body potential. First one observes the existence of three basic "building blocks" (triangle, square, and pentagon) in the ground-state energy structures for each cluster Si_{3-14}. These not only form the lowest-energy configurations for Si_3, Si_4, and Si_5 but also by trivially stacking these three planer figures one obtains the ground-state configurations of Si_6, Si_8, and Si_{10}. All metastable configurations for large clusters are formed by 3-, 4-, and 5-membered rings which are the distorted ground-state configurations for Si_3, Si_4, and Si_5.

The plot of ground-state energy per atom (E_0) vs number (N) in Fig. 3 does not show any indication for the existence of the magic numbers (N=4,6,10) that appeared in the E vs T curves of the previous quantum-chemical calcuations of Raghavachari and Logovinsky[8] (RL) and Tománek and Schlüter[9] (TS). In the calculations of RL and TS, a limited search of possible ground-state geometries resulted in

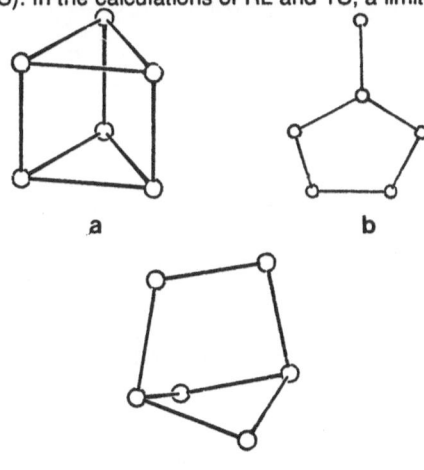

FIG. 2 Example of a few Si_6 structures: (a) Ground-state geometry can be formed by symmetrically stacking two equilateral triangles; (b) Corner-capped pentagon; (c) Reconstructed face-capped pentagon.

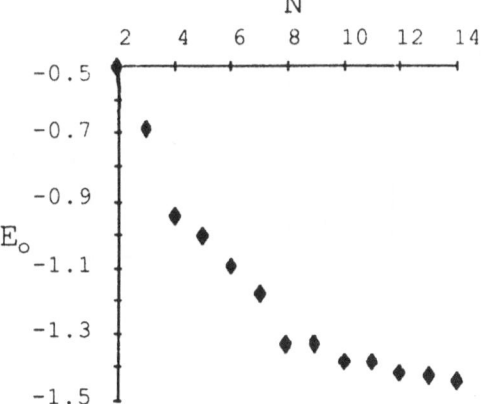

FIG. 3 Ground-state energy per atom (E_0) versus Number (N) . In the present calculation the presence of magic numbers can not be explained by ground-state energies or structures but only by the topology and energetics of finite-temperature clusters.

relatively stable structures for $Si_{4,6,10}$. The unusually common occurrence of these clusters in the photofragmentation spectra of silicon was then explained by their particularly stable ground-state configurations. We have found structures topologically equivalent to the ground-state structures of RL and TS, but we are unable from the zero-temperature ground-state energies and structures to make any claims to the presence of magic numbers. The present static zero-temperature results indicate all even-numbered clusters (particularly Si_8) are relatively stable with the exception of Si_6. The ground-state configuration for Si_6 contains a large contribution to the 3-body potential energy due to the two equilateral triangles. Since fragmentation occurs at such high temperatures, the structures

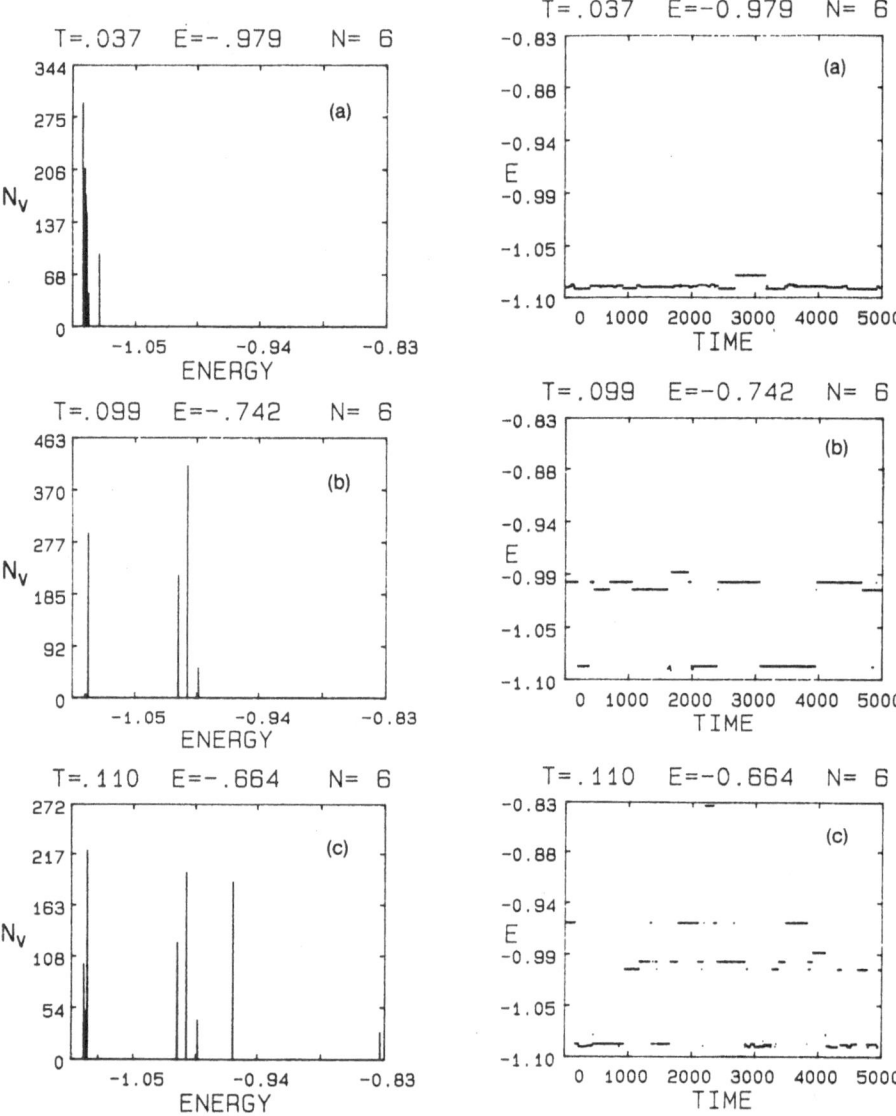

FIG. 4 Histograms of the visited potential minima for Si_6 at: (a) T^*=.037, (b) T^*=.099 and (c) T^*=.110. The structures accessible to Si_6 can be grouped into sets of configurations of nearly equal energies. Transitions between subgroups occur with increasing temperature, indicated by the small jumps in E vs T (Fig 6). A reduced temperature (T^*) of .1 corresponds to 2500 K.

FIG. 5 The underlying potential minimum is plotted against the number of MD steps for the same three simulations depicted in Fig. 4. Not only does the majority of underlying state shift towards higher-energy configurations but the frequency of transitions also increase with increasing temperature.

underlying the high-temperature cluster must play an important role in the determination of the fragmentation spectra and they can only be studied by investigating finite-temperature systems.

CLUSTERS AT HIGH TEMPERATURES

As an example of our investigation into the energetics and topology of each cluster Si_{2-14}, a discussion Si_6 clusters will be given. Sixteen different potential-energy minima were found for Si_6 through the SDQ. The lowest-energy structure (Fig. 2a) was not observed in systems near the fragmentation energy where more open structures were commonly found. A distorted octahedron, similar to the ground-state structure of Si_6 determined by RL and TS, is found though a corner-capped pentagon (Fig. 2b) is closer in energy to the ground-state and more frequently visited. Even at relatively low temperatures $T^* \sim .037$ the cluster undergoes many transitions between configurations of nearly equal energies. Two notable intermediate-energy structures are the reconstructed face-capped pentagon (Fig. 2c) and two edge-sharing squares forming a tetrahedral angle. The most commonly found geometries near E_f were the corner-capped pentagon, a dimer corner-capped to a square, and the distorted tetragonal bipyramid. In Fig. 4 the number of visitations (N_v) versus the underlying potential minima (E) is plotted for three finite-temperature systems. The local potential energy minimum resulting from the SDQ performed in parallel to an MD run is plotted against the number of time steps to show the frequency of transitions between allowed structures in Fig. 5. From these two figures one sees that not only does the majority of underlying states shift towards higher-energy configurations but the frequency of transitions also increases with increasing temperature. The two small jumps in E versus T in Fig. 6 can be explained by the transitions between three groups of structures, shown in Fig. 4c and not just between single structures.

MAGIC NUMBERS

The results of the present MD simulation indicate that 4-, 6- and 10-atom clusters are relatively stable structures at temperatures on the order of the melting temperature. Curves of fragmentation energy per atom (E_f) versus number (N) and fragmentation temperature (T_f) versus number (N) are shown in Fig. 7. The fragmentation energy, the highest attainable energy of a bound system, was determined from approximately 60 systems generated for each cluster Si_{3-14}. The fragmentation temperature was determined from the average kinetic energy of the system with energy E_f. Dips in the curve for E_f vs N at N=6,10 and peaks in the T_f vs N curve at N=4, 6, and 10, clearly indicate Si_6, Si_{10} and probably Si_4 microclusters form particularly stable high-energy configurations.

Fragmentation occurs when a subset of atoms no longer interacts with the remaining atoms in the parent cluster. This definition restricts our MD simulation of fragmentation to events where only single bonds are broken. The dominant fragmentation channel for all clusters Si_N (N=2-14), is the dissociation of single atom from the parent cluster, $Si + Si_{N-1}$. The relative probability of the cluster Si_N dissociating into the fragments $Si_{N-M} + Si_M$ (M=1,2,3) is given in Table 1. Evidence for the stability of the magic numbers is found in the nontrivial fragmentation channels N-->2+(N-2) and N-->3+(N-3). The $Si_{6,8,12}$ clusters were observed to have the highest probability of nontrivial fragmentation with at least 25% of the observed events resulting in $Si_2 + Si_{4,6,10}$, respectively (see Table 1). Seven and nine

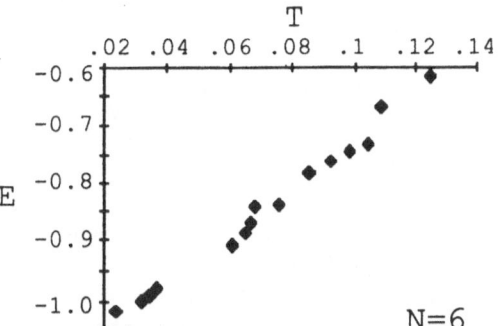

FIG. 6 The total internal energy per atom versus temperature curve is shown for Si_6. The cluster was started in its ground-state configuration and monotonically heated by multipling the velocities by 1.1. After each scaling the temperature was determined by averaging the kinetic energy over 11,000 MD steps. Energy and Temperature are expressed in reduced units (ε=2.17 eV) with a temperature of .1 corresponding to 2500 K.

atom clusters, $Si_{7,9}$, were also found to have a large probability of dissociating into Si_3 and $Si_{4,6}$. These relatively high probabilities for nontrivial fragmentation into 4-, 6- and 10-atom clusters are in good agreement with the experimental fragmentation spectra and support the claim for unusually stable $Si_{4,6,10}$ microclusters.

Our results indicate that the knowledge of the zero-temperature ground-state structure and energies is not sufficient to explain the existence of unusually stable clusters for covalently bonded systems. Rather, it is the topology and energetics of these high-temperature clusters which determine its relative stability with respect to the number of atoms. The fragmentation energy (E_f) and temperature (T_f) indicate $Si_{4,6,10}$ are particularly stable clusters at high temperatures, the order of the melting temperature (T ~ 2000 K). Though we have only investigated the primary channel of fragmentation, the fragmentation spectra N-->2+(N-2) and N-->3+(N-3) indicates the unusually common occurrence of $Si_{4,6,10}$ fragments in the dissociation of silicon microclusters. The results of this MD simulation agree quite well with the BFB photofragmentation experiment and gives us good reason to believe that N=4,6,10 are the magic numbers for silicon. We believe that the SW 3-body potential describes the essential features of the energetics and fragmentation of silicon microclusters. It can also be argued that many of the results described above will be common to Group IV semiconductors, in particular to germanium clusters. There will certainly be details which will differ when a more accurate interaction potential becomes available or a fully self-consistent finite-temperature first-principles electronic structure calculation becomes possible.   *Work supported by U.S. D.O.E, BES-Materials Sciences.

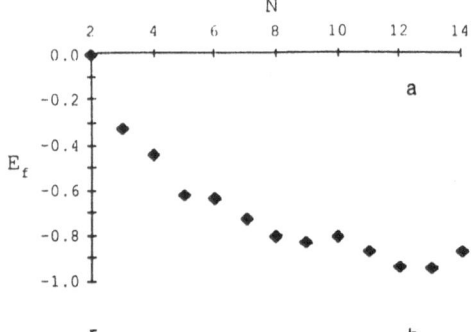

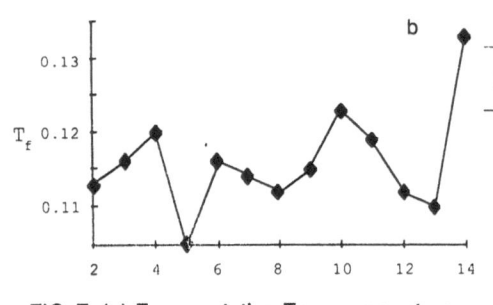

FIG. 7 (a) Fragmentation Energy per atom (E_f) vs Number (N); (b) Fragmentation Temperature (T_f) vs Number (N). Peaks in these curves at N=4,6,10 (magic numbers) clearly indicate the particular stability of the $Si_{4,6,10}$ microclusters at high energies and temperatures.

TABLE 1   Primary channels of fragmentation for Si_{2-14} as determined from 20-30 events for each size cluster. Fragmentation from Si_N to $Si_{(N-2)}+Si_2$ indicates $Si_{4,6,10}$ are particularly common fragments while the fragmentation from Si_N to $Si_{(N-3)}+Si_3$ support the common occurrence of Si_4 and Si_6 fragments.

N	(N-1)+1	(N-2)+2	(N-3)+3
2	1.00		
3	1.00		
4	0.90	0.10	
5	0.82	0.18	
6	0.64	0.25	0.11
7	0.62	0.19	0.19
8	0.55	0.35	0.05
9	0.64	0.12	0.16
10	0.74	0.22	0.00
11	0.65	0.15	0.10
12	0.70	0.25	0.05
13	0.76	0.12	0.04
14	0.76	0.14	0.00

[1]B.K. Rao and P. Jena, Phys. Rev. B 32, 2058 (1985). [2]P.A. Montano, G.K. Shenoy, E.E. Alp, W. Schulze, and J. Urban, Phys. Rev. Lett. 56, 2076 (1986). [3]F.F. Abraham, Rep. Prog. Phys. 45, 1113 (1982). [4]I.A. Harris, R.S. Kidwell, and J.A. Northby, Phys. Rev. Lett. 53, 2390 (1984). [5]J. Farges, M.F. de Feraudy, B. Raoult, and G. Torchet, J. Chem. Phys. 78, 5067 (1983). [6]L.A. Bloomfield, R.R. Freeman, and W.L. Brown, Phys. Rev. Lett. 54, 2246 (1985). [7]T. P. Martin and H. Schaber, J. Chem. Phys. 83, 855 (1985). [8]K. Raghavachari and V. Logovinsky, Phy. Rev. Lett. 55, 2853 (1985). [9]D. Tomanek and M.A. Schluter, Phys. Rev Lett. 56, 1055 (1986). [10]F.H. Stillinger and T.A. Weber, Phys.Rev. B 31, 5262 (1985). [11]F H. Stillinger and T.A. Weber, Science 225, 983 (1984).

SPECTROSCOPIC INVESTIGATION OF Ag-METAL CLUSTERS

F.W. Froben[1], K. Möller[1], W. Schulze[2] and B. Winter[2]

[1] Dept. Physics, Free University Berlin, Arnimallee 14
1000 Berlin 33, West Germany
[2] Fritz-Haber-Institut, Faradayweg 4, 1000 Berlin 33

ABSTRACT

The investigation of silver clusters, prepared by the gas aggregation technique, sputtering or laser vaporization, using mass-or fluorescence spectroscopy for detection, provides new information of the influence of experimental parameters on the cluster size distribution. Especially the gas aggregation technique proves to be a very intense source generating metal clusters in narrow size distributions. This high intensity allows to measure multiple charged clusters below the calculated size limit. The observation of time resolved spectra of the cluster beam produced by laser vaporization allows to distinguish spectroscopically between different species. In the spectral range of 200 to 500 nm spectra of neutral and ionized atoms, dimers and larger molecules have been detected.

INTRODUCTION

During recent years the investigation of metal clusters has increased dramatically mainly because of the important applications of cluster research but also due to the invention of the laser vaporization method. Most of the results have been obtained using mass spectroscopy neglecting more or less the problems of fragmentation, e.g. the question of the difference between cluster distribution after ionization and the size distribution for the neutral precourser. In this paper we try to combine mass spectroscopical measurements with data from optical spectra using different techniques of production and varying experimental parameters.

EXPERIMENTAL

For the production of clusters three different experimental techniques have been used, the gas aggregation technique described previously (1), sputtering and laser vaporization described by Smalley (2). The main advantage of the gas aggre-

gation method is the continuous formation of neutral clusters
of a size distribution adjustable by the metal supersaturation
in the nucleation regime. The clusters/molecules can be measured by a magnetic mass spectrometer (3) or for small molecules by their optical spectra. For sputtering a simple continuous discharge was used and the optical spectra recorded in
the gas phase. The laser vaporization device is modified from
(2) to allow for a change of nozzles during the experiment and
to observe the emission in front of the nozzle, perpendicular
to the cluster beam. By moving the detection window on the
time axis it is possible to register time resolved spectra.
For further details see (4). The laser vaporization source has
the advantage of high rates but only in short pulses and the
difficulties of very rich optical spectra not trivial to
assign and a production under nonequilibrium conditions.

RESULTS AND DISCUSSION

Mass spectroscopic data have so far been obtained only in
the gas aggregation experiment. Fig. 1 shows an example for
multiple charged clusters. It is obvious from the figure that
double charged clusters are stabilized below the critical
values of 30 resp. 40 for triple charged ones (5,6).
Optical spectra in the gas phase have been measured for all
three experimental setups. Neutral atoms and atomic ions are
assigned according to (7,8), the He dimer after (9) and Ag

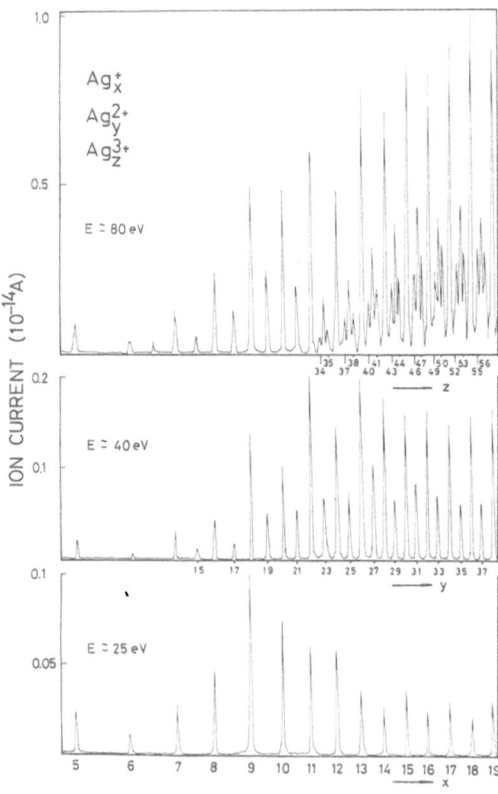

Fig. 1   Influence of excitation energy on multiple
charged Ag-clusters

dimers using (10). There is no definitive identification of
Ag_x, x > 2 reported, except in low temperature matrices (11).
But the interaction in the matrix will cause a shift of some
hundred wave numbers and make the comparison very doubt-
ful.Except the well known spectral features from atoms and
dimers there are 3 fact to remark.
First, under high gas flow in the gas aggregation experiment
some emission between 360 and 380 nm and at 395 nm and 310 nm
are observed most probably due to larger species.
Second, in the beam of laser vaporized species there are some
spectral features remaining after subtraction of all the known
atomic and molecular bands. The most intense band systems
appear around 350 and between 420 and 428 nm, in the same
regime where Ag_3 in the matrix has been reported. The measurement of time resolved spectra - Fig. 2 shows an example -
allows to assign these systems to a species containing Ag,
most probably the Ag trimer.
Third, in addition to the metal species, there is also a rare
gas molecule formed in the beam after laser vaporization. For
high gas flow, high laser power and short nozzles the He_2
systems: $e^3 - > a^3$ around 465 nm, $h^3 > b^3$ around 455 nm and
the j^3 , ^{3} .^{3} > b^3 systems around 440 nm have been observed.
The relative strong emission of these systems shows, that the
atoms are highly excited when they combine in the upper electronic state to form the diatomic molecule.

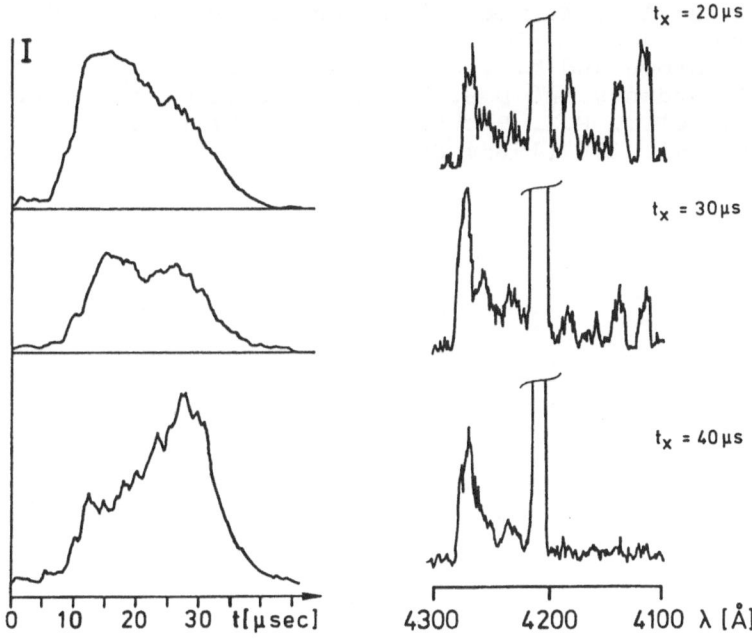

Fig. 2   left part: time resolved spectra at 447.4 nm with HeI
          at 15 μsec and AgI at 30 μsec after the laser shot.
          Laser power decreasing from top to bottom.
          right part: spectra between 410 and 430 nm measured
          at 3 different delay times after the laser shot.
          Three different species are to be seen: at 428 nm Ag_x,
          x > 2, at 421 nm AgI and a third one between 410 and
          420 nm.

ACKNOWLEDGMENT

This investigation was supported by the "Deutsche Forschungsgemeinschaft" through SFB 6 and 161 which is gratefully acknowledged.

REFERENCES

1. F. Frank, W. Schulze, B. Tesche, J. Urban and B. Winter, Surf.Sci. 156 (1985) 90
2. J.B. Hopkins, P.R.R. Langridge-Smith, M.D. Morse and R.E. Smalley, J.Chem.Phys. 78 (1983) 1627
3. I. Goldenfeld, F. Frank, W. Schulze and B. Winter, Int.J.Mass.Spec. and Ion Proc. 71 (1986) 103
4. F.W. Froben and K. Möller, in preparation
5. W. Schulze, B. Winter, J. Urban and I. Goldenfeld, Z.Phys.D (in press)
6. D. Tomanek, S. Mukherjee and K.H. Bennemann, Phys.Rev.B 28 (1983) 665
7. A.N. Zaidel, V.K. Prokot'ev, S.M. Raiskii, V.A. Slavnyi and E.Ya. Shreider: Tables of spectral lines. IFI/Plenum, New York, London 1970
8. A.G. Shenstone, Phys.Rev. 57 (1940) 894 H.U. Johannsen and R. Lincke, Z.Phys.A 272 (1975) 147
9. C.M. Brown and M.L. Ginter, J.Mol.Spectr. 40 (1971) 302
10. B. Kleman and S. Lindkvist, Arkiv Fysik 9 (1955) 385, C.Shin-Piaw, W. Loong-Seng and L. Yoke-Seng, Nature 209 (1966) 1300 C.M. Brown and M.L. Ginter, J.Mol.Spectr. 69 (1978) 25 V.I. Srdanov and D.S. Pesic, J.Mol.Spectr. 90 (1981) 27
11. U. Kettler, P.S. Bechthold and W. Krasser, Surf.Sci. 156 (1985) 867.

PHOTOFRAGMENTATION OF MASS-SELECTED MAIN GROUP ELEMENT CATIONS

M. Geusic, M. Jarrold, R. Freeman, W. Brown,
T. McIlrath* and M. Duncan*

AT&T Bell Laboratories
Murray Hill, N.J.

INTRODUCTION

The recent growth in the field of clusters has brought about the development and application of a number of techniques.[1] These are concerned with both the production and study of physical properties of clusters. One technique that has been used in our lab is photofragmentation of mass-selected cluster cations.[2] The primary objective in using this technique has been to obtain datum concerning dissociation branching ratios, thresholds and cross sections as well as determining the mechanism for dissociation.

EXPERIMENTAL

The overall experimental set-up has been described in detail elsewhere.[3] In brief, the experimental sequence is as follows: A pulsed Nd:YAG laser (second harmonic) is used to vaporize a rotating sample target within the throat of a pulsed supersonic nozzle. The firing of the laser is timed, such that, as the helium pulse flows over the sample the laser pulse vaporizes some material from the rod, which is entrained within the helium. The helium plus cluster cations, anions and neutrals are then expanded under supersonic conditions. From here, the cluster ions enter a pulsed (2.25 Kv) acceleration region of a Wiley-Mclaren[4] time of flight. The cations are turned 90 degrees and proceed down the beam line through a field free region. Within this drift region the ions encounter deflection plates and an einzel lense system for enhanced transmission efficiency throughout the machine. Next, the ions enter a mass selector region which allows the selection of a particular size cluster packet. The packet is decceler-ated to approximately 1 Kv and crossed at 90 degrees by a pulsed excimer laser beam which is focused to .4 cm^2. The unfragmented cations and product cations then enter a single stage acceleration of a time of flight mass spectrometer. The ions are detected using dual microchannel plates. The signal is then amplified and analyzed by a 200 MHz transient

---

* Visiting from U. of Maryland, U. of Georgia

recorder (Model TR8828.B, Lecroy). The digitized signal is then transferred along a Camac dataway to a PDP 11/34 computer for analysis. The photodissociating laser intensity is monitored by a UV diode and boxcar averager and is fed to the computer through a Camac based A to D converter.

RESULTS

The cluster ions for these studies were derived directly from the source[2] in an attempt to produce internally cold ions. It has been shown that elements like Si, Ge, C, Bi and Sb all produce cluster ions directly from the source without the use of an ionizing technique (i.e. electron bombardment, photoionization, etc.). Although the formation process of the ion clusters is not distinguishable between either ion molecule reactions or direct production off the sample, it is reasonably argued that due to approximately 10^4-10^5 collisions of helium with the cluster ions the ion will internally cool to~10K rotational ~125K vibrational and ~ 5K translational.

In our lab we have carried out mass-selected photodissociation experiments on C_n^+ (n=3-20), Bi_x^+ (n=2-8) and Sb_n^+ (n=2-8). In the case of Carbon, the most extensive study has been carried out enabling us to obtain branching rations, limits on dissociation thresholds and cross sections.

For photofragmentation with 532 nm, 351 nm and 248 nm light the branching ratios for C_n^+ (n=3-20) are the same. The dominant pathway for fragmentation is seen to be $C_n^+ \xrightarrow{h\nu} C_{n-3} + C_3$ as exemplified for 248 nm in Table 1. The data was taken at 2 mjcm^{-2} so that only the primary products (single photon) are seen. Although, for the products which have intensities less than 5%, the signal to noise ratio is insufficient for us to positively determine the linearity of the fluence dependence of the signal.

The photodissociation thresholds of C_n^+ (n=3-20) were bracketed using both 351 nm and 248 nm light. These results are summarized in Table 2. At 248 nm (4.98 eV) clusters with n=4 and n=6-20 photodissociate with one photon while for n=5 the photodissociation is multiphoton. At 351 nm (3.53 eV) clusters with n=6-20 are one photon. For n=3 we were unable to obtain sufficient depletion with either 351 nm or 248 nm (at low fluences) to determine the laser fluence dependence. We observed the same result with n=4 at 351 nm. The rational for the result is that the depletion is multiphoton for these particular clusters at the specified wavelengths.

Photodissociation cross sections were also obtained for C_n^+ (n=3-20) for both 351 nm and 248 nm light. With 248 nm light, the cross sections are seen to rise rapidly to n=9 then fall dramatically at n=10 and rise again out to n=20. While the cross sections with 351 nm rise rapidly to n=9 and fall dramtically at n=10, they rise again only slightly out to n=20.

Preliminary results concerning photodissociation have also been obtained for Bismuth and Antimony[5] clusters (n=2-8) using 248 nm (4.98 eV) light. Although these elements are metallic, the primary fragmentation pathway is different from that observed for the transition metal clusters.[6] For

Table 1. Fragmentation patterns for carbon clusters cations at 248 nm. The numbers shown represent the relative per cent of cation fragments formed.

	1	2	3	4	5	6	7	8	9	10	11	12	13	14	15	16	17
C_4^+	44	-	56	-	-	-	-	-	-	-	-	-	-	-	-	-	-
C_5^+	-	-	100	-	-	-	-	-	-	-	-	-	-	-	-	-	-
C_6^+	-	-	94	1	5	-	-	-	-	-	-	-	-	-	-	-	-
C_7^+	-	-	3	78	18	-	-	-	-	-	-	-	-	-	-	-	-
C_8^+	-	-	1	-	94	-	5	-	-	-	-	-	-	-	-	-	-
C_9^+	-	-	7	-	-	84	8	1	-	-	-	-	-	-	-	-	-
C_{10}^+	-	-	2	1	-	97	-	-	-	-	-	-	-	-	-	-	-
C_{11}^+	-	-	-	9	-	-	71	-	21	-	-	-	-	-	-	-	-
C_{12}^+	-	-	-	-	-	4	-	53	12	31	-	-	-	-	-	-	-
C_{13}^+	-	-	-	-	-	-	-	-	98	2	-	-	-	-	-	-	-
C_{14}^+	-	-	-	-	-	-	-	3	3	94	-	-	-	-	-	-	-
C_{15}^+	-	-	-	-	-	-	2	10	5	63	18	2	-	-	-	-	-
C_{16}^+	-	-	-	-	-	-	-	12	37	2	43	4	3	-	-	-	-
C_{17}^+	-	-	-	-	-	-	-	-	4	3	-	84	8	-	-	-	-
C_{18}^+	-	-	-	-	-	-	-	2	3	3	3	5	84	-	-	-	-
C_{19}^+	-	-	-	-	-	-	-	-	6	-	2	26	12	55	-	-	-
C_{20}^+	-	-	-	-	-	-	-	2	4	12	3	15	30	10	31		

transition metal clusters the dominant dissociation pathway is the loss of a single atom, whereas, for both Bismuth and Antimony the dominant dissociation pathway involves the loss of a molecular species.

The tabulated data for photodissociation of various Antimony and Bismuth clusters (n=2-8) is shown in Table 3. As demonstrated, all cluster cations except Bi_2^+ produce more than one primary fragment. In every case however, there is one fragment which dominants the dissociation process. Corresponding clusters for Sb and Bi have quite different branching ratios in the size range of n=3-5, while in the n=6-8 range the fragmentation channels are similar for both metals.

CONCLUSION

Mass-selected photofragmentation of cluster cations of Carbon, Bismuth and Antimony have been presented. In the case of Carbon a detailed study has led to dissociation branching ratios, thresholds and cross sections. The dominant feature seen in the dissociation pathway for C_n^+ is the loss of a neutral C_3 for clusters in the n=6-20 range. For photodissociation thresholds and cross sections, the most interesting feature is the change in these properties as a function of cluster size. Wavelength independence of the dissociation products coupled with collisional dissociation data suggests that the mechanism is most likely a statistical dissociation (i.e. unimolecular decay).

For Bismuth and Antimony our preliminary results have yielded dissociation branching ratios. In all cases the primary dissociation pathway is seen to be the loss of a molecular species unlike the single atom loss for transition metal clusters. Due to the fact that the photodissociation pathways tend to favor the production of a stable product molecule (i.e. stable cations or neutrals), the same statistical dissociation mechanism as proposed for Carbon seems likely to occur in Bismuth and Antimony. Further wavelength dependent studies and collisional dissociation experiments need to be carried out to confirm this mechanism.

Table 2. Dissociation Energies for the Carbon Cluster Ions

Carbon Cluster Ion	Dissociation Energies, eV	
	Laser Fluence Dependence	Literature[a]
C_3^+	(> 4.98)	7.0
C_4^+	< 4.98	4.3
C_5^+	> 4.98	5.9
C_6^+	< 3.53	< 3.8
C_7^+	< 3.53	< 6.8 [b]
$C_8^+ - C_{20}^+$	< 3.53	

a) These values can only be considered reliable to $\pm$ 1 eV.
b) Ionization energy for this cluster estimated from IE (C_6) and IE (C_5).

296

Table 3. Fragmentation patterns for bismuth and antimony cluster cations at 248 nm. The numbers shown represent the relative per cent of cation fragments formed.

	1	2	3	4	5	6	7
Bi_2^+	100	-	-	-	-	-	-
Bi_3^+	73	27	-	-	-	-	-
Bi_4^+	29	65	6	-	-	-	-
Bi_5^+	-	-	100	-	-	-	-
Bi_6^+	-	-	63	37	-	-	-
Bi_7^+	-	-	38	41.5	13	7.5	-
Bi_8^+	-	-	27	39	33	-	-
Sb_2^+	100	-	-	-	-	-	-
Sb_3^+	17	83	-	-	-	-	-
Sb_4^+	-	25	75	-	-	-	-
Sb_5^+	-	-	61	39	-	-	-
Sb_6^+	-	-	69	31	-	-	-
Sb_7^+	-	-	46	28	26	-	-
Sb_8^+	-	-	18	59	23	-	-

REFERENCES

1) M. D. Morse, Chem. Rev., in press.

2) a) L. A. Bloomfield, R. R. Freeman and W. L. Brown, Phys. Rev. Lett. 50 (1985) 2246.

   b) L. A. Bloomfield, M. E. Geusic, R. R. Freeman and W. L. Brown, Chem. Phys. Lett. 121(1985)33.

   c) M. E. Geusic, T. J. McIlrath, M. F. Jarrold, L. A. Bloomfield, R. R. Freeman and W. L. Brown, J. Chem. Phys. 84(1986)2421.

3) M. E. Geusic, M. F. Jarrold, T. J. McIlrath, R. R. Freeman and W. L. Brown, J. Chem. Phys., submitted.

4) W. C. Wiley and I. H. McLaren, Rev. Sci. Instrum. 26(12), 1150(1955).

5) M. E. Geusic, R. R. Freeman and M. A. Duncan, Chem. Phys. Lett., submitted.

6) a) L. S. Zheng, P. J. Brucat, C. L. Pettiette, S. Yang and R. E. Smalley, J. Chem. Phys. 83(1985)4274.

b) P. J. Brucat, L. S. Zheng, C. L. Pettiette, S. Yang and R. E. Smalley, J. Chem. Phys. 84(1986).

7) S. W. McElvany, B. I. Dunlap and A. O'Keefe, J. Chem. Phys., submitted.

DESORPTION AND DECOMPOSITION OF MULTIPLY CHARGED CLUSTERS FROM

THE CRYSTAL OF GROUP IV ELEMENTS BY LASER EXCITATION

A. Kasuya and Y. Nishina

The Research Institute for Iron, Steel and Other Metals
Tohoku University, Sendai 980 Japan

Dynamical aspects of ion desorption from the solid target of group IV elements by laser excitation have been investigated by means of a time-of-flight spectrometer, a 127° electrostatic energy analyzer and of a quadrupole mass spectrometer. The mass distribution of emitted clusters and their decomposition times depend considerably on the wavelength and intensity of exciting laser beam. Our TOF spectrum on graphite by N_2 laser excitation shows a series of sharp lines at the flight times for C_{3i}^{2+} clusters (i= 1 to 7). The results on the excitation by excimer lasers of KrF and ArF, on the other hand, show band spectra for the peaks with i = 1 and 3. The spectral profile is quite similar to that observed in Si by N_2 laser excitation. Hence, the spectral broadening for peaks with i = 1 and 3 is caused by their decompositions in the time scale of 100 ns. For quadrupole mass analysis, the dominant species is identified with C_3^+ for N_2 and KrF and C^+ for ArF. The measurements of photo-electron energy distribution gives the average kinetic energy in the order of 1 eV for the excited electrons in the solid. This kinetic energy becomes higher for the laser excitation in order of increasing photon energy from 3.68 eV (NL) to 5.0 eV (KrF) to 6.4 eV (ArF). This result suggests that, at the time of desorption, the surface electronic system is excited toward higher levels in this order so that the emitted clusters decompose more readily to their ultimate form of atomic ions.

I. INTRODUCTION

Laser desorption technique has been applied for vaporizing non-volatile materials such as refractory metals and semiconductors.[1] This desorption process has been described in terms of the thermal evaporation process in which atoms are thermally ejected independently of each other from their isolated lattice sites on the surface. Our time-of-flight (TOF) mass analysis of desorbed ions,[2] on the other hand, shows that the atoms/ions are desorbed from the surface in coagulated forms as a consequence of the critical excitation of electronic system. This experimental result suggests that time resolving analysis is essential in identifying the microscopic mechanism of desorption process and in predicting the charge, mass and kinetic energy distribution of emitted ions in steady states. This paper describes our experimental investigation on the wavelength dependence of desorption and subsequent decomposition process of ionic clusters of group IV elements. The TOF spectrometer serves as analyzing the decomposition process, while a quadrupole (QM) mass analyzer determines the quasi-thermal equilibriumm state of mass distribution. The kinetic energy distribution of photoelectron is measured by a 127° electrostatic analyzer, in order to investigate the degree of excitation in the electronic system. The beam sources are N_2 laser (NL) with the photon energy of 3.7 eV, KrF-excimer laser (KL) of 5.0 eV and ArF (AL) of 6.4 eV, respectively. The pulse width is approximately 10 ns in all of the lasers.

II. TIME-OF-FLIGHT MASS ANALYSIS

Figures 1-a and 1-b show TOF spectra (acceleration voltage: 300 V, flight path length: 62 cm) obtained by excitation of graphite (highly oriented pyrolitic graphite) by AL and KL, respectively. For comparison, Fig. 1-c gives a similar spectrum by NL. The results show that the spectrum depends definitely on the photon energy of the incident laser beam. The spectral peaks corresponding to the flight time of C_3^{2+} (11 μs) and C_9^{2+} (19 μs) are broadened as shown in Figs. 1-a and 1-b in comparison to Fig. 1-c. The line-profile of these peaks is quite similar to that of TOF specra of Si excited by NL.[3] Namely, it has a sharp leading edge as well as a trailing, and both edges are connected by a relatively smooth curve. As found in the case of Si by NL excitation, the profile fluctuates considerably from shot to shot of laser pulses even if their peak intensity is nearly constant. The spectral leading edge and the trailing one appear at flight times close to each other or far apart in out-of-phase displacements as shown in Fig. 2. These spectra are a few examples recorded for each laser shot of a nearly equal intensity of AL. The spectra are displayed in increasing order of the spectral width as one goes from Figs. 2-a through 2-d. In Fig. 2-a, the peak at the flight time for C_3^{2+} exhibits

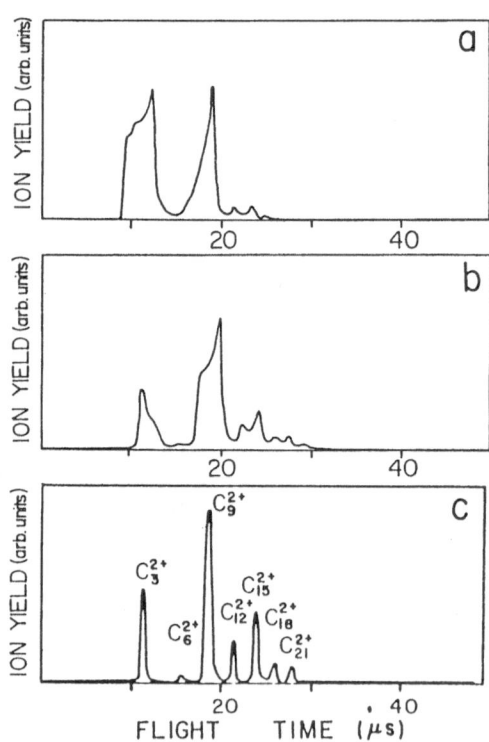

Figure 1. Time-of-flight spectra of ions emitted from graphite surface by (a) ArF laser (b) KrF laser and (c) N_2 laser for the peak intensity of 50 MW/cm^2. The pulse width of these lasers is $\sim$ 10 ns. The acceleration voltage is 300 V.

an appreciable broadening but that for C_9^{2+} shows a relatively sharp peak. In Fig. 2-d, on the other hand, both peaks are broadened, and show a line-profile different from Fig. 2-a. These spectral characteristics are less pronounced in the case of KL in comparison to AL, and is absent in NL. In the case of KL, the line profile is narrower than that of AL and is nearly constant for a given laser intensity. The spectral width tends to increase with increase in intensity for both KL and AL. According to our interpretation of previous measurements in Si, this type of spectral broadening results from decomposition of ions during acceleration time (1 μs) of the TOF spectrometer. The spectral width depends on how rapidly the decomposition proceeds through successive fragmentation of the form, $C_k^{a+} \rightarrow C_m^{b+} + C_n^{c+}$ with a = b + c, k = m + n. The more readily the fragmentation proceeds, the wider the spectral width becomes. Hence ions produced by AL decompose more readily than those by KL with the broad spectra mentioned above, and far more than those by NL which give in remarkable contrast a series of sharp lines as shown in Fig. 1-c. In other words, the ions produced by AL and KL excitation are much less stable in comparison to those produced by NL. From the flight time position of both edges in the line-profile, one can deduce the mass-to-charge ratio of the parent ions, k/a, as well as fragment ions, m/b and n/c. The value of k/a is obtained from the plot for the flight times of both edges vs. spectral width for a number of spectra such as those in Figs. 2-a through 2-d. The details of this procdure is given in ref. 2. The result gives 3/2 and 9/2 for the peaks near 11 μs and 19 μs, respectively. From the spectral position of both edges, one finds possible fragmentation processes for C_3^{2+} as $C_3^{2+} \rightarrow C^+ + C_2^+$, and C_9^{2+} as $C_9^{2+} \rightarrow C_4^+ + C_5^+$, respectively.

The spectral peaks of larger clusters are not broadened appreciably as seen in Figs. 1-a and 1-b but their intensities relative to the first peak are lower in comparison with those of Fig. 1-c. The ratio of integral intensity for the first peak with respect to the rest of the peaks decreases in going from Figs. 1-a to 1-b, and then to 1-c. These results indicate that chemical stabilities of these larger clusters are similar to those produced by NL, but the efficiency of producing large clusters at the initial stage of formation on or near the surface increases in going from AL to KL, and then to NL.

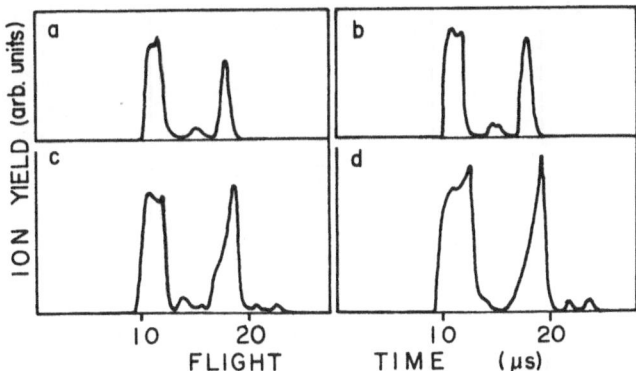

Figure 2. Fluctuation in Time-of-flight spectra of ions emitted from graphite surface by ArF laser. The spectra are few examples recorded for each laser shot and displayed in an increasing order of the spectral width.

Figure 3 shows QM spectra obtained by excitation with AL, KL and NL under excitation conditions similar to those for TOF. Spectra also show a definite photon energy dependence of laser beam. Figure 3-a for AL shows spectrum quite different from those by KL and NL. The intensity of C^+ is the dominant chemical species for AL, while C_3^+ is dominant for KL (3-b) and NL (3-c). The spectrum of Fig. 1-b is similar to Fig. 1-c in that it tends to show the intensity alternation of odd species with even. It is also found that the intensity of larger clusters with respect to C^+ increase in going from AL to KL and to NL. These results are in accordance with mass number distributions deduced from TOF measurements as shown in Fig. 1. It should be noted that the intensity ratio of C_3^+ to C^+ excited by AL is more than two orders of magnitudes smaller than that by KL. This QM result is difficult to explain in terms of the thermal model of ion desorption. Many experimental[4] as well as theoretical[5] studies show that C_3^+ is the most stable cluster of carbon in the linear chain form. In our results, the intensity of C_3^+ relative to C^+ is the lowest for the spectrum by AL but increases in going to KL, and then to NL. If the laser energy is merely converted into heat in the lattice system before desorption of surface atoms, one should observe similar mass spectra with little dependence, if at all, on the incident photon energy. Our results clearly show that the observed mass distribution depends on both initial formation process and subsequent decomposition process toward thermal equilibrium.

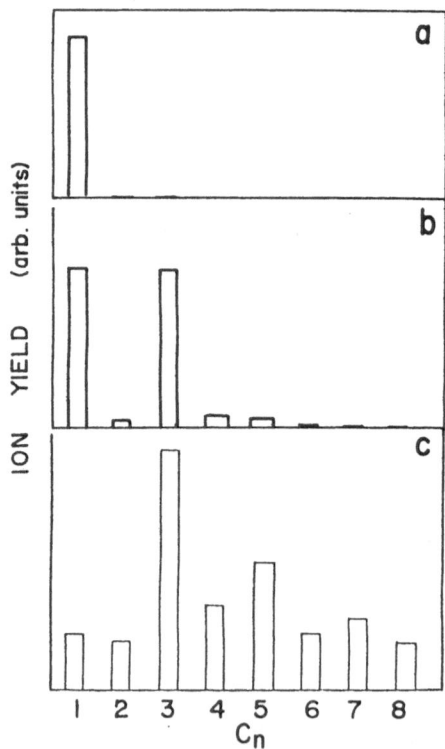

Figure 3. Quadrupole mass spectra of ions emitted from graphite surface by (a) ArF laser (b) KrF laser and (c) N_2 laser for the peak intensity of $\sim$50 MW/cm^2.

Hintenberger et al.[6,7] find in their measurement of ions from graphite spark source that the C^+ is the dominant species and its intensity is about two orders of magnitude larger than that of C_3^+ which is the second highest in intensity in their measurements. The difference in the spectrum indicates that the process of excitation by laser beam is quite different from that in the spark source.

IV. ELECTRON ENERGY DISTRIBUTION

In order to obtain information on the excitation level of electronic system in the vicinity of ion emission threshold of the sample, we have measured the electron energy distribution by means of a 127° electrostatic energy analyzer. Figure 4 shows the result by NL excitation with pulse intensities up to 30 MW/cm^2. The electron energy is expressed as the sum of kinetic energy and the work function of graphite, 4.6 eV.[8] Since the work function of graphite is larger than the photon energy of NL, there is no practical contribution from ordinary one-photon photoemission process. The total electron yield is found to increase super-linearly with respect to the laser intensity. Hence the two-photon photoemission is one of the dominant mechanisms of emission. In this case, the high energy edge of electron energy spectrum coincides with twice the photon energy of NL, 7.6 eV, as measured from the valence band maximum. Any contribution to the electron energy distribution in the higher energy comes from the photo-electrons which gain energies in various channels of collision processes such as electron-elecron, electron-plasmon and electron-phonon. The energy

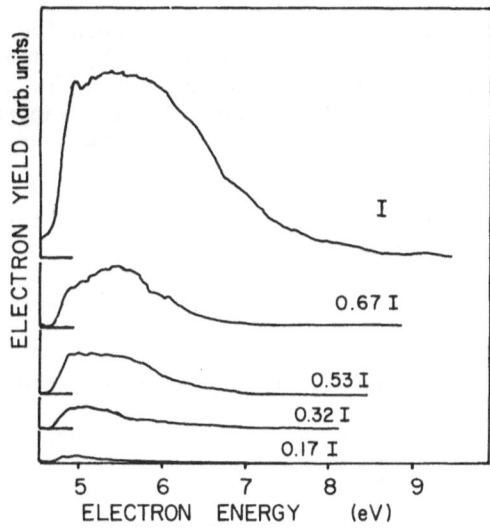

Figure 4. Electron energy distribution of photoelectons emitted from graphite surface by nitrigen laser excitation of intensities up to I = $\sim$30 MW/cm^2. The electron energy is expressed as the sum of kinetic energy and work function of graphite, 4.6 eV. The threshold for ion emission is $\sim$50 MW/cm^2.

distribution in Fig. 4 shows a relatively sharp rise at the work function of graphite and a broad maximum followed by a long tail toward higher energies. This tail extends more toward higher energy as the intensity approaches the threshold of $\sim$50 MW/cm^2 for ion emission. Since the decay of the spectral intensity begins to appear in the energy range much lower (<1.5 eV) than twice the photon energy, the contribution from the two-photon process without collision is small in the electron energy range from 6.5 to 7.6 eV. Assuming that the slope in the log-log plot of the tail represents the average kinetic energy, E_{av} of photo-carriers in the sample, one can estimate the degree of excitation in the electronic system near the ion threshold. One finds that E_{av} is less than 1 eV below the threshold. For spectra obtained by excitation with KL and AL, similar profile is found with their tails extending much more toward higher energies in going from NL to KL and then to AL. These results indicate that E_{av} of photo-carrier in the sample near the threshold of ions increases in this order. Hence at the threshold intensity, the electronic sytem of the surface atoms is excited to higher energies in this order. This observation suggests that the difference in the photon energy dependence of the TOF spectrum is related to the degree of excitation in electronic system. The higher is the degree of excitation in the electronic system, the shorter the decomposition time of ionic clusters produced in laser desorption process.

V. CONCLUSION

The process of ion desorption from the solid target of group IV elements by laser excitation has been investigated by TOF and QM at the photon energy of 3.7, 5.0 and 6.4 eV. The spectra of both TOF and QM show definite photon energy dependences. The results suggest the presence of an photon energy threshold for fragmentation of C_3^{2+} and C_9^{2+} somewhere between 3.7 and 5.0 eV. The desorption process does not take place in thermal equilibrium so that the mass, charge and kinetic energy distribution of emitted ions depend on their initial process of formation and subsequent fragmentation. Thus the chemical species of ionic clusters have been identified as functions of photon energy as well as beam intensity. The transition in their physical and chemical states in the time scale of 100 ns to 10 µs are found entirely different from those in the ordinary evaporation process as deduced from the mass analysis of ions as well as the energy analysis of electrons.

The authors would like to thank Professor T. Goto for his interest in this work and for valuable discussion. This work was supported in part by the Mitsubishi Foundation.

REFERENCES

1. J.F. Ready, Effects of High Power Laser Radiation (Academic Press, New York, 1971) Chap 4.
2. A. Kasuya and Y. Nishina, Phys. Rev. B28, 6571 (1983).
3. A. Kasuya and Y. Nishina, Phys. Rev. Lett. 57, 755 (1986).
4. K. S. Pitzer and E. Clementi, J. Am. Soc. 81, 4477 (1959)
5. J. Berkowitz and W. A. Chapka, J. Chem Phys. 40, 2735 (1964).
6. E. Dornenberg and H. Hintenberger, Z. Naturforsch. 14a, 767 (1959).
7. E. Dornenberg, H. Hintenberger and J. Franzen, Z, Naturforsch. 16a, 532 (1961).
8. T. Takahashi, H. Tokailin and T. Sagawa, Phys. Rev. B32, 8317 (1985).

ENERGY OF FORMATION FOR SMALL MOLECULAR CLUSTERS: $(H_2O)_n$

C. K. Lutrus and S. H. Suck Salk

Department of Physics and
Graduate Center for Cloud Physics Research
University of Missouri-Rolla
Rolla, Missouri 65401

I. INTRODUCTION

As is well known, there exists a great discrepency between the classical capillarity approximation and statistical mechanical treatment in defining the energy of formation of clusters. The objective of the present paper is to evaluate the energy of formation of small molecular clusters, namely $(H_2O)_n$, based on the statistical mechanical treatment.

Conventionally the energy of formation $(\Delta\Phi_i)$ of droplets or clusters of size i is defined through the relation[1]

$$n_i = n_1 \exp(-\Delta\Phi_i/kT) \quad , \tag{I.1}$$

where n_i is the concentration of clusters of size i, k, the Boltzmann constant and T, the absolute temperature. According to the capillarity approximation, the energy of formation $\Delta\Phi_i$ above is given by[1]

$$\Delta\Phi_i = -i\,kT\,\ln S + \sigma a_o i^{2/3} \quad , \tag{I.2}$$

where

$$a_o = 4\pi r^2 / i^{2/3} \quad . \tag{I.3}$$

Here i denotes the total number of monomers in the droplet. S is the saturation ratio; σ, the surface tension; a_o, the surface area per surface molecule; and r, the radius of the droplet.

In the present paper we employ a statistical mechanical treatment to express the energy of formation. The newly derived energy of formation differs from other approaches in that it is casted into a form for which a quantum mechanical treatment can be made.

II. A STATISTICAL MECHANICAL DESCRIPTION OF THE ENERGY OF FORMATION FOR HOMOMOLECULAR CLUSTERS

We consider the formation of homomolecular clusters in an environment where monomer (that is, i=1) population is dominant. They are subject to the bi-molecular association (forward process) and dissociation (reverse process) reactions of the type,

$$A_{i-1} + A_1 \overset{\rightarrow}{\leftarrow} A_i \quad . \tag{II.1}$$

Here A_i represents the homomolecular cluster of size i. The size i here refers to the total number of monomers present in the cluster.

The law of mass action for the monomer dominating system above is,

$$N_i \;/\; (N_{i-1} N_1) = q_i \;/\; (q_{i-1} q_1) \quad , \tag{II.2}$$

or

$$N_i = (N_{i-1}/q_{i-1}) \; (N_1/q_1) \; q_i \quad . \tag{II.3}$$

Here N_i is the total number of the homomolecular cluster of size i and q_i, the partition function of the molecular cluster. N_1 is the total number of monomers and q_1, the partition function of the monomer. Applying successive substitution, the expression (II.3) becomes

$$N_i = (N_1/q_1)^i \; q_i \quad . \tag{II.4}$$

The partition function of the i-cluster A_i is written in the following product form,

$$q_i = \xi_i \; \xi_i^e \quad . \tag{II.5}$$

Here ξ_i is the partial partition function which can be represented as the product of translational (ξ^t), rotational (ξ^r) and vibrational (ξ^v) parts of the partition function,

$$\xi_i = \xi_i^t \; \xi_i^r \; \xi_i^v \quad , \tag{II.6}$$

for the case of weak coupling. ξ_i^e is the electronic part of the partition function.

Denoting E_i as the electronic energy of the cluster of size i, the substitution of (II.5) and (II.6) above into (II.4) yields

$$N_i = N_1^i \; (\xi_i \;/\; \xi_1^i) \; \exp(-\Delta E_i/kT) \quad , \tag{II.7}$$

where ΔE_i is the total binding energy of the cluster of size i,

$$\Delta E_i = E_i - i \; E_1 \quad . \tag{II.8}$$

The expression (II.8) can be rewritten

$$\Delta E_i = \sum_{j=1}^{i} \Delta E_{j-1,j} \quad , \tag{II.9}$$

where $\Delta E_{j-1,j}$ is the difference in binding energy between the (j-1)-cluster and j-cluster,

$$\Delta E_{j-1,j} = \Delta E_j - \Delta E_{j-1} \quad . \tag{II.10}$$

Using (II.8) we now rearrange (II.7) to obtain

$$N_i = N_1 \exp\{-(\Delta E_i - kT \ln \xi_i + i \, kT \ln \xi_1 - (i-1) \, kT \ln N_1)/kT\} \quad .$$

$$(II.11)$$

Dividing both sides of (II.11) by the volume V and rearranging the resulting expression, we obtain the number concentration of the i-cluster,

$$n_i = N_i/V = n_1 \exp\{-(\Delta E_i - kT \ln(\xi_i/V) + i \, kT \ln(\xi_1/V)$$
$$- (i-1) \, kT \ln(N_1/V) \,) \, / \, kT\} \quad . \qquad (II.12)$$

We introduce the equilibrium concentration n_1^o of monomer at the saturation ratio, S = 1 and the equation of state for an ideal gas, in order to write

$$\ln(N_1/V) = \ln(N_1/N_1^o) + \ln(N_1^o/V)$$
$$= \ln S \quad + \ln n_1^o \quad , \qquad (II.13)$$

where N_1^o is the total number of monomers at S=1. The substitution of (II.13) into (II.12) yields

$$n_i = n_1 \exp\{-(\Delta E_i - kT(\ln \xi_i' - i \ln \xi_1')$$
$$- (i-1) \, kT \ln S - (i-1) \, kT \ln n_1^o) \, / \, kT \} \, , \quad (II.14)$$

where

$$\xi_i' = \xi_i / V \quad . \qquad (II.15)$$

We now rewrite (II.14)

$$n_i = n_1 \exp(-\Delta \Phi_i/kT) \quad , \qquad (II.16)$$

in order to difine the energy of formation,

$$\Delta \Phi_i = \Delta E_i - kT(\ln \xi_i' - i \ln \xi_1') - (i-1) \, kT \ln S - (i-1) \, kT \ln n_1^o \quad .$$
$$(II.17)$$

It is now to be noted that this expression is markedly different from the energy of formation given by the classical capillarity approximation shown in (I.2).

III. STABILIZATION ENERGIES AND ENERGIES OF FORMATION FOR SMALL WATER
     CLUSTERS

The total binding energy ΔE_i represents the stabilization energy of clusters of size i at 0^oK. Thus this energy excludes the effects of entropy. Consequently it represents the intrinsic property of molecular clusters. On the other hand, the 'energy of formation' $\Delta \Phi_i$ includes the entropy effects. Here we discuss the computed stabilization energy and the energy of formation $\Delta \Phi_i$ for small water clusters.

For the computation of stabilization energies we used a highly success-ful semi-empirical effective Hamiltonian method called 'I-MNDO'[2,3], that we developed earlier. As shown in Fig. 1, the stabilization or binding energy ΔE_i decreases with cluster size. This implies that the clusters become increasingly stable with increasing cluster size at 0^oK. Note that the positive y axis represents the negative values of ΔE_i, that is, $y=-\Delta E_i$. The computed results are in good agreement with a recent ab initio study of Tomoda and Kimura[4]. The solid curve represents the best analytic fit to the computed results. The analytic values of the stabilization energy plotted in Fig. 1 are obtained by the following procedure.

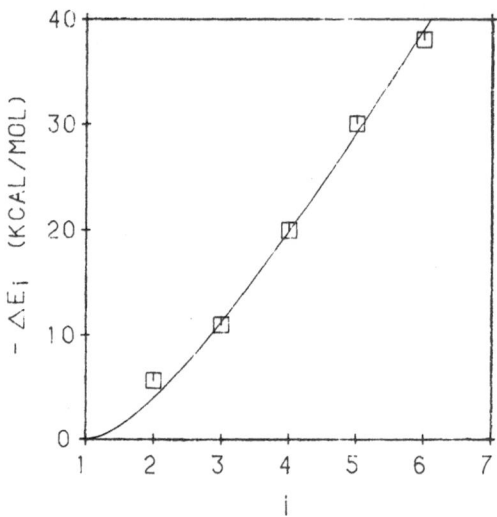

Fig. 1. Stabilization energy ΔE_i as a function of cluster size i. The computed results are based on various open structures of water clusters. See Ref. 4 for the details of the geometric structures. Only the 'Group A' structures in the reference are considered here.

The binding energy per bond is given by

$$B_i = \Delta E_i \, / \, n \qquad\qquad (III.1)$$

where n is the total number of bonds in the cluster. Earlier we found that the best fit to the binding energy per bond is given by the relation[3]

$$B_i = B_\infty \, (1 - 1/i^\alpha) \qquad\qquad (III.2)$$

where $B_\infty = -12.7$ Kcal/mol or -0.55 eV and $\alpha = 0.531$. It is interesting to note that $B_\infty = -12.7$ Kcal/mol at $i = \infty$ is close to the observed lattice energy (-13.4 Kcal/mol) of ice. Using (III.2), the stabilization energy ΔE_i in (III.1) is then

$$\Delta E_i = n \, B_\infty \, (1 - 1/i^\alpha) \qquad\qquad (III.3)$$

The expression (III.3) indicates nonlinear relationship between the stabilization energy and cluster size i for small clusters.

Assuming that the intramolecular vibrational frequencies of the molecular clusters are not considerably different from those of monomer (an isolated molecule), we write

$$- kT(\ell n \, \xi_i^v - i \, \ell n \, \xi_1^v) \doteq - kT \, \ell n \, \xi_i^{v'} \qquad\qquad (III.4)$$

where v' is to denote only the intermolecular part of normal mode vibrational frequencies. Now for the computation of intermolecular normal mode vibrational frequencies for the expression (III.4) and thus (II.17), we used the aforementioned Hamiltonian method of I-MNDO[2,3]. For example, see ref. 2 for the predicted intermolecular normal mode vibrational frequencies of $(H_2O)_2$. The normal mode vibrational frequencies are obtained by using the potential energy surface of the water cluster extracted from this method.

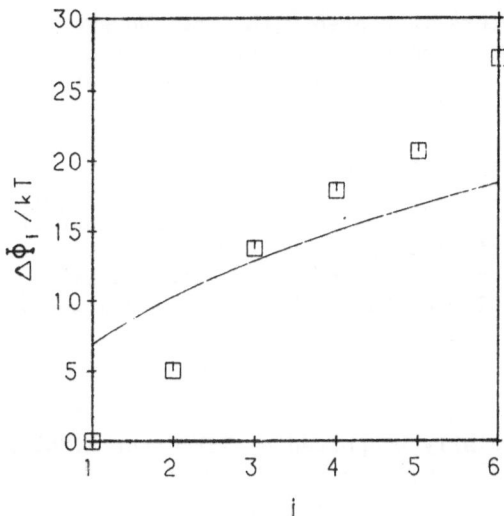

Fig. 2. The computed dimensionless energy of formation, $\Delta\Phi_i/kT$, as a function of cluster size i. The solid line represents the capillarity approximation.

The rotational part of the partition function is obtained using the well known relation

$$\xi_i^r = (8\pi^2 kT/h^2)^{3/2} (I_a I_b I_c)^{1/2} / \eta \qquad (III.5)$$

h is the Planck constant and η, the symmetry number. I_α with α = a, b and c are the principal moments of inertia. The principal moments of inertia are computed by using the predicted equilibrium geometries of the small clusters from our quantum calculations (I-MNDO[2]).

Fig. 2 shows the computed energy of formation (denoted by □) as a function of cluster size i at T = 300°K and S = 5. It is of note that the energy of formation $\Delta\Phi_i$ increases with cluster size, in sharp contrast to the case of stabilization energy ΔE_i discussed above. Although not shown here, we find that the same trend persists at other temperatures. This indicates that the clusters at finite temperatures are thermodynamically unstable for their growth. This is attributed to the increasing importance of entropy effects at finite temperatures. As expected, the energy of formation at small cluster sizes does not smoothly vary as a function of cluster size unlike the case of of the classical capillarity approximation (denoted by a solid curve). The capillarity approximation assumes the shape of a spherical droplet even for small size clusters which are molecular (rather than bulk) in dimension. Thus computed results for the small water clusters based on this liquid droplet theory should not be considered too seriously.

IV. SUMMARY

In this work, a statistical mechanical expression for the energy of formation of molecular clusters was presented. Unlike other methods[5,6,7], the derived energy of formation was casted into a form which can be computed by quantum mechanical means. Considering only the stabilization energy which measures the intrinsic stability of clusters, we noted that the stability of the clusters increases with size. However, the predicted energy of formation showed the importance of entropy effects in causing instability, that is, $\Delta\Phi_i - \Delta\Phi_{i-1} > 0$ for the selected water clusters of small size. This trend is expected to remain true despite anharmonicity correction to be made for the accurate evaluation of the partition functions. Extension of the present statistical mechanical approach to include larger clusters is currently in progress. Such a study will be of great value to understand the differences between the classical capillarity approximation which depends on the bulk physical properties of macroscopic systems and the present molecular approach which combines the statistical mechanical treatment with the first principle (quantum mechanical) calculations.

ACKNOWLEDGEMENT

This work is partially supported by NSF Grant ATM82-12328

REFERENCES

1. F. F. Abraham, 'Homogeneous Nucleation Theory' (Academic Press, New York, 1974).
2. S. H. Suck Salk, T. S. Chen, D. E. Hagen and C. K. Lutrus, Theoretica Chim. Acta $\underline{70}$, 3 (1986).
3. S. H. Suck Salk and C. K. Lutrus, Phys. Rev. A, submitted.
4. S. Tomoda and K. Kimura, Chem. Phys. Lett. $\underline{102}$, 560 (1983).
5. H. Reiss, Adv. Colloid and Interface Sci. $\underline{7}$, 1 (1977); H. P. Gillis, D. C. Marvin and H. Reiss, J. Chem. Phys. $\underline{66}$, 223 (1977).
6. P. L. M. Plummer and B. N. Hale, J. Chem. Phys. $\underline{56}$, 4329 (1972).
7. J. J. Burton and C. L. Briant, Adv. Colloid and Interface Sci. $\underline{7}$, 131 (1977).

ENERGY REDISTRIBUTION AND CLUSTER STABILITY[1]

Patricia L. Moore Plummer and T.S. Chen

Departments of Physics and Chemistry
University of Missouri-Columbia
Columbia, MO   65211   U.S.A.

ABSTRACT

The stability of small molecular clusters is not independent of the rate
or the mode of energy transfer to the cluster. Either can lead to structure
disruption at total energies at which the cluster would otherwise be stable.
Molecular dynamics calculations are used to monitor the changes in cluster
structure as energy is added or removed from the system. The energy was added in different ways in the attempt to simulate: 1) an isolated collision,
2) interaction with a bath at a temperature different from that of the cluster and 3) a transition to a single excited vibrational state. The clusters
used in these experiments were water clusters containing from three to twenty
molecules. The potential used was the central force pair potential proposed
by Stillinger and Rahman[2].

INTRODUCTION

Before condensation and/or aerosol formation can take place, small
embryonic clusters are formed from single molecules. In order to grow, the
cluster should neither decay nor significantly change its structure in a time
shorter than the time between collisions with other molecules in the system.
Thus knowledge of the stability of the initial embryos of the new phase is of
major importance in determining the mechanisms by which prenucleation clusters and aerosol particles form and grow. The stability of the clusters will
depend not only on the total energy of the cluster, but how energy is absorbed by the cluster and how it is distributed within the cluster.

The detailed mechanisms for energy absorption and redistribution by
clusters will depend on the chemical nature of the cluster. Thus we expect
atomic and molecular clusters to differ widely in both the magnitude and
variety of processes available for energy absorption and distribution. Water
clusters were chosen for this study because the interactions present range
over a wide energy spectrum. For example, the bonds between the molecules
are highly directional but more than an order of magnitude smaller in energy
than the bonds within the molecules.

Molecular dynamics calculations were used to simulate the introduction
of energy into a water cluster and follow the time evolution of the excited
cluster. The simulations are described in the next section and the results
are discussed in the final section.

CALCULATIONS

In molecular dynamics the classical equations of motion for the molecules are numerically solved to produce the position and velocity of the particles as functions of discrete time steps. The central force pair potential model of Stillinger and Rahman[2] is used to describe the interaction between each pair of atoms. This potential provides for stable water monomers and dimers and has been used successfully in simulations of larger water clusters[3] and bulk water[2]. A central difference method due to Verlet[4] is used to integrate the resulting equations of motion. A time step of $\Delta t = 0.00025$ ps was found in previous studies[3] to provide energy conservation to within 0.001 kcal per molecule.

Most of the simulations in this study were carried out for clusters containing 20 water molecules. However clusters containing 3, 5, 6 and 21 molecules were also examined.

In the first set of experiments, a highly artificial elastic collision between a water monomer and a 20 molecule cluster is simulated. Prior to the collision a 20 molecule cluster, initially in a pentagonal dodecahedral configuration, is heated[3] to ca. 225 K. At this temperature, much of the second and third neighbor orientational order has been lost but first neighbor order is essentially unchanged from that at 20 K. The motion of the 20 molecule cluster is then frozen and a monomer with translational energy ca. 225 K is introduced at impact parameters of 0, 5 and 10 A. The initial monomer energy was allowed to redistribute between translation, rotation and internal angle deformation but, since the collision was elastic and the cluster held rigid, not to transfer to the 20 molecule cluster.

The second set of experiments was similar to the first except that it was inelastic. Energy transfer between the monomer and the cluster molecules was allowed so that the 20 molecule cluster was no longer rigid, but participated fully in the dynamics.

In the third experiment 120 kcal/mole was added to the 20 molecule cluster initially in the dodecahedral clathrate structure equilibrated at 20 K. All of the added energy was introduced into the symmetric stretching mode of the cluster (the only singly degenerate intermolecular mode of the cluster).

The next set of simulations was carried out on clusters containing five water molecules. These simulations were used to examine the rate and temperature dependence of structure interconversion. First, clusters having different hydrogen bonding configurations were cooled to determine which structures were the more stable at low temperatures (1-5 K). Initially, six different bonding configurations were considered. The four more stable structures were then subjected to a series of heating and equilibrium runs. The heating rate was ca. 10 K/ps with 500 time steps allowed for equilibration. Properties of the cluster at each energy were obtained by time averaging over 32768 time steps or approximately 8 ps.

The temperature of the cluster[5] is calculated from the time average of the kinetic energy:

$$T_{cluster} = (3N/3N-6)(2/3k)\langle(1/2N)\sum_i m_i v_i^2\rangle \quad (1)$$

and the power spectrum[6] from the Fourier transformation of the velocities:

$$\breve{E}(\omega_k) = \sum_i^N \sum_j^3 \frac{1}{2} m_i |\breve{v}_{ij}(\omega_k)|^2 . \quad (2)$$

RESULTS AND DISCUSSION

In the collisions between a monomer and a rigid 20 molecule cluster, for all impact parameters, there was an increase in the librational energy and a decrease in the translational energy of the monomer as the monomer neared the cluster and the interaction energy increased. This redistribution of monomer energy reflects the attempt of the monomer to accomodate to the cluster. The trajectories for impact parameters of 0 and 10 A are shown in Fig. 1-2. The numbers alongside the trajectory show the elapsed time in units of 1000 Δt. The lines along the trajectory indicate the changing monomer orientations. For impact parameters of 0 and 10 A, the interaction energy was in excess of that required for bond formation. However, large oscillations in the librational energy of the monomer produced a configuration having a weaker interaction (see Fig. 3 for b=10A) and in all cases the monomer eventually escaped.

In contrast, when the cluster was not held rigid and energy transfer between the monomer and the cluster was allowed, the monomer was incorporated in the cluster. The interaction energy between the monomer and cluster ranged from 7 to 16 kcal/mol or about 1 to 3 hydrogen bonds. The cluster structure was altered under the influence of the monomer even when the interaction energy was small (less than one kcal/mole). In the addition process, the cluster absorbs about 10 to 15% of the initial monomer energy. Figs. 4 and 5 indicate two "snapshot" views of the monomer-cluster absorption process for the 0 A impact parameter case. Instead of the 'bounce' occuring in the frozen cluster collision, at the time of the monomer's first contact with the cluster, the molecules in the cluster reorient so as to increase the interaction and prevent the escape of the monomer. In fact, for all impact

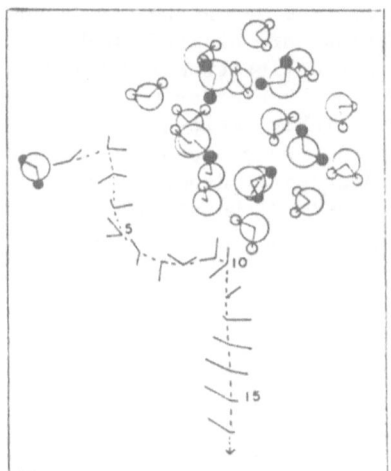

Fig. 1. Collision sequence of a water monomer with a rigid cluster Impact parameter is 0 A, T = 225 K

Fig. 3. Interaction energy.

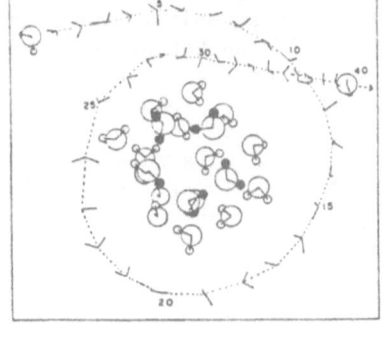

Fig. 2. Impact parameter is 10 A

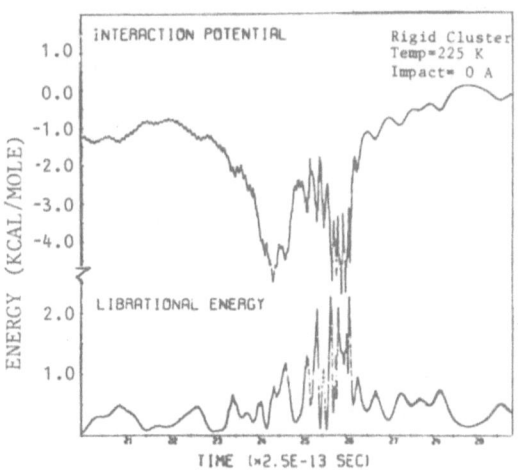

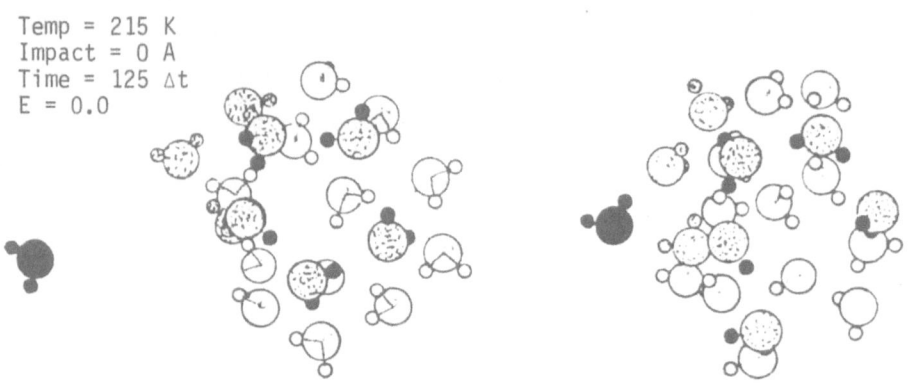

Temp = 215 K
Impact = 0 A
Time = 125 Δt
E = 0.0

Fig. 4. t=0, b=0 Å, E=0 Kcal/mole. Fig. 5. t=1 ps, b=0 A, E=10.1 Kcal/mole.

parameters investigated, the results indicate that once a bond formed (when the average interaction potential exceeds 4 kcal/mole) the monomer and cluster molecules reorient within about 100 intramolecular vibrations to form a second bond with the cluster. The interaction energy and the monomer librational energy for the collision illustrated in Fig. 4-5 are shown in Fig. 6.

When a large amount of energy is suddenly added to a single vibrational mode, initially all of the energy is in translation of the individual molecules toward the center of mass of the cluster. However as the molecules pass their equilibrium positions (around 500 Δt) the energy begins to build in the librational modes. The internal vibrations are excited 1000Δt later. Observation of a number of cases suggests that the librational motions serve as a 'bridge' for the transfer of energy between the cluster modes and the internal vibrations. As the librations become highly excited, the preferred alignment of hydrogens along a bond is disturbed. Once a bond is disrupted by an exceptionally large oscillation, the dodecahedral geometry is likely to deteriorate too quickly for the more highly ordered structure to be reestablished. After the structure disruption occurs, energy fluctuations in the modes are considerable reduced and the energy rapidly becomes equally distributed among the available modes.

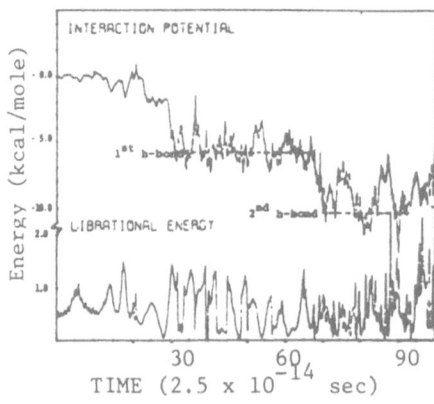

Fig. 6. Interaction energy and monomer librational energy collision at b=0 A.

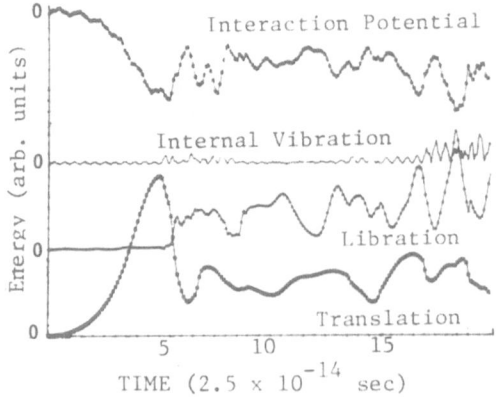

Fig. 7. Analysis of average molecular contributions to cluster for energy on vibrational excitation.

Two other experiments were conducted on the 20 molecule cluster. For both of these, the molecules were in the clathrate configuration and had been equilibrated at ca. 65 K. In the first, an inelastic virtual collision was simulated by giving one cluster molecule extra translational energy—on the order of 2.5 kcal/mole. In this process the injected energy was directed toward the center of mass of the cluster. The sudden transfer of energy to a single molecule did not, as has been speculated, disrupt the cluster structure significantly nor did it result in the loss of a molecule. The energy was rapidly transfered to adjacent molecules and distributed throughout the cluster. Subsequent observations revealed radial distribution functions were relatively unchanged and there was no evidence for any build-up of energy in a single bond or mode which could result in the evaporation of a molecule. In the second of these experiments, a virtual evaporation was simulated by instantaneously removing one molecule from the cluster, leaving the remaining molecules intact. Surprisingly, the cage-like structure did not immediately collapse and there was no extensive rearrangement of the remaining bonds. The cage expanded somewhat and the radial distribution showed loss of definition in the third and subsequent peaks. However, much of the nearest neighbor and five membered ring structure of the original cluster remained intact.

Simulations were also performed on smaller clusters of water molecules, in part to see if the trends observed for the 20 molecule cluster, such as the importance of the librational modes in energy redistribution and structure disruption, were present in clusters containing 3 or 5 molecules. The details of these simulations will be reported elsewhere[7] but some results will be described here.

For the five molecule cluster, a ring structure was found to be stable over a wide temperature range. In addition, in the energy optimization calculations, several of the more open structures convert to the ring as the temperature is lowered. A second structure referred to as C2 is obtained from optimizing a tetrahedral pentamer or a branched chain pentamer. At 5 K the total energy of the C2 structure is 2-3 kcal/mole higher than the ring but the barrier for interconversion of the structure is sufficiently high that no conversion can take place below about 70 K.

The total energy of the ring pentamer and of the C2 pentamer are plotted as a function of temperature in Fig. 8. This plot shows the transition of the

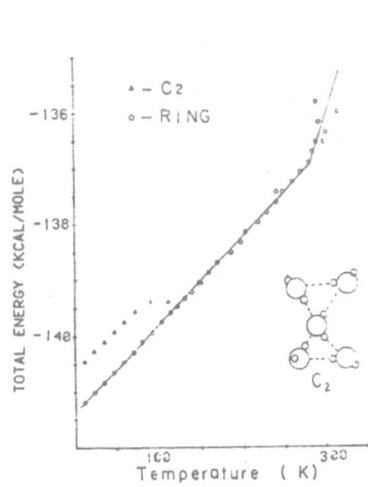

Fig. 8. Energy as a function of temperature for 2 configurations of water pentamers.

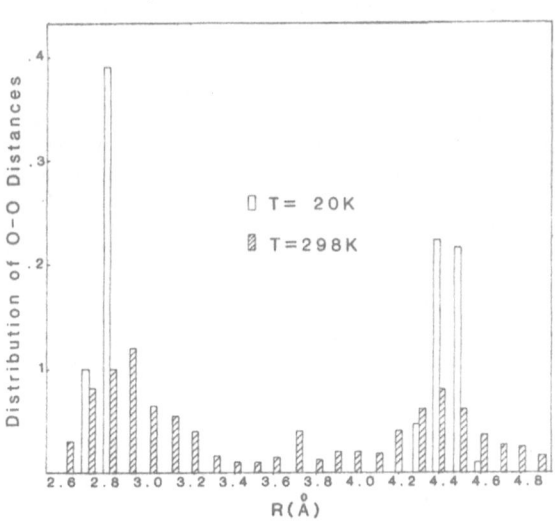

Fig. 9. O-O distances for ring water pentamer at 20 K and 298 K.

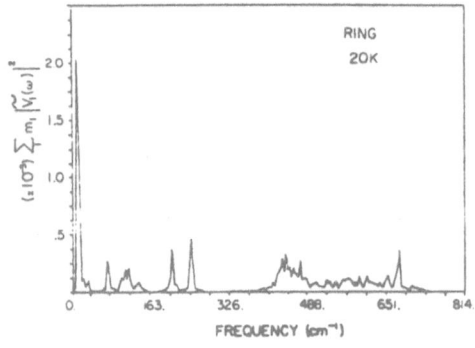 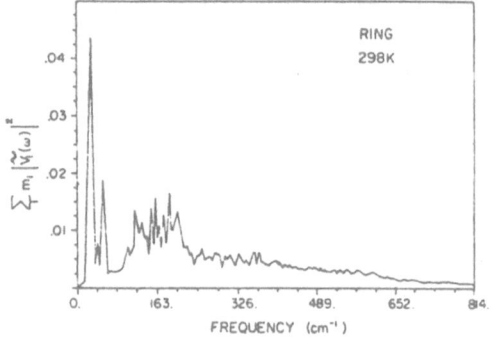

Fig. 10. Power spectrum for ring   Fig. 11. Same as 9, with T=298 K.

C2 structure to the ring at approximately 100 K. (At higher heating rates-
-40 kcal/ps the transition takes place at ca. 75 K but lower heating rates do
not raise the transition temperature). The change in slope around 300 K
corresponds to fragmentation of the ring and possibly can be considered as a
transition from dense to gas phase. Even at 298 K the average structure is
still a ring, but the hydrogen bonds are becoming very flexible and an
increasing amount of energy is containd in the librational motions. As was
seen for the case of the larger clusters, a build-up of energy in the
librational modes leads to a disruption of the hydrogen bonding structure.

For the ring pentamer the structure or bonding pattern remains
essentially unchanged from 2 to 280 K. Above 290 K structures having fewer
total hydrogen bonds or more 'dangling' molecules are observed with increasing frequency. The ring is also expanding somewhat at the higher temperature
and a greater variety of 0-0 separations are observed (see Fig. 9). The power spectrum associated with group cluster motions at 20 K and 298 K are shown
in Figs. 10-11. The loss of ring structure is evidenced by the shift of the
librational molds to lower frequencies as the temperature increases.

Even though these studies simulate highly simplified energy transfer
processes, they can provide significant insight into the complex mechanisms
present in small clusters, especially molecular clusters. We see that even
for the clusters containing less than 10 molecules, the most stable structure
depends not only on the temperature of the cluster but the way in which the
cluster reaches that temperature. Too rapid 'heating' tends to concentrate
the energy in modes which lead to structure disruption. On the other hand,
the cluster is capable of absorbing substantial amounts of energy if
sufficient time is allowed for the energy to distribute among the internal
degrees of freedom. The efficiency of energy transfer is further evidenced
by the collision studies which support the assumption of an accommodation
coefficient of near one for even the small water clusters.

REFERENCES

1. This research supported in part by Atm. Sci. Section, NSF under grants
   NSF ATM77-72614 and NSF ATM80-19752.
2. F.H. Stillinger and A. Rahman, J. Chem. Phys. 68, 666 (1978).
3. P.L.M. Plummer and T.S. Chen, J. Phys. Chem. 87, 4190 (1983) and T.S.
   Chen, Ph.D. Dissertation, University of Missouri-Rolla, Dec. (1980).
4. L. Verlet, Phys. Rev. 159, 98 (1967).
5. D.J. McGinty, J. Chem. Phys. 58, 4733 (1973).
6. P.O. Esbjorn, E.J. Jensen, W.D. Kristensen, J.W. Martin, and L.B.
   Pedersen, J. Comput. Phys. 12, 289 (1973).
7. P.L.M. Plummer and T.S. Chen, J. Chem. Phys. (submitted).

FRAGMENTATION OF NEUTRAL AND IONIC CARBON CLUSTERS

Krishnan Raghavachari

AT&T Bell Laboratories
Murray Hill, NJ 07974

J. Stephen Binkley

Sandia National Laboratories
Livermore, CA 94550

ABSTRACT

The structures and energies of the neutral and ionic forms of small carbon clusters (C_2-C_{10}) have been calculated by means of accurate ab initio molecular orbital techniques. The effects of polarization functions and electron correlation have been included in these calculations. The ground state geometries are calculated to change from linear structures toward cyclic structures as the cluster size increases, but with significant odd-even differences. Harmonic vibrational frequencies have been reported for all the clusters and may be of aid in the assignment of matrix isolation infrared spectra for such species. The fragmentation behavior of both the neutral and ionic clusters has also been carefully investigated. The lowest energy channel for the fragmentation of C_n^+ corresponds to the loss of neutral C_3 to give C_{n-3}^+. This observed fragmentation pattern is not only due to the extra stability associated with C_3 but also has a significant contribution from the ionization potential of the cluster C_{n-3}. The results are in good agreement with the recent experimental results on the photofragmentation of small carbon cluster ions.

INTRODUCTION

Carbon clusters have been the focus of many different experimental[1-11] and theoretical[12-26] studies during the past 30 years. The experimental works range from the early investigations of Honig[1] and Drowart et al.[2] on the small clusters (2-7 atoms) to the recent studies by Smalley and coworkers[8,9] on the larger clusters (40-80 atoms) suggesting an unusually stable spheroidal structure for C_{60}. The theoretical works consist of several ab initio molecular orbital studies[12-18] on the smaller clusters (2-4 atoms) but only empirical and semi-empirical calculations on the larger clusters.[19-26] In this work, we have carried out a systematic theoretical study using accurate quantum chemical techniques to determine the ground state structures and stabilities of neutral and ionic carbon clusters (C_2-C_{10}). We have then used the results to calculate the energetics of fragmentation of these clusters in order to interpret the unusual behavior seen by Geusic et al.[6,7] in photofragmentation experiments.

THEORETICAL METHODS

All electron _Ab initio_ molecular orbital techniques have been used throughout.[27] The Hartree-Fock method was used with the large 6-31G* basis set[28] (double-zeta sp- plus a set of d-type polarization functions on each carbon) in the optimization of molecular geometries for each cluster. Harmonic force constants and the associated vibrational frequencies were then calculated[29] for all the ground state structures with the same basis set. Electron correlation effects were included by means of an augmented coupled cluster scheme[30] which is capable of yielding reliable energies even in cases where there are low-lying excited states. At these levels of theory, the calculated bond lengths can be expected to be reliable to within 1-2%, the vibrational frequencies typically too high by about 10%, and the binding energies typically too low by about 10%.[27]

GROUND STATE GEOMETRIES

Linear, cyclic and other three-dimensional structures were considered initially for each cluster. However, only linear or cyclic structures were found to be the ground states in all cases. This is clearly due to the importance of strong π-bonding involving carbon which is in complete contrast to the behavior seen previously for silicon clusters.[31-33] The calculated ground state geometries for all the neutral carbon clusters are shown in Figure 1. The nature of the bonding in such structures has been discussed in detail elsewhere.[34,35] A surprising observation, not suggested by the previous calculations, is the strong odd-even alternation with all the even-clusters having cyclic ground states. It should be noted, however, that there are alternative linear structures for C_4 and C_8 which are only a few kcal/mol higher in energy than the cyclic structures shown in Figure 1.[16-18] The energy difference is larger for C_6 where the cyclic structure is about 10-15 kcal/mol more stable than the linear form.[34] The first cluster where the cyclic structure is substantially more stable than the linear form is C_{10} (by about 50 kcal/mol).[35]

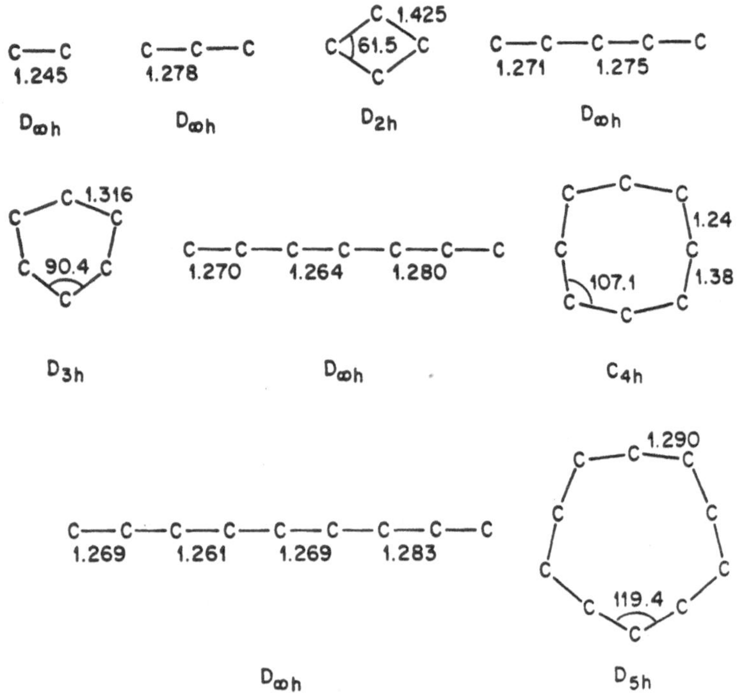

Fig. 1.   Calculated ground state geometries (Å and degrees) of neutral clusters . Note that there are also low-lying linear forms of C_4 and C_8 (see text).

Ionization changes these results somewhat. The odd-clusters still remain linear. For C_4^+, the cyclic and the linear forms are still very close in energy. However, for the larger even-clusters where there is a linear form (triplet) and a cyclic form (singlet), the linear structure has a lower ionization potential and thus the ground state of the ion may be different from that of the neutral. Such an analysis[35] indicates that the ground state of C_{10}^+ is definitely cyclic but C_8^+ and perhaps C_6^+ may have linear ground states. This suggests that previous simple ideas[19,20] that the ions up to C_9^+ may be linear appear to be reasonable.

VIBRATIONAL FREQUENCIES

As mentioned earlier, harmonic vibrational frequencies have been computed for all the ground state structures and are listed in Table I. The frequencies for the linear form of C_4 have also been included since there is some evidence for its observation in matrix-isolation experiments.[36] As expected, the frequencies of C_2 and C_3 are about 10% larger than the experimental values though there is a larger deviation for the low bending frequency in C_3.[37] The calculated frequencies for linear C_4 and C_5 are in good agreement with those derived by Sanborn by means of empirical correlations.[38] These results can be used in the calculation of the thermodynamic properties of these clusters and may also be useful in the assignment of frequencies observed in matrix isolation infrared spectroscopy.

Table I.   Harmonic Vibrational Frequencies Calculated for Neutral Carbon Clusters

Mol[a]	Vibrational frequencies (cm^{-1})
C_2	$1940(\sigma_g)$
C_3	$154(\pi_u),1367(\sigma_g),2311(\sigma_u)$
C_4	$350(b_{2u}),450(b_{3u}),1088(a_g),1103(b_{3g}),1431(a_g),1568(b_{1u})$
$C_4{}^b$	$209(\pi_u),408(\pi_u),1022(\sigma_g),1741(\sigma_u),2345(\sigma_g)$
C_5	$112(\pi_u),222(\pi_g),648(\pi_u),863(\sigma_g),1632(\sigma_u),2220(\sigma_g),2344(\sigma_u)$
C_6	$458(a_2''),575(e''),738(e'),888(a_1'),1269(a_1'),1285(e'),1349(a_2'),$ $1764(e')$
C_7	$73(\pi_u),157(\pi_g),240(\pi_u),598(\pi_g),631(\sigma_g),710(\pi_u),1206(\sigma_u),$ $1745(\sigma_g),2132(\sigma_u),2281(\sigma_u),2376(\sigma_g)$
C_8	$190(b_g),294(b_u),344(b_u),371(a_u),482(e_g),512(b_g),677(e_u),$ $738(a_g),1014(a_g),1097(e_u),1276(b_g),1957(b_g),2007(e_u),2050(a_g)$
C_9	$49(\pi_u),114(\pi_g),187(\pi_u),252(\pi_g),496(\sigma_g),567(\pi_u),658(\pi_g),783(\pi_u),$ $960(\sigma_u),1393(\sigma_g),1803(\sigma_u),2084(\sigma_u),2134(\sigma_g),2338(\sigma_u),2415(\sigma_g)$
C_{10}	$184(e_2'),253(e_2''),419(a_2''),497(e_2'),555(e_1''),568(e_2'),577(a_2'),$ $661(a_1'),690(e_1'),946(a_1'),1118(e_1'),1522(e_2'),1971(e_2'),2013(e_1')$

a   Unless otherwise indicated, the structure corresponds to the ground state geometry as shown in Figure 1.
b   The structure corresponds to a linear geometry which is not the ground state (see text).

CLUSTER FRAGMENTATION

Geusic et al.[6,7] have carried out detailed photofragmentation experiments on $C_3^+ - C_{20}^+$ and have found that the dominant fragmentation process for C_n^+ corresponds to the loss of a C_3 unit to give C_{n-3}^+. In addition, they have used the laser fluence dependence of the cluster ion depletion to get an estimate of the fragmentation energies of the cluster ions. Their data suggests that $C_6^+ - C_{20}^+$ require less than 3.53 eV, C_5^+ requires more than 4.98 eV and C_4^+ requires less than 4.98 eV for fragmentation.

We have used our calculated binding energies and ionization potentials[35] to calculate the energetics of different fragmentation processes.

Table II lists the energy required to fragment a given neutral cluster into different possible products and Table III lists the corresponding information for ionic clusters. Detailed analysis of Tables II and III shows that two factors are responsible for the observed fragmentation behavior. The first factor is obviously the stability of the neutral cluster C_3. This has been pointed out previously[18,35] and can also be seen from the known experimental binding energies of the small clusters.[37] Table II shows that even with the neglect of ionization effects, the fragmentation of C_n in most cases leads to C_3 and C_{n-3}. Though this is a dominant effect, subtle differences are found for the larger neutral clusters. For example, the energy required to fragment C_{10} into C_7 and C_3 is almost the same as that required to give $2C_5$, though the latter is not a prominent observed channel. Similarly, the energy required to fragment C_{12} into C_9 and C_3 is very similar to that required for the channel leading to C_7 and C_5. This effect is due to the fact that the bonding in all odd-atom clusters (excluding the atom) is very similar and hence fragmentation of an even cluster to yield different combinations of odd products are all similar in energy.

Table II. Binding and Fragmentation Energies (eV) for Neutral Carbon Clusters

Number of atoms (n)	Binding energy for C_n	Fragmentation energy[a]				
		$C_{n-1}+C$	$C_{n-2}+C_2$	$C_{n-3}+C_3$	$C_{n-4}+C_4$	$C_{n-5}+C_5$
2	5.8	5.8				
3	12.7	6.9				
4	17.3	4.6	5.7			
5	23.9	6.6	5.4			
6	28.9	5.0	5.8	3.5		
7	35.2	6.3	5.5	5.2		
8	40.3	5.1	5.6	3.7	5.7	
9	46.4	6.1	5.4	4.8	5.2	
10	53.9	7.5	7.8	6.0	7.7	6.0

[a] The lowest fragmentation energy is underlined.

Table III. Binding and Fragmentation Energies (eV) for Ionic Carbon Clusters

Number of atoms (n)	Binding energy for C_n^+	Fragmentation energy[a]				
		$C_{n-1}^+ +C$	$C_{n-2}^+ +C_2$	$C_{n-3}^+ +C_3$	$C_{n-4}^+ +C_4$	$C_{n-5}^+ +C_5$
2	4.7	4.7				
3	12.3	6.5[b]				
4	17.8	5.1[b]	7.3			
5	24.2	6.4	6.1			
6	30.0	5.8	6.4	5.0		
7	36.2	6.2	6.2	5.7		
8	42.0	5.8	6.2	5.1	6.9	
9	48.0	6.0	6.0	5.3	6.3[b]	
10	54.7	6.7	6.9	5.8	7.4	6.6

[a] The lowest fragmentation energy is underlined.
[b] For these cases, the smaller cluster fragment has a lower ionization potential than the larger fragment and hence is assumed to carry the charge.

The inclusion of ionization in Table III has an important effect on the fragmentation energies. This is principally due to the fact that, in general, larger clusters are easier to ionize than smaller clusters. Since C_{n-3} is one of the larger possible products, it is much easier to ionize than, for example, C_{n-5}. Thus, for the fragmentation of C_{10}^+, the channel leading to the formation of C_7^+ and C_3 is more favorable than the one leading to C_5^+ and C_5 mainly because C_7 has a lower ionization potential than C_5. Similar explanations appear to be valid for the larger clusters also.

Experimentally, there are some exceptions to this fragmentation pattern. For example, C_5^+ gives neutral C_2 and C_3^+. This is consistent with the calculations since C_3 has a lower ionization potential than C_2.[35] There are also instances where the smaller cluster appears to carry the charge. For example, in the fragmentation of C_4^+, a significant channel is found where the charge appears on the carbon atom.[6] This can also be explained easily, since the atom is calculated to be easier to ionize than C_3.[35]

The trends in the observed fragmentation energies,[6,7] e.g., C_5^+ requiring more energy to fragment than both C_4^+ and C_6^+, are also reproduced in the calculations. However, the absolute value of the experimentally derived fragmentation energies[6,7] are found to be significantly smaller than the corresponding theoretical estimates. This is particularly true for the larger clusters ($\geq$ 7 atoms) where the experimentally determined[6,7] upper limit is 3.5 eV whereas the calculations indicate fragmentation energies greater than 5 eV. From the known systematic deficiency of the calculations[35] which uniformly yield about 90% of the true binding energy, it is not clear how the true fragmentation energies can be significantly smaller than the calculated values. In fact, the true fragmentation energies are likely to be somewhat greater than the calculated values. Of course, the calculations assume that the initial cluster ion is in its ground state. However, if the ions have considerable internal energy in the experiments, it can lead to an underestimation of its determined fragmentation energy. Another possibility, as pointed out by Geusic et al.,[6,7] is that the apparent one-photon dissociation process with the 3.5 eV photon as seen by its laser fluence dependence could really be a two-photon process if one of the cross sections is considerably greater than the other. Further experiments may be necessary to resolve this discrepancy.

The calculated fragmentation energies illustrate the nature of the stabilities of the cluster ions. Figure 2 plots the fragmentation energies as a function of the size of the cluster. It is clear that C_3^+, C_5^+, and C_7^+ appear as stable clusters from this Figure. This appears to be consistent with many experimental observations[4,6,10] which show dominant peaks for odd clusters up to C_7^+. Thus, energetic considerations appear to explain many of the interesting aspects of the behavior of these clusters including the detailed fragmentation patterns.

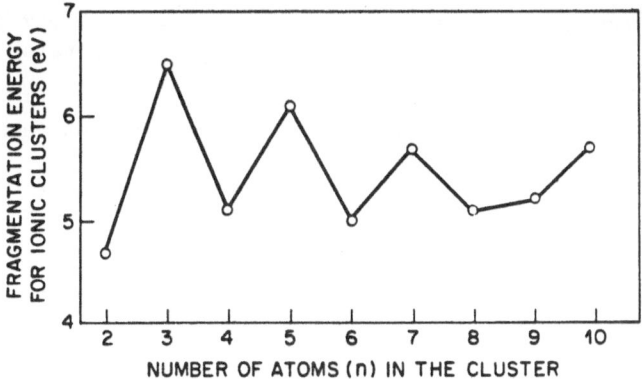

Fig. 2.   Fragmentation energy for the most favorable channel for ionic clusters C_n^+ vs. the number of atoms n.

REFERENCES

1. R. E. Honig, J. Chem. Phys. 22:126 (1954).
2. J. Drowart, R. P. Burns, G. DeMaria and M. G. Inghram, J. Chem. Phys. 31:1131 (1959).
3. N. Fürstenau and F. Hillenkamp, Int. J. Mass Spectrom. Ion. Phys. 37:135 (1981).
4. E. A. Rohlfing, D. Cox, and A. Kaldor, J. Chem. Phys. 81:3322 (1984).
5. L. A. Bloomfield, M. E. Geusic, R. R. Freeman and W. L. Brown, Chem. Phys. Lett. 121:33 (1985).
6. M. E. Geusic, T. J. McIlrath, M. F. Jarrold, L. A. Bloomfield, R. R. Freeman and W. L. Brown, J. Chem. Phys. 84:2421 (1986).
7. M. E. Geusic, M. F. Jarrold, T. J. McIlrath, L. A. Bloomfield, R. R. Freeman and W. L. Brown, Z. Phys. D. (in press).
8. H. W. Kroto, J. R. Heath, S. C. O'Brien, R. F. Curl and R. E. Smalley, Nature (London) 318:162 (1985).
9. Q. L. Zhang, S. C. O'Brien, J. R. Heath, Y. Liu, R. F. Curl, H. W. Kroto and R. E. Smalley, J. Phys. Chem. 90:525 (1986).
10. R. D. Knight, R. A. Walch, S. C. Foster, T. A. Miller, S. L. Mullen and A. G. Marshall, Chem. Phys. Lett. 129:331 (1986).
11. S. W. McElvany, W. R. Creasy and A. O'Keefe, J. Chem. Phys. 85:632 (1986).
12. K. Kirby and B. Liu, J. Chem. Phys. 70:893 (1979).
13. P. J. Bruna, S. D. Peyerimhoff and R. J. Buenker, J. Chem. Phys. 72:5437 (1980).
14. D. H. Liskow, C. F. Bender and H. F. Schaefer, J. Chem. Phys. 56:5075 (1973).
15. R. A. Whiteside, R. Krishnan, M. J. Frisch, J. A. Pople and P. v. R. Schleyer, Chem. Phys. Lett. 80:547 (1981).
16. R. A. Whiteside, R. Krishnan, D. J. DeFrees, J. A. Pople and P. v. R. Schleyer, Chem. Phys. Lett. 78:538 (1981).
17. D. H. Magers, R. J. Harrison and R. J. Bartlett, J. Chem. Phys. 84:3284 (1986).
18. B. K. Rao, S. N. Khanna and P. Jena, Solid State Commun. 58:53 (1986).
19. K. S. Pitzer and E. Clementi, J. Am. Chem. Soc. 81:4477 (1959).
20. S. J. Strickler and K. S. Pitzer, in "Molecular orbitals in chemistry, physics and biology", B. Pullman and P.-O. Lowdin, eds., Academic Press, New York, 1964, p. 281.
21. R. Hoffmann, Tetrahedron 22:521 (1966).
22. Z. Slanina and R. Zahradnik, J. Phys. Chem. 81:2252 (1977).
23. D. W. Ewing and G. V. Pfeiffer, Chem. Phys. Lett. 86:365 (1982).
24. P. Joyes and M. Leleyter, J. Phys. 45:1681 (1984).
25. J. Bernholc and J. C. Phillips, Phys. Rev. B 33:7395 (1986).
26. S. W. McElvany, B. I. Dunlap and A. O'Keefe, (to be published).
27. W. J. Hehre, L. Radom, P. v. R. Schleyer and J. A. Pople, "Ab Initio Molecular Orbital Theory", John Wiley, New York, 1986.
28. P. C. Hariharan and J. A. Pople, Chem. Phys. Lett. 66:217 (1972).
29. J. A. Pople, R. Krishnan, H. B. Schlegel and J. S. Binkley, Int. J. Quant. Chem. Symp. 13:255 (1979).
30. K. Raghavachari, J. Chem. Phys. 82:4607 (1985).
31. K. Raghavachari and V. Logovinsky, Phys. Rev. Lett. 55:2853 (1985).
32. K. Raghavachari, J. Chem. Phys. 84:5672 (1986).
33. D. Tománek and M. Schlüter, Phys. Rev. Lett. 56:1055 (1986).
34. K. Raghavachari, R. A. Whiteside and J. A. Pople, J. Chem. Phys. (in press).
35. K. Raghavachari and J. S. Binkley, J. Chem. Phys. (submitted).
36. W. R. M. Graham, K. I. Dismuke and W. Weltner, Astrophys. J. 204:301 (1976).
37. D. R. Stull and H. Prophet, "JANAF Thermochemical Tables", NSRDS-NBS-37, U. S. GPO, Washington, D. C., 1971.
38. R. H. Sanborn, J. Chem. Phys. 49:4219 (1968).

PRODUCTION OF CLUSTER IONS BY LASER VAPORIZATION

M.M. Ross, A. O'Keefe, and A.P. Baronavski

Chemistry Division
Naval Research Laboratory
Washington, D.C. 20375-5000

INTRODUCTION

There have been numerous reports of the production of cluster ions by the method of laser vaporization of solid samples, entrainment of the vaporized species in a supersonic molecular beam, and photoionization of the neutral clusters [1-3]. The exact cluster growth mechanism(s) are not completely understood, and the observed cluster ion distribution may depend upon numerous factors including instrumental parameters. Although the laser vaporization/molecular beam/photoionization technique has been successful in the production of high-mass clusters of a wide variety of materials, other methods of cluster production have been used. In particular, carbon cluster ions have been generated directly by particle bombardment and laser vaporization combined with mass spectrometric detection [4-6]. These direct cluster ion production methods have allowed detection of only relatively small carbon cluster ions, $[C_n]^+$ with n<30. The laser/molecular beam method has allowed detection of a bimodal distribution of carbon cluster ions; both the low-mass distribution, n<30, and a high-mass distribution, 30<n<150. The low-mass distribution with the characteristic cluster interval of n=1 is essentially the same in both methods, despite the significant differences in the experimental conditions. To this point, the high-mass carbon cluster ion distribution, with the cluster interval of n=2 and a maximum ion abundance at n=60, has been observed only using a molecular beam.

We report herein the observation of high-mass carbon cluster positive ions produced directly by laser vaporization of graphite with mass spectrometric detection [7]. The cluster ion abundance distribution was found to be very dependent on experimental parameters. With particular conditions of the laser power, sample irradiation time, and delay time between the laser pulse and the ion extraction pulse, the distribution of cluster ions was found to be qualitatively similar to that from the molecular beam method.

In addition, cluster ions were produced directly from the laser vaporization of single- and mixed-metal solid samples. The cluster ions obtained from mixed-element samples were indicative of chemistry occurring in the laser-induced plasma. Finally, the kinetic energies of the cluster ions produced from different elements, considered along with the cluster ion distributions, suggested significant differences in cluster formation processes.

EXPERIMENTAL

The mass spectrometer used in these studies was a modified CVC MA-2 time-of-flight instrument with a Galileo channeltron electron multiplier. The ion signals were amplified externally, averaged, and displayed on a Le Croy 125 MHz digital oscilloscope. The samples were mounted as pressed pellets between the backing plate and the first extraction plate of the ion source. In contrast to the laser/molecular beam experiments, the sample was stationary so that the laser irradiated a fixed spot. The samples were irradiated by the focused (1-m lens) output (30 mJ per pulse, 10-ns pulsewidth) from a frequency-doubled Nd^{3+}:YAG laser (Quantel 581c) operated at a repetition rate of 10 Hz. The laser beam was directed along the normal to the sample surface, which was orthogonal to the ion flight tube of the mass spectrometer.

The delay time between the laser pulse and the ion extraction pulse was set by a series of pulse generators. This timing was found to have a large effect on the distribution of cluster ions observed because of the differences in the velocities of the cluster ions. The distributions of times required for different mass ions to move from the sample to the ion extraction region (approximately 1 cm) were determined by setting the mass spectrometer detector to monitor a narrow time-of-flight window (one mass-to-charge ratio) and varying the laser/extraction pulse delay time. These cluster ion arrival times were on the order of 5 to 10 microseconds for $[C_n]^+$ with 3<n<20, depending on n.

RESULTS

Carbon Clusters

In an effort to obtain more detailed information on the cluster formation process, our experiment was designed to observe the cluster ions that are formed directly in the laser-induced plasma, without additional "cooling" or entrainment of the vaporized species in a molecular beam. Figures 1 and 2 present the mass spectra of the carbon cluster ions that were observed using two different laser pulse/extraction pulse delay times. In Fig. 1 a delay time of 16 microseconds was used and only the low-mass distribution of carbon cluster ions was observed. This distribution has the characteristic cluster interval of n=1, a maximum ion abundance at $[C_{11}]^+$, and is qualitatively similar to the distributions as measured by others [6]. The short delay time resulted in the observation of only the high-velocity, low-mass species.

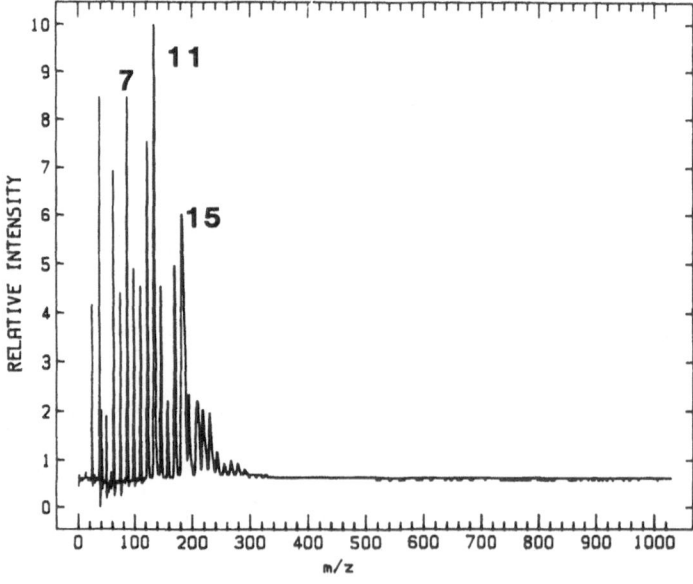

Figure 1. The laser vaporization/time-of-flight mass spectrum of carbon cluster ions, $[C_n]^+$ with n labelled, with a 16-microsecond delay time.

In Fig. 2 a delay time of 30 microseconds was used and both low- and high-mass cluster ion distributions were observed. The high-mass distribution has the characteristic interval of n=2 and enhanced ion abundances of $[C_n]^+$ at n= 44, 50, 60, and 70. It should be noted that the maximum of the low-mass distribution has shifted to $[C_{15}]^+$ and other cluster ions larger than $[C_{15}]^+$ have increased abundances. The longer delay time enhanced the observation of the higher mass or lower velocity cluster ions.

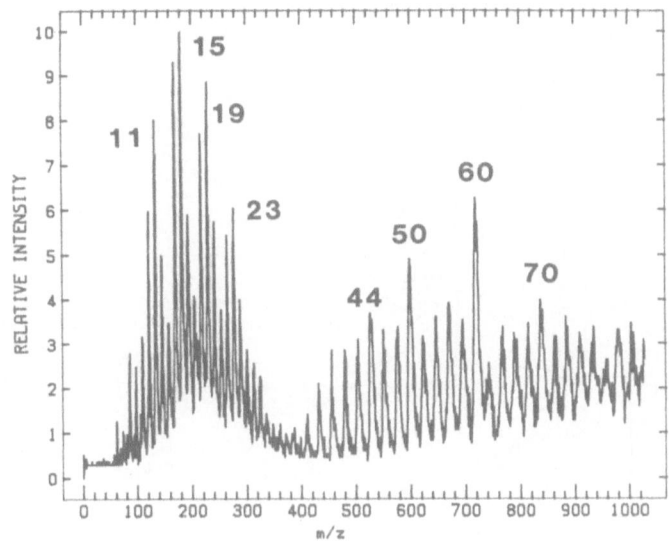

Figure 2. The laser vaporization/time-of-flight mass spectrum of carbon cluster ions, $[C_n]^+$ with n labelled, with a 30-microsecond delay time.

The relative abundances of different carbon cluster ions could be varied also by changing the laser irradiation time. At the beginning of an experiment (sample exposure time < 1-2 min.) only the low-mass ions were observed. Upon prolonged laser irradiation and with a delay time suitable for the observation of larger clusters, the high-mass cluster ions increased in abundance dramatically. The surface of the graphite pellet was examined after such an experiment and a hole approximately 0.5-mm diameter and 1- to 2-mm depth was found at the site of laser irradiation. Through several experiments we found that this laser-produced "channel" was necessary to promote the formation of the high-mass carbon cluster ions. We estimate that the density of vaporized species in this channel could be on the order of 10^{18} cm^{-3}. This condition could give rise to numerous reactions within the channel, which could result in the formation of the observed cluster ions. Further evidence for this mechanism is the observation that the production of the high-mass cluster ions was dependent upon the laser power as well. If the laser power was below 2-5 mJ then only the low-mass ions were observed. The high laser power was required to create the channel in the graphite pellet that provides a confined space and higher particle densities for the plasma in which the cluster formation reactions occur. It should be noted that laser powers greater than 10-20 mJ had the effect of decreasing the abundances of the cluster ions, which could be the result of dissociation processes.

Ion Velocity Measurements

By setting the mass spectrometer detector to monitor a narrow time window, or essentially one cluster ion m/z, and varying the laser pulse/extraction pulse delay time the distribution of cluster ion arrival times (sample to extraction region) could be measured. From this distribution of ion arrival times an average ion velocity was calculated. For the carbon cluster ions, a plot of the square of the ion velocity versus the inverse of the cluster ion mass did not yield a straight line. Therefore, the distribution of carbon cluster ions could not be characterized by a single kinetic energy. Comparison of this result with similar measurements on different elemental systems will be given in the next section.

In addition to carbon, cluster ions from other elements have been produced directly by laser vaporization. For example, a pressed pellet mixture of tungsten and carbon was used to obtain the mass spectrum shown in Figure 3.

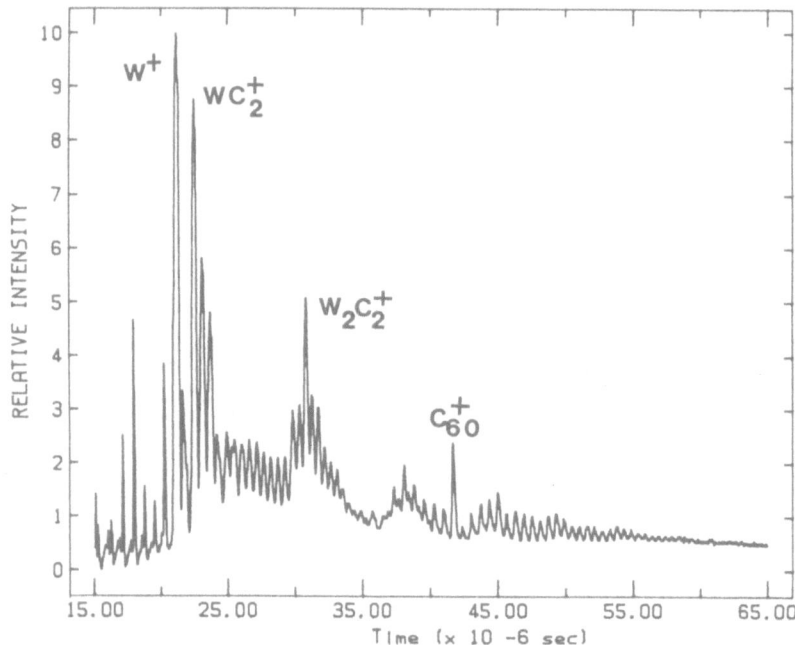

Figure 3. The laser vaporization/time-of-flight mass spectrum of cluster ions from a tungsten/carbon sample.

In addition to the cluster ions of tungsten alone, $[W_n]^+$, mixed cluster ions of tungsten and carbon were observed as $[W_nC_m]^+$ species with $n=1$ to 4 and $m=1$ to 14. The enhanced ion abundances of $[W_nC_2]^+$ species could indicate enhanced relative stabilities of these clusters but the reason for this is unknown at the present time. The relative abundances of tungsten and tungsten carbide clusters could be changed by variation of the relative amounts of tungsten and carbon in the sample pellet. Increased amounts of tungsten in the sample resulted in increased abundances of tungsten cluster ions relative to the tungsten carbide ions such that the highest tungsten cluster observed increased from $n=4$ to 5 to $n=10$ to 11. Increased amounts of carbon resulted in spectra such as the one in Figure 3 where tungsten carbide and carbon cluster ions predominate, to the point that abundant $[C_{60}]^+$ species are readily observed.

Mixtures of two metals have been investigated, also. Figure 4 shows the laser vaporization mass spectrum of the cluster ions obtained from a mixture of bismuth and antimony. In addition to the single-element cluster ions, $[Bi_n]^+$ and $[Sb_m]^+$, the variety and high abundances of the mixed-element cluster ions; $[BiSb_m]^+$, $n=1$-4, and $[Bi_nSb]^+$, $n=1$-4, are indicative of the reactions that must occur in the plasma above the sample surface.

Ion Velocity Measurements

Ion arrival time measurements were made on cluster ions from both single- and mixed-element samples. In all cases, the plots of the square of the ion velocity versus the inverse of the cluster ion mass yielded straight lines. The slope of such a plot is a kinetic energy by which the cluster ion distribution can be characterized. These energies were typically in the range of 1 to 10 eV.

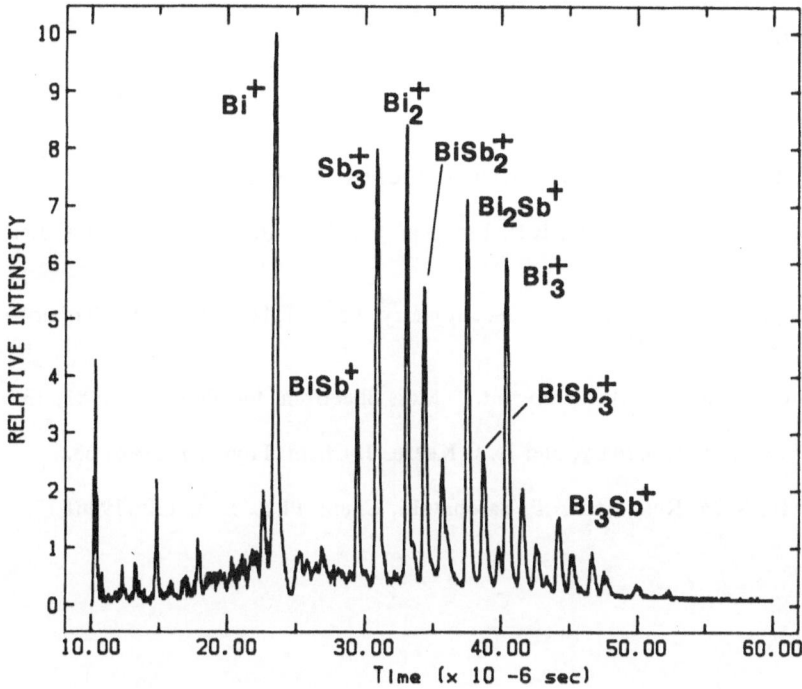

Figure 4. The laser vaporization/time-of-flight mass spectrum of the cluster ions from a bismuth/antimony sample.

As previously discussed, the distribution of carbon cluster ions could not be characterized by a kinetic energy. To date, carbon is the only sample for which this is true. In addition, carbon displays the unique bimodal cluster ion distribution, which is in sharp contrast to the cluster ion distributions observed for all other samples. These differences must point to differences in the cluster formation processes between carbon and other elements. For reasons as yet unknown, with carbon, under laser irradiation, a plasma is established in which ion/molecule reactions can occur that produce stable and abundant high-mass cluster ions.

CONCLUSIONS

We have demonstated that high-mass carbon cluster ions can be generated directly by laser vaporization of graphite. It is significant that the distribution of carbon cluster ions observed is very similar to that observed using a molecular beam and photoionization. Considering the large differences between these two methods this agreement would not be expected a priori. The distribution of carbon cluster ions was found to depend upon the laser power, sample irradiation time, and the time delay between the laser pulse and ion extraction pulse. The plasma set up in the laser-formed channel with the relatively high laser power is required to generate the high-mass carbon cluster ions (n>30).

Cluster ions produced by laser vaporization of single- and mixed-element samples indicate that there are extensive reactions that occur in the laser-induced plasma. Except for carbon, the distributions of cluster ions from all samples could be characterized by an ion kinetic energy. Carbon is the exception with respect to this property and because of the unique bimodal cluster ion distribution. The reason(s) for these observations are under investigation.

REFERENCES

1. H.W. Kroto, J.R. Heath, S.C. O'Brien, R.F. Curl, and R.E. Smalley, Nature 318 (1985) 162.

2. E.A. Rohlfing, D.M. Cox, and A. Kaldor, J. Chem. Phys. 81 (1984) 3322.

3. L.A. Bloomfield, M.E. Geusic, R.R. Freeman, and W.L. Brown, Chem. Phys. Lett. 121 (1985) 33.

4. R.E. Honig, in Advances in Mass Spectrometry, vol.2, R.M. Elliot, ed. (Pergamon Press, Oxford, 1963), p.25.

5. N. Fuerstenau and F. Hillenkamp, Int. J. Mass Spectrom. Ion Phys. 35 (1981) 135.

6. S.W. McElvany, W.R. Creasy, and A. O'Keefe, J. Chem. Phys. 85 (1986) 632.

7. A. O'Keefe, M.M. Ross, and A.P. Baronavski, Chem. Phys. Lett. 130 (1986) 17.

FROM METAL CLUSTERS TO CLUSTER METALS

Günter Schmid and Norbert Klein
Institut für Anorganische Chemie
Universität Essen

4300 Essen 1, W.-Germany

ABSTRACT

Ligand stabilized full-shell clusters show a special stability and have been described for the magic numbers 13 and 55[1]. On the other hand, there is no experimental information on the stability of naked full-shell clusters, presumably because there was no possibility to generate those clusters. But numerous papers occupy with the formation and stability of naked full-shell clusters from the theoretical point of view[2-12]. Two-shell clusters of the type $M_{55}L_{12}Cl_n$ (M = Au, Rh, L = $P(C_6H_5)_3$, n = 6; M = Rh, Ru, L = $P(tert-C_4H_9)_3$, n = 20; M = Pt, L = $As(tert-C_4H_9)_3$, n = 20) for the first time present the occasion of generating naked M_{13} clusters by peeling off the outer metal shell together with the ligands. The destiny of these micro-clusters should easily been followed up, as only two routes for their stabilization seem open: either they degrade to form normal bulk metal or they aggregate to novel metal modifications with M_{13} clusters as building blocks.

INTRODUCTION

The successful synthesis of the first two-shell clusters, built up by 55 cubic closest-packed metal atoms, enabled us to investigate for the first time very small particles of uniform size[1,13-17]. Besides very surprising chemical properties, M_{55} clusters show fascinating properties in a physical sense. One of the most exciting property was recently found by us[17]: the clusters $Au_{55}[P(C_6H_5)_3]_{12}Cl_6$, $Rh_{55}[P(C_6H_5)_3]_{12}Cl_6$, $Rh_{55}[P(tert-C_4H_9)_3]_{12}Cl_{20}$, $Ru_{55}[P(tert-C_4H_9)_3]_{12}Cl_{20}$, and $Pt_{55}[As(tert-C_4H_9)_3]_{12}Cl_{20}$, dissolved in dichlormethane, can be degraded by 20 V d.c. on platinum or graphite electrodes. The ligands, together with the metal atoms of the outer shell, are removed, whereby the inner M_{13} nuclei are generated. The destiny of the ligands and the outer metal atoms is not completely clear. In the case of the Au_{55} cluster, we know that $[(C_6H_5)_3P]_2AuCl$ is formed. It remains in solution, whereas the rest of the outer-shell gold atoms is precipitated as bulk metal. It seems reasonable to suppose a comparable behaviour in the other cases, though the corresponding complexes could not be fully characterized up to now. First of all our interest was focussed upon the behaviour of the M_{13} clusters.

If M_{13} nuclei, the smallest full-shell clusters, are relatively long-lived species, they should try to organize themselves. If their life-time is not long enough, the clusters should degrade and give either normal metal particles or possibly form polytetrahedral structures. As the M_{13} clusters consist of cubic colsest-packed atoms, there is no conspicuous reason why they should decompose.

RESULTS

Representative for all degradation processes, the experiment with $Ru_{55}[P(tert-C_4H_9)_3]_{12}Cl_{20}$ is described here in more detail. Two platinum electrodes or a graphite cathode and a platinum anode are dipped into a stirred solution of 80 mg $Ru_{55}[P(tert-C_4H_9)_3]_{12}Cl_{20}$ in 40 ml of dichlor-methane, and 20 Volt d.c. are applied. After 24 hours the reaction has finished and a black precipitate can be collected from the bottom of the reaction vessel and partially also from the electrodes. This degradation is not caused by an electrolysis, as no discharging is observed. Rather it could be a matter of a catalytic process on the surface of the electrodes, as the metallic degradation products are found at the cathode as well as at the anode. It should be mentioned that, if the electrodes are not dipped into the cluster solution, but into layers of water covering the dichlormethane solution in a U-formed tube as a reaction vessel, electrophoresis instead of degradation is observed. The collected black precipitate contains, after drying under vacuum, small amounts of organic materials (solvent, ligands). These can be eliminated by warming up the samples under vacuum. The evolution of impurities can be followed by DMC measurements (see below). After this, the elemental analysis results 99 - 100% of metal. The solutions normally contain some colloidal metal which is disposed together with the solutions. Debye-Scherrer diffraction patterns ($Cu_{K\alpha}$, $\lambda = 1.5418\overset{\circ}{A}$) of the precipitated samples prove clearly a novel metallic state: together with some 'normal' metal powder, there exist microcrystals of $[(M_{13})_{13}]_n$. Table 1 shows the X-ray data leading to this formula.

Table 1. 2θ-values, interplanar spacings $d(\overset{\circ}{A})$ of
$[(Ru_{13})_{13}]_n$ and, for comparison, hkl and
d-values of Rh_x.

$2\theta(°)$	$d_{obs}[(Ru_{13})_{13}]_n$		hkl (Rh_∞)	d (Rh_∞)	Occupation
5.8	15.3	d^1	200	1.902	$8 \times d(Rh_\infty) = 15.216$
14.5	6.11	d^2	331	0.8724	$7 \times d(Rh_\infty) = 6.1075$
27.5	3.24	d^3	222	1.0979	$3 \times d(Rh_\infty) = 3.2937$
30.75	2.908	d^4	400	0.9508	$3 \times d(Rh_\infty) = 2.8524$
34.75	2.582	d^5	331	0.8724	$3 \times d(Rh_\infty) = 2.6172$
40.25	2.241	d^6	111	2.1960	$1 \times d(Rh_\infty) = 2.1960$
(43.5	2.080)*				
45.75	1.983	d^7	200	1.902	$1 \times d(Rh_\infty) = 1.902$
54.25	1.691	d^8	420	0.8504	$2 \times d(Rh_\infty) = 1.7008$
(58.5	1.578)*				
(67.5	1.388)*				
71	1.33	d^9	220	1.345	$1 \times d(Rh_\infty) = 1.345$
(83	1.16)*				
84.5	1.15	d^{10}	311	1.1468	$1 \times d(Rh_\infty) = 1.1468$

*The values in parantheses belong to bulk hexagonal Ru_∞. Therefore they are not to be compared with those of $[(Ru_{13})_{13}]_n$. The values of the other $[(M_{13})_{13}]_n$ species are not given here but correspond to those of $[(Ru_{13})_{13}]_n$. The weaker X-ray reflexes d^5 - d^{10} are not always observed. This depends mainly on the yield of 'normal' metals M_∞ and on the size of the microcrystals.

Two facts following from Table 1 have to be pointed out: the observed interplanar spacings in the novel ruthenium d^1 - d^{10} are multiples of those existing in the normal rhodium metal (Rh_∞): rhodium crystallizes in a cubic densest packing. It's atomic radius is the same as that of ruthenium. As normal Ru_∞ has a hexagonal structure the X-ray pattern of the novel Ru must be related to that of Rh_∞. The superstructures are mainly characterized by the most intensive d-values d^1, d^2 and d^3. Four very weak reflexions belong to the hexagonal Ru_∞ which originates from outer-shell atoms. The course of d^1, d^2 and d^3 is shown in Fig. 1. The other d-values are difficult to be shown, but the fact that they are also multiples of Rh_∞-values proves the new structures definitely. Consequently the d-values for $[(Rh_{13})_{13}]_n$ are identical with those for $[(Ru_{13})_{13}]_n$. Scheme 1 summarizes the formation of the novel metal structures.

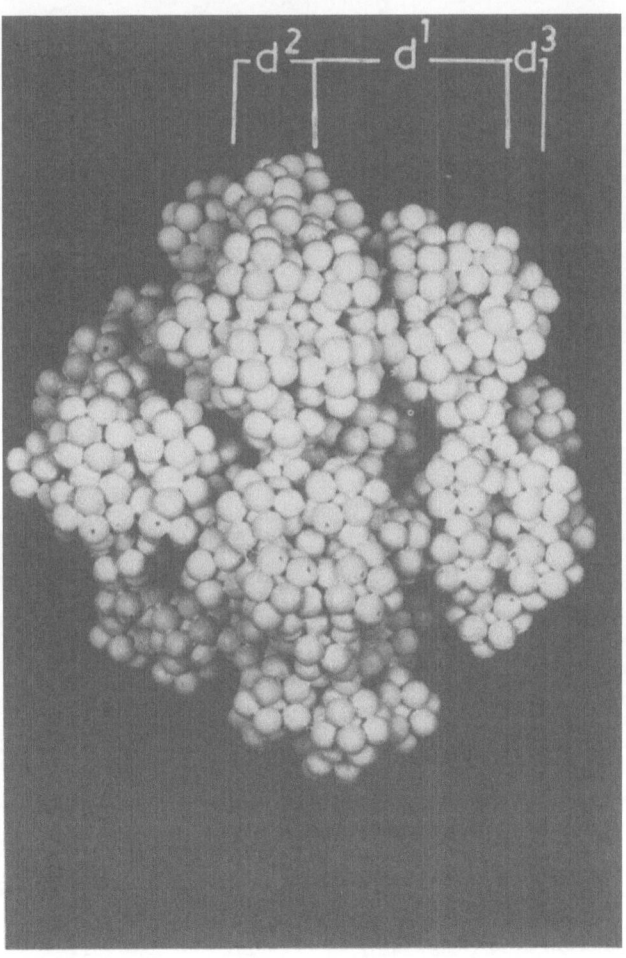

Fig. 1. Model of $[(M_{13})_{13}]_{13}$ showing the course of the 3 most intensive interplanar spacings.

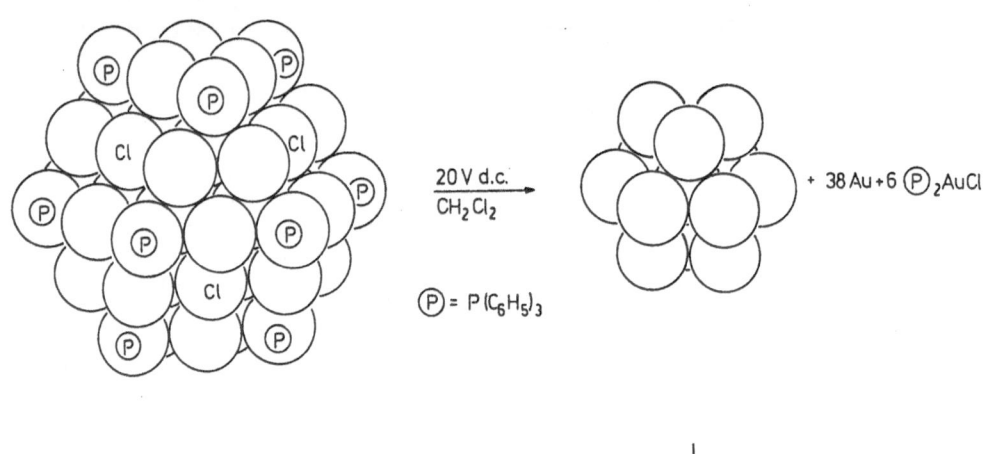

20 V d.c.
————————
CH₂Cl₂

+ 38 Au + 6 (P)₂AuCl

(P) = P(C₆H₅)₃

$$\frac{20\,\text{V d.c.}}{CH_2Cl_2}$$

$+\ 38\,Au + 6\ \textcircled{P}_2AuCl$

$\textcircled{P} = P(C_6H_5)_3$

| × 13

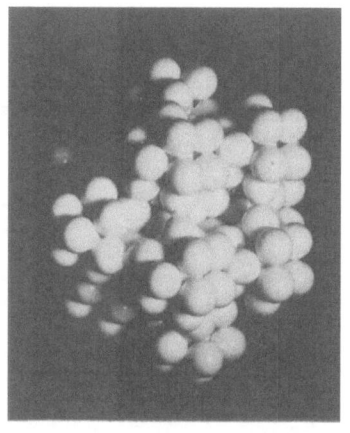

| × 13

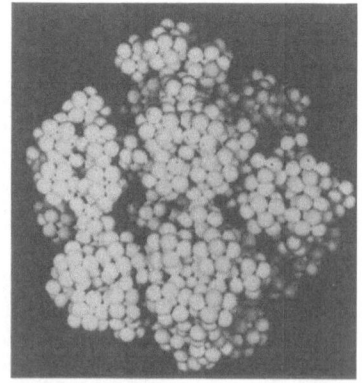

Scheme 1: Formation of $[(M_{13})_{13}]_n$ starting with the M_{55} Clusters, shown for means of $Au_{55}[P(C_6H_5)_3]_{12}Cl_6$

One of the most interesting properties of the $[(M_{13})_{13}]_n$-modification is its thermal stability. As $[(Au_{13})_{13}]_n$ is the only air-stable modification, we studied its stability by means of differential microcalorimetry (DMC) measurements. At 437, 441, 447, 454, 464, 472 and 480°C there are very sharp phase transitions showing the stepwise break-down of $[(Au_{13})_{13}]_n$. The X-ray pattern after heating agrees with that of normal gold, Au_∞.

REFERENCES

1. G. Schmid, Structure and Bonding 62:51 (1985)
2. B. G. Bagley, Nature 208:674 (1965)
3. B. G. Bagley, ibid 225:1040 (1970)
4. B. G. Bagley, J. Cryst. Growth 6:323 (1970)
5. J. J. Burton, Cat. Rev. Sci. Eng. 9:209 (1974)
6. A. L. Macky, Acta Cryst. 15:916 (1962)
7. M. R. Hoare and P. Pal, Nature, Phys. Sci. 236:35 (1972)
8. J. J. Burton, Phys. Lett. 3:594 (1970)
9. J. J. Burton, Nature 229:335 (1971)
10. M. R. Hoare and P. Pal, Adv. Phys. 24:645 (1975)
11. M. R. Hoare, Adv. Chem. Phys. 40:49 (1979)
12. B. Mutaftschiev, J. de Phys. Colloq. 41:C3 (1980)
13. G. Schmid, R. Pfeil, R. Boese, F. Bandermann, S. Meyer,
    G. H. M. Calis, and J. W. A. van der Velden,
    Chem. Ber. 114:3634 (1981)
14. G. Schmid, U. Giebel, W. Huster, and A. Schwenk,
    Inorg. Chim. Acta 85:97 (1984)
15. L. R. Wallenberg, J.-O. Bovin, and G. Schmid,
    Srfce. Sce. 156:256 (1985)
16. G. Schmid and W. Huster, Z. Naturforsch. 41B:1028 (1986)
17. G. Schmid and N. Klein, Angew. Chem. 98:910 (1986),
    Angew. Chem. Int. Ed. Engl. 25:922 (1986)

THERMAL STABILITY OF MICRO-CLUSTERS

S. Shimamura

Department of Applied Science, Faculty of Engineering
Yamaguchi University, Ube 755, Japan

ABSTRACT

Anharmonic vibrations of atoms in a micro-cluster are theoretically studied by using the self-consistent phonon formalism. Atoms are assumed to interact on each other through a pairwise interatomic potential. The free energy of a cluster is minimized by using a model reference system in which a central atom is vibrating with a frequency ω_0 and the surrounding atoms with ω_1. We investigate the thermal softening of atomic vibrations, the thermal expansion and the vibrational instability for different micro-clusters. As for N=13 clusters, Cuboctahedron becomes instable at lower temperature than Icosahedron. The thermal stability of micro-clusters is discussed in terms of the instability temperature.

INTRODUCTION

There has been a growing interest in the stability of small clusters. Since the " magic numbers " were found experimentally in the mass spectra of alkali-metal clusters[1], the stability of clusters has been studied theoretically by a number of authors. Recent experiments on the photo-fragmentation[2] and the Coulomb explosion of inonized clusters[3] have also stimulated theoretical investigations of the stability of clusters.

Many theoretical studies are based on the calculation of electronic energy levels in a cluster. These studies have shown that the stability of simple clusters such as alkali-metal clusters is associated with the filling of electronic orbital shells. On the other hand the investigation of the stability from the thermal point of view is also of considerable interest. Berry et al.[4] investigated the melting and freezing of clusters ; they discussed the stability of the solidlike and the liquidlike forms using a model based on the energy level correlation diagram of vibrational frequencies.

The phase transformation such melting and the thermal expansion are caused by the anharmonic effect of atomic vibrations. The self-consistent phonon formalism[5] has been successfully applied to the investigation on the anharmonic vibrations of atoms. However the self-consistent calculation based on the general version of this formalism is not easy to be performed. Matsubara et al.[6,7] have developed a simple version based on the Einstein model for atomic vibrations and have applied it to the lattice vibration in metallic small particles. They have investigated the particle size dependence of the melting temperature on the basis of the instability condition of the self-consistent Einstein phonon. The model has been also used by Hasegawa et al.[8] to study more extensively the melting of metallic small particles ; they used the thermodynamic criterion for melting by comparing the free energy of a solid phase with that of a liquid phase.

In the present paper we will apply the self-consistent phonon formalism in a simple version developed by Matsubara et al.[6,7] to micro-clusters consisting of about ten atoms. The main purpose of this paper is to investigate the thermal softening of vibrations and the thermal expansion of different micro-clusters. We will also discuss the thermal stability of micro-clusters in terms of the vibrational instability in the present model.

MODEL CLUSTER

We consider a cluster composed of atoms which interact on each other through a pairwise interatomic potential $v(r)$. The Hamiltonian of the cluster is given by

$$H = \frac{M}{2} \sum_n \dot{u}_n^2 + \frac{1}{2} \sum_{n \neq n'} v(R_n - R_{n'} + u_n - u_{n'}) , \tag{1}$$

where M is the atomic mass, and R_n and u_n are, respectively, the equilibrium position and the displacement vectors of the n-th atom in a cluster. In order to calculate the free energy of the cluster, we use the variational method based on the Gibbs-Bogoliubov inequality

$$F_{exact} \leq F \equiv F_0 + < H - H_0 > , \tag{2}$$

where F_0 and H_0 are, respectively, the free energy and the Hamiltonian for a reference system, and $<\cdots>$ represents the thermal average over the reference system. We choose as the reference system the harmonic system with the frequency ω_n for the n-th atom :

$$H_0 = \frac{M}{2} \sum_n \dot{u}_n^2 + \frac{M}{2} \sum_n \omega_n^2 u_n^2 . \tag{3}$$

Using the Fourier transform of $v(r)$, $V(q)$ defined by

$$v(r) = \int dq \, V(q) \, \exp(iq \cdot r) , \tag{4}$$

we have the following expression for the free energy of our cluster[7,8] :

$$F = \frac{1}{2} \sum_{n \neq n'} \int dq \, V(q) \, \exp[iq \cdot (R_n - R_{n'})] \, \exp[-q^2 (W_n + W_{n'})]$$

$$+ 3k_B T \sum_n \{ \ln[2 \sinh(\frac{\hbar}{2k_B T} \sqrt{\frac{\phi_n}{M}})] - \frac{\phi_n}{k_B T} W_n \} , \tag{5}$$

where

$$W_n = <u_n^2>/6 = \frac{\hbar}{4M} \sqrt{\frac{M}{\phi_n}} \coth(\frac{\hbar}{2k_B T} \sqrt{\frac{\phi_n}{M}}) \tag{6}$$

and $\phi_n = M\omega_n^2$.

Now we minimize the right-hand side of Eq.(5) with respect to ϕ_n. From the condition $\delta F/\delta\phi_n = 0$, we have a set of n equations[7,8]

$$\omega_n^2 = \sum_{n'(\neq n)} \frac{1}{3M} \sum_{\alpha=x,y,z} <v_{\alpha\alpha}(R_n-R_{n'} + u_n-u_{n'}) > \qquad (7)$$

$$= -\frac{1}{3M} \sum_{n'(\neq n)} \int dq \; q^2 V(q) \; \exp[iq\cdot(R_n-R_{n'})] \; \exp[-q^2(W_n+W_{n'})], \qquad (8)$$

where $v_{\alpha\alpha}(r) = \partial^2 v/\partial r_\alpha^2$. Equations (8) are self-consistent equations for $\{\omega_n\}$ under the fixed positions $\{R_n\}$ of atoms in a cluster.

As the next step of our variational method we minimize the free energy with respect to atomic positions. For this purpose we fix the symmetry of atomic configulation and then calculate the free energy as a function of the nearest-neighbor distance under the fixed symmetry. Different micro-clusters considered in our calculation are shown in Fig.1. In these clusters there are two frequencies, ω_0 for a central atom and ω_1 for the surrounding atoms ; from symmetry consideration all surrounding atoms vibrate with the same frequency. Thus we take the following steps in the calculation.
[1] For a given nearest-neighbor distance a, we obtain ω_0 and ω_1 by solving Eqs.(8) self-consistently.
[2] Using the values of a, ω_0 and ω_1, we calculate the free energy given by Eqs.(5) and (6).
[3] For different values of a, we repeat the steps [1] and [2]. Thus we obtain a set of values of a, ω_0 and ω_1 for which the free energy has the minimum value.
In the calculation we have used a pairwise potential of the Gaussian type given by

$$v(r) = A \exp(-\alpha r^2) - B \exp(-\beta r^2) \qquad (9)$$

with the potential constants A, B, α and β determined by Matsubara et al.[7]

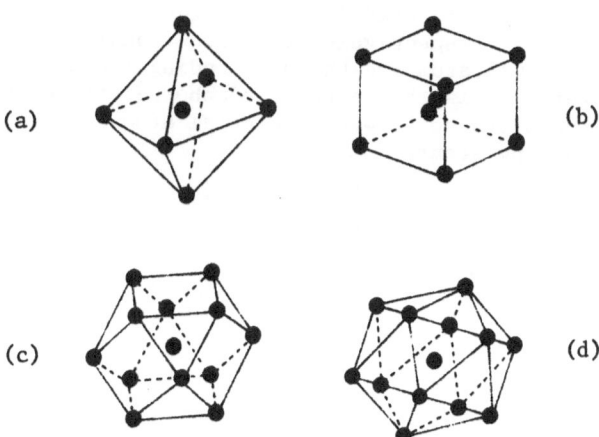

Fig. 1. Different micro-clusters : (a) Octahedron (N=7), (b) Hexahedron (N=9), (c) Cuboctahedron (N=13), and (d) Icosahedron (N=13).

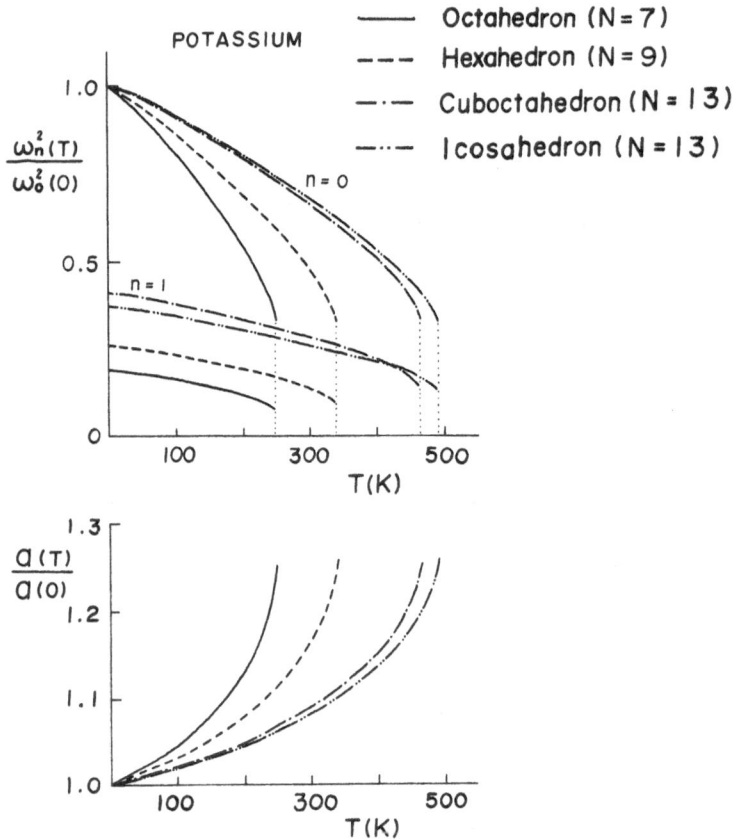

Fig. 2. The temperature dependences of $\omega_n{}^2(T)/\omega_0{}^2(T=0)$ and
$a(T)/a(T=0)$ for potassium micro-clusters. Here ω_0
and ω_1 are, respectively, the vibrational frequencies
of a central atom and the surrounding atoms in a
cluster ; a is the nearest-neighbor distance, that
is, the distance between a central atom and one of the
surrounding atoms.

RESULTS AND DISCUSSION

The calculation has been performed for some alkali-metal clusters. We will here show the results for potassium clusters ; the results for other alkali-metal clusters are qualitatively similar to those for potassium clusters. The calculated values of the equilibrium nearest-neighbor distance at T=0 are 5.55 A, 5.64 A, 5.97 A and 5.87 A for Octahedron (N=7), Hexahedron (N=9), Cuboctahedron(N=13) and Icosahedron (N=13), respectively ; these values are compared with 4.74 A calculated[7] for bulk potassium using the same interatomic potential. The frequency ratios squared ω_1^2/ω_0^2 at T=0 are 0.19, 0.26, 0.41 and 0.37 for the clusters in the order described above. In N=13 clusters the surrounding atoms in Cuboctahedron are bound more tightly, relatively to a central atom, than in Icosahedron. This is attributed to the fact that Cuboctahedron has more number of nearest-neighbor bonds than Icosahedron.

Figure 2 shows the temperature dependences of $\omega_n^2(T)/\omega_0^2(T=0)$ and $a(T)/a(T=0)$ for four clusters shown in Fig. 1. As a result of anharmonic vibrations, the vibrational frequencies decreases and the nearest-neighbor distance expands with increasing temperature. Since the thermal expansion causes the softening of atomic vibrations and *vice versa*, these effects are enhanced cooperatively by an increase of temperature. As for N=13 clusters, the ratio $\omega_1^2(T)/\omega_0^2(T=0)$ at lower temperatures is larger for Cuboctahedron than for Icosahedron ; at higher temperatures, however, the ratio for Cuboctahedron becomes smaller than that for Icosahedron. A cluster falls into instability at a certain temperature at which a local minimum of the free energy as a function of the nearest-neighbor distance disappears. Cuboctahedron becomes instable at lower temperature than Icosahedron.

The instability temperatures in the present model are about 250 K, 340 K, 470 K and 490 K for Octahedron (N=7), Hexahedron (N=9), Cubocta-hedron (N=13) and Icosahedron (N=13), respectively. These values are compared with 336 K, the melting temperature of bulk potassium. It should be noted that the instability temperature in the present model clusters does not correspond to the melting temperature because the melting point represents a temperature at which the solid phase coexists with the liquid phase. Hasegawa et al.[8] , in their study of the size dependence of the melting temperature of small metallic particles, have investigated two criterions for melting : the thermodynamic melting temperature defined by the temperature at which the value of the free energy of the solid phase is equal to that of the liquid phase, and the instability temperature in a similar model to the present one. They have shown that the instability temperature calculated for bulk metals is much larger than the bulk melting temperature. This is partly ascribed to the fact that they have neglected the effect of the thermal expansion on the softening of atomic vibrations ; if we neglect the effect of the thermal expansion in the present calculations, we have the instability temperature which is two to three times as large as the present instability temperature. On the other hand we should also note the following : for a micro-cluster composed of about ten atoms, it is not obvious that the well-defined " liquid phase " exists. In order to study extensively the thermal stability of micro-clusters, it is necessary to investigate the existence of any possible phase carefully. The present instability temperature, therefore, should be considered a measure of the thermal stability of micro-clusters.

REFERENCES

1. W. D. Knight, K. Clemenger, W. A. de Heer, W. A. Sounders, M. Y. Chou and M. L. Cohen, Phys. Rev. Lett. 52 (1984) 2141.
2. L. A. Bloomfield, R. R. Freeman and W. L. Brown, Phys. Rev. Lett. 54 (1985) 2246.
3. A. Howie, Nature 320 (1986) 684.
4. R. S. Berry, J. Jellinik and G. Natanson, Phys. Rev. A 30 (1984) 919.
5. T. R. Koehler, *Dynamical Properties of Solids*, ed. G. K. Horton and A. A. Maradudin, Amsterdam, North-Holland.
6. T. Matsubara and K. Kamiya, Prog. Theor. Phys. 58 (1977) 767.
7. T. Matsubara, Y. Iwase and A. Momokita, Prog. Theor. Phys. 58 (1977) 1102.
8. M. Hasegawa, K. Hoshino and M. Watabe, J. Phys. F : Metal Phys. 10 (1980) 619.

MEASUREMENT OF Bi_2 AND Bi_4 ELECTRON IMPACT FRAGMENTATION

USING MAGNETIC POLARIZATION

K.P. Ziock[†] and W.A. Little

Physics Department
Stanford University
Stanford, Ca. 94305

ABSTRACT

We have measured the fragmentation of Bi_2 under electron impact ionization (45 eV) by measuring the relative intensity shift produced by a quadrupole magnet on an electrically neutral beam of atomic bismuth (magnetic moment ~ one Bohr magneton) and Bi_2 (magnetic moment zero). The large discrepancy between the theoretically predicted and the observed shifts was ascribed to fragmentation in the ionization process. We obtain a fragmentation as defined by Kohl et al.[1] of 2.02 ± .35. This is slightly higher than their estimates of 1.54 ± .20 and exemplifies the importance of fragmentation processes to cluster mass spectroscopy. The same measurement performed on a beam of Bi_3 and Bi_4 showed no intensity shift for the bismuth trimer. This result is ambiguous due to the unknown magnetic moment of Bi_3. However, assuming a single Bohr magneton moment, to within our statistical accuracy, we could ascribe all of the Bi_3 detected (Knudsen source) to fragmentation of Bi_4 (the observed peak heights for Bi_3 and Bi_4 were identical.)

INTRODUCTION

Much of the recent work in the field of small metal particles in molecular beam environments is based on a "generic" apparatus which consists of a cluster source, some form of ionizing radiation, and a mass spectrometer. Typically, cluster properties are inferred by measuring changes in the mass spectrum due to variations in either the source conditions, the ionizing radiation, or both.[2]

One of the problems with this technique is the possibility that the observed ion spectrum may not be a direct measure of the neutral clusters present before ionization. In general, the ionization energy is greater than the bond strengths of the particle such that one may fragment the neutral cluster into several smaller, charged pieces detected by the mass spectrometer. When one alters the ionizing radiation, one changes both the ionization (desired quantity) and the fragmentation cross sections. The intensity changes in the mass spectrum are a convolution of both these processes and difficult to separate.

[†]Current address: Lawrence Livermore National Laboratory
Livermore, CA 94550

There has been some recent work to avoid this problem through the inclusion of a variable experimental probe (dipole magnet,[3,4] electric field[5]) between the cluster source and the ionization region. Measurement of the mass spectrum as a function of this probe with constant ionizing radiation allows for conditions of constant fragmentation. If the fragmentation cross sections were known, then one could theoretically correct the spectra. Unfortunately, these values are generally unknown.

In this paper we present the results of an experiment which uses a quadrupole magnet as the variable probe. We have taken the novel step of reversing the analysis to calculate the fragmentation during ionization from a comparison of the the calculated and measured effect of the magnet on the cluster beam.

THEORY

One of the earliest predictions for anomalous behavior of clusters is variation of the magnetic moment with size[6]. Specifically, for odd atomic number elements, clusters of odd numbers of atoms are expected to have a magnetic moment of at least one Bohr magneton due to the unpaired spin of the odd electron; whereas, the clusters with even numbers of atoms might show no magnetic moment due to pairing of the spins. One may take advantage of this to differentially manipulate the clusters in an electrically neutral beam by passing it through a region of inhomogeneous magnetic field as described below.

An electrically neutral particle in a magnetic induction, $\mathbf{B}$, with magnetic moment, μ, will experience a force

$$\mathbf{F} = -\nabla (\mu \bullet \mathbf{B}) = \pm \nabla (|\mu||\mathbf{B}|) \tag{1}$$

where the last step relies on the alignment of the particles z-axis with the magnetic field direction. The magnetic induction of a quadrupole magnet of pole tip field, B_t, and radius, r_m, has a magnitude

$$|\mathbf{B}| = \frac{B_t}{r_m} |\mathbf{r}| \tag{2}$$

where $\mathbf{r}$ is the radial displacement. Combining (1) and (2) with the fact that the magnitude of the axial velocity (v) is a constant, allows one to interchange position and time[7] yielding the following particle trajectory

$$r'' = \frac{L_0^2}{m^2 v^2 r^3} \pm \frac{\mu B_t}{m r_m v^2} \tag{3}$$

Here the prime refers to differentiation with respect to the z-direction, μ is the magnitude of the magnetic moment, and L_0 is the angular momentum of the particle on entering the magnet. The two trajectories described by (3) are: for particles with magnetic moment anti-parallel to the field, oscillation about the axis of the magnet; for particles with magnetic moment parallel to the field, an off-axis motion to infinity.

The effect of the magnet on an equal mix beam of spin-up and spin-down particles is to increase the observed intensity downstream of the magnet when it is turned on. This seems counter-intuitive since the magnet removes particles from the beam. Consider, however, that all particles of the

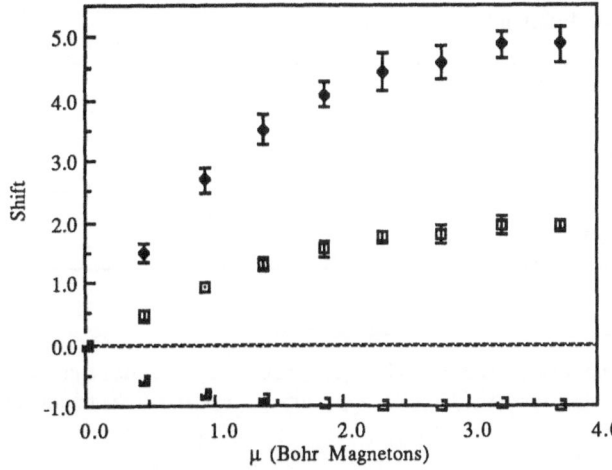

net · focused ⌐ defocused

Figure 1: Shift induced
in the neutral cluster beam
as calculated by Monte
Carlo technique for the
two spin configurations
(up, down) and for a
mixed beam (net).

proper orientation which enter the magnet within a given phase space are
transported through the magnet. This phase space is sufficiently larger
then that of geometric transport through the same limiting apertures to make
up for the opposite spin particles which are lost.

To calculate the magnet's effect we used a Monte Carlo simulation.[8] A
thermal distribution of velocities was assumed for the cluster beam.

$$\Gamma(v)\,dv \;=\; 2\Gamma_0 \left[\frac{v}{\alpha}\right]^3 e^{-(v/\alpha)^2}\frac{dv}{\alpha}\;\frac{mol}{cm^2\text{-}sec} \qquad where \qquad \alpha = \sqrt{\frac{2\,kT}{m}} \qquad (4)$$

and Γ_0 is selected so the integral over all velocities is normalized. The
effect of the magnet for such a velocity distribution is independent of the
particle mass since both L_0 and μ in equation (3) contain a mass which can-
cels those in the denominator. Physically, the smaller acceleration experienced by the more massive clusters is just enough smaller to maintain the
same trajectory as lighter more rapidly moving particles.

The trajectory of the particles from the source through the magnet was
determined for the magnet off (I_0) and for the magnet on both for the
focused (I_f) and defocused particles (I_d). The shift in intensity, ΔI, of
the magnet on versus the magnet off was calculated as

$$\Delta I = \frac{I_f + I_d}{2I_0} - 1 \qquad (5)$$

for several different values of μ to arrive at the results shown in Figure
1. From the plot, the effect of the magnet on a bismuth beam is determined
as shown in Table 1.

EXPERIMENTAL

The experimental apparatus is depicted in Figure 2. The source of
bismuth clusters was an inert gas condensation cell similar to that reported
by Sattler[9] but run without the inert gas, and differential pumping, such
that Knudsen conditions existed (Bi_n with $1 \le n \le 4$ were observed in the
ratios given in Table 1). The experimental probe was provided by an 1.20 ±
.06 Tesla pole tip field, permanent, quadrupole magnet described elsewhere.[10]
This could be turned "on" or "off" by interchanging it with an aluminum

Table I: Effect of Quadrupole Magnet

Species	Relative Intensity	μ (Bohr Magnetons)	Predicted Shift	Observed Shift	α
Bi $_1$	1.00	1,3	$1.43 \pm .17$	$0.346 \pm .088$	--
Bi $_2$	0.37	0	0	$-0.028 \pm .061$	$2.02 \pm .35$
Bi $_3$	0.02	1*	$1.00 \pm .16$	$0.012 \pm .090$	--
Bi $_4$	0.02	0*	0	$0.002 \pm .045$	$0.99 \pm .08$

*Hypothesized minimum value.

tube having the same geometric apertures as the magnet. The clusters were ionized by electron impact ionization (45 eV) and analysed with a time-of-flight mass spectrometer.

The effect of the magnet on the clusters was determined by collecting mass spectra for fixed time intervals -- alternating either the magnet or the tube in the beam. In addition, spectra of the background were regularly taken by rotating either the tube or the magnet such that the beam was blocked. The counts under each of the observed peaks for a single run were integrated (with correction for dead-time in the electronics). The results for a series of such runs were plotted as a function of time (see Figure 3) and a least squares' fit of intensity versus time performed for each of the three conditions (magnet, tube, and block). The areas under the plots were used to calculate the shift in intensity, ΔI, due to the magnet as

$$\Delta I = \frac{A_m - A_t}{A_t - A_b} \qquad (6)$$

where A_m, A_t, and A_b are the areas under the magnet, tube, and blocked curves respectively. The results are presented in Table 1.

We observe a larger then expected scatter in the day-to-day results which is not clearly understood. The most likely cause of the scatter is changes in the source position due to thermal expansion on heating. Such an alignment shift could explain the scatter if the different species do not have the same radial distribution in the beam. In light of this, our quoted error is the standard deviation of several different days runs. The actual propagated error would be considerably smaller.

Figure 2: Experimental apparatus. The clusters are generated in the source (a), pass through the tube or magnet (b), into the ionization region for ionization by the electron beam (behind the ion optics), and are drawn into the time- of-flight tube (c) by the ion optics (d).

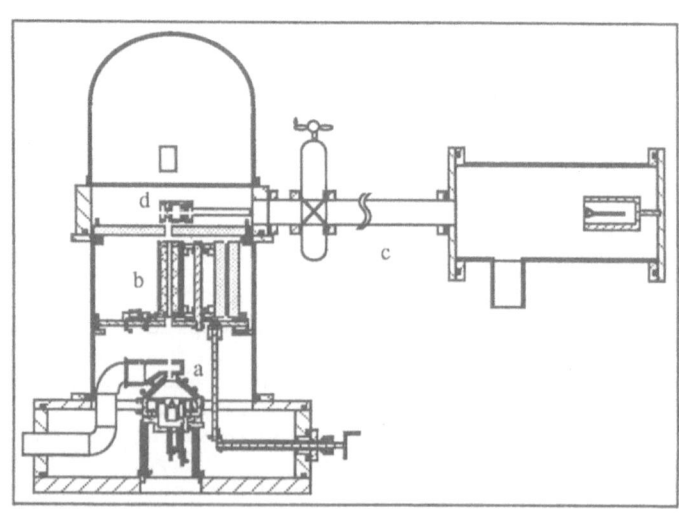

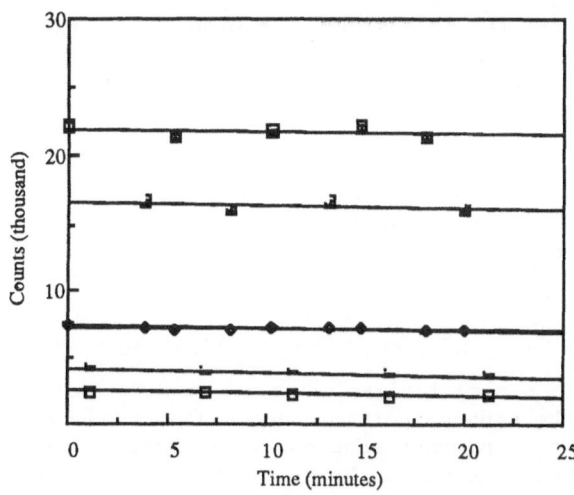

Figure 3: Typical Bi_1 -- Bi_2 data. Note the difference in effect of the magnet on the two species.

Magnet	Tube	Block
▫ Bi_1	◻ Bi_1	' Bi_1
◆ Bi_2	◇ Bi_2	▫ Bi_2

DISCUSSION

A comparison between the theoretical and experimental intensity shifts shown in Table I for Bi_1 shows a significant discrepancy. One may obtain a measure of the fragmentation by invoking fragmentation processes to explain the difference. We proceed with an analysis similar to Kohl et al.[1] who interpret the observed mass intensity for the monomer as having contributions from both atomic Bi_1 from the source (I_1) and fragmentation products form the dimer (αI_2). (It is assumed that the contribution to Bi_1 from fragmentation of Bi_3 and Bi_4 are negligible due to the significantly lower observed intensities of these species.)

$$I_t = I_1 + \alpha I_2$$
$$I_m = \kappa I_1 + \alpha I_2$$

(6)

where κ is the enhancement of the beam due to the magnet, and the subscript t (m) refers to the tube (magnet) in the beam. We have assumed that the magnet has no effect on the dimer (experimentally verified -- see Table I) and have folded all of the various detection efficiencies which remain constant (such as the ionization cross section) into I_1 and I_2. A value for κ (2.43) may be derived from the predicted shifts given in Table I which allows one to obtain $\alpha = 2.02 \pm .35$. With the interpretation of α as the fraction of the observed monomer peak due to fragmentation of the dimer and since $I_2/I_t = .37$, we see that this is a significant correction to the mass spectrum.

This result may be compared with that of Kohl et al.[1] who report $\alpha = 1.54 \pm .20$ at an electron beam energy of 60 Volts. From the data in their paper one may estimate that this value would increase slightly (~ .1 to .2) at 45 eV. Our result remains somewhat higher then theirs, although the error intervals overlap. It is important to note that the measurement of Kohl et al. relies on the assumption that the ratio of ionization cross sections of Bi_1 to Bi_2, is a constant on a relative energy scale, E − A where E is the electron energy and A is the appearance potential of the clusters. This is not an exact procedure, rather it is "speculated that this is roughly valid for practical use in the absence of any other known method."[1] Our result is a direct measurement relying on the known magnetic moments of the two species involved.

If one assumes that no species larger then Bi_4 are produced in the source, then one may treat Bi_3 and Bi_4 the same as Bi_1 and Bi_2. Unfortunately, the magnetic moments of these species are not known and an analysis is tenuous. However, it is interesting to speculate that the moment of the quadrumer is zero, while the smallest moment likely for the trimer is a single Bohr magneton. Using these values and the null effect of the magnet on the trimer allows us to calculate a reasonable minimum $\alpha = 1.00 \pm .08$ i.e., almost all of the Bi_3 observed in the experiment was due to fragmentation of Bi_4. This agrees with other reported results.[1,11,12]

SUMMARY

The process of fragmentation during ionization makes it difficult to relate mass spectra obtained to the neutral particle spectra emitted by a source. Knowledge of the fragmentation cross sections would facilitate work in the field and makes this an important goal. Measurements of fragmentation cross sections are possible in simple systems (few cluster species present) if one finds a means of changing the relative populations in a known manner -- such as with a magnetic field. By working with progressively more complex systems, one might hope to measure the fragmentation of larger clusters by eliminating the previously measured values for smaller species.

ACKNOWLEDGMENT

We are indebted to the National Science Foundation, Grant No. DMR-80-02437 for support of this work.

REFERENCES

1. F.J. Kohl, O.M. Uy, K.D. Carlson, J. Chem. Phys. **47**, 2667 (1967).
2. see e.g. "Proc. of the 3rd Int. Meeting on Small Part. and Inorg. Clust.," Surf. Sci. **156** (1985).
3. W.D. Knight, Helv. Phys. Acta. **56**, 521 (1983).
4. D.M. Cox, D.J. Trevor, R.L. Whetten, E.A. Rohlfing, A. Kaldor, Phys. Rev. B **32**, 7290 (1985).
5. R.G. Keesee, R. Sievert, A.W. Castleman Jr., Ber. Bunsenges. Physik. Chem. **88**, 273 (1984).
6. D.A. Greenwood, R. Brout, J.A. Krumhansl, Bull. Amer. Phys. Soc. **5**, 297 (1960).
7. J.M. Dickson, Prog. Nucl. Tech. Inst. **1**, 103 (1965).
8. K.P. Ziock, "Molecular Beam Studies of the Magnetic Moments of Small Metal Particles," thesis, Stanford Univ. (1985).
9. K. Sattler, J. Mühlbach, E. Recknagel, Phys. Rev. Lett. **45**, 821 (1980).
10. K.P. Ziock, W.A. Little, Rev. Sci. Inst., to be published.
11. L. Rovner, A. Drowart, J. Drowart, Trans. Faraday Soc. **63**, 2906 (1967).
12. B. Eberle, H. Sontag, R. Weber, Surf. Sci. **156**, 751 (1985).

METASTABLE DECAY OF PHOTOIONIZED NIOBIUM CLUSTERS:

EVAPORATION VS. FISSION FRAGMENTATION

S. K. Cole, K. Liu, and S. J. Riley

Chemistry Division
Argonne National Laboratory
Argonne, IL 60439 USA

ABSTRACT

The metastable decay of photoionized niobium clusters (Nb_n^+) has been observed in a newly constructed cluster beam machine. The decay manifests itself in the time-of-flight (TOF) mass spectrum as an asymmetric broadening of daughter ion peaks. Pulsed ion extraction has been used to measure the decay rate constants and to establish the mechanism of the fragmentation, evaporation and/or fission of the photoionized clusters. It is found that within the experimental time window evaporation dominates for the smaller clusters ($n < 15$), whereas fission fragmentation becomes facile for $15 < n < 30$. The decay rate constants obtained all fall within the range $0.5-2.0 \times 10^6$ sec^{-1}. The average kinetic energy release is also determined and is found to be on the order of 5 meV.

I. INTRODUCTION

Fragmentation of atomic cluster ions has been studied in a number of recent experiments.[1-7] The observed fragmentation patterns appear to fall into distinct categories depending on the types of clusters. In general, metal clusters tend to lose one atom at a time (evaporation), whereas semiconductor clusters exhibit both evaporation and fission fragmentation (cleavage of a large cluster). From the pieces of information available, it seems that fragmentation patterns generally reflect the relative stability of the fragments. However, this conjecture still does not explain why the fragmentation pattern should be different for metal and semiconductor clusters. In this study, we report the first observation of fission fragmentation for a transition metal cluster. It is found that there is a critical cluster size above which fission fragmentation becomes facile, analogous to the nuclear fission process.

II. EXPERIMENTAL

The apparatus consists of an ion time-of-flight (TOF) mass spectrometer and an electron TOF spectrometer coupled with a laser vaporization source for generation of the metal clusters. The second harmonic of a 20 Hz Nd:YAG laser is weakly focused onto a niobium rod resulting in vaporization of the metal. A pulsed flow of He cools the plasma and stimulates cluster growth. For the study of neutral clusters various excimer lasers are used for ionization prior to extraction into the TOF apparatus. TOF mass spectra have been obtained at four photon energies: 6.4 eV (ArF excimer laser), 5.0 eV (KrF excimer laser), 4.0 eV (XeCl excimer laser), and 3.5 eV (XeF excimer laser).

III. PHOTOFRAGMENTATION PATHWAY

Metastable decay of molecular ions in a TOF mass spectrometer is a well understood phenomenon. The signature of such a process, asymmetric broadening of daughter ion peaks, is a consequence of parent ion fragmentation on a time-scale comparable to the ion extraction time. If the dissociation takes place after the parent ion has experienced significant acceleration, but before the acceleration is complete, the daughter ion time of flight will be between that of the parent ion and an unfragmented daughter ion. The width and shape of the asymmetric broadening will depend upon the time-of-flight parameters, i.e., voltages and distances, and the nature of the fragmentation. In a typical experiment of this type, all parameters are known except the parent ion dissociation rate for a particular daughter ion and the kinetic energy released in the dissociation. These two unknowns can be determined from the TOF spectrum. However, in the case of a distribution of cluster sizes, the parent ion is also unknown, thereby complicating the analysis. To sort out what processes are occurring, a delayed ion extraction technique has been used in this study to determine the dissociation rates for niobium clusters in the size range 5 to 30.

Though the asymmetric broadening of the ion peaks in the KrF spectrum might be due to ion metastable decay, other possible mechanisms must be considered and tested. For example, a delayed ionization process such as dissociative ionization, field ionization, or ion pair formation followed by autodetachment will also result in asymmetric broadening of the TOF peaks. These processes will also broaden the electron TOF spectrum. The observed electron TOF spectrum shows but a single sharp peak (FWHM < 30 ns), which rules out any delayed ionization process. Ion-molecule reactions during ion extraction might also result in a metastable peak shape. To test this possibility the delayed ion extraction technique was used. If an ion-molecule reaction is occurring the peak shape will be independent of the extraction time. As can be seen in Fig. 1 this is not the case. With increasing delay between the laser pulse and the ion extraction pulse the ion peaks become sharper, indicating that an ion-molecule reaction is not responsible for the observed peak shapes.

More importantly, since the delayed ion extraction technique is, in essence, a mass spectrometric version of the pump-and-probe method, it should provide a <u>direct</u> measurement of the absolute decay rate constant.

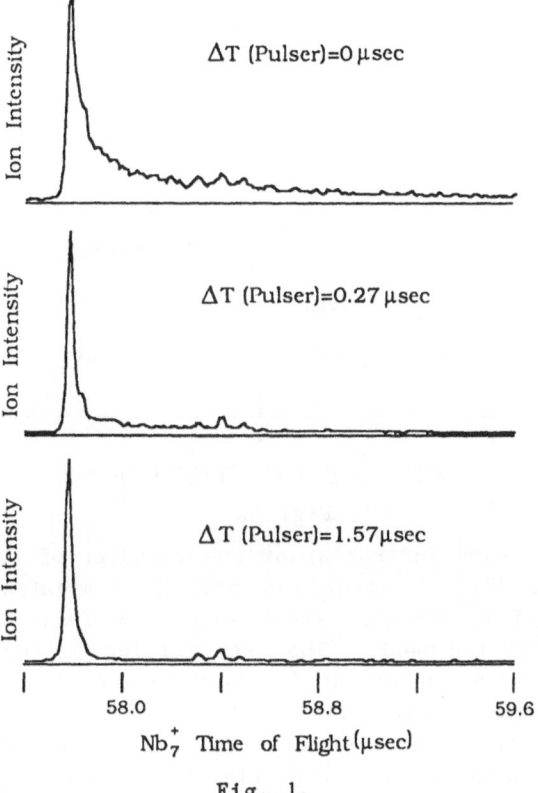

$\Delta T \text{ (Pulser)} = 0 \, \mu\text{sec}$

$\Delta T \text{ (Pulser)} = 0.27 \, \mu\text{sec}$

$\Delta T \text{ (Pulser)} = 1.57 \, \mu\text{sec}$

Nb_7^+ Time of Flight (μsec)

Fig. 1.

This may be seen by considering a single exponential decay of a parent ion cluster of size m to a daughter ion i.

$$Nb_m^+ \rightarrow Ni_i^+ + Nb_{m-i} \,. \tag{1}$$

The number of daughter ions formed per unit time at time τ can be expressed as:

$$dn_i/d\tau = k_{m \rightarrow i} \, n_m^0 \, \exp(-k_m \tau) \tag{2}$$

where n_m^0 is the initial parent ion number density, k_m is the total decay rate constant of the parent ion, and τ is the time after photo-excitation.

Since the actual data is displayed in the TOF time domain, one has to consider the mapping of the formation times τ of a daughter ion in the ion source into the arrival times t at the detector. For each parent and daughter mass in a given electric field there is a unique transformation from the TOF time domain into the real decay time domain. Furthermore, it is found that under space focusing conditions the transformation is approximately a linear function, i.e., $\tau = \alpha t$. Therefore the rate equation becomes

$$dn_i/dt = k'_{m \rightarrow i} \, n_m^0 \, \exp(-k'_m t) \tag{3}$$

where $k'_m = \alpha k_m$ is the "apparent" decay constant in the TOF time domain and α is the transformation coefficient. Thus, the signal in the TOF spectrum should also behave as an exponential decay. One can readily extend the above analysis to the case of several parent ions or isomers.

349

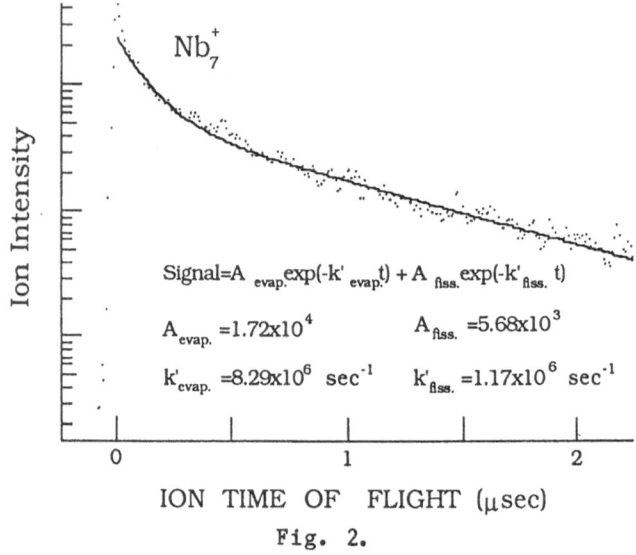

Fig. 2.

Figure 2 gives some indication of the validity of this approximation, in which the Nb_7^+ daughter ion peak is simulated by a double exponential decay of metastable parent ions. At least two components are necessary to fit the peak. The data fitting is conducted in two steps. First, the "slow" component (tailing to long time of flight) is fit by a single exponential. The "fast" component is then fit by a single exponential after subtracting the TOF signal form the "slow" component fit. The ratio of the preexponential factor divided by the apparent rate constant to the total peak intensity then gives the fraction of ions resulting from each metastable decay component.

In the case of pulsed ion extraction with a time delay ΔT between the photon pulse and the extraction pulse, Eq. (3) becomes:

$$dn_i/dt = [k'_{m \to i} \, n_m^0 \, \exp(-k_m \Delta T)] \, \exp(-k'_m t) \qquad (4)$$

where t is still in the TOF time domain but starts from the onset of the delayed extraction pulse. The idea of the delayed extraction technique is to follow the time evolution of the amount of metastable decay with the variable time delay ΔT as illustrated in Fig. 1. Quantitatively, it can readily be shown from Eq. (4) that the fraction of the total signal in the metastable tail should be proportional to $\exp(-k_m \Delta T)$. Thus, the slope of a plot of ln(fraction) vs. ΔT yields the decay rate constants k_m directly. Once the rate constants k_m and k'_m are known for a given daughter ion, the transformation coefficient α, and hence the parent ion mass, can be determined.

Table I summarizes the results for Nb_i^+ daughter ions $5 \le i \le 15$. Note that the decay rate constants are nearly all identical. The evaporation component appears "fast" simply as a result of the time domain transformation (α is large). The identification of the parent ion for the evaporation component is in fact simplified by the transformation, because α is a rapidly varying function of the parent ion and daughter ion masses in this mass range. On the other hand, as the ratio of the parent to daughter ion masses increases, α becomes less sensitive to the exact parent mass. Therefore, there is a range of parent ion masses for the fission fragmentation.

Table 1. Fragmentation Decay Rates for $Nb_m^+ \rightarrow N_i^+ + Nb_{m-i}$

	EVAPORATION ("fast")			FISSION ("slow")			
i	$k_{evap.}$ $(10^6 s^{-1})$	Parent Ion (m)	Fraction[a]	$k_{fiss.}$ $(10^6 s^{-1})$	Parent Ion (m)	Fraction[a]	Total Fraction[b]
5	0.90	6	0.20	–	–	–	0.20
6	1.41	8	0.27	–	–	–	0.27
7	2.04	9	0.29	0.85	17–30	0.59	0.88
8	1.82	10	0.24	0.70	14–17	0.32	0.56
9	1.81	10,11	0.35	0.61	16–20	0.67	1.02
10	0.82	11	0.26	0.56	16–19	0.33	0.59
11	1.06	12	0.24	0.78	21.24	0.39	0.63
12	0.74	13	0.26	0.64	20.24	0.39	0.65
13	0.94	14	0.22	0.75	25–30	0.50	0.72
14	0.69	15	0.16	0.62	24–27	0.47	0.63
15	0.64	16	0.15	0.61	23–28	0.51	0.66

[a]Refers to the ratio of the integrated intensity of the particular component relative to the total integrated intensity of the ion peak.
[b]Refers to the fraction of the total integrated peak intensity due to fragmentation (evaporation and fission).

It is worthwhile to point out several interesting findings from these experiments. (1) For cluster ions Nb_n^+ in the size range n=7 to 15, all TOF peaks exhibit two distinct components to their asymmetric broadening. As shown in Table I, one results from the fission fragmentation of the larger cluster ions Nb_{15}^+ to Nb_{30}^+, and the other from evaporation of monomers and dimers from clusters having one or two more niobium atoms than the observed daughters. However, the daughter ions 5 and 6 result only from evaporation pathways. (2) For decay processes occurring within the experimental time window (0.1 to 60 μs), evaporative fragmentation dominates for the smaller parent clusters (6≤n≤15), while fission fragmentation becomes facile for 15<n<30. (3) The rate constants for fragmentation of all parent clusters except Nb_7^+ are within the range 0.5 to 2 × 10^6 sec^{-1}. The Nb_7^+ cluster is surprisingly stable and undergoes no fragmentation once it is formed. Other evidence for a stable Nb_7^+ ion is seen in the spectrum of cluster ions coming directly from the source, where once again Nb_7^+ is the most prominent species. In fact, there are other similarities between the spectra of ions from the source and from metastable decay. (4) The average kinetic energy imparted to the daughter ions has been measured and found to be on the order of 5 meV. This low energy, together with the observation that fragmentation tends to lead to the more stable ionic clusters, suggest that the fragmentation processes are statistical in nature and have no substantial reverse activation barriers.

IV. PHOTOEXCITATION PATHWAY

To better understand the fragmentation pathways observed here, a series of experiments has been conducted to elucidate the photoexcitation pathway and the energetics involved. The dependence of the ion signal on laser fluence shows a slope of 1.2 for the KrF laser excitation and 0.9 for the 6.4 eV ArF excitation. In the size range n=5 to 30, niobium clusters have ionization potentials ranging from 4.5 to 5.5 eV.[8] Thus it appears KrF laser ionization is a two-photon process. Interestingly, preliminary photoelectron energy measurements indicate that the average kinetic energies of photoelectrons from KrF and ArF laser ionization are not very different, being 0.8 eV for KrF and 1.3 eV for ArF. An additional observation is that when the KrF laser is fired at the cluster ions coming from the source, no asymmetric broadening appears.

This complex series of observations points to the following picture of the photophysical and photochemical processes in Nb clusters. When parent neutral clusters absorb one KrF photon, even if the cluster ionization potential is lower than 5 eV, only a small fraction of them ionize. This is consistent with the TOF spectrum in which the parent ion signals are much smaller than the daughter ion signals. The more probable ionization pathway is the absorption of an additional photon followed by ionization and fragmentation into daughter ions. This tentative scenario may have important implications for the study of transition metal clusters because it requires the existence of some isolated excited states which have little coupling with the rest of the "vibronic soup" that is thought to characterize transition metal clusters. A direct experimental verification of this photoexcitation pathway is currently under way.

Acknowledgments — KL would like to thank Dr. E. K. Parks for his help in construction of the apparatus and Prof. W. R. Gentry and C. F. Giese for the gift of the pulsed valve. This work is supported by the U. S. Department of Energy under contract W-31-109-Eng-38.

REFERENCES

[1] P. J. Brucat, L.-S. Zheng, C. L. Pettiette, S. Yang, and R. E. Smalley, J. Chem. Phys. **84**, 3078 (1986).

[2] J. R. Heath, Y. Liu, S. C. O'Brien, Q.-L. Zhang, R. F. Curl, F. K. Tittle, and R. E. Smalley, J. Chem. Phys. **83**, 5520 (1985).

[3] L. A. Bloomfield, R. R. Freeman, and W. L. Brown, Phys. Rev. Lett. **54**, 2246 (1985).

[4] W. Begemann, K. H. Meiwes-Broer, and H. O. Lutz, Phys. Rev. Lett. **56**, 2248 (1986).

[5] M. E. Geusic, T. J. McIlrath, M. F. Jarrold, L. A. Bloomfield, R. R. Freeman, and W. L. Brown, J. Chem. Phys. **84**, 2421 (1986).

[6] C. Brechignac, Ph. Cahuzac, and J. Ph. Roux, Chem. Phys. Lett. **127**, 445 (1986).

[7] C. Brechignac and Ph. Cahuzac, Chem. Phys. Lett. **112**, 20 (1984).

[8] R. L. Whetten, M. R. Zakin, D. M. Cox, D. J. Trevor, and A. Kaldor, J. Chem. Phys. **85**, 1697 (1986).

ESR OF SMALL CLUSTERS

William Weltner, Jr. and Richard J. Van Zee

Department of Chemistry
University of Florida
Gainesville, FL 32611, USA

INTRODUCTION

It is not straightforward to determine the detailed electronic and magnetic properties of individual clusters. Optical spectroscopy has been applied successfully to conventional[1] and supersonic sources,[2] but particularly for clusters larger than diatomics, the complexity of the spectra may make the analysis very difficult.[3] Electron-spin-resonance (ESR) spectroscopy at 4 K offers a means of determining a number of important properties of small clusters, such as:
1. Multiplicity; i.e., total spin S.
2. Spin populations and s, p, and/or d character. This is obtained from the hyperfine splittings (hfs) due to the interaction of the unpaired electrons with the nuclear moments.
3. Geometry of the molecule (linear, equilateral triangle, number of equivalent nuclei, etc.). This may also be obtained from hfs.
4. Zero-field splitting, D (and E). Arises from magnetic interaction between unpaired spins (spin-spin interaction) or through second-order spin orbit interaction, when $S > 1$.
5. g tensor components. Shifts from the free electron value, $g_e = 2.0023$, indicate the degree of spin-orbit coupling with excited electronic states and can indicate the presence of low-lying states.
6. Spin-rotation constants. Derived from g shifts using Curl's equation.[4]
7. Electric quadrupole coupling constants. Arise from interaction of the electric field gradient with a nucleus having a quadrupole moment.
Since all the measurements on clusters are made in matrices (either solid rare-gases, nitrogen, or hydrocarbons) at low temperatures, the detailed moment of inertia data obtained from gas phase ESR[5] cannot be obtained for these species, which are usually non-rotating or at most undergoing perturbed rotation in sites in the solid matrix. Other than this effect upon their rotational properties, the matrix ESR parameters are usually insignificantly different from those of gas-phase molecules.

The intention here is to review all of the ESR measurements that have been made on small clusters, and where possible, to compare experiment with theory. The clusters that have been studied are the Group IA and IB metals, the transition-metals, and one carbon molecule, C_4.

Na_3 and K_3

Lindsay, Herschbach, and Kwiram,[6] prior to the intense interest by theoreticians, showed by measurements of the ESR spectrum of sodium vapor trapped in solid argon at temperatures near 4 K that Na_3 is chemically bonded but does not have an equilateral triangular geometry. They found from hyperfine structure (^{23}Na has a nuclear moment with spin I = 3/2) that the single unpaired electron spin was almost entirely distributed between two equivalent Na atoms in the molecule with only small density on the remaining atom.

If the unpaired spin is interacting with two equivalent nuclei, each with I = 3/2, one expects to observe, if the interaction is not too large, 2(2I) + 1 = 7 hyperfine lines having relative intensities 1:2:3:4:3:2:1 corresponding to M_I = ±3, ±2, ±1, 0. However, if the hyperfine interaction is strong enough, the degeneracy in M_I will be removed completely, and one expects to observe $(2I + 1)^2 = 16$ lines for S = 1/2, all with the same intensity. This is the case in $^{23}Na_3$, indicating that two Na atoms are equivalent, and the spin densities at those nuclei are relatively large. A further small splitting of each of these 16 lines into quartets, due to interaction of the electron spin with a third unique nucleus, was also detected. Hence the hyperfine pattern identifies a trimer and one that does not have an equilateral triangle structure.

The lines in the ESR spectrum appear isotropic in shape. Indeed, calculation of the spin densities on the basis that the unpaired electron is pure 3s, as in the atom, accounts for almost all of the spin. Thus Lindsay et al.,[6,7] found $\rho_{3s}^1 = \rho_{3s}^3 = +0.47$ and $\rho_{3s}^2 = -0.07$. The small negative spin on the middle atom was attributed to spin polarization; this was confirmed later from observation of "dynamic" spectra.[7] From these approximate spin densities, one then concludes that the unaccounted-for spin, about 10%, is due to p character, i.e., s-p hybridization, in the wavefunction. As is well known, the hyperfine interaction due to p orbitals is more than an order of magnitude less than for an s electron, and therefore its effect upon the hyperfine splittings would be small.

An important deduction from these properties of the wavefunction is that it has a node, or near-node, at the middle atom. Although this finding does not exclude a symmetrical linear structure ($^2\Sigma_u^+$) for Na_3 or a bent C_{2v} structure with an apical angle greater than 60^0, it does appear to exclude a more acute-angled structure and, of course, an equilateral triangle. Simple molecular-orbital considerations[8] indicate that the observed spin density corresponds to the odd electron being in an anti-bonding orbital between the terminal atoms with a node at the central atom. This orbital has b_2 symmetry for a C_{2v} molecule and results in a $^2B_2$ ground state for Na_3. The alternative C_{2v} case, occurring for a small apical angle, would place the unpaired spin in a bonding orbital (a_1) between the end atoms, which however places about 50% of the spin on the central atom, clearly not in accord with observation. Thus, from straightforward consideration of the ESR spectra alone, the ground state of Na_3 is determined to be either $^2\Sigma_u^+$ or $^2B_2$. This is the static picture both for Na_3 and, from very similar findings, for K_3.[8,9] In fact, the spin densities are essentially the same for K_3, +0.47 and -0.06.

However, this is not the whole story for Na_3 and K_3 in argon matrices since for both trimers at T < 20 K there is an additional ESR spectrum overlapping the static one.[8-10] Whereas the static spectra reach a maximum intensity at ~10 and 19 K, respectively, and then decrease to zero intensity, the second spectrum in each case reaches a maximum near 35 K. This second dynamic spectrum exhibits the hyperfine splittings of a trimer with three equivalent I = 3/2 nuclei, i.e., it consists of ten equally spaced lines with relative intensities 1:3:6:10:12:12:10:6:3:1. For Na_3, the splitting $\bar{a}$ = 94(1)G and g_0 = 2.0016(3), which is in close agreement with the average of the static hyperfine constants (if that of the middle atom is negative) and g_0 = 2.0012 (12). A similar comparison occurs for K_3.

The dynamic spectra are the result of rapid exchange between the three equivalent obtuse-angled geometries of the static trimer. Raising the temperatures of the matrices slightly, to only about 30 K, causes the Na_3 and K_3 molecules to reach this dynamic state, indicating that the barriers to this "pseudorotation" are quite low. These three equivalent minima in the potential energy surface can be thought of as resulting from the Jahn-Teller distortion of the degenerate ^2E electronic state of the (D_{3h}) equilateral structured molecule.[11] As the temperature is raised, vibronic coupling allows the molecule to surmount the barriers between these three minima and pseudorotate. At the highest temperatures the molecule becomes completely "fluxional" and the spectrum becomes motionally averaged with a corresponding averaging of the hyperfine splittings. The pseudorotational barrier, which is low in these molecules (calculated as 0.39 and 0.52 kcal/mole for Na_3 and K_3, respectively[12]) and the connected motional processes, could be affected by the matrix environment and its dynamics.

Of interest is the unexpected (and unexplained) fact that both the static and dynamic spectra of Na_3 and K_3 can be observed simultaneously in argon matrices in the lowest temperature range. Presumably this is due to the trapping of the trimers in two pseudorotationally-sensitive sites (to be distinguished from energetically different sites for the static or for the motional molecules) where exchange is more inhibited in one than the other. These two distinct sites might, for example, occur on the surface of argon microcrystals and at distorted substitutional positions within the argon lattice.[13]

Li_3

Li_3 has been prepared in solid argon at 4 K and also in a hydrocarbon matrix (adamantane) at 77 K.[15] [The latter was an extension by Howard, Sutcliffe, and Mile of ESR studies of Cu, Ag, and Au clusters in hydrocarbon matrices (see below)]. The spectrum obtained in both cases was that of three equivalent nuclei, and variations in temperature, even over the range of 4 to 298 K had little effect upon the spectrum. Thus the ESR spectrum of $^6Li_3$ exhibits a pattern of seven equally spaced hyperfine transitions with relative intensities corresponding to the values expected for three equivalent I = 1 nuclei. The $\rho_{2s} \cong 0.23$ on each atom, which is lower than in the heavier trimers, indicated, not unexpectedly, a larger s-p hybridization (~30% p) of the unpaired spin. These findings clearly concurred with theory[16,17] which is in quantitative agreement with the increased p contribution and also yields a barrier to pseudorotation which is very small (<0.05 kcal according to Thompson et al.[12]). The latter means that the molecule should be completely fluxional even at very low temperatures; the observations indicate even at 4 K. Of interest also is a comparision of the magnetic parameters in the two matrices: a(^7Li) = 32.2 G and g = 2.0026 in argon; a = 33.1 G and g = 2.001 in adamantane, demonstrating the essential independence of the data to the matrix solid.

These larger clusters were also prepared in argon matrices by Lindsay's group[18,19] and characterized by the hyperfine pattern in each spectrum. In each case the hfs pattern of seven lines showed that there were two equivalent nuclei and associated with each of the seven lines was a further smaller splitting into 16 lines corresponding to five equivalent nuclei. Care was taken to signal-average these superhyperfine patterns to establish that they contained (at least) 16 lines since those in the wings drop rapidly in intensity. A pentagonal bipyramid structure, later also supported by theory,[20] was proposed. These are then $^2A_2''$ (D_{5h}) ground-state molecules. The unpaired spin is largely on the two apical atoms ($\rho_s \cong 35\%$ each), and much smaller density ($\rho_s \cong \pm 2\%$) on the ring atoms. Again, considerable p character must be invoked to account for all of the spin. This is obtained approximately from the difference of the isotropic spin population from unity and leads, not unexpectedly, to about 40% p character for Na$_7$ and K$_7$ but 60% for Li$_7$.[18]

The discovery of this geometry for these clusters has been an important contribution since it shows that one cannot visualize the "aufbau" process as simply the binding of atoms to triangular faces to grow stable clusters by the successive addition of tetrahedra.[21] These seven-atom structures are sections of an icosahedron and can be visualized as five slightly distorted tetrahedra meeting at an edge to form a decahedron, thereby having a pentagonal axis.

Pentamers, Where Are They?

One notes that Li$_5$, Na$_5$, and K$_5$ appear to have been bipassed in the experimental observations, although it is certainly logical to assume that they are present in matrices where the trimers and heptamers are present. We believe this is probably an ESR detection problem and not due to any anomaly in relative concentrations. If theory may be assumed reliable then the most stable form of the pentamer molecule is as indicated in Fig. 1. If planar, it has C_{2v} symmetry and a ^2A or ^2B ground state. It would have one unique atom and two sets of 2 equivalent atoms and the g and A tensors would presumably be completely anisotropic. Thus, the spectrum could be spread out and weakened around g = 2.0 due to the complex hyperfine structure on top of the anisotropy in g. The inability to detect the molecule may, in fact, be taken as evidence for a low symmetry structure since a higher symmetry, as in an axial trigonal bipyramid or a square-based pyramid, would yield a simpler and more easily detected spectrum. It is worth noting at this point that Howard, et al., have presumably observed the ESR spectra of Cu$_5$ and Ag$_5$ molecules in hydrocarbon matrices and from the hyperfine pattern they give it a distorted (nonaxial) trigonal bipyramid structure, but there may be doubt about these assignments. This will be discussed more thoroughly in the next section.

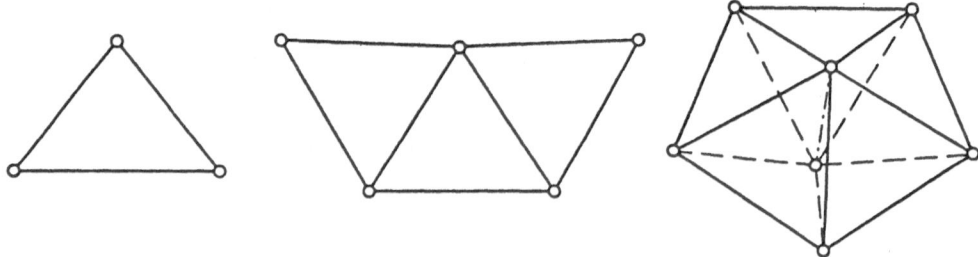

Fig. 1. Clusters containing an odd number of atoms. These are their most stable structures according to theory.

Similar to Lindsay, et al´s work on the alkali metals, ESR spectra of Cu, Ag, and Au molecules have been investigated largely by one group of researchers, Howard, Sutcliffe, Mile, Preston, and their coworkers at the National Research Council of Canada.[22] Recently Lindsay´s group has also studied Ag_3 and Cu_3 in nitrogen matrices with some unexpected results.

These metal atoms are the transition-metal counterparts of the alkali metals in that their lowest electronic states have an s electron outside of a closed shell. Thus one expects, and finds, that the small clusters formed by these Group IA and IB metals are similar.

A unique aspect of the Canadian research, which is at first worrisome, is their use of organic compounds, solid benzene, cyclohexane, and adamantane, as matrices at 77 K. Indeed, complexes or reactants are often formed, providing background spectra.[23] These apparently are not dominant species and are either recognizable from other ESR studies or can be preferentially annealed away. Deuterated matrix compounds may also be used and serve as tests for whether cluster hydrides are formed. In the one case[15] where there is a direct comparison with argon matrix data, that of Li_3, the agreement between the measured magnetic parameters was excellent (see above). To simplify the spectra it has been expedient to use metals enriched in one isotope, such as $^{63}Cu(I = 3/2)$ and $^{109}Ag(I = 1/2)$. Gold presents no problem since it contains 100% natural abundance of $^{197}Au(I = 3/2)$.

The distinct advantage of hydrocarbon matrices is the large temperature range over which they are rigid enough that solid-state diffusion does not occur, and ESR measurements can even be made up to room temperature.

Cu_3, Ag_3, Au_3 and Cu_2Ag

In each of these cases, isotropic spectra were measured which were basically similar to those of the alkali-metal trimers.[24-27] Very large hfs was found at two equivalent metal atoms and a much smaller splitting at a third atom. For example, for $^{63}Cu_3(I = 3/2)$ sixteen lines were observed due to the strong interaction of the one unpaired spin with the two equivalent nuclei, and each of these lines is then split into four lines by the much weaker interaction with a unique third atom. The analogy with the alkali metals is complete in that spin density distributions are also comparable, as indicated in Table 1, so that again obtuse-angled $^2B_2$ ground states are assigned.

Table 1. Spin Populations of "s" Character in Group IA and IB Trimer Molecules

	ρ(basal atom)	ρ(apical atom)
$Li_3{}^a$	0.23	0.23
Na_3	0.47	−0.07
K_3	0.47	−0.06
Cu_3	0.29	±0.026
Ag_3	0.44	±0.06
CuAgCu	0.41	+0.04
Au_3	0.39	±0.056

[a] pseudorotating.

Cu$_3$ and Ag$_3$ molecules have not been observed in rare-gas matrices,[28] but recently Lindsay and coworkers have detected their ESR spectra in nitrogen matrices at 4 K.[29,30] The perhaps surprising result in each case is that the spin distributions are quite different than when trapped in the solid hydrocarbons. For Ag$_3$ in N$_2$, the 5s spin population on the apical atom is 0.51 relative to 0.12 on each basal atom, whereas in a C$_6$D$_6$ matrix,[25] the corresponding populations are 0.06 and 0.44, respectively. This complete reversal of spin densities indicates that the character of the molecular orbital occupied by the unpaired spin has changed from essentially antibonding (b$_2$) to bonding (a$_1$) between the basal atoms in going from a solid hydrocarbon to a nitrogen environment and suggests that the bending force constant in the molecule is low, i.e., the potential surface is very flat. Theory supports this finding in showing that the energy difference between the "obtuse" and "acute" geometries of Ag$_3$ is only a few hundred wave numbers.[30,31] Thus, it is understandable that the electronic properties of the molecule, which are so sensitive to apical angle, are affected by the surrounding matrix. Here matrix effects, as is sometimes the case,[32] can clearly reveal the "soft", or energetically-sensitive, features of a molecule.

A sharp divergence from the behavior of the alkali-metal trimers does arise in the lack of evidence for pseudorotation in the Group IB series, even in hydrocarbon matrices at temperatures up to 150 K. In view of the relative ease of conversion from obtuse to acute-angled species in the latter series, it would seem that the effective high barrier to pseudorotation must be attributed largely to their heavier masses.

Cu$_5$, Ag$_5$, Cu$_2$Ag$_3$ and CuAg$_4$

In discussing these molecules, it should perhaps be pointed out that their detection, (if positive) marks the first departure from the analogy between Group IA and IB metal clusters arising in this review. As noted earlier, pentamer species have not yet been detected among the alkali metals via ESR, although it is logical to assume that they are present if the trimers and heptamers are. Then it is important to examine the evidence for observation of Cu$_5$, etc. and to ask whether there is an alternative explanation of the ESR spectra. At the time of writing this review, we believe that we have detected the ESR spectrum of the Ag$_7$ molecule in a neon matrix at 4 K. It appears to be very similar to the spectrum of Ag$_5$ published by Howard, et al.[34] The question then arises as to how there can be ambiguity in the interpretation of that spectrum. So we will discuss the analysis of the spectrum of Ag$_5$ in a C$_6$D$_{12}$ matrix first:

Three sets of lines are observed separated by about 200 G but the middle one is split by 13 G. This in accord with the presence in the molecule of two equivalent ^{107}Ag nuclei with which the one unpaired electron is interacting strongly. Each of these four lines in turn exhibit superhyperfine splitting into a quintet with a much smaller hfs of 5.5 G. These five lines are attributed to two overlapping sets of triplets with fortuitous spacings such that only five lines can be observed. Thus the spectrum is assigned to a molecule containing two sets of 2 equivalent atoms and one unique atom, i.e., five total atoms. The structure is then suggested to be a distorted trigonal bipyramid with the spin predominantly on the two apical atoms, and the triangular base distorted to yield 2 equivalent and one unique atom. One notes that this analysis of the spectrum depends directly on the resolution (and intensity distribution) of the quintet structure in each of the three dominant lines.

If indeed, as we believe, this quintet structure really contains six lines, then this superhyperfine structure can be attributed to five equivalent Ag nuclei and the interpretation is the same as given above for Li_7, etc., i.e., the structure would be a pentagonal bipyramid. The spectrum that we have recently observed in neon matrices is highly resolved and has such a six line pattern; its magnetic parameters are otherwise almost the same[36] as those of Howard, et al.

The ESR spectrum of $^{63}Cu_5$ has been interpreted in an analogous way by these authors.[33] Again there is definite evidence of large spin density at two equivalent atoms from the observed 16 sets of multiplets. Each of these multiplets consist of 13 lines attributed to two overlapping sets of seven lines and a proposed structure for Cu_5 as a distorted trigonal bipyramid. If this 13-line pattern were really 16 lines, then one would again attribute this spectrum instead to a septamer with a pentagonal bipyramid structure. The resolution is much better than in the Ag_5 spectrum so that perhaps all of the superhyperfine structure has been accounted for here, but the expected analogy with Group IA molecules again makes the assignment to a pentamer suspect.

TRANSITION METALS

ESR has also been a prime source, in fact the predominant experimental source, of information on transition-metal molecules. These, beginning with the diatomics, have been discussed in two relatively recent reviews,[37,38] one specifically concerned with ESR contributions.[38] Although there have apparently been no significant recent advances, a few points are worth emphasizing:

Dimers

Among the 45 possible diatomic molecules, both homo- and heteronuclear, that can be formed from the nine first-row transition metal atoms (Sc,...,Cu) only ten diatomics have been characterized experimentally, and among those, seven by ESR studies in matrices. Thus, much more research is needed. One inducement for further studies may be a comparison of isoelectronic molecules, i.e., those containing the same number of electrons, which form subsets of the array. For example, Cr_2, MnV, FeTi, and CoSc are isoelectronic. Cr_2 is known to have a $^1\Sigma$ ground state,[39] and FeTi and CoSc have also been inferred to be $^1\Sigma$;[38] therefore, is MnV also $^1\Sigma$? It would be surprising if such a simple molecular orbital principle could apply to these molecules where many low-lying electronic states are probably the rule.

Only a few scattered diatomics, such as $Mo_2(^1\Sigma)$, involving second- and third-row transition metals have been characterized. In ESR spectroscopy the studies of these heavier molecules is made more difficult because of their large spin-orbit coupling. This leads to large zero-field (second-order spin orbit interaction) splittings which often prevent their detection by X-band ESR, particularly for odd multiplicities where Kramers doublets do not arise.

As emphasized previously,[37] beyond diatomics experimental information is meager and most of it comes from ESR investigations.

Trimers

The hyperfine splitting pattern in the Sc_3 ESR spectrum shows clearly that the molecule contains three equivalent nuclei and is therefore an equilateral triangle of D_{3h} symmetry.[40] Its one unpaired spin was assigned to an a_1^- orbital on the basis of its observed hyperfine. An electron in an a_1' orbital derived from a $d\sigma$ or $d\delta$ atomic orbital was calculated to display the observed A_{dip}. This is in contrast to Walch and Bauschlicher[41] who calculate the electronic ground state to be $^2A_2''$ derived from an electron in an a_2'' orbital which results from a $3d\pi$ atomic orbital.

The most important result of the theoretical calculation is that the ground state is ^2A and not a Jahn-Teller distortion of a $^2E'$ state. Thus the observed equilateral triangle geometry is the real ground state geometry and not a fluctuating pseudo-rotating geometry.

The Y_3 molecule, however, is not equilateral, based on the observed hyperfine splittings, and is probably a bent molecule.[40] Again, the spin is distributed in 3d orbitals with a $^2B_2$ ground state.

Mn_5

Mn_5 is perhaps the most intriguing transition-metal molecule detected because of its ferromagnetic character.[42] It has a total spin of S = 25/2 so that it is almost the equivalent molecularly of a single domain in a ferromagnetic solid; i.e., it can be considered as "superparamagnetic".[43] Presumably the bonding between the five Mn atoms, thought to probably form a ring, occurs through s-p hybridized bonds, and the 25 unpaired spins are essentially in d orbitals coupled ferromagnetically through these bonds. Unresolved hyperfine structure in the ESR spectrum supports such a picture. The molecule was found to be relatively highly oriented in argon and krypton matrices, which usually suggests a prolate or, in this case, oblate shaped molecule rather than a compact cluster such as a trigonal bipyramid. Thus, considerable information has been obtained or can be inferred from the ESR data.

Sc_x

Finally, there is what we have referred to as a "real" cluster, Sc_x where x > 9.[40] A broad Lorentzian line at g = 2.0 with extensive hyperfine structures disappearing in the wings must be a large scandium molecule containing at least 9 equivalent Sc atoms. Establishing the upper limit to this number is difficult because the hyperfine lines are dropping rapidly in intensity on the shoulders of the broad signal. Further signal averaging of these regions might settle that question. However, lacking firm experimental evidence, it was assumed that the molecule being observed was Sc_{13}, where one atom was surrounded by a shell of twelve equivalent atoms. This structure was chosen because of the expected stability of 13-atoms in a close-packed cubic or hexagonal array or in an icosahedron. Also, an Xα calculation by Salahub[44] indicated that hexagonal close-packed Sc_{13} should have the observed S = 1/2 ground state. The small hyperfine splitting implies that the single unpaired spin is in a delocalized molecular orbital of predominantly d character.

Transition-metal molecules and clusters are difficult species to study, both experimentally and theoretically. Because of their unfilled d shells one can expect them generally to be magnetic and therefore susceptible to ESR investigation; however, their reactivity and relatively large spin-orbit constants are two reasons why ESR may not always be a successful tool. Variable frequency ESR would be a distinct advantage in their study.

CARBON

Pure carbon molecules may be involved in interstellar and

circumstellar processes and perhaps in the formation of soot. Besides that, carbon containing compounds are intrinsically interesting to chemists, and here they are in their most basic form. The C_2 and C_3 molecules are well known in astronomical sources and are often detected during the combustion of hydrocarbons. Ballik and Ramsay[45] first established that the ground state of C_2 is $^1\Sigma_g$ and that the $^3\Pi_u$ state, thought for many years to be the lowest state, lies 700 cm^{-1} higher. The observation of a more highly resolved spectrum of C_3 made it possible for Gausset, Herzberg, Lagerquist and Rosen[16] to analyze its optical spectrum and establish that its ground state is also $^1\Sigma$. Its spectrum is complicated by a low bending frequency ($\sim$60 cm^{-1}) in the ground state and Renner-Teller splittings in the upper $^1\Pi$ state.

C_4 was first detected mass spectrometrically, but there are not yet any observations of its gas phase spectra. It has been trapped in neon and argon matrices at 4 K[47] and a characteristic vibrational frequency has been assigned as 2170 cm^{-1} in the ground state.[48] Its ESR spectrum has also been observed in argon matrices at low temperatures. It is very likely that the spectrum was due to the C_4 molecule since the same ESR spectrum was obtained when carbon vapor was trapped as when diacetylene, C_4H_2, was mixed with the argon and photolyzed. (The vapor over graphite at about 2500 C contains only about 0.1% C_4, but the C, C_2, and C_3 species are very reactive, and it can be expected that the C_4 concentration in the matrices would be considerably higher.)

The ESR spectrum of $^{12}C_4$ and $^{13}C_4$ are those of a linear $^3\Sigma$ molecule with a zero-field splitting $|D| = 0.256$ cm^{-1}. The ^{13}C hyperfine splitting is small and shows clearly that the molecule is not an acetylenic diradical $\cdot C\equiv C-C\equiv C\cdot$ where sp hybridization would imply large hfs. The shift in the measured g tensor component indicates that there is a low-lying $^3\Pi$ state, approximately at 6000 cm^{-1}. Optical spectra of the trapped molecules yield a 0,0 transition at 19,600 cm^{-1} and a vibrational frequency in the upper state of 2100 cm^{-1}. At that time these matrix observations were in accord with approximate theory[49] indicating that C_n molecules could be expected to be linear; those with n odd would be singlet and with n even would be triplet, as found for C_3 and C_4. This was expected to prevail until the chains became long enough (n $\cong$ 10) where strain energy would not prevent a cyclic molecule to form.

However, beginning with Slanina and Zahradnik[50] and the more recent theoretical work of Whiteside, et al.,[51] the assumption of a linear structure for C_4 in the ground state has been more thoroughly examined. Those authors, and now several other groups performing _ab initio_

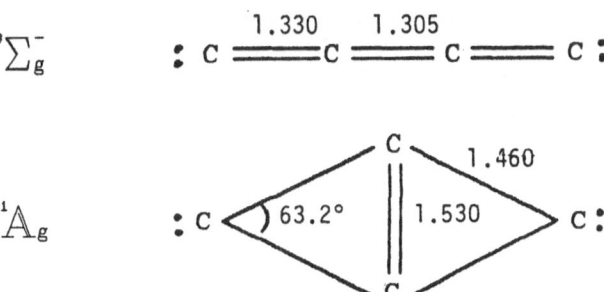

Fig. 2. Computed structures of C_4 (Magers, Harrison and Bartlett[55]).

calculations,[52-55] have found, as Fig. 2 indicates, that a planar rhombic singlet form of C_4 is slightly more stable than the linear triplet molecule. For example, Bartlett, et al.,[55] have calculated using MBPT theory that this energy difference is only about ~5 kcal/mole = ~2 K. This small ΔE is difficult to defend as completely certain even at this level of sophistication of theory, so that it is possible that the triplet state is lowest, but all recent ab initio results agree in placing the singlet lowest. The matrix work, however, admits only to the observation of the $^3\Sigma$ molecule, since observed positions of the excited electronic states are only in agreement with those calculated for the linear triplet molecule. Thus, there is no experimental evidence of the ^1A molecule in the matrix at these low temperatures. But since the rhombic molecule is non-magnetic and its detection only depends upon observation of its electronic transitions, perhaps a more thorough investigation via optical spectroscopy might lead to its discovery. (Note that variations in temperature in these matrices are limited to less than about 35 K.)

It is also possible that, with these methods of preparation (vaporization of graphite and photolysis of diacetylene), both molecules are simultaneously present in the matrices. Then since the activation energy required to convert one form into the other is probably high, the triplet excited molecule could have a very long lifetime in the matrix at 4 K. However, the singlet state should also be detectable optically, unless, improbably, only the triplet is indeed produced by these generation procedures.

These studies of C_4 suggest the possibility that the larger carbon molecules could be completely or partially cyclic, contrary to early theoretical expectations.[49] The vaporization of graphite produces many carbon species, some of which appear in matrix ESR spectra, but are not yet assigned. The infrared spectra of such matrices contain bands due to even more carbon molecules (including, of course, C_3) and an attempt has been made to assign them to specific molecules based on ^{13}C substitution and on the changes in the spectra with successive warm-ups, diffusion, and quenching.[48] This approach is fraught with difficulties.

Thus, the present situation is that although an extensive series of carbon clusters have been detected in supersonically-expanded beams[56] those larger than C_4 (up to C_{60}![57]) remain essentially uncharacterized.

ESR IN MATRICES

Perhaps it would be worthwhile to discuss briefly the accuracy of the data obtained from spectroscopic measurements of molecules trapped in solid matrices since they are sometimes questioned, particularly if there is disagreement with theory. Matrix-isolation in rare-gas solids has been in use now for more than 30 years, and a great deal of data has been obtained where direct comparison can be made with gas-phase measurements. The perturbations due to the matrix are generally too small to affect the interpretation and significance of the data, but understandably, they do vary with the matrix gas. Most notably the perturbations generally increase with increasing polarizability of the matrix atoms so that neon matrices are expected to, and usually do, yield the most "gas-like" data.[58] [Nitrogen matrices are generally more perturbing, presumably because of their more unsymmetrical sites and perhaps because of the added interaction with a trapped molecule due to the quadrupole moment of N_2]. Jacox has recently demonstrated this in her extensive comparison of vibrational frequencies of molecules trapped in various matrices and in the gas phase.[59] Also understandably, highly-ionic matrix-isolated molecules (such as LiF) can be expected to interact more strongly with the surrounding solid, causing larger shifts in the gas-phase vibrational frequencies.

In ESR similar effects are found: for example, hyperfine interaction constants and zero-field splittings (due to spin-spin and spin-orbit interactions) can be altered in the solid. Even for ionic species, however, a comparison shows that the magnetic data in solid neon are in excellent agreement with gas phase data. For example, from Childs, et al's gas data[60] for the CaF ($^2\Sigma$) molecule, one may derive $A_{\parallel}(^{19}F) = 149.302$ MHz and $A_{\perp}(^{19}F) = 109.183$ MHz. The values found via ESR in a neon matrix[61] (prior to the gas-phase work) are 149 and 106 MHz, respectively, where the uncertainty is ±3 MHz. Thus, the discrepancy is small and insignificant from the point of view of interpretation of spin population.

Comparisons of zero-field splitting parameters, $D(= b_2^0)$, are less abundant but a well-studied case is the CCO radical. The gas value is 0.772 cm^{-1} [62] and in matrices it is 0.739 (neon), 0.712 (argon) and 0.57 (krypton).[63] The theory of such shifts is rather complex but involves enhancement by the matrix of spin-orbit mixing in the trapped molecule.[63]

Of most concern to those not familiar with matrix-isolation is the possibility of an alteration in the energies of the lowest electronic states such that the ground state of the free molecule is different in the matrix. There is presently no evidence of any such flipping of states in any diatomic molecule trapped in a rare-gas matrix. Atoms are known to be more strongly perturbed in matrices by crystal-field and Jahn-Teller effects,[58] but there is only one case where a juxtaposition of states occurs in Ar, Kr, and Xe (but not in Ne) matrices.[64] In those matrices the $^3D_3$ state of Ni atom, lying 205 cm^{-1} above the $^3F_4$ ground state in the gas, has been found to be the lowest state.

In molecules there is, however, evidence of bending of a normally linear molecule in some matrices if its bending force constant is low, i.e., if it has a small bending vibrational frequency. For example, ESR evidence indicates that the molecule SiCO is slightly bent in argon matrices.[65] Similar bending of the MgOH molecule in argon matrices was also suggested;[66] this would be expected to be particularly easy for such highly ionic species. Therefore slight bending distortions in the confined sites in matrices would not seem to be extraordinary, and indeed, as was discussed above, the nonlinear triatomic Ag_3 and Cu_3 molecules appear to exhibit such an effect.

Perhaps the most pervasive problem in the observations in matrices is the identification of the particular molecule producing the signal. This is particularly true for metal clusters since there are always several cluster sizes present in the matrix at the same time. (Mass separation could in principle allow the trapping of a single cluster species, as discussed in this volume by Wöste.) It is evident that this problem can easily arise in the assignment of observed electronic absorption bands to particular clusters, although the "photoaggregation" technique may be of value here.[67] The beauty of ESR is that it can often identify the molecule being observed from the hyperfine structure in the spectra, as has been amply demonstrated in this review. Thus, as in infrared spectroscopy where isotopic substitution has been used so rewardingly by Milligan and Jacox and others, stable isotopes with known nuclear spins are of invaluable assistance in ESR studies in matrices.

CONCLUSIONS

Groups IA (Li, Na, K) and IB (Cu, Ag, Au)

The evidence indicates that the odd-atom clusters (S = 1/2) up to n = 7 are similar within the two series:

M_3 clusters can be expected to have C_{2v} symmetry in their lowest states in the gas phase and probably bond angles greater than 60^0 but with low bending force constants, i.e., flat bending potential energy surfaces. Unpaired spin is almost all on the two basal atoms and increases in p character in Li_3 and Cu_3. Pseudorotation barriers are low in the alkali-metal trimers, but such vibronic effects have not been observed for Group IB clusters in matrices.

M_5 clusters have not been observed for Group IA elements and Ag, Au, perhaps because of intrinsic broad, weak ESR spectra. Although Cu_5 may have been observed as a distorted trigonal bipyramid, confirmation is needed because it would be unique in the series.

M_7 clusters have a pentagonal bipyramid structure (D_{5h} symmetry) with almost all of the unpaired spin on the two axial atoms. These septamers have been detected for Li, Na, K, and Ag.

Transition Metals

M_2: Ten of a possible 45 first-row diatomics have been characterized, but only a few containing second- and third-row atoms. There is some evidence that their ground states may follow an isoelectronic principle.

M_3 clusters: Sc_3 has S = 1/2 and a true (non-fluxional) equilateral triangle structure, but Y_3 is bent and has a $^2B_2$ ground state. Evidence of pseudorotation was not observed in the latter case.

Mn_5 has S = 25/2 and is probably a near pentagonal ring. Small hfs implies that the unpaired spins are ferromagnetically-coupled d orbitals.

Sc_{13} may have been observed as an S = 1/2 molecule. Hfs is observed corresponding to at least 9 equivalent atoms. The unpaired electron is delocalized in an orbital of largely d character.

Carbon

Other than C_2 and C_3, the only observed small cluster is C_4. ESR at 4 K indicates that its ground state is $^3\Sigma$, and its magnetic and optical parameters have been determined. However, ab initio theory presently suggests that a ^1A_g rhombic structure is slightly lower in energy.

REFERENCES

1. Y. M. Efremov, A. N. Samoilova, V. B. Kozhubhovsky and L. V. Gurvich, J. Mol. Spectr. 73:430 (1978).
2. D. L. Michalopoulos, M. E. Geusic, S. G. Hansen, D. E. Powers, R. E. Smalley, J. Phys. Chem. 86:3914 (1982), V. E. Bondybey, and J. H. English, Chem. Phys. Let. 94:443 (1983); P. R. R. Langridge-Smith, M. D. Morse, G. P. Hansen, R. E. Smalley, A. J. Merer, J. Chem. Phys. 80:593 (1984).
3. Even for cold Cu_3 the analysis is difficult: M. D. Morse, J. B. Hopkins, P. R. R. Langridge-Smith, R. E. Smalley, J. Chem. Phys. 79:5316 (1983); T. C. Thompson, D. G. Truhlar, and C. A. Mead, J. Chem. Phys. 82:2392 (1985); E. A. Rohlfing, and J. J. Valentini, Chem. Phys. Lett. 126:113 (1986); J. Chem. Phys. 84:6560 (1986).
4. R. F. Curl, Mol. Phys. 9:585 (1965).
5. H. E. Radford, Phys. Rev. 122:114 (1961); ibid 136A:15 (1964); J.

Chem. Phys. 40:2732 (1964); A. Carrington, D. H. Levy, and T. A. Miller, Adv. Chem. Phys. 18:149 (1970).

6. D. M. Lindsay, D. R. Herschbach, and A. L. Kwiram, Mol. Phys. 32:1199 (1976) and 39:529 (1980).

7. D. M. Lindsay and G. A. Thompson, J. Chem. Phys. 77:1114 (1982).

8. G. A. Thompson and D. M. Lindsay, J. Chem. Phys. 74:959 (1981).

9. G. A. Thompson, F. Tischler, D. Garland, and D. M. Lindsay, Surf. Sci. 106:408 (1981).

10. D. M. Lindsay, D. Garland, F. Tischler, and G. A. Thompson, Am. Chem. Soc. Symp. Ser. (J. L. Gole and W. C. Stwalley, Editors) No. 179, 69 (1982).

11. R. L. Martin and E. R. Davidson, Mol. Phys. 35:1713 (1978).

12. T. C. Thompson, G. Izmirlian, Jr., S. J. Lemon, D. G. Truhlar, and C. A. Mead, J. Chem. Phys. 82:5597 (1985).

13. P. H. Kasai and D. McLeod, Jr., J. Chem. Phys. 55:1566 (1971).

14. D. A. Garland and D. M. Lindsay, J. Chem. Phys. 78:2813 (1983).

15. J. A. Howard, R. Sutcliffe, and B. Mile, Chem. Phys. Lett. 112:84 (1984).

16. J. Kendrick and J. H. Hillier, Mol. Phys. 33:635 (1977); P. S. Bagus, G. del Conde, and D. W. Davies, J. Chem. Soc. Faraday Trans. II, 62:321 (1977)

17. W. H. Gerber and E. Schumacher, J. Chem. Phys. 69:1692 (1978); E. Schumacher, W. H. Gerber, H. P. Harri, M. Hoffmann, and E. Scholl, Am. Chem. Soc. Symp. Series 179:83 (1982).

18. D. A. Garland and D. M. Lindsay, J. Chem. Phys. 80:4761 (1984).

19. G. A. Thompson, F. Tischler, and D. M. Lindsay, J. Chem. Phys. 78:5946 (1983).

20. P. Fantucci, J. Koutecky, and G. Pacchioni, J. Chem. Phys. 80:325 (1984).

21. J. J. Burton, Catal. Rev. 9:209 (1974).

22. J. A. Howard, R. Sutcliffe, and B. Mile, Surf. Sci. 156:214 (1985).

23. See, for example, Ag atom reactions with hydrocarbons: P. H. Kasai and D. McLeod, Jr., J. Am. Chem. Soc. 97:6602 (1975); 100:625 (1978); P. H. Kasai, D. McLeod, Jr. and T. Watanabe, J. Am. Chem. Soc. 102:179 (1980).

24. J. A. Howard, K. F. Preston, and R. Sutcliffe, J. Phys. Chem. 87:536 (1983).

25. J. A. Howard, K. F. Preston, and B. Mile, J. Am. Chem. Soc. 103:6226 (1981).

26. J. A. Howard, R. Sutcliffe, and B. Mile, J. Chem. Soc., Chem. Comm. 1449 (1983).

27. J. A. Howard, R. Sutcliffe, and B. Mile, J. Am. Chem. Soc. 105:1394 (1983).

28. In recent work in our laboratory the ESR spectrum of Ag_3 in a neon matrix has been observed.

29. D. M. Lindsay, G. A. Thompson, and Y. Wang (to be published).

30. K. Kernisant, G. A. Thompson, and D. M. Lindsay, J. Chem. Phys. 82:4739 (1985).

31. T. C. Thompson, D. G. Truhlar, and C. A. Mead, J. Chem. Phys. 82:2392 (1985).

32. Matrix effects associated with low bending force constants have been observed in the ESR spectra of the SiCO molecule [Lembke, Ferrante, and Weltner, J. Am. Chem. Soc. 99:416 (1977)] and the VCO molecule [Van Zee, Bach, and Weltner, J. Phys. Chem. 90:583 (1986)] and references cited there. See the section in this review on ESR in Matrices.

33. J. A. Howard, R. Sutcliffe, J. S. Tse, and B. Mile, Chem. Phys. Lett. 94:561 (1983).

34. J. A. Howard, R. Sutcliffe, and B. Mile, J. Phys. Chem. 87:2268 (1983).

35. J. A. Howard, R. Sutcliffe and B. Mile, J. Phys. Chem. 88:2183 (1984).

36. Tentative parameters for $^{107}Ag_7$ in a neon matrix are: $g_\parallel = 2.094$, $|A_\parallel|_2 = 204$ G, $|A|_5 = 7.4$ G, as compared to $g_\perp = 2.085$, $|A_\perp|_2 = 201$ G, $|A|_5 = 5.5$ G in the C_6D_{12} matrix spectrum[134].

37. W. Weltner, Jr. and R. J. Van Zee, Ann. Rev. Phys. Chem. 35:291 (1984).

38. W. Weltner, Jr. and R. J. Van Zee in "Comparison of Ab Initio Quantum Chemistry with Experiment for Small Molecules", R. J. Bartlett, Ed., Reidel, Dordrecht (1985) pages 1-16.

39. Y. M. Efremov, A. N. Samoilova, V. B. Kozhubhovsky, L. V. Gurvich, J. Mol. Spectr. 73:430 (1978); D. L. Michalopoulos, M. E. Geusic, S. G. Hansen, D. E. Powers, R. E. Smalley, J. Phys. Chem. 86:3914 (1982); S. J. Riley, E. K. Parks, L. G. Pobo, S. Wexler, J. Chem. Phys. 79:2577 (1983); V. E. Bondybey and J. H. English, Chem. Phys. Lett. 94:443 (1983).

40. L. B. Knight, Jr., R. W. Woodward, R. J. Van Zee, W. Weltner, Jr., J. Chem. Phys. 79:5820 (1983).

41. S. P. Walch and C. W. Bauschlicher, Jr., J. Chem. Phys. 83:5735 (1985).

42. C. A. Baumann, R. J. Van Zee, S. V. Bhat, and W. Weltner, Jr., J. Chem. Phys. 78:190 (1983).

43. J. Frenkel and J. Dorfman, Nature (London) 126:274 (1930); E. C. Stoner, Phil. Trans. Roy. Soc. London Ser. A 235:165 (1936); J. J. Becker, Trans. AIME 209:59 (1957); C. P. Bean and J. D. Livingston, J. Appl. Phys. 30:1265 (1959).

44. D. R. Salahub, "Impact of Cluster Physics in Material Science and Technology", J. Davenas, Ed., The Hague, Nijhoff, (1983).

45. E. A. Ballik and D. A. Ramsay, J. Chem. Phys. 31:1128 (1959).

46. L. Gausset, G. Herzberg, A. Lagerqvist and B. Rosen, Discuss. Faraday Soc. 35:113 (1963).

47. W. R. M. Graham, K. I. Dismuke, and W. Weltner, Jr., Astrophys. J. 204:301-10 (1976).

48. K. R. Thompson, R. L. DeKock, and W. Weltner, Jr., J. Am. Chem. Soc. 93:4688 (1971).

49. K. S. Pitzer and E. Clementi, J. Am. Chem. Soc. 81:4477 (1959); E. Clementi, J. Am. Chem. Soc. 83:4501 (1961); S. J. Strickler and K. S. Pitzer, "Molecular Orbitals in Chemistry, Physics and Biology", B. Pullman and P.-O. Löwdin, Eds., Academic, NY (1964); R. Hoffman, Tetrahedron 22:521 (1966).

50. Z. Slanina and R. Zahradnik, J. Phys. Chem. 81:2252 (1977).

51. R. A. Whiteside, R. Krishnan, D. J. DeFrees, and J. A. Pople, Chem. Phys. Lett. 78:538 (1981).

52. Z. Z. Wang, R. N. Diffenderfer, and I. Shavitt (to be published).

53. J. P. Ritchie, H. F. King, and W. S. Young, Proc. NATO/ASI in "Fast Reaction Kinetics", Aug. 25 to Sept. 7, (1985), Crete, Greece.

54. J. Koutecky, G. Parchioni, G.-H. Jeung, and E.-C. Hass, Surf. Sci. 156:650 (1985).

55. D. H. Magers, R. J. Harrison, and R. J. Bartlett, J. Chem. Phys. 84:3284 (1986).

56. E. A. Rohlfing, D. M. Cox and A. Kaldor, J. Chem. Phys. 81:3322 (1984).

57. H. W. Kroto, J. R. Heath, S. C. O'Brien, R. F. Curl, R. E. Smalley, Nature (London) 318:162 (1985).

58. W. Weltner, Jr., "Magnetic Atoms and Molecules", Van Nostrand Reinhold, New York, (1983).

58a. M. Vala, K. Zeringue, J. ShakhsEmampour, J.-C. Rivoal, and R. Pyzalski, J. Chem. Phys. 80:2401 (1984); C. Grisolia, J. C. Rivoal, J. Pyka, and M. Vala, J. Chem. Phys. (submitted).

59. M. E. Jacox, J. Mol. Spectr. 113:286 (1985).

60. W. J. Childs, G. L. Goodman and L. S. Goodman, J. Mol. Spectr. 86:365 (1981).

61. L. B. Knight, Jr., W. C. Easley, and W. Weltner, Jr., J. Chem. Phys. 54:322 (1971).

62. C. Devillers and D. A. Ramsay, Can. J. Phys. 49:2839 (1971).

63. G. R. Smith and W. Weltner, Jr., J. Chem. Phys. 62:4592 (1975).

64. C. P. Barrett, R. G. Graham and R. Grinter, Chem. Phys. 86:199 (1984); W. Schrittenlacher, W. Schroeder, H. H. Rotermund, and D. M. Kolb, Chem. Phys. Lett. 109:7 (1984).

65. R. R. Lembke, R. F. Ferrante, and W. Weltner, Jr., J. Am. Chem. Soc. 99:416 (1977).

66. J. M. Brom, Jr. and W. Weltner, Jr., J. Chem. Phys. 58:5322 (1973).

67. H. Huber and G. A. Ozin, Inorg. Chem. 17:155 (1978); W. Klotzbücher and G. A. Ozin, J. Am. Chem. Soc. 100:2262 (1978), J. Mol. Catalysis 3:195 (1977).

STRUCTURE OF HOMO- AND HETERO-NUCLEAR MICRO-CLUSTERS

B. K. Rao, S. N. Khanna, and P. Jena

Virginia Commonwealth University
Richmond, VA 23284

ABSTRACT

Using ab initio SCF-LCAO-MO theory, studies have been made on small clusters of lithium, beryllium, magnesium, and carbon atoms. These studies include the investigation of the equilibrium geometries, relative stabilities, and the evolution of various structural and electronic properties. The study of the electronic properties shows that the geometries of the clusters have a direct relationship with their spin configurations. The effects of foreign atoms on the properties of clusters have been investigated using clusters of Li_NNa, Li_NMg, and Li_NAl. The effect of ionization on cluster stability and the dissociation channels of multiply ionized clusters are also discussed.

INTRODUCTION

Although the first discovery[1] of cluster formation in supersonic expansions was made almost thirty years ago, only recently[2] experimental techniques have advanced enough to enable researchers to produce and characterize small atomic clusters ranging from two to a few hundred atoms per cluster. A large variety of clusters consisting of rare gas,[3-5] metal,[6-9] semi-conductor,[10-12] and binary mixtures of atoms[13,14] have been investigated by now. The interest in small clusters is sparked both by challenging problems the field poses at a fundamental level and by its potential for industrial applications.

At a fundamental level, the study of small clusters is aimed at understanding the evolution of both structural and electronic properties of the solid state starting from the atoms and molecules. Among these are studies of equilibrium structures, nucleation, abundances, crystal growth, electronic structure, magnetism, probability of formation (magic numbers), chemisorption, and reactivity as a function of cluster size. Since the surface to volume ration of clusters can be varied with cluster size, the evolution of surface properties can also be observed and understood.

The potential for industrial applications arises mainly due to the fact that many of the properties of clusters are size-specific. The properties do not change monotonically with the increase of cluster size. The rate of change also depends upon the property itself. Because of this, the behavior of the clusters usually differs from that of the bulk material. This suggests the possibility of manufacturing "cluster materials" with specific tailored characteristics.

The observation of magic numbers (strong peaks in the intensities of mass spectra) for clusters having 8, 20, 40, ... atoms in alkali atom systems[6,8] has already generated a lot of excitement in the cluster community. Photofragmentation studies on semiconductor clusters[10] also exhibit similar behavior. In this case, clusters with specific sizes are found to be very stable while the others fragment easily with "magic" clusters as one of the predominant byproducts. In compound clusters (clusters consisting of hetero-nuclear atoms) the magic numbers[13] differ depending on the constituent atoms. Electronic properties associated with chemisorption of H_2, N_2, and CO on transition metal clusters[15-17] have been observed to depend dramatically on the cluster size.

Once the cluster size begins to grow, they can exist in different geometrical arrangements[18-20] (structural isomers) even though the same number of atoms are involved. The structural isomers sometimes exhibit different properties. For example, the ground state spin configuration of an alkali atom tetramer is a singlet and when it is forced to assume a three-dimensional structure, it becomes a spin triplet.[21] The sensitivity of some cluster properties to the geometry indicates that a theoretical understanding of small clusters has to proceed from a fairly precise determination of their equilibrium geometries. One must start with a single atom and add one extra atom at a time to see how the geometries and the properties change.

Theories for studying molecular clusters have been around for a long time. While new methods are being developed the old methods are also being improved. To cite a few examples, one can mention the works of Koutecky,[20] Ellis,[22] Bagus and Bauschlicher,[23] Salahub,[24] Johnson,[25] and Pople.[26] Because of the exhaustive amount of literature available, the theoretical procedure used in the present work is presented very briefly in the next section. In the following sections, results from the study of Li, Be, Na, C, Be, Mg, Al clusters and some binary mixtures of these are presented. The studies are concerned with the investigation of the equilibrium geometries, energies, and electronic properties of homo-nuclear and hetero-nuclear clusters. The effect of ionizing the clusters is also discussed.

THEORETICAL TECHNIQUE

The conventional approaches for studying molecular clusters can either be semi-empirical or ab initio in nature. In the present case, ab initio self-consistent field (SCF) molecular orbital (MO) method[26] has been used to calculate the ground state energy of the clusters in a given configuration. The cluster wave function is a Slater determinant formed out of molecular spin orbitals which are, in turn, expressed as a linear combination of atomic orbitals. Thus, the MO is written down as

$$\psi_i^\sigma = \sum_\mu C_{\mu i}^\sigma \phi_\mu \tag{1}$$

where σ denotes the spin state of the MO, ψ_i and $C_{\mu i}$ are combination coefficients for the atomic orbitals ϕ_μ. One obtains the combination coefficients $C_{\mu i}$ by solving the SCF equation

$$[-\tfrac{1}{2} \nabla_i^2 + V_i] \psi_i^\sigma = \varepsilon_i^\sigma \psi_i^\sigma \tag{2}$$

using a variational technique. In Eq. (2) V_i is the potential experienced by the ith electron and ε_i^σ is the single particle eigenvalue. V_i has the following components:

$$V_i = V_{es} + V_{ex} + V_{corr} \quad . \tag{3}$$

V_{es}, V_{ex}, and V_{corr} arise from the electrostatic, exchange, and the correlation effects respectively. Since V_i is basically a many-electron potential, the solution of eq. (2) is not a trivial exercise. In actual calculations one uses various types of approximations for V_{ex} and V_{corr}. In our calculations, V_{ex} has been treated by the unrestricted Hartree-Fock (UHF) method.[26] V_{corr} has been treated either by Möller-Plesset[27] type perturbation procedure or by configuration interaction (CI)[28] method. In the density functional approach, one treats the exchange and correlation together as a functional of the local density.[24,25]

In all these cases, the MO is being represented as a linear combination of atomic orbitals. Therefore, the results may depend on the choice of the atomic basis set and this problem should be carefully addressed. For our work, we have chosen to represent the atomic orbitals as linear combinations of gaussian functions. This facilitates analytic determination of all the integrals and removes the inaccuracy normally associated with numerical basis. In the discrete variational method (DVM)[22] the hamiltonian matrix elements are computed as discrete sums and this eliminates any errors arising due to fitting of a wavefunction by a sum of primitive gaussian functions. However, the energy integrals have to be computed numerically and this is dependent upon the choice of grid points. Thus, no matter what procedure one chooses, one must face a certain amount of inexactness. One eliminates this as much as possible through a suitable choice of the atomic basis set.

All the above prescribe the technique used to obtain the ground state wavefunction and the energy of a particular configuration of atoms in a cluster. However, since one is interested in the determination of the global equilibrium geometry of a cluster, and there is a possibility of the clusters to existing in one of the many nearly equal energy states with different geometries, one must determine the ultimate ground state geometry by optimizing the total energy of the cluster with respect to all possible bond lengths and angles. We begin with the constituent atoms placed at arbitrary positions and calculate the total energy self-consistently as mentioned. Then the force at every atomic site is calculated by the gradient technique. The atoms are moved along the direction of the forces and the total energy is again calculated. These steps are repeated till the forces at the atomic sites vanish. Since it is possible to reach a local minimum in the energy surface, one must be careful to search for the global minimum. Therefore, after a minimum is reached, we restart the complete procedure with another random distribution of the atoms. When several initial choices of distributions lead to the same "minimum" in energy, one can "safely" assume that the actual equilibrium geometry has been obtained.

While looking for the equilibrium geometry, one must not forget the fact that the spin distribution in the cluster also affects[21] the final arrangement of the atoms in the cluster. Therefore, the above procedures are repeated with different spin configurations to obtain the lowest energy.

With this geometry, one then proceeds to look at various properties of the cluster.

NEUTRAL HOMO-NUCLEAR CLUSTERS

In this section, we present the results from our UHF-CI study of neutral clusters of Li, Be, Mg, and C atoms in terms of their equilibrium geometries, relative stabilities, evolution of different properties, and magnetism. For lithium atoms, we have used STO-6G basis set. For C, we have used the split valence 6-31G functions. For Be, an extensive 6-311G** basis has been used. For Mg the STO-6G* basis sets have been employed. For the CI procedure, all the single and double excitations have been used for the clusters of Li, Be, and C atoms. For the Mg clusters, only the excitations of the valence electrons have been considered.

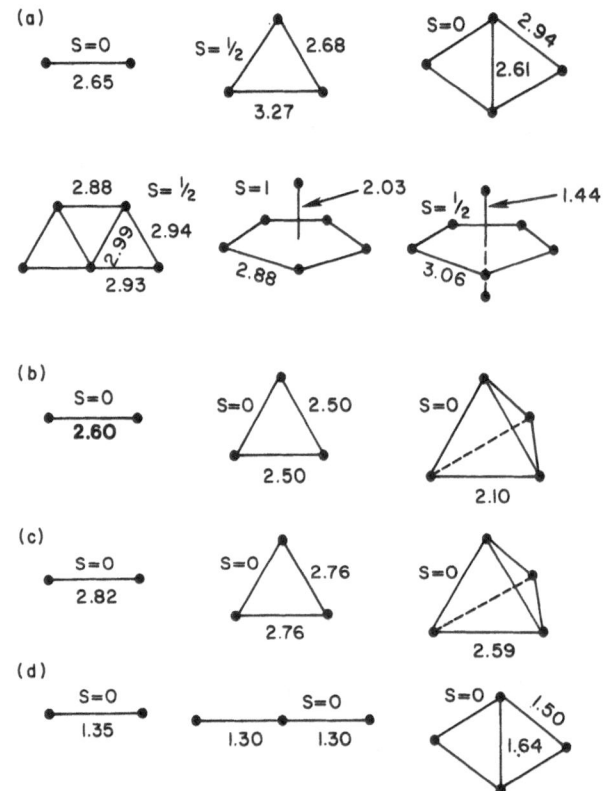

Fig. 1   Optimized geometries and spin of neutral
(a) Li, (b) Be, (c) Mg, and (d) C
clusters. The bond lengths are in Å.

The optimum geometries for neutral clusters of Li, Be, Mg, and C atoms are presented in Fig. 1 along with their spin configurtions. Clusters of Li_N (up to N<5) are planar as is observed for Na_N clusters in local density calculations.[19] This is not what would be expected from a jellium-type calculation.[29] However, this can be explained easily in a simple bonding-antibonding type picture[30] using only the valence electrons. This also predicts that Be and Mg being closed shell atoms, Be_3 and Mg_3 would be equilateral traingles as observed in our ab initio results. Be_4 and Mg_4 are expected to be tetrahedral in shape which is also borne out (see Fig.1). These arguments predicting the symmetries of clusters also match with the results of Al clusters from local density calculations.[31] The optimum geometries of the carbon clusters (for N≤4) agree with the results of Raghavachari[32] who has also found that small carbon clusters have linear geometry for odd values of N and cyclic structures when N is even.

The magic numbers guiding the stabilities of clusters have been explained[8] in terms of filling of electronic shells. For example, in the case of alkali atom clusters, one can assume that the effective potential experienced by the valence electrons due to the positive ion cores is spherically symmetric. For such a potential, orbital angular momentum is a good quantum number and the energy levels of the electrons in the cluster can be characterized by their s, p, d, ... characters. When one of the shells gets filled, extra energy is needed to raise the next electron to the higher level. This produces a jump in the ground state energy of the cluster and the cluster is less stable than the previous smaller cluster. This, obviously explains the magic numbers in the case of alkali atom clusters (i.e. 2, 8, 20 ... etc.). However, there are additional features in the same mass spectra that cannot be successfully explained by the simplistic shell model above. One such feature is the odd-even alternation in the stability of the clusters. Even numbered neutral alkali atom clusters have been observed[6,8] to be more stable than the odd-numbered ones. Similar odd-even alternation has also been observed in noble metal clusters.

The odd-even alternation in the mass spectra can be studied in more detail when one considers the energetics of formation of the clusters. Let us define the energy needed to add an extra atom to an existing cluster of N-1 atoms to produce a cluster of N atoms as

$$\Delta E_N = E_N - E_{N-1} - E_1 \tag{4}$$

where E_N denotes the ground state energy of an N-atom cluster. In a plot of N against ΔE_N, the dips would indicate that for the particular value of N, the formation of the cluster is more preferred than the adjacent values of N. In Fig. 2 such a plot is given for Li_N and Na_N clusters which show the odd-even alternation nicely. Here the results for Na_N clusters are taken from the local density work of Martins et al.[19]

The evolution of bulk materials starting from atoms must go through the intermediate step of cluster formation. This can be observed by examining the variation of the nearest neighbor distance of atoms as the size of the cluster changes. We have calculated the root-mean square distance, <R>,

$$< R > = [\ \Sigma_{i<j}^{N} \ R_{ij}^2 \ / \ N \ ]^{\frac{1}{2}} \tag{5}$$

where R_{ij} is the distance between two nearest neighbor atoms. The results, normalized to bulk value, are plotted in Fig. 3. It is noticeable that for clusters with only 7 atoms, <R> is within 2% of the bulk value.

The evolution of the electronic structure can be observed through the study of the changes in the electron density distribution, binding energy per atom, band gap, etc. Using the lithium clusters as an example again, the evolution of these properties has been observed. Defining band gap as the energy difference between the lowest unoccupied MO and highest occupied MO, one obtains this gap as 7.7 eV for a lithium atom while in the case of Li_7 it decreases to 5.18 eV. The binding energy per atom is approximately 0.9 eV for the 7-atom cluster as compared with 1.58 eV for bulk lithium. These show that even though the structural parameters seem to be close to the bulk value, in a 7-atom cluster the electronic properties are no where close to that of the extended solid. As a final example, the valence electron density in the interstitial region of Li_7 has been obtained to be $0.0068a_o^{-3}$ which compares favorably with interstitial density ($0.0071la_o^{-3}$) of bulk Li metal. This indicates the delocalised nature of the electrons.

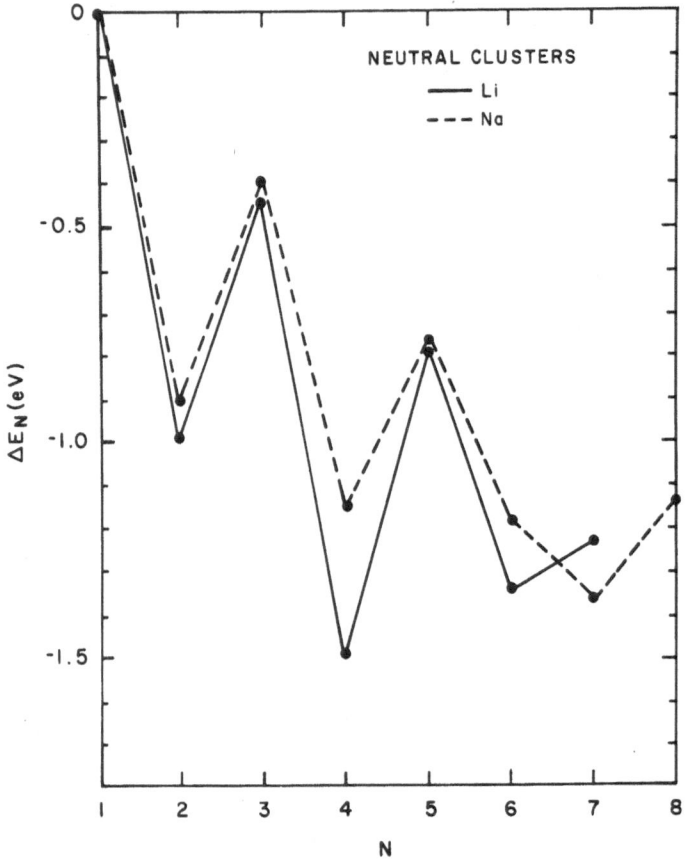

Fig. 2 Plot of ΔE_N versus N for Na_N and Li_N clusters.

As already discussed, one must remember to optimize the spin configuration while optimizing the geometries of a cluster. When a cluster is highly symmetric in structure, the energy levels are usually degenerate. In that case, due to Hund's rule, the spins try to be parallel and occupy the different degenerate states, thus producing a configuration with high spin multiplicity. When this degeneracy is lifted due to Jahn-Teller distortion of the geometry, the electrons choose the lower levels and pair up to produce a low spin configuration. These two mechanisms guide the geometrical and spin structures of small clusters.

The competition of these two mechanisms can be illustrated in calculations with a model cluster. For this, we started with a Li_4 cluster in a planar square configuration. As this is a symmetric ase, Hund's rule is expected to be predominant and the spin triplet (the two outermost valence electrons being with parallel spins) should have lower energy than the spin singlet. This was verified to be the case by actual calculation where we optimized the bond length (keeping the square structure intact) for both spin configurations. To verify whether the effect of Jahn-Teller distortion could overtake that of the Hund's rule, the Li_4 cluster was distorted from the square geometry by only changing the apex angle of the

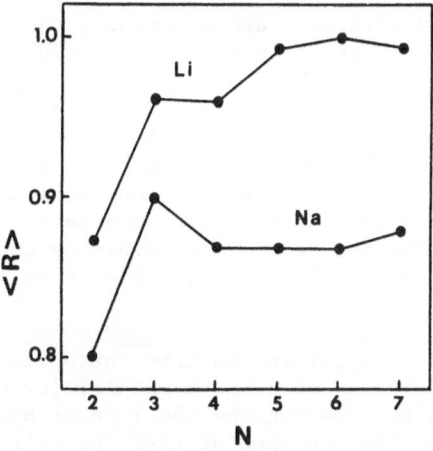

Fig.3 Root mean square nearest neighbor distance
for Li_N and Na_N clusters. These have been
normalized to their bulk values.

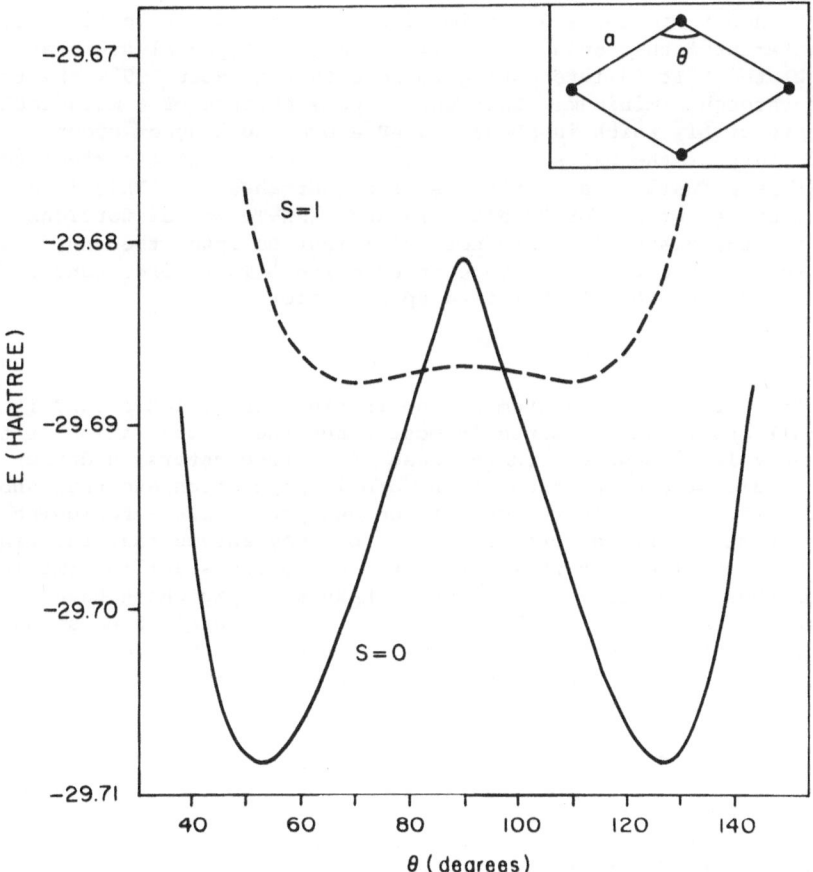

Fig. 4 Variation of energy as a function of apex
angle of the Li_4 rhombus geometry.

square to convert it into a rhombus and optimizing the bond length again. Energies for the singlet and the triplet cases were calculated for different values of the apex angle. These have been shown in a plot in Fig. 4. When the apex angle decreases from 90°, the energy gap between the singlet and triplet decreases and at about 82° the singlet begins to be the preferred state. It is interesting to notice that when the angle reaches approximately 52°, it goes through the deepest minimum in the singlet configuration. This is the optimum geometry[18] for Li_4. In this case, obviously, the Jahn-Teller distortion has split the degenerate levels by such an amount that filling the lower state by two electrons (with spin up and down) is energetically more favorable.

One can also extend this discussion into three dimensions since Li_4 cluster is the smallest cluster that has a possibility of existing in a three dimensional structure. To examine the role of Hund's rule coupling and Jahn-Teller distortion in the case of Li_4, the following comparison was done. Starting with the planar rhombus configuration, one of the atoms was slowly lifted off the plane and the energies were calculated for several values of the dihedral angle. For each value of dihedral angle, the length of the arms of the rhombus and the bond angles were optimized to get the most relaxed form. This calculation was done both for S=0 and S=1 states. The results are shown in Fig. 5. As expected, S=0 planar rhombus state has the lowest energy. However, when S=1 (i.e. when Hund's rule is forced) the cluster has much higher energy when it is planar (i.e. the dihedral angle is 180°). When the dihedral angle is decreased, it finally crosses the singlet curve and becomes the lower energy state. It is needless to mention that now it is three dimensional in geometry which confirms our earlier statement[21] that spin and atomic arrangements in clusters are intimately related. It is interesting to note that at about 90°, the triplet curve goes through a minimum. This shows the existence of a metastable triplet state of Li_4 which is about 0.4 eV above the singlet ground state. The ground state of the S=1 state of Li_4 is, however, not a perfect tetrahedron. It is actually a slightly distorted tetrahedron. This is an example of simultaneous existence of Hund's rule and Jahn-Teller distortions. In this case, the distortions are not sufficient to lower the energies enough as to switch off the exchange interaction. Therefore, Hund's rule still holds good and the cluster is a spin triplet.

NEUTRAL HETERO-NUCLEAR CLUSTERS

The effect of a foreign atom on the lattice distortion around it is usually small and localized since in most cases the concentration of the defect is very low. However, in the case of micro-clusters, a defect atom represents a significant portion of the atomic population and thus should have major influence on cluster geometries and properties. Following the reasonings of the electronic shell model,[8] one may assume that the final number of valence electrons will actually dictate the stability (magic numbers) of these clusters. For example, K_6Zn and K_6Mg which have 8 valence electrons each, should be "magic" clusters. Experiments[13] have shown that, on the contrary, K_8Mg and K_8Zn are the magic numbers. This may suggest that Mg or Zn do not interact with the potassium atoms because their valence electron energies are not close.

To test this and to examine the details of the effects of a "foreign" atom, optimized geometries for Li_2Na, Li_2Mg, and Li_2Al were obtained. Obviously Na, Mg, and Al were chosen as the "defect" atoms because of the different number of valence electrons available in each case. As before, STO-6G basis was used for the lithium atoms. For Mg and Al, STO-6G* wave functions were used. In this section, all the CI calculations include only the excitations of the valence electrons.

376

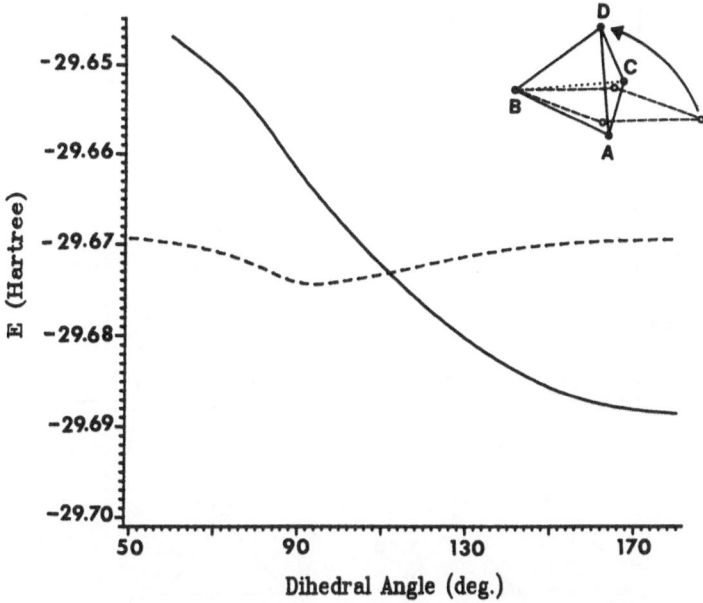

Fig. 5 Energy of Li$_4$ clusters as a function of the
dihedral angle. The inset shows the geometries.
Soild and dashed curves stand for S=0 and S=1
respectively.

As shown in Fig. 6(a) only the valence states of Li, Na and Al are
reasonably close. Therefore, on this ground, only Li$_2$Na and Li$_2$Al should
show geometries which are similar. Li$_2$Mg must show a geometry that would
be different from these two. Fig. 6(b) immediately disproves this. It
is clear that the number of valence electrons seems to play a major role
here. As Na has the same number of valence electrons, Li$_3$ and Li$_2$Na geometries are almost difficult to distinguish from each other. Mg and Al
having 2 and 3 valence electrons respectively, Li$_2$Mg and Li$_2$Al geometries
are completely different (see Fig. 6(b)) from that of Li$_3$ or Li$_2$Na. Three
questions now arise. Do the defect atoms interact very strongly with the
host atom ? Do they change the geometry ? Do they affect the stability?

To answer the second question first, Fig. 7 shows the comparison of
the equilibrium geometries of Li$_N$ clusters and Li$_{N-1}$ Mg clusters (obtained
by replacing one Li atom with Mg). It is obvious that in the case of Li$_{N-1}$Mg
clusters the Mg atom always tries to be in the middle of the cluster and
wants to be bonded with as many Li atoms as possible. This, in fact, increases the symmetry of the clusters in most cases. Upon examination of
the molecular orbitals, it is noticed that the interaction of the Mg atom
with the Li atoms is rather strong. A similar situation is also observed
when one investigates the case of Li$_N$Al clusters.[33] While testing the
stability of the Li$_N$Mg clusters using Eq. (4), one again obtains the odd-even
alternation[33] very similar to that of Fig. 2 for Na$_N$ and Li$_N$ clusters.
Therefore, one can conclude that in small clusters the electronic structure
of constituent atoms plays a decisive role in determining the geometries
and stabilities.

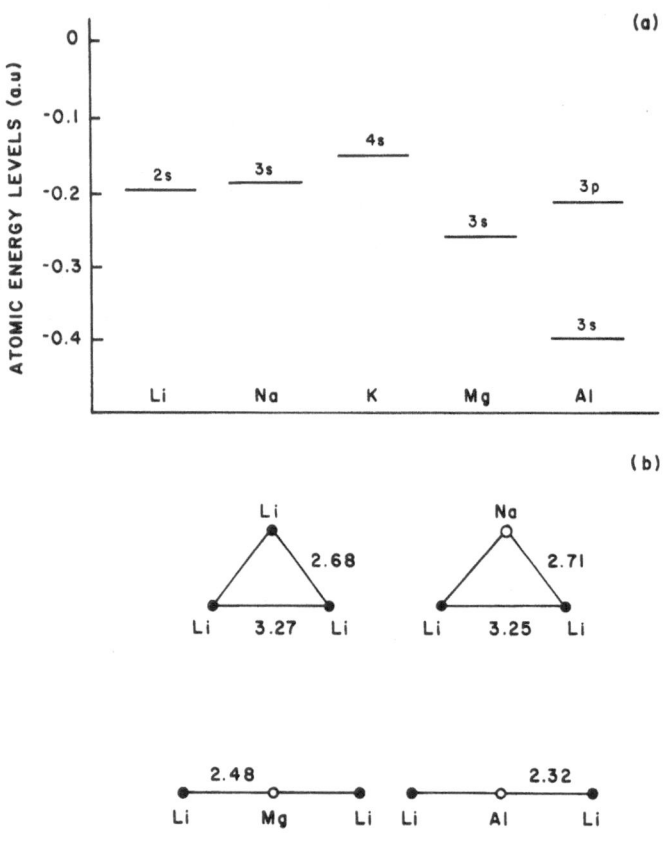

Fig.6 (a) One electron energy levels of Li, Na, K, Mg, and Al atoms.

(b) Equilibrium geometries of Li_2 interacting with Li, Na, Mg, and Al .

IONIZED CLUSTERS

Study of clusters using mass spectroscopy requires that the clusters be ionized. This is achieved by subjecting them to ionizing radiation. Depending upon the stability of the clusters, some of them may undergo fragmentation during this process. Therefore, the issue of magic numbers in the intensity of mass spectra gets uncertain. Moreover, other techniques like laser vaporization of materials[10] to produce clusters also can produce pre-ionized clusters which get sufficient time to relax during the flight. One must, therefore, examine the question of adiabatic ionization versus vertical ionization, ultimate equilibrium geometries, and stabilities of ionized clusters.

To address these questions, calculations have been performed on Li_N^+ and Li_N^- clusters with and without relaxation from the corresponding neutral geometries. The difference between vertical and adiabatic ionization potentials have been observed to be as much as 0.4 eV(for Li_5^+). However, the most interesting observation can be made from Fig. 8 when one compares the optimized geometries of Li_N^+ and Li_N^- with that of Li_N and $Li_{N-1}Mg$ respectively. It is easy to realize that in each case of Li_N^+, the removal of an electron has opened up the geometry a bit; as compared to Li_N geometries given in Fig. 1. This is to be expected because of electrostatic reasons. The minor changes of symmetry can be explained using the bonding-antibonding arguments,[30] as already mentioned. Curiously, the geometries of Li_N^- and Li_{N-1} Mg are very much similar in symmetry. One can point out here that both the clusters have the same number of valence electrons. As compared with the case of Li_2Na, it is clear that the number of valence electrons seems to guide the equilibrium geometry.

To examine the stability of ionized clusters, one can define the energy gain from dissociation as

$$\Delta E_{NM}^+ = E^+(N) - E^+(N-M) - E(M) \ , \quad N > M \geq 1 \tag{6}$$

where the numbers in the parentheses denote the number of atoms in that particular cluster. Depending upon the value of N and M, one can get the fragmentation to proceed in many different channels. When M=1, this is the same quantity as defined in Eq. (4) except that the present case defines the gain of energy in adding a neutral atom to an existing singly charged cluster. For Li_N^+ and Na_N^+ atoms, such plots have been made[34] and these show dips at 3, 5, 7 ... instead of 2, 4, 6 ... etc. of Fig. 2. Obviously, in pre-ionized clusters with a single positive charge, the odd-numbered clusters (with even number of valence electrons) are more stable. Calculation of the ΔE_{NM}^+ for different values of N and M for Li_N^+ and Na_N^+ clusters gives us the best possible channel for fragmentation. In each case, one of the fragmentation products seems to be the Li_3^+ (or the Na_3^+) cluster. This shows the stability of Li_3^+ clusters in the fragmentation of pre-ionized clusters. On the other hand, the study of fragmentation channels of neutral Li_N and Na_N yield the best values of ΔE_{NM} (same as Eq. (6), but without the charge) when one of the clusters is a dimer. Thus, the study of fragmentation products for determining magic numbers[10] shows that for neutral clusters or pre-ionized clusters, the stable species differ.

As already mentioned, all singly ionized clusters do not necessarily fragment. In Eq. (6), when ΔE_{NM}^+ is negative, it means that the cluster E_N^+ needs that much energy to go through dissociation in that channel. For all clusters of Li_N^+ and Na_N^+ studied, ΔE_{NM}^+ has always been negative.[34] The value of ΔE_{NM}^+ has only demonstrated the relative stabilities of these clusters. However, when the clusters get multiply ionized, due to Coulomb repulsion, this energy can become positive. This would mean that the cluster would fragment spontaneously - the so called "Coulomb explosion" would occur. This has been investigated[35] by us on Li_N^+, Li_N^{++}, and Li_N^{+++} (N<7) clusters using the ab initio UHF-CI method. The geometries of these clusters have been left at that of the neutral cluster because the values of ΔE involved here are much more than the difference of the adiabatic and vertical ionization potentials of these clusters. Moreover, upon optimization of geometry, these clusters would fly apart as these clusters are not stable. Allowing all possible channels of dissociation, it has been observed that all the Li_N^+ clusters have negative ΔE and therefore, all of these are stable. In the case of Li_N^{++}, all the clusters automatically dissociate into two fragments, each with one positive charge. The Li_N^{+++} clusters prefer spontaneous fragmentation into three fragments, each with one positive charge. These show that the effect of Coulomb repulsion in these cases is rather predominant. Also, it is noticed that whenever a cluster of 9>N>3 fragments, one of the fragments is always Li_3^+. As mentioned before, this preponderance of Li_3^+ is a hint at the "magic" nature of clusters with two valence electrons. One must, however, be careful about mentioning the spontaneity of fragmentation in the case of multiply ionized clusters. The positive value of ΔE always may not indicate that the dissociation would be automatic. There may be cases (e.g. Be_2^{++} ions[36]) where the clusters need to overcome a barrier before disintegration. In the case of the Li_N clusters studied, these kind of metastable states have not been observed.

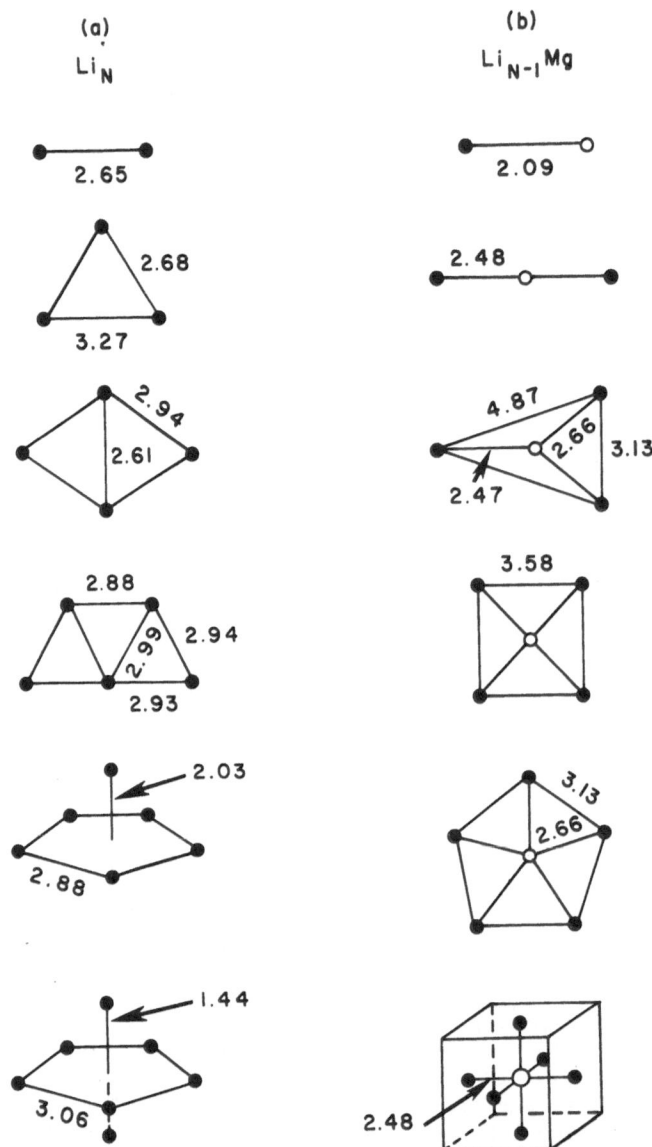

Fig. 7 Equilibrium geometries of Li_N and $Li_{N-1}Mg$ clusters. All bond lengths are in Å.

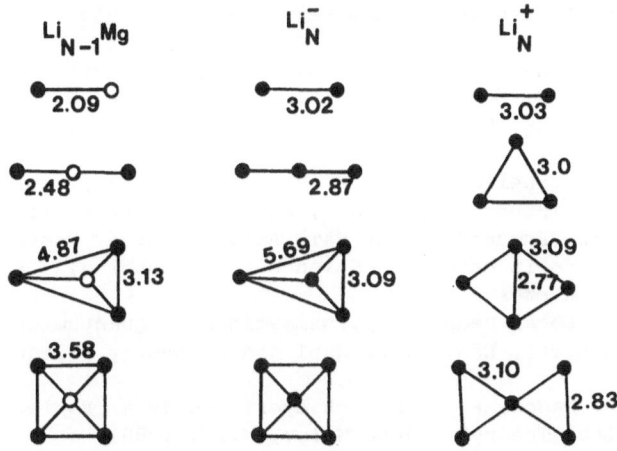

Fig. 8 Equilibrium geometries of ionized Li_N compared with those of $Li_{N-1}Mg$. All bond lengths are in Å units. Li_5^- is not completely optimized.

ACKNOWLEDGEMENT

    This work was supported in part by the Army Research Office (DAAG 29-85-K-0244) and the Jeffress Trust.

REFERENCES

1. E. W. Becker, K. Bier, and W. Henkes, Z. Phys. 146, 333 (1956).
2. T. D. Mark and A. W. Castleman, Jr., Adv. At. and Mol. Phys. 20, 65 (1985). Also see the references therein.
3. I. A. Harris, R. S. Kidwell, and J. A. Northby, Phys. Rev. Lett. 53, 2390 (1984).
4. O. Echt, K. Sattler, and E. Recknagel, Phys. Rev. Lett. 47, 1121 (1981).
5. J. Farges, M. F. DeFeraudy, B. Raoult, and P. Torchet, J. Chem. Phys. 78, 5067 (1983).
6. K. Kimoto, I. Nishida, H. Takahishi, and H. Kato, Jpn. J. Appl. Phys. 19, 1821 (1980).
7. M. M. Kappes, R. W. Kunz, and E. Schumacher, Chem. Phys. Lett. 91, 413 (1982).
8. W. D. Knight, K. Clemenger, W. A. deHeer, W. A. Saunders, M. Y. Chou, and M. L. Cohen, Phys. Rev. Lett. 52, 2141 (1984).
9. F. W. Froben and W. Schulze, Ber. Bunsenges, Phys. Chem. 88, 312 (1984).
10. L. A. Bloomfield, R. R. Freeman, and W. L. Brown, Phys. Rev. Lett. 54, 2246 (1985).
11. A Kasuya and Y. Nishina, Phys. Rev. Lett. 57, 755 (1986).
12. E. A. Rohlfing, D. M. Cox, and A. Kaldor, J. Chem. Phys. 81, 3322 (1984).
13. M. M. Kappes, P. Radi, M. Schar, and E. Schumacher, Chem. Phys. Lett. 119, 11 (1985).
14. T. P. Martin, J. Chem. Phys. 81, 4426 (1984).
15. M. D. Morse, M. E. Geusic, J. R. Heath, and R. E. Smalley, J. Chem. Phys. 83, 2293 (1985).
16. S. C. Richtsmeier, E. K. Parks, K. Liu, L. G. Pobo, and S. J. Riley, J. Chem. Phys. 82, 3659 (1985).
17. R. L. Whetten, D. M. Cox, D. J. Trevor, and A. Kaldor, Phys. Rev. Lett. 54, 1494 (1985).
18. B. K. Rao and P. Jena, Phys. Rev. B. 32, 2058 (1985).
19. J. L. Martins, J. Buttet, and R. Car, Phys. Rev. B 31, 1804 (1985).

20. J. Koutecky and P. Fantucci, Chem. Rev. $\underline{86}$, 539 (1984). See references therein.
21. B. K. Rao, S. N. Khanna, and P. Jena, Chem. Phys. Lett. $\underline{121}$, 202 (1985).
22. D. E. Ellis, Int. J. Quant. Chem. $\underline{18}$, 183 (1984); B. Delley, D. E. Ellis, A. J. Freeman, E. J. Baerends, and D. Post, Phys. Rev. $\underline{27}$, 2132 (1983).
23. P. S. Bagus, C. J. Nelin, and C. W. Bauschlicher, Jr., Suf. Sci. $\underline{156}$, 615 (1985); C. W. Bauschlicher, Jr., Chem. Phys. Lett. $\underline{117}$, 33 (1985).
24. D. R. Salahub in "Proceedings of NATO Advanced Study Institute on Impact of Cluster Physics in Materials Science and Technology", Ed. J. Davenas (Plenum, N.Y., in press).
25. K. H. Johnson, "Local Density Approximations in Quantum Chemistry and Solid State Physics", Eds. J. P. Dahl and J. Averry (Plenum, N.Y., 1984).
26. W. J. Hehre, L. Radom, P. v. R. Schleyer, and J. A. Pople, "Ab Initio Molecular Orbital Theory", (John Wiley, N.Y., 1986).
27. C. Möller and M. S. Plesset, Phys. Rev. $\underline{46}$, 618 (1934); J. S. Binkley and J. A. Pople, Int. J. Quant. Chem. $\underline{9}$, 229 (1975).
28. A Szabo and N. S. Ostlund, "Modern Quantum Chemistry: Introduction to Advanced Electronic Structure Theory", (MacMillan, N.Y., 1982).
29. B. K. Rao, P. Jena, and M. Manninen, Phys. Rev. B. $\underline{32}$, 477 (1985).
30. B. K. Rao, S. N. Khanna, and P. Jena, Solid State Commun. $\underline{56}$, 731 (1985).
31. T. H. Upton, Phys. Rev. Lett. $\underline{56}$, 2168 (1986).
32. K. Raghavachari and V. Logovinski, Phys. Rev. Lett. $\underline{55}$, 2852 (1985).
33. B. K. Rao, S. N. Khanna, and P. Jena, Z. Phys. D $\underline{3}$, 219 (1986).
34. P. Jena, B. K. Rao, and R. M. Nieminen, Solid St. Commun. $\underline{59}$, 509 (1986).
35. B. K. Rao, P. Jena, M. Manninen, and R. M. Nieminen (to be published).
36. M. R. Press, B. K. Rao, S. N. Khanna, and P. Jena; see page 431 in this book.

STABILITY OF SMALL METALLIC CLUSTERS :

CONCEPTUAL AND COMPUTATIONAL CHALLENGES

Jean-Paul Malrieu

Laboratoire de Physique Quantique (U.A. 505 du C.N.R.S.)
Université Paul Sabatier, 118, route de Narbonne
31062 Toulouse Cedex (France)

ABSTRACT

The flatness of the potential energy surfaces, the possible existence of several local minima, of different spin and space symmetries make especially difficult the theoretical study of atomic clusters. One recalls the Valence-Bond distinction between neutral and ionic states and the various components of binding energies - delocalisation and charge and spin correlation within the valence band - hybridization, dispersion and dynamical correlation effects going out of the valence band. The extent and qualitative effects of these various components is first illustrated on filled bands and nearly filled band problems (neutral and positively charged clusters of alkaline earth and rare gas atoms) and on half-filled band situations (alkali and transition metal clusters). The presently proposed lithium cluster conformations correspond to <u>ionic</u> states, but possible neutral structures with ordered spin waves should be searched. The intra-atomic ferromagnetism of the transition metals introduces a constraint neglected by ab initio RHF approaches and taken into account in unrestricted HF and LSD formalisms.

PRACTICAL CONSTRAINTS OF QUANTUM CHEMICAL CALCULATIONS

In the rapidly increasing field of ab initio quantum mechanical calculations concerning metallic clusters, a recent review [1] gives a nearly complete survey of the presently available theoretical results. Its interest is concentrated on the structure and stability of the very small clusters (n<10) which may actually be explored through the ab initio techniques. One should mention immediately that the contribution of quantum chemistry calculations to this field is severely limited by the rapid increase of the computational cost. The limits are of two types :

- the dependance on the number of (valence) electrons (n_e) ; the SCF processes go as n_e^4, the crudest explicit calculation of correlation energy (2nd order Moller-Plesset perturbation) requires a n_e^5 step ... In that sense the <u>practical</u> superiority of density functionnal approaches is evident.

- the dependance on the number of atoms n_a resulting in $3n_a-6$ degrees of freedom is perhaps more crucial. Since the clusters are rather labile entities, one cannot rest on any intuition or rule concerning the most likely conformations of the cluster, which one had in chemistry through the rules concerning the valency of the atoms and bond building. The dramatic features in this field are the flatness of the potential energy surface and the possible occurance of numerous minima. The flatness results in rather slow convergence of gradient procedures (which are only available for SCF and a few CI approximations). The multiplicity of the minima cannot be established a priori. In practice certain initial conformations are introduced at the beginning of the calculation, assuming for instance a certain symmetry. The symmetry is usually maintained and the optimized geometry, although a singular point on the potential energy surface, may not be a real minimum. Calculations of the full hessian are never reported, so that a certain uncertainty remains on the nature of the so called geometry optima and on the lifetime of the cluster at finite temperature. Another question concerns the completeness of the exploration ; for the future stochastic approaches will be necessary (when the computation time of the energy for each conformation will be low enough) in order to be sure that no important region of the (3n-6) dimensional potential surface has been omitted. As will be discussed later on, the question is made even more complex when one remembers that a single surface is not sufficient ; the lowest state may be of different space and spin symmetries in different regions of the configuration space.

The present contribution is inspired by the idea that an effort must be done to understand the physical content of the wave-functions, and that this understanding, which provides qualitative pictures and ways of thinking, in turn may help to research prefered architectures of medium-sized clusters.

BASIC ELECTRONIC EFFECTS AND FUNDAMENTAL CONCEPTS : A GLOSSARY

There exist two main modes of thinking in Quantum Chemistry, which are both valid and useful even if the second one is less popular and of little help in practical computations : the MO-CI (Molecular Orbital plus (eventually) Configuration Interaction) and Valence Bond approaches. These architectures introduce some essential distinctions, to which one should refer in order to understand and criticize theoretical results.

Valence Bond is a convenient tool to understand what happens within the valence shell (minimal basis set). It essentially introduces a basic distinction between neutral and ionic structures. In neutral structures each atom is neutral ; it may be in its ground state, or promoted to an excited state (hybridized). In ionic structures one or several pairs of atoms are ionized with positive and negative charges. The higher the number of ionized (or excited) atoms in a VB structure, the higher its mean energy and the lower - in principle - its contribution to the ground state wave-function. This is excellently pictured through the well-known Hubbard Hamiltonian [2] of a single band

$$H = \sum_{p,q} \beta_{pq} a_p^+ a_q + U \sum_p a_p^+ a_{\overline{p}}^+ a_{\overline{p}} a_p$$

(where a_p^+ and a_p are creation and annihilation operators in the AO p). The VB components are energetically hierarchized according to their ionicity, the neutral ones being the lowest (at zero) the singly-ionic ones lying at energy U, the doubly ionic at 2U and so on, when the hopping integrals β_{pq} increases and mixes neutral and ionic states, the degeneracies are removed, splitting the eigenstates into families of (predominantly) neutral, singly ionic ... states. The mixing increases with the ratio β/U but it remains possible to distinguish the dominant VB nature of the eigenstates.

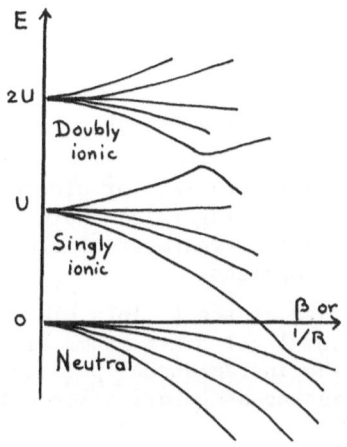

Notice that for weak β/U ratio one may built an effective Hamiltonian restricted to the neutral determinants, and giving the exact energies of the neutral eigenstates. This effective Hamiltonian, established by means of the Quasi Degenerate Perturbation Theory reduces (at the 2nd order expansion) to the well-known Heisenberg Hamiltonian for a half-filled band, as shown by Anderson [3]. For half-filled bands one expect that the lowest eigenstates belong to the family of neutral states (i.e. that their main components are on neutral VB components) and that among the neutral components, those presenting the largest number of spin-alternations between adjacent atoms will be of largest amplitudes (antiferromagnetic spin waves).

If ones moves then to a MO-CI approach, one may consider first a (formal) problem restricted to a single band (for instance valence s electrons for alkali and alkine earth clusters). A purely monoelectronic model (the Extended Hückel (EHT) hamiltonian, or the hight binding band model) takes simply into account the electronic delocalization and mainly the kinetic energy (EHT also includes overlap repulsive effects).

The mean electrostatic effects of nuclei and electrons may be introduced through the Hartree-Fock Self Consistent Field (SCF) approximation ; they do not result in significant changes from the EHT model for neutral homogeneous clusters (except for charge instabilities, vide infra), due to the local electrostatic neutrality. When several holes will be present in the band, the repulsion between the holes will result in qualitative deviations from the simple monoelectronic picture.

At a third step the CI within the valence shell will introduce the so called non-dynamical correlation effects : they essentially result in a diminution of the charge fluctuation which is grossly overestimated by the single determinantal approximation. Using the VB language one says that CI diminishes the weight of ionic structures. When the atoms bear several electrons in open-shell(s) the second task of the valence CI is to diminish the spin disorder by increasing the weight of the VB structures in which the atoms have a maximum spin (i.e. are in their ground state, according to Hund's rule). A correct inclusion of these two qualitative effects [4] (diminution of the local charge fluctuation, increase of the local spin fluctuation) are necessary to obtain reliable dissociation energies (i.e. binding energies) and should require in principle a full-valence CI, which is already a non trivial work for 10 atoms.

One may then introduce AOs not involved in the atomic ground state (and non minimal basis sets). At the monoelectronic level of description, this inclusion may result in either

- quantitative effects, through contractions and distorsions of the ground state AOs of the atom, contributing to a decrease of the cluster energy or

- qualitative effects, when the cluster can no longer be viewed as built from ground state atoms, and when one must consider that electrons have been promoted to other AOs (and their MO combinations). This second effect is named band mixing in Solid State Physics ; in Molecular Physics one speaks of promoted atoms, but hybridization is unhappily unployed for both small mixtures (first type of effect) and promotion.

The non minimal basis set will also introduce same <u>static</u> (i.e. mean field) <u>polarization</u> effects when the cluster is charged.

Going to the CI level, the use of non minimal basis sets permits in principle to treat three different effects

- the <u>dispersion energy</u> between atoms, which is crucial for closed shell atoms ; in practice the ab initio calculation of dispersion energies is <u>very</u> difficult even for diatoms. The correct ab initio treatment of clusters of closed shell atoms is for the present stage hopeless.

- the <u>dynamical polarization</u> effect in charged clusters. In a cluster X_n of n atoms, when the electron-defect is on atom A all other atoms are polarized toward A, when the hole jumps on atom B, the instantaneous polarization of the other atoms is changed, and this instantaneous polarization effect cannot be dealt with at the SCF level.

- the <u>atomic angular and radial correlation</u> effects essentially result in a lowering of the effective energy of the ionic VB structures,[5,6] they therefore permit slightly more charge (and spin) disorder than would result from a full-CI in a minimal basis set.

The next sections will illustrate on two limit cases (filled and half-filled bands) the possible role of these various effects.

FILLED BANDS AND NEARLY FILLED BANDS

As examples of filled band problems one may quote the (small) clusters of alkaline-earth atoms (s^2) or rare gas atoms $(s^2 p^6)$. When the number of atoms is small enough, the cohesion of the cluster is essentially insured by dispersion forces ; as already mentionned, accurate ab initio calculations of these energies are impossible. Empirical two-body potentials may be used but they lack important attractive three-body interactions ($\cong$ 20% of the two body ones) and repulsive four-body contributions.[7] The bulk is known to have a cohesive energy which cannot be explained only by dispersion correction and involves the mixing with upper "valence" band, p for alkaline earth and d for rare gas. One should notice that Be_2 itself is not a Van der Waals molecule,[8] it presents a weak (R_e=2.2 kcal/mole) short distance minimum (r_e=4.6 a.u.)[9] which is diabatically correlated to an $Be(^3P$ sp$)+Be(^3P$ $\overline{sp})$ limit,[10] 5.5 eV above the ground state $Be(^1S$ $s^2)+Be(^1S$ $s^2)$ limit. Two bonds s_a+s_b and p_a-p_b may be built from the upper limit, but their treatment requires a correct treatment of the <u>charge</u> and <u>spin</u> correlations. (The singlet Be ^1P sp configuration is 2.5 eV above the triplet one). The obtention of the short range minimum - especially when starting from $s_A^2 s_B^2$ MOs - has been difficult and controversial.[8] The trimer of Be atoms should be weakly bond (20 kcal/mole).[11]

Tetramers of Be, Mg and Ca have been studied quite extensively.[11,12,13] As a first qualitative feature one must mention that Mg and Ca exhibit a low tendency to involve p AOs [13] : the tendency to hybridization is maximum in the first line of the periodic table, due to similar spatial extents of s and p AOs. The calculated binding energies of Mg_4 and Ca_4 remain weak.[12,13] In tetrahedral Be_4 geometry the p AOs play a large role with large (.3) coefficients in the t_2 MOs.[12] The binding energy per atom is quite large from .33 to .37 eV with s+p basis sets [12] to .7 eV when d AOs are included.[11,13]

Be$_7$, B$_{10}$ and Be$_{13}$ fragments of the hcp lattice have been calculated.[12]
The most interesting conclusions concern the near degeneracies of states and
their inversions under CI inclusion. For planar Be$_7$ (centered hexagon) triplet states are the lowest at both SCF ($^3A_{2n}$, $^3B_{1g}$) and CI levels ($^3B_{1g}$,$^3B_{2u}$)
but of different nature. The occurence of openshell configurations, invol-
ving the " π " fully symmetric a$_{2u}$ MO (and a σ b$_{2u}$ MO essentially built on p
AOs), is a dramatic proof of the extent of hybridization to band mixing. The
same is true for Be$_{10}$[12] and Be$_{13}$ fragments ; even in the ^1A_g SCF configuration [14] the MO content is in contradiction with an s^2 nature, and a recent
openshell SCF study [15] proposes a quintet $^5A''_1$ ground state, with 14 states
below 2 eV.

Singly and doubly ionized clusters of Mg have been studied through both
UHF ab initio and DIM-type models.[16] A singly ionized cluster is governed by
the delocalization of one hole in a filled band : the electron leaves the
highest MO of the band and the cluster should keep the conformation leading
to an MO of __highest__ energy. Mg$_3^+$ is actually linear (ε_{max}=-β). For Mg$_4^+$ the
topology should prefer a square (ε_{max}=-2β) for which the charge is equally
spread on the four atoms. The SCF calculation predict a centered triangle

(ε_{max}=-$\sqrt{3}\beta$) with a charge highly concentrated (-.66 e$^-$) on the

central atom, .25 eV below the square cluster for which all charges are
equal (-.25 e$^-$). This difference is due to the inclusion of the __static__ polarization in the SCF calculation, the central charge Mg$^+$ polarizing the three
surrounders. When dynamical polarization is included, as occurs in the DIM
type model, the energy difference between the square (delocalized) and centered triangle (localized) is of course reduced, the square appearing slightly lower (by .07 eV). Extensive CI calculations have been performed[17] to
clarify this point, and they confirm the DIM-model conclusions = SCF calculations overstabilize centered structures where the charge is localized on
a central atom, the inclusion of dynamical polarization makes competitive
more delocalized architectures. For Mg$_5^+$ the centered tetrahedron and centered square are nearly degenerate, with a central charge of -.71 e$^-$ in both
structures.

The Ar$_3^+$ cluster is linear according to both ab initio[18] and DIM type
calculations.[19] While in s^2-atoms the hole was isotropic, in the rare-gases
the hole concerns the p shell ; p AOs are directional and the hopping integral is much larger between σ p$_z$ AOs than between π p$_x$ AOs.[19] The delocalization is therefore essentially monodimensional in rare gas.[19] Ar$_4^+$ would be
linear from this criterion, but the gain in delocalization energy is smaller
than the benefit of polarizing a neutral Ar by Ar$_3^+$, and Kuntz[19] claims that
Ar$_n^+$ clusters are essentially Ar$_3^+$ clusters surrounded by n-3 neutral polarized atoms.

In __doubly ionized__ clusters of filled-band clusters, the two holes
introduce correlation effects through their r^{-1} repulsion. In naive electrostatic pictures the repulsion between the two holes should lead to an
explosion of small clusters, while Hg$_n^{++}$ clusters are observed for n$\geqslant$5.[20]
Here again static and dynamic polarization effects play a key role in the
stability of these cluster.[16] Mg$_5^{++}$ would take a centered tetrahedron structure, with the two holes having a large probability to be located on the
same central atom (the mean charge of which is -1. e$^-$). The energy being
1.72 eV above the Mg$_2^+$ + Mg$_3^+$ products of the prefered dissociation channel,
but the Mg$_5^{++}$ structure is a stable minimum protected by a .21 eV barrier.
Mg$_6^{++}$ has a square bipyramid conformation,[21] with charges essentially located
on the two summits (q=-.65 e$^-$), but the distance between the two summits is
large (7.0 a.u.) and on cannot regard the cluster is Mg$_2^{++}$ surrounded by from
polarized neutral atoms. A triplet minimum corresponding to an icosaedral
geometry and a perfectly delocalized charge lies only .2 eV above. Mg$_7^{++}$ in
s pentagonal bipyramid structure is already of lower energy than Mg$_3^+$ + Mg$_4^+$;

it is of singlet spin symmetry but a triplet state is nearly degenerate (.02 eV above !).

HALF FILLED BANDS WITH s e⁻ PER CENTRE

Alkali clusters and noble metals, with their single valence electron in a s type AO, represent the priviledged problem for theory : the computational cost is minimal and their study should give crucial informations on the appearance of metallic properties. Calculations are actually numerous, at least for Li_n. For Na and K clusters the core-valence correlation effects, which are crucial at least for interatomic distances, make the calculations more difficult (and more rare or less accurate). The same is true for noble metals (Cu, Ag, Au) for which the d^{10} shell is polarizable.

Li_2 is a typical single band, its $^1\Sigma_g^+$ ground state is dominated by neutral VB components. Li_3 ground state conformation is an open triangle[22] ($\theta \cong 75°$) ; its ^2B_u wave function corresponds to a $|a_g^2 b_{13}|$ MO filling and is essentially neutral according to VB criteria dominated by a spinwave

↑⁀⁀↑ . The nearly degenerate ^2A_g minimum ⌈⁻⁻⁻⁀➤ is of intermolecular nature. The first ionic state would be of $|a_g b_u^2|$ content, it would be linear and corresponds to a charge density wave. The central atom p_z AO

would stabilize the b_u MO. One may wonder why the ground state conformation is so largely bent, although only a few kcal/mole below the linear structure. Maynau[23] has shown that this bonding is due to the three-body overlap effects, which are attractive : when fixed interatomic $r_{AB} = r_{BC}$ are imposed to the ↑A⁀ ⁀C↑ quartet state (which ignores delocalization), these three-body forces favors the equilateral geometry. Models ignoring these three-body effects (such as Hubbard Hamiltonian, Heisenberg[24] on DIM models[25]) cannot reproduce the closed triangle geometry.

Li_3^+ has a regular triangular geometry and a great stability.

Li_4 has been calculated to have a rhombic singlet ground state below Li_2.[26-28] One should point out that this closed shell state belongs to the family of ionic states in VB terminology. The net charges reduced to the s band are positive on the short diagonal, negative on the long one.

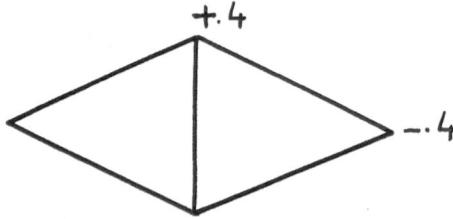

The stability of this cluster may be understood as due to i) the great stability of Li_3^+ triangle ii) the Li_3^+ ...Li^- electrostatic interaction iii) the resonance between the two ionic forms, which proceeds through a jump of one electron from one side of the long diagonal to the other side through the antisymmetric p_σ AO of the central Li atoms.

One might wonder whether <u>neutral</u> states could not occur as well. The DIM method,[25] the Heisenberg Hamiltonian[24] and X_α-SW method[29] had predicted a planar square, which actually would be neutral, and supported by spinwaves. Four-body effects destabilize[30] the corresponding openshell singlet, which is above the ionic one even for square geometries.

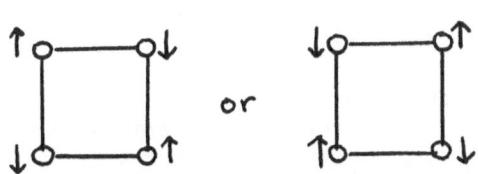

As suggested by Beckmann et al.[26] and later on more extensively confirmed by Rao et al.[27], the triplet state potential surface has a real minimum for a D_{2h} distorted tetrahedron geometry; for these conformations the singlet is at higher energies, and this is a case of a secondary minimum, lying above the ground state rhombic singlet.

Li_5[27,31] is again of ionic nature, the positive charge tends to concentrate on the most central atom. Notice that this may be viewed as well as the addition of the ionic state $^2\Sigma_g^+$ of Li_3 on a Li_2 $^1\Sigma_g^+$ molecule.

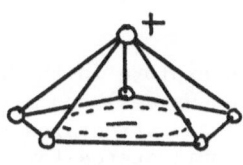

Li_6 has been calculated to have two nearly degenerate minima one being a C_{5v} pentagonal bipyramid, the other being a D_{3h} planar addition of three atoms on a regular triangle

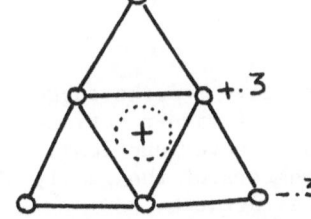

It is evident from the MO requirements (avoidment of Jahn-Teller distorsions) that these states must be ionic (or of high spin).

One may wonder way neutral states do not appear for these low number of atoms. Since neutral states would prefer spin waves, they dislike triangles

and would accept structures of the type

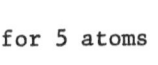 for 5 atoms 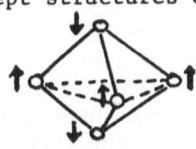 ($S_z=1/2$) or 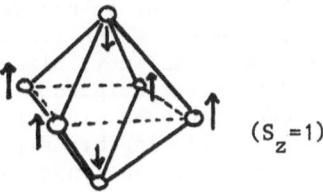 ($S_z=1$)

for 6 atoms. Such low space and/or spin-symmetry species have perhaps been missed.

The first occurence of a neutral state is reported for cubic Li_8 through an UHF calculation which gives a symmetry broken perfect spin wave

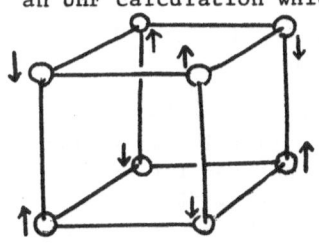

The binding energy per atom is only .11 eV but CI is lacking. A X_α-SW calculation[29] reports a .78 eV binding energy for the same structure.

One should notice at this stage that the bcc lattice of the bulk involves only square four-body graphs between adjacent atoms, and that its cohesive energy, lattice parameter and bulk modulus are perfectly reproduced by an Heisenberg Hamiltonian extracted from ab initio or experimental potential curves of Li_2[24]. This is linked to the fact that perfectly alternant spin waves may be put on the bcc lattice. From this remark one may guess that Li_{15} bcc with 6 atoms of β spin above the 6 faces of a cube of α spins cen-

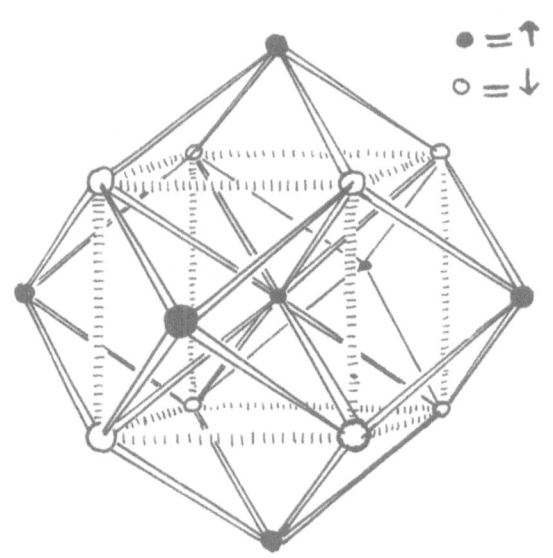

tered on a β spin atom will support a low energy spin wave and actually a strong spin localization occurs when one calculates this cluster in an UHF approach.[34] We played with an analogous Li_{24} cluster of bcc structure and found that the UHF solution, i.e. a similar spin wave, is significantly below the RHF solution, which looks more like a charge wave.[34]

● = ↑
○ = ↓

These tentative remarks suggest that one should explore in the future different kinds of wave-functions (including high S_z values) and structures adapted to these wave-functions, the presently proposed structures being essentially members of the family of the low-spin ionic states for compact (triangular) structures.

HALF-FILLED BANDS WITH SEVERAL ELECTRONS PER ATOM ; INTRODUCTION OF THE LOCAL EXCHANGE CONSTRAINT

Moving to the atoms of group V (s^2p^3 atomic ground state, i.e. Nitrogen, Phosphorus) leads to non-metallic elements (which may be understood as follows ; elements which (in homogeneous buildings at least) never deviate from their normal valency, always being triply bonded to one atom or singly bonded to three neighbours (no hypervalency). Nevertheless metallic phases may occur for such elements as well, and it is worth analysing this problem, before considering briefly the transition metal problem. The ground state of the atoms is governed by the Hund's rule, i.e. the three p electrons are spread into the three p AOs with a quartet (^4S) spin arrangement (all spins are parallel). It is costly to violate this intraatomic ferromagnetism, the lowest excited states of N and P lie at 2.4 and 1.4 eV respectively above the ground state. The atoms are reluctant to leave this intraatomic spin ordering and a correct treatment of a cluster or molecule requires a CI which diminishes both the ionicity and the spin-disorder, which are intrinsic to the monodeterminantal approximation : P_4 is more stable than $2P_2$, while N_4 is highly unstable, a cubic cluster P_8 has been suggested[35] and theory predicts that it is a stable structure (although slightly less stable than $2P_4$). Hartree-Fock spin instabilities happen in that system,[36] leading to a perfectly alternating spin-wave, which keeps in a collective mode both the tendency to neutrality, the ferromagnetism of the atoms and the anti-ferromagnetic interatomic coupling, as pictured for cubic Li_8. Starting from a Restricted H.F. calculation and correctly treating these collective effects is already a hard task for usual CI algorithms.

These remarks take their full strength when real transition metals ($d^n s^1$, $n<10$) are considered. For Cr for instance the d shell is half filled, as is the s shell, the ground state is a ^7S, i.e. with all spins parallel, and reversing the spin of the s electron costs .95 eV, while reversing a spin in the d shell costs 2.5 eV. One understands first why the Restricted HF approximation gives a negative binding evergy of Cr_2 (De=-19 eV) through its sextuple bond formation which introduces absurd charge and spin fluctuations. The UHF solution[28] $Cr(d^5s).Cr(\overline{d^5}s)$ is much more stable since it preserves the ground state of the atoms and an interatomic ferromagnetic coupling, but the UHF solution lacks a left-right resonance and remains unbound at the observed[39] interatomic distance of Cr_2 (r_e=1.68 Å) which has a D_e=1.6 eV dissociation energy.[40] So far and despite tremendous efforts quantum chemists have failed to correctly reproduce this molecular well in an ab initio explicitly correlated technique. In view of this basic failure there is little hope for reliable treatments of Cr_n clusters, and moderate hope for clusters of metals having $d^n s^1$ structures when $2<n<8$ (several open shells). One should notive that

i) For the bulk interatomic distance, the UHF solution - or better its 2nd order correlated energy[41] - becomes reasonable (the left-right resonance becoming of lesser importance) and we may suggest that UHF approaches are the convenient starting point for the study of such systems.

ii) The Local Density functionnal, and especially the LSD[42] version (i.e. an UHF approach), succeed to reproduce the Cr_2 potential curve. This success may be due to the positive interference between the use of a spin-unrestricted formalism - which favors both the neutrality of the atoms and their high-spin ordering - and of an attractive correlation potential - which lowers the energy and accept larger ionic components. It seems actually that the spin polarization or magnetization $n_\uparrow - n_\downarrow$ of the atoms is lesser in LSD calculations than in UHF ab initio calculations. (Compare refs. 38 and 42). For a Fe_{13} bcc cluster studied in LSD formalism[43] $n_\uparrow - n_\downarrow$=2.7 is however.

A careful comparison of UHF, RHF and LSD wave functions would be fruitful and it would be worthwhile to compare the LSD density matrix with that resulting from an explicit treatment of correlation.

CONCLUDING REMARKS

The necessary research of equilibrium structures and energies of small clusters is a painfull task. We stressed on the fact that these explorations should be more extensive and concern several space and spin symmetries, considering low and high spin states, ionic and neutral states in a more systematic manner.

The wave functions should be analyzed more explicitly and qualitative pictures should be proposed in order to understand both the origin of the stability of a given cluster and the possible aggregation or evaporation on it or from it, in a genealogic scheme.

New qualitative pictures of the bulk itself might arise from these efforts. As an example one may quote the recent proposal by Goddard[44] from studies on Li_4, Li_{10} and Li_{13}^+ ; the wave functions resulting from so called GVB calculations (where correlation is introduced within separated pairs of electrons, which is reasonnable picture for Lewis molecules, not for hypervalent buildings) suggest that the electrons might be distributed in singly occupied MOs in different regions of space. In that sense the description is more similar to a magnetic description than to the classical MO picture, but these singly occupied MOs are no longer AOs they are regional orbitals, spread on one bond, a triangle or a tetrahedron, i.e. involving a local delocalization (and hybridization to maintain orthogonality between adjacent MOs). This description, and their extrapolations to the bulk, are somewhat tentative but very stimulating and illustrate the major interest of that field, where Molecular Physics, Quantum Chemistry and Solid State Physics converge, in a perhaps contradictory manner, to question conventional representations and rise new questions.

REFERENCES

1. J. Koutecky and P. Fantucci, Chem. Rev. 86, 539 (1986).
2. T. Hubbard, Proc. Roy. Soc. London, Ser. A. 276, 283 (1986).
3. P.W. Anderson, Phys. Rev. 115, 2 (1959) ; in Magnetism, edited by C.T. Rado and H. Juhl (Academic, New York, 1963) p. 25.
4. P. Karafiloglou and J.P. Malrieu, Chem. Phys. 104, 383 (1986).
5. M.M. Goodgame and W.A. Goddard III, Phys. Rev. Letters 54, 661 (1985).
6. F. Spiegelmann, J.P. Malrieu, D. Maynau and J.P. Zurru, J. Chim. Phys. 83, 69 (1986).
7. J.P. Daudey, O. Novaro and M. Berrondo, J. Chem. Phys. 71, 4297 (1979).
8. B. Liu and A.D. McLean, J. Chem. Phys. 72, 3418 (1980) ; R.J. Harrison and N.C. Handy, Chem. Phys. Letters 98, 97 (1983) ; B.H. Lengsfield, A.D. McLean, M. Yoshimire and B. Liu, J. Chem. Phys. 79, 1891 (1983).
9. V.E. Bondybey and J.H. English, J. Chem. Phys. 80, 568 (1984).
10. J. Gerratt, in ref. 11.
11. R.J. Harrison and N.C. Handy, Chem. Phys. Letters 123, 321 (1986).
12. G. Pacchioni and J. Koutecky, Chem. Phys. 71, 181 (1982) ; G. Pacchioni and J. Koutecky, J. Chem. Phys. 77, 5850 (1982).
13. C.W. Bauschlicher, P.S. Bagus, B.N. Cox, J. Chem. Phys. 77, 4032 (1982).
14. C.W. Bauschlicher and L.G.M. Petterson, J. Chem. Phys. 84, 2226 (1986).
15. W.C. Ermler, C.W. Kern, R.M. Pitzer, N.W. Winter, J. Chem. Phys. 84, 3937 (1986).
16. G. Durand, J.P. Daudey, J.P. Malrieu, J. de Phys. 47, 1335 (1986).
17. G. Durand, J.P. Daudey, J.P. Malrieu, to be published.
18. W.R. Wadt, Appl. Phys. Letters 38, 1030 (1981). This work also includes DIM calculations ; H.U. Böhmer, S.D. Peyerimhoff, in press.
19. J. Hesslich, P.J. Kuntz, Z. Phys. D 2, 251 (1986).
20. C. Bréchignac, M. Broyer, Ph. Cahuzac, G. Delacretaz, P. Labastie and L. Wöste, Chem. Phys. Letters 118, 174 (1985).
21. G. Durand, to be published.
22. J. Kendrick and I.H. Hillier, Mol. Phys. 33, 635 (1977) ; J.W. Kress, J.J. Carberry and G.C. Kuczinski, Mol. Phys. 36, 717 (1978) ; J.L. Gole, R.H. Childs, D.A. Dixon and R.A. Eades, J. Chem. Phys. 72, 6368 (1980).
23. D. Maynau, private communication.
24. J.P. Malrieu, D. Maynau, J.P. Daudey, Phys. Rev. B 30, 1817 (1984).
25. A.L. Companion, Chem. Phys. Letters 56, 500 (1978) ; S.C. Richtsmeier, D.A. Dixon, J.L. Gole, J. Chem. Phys. 86, 3942 (1982).
26. H.O. Beckmann, J. Koutecky, V. Bonacic-Koutecky, J. Chem. Phys. 73, 119 (1979) ; G. Pacchioni, H.O. Beckmann, J. Koutecky, Chem. Phys. Letters 87, 151 (1982).
27. B.K. Rao, P. Jena, Phys. Rev. B 32, 2058 (1985).
28. A.K. Ray, J.L. Fry, C.W. Myles, J. Phys. B 18, 381 (1985).
29. J.G. Fripiat, K.T. Chow, M. Boudart, J.B. Diamond, K.H. Johnson, J. Mol. Catal. 1, 59 (1975).
30. J.P. Malrieu, D. Maynau, J. Am. Chem. Soc. 104, 3021 (1982) ; D. Maynau, J.P. Malrieu, ibid. 104, 3029 (1982) ; D. Maynau, Ph. Durand, J.P. Daudey, J.P. Malrieu, Phys. Rev. A 28, 3193 (1983).
31. P. Fantucci, J. Pacchioni, J. Koutecky, J. Chem. Phys. 80, 325 (1984) ; G. Pacchioni, J. Koutecky, Ber. Bunsenges Phys. Chem. 88, 242 (1984).

32. P. Fantucci, V. Bonacic-Koutecky, J. Koutecky, J. Phys. Chem. <u>87</u>, 1096 (1983), J. Comput. Chem., <u>6</u>, 462 (1985).
33. R.F. Marshall, R.J. Blint, A.B. Kunz, Phys. Rev. B <u>13</u>, 333 (1976) ; Solid State Commun. <u>18</u>, 371 (1976).
34. D. Maynau, J.P. Daudey, J.P. Malrieu, to be published.
35. G. Trinquier, J.P. Daudey, N. Komiha, J. Am. Chem. Soc. <u>107</u>, 7210 (1985).
36. G. Trinquier, private communication.
37. A.D. McLean, B. Liu, Chem. Phys. Letters <u>101</u>, 144 (1983).
38. M.M. Goodgame, W.A. Goddard, Phys. Rev. Letters <u>48</u>, 135 (1982).
39. Y.M. Efrenov, A.N. Samoilova, V.B. Kozhukosky, L.V. Gurvich, J. Mol. Spectrosc. <u>73</u>, 430 (1978) ; V.E. Bondibey, J.H. English, Chem. Phys. Letters <u>94</u>, 443 (1983).
40. A. Kant, B.H. Strauss, J. Chem. Phys. <u>45</u>, 3161 (1966).
41. M. Pélissier, J.P. Daudey, private communication.
42. B. Delley, A.J. Freeman and D.E. Ellis, Phys. Rev. Letters <u>50</u>, 488 (1983).
43. K. Lee, J. Callaway, K. Kwong, R. Tang, A. Ziegler, Phys. Rev. B <u>31</u>, 1796 (1985).
44. M.H. McAdon, W.A. Goddard, Phys. Rev. Letters <u>55</u>, 2563 (1985).

SELF-CONSISTENT CALCULATION OF THE COLLECTIVE EXCITATIONS OF SMALL

JELLIUM SPHERES

D. E. Beck

Department of Physics and Laboratory for Surface Studies
University of Wisconsin-Milwaukee
Milwaukee, Wisconsin 53201

The time-dependent local density approximation is used to compute the complex eigenfrequencies for the dipolar excitations of small jellium spheres. The dynamic response of the sphere to a time-dependent field is continued to complex frequencies in order to isolate the contributions of the individual eigenmodes of the system. The computations are reported for closed-shell configurations containing from 8 to 198 electrons.

INTRODUCTION

The density-functional formalism in combination with a linear response approximation provides for a quantum-mechanical calculation of the static polarizability of atoms[1] and small atomic clusters.[2,3] The complexity of the self-consistent density-functional calculation of the electronic structure and linear response of the system is reduced by introducing a local density approximation for the exchange-correlation energy.[4] The time-dependent local density approximation (TDLDA) is the dynamic extension of this self-consistent procedure.[1,5] The jellium model for the conduction electrons in a small metal sphere has N electrons confined by a rigid uniform spherical charge density, $n^+ = 3/4\pi\, r_s^3$, of radius R. This model for a metal sphere has been used successfully in interpreting the size distribution of alkali clusters in molecular beam experiments,[6] and it provides a qualitative agreement between the computed static polarizability of these spheres[2,3,7] and the experimentally determined polarizability.[8] In order to avoid the additional computational complexity presented by the lack of spherical symmetry, we confine our calculations to ground state systems which are closed shell configurations.[9]

In considering the dynamic response of this electronic system there are two classes of excitations. In a "single-particle" excitation an electron is excited from an occupied ground state energy level into an unoccupied level (electron-hole transition) or a continuum state (ionizing transition). These single-particle transitions are responsible for most of the prominent features in the computed response of the system to a time-dependent perturbation,[3,10] Fig. 1. However, we should not expect the TDLDA to accurately predict the excitation energies of these modes. For example, the continuum threshold in the imaginary part of the dynamic dipolar polarizability in Fig. 1 occurs at a frequency corresponding to the energy of the highest occupied level of the ground state system and

not at a frequency determined by the ionization potential for this system.[11]
The TDLDA gives the linear response of the system to an applied stimulus,
hence, the induced density response, $\delta n(\vec{r},t)$, is treated as a "small perturbation" on the ground state density. An additional weakness of this
approximation for the description of these excitations is that the change
in the exchange-correlation energy is treated by expanding the local density
expression and retaining only the term proportional to $\delta n(\vec{r},t)$ in the
effective potential for the electrons.

The other class of excitations is the "collective or plasma" excitations where the electronic charge as a whole is excited and oscillates as
a result of the restoring force provided by the rigid positive background.
For these excitations the TDLDA provides an adequate description, since the
induced density associated with the excitation involves only a small perturbation of the ground state system. These collective excitations show
only a small dependence on particle size so they should account for the
overall structure in the experimental data obtained using samples with a
distribution of particle size. However, the features associated with these
excitations tend to be less prominent in the computed dipolar polarizability
than the single particle features and, consequently, it is difficult to
determine their excitation energies.

In Fig. 1 the imaginary part of the dynamic dipolar polarizability
for a sphere containing $N = 18$ electrons and a positive charge density
corresponding to a metal with $r_s = 4.0$ (a.u.) is presented [we use atomic
units (a.u.) where $e=\hbar=m=1$]. Below the continuum threshold $\text{Im } \alpha_1(\omega)$ is
zero, since this model has no mechanism for an excitation with an energy
below the threshold to dissipate its energy. The vertical lines below the
threshold indicate the energies of the low-lying excitations. The response
even for this small highly symmetric system is quite complicated. There
are Fano resonances of single-particle modes with the continuum states

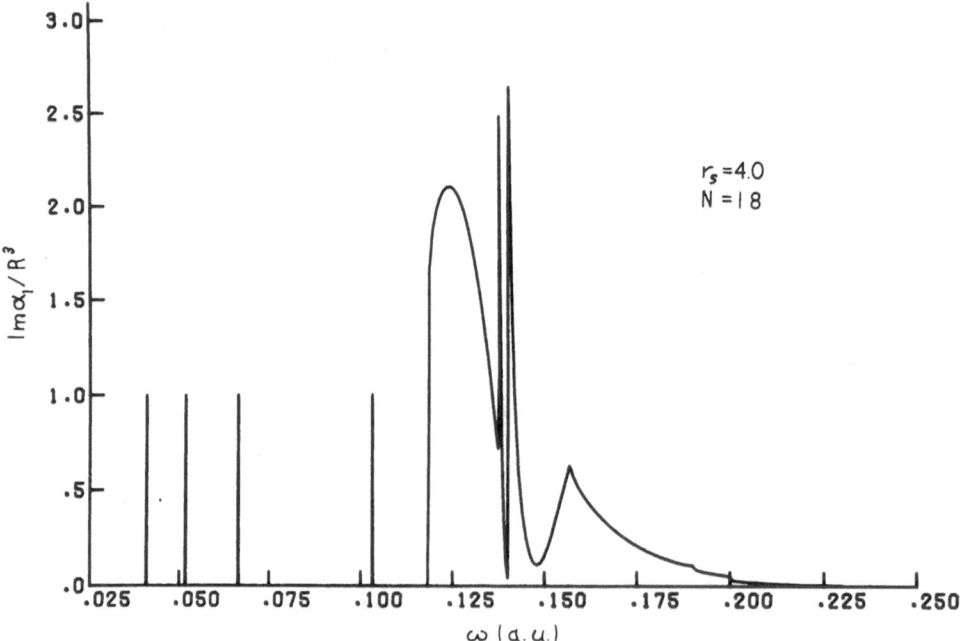

Fig. 1   The imaginary part of the dynamic dipolar polarizability for a
jellium sphere with $r_s = 4.0$ (a.u.) and containing 18 electrons.
The vertical lines, below the continuum threshold at 0.118 (a.u.),
indicate the frequencies at the low-lying excitations.

and sharp features one would associate with single-particle excitations. However the identification of the features to be associated with the collective modes requires a more extensive analysis.[3,10]

In order to isolate the contributions of the various excitations of the system to the dynamical response, we have formulated the TDLDA as an eigenvalue problem and computed the complex eigenfrequencies of the spheres [our computation requires the continuation of the dynamic response into the complex frequency plane which complicates our search for the eigenmodes]. The identification of the character of the excitation is inferred from the induced electronic density perturbation. The features of the induced densities for most of the single-particle modes are quite regular and they are principally confined to the interior of the sphere. The induced density for the collective modes have more prominent surface features which extend outside of the positive charge density.

Our calculation of the static polarizability of these spheres contains a description of the procedures and numerical methods,[2] and the formulation of the TDLDA have been thoroughly detailed.[1,3,5] Hence, only a brief description of our notation and the formulation of the eigenvalue problem for the spherically symmetric system are provided below. A summary of our calculated eigenfrequencies for the collective modes of these spheres is presented in the concluding section of this paper.

FORMALISM

We consider a spherically symmetric system subjected to an infinitesimal time-dependent perturbation,

$$\delta V_M(\vec{r},t) = \lambda^o \, r^M \, Y_M^o(\hat{r}) \, e^{-i\omega t}$$

which develops an Mth order moment, $P_M = \lambda^o \alpha_M(\omega)$. For the closed shell (spherically symmetric) system, the only non-zero term in the expansion of the linear density response in spherical harmonics is the term with the same symmetry as the perturbation, and $\delta n(\vec{r},\omega) = \lambda \, Y_1^o(\hat{r}) \, n_1(r,\omega)$ for the dipole response. This density response is related to the self-consistent field potential, $\lambda V_1(r,\omega)$ by the random-phase approximation for the generalized susceptibility, $\chi_1^o(r,r';\omega)$, of the Kohn-Sham independent particle model of the ground state system,[1,3]

$$n_1(r,\omega) = \int_0^\infty r'^2 dr' \, \chi_1^o(r,r';\omega) V_1(r',\omega). \tag{1}$$

Inserting this into the expression relating $V_1(r,\omega)$ to the charge density response of the system, we obtain the integral equation,

$$V_1(r,\omega) = r[1 + \int_0^\infty r'^2 dr' \, K_1(r,r';\omega) V_1(r',\omega)] \tag{2}$$

with the kernel

$$K_1(r,r';\omega) = \int_0^\infty dr''[(4\pi/3)\theta(r-r'')((r''/r)^3-1) +$$
$$+ \delta(r-r'')r''^{-1}(d^2 n \varepsilon_{xc}(n)/dn^2)] \, \chi_1^o(r'',r';\omega). \tag{3}$$

Here $\varepsilon_{xc}(n)$ is the local density expression for the exchange-correlation energy.[4] We have set

$$\lambda = \lambda^o [1 - (4\pi/3)\int_0^\infty dr \, n_1(r,\omega)]^{-1}, \tag{4}$$

and the dipole polarizability of the system is given by

$$\alpha_1(\omega) = (4\pi\lambda/3\lambda_o)\int_0^\infty dr\ r^3\ n_1(r,\omega). \tag{5}$$

The self-sustained excitations or eigenmodes of this system are those modes where one obtains a finite response even for an infinitesimal perturbation. Hence, we require solutions for which λ, Eq. 4, remains finite for $\lambda^o \to 0$; i.e., the complex eigenfrequencies, z_j, are solutions of

$$\xi(z_j) = (4\pi/3)\int_0^\infty dr\ n_1(r,z_j) = 1. \tag{6}$$

The excitations with frequencies above the continuum threshold are damped and the generalized susceptibility must be analytically continued into the lower-half of the complex frequency plane in order to locate these eigenfrequencies.[12]

COLLECTIVE EXCITATIONS

A classical model for a metal sphere where the positive and negative charge densities are uniform will have a multipole surface plasma mode (non-retarded) at a frequency given by $\omega_M^c = \omega_p \sqrt{M/2M+1}$ where the bulk plasma frequency is given by $\omega_p^2 = 4\pi\ ne^2/m = 3/r_s^3$ (a.u.). The induced density associated with the mode is localized at the surface $\delta n(r) \propto \delta(r-R)$. In the jellium model the electronic charge density can "spill-out" from the rigid positive background charge and the hydrodynamic calculations using a diffuse electronic surface density predict a size-dependent "red-shift" of this surface plasma mode with decreasing sphere size; i.e. Re $z_j < \omega_M^c$. We also expect to find additional surface plasma modes which for the classical model would be degenerate with ω_M^c. In addition, there are theoretical predictions[13] and experimental[14] results which indicate the existence of "bulk" plasma modes at frequencies above ω_p.

Because of the complexity of the solution spectrum, $\xi(z)$, Eq. 6, one cannot guarantee that all of the eigenfrequencies in a given region of the complex z-plane have been located. In addition it is often difficult to establish an unambiguous characterization of a particular eigenmode. Nonetheless, we attribute a number of the eigenfrequencies to dipolar plasma modes for the closed shell configurations of jellium spheres containing from N=8 to 198 electrons. The complex eigenfrequencies for some of these collective modes are presented in Fig. 2. The positive charge densities are approximately the same as the conduction electron density in bulk aluminum and sodium (2.07 and 3.99, respectively). The bar associated with the datum point denote the magnitude of the imaginary part of the eigenfrequency. The low-lying mode for the $r_s = 4.0$ case occurs near or below the continuum threshold, and consequently, the imaginary parts of these frequencies are infinitesimal.

The eigenfrequencies for the low-lying surface mode is "red-shifted" from ω_1^c, but for the spheres with $r_s = 4.0$ display only a slight radial dependence. The size-dependent red-shift is more pronounced for the spheres with $r_s = 2.0$. All of the bulk mode eigenfrequencies [two upper sets of data points] show the "blue-shift" with decreasing sphere size which was reported by Ekardt.[3] There is also a decrease in the imaginary part of these frequencies with increasing sphere radius which accounts for the prominence of these bulk features in the dynamic polarizability of the larger spheres.[3]

There are eigenmodes with prominent surface features in the range where $\omega_1^c < $ Re $z < \omega_p$. However, the cuts, introduced to define the analytic continuation of $\chi_1^o(r,r';\omega)$ and the single-particle excitations complicate the analytic structure of $\xi(z)$ and prevent the unique characterization of

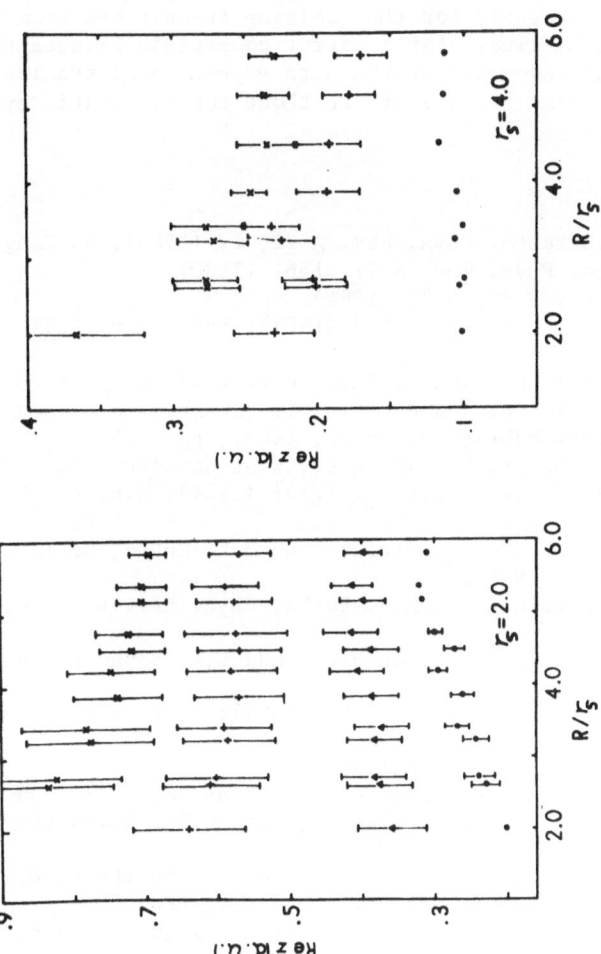

Fig. 2 The eigenfrequencies for plasma modes in closed shell configuration of electrons in jellium spheres. The data points indicate the real part of the frequencies and the magnitude of the imaginary part is denoted by the bar accompanying the datum point. The upper two sets of points are associated with "bulk" modes and the lower sets with "surface" modes. The bulk and classical surface plasma frequencies for $r_s = 2.0$ and 4.0 are $\omega_p = 0.6124$ and 0.2165 (a.u.) and $\omega_1^c = 0.3536$ and 0.1250 (a.u.), respectively.

399

many of these modes. For the spheres with $r_s = 2.0$ (a.u.) it was possible to isolate an additional surface mode and to trace its dependence on sphere radius, and the frequencies for this "surface" mode are also plotted in Fig. 2.

The experimental results for the dynamic response of metal particles[14] have been obtained using samples which contain a distribution of particle sizes, and the median radius for the sample is usually much larger than the radii of our spheres. Duthler et al.[14] using sodium particles, of less than 100 Å in diameter, reported a blue shift of 15% from ω_1^c for the low-lying peak in their data which also has a peak near the wavelength corresponding to ω_p. The surface mode for our largest sphere with $r_s = 4.0$ is red-shifted 11%; however, this eigenmode lies just below the continuum threshold which is the threshold for the ionizing transitions from the conduction band. It seems clear that a direct comparison between these theoretical calculations and experiment awaits experimental results for the dynamic response of the same specificity as those for the static system.[8]

REFERENCES

1. M.J. Stott and E. Zaramba, Phys. Rev. A 21, 12 (1980); A. Zangwill and Paul Soven, Phys. Rev. A 21, 1561 (1980).
2. D.E. Beck, Phys. Rev. B 30, 6935 (1984).
3. W. Ekardt, Phys. Rev. Lett. 52, 1925 (1984); Phys. Rev. B 31, 6360 (1985).
4. O. Gunnarsson and B.I. Lundqvist, Phys. Rev. B 13, 4274 (1976).
5. A. Zangwill in "Atomic Physics 8", Eds. I. Lindgren, A. Rosen and S. Svanberg (Plenum Publishing Corp., 1983), pg. 339.
6. W.D. Knight, K. Clemenger, W.A. de Heer, W.A. Saunders, M.Y. Chou and M.L. Cohen, Phys. Rev. Lett. 52, 2141 (1984); W.D. Knight, W.A. de Heer, K. Clemenger and W.A. Saunders, Solid State Commun. 53, 445 (1985); M.Y. Chou, A. Cleland and M.L. Cohen, Solid State Commun. 52, 645 (1984).
7. M. Manninen, R.M. Nieminen and M.J. Puska, Phys. Rev. B 33, 4289 (1986).
8. W.D. Knight, K. Clemenger, W.A. de Heer, and W.A. Saunders, Phys. Rev. B 31, 2539 (1985).
9. D.E. Beck, Solid State Commun. 49, 381 (1984); W. Ekardt, Phys. Rev. B 29, 1558 (1984).
10. W. Ekardt, Phys. Rev. B 32, 1961 (1985).
11. Zangwill and Soven in their atomic calculations (Ref. 1) found that their thresholds typically began several volts below the observed thresholds.
12. Details of this continuation as well as a more comprehensive report of this calculation will be published elsewhere.
13. R. Ruppin, Phys. Rev. B 11, 2871 (1975); B.B. Dasgupta and R. Fuchs, Phys. Rev. B 24, 554 (1981).
14. C.J. Duthler, S.E. Johnson and H.P. Broida, Phys. Rev. Lett. 26, 1236 (1971); a recent review of optical absorption of small metal particles [U. Kreibig and L. Genzel, Surface Sci. 156, 678 (1985)] presents the experimental and theoretical results for their plasma modes.

MAGNETIC PROPERTIES OF MOLECULAR METAL CLUSTERS

Robert E. Benfield

Chemical Laboratory, University of Kent at Canterbury
Canterbury CT2 7NH, Kent, U.K.

Relatively little is known of the physical properties of molecular metal cluster compounds. Magnetic susceptibility and electron paramagnetic resonance (EPR) experiments have shown that osmium carbonyl clusters of sufficiently high nuclearity have unusual "metametallic" magnetic properties. These represent a molecular analogue of the types of magnetism observed in small particles and colloids of several metals, but are beyond the ability of current theories to explain quantitatively. Results of low-temperature EPR spectroscopic measurements on several molecular cluster compounds such as $H_2Os_{10}C(CO)_{24}$ and $Os_{10}S_2(CO)_{23}$ are presented, and parallels between their properties and those of non-molecular metal clusters are discussed.

INTRODUCTION

Molecular carbonyl clusters of the transition metals are good structural models for small metal particles. They comprise discrete clusters of up to 55 metal atoms enveloped by ligands (1-4), with complete homogeneity of size and shape in experimental samples. Their geometries, determined by X-ray crystallography, include (Figure 1) fragments of bulk lattices, polytetra-hedral and pentagonal forms.

There are enormous electronic differences between covalently bonded, low-nuclearity molecular clusters and bulk metals (5,6); however, little is known of the physical properties of the larger clusters, and a smooth trend towards metallic properties with increasing nuclearity has been implicitly anticipated. It has now been found that high-nuclearity molecular clusters show instead unusual magnetic properties. These properties appear to represent a molecular analogue of the quantum-size effects (7-9) shown by small metal particles, and are termed "metametallic" to emphasise their qualitative difference from both the molecular and bulk metallic limits.

MAGNETIC PROPERTIES

Magnetic susceptibility measurements (10) have shown that even-electron osmium carbonyls with fewer than 10 metal atoms are diamagnetic, with a temperature-independent paramagnetic (TIP) susceptibility component which increases with cluster nuclearity. This directly demonstrates the smaller frontier orbital separation in the larger clusters, complementing electronic spectroscopy as the selection rules for orbital interactions differ.

A fairly complete quantitative picture of the frontier orbital structure in $Os_3(CO)_{12}$ can be built up from several calculational and experimental methods, indicating a 3 eV separation between occupied and unoccupied levels. This energy gap is comparable with that in octahedral low-spin complexes of the Co(III) ion (11). These are also diamagnetic, with a TIP component which perturbs the NMR chemical shift of the central cobalt nucleus.

The decanuclear cluster $H_2Os_{10}C(CO)_{24}$ (Fig. 1) behaves differently. It is paramagnetic below 70K, following the Curie law below 30K with a nominal magnetic moment of 0.6 Bohr Magnetons per cluster molecule (12). This paramagnetism is anomalous for an even-electron molecule. It is quite different from the diamagnetism of lower-nuclearity osmium clusters and from the Pauli paramagnetism (also ideally temperature-independent) of bulk metals.

The same type of weak paramagnetism has subsequently been found in carbonyl clusters of nickel and platinum (13, 14). Its origin most probably lies in the small frontier orbital separation in these compounds, which is less than 1 eV in $H_2Os_{10}C(CO)_{24}$. To pursue the analogy with octahedral complexes of transition metal ions, a high-spin electronic configuration (15) can be envisaged. The consequent magnetic properties are difficult to predict, because the metal atom hybridisation and the extent of electron delocalisation in clusters of this size are not known.

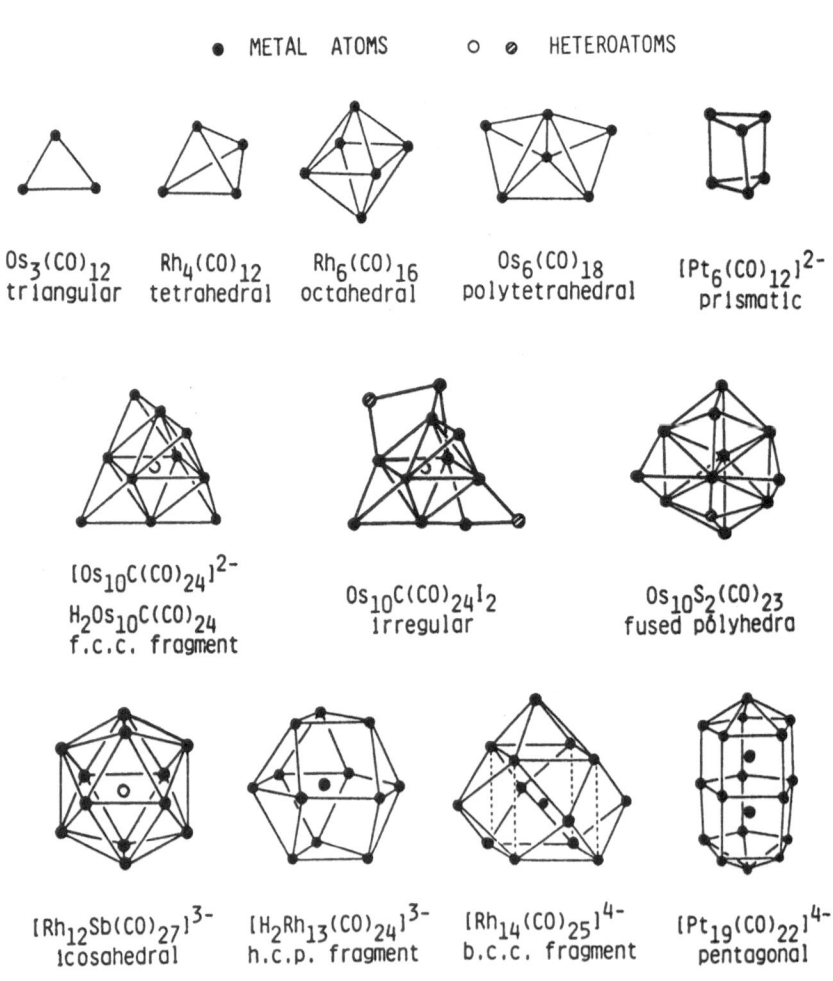

● METAL ATOMS   ○ ⊘ HETEROATOMS

$Os_3(CO)_{12}$
triangular

$Rh_4(CO)_{12}$
tetrahedral

$Rh_6(CO)_{16}$
octahedral

$Os_6(CO)_{18}$
polytetrahedral

$[Pt_6(CO)_{12}]^{2-}$
prismatic

$[Os_{10}C(CO)_{24}]^{2-}$
$H_2Os_{10}C(CO)_{24}$
f.c.c. fragment

$Os_{10}C(CO)_{24}I_2$
irregular

$Os_{10}S_2(CO)_{23}$
fused polyhedra

$[Rh_{12}Sb(CO)_{27}]^{3-}$
icosahedral

$[H_2Rh_{13}(CO)_{24}]^{3-}$
h.c.p. fragment

$[Rh_{14}(CO)_{25}]^{4-}$
b.c.c. fragment

$[Pt_{19}(CO)_{22}]^{4-}$
pentagonal

Figure 1. Typical metal atom geometries in cluster carbonyls.

The paramagnetism of $H_2Os_{10}C(CO)_{24}$ has been studied by EPR spectroscopy (Figure 2) (16). Below 100K, a single Lorentzian resonance without hyperfine splitting was observed; its g-factor of 2.280 ± 0.005 confirmed the strong orbital component expected in a cluster of low-valent 5d metal atoms. Resonance intensity followed the Curie law between 20K and 100K. The line-width varied linearly from 36 ± 1 Gauss at 20K to 94 ± 2 Gauss at 120K, corresponding to an electronic spin-spin relaxation time T_2 of the order of 10^{-9} seconds. A spin-lattice relaxation time T_1 of about 10^{-3} seconds was shown by power saturation measurements at 12K. Note particularly that in metallic cubic-symmetry systems, these two relaxation times are equal.

A progressive development of EPR lineshape asymmetry above 100K is associated with mobility of the interstitial hydrogen ligands within the $H_2Os_{10}C(CO)_{24}$ cluster. This type of stereodynamic behaviour is well-known in molecular clusters (17). Here, it represents localised electrical charge transport which perturbs the dielectric properties of the solid, mixing microwave absorption and dispersion to give a Dysonian-type lineshape (18).

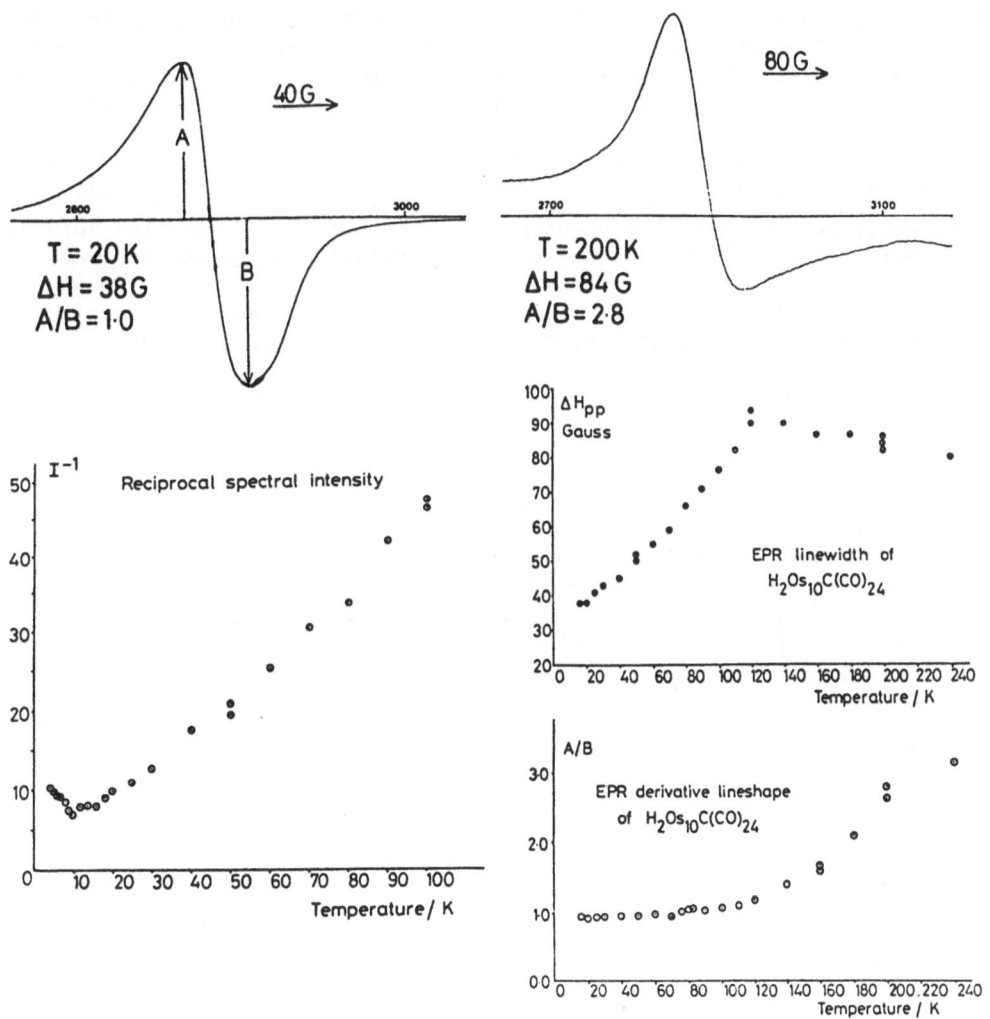

Figure 2. EPR of $H_2Os_{10}C(CO)_{24}$. Representative spectra, and temperature dependence of spectral intensity, linewidth and lineshape.

We have observed EPR in several other (3) decanuclear osmium clusters. Polycrystalline $[Os_{10}C(CO)_{24}]^{2-}[N(PPh_3)_2]_2^+$ (Fig. 1) showed Curie-law EPR intensity variation up to 100K. In contrast to the single resonance of $H_2Os_{10}C(CO)_{24}$, the lower crystal symmetry gave rise to three principal g-values (Figure 3). The central g-value of 1.92 was enormously shifted from the 2.280 of the isoelectronic parent compound. Removal of the two protons from the cluster may greatly perturb its frontier orbital pattern. One other feature of the EPR of this anionic cluster is easily understood: no lineshape asymmetry developed above 100K, because there are no interstitial hydrogen ligands to become mobile.

The axially symmetric substituted cluster $[Os_{10}C(CO)_{24}AuPPh_3]^-[PPh_3Me]^+$ showed EPR below 50K with $g_\perp$ = 2.28 and $g_{\parallel}$ = 2.44. $[Os_{10}C(CO)_{24}I]^-[N(PPh_3)_2]^+$ gave a resonance at g= 2.14 with evidence of a 96 Gauss hyperfine coupling to the iodine nucleus, implying spin density localisation on this atom.

Very interesting EPR spectra have been obtained from $Os_{10}S_2(CO)_{23}$. At 150K, a symmetric resonance 110 Gauss wide was observed at g = 2.23 (Fig. 3). Divergence of linewidth and a sharp change in spectral intensity in the temperature range 100-120K were evidence for a phase change. This is an almost unexplored aspect of the crystal chemistry of carbonyl clusters. On cooling below 100K, the resonance moved to lower field, reaching g = 3.0 at 25K.

This behaviour resembles that of some antiferromagnetic compounds (19). It can be related to the structure (20) of $Os_{10}S_2(CO)_{23}$ (Fig. 1). Crystals of this molecule contain a high concentration of osmium atoms. The two sulphur atoms bonded to the cluster do not bear any carbonyl groups, and so create "holes" in the ligand envelope which would otherwise completely surround each cluster (2,21). This promotes a potential pathway for magnetic interactions between clusters, which are not possible in the other Os_{10} compounds.

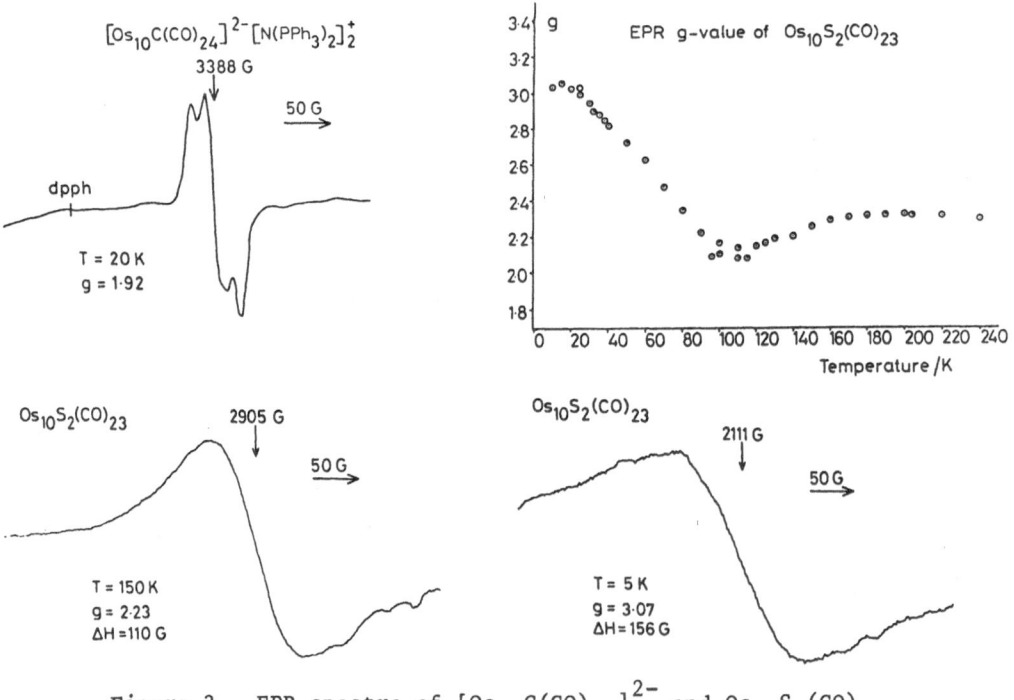

Figure 3. EPR spectra of $[Os_{10}C(CO)_{24}]^{2-}$ and $Os_{10}S_2(CO)_{23}$.

In contrast to these results, $Os_{10}C(CO)_{24}I_2$ (Fig. 1) gave no EPR; this compound is diamagnetic (10). Its structure has a central cluster of only eight osmium atoms. Similarly, $[Os_{11}C(CO)_{27}]^{2-}[PPh_3Me]^+_2$, which has an irregular structure (Fig. 1), was EPR silent. Lower nuclearity clusters such as $Os_6(CO)_{18}$ are also diamagnetic. For even-electron osmium carbonyl clusters to show Curie paramagnetism and EPR, it therefore appears that at least ten metal atoms must participate fully in cluster bonding. This condition does not seem to apply to platinum cluster carbonyl anions. Curie paramagnetism has been observed (14) in the trigonal prismatic $[Pt_6(CO)_{12}]^{2-}$ (Fig. 1).

Exchange interactions and spin-orbit coupling must strongly influence the magnetic properties of molecular clusters of these 5d metals. The paramagnetism of even-electron carbonyl clusters of nickel (13) can be related to the bulk ferromagnetism of this 3d metal. No examples of paramagnetism have yet been found in the corresponding even-electron compounds of 4d metals such as ruthenium and rhodium.

COMPARISON WITH METAL PARTICLES

As accurate electron wavefunctions and orbital energies for these large liganded molecular clusters cannot be calculated, quantitative understanding of their magnetic properties is still beyond us. However, there are strong similarities in magnetic behaviour between these molecular clusters and ligand-free metal particles small enough to show quantum-size effects (7-9).

Low-temperature Curie paramagnetism has been reported for microcrystals of even-electron metals such as platinum (22) and magnesium (23). These small clusters are believed to maintain charge neutrality (7,8,24). Susceptibility enhancement in even-electron metal particles can be predicted if level separation and exchange energy satisfy certain conditions (25, 26).

Lorentzian EPR spectra have been observed from small particles and colloids of several even-electron metals (27). Examples include magnesium and calcium evaporated into xenon matrices (28, 29); cadmium (30) and palladium (31) particles generated by annealing electrolytically coloured KCl; and 22 Å platinum particles formed by reduction of chloroplatinic acid in gelatin (32). These EPR spectra are consistent with "anomalous" paramagnetism in the metal particles, of the same type as observed in the molecular clusters.

In these particles, and in those of odd-electron metals such as silver (33), a well-established quantum-size effect is the inhibition of electron spin relaxation by the separation of the energy levels, so that EPR lines are narrower than conduction electron spin resonances of bulk metals. Exactly the same effect is observed in the decanuclear molecular osmium clusters. The relaxation time T_2 is longer than that in bulk osmium by a factor of 10^5.

CONCLUSION

The quantum-size effects which modify the properties of small metal particles are still poorly understood. Theory is extremely complex, and particle size distribution problems make many experimental observations of properties as a function of nuclearity difficult to interpret.

We have found that decanuclear molecular clusters of osmium have anomalous magnetic properties, showing several features in common with those of metal particles. The "high-nuclearity" molecular clusters and the "small" metal particles constitute a distinct size range in which quantum-size effects are important. We term their properties "metametallic" to emphasise their qualitative difference from bulk metallic behaviour.

The metametallic size regime is thus defined as comprising metal particles and molecular clusters with properties influenced by quantum-size effects. Molecular clusters are an exceptionally promising system for the experimental study of these effects.

ACKNOWLEDGEMENTS

This work was supported by the Royal Society. The U.K. Science and Engineering Research Council is also thanked for funding.

REFERENCES

1. P. Chini, J. Organomet. Chem. 200:37 (1980).
2. G. Schmid, Structure and Bonding (Berlin) 62:51 (1985).
3. J. Lewis and B.F.G. Johnson, Pure and Applied Chem. 54:97 (1982).
4. B.F.G. Johnson, ed. "Transition Metal Clusters", Wiley, New York (1980).
5. D.M.P. Mingos, Chem. Soc. Rev. 15:31 (1986).
6. E.L. Muetterties and R.M. Wexler, Surv. Prog. Chem. 10:61 (1983).
7. R. Kubo, J. Phys. Soc. Japan 17:975 (1962).
8. A. Kawabata, J. Phys. Soc. Japan 29:902 (1970).
9. R.F. Marzke, Catal. Rev. Sci. Eng. 19:43 (1979).
10. D.C. Johnson, R.E. Benfield, P.P. Edwards, W.J.H. Nelson and M.D. Vargas, Nature (London) 314:231 (1985).
11. C.J. Ballhausen, "Introduction to Ligand Field Theory", McGraw Hill, New York (1962), p.147.
12. R.E. Benfield, P.P. Edwards and A.M. Stacy, J. Chem. Soc. Chem. Commun. 525 (1982).
13. B.J. Pronk, H.B. Brom, L.J. de Jongh, G. Longoni and A. Ceriotti, Solid State Commun. 59:349 (1986).
14. B.K. Teo, F.J. DiSalvo, J.V. Waszczak, G.Longoni and A. Ceriotti, Inorg. Chem. 25:2262 (1986).
15. L.E. Orgel, "Introduction to Transition Metal Chemistry" 2nd edn., Methuen, London (1966), Chapter 3.
16. R.E. Benfield, J. Phys. Chem. in the press (1986).
17. B.F.G. Johnson and R.E. Benfield, Ref.4 , Chapter 7.
18. F.J. Dyson, Phys. Rev. 98:349 (1955).
19. P.R. Elliston, J. Phys. C7:425 (1974).
20. J.P. Attard, B.F.G. Johnson, J. Lewis, J.M. Mace, M. McPartlin and A. Sironi, J. Chem. Soc. Chem. Commun. 595 (1984).
21. R.E. Benfield and B.F.G. Johnson, J. Chem. Soc. Dalton 1743 (1980).
22. R.F. Marzke, W.S. Glaunsinger and M. Bayard, Solid State Commun. 18:1025 (1976).
23. J-L. Millet and J-P. Borel, Surface Science 106:403 (1981).
24. R. Kubo, A. Kawabata and S. Kobayashi, Ann. Rev. Mat. Sci. 14:49 (1984).
25. R. Denton, B. Mühlschlegel and D.J.Scalapino, Phys. Rev. B7:3589 (1973).
26. J. Sone, J. Phys. Soc. Japan 42:1457 (1977).
27. A.E. Hughes and S.C. Jain, Adv. Phys. 28:717 (1979).
28. J-P. Borel and J-L. Millet, J. de Physique 38:C2-115 (1977).
29. S. Sako and K. Kimura, Surface Science 156:511 (1985).
30. S.C. Jain, G.D. Sootha and R.K. Jain, J. Phys. C1:1220 (1968).
31. S.C. Jain, S.K. Agarwal, G.D. Sootha and R. Chander, J. Phys. C3:1343 (1970).
32. D.A. Gordon, R.F. Marzke and W.S. Glaunsinger, J. de Physique 38:C2-87 (1977).
33. G.A. Ozin, J. Amer. Chem. Soc. 102:3301 (1980).

ELECTRONIC PROPERTIES OF WATER CLUSTERS BY AN EFFECTIVE

HAMILTONIAN TREATMENT

D.E. Hagen, L. Jin, M.S. Choe, C.K. Lutrus
T. Oshiro, and S.H. Suck Salk

Department of Physics and Graduate Center for Cloud
Physics Research
University of Missouri-Rolla
Rolla, MO

INTRODUCTION

A computationally efficient semi-empirical effective Hamiltonian
method[1-3] for treating molecular clusters was recently developed. This
treatment led to reliable intermolecular binding energies and electronic
properties for hydrogen-bonded water clusters in agreement with ab initio
calculations[4], while other semi-empirical theories failed[5].
Encouragingly, physical properties such as ionization potentials, dipole
moments and normal mode vibrational frequencies were found to be in good
agreement with experiments[6-9] and theory[4-10]. In the present study we
focus our attention only on the ionization potentials and dipole moments
of the water clusters.

Currently most of the ab initio studies of molecular water clusters
$(H_2O)_n$ are limited to small sizes, usually to the n = 2 water dimer[10-11].
This is due to the excessive computation time. To the best of our
knowledge, there are no ab initio H-F (Hartree Fock) studies which pay
attention to the optimization (minimum energy configuration search) of
equilibrium geometries in association with water clusters. This study
serves as an extension to a more comprehensive and systematic
investigation of even larger clusters that were not reported previously.

PREDICTED FIRST IONIZATION POTENTIALS (ENERGIES) AND DIPOLE MOMENTS

To save space, in Fig. 1 we show only the selected geometric
structures of water cluster corresponding to various parts of the

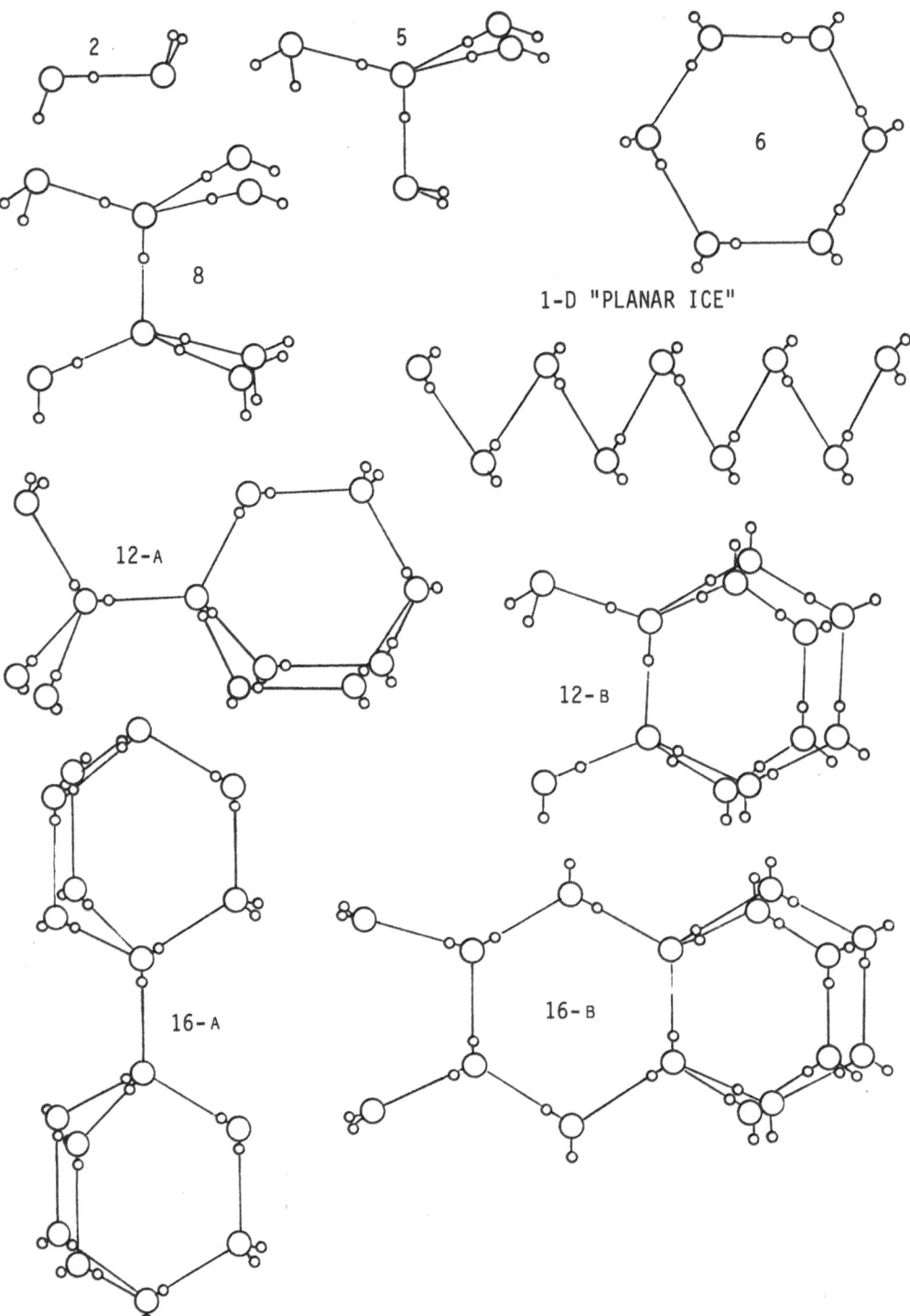

Fig. 1 Geometric configurations of selected
water clusters.

ordinary hexagonal ice (ice I) lattice, including an infinite
1-dimensional "planar ice". Geometry optimization was made in
association with self-consistent field calculations. To save space we
report only the general characteristics of predicted geometries in the
following. Complete information on the optimized geometries of $(H_2O)_2$
was reported elsewhere[1]. The predicted hydrogen bond length O-H--O
varied from 2.88 Å to 3.03 Å. For the tetrahedral pentamer, $(H_2O)_5$, we
found the hydrogen bond length of ~ 2.9 Å, the non-bonded O-H distance of
0.958 Å, the bonded O-H distance of 0.977 Å, the O--O--O angles of ~ 107°
to ~ 114°, the intramolecular HOH angles of ~ 105° and ~ 108°, the
intermolecular HOH angles of ~ 110° and ~ 115°, and the intermolecular
O-H--O angles of ~ 175° and ~ 178°. The rest of the clusters showed
similar geometric configurations in both the O-H bond lengths and the HOH
angles, while showing greater differences in the intermolecular O-H--O
bond angles, as expected.

It is well known that some of the semi-empirical methods[12-14]
reproduce observed physical and chemical properties reasonably well
despite parameterization with respect to only a small number of selected
molecules. Indeed, our past studies[15] using the MNDO (modified neglect
of diatomic-differential overlap)[14] methods have yielded the reliable
predictions of various physical properties of molecules including the
extended systems of organic crystals.

The new method of I-MNDO[1] which introduces the additional
intermolecular set of parameters into MNDO was proven to be highly
successful in obtaining reliable physical and chemical properties of the
hydrogen-bonded molecular water clusters for which other semi-empirical
theories[5] failed. Among some of the reliably predicted physical
properties are the stabilization energies or binding energies, ionization
potentials, dipole moments and normal mode vibrational frequencies as
mentioned earlier.

Figure 2 displays the predicted first ionization energies of $(H_2O)_n$
ranging n = 1 to n = 16. Our computed first ionization energies are
based on Koopman's theorem. That is, the negative of the highest
occupied orbital energy ε is the first ionization energy, $- \varepsilon = E^{N-1} - E^N$, where E^N and E^{N-1} are the Hartree Fock expectation values of the
energy of two single determinants corresponding to the N- and
(N-1)-electron systems respectively. Thus in our calculations of the
first ionization energies, correction for electron relaxation effects is
not introduced. Due to the parameterization, the electron correlation
effects are partially introduced in our semi-empirical effective

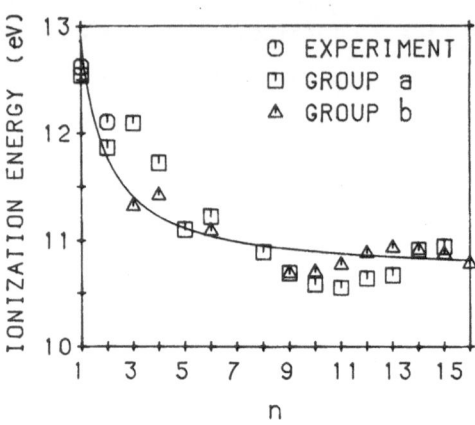

Fig. 2   Ionization energy (eV) vs.
cluster size n.

Hamiltonian calculations.   However, in general for systems containing
large numbers of electrons, the electron relaxation and correlation
effects tend to cancel each other.   Thus our predicted results for the
first ionization potentials, particularly for large clusters are expected
to be highly reliable.

Interestingly the predicted ionization potentials not only agree
well with the ab initio results of Tomoda and Kimura but also rapidly
converge to the bulk value for ice (11.0 eV[7] or 11.1 eV[8]), as shown in
Fig. 2.   Some of the predicted first ionization potentials are listed in
Table 1.   n = ∞ represents the infinite linear chain of $(H_2O)_n$.   Earlier
we[2] found that the predicted ionization potentials are scattered

Table 1.   First ionization energies ε in eV as a function of cluster size
n. ε(T-K) below stands for the ab initio results of Tomoda and
Kimura (See ref. 4 for n = 8a below.   The geometric structure
for 16a is shown in Fig. 1.)

n	ε(ours)	ε(T-K)	ε(expt.)
1	12.54	12.81	12.62 (ref. 9)
2	11.86	11.88	12.1 (ref. 10)
5	11.10	11.31	
8a	10.89	11.08	
12	10.88		
16a	10.78		
∞(ice)	11.0		11.1 (ref. 11)

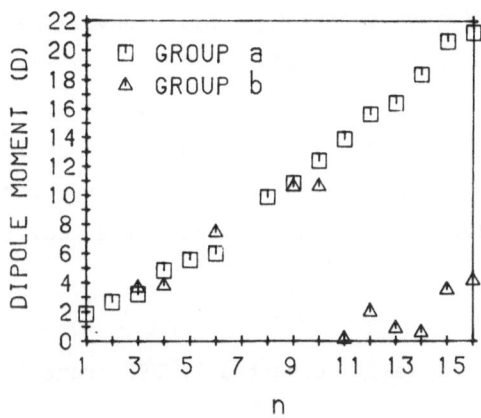

Fig. 3 Dipole moment (Debye) vs. cluster size n.

around the line, $\varepsilon = \alpha(1/n) + \beta$, with the slope of $\alpha = 2.19$ and the intercept $\beta = 10.7$.

Figure 3 shows the predicted dipole moments of various water clusters of size up to n = 16. For the clusters of Group a shown in Fig. 1, the dipole moments are seen to generally increase with cluster size. This implies that the dipole moments tend to cooperatively align themselves in the clusters as long as they are not heavily dominated by ring type structures. Indeed, the predicted dipole moment for the structure 16b is very small with the value of m = 4.13. Some of the predicted dipole moments are m = 2.61 D for the water dimer in good agreement with the experimental value of 2.60 D[16] and m = 5.52 D for the tetrahedral water cluster $(H_2O)_5$. To the best of our knowledge, there exists no measurements for water clusters beyond n = 2. Finally it is of note that the predicted dipole moments of 'Group a' clusters are linear in cluster size n.

SUMMARY

In the present study, we investigated the molecular water clusters of large size based on the optimized geometries. Presently such studies are not feasible by ab initio H-F calculations. The selected structures of the water clusters were taken from parts of the hexagonal ice lattice. Thus the physical properties (particularly ionization energies and dipole moments) studied here are by no means complete. However, general propensity rules are discovered through this study. They are now in order; 1) the first ionization energies rapidly converge to the bulk value at surprisingly small water cluster sizes, and 2) the dipole moments of water clusters tend to linearly increase as long as the structures of clusters are not dominated by ring type structures, indicating a cooperative phenomena of constructive addition.

ACKNOWLEDGEMENT

This material is partially supported by NSF grant ATM82-12328.

REFERENCES

1. S.H. Suck Salk, T.S. Chen, D.E. Hagen, and C.K. Lutrus, Theor. Chim. Acta 70, 3 (1986).
2. S.H. Suck Salk, and C.K. Lutrus, Phys. Rev. A., submitted.
3. S.H. Suck Salk, and C.K. Lutrus, in this issue.
4. S. Tomoda, and K. Kimura, Chem. Phys. Lett. 102, 560 (1983).
5. W. Thiel, Theor. Chim. Acta 48, 357 (1978); M. Gordon, D. Tallman, C. Monroe, M. Steinbach, and J. Ambrust, J. Am. Chem. Soc. 97, 1326 (1975); T. Zielenski, D. Breen, and R. Rein, J. Am. Chem. Soc. 100, 6266 (1978); G. Klopman, P. Andreozzi, A. Hoffinger, O. Kibuchi, and M. Dewar, J. Am. Chem. Soc. 100, 6268 (1978).
6. K. Kimura, S. Katzumata, Y. Achiba, T. Yamazaki, and S. Iwata, in 'Handbook of Photoelectron Spectra of Fundamental Organic Molecules' (Halsted Press, New York, 1981).
7. S. Tomoda, Y. Achiba, and K. Kimura, Chem. Phys. Lett. 87, 197 (1982).
8 A. Yencha, H. Kubota, T. Fukuyma, T. Kondow, and K. Kuuchitsu, J. Electron Spectry 23, 431 (1981).
9. D. Eisenberg, and W. Kauzmann, 'The Structure and Properties of Water' (Oxford Press, London 1969).
10. L.A. Curtiss, and J.A. Pople, J. Mol. Spectro. 55, 1 (1975).
11. O. Matsuoka, E. Clementi, and M. Yoshimmine, J. Chem. Phys. 64, 1315 (1976); K. Morokuma, and L. Pederson, J. Chem. Phys. 48, 3275 (1968); K. Morokuma, and J.R. Winich, J. Chem. Phys. 52, 1301 (19770); D. Hankins, J.W. Moskowitz, and F.H. Stillinger, J. Chem. Phys. 53, 4544 (1970); J. Del Bene, and J.A. Pople, J. Chem. Phys. 52, 4858 (1970); H. Popkie, H. Kistenmacher, and E. Clementi, J. Chem. Phys. 58, 5296 (1973); G.H.F. Diercksen, Theor. Chim. Acta 21, 335 (1971); 26, 249 (1975); P.A. Kollman, and L.C. Allen, J. Chem. Phys. 51, 3286 (1969); E. Clementi, 'Lecture Notes in Chemistry 2' (Springer, Berlin, Heidelberg, New York, 1976).
12. M.J.S. Dewar, 'The Molecular Orbital Theory of Organic Chemistry' (McGraw Hill, New York, 1967).
13. R. Bingham, M. Dewar, and D. Lo, J. Am. Chem. Soc. 97, 1285 (1975).
14. M.J.S. Dewar, and W. Thiel, J. Am. Chem. Soc. 99, 4899 (1977); ibid 4907 (1977); M.J.S. Dewar, E.G. Zoebisch, E.F. Healy, J.J.P. Stewart, J. Am. Chem. Soc. 107, 3904 (1985).
15. S.H. Suck Salk, J.L. Kassner, and Y. Yamaguchi, Appl. Opt. 18, 2609 (1979); S.H. Suck Salk, A.E. Wetmore, T.S. Chen, and J.L. Kassner, Appl. Opt. 21, 1610 (1982); V. Young, S.H. Suck Salk, and E.W. Hellmuth, J. Appl. Phys. 50, 6088 (1979); M.J.S. Dewar, Y. Yamaguchi, and S.H. Suck Salk, Chem. Phys. 43, 145 (1979).
16. J. Reimmers, R. Waats, and M. Klein, Chem. Phys. 64, 95 (1982).

AN AB-INITIO STUDY OF THE INTERACTION OF ATOMIC HYDROGEN WITH LITHIUM CLUSTERS

A. S. Hira, A. K. Ray and J. L. Fry

Department of Physics, University of Texas at Arlington
Arlington, Texas 76019

INTRODUCTION

As is well known, the interaction of the atoms or molecules of the different gases with the various metallic surfaces is important in the investigation of chemisorption and catalytic processes. In recent years, there has been significant scientific work on the interaction of gases such as hydrogen, oxygen and nitrogen, and some of the rare gases, with transition and other metals, including tungsten, copper, silver and platinum, and this work has contributed to the understanding of the energetics of chemisorption (Gasser, 1985; Tompkins, 1978). From a computational point of view, for obvious reasons, lithium is particularly convenient to work with, allowing the use of sophisticated techniques such as the Unrestricted Hartree-Fock Method with correlation included at various levels of approximation. This paper builds on the previous work of the authors (Ray, 1986; references therein) and other researchers (Beckmann and Koutecky, 1982; Pacchioni et al. 1984), and extends it to an ab-initio study of the interactions of atomic hydrogen with lithium clusters.

GEOMETRY AND COMPUTATIONAL METHOD

The work reported here is done within the context of the Unrestricted Hartree-Fock formalism. All the computations were carried out on an IBM 4381 computer, using the program GAMESS, written by Dupuis et al (1979). The clusters considered were Li_2 through Li_{10} in the bcc(100) and bcc(110) crystal structures. For each of the clusters, we varied the "lattice constant" a i.e. the second nearest neighbor distance. The minimum of the total cluster energy was used to calculate the optimum lattice constant and the maximum binding energy per atom in kilocalorie per mole. We tried calculations with 3-21G, 6-21G and 6-31G basis sets, for the bcc(100) clusters, and selected 6-21G as the best basis set for our purpose since it gave the lowest total energy for the Li_2, Li_4 (4,0,0) and Li_4 (2,2,0) clusters. This is in agreement with the discussion in the paper of Binkley et al (1980), where 6-21G is recommended for lithium over 6-31G. It was pointed out there that the single valence electron in lithium has a tendency to "fall inward" with the 6-31G basis set, because energy minimization dictates an inner shell function for this electron without regard to a proper description of the valence region. For the rest of our calculations we used the 6-21G basis set exclusively.

Next, we made runs for the interaction of a single hydrogen atom with some of the bcc(100) and bcc(110) clusters selected from among those in the first part. In this selection we followed the example of Beckmann and Koutecky (1982). From the point of view of the interaction with a crystalline cluster, the approach position of an atom is classified into three categories: the on top position, the open site position and the bridge position. In the on top position, the hydrogen atom takes a line of approach passing through the center atom among the atoms at the first level in the cluster. In the open site position, the line of approach is clear of all the atoms at the first level and is symmetrically enclosed by the lithium atoms. In the bridge position, the approaching hydrogen atom straddles the line joining two lithium atoms close together, in the first level of the cluster. The idea here is to model the peaks, valleys and ridges among the irregularities on a lithium surface. The largest number of cases was for the on top position, including both planar and non-planar clusters, while the smallest number was for the bridge position.

In this part, the variation in the total energy of the system as a function of the variation in the separation distance R of the hydrogen from the cluster was studied, R being allowed negative values for positions inside the non-planar clusters. The minimum of energy found here was used, together with the results from part one, to compute the energy of chemisorption and the equilibrium position of the hydrogen atom for each cluster.

RESULTS AND DISCUSSION

All the clusters studied in this paper are shown in figures 1 and 2. The binding energy per atom for the cluster Li_n is defined by

$$E_b(Li_n) = (nE(Li_1) - E(Li_n))/n \qquad (1)$$

while the chemisorption energy is defined by

$$E_c(Li_n) = E(H) + E(Li_n) - E(H - Li_n) \qquad (2)$$

The single atom energies $E(H)$ and $E(Li_1)$ were taken from the paper by Binkley et al. The notation used for the lithium clusters is $Li_n(n_1, n_2, n_3; 1m0)$, where n is the total number of atoms in the cluster; n_1 = number of atoms in first layer; n_2 = number of atoms in second layer; n_3 = number of atoms in third layer and m = 0 or 1, for bcc(100) or bcc(110).

The results from the first part of our calculations are summarized in Tables 1 and 3. The optimum value, a_{OPT}, of the lattice constant for each cluster was determined from a plot of the total energy versus a and then a separate run was made for a_{OPT} to get the minimum of energy. In all cases we obtained larger binding energies than those of Beckmann and Koutecky from their SCF runs, but smaller than those from their MRD-CI runs. Our UHF values are often approximately half their MRD-CI values. For our results, the bcc(110) clusters have larger binding energies than the corresponding bcc(100) clusters and, hence, are the more stable structures. Our binding energies for Li_3, Li_6 and Li_7 show considerable increase over the SCF values of Beckmann and Koutecky. We found Li_4 (2,2,0: 100), Li_5 (5,0,0: 110), Li_6 (6,0,0: 110) and Li_{10} (5,4,1: 110) to be the most stable structures for Li_4, Li_5, Li_6 and Li_{10}, in disagreement with their results. According to their work, the binding energy per atom, E_b, should approach the experimental value of 36.4 kcal per mole per atom for an infinite crystal as the number of nearest neighbours, n_b, and the number of next nearest neighbors, n_c, tend to 8 and 6 respectively. We plotted E_b against the average value of n_b, for the most stable Li_n cluster for a given n, and extrapolated to get a value of 25.3 kcal per mole per atom at

$n_b = 8$. The total range for a_{OPT} is found to be from 5.25 bohr to 6.61 bohr for the bcc(100) cluster and from 6.13 bohr to 6.75 bohr for bcc(110).

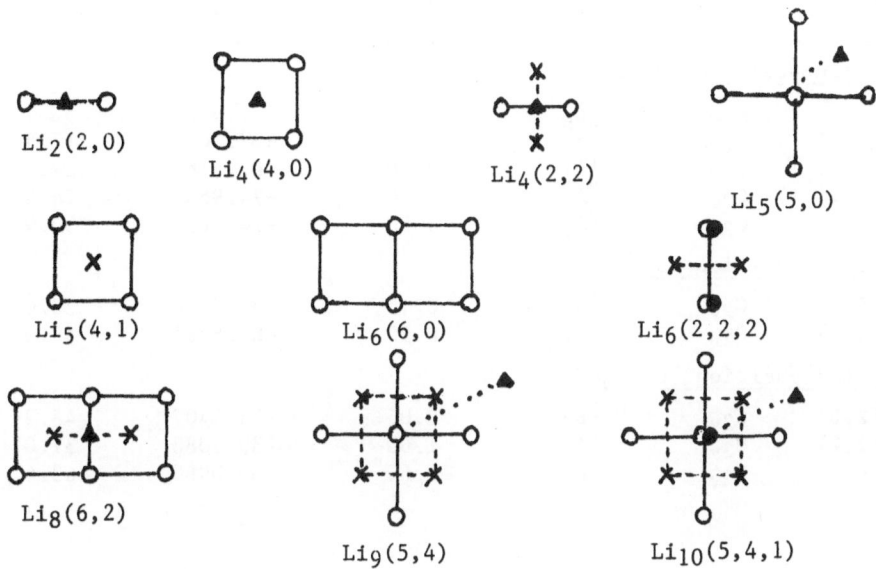

Fig. 1. Clusters modelling bcc(100) surface of the Li lattice: (O) show the first layer atoms; (✕) show the second layer; (●) show the third layer; (▲) represent the approach position of the hydrogen atom.

Table 1. Ground States and Energies of Li_n bcc(100) Clusters

Cluster	Geometry	Ground State	a_{min} (au)	Energy (au)	Binding Energy per atom (kcal/mole)
$Li_2(2,0)$	$D_{\infty v}$	$^1\Sigma_g$	5.32	-14.86674	1.72
$Li_4(4,0)$	D_{2h}	$^3B_{2g}$	5.50	-29.74474	3.49
$Li_4(2.2)$	C_{2v}	$^3A_2$	6.45	-29.75281	4.75
$Li_5(5,0)$	D_{2h}	$^2B_{1u}$	5.25	-37.17732	3.03
$Li_5(4,1)$	C_{2v}	$^2B_2$	6.20	-37.19054	4.69
$Li_6(6,0)$	D_{2h}	^1A_g	5.50	-44.60313	2.02
$Li_6(2,2,)$	C_{2v}	$^1A_1$	6.40	-44.63194	5.04
$Li_8(6,2)$	C_{2v}	$^1A_1$	6.30	-59.49755	4.12
$Li_9(9,0)$	D_{2h}	$^2B_{2g}$	5.60	-66.94300	4.69
$Li_9(5,4)$	C_{2v}	$^2A_1$	6.54	-66.97537	6.95
$Li_{10}(5,4,1)$	C_{2v}	$^1A_1$	6.61	-74.41940	7.10

The chemisorption energies are summarized in Tables 2 and 4. The multiplicities for the ground states were taken from the paper of Beckman and Koutecky, but were checked in most cases. For hydrogen interaction with Li_7 (7,0,0: 110), Li_9 (5,4,0: 110) and $Li_3(2,1,0: 110)$, we found the ground states to have multiplicity 3 instead of 1, in disagreement with their results.

We obtained larger chemisorption energies for the bcc(110) clusters

Table 2.   Ground State Energies and Equilibrium Distances R for H-Li$_n$ bcc(100) Systems

Cluster	Geometry	Ground State	R_{min} (au)	Energy (au)	E_c (kcal/mole)
Adsorption position "on top"					
H-Li$_1$	$C_{\infty v}$	$^1\Sigma_g$	3.15	$-$ 7.9787	32.1
H-Li$_5$(5,0)	C_{2v}	$^3A_2$	3.25	-37.7023	17.6
H-Li$_9$(5,4)	C_{2v}	$^1A_1$	3.30	-67.5167	27.8
H-Li$_{10}$(5,4,1)	C_{2v}	$^2A_1$	3.50	-74.9559	24.8
H-Li$_{10}$(1,4,5)	C_{2v}	$^2A_1$	3.50	-74.9656	30.9
Adsorption position "open"					
H-Li$_4$(4,0)	C_{2v}	$^4A_2$	0.45	-30.2784	23.0
H-Li$_9$(4,5)	C_{2v}	$^1A_1$	0.50	-67.5497	49.3
Adsorption position "bridge"					
H-Li$_2$(2,0)	C_{2v}	$^2B_2$	2.05	-15.4407	48.3
H-Li$_4$(2,2)	C_{2v}	$^4A_2$	1.60	-30.3088	37.0
H-Li$_8$(6,2)	C_{2v}	$^2A_2$	-2.60	-60.0960	63.6

than for the bcc(100) clusters, except in the case of Li$_5$ (5,0,0), Li$_9$ (5,4,0) and Li$_8$ (6,2,0). They are approximately equal for Li$_9$ (5,4,0) and Li$_9$ (4,5,0). The minimum energy for each H-Li$_n$ systems was determined from a UHF calculation for R_{OPT}, which is the optimum R value obtained from a total cluster energy versus R plot. According to Gasser, a typical value for chemisorption energy is 400 kJ/mole, while Tompkins give sthe range 50 to 400 kJ/mole for gas/metal systems. Tompkins' range is 11.95 to 95.6 in kcal/mole and all our energies lie in this range, except for H-Li$_5$ (5,0,0; 110) and H-Li$_7$ (7,0,0; 110).

In the on-top approach of the hydrogen atom, there is a sharp decrease in the chemisorption energies for the planar clusters as the number of lithium atoms is increased: from 32.1 kcal/mole for H-Li$_1$ to 0.845 kcal/mole for H-Li$_7$ (7,0; 110). The energies vary little for systems with four atoms in the second layer, including Li$_5$ (1,4), Li$_9$ and Li$_{10}$, even though the number of atoms in the first and third layers vary. These energies average at about 30 kcal/mole. Apparently, systems with atoms in the second and third layers are energetically favored and, even among these, the ones with a small number of atoms in the first layer have larger chemisorption energies. The chemisorption energies for the open-site and bridge approach positions are, predominantly, higher than for the on-top positions, which is to be expected because of the coulomb repulsion between the hydrogen atom and the lithium atom directly underneath it. These energies average about 43 kcal/mole. For the open site position, the presence of the second layer atoms increases the chemisorption energy, while the presence of atoms in the first layer decreases it. No conclusions of this type can be drawn from the data for the bridge position.

The optimum site for the hydrogen atom with respect to the cluster averages at about 3.20 bohr in the on-top approach, with small variations. R_{OPT} varies much more for the open site and the bridge positions: 2.05 bohr outside the surface for Li$_2$ to 2.60 bohr inside the Li$_8$ (6,2,0; 100) cluster. For Li$_4$ (4,0; 100) and Li$_9$ (4,5; 100), the hydrogen sits close to the surface at 0.45 bohr and 0.50 bohr. There are also local minimums at additional sites for Li$_{10}$ (5,4,1), Li$_{10}$ (1,4,5), Li$_4$ (2,2: 100), Li$_5$ (4,1; 110) and Li$_9$ (4,5: 110).

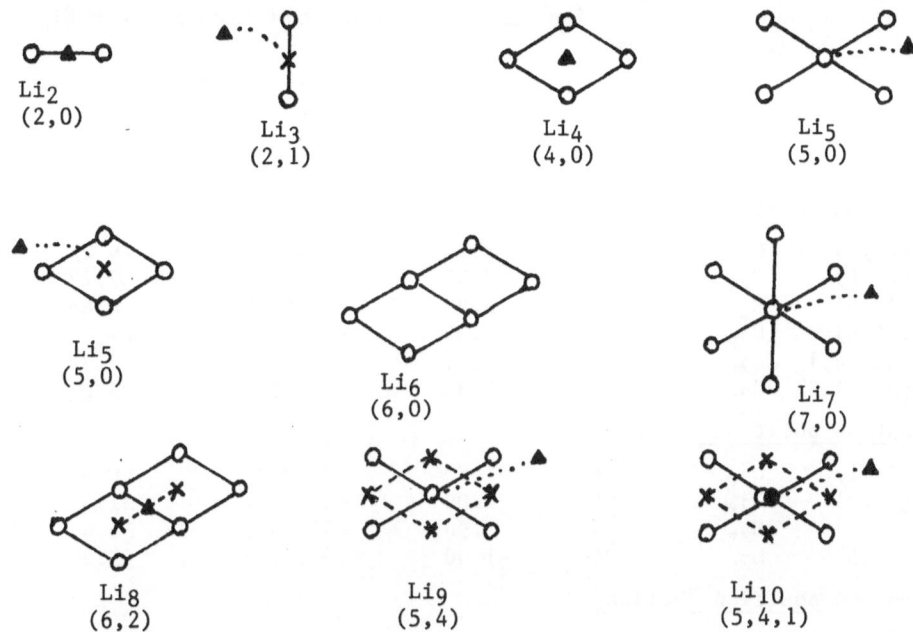

Li$_2$
(2,0)

Li$_3$
(2,1)

Li$_4$
(4,0)

Li$_5$
(5,0)

Li$_5$
(5,0)

Li$_6$
(6,0)

Li$_7$
(7,0)

Li$_8$
(6,2)

Li$_9$
(5,4)

Li$_{10}$
(5,4,1)

Fig. 2. Clusters modelling bcc(110) surface of the Li lattice. Symbols are the same as for Fig. 1.

Table 3. Ground States and Energies of Li$_n$ bcc(110) Clusters

Cluster	Geometry	Ground State	a_{min} (au)	Energy (au)	Binding Energy per Atom (kcal/mole)
Li$_2$(2,0)	D$_{\infty v}$	$^1\Sigma_g$	6.13	−14.86674	1.72
Li$_3$(2,1)	C$_{2v}$	^{2}B$_2$	6.13	−22.31078	3.95
Li$_4$(4,0)	D$_{2h}$	^{1}A$_g$	6.72	−29.73634	2.17
Li$_5$(5,0)	D$_{2h}$	^{2}B$_{3u}$	6.33	−37.19472	5.22
Li$_5$(4,1)	C$_{2v}$	^{4}A$_2$	6.36	−37.18872	4.46
Li$_6$(6,0)	C$_{2v}$	^{3}B$_2$	6.68	−44.65140	7.07
Li$_7$(7,0)	D$_{2h}$	^{2}A$_g$	6.65	−52.09355	7.09
Li$_8$(6,2)	C$_{2v}$	^{3}B	6.75	−59.52588	6.34
Li$_9$(9,0)	D$_{2h}$	^{2}G$_{2g}$	6.65	−66.95728	5.69
Li$_9$(5,4)	C$_{2v}$	^{2}A$_2$	6.54	−66.99478	8.30
Li$_9$(5,4,1)	C$_{2v}$	^{3}A$_2$	6.61	−74.43856	8.30

Some broad conclusions may be drawn from our work. For lithium clusters, the bcc(110) clusters are more stable, in general, than the bcc(100) clusters, though there are notable exceptions like Li$_4$ and Li$_5$. The optimum lattice constant is larger for the bcc(110) clusters than for the bcc(100) clusters, but stabilizes at about 6.50 bohr as the number of atoms reaches 10. The trend in the binding energy per atom with increasing n_b, the average number of the nearest neighbors in the cluster, is upward as expected and, in our calculations, extrapolates to 25.3 kcal per mole per atom for n_b equal to 8.

Table 4. Ground State Energies and Equilibrium Distances R for H-Li$_n$ bcc(110) Systems

Cluster	Geometry	Ground State	R_{min} (au)	Energy (au)	E_c (kcal/mole)
Adsorption position "on top"					
H-Li$_5$(1,4)	C_{2v}	$^1A_1$	3.20	-37.7292	29.2
H-Li$_5$(5,0)	C_{2v}	$^1A_1$	3.30	-37.7032	7.23
H-Li$_7$(7,0)	C_{2v}	$^3A_1$	3.20	-52.5919	0.845
H-Li$_9$(5,4)	C_{2v}	$^3A_1$	3.25	-67.5359	27.7
H-Li$_{10}$(5,4,1)	C_{2v}	$^2A_1$	3.20	-74.9823	29.3
H-Li$_{10}$(1,4,5)	C_{2v}	$^2A_1$	3.35	-74.9884	33.2
Adsorption position "open"					
H-Li$_3$(2,1)	C_{2v}	$^3A_1$	1.80	-22.8884	50.9
H-Li$_4$(4,0)	C_{2v}	$^4A_2$	1.00	-30.2834	31.4
H-Li$_5$(4,1)	C_{2v}	$^1A_1$	1.90	-37.7407	36.4
H-Li$_9$(4,5)	C_{2v}	$^3A_2$	-1.00	-67.5707	49.5
Adsorption position "bridge"					
H-Li$_2$(2,0)	C_{2v}	$^2A_1$	2.05	-15.4407	48.3
H-Li$_8$(6,2)	C_{2v}	^2A	1.65	-60.0923	43.6

Chemisorption occurs for all the clusters considered, with the chemisorption energy varying from 0.845 kcal/mole for H-Li$_7$ to 63.6 kcal/mole for kcal/mole for H-Li$_8$ (6,2; 100). The energies average about 35 kcal/mole, which compares favorably with other gas/metal systems. Chemisorption is stronger for the bcc(110) lithium surface. The presence of the second layer atoms facilitates chemisorption, while the presence of the third layer atoms hinders it. Chemisorption is favored for the valleys and ridges (the open-site and bridge positions) in the lithium surface over the hills or regular planes in the surface (the on top positions). In some cases, the hydrogen will dissolve inside a cluster or lattice, after overcoming an energy barrier, while other clusters have dissolution states which are transient.

Acknowledgements

We thank the Robert A. Welch Foundation (Grant No. Y707) for partial support of this work.

REFERENCES

Bauschlicher C. W. Jr., Bagus P.S., and Schaefer H. F. III, 1978, Model study in chemisorption: molecular orbital cluster theory for atomic hydrogen on Be(0001), IBM. J. Res. Develop., 22:213.

Bauschlicher C. W. Jr., Bender C. F., Schaefer H.F. III, and Bagus P. S., 1976, Chemisorption and the properties of metal clusters, Chem. Phys., 15:227.

Bauschlicher C. W. Jr., Liskow D. H., Bender C. F., and Schaefer H. F. III, 1975, Model studies of chemisorption. Interaction between atomic hydrogen and beryllium clusters, J. Chem. Phys., 62:12:4815.

Beckmann H-O, 1983, Ab initio investigation of hydrogen atom adsorption on Li clusters: embedded cluster model, J. Elec. Spect. and Related Phenom., 29:77.

Beckmann H-O., and Koutecky J., 1982, Ab-initio CI investigation of the interaction of a hydrogen atom with clusters which model the (100) and (110) surfaces of bcc Li-lattice, Surface Sci.. 120:127.

Binkley J.S., Pople J. A., and Hehre W. J., 1980, Self-consistent molecular orbital methods 21. Small split-valence basis sets for first-row elements, J. Am. Chem. Soc., 102:3:939.

Companion A. L., 1976, Potential energy surfaces for the interaction of atomic and diatomic hydrogen with lithium metal clusters, Chem. Phys., 14:1.

Dupuis M., Spangler D., Wendoloski J. J., 1983, GAMESS version 1.02 - revision 1 Oct. 80 - QG01.2, NRCC staff. IBM version - Quantum chemistry group - North Dakota State University.

Fantucci P., Pacchioni G., and Fernandez - Rico J., 1981, The electronic structure of small metallic clusters. Part III. Ab initio calculations of the interaction of atomic hydrogen with small Li_n (n = 3, 4, 5) clusters, J. Mol. Catal., 12:245.

Gasser R. P. H., 1985, "An Introduction to Chemisorption and Catalysis by Metals", Oxford University Press, New York.

Isett I. C., and Blakely J. M., 1975, Binding energies of Carbon to Ni (100) from equilibrium segregation studies, Surface Sci., 47:645.

Kunz A. B., Mickish D. J., and Deutsch P. W. , 1973, On the interaction of a hydrogen atom with a lithium metal surface, Solid State Commun., 13:35.

Lavery R., and Hillier I. H., 1978, Cluster and crystal orbital approaches to the adsorption of hydrogen on lithium metal, J. Mol. Catal., 4:9.

Pacchioni G., Koutecky J., and Beckmann H-O., 1984, Mixed-basis approach to the chemisorption problem; example of H chemisorption on Li_n clusters which the Li (100) surface, Surface Sci., 144:602.

Ray A. K., 1986, Minimal basis-set SCF-MO calculations on aggregates of lithium and beryllium atoms, J. Phys. B: At. Mol. Phys., 19:1253.

Stoll H., and Preuss H., 1977, On the chemisorption of hydrogen on lithium clusters, Surface Sci., 65:229.

Tompkins F. C., 1978, "Chemisorption of Gases on Metals", Academic Press Inc., New York.

Votjik J., and Fiser J., 1982, Interaction of atomic hydrogen with lithium metal clusters: breakdown of the adiabatic approximation, Surface Sci., 121:111.

THE EPR SPECTRUM OF Na_3 IN ADAMANTANE[1]

J.A. Howard, C.A. Hampson*, M. Histed*, and H. Morris* and
B. Mile

National Research Council of Canada
Division of Chemistry
Montreal Road
Ottawa, Ontario, Canada K1A 0R9

ABSTRACT

The sodium trimer, Na_3, has been prepared in adamantane and its EPR
spectrum recorded from 4 to 170 K. It has three magnetically equivalent
Na nuclei with magnetic parameters $\underline{a}_{23}$ = 93.6 ± 0.3 G and
$\underline{g}$ = 2.0027 ± 0.0003. These parameters are similar to those for
pseudorotating Na_3 in argon. It has a "normal" spectrum with no evidence
for an alternating linewidth effect or dynamic frequency shifts. This
suggests that the correlation time, τ_c, is faster in adamantane than it is
in argon.

INTRODUCTION

The paramagnetic alkali-metal trimers Li_3, Na_3, and K_3 isolated in
inert gas matrices have been thoroughly studied by Lindsay and
coworkers[2-7] by electron paramagnetic resonance spectroscopy (EPR). They
found that Li_3[7] has three magnetically equivalent nuclei even at 4 K in
solid argon, i.e., it has the unpaired spin population equally distributed
between three nuclei and must be a fluxional molecule because a static
equilateral triangular array of nuclei is unstable relative to the obtuse
($^2B_2$) and acute ($^2A_1$) triangular structures. Pseudorotation is believed
to occur between the three possible obtuse forms with the acute form at
the saddle points of the energy barrier. Pseudorotating Na_3 and K_3 also
exist in solid argon down to 4 K and are isostructural with Li_3.
Interestingly, rigid obtuse triangular forms of these trimers are also
found in this solid matrix but interconversion of rigid and pseudorotating
forms has not been observed.

---

*Department of Chemistry and Biochemistry, Liverpool Polytechnic,
 Liverpool, England L3 3AF

Pseudorotating Na_3 and K_3 are of further interest because they exhibit an "alternating linewidth" effect which is characteristic of a radical that pseudorotates with a frequency similar to the hyperfine interaction. In addition the EPR spectrum of pseudorotating Na_3 in argon is complicated by a dynamic frequecy shift which like the second order effect moves certain EPR transitions to lower fields.

The only group 1 trimer that has been reported in inert hydrocarbon matrices at 77 K is Li_3[8] and it has the same pseudorotating structure in this matrix as it does in argon with similar magnetic parameters in the two matrices.

In the present paper we describe the preparation, magnetic properties, and structure of a second alkali-metal trimer, Na_3, in adamantane.

EXPERIMENTAL

Na_3 was prepared by agglomeration of Na atoms in adamantane at 77 K on a rotating cryostat.[9,10] The Na atoms were prepared by vapourizing metallic sodium in a resistively heated furnace at ~800 K.

EPR spectra were recorded at X-band on Varian E-4 and E-12 spectrometers equipped with accessories for measurement of the magnetic field and microwave frequency. Temperature control of the sample was achieved with a homemade cold N_2 gas thermostat or with an Oxford Instruments He cryostat.

RESULTS

EPR spectra at 77 K of ^{23}Na (I=3/2) atoms deposited in adamantane are shown in Figure 1. Figure 1a was taken at a power level of 2mW and consists of two quartets from Na atoms in two trapping sites I and II with the magnetic parameters a_{23}^I = 299.4 G and g = 2.0017 and a_{23}^{II} = 226.4 G and g = 2.0027. In addition there is a ten-line spectrum exhibiting further splittings on all the lines but the outer two from a species with three magnetically equivalent Na nuclei. The outer transitions, M_I = ±9/2, J = 9/2; M_I = ±7/2, J = 7/2, 5/2; M_I = ±5/2, J = 9/2, 7/2, 5/2 had the relative intensities 1:1:2:1:2:3, expected for a second-order shift, i.e., there was no broadening of the M_I = ±7/2, and ±5/2 components relative to M_I = ±9/2 components. Exact solutions of an isotropic spin Hamiltonian for the J = 9/2, 7/2, 5/2, and 3/2 spin states using the ±M_I lines in pairs gave the magnetic parameters listed in Table 1. A statistical analysis gave a_{23} = 93.6 ± 0.1 and g = 2.0017 ± 0.0003. The small errors for a_{23} and g suggest that the spectrum of Na_3 is not affected by dynamic frequency shifts at 77 K. The field positions of the 30 lines of Na_3 are shown as a "stick diagram" in Figure 1a.

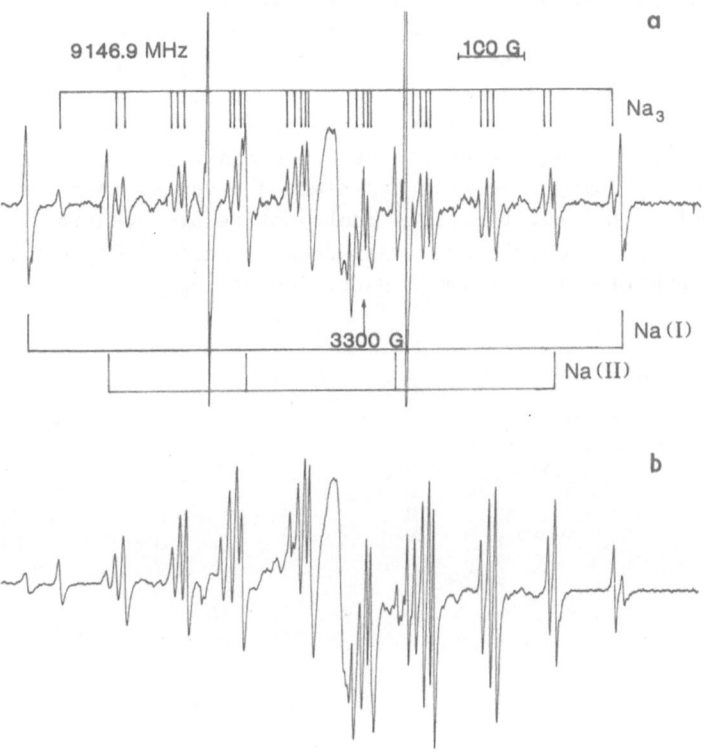

Figure 1.   EPR spectrum of Na₃ in adamantance at 77 K and power levels
of 2 mW a) and 50 mW b)

Table 1   Magnetic parameters of Na₃
in adamantane at 77 K.

J	$\underline{M}_I$	$\underline{a}_{23}$/G	g
4.5	±4.5	93.6	2.0012
	±3.5	93.6	2.0007
	±2.5	92.8	2.0007
	±1.5	93.3	2.0012
3.5	±3.5	93.4	2.0016
	±1.5	93.4	2.0011
	±2.5	93.6	2.0015
2.5	±2.5	93.8	2.0014
	±1.5	93.1	2.0009
1.5	±1.5	93.7	2.0015

When the spectrum was recorded at a power setting of 50 mW the intensity of the atom lines was drastically reduced and the spectrum from Na_3 dominated the overall spectrum (Figure 1b). The spectrum of Na_3 was recorded at ∼4 K (Figure 2a) and at 170 K (Figure 2b). At ∼4 K there was some loss of intensity of the lines from $J < 9/2$ while the $J = 9/2$ lines remained unaffected. Na_3 was stable at 170 K and was in fact still present at 270 K. These annealing experiments revealed that Na^{II} disappeared before Na^I but the atoms could not be preferentially removed from adamantane to leave only Na_3. All these species disappeared at 300 K leaving the conduction EPR of small particles of Na.

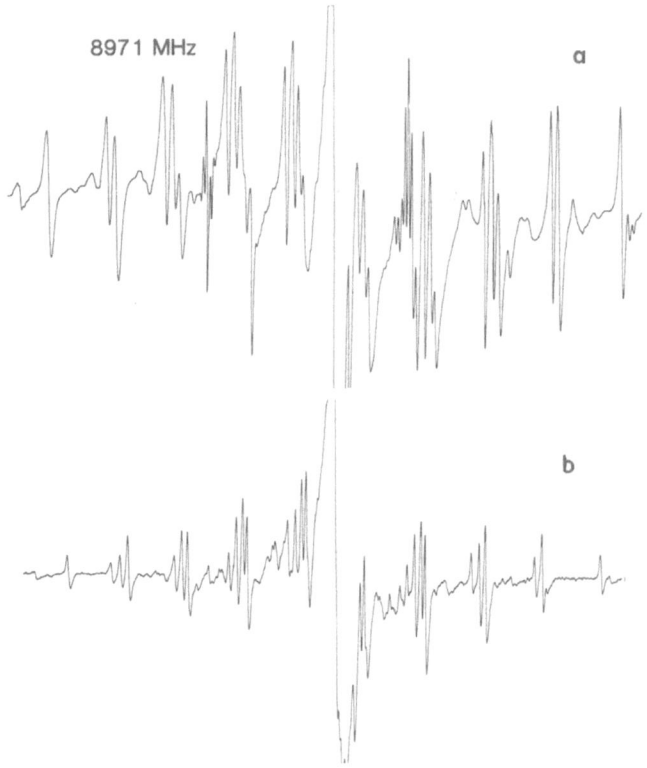

Figure 2. EPR spectrum of Na_3 in adamantane at ∼4 K a) and 170 K b)

Although there were other unassigned lines in the spectrum there was no evidence for a septet of quartets from obtuse Na_3 or a quartet of septets from acute Na_3.

DISCUSSION

It is apparent from the present work that the only trimer that is formed by condensing sodium atoms in adamantane at 77 K is Na_3 with three magnetically equivalent sodium nuclei. The Jahn-Teller theorem precludes a stationary equilateral triangular array of nuclei (point group D_{3h}) because the orbital degeneracy of the resulting $^2E'$ state must be lifted to give either an obtuse ($^2B_2$) or acute ($^2A_1$) angled structure of C_{2v} symmetry. The Na_3 produced in adamantane must therefore be a fluxional molecule which pseudorotates between the three obtuse or acute triangular forms e.g.

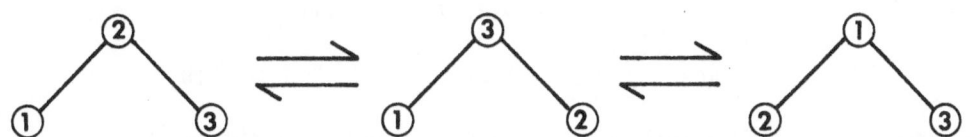

Theoretical calculations[11] predict the $^2B_2$ state to have the lowest energy with three equivalent potential wells distinguished by a different Na atom at the apical position and separated from each other by energy barriers whose saddle points correspond to the acute structures. The surface has three-fold symmetry about the central high energy equilateral configuration and is described as a "Mexican hat". Clearly the energy barrier between these wells in Na_3 is sufficiently small that the molecule is completely fluxional within the time scale of the EPR experiment ($\sim 10^{-7}$s). Na_3 is, therefore, like Li_3, pseudorotating in adamantane but differ from Cu_3[12], Ag_3[13], and Au_3[14] which give only obtuse trimers in hydrocarbon matrices and acute trimers in solid nitrogen.[15]

The EPR parameters for Na atoms and Na_3 in adamantane are collected in Table 2 along with some of the values obtained in powdered argon and the gas phase. The hyperfine interaction for Na atoms in site I of adamantane is close to the gas phase value ($a_{adam}/a_{gas} = 0.95$) whereas it is well below the gas phase value in site II ($a_{adam}/a_{gas} = 0.71$) indicating significant loss of unpaired spin population to the matrix. There is a similar spread of hfi in argon for Na atoms with at least five trapping sites.[16]

There is only one trapping site for Na_3 in adamantane compared with three in argon. Interestingly, the magnetic parameters for Na_3 in adamantane and argon are much closer than those for the atoms in these two matrices. Dividing a_{23} for Na_3 by a_{gas} gives an unpaired s spin population of 0.3 per nucleus and a total s spin population of 0.9. The singly occupied molecular orbital for Na_3 has, therefore, at the most 10% p character. This can be compared with Li_3 which has a SOMO with about 30% p character.

Table 2   EPR parameters of ^{23}Na and ^{23}Na$_3$ in inert matrices

		Matrix	$\underline{a}_{23}$/G	g-factor	Ref
Na	site I	adamantane	299.4	2.0017	This work
	site II	adamantane	226.4	2.0027	This work
		argon	331.0	1.9995	17
		argon	248.5 -331.7	1.9997 - 2.0000	16
		gas	316.1		18
Na$_3$		adamantane	93.6	2.0016	This work
	site I	argon	93.2	2.0012	4
	site II	argon	95.6	2.0016	4
	site III	argon	94.7	2.0018	4

DIFFERENCES BETWEEN Na$_3$ IN ADAMANTANE AND ARGON:

Despite the similarity of the magnetic parameters for pseudorotating Na$_3$ in adamantane and argon there are two major and important differences between the EPR spectra in the two matrices. (i) In argon only the M_I = ±9/2 and ±3/2 transitions are observed, the other transitions being broadened by the switching between the three $^2B_2$ forms which though of identical energy have a different m_I value at one of the three sodium nuclei. In adamantane all transitions are seen and M_I = ±7/2, ±5/2, and ±1/2 lines are not appreciably broadened and have the intensities expected for a "normal" spectrum with no motional broadening or narrowing. (ii) There is no detectable dynamic frequency shift in adamantane although a shift of 5-10 G was observed in argon. Both these differences are explicable if the correlation time, τ_c, in adamantane is much shorter (1/10th) than that in argon; τ_c being a measure of the residence time for each $^2B_2$ state in its shallow potential well. Considering first the motional broadening of the M_I = ±7/2, ±5/2, and ±1/2 lines, we can use the equation quoted by Lindsay for the dynamic linewidth $\Gamma(\underline{M},k)$

$$\Gamma(\underline{M},k) = \frac{2}{9}(\underline{a}_0 - \underline{a}_3)\left[M_I^2 - 3(\underline{m}_1\underline{m}_2 + \underline{m}_1\underline{m}_3 + \underline{m}_2\underline{m}_3)\right]\tau_c \qquad [1]$$

where $\underline{a}_0$ and $\underline{a}_3$ are the terminal and apical hfi, respectively. It is easily seen that the second term in parenthesis is zero for M_I = ±9/2 and

±3/2 and hence these four transitions are always seen. For the remaining values of M_I this term is not zero and if τ_c is sufficiently long the linewidths of these transitions can be broadened beyond detection. This occurs in an argon matrix. The observation of all these transitions in adamantane requires that τ_c be considerably shorter in this matrix than in argon.

Turning now to the dynamic frequency shift, H_1, which is 5-10 G for Na_3 in argon and virtually zero in adamantane. We again use the expression given by Lindsay based on Redfields perturbation theory.[19]

$$H_1 = -\frac{<\delta \underline{a}^2>}{2H} \left[3I(I+1) - \sum_q \underline{m}_q^2 \right] \frac{\omega_0^2 \tau_c^2}{1+\omega_0^2 \tau_c^2} \qquad [2]$$

where $<\delta \underline{a}^2> = \frac{2}{9}(\underline{a}_0 - \underline{a}_3)^2$, ω_0 is the Larmor frequency and $\underline{m}_q = \pm 3/2, \pm 1/2$. This expression when combined with the eigenvalues of the static Hamiltonian (H_0) predicts, with the selection rules $\Delta \underline{m}_s = \pm 1$, $\Delta \underline{m}_q = 0$, transitions (in G) at the following field positions.

$$H = H_0 + H_1 = \frac{g_e H_e}{g_0} - \bar{\underline{a}} M - \frac{a^2}{2H} \left[3I(I+1) - \sum_q \underline{m}_q^2 \right]$$

$$- \frac{<\Delta \underline{a}^2>}{2H} \left[3I(I+1) - \sum_q \underline{m}_q^2 \right] \frac{\omega_0^2 \tau_c^2}{1+\omega_0^2 \tau_c^2} \qquad [3]$$

where $\bar{a} = (2\underline{a}_0 + \underline{a}_3)/3$. The last term of equation [3], the dynamic frequency shift causes the greatest downfield movement for the $M_I = \pm 3/2$ lines and the lowest downfield shift for $M_I = \pm 9/2$, the magnitude of the shift depending critically on the magnitude of $\omega_0^2 \tau_c^2$ compared to unity. When $\omega_0^2 \tau_c^2 \gg 1$, as is the case of Na_3 in argon, the term becomes independent of τ_c and has the value

$$-\frac{<\delta \underline{a}^2>}{2H} \left[3I(I+1) - \sum_q \underline{m}_q^2 \right]$$

However, when $\omega_0^2 \tau_c^2 \ll 1$ the term becomes

$$-\frac{<\delta \underline{a}^2>}{2H} \left[3I(I+1) - \sum_q \underline{m}_q^2 \right] \omega_0 \tau_c^2$$

This appears to be the case for Na_3 in adamantane where the dynamic frequency shift is < 0.5 G. Inserting the values for $\delta \underline{a}^2$, I, $\underline{m}_q$, and H leads to an estimate of $\omega_0^2 \tau_0^2 < 0.15$ and hence $\tau_c < 4 \times 10^{-11}$ sec., if $\omega_0 = 9.12 \times 10^9$ sec.$^{-1}$ at 77 K, with only a slight increase at 4 K. This compares with $\tau_c \sim 2 \times 10^{-10}$ sec. at 30 K for Na_3 and K_3 in argon. The residence of the Na_3 trimer in any obtuse geometry in adamantane is therefore about five times shorter than in argon. This, in turn, indicates that the small energy barriers surrounding these vibronic states are lower by 0.5 to 1 kJ mol^{-1} in adamantane assuming similar frequency terms for both matrices. We believe that this suggests that there is less perturbation of these vibronic states in adamantane and that they more closely resemble the free "gas-phase" state. This is probably because the substitutional site in adamantane is larger than in argon (5.6 Å compared with 3.8 Å).

Two puzzling features remain. First, although EPR spectra of alkali and coinage metal trimers have been observed in adamantane only alkali metal trimers have been observed in argon. There is no obvious explanation for this anomoly but it may be pertinent that on cooling below 30 K the EPR intensity of rigid Na_3 in argon first increased to a maximum at 10 K and then fell to zero[3]. It is possible that the temperature for maximum intensity of the coinage metal trimers may be much higher than 10 K although we, like Lindsay, cannot offer any explanation for such behavior.

Second, unlike the alkali metal trimers, EPR spectra of the coinage metal trimers in adamantane indicate static obtuse molecules and hence appreciably higher barriers at the saddle points. The virtual absence of such barriers for alkali metal trimers in the same matrix suggests that the barriers in the coinage metal trimers are molecular and intrinsic in origin rather than matrix effects. This is not entirely in accord with the laser Raman studies of Cu_3 in argon which indicate a pseudorotating trimer.[20]

To end on a cautionary note it is conceivable that the spectrum we have assigned to pseudorotating Na_3 could arise from Na_3^{2+} which would have the unpaired electron in the bonding σ orbital. We do, however, consider this to be unlikely because there is no obvious mechanism for formation of a doubly charged cation in hydrocarbon matrices.

ACKNOWLEDGEMENT: We thank NATO for financial support in the form of a travel grant. (No.442/82).

REFERENCES

1. Issued as NRCC No. 26673.

2. D.M. Lindsay, D.R. Herschbach, and A.L. Kwiram. Mol. Phys. 32, 1199 (1976); 39, 529 (1980).

3. G.A. Thompson, F. Tischler, D. Garland and D.M. Lindsay. Surf. Sci. 106, 408 (1981).

4. D.M. Lindsay and G.A. Thompson. J. Chem. Phys. 77, 1114 (1982).

5. D.M. Lindsay, D. Garland, F. Tischler and G.A. Thompson. Am. Chem. Soc. Symp. Ser. 179, 69 (1982).

6. G.A. Thompson and D.M. Lindsay. J. Chem. Phys. 74, 959 (1981).

7. D.A. Garland and D.M. Lindsay. J. Chem. Phys. 78, 2813 (1983).

8. J.A. Howard, B. Mile and R.Sutcliffe. Chem. Phys. Lett. 112, 84 (1984).

9. J.E. Bennett, B. Mile, A. Thomas and B. Ward. Adv. Phys. Org. Chem. 8, 1 (1970).

10. A.J. Buck, B. Mile and J.A. Howard. <u>J. Am. Chem.Soc.</u> <u>105</u>, 3381 (1983).

11. R.L. Martin and E.R.Davidson. <u>Mol. Phys.</u> 36, 1713 (1978).

12. J.A. Howard, K.F. Preston, R. Sutcliffe and B. Mile. <u>J. Phys. Chem.</u> 87, 536 (1983).

13. J.A. Howard, K. Preston and B. Mile. <u>J. Am. Chem. Soc.</u> 103, 6226 (1981).

14. J.A. Howard, R. Sutcliffe and B. Mile, <u>J. Chem. Soc. Chem. Commun.</u> 1449 (1983).

15. K. Kernisant, G.A. Thompson, and D.M. Lindsay. <u>J. Chem. Phys.</u> 82, 4739 (1985); D.M. Lindsay, G.A. Thompson, and Y. Wang. In press.

16. J.P. Goldsborough and T.R. Koehler. <u>Phys. Rev. A.</u> 133, 135 (1964).

17. G.A. Thompson, F. Tischler, and D.M. Lindsay. <u>J. Chem. Phys.</u> 78, 5946 (1983).

18. J.R. Morton and K.F Preston. <u>Landolt-Börnstein, New Series;</u> H. Fischer and K.-H. Hellwege, Eds.; Springer-Verlag: West Berlin, Vol 9, Part a (1979).

19. A.G. Redfield. <u>Adv. Mag. Res.</u> 1, 1 (1965).

20. M. Moskovits. <u>Chem. Phys. Lett.</u> 118, 111 (1985).

ELECTRONIC STRUCTURE AND PROPERTIES OF SMALL Be-CLUSTERS

M. R. Press, B. K. Rao, S. N. Khanna, and P. Jena

Physics Department
Virginia Commonwealth University
Richmond, VA 23284-0001

ABSTRACT

Equilibrium geometries and electronic structure of Be-clusters up to four atoms per cluster are obtained by complete geometry optimization using selfconsistent ab initio theories. The exchange and correlation contribution to total energy are obtained both from Hartree-Fock configuration interaction and local spin density schemes. The interaction between atoms in charged Be_2 clusters is also studied as a function of inter-nuclear distance to illustrate the role of charge polarization. The results from both theories are compared.

INTRODUCTION

Atoms of alkaline earth series, such as Be, Mg have closed outermost electronic shell. Consequently, dimers of these atoms interact weakly through a Van der Waals type mechanism. The experimental binding energy of Be_2 and Mg_2 are respectively 0.12 eV and 1.15 eV. Yet, Mg and Be form good metals[2] in solid phase and their bulk cohesive energies are 1.51 eV and 3.32 eV respectively. Thus, it is interesting to see how these weakly bound dimers evolve into a strongly cohesive solid with increasng size. It is also interesting to study the energetics of charged clusters. Consider Be_2 as an example. The dimer has four valence electrons. Two of these form a bonding orbital whereas the other two go to the antibonding state. In a Be_2^+ cluster, it is an antibonding electron that is removed, thus making Be_2^+ a strongly bonded ion compared to Be_2 that is very weakly bound. This is, of course, quite contrary to what one observes in alkali atom clusters.

CALCULATION

We have calculated the energetics of Be clusters using self-consistent field linear combination of atomic orbitals-molecular orbital (SCF-LCAO-MO) method. The total energies of these clusters and the corresponding geometries have been obtained using two different approximations. In one,[3] we calculate the exchange energy using unrestricted Hartree-Fock method. The correlation contribution is incorporated through a configuration interaction scheme involving pair-excitations of all the valence and core-electrons. We have used the Gaussian 6-311G** basis set[4] for Be for this calculation. In the other SCF-LCAO-MO method, the exchange-correlation energy is calculated using the local spin density approximation.[5] The density functional equation is solved through the discrete variational method[6] (DVM) where we use numerical basis sets. The advantage of this method over the Gaussian one is that here the atomic wave functions are not fitted to a set of Gaussian functions. Consequently, any errors in the total energy due to fitting procedure are eliminated in the DVM procedure. One, however, pays a price for this advantage. The total energy in the DVM procedure are obtained through numerical integration, whereas in the Gaussian procedure, the integration is analytic. Consequently, the later method can yield more accurate total energy than the former. In order to bridge the gap between these two methods, we have also carried out test calculations using Gaussian basis sets in the local density functional approximation[7] (LCGTO). Many calculations[8,9] using these procedures are available for small clusters in literature. A quantitative comparison between these is often difficult since various authors use different numerical procedures and their effect on computed energetics may be comparable to that due to varying approximations in exchange-correlation energy and choice of basis sets. It is, therefore, meaningful to carry out these different calculations in one laboratory to minimize ambiguities otherwise inherent in these types of calculations. We hope that the following comparison is a step in the right direction. Due to space limitations, it is not possible to provide details of all of our calculations. We refer the reader to the existing literature and to our forthcoming paper.[8]

Table 1. Properties of Be clusters. E_b is in eV and the rest are in atomic units.

Cluster	Property	CI	LSD–DVM
Be_2	E_b	−0.03	−0.36
	R_{eq}	4.91	5.04
	efg		0.0056
Be_3	E_b	−0.06	−1.38
	R_{eq}	4.73	4.33
	efg		−0.011
Be_4	E_b	−0.72	−3.66
	R_{eq}	3.95	4.20
	efg		0.00

RESULTS

In table 1 we give the equilibrium geometries, binding energies, and equilibrium bond lengths of Be clusters having upto four atoms. Note that the bond lengths obtained from these two different theories are quite comparable to each other. However, the binding energies are very different.

It has been shown[9] in the past that an appropriate choice of basis sets for Be is very important to make Be_2 bound. We have repeated these calculations in the UHF-CI formalism by taking 6-31G and 6-311G** basis set[4] for Be. The former, indeed, does not give rise to a bound Be dimer. The later basis set gives a very small binding for Be_2. There are, however, more extended basis sets available in literature that give rise to better binding. In addition, the CI procedure can also be extended to include higher order excitations and corrections to minimize the errors due to finite basis sets. Our aim of the present work, however, has not been to improve upon existing calculations for Be_2 but to compare the results obtained from different approaches such as the DVM and LCGTO procedures.

In order to see if the large difference between the binding energies of Be_2 in the UHF-CI and LSD-DVM is primarily due to basis set limitations or inherent approximations in the exchange-correlation energy, we have carried out LSD calculations using Gaussian[10] (9s3p) type functions fitted to atomic Be wavefunctions. The corresponding bond length and binding energy for Be_2 in the LCGTO method is 4.8 a_0 and 0.37 eV respectively. These results agree very well with the LSD-DVM method indicating that the large difference in Be_2 binding energy in Table 1 is primarily due to the role of exchange-correlation energy functional.

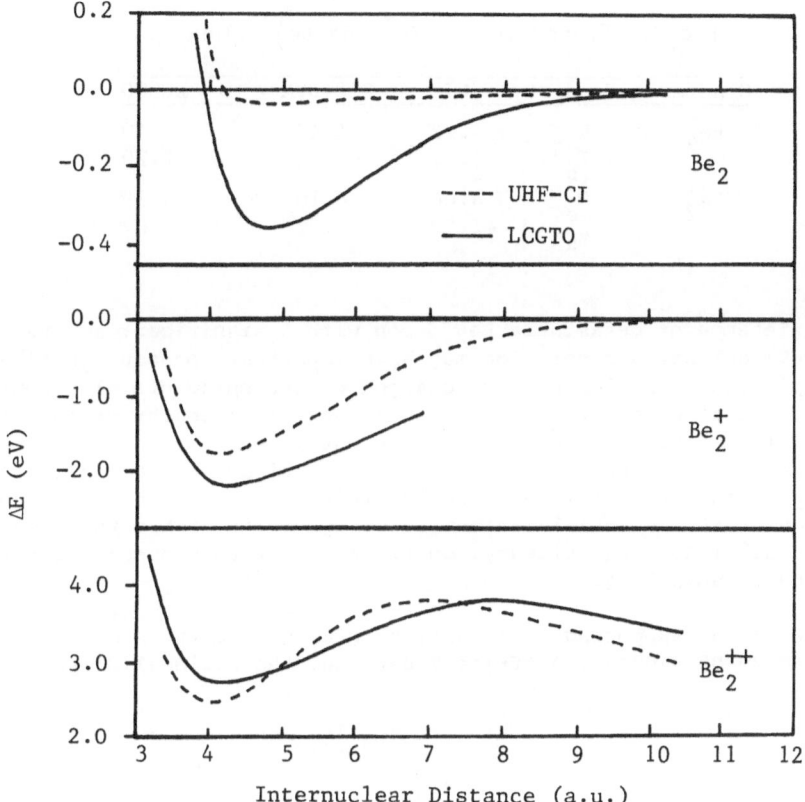

Fig. 1. Potential energy curves for Be_2, Be_2^+, and Be_2^{++}.

To further illustrate this comparison, we plot in Fig. 1 the potential energy curves for Be_2, Be_2^+, and Be_2^{++} clusters using UHF-CI and LCGTO method. Except for Be_2, the various schemes compare favorably. In Table 2 we compare the bond-lengths, binding energies, and total energies of the neutral and charged dimers obtained from three methods. Note again that the bond lengths are insensitive to details in the theoretical procedure. The percentage disagreement between binding energies of Be_2^+ obtained from three methods is much less than in Be_2. This indicates that in weakly bound systems, an incomplete description of configuration interaction is inadequate in achieving proper binding. That local density approximations give rise to enhanced binding is apparent from Table 2.

The potential energy curve for Be_2^{++} in Fig. 1 deserves special attention. Classically, two equal charges repel. Thus, if quantum mechanics did not play a role, the total energy of Be_2^{++} molecule should monotonically decrease as a function of the interatomic distance. An analysis of Fig. 1 suggests that this classical behavior does not become effective until the internuclear distance is greater than 10 a_0. At short distances, however, these two positively charged atoms attract ! This behavior has recently been seen by Durand et al.[11] for Mg_2^{++} cluster. As the charged atoms come close together, the electron-electron interaction dominates over the nuclear-nuclear term, and gives rise to a quasi-bound state. This, however, is not a universal phenomenon in all doubly charged dimers. In Li_2^{++} cluster, for example, the potential energy is always repulsive. Thus, the short range attraction in some charged clusters may be intimately linked to the electronic structure of the atom.

Table 2. Properties of Be_2^+ and Be_2^{++} clusters.

Cluster	Property	CI	LSD-DVM	LCGTO
Be_2^+	R_{eq}	4.22	4.40	4.40
	E_b	-1.77	-2.24	-2.20
Be_2^{++}	R_{eq}	4.04	4.10	4.19
	E_b	-1.32	-1.32	-1.07

The existence of metastable Be_2^{++} ion with a significant activation barrier (1.3 eV) against dissociation may have important consequence in studying coulomb explosion. As clusters are charged by stripping off electrons, the stored electrostatic energy may be large enough to cause the parent cluster to fragment into smaller pieces. If no potential barriers exist, such fragmentation may be spontaneous. However, in situations such as in Be_2^{++}, the cluster explosion may be delayed or even hindered even if the total energy of Be_2^{++} ion at its potential minimum (see Fig. 1) is higher than two fully dissociated Be^+ ions. Experimental confirmation of this observation would certainly be illuminating.

This work was supported , in part, by grants from the Army Research Office (DAAG 29-85-K-0244), Jeffress Trust, and the National Science Foundation.

REFERENCES

1. V. E. Bondybey and J. H. English, J. Chem. Phys. 80, 568 (1984).
2. W. J. Balfour and A. E. Douglas, Can. J. Phys. 48, 901 (1970); C. R. Vidal and H. Scheingraber, J. Mol. Spectrosc. 65, 46 (1977).
3. B. K. Rao and P. Jena, Phys. Rev. B 32, 2058 (1985).
4. Gaussian - 80 program (Quantum Chemistry Program Exchange, QCPE-437) by J. S. Binkley, R. A. Whiteside, R. Krishnan, R. Seeger, D. J. DeFrees, N. B. Schlegel, S. Topiol, L. R. Kahn and J. A. Pople.
5. P. Hohenberg and W. Kohn, Phys. Rev. 136 B, 864 (1964); W. Kohn and L. J. Sham, Phys. Rev. 140 A, 1133 (1965).
6. D. E. Ellis, Intl. J. of Quantum Chem. 2S, 35 (1968); D. E. Ellis and G. S. Painter, Phys. Rev. B 2, 2887 (1970), E. J. Baerendo, D. E. Ellis and P. Ros, Chem. Phys. 2, 41 (1973), B. Delley and D. E. Ellis, J. Chem. Phys. 76 (4), 1949 (1982).
7. J. L. Martins, J. Buttet and R. Car, Phys. Rev. B 31, 1804 (1985).
8. B. K. Rao, S. N. Khanna, and P. Jena (Unpublished).
9. R. J. Harrison and N. C. Handy, Chem. Phys. Lett. 98, 97 (1983). See the references therein.
10. "Gaussian Basis Sets for molecular calculations", Ed. S. Huzinaga (Elsevier, 1984).
11. G. Durand, J.-P. Daudey and J.-P. Malrieu, J. Physique 47, 1335 (1986).

JAHN-TELLER DISTORTION, HUND'S COUPLING AND METASTABILITY IN

ALKALI TETRAMERS

S. N. Khanna, B. K. Rao, P. Jena, and J. L. Martins[*]

Virginia Commonwealth University, Richmond, Virginia 23284
[*]University of California, Berkeley, California 94720

ABSTRACT

The equilibrium geometries of small metal clusters are strongly influenced by the number of valence electrons available. These electrons occupy the molecular energy levels in different ways depending upon the degeneracies of the one-electron states. The symmetry of the atomic arrangements guide the degeneracies and the consequent spin configuration of the clusters. The interplay of the geometry and the spin configuration is demonstrated using model clusters of Na_4 and Li_4.

INTRODUCTION

Recently, there has been considerble interest in the electronic structure and geometries of small metal clusters. Ab-initio theoretical studies on small simple metal clusters show that their ground state structures are very different from their bulk counterparts. In addition to the true ground state, the clusters can also exist in metastable states corresponding to various geometric or spin structures. There are two effects which seem to govern the equilibrium geometries of small clusters. First, the Jahn-Teller effect tries to lower the total energy of the system by breaking any electronic state degeneracies arising due to spatial symmetry, thus giving rise to a less symmetric structure. The ground state of alkali trimers as obtuse angled triangular structures instead of equilateral triangle, and the rhombus instead of square or tetrahedral structure for a tetramer are the classic examples.[1] Secondly, high symmetric arrangement of atoms in a cluster can lead to degeneracies in electronic levels. In such situations, the cluster can lower its energy by maximizing its spin multiplicity through the exchange or Hund's rule coupling. Thus, the Jahn Teller effect and Hund's rule appear to play competing roles in determining the ground state geometries of microclusters. We illustrate these rules by considering Na_4 and Li_4 clusters as model systems and where the atoms are confined to lie on a plane. We further show that in three dimensional structures of alkali tetramers, the Jahn-Teller effect and Hund's rule work together to lower the energy where the ground state is a distorted tetrahedron with spin triplet configuration.

Our studies on Na_4 were based on the density-functional approach. The effect of core electrons was treated through a non-local, norm-conserving pseudopotential proposed by Bachelet et al.[2] The exchange-correlation corrections were incorporated in the local-spin-density approximation. The Kohn-Sham equations were solved by expanding the molecular orbitals as a sum of Gaussian atomic orbitals. For other details of the method, the readers are referred to articles[3] available in the literature. For Li_4, we have used unrestricted Hartree-Fock method with configuration interaction. One can refer to a recent article[1] for details of the calculational procedure.

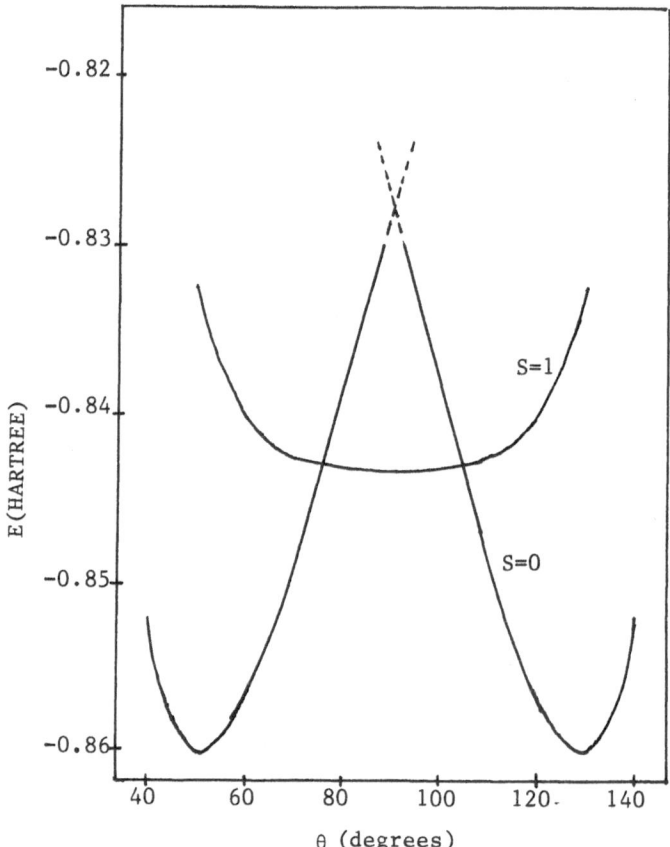

Fig.1 Variation of the total energy of a
       sodium tetramer, in singlet and triplet
       state, as a function of the angle θ for
       plane rhombus configurations. (For each
       θ, the length ,a, of each side of rhombus
       has been optimized)

In Fig. 1 we show the total energy of Na_4 as a function of the apex angle θ for the planar rhombus configurations corresponding to singlet and triplet states. For a fixed value of θ , the bond lengths have been optimized. It is seen that as the apex angle is increased from θ = 52°, the total energy of the singlet state rises and at an angle of θ = 76°, the triplet becomes lower than the singlet. As θ is further increased towards

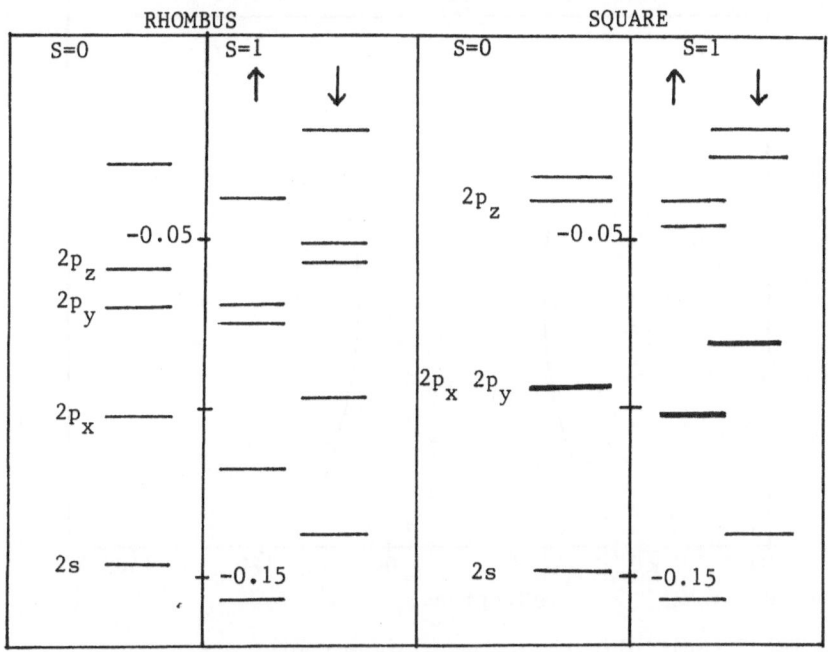

Fig.2   One electron energy levels (in Hartree)
corresponding to singlet and triplet state
for rhombus(θ=52°) and square geometries.

θ = $\pi/_2$, the triplet energy decreases towards its minimum value at θ =
$\pi/_2$. In order to understand this transition, we show in Fig. 2 the valence
electron levels for θ = 52° and θ = 90° for the singlet and triplet configurations of the Na_4 cluster. Using the terminology of the jellium model,
we have labeled the valence states as 2s, $2p_x$, $2p_y$, and $2p_z$. As θ tends
toward $\pi/_2$, the difference between p_x and p_y levels decreases until they
become degenerate at $\theta = \pi/_2$. There is a range of θ values around $\theta = \pi/_2$
for which the difference in the one electron levels is smaller than the
exchange energy and the cluster prefers a triplet configuration with an electron in both p_x and p_y levels instead of a singlet with both electrons in
lowest p_x levels. The triplet state at $\theta = \pi/_2$ has a non-degenerate electronic configuration. As θ decreases towards 52°, the degeneracy in the
p_x, p_y-like orbital is lifted. The gap in the energy of these orbitals
becomes larger than the exchange energy gained by putting two electrons of
parallel spin in p_x and p_y orbital. Consequently, as seen in Fig. 1, the
cluster assumes a spin singlet configuration. The Li_4 clusters exhibit
exactly the same trend.

To understand if the preferred spin structure of the tetramer is
symmetry related (ie. dependent only on θ) or also dependent on spatial
parameters (ie. bond lengths), we plot in Fig. 3 a phase diagram showing
regions of spin singlet and triplet state for a given value of the bond
length a and bond angle θ. The fact that the plot of a vs θ has a weak θ-
dependence clearly suggests that the magnetism of clusters is a symmetry
driven problem.

One can also study the spin-problem by inducing degeneracy in p_x, p_y,
and p_z orbitals by going to a tetrahedral configuration. Elsewhere in this
book[4] we show the total energy of the Li_4 as a function of the dihedral
angle (shown in the insert). One notices a transition as a function of the

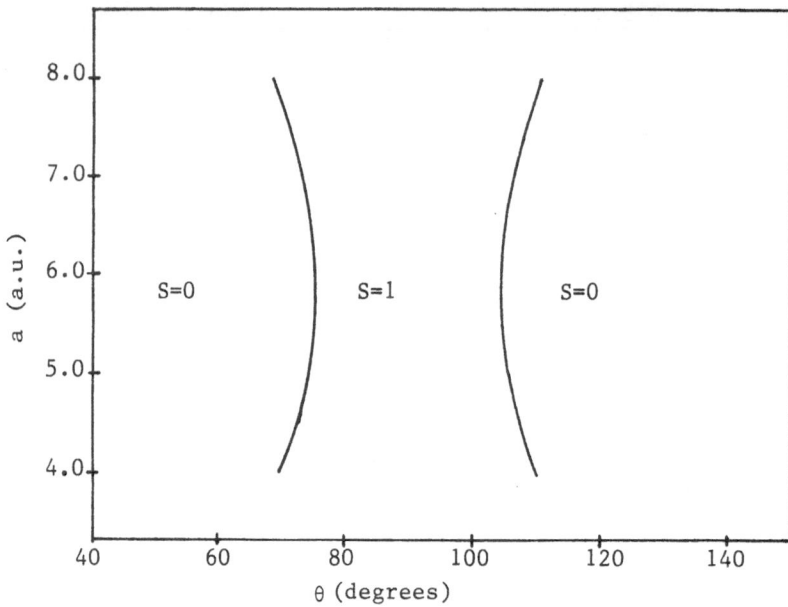

Fig.3 Multiplicity of the minimum energy state for rhombus configurations of a sodium tetramer.

dihedral angle and the existence of a metastable triplet minimum. It is interesting to note that the triplet minimum does not correspond to a perfect tetrahedron which would occur at $\theta \cong 70°$. A perfect tetrahedron leads to a degeneracy of p_x, p_y and p_z. Since the electronic levels are degenerate, the tetrahedral structure has triplet lower than the singlet. However, contrary to the case of square (2 dimensions), parallel arrangement of spins does not remove the electronic state degeneracy. The tetrahedral triplet has a degenerate electronic ground state and consequently undergoes a Jahn-Teller distortion lowering its energy through a deformation of the perfect tetrahedron. Thus, the Hund's rule coupling and the Jahn-Teller distortion now act together to lower energy, resulting in a distorted tetrahedron in the spin triplet configuration.

In conclusion, we have demonstrated through model calculations the relative roles played by Jahn-Teller effect and Hund's rule in determining the ground state geometry and spin structure of clusters. We have also demonstrated that the nature of these interactions are qualitatively different in two and three dimensions. Thus, it is essential to study the possible spin multiplicity of clusters before deciding on their true equilibrium geometry, for these two are intimately related.

This work was supported, in part, by the Army Research Office (DAAG 29-85-K-0244) and the Jeffress Trust.

REFERENCES

1. B. K. Rao and P.Jena, Phys. Rev. B **32**, 2058 (1985).
2. G. B. Bachelet, D. R. Hamann and M. Schluter, Phys. Rev. B **26**, 4199 (1982).
3. J. L. Martins, J. Buttet, and R. Car, Phys. Rev. B **31**, 1804 (1985).
4. B. K. Rao, S. N. Khanna, and P. Jena, page 369 in this book. See Fig.5.

ELECTRONIC STRUCTURE OF SMALL CARBON, SILICON, AND GERMANIUM CLUSTERS

Gianfranco Pacchioni

Dipartimento di Chimica Inorganica e Metallorganica, Centro CNR, Università di Milano, I-20133 Milano, Italy

Jaroslav Koutecký

Institut für Physikalische Chemie, Freie Universität Berlin 1000 Berlin 33, West Germany

ABSTRACT

C_n (n = 3-5), Si_m and Ge_m (m = 3-7) clusters were investigated by means of all electron (C) and effective core potential (Si and Ge) configuration interaction calculations. The optical spectra of linear C_3, C_4, and C_5 and rhombic C_4 clusters were determined. The nature of the bonding and the electronic properties of C, Si, and Ge clusters are discussed.

I. INTRODUCTION

The general interest in semiconductor materials has recently stimulated several experimental and theoretical investigations on clusters of Group IVa atoms (in particular C, Si, and Ge).[1]

There are at least two basic question of general interest for the theoreticians concerning this class of clusters: i) the different properties and electronic structures of carbon with respect to silicon and germanium clusters, and ii) the different bonding nature · in small clusters and respective crystal lattices and, therefore, the changes accompanying the transition from a microcluster to the solid.

In this paper we report preliminary results of all electron configuration interaction (CI) calculations on small C clusters. The optical spectra of C_3, C_4 and C_5 molecules have been theoretically determined and compared with the experimental results.[2] The characteristic features of Si and Ge clusters as derived from effective core potential (ECP) CI calculations are then briefly reported and compared with those of C clusters.

II. COMPUTATIONAL DETAILS

The electronic structure of C clusters has been investigated by means of all electron MRD CI calculations.[3] The AO basis set adopted is (9s5p1d/4s2p1d). The reliability of the MRD CI procedure has been

Table 1 - Results of MRD CI calculations for rhombic and linear C_4 clusters

State	nM/nR	E_{CI} (au)	$\sum_i c_i^2$	vertical T_e (eV)
Rhombic C_4				
^1A_g	3M/1R	-151.5913	0.87	0.0
$^1B_{1u}$	4M/1R	-151.5034	0.86	2.39
$^1B_{2g}$	5M/1R	-151.5029	0.87	2.41
$^1B_{3u}$	3M/1R	-151.3385	0.86	6.88
$^1B_{2u}$	4M/1R	-151.3327	0.86	7.03
Linear C_4				
$^3\Sigma_g^-$	11M/1R	-151.5952	0.89	0.0
$^1\Delta_g$	10M/2R	-151.5800	0.88	0.41
$^1\Sigma_g^+$	10M/2R	-151.5701	0.88	0.68
$^3\Sigma_u^+$	10M/2R	-151.5138	0.87	2.21
$^3\Pi_g$	7M/1R	-151.5583	0.86	1.00
$^3\Pi_u$	11M/3R	-151.5385	0.87	1.54
$^1\Pi_g$	6M/1R	-151.5315	0.86	1.73
$^1\Pi_u$	6M/1R	-151.5199	0.87	2.05
$^3\Sigma_u^-$	12M/3R	-151.4902	0.88	2.86

carefully analyzed for the case of C_2 and C_3 molecules for which accurate theoretical and experimental values are known. The overall agreement found is very satisfactory since the experimental transition energies (T_e) are reproduced within an error of one tenth of eV.[4]

The calculations have been performed at both single reference and multireference CI levels (nM/nR). In general, the SR CI is inadequate for the description of some excited states where the leading configuration is strongly mixing with other excited configurations.[4] In the following we report only the results of the MRD CI calculations. Some details of the procedure are given in Tables 1 and 2.

The study of Si and Ge clusters was performed with the ECP[5] version of the MRD CI program adopting a double zeta basis set. Further computational details related to this part of the work are given in Ref. 6.

III. CARBON CLUSTERS

Both theory and experiments agree about the fact that C_3 is a π-bonded linear molecule while in the case of C_4 it has been proposed that linear and rhombic forms are equally favorable.[7-9]

We have performed an independent optimization of angle and bond lengths in linear and rhombic C_4 clusters at CI level. The equilibrium bond distances in linear C_4 (1.30 Å) are much shorter than in the rhombic structure (1.46 Å). The rhombus has an internal angle of 118°.

The total energies of the two structures were computed at SCF, SR CI, MRD CI and full CI (Davidson's correction) levels. The two forms are both energetically favorable, but the relative ordering of stabilities depends on the theoretical treatment. At the highest level of treatment, i.e. full CI, the rhombic form is more stable than the linear one by 0.17 eV (4 kcal/mol), in close agreement with the results of Whiteside et al.[7] and of Magers et al. [8]. However, the energies at sake are too small to draw definitive conclusions about the most stable form of carbon tetramer.

More important is that, on the basis of these results, one cannot rule out the possibility of the existence of a dynamical equilibrium between the two structures, and therefore the interpretation of the optical spectra of carbon clusters in gas-phase should also consider the possible existence of cyclic carbon species.

Rhombic C_4 cluster has a ^1A_g ground state. The first allowed transition is $^1B_{1u} \leftarrow {}^1A_g$ with a $T_e = 2.39$ eV. This transition occurs nearly at the same energy of the vibronically allowed $^1B_{2g} \leftarrow {}^1A_g$ transition (2.41 eV). The two other excited states considered, namely $^1B_{3u}$ and $^1B_{2u}$, lie at much higher energies (Table 1).

There is general agreement about the ground state of linear C_4 which is $^3\Sigma_g^-$ according to both theoretical[7-9] and experimental [2] results. The valence configuration and the MO spectrum of linear C_4 are shown in Fig. 1. From this MO scheme the different electronic structure of linear C_4 with respect to linear C_3 is apparent. In this latter molecule, in fact, the $1\pi_u$ HOMO is fully occupied, while in C_4 the $1\pi_g$ HOMO is partially unfilled. Therefore, the electronic transitions in linear C_4 involve a set of rather close-lying energy levels giving rise to small excitation energies, differently from the C_3 case where

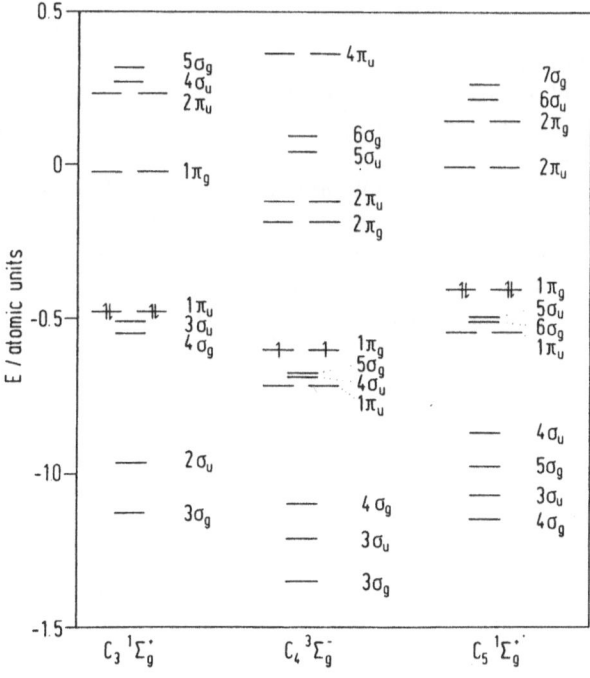

Fig. 1 - MO spectra of linear C_3, C_4, and C_5 clusters. The highest occupied molecular orbitals are indicated.

Table 2 - Results of MRD CI calculations for linear C_5 cluster

State	nM/nR	E_{CI} (au)	$\sum_i c_i^2$	vertical T_e (eV)
$^1\Sigma_g^+$	5M/1R	-189.5616	0.86	0.0
$^3\Sigma_u^+$	6M/2R	-189.4681	0.86	2.54
$^3\Pi_g$	5M/1R	-189.4675	0.85	2.56
$^3\Pi_u$	5M/1R	-189.4668	0.85	2.58
$^3\Delta_u$	6M/2R	-189.4675	0.86	2.56
$^3\Sigma_u^-$	6M/2R	-189.4647	0.86	2.64
$^1\Delta_u$	6M/2R	-189.4601	0.85	2.76
$^1\Sigma_u^-$	6M/2R	-189.4592	0.85	2.79
$^1\Pi_g$	3M/1R	-189.4453	0.85	3.16
$^1\Pi_u$	4M/1R	-189.4446	0.85	3.18
$^1\Sigma_u^+$	10M/2R	-189.3133	0.85	6.76

the first allowed transitions are found around 3 eV. The electronically allowed transitions in linear C_4 are the $^3\Pi_u \longleftarrow {}^3\Sigma_g^-$ and $^3\Sigma_u^- \longleftarrow {}^3\Sigma_g^-$. In addition, two other transitions, $^3\Pi_g \longleftarrow {}^3\Sigma_g^-$ and $^3\Sigma_u^+ \longleftarrow {}^3\Sigma_g^-$, are vibronically allowed. These states, toghether with some spin or symmetry forbidden transitions, have been considered in our study (Table 1).

The $^1\Delta_g$ ($1\pi_g^2$) and $^1\Sigma_g^+$ ($1\pi_g^2$) states are very close in energy to the $^3\Sigma_g^-$ ground state (Table 1). The first state which could be important for the assignment of the absorption lines of C_4 is the $^3\Pi_g$ ($5\sigma_g^1 1\pi_g^3$) which lies 1.0 eV above the ground state.

The lowest electronically allowed transition is the $^3\Pi_u \longleftarrow {}^3\Sigma_g^-$ ($4\sigma_u^1 1\pi_g^3$) with T_e = 1.54 eV. The spin forbidden $^1\Pi_g$ ($5\sigma_g^1 1\pi_g^3$) and $^1\Pi_u (4\sigma_u^1 1\pi_g^3)$ states are 0.5-0.7 eV higher than the corresponding triplet states. The electronically allowed transition $^3\Sigma_u^- \longleftarrow {}^3\Sigma_g^-$ ($1\pi_u^3 1\pi_g^3$) lies 2.86 eV above the ground state.

The geometry of linear C_5 has not been optimized and all C-C bond lengths were fixed at the value 1.31 Å. The molecular ground state of linear C_5 is $^1\Sigma_g^+$. Similarily as in linear C_3, also in linear C_5 the HOMO is a fully occupied MO of π symmetry (Fig. 1). Therefore, the lowest transitions in linear C_5 are expected to be higher than in the corresponding C_4 cluster and to be comparable with those of linear C_3.

The two electronically allowed transitions in linear C_5 are the $^1\Pi_u$ ($6\sigma_g^1 2\pi_u^1$) $\longleftarrow {}^1\Sigma_g^+$ and the $^1\Sigma_u^+$ ($1\pi_g^3 2\pi_u^1$) $\longleftarrow {}^1\Sigma_g^+$. The former has a T_e = 3.18 eV whereas the $^1\Sigma_u^+$ state lies at higher energies (T_e = 6.76 eV). Two further singlet states, $^1\Sigma_u^-$ ($1\pi_g^3 2\pi_u^1$) and $^1\Pi_g (5\sigma_g^1 2\pi_u^1)$ are found in the region around 3 eV (Table 2).

It is interesting to note the similarity existing between the optical spectra of linear C_3 and linear C_5 clusters. This is particularly evident for the electronically allowed vertical transition energies. The $^1\Pi_u \longleftarrow {}^1\Sigma_g^+$ transitions are located at about 3 eV above the ground state and the

$^1\Sigma_u^+ \longleftarrow {}^1\Sigma_g^+$ transitions fall in the near ultraviolet region in both C_3 and C_4 systems.

In conclusion, the results of the present calculations support the assignements done by Graham et al. [2] for linear C_4 on the basis of spectroscopic data. The $^3\Pi_g$ state, which according to the experiment should not be higher than 6000 cm^{-1} (about 0.75 eV) is predicted to lie about 1 eV above the ground state by the present theoretical calculations. The second transition occuring at 19564 cm^{-1} (2.42 eV), attributed in the experiment to either $^3\Sigma_u^-$ or $^3\Pi_{u_3}$ states, must be assigned to the $^3\Sigma_u^- \longleftarrow {}^3\Sigma_g^-$ transition, since the $^3\Pi_u \longleftarrow {}^3\Sigma_g^-$ excitation occurs at much lower energies (about 1.5 eV).

The comparable stability of the linear and rhombic forms of carbon tetramer suggests that also the cyclic structure should be considered for the interpretation of the optical spectrum of the C_4 cluster.

The optical spectrum of linear C_5, as determined according to MRD CI calculations, is basically different from that of linear C_4 since in the former the lowest electronically allowed transitions occur at energies of 3 eV and more in full analogy with the case of linear C_3. The similarity in the optical spectra of C_3 and C_5 molecules can be qualitatively rationalized on the basis of the MO energy scheme.

As found for the case of carbon tetramer, however, also for the C_5 cluster it is possible that favorable cyclic structures exist. Since the spectra of this kind of structures could be in principle substantially different from that of the linear form, further work is necessary in order to determine the optical spectrum of C_5.

IV. SILICON AND GERMANIUM CLUSTERS

Recently reported theoretical investigations have contributed to elucidate the electronic and geometric structures of Si and Ge clusters[6,10-12]. Some general features have been established. Here we report the most important of these features as determined according to our ECP MRD CI calculations.[6]

Si and Ge clusters have very similar electronic structures and phisico-chemical properties. The bonding in these clusters has the same character and does not involve any important sp hybridization. The Si 3s and Ge 4s valence orbital populations (1.75 - 1.95 electrons) indicate that the electronic configuration of Si and Ge atoms in clusters is close to that of the free atoms. The electronic redistribution (hybridization) occuring with the formation of the cluster is therefore rather small. Also in tetrahedral Ge$_5$ the central atom, which has the same coordination as a bulk atom, shows an orbital population $4s^{1.74}4p^{2.3}$ markedly different from that of a sp^3 hybrid.

In this respect Si and Ge clusters are substantially different from analogous small carbon clusters. In fact, linear π-bonded Si and Ge clusters are clearly less favorable than more compact geometries. On the other hand, some planar Si and Ge structures possess considerable stability. This is the case, for instance, of the rhombus which is the most stable Si and Ge tetramer.

Very stable Si and Ge clusters (e.g. rhombus, trigonal bipyramid, square pyramid, tripyramid, octahedron, pentagonal bipyramid)[6] do not represent neither regular nor deformed sections of the corresponding crystal lattices. On the contrary, they can be considered as segments of closed-packed lattices or as steps in pentagonal crystal growth.

The clear tendency of small Si and Ge clusters to assume geometrical and electronic structures different from those of the respective bulk crystals is due to the fact that the majority of atoms in small clusters are surface atoms. The valence of Si and Ge atoms in the respective crystal lattices is saturated by bonding with four nearest neighbors located at the vertices of a regular tetrahedron. In order to reproduce such kind of bonding a very large number of atoms is necessary to minimize the "edge" effects resulting from the unsaturation of the surface atoms.

The appearance of typical bulk properties is therefore expected only for quite big Si or Ge clusters (better to say small particles) with small surface atoms/bulk atoms ratio. This explains also the observed slow convergence of the binding energy of clusters of increasing size to the cohesive energy of the crystal.[6]

ACKNOWLEDGMENT - This work was supported in part by the DFG-SFB6 "Structure and Dynamics of Interfaces" and in part by the Italian CNR.

REFERENCES

1. J. Koutecký and P. Fantucci, Chem. Rev., **86**, 539 (1986).

2. W.R.M. Graham, K.I. Dismuke, and W. Weltner Jr., The Astrophys. J., **204**, 301 (1976).

3. R.J. Buenker and S.D. Peyerimhoff, Theoret. Chim. Acta, **35**, 33 (1974); ibidem, **39**, 217 (1975).

4. G. Pacchioni and J. Koutecký, to be published.

5. J.C. Barthelat, Ph. Durand, and A. Serafini, Mol. Phys., **33**, 159 (1977).

6. G. Pacchioni and J. Koutecký, J. Chem. Phys., **84**, 3301 (1986).

7. R.A. Whiteside, R. Krishnan, D.J. Defrees, J.A. Pople, and P.v.R. Schleyer, Chem. Phys. Letters, **78**, 538 (1981).

8. D.H. Magers, R.J. Harrison, and R.J. Bartlett, J. Chem. Phys., **84**, 3284 (1986).

9. J. Koutecký, G. Pacchioni, G.H. Jeung, and H.C. Hass, Surf. Sci., **156**, 650 (1985).

10. D. Tománek and M.A. Schlüter, Phys. Rev. Letters, **56**, 1055 (1986).

11. K. Raghavachari, J. Chem. Phys., **84**, 5672 (1986).

12. S. Saito, S. Ohnishi, and S. Sugano, Phys. Rev. B, **33**, 7036 (1986).

KEYWORDS: Carbon / Silicon / Germanium / Electronic structure
Stability / Excited states / Configuration interaction

PERIODIC ANDERSON MODEL FOR SMALL CLUSTERS

P. K. Misra[1], D. G. Kanhere[2] and Joseph Callaway[3]

[1]Dept. of Physics, Univ. of Rhode Island, Kingston, RI 02881
[2]Dept. of Physics, Univ. of Poona, Pune 41107, India
[3]Dept. of Physics, Louisiana St. Univ., Baton Rouge, LA 70803

ABSTRACT

The periodic Anderson model is applied to three different four-site clusters to study the f-site occupation (n_f), specific heat (C_v) and magnetic susceptibility (χ_f) of rare-earth and actinide compounds. We consider one localized f orbital per site per spin with energy E_f with a Coulomb repulsion U between two electrons in 'f' orbitals in the same site, one extended orbital per site per spin with an interatomic transfer integral t and a positive hybridization term V between the localized and extended orbitals of same spin in different sites. The interaction between different sites is restricted to nearest neighbors and the number of electrons is taken to be one per site. The many-body eigenvalues and eigenstates are calculated exactly by diagonalizing the Hamiltonian. The principal features of our results for various choices of parameters are (i) observation of heavy-fermion behavior when the non-magnetic ground state is nearly degenerate with two excited magnetically ordered states; (ii) transition from a Kondo-lattice to a magnetic regime and subsequent re-entry to either a Kondo-lattice or a mixed-valence regime as E_f is increased; (iii) coexistence of magnetic order and mixed-valence for a narrow range of E_f; (iv) the 'benchmark' results of C_v and χ_f of single-impurity Anderson model are reproduced only at high temperatures.

INTRODUCTION

The rare-earth and actinide compounds can be broadly classified into three categories. The majority of these compounds have local moments and magnetic transitions occur at low temperatures. These can be classified as systems in the magnetic regime. The second category exhibit mixed-valence (MV) phenomena[1]. The valence fluctuations in such systems are at the root of many remarkable physical properties. The third category has negligible valence fluctuations but non-magnetic ground states and are identified as systems in the Kondo-lattice regime. A restricted class of this category has anomalously large values of specific heat (C_v) and magnetic susceptibility (χ_f) below a characteristic coherence temperature and are identified as heavy-fermion (HF) systems[2]. There has been extensive theoretical work on mixed-valence and Kondo systems[3]. The thermodynamic (C_v and χ_f), transport and excitation properties of the Anderson model[4] have been extensively studied and 'benchmark' results have been obtained using the renormalization group technique[5] or the Bethe Ansatz method[3]. However, the lattice problem is neither well understood nor solved although significant advances have been made by using 1/N expansion, fermi-liquid, Hubbard-Gutzwiller model or variational theories[3]. In this paper, we use the periodic Anderson model[1,2] to calculate the f-state occupation (n_f), C_v and χ_f of three different four-site clusters (square, rhombus and tetrahedron) with periodic boundary conditions. The first such calculation for MV systems was made by Arnold and Stevens[6] who considered an isolated three-atom cluster. However, their calculation implicitly assumes an infinite value of U. Parlebas et al.[7] have applied the periodic Anderson model to a four-site tetrahedral cluster with periodic boundary conditions and calculated the variation of n_f for different parameters. Hirsch and Fye[8] have used a Monte Carlo method to treat a small number of magnetic impurities interacting with the conduction electrons in a metal but find systematic errors when U is increased.

MODEL

The periodic Anderson model[1] is given by the Hamiltonian

$$H = t\sum_{i \neq j, \sigma} C^+_{i\sigma}C_{j\sigma} + E_f\sum_{i, \sigma} f^+_{i\sigma}f_{i\sigma} + U\sum_i f^+_{i\uparrow}f_{i\uparrow}f^+_{i\downarrow}f_{i\downarrow}$$

$$+ V\sum_{i \neq j, \sigma}(C^+_{i\sigma}f_{j\sigma} + f^+_{j\sigma}C_{i\sigma}). \tag{1}$$

Here t is the hopping integral of the extended orthogonal orbitals between sites i and j (restricted to nearest neighbors in our model), $C^+_{i\sigma}$ and $C_{i\sigma}$ are the creation and annihilation operators for these extended orbitals with spin σ. There is one extended orbital per site per spin with a mean energy which is the origin of the energy scale. $f^+_{i\sigma}$ and $f_{i\sigma}$ are the creation and annihilation operators for the localized orthogonal non-degenerate 'f' orbital with energy E_f. U is the Coulomb repulsion between two electrons of opposite spin in the f orbital and describes a short-range interaction between them. V is a positive hybridization parameter between the localized and band orbitals in the neighboring sites. U is positive while t and E_f can have either sign. The total number of electrons is taken to be one per site. We consider the application of the Hamiltonian (1) to small clusters of four sites (square, rhombus and tetrahedron) with periodic boundary conditions. These systems differ principally in the number of nearest neighbors (4, 5 and 6). By using various geometries, we explicitly include the importance of band structure effects. For example, our model Hamiltonian for a tetrahedron is identical to the Hamiltonian of the infinite face-centered-cubic lattice if the Brillouin zone sampling is restricted to four reciprocal lattice points, the zone center Γ, and the three square-face-center points X[7]. The many-body eigenvalues and eigenstates are calculated exactly by constructing a computer program to diagonalize the Hamiltonian (1) within subspaces of fixed values of spin which is a good quantum number, and the states are classified as singlets, triplets and quintets. We did not form symmetric combinations of the basis states since we study the different geometric structures within the same program. n_f, C_v and χ_f are calculated from the many-body eigenvalues and eigenstates by using standard statistical mechanics procedure in a canonical ensemble. We calculate n_f, C_v and χ_f for a wide range of parameters $U/|t|$, $V/|t|$ and $E_f/|t|$ for both positive and negative values of t.

RESULTS AND DISCUSSION

In Fig. 1, we show the first few eigenvalue spectra of the many-body states for a typical value of U=50, V=0.1, E_f=-3 for both t=±1 for the three geometries. This picture implies a strong dependence of low temperature

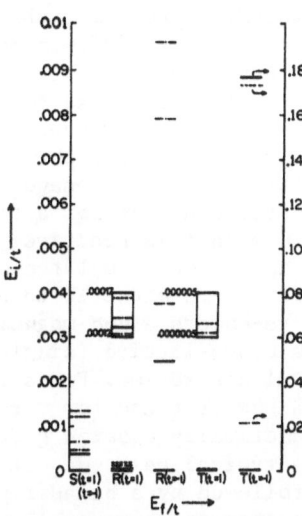

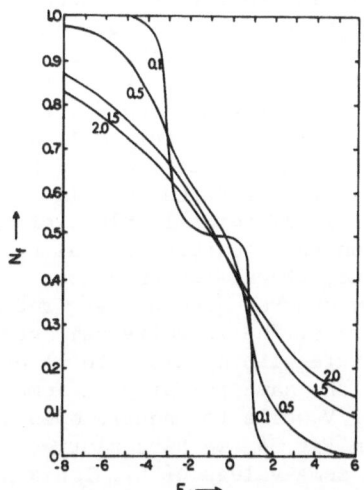

Fig. 1. Eigenvalue spectrum of the first few many-body states for U=50, V=0.1, E_f=-3 for both t=±1 for the square, rhombus and the tetrahedron. Singlets are denoted by solid lines, triplets by dashed-dot lines (-·-·-·) and quintets by dashed lines.

Fig. 2. The f-occupation number per site, n_f, in the four-electron ground state in terms of E_f and various hybridization energies V for t=-1, U=50 for a tetrahedron.

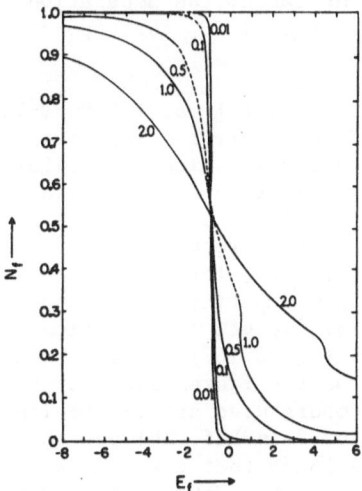

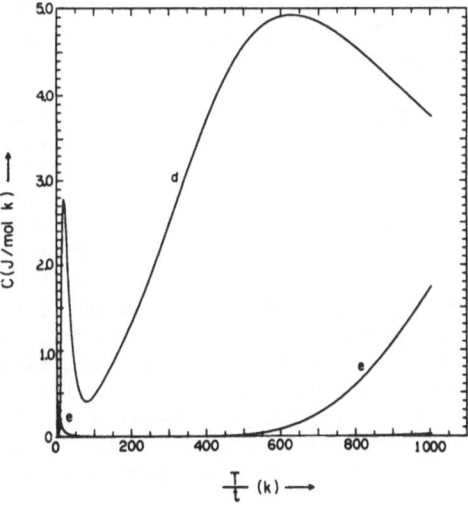

Fig. 3. The f-occupation number per site, n_f, in the four-electron ground state in terms of E_f and various hybridization energies V for t=1, U=50 for a tetrahedron. The singlets are represented by solid lines, triplets by dashed-dot lines (-·-·-·) and quintets by dashed lines.

Fig. 4. C_v vs. T/t for various E_f for t positive, U=50, V=0.1 for a tetrahedron.
c. E_f = -2.0.
d. E_f = -1.0.
e. E_f = -0.5.
(All parameters in units of |t|.)

447

thermodynamic properties on geometry. In mixed-valence materials, the charge
fluctuations tend to destroy magnetic order while the sf (or df)-exchange
interaction tends to stabilize it. In order to study such behavior, we have
plotted in Figs. 2-3 n_f vs. E_f for various hybridization energies for
$U/|t|=50$ for a tetrahedron for $t=\pm 1$. From Fig. 2, we note that there is no
magnetic ground state for $t=-1$. Here, and in the following, the phrase
"magnetic ground state" means $S>0$. The possibility for antiferromagnetic
order is not considered. We find regimes of magnetic ordering for a tetra-
hedron (Fig. 3) for all values of $V/|t|$ below 1.0 when t is positive. We
find an unusual feature that as E_f is gradually increased from large negative value, the system undergoes a transition from a non-magnetic to a magnetically ordered ground state and subsequently re-enters a non-magnetic
ground state. The re-entry can occur either to a Kondo-lattice (singlet
ground state with n_f close to 1) or to a mixed-valence regime. For a rhombus
we found the same re-entry phenomenon for both signs of t and for very small
values of V while the square does not have a magnetically ordered ground
state. In Fig. 4, we have plotted C_v vs. T for a typical case. We notice
that for some values of E_f, C_v has a sharp peak followed by a broader peak.
The first peak is due to the low-lying nearly degenerate states well separated from higher excited states. The second peak, which is much broader,
is generally of the Schottky shape[7] and is similar to the specific heat of
a Kondo system[9]. The low-temperature rise in specific heat indicates the
onset of heavy-fermion behavior.

In order to study this in detail, we present a typical example and plot
(Fig. 5) C_v/T vs. T for E_f ranging from -5.0 to -4.0 (n_f varies from 0.9943
to 0.9788). We notice that the rate of increase of C_v/T at very low temperatures is maximum for $E_f=-5.0$ but gradually decreases as E_f is increased.
This heavy-fermion feature practically disappears when $E_f=-4.2$. In Fig. 6,
we plot the energy-level diagram of the first few many-body states for each
of these E_f as well as for $E_f=-3.0$. As E_f is increased, the separation between the lowest three levels increases and the rise in C_v correspondingly
decreases.

In Fig. 7, we plot $k_B \chi_f T/(g\mu_B)^2 (\equiv \chi_f T)$ vs. log $T/|t|$ for $\Gamma \sim U(\Gamma = \pi V^2)$.
For both $E_f=-5$ and -3, there is a transition from the frozen-impurity
($\chi_f T=0$) to a free-orbital ($\chi_f T=0.125$) regime, a result which is similar to
Fig. 9B of Ref. 5b. For higher E_f, there is a transition from the frozen-
impurity to an intermediate ($0<\chi_f T<0.125$) regime. There are no corresponding
$\chi_f T$ curves for single-impurity Anderson model[5] where results only for $\Gamma=0$,
$\Gamma<<U$ and $\Gamma>>U$ have been evaluated. When $\Gamma<<U$ (Fig. 8), for $E_f=-5.0$
($n_f=0.994$), there is a transition from the frozen-impurity to the local
moment ($\chi_f T \tilde{=} 0.25$) regime. For $E_f=-3.0$ ($n_f=0.725$), we get a transition from
the frozen-impurity to the valence-fluctuation ($\chi_f T \tilde{=} 0.167$) regime. When E_f
is further increased, the transition occurs from the frozen-impurity to the
free-orbital regime. Our results for higher temperatures are in excellent
agreement with the single-impurity results (Fig. 8 in Ref. 3b) but the shape
of curves differs at low temperatures. We have found that there is a gradual
shift in the χT curves towards lower temperatures when V=0.01. Thus V plays
a crucial role in the highly correlated regime. We note that for the same
parameters, the specific heat maxima generally occur below the temperature
at which χT reaches its high-temperature value. The same broad features have
been recently observed experimentally[10]. Finally, we note that the low-
temperature χ values of all Kondo-like systems cannot be explained by a
Fermi-liquid theory. There will be correlation between the spins in different sites brought about by RKKY-type interactions at low temperatures. The
same conclusion has been recently inferred from analysis of neutron-scatter-
ing measurements[11] on heavy-fermion system $CeCu_6$.

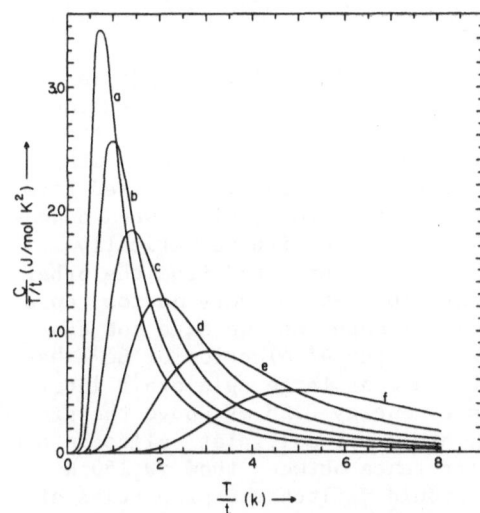

Fig. 5. C_V/T vs. T for various E_f for t negative, U=50, V=0.1 for a tetrahedron. a. E_f=-5.0; b. E_f=-4.8; c. E_f=-4.6; d. E_f=-4.4; e. E_f=-4.2; f. E_f=-4.0. (All parameters are in units of $|t|$.)

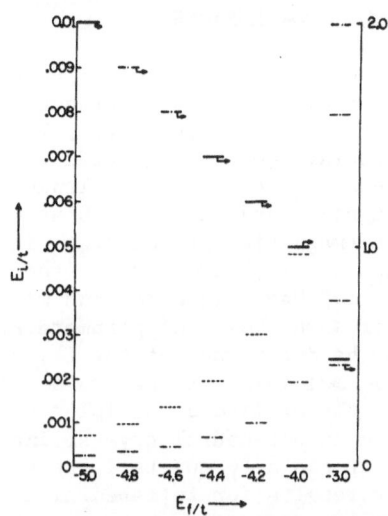

Fig. 6. Energy-level diagram of the first few many-body states for various E_f for t negative, U=50, V=0.1 for a tetrahedron. Singlets are denoted by solid lines, triplets by dashed-dot lines (−·−·−.) and quintets by dashed lines.

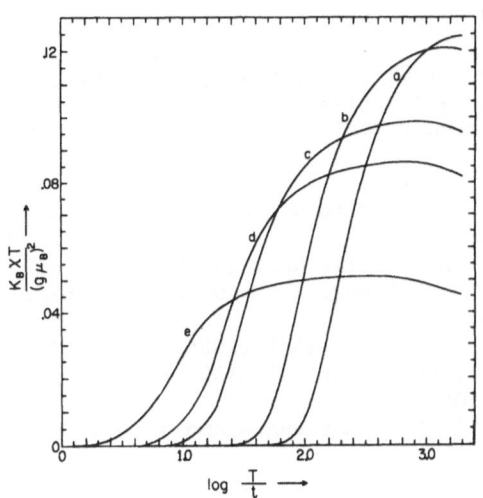

Fig. 7. $k_B \chi_f T/(g\mu_B)^2$ vs. log $T/|t|$ for t=-1, U=1 and V=1 for various E_f for a tetrahedron. a. E_f=-5.0; b. E_f=-3.0; c. E_f=-1.0; d. E_f=-0.5; e. E_f=0.5. (All parameters are in units of $|t|$.)

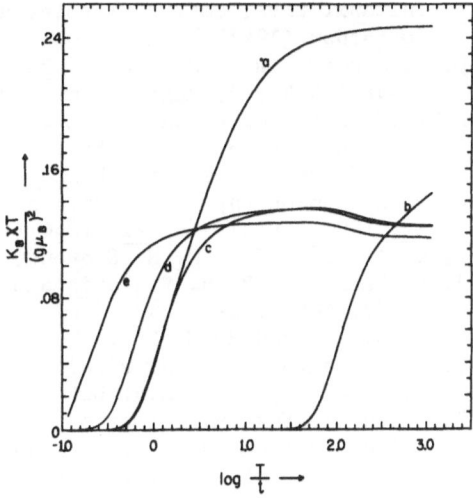

Fig. 8. $k_B \chi_f T/(g\mu_B)^2$ vs. log $T/|t|$ for t=-1, U=50 and V=0.1 for various E_f for a tetrahedron. a. E_f=-5.0; b. E_f=-3.0; c. E_f=-1.0; d. E_f=-0.5; e. E_f=0.5. (All parameters are in units of $|t|$.)

SUMMARY AND CONCLUSIONS

Our principal results can be summarized as follows: (i) The simple cluster model is able to reproduce the basic features of the mixed-valence and heavy-fermion systems. (ii) In some cases, there is a transition from the Kondo-lattice to a magnetic regime and subsequent re-entry into either a Kondo-lattice or a mixed-valence regime as E_f is gradually increased from large negative values. (iii) Heavy-fermion features are obtained when a non-magnetic (magnetic) ground state is nearly degenerate with magnetically (non-magnetically) ordered excited states but well separated from the other states. (iv) Magnetic order and mixed-valence coexist for very narrow range of E_f for some choice of parameters. Finally we note that we have not considered the degeneracy of the f level. However, the 4f electron of Ce^{3+} has the $^2F_{5/2}$ multiplet as the ground state. Because of large spin-orbit interaction, the excited J multiplet $^2F_{7/2}$ has an energy 0.28 eV above the ground state. In an octahedral crystalline field, the J=5/2 multiplet splits into a Γ_7 doublet and a Γ_8 quartet, the energy difference between them is 150 K. Thus our results for tetrahedral clusters should reflect the properties of cerium alloys.

ACKNOWLEDGMENT

This research was supported in part by the Division of Materials Research of the U. S. National Science Foundation under grant DMR-8504259.

REFERENCES

1. J. M. Lawrence, P. S. Riseborough and R. D. Parks, Rep. Prog. Phys. 44, 1 (1981); Valence Fluctuations in Solids, eds. L. M. Falicov, W. Hanke and M. B. Maple, North-Holland, Amsterdam (1981); Valence Instabilities, eds. P. Wachter and H. Boppart, North-Holland, Amsterdam (1982).
2. G. R. Stewart, Rev. Mod. Phys. 56, 755 (1984); J. Magn. Magn. Mater. 47-48 (1985); J. Magn. Magn. Mater. 54-57 (1986).
3. A brief review of these theories has been recently made by B. H. Brandow, Phys. Rev. B 33, 215 (1986).
4. P. W. Anderson, Phys. Rev. B 124, 41 (1961).
5. H. R. Krishnamurthy, J. W. Wilkins and K. G. Wilson, (a) Phys. Rev. B 21, 1003 (1980); (b) 21, 1044 (1980).
6. R. G. Arnold and K. W. H. Stevens, J. Phys. C 12, 5037 (1979).
7. J. C. Parlebas, R. H. Victoria and L. M. Falicov, J. Magn. Magn. Mater. 54-57, 405 (1986).
8. J. E. Hirsch and R. M. Fye, Phys. Rev. Lett. 56, 2521 (1986).
9. L. N. Oliveira and J. W. Wilkins, Phys. Rev. Lett. 47, 1553 (1981); V. T. Rajan, Phys. Rev. Lett. 51, 308 (1983).
10. K. Satoh, T. Fukita, Y. Maeno, Y. Onuki, T. Komatsubara and T. Onuki, Solid State Commun. 56, 327 (1985).
11. G. Aeppli, H. Yoshisawa, Y. Endoh, E. Bucher, J. Hufnagl, Y. Onuki and T. Komatsubara, Phys. Rev. Lett. 57, 122 (1986).

ELECTRONIC THEORY OF SEGREGATION IN SMALL BIMETALLIC CLUSTERS

S. Mukherjee

Institut de Physique Expérimentale
Ecole Polytechnique Fédérale de Lausanne
CH-1015 Lausanne, Switzerland

J.G. Pérez-Ramírez[1] and J.L. Morán-López[1,2]

Escuela Superior de Física y Matemáticas[1]
Departamento de Fisica[2]
Centro de Investigacion del I.P.N., 07000 Mexico

ABSTRACT

Surface segregation in bimetallic cubooctahedral and icosahedral clusters is studied within a simple tight-binding model. Results for equilibrium concentrations on different inequivalent sites in the cluster are given for Cu-Ni and Cu-Ru alloy clusters at various temperatures.

INTRODUCTION

Metal alloys have a widespread application in catalytic processes[1,2]. Alloy catalysts used in industrial applications are normally in the form of small particles, which has a large surface area per unit mass of the material[1]. The catalytic properties of these catalysts are determined largely by the composition of their surfaces, therefore the segregation of one or other constituents to the surface will have a direct effect on its catalytic property. Furthermore, the study of segregation in bimetallic alloy clusters like copper-ruthenium or copper-osmium is important for the understanding of the size dependent miscibility properties of these alloys, which are found to be not miscible in the form of larger aggregates[1,3].

In this paper we present a theory for the calculation of the surface composition of bimetallic alloy clusters at different temperatures, which can be applied to the clusters of any transition-metal binary alloy. It has been observed that small metal clusters are formed in close-packed structures[4], cubooctahedron and icosahedron being most common. We therefore apply our theory to calculate the surface composition of 55 and 147-atom cubooctahedral and icosahedral clusters of Cu-Ni and Cu-Ru alloys.

THEORY

In the present theory the internal energy of the bimetallic alloy cluster of N atoms with N_A A-atoms and $N-N_A$ B-atoms is calculated in a tight-binding model. The local density-of-states of the d-electrons are assumed to be of rectangular shape with bandwidth, band center and band filling as the given parameters. This approximation has been successfully applied to calculate the cohesive energy and surface tension of metals[5,6], heat of formation of alloys[7] and the surface segregation in semi-infinite binary transition-metal alloys[8]. Tight-binding models have been used previously for the calculation of cohesive and structural properties of small metal clusters[9-11].

In small metal clusters the different atoms at the surface have different atomic environment and therefore the binding energies at inequivalent sites will be different. In binary alloy clusters this would lead to site dependent heat of segregation at the surface[12]. In the present model we divide the cluster in concentric spherical shells such that the atoms lying on each shell have same number of nearest neighbor atoms.

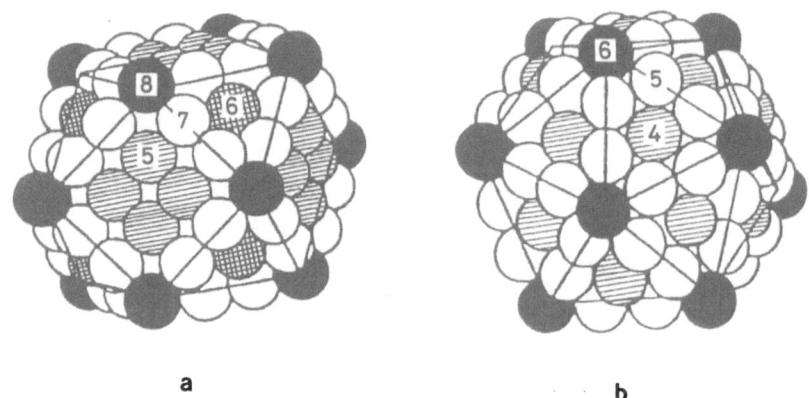

a   b

Fig. 1 : Cubooctahedron (a) and icosahedron (b) with 147-atoms. In each cluster, equivalent sites on the surface have same shading. The numbers represent the shells on which they lie.

A. Cubooctahedral clusters

In a cubooctahedral cluster of 147 atoms there will be eight such shells of sites around the central atom. The number of sites in the first, second, ... etc. shells are 12, 6, 24, 12, 24, 8, 48 and 12 respectively. The surface of this cluster consists of atoms in the 5th, 6th, 7th and 8th-shells (Fig. 1a). Each site in a given shell has a characteristic set of nearest neighbors, which is summarized in Table 1. Table 1 also contains similar information for a 55 atom cubooctahedron containing only four shells around the central atom.

The band energy of the cluster is calculated from its effective band-width W_{AB} of its d-band,

$$E = - \frac{W_{AB} \, n \, (10-n)}{20} + n \, \varepsilon, \tag{1}$$

where n is the occupation of the d-band of the alloy and ε is the average mean energy (center of the band) given by,

$$n = x \, n_A + y \, n_B \tag{2}$$

$$\varepsilon = x \, \varepsilon_A + y \, \varepsilon_B . \tag{3}$$

Here $n_A(n_B)$ and $\varepsilon_A(\varepsilon_B)$ are the occupation and the mean energy of the d-band of the pure constituent A(B) respectively and $x = N_A/N$ $(y = 1-x)$ is the average concentration of A(B) in the alloy.

Table 1 : Number of nearest neighbors Z_{ij} of an atom in shell i located in shell j, in a cubooctahedron with 147 atoms.

i \ j	0	1	2	3	4	5	6	7	8	Total
0	0	12	0	0	0	0	0	0	0	12
1	1	4	2	4	1	0	0	0	0	12
2	0	4	0	4	0	4	0	0	0	12
3	0	2	1	2	2	2	1	2	0	12
4	0	1	0	4	0	2	0	4	1	12
5	0	0	1	2	1	2	0	2	0	8
6	0	0	0	3	0	0	0	6	0	9
7	0	0	0	1	1	1	1	2	1	7
8	0	0	0	0	1	0	0	4	0	5

The effective bandwidth W_{AB} of the alloy cluster will depend on the concentrations x_1, x_2, ... etc. of each shell, and is calculated from the second moment[5] μ associated to a site p in any shell, written as,

$$\mu_p = <p|H^2|p> \tag{4}$$

where H represents the tight-binding Hamiltonian. Taking the nearest neighbor interactions and assuming a random distribution of the atoms in each shell, the second moment for any site in the shell i can be written from eq. (4) as,

$$\mu_i = x_i \epsilon_A^2 + y_i \epsilon_B^2 +$$
$$(1/Z^2) \left[(x_i W_A + y_i W_B) \{ \sum_j Z_{ij} (W_A x_j + W_B y_j) \} \right] \tag{5}$$

where Z is the bulk coordination number, Z_{ij} is the number of nearest neighbors of an atom in shell i, in shell j. $W_A (W_B)$ is the bandwidth of the pure constituent A(B) of the alloy. The average second moment of the cluster can be written as,

$$\mu_{av} = (1/N) \sum_i N_i \mu_i \tag{6}$$

The effective bandwidth W_{AB} of the cluster is related to the second moment μ_{av},

$$\mu_{av} = \epsilon^2 + W_{AB}^2 / 12 \tag{7}$$

The band energy is calculated from eqs. (1) and (7). The configurational entropy of the cluster is given by,

$$S = \sum_i \ln \left[N_i ! / \{ N_{A,i} ! (N_i - N_{A,i})! \} \right] \tag{8}$$

where N_i is the total number of sites in i-th shell occupied by $N_{A,i}$ A-atoms and the summation is carried over all different shells. The equilibrium values for the x_i at any temperature T are obtained by minimizing the free energy,

$$F = E - TS \tag{9}$$

The minimization is done with the constraint that the average concentration,

$$x = (1/N) \sum_i N_{A,i}$$

remains constant.

B. Icosahedral clusters

In a regular icosahedron of 147 atoms there are six equivalent shells around the central site, the surface atoms being confined in 4th, 5th and 6th-shell (Fig. 1b). However, in icosahedron there are two characteristic nearest neighbor distances, one being about 1.05 times larger than the other. We therefore take a distance dependent electron hopping parameter and the second moment at any shell i can be written in a similar way,

$$\mu_i = x_i \varepsilon_A^2 + y_i \varepsilon_B^2 + (1/Z^2) \left[(x_i W_A + y_i W_B) \right.$$

$$\left. \{ \sum_k Z_{ik} (W_A x_k + W_B y_k) + \sum_\ell \alpha \, Z_{i\ell} (W_A x_\ell + W_B y_\ell) \} \right] \qquad (10)$$

Here the summation over k goes through all those neighbors whose distances are same as in bulk and that over ℓ through those which are 1.05 times farther. The factor α is evaluated from the distance dependent electron hopping parameter allowing a $1/r^5$ fall off with distance[13]. The rest of the calculation is done as above.

RESULTS AND DISCUSSIONS

The essential input parameters – bandwith, band center and the band filling of the d-band for the pure elements are taken from ref. 14. In Fig. 2 results are shown for the concentration of Cu atoms in $Cu_{28}Ni_{27}$ 55-atom binary alloy cluster. This corresponds to an average Cu-concentration of about 0.5.

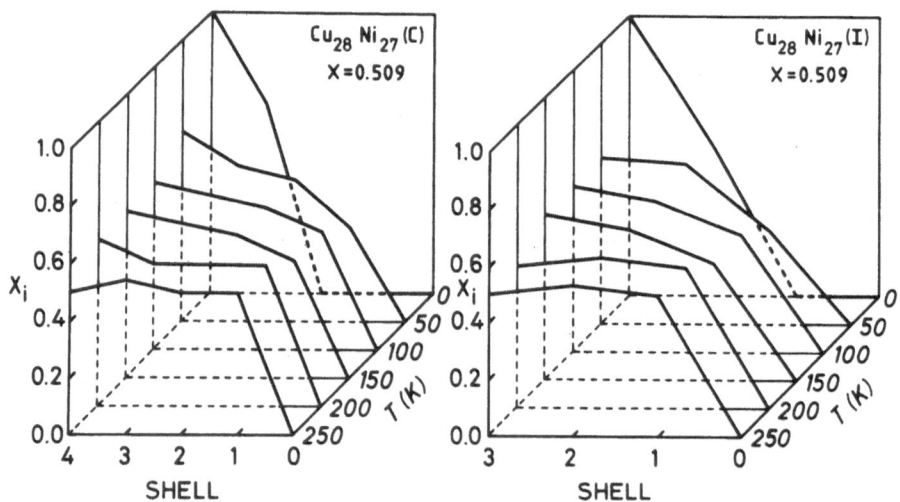

Fig. 2 : The copper concentration profile at different temperatures for the $Cu_{28}Ni_{27}$ cubooctahedral (C) and icosahedral (I) clusters of 55 atoms

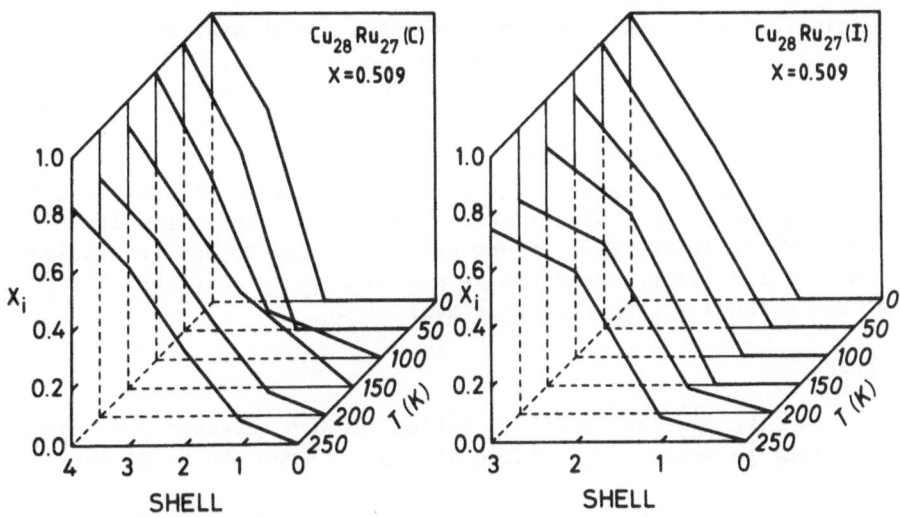

Fig. 3 : Same as Fig. 2 but for $Cu_{28}Ru_{27}$.

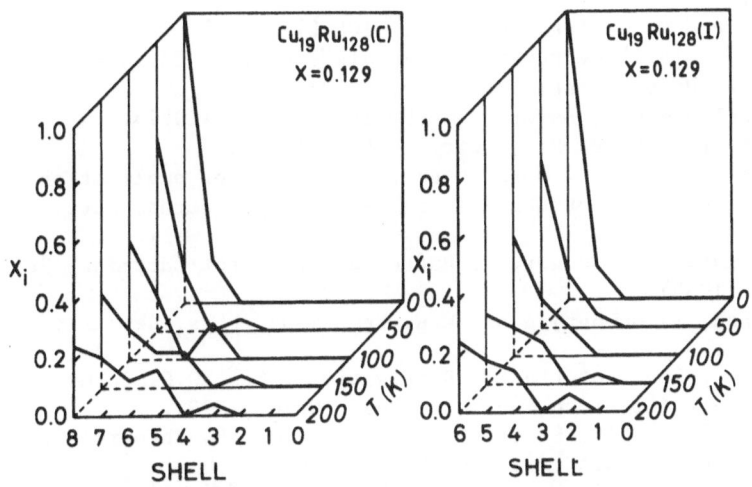

Fig. 4 : Same as Fig. 2 but for 147 atom Cu-Ru cluster with
19 Cu-atoms.

From Fig. 2 we observe that the concentration of Cu at surface sites are
nearly same, except for cubooctahedron in which the concentration in corner
sites (4th shell) is slightly larger. This happens due to less coordination
number of these sites compared to those in icosahedron. In Fig. 3 results
are shown for 55-atom Cu-Ru clusters with 28 Cu-atoms. We find a larger
segregation of Cu-atoms onto the corner sites and the stronger decrease of
segregation trend in successive inner lying shells. Thus for a 55-atom
cluster at 250 K we expect more Cu-atoms on corner sites and more Ru-atoms
on the surface planes. These segregation trends decrease with increasing
temperature. Similar trends have been reported[15] for Cu-Ni clusters in a

regular solution type model. In Fig. 4 results are shown for 147-atom Cu-Ru clusters with 19 Cu-atoms with average concentration 0.129. In this case we observe a stronger concentration fluctuation in different shells, owing to less number of Cu-atoms. Moreover, the temperature dependent changes are also large due to entropy effects.

In conclusion, we present a theory for calculation of the spatial distribution of atoms in a binary metal cluster and report results for Cu-Ru and Cu-Ni clusters. This theory can be applied to any transition-metal binary alloy clusters and can be extended to include bond relaxation effects[11], not taken in the present formulation.

ACKNOWLEDGMENTS

We would like to thank Prof. J. Buttet for his interest and support. S.M. would like to acknowledge financial support from the Swiss National Funds.

REFERENCES

1. J.H. Sinfelt, Rev. Mod. Phys. 51:569 (1979).
2. V. Ponec, Adv. Catal. 32:149 (1983).
3. J.H. Sinfelt, Y.L. Lam, J.A. Cusumano and A.E. Barnett. J. Catal. 42: 227 (1976).
4. A. Renou and M. Gillet, Surf. Sci. 106:27 (1981).
5. J. Friedel, in: "The Physics of Metals", vol. 1, Ed. J.M. Ziman (Cambridge Univ. Press, Cambridge, 1969).
6. J. Friedel and C.M. Sayers, J. Physique 38:697 (1977).
7. D.G. Pettifor, Phys. Rev. Lett. 42:846 (1979).
8. S. Mukherjee and J.L. Morán-López, submitted for publication.
9. M.B. Gordon, F. Cyrot-Lackmann and M.C. Desjonquères, Surf. Sci. 80: 159 (1979).
10. S.N. Khanna, J.P. Bucher, J. Buttet and F. Cyrot-Lackmann, Surf. Sci. 127:165 (1983).
11. D. Tománek, S. Mukherjee and K.H. Bennemann, Phys. Rev. B 28:665 (1983).
12. C.R. Helms, in: "Interfacial Segregation", Eds. W.C. Johnson and J.M. Blakely (American Soc. of Metals, Ohio, 1979).
13. V. Heine, in: "Solid State Physics", 35, Eds. H. Ehrenreich, F. Seitz and D. Turnbull (Academic, New York, 1980).
14. O.K. Andersen, O. Jepsen and D. Glötzel, in: "Highlights of Condensed Matter Theory", Eds. F. Bassani, F. Fumi and M.P. Tosi (Soc. Italiana di Fisica, Bologna, 1985).
15. J.L. Morán-López and C.A. Balseiro, Phys. Rev. 33:4849 (1986).

THEORETICAL CLUSTER MODEL OF A Zn-C REFORMATSKY INTERMEDIATE

F. Orsini[1], F. Pelizzoni[1], D.D. Shillady[2] and L.M. Vallarino[2]

[1]Centro Studi Sostanze Organiche Naturali del C.N.R.
Milano, Italy
[2]Department of Chemistry, Virginia Commonwealth University
Richmond, VA 23284, USA

ABSTRACT

A large body of experimental data exists for the reactions of $BrZnCH_2COOC(CH_3)_3$ (I) and related organo-zinc compounds. Orsini et al. isolated a crystalline form of (I) from tetrahydrofuran; the X-ray structure determined by Boersma et al. showed a cyclic dimer of 68 atoms. In the present work, relativistic extended Huckel (REX) calculations using the method of Pyykko and Lohr have been carried out for the approach of CH_3I and CH_3Br to the Zn-C bond of (I). Several mechanisms have been modeled, showing that the Br or I of an alkyl halide approaching Zn(II) plays a very important role in the cleavage of the Zn-C bond. The results of model cluster calculations on $Zn(CH_3)_2$, $ZnCl_2$, ZnH_2, and CH_3Cl using HONDO/5 and STO 6/4G basis have been compared to those obtained with REX and with the ab initio pseudopotential results of Ratner et al. with respect to Mulliken atomic charges and optimized geometries. The REX model agrees qualitatively with the experimental data and with the atomic charges of pseudopotential calculations, but the nonrelativistic ab initio HONDO/5 atomic charges in a minimum basis give a less clear description of the Zn-C bond when used with REX screening values. The REX model appears to be useful for clusters up to 100 atoms and 400 spin-orbitals and for large zinc-containing systems of biological interest.

INTRODUCTION

Zinc is important in biological systems and in synthetic organic chemistry, but many of the reaction mechanisms have not been fully elucidated. Recently, a body of experimental evidence has been presented proving the existence in solution of C-metallated rather than O-metallated species for the typical Reformatsky reagents $BrZnCH_2COOC(CH_3)_3$, $BrZnCH(CH_3)COOC(CH_3)_3$, and $BrZnC(CH_3)_2COOC(CH_3)_3$.[1] The reactivity of these compounds as selective nucleophiles has also been extensively investigated. New insight on this interesting class of organo-zinc reagents has been provided by the single crystal X-ray analysis[2] of $BrZnCH_2COOC(CH_3)_3$. In the solid state, this was shown to be a dimer in which each of the two equivalent Zn atoms is tetrahedrally surrounded by two oxygen, one bromine, and one carbon atom. The resulting dimeric structure involves an 8-membered non-planar ring, $(ZnCCO)_2$ (see schematic framework in Figure 1). The dimer persists in tetrahydrofuran (THF) but breaks down in strongly polar solvents to give solvated monomers.[1,2]

RESULTS AND DISCUSSION

The knowledge of the heavy-atom coordinates of $[BrZnCH_2COOC(CH_3)_3]_2$ from X-ray crystallography provided a useful initial framework for the study of the interaction of this compound with alkyl halides in solution, as a model for organo-zinc reactions. In order to investigate the possible electronic mechanism of the reaction between alkyl halides and a Reformatsky reagent from a theoretical viewpoint, one must use a method capable of treating a multi-atom system including Zn, Br, and I. For this study the relativistic extended Huckel method of Pyykko and Lohr[3-6] was selected and enlarged to treat up to 100 atoms and 400 spin-orbitals on the VCU IBM 3081D computer. The REX method treats all elements up to atomic number 116 using Desclaux's[7] relativistic calculations for free atoms. Of special interest is the use of separate radial functions (Slater-type orbitals) for the spin-up and spin-down wavefunctions; this produces a /J,Jz) orbital basis which incorporates one-center spin-orbit coupling in a natural way for the heavier elements while retaining the same quality as Hoffmann's method[8] for the lighter elements. Even though there are well known limitations to extended-Huckel type calculations, such as exaggerated charges and lack of electron repulsion, the relativistic parameters in REX give a very pleasing uniform treatment of the entire Periodic Chart. This substantially contributes to the ability to study organometallic chemistry in a qualitative way.

To assess the validity of the REX method when applied to the Reformatsky system of interest here, model cluster calculations were optimized for several simple Zn(II) compounds by HONDO/5 in a STO-6/4G basis. The results obtained were compared to those from REX as well as to the literature data for atomic charges (Table I). The Mulliken atomic charge is the only available common feature of the three methods compared.

Considering that the extended Huckel method is known to give exaggerated charges, the results obtained by REX for $Zn(CH_3)_2$, ZnH_2, and $ZnCl_2$ may be considered to be in good agreement with the ab initio pseudopotential results in the literature[9] and with HONDO/5. A discrepancy is observed for CH_3Cl; this arises from the use in REX of diffuse screening constants fitted to relativistic atomic functions. In HONDO/5[10], we used core STO-6G functions scaled to Clementi's values[11] with STO-4G valence orbitals scaled to the diffuse REX values (Table II). It should be pointed out that the atomic charge values obtained from REX are qualitatively consistent with the experimental behavior of organozinc compounds. The simple formula in equation 1 leads to qualitative agreement with experiments, even though the same basis set in the nonrelativistic HONDO/5 program yields an opposite carbon charge in CH_3Cl.

$$H_{ij} = 0.875(H_{ii} + H_{jj})S_{ij}; \qquad HC = ESC \qquad (1)$$

The starting point in the present study was the use of the X-ray structural data to provide the coordinates of the heavy atoms. Idealized angles and a C-H bond length of 1.08 A were then employed to include the hydrogen atoms in the structure of the Reformatsky dimer. Methyl bromide and iodide were chosen as the reagents, to allow for a sufficiently small model while maintaining a chemically meaningful relationship with the

Table I. Atomic Mulliken Charges from REX and Other Methods

Molecule/Atom	REX	Pseudopotential	HONDO/5
ZnCl$_2$	r(Zn–Cl)=2.073 A	r(Zn–Cl)=2.049 A	r(Zn–Cl)=2.073 A
Zn	+1.60	+0.92	+1.04
Cl	–0.80	–0.46	–0.52
			E=–2686.556917 au
Zn(CH$_3$)$_2$	r(Zn–C)=1.784 A	r(Zn–C)=1.83 A	r(Zn–C)=1.784 A
Zn	+1.93	+0.65	+1.38
C	–0.98	–0.73	–1.03
H	+0.04	+0.14	+0.11
			E=–1841.450252 au
ZnH$_2$	r(Zn–H)=1.465 A		r(Zn–H)=1.465 A
Zn	+1.15		+1.73
H	–0.58		–0.37
			E=–1770.776415 au
CH$_3$Cl	r(C–Cl)=1.843 A		r(C–Cl)=1.843 A
C	+0.20		–0.52
H	+0.04		+0.21
Cl	–0.34		–0.12
			E=–494.206053 4au

higher molecular weight organic halides of practical importance.

Several model reactions of CH$_3$I with the dimeric zinc complex were simulated on the basis of space-filling models and of mechanistic pathways proposed in the literature[12] for organometallic compounds of this type. Since preliminary studies of the REX wavefunctions for the dimer showed that direct analysis of the LCAO real and complex matrices was impractical, bond orders from the Mulliken overlap population analysis were chosen and attention was focused on changes in the Zn–C bond order.

The results obtained are illustrated in Figure 1. When the approach of CH$_3$I to Zn(II) was modeled as required for a bimolecular nucleophilic substitution (SN$_2$: path d) or for a bimolecular elecrophilic substitution (SE$_2$: paths a and d), no breaking of the Zn–C bond was observed. Accordingly, these mechanisms could reasonably be ruled out.

We next simulated the approach of CH$_3$I and CH$_3$Br along directions opposite to each of the four bonds of Zn(II), with the iodine or bromine

Table II. Valence Shell Screening Parameters for HONDO/5 from REX Values

Zn		C		Cl		H	
3d	3.980285	2s	1.57735	3s	2.25831	1s	1.000018
4s	1.579840	2p	1.43472	3p	1.902267		
4p	1.259410						

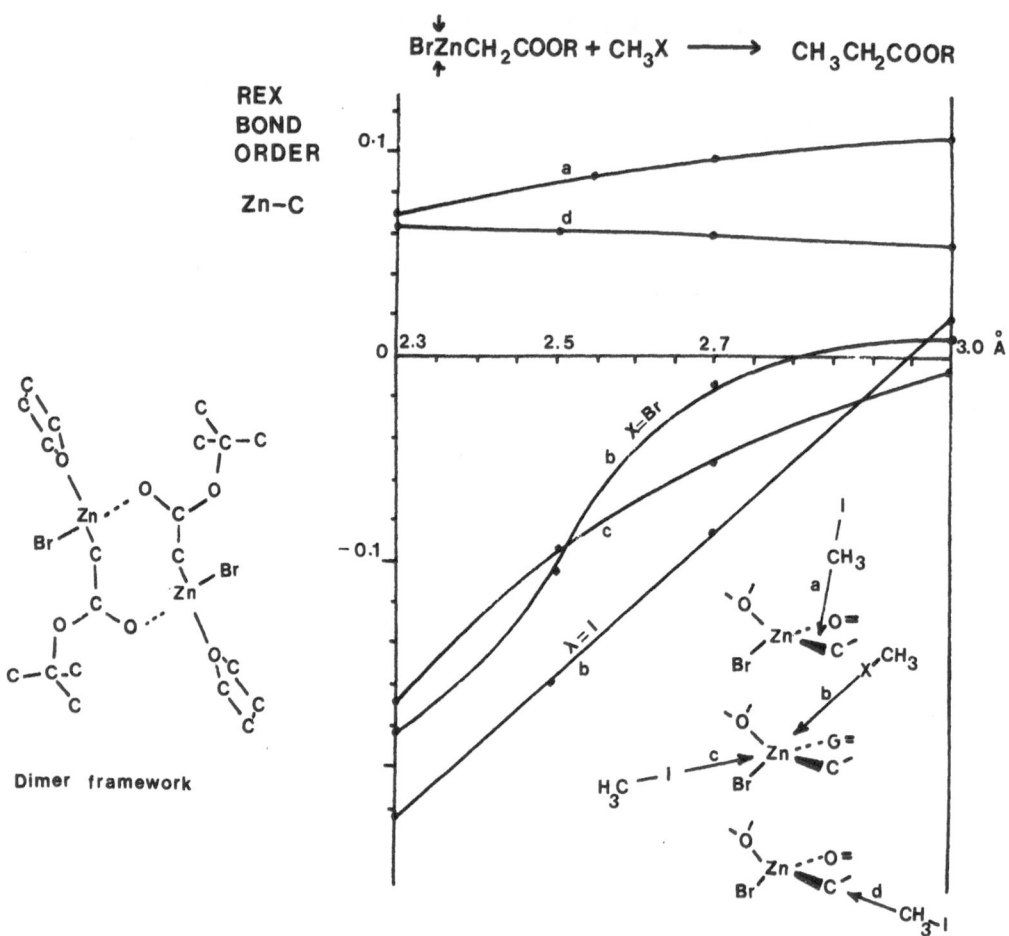

Figure 1. Simulated approaches of methyl halides to dimeric Reformatsky cluster.

atom of the alkyl halide pointing toward the zinc atom. Of these approaches, only two (one opposite to the Zn-Br bond and the other opposite to the Zn-O=C bond) were found to result in cleavage of the Zn-C bond, as indicated by its increasing antibonding (negative bond order) character. In both approaches, the Zn-keto and Zn-THF bond orders (not shown in Figure 1) change somewhat as the halogen approaches the Zn atom. In the free dimer the Zn-keto interaction is slightly bonding (+0.002) and the Zn-THF interaction is slightly antibonding (-0.012); as the halogen approaches the zinc atom, however, both Zn-O coordinate bonds become antibonding, with the Zn-keto bond order falling more rapidly. These changes are consistent with the experimental observation that the formation of the dimer is solvent dependent and that the dimeric structure is lost as the reaction proceeds. The results obtained with REX point to an important initial cleavage of the Zn-C bond by the Zn-halogen interaction. Indeed, in the case of CH_3I, coordination of iodine to zinc prior to cleavage of the zinc-carbon bond may reasonably be postulated. It is noteworthy that the approach of CH_3Br is not equally effective in bringing about the cleavage of the Zn-C bond (path b: X = Br); this is a key agreement with experimental data, for alkyl iodides are known to react with Zn(II) Reformatsky reagents much more effectively than do alkyl bromides.[1]

A four-center concerted mechanism also appears to be ruled out by calculations. Changing the position of the methyl group so that the CH_3-I and the Zn-CH_2 bonds are aligned produces a stronger Zn-I bonding interaction but an increasingly antibonding I-CH_3••••CH_2-Zn interaction.

CONCLUSIONS

The work presented here provides an understanding of the initial phase of the reaction between an alkyl halide and the dimeric Reformatsky reagent. The results obtained using the REX model agree with two key experimental facts: the breaking of the dimer in polar solvents and the higher reactivity of alkyl iodides relative to bromides. This agreement implies rejection of the concerted four-center, SN_2 and SE_2 mechanisms. Elucidation of the mode of formation of the CH_2-CH_3 bond requires treatment of a transition state using a realistic set of force constants at a saddle point geometry.

Acknowledgement

This work has been supported as a bilateral project by the Italian Research Council and Virginia Commonwealth University.

REFERENCES

1. F. Orsini, F. Pelizzoni, G. Ricca, C-Metallated Reformatsky Intermediates. Structure and Reactivity, Tetrahedron, 40: 2781 (1984).
2. Y. Dekker, P.H.M. Budzelaar, Y. Boersma, G. Van der Kerk, The Nature of the Reformatsky Reagent. Crystal Structure of (BrZnCH₂COO-t-Bu•THF)₂, Organometallics, 3: 1403 (1984).
3. L. Lohr, P. Pyykko, Relativistic Parametrized Extended Huckel Theory, Chem. Phys. Letters, 62: 333 (1979).
4. P. Pyykko, L. Weisenfeld, Relativistic Parametrized Extended Huckel Calculations IV. Nuclear Spin-spin Coupling Tensors for Main Group Elements, Mole. Phys., 43: 557 (1981).
5. P. Pyykko, L. Lohr, Relativistic Parametrized Extended Huckel Calculations. 3. Structure and Bonding for Some Compounds of Uranium and Other Heavy Elements, Inorg. Chem., 20: 1950 (1981).
6. L. Lohr, M. Hotokka, P. Pyykko, Relativistic Parametrized Extended Huckel Program, Quantum Chemistry Program Exchange No. 387, Indiana University, U.S.A.
7. Y. P. Desclaux, Atomic Data and Nuclear Data Tables, in: "Relativistic Dirac-Fock Expectation Values for Atoms with Z = 1 to Z = 120", Academic Press, New York (1973).
8. R. Hoffmann, An Extended Huckel Theory. I. Hydrocarbons, J. Chem. Phys., 39: 1397 (1963).
9. M. A. Ratner, Y. W. Moskowitz, S. Topiol, Pseudopotential Calculations. 5. Results for Group 2A and 2B Dimethyls and Chlorides, Y. Am. Chem. Soc., 100: 2329 (1978).
10. M. Dupuis, J. Rys and H. F. King, Ab Initio Hartree-Fock Self-Consistent Field Program, Quantum Chem. Program Exchange No. 403, Indiana University, U.S.A.
11. E. Clementi, D. L. Raimondi, Atomic Screening Constants from SCF Functions, G. Chem. Phys., 38: 2686 (1983).
12. Ei-ichi Negishi, General Patterns of Organometallics Reaction, in "Organometallics in Organic Synthesis", John Wiley and Sons, New York (1980).

CALCULATION OF THE MAGNETIC PROPERTIES OF SMALL Fe-CLUSTERS

G.M. Pastor, J. Dorantes-Dávila* and K.H. Bennemann

Freie Universität Berlin, FB 20, WE 5
1000-Berlin 33, W.-Germany

ABSTRACT

We report calculations of the size dependence of the magnetic moments of small Fe-clusters. We use the d-band tight-binding Hamiltonian, and treat electron correlations within the Hubbard model. The clusters considered are b.c.c.-like Fe_9, Fe_{15} and Fe_{27}. We obtain for both Fe_9 and Fe_{15} a total magnetic moment per atom $\mu = 2.79$ μ_B, in agreement with experiment and previous calculations. For Fe_{27} we get $\mu = 2.64$ μ_B as convergence to bulk magnetic moment $\mu_b = 2.2$ μ_B starts to take place. Temperature dependent effects and structural stability are discussed.

I. INTRODUCTION

The study of the magnetic properties of small clusters has become a subject of considerable interest /1,4/. Recently /1/, non-resonant magnetic deflection experiments were for the first time performed on Fe-clusters composed of 2 to 17 atoms. One observes that small Fe-clusters are already magnetic with a magnetic moment per atom μ_n approximately independent of the number of cluster atoms n. In addition, μ_n was seen to be larger or equal than bulk magnetic moment μ_b. We present here preliminary results for the magnetic moments of Fe_9, Fe_{15} and Fe_{27}, and compare them with those obtained by other calculational procedures. Convergence to bulk behaviour, structural stability and finite temperature effects are briefly discussed.

II. THE MODEL

We consider the Hubbard-model:

$$H = H_{TB} + H_{corr} \quad ,$$

where H_{TB} refers to the usual d-band tight-binding Hamiltonian. The many electron term H_{corr}, treated in the Hartree-Fock approximation, is site-diagonal and is given by:

$$H_{corr} = \sum_{l\sigma} \epsilon_{l\sigma} n_{l\sigma} - E_{corr} \quad , \quad \epsilon_{l\sigma} = U\, n(1) - \sigma J \mu(1) \quad .$$

Here $n_{1\sigma}$ refers to the number operator of spin σ at cluster site 1, and E_{corr} to the correction term due to double counting. J and U stand respectively for the exchange and direct Coulomb integrals, and are taken to be independent of orbital, l and cluster size. $\mu(1)$ and $n(1)$ are the magnetic moment and number of d-electrons on atom 1, and are given by

$$n(1) = <n_{1\uparrow} + n_{1\downarrow}> \quad , \quad \mu(1) = <n_{1\uparrow} - n_{1\downarrow}> \quad .$$

Self-consistency is achieved by requiring

$$<n_{1\sigma}> = \int_{-\infty}^{\epsilon_F} N_{1\sigma}(\epsilon)\ d\epsilon \quad ,$$

where $N_{1\sigma}(\epsilon)$ refers to the local density of states which we calculate using the recursion method technique /5/. Note, all orbitals were taken here to be of the same kind (d-orbitals), but there is no difficulty in extending the model to include s and p orbitals.

III. RESULTS AND DISCUSSION

We present here results obtained considering only d-orbitals with 1st and 2nd neighbor hopping integrals $dd\sigma$, $dd\pi$ and $dd\delta$ appropriate for bulk α-Fe /6/. The value of J was determined to be J = 0.83 eV by fitting to the experimental bulk magnetic moment at T = 0 (μ_b = 2.2 μ_B). We assume for the d-band filling n_d = 7.2 electrons per atom, usually taken for bulk Fe /7/, and calculate self-consistently charge redistribution throughout the cluster. Note, the lack of translational symmetry in the cluster will result in general in charge transfer between non-equivalent sites which, however, tends to be compensated by the direct Coulomb repulsion U. We consider here 2 limiting cases: 1) U = 0 which physically corresponds to assume that sp electrons screen perfectly any charge imbalance among d-electrons /6/; and 2) U = 6.1 eV estimated from experimental atomic data, which in practice is equivalent to the strong correlation limit (U → +∞) and implies approximately local charge neutrality.

Calculations were performed for Fe_9, Fe_{15} and Fe_{27}. We assumed the geometrical structure obtained by adding to a central atom the successive closed shells of its 1st, 2nd and 3rd neighbors in a b.c.c.-lattice (see Fig.1). A summary of the results for the size dependence of the magnetic moments is shown in Table 1. We obtain for Fe_9 and Fe_{15} approximately the same total magnetic moment per atom $\mu_9 \cong \mu_{15}$ = 2.79 μ_B. This is in agreement with experimental results /1/ showing that the total magnetic moment increases linearly as a function of cluster size, and also compares with previous calculations /2,8/. This can be understood noting that most of the atoms of Fe_9 and Fe_{15} are surface atoms and therefore have a reduced local coordination number and effective band-width W_n. This causes the effective exchange integral J/W_n to be larger and leads to larger magnetic moments. Note, the same effect was already reported for transition-metal surfaces /9/. For Fe_{27} we obtain μ_{27} = 2.64 μ_B which, although still reflects reduced coordination number effects, starts to approach bulk magnetic moment (μ_b = 2.21 μ_B). It is important to note that not the central atom (denoted by 1) but its nearest neighbors (denoted by 2) show a magnetic moment that converges most rapidly to the bulk value. This is due to the fact that the perturbation introduced by the cluster surface is symmetric around the central atom and therefore produces very strong changes in its local density of states. Remarkably, the results are almost independent of the direct Coulomb repulsion U, since charge transfer is not strong (⩽ 0.1 el.) even for U = 0.

The ferromagnetic order (all moments pointing in the same direction) was in all cases the solution of the Hartree-Fock equations with the lowest energy. This is in contrast with the results of Lee et al. /2/ which show

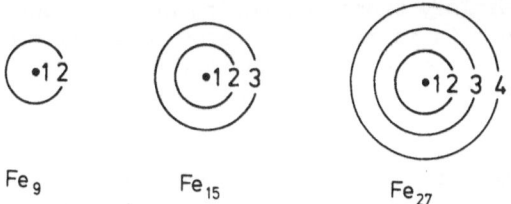

Fig. 1. Illustration of the geometrical structure of the
clusters. The numbers label different shells
composed of symmetry related atoms.

Table 1. Summary of the results for the magnetic moments
(in units of μ_B) of small Fe-clusters. Parameters: n_d=7.2 el., J=.83 eV, U=6.1 eV. In brackets
results for U=0. Shell numbers as in Fig. 1.

	Fe_9	Fe_{15}	Fe_{27}	Bulk
$\mu_{av.}$	2.79 (2.79)	2.79 (2.79)	2.64 (2.62)	2.21
$\mu(1)$	2.77 (2.70)	2.78 (2.79)	2.81 (2.91)	
$\mu(2)$	2.79 (2.80)	2.79 (2.81)	2.50 (2.48)	
$\mu(3)$		2.79 (2.76)	2.63 (2.56)	
$\mu(4)$			2.72 (2.72)	

a spin density around the central atom of Fe_9 and Fe_{15} dominated by minority spins. This disagreement is probably due to the rather large, cluster
size independent number of d-electrons considered here. One expects namely
that transition-metal clusters have less d-electrons than bulk /10/, what
would imply a tendency towards antiferromagnetic order since we approach
half-band filling. We would therefore like to emphasize that the present
results should be taken as preliminary since n_d is not known very well,
s-d charge transfer might play a significant role as a function of n and
then should be included more carefully. Also changes in the hopping integrals due to relaxation should be taken into account (it might play a role
for μ_n as well as for the magnetic order). Also ε_0 = const. should be improved by using $\varepsilon_1 = \varepsilon_0 - az_1$ /11/. Calculations are in progress to study
these effects.

In order to study possible structural transitions (e.g. b.c.c.-f.c.c.)
we performed calculations for f.c.c.-like clusters. We obtained that small
f.c.c. Fe-clusters possess strong magnetic moments in contrast with bulk

γ-Fe which does not satisfy Stoner criterion and is probably a weak anti-ferromagnet /12/. The gain in magnetic energy of small f.c.c. Fe-clusters with respect to bulk γ-Fe might cause reconstruction from the rather open surface of b.c.c.-like Fe_{27} to a more close-packed one resembling f.c.c.-like structure.

Concerning the temperature dependence of the magnetization $M_n(T)$ we feel that reliable conclusions need studies taking into account short-range spin correlations, which are important for surface magnetism /13/, in particular when the cluster size is of the same order as the correlation length.

Helpful discussions with Dr. A. Aligia are greatfully acknowledged.

REFERENCES

* Present address: Departamento de Física, CINVESTAV-IPN, Apdo. postal 14-740, 07000 México D.F., México.

1. D.M. Cox, D.J. Trevor, R.L. Whetten, E.A. Rohlfing, and A. Kaldor, Phys. Rev. B 32, 7290 (1985).
2. K. Lee, J. Callaway, and S. Dhar, Phys. Rev. B 30, 1724 (1984).
3. J. Callaway, D.P. Chen, and R. Tang, Z. Phys. D 3, 91 (1986).
4. Y. Ishii and S. Sugano, submitted to J. Phys. Soc. Japan (1986).
5. R. Haydock, Solid State Phys. 35, 216, New York, Academic Press, 1980.
6. V. Heine, J.H. Samson, and C.M.M. Nex, J. Phys. F: Met. Phys. 11, 2645 (1981).
7. H. Hasegawa, submitted to J. Phys. F: Met. Phys. (1986).
8. C.Y. Yang, K.H. Johnson, D.R. Salahub, J. Kaspar, and R.P. Messmer, Phys. Rev. B 24, 5673 (1981).
9. R.H. Victora, L.M. Falicov, and S. Ishida, Phys. Rev. B 30, 3896 (1984).
10. M.G. Mason, L.J. Gerenser, and S.T. Lee, Phys. Rev. Lett. 39, 288 (1977).
11. G.M. Pastor, J. Dorantes-Dávila, and K.H. Bennemann, in this volume.
12. G.J. Johanson, M.B. Mc Girr, and D.A. Wheeler, Phys. Rev. B 1, 3208 (1970); J. Kübler, Phys. Lett. 81A, 81 (1981).
13. J. Dorantes-Dávila, G.M. Pastor, and K.H. Bennemann, Solid State Commun. (1986), in press.

DISCUSSION OF THE SIZE DEPENDENCE OF IONIZATION THRESHOLD ENERGY OF

Fe-CLUSTERS

G.M. Pastor, J. Dorantes-Dávila* and K.H. Bennemann

Freie Universität Berlin, FB 20, WE 5
Arnimallee 14
D-1000 Berlin 33, Germany

ABSTRACT

We discuss the size dependence of the ionization threshold energy I_n of small Fe clusters using a simple tight-binding type electronic theory. In agreement with recent experimental results we obtain characteristic structure in I_n as a function of the number n of cluster atoms. From comparison with experiment we conclude that small Fe-clusters have b.c.c.-like structure.

The properties of small clusters are at present a subject of considerable interest /1-5/. Much of the experimental and theoretical work has been focused on metal clusters because of its fundamental importance and practical applications. For example, measurements of the photoionization threshold energy /3,4/ for small Fe clusters have revealed interesting structure as a function of cluster size. One observes that the ionization energy of Fe_n increases for n = 2 to 4 and n = 10 to 15, and decreases for n = 4 to 10 and n = 15 to 25. In order to explain these experimental results we calculate the ionization energy using a simple tight-binding type electronic theory. We assume for n < 9 the structure geometries shown in the inset of Fig. 1 and for n ⩾ 9 b.c.c.-like cluster structure.

The ionization-threshold energy is given by:

$$I_n = E_n^{i-1} - E_n^i + \Delta I_n \ ,$$

where $E_n^i = \int_{-\infty}^{\varepsilon_{max}} d\varepsilon \, \varepsilon \, N_n(\varepsilon)$ is the total electronic energy of a cluster of n atoms and i electrons. ε_{max} refers to the highest occupied state and $N_n(\varepsilon)$ to the total density of states of a n atom cluster. ΔI_n is the many electron contribution resulting from the interaction between the photoelectron and the hole produced in the cluster, which is neglected in our calculation of E_n^i. Using a classical approximation ΔI_n is given by /6/:

$$\Delta I_n = const. + (3/8)(e^2/R) \ .$$

Here, $R = (\frac{3V}{4\pi} n)^{1/3}$ is the average radius of the cluster, V refers to the atomic volume. We consider the Hubbard Hamiltonian in the Hartree-Fock approximation:

$$H = \sum_{l\sigma} (\varepsilon_l + U < n_{l\bar{\sigma}}>) \, c^+_{l\sigma} c_{l\sigma} + \sum_{ij} t_{ij} c^+_{i\sigma} c_{j\sigma} - \sum_l U < n_{l\uparrow}>< n_{l\downarrow}> \quad ,$$

where $c^+_{l\sigma}$, $c_{l\sigma}$, $n_{l\sigma}$ are the usual creation, destruction and number operators of spin σ at site l, and t_{ij} and U refer to the hopping and Coulomb
exchange integrals. The d-level energy ε_l is approximately given by /7/:

$$\varepsilon_l = \varepsilon_o + \Delta\varepsilon_l \quad .$$

Here, ε_o results from the potential of atom l and is taken to be independent of l and cluster size. The shift $\Delta\varepsilon_l$, essentially due to the nearest
neighbours of l, is approximately given by $\Delta\varepsilon_l = -az_l$, z_l is the coordination
number of the cluster site l. The constant a is a measure of the strength
of the screened potential of the nuclear charge on a neighbour atom and is
expected to depend smoothly on n through the cluster size dependent interatomic distance and electron charge distribution. We determine $N_n(\varepsilon)$ using
the second moment approximation which was shown to give good qualitative
results for energy calculations /8/. In this way one has

$$N_n(\varepsilon) = N_{n\uparrow}(\varepsilon) + N_{n\downarrow}(\varepsilon) \quad ,$$

$$N_{n\sigma}(\varepsilon) = \frac{2}{\pi W_n^2} \left[W_n^2 - (\varepsilon_n + < n_{\bar{\sigma}} > U - \varepsilon)^2 \right]^{1/2}$$

Here, $\varepsilon_n = \varepsilon_o - az_n$ is the average d-level energy and z_n the average coordination number of a n atom cluster. The band width W_n is given by:

$$W_n = \left[\frac{z'_n}{z'_b} W_b^2 + <\varepsilon_l^2>_n - \varepsilon_n \right]^{1/2}$$

where z'_n is the average effective coordination number (b refers to bulk),
and $<\varepsilon_l^2>_n$ means average over the cluster atoms denoted by l . For clusters with b.c.c.-like structure z'_l is given by /8/ $z'_l = z_l + (1/2)z_{2l}$
(z_{2l} is the number of second neighbours). The small correction $(<\varepsilon_l^2> - \varepsilon_n^2)$
results in an increase of W_n since ε_l is not constant throughout the cluster.

Using the parameters $U = 1.2$ eV, $W_b = 4$ eV and $N_{el} = 7.2$ electrons,
usually taken for bulk Fe, and for the atomic volume the bulk one, we obtain results for I_n shown in Fig. 1. For a we use $a = 0.2$ eV which is close
to the value $a = 0.16$ eV determined from recent density functional calculations /5/ for Fe_9 and Fe_{15} by integrating the local d-electron density
of states /9/. We can also estimate a and its dependence on the interatomic distance r assuming that d-electrons localized on neighbouring sites
screen perfectly the nuclear charge and that s-electron screening can be
characterized by a Thomas-Fermi wave vector appropriate for bulk Fe
($k_{TF} \cong 1$ au) /7/. Then one has approximately $\Delta\varepsilon_l = - \frac{z_l}{r} \exp(-k_{TF}r)$.
Using the dimer interatomic distance ($r = 3.77$ au) /4/ we obtain $a = 0.17$eV;
and using the bulk one ($r = 4.71$ au) /7/ $a = 0.05$ eV. This shows that the
shift $\Delta\varepsilon_l$ appears to be more important for small clusters than what one
would expect in terms of estimates for bulk. In addition k_{TF} should decrease with decreasing cluster size since screening in small clusters
seems to be less effective than in bulk metal.

It is remarkable that the simple theory presented here reproduces
quite closely the experimentally observed structure in the ionization-ener-
gy as a function of cluster size. We conclude that a simple interpretation
of the experimental results in terms of the changes of the average coordination number z_n as a function of n is possible. On one hand we have
$\varepsilon_n = \varepsilon_o - a \, z_n$, what causes I_n to increase with z_n. On the other hand it
holds approximately $W_n \alpha \sqrt{z_{eff}}$, this produces an increase of I_n if

$\varepsilon_{max} > \varepsilon_n$ or a decrease of I_n if $\varepsilon_{max} < \varepsilon_n$, with increasing z_{eff} (see Fig.2). Hence, depending on the value of a, W_b and the filling of the d-band, one can expect different behaviours. If $\Delta\varepsilon_n$ dominates over ΔW_n (as is the case for small Fe-clusters) I_n increases with coordination number. If, however, ΔW_n dominates over ε_n (and $\varepsilon_{max} < \varepsilon_n$) I_n decreases as the coordination number increases. In addition, the contribution ΔI_n causes I_n to decrease monotonically with n, and rules the dependence of I_n on cluster size when ΔW_n and $\Delta\varepsilon_n$ cancel each other and when the coordination number does not depend strongly on cluster size as it occurs, for example, when the cluster is large enough so that bulk like coordination is approached.

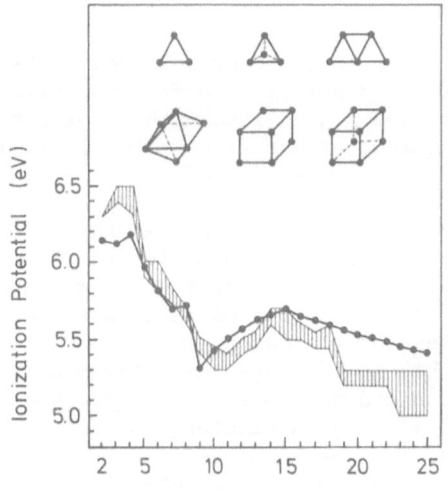

Fig. 1: The ionization threshold energy of Fe_n as a function of cluster size. The calculation refers for $n \geqslant 9$ to b.c.c.-like clusters. The inset shows the cluster geometry assumed for $n < 9$. The dashed band reflects the uncertainties of experimental results (Ref. 3).

It is important to note that the structure in I_n also reflects the b.c.c.-like geometry of small Fe_n clusters. Calculations performed assuming f.c.c.-like geometry show namely no oscillation in I_n for $n = 13$ to 19, as the 3rd shell of the cluster is filled up, in disagreement with experiment.

It is possible to improve our theory taking into account relaxation and the resulting change in a and W_n with cluster size. This would increase the value of a for Fe_2-Fe_4 and therefore the calculated I_2-I_4, improving the agreement with experiment. Note, cluster contraction in Fe_9 with respect to Fe_{10} would increase I_9 with respect to I_{10} and shift the

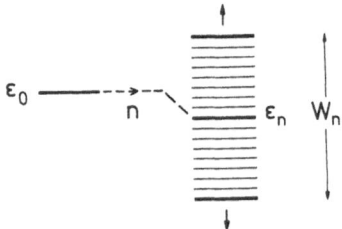

Fig. 2: Illustration of the size dependence of I_n due
to changes in ε_n and W_n. Approximately
$W_n \alpha \sqrt{z_{eff}}$, and $\varepsilon_n = \varepsilon_o - az_n$.

maximum to clusters with n = 10 as observed experimentally. Also we expect
for larger cluster (n $\geqslant$ 19) possibly a surface reconstruction from b.c.c.-
like to a more closed-packed f.c.c.-like with a lower surface energy. This
could give a better agreement of our results for I_n (n $\geqslant$ 19) with experiment.

So far we have not considered the changes in the center of gravity of
the d-band coming from the size dependent electron charge distribution and
e-e interactions. For example, as the cluster grows the s-charge density
around the nuclei is gradually increased as a consequence of the overlap
of the atomic s-densities and the change in the boundary conditions on the
s-wave function /10/. This would cause the d-level energy to increase with
n because of the enhanced s-d coulombic repulsion. However, we expect this
contribution to depend smoothly on cluster size and to be not very structure sensitive, since s-wave functions are quite spread out. Also size dependent screening should be important in determining d-level shifts induced
by e-e interactions.

A more detailed discussion of the size dependence of the ionization-
energy and also of the cohesive-energy of Fe-clusters will be presented
elsewhere /11/.

Helpful discussions with Dr. P. Stampfli are greatfully acknowledged.

REFERENCES

*   Present address: Departamento de Física, CINVESTAV-IPN, Apdo. postal
    14-740, 07000 México D.F., México.
1. W.D. Knight, K. Clemenger, W.A. de Heer, W.A. Saunders, M.Y. Chou, and
    M. Cohen, Phys. Rev. Lett. 52, 2141 (1984).

2. J.L. Martins, J. Buttet, and R. Car, Phys. Rev. Lett. $\underline{53}$, 655 (1984).
3. R.L. Whetten, D.M. Cox, D.J. Trevor, and A. Kaldor, Phys. Rev. Lett. $\underline{54}$, 1494 (1985).
4. E.A. Rohlfing, D.M. Cox, A. Kaldor, and K.H. Johnson, J. Chem. Phys. $\underline{81}$, 3846 (1984).
5. K. Lee, J. Callaway, and S. Dhar, Phys. Rev. B $\underline{30}$, 1724 (1984).
6. D.M. Wood, Phys. Rev. Lett. $\underline{46}$, 749 (1981).
7. N.W. Ashcroft and N.D. Mermin, Solid State Physics (Holt, Rinehart and Winston, New York, 1976), chapters 10 and 17.
8. D. Tománek, S. Mukherjee, and K.H. Bennemann, Phys. Rev. B $\underline{28}$, 665 (1983).
9. $a = (\varepsilon_{15} - \varepsilon_9) / (z_9 - z_{15})$, where $n = \frac{1}{10} \int_{-\infty}^{\varepsilon_{max}} d\varepsilon \, \varepsilon \, N_{nd}(\varepsilon)$ and z_n is the average coordination number in Fe_n.
10. See for instance M.O. Robbins and L.M. Falicov, Phys. Rev. B $\underline{29}$, 1333 (1984) and references therein.
11. G.M. Pastor, J. Dorantes-Dávila, and K.H. Bennemann, to be published.

SELF-INTERACTION-CORRECTED CALCULATION OF THE STATIC POLARIZABILITY OF

SMALL METAL SPHERES

P. Stampfli and K.H. Bennemann

Freie Universität Berlin, FB 20, WE 5
Arnimallee 14
D-1000 Berlin 33, Germany

ABSTRACT

   We calculate selfconsistently the static electric polarizability of
small metal spheres using density-functional theory in the local
density approximation and the spherical jellium approximation. The
erroneous interactions of the electron with itself contained in the
usual density-functional theory are corrected. We obtain good quantitative agreement with the polarizabilities of Na-cluster. Previous
calculations have given values which were about 20 % lower than our
results. The improvement is mainly due to the elimination of the erroneous screening of the electron by its own induced charge contained
in previous calculations.

INTRODUCTION

     Previous calculations of the static electric polarizability of small
metal spheres /1,2,3/ have given values which are about 20 % too small compared to the polarizabilities of small Na-clusters as observed in experiment /4/. These calculations used a self-consistent density functional
scheme with a local density exchange-correlation potential for the electrons and a spherical jellium approximation for the positive metal ions.
They did not correct the erroneous interaction of the electron with itself contained in the electrostatic and the exchange-correlation potentials in the one electron equations by Kohn and Sham /6/. A recent calculation /5/ includes elastic deformations of the positive background, but
this increases the polarizability only by about 1 % and does not resolve
the disagreement with experiment. Improving on the jellium approximation
and using atomic pseudopotentials one does not obtain a more diffuse density profile for the conduction electrons at the surface of alkalime-
tals /13/. Thus, we expect that the polarizabilities of Na-clusters would
not be significantly increased if one includes the atomic structure in
the calculation.

The ground state electronic structure is calculated using the Kohn-Sham one-electron equations with an approximate correction of the self-interaction. This has already been discussed extensively /7/. The density of conduction electrons of the bulk metal is $n_o = (4\pi r_s^3/3)^{-1}$ and the number of conduction electrons of the metal cluster is N. The positive charge of the metal ions is approximated with a jellium charge density $n_+(\vec{r})$ which is uniform $n_+ = n_o$ inside a sphere of radius $R = N^{1/3} r_s$ and $n_+ = 0$ outside. The electronic density is

$$\rho(\vec{r}) = \sum_{i=1}^{N} \rho_i(\vec{r}) \quad , \quad \rho_i(\vec{r}) = |\psi_i(\vec{r})|$$

where ψ_i are the one-electron wave functions. Using atomic units $(e = m = \hbar = 1)$ the Kohn-Sham equations are

$$(-\frac{1}{2} \vec{\nabla}^2 + V_{es}(\vec{r}) + V_{xc}(\vec{r}))\psi_i = \varepsilon_i \psi_i \tag{1}$$

The Coulomb potential is

$$V_{es}(\vec{r}) = -\int d^3r' \; \frac{n_+(\vec{r}') - \rho(\vec{r}')}{|\vec{r} - \vec{r}'|} \tag{2}$$

and the exchange-correlation potential is in the local density approximation

$$V_{xc}(\vec{r}) = V_{LD}(\rho(\vec{r})) \tag{3}$$

where

$$V_{LD}(\rho) = \frac{\partial}{\partial \rho} (\rho \varepsilon_{xc}(\rho))$$

We use Wigner's interpolation formula /3,8/

$$\varepsilon_{xc}(\rho) = -0.458 \; (4\pi\rho/3)^{1/3} - 0.44/ \; (7.8 + (4\pi\rho/3)^{-1/3})) \tag{4}$$

According to (1) and (2) an electron occupying orbital i moves in an electrostatic potential V_{es} due to all electron charges including its own charge distribution $-\rho_i(\vec{r})$. This falsely implies that each electron interacts with itself electrostatically. Similarly (1) and (3) approximately describe an exchange and correlation interaction of an electron with all electrons of the system including itself. These errors in V_{es} and V_{xc} partially cancel each other in the total potential $V_{tot} = V_{es} + V_{xc}$. But there remain large errors except for systems with delocalized electrons such as bulk-metal. The potential V_{es} is straightforward to correct, we obtain

$$V_{es}^{(i)}(\vec{r}) = -\int d^3r' \; \frac{n_+(\vec{r}') - (\rho(\vec{r}') - \rho_i(\vec{r}'))}{|\vec{r} - \vec{r}'|} \tag{5}$$

The index i indicates that the self-interaction corrected potential is different for electrons occupying different orbitals. For a system with only one electron the corrected V_{xc} should be exactly zero. For a system with N electrons there is a positively charged hole around each electron because of exchange and correlation with the other N-1 electrons. The charge of the hole has to be compensated by a negative charge distributed over the whole electronic system. The exchange-correlation hole is localized near the position of the electron and it is reasonable to use the local density approximation (3). For the interaction with the compensating charge we really should use a nonlocal description. But this would make the calculation very complicated, and we need a rough local approximation. The appropriate correction is /7/

$$V_{xc}^{(i)}(\vec{r}) = V_{LD}(\rho(\vec{r})) - V_{LD}(\rho_i(\vec{r})) \tag{6}$$

There $- V_{LD}(\rho_i(\vec{r}))$ very roughly simulates a potential of a negative charge distributed over the whole system because $\rho_i(\vec{r})$ simply is an orbital density of one electron localized on the cluster. Note that this approximation is not good near the surface of the system where the wave functions $\psi_i(\vec{r})$ and thus $\rho_i(\vec{r})$ become small and vanish. For the calculation of ground state properties this approximation may be quite good because the errors near the surface are then not too important. The one-electron equation is now

$$(- \frac{1}{2} \vec{\nabla}^2 + V_{es}^{(i)}(\vec{r}) + V_{xc}^{(i)}(\vec{r}))\psi_i = \epsilon_i \psi_i \tag{7}$$

In order to obtain spherical symmetry we take the spherical averages of the electron densities to calculate the potentials $V_{es}^{(i)}$ and $V_{xc}^{(i)}$. Equations (5), (6) and (7) define the electronic structure. We have to achieve self-consistency simultaneously in each of the potentials.

THE POLARIZABILITY

First, we present the modified Sternheimer equations used to calculate the polarizability of metal spheres neglecting the self-interaction corrections. They have been discussed extensively by other authors /9,10/. Then we shall introduce the appropriate self-interaction corrections.

The external perturbation, such as an applied electric field, is assumed to be small. Thus only the changes in (1), (2) and (3) of first order in the perturbation are considered. The polarizing electric field $\vec{E}$ is taken to be parallel to the z-axis. We use $\vec{E} = (0,0,1)$ and the external perturbation is then $\delta V_{ext}(\vec{r}) = r \cos \theta$. From (1) we obtain the perturbation $\delta \psi_i$ of the one-electron wave function ψ_i. This induces a change $\delta \rho_i = 2 \, \mathrm{Re}(\psi_i^* \delta \psi_i)$ of the electron density ψ_i. The total induced charge density is

$$- \delta \rho = - \sum_{i=1}^{N} \delta \rho_i$$

with the spatial dependence $\delta \rho(\vec{r}) = \delta \rho(r) \cos \theta$. This gives rise to perturbations δV_{es} and δV_{xc} in V_{es} and V_{xc}. The net perturbation is

$$\delta V = \delta V_{ext} + \delta V_{es} + \delta V_{xc}$$

with the spatial dependence $\delta V(r) \cos \theta$. The induced dipole moment and the polarizability is

$$\alpha = - \int d^3r \; z \; \delta \rho(\vec{r})$$

Keeping only first order terms, the perturbation $\delta \psi_i$ is given by (1)

$$(- \frac{1}{2} \vec{\nabla}^2 + V_{es} + V_{xc} - \epsilon_i) \delta \psi_i = - (\delta V_{ext} + \delta V_{es} + \delta V_{xc}) \psi_i \tag{8}$$

where ψ_i is the unperturbed wave function and

$$\delta V_{es}(\vec{r}) = \int d^3r' \; \delta \rho(\vec{r}') \, / \, |\vec{r} - \vec{r}'| \tag{9}$$

$$\delta V_{xc}(\vec{r}) = \delta V_{LD}(\vec{r}) = \delta \rho(\vec{r}) \; (\frac{\partial}{\partial \rho} V_{LD}(\rho(\vec{r}))) \tag{10}$$

The equations (8), (9) and (19) again contain spurious self-interactions which have to be corrected. An electron in orbital i does not react to its own induced charge $\delta\rho_i$, therefore

$$\delta V_{es}^{(i)}(\vec{r}) = \int d^3r' \frac{\delta\rho(\vec{r}') - \delta\rho_i(\vec{r}')}{|\vec{r} - \vec{r}'|} \tag{11}$$

is the self-interaction corrected perturbation of the electrostatic potential applying to this electron. It would not be appropriate to derive the correction of the perturbation δV_{xc} of the exchange-correlation potential directly from (6) because (6) is a very rough local approximation. Note that the charges $\delta\rho_i$ are induced near the surface of the metal sphere where the approximation (6) essentially fails. For a system with only one electron the self-interaction corrected $\delta V_{xc}^{(i)}$ should be zero exactly. Similarly for a many electron system, whenever only one electron in a certain orbital i contributes to the change in the total electron density $\delta\rho = \delta\rho_i$ then the self-interaction corrected $\delta V_{xc}^{(i)}$ should vanish too for this electron. The correction should thus be

$$\delta V_{xc}^{(i)}(\vec{r}) = (\delta\rho(\vec{r}) - \delta\rho_i(\vec{r})) \frac{\partial}{\partial\rho} V_{LD}(\rho(\vec{r})) \tag{12}$$

The equation (8) is then changed to

$$(-\frac{1}{2}\vec{\nabla}^2 + V_{es}^{(i)} + V_{xc}^{(i)} - \varepsilon_i)\delta\psi_i = -(\delta V_{ext} + \delta V_{es}^{(i)} + \delta V_{xc}^{(i)})\psi_i \tag{13}$$

Note that the corrections (6) and (12) of the self-exchange and self-correlation are formally not consistent with each other because (12) is not the variation of (6) with respect to small changes $\delta\rho_i$ of the electronic densities. Thus, the force-sum rule /12/ will not be satisfied. But this is only a minor inconvenience compared to the advantages of using local approximations for exchange and correlation. We solve eqn. (13) using an expansion of $\delta\psi_i$ in unoccupied electron states. This makes sure that only true electron-hole excitations contribute to the polarizability.

DISCUSSIONS OF RESULTS

For our calculations we have used a bulk electron density with r_s= 4 which simulates Na. We have done our calculation only for metal spheres with full shells of degenerate electron states because for partially filled shells aspherical deformations should become important /11/. We stopped with our calculation at a cluster size of 68 conduction electrons but the calculation could easily be continued to larger clusters. We have checked the numerical accuracy of our program using the force sum rule/12/. The condition that at the surface of the jellium $\delta V_{es}(R) = -\delta V_{ext}(R)$ is fulfilled within less than 1 % error for our calculation without self-interaction correction. This condition does not apply for the self-interaction corrected calculation because of the inconsistency of the self-exchange and self-correlation correction. But the numerical accuracy is not impaired and should remain the same.

Our results for the ground-state electronic structure of a metal sphere agree well with the results of other authors /1,2,3,8/. The self-interaction correction changes the electron density profile only slightly. The electron density extends a bit further out from the surface of the jellium background. But this alone would have a negligible effect on the polarizability.

In tab. 1 we present the values for the polarizability α of metal spheres calculated with and without self-interaction correction together with the polarizabilities of Na-clusters as measured by Knight et al /4/. For convenience they are given in units of the classical polarizability R^3.

TABLE 1. Static polarizabilities α of Na$_n$-clusters in comparison with the classical polarizability R^3 (Values given as αR^{-3})

Size n of the cluster	Previous calculations[a]	New calculation[b]	Measured polarizability[c]
2	1.56	1.74	2.02 ± 0.06
8	1.40	1.76	1.73 ± 0.03
18	1.31	1.55	1.63 ± 0.05
20	1.34	1.61	1.58 ± 0.04
34	1.24	1.43	1.58 ± 0.03
40	1.30	1.61	1.56 ± 0.03
58	1.21	1.38	---
68	1.25	1.44	---

a) Our own values, reproducing previous calculations /1,2,3/
b) With correction of the self-interaction errors
c) Experiment by Knight et al /4/

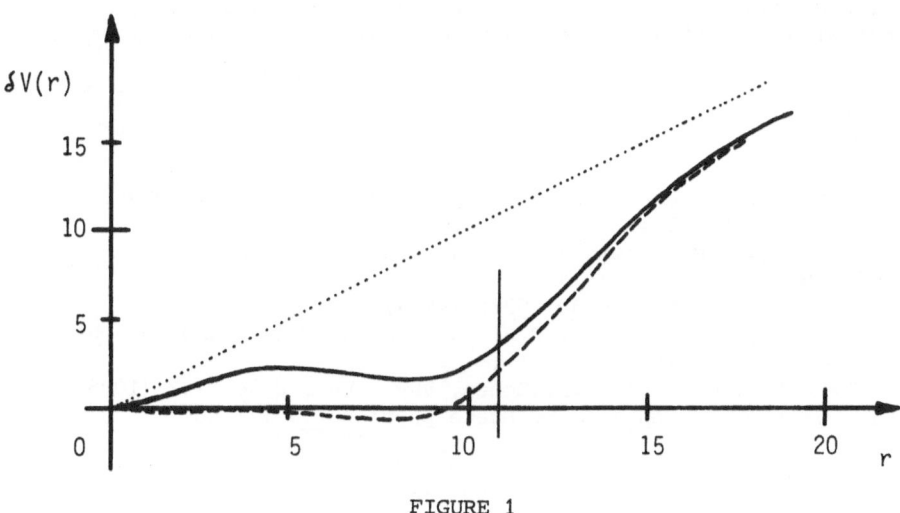

FIGURE 1

The perturbation $\delta v^{(i)}$ of the effective one-electron potential $v^{(i)}$ for an electron i in the highest occupied orbital (2s) of a metal sphere with 20 electrons and $r_s = 4$ (atomic units). The vertical line marks the jellium edge (R = 10.9).

    Continuous line: Calculation with correction of the self-interaction
    Dashed line   : Calculation without correction
    Dotted line   : External perturbation v_{ext} corresponding to the polarizing electric field

The self-interaction correction increases the polarizabilities by about 20 % compared to the calculation without correction. Good agreement with the polarizabilities of Na-clusters is achieved except for the cluster with 34 electrons. The self-interaction correction mainly seems to increase the induced charge density $\delta\rho(\vec{r})$ but its profile remains essentially the same.

In fig. 1 we show the total perturbation $\delta V_{tot} = \delta V_{ext} + \delta V_{es} + \delta V_{xc}$ of the effective one-electron potential V_{tot} applying to an electron i in the highest occupied orbital of a sphere with 20 electrons. Without self-interaction correction we have approximately $\delta V = 0$ inside the sphere. With the correction the perturbation $\delta V_{tot}^{(i)}$ does not vanish. This is due to the fact that essentially only the electrons in the highest occupied orbital contribute to the screening of the external electric field. For each of these electrons only the screening charge due to the other electrons is relevant because it should not be influenced by its own change. Thus, screening is not complete for these electrons. In our special case there are two electrons in the highest occupied level and the perturbation $\delta V_{tot}^{(i)}$ is thus reduced only to approximately one half of δV_{ext} rather than zero near the center of the sphere. Thus, the self-interaction correction increases the polarizability because the spurious self-screening of the electrons contained in previous calculations is removed. This reduces the effective screening and makes the interaction with the polarizing electric fields stronger.

CONCLUSIONS

We obtain increased polarizabilities for metal spheres because we eliminate the spurious self-screening of the electron contained in previous calculations. We achieve good agreement with experiment. However, the local approximations used for the self-interaction correction in the calculation of the ground state structure and the polarizability are not consistent with each other. This could be resolved in a nonlocal calculation.

REFERENCES

/1/   W. Ekardt, Ber. Bunsenges. phys. Chem. 88, 289 (1984); Phys. Rev. Lett. 52, 1925 (1984)
/2/   M. Manninen, R.M. Nieminen and M.J. Puska, Phys. Rev. B 33, 4289 (1986)
/3/   D.E. Beck, Phys. Rev. B 30, 6935 (1984)
/4/   W.D. Knight, K. Clemenger, W.A. de Heer and W.A. Saunders, Phys. Rev. B 31, 2539 (1985)
/5/   P. Shen, M.Y. Chou and M.L. Cohen, Phys. Rev. B 34, 732 (1986)
/6/   W.E. Pickett, Comments on S. State physics 21, 1 (1985)
/7/   J.P. Perdew and A. Zunger, Phys. Rev. B 23, 5048 (1981)
/8/   D.E. Beck, S. State Commun. 49, 381 (1984)
/9/   G.D. Mahan, Phys. Rev. A 22, 1780 (1980)
/10/ D.E. Beck, Phys. Rev. B 30, 6935 (1984)
/11/ K. Clemenger, Phys. Rev. B 32, 1359 (1985)
/12/ R.S. Sorbello, S. State Commun. 48, 989 (1983)
/13/ G.P. Alldredge and L. Kleinmann, Phys. Rev. B 10, 559 (1974)

KEYWORDS: polarizability/ metal cluster/ self-interaction-correction density functional theory/ jellium approximation

ESR AND ELECTRON SPIN RELAXATION TIME T_1

OF Be SMALL PARTICLES

Sanshiro Sako

Department of Physics
Faculty of Education, Mie University
Tsu, Mie 514, Japan

ABSTRACT

The ESR spectra in the Be small particles (size 28Å~70Å) were investigated in the temperature range from 3K to room temperature at 9.1GHz and
at room temperature at 35GHz. In any spectrum, the Curie-like paramagnetic
component was, more or less, detected in addition to the Pauli paramagnetic
one with the g value of 2.0033±0.0002. Those spin species are discussed.
The size dependence of the spin relaxation time T_1 was measured in the above
temperature range. The increasing tendency of T_1 with the decrease of size
was observed. It is indicated from the experimental result of T_1 that those
spectra reflect the softening effect of lattice vibration or the quantum
size effect.

INTRODUCTION

In general, the conduction electron spin resonance (CESR) is difficult
to detect because of the strong spin-orbit interaction except for alkaline
metals. However, in such divalent metals as Mg and Be, the ESR signal is
easily obtained. The small particles of such metals are the even particles[1] which are interesting for the examination of the quantum size
effect[1].

As for Mg small particles, the characteristic ESR spectrum with the
asymmetrical line-shape and the low field shoulder in addition to the broad
line[2] due to the normal conduction electron spins had been detected[2,3] and
discussed comparing with the theories of the quantum size effect[1,4,5,6]
However the effect of surface cannot be neglected in many cases because
the surface to volume ratio is large in a particle. The adsorbed radical
molecules or atoms on the surface oxide layer will affect the ESR spectrum
in many cases. Especially, in Mg small particles, we think that the influence of O_2^- molecules[7] adsorbed on the surface MgO layer will be large.

In this paper, we report about the ESR of small particles of Be which
have the same h.c.p. structure as Mg. The resonance spectrum has the line-
shape of Lorentzian type and shows that Pauli paramagnetic behaviour. However, the absorption intensity increased at low temperature in accordance
with the Curie law. This suggests the addition of the Curie-like paramagnetic spins to the Pauli paramagnetic ones. As a clue to elucidate the spin

species of the spectrum, the size dependence of spin relaxation time T_1 was measured by the saturation method. As a result, we found the tendency that T_1 becomes large as the size becomes smaller. It was the same result as that obtained by Borel *et al.*[8] in Li small particles. The spin species of the Curie-like component and the spin relaxation mechanism of Be small particles are discussed in relation to the softening effect of lattice vibration and the quantum size effect.

EXPERIMENTAL PROCEDURE

Be small particles were produced under the atmosphere of high purity argon or helium gas (both 99.999% pure) by the gas evaporation method. The purity of the original metal was more than 99.99% (more than 99.999% on magnetic impurities). The particles evaporated on the substrate cooled by liquid nitrogen were collected under air and put into a quartz tube for the ESR measurement and then sealed together with helium gas. The strong BeO pattern in the electron diffraction was observed overlapping with that of Be metal because of the surface oxidation. The thermogravimetry analysis of Be small particles was carried out in order to know the degree of oxidation just after the ESR measurement The mass of particles decreased at first with increasing temperature in the stream of dry air for the evaporation of H_2O on the surface, and then increased with the progress of oxidation as shown in Fig. 1. It reached constant value when particles were perfectly oxidized. From the change of the mass, we can know the mass of unoxidized core Be. The ratios of core Be to all Be atoms in a particle were 30~50% in the samples used in this experiment. The size distributions of particles were observed by means of the transmission electron microscope (TEM). In an example shown in Fig 2, the broken line shows the size distribution of core Be which was estimated from the size distribution gained by TEM observation (solid line) under the assumption that the thickness of surface oxidized layer is constant independent of particle size. The size with a maximum population in core size distribution was used for the size of a sample as a measure. The sample size used for the measurement of T_1 was in the range from 28Å to 60Å.

The ESR signals were measured by means of a reflection spectrometer

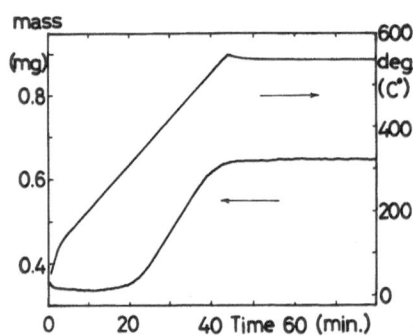

Fig. 1. Change of sample mass with temperature and time by the thermogravimetry. Increase of mass is caused by oxidation. The perpendicular axis is mass of sample (left side) and temperature of sample (right side).

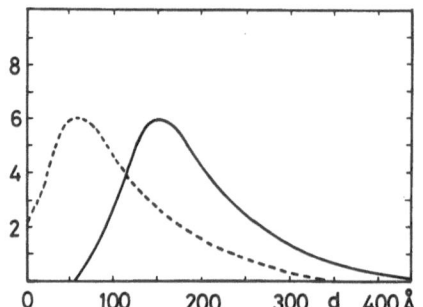

Fig. 2. Size distribution of a sample of Be small particles. The solid line shows the original size distribution. The broken line shows the core Be size distribution corrected for surface oxide. The perpendicular scale is arbitrary.

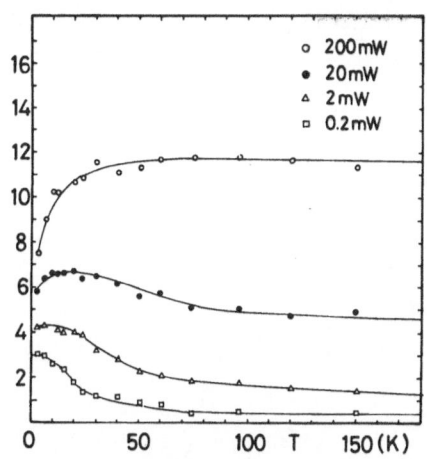

Fig. 3. Absorption intensity vs temperature in each power. The sample size is 70Å.

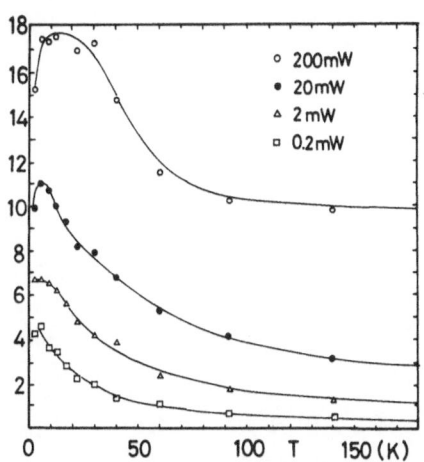

Fig. 4. Absorption intensity vs temperature in each power. The sample size is 45Å.

with magnetic field modulation at 100kHz in the temperature range from 3K to room temperature at 9.1GHz and at room temperature at 35GHz. The temperature was controlled by the gas flow type controller (Oxford ESR-9 cryostat).

EXPERIMENTAL RESULT

The g value of the spectrum of Be small particles was 2.0033±0.0002 which was independent of temperature. The value is between the anisotropic bulk g values of $g_\perp$=2.00325 and $g_{//}$=2.00374 given by Cousins and Dupree[9]. It agreed with the result in the polycrystalline metal by Feher and Kip[10] (the g shift is +(9±1)×10^{-4}). The derivative peak to peak linewidths of the range 5~6 gauss at room temperature at 9.1GHz were broadened to the range 0.7~0.9 gauss at 35GHz. The broadening due to the frequency difference is approximately equal to that expected from the anisotropy of g value. The slight Dysonian asymmetry in line-shape owing to the skin-depth appeared in the large size sample. The ratio A/B of the height A and B of the derivative absorption line (A is the peak height in the lower field) increased with decreasing temperature because of the increase of the skin-depth. The value of A/B was about 1.1 at the maximum. These are the Pauli paramagnetic behaviour due to the conduction electrons.

However, the absorption intensity increased at low temperature in any sample, as shown also in the example of Fig. 3, Fig. 4. The power change reflects the saturation phenomenon. The difference between Fig. 3 and Fig. 4 is the sample size. The size in Fig. 3 is 70Å and the one in Fig. 4 is 45Å. Those figures show the overlap of the Curie-like paramagnetic component with the same g value judging from the symmetrical line-shape, on the Pauli paramagnetic one. The Curie-like component has a tendency to become larger as the size becomes smaller.

The spin lattice relaxation time T_1 at a certain temperature was measured from the power dependence of intensity in the power range from 0.01mW to 200mW by the saturation method using DPPH as the reference. The temperature dependence of T_1 is obtained by measuring T_1 at each temperature as seen in Fig. 5. The linear dependence in logarithmic scale means the pro-

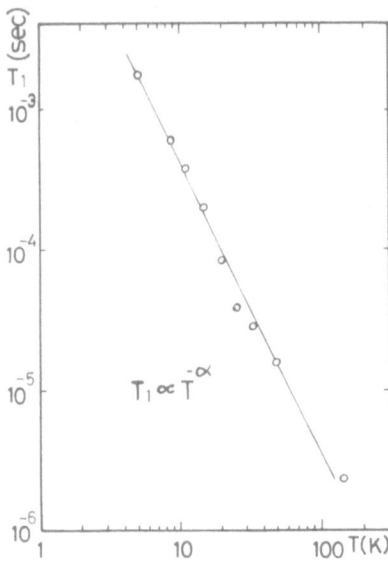

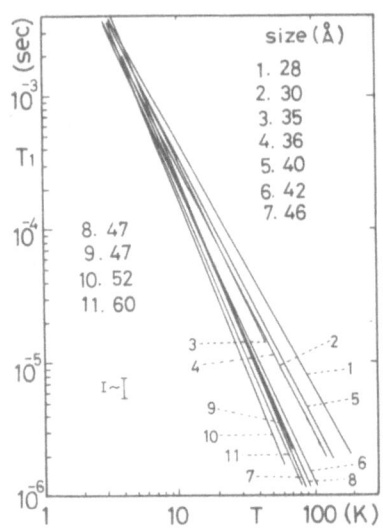

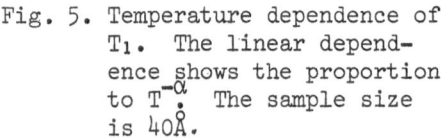

Fig. 5. Temperature dependence of T_1. The linear dependence shows the proportion to $T^{-\alpha}$. The sample size is 40Å.

Fig. 6. Temperature dependence of T_1 in 11 samples with different size. Smaller sample number in the figure shows smaller sample size.

portion of T_1 to $T^{-\alpha}(\alpha>0)$. Such temperature dependence of T_1 was measured in 11 samples with different size. Those results are shown in Fig. 6. T_1 become longer as the size becomes smaller at any temperature. Fig. 7 shows the size dependence of T_1 at 30K. The size dependence of α is shown in Fig. 8. Those are estimated from the gradient in Fig. 6. The value of α becomes larger as the size becomes larger.

DISCUSSION

The resonance spectrum includes the Pauli paramagnetic component which is dominant at high temperature and the Curie-like paramagnetic one which is dominant at low temperature. We can, moreover, estimate that the smaller particles in a sample contribute to the Curie-like component and the larger ones in the same sample contribute to the Pauli paramagnetic one. A smaller particle has the larger surface and interface ratio to volume. Therefore, the effect of adsorbed molecules or atoms on the surface oxide, or the effect of the interface with the unstable bond between oxide and core metal may contribute to the Curie-like component. As another possibility of the component, we can consider the breakdown of the electrical neutrality[1] in such an even particle as Be particle if the quantum size effect takes place.

We can consider the following three mechanisms as the spin relaxation mechanism which makes T_1 longer with decreasing size. The first one is the cut-off of the phonon energy smaller than hc/2d (c is the light velocity). It originates in the fact that the phonon with the longer wavelength than the size doesn't exist. This mechanism doesn't work at high temperature as the cut-off energy for the size of 50Å corresponds to 10K. The second one is the decrease of the number of the proper lattice vibration[11] which

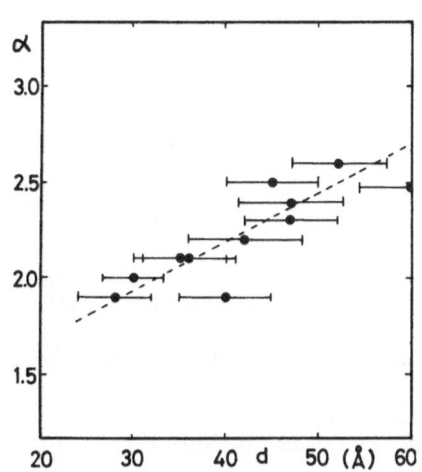

Fig. 7. Size dependence of T_1 at 30K.

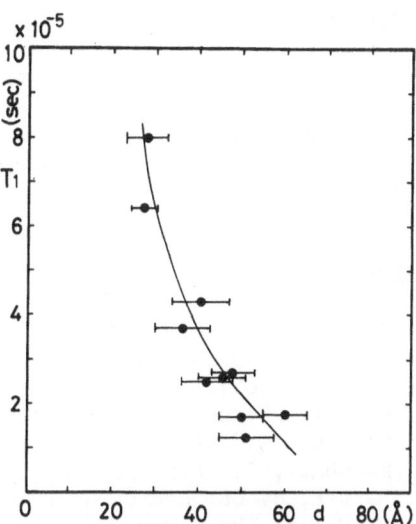

Fig. 8. Size dependence of the value of α.

is caused by the mobility of atoms toward the direction perpendicular to the surface (the softening effect of lattice vibration) This mechanism can explain our experimental results as it works at high temperature. The third one is the effect of the separation of the conduction electron energy level (the quantum size effect). The spin relaxation mechanism by Elliott[12] should not work when the energy level spacing becomes larger than Zeeman energy. Kawabata presented the conditions[4] under which the quantum size effect takes place. It is estimated that those conditions begin to be satisfied when d<140Å at 9.1GHz for Be. The samples used in this experiment satisfy the above condition. We think that the above second and third mechanism can be adapted for our experiment.

The dominant relaxation mechanism depends on the spin species. If the radical unpaired spins on the oxide surface contribute to the Curie-like component, the size dependence of T_1 in our experiment should reflect the relaxation mechanism due to the softening effect of lattice vibration. If the unpaired spins on the interface do, it should be attributed to the quantum size effect or the additional effect of the softening of lattice vibration. In the latter case, the magnetic susceptibility of the spin system of unpaired spins on the interface and conduction electrons is easily calculated[13] using the equations by Denton *et al.*[14] when those local spins have the strong interaction with conduction electrons and the energy level near Fermi level.

According to the spin relaxation mechanism through the spin-orbit interaction by Elliott[12], T_1 is proportional to T^{-3} at much lower temperature than Debye temperature (1160K in Be). The value of α approaches 3.0 in this experiment as the size of sample becomes larger. The sample with a larger size has the stronger Pauli paramagnetic component. This suggests the validity of the above Elliott relation.

In conclusion, we could detect the softening effect of lattice vibration or the quantum size effect, or both additional effects in the T_1 measurement of the ESR spectra of Be small particles though the origin of the Curie-like paramagnetic component could not be discriminated in this experiment.

ACKNOWLEDGEMENT

The author wishes to express thanks to Dr. K. Kimura and Mr. S. Bandow of the Institute for Molecular Science for useful discussion. Thanks are due to the Instrument Center, the Institute for Molecular Science, for assistance in obtaining the low temperature ESR spectra and analysis by the thermogravimetry.

REFERENCES

1. R. Kubo, J. Phys. Soc. Jpn. 17:975(1962).
2. S. Sako and K. Kimura, J. Phys. Soc. Jpn. 53:1495(1984), Surf. Scie. 156:511(1985).
3. J. -L. Millet and J. -P, Borel, Surf. Scie. 106:403(1981), Solid State Commun. 43:212(1982).
4. A. Kawabata, J. Phys. Soc. Jpn. 29:902(1970).
5. J. Buttet, R. Car and C. W. Myles, Phys. Rev. B26:2414(1982).
6. C. W. Myles, Phys. Rev. 26:2648(1982).
7. J. H. Lunsford and J. P. Jane, J. Chem. Phys. 44:1487(1966).
8. J. -P. Borel, C. Borel-Narbel and R. Monot, Helv. Phys. Acta 47:537 (1974).
9. J. E. Cousins and R. Dupree, Phys. Lett. 19:464(1965).
10. G. Feher and A. F. Kip, Phys. Rev. 98:337(1955).
11. J. M. Dicky and A. Paskin, Phys. Rev. B1:851(1970).
12. R. J. Elliott, Phys. Rev. 96:266(1954).
13. K. Kimura and H. Hayashi, private commun.
14. R. Denton, B. Mühlschlegel and D. J. Scalapino, Phys. Rev. B7:3589 (1973).

ELECTRONIC STATES AND GEOMETRICAL

STRUCTURES OF HUBBARD CLUSTERS WITH AN IMPURITY

N. L. Sharma

Physics Department

Virginia Commonwealth University, Richmond, VA 23284

ABSTRACT

The relative stability and electronic excitation spectrum of three, four, and five site clusters is studied in the context of Hubbard model. Electron correlation is found to decrease the cohesive energy appreciably in the half-filled case. Planar geometries with a lower symmetry (pure cluster) or a large number of impurity bonds (impurity clusters) are found to be relatively stable. Noninteracting ground state is found to differ from the single particle ground state obtained by filling single particle states using the Pauli principle.

INTRODUCTION

Traditionally finite atomic systems have been studied in order to understand or mimic the bulk properties of matter. Today, with the availability of experimental techniques[1,2] to produce and analyse small clusters and with the future possibility of their technological[3] application, the study of atomic clusters has grown to an interesting field in its own right. There are two kinds of methods which have been used in the study of bulk many-body problems. In one we start with the actual many-body Hamiltonian and then try to solve it under different approximations either of perturbative (Ring or Ladder diagrams) or of selfconsistant (Hartree-Fock) nature. In the other, we start with a model Hamiltonian and try to solve it exactly. Ising, Hubbard, and Anderson models are examples of the latter. While the Ising model has been solved exactly in one and two dimensions, the Hubbard model has been solved exactly only in one dimension[4] and only for the ground state. From approximate solutions of the Hubbard model, however, we have been able to understand several interesting phenomena such as the metal insulator transition, and charge and spin density waves. Thus initially introduced to make a many-body problem tractable, the Hubbard Hamilton has proved to be an interesting model for problems involving short range interaction.

For the cluster problem, of course, the solution of the full Hamiltonian, i.e., the many-particle Schrodinger equation, under appropriate approximations (exact solutions are known only for hydrogen and helium) is the legitimate way. The use of the Hubbard model can nevertheless provide interesting clues to the real problem and sometime much more. Cluster calculations based on Hubbard model have been reported earlier. Falicov and Victora[5] have studied a tetrahedral cluster analytically with the use of the full tetrahedral point symmetry group. Ishii and Sugano[6] have considered the relative stability of various 4-site geometrical structures as a function of electron occupation number. Callaway et al[7] have given more attention to spin and magnetic properties of the ground state. Here we have used the Hubbard model to study the relative stability and excitation spectra of triatomic clusters. The variation of ground state energy with number of electrons is studied including on-site correlation between electrons. We have also studied three-site and four-site heteronuclear clusters (clusters with one impurity site) with different geometries to mimic band structure effects in the ground state energy. For example a tetrahedral geometry mimics a fcc lattice[5].

THE MODEL

The Hubbard model is defined by the Hamiltonian

$$H = - \sum_{ij\sigma} t_{ij} \, c^{\dagger}_{i\sigma} c_{i\sigma} + U \sum_{i} n_{i\uparrow} n_{i\uparrow} \quad ,$$

where $C_{i\sigma}$ ($C^{\dagger}_{i\sigma}$) is the annihilation (creation) operator for an electron

with spin σ at the i-th site, and n_i is the corresponding number operator. The first term in the Hamiltonian represents the hopping of electron between nearest neighbor sites. For pure clusters of equal side lengths we have taken the hopping parameter $t_{ij} = t$ for all i and j. For impurity clusters, we have $t_{ij} = s$ for the hopping integral involving an impurity site. The second term in the Hamiltonian expresses the increase in the energy by U when two electrons with opposite spins occupy the same site. For simplicity we have restricted our study to non-negative values of t, s, and U. Also, we only consider geometries with sides of equal length. Different geometries with same number of sites differ only in the number of distinct nearest neighbor pairs. This is to minimize the number of parameters in the model which is two in our case.

THE METHOD

(a) Single Electron Approximation

For a single electron the on-site correlation term reduces to zero and the eigenvalues of the Hamiltonian for a general three-site cluster shown in Fig.(1a) are

$$\varepsilon_0 = -0.5 - 0.5 \, [8t^2+1]^{\frac{1}{2}}$$

$$\varepsilon_1 = 1$$

$$\varepsilon_3 = -0.5 + 0.5 [\, 8t^2+1 \,]^{\frac{1}{2}} \quad .$$

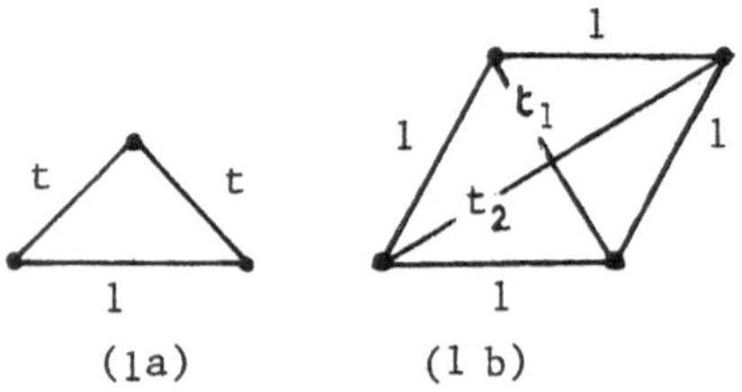

Fig.1 3-site and 4-site clusters with hopping parameters specified along the respective bonds.

For a 4-site cluster, shown in Fig. (1b), the eigenvalues are

$$\varepsilon_1 = t_1 \quad , \quad \varepsilon_2 = t_2$$

$$\varepsilon_{0,3} = -0.5(t_1 + t_2) \mp 0.5 \left[(t_1 - t_2)^2 + 16 \right]^{\frac{1}{2}} \quad .$$

486

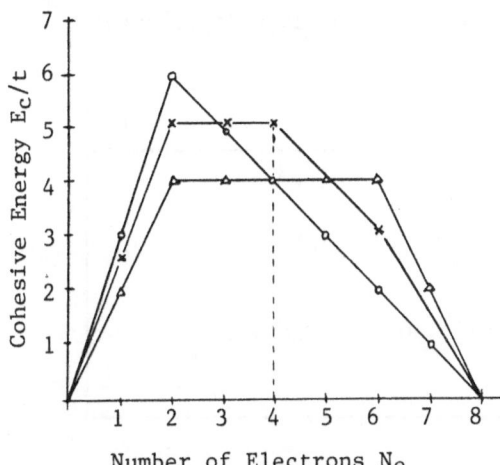

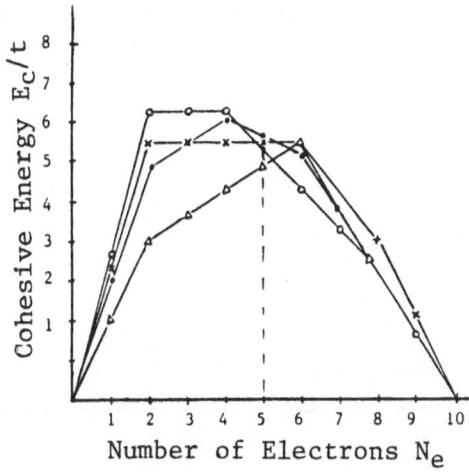

Fig.2(a). The cohesive energies of tetrahedron (o), rhombus (x), and square(Δ). The vertical line marks the half-filled case.

Fig.2(b). The cohesive energies of pentagon(Δ), Truss(·), bipyramid(o), and square pyramid(x). The shape of these clusters is shown in the table 2 in that order.

Many-electron states are obtained by accommodating no more than two electrons in each level in order that the Pauli principle is satisfied. The energies of many-electron states are calculated by adding energies of all the electrons.

(b) Interacting Case

Here the base states are the configurations obtained by distributing N_e electrons on N sites, putting no more than two electrons of opposite spin on the same site. Thus, the size of the matrix to be diagonalized is equal to $(2n)!/\left[N_e!(2N-N_e)!\right]$. This matrix can be block-diagonalized by using the fact that states with different S_z do not mix. The diagonalization was done numerically for each value of N_e. However, due to an ambiguity in the interacting ground which we shall mention below, we have not included the electron-electron correlation in the calculation of four and five-site clusters.

THE RESULTS

(a) Single-particle Calculation

For 4-site and 5-site cluster geometries with equal side lengths we find, see Figs. 3(a) and 3(b), that structures with higher symmetry, i.e., closed packed structures such as tetrahedron and bipyramid, are stable for the less than half-filled case and planar geometries with a lower symmetry, rhombus and truss, are stable for the half-filled case. This seems to be in accordance with the Jahn-Teller effect.

For impurity clusters as well, planar geometries with a large number of impurity bonds seems to be preferred if the impurity bond is stronger (s/t > 1) than the normal bond. This can be seen from Table (1), for 3-site clusters, and the last column of Table (2) for others. In fact if we

Table 1. Ground state energy of two atom plus one impurity clusters with three ($N_e=3$) electrons.

Cluster Shape	Eigen Values	Ground State Energy
	$\pm[1 + s^2]^{\frac{1}{2}},\ 0$	$-[4s^2 + 4]^{\frac{1}{2}}$
	$\pm\ 2^{\frac{1}{2}}s\ \ ,\ 0$	$-[8s^2]^{\frac{1}{2}}$
	$0.5 \pm 0.5[8s^2 + 1]^{\frac{1}{2}},\ 0$	$-[8s^2 + 1]^{\frac{1}{2}}$

Table 2. Ground state energy of clusters with one impurity and number of electrons equal to the number of sites. For pure case $s = t$.

Cluster Shape	Total Bonds	Impurity Bonds	S/t Values					
			0	0.1	0.5	1.0	2.0	10.0
	4	2	2.83	2.84	3.16	4.0	6.32	28.4
	5	3	2.83	2.87	3.63	5.12	8.44	36.0
	5	2	4.0	4.01	4.32	5.12	7.40	29.4
	6	3	4.0	4.0	4.0	4.0	7.21	34.7
	5	2	4.47	4.49	4.94	5.85	8.53	30.9
	7	4	4.47	4.49	5.07	6.64	10.3	42.3
	7	3	4.96	5.0	5.49	6.64	9.68	37.0
	7	2	5.12	5.20	5.69	6.64	9.05	31.1
	9	3	5.00	5.01	5.36	6.29	9.00	35.8
	9	4	5.12	5.14	5.47	6.29	9.93	41.6
	8	4	4.0	4.04	4.83	6.47	10.2	42.1
	8	3	5.12	5.14	5.5	6.47	9.24	36.2

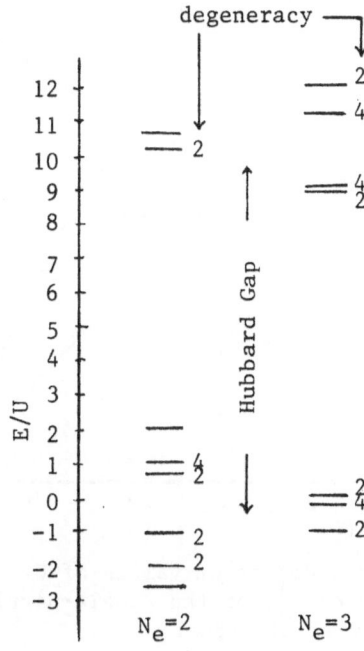

Fig.3   Energy level diagram of 3-site clusters
for U=10 and N_e=2 and 3.

look closely, then for 4-site impurity clusters the tetrahedron is next in
stability to the rhombus. Similarly, for 5-site clusters the square pyramid
with the impurity atom at its apex is almost degenerate in energy with the
truss which has got the same number of impurity bonds as the square pyramid.
This suggests that if there were freedom, which does not exist in our case
of equal side lengths, the impurity would have liked to sit at the center
of a symmetric planar structure. This indeed is found by Rao et al[8] through
a realistic calculation.

(b) Interacting Case

     In Fig.3 we show the excitation spectra for U=10 for two different
cases. We see that the states divide clearly into two sets, the lower set
not involving (approximately) double occupancy of any site, and a higher
set involving one doubly occupied site and hence higher in energy by about U.
Thus there exists a Hubbard gap separating the two sets of states.

     Through exact diagonalization of the full Hamiltonian containing the
on-site correlation term we find that the on-site correlation decreases
the cohesive enrgy markedly in the half-filled case. We have done this
calculation only for a 3-site cluster. This is shown in Fig.4. Here we
also notice a surprising phenomenon. When we reduce the on-site interaction U to zero we do not recover the single-particle ground state for N_e
electrons. This is not due to any programming error as we have verified
it analytically for N_e=5 where the size of the matrix can be reduced
to a 3x3 matrix. This means that the noninteracting ground state is not
the same as the single-particle ground state obtained by filling single-
particle states using Pauli principle. Also the non-interacting ground
state is lower in energy than the single-particle ground state. Which one
is the true ground state? We do not have a conclusive answer to this

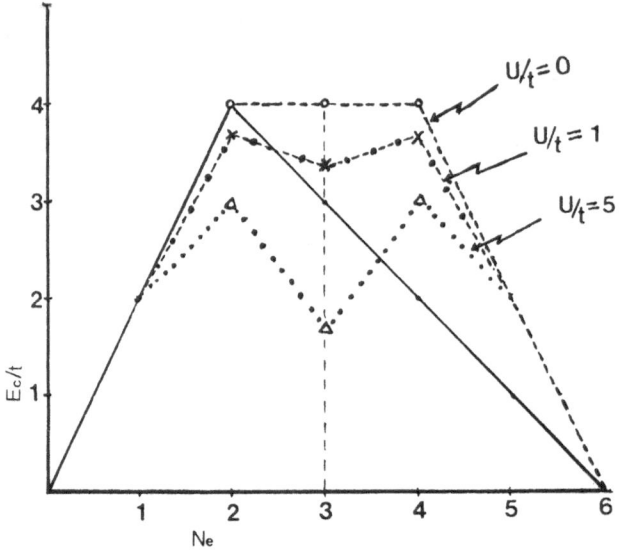

Fig.4 Cohesive energy for different values of U. The solid line depicts the cohesive energy of the single-particle ground state for different number of electrns.

question. However, we suggest the following explanation.

For the interacting many-body case the dimensionality of the Hilbert space (basis set) is quite different than that for the single particle case. Since we are solving the problem exactly, and not perturbatively, the inter-acting ground state, in general, can not be expanded as a perturbation series in the interaction parameter. So when we reduce the parameter U to zero we could not always expect to end up in the single-particle ground state. This is often encountered when there is a broken symmetry. We see here a situa-tion opposite to the Gell-Mann and Low theorem,[9] where even an adiabatic switching on (in our case it is switching off) of the interaction does not necessarily lead the system to its interacting ground state. So in this situation, we think, the noninteracting ground state is the true one.

The other possibility is that it is an artifact of the model. If that is the case then we must treat the single-particle ground state as the true ground state. We plan to explore further on this issue.

ACKNOWLEDGEMENTS

The author wishes to thank Dr. P. Jena for bringing my attention to this problem, and to Drs. S. Mahanti, B. K. Rao, and M. F. Bishop for helpful discussions.

REFERENCES

1. W.D. Knight, K. Clemenger, W.A. deHeer, W.A. Saunders, M.V. Chou, and M.L. Cohen, Phys. Rev. Letts. 52, 2141 (1984).
2. A. Hermann, E. Schumacher, and L. Woste, J. Chem. Phys. 68, 2327 (1978)
3. B.K. Rao. S.N. Khanna, and P. Jena, Chem. Phys. Letts. 121 , 202 (1985)
4. E.H. Lieb and F.Y. Wu, Phys. Rev. Letts. 20 , 1445 (1968)

5. L.M. Falicov and R.H. Victora, Phys. Rev. <u>B30</u> ,1695 (1984).
6. Y. Ishii and S. Sugano, J. Phys. Soc. Jpn. <u>53</u> , 3895 (1984).
7. J. Callaway, D.P. Chen, and R. Tang, Private Commucation.
8. B.K. Rao, S.N. Khanna, P. Jena, Z. Phys. <u>D3</u> , 219 (1986)
9. A.L. Fetter and J.D. Walecka in " Quantum Theory of Many- Particle Systems " (McGraw-Hill, New York, 1971), p.61.

THEORETICAL PREDICTIONS OF THE VIBRATIONAL FREQUENCIES OF SINGLET Li_4

D.D. Shillady, B.K. Rao, and P. Jena

Departments of Chemistry and Physics
Virginia Commonwealth University
Richmond, VA 23284 U.S.A.

ABSTRACT

The geometry of the singlet ground state of Li_4 has been optimized using gradient methods in a 6-21G and 6-21G* basis using MICROMOLE and HONDO/5 computer programs respectively. The geometries were exhaustively optimized to provide the best energy minimum point at which to calculate the force constant matrix from a finite gradient difference and thus estimate the vibrational frequencies of the cluster. The lowest energy vibration is estimated as 100 to 106 cm^{-1} in energy and offers a vibronic mechanism to a quasi-tetrahedral triplet state. The triplet geometry was optimized in an STO-6G basis and both singlet and triplet geometries were compared in a Van Duijneveldt (10s, 2p/4s, 1p) basis with configuration interaction (C.I.) treatment including valence shell single, double, and selected quadruple excitations. The 99 C.I. singlet energy at the 6-21G geometry is -29.782356 au and the 100 C.I. triplet energy at the STO-6G geometry is -29.754456 au. Three quadruple excitations contributed only 0.064 ev to the singlet, and split-double (11) excitations were more important than quadruple excitations in the triplet C.I. This study suggests an electronic singlet-triplet transition in Li_4 at about 0.8 ev (1550 nm wavelength) which may couple with a vibrational mode of about 100 cm^{-1}.

INTRODUCTION

The previous study of small lithium clusters by Rao, Khanna and Jena[1] predicted a planar rhombus for Li_4 in the ground state and a metastable quasi-tetrahedral triplet state (Figure 1). Earlier work by Rao, Jena and Shillady[2] showed that lithium clusters can be successfully treated at a qualitative level with a Roothaan-Hartree-Fock[3] basis expansion method in a basis set which includes 2p orbitals on each Li atom. In this work we have taken advantage of the recent development of energy-gradient optimizing computer programs such as MICROMOLE[4] for the IBM PC and the HONDO/5[5] program for mainframe computers to obtain both optimized geometries and vibrational frequencies for Li_4.[6] The geometries found were then treated using a Boys-type[7] configuration interaction program we have developed previously[7] (LOBE140) to compute the lowest electronic singlet-triplet transition.

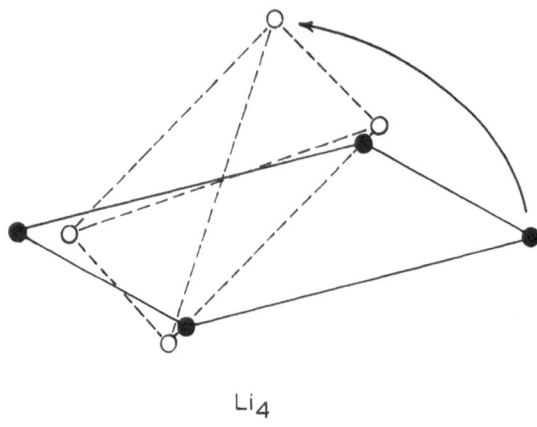

Li_4

●—PLANAR SINGLET
o--QUASI-Td TRIPLET

Figure 1. Geometries of Li_4

COMPUTATIONAL METHODS

In order to determine the vibrational frequencies of a molecular cluster using MICROMOLE or HONDO/5 one must first optimize the geometry of the cluster. Both of the programs vary all 3N atomic coordinates during the energy gradient optimization so that 6 false frequencies will be found for translational and rotational degrees of freedom unless the geometry is highly optimized. The first results (1 in Table I) were obtained using MICROMOLE on an IBM PC/XT with an 8087 arithmetic coprocessor and a small basis set. Optimizing to a maximum energy gradient of 0.0001 au/Bohr and using a finite step size of 0.005 au for the gradient-difference approximation of the second energy derivative produced a set of frequencies with a maximum false frequency of 7.3 cm^{-1}. The initial optimized geometry was then used as a starting point for the calculations based on Pople's 6-21G and 6-21G* basis sets. Result No. 4 used a 6-21G* basis set including a set of d-orbitals on each lithium atom with gaussian exponents of 0.8. Calculation No. 4 was the result of exhaustive optimization to 2.0×10^{-7} maximum gradient and use of a step size of 0.010 au in the gradient difference and should be the most accurate calculation at the single-determinental level with all six false frequencies less than 3.8 cm^{-1}. Calculation No. 3 is shown to illustrate the effect of a large step size (1.0 au) on the computed frequencies; to a certain extent the larger false frequencies reflect anharmonicity of the vibrational potentials. Calculations 3 and 4 were performed on the VCU IBM 3081D in about six hours and twenty minutes each, while the largest calculation on the IBM PC/XT (No. 2) required almost eight days.

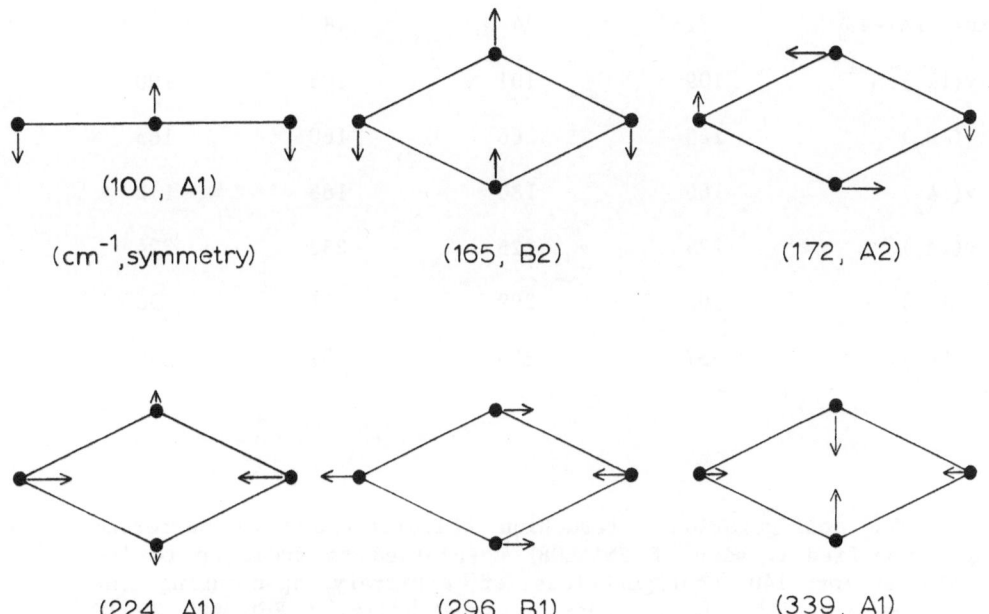

Figure 2. Normal modes of singlet Li_4 in C_{2v}

Table I. Computed Vibrational Frequencies of Li_4 (C_{2v})

Calculation No.	1	2	3	4
Program	MICROMOLE	MICROMOLE	HONDO/5	HONDO/5
Basis Set	(5,1,s/2p)	6-21G	6-21G*	6-21G*
Energy (au)	-29.724119	-29.752083	-29.753575	-29.753575
Max. Grad.	0.0001	0.0002	0.000003	0.0000002
Step Size (au)	0.005	0.010	1.000	0.010
Max. False ν	7.3	24.0	48.7	3.8
$\nu(1A_1)cm^{-1}$	109	101	105	100
$\nu(1B_2)$	165	166	160	165
$\nu(1A_2)$	166	180	188	172
$\nu(2A_1)$	229	225	232	224
$\nu(1B_1)$	308	299	311	296
$\nu(3A_1)$	337	336	357	339

The configuration interaction[10] calculations were performed using a modified version of SMALLOBE enlarged to treat up to 140 orbitals and/or 140 configurations of arbitrary[7] spin using the Boys-Reeves algorithm in a gaussian lobe basis. For the 6-21G geometry of the singlet (Li at X = ±5.401857 and Y = ±2.554030 Bohrs) a 99x99 C.I. treatment was carried out using all single and double valence electron excitations with 1s core electrons frozen. A survey of low-lying quadruple excitations led to only three important quadruply excited configurations in the final C.I. The C.I. energy changed less than 0.064 ev when these quadruple excitations were included. The triplet geometry was optimized at the STO-6G level and a C.I. treatment was carried out using the same Van Duijneveldt[11] 10/7,1,1,1)s basis with a four-gaussian-sphere (2G) hydrogenic 2p orbital scaled to 0.790 as optimized for Li_2. This C.I. program is the only computer code available to us in which quadruple excitations could be tested for contributions to both singlet and triplet states. A survey of quadruple excitations and split-double excitations indicated that quadruple excitations are negligible compared to split-double excitations for the triplet state. The final 100x100 C.I. for the triplet thus included all valence single and pair-double excitations with eleven selected split-double excitations.

RESULTS AND DISCUSSION

It seems clear from Table II that at a comparable level of C.I. the ground state is a singlet and that the quasi-tetrahedral triplet is some 0.0279 Hartrees=0.76 ev higher in energy (1,633 nm). A _vertical_ transition of the planar singlet to a planar triplet at the same 6-21G geometry would require 0.815 ev (1,522 nm). Thus the first

Table II. Configuration Interaction Energies in a $(7,1,1,1,s/2,p)$
Lobe Basis.

| | Rhombus Singlet | | Tetrahedral Singlet |
	STO-6G Geometry	6-21 Geometry	Perfect Td Geometry
E(SCF)*	−29.741597	−29.757064	−29.716813
E(99CI)	−29.771362	−29.782356	−29.747137

| | Rhombus Triplet | | Tetrahedral Triplet |
	STO-6G Geometry	6-21 Geometry	C2v STO-6G Geometry
E(SCF)*	−29.741597	−29.757064	−29.715596
E(100CI)	−29.746917	−29.752408	−29.754456

(*Triplet C.I. from singlet SCF wavefunction.)

singlet-triplet absorption of Li_4 would probably be a weak broad band between 1650 and 1500 nm with vibronic fine structure related to the A_1 vibrational modes of 100 cm^{-1}, 224 cm^{-1} and 339 cm^{-1}. Note that only $1A_1$ at 100 cm^{-1} directly assists the nonplanar singlet-triplet transition. When d orbitals were added to the 6-21G basis the geometry did relax further to (Li at X = ±5.412815 and Y = ±2.537562 Bohrs), but the 6-21G geometry was used for the C.I. step since the C.I. did not include d orbitals. It is believed that this work is sufficiently accurate to guide assignments of the spectra of Li_4 in the near infrared when such data becomes available.

ACKNOWLEDGEMENTS

This work was supported by a grant from the Army Research Office Durham: DAAG 29-85-K-0244. B.K.R. also wishes to thank Virginia Commonwealth University for a Faculty Grant-in-Aid.

REFERENCES

1. B. K. Rao, S. N. Khanna and P. Jena, Equilibrium Geometries of Small Metal Clusters and Their Relationship to Crystalline Structure, Solid State Commun. 56:731 (1985).
2. B. K. Rao, P. Jena and D. D. Shillady, Structural Properties of Microcrystallites, Phys. Rev. 30: 7293 (1984).
3. C. C. J. Roothaan, New Developments in Molecular Orbital Theory, Rev. Mod. Phys. 23:69 (1951).
4. S. M. Colwell, A. R. Marshall, R. D. Amos and N. C. Handy, Quantum Chemistry on Microcomputers, Chem. in Britain, p 657, July (1985).
5. M. Dupuis, J. Rys and H. F. King, Ab Initio Hartree-Fock Self-Consistent Field Program, Quantum Chemistry Program Exchange No. 403, Indiana University U.S.A.
6. S. F. Boys and G. B. Cook, Mathematical Problems in the Complete Quantum Predictions of Chemical Phenomena, Rev. Mod. Phys. 32:285 (1960).
7. D. D. Shillady and S. Baldwin-Boisclair, Dipole-Optimized Gaussian Orbitals for Rapid Computation of Electrostatic Molecular Potential Contour Maps, Int. J. Quantum Chem., Quantum Biology Symposium No. 6, 105 (1979).
8. M. Dupuis and H. F. King, Molecular Symmetry II. Gradient of Electronic Energy with Respect to Nuclear Coordinates, J. Chem. Phys. 68:3998 (1978).

9. J. S. Binkley, J. A. Pople and W. J. Hehre, Self-Consistent Molecular Orbital Methods 21. Small Split-Valence Basis Sets for First-Row Elements, J. Am. Chem. Soc. 102:939 (1980).

10. D. D. Shillady, Ab Initio Gaussian Lobe SCF Closed Shell Program, Quantum Chemistry Program Exchange No. 239, Indiana University, U.S.A.

11. F. B. van Duijneveldt, Gaussian Basis Sets for the Atoms H-Ne for use in Molecular Calculations, IBM Publication RJ 945 (No. J6437).

EVAPORATION OF METAL CLUSTERS

Michael Vollmer and Frank Träger

Physikalisches Institut der Universität Heidelberg
Philosophenweg 12, D-6900 Heidelberg
Federal Republic of Germany

ABSTRACT

The evaporation of sodium clusters on an insulator single crystal surface has been investigated. For this purpose the substrate was heated slowly, and sodium atoms desorbing from the cluster surface were detected with a mass spectrometer. We demonstrate that the resulting spectra can be used to extract cluster binding energies as a function of cluster size. They approach the bulk value for clusters containing as many as 10^6 to 10^7 atoms. This is consistent with theoretical calculations.

INTRODUCTION

Recently, an increasing number of experiments have been undertaken to measure the size dependences of cluster properties. Among others, the thermodynamic features of clusters seem to be of particular interest. However, only few experiments have been carried out to investigate these characteristics in detail. For example, the evaporation of lead clusters was observed directly by electron microscopy as a test of the Kelvin equation for small particle radii[1]. Also, the melting temperature of gold and other metal clusters has been determined as a function of size[2]. The present paper reports new work on the thermodynamic properties of metal clusters and describes how the evaporation of such particles takes place. The measurements do not only give insight into the evaporation in a microscopic picture but also permit to extract the binding energies of individual atoms on the cluster surface and their convergence to the bulk value. In addition, scattering experiments with thermal atoms from a surface partly covered with clusters have been carried out. They yield information on the cluster nucleation associated with atom diffusion on surfaces and permit to determine the average size of the clusters. The measurements have been performed with Na clusters on a LiF(100) single crystal surface under ultrahigh vacuum conditions.

The vapor deposition of atoms on substrate surfaces and the subsequent thin film formation have been investigated in detail, both experimentally and theoretically[3]. From the three different growth modes in epitaxy the Volmer-Weber mode is of particular interest here since it describes the growth of threedimensional particles on supports. In general, impinging atoms may either be scattered elastically or adsorbed for a certain residence time. For the present case of Na atoms directed onto a LiF surface at liquid nitrogen temperature elastic scattering processes are negligible because the adsorption probability is practically 100%. Based on statistical considerations Frenkel[4] showed that the mean residence time of adsorbed atoms is given by

$$\tau_{res} = \tau_0 \times e^{E/kT}. \tag{1}$$

τ_0 denotes the oscillation period of an atom in the surface potential, T the substrate temperature and E the adsorption energy. Since the potential barrier for surface diffusion is less than one half of the adsorption energy[5] the atoms can perform a large number of diffusion hops to neighboring adsorption sites during their residence time. Each atom therefore covers a large area during its random walk diffusion on the substrate. If it occasionally meets defects or already grown clusters, the binding energy is increased and it can no longer desorb. During the deposition of atoms, the concentration of diffusing single adatoms should therefore decrease with time as the clusters grow and the cluster free area of the surface becomes smaller. Consequently, fewer and fewer atoms can desorb and the rate for inelastic scattering processes decreases. With a model based on nucleation kinetics and surface diffusion we have calculated the scattering signal as a function of time for a given flux of impinging atoms[6]. A comparison with the experimental data then permits to determine the number density of clusters and the mean cluster size for a given total number of sodium atoms on the surface.

For the nucleation of clusters defects on the surface play an important role. They act as nucleation centers and may consist of vacancies, impurities, steps or dislocations. Defect densities typically range from 10^9 to 10^{11}/cm^2. After cleaning and annealing of the surface by heat treatment the number of defects may be reduced considerably and is of the order of several 10^8/cm^2 for alkali halide surfaces. In an ultrahigh vacuum environment such a surface remains clean for hours[7]. Because the defects serve as nucleation centers their number roughly equals the number of clusters on the surface. We find from our scattering experiments that the density of clusters is 5×10^8/cm^2 and remains practically constant during the deposition. The average cluster size is therefore proportional to $n^{1/3}$ where n denotes the total number of atoms on the surface. The scattering experiments also give the result that clusters with radii between about 100 Å and 1500 Å were investigated.

CLUSTER EVAPORATION

Evaporation of clusters can be accomplished by rising the temperature of the substrate and monitoring the desorbing species with a mass spectrometer (see Fig. 1). Since the binding of a cluster to the surface is quite strong it cannot desorb as an intact particle. Rather, evaporation of single atoms from the cluster surface will take place. The rate dn/dt of desorbing particles is given by the following Arrhenius equation

$$dn(T)/dt = n(T)^x \ \nu \ e^{-E/kT} \tag{2}$$

The number of particles $n_{des}(T)$ that can participate in the desorption process is expressed as $n(T)^x$ where n denotes the total number of atoms on the surface. x is introduced as the formal order of the reaction. ν is the frequency of vibration in the surface potential. Since evaporation is only possible for atoms at the cluster surface, it follows that x<1, implying a *fractional* order desorption process.

The method outlined above and described theoretically with Equ.(2) is well known in surface science as thermal desorption or temperature programmed desorption[8]. A spectrum

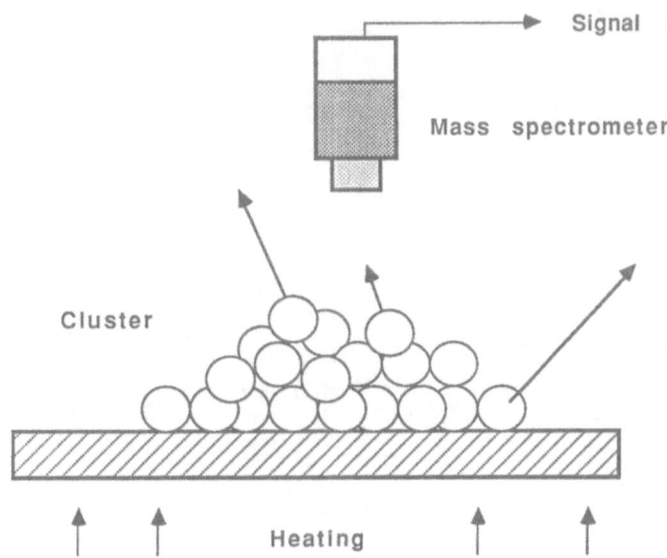

Fig. 1: Schematic view of a cluster on a surface. By rising the temperature of the substrate atoms from the cluster surface are evaporated. The desorption process is monitored with a quadrupole mass spectrometer.

of dn(T)/dt versus T shows a maximum at that temperature where the rise of the desorption signal due to the increasing temperature equals the decrease resulting from the vanishing coverage. The upper part of Fig. 2 displays two spectra for Na/LiF with different initial coverages, i.e. different cluster sizes. The analysis of the spectra according to Equ.(2) is made with logarithmic plots of the desorption rate versus the reciprocal temperature. The lower part of Fig. 2 depicts such Arrhenius diagrams of the spectra. They have been used to extract the binding energies and the order of desorption. It should be mentioned that a more detailed analysis of the desorption spectra has also been made. It takes into account that the cluster size decreases during the evaporation[9].

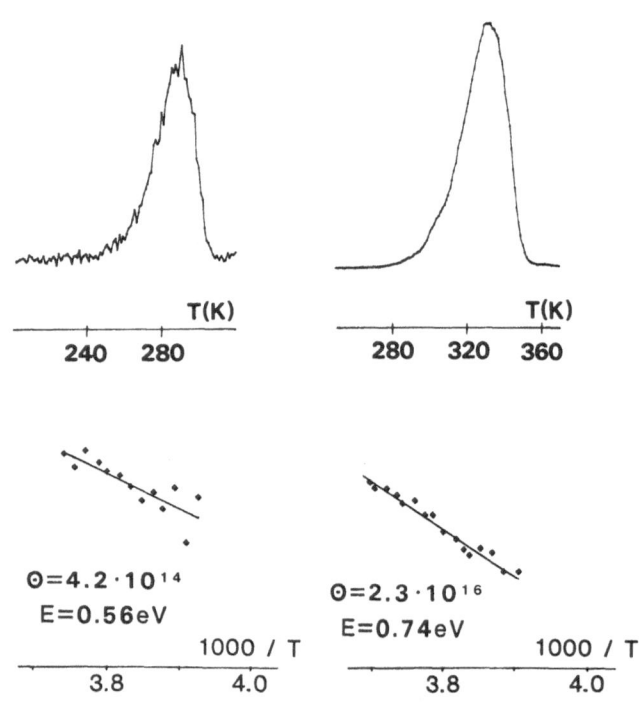

Fig. 2: Results of thermal desorption experiments for cluster sizes of 315 Å (left hand side) and 1200 Å (right hand side) corresponding to coverages of 4.2×10^{14} and 2.3×10^{16} atoms/cm^2, respectively. In the upper part, the desorption rates are plotted in arbitrary units as a function of temperature. In the lower part the logarithm of the desorption rates is plotted as a function of 1/T for a decrease of the initial coverage of several percent.

DISCUSSION

The results of our experiments show that the desorption energy per atom varies between about 0.55 eV and 0.8 eV and follows the dependence displayed in Fig. 3. Indeed, one would expect that the binding energy of an atom on or in a cluster increases with size until its average coordination number does not change any further. In the present study, this saturation value is reached for clusters as large as 1000 Å, i.e. with as many as 10^7 atoms. This is consistent with theoretical calculations which show that this limit should occur for clusters with approximately 10^6 to 10^7 atoms[10,11]. On the other hand, the binding energy of the sodium dimer should impose the lower limit for the cohesive energy. Presently further experiments are under way in our laboratory with the goal to investigate the evaporation of much smaller clusters than reported here. Even though it seems unlikely that particles as small as the dimer or even ten membered clusters will give a sufficiently large signal-to-noise ratio at the mass spectrometer, we are confident that the dependence of the desorption energies can be followed down into the size range of microclusters, i.e. with radii below 10 Å.

The maximum value of the binding energy is around 0.8 eV. This is considerably smaller than the heat of sublimation of sodium of 1.113 eV. This apparent discrepency can be explained in a simple picture where the small crystalline sodium particles are e.g. regarded as cubes. As already stated above, the number of nearest neighbors influences the binding energies, an effect which is particularly pronounced at the edges of the cubes. The number of

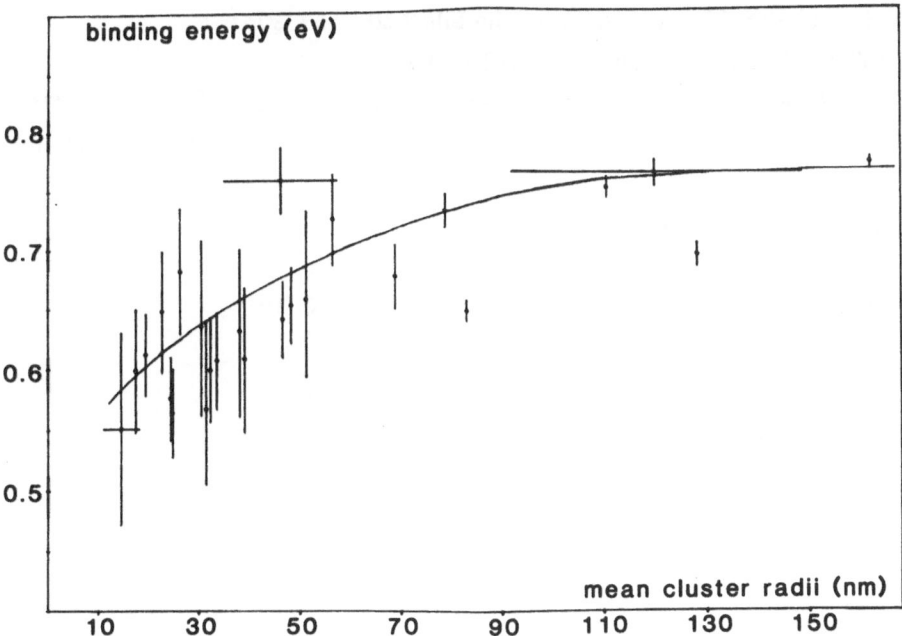

Fig. 3: Binding (desorption) energies of Na-atoms on the surface of sodium clusters as a function of size. The solid line is only to guide the eye.

nearest neighbors of *edge* atoms is only about 0.70 of the corresponding number for atoms on the flat surfaces. Thus, the bulk sublimation energy of 1.113 eV should decrease to 0.78 eV for atoms located at the edges of such a cluster, a value which almost perfectly agrees with our maximum desorption energy. It is interesting to note that this interpretation of the cluster binding energies is consistent with a microscopic picture[6,9] of the evaporation process that also takes into account the experimental value for the order of desorption of x=0.79(8). These arguments are certainly very crude and qualitative. Nevertheless, they explain the experimental findings in a way which is also consistent with other results and considerations. The coordination number is certainly of major importance for the binding energies. There is no doubt, however, that further experimental and theoretical work is needed to obtain a more sophisticated picture of cluster evaporation and of the dependence of the cohesive energies on the cluster size.

REFERENCES

1. J.R. Sambles, L.M. Skinner, N.D. Lisgarten, Proc. Roy. Soc. London **A318**, 507 (1970)
2. Ph. Buffat, J.-P. Borel, Phys. Rev. **A13**, 2287 (1976)
3. J.A. Venables, G.D.T. Spiller, M. Hanbücken, Rep. Progr. Phys. **47**, 399 (1984)
4. J. Frenkel, Z. Phys. **26**, 117 (1924)
5. V.N.E. Robinson, J.L. Robins, Thin Solid Films **20**, 155 (1974)
6. M. Vollmer, F. Träger, Z. Phys. **D 3**, 291 (1986)
7. J. Estel, H. Hoinkes, H. Kaarmann, H. Nahr, H. Wilsch, Surf. Sci. **54**, 393 (1976)
8. D. Menzel, Thermal Desorption, in Chem. Phys. Sol. Surf. IV, Springer Ser. Chem. Phys. **20**, (1982)
9. M. Vollmer, F. Träger, to be published
10. P. Kadura, L. Künne, Phys. Stat. Sol. **B 88**, 537 (1978)
11. L. Künne, L. Skála, O. Bílek, Czech. J. Phys. **B29**, 1030 (1979)

ANALYSIS OF THE BASIS SET DEPENDENCY ON THE ELECTRONIC STRUCTURE OF Nb_2

Arne Rosén and Tomas Wahnström

Department of Physics
Chalmers University of Technology and University of Göteborg
S-412 96 Göteborg, Sweden

ABSTRACT

The electronic structures of the valence and first excited states for Nb_2 have been calculated within the local spin density approximation using the LCAO method. Calculations were done with basis sets of the neutral atom consisting of occupied plus virtual orbitals generated with different occupation numbers of the 4d and 5s orbitals. In addition polarization functions have been included by using the 4d, 5s and 5p functions for ionized atom. Inclusion of the ionized basis is found to be important while the addition of basis functions with changes in occupation numbers for the 4d and 5s electrons do not give any further improvement.

INTRODUCTION

The electronic structure of transition elements are characterized by having low lying open d-shell configurations of the type $nd^N (n+1)s^2$, $nd^{N+1}(n+1)s$ and nd^{N+2}. These configurations which are close in energy and of the same parity give rise to many states[1] which in some cases perturb each other very strongly[2]. The presence of all these manifold of states implies also complicated spectra, which however can be a benefit in detailed studies of electronic interaction of electrostatic and magnetic origin[3,4]. A still more sensitive probe of the electronic structure for these elements can be achieved from experimental and theoretical investigations of the hyperfine structure (hfs)[5,6], which is one of the most sensitive probes of the relative contributions of d and s electrons in different states[7,8].

The presence of all these manifold of low lying states for the transition elements will also influence the electronic structure of the bonding properties and energies of the low lying states for molecules created of the transition elements. Further, quite different activation barriers for hydrogen chemisorption have been calculated depending on which atomic configuration and state is involved in the interaction [9]. For Nb there are 46 LSJ states of even parity below 2 eV, where one finds the first state of odd parity[1]. The question is how many of all these manifold of states of even parity will be involved in the formation of clusters from transition elements using the laser evaporation technique[10] It is for example known that highly excited metastable states can be populated in atoms by electron bombardment of a molten ball of the transition element [11]. Detailed high resolution spectroscopic investigations of the assignments of the low lying states of molecules of the transition elements [12,13] are therefore of great basic interest.

In the present work molecular cluster calculations are performed for Nb_2 within the local density approximation (LDA) [14-16]. A coupling of the one electron momenta to a well defined many electron state is not well defined in LDA theory and requires detailed analysis [17,18]. In order to at least include some angular momentum coupling spin polarized local density calculations are normally performed in which the multiplicity of the ground state is automatically obtained. The calculations in this work were done with the exchange correlation potential of Barth and Hedin [19] using the LCAO method with different type of basis sets.

COMPUTATIONAL METHOD

The important feature of the self-consistent local density model is to replace the many-body Hamiltonian by an effective single particle Hamiltonian which in atomic units is given as

$$h(r) = -1/2 \nabla^2 + v_{eff}(r)$$

where the first term on the right hand side represents the kinetic energy operator of an electron in an effective potential $v_{eff}(r)$, which includes the nuclear and electronic Coulomb potentials and the exchange and correlation contribution.

The Schrödinger equation is solved variationally to determine the molecular wavefunctions which are given as

$$\Psi_i(r) = \sum_j X_j(r) C_{ji}$$

in terms of an expansion of symmetry-adapted basis functions. These symmetry functions are composed of free atom and of free ion radial wavefunctions $u_{nl}(r)$ in numerical form and the spherical harmonics $Y_{lm}(r)$ as

$$X_j(r) = \sum_{\nu nlm} W_{nlm}^{j\nu} u_{nl}(r_\nu) Y_{lm}(\hat{r}_\nu)$$

where W are coefficients chosen to generate the symmetry functions which transform according to the irreducible representation of the point group of the Hamiltonian.

Matrix elements of the Hamiltonian and overlap matrices in the secular equation are evaluated by applying the DVM numerical integration technique[20]. The one-electron equations are solved by expansion of the true molecular and spin charge densities as a sum of spherical density components[21] and expansion of multipole functions with $l \leqslant 2$ around each atom in the molecule[22]. The charge and spin densities are then evaluated from the molecular wavefunctions and determined to obtain self consistency in the expansion coefficients. Core functions were frozen and the valence symmetry orbitals were orthogonalized against the core functions. The different basis sets used in the construction of the molecular wavefunctions were generated with an atomic self consistent program using different atomic and ionic configurations.

RESULTS AND DISCUSSION

   The ground state term in the Nb atom originates from the $4d^4 5s^1$ configuration [1]. A suitable minimal basis set (denoted with a) in molecular calculations would then be wave functions generated for the $4d^4 5s^1 5p^0$ configuration, where the 5p virtual orbital has been added to get more variational freedom. This basis set has then been extended by inclusion of 4d, 5s, and 5p polarization functions (denoted with b) for a Nb 4+ ion with the configuration $4d^1 5s^0 5p^0$. To get still more extended basis sets 4d, 5s and 5p wavefunctions were added from the configurations $4d^3 5s^2 5p^0$ (denoted with c) and $4d^5 5s^0 5p^0$ (denoted with d), respectively. Core wavefunctions have only been included from the $4d^4 5s^1 5p^0$ configurations since core functions for the other configurations will be linearly dependent. Features of the basis sets can be examined by evaluation of the expectation value of r for the different wavefunctions as for example reviewed by Walch and Bauschlicher [23].

   Using these different basis sets, molecular calculations were done for Nb_2 at different bond lengths around the bond length calculated by Walch and Bauschlicher[23] in extensive CASCF calculations. The molecular eigenvalues calculated for these bond lengths by using the basis sets denoted by a, b and c are shown in Fig. 1.

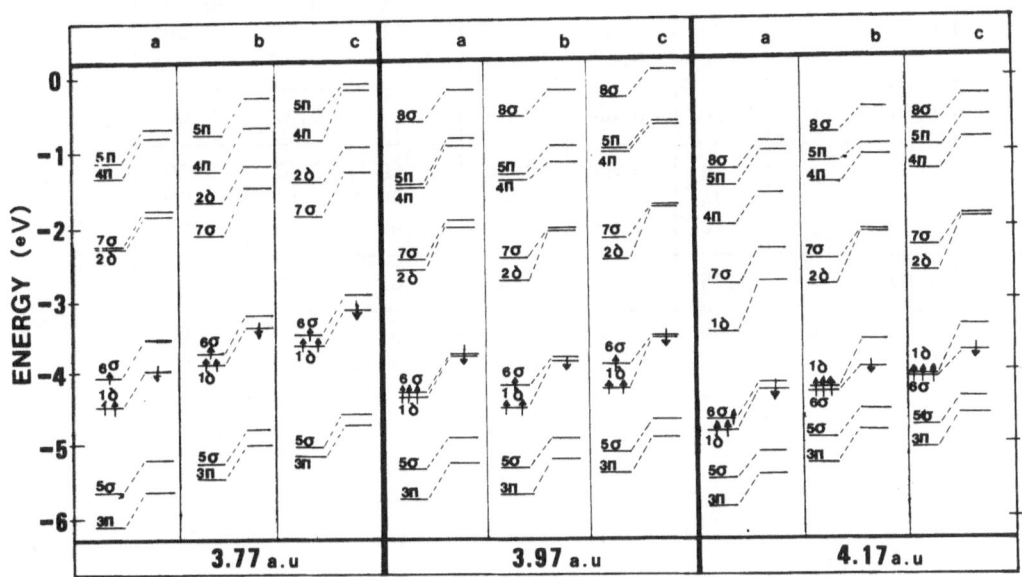

Fig. 1. Molecular eigenvalues for Nb_2 calculated at different bond lengths and basis sets as described in the text.

   Calculations by inclusion of basis set d failed due to a very strong linear dependence with big occupation numbers of different signs for the 4d, 5s, and 5p basis functions from the different configurations. There is a significant change in the spacing between the different levels and the absolute values in using neutral atom basis set denoted by a compared with the inclusion of the polarization functions in basis set b. Addition of the wavefunctions from the $4d^3 5s^2 5p^0$ configuration did not give any significant change in the eigenvalues. We notice how the ordering of the last occupied states change as function of bond length for all three basis sets. Some of the levels as 6σ↑ 1δ↑ 6σ↓ 1δ↓ are very close in energy and the assignment of the ground state configuration is difficult. Summation of the spin components for the different electrons shows that the total $M_S=1$. The configuration at the bond length of 3.97 a.u. is

$$(3\pi\uparrow)^2(5\sigma\uparrow)^1(3\pi\downarrow)^2(5\sigma\downarrow)^1(1\delta\uparrow)^2(6\sigma\uparrow)^1(6\sigma\downarrow)^1)$$

Here $\uparrow\downarrow$ refers to the spin direction. This should be compared with the calculations by Walch and Bauschlicher[23] who found that the ground state would be $^3\Sigma_g^-$ with close lying $^3\Delta_g$, $^1\Gamma_g$ states.

In order to investigate the relative importance of 4d and 5s electrons in the bonding of this dimer the molecular orbitals were examined through a Mulliken orbital and spin analysis in terms of the basis functions. The contribution from the different wavefunctions of a certain orbital symmetry have been added together and are presented in Table 1.

Table 1   Mulliken orbital and spin populations at different bond lengths and with different basis sets a,b,c as described in the text.

R	basis set	$n\uparrow + n\downarrow$			$n\uparrow - n\downarrow$		
		4d	5s	5p	4d	5s	5p
3.77	a	4.38	0.60	0.03	0.59	0.41	−0.00
	b	4.32	0.61	0.08	0.60	0.39	0.00
	c	4.30	0.67	0.16	0.59	0.40	0.01
3.87	b	4.26	0.68	0.08	0.67	0.32	0.00
3.97	a	4.29	0.68	0.04	0.68	0.31	0.00
	b	3.94	0.99	0.08	1.00	−0.00	0.00
	c	3.90	1.05	0.15	1.00	−0.00	−0.01
4.07	b	3.95	0.99	0.07	1.00	−0.00	0.00
4.17	a	3.99	0.98	0.04	1.00	−0.00	0.00
	b	3.95	0.99	0.07	1.00	−0.00	0.00
	c	3.92	1.04	0.13	1.00	−0.00	0.01

Inspection of the values in the table shows that the contribution of 4d electrons increases for smaller bond lengths with a minimum at about 4.0 a.u.,while the contributions of 5s electrons seems to approach a value of 1.0 at this bond length. Use of bigger basis sets seem to increase the contribution from the 5p wavefunctions. Further, a net spin contribution is only obtained from the 4d electrons at this bond length, while the 5s electrons seem to pair off.

Calculations are now in progress to extend these calculations for evaluation of the total energy, ionization energies for comparison with calculations within the MSXα - method[24,25],CASCF[23] and experimental data[26].

ACKNOWLEDGEMENTS

The authors wish to thank Tapio Rantala, Bengt Lindgren, Per Siegbahn and various colleagues in the molecular physics group at Chalmers for enlightful discussions. This work has been supported by the Swedish National Research Council (NFR).

REFERENCES

1. C.E. Moore, Atomic Energy Levels. NBS Circ. 467 Washington DC (1971).
2. R.E. Trees, Phys. Rev. $\underline{83}$, 756 (1951); $\underline{84}$, 1089 (1951); $\underline{85}$, 382 (1952).
    R.E. Trees and M.M. Harvey, J. Res. NBS $\underline{49}$, 397 (1952).
3. J.E. Hansen and B.R. Judd, Comments At. Mol. Phys. $\underline{18}$, 125 (1986).
4. B.R. Judd, Rep. Prog. Phys. $\underline{48}$, 907 (1985).
5. L. Armstrong, Theory of the Hyperfine Structure of Free Atoms, Wiley-Interscience, New York (1971).
6. I. Lindgren and A. Rosén, Case Stud. Atomic Phys. $\underline{4}$, 93 (1974).
7. G. Olsson and A. Rosén, Physica Scripta, $\underline{26}$, 168 (1982); Phys. Rev. $\underline{A25}$, 658 (1982).
8. S. Büttgenbach, Hyperfine Structure in 4d and 5d Shell Atoms, Springer Tracts in Modern Physics, 96 Berlin (1982).
9. P.E.M. Siegbahn, M.R.A. Blomberg, C.W. Bauschlicher Jr, J. Chem. Phys. $\underline{81}$, 1373 (1984).
10. R.E. Smalley, in Comparison of Ab Initio Quantum Chemistry with Experiment for Small Molecules, The State of the Art, Ed., R.J. Bartlett, D. Reidel Publ. Comp. Dordrecht, Holland 1985. In this review references are given to experimental works on clusters from a number of groups.
11. G. Olsson, T. Olsson, L. Robertsson and A. Rosén, Physica Scripta, $\underline{29}$, 1 (1984).
12. J.B. Hopkins, P.R.R. Langridge-Smith, M.D. Morse and R.E. Smalley, J. Chem. Phys. $\underline{78}$, 1627 (1983).
13. S.J. Riley, E.K. Parks, L.G. Pobo and S. Wexler, J. Chem. Phys. $\underline{79}$, 2577 (1983).
14. P. Hohenberg, W. Kohn, Phys. Rev. $\underline{136}$, B864 (1964).
15. W. Kohn, L.J. Sham, Phys. Rev. $\underline{A140}$, 1133 (1970).
16. J.P. Dahl and J. Avery, Eds. Local density approximations in quantum chemistry and solid state physics", Plenum Press, New York (1984).
17. U. von Barth, Phys. Rev. $\underline{A20}$, 1693 (1979).
18. T. Ziegler and A. Rauk, Theo Chim. Acta. $\underline{43}$, 261 (1977).
19. U. von Barth, L. Hedin, J. Phys. $\underline{C5}$, 1629 (1972).
20. D.E. Ellis and G. Painter, Phys. Rev. $\underline{B2}$, 2887 (1973).
21. A. Rosén, D.E. Ellis, H.Adachi and F.W. Averill, J. Chem. Phys. $\underline{65}$, 3629 (1976).
22. B. Delley, D.E. Ellis, J. Chem. Phys. $\underline{76}$, 1949 (1982).
23. S.P. Walch and C.W. Bauschlicher, Jr., in Comparison of Ab initio Quantum Chemistry with Experiment for Small Molecules, The State of the Art, Ed. R.J. Bartlett, D. Reidel Publ. Comp., Dortrecht, Holland (1985).
24. G. Seifert, E. Mrosan, H. Müller and P. Ziesche, Phys. Status Solidi $\underline{89}$, K175 (1978).
25. H. Müller, Ch. Optiz, G. Seifert, Z. Phys. Chem. Leipzig $\underline{263}$, 1005 (1982).
26. R.E. Smalley as quoted in Ref. 23.

ANALYSIS OF THE ELECTRONIC PROPERTIES OF SMALL NIOBIUM CLUSTERS

Tomas Wahnström, Arne Rosén, and Tapio T. Rantala*

Department of Physics, Chalmers University of Technology
and University of Göteborg, S-412 96 Göteborg, Sweden
*Department of Physics, University of Oulu
SF-90570 Oulu 57, Finland

ABSTRACT

The electronic structures of small niobium clusters have been calculated within the local spin density approximation using the LCAO method.
The calculations were done for optimized number of interaction bonds at
the bond length in the bulk and a bond length assuming to be valid for the
dimer. The changes in the Fermi energy is found to be more smooth as a
function of cluster size compared with similar calculations for cobalt
clusters. This may indicate that the change in reactivity for hydrogen
chemisorption on niobium clusters as a function of cluster size is not a
pure electronic structure effect but geometrical.

INTRODUCTION

After the access to modern computers molecular cluster calculations
within different approximations have been performed to evaluate the
electronic structures of bulk materials and different well characterized
surfaces [1-3]. In these works much efforts have been directed to extend the
calculations to bigger clusters in order to better model the electronic
properties of the bulk material or an extended surface. The new technique
of producing small clusters of virtually any element [4] has however resulted
in a growing interest also for theoretical studies of molecular clusters of
small and medium size [5-7]. The new experimental data open a possibility for
detailed studies of the gap between atoms or small molecules and the bulk by
the successive addition of atoms to the clusters [8-11]. Further, the experimentally observed irregular behaviour of abundances in photoionization mass
spectra [12], ionization potentials [13], stability etc. as a function of cluster
size have made the research on clusters a real challenge for the theoreticians.

Studies on transition metal clusters open another dimension of research
due to the technological use of these elements in commercial catalysts and
as model systems for applied and basic surface research [14-17]. Of particular
interest and very challenging for the theoreticians are the recent observations [4,18-22] of the strong variation in reactivity as a function of cluster
size for adsorption of D_2, O_2, H_2S and N_2 on some transition element clusters.
Copper clusters, for example, were found to be quite inert to D_2, while the
activity for hydrogen chemisorption changed quite dramatically as a function
of the number of atoms for clusters of cobalt and niobium. It is for example

known from surface studies [23] that adsorption of tungsten on different tungsten surfaces give a change of the workfunction which could be compared with the change of ionization thresholds observed between neighbouring metal clusters[13]. Selective behaviour of reactivity of species on different surfaces of the same type of element have also been found in surface studies [24]. If however, this dramatic change in reactivity for clusters of Nb is due to electronic or geometric structure is still an open question [25]. However, the trends in reactivity for the transition elements are quite similar to the trends in hydrogen chemisorption on transition metal surfaces as recently summarized by Scillard and Hoffmann[26] and Nordlander et al.[27]. One would therefore expect that the different models for chemisorption, which with great success have been used in surface physics would be applicable for reaction studies of metal clusters[17,28-31]. Recently, we evaluated the electronic structures of small cobalt clusters [32] and found a striking correlation between the cluster size dependence of the Fermi energy, and the magnetic moments to the experimentally observed reactivity of hydrogen chemisorption. We have now extended these type of calculations to clusters of niobium for which also ionization thresholds have been measured quite recently [33].

COMPUTATIONAL APPROACH

The electronic structures of the small niobium clusters presented in this work were calculated within the local-density approximation [34,35] with the spin-polarized exchange correlation potential by von Barth and Hedin [36]. A variational method is used to find the molecular wave functions which are determined within the LCAO method with numerical basis functions. The one electron equations are solved within the self-consistent multipolar (SCM) method [37] with a multipole expansion in $\ell \leqslant 2$ in addition to the occupation charges calculated in the self consistent charge method [38]. The basis functions have been generated from a free atom Nb $4d^4 5s^1 5p^0$ configuration with the addition of 4d, 5s and 5p wave functions generated from a free ion Nb^{4+} with the $4d^1 5s^0 5p^0$ configuration. This type of basis has in calculations for Nb_2 [39] been found to be varitionally better compared with the use of a neutral atom basis consisting of occupied and virtual orbitals.

The conformation or the geometry of small Nb clusters is not known and the calculations of all the clusters were performed with the bond-distance of the bulk [40], d(Nb-Nb) = 5.4 a.u. However, in order to test the sensitivity of the electronic structure and particularly the properties of levels close to the Fermi level calculations were also done for a bond-distance of d(Nb-Nb) = 3.97 a.u. which is close to the equilibrium value found by Walch and Bausclicher [6] in extensive CASCF calculations. Calculations for bond distances of the two extremes are therefore expected to show the general trends of the electronic structure for these niobium clusters.

Optimization of the geometry by minimization of the total energy would be highly desirable but would lead to too extensive calculations for the series of clusters treated in this work. Nb form in the bulk a bcc structure which in a molecular cluster calculation can be treated with Nb_9 and Nb_{15} clusters including the nearest and second nearest neighbor atoms. To investigate the electronic structure for the small clusters calculations were done for clusters with the optimum number of bonds of equal length which can be done within the same symmetry group C_{3v}. Nb_3 is in our calculations treated as a triangle with extension to a pyramid and double pyramid for Nb_4 and Nb_5, respectively. Nb_6 has been calculated as a double pyramid with four atoms in a plane, one below and one above the plane which geometry can also be oriented to C_{3v} symmetry. This geometry has then been extended to the bigger clusters up to Nb_7 and Nb_8 and Nb_{10}. The calculation for the Nb_9 was done in C_{4v} symmetry in bcc structure since we found it difficult to construct, such a cluster in C_{3v} symmetry. The levels close to the Fermi level have been occupied with a Fermi-Dirac distribution corresponding to a temperature of about 160 K or a thermal energy of about 14 meV.

RESULTS AND DISCUSSION

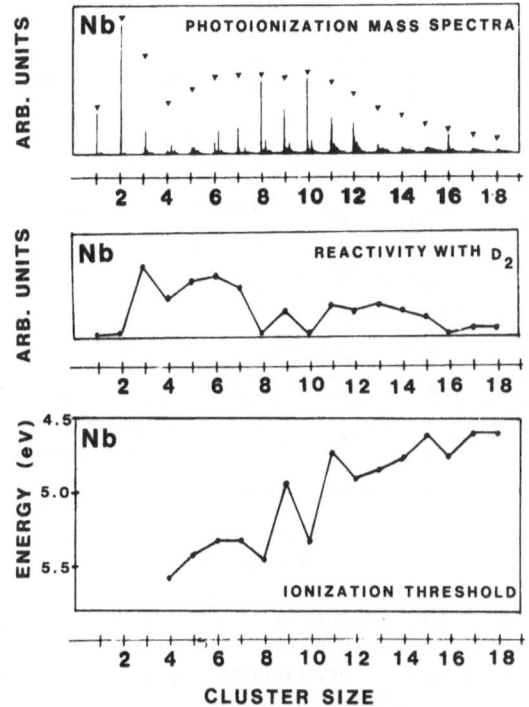

Fig. 1. A schematic overview of reactivity[21] and ionization threshold [33] for small Nb_N clusters.

The main intention with these calculations is to find some model which can explain the dramatic change of observed cluster size dependence of hydrogen chemisorption on small Nb microclusters which for convenience is displayed in the upper part of Fig. 1. The activity for hydrogen chemisorption is given by the difference in intensity of the photoionization mass spectra between the beam of pure Nb and the beam seeded with D_2 as visualized in the middle of Fig. 1. The recently measured photoionization thresholds [33] for these small clusters are displayed in the lower part of Fig. 1. We notice how the clusters Nb_8 and Nb_{10} which are inert for hydrogen chemisorption have higher ionization thresholds compared with the neighbouring reactive ones. This does not hold for Nb_4 which is reactive but has also a high ionization threshold. Whether, the changes in ionization thresholds and reactivity as a function of cluster size is due to changes of the electronic or in geometrical structure requires extensive minimization of the total energy for the different clusters. It is however from the experimental values difficult to derive some absolute number of the relative reactivity for the different clusters.

Recently [39] we investigated changes in the electronic structure of Nb_2 for different bond lengths and found that the energy of the last occupied level changed of the order of 0.4 eV, when the bond length was changed by ± 0.2 a.u. Calculations for reasonable geometries and a fixed bond length is therefore expected to show some indication of abrupt changes of the electronic structure by the addition of an atom to a smaller cluster. Molecular one electron eigenvalues for the Nb_N clusters for N=1-10 calculated for a bond length of the dimer[6] is shown in Fig. 2. We notice how the position of the last occupied levels i.e. the Fermi level generally shifts up for the bigger clusters except for the Nb_4, Nb_8 and Nb_{10} clusters for which the trend is broken. The general trends of the experimental values seem therefore to be reproduced by these theoretical values. In order to get some indication of the sensitivity of the electronic structure as a function of bond length calculations were done for the bond distance in the bulk. In this case a more smooth change take place as a function of cluster size although there is some indication of small changes for the Nb_4, Nb_8 and Nb_{10} clusters.

In the recent calculations for Co clusters the majority up-spin was found to become stabilized and the minority down-spin to be destabilized for clusters up to the size of five atoms. For Co_6, which was the first big

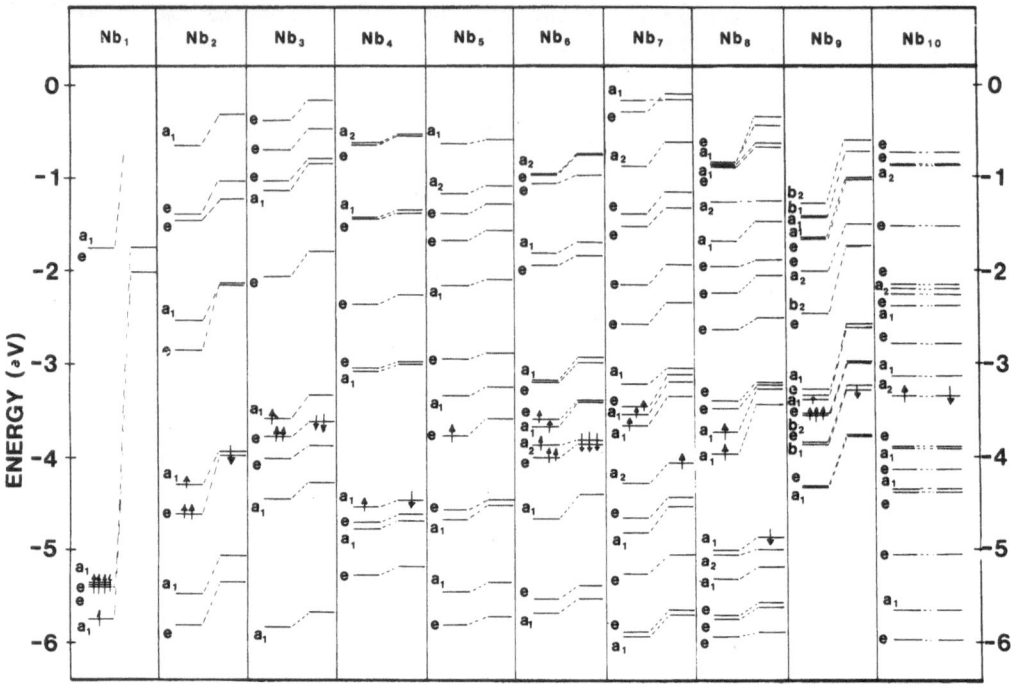

Fig. 2. Energy level diagram of the molecular one electron levels in
the Nb_N clusters for N = 1-10 calculated for a bond length of
3.97 a.u. The labeling of the levels refer to C_{3v} symmetry
except for Nb_9 which is treated in C_{4v} symmetry. The levels
to the left are those with majority up-spin while those to the
right are those with minority down-spin. Occupancies of the
last occupied levels are marked with arrows.

cluster inert for hydrogen chemisorption, the position of the last occupied levels of up and down spin were located at the same energy. The
exchange splitting for the Nb clusters seem to be considerably smaller
and very close to zero for some of the clusters which are inert to
hydrogen chemisorption as Nb_4, Nb_{10}.

A commonly discussed feature in these type of calculations has been
the relative importance of d and s electrons. In order to investigate
this further for the Nb clusters the molecular orbitals were analyzed
through a Mulliken orbital and spin analysis in terms of our basis
functions. In this analysis the contribution of the same orbital symmetry for the neutral and ionized basis have been added together. The net
spindensity from the 4d wavefunctions are reduced to less than one electron for clusters bigger than three while the contribution from the 5s
wavefunctions vary between 0.4-0.0 electrons for these clusters. The
net spin density contribution from all basis functions add up to zero
for the Nb_4 and Nb_{10} clusters.

The models for chemisorption[17,28-31] of H_2 on transition metal
clusters require participation of an unfilled d-shell and a reduction
of the 5s contribution. Our spin-polarized calculations seem to indicate that chemisorption of molecules on small transition metal clusters
require a net but not too big spin contribution which holds for Nb_3,
Nb_{5-7}, and Nb_9. Breaking of the $1\sigma_g^2$ bond in H_2 requires a transfer of

514

electrons to the metal cluster followed by a backdonation to the anti-bonding σ_u levels of H_2. The possibility of this charge transfer to take place needs the opening of channels for both spin directions in the HOMO as well as LUMO levels of the substrate. In the case that the HOMO levels of the substrate were filled as in the case of Nb_4 and Nb_{10} would mean that the electrons have to be transferred to the LUMO levels. The efficiency of the different charge transfers will depend on the energy spacing of the HOMO and LUMO levels of the substrate as well as on the symmetries of the HOMO and LUMO levels of the substrate and adsorbate which with great success has been used in the interpretation of CO chemisorption on transition metal complexes and surfaces.[41-43] These require more detailed calculations on the combined systems which are now in progress.

ACKNOWLEDGEMENTS

The authors wish to thank P. Siegbahn and various colleagues in the molecular and surface physics groups at Chalmers for enlightful discussions. This work has been supported by the Swedish National Research Council (NFR), Joint Committee of the Nordic National Science Research Council (NOS-N) and Nordic Research Courses (Nordic Council of Ministers).

REFERENCES

1.

2. R.P. Messmer in "The Physical Basis for Heterogeneous Catalysis", Eds. E. Drauglis and R.I. Jaffee, Plenum, New York (1975).
3. T.N. Rhodin, G. Ertl, "The Nature of the Surface Chemical Bond", North Hollan, Amsterdam (1979).
4. R.E. Smalley, in "Comparison of Ab Initio Quantum Chemistry with Experiment for Small Molecules, the State of the Art", Ed. R.J. Bartlett, D. Reidel Publ. Comp. Dordrecht, Holland (1985). In this review references are given to experimental works on clusters from a number of groups.
5. W. Weltner, Jr. and R.J. Van Zee, Ann. Rev. Phys. Chem. 35, 291 (1984). References are given to a number of calculations.
6. S.P. Walch and C.W. Bauschlicher, Jr., in "Comparison of Ab Initio Quantum Chemistry with Experiment for Small Molecules, the State of the Art", D. Reidel Publ. Comp. Dordrecht, Holland (1985).
7. B. Delley, D. Ellis, A. Freeman, E. Baerends and D. Post, Phys. Rev. B27, 2132 (1983).
8. D.E. Ellis, B. Delley, in "Local Density Approximations in Quantum Chemistry and Solid State Physics", Eds. J.P. Dahl and J. Avery, Plenum Publ. Corp. (1984).
9. B.K. Rao and P. Jena, J. Phys. F: Met. Phys. 16, 461 (1986).
10. J. Koutecky, G. Pacchioni, G.H. Jeung, E.C. Hass, Surf. Sci. 156, 650 (1984).
11. J.L. Martins, J. Buttet, R. Car, Phys. Rev. B31, 1804 (1985).
12. W.D. Knight, K. Clemenger, W.A. de Heer, W.A. Saunders, M.Y. Chou, and M.L. Cohen, Phys. Rev. Lett. 52, 2141 (1984).
13. R.L. Whetten, D.M. Cos, D.J. Trevor, A. Kaldor, Phys. Rev. Lett. 54, 1494 (1985); J. Phys. Chem. 89, 566 (1985).
14. D.A. King, D.P. Woodruff, "The Chemical Physics of Solid Surfaces and Heterogeneous Catalysis, Vol. 1. Clean Solid Surfaces, Vol. 2. Adsorption at Solid Surfaces, Vol. 3. Chemisorption Systems, Vol. 4. Fundamental Studies of Heterogeneous Catalysis". Elsevier, Amsterdam (1983).
15. G.A. Somorjai, Science 227, 902 (1985).
16. E. Shustorovich, R.C. Baetzold, Science 227, 876 (1985).

17. E. Shustorovich, R.C. Baetzold, E.L. Muetterties, J. of Phys. Chem. 87, 1100 (1983).

18. S.J. Riley, E.K. Parks, G.C. Nieman, L.G. Pobo, S. Wexler, J. Chem. Phys. 80, 1360 (1984).

19. S.C. Richtsmeier, E.K. Parks, K. Lin, L.G. Pobo, S.J. Riley, J. Chem. Phys. 82, 3659 (1985).

20. E.K. Parks, K. Lin, S.C. Richtsmeier, L.G. Pobo, S.J. Riley, J. Chem. Phys. 82, 5470 (1985).

21. M.E. Geusic, M.D. Morse, R.E. Smalley, J. Chem. Phys. 82, 590 (1985).

22. R.L. Whetten, D.M. Cox, D.J. Trevor, A. Kaldor, Phys. Rev. Lett. 54, 1494 (1985); J. Phys. Chem. 89, 566 (1985).

23. K. Besocke and H. Wagner, Phys. Rev. B8, 4597 (1973); Surf. Sci. 53, 351 (1975).

24. G.A. Samorjai, N.D. Spencer, and R.C. Schoonmaker, J. Catal. 74, 129 (1982).

25. J.C. Philips, J. Chem. Phys. 84, 1951 (1986).

26. J.E. Scillard, R. Hoffmann, J. Am. Chem. Soc. 106, 2006 (1984). This work gives references to experimental and theoretical works on metallorganic compounds and surfaces.

27. P. Nordlander, S. Holloway, J.K. Norskov, Surf. Sci. 136, 59 (1984). This work gives references to earlier studies of hydrogen chemisorption on surfaces.

28. J.K. Norskov, A. Houmøller, P.K. Johanssen, and B.I. Lundqvist, Phys. Rev. Lett. 46, 257 (1981).

29. P. Madhavan and J.L. Whitten, J. Chem. Phys. 77, 2673 (1982).

30. P.E.M. Siegbahn, M.R.A. Blomberg, and C.W. Bauschlicher, J. Chem. Phys. 81, 1373 (1984).

31. J. Harris, S. Andersson, Phys. Rev. Lett. 55, 1583 (1985).

32. A. Rosén and T.T. Rantala, Z. Phys.   (1986).

33. R.L. Whetten, M.R. Zakin, D.M. Cox, D.J. Trevor and A. Kaldor, J. Chem. Phys. 85, 1697 (1986).

34. P. Hohenberg, W. Kohn, Phys. Rev. 136, B864 (1964).

35. W. Kohn, L.J. Sham, Phys. Rev. A140, 1133 (1970).

36. U. von Barth, L. Hedin, J. Phys. C5, 1629 (1972).

37. B. Delley, D.E. Ellis, J. Chem. Phys. 76, 1949 (1982).

38. A. Rosén, D.E. Ellis, H. Adachi and F.W. Averill, J. Chem. Phys. 65, 3629 (1976).

39. A. Rosén and T. Wahnström, this volume.

40. J.C. Slater, "Quantum Theory of Molecules and Solids", Vol. 2, McGraw-Hill Book Comp., New York (1965).

41. G. Blyholder, J. Phys. Chem. 68, 2772 (1964).

42. E.W. Plummer, W. Eberhardt, Adv. Chem. Phys. 49, 533 (1982).

43. Sung Shen-Shu and R. Hoffmann, J. Am. Chem. Soc. 107, 578 (1985).

LINEARIZED AUGMENTED PLANE WAVE (LAPW) METHOD

FOR ISOLATED CLUSTERS AND MOLECULES

Henry Krakauer[*] and Michael J. Mehl[**]

* Department of Physics
  College of William and Mary
  Williamsburg, VA 23185

** Condensed Matter Physics Branch
   Naval Research Laboratory
   Washington, D.C. 20375-5000

ABSTRACT

A linearized augmented plane wave (LAPW) method for isolated clusters or molecules is presented. This method avoids the use of a supercell geometry and is relatively easy to implement into an existing bulk LAPW computer code. As in the bulk LAPW method, the potential and charge density are represented by shape-unrestricted expansions, i.e. the potential and charge density are permitted to have full variation in all regions of space. Local density-functional-theory calculations have been performed for two test systems. Results obtained for a Cu atom are in excellent agreement with that of a standard atomic program. Calculations were also performed for the O_2 molecule and are in excellent agreement with results obtained using standard molecular methods.

Many of the theoretical techniques which had been specifically designed for obtaining the band structure of infinitely periodic solids have been successfully adapted for use in treating isolated clusters and molecules. These include partial-wave methods like the Kohn, Korringa, and Rostoker (KKR) method[1] adapted by Johnson et al.[2] and the Muffin-Tin Orbital (MTO) method[3] adapted by Andersen and Wooley[4] and applied by Gunnarsson et al.[5]. Of course, other methods based on the linear combination of atomic (or other) orbitals (LCAO methods) can be applied with equal ease to both periodic and molecular systems. In the case of plane wave based methods, however, the few reported molecular calculations have always been performed using a supercell to model the isolated cluster (see for example Refs. 6 and 7). In this paper we present an adaptation of the LAPW method which can be applied to a truly isolated cluster or molecule without resorting to the use of a supercell geometry. This capability is relatively easy to implement into an existing bulk LAPW computer code.

In a supercell approach, the cluster or molecule to be studied is placed on each lattice site of an infinite three-dimensional[6] or two-dimensional[7] crystal. If the nearest-neighbor distance between two adjacent clusters is large enough, the calculated results can be used to approximate the truly isolated cluster or molecule. In such calculations for the O_2 molecule,[6,7] neighboring molecules were separated by about three bond lengths. There are several disadvantages to this approach. Since the size of the variational basis employed depends on the size of the unit cell, there will always be spurious inter-molecular interactions in any practical application, especially for the more extended molecular states. This is discussed further below. This is an especially undesirable feature if one is attempting to study a cluster or molecule which has a dipole moment, since inter-molecular interactions will be of long range. Furthermore, in a plane-wave based method, one relies on a linear combination of plane-waves to reproduce the exponential fall-off of the charge density away from the cluster or molecule. This means that very large basis-sets must be employed, increasing the computational resources required. For these reasons, we have developed an approach which avoids the supercell geometry and instead treats an isolated cluster.

In the cluster LAPW method, all space in partitioned into three types of regions as in the KKR multiple scattering method[2]: (1) nonoverlapping muffin-tin (MT) spheres centered on each atom, (2) the vacuum region, defined as all space outside of a Watson[2] sphere which encloses the smaller MT spheres, and (3) an interstitial region, defined as the space between the small MT spheres and the Watson sphere. The vacuum region is simply treated as an exterior MT sphere when defining the LAPW basis functions, Φ:

$$
\Phi_G(r) = \begin{cases}
\exp(i\, G \cdot r) & r \in \text{Inter.} \\[2mm]
\Sigma_{l,m}[a^{\alpha}_{lm} u^{\alpha}_l(E_1,r) + b^{\alpha}_{lm} ue^{\alpha}_l(E_1,r)]\, Y_{lm}(\theta,\Phi) & r \in \alpha\text{-MT} \\[2mm]
\Sigma_{l,m}[a^{V}_{lm} u^{V}_l(E_1,r) + b^{V}_{lm} ue^{V}_l(E_1,r)]\, Y_{lm}(\theta,\Phi) & r \in \text{Vacuum}
\end{cases}
\tag{1}
$$

The LAPW wave function is given by

$$
\psi = \Sigma_G\, c_G\, \Phi_G(r)
\tag{2}
$$

Notationally, this looks just like the usual definition of the bulk LAPW basis functions for Bloch momentum k=0. Here, u_1 is a radial wave function and ue_1 is its energy derivative, and Y_{lm} is a spherical harmonic. In the bulk, however, the G-vectors are just the reciprocal lattice vectors. In the cluster case, we introduce a fictitious reciprocal lattice by enclosing the entire cluster by a box. For example, in the calculations reported here we enclosed the Watson sphere in an fcc unit-cell shaped box. A similar trick is used in the slab-LAPW method[7] to introduce a component of the reciprocal lattice vector normal to the surface. As in the slab case[7], we emphasize that the plane waves thus defined are employed only in the interstitial region. The correct boundary conditions at infinity are thus ensured by the representation given in Eq. (1). We note that as defined in Eq. (1), the vacuum region can be handled by the method as just another MT sphere. This means that it will be transparent to most parts of a bulk LAPW program that a cluster or molecule is in fact being treated.

The major change that must be made in adapting a bulk LAPW program to treat an isolated cluster or molecule is in the solution of Poisson's equation to obtain the Coulomb potential. This is now briefly described. (1) We first solve for the Coulomb potential in the vacuum region. To do this

518

we only need to construct the multipole moments of the charge density inside the Watson sphere. Together with the vacuum charge density, this constitutes enough information to determine the Coulomb potential in the vacuum region. In spherical coordinates, we then can perform a simple one-dimensional radial integration involving the vacuum charge density and obtain the vacuum potential as well as the boundary condition on the Watson sphere surface. (2) Using the LAPW pseudocharge density (defined in Ref. 7) and the boundary condition on the Watson sphere surface, the Coulomb potential can be obtained everywhere inside the Watson sphere but outside the smaller atomic MT spheres

Table I.   Cluster LAPW Results for a Cu atom

Eigenvalues (in Ry):

Core-Level Eigenstates

n	l	j	Cluster LAPW	Atomic Prog.	Occup.
1	0	1/2	-649.0675	-649.0652	2
2	0	1/2	-77.9158	-77.9131	2
2	1	1/2	-68.3440	-68.3413	2
2	1	3/2	-66.8411	-66.8385	4
3	0	1/2	-8.3941	-8.3924	2
3	1	1/2	-5.4276	-5.4260	2
3	1	3/2	-5.2370	-5.2353	4

Valence-Level Eigenstates

3	2	--	-0.3970	-0.3981	10
4	0	--	-0.3639	-0.3641	1
4	1	--	-0.0628	-0.0629	0

Total Energy (in Ry):

-3304.520716   -3304.520417

using the sphere Green's function.   (3) This gives us the boundary condition on the atomic MT spheres and using the true charge density inside these spheres and the sphere Green's function again, the Coulomb potential is finally obtained everywhere inside the atomic MT spheres. In a final step, the potential in the interstitial is re-expressed in terms of a plane wave expansion for subsequent eigenvalue determination.

With a few exceptions, the remainder of the program, including the total energy evaluation, remains unchanged. As in our general potential bulk LAPW method,[8] the potential and charge density are represented by shape-unrestricted expansions, i.e. we do not require spherical symmetry inside the MT spheres and allow full variation in the interstitial region. Band states are treated scalar-relativistically (i.e. without spin-orbit coupling). The core electron states are treated fully relativistically with use of an atomiclike program which employs only the self-consistent[9] spherical part of the potential in the atomic MT sphere.

We now present results obtained for two test systems, the Cu atom and the O_2 molecule.

For the Cu atom we used an atomic MT radius of 2.12 a.u. and a Watson sphere radius of 3.2 a.u. With this choice there are about 9.2 d-electrons inside the atomic MT sphere, a total of 0.96 electrons in the interstitial region and about 0.5 electrons in the vacuum region. An fcc unit-cell corresponding to a = 12.4 a.u. was used as the "box" which generates the reciprocal-space-like G-vectors. The basis consisted of 531 LAPWs which corresponds to a cutoff of $G_{max} R_{MT}$ = 8.4 (where R_{MT} is 2.12 a.u.). Our results are presented in Table I and compared to those obtained with a standard atomic program. The atomic program was modified so as to calculate scalar-relativistic valence states for comparison with the present calculations. The Hedin-Lundqvist[10] exchange-correlation potential was used in both programs. Eigenvalues in both cases are unshifted and with respect to vacuum zero. As seen in Table I excellent agreement is obtained for this simple test case.

We performed another test calculation for the Oxygen molecule. An atomic MT radius of 1.05 a.u. and a Watson sphere radius of 2.5 a.u. were used, and an fcc "box" with a = 8.46 a.u. was used to generate the G-vectors. Calculations were performed at the experimental bond length of 2.279 a.u. A basis of 531 LAPWs was used which corresponds to a cutoff of $G_{max} R_{MT}$ = 6.0 (where R_{MT} is 1.05 a.u. This is identical to the cutoff used in the supercell calculation in Ref. 7). For comparison to previous work the exchange-correlation potential is approximated by the simple $X\alpha$ potential using α = 0.7. Our results are presented in Table II and compared with some of the more recent self-consistent calculations.[7,11,12]

Table II. Cluster LAPW Results for the O_2 Molecule

Eigenvalues ($-\epsilon_i$ in Ry):

Orbital	Cluster LAPW	LCAO-Gaussian[a]	DVM-SCM[b]	FLAPW[c]
1s	37.586	--	37.599	37.578
$2\sigma_g$	2.329	2.331	2.334	2.309
$1\sigma_u$	1.371	1.378	1.379	1.381
$3\sigma_g$	0.917	0.919	0.921	0.923
$1\pi_u$	0.889	0.889	0.895	0.859
$1\pi_g$	0.376	0.374	0.383	0.350

[a] Reference 11. Pseudopotential linear combination of Gaussian orbitals.

[b] Reference 12. All-electron LCAO Discrete Variational Method.

[c] Reference 7. All-electron slab LAPW method using a two-dimensional supercell geometry.

As can be seen in Table II, excellent agreement is obtained with conventional cluster methods[11,12] as well as with the supercell[7] calculation for the lower lying orbitals. For the higher lying $1\pi_u$ and $1\pi_g$ orbitals, the supercell results are consistently higher by a significantly large value. This reflects the fact that for these more delocalized states spurious inter-molecular interactions are artificially raising the energy of these states.

In conclusion, we have presented an LAPW method which can determine the electronic structure and total energy of clusters and molecules to a high degree of accuracy. For somewhat larger clusters than have been considered here, the computer resources required are roughly equivalent to that required for a bulk calculation with an equivalent number of atoms in the unit cell. Since only a single diagonalization is required the method is practical for moderately sized clusters.

ACKNOWLEDGEMENTS

Work at the College of William and Mary was supported by National Science Foundation Grant DMR-84-16046. One of us (HK) acknowledges NSF support for computations carried out at the National Center for Supercomputing Applications at the University of Illinois.

REFERENCES

1. J. Korringa, Physica 13, 392 (1947); W. Kohn and N. Rostoker, Phys. Rev. 94, 1111 (1954).
2. K. H. Johnson, J. Chem. Phys. 45, 3085 (1966); K. H. Johnson and F. C. Smith, Phys. Rev. Lett. 22, 1168 (1969).
3. O. K. Andersen, Phys. Rev. B 12, 3060 (1975).
4. O. K. Andersen and R. G. Woolley, Mol. Phys. 26, 905 (1973).
5. O. Gunnarsson, J. Harris, and R. O. Jones, Phys. Rev. B 15, 3027 (1977).
6. G. P. Kerker, A. Zunger, M. L. Cohen, and M. Schluter, Solid State Commun. 32, 309 (1979).
7. E. Wimmer, H. Krakauer, M. Weinert, and A. J. Freeman, Phys. Rev. B 24, 864 (1981).
8. S.-H. Wei and H. Krakauer, Phys. Rev. Lett. 55, 1200 (1985).
9. S.-H. Wei, H. Krakauer, and M. Weinert, Phys. Rev. B 32, 7792 (1985).
10. L. Hedin and B. I. Lundqvist, J. Phys. C 4, 2064 (1971).
11. J. Bernholc and N. A. W. Holzwarth, J. Chem Phys. 81, 3987 (1984).
12. B. Delly and D. E. Ellis, J. Chem. Phys. 76, 1949 (1982).

THE NICKEL DIMER AND TRIMER

Irene Shim

Chemical Physics, Chemistry Department B
The Technical University of Denmark, DTH 301
DK-2800 Lyngby, Denmark

Karl A. Gingerich

Department of Chemistry
Texas A&M University
College Station, Tx 77843, U.S.A.

ABSTRACT
The chemical bonds and the low-lying electronic states of the
molecules Ni_2 and Ni_3 have been investigated by performing all electron
ab initio Hartree-Fock and configuration interaction calculations. The
results reveal that the chemical bonds in Ni_2 and Ni_3 are of similar natures. Thus, in each molecule the bond is mainly due to a delocalized molecular orbital composed of the Ni 4s orbitals. The 3d electrons localize
around the nuclei, and their weak interactions give rise to "bands" of
low-lying electronic states. As for the Ni_2 molecule, the lowest lying
states of the linear Ni_3 molecule are due to a localized hole in the $3d\delta$
subshell of each Ni atom. The ground state of the Ni_3 molecule has been
derived as $^5A_2[...(18a_1)^2....(3a_2)^2(4a_2)^1(5a_2)^1....(13b_1)^2(14b_1)^1....(6b_2)^2$
$(7b_2)^1]$. With the Ni-Ni bond distances fixed to 4.709 a.u., the bent Ni_3
molecule with a Ni-Ni-Ni angle of $90°$ is found to be 0.08eV more stable than
the linear molecule and 0.29eV more stable than the triangular molecule.

INTRODUCTION

As is reflected by the present International Symposium on the Physics
and Chemistry of Small Clusters, as well as by the Faraday Symposium 1980,
Diatomic Metals and Metallic Clusters,[1] there has been growing interest,
both experimentally and theoretically, for small molecular systems involving coordinatively unsaturated transition metal atoms. From several pertinent reviews,[2-4] it is in particular noted that it has now become possible
to perform experimental as well as theoretical investigations on well-
defined "naked" transition metal clusters, and the numbers of such studies
are increasing rapidly. The current interest in the transition metal clusters is of course influenced by their relevance to surface science and
heterogeneous catalysis. Thus, fundamental understanding of the well-defined transition metal clusters is essential for gaining deeper insight
into the mechanism of the catalytic reactions at the atomic and molecular
levels.

The present paper has been devoted to the smallest possible clusters consisting of Ni atoms, namely the molecules Ni_2 and Ni_3. The electronic structures and the natures of the chemical bonds in these molecules have been elucidated by performing all electron *ab initio* Hartree-Fock (HF) and configuration interaction (CI) calculations. For the Ni_3 molecule, the computations have been performed for four different geometric configurations, namely a linear, a triangular, as well as two bent configurations. The results obtained compare favourably with available experimental data.

COMPUTATIONAL DETAILS

The HF calculations have been carried out in the open shell Hartree-Fock-Roothaan formalism.[5] The integrals have been evaluated by using the program MOLECULE,[6] and the ALCHEMY program system[7] was utilized for the HF calculations. The CI calculations have been performed by using ALCHEMY in conjunction with the program ENERGY[8] for generating the symbolic energy expressions. The deformation density maps have been produced by means of the program MOLPLOT.[9]

The wave functions for the molecules Ni_2 and Ni_3 have been expanded in Gaussian type functions. The basis set employed for the Ni atom is basicly Wachters',[10] but modified and extended as described in detail in Ref. 3. The primitive basis set (14s,11p,6d) has been contracted to (8s, 6p,3d). Thus, in the contracted basis the 3d orbital is represented by a triple zeta function, while all core orbitals as well as the valence 4s and 4p orbitals are represented by double zeta functions. At this point, we would like to emphasize that the splittings between the low-lying terms of the Ni atom derived from HF calculations based on the present basis set are in excellent agreement with those derived from numerical HF calculations.[3,11]

The partly filled 3d shell of the Ni atom gives rise to uncertainties as to electronic ground state assignments of the molecules Ni_2 and Ni_3. Therefore, we have performed HF calculations to determine unique sets of molecular orbitals, which are suitable for utilization in CI calculations to describe all the low-lying electronic states of the molecules. In the HF calculations we have searched for wave functions capable of describing bound molecules, i.e., the total energies of the molecules should be lower than the sum of the energies of the free atoms in their ground term.

For the Ni_2 molecule the optimized molecular orbitals have been utilized in CI calculations allowing full reorganization within the 3d and 4s shells. In addition, the lowest lying electronic states have been investigated further by performing CI calculations that included additional configurations derived from single and double excitations into the 4p and 5s orbitals. For the Ni_3 molecule the CI calculations allowed full reorganization within the 3d shells, but for the linear Ni_3 molecule we also carried out CI calculations allowing full reorganization within both the 3d and the 4s shells.

The chosen procedure, of course, does not provide the optimum description of each state individually, but we expect it results in balanced descriptions of the many low-lying electronic states. In particular, it assures a correct description of the d electron part of the wave functions in the dissociation limits, and it allows for localization of the d electrons around the nuclei.

For the linear molecules the calculations have been performed in the subgroup D_{2h} of the full symmetry group $D_{\infty h}$. For triangular Ni_3 as well as for the bent configurations the calculations have been carried out in the ·

symmetry group C_{2v}. The number of configurations included within each space and spin symmetry is within the ranges 208-3219 and 46-1158 for the Ni_2 and Ni_3 molecules, respectively.

THE NICKEL DIMER

According to the spectroscopic data in Moore's Tables,[12] the two lowest lying terms of the Ni atom, $^3D(3d)^9(4s)^1$ and $^3F(3d)^8(4s)^2$, are practically degenerate. In a previous work[13] we have shown that the interaction between two Ni atoms is strongly dependent on the initial atomic configuration. Thus, the interaction between two Ni atoms, both in the $^3F(3d)^8(4s)^2$ term, does not lead to the formation of a stable molecule. Their interaction is of a Van der Waals' type, because the partly occupied d orbitals are very well localized, and completely screened by the fully occupied 4s orbitals.

However, two Ni atoms, both in the $^3D(3d)^9(4s)^1$ configuration, interact by forming a stable molecule [13,3]. As shown in Fig. 1, the *ab initio* calculations have revealed the existence of a "band" consisting of 30 low-lying electronic states within an energy range of 0.6eV at the internuclear distance 4.709 a.u. In spite of the numerous low-lying states, a simple picture of the bonding interaction between the two Ni atoms is emerging from the calculational results: The 3d electrons localize around the nuclei, and the chemical bond is almost entirely due to the delocalized $4s\sigma_g$ molecular orbital. As the two Ni atoms approach each other, the axial symmetry introduced

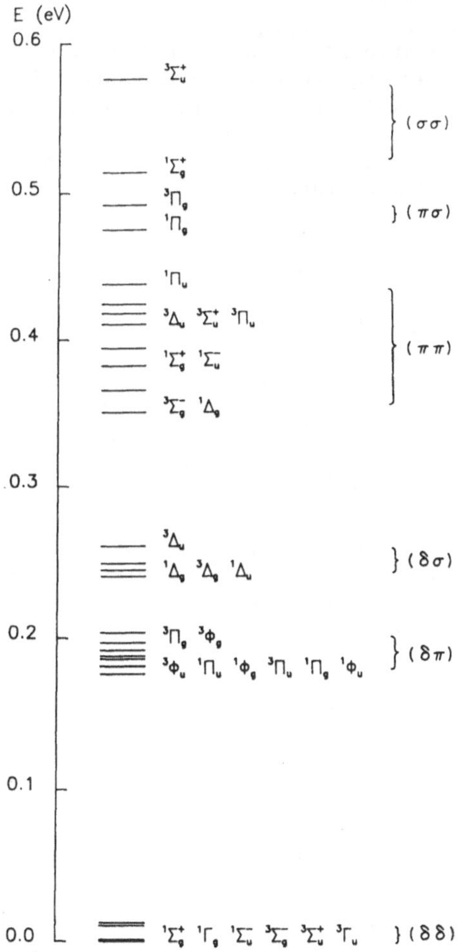

Figure 1. Relative energies in eV of the 30 low-lying electronic states of the Ni_2 molecule at the internuclear distance 4.709 a.u. The states are listed in order of increasing energy.

525

causes the 3d orbitals to split into σ, π, and δ orbitals with slightly different energies. The weak interactions between the split 3d orbitals are describable in terms of Heisenberg exchange couplings,[14] and this gives rise to the small energy splittings between the many low-lying potential energy curves.

From Fig. 1 it is noted that all the 30 low-lying electronic states can be characterized by the symmetry of the hole in the 3d shell of each Ni atom. Thus, the six lowest lying states all have a hole in the 3dδ subshell of each Ni atom. The following eight states have a hole in the 3dδ subshell of one atom and in the 3dπ subshell of the other atom. Thereafter follows four (δσ) hole states, six (ππ) hole states, and four (πσ) hole states. The two highest lying states in Fig. 1 are the (σσ) hole states.

It is interesting to note that the lowest lying states of the Ni_2 molecule are due to localized holes in the 3dδ subshell of each atom. A priori, a hole in the 3dσ subshell seems more favourable, because this could give rise to the formation of a 3dσ bond at closer approach. However, the preference of the 3dδ hole indicates that each atom experiences the other as a positive unit. Therefore, the Ni_2 molecule can be regarded as two Ni^+ ions kept together by the delocalized charge cloud originating from the 4s electrons. This point of view is supported further by noting that the 30 low-lying electronic states are exactly those expected from coupling of the angular momenta of the two Ni^+ ions in the $^2D(3d)^9$ term.

As evident from Refs. 2-4, the Ni_2 molecule has been studied by various experimental and theoretical techniques. At this point we would like to emphasize the very good agreements between the results of the all electron calculations[3,13] and those derived in the pseudopotential calculations.[15,16] Furthermore, recent spectroscopic investigations of the Ni_2 molecule in the gas phase[17] have revealed a dense manyfold of highly perturbed levels. This is consistent with the theoretical predictions of the nature of the chemical bond in the Ni_2 molecule, including the existence of the many low-lying electronic states.

THE NICKEL TRIMER

The Ni_3 molecule has been treated in *ab initio* calculations using similar methods to those employed for the Ni_2 molecule. The calculations have been performed for four different geometries, namely a linear, a triangular as well as two bent configurations with Ni-Ni-Ni angles of 120° and 90°, respectively. For all four geometries the bond distances have been fixed to 4.709 a.u., which is the nearest neighbour internuclear distance in the bulk metal.

In Fig. 2 we present the lowest lying electronic states as derived in CI calculations for the four different geometries of the Ni_3 molecule considered. Table I shows the correlations between the symmetries of the three point groups, D_{3h}, C_{2v}, and $D_{\infty h}$, associated with the different geometries of Ni_3.

The sixteen lowest lying electronic states of the linear Ni_3 molecule, all due to a localized hole in the 3dδ subshell of each Ni atom, appear in four separate groups, and each group consists of four practically degenerate states (Fig. 2). The two lowest lying groups of states, including the quintet states $^5\Delta_g$, $^5\Delta_u$, $^5\Delta_g$, and ^5I_g, respectively the triplet states, $^3\Delta_g$, ^3I_g, $^3\Delta_u$, and $^3\Delta_g$, can be considered as arising from the three lowest lying triplet states of Ni_2, $^3\Sigma_g^-$, $^3\Sigma_u^+$, $^3\Gamma_u$, by coupling with the third Ni atom in a $^3\Delta_u$ state. Likewise, the following two groups of higher lying states, i.e. the triplet states $^3\Delta_u$, $^3\Delta_g$, $^3\Delta_u$, ^3I_u and the singlet states $^1\Delta_u$, $^1\Delta_g$, $^1\Delta_u$, ^1I_u,

Table I. Correlation between symmetries of the point groups D_{3h}, C_{2v}, and $D_{\infty h}$

D_{3h}	C_{2v}	$D_{\infty h}$
A_1'	A_1	Σ_g^+
A_2'	B_1	Σ_u^+
A_1''	A_2	Σ_u^-
A_2''	B_2	Σ_g^-
E'	A_1+B_1	
E''	A_2+B_2	
	A_1+B_2	$\Pi_u, \Delta_g, \Phi_u \ldots$
	A_2+B_1	$\Pi_g, \Delta_u, \Phi_g \ldots$

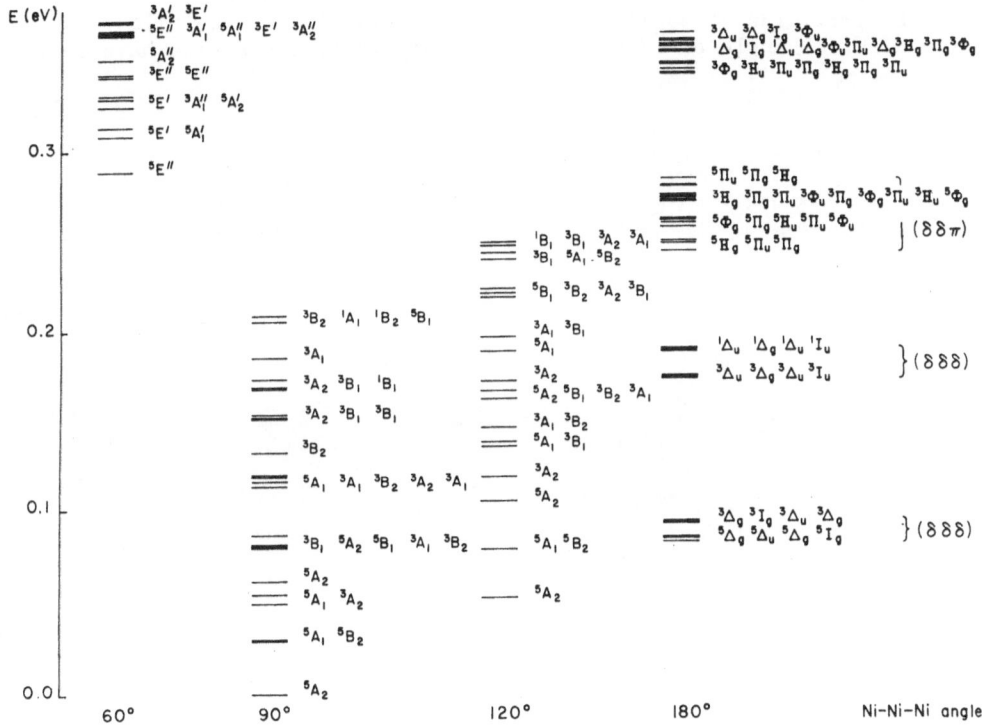

Figure 2. Relative energies in eV of the low-lying electronic states of the Ni$_3$ molecule in a linear, a triangular as well as two bent geometric configurations with bond angles of 90° and 120°, respectively. In all cases the bond distances have been kept fixed at the nearest neighbour internuclear distance in the bulk metal, 4.709 a.u. The electronic states are listed in order of increasing energy.

arise when the three lowest lying singlet states of Ni_2, $^1\Sigma_g^+$, $^1\Sigma_u^-$, and $^1\Gamma_g$, are coupled with the third Ni atom in the $^3\Delta_u$ and $^1\Delta_u$ states, respectively. After these states follows a manifold of $(\delta\delta\pi)$ hole states starting with the ^5H_g state 0.16eV above the $^5\Delta_g$ ground state.

The calculated atomization energy of the linear Ni_3 molecule is 0.98eV. This is only slightly higher than the dissociation energy of the Ni_2 molecule, 0.95eV, derived in calculations equivalent to those performed for Ni_3.[3] This is consistent with our findings that the linear Ni_3 molecule has a localized hole in the $3d\delta$ subshell of each Ni atom. The chemical bond is mainly due to the 4s electrons, and relative to the Ni_2 molecule the additional 4s electron enters into an almost non-bonding $4s\sigma_u$ orbital.

In Figs. 3 and 4 we present the deformation density maps of the Ni_2 and the linear Ni_3 molecules, respectively. Although the similarity between the deformation density maps of Ni_2 and Ni_3 is striking, it is observed that the deformation density map of Ni_3 exhibits "edge" effects. Thus, charge is pushed away from the center atom toward the two end atoms.

The ground state of the triangular Ni_3 molecule has been determined as $^5E''(6a_1')^2(7a_1')^2(8a_1')^2(1a_1'')^1(3a_2')^2(3a_2'')^2(8e')^4(9e')^4(10e')^4(11e')^1(3e'')^4(4e'')^2$, but as noted from Fig. 2, numerous other electronic states have been located at only slightly higher energies than the $^5E''$ ground state. The triangular molecule is found to be 0.21eV less stable than the linear molecule, and its atomization energy has been derived as 0.78eV.

The ground states of the bent Ni_3 molecules have been determined to be $^5A_2[\ldots(18a_1)^2\ldots(3a_2)^2(4a_2)^1(5a_2)^1\ldots(13b_1)^2(14b_1)^1\ldots(6b_2)^2(7b_2)^1]$. With the Ni-Ni-Ni angle of 120° and 90° the atomization energies have been

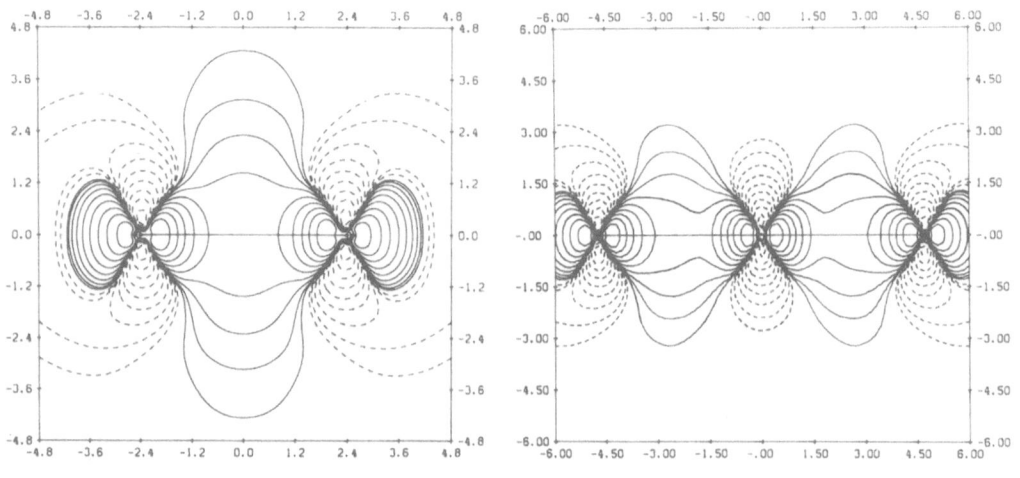

Figure 3   Figure 4

Figures 3 and 4.   Deformation density maps of the Ni_2 (Fig. 3) and the linear Ni_3 (Fig. 4) molecules. The superpositioned atomic charge densities have been subtracted from the molecular charge densities as derived in CI calculations for Ni_2 and in HF calculations for Ni_3. Solid contours show enhanced electron charge relative to the superpositioned atoms, dashed contours show diminished charge. The smallest contour value is 0.000625 e/a.u.[3] Adjacent contours differ by a factor of 2.

derived as 1.01 and 1.07eV, respectively. As for linear and triangular Ni_3, the electronic states of the bent Ni_3 molecules are closely spaced (Fig. 2). Thus, for the bent Ni_3 molecule with the Ni-Ni-Ni angle of $90°$, the fifth electronic state, i.e. the lowest lying triplet state $^3A_2$, is only 0.06eV above the $^5A_2$ ground state, and the lowest lying singlet state, $^1B_1$, is 0.17eV above the $^5A_2$ ground state.

As for the Ni_2 molecule, the total overlap populations in the different geometries of the Ni_3 molecule considered are almost entirely due to the 4s electrons. Each Ni atom is associated with the atomic configuration $(3d)^9 (4s)^1$, and the hole in the 3d shell of each Ni atom has been found in a localized δ or δ-like orbital.

Ab initio calculations have previously been applied to the Ni_3 molecule.[18,19] Thus, the Ni_3 molecule in the linear and the triangular configurations has been investigated in the HF approximation by means of pseudopotential calculations,[18] and recently all electron HF calculations have been applied to the triangular Ni_3 molecule.[19] For the linear molecule, the results pf the present work are in agreement with the results of the pseudopotential calculations, but for the triangular molecule the ground state in the pseudopotential calculations was derived as $^5E'(6a_1')^2(7a_1')^2(8a_1')^1(1a_2'')^2(3a_2')^2(3a_2'')^2(8e')^4(9e')^4(10e')^2(11e')^1(3e'')^4(4e'')^4$. In our HF calculations this $^5E'$ state is found to be 0.25eV less stable than the $^5E''$ ground state of the triangular Ni_3 molecule. Our findings are consistent with the characters of the holes in the 3d shells of the Ni atoms in the $^5E'$ and $^5E''$ states, respectively. Thus, in the $^5E'$ state with the above mentioned configuration the holes in the 3d shells are σ-like, while the $^5E''$ ground state is associated with δ-like holes.

The Ni_3 molecule has been observed in Ar matrices, and its resonance Raman spectrum has been measured.[20,21] The spectrum has been interpreted as arising from a bent Ni_3 molecule with a Ni-Ni-Ni angle in between $90°$ and $100°$. Furthermore, it is suggested that the bent geometry is caused by distortion from a linear Δ_u ground state due to matrix site or Renner-Teller effects.[21] Although the interpretation of the spectrum is questionable,[2,4] it agrees well with our calculational results. Thus, the potential as function of the bending coordinate has been found to be flat with a minimum around the bond angle of $90°$. In addition, it is noted that the predicted $^5A_2$ ground state of the bent Ni_3 molecule correlates with the low-lying $^5\Delta_u$ state of the linear molecule.

CONCLUSIONS

The present work is concerned with all electron *ab initio* HF-CI calculations on the molecules Ni_2 and Ni_3. It has been pointed out that the results obtained are consistent with recent experimental data. As part of the Ni_2 and the Ni_3 molecules, the Ni atom essentially retains the $(3d)^9(4s)^1$ configuration, and the chemical bonds are almost entirely due to the delocalized 4s electrons. The 3d electrons localize around the nuclei, and the lowest lying molecular states are due to a hole in the δ or δ-like orbital of each Ni atom. The weak interactions between the 3d electrons associated with the different nuclei give rise to the numerous low-lying electronic states.

For the Ni_3 molecule, the potential energies as functions of the Ni-Ni-Ni angle reveal shallow minima around the bond angle $90°$. This indicates that the Ni_3 molecule, most likely has the symmetry C_{2v} in the electronic ground state. The ground state of the Ni_3 molecule has been derived as $^5A_2[\ldots(18a_1)^2\ldots(3a_2)^2(4a_1)^1(5a_2)^1\ldots.(13b_1)^2(14b_1)^1\ldots.(6b_2)^2(7b_2)^1]$ with the atomization energy 1.07eV for the bond angle $90°$.

ACKNOWLEDGEMENTS

The computations have been performed at the Computing Service Center at Texas A&M University, and at Uni·C, Lyngby. The computational work in Denmark has been supported by the Danish Natural Science Research Council. I.S. acknowledges the Royal Danish Academy of Sciences and Letters for awarding the Niels Bohr fellowship. The work at Texas A&M University has been supported by Texas A&M University fund for Material Science. The authors also appreciate the support by a NATO grant No. RG 85/0448 for international collaboration in research.

REFERENCES

1. Faraday Symposia of the Royal Society of Chemistry No. 14, Diatomic Metals and Metallic Clusters (1980).
2. W. Weltner, Jr. and R.J. Van Zee, Ann. Rev. Phys. Chem. 35: 291 (1984).
3. I. Shim, Kgl. Danske Vid. Selsk. Matt.-Fys. Medd. 41: 147 (1985).
4. M.D. Morse, Chem. Rev. in press.
5. C.C.J. Roothaan, Rev. Mod. Phys. 32: 179 (1960).
6. J. Almlöf in "Proceedings of the Second Seminar on Computational Problems in Quantum Chemistry, Max-Planck Institut, München (1973).
7. The ALCHEMY program system is written at IBM Research Laboratory in San José, Ca., by P.S. Bagus, B. Liu, M. Yoshimine, and A.D. McLean.
8. C.R. Sarma and S. Rettrup, Theor. Chim. Acta (Berlin) 46: 63 (1977); S. Rettrup and C.R. Sarma, Theor. Chim. Acta (Berlin) 46: 73 (1977).
9. H. Johansen, private communication.
10. A.J.H. Wachters, J. Chem. Phys. 52: 1033 (1970).
11. R.L. Martin and P.J. Hay, J. Chem. Phys. 75: 4539 (1981).
12. C.E. Moore, Nat. Bur. Stand. Circ. No. 467 (U.S. GPO, Washington, D.C. 1952) vol. 2.
13. I. Shim, J.P. Dahl, and H. Johansen, Int. J. Quantum Chem. 15: 311 (1979).
14. I. Shim, Mol. Phys. 39: 185 (1980).
15. T.H. Upton and W.A. Goddard III, J. Am. Chem. Soc. 100: 5659 (1978).
16. J.O. Noell, M.D. Newton, P.J. Hay, R.L. Martin, and F.W. Bobrowicz, J. Chem. Phys. 73: 2360 (1980).
17. M.D. Morse, G.P. Hansen, P.R.R. Langridge-Smith, L.-S. Zheng, M.E. Geusic, D.L. Michalopoulos, and R.E. Smalley, J. Chem. Phys. 80: 5400 (1984).
18. H. Basch, M.D. Newton, and J.W. Moskowitz, J. Chem. Phys. 73: 4492 (1980).
19. M. Tomonari, H. Tatewaki, and T. Nakamura, J. Chem. Phys. 85: 2875 (1986).
20. M. Moskovits and J.E. Hulse, J. Chem. Phys. 66: 3988 (1977).
21. M. Moskovits and D.P. DiLella, J. Chem. Phys. 72: 2267 (1980).

INELASTIC LIGHT SCATTERING FROM METAL CLUSTERS:

SOME RECENT DEVELOPMENTS

Paul S. Bechthold*

Institut für Festkörperforschung der Kernforschungsanlage Jülich
Postfach 1913, D-5170 Jülich, FRG

ABSTRACT

Most Raman and resonance Raman experiments on small metal aggregates have been performed with the scattering particles isolated in rare gas matrices. The matrices permit a high cluster density, but they also influence their spectroscopic properties. Cluster-support interactions become important and may cause a multiplication of vibrational modes, e.g., due to trapping site effects or isomerization processes. The least perturbed vibrational mode usually will be the totally symmetric one. External modes of the particle may appear which directly probe the interaction of the cluster with the matrix support at the particular trapping site. For silver dimers these were used to identify the geometries of various trapping sites by means of a computer simulation.

Clusters deposited on solid surfaces have been little studied by light scattering techniques, despite their importance in catalysis and surface enhanced Raman scattering. Germanium clusters have been probed by Raman scattering. Elastic forces between gold clusters on NaCl have been studied by surface Brillouin scattering.

A few light scattering experiments have been performed with van der Waals and molecular clusters in free jet expansions.

INTRODUCTION

Small metallic particles and clusters are of great importance to a variety of fields, the most obvious being perhaps heterogeneous catalysis. When trapped in a solid they influence the properties of the material. In laser materials clusters cause concentration quenching, and in phototropic glasses they are responsible for the reversible light-induced changes of optical absorption. In metals they form precipitates which have a pronounced influence on elastic properties and the brittleness of the material. Thus, clusters show a multitude of chemical and physical effects in a variety of environments. To get a more detailed understanding of such effects we need to study the geometric and electronic structures of the clusters themselves and their interactions with their supports. Inelastic light scattering, namely Raman- and Brillouin-scattering, can provide a great deal of such information.

Presently, our spectroscopic knowledge of metallic particles and clusters is still limited and either confined to the smallest particles[1] or to collective phenomena of the very big ones.[2] Here I will review Raman scattering from small metal particles isolated in rare gas matrices and Raman and Brillouin scattering from larger particles deposited on a solid surface. Prospects of Raman scattering experiments in free cluster beams will briefly be discussed.

Since clusters are always produced with a certain size distribution, it is difficult to identify the carrier of an observed vibrational mode. Therefore, the correlation with other spectroscopic data due to UV-, VIS-, and IR-absorption, fluorescent emission, magnetic circular dichroism, and electron spin resonance is of fundamental importance. For free particles the most important correlation is of course with mass spectrometry.

CLUSTERS IN MATRICES

The particle density in a free jet expansion is too small for conventional Raman techniques. Therefore, the particles are accumulated and stabilized in matrices, usually rare gas matrices, where they suffer the least perturbations as compared to the gas phase. Matrices are produced by codeposition of a sufficient flux of vaporized metal together with the rare gas on a cooled substrate at temperatures typically between 4 and 25 K depending on the rare gas used. The metal content is usually in the range of 0.5-2%. The clustering occurs during matrix production by diffusion of the metal atoms in the condensation zone and can be further enhanced by thermal annealing or photoaggregation.

In cases where the internal vibrational frequencies of the particles strongly exceed the phonon frequencies of the matrix itself, the observed molecular frequencies will agree with the gas-phase values within a few percent. Low energetic vibrations like bending modes or vibrations of weakly bound species like van der Waals clusters may be more strongly perturbed, depending on the compressing forces exerted by the surrounding matrix cage, and cannot be considered representative of the isolated gas phase species. For example, the van der Waals dimer Cd_2 shows a vibration at 58 cm^{-1} in a krypton matrix,[3] whereas the corresponding gas-phase value is reported to be 22 cm^{-1}.[4]

The occupation of various trapping sites in the same matrix[5] may give rise to a multiplication of observed spectral features. Matrix effects may lift degeneracies of modes. For particles with several low lying electronic states the matrix may even change the electronic ground state, thus giving rise to isomerization processes.[6,7] Diverse isomers may get stabilized in different environments. In addition, external modes may appear in the matrix spectra.[5] These probe directly the interaction of the particle with the surrounding matrix cage and provide information about the local geometry. Thus while matrices perturb and multiply the vibrational modes of the pure clusters, they also provide us with the most simple systems for fundamental studies of cluster-support interactions.

In conventional Raman scattering experiments from matrices one attempts to deduce assignments to a specific particle size mainly from concentration dependences.[8-10] Such techniques are usually limited to dimers or trimers and will reliably work for selected cases only. The multitude of lines generally remains uninterpretable and is not even correlated with other spectroscopic features. Weak unknown resonance effects may further increase the confusion. Nevertheless, it has been possible to identify a complete set of vibrational modes for Sb_4, indicating a tetrahedral geometry for this cluster.[10] In general, however, only the correlation of different experimental data can finally lead to an unambiguous identification of the carrier of an observed transition. In Raman spectroscopy such a correlation may be obtained from resonance phenomena.[11,12]

In resonance Raman scattering the exciting laser is tuned to an absorption band of the particle studied. Certain vibrational modes of this particle are then strongly enhanced, normally the totally symmetric ones. The correlation with the particular absorption band, thereby drastically increases the possibilities of identification. Since resonance Raman scattering is associated with a single absorption band it gives information also on the resonant electronic state and its coupling to vibrations. This is in contrast to conventional Raman scattering where the signal intensity is associated with a sum over products of transition moments of _all_ molecular electronic states.[11] Overtones are often seen in resonance Raman spectra. Therefore, with favorable conditions anharmonicities can be determined and bond dissociation energies estimated. The isotopic fine structure of higher harmonics, which can be better resolved, may allow the identification of the carrier and simultaneously provide structural information on this species.[12,13] Resonance Raman scattering within electronically excited states[12,14] and electronic resonance Raman scattering[12,15] have been reported for small metal particles.

Matrix induced perturbations have been studied in detail by resonance Raman scattering from Ag_2[5,16-18] and Ag_3.[7,17,19] It is found experimentally that dimers can occupy one, two, and three different trapping sites in xenon, krypton, and argon matrices, respectively. As an example, Fig. la shows the resonance Raman spectrum of Ag_2 in krypton matrices obtained at 20 K with 406.74 nm excitation of a krypton ion laser. The spectrum is correlated with both trapping sites of krypton matrices simultaneously, because the two associated Ag_2 absorption bands overlap strongly.[16,20] The absorption peaks are separated by less than 10 nm. The Raman lines at 194 and 203 cm^{-1} correspond to the internal stretching vibrations of the Ag_2 in the two trapping sites, respectively. The line at 203 cm^{-1} forms a combination band at 260 cm^{-1} with the low energy excitation at 57 cm^{-1}. Therefore, all three lines belong to the same trapping site. Similarly, the line at 194 cm^{-1} forms a combination band with the broad structure centered at 40 cm^{-1}. An overtone of the line at 194 cm^{-1} is seen at 387 cm^{-1}. The correlation of the lines as described can be further verified after the matrix is annealed at 50 K for a short time and then cooled to 20 K again (Fig. 1b). One observes a simultaneous increase of the lines at 57, 203, and 260 cm^{-1} with respect to the broad structures at 40 and 234 cm^{-1} and the line at 194 cm^{-1}.

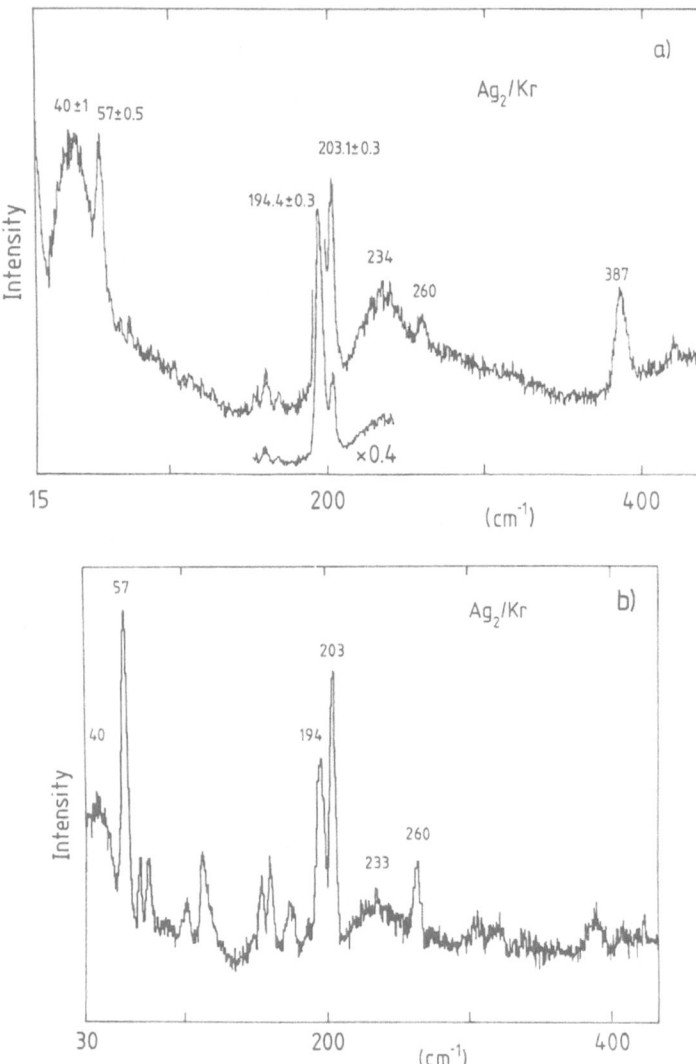

Fig. 1. Resonance Raman spectra of Ag_2 in krypton matrices demonstrating the presence of two trapping sites and correlation of modes as described in the text. a) freshly prepared matrix. b) after a short annealing at 50 K. The new modes coming up are due to larger clusters. (The small peak at 418 cm^{-1} is due to the sapphire substrate).[16,17]

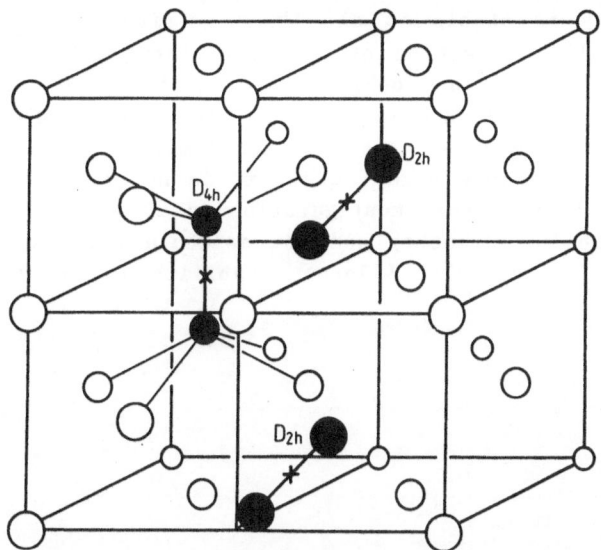

Fig. 2.   Geometries of realized trapping sites of Ag_2 in fcc-rare
gas host lattices (see text). The monovacancy site with
D_{4h} symmetry is shown on the left. The upper right side
shows the monovacancy site with D_{2h} symmetry; the lower
right side shows the double vacancy site.

Since the low energy excitations at 40 and 57 cm^{-1} form combination bands
with the respective Ag_2 stretching vibrations they must be attributed to
the dimers, too. They have to be assigned to external modes of the molecule because the dimer itself has a single internal frequency only.
Therefore, they directly probe the interaction with the host lattice at
the corresponding trapping sites and provide information on the local
environments of the molecules. The line at 57 cm^{-1} lies above the maximum phonon energy (50 cm^{-1})[21] of crystalline krypton. It is, therefore,
sharp and localized. The external mode of the other trapping site is
within the range of host lattice phonons. Therefore, it couples to a
multitude of lattice vibrations which causes the broad resonant structure
observed. Respective spectra were obtained for the various trapping
sites in the other matrices.[5,16,17] The experimental data have been
quite precisely reproduced by a computer simulation.[18] The calculations
were done without any adjustable parameters, taking reasonably chosen
interaction potentials as the only input. The results are summarized together with experimental values in Ref. 18. (The calculated frequency
distributions of Ag_2 in krypton matrices are shown for comparison in
Figs. 3 and 4 of that paper.) As the calculations demonstrate only the
three geometries shown in Fig. 2 form (meta)stable dimer configurations
in the three fcc-rare gas host lattices. The monovacancy configuration

with D_{4h} symmetry is stable in all three matrices. The molecule is aligned in the <100> direction of the host lattice and centered with respect to the vacancy. The double vacancy configuration with D_{2h} symmetry and the molecule aligned in the <110> direction is the most stable one for krypton and argon matrices. In xenon matrices it is metastable, but due to its unfavorably high energy, it is not formed in measurable quantities. The third stable configuration in argon is the monovacancy configuration with D_{2h} symmetry and the molecule aligned in the <110> direction. From the calculated larger stability of the double vacancy

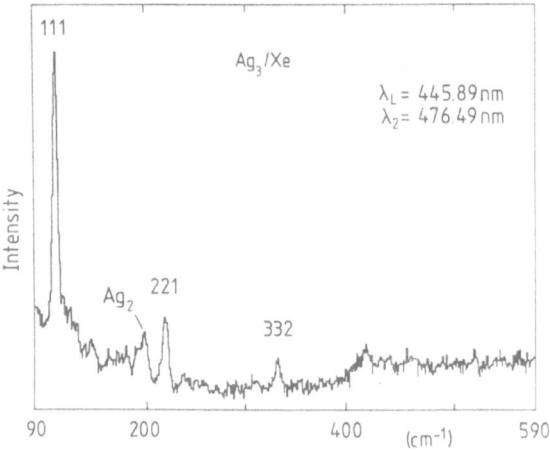

Fig. 3. Dual-beam resonance Raman spectrum of the silver trimer with the 442 nm absorption band in solid xenon (λ_L: wavelength of exciting laser; λ_2: wavelength of auxiliary laser). The nonresonant line of Ag_2 at 198 cm^{-1} is also seen.[7]

configurations in krypton and argon matrices one is tempted to assume that the abundances of dimers with these configurations increase on thermal annealing of the matrix with respect to those which populate the monovacancy sites with D_{4h} symmetry. In the experiment the opposite is observed.[16] The reason lies in the heterogenous composition of the matrices and in the large atomic radii of the rare gas atoms. Due to these large radii the double vacancy sites can accept a third silver atom so that a trimer can be formed. During the annealing process silver atoms start to diffuse so that dimers in double vacancy sites are preferentially transformed to trimers which drastically reduces the relative abundances of the respective dimers.

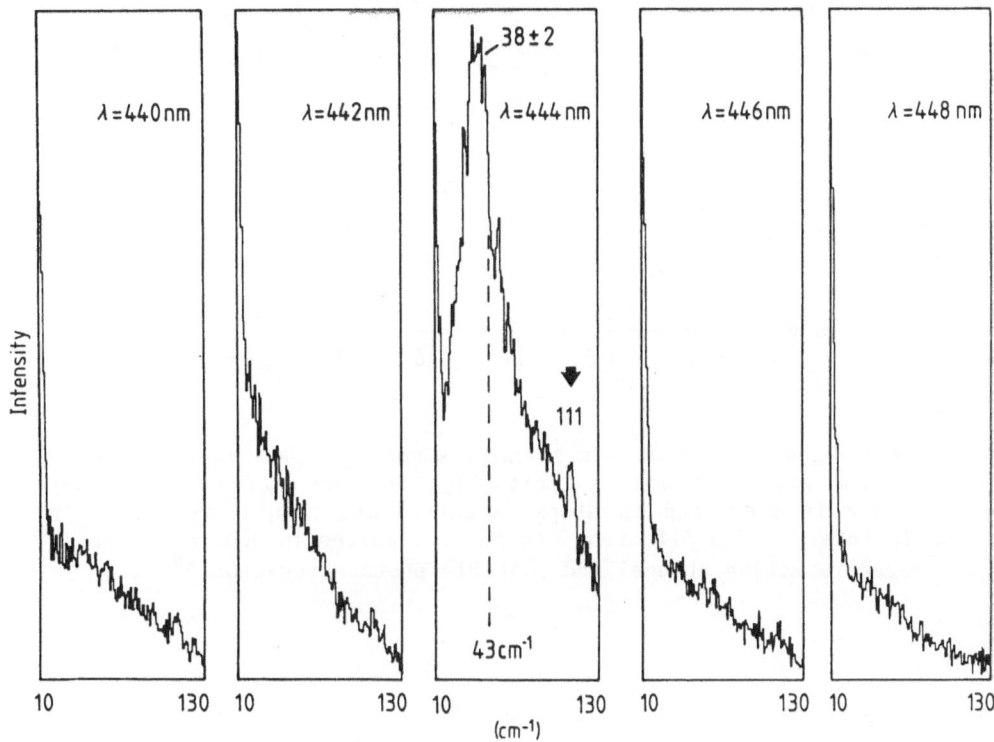

Fig. 4.   Dependence of the strong resonant mode at 38 cm^{-1} on the excitation wavelength.   The maximum phonon energy of the host lattice is 43 cm^{-1}.   The arrow marks the internal mode of the trimer.[19]

In freshly prepared xenon matrices with silver concentrations of about 2% two absorption bands are observed at 425 and 442 nm which were attributed to Ag$_3$ from concentration dependent measurements.[6,22]   Their strengths can be enhanced by an annealing process.   The corresponding lines in krypton matrices are centered at 417 and 423 nm and show a considerable overlap, so that an asymmetric structure results.[7,17]   The respective trimer bands in argon matrices overlap unfortunately with the dimer absorption bands so that they cannot be identified unambiguously.[23] Therefore, the following considerations are confined to xenon and krypton matrices.   The Ag$_3$ species are particularly photosensitive.   During irradiation into one of the absorption bands this band will disappear while a third absorption band is created at 475 nm in xenon matrices and at 448 nm in krypton matrices, respectively.[6,7,17]   Irradiation at the wavelengths of the new lines almost re-establishes the original spectra. Due to this reversibility the phototransformations can be cycled several times.   Ozin et al.[6] first suggested that these phenomena are photoisomerization processes of the Ag$_3$/matrix cage units.

Since the phototransformations were too fast for conventional resonance Raman experiments, the reversibility of the processes was utilized to get resonance Raman spectra in dual beam experiments.[7,17]   The overall light intensity was kept below 170 mW/cm^2 during the measurements to avoid overheating of the matrices and subsequent aggregation.   Figure 3 shows the resonance Raman spectrum of the species with the 442 nm

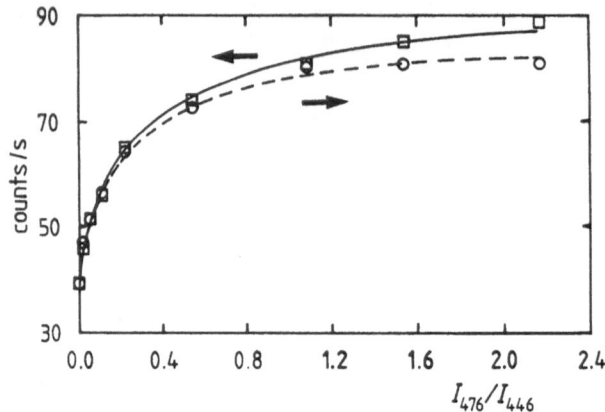

Fig. 5. Dependence of the 111 cm^{-1} Raman signal on the intensity ratio of the two laser beams. Initially, the power of the auxiliary laser is decreased in steps (squares) and then increased again (circles). The difference in the two curves is due to the third transformation channel and possible photoaggregation.[19]

absorption band in xenon. The exciting dye laser was tuned to ~446 nm while the auxiliary argon laser was operated on the 476 nm line to drive the back transformation process and stabilize the concentration of the species studied. A fundamental vibration at 111 cm^{-1} and two overtones can be identified. The nonresonant mode at 198 cm^{-1} of dimers which are present in much higher concentration can also be seen. When the exciting dye laser is tuned across the trimer absorption band at 442 nm one observes in a very narrow spectral range of excitation wavelengths a broad resonant structure in the Raman spectra with the trimer line at 111 cm^{-1} superimposed at the slope (Fig. 4). Since the peak of this broad structure lies below the maximum energy of the host lattice phonons (43 cm^{-1})[21] it is interpreted as an external mode of the trimer in analogy with the observations for dimers. All attempts to obtain Raman spectra corresponding to the other two absorption bands at 425 and 475 nm in xenon matrices failed, so far. A photokinetic study[19] was undertaken to explain this phenomenon. The dual beam stabilized stationary concentration of the trimers with the 442 nm absorption band depends on the intensity ratio of the two laser beams. Therefore the intensity of the 111 cm^{-1} Raman line was measured as a function of this ratio (Fig. 5). To reduce laser induced aggregation the overall irradiance was kept at a low level of 54 mW/cm^2 during these measurements. The power of the blue dye laser was fixed. Starting at a ratio of 2.2 the intensity of the

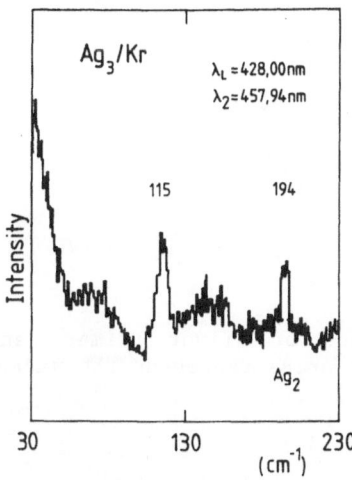

Fig. 6. Preresonant Raman spectrum of Ag_3/Kr obtained with the dual beam technique. The line at 115 cm^{-1} is due to the trimer with the 423 nm absorption band. The line at 194 cm^{-1} is due to a dimer.[17]

Raman signal decreases as the power of the auxiliary laser decreases and increases again with increasing power. However, the initial Raman intensity cannot be reached due to the third transformation channel and possible aggregation to larger clusters. The experimental data were fit to a simplified rate equation[19] and it was found that the rate constant for generation of the 442 nm species is at least 3 times larger than that of the back transformation. This explains why a Raman spectrum of the 475 nm species could not yet be detected. The dual beam technique was applied to the trimers in krypton matrices, too.[17] Only a preresonant spectrum could be obtained (Fig. 6) because the wavelength of the auxiliary argon laser at 458 nm is only at the edge of the 448 nm absorption band of the photoinduced species. A trimer mode at 115 cm^{-1} and the dimer mode at 194 cm^{-1} can be identified. The mode at 115 cm^{-1} is not identical with the Raman line at 120 cm^{-1} reported by Schulze et al.[24] The latter line was observed, too,[17,23,25] but only at excitation with

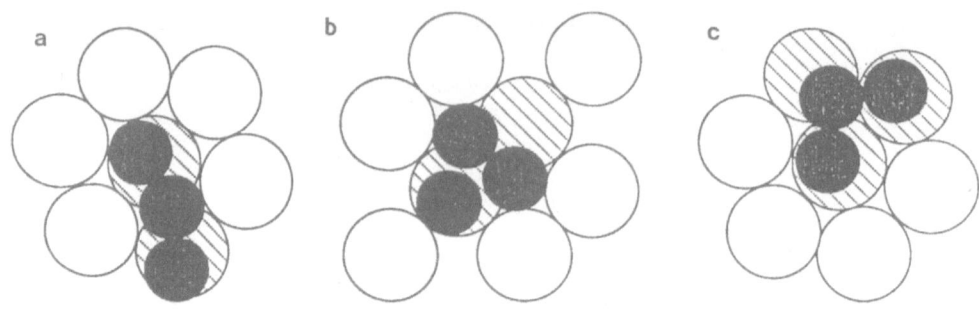

Fig. 7. Proposed geometries of silver trimers and associated matrix sites. The shaded areas represent the vacancies.

the 514.5 nm line of the argon laser. Therefore, it is believed that this line is rather due to a preresonant Raman excitation associated with the 510 or 524 nm absorption bands in krypton matrices. So far, the experimental data do not allow a final decision whether a true photoisomerization of the Ag_3 takes place. There is, however, some evidence mainly from theoretical data and also from ESR measurements in favor of this interpretation.

The D_{3h} equilateral triangular geometry of Ag_3 is unstable due to the Jahn-Teller effect.[26] The symmetric degenerate $^2E'$ electronic ground state will split by a distortion to C_{2v} symmetry into $^2A_1$ and $^2B_2$ states with configurations $6a_1^2 7a_1^1$ and $6a_1^2 5b_2^1$, respectively. Ab initio calculations[26-28] show that the $^2B_2$ state (apex angle >60°) is lower in energy. At apex angles below 60° the $^2A_1$ state becomes the ground state. Basch[26] calculated its potential minimum to be about 0.14 eV above the $^2B_2$ potential minimum. In a more recent calculation taking more configurations into account and allowing for more free parameters Andreoni and Upton[28] showed that the separation is even smaller, ~30 meV, not much exceeding the corresponding value of Cu_3.[29] Such an energy difference can be supplied by the matrix cage, so that the matrix can invert the sequence of the $^2B_2$ and $^2A_1$ states. This was verified in ESR measurements which show a change of the Ag_3 electronic ground state from $^2B_2$ in a C_6D_6 matrix to $^2A_1$ in a N_2 matrix.[30,31] The ESR data prove that different isomers can get stabilized in different environments. It seems that after photoexcitation of Ag_3 the cluster/cage unit can relax to a new geometry which is associated with an isomerization of the molecule. Moreover, matrix effects may even further decrease the energy separation of the $^2A_1$ and $^2B_2$ potential minima. In such a case a pseudorotation of the cluster may result as a consequence of the dynamic Jahn-Teller effect, as was found recently for Cu_3.[32-34]

At this point it might be interesting to suggest morphologies for silver trimers in their trapping sites. If for the silver atomic diameter we take the mean value between the atomic separation of the dimer and that of bulk silver we get the proportions for Ag_3 in krypton matrices as shown in Fig. 7. A more or less linear trimer may occupy a double vacancy site in the krypton matrix (Fig. 7a). This picture is in accordance with what was said above about the behavior of dimers on annealing of the matrix. A strictly linear configuration is unlikely due to Andreoni and Upton[28] because of an energy increase of 100 meV with respect to the $^2B_2$ potential minimum. One might expect that this is the most stable configuration, some relaxation of the lattice included. One metastable site might have the geometry of Fig. 7b with a $^2A_1$ ground state of the cluster. The molecule is located in a divacancy configuration, too, but the plane of the molecule now is in the (100) plane of the lattice. Two atoms are clamped in a position perpendicular to the axis of the double vacancy. These atoms hinder themselves from relaxation to the geometry of Fig. 7a. Some lattice relaxation will also be necessary due to the overlap with lattice atoms in the planes below and above the plane of the figure. The third configuration might be that of a trimer in a trivacancy in the (111) plane (Fig. 7c). Due to the larger size of the vacancy this configuration is expected to be metastable, too. Perhaps as in the case of Cu_3 a pseudorotating form of Ag_3 may occur, although no experimental evidence has, so far, been found.

CLUSTERS ON SURFACES

Clusters on surfaces are of general importance, because they are widely used systems in catalysis. They are of special interest to light scattering, because they seem to play a prominent role in surface enhanced Raman scattering (SERS).[35] They are believed to be responsible for the "chemical enhancement". Co-adsorbed molecules may show enhanced scattering under conditions of resonance for the clusters. These are, however, extremely difficult to identify because the substrate consists of the same material as the clusters. It was proposed recently that silver tetramers, possibly positively charged, are a dominant species on SERS-active silver electrodes.[36] They were said to be responsible for some low energy excitations observed in SERS-experiments. These results are compared to a matrix isolation study in Ref. 25.

Despite their importance, few other experiments are reported on clusters on surfaces or cluster-substrate interactions. One considers soot-like deposits of germanium smoke particles on glass plates.[37] By Raman scattering and electron diffraction it was found that germanium particles of ~80 Å diameter have an amorphous structure, whereas particles with > 180 Å diameter have the diamond structure. The bigger particles show bulk-like Raman spectra with size dependent shifts and broadening of spectral lines.

Another experiment treats Brillouin scattering from Au particles deposited on NaCl,[38] a simple and easy to handle model system. Brillouin scattering was chosen as the spectroscopic tool because the clusters are bigger in this case and, therefore, the expected frequencies smaller. Gold clusters produced by vacuum deposition on freshly cleaved or annealed alkali halide surfaces have long been used by crystallographers to visualize slip and cleavage steps, point defects, and screw dislocations by electron microscopy. More recently nucleation phenomena have been studied and other metals and substrates have also been employed.[39] The clusters can be produced with varying density and sizes between 10 Å and several hundred Å depending on evaporation time and substrate temperature. However, with favorable preparation conditions individual samples show a very narrow size distribution. This property makes them attractive for spectroscopic studies in photoemission, Auger, optical absorption, and light scattering experiments. The particles can be

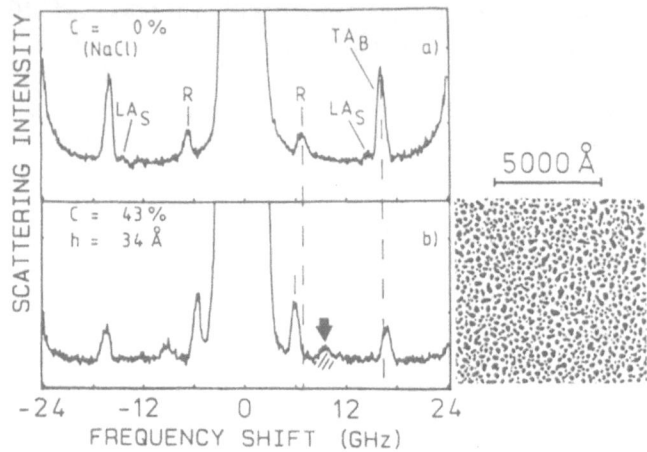

Fig. 8. Brillouin spectra of pure NaCl and of gold clusters on
NaCl. An electron micrograph of the sample is shown on
the right. The scattering intensities are normalized
with respect to a constant photon flux. The angle of
incidence of the laser beam is 45°. C denotes the degree
of coverage of the NaCl surface by the gold layer as
obtained by Rutherford ion backscattering. The arrow
marks the new "cluster mode". The other modes are: R:
Rayleigh mode, LA_s: longitudinal acoustic surface mode
(mixed mode), TA_B: transverse acoustic bulk mode. The
increasing scattering intensity at 24 GHz is due to the
onset of the longitudinal acoustic bulk mode. Its
intensity is 2-3 orders of magnitude larger than that of
the TA_B phonon which is symmetry forbidden in the backscattering geometry used.[38]

characterized by several means: electron microscopy allows to measure
the coverage and lateral size distribution. The mass deposited per unit
area is measured by Rutherford ion backscattering.[40] Both techniques
together provide a mean particle volume. X-ray and electron diffraction
techniques allow one to determine the orientation of the particles with
respect to the substrate. It was found that cube-octahedral gold clusters can orient in 17 different ways with respect to the NaCl substrate,
having either the (100), (111) or (115) planes parallel to the substrate
surface.[41]

Brillouin spectra of gold clusters on NaCl (Fig. 8) obtained with a
tandem Fabry-Perot spectrometer show a new mode besides transverse and
longitudinal bulk modes and the Rayleigh surface mode of the substrate.
The frequency of this new mode depends on the coverage but also on the
wave vector component of the incident laser beam parallel to the substrate surface. Therefore, the new mode is not simply an eigenmode of
the clusters involved, but also probes the cluster-substrate interaction.

From the coverage dependence of the mode frequency the percolation threshold can be determined.[38] As the amount of deposited gold increases the clusters grow together and finally form a closed gold film. Simultaneously, the new mode grows into a first order Sezawa mode. The peristaltic character of this surface acoustic mode indicates that internal vibrations of the clusters are involved in the cluster regime. The new mode can most probably be described by the coupling of internal cluster modes to a surface acoustic distortion.

It has been shown by analysis of the radial distribution of gold clusters on alkali halides, that repulsive elastic forces exist between pairs of short distant clusters.[42] The repulsive interaction originates from the overlap of strain fields which are induced in the substrate because of the appreciable misfit between Au and NaCl lattice parameters. It seems that Brillouin scattering provides an alternative probe of these interactions although the detailed nature of the new mode still has to be evaluated.

CLUSTERS IN A BEAM

Light scattering experiments on cluster beams rely on nonlinear coherent Raman scattering mechanisms, because these provide up to 5 orders of magnitude larger scattering intensities. Only van der Waals and molecular dimers and clusters have, so far, been studied in a beam.[43-47] Structural information on molecular clusters could be obtained.[45,46] The ion-dip technique[47] seems to be particularly promising, because it is capable of generating Raman signals at a particle density of $10^{11}/cm^3$, a value that can be reached in a metal cluster beam under favorable conditions.[34] A particular advantage of the technique is also the correlation with mass-spectrometry.

CONCLUSION

Recent light scattering experiments on metal particles and clusters have been reviewed. It was shown that particle support interactions are important for clusters in matrices and on surfaces. For metal dimers in rare gas matrices the experimental results can be quantitatively interpreted by computer simulations. This is encouraging and might put matrix spectroscopy on a more quantitative basis. Such investigations can be extended to larger clusters once reliable potential energy surfaces are available. Actually, what is needed are the energy surfaces in the vicinity of the local minima. Matrix spectra might be better interpreted. Perhaps even more important, matrix spectroscopy may be developed to reliably test ab initio calculated energy surfaces, even for isomers that might get stabilized in the matrices, but do not exist in the gas phase.

Light scattering experiments from surface deposited clusters should be given more attention, because of the general importance of these systems.

Light scattering experiments on free jet expansions will become increasingly important in the near future, although at the large light intensities needed predissociation and fragmentation phenomena due to multiphoton processes might cause some complications.

Acknowledgments: It is a pleasure to thank M. Campagna for his continuous support and U. L. Kettler, H. R. Schober, W. Krasser, and N. Herres in Jülich and B. Hillebrands, R. Mock, and G. Güntherodt at the "Universität zu Köln" for their collaboration. I thank the Chemistry Division of Argonne National Laboratory and in particular S. J. Riley, E. K. Parks, K. Liu, J. Jellinek, L. G. Pobo, W. F. Hoffman, and S. K. Cole for their generous hospitality.

REFERENCES

*Current address: Chemistry Division, Argonne National Laboratory, Argonne, IL, 60439, USA.

[1] M. D. Morse, Chem. Rev. 86 (6), (1986) (in press).

[2] U. Kreibig and L. Genzel, Surf. Sci. 156, 678 (1985).

[3] A. Givan and A. Loewenschuss, Chem. Phys. Lett. 62, 592 (1979).

[4] A. Kowalski, M. Czajkowski, and W. H. Breckenridge, Chem. Phys. Lett. 119, 368 (1985).

[5] P. S. Bechthold, U. Kettler, and W. Krasser, Solid State Comm. 52, 347 (1984).

[6] G. A. Ozin, H. Huber, and S. A. Mitchell, Inorg. Chem. 18, 2932 (1979).

[7] U. Kettler, P. S. Bechthold, and W. Krasser, Surf. Sci. 156, 867 (1985).

[8] W. Schulze and H. Abe, Faraday Symp. Chem. Soc. 14, 87 (1980).

[9] F. W. Froben and W. Schulze, Surf. Sci. 156, 765 (1985).

[10] H. Sontag and R. Weber, Chem. Phys. 70, 23 (1982).

[11] D. L. Rousseau, J. M. Friedman, and P. F. Williams in: Raman Spectroscopy, A. Weber (Ed.), Springer Verlag, Berlin, Heidelberg, New York (1979).

[12] M. Moskovits and D. P. DiLella, ACS Symp. Ser. 179, 153 (1982).

[13] H. Schnöckel, H. J. Göcke, and R. Elsper, Z. anorg. allg. Chem. 494, 78 (1982); H. Schnöckel, Z. anorg. allg. Chem. 510, 72 (1984).

[14] H. Sontag, B. Eberle, and R. Weber, Chem. Phys. 80, 279 (1983); B. Eberle, H. Sontag, and R. Weber, Chem. Phys. 92, 417 (1985); B. Eberle, H. Sontag, and R. Weber, Surf. Sci. 156, 751 (1985).

[15] M. Moskovits, D. P. DiLella, and D. P. Limm, J. Chem. Phys. 80, 626 (1984).

[16] P. S. Bechthold, U. Kettler, and W. Krasser, Surf. Sci. 156, 875 (1985).

[17] P. S. Bechthold, U. Kettler, H. R. Schober, and W. Krasser, Z. Phys. D3, 263 (1986).

[18] P. S. Bechthold and H. R. Schober (this conference).

[19] U. Kettler, P. S. Bechthold, and W. Krasser, Chem. Phys. Lett. (in press).

[20] S. A. Mitchell, G. A. Kenney-Wallace, and G. A. Ozin, J. Am. Chem. Soc. 103, 6030 (1981).

[21] Y. Endoh, G. Shirane, and J. Skalya, Phys. Rev. B 11, 1681 (1975).

[22] W. Schulze, H.-U. Becker, and H. Abe, Chem. Phys. 35, 177 (1978).

[23] U. Kettler and Ph.D. Thesis, Universität zu Köln, (1984), (Berichte der Kernforschungsanlage Jülich No 1980 (1985)).

[24] W. Schulze, H.-U. Becker, R. Minkwitz, and K. Manzel, Chem. Phys. Lett. 55, 59 (1978).

[25] U. L. Kettler, P. S. Bechthold, and W. Krasser, (this conference).

[26] H. Basch, J. Am. Chem. Soc. 103, 4657 (1981).

[27] J. Flad, G. Igel-Mann, H. Preuss, and H. Stoll, Chem. Phys. 90, 257 (1984).

[28] W. Andreoni and T. H. Upton (unpublished).

[29] S. P. Walch and B. C. Laskowski, J. Chem. Phys. 84, 2734 (1986).

[30] J. A. Howard, K. F. Preston, and B. Mile, J. Am. Chem. Soc. 103, 6226 (1981).

[31] K. Kernisant, G. A. Thompson, and D. M. Lindsay, J. Chem. Phys. _82_, 4739 (1985).

[32] M. Moskovits, Chem. Phys. Lett. _118_, 111 (1985).

[33] M. D. Morse, J. B. Hopkins, P. R. Langridge-Smith, and R. E. Smalley, J. Chem. Phys. _79_, 5316 (1983).

[34] E. A. Rohlfing and J. J. Valentini, Chem. Phys. Lett. _126_, 113 (1986).

[35] A. Otto, in: Light scattering in solids, Vol. 4, M. Cardona and G. Güntherodt (Eds.), Springer Verlag, Berlin, Heidelberg, New York (1984).

[36] D. Roy and T. E. Furtak, Chem. Phys. Lett. _124_, 299 (1986).

[37] S. Hayashi, M. Ito, and H. Kanamori, Solid State Comm. _44_, 75 (1982)

[38] B. Hillebrands, R. Mock, G. Güntherodt, P. S. Bechthold, and N. Herres, to be published.

[39] K. Reichelt and B. Lampert, Vaccuum _30_, 383 (1980).

[40] K. Reichelt, B. Lampert, and H.-P. Siegers, Surf. Sci., _93_ 159 (1980).

[41] N. Herres, Ph.D. Thesis, Rheinisch Westfälische Technische Hochschule Aachen (1986).

[42] J. C. Zanghi, J. J. Metois, and R. Kern, Surf. Sci. _52_, 556 (1975), erratum _55_, 761 (1976).

[43] F. König, P. Oesterlin, and R. L. Byer, Chem. Phys. Lett., _88_, 477 (1982).

[44] H. P. Godfried and I. F. Silvera, Phys. Rev. Lett., _48_, 1337 (1982).

[45] G. A. Pubanz, M. Maroncelli, and J. W. Nibler, Chem. Phys. Lett. _120_, 313 (1985).

[46] M. Maroncelli, G. A. Hopkins, J. W. Nibler, and T. R. Dyke, J. Chem. Phys. _83_, 2129 (1985).

[47] W. Bronner, P. Oesterlin, and M. Schellhorn, Appl. Phys. B 34, 11 (1984).

OPTICAL AND INFRARED PROPERTIES OF SMALL METAL PARTICLES

D.G. Stroud and P.M. Hui

Department of Physics
The Ohio State University
Columbus, Ohio 43210

The optical and infrared properties of small metal particles are
reviewed. Among the properties discussed are surface plasmon resonances
in small particles, electric and magnetic dipole absorption in the far
infrared region, changes in small particle optical properties between the
normal and superconducting state, and the influence of clustering on the
a.c. properties of small particles. Illustrations will be drawn from
experiments, and from calculations by the authors and by other workers.

INTRODUCTION

This paper will emphasize "clusters" somewhat larger than those

which are the principal focus of this Workshop – metal particles of

linear dimensions of order 30 Å or more. The properties of such parti-

cles are difficult to treat by focusing on individual atoms and calcu-

lating, for example, the individual electronic eigenstates – there are

too many of them. Instead, a more natural starting-point is to imagine

the particles as small chunks of bulk material, with intensive proper-

ties, such as conductivity or dielectric function, approximating those of

the bulk. Of course, this starting point must fail at some level –

surface scattering, for example, or the effects of disorder which inevi-

tably appears in small-particle systems. will certainly affect the bulk

properties in sufficiently small particles. But these may well be

effects of secondary importance for particles of the size considered.

We will be concerned with the infrared and optical properties of

small metal particles. These properties are determined by the complex

frequency-dependent dielectric function, $\epsilon(\omega)$, of the small metal par-

ticle. We will consider, in particular, an anomaly first discovered by

Tanner, Sievers, and Buhrman (1) about ten years ago – the enormous

enhancement of far-infrared absorption by small metal particles. It will

be shown that this enhancement need no longer be considered anomalous, but can instead by attributed to a variety of extrinsic factors, such as clustering of small metal particles.

QUASISTATIC APPROXIMATION

The simplest classical approach to electromagnetic absorption by small metal particles is the quasistatic approximation. In this approach, one imagines a small metal particle (assumed for convenience here to be spherical), centered at the origin, and of radius a, with complex frequency-dependent dielectric function $\varepsilon_m(\omega)$, immersed in a homogeneous host medium of dielectric constant ε_i, which we choose here to equal unity, and subjected to a plane incident electromagnetic wave of electric field

$$\vec{E}_1\,(\vec{x},t) = [\text{Re }\vec{E}_{inc}\,(\vec{x},t)]$$

$$\vec{E}_{inc} = \vec{E}_0 \cdot \exp(i\vec{k}\cdot\vec{x}-i\omega t) \tag{1}$$

where $k = \sqrt{\varepsilon_i}\ \omega/c = \omega/c$. If the particle radius satisfies ka<<1, then $\vec{E}_{inc} \sim \text{Re}[\vec{E}o\ \exp(-i\omega t)]$ and the field inside the small particle is uniform and given by

$$\vec{E}_2 = \text{Re }\vec{E}_{in}$$

$$\vec{E}_{in} = [3\vec{E}_0/(\varepsilon_m+2)]\exp(-i\omega t). \tag{2}$$

The factor (ε_m+2) comes from the spherical shape of the particle; other ellipsoidal shapes still lead to a uniform field within the particle but a different denominator.

The time-averaged power absorbed by the small particle is

$$<P>_{av} = (4\pi/3)a\ \cdot(1/2)\text{Re}(\vec{J}\cdot\vec{E}_{in}^*) \tag{3}$$

where $\vec{J}$ is the current density in the particle, given by

$$\vec{J} = \sigma\vec{E}_{in}$$

$$\sigma = \omega[\varepsilon_m(\omega)-1]/(4\pi i). \tag{4}$$

It is useful to convert this expression to an <u>absorption coefficient</u> α_e in the quasistatic approximation, defined as

$$\alpha_e = \frac{\text{power absorbed per unit volume}}{\text{incident power per unit area}} \tag{5}$$

Since the incident power density is $c|E_0|^2/(8\pi)$, we find

$$\alpha_e = (9\omega f/c)\,\text{Im}\{[\varepsilon_m(\omega)-1]/|\varepsilon_m(\omega)+2|^2\}. \tag{6}$$

548

This expression is valid for a suspension of spheres of radius a, dielectric function $\varepsilon_m(\omega)$, embedded in a matrix of dielectric constant unity, such that the total volume fraction occupied by the spheres is f. The derivation clearly assumes that there are no interactions between the spheres, i.e. that the volume fraction of spheres is small, $f \ll 1$. Note that in this quasistatic approximation the sphere radius drops out (though the shape still plays a role).

To illustrate the predictions of this simple model, we apply it to a Drude metal with dielectric function

$$\varepsilon_m(\omega) = 1 - \omega_p^2 / [\omega(\omega + i/\tau)] \tag{7}$$

Here ω_p is the plasma frequency of the metal, and τ is a characteristic relaxation time. For most metals in bulk at room temperature, $\omega_p \tau$ is of order 100. In small particles τ may be appreciably reduced by surface scattering, which may even dominate the scattering processes for sufficiently small particles.

The dielectric function (7) leads to two characteristic electromagnetic responses in small metal particles. The first of these is the so-called surface plasmon peak. This occurs, according to eq. (6), when the denominator, $|\varepsilon_m + 2|^2$, is very small. For a Drude metal in the limit $\omega_p \tau \gg 1$, the resonance occurs when $1 - \omega_p^2/\omega^2 \sim -2$ or

$$\omega \sim \omega_p/\sqrt{3} \tag{8}$$

which is the surface-plasmon frequency for small spherical metal particles. The corresponding coefficient of absorption, from eq. (6), is very large near this frequency. The absorption represents the resonant transfer of energy from the incident electromagnetic wave to a natural oscillation of the electron gas in the small metal particle at frequency $\omega_p/\sqrt{3}$. As is well known, small metal particles have numerous other oscillations [the electric ℓ-pole oscillation, for example, has a frequency $\omega = \omega_p/\sqrt{2\ell+1}$. However, only the lowest mode has a net electric dipole moment and can be excited by a uniform incident electric field.

The second phenomenon arising in small particles of Drude metal is a far infrared absorption varying as the square of the frequency. This absorption can readily be calculated in the quasistatic approximation by substituting the Drude form (7) into (6). In the low-frequency limit ($\omega\tau \ll 1$), one finds

$$\alpha_e = 9\omega^2 f/[4\pi\sigma c) = C_e \omega^2 f \tag{9}$$

where

$$\sigma = \omega_p^2 \tau/4\pi \tag{10}$$

is the static conductivity. As will be illustrated later, this absorption exhibits the same dependence on frequency and on filling fraction f, as experiment, but the experimental behavior,

$$\alpha_{expt} = C_{expt}\omega^2 f, \tag{11}$$

is much stronger than predicted by the quasistatic approximation: $C_{expt} \gg C_e$. The discrepancy is of the order of $10^4 - 10^6$ or even more.

Over a period of several years, numerous possible explanations of this excess absorption have been examined (2-6). Among these possibilities are

- Magnetic dipole absorption;
- Oxide coatings;
- Clustering;
- Quantum size effects.

In the remainder of this paper, we will discuss each of these in turn.

BEYOND THE QUASISTATIC APPROXIMATION: MAGNETIC DIPOLE ABSORPTION

The extinction coefficient for a sphere of dielectric function $\varepsilon_m(\omega)$ can be calculated exactly, not just in the long-wavelength limit. The problem was first solved by Mie in 1903, by expanding the incoming and outgoing radiation in a partial wave series. The result for the total extinction coefficient α_{tot} is

$$\alpha_{tot} = (4\pi/k^2) \text{ Re } S(0) \tag{12}$$

$$S(0) = (1/2) \sum_\ell (a_\ell + b_\ell) \tag{13}$$

where $S(0)$ is the forward scattering amplitude and a_ℓ and b_ℓ are the electric and magnetic multipole coefficients; they are given by

$$a_\ell = \frac{\psi'_\ell(y)\psi_\ell(x) - \mu\psi_\ell(y)\psi'_\ell(x)}{\psi'_\ell(y)\delta_\ell(x) - \mu\psi_\ell(y)\delta'_\ell(x)} \; ; \tag{14}$$

$$b_\ell = \frac{\mu\psi'_\ell(y)\psi_\ell(x) - \psi_\ell(y)\psi'_\ell(x)}{\mu\psi'_\ell(y)\delta_\ell(x) - \psi_\ell(y)\delta'_\ell(x)} \tag{15}$$

with $x = \omega a\sqrt{\varepsilon_m}/c$, $y = \omega a/c$, $\mu = y/x$, $\psi_\ell(x) = xj_\ell(x)$, $\delta_\ell(x) = xh_\ell^{(1)}(x)$, and $j_\ell(x)$ and $h_\ell^{(1)}(x)$ the standard spherical Bessel and Hankel functions.

The coefficient α_{tot} is the sum of losses due to scattering and absorption, but at long wavelengths the scattering losses are negligible. For $ka \ll 1$, there are _two_ comparable contributions, arising from a_1 and b_1. These can be calculated by expanding the spherical Bessel and Hankel functions in powers of ka; the result is (2)

$$\alpha_{tot} = \alpha_e + \alpha_m$$

$$\alpha_e = C_e \omega^2 f$$

$$\alpha_m = C_m \omega^2 f \qquad\qquad (16)$$

$$C_e = 9/(4\pi\sigma c)$$

$$C_m = (2\pi/5)\sigma a^2/c^3.$$

The electric dipole contribution α_e is as given in eq. (9). The magnetic dipole contribution α_m also varies as $\omega^2 f$, but unlike α_e increases with particle radius; this is because the induced eddy currents dissipate more energy in large particles.

For particles of conductivity comparable to Al at room temperature, C_m starts to exceed C_e at particle radius as small as 30 Å. For 100 Å particles, inclusion of the magnetic dipole absorption gives an enhancement of far infrared absorption of order 10^2 over the quasistatic approximation. However, this enhancement is still inadequate to explain the factor of $10^4 - 10^6$ seen in many experiments.

OXIDE COATINGS

Several authors [3,7] have proposed that the anomalous far-infrared enhancement stems, in part, from oxide coatings on the small metal particles. An oxide coating, if present, would indeed enhance the electric the electric dipole contribution to the absorption. α_e varies as the inverse of the metal particle conductivity. If the particle were coated with poorly conducting oxide, the effective conductivity of the particle would be much reduced, greatly increasing the far infrared absorption. This mechanism may be of importance in oxide-coated small metal particles. A representative calculation of this enhancement is shown in Fig. 1, for small particles of Al coated with Al_2O_3.

CLUSTERING

Even if the volume fraction of metal in a composite is small, the metal particles may be clumped in a non-random fashion, to form much larger clusters. It is intuitively plausible that magnetic dipole absorption will be enhanced by such clumping: the presence of large clusters gives an effectively bigger particle radius, and potentially much larger current loops, through which induced eddy currents may pass to give absorption. We will give below similar arguments for enhancement of the electric dipole contribution by clustering.

A variety of experimental evidence suggests that clumping is a major factor in far infrared enhancement. Fig. 2, due to Curtin, Spitzer,

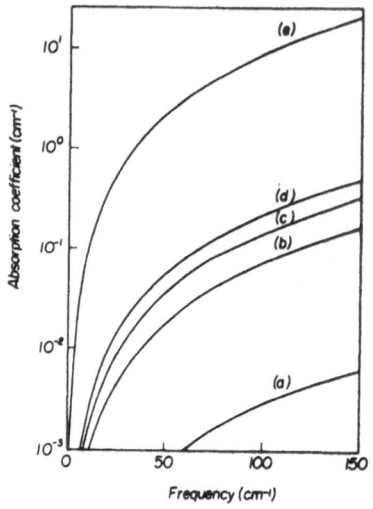

Fig. 1

Absorption coefficient of a composite of isolated
spheres of diameter 50Å and volume fraction 0.015.
(a) Unoxidized Al, (b) 5Å oxide layer, (c) 10Å oxide
layer, (d) 25Å oxide layer, (e) experimental results,
[After ref. (7)].

Ashcroft, and Sievers (8), shows the effects of heat-treating composites
of small metal particles which are oxide coated or oxide free. In the
oxide-free samples, heat treatment melts the weak electrical contacts
between the metal particles in the cluster. This causes the cluster to
have higher conductivity, and hence, according to eq. (9), lower
electric dipole and higher magnetic dipole absorption. The absorption

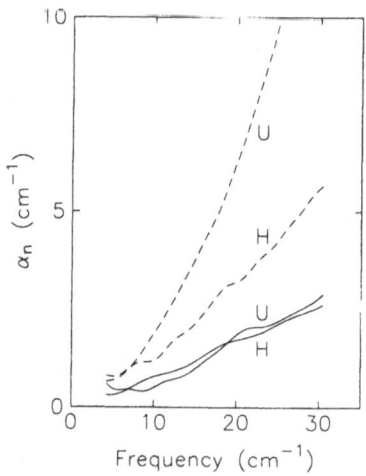

Fig. 2

Experimental measurements of absorption coefficients
for oxide coated (solid lines) and oxide-free (dashed
lines) samples before (U) and after (H) heat-treat-
ment. [After ref. (8)].

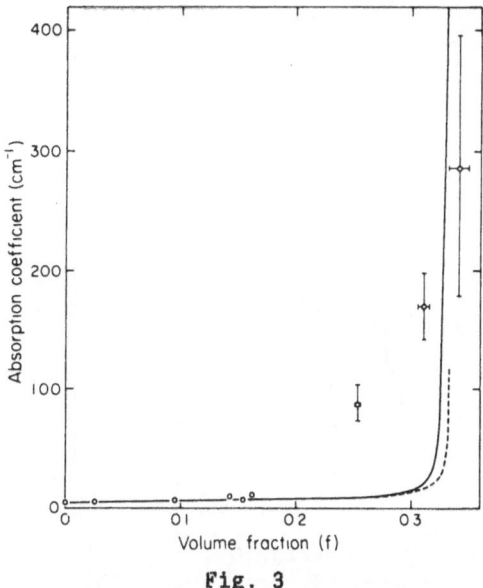

Fig. 3

Absorption coefficient of acetone-dried Ag in gelatin. Circles indicate experimental data. The solid line shows the prediction of the effective medium approximation for 100Å-diameter Drude Ag particles in gelatin. [After ref. (10)].

of the heat-treated cluster drops and its absorption becomes dominated by the magnetic dipole term. The oxice-coated particles show a much smaller change on heat-treating, probably because the conductivity of the barriers shows a much smaller change on melting. Experiments by Lee et al (9) show a similar change between clustered and unclustered metal particles. Curtin and Ashcroft have developed a cluster-percolation model (5) to account for these experiments; in this model, some of the clusters are near the metal percolation threshold, and therefore have the low conductivity to account for the electric dipole dominance of the absorption.

Fig. 3 shows the experimental results of Devaty and Sievers (10) for far infrared absorption by small Ag particles embedded in gelatin, at a frequency of 15 cm^{-1}. Here the clustering is due not to a clumping process driven by some attractive force between the metal particles, but to the purely statistical likelihood of forming larger clusters of metal particles as the volume fraction of metal increases. In a randomly prepared metal-insulator composite, as is well known, the metallic component will form an "infinite cluster," extending across the sample, and having finite d.c. conductivity, above a percolation threshold f_c which is about 1/3 in three dimensions. Slightly below f_c one expects large but finite clusters. The measured far-infrared absorption is indeed diverging as f_c is approached from below; this shows that the absorp-

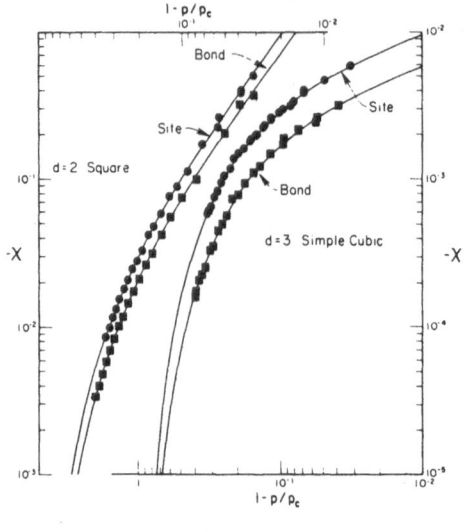

Fig. 4

Calculated susceptibility for square and simple cubic
superconducting lattices below the percolation threshold.
[After ref. (11)].

tion <u>per unit volume of metal</u> is indeed increased by the formation of
large clusters.

FRACTAL CLUSTER MODEL FOR FAR-INFRARED ABSORPTION

Several models have been discussed in the literature to explain
enhancement of anomalous far-infrared absorption by the formation of
clusters. Curtin ans Ashcroft (5) have given a detailed discussion of
how cluster formation enhances both electric and magnetic dipole
absorption, in both normal and superconducting small metal particles.
Bowman and Stroud (11) [Fig 4] have calculated the diamagnetic susceptibility of metal clusters near the percolation threshold, using numerical
simulation in both two and three dimensions. They find that the susceptibility and hence the far infrared absorption per unit volume of metal
do indeed diverge as the percolation threshold is approached. The divergence is stronger in two than in three dimensions.

In order to make these considerations more explicit, we now review in
detail a model of Hui and Stroud (12) for treating electromagnetic absorption due to <u>fractal clusters</u>. A fractal object is one in which the
mass M is related to a linear dimension L by

$$M = AL^{d_f} \tag{17}$$

where A is a constant and d_f is a non-integer. An example of a fractal
object (embedded in a two-dimensional Euclidean space) is shown in
Fig. 5. Here we see that $L = a \cdot 2^{n-1}$ and $M = \rho a^2 \cdot 3^{n-1}$ at the Nth stage of
development of the fractal, where ρ is the mass density of a metal square

554

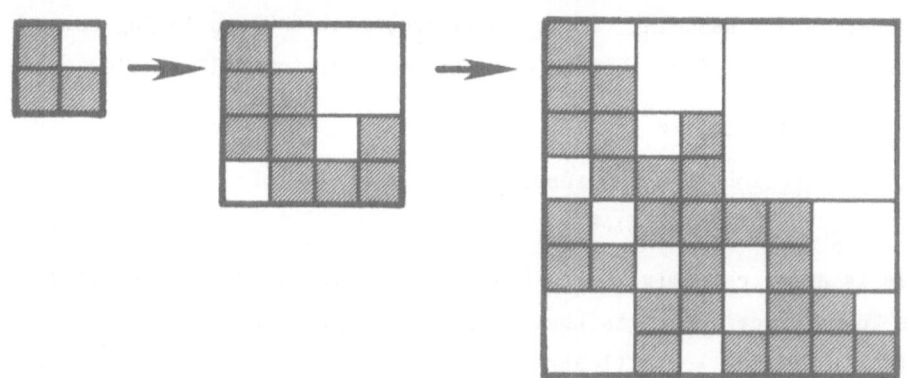

Fig. 5

Schematic diagram showing one possible way of building
up a fractal. [After ref. (12)].

square and a is its edge. Thus in this case d_f = dln M/dln L = ln3/ln2
= 1.585.

A number of workers have reported that metal particles prepared in
colloidal suspensions do indeed form fractal clusters, the fractal dimensions of which may depend on the method of preparation [see, for example,
Weitz et al (13)]. Thus, it is not unreasonable to model the electromagnetic absorption of such fractal clusters.

The model of Hui and Stroud (12) is based on the so-called differential effective-medium approximation. According to this picture, we
imagine we have a cluster of radius R (embedded in a three-dimensional
space), volume V, and mass M, and we enlarge this cluster by infinitesimal increments. Thus the radius, volume, and mass change according
to the formulas

$$R \rightarrow R' = R + \delta R$$

$$V \rightarrow V' = A(R + \delta R)^3 \tag{18}$$

$$M \rightarrow N' = \rho f(R+\delta R)V'$$

where $A = 4\pi/3$, ρ is the mass density, and $f(R)$ is the metal volume
fraction of a particle of radius R. These equations give

$$\delta V = 3A\rho^2 \delta R \tag{19}$$

$$\delta M = \rho[3AR^2 f(R) + AR^3 f'(R)\delta R. \tag{20}$$

Now we conceptually divide δV into two parts,

$$\delta V = \delta V_1 + \delta V_2 \tag{21}$$

and assume that <u>all</u> the extra metal is added to δV_1, which is assumed to
have the same <u>volume fraction</u>, $f(R)$, of metal as the bulk of the
cluster. Thus we have

$$\delta M/\delta V_1 = \rho f(R) \tag{22}$$

[Note that we are assuming, for convenience, that the insulating component is massless.] Substitution of (21) and (20) into (22) yields

$$\delta V_2 = -f'(R)/f(R) \cdot AR^3 \delta R. \tag{23}$$

The _volume fraction_ of added insulator is

$$\eta = \delta V_2/V = -f'(R)/f(R)\delta R. \tag{24}$$

If this is added _randomly_ to the host particle (of conductivity $\sigma(R)$), then a known _exact result_ is that there is a change in conductivity,

$$\sigma(R') = \sigma(R+\delta R) = \sigma(R)(1-3\eta/2) \tag{25}$$

so that (by using (24)),

$$d\sigma/dR = (3/2)\sigma(R)f'(R)/f(R), \tag{26}$$

which can be integrated to give

$$\sigma(R) = \sigma(a)[f(R)/f(a)]^{3/2} \tag{27}$$

which expresses the conductivity of a cluster of radius R in terms of that of radius a.

The same kind of argument can readily be extended to finite frequencies, and expressed in terms of complex dielectric function $\varepsilon(R)$ rather than a zero-frequency conductivity, with the result that $\varepsilon(R)$ is given as the solution to a cubic equation,

$$\frac{\varepsilon(R)}{\varepsilon(a)}\left[\frac{\varepsilon_i-\varepsilon(a)}{\varepsilon_i-\varepsilon(R)}\right]^3 = \frac{1}{[f(R)]^3} \tag{28}$$

which expresses $\varepsilon(R)$ in terms of the dielectric constant of a pure metal particle of radius a, and that of an insulator of dielectric constant ε_i. [Here, we assume that radius a is the dimension of a particle of pure metal, i.e., $f(a) = 1$.] Eq. (28) has been derived previously by Sen, Scala, and Cohen (14), but in a rather different context, that of the conductivity of porous rocks.

It is now easy to see how the far-infrared absorption is enhanced if the cluster has a fractal dimension d_f. In this case, we have

$$f(R) = (R/a)^{d_f-3}, \tag{29}$$

and the corresponding conductivity is

$$\sigma(R) = \sigma(a)(R/a)^{-y}; \quad y=(3/2)(3-d)_f. \tag{30}$$

The electric dipole absorption per unit volume of _cluster_ is $\alpha_e \sim \omega^2/\sigma(R)$. That per unit volume of _metal_ must be multiplied by an additional factor $(R/a)^{3-d_f}$. Combining this factor with that implied by eq. (30), we find that the electric dipole absorption per unit volume of metal is enhanced relative to that of isolated particles according to the rule

556

$$\left(\alpha_e\right)_{cl} \sim \alpha_{e_{ip}} \cdot (R/a)^{5/2(3-d_f)} \qquad (31)$$

which can amount to a huge enhancement for clusters of low fractal
dimensionality.

A similar argument can be made for the effect of clustering on mag-
netic dipole absorption. According to eq. (16), α_m varies as $\sigma \cdot a^2$ per
unit volume of metal, for a particle of radius a, conductivity σ.
Taking into account how the conductivity varies with particle radius, in
the hypothetical fractal cluster, we find that the magnetic dipole
absorption per unit volume of metal is enhanced, relative to that of an
isolated particle of radius a, according to the law

$$\left(\alpha_m\right)_{cl} = \alpha_{m_{ip}} \cdot (R/a)^{1/2(1+d_f)} . \qquad (32)$$

Thus both electric and magnetic dipole absorption are predicted to be
enhanced by clustering, the degree of enhancement depending on the
fractal dimension of the cluster.

The physical origin of the enhancement lies in the decrease of
cluster conductivity with increasing radius. Such a decrease is clearly
suggested by Fig. 5. In this figure, we move from one stage to the next
by putting several clusters of the previous stage together to form a
larger cluster. Since the larger cluster includes insulator as well as
smaller clusters, it must have a lower conductivity. Thus the conductivity will clearly decrease with radius, though the exact manner of the
decrease may differ from the predictions of the differential effective-
medium approximation. This increasing impedance is responsible for the
increased absorption.

Figs. 6a and 6b show the effects of clustering on the far-infrared
absorption of a composite of 1% metal (by volume) as calculated by Hui
and Stroud. Both electric and magnetic dipole absorptions are evidently
enhanced by clustering, and the enhancement is greater for lower fractal
dimension, particularly for the electric dipole absorption. Putting all
factors together, one can easily obtain enhancements of 10^3-10^4 relative
to the unclustered, quasistatic approximation, for parameters appropriate
to Al in the far infrared.

If clustering has such a large effect in the far-infrared, what does
it do in the visible, wherein the surface plasmon resonance lies? Model
calculations of the effect of clustering in this regime are shown in
Fig. 7. Evidently, fractal clustering both broadens and red-shifts the
surface plasmon line, relative to the unclustered (isolated particle)
limit, while non-fractal clustering simply broadens this line but does

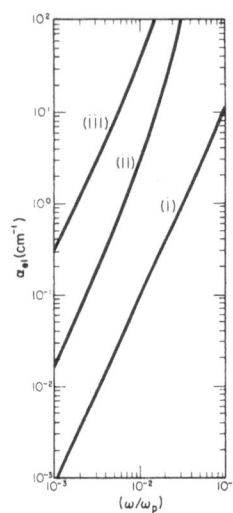 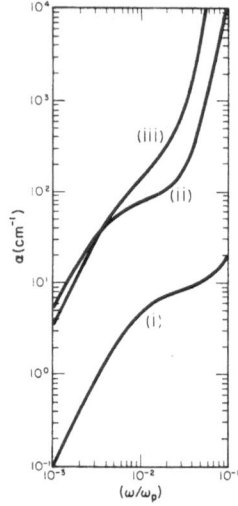

Fig. 6

(a)   (b)

Calculated absorption coefficient in the presence of
fractal clustering. In (b) both electric and magnetic
dipole absorption are included and in (a) only the
electric dipole absorption is included. (i) no
clustering, (ii) R=10a, d_f=2.5, (iii) R=10a and
d_f=2.0. After ref. [(12)].

not notably red-shift it. Experimentally, the surface plasmon line seems
to be only slightly red-shifted and broadened relative to
the predictions of the unclustered, quasistatic approximation.

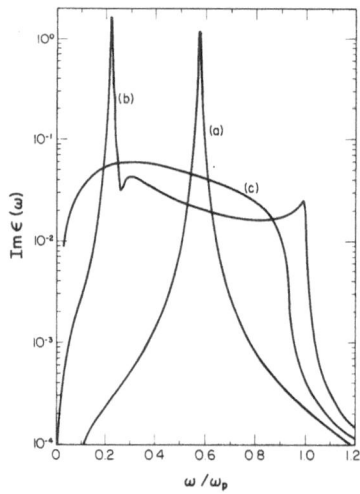

Fig. 7

Imaginary part of the composite dielectri constant:
(a) Maxwell-Garnett approximation with no clustering
(b) fractal clustering with R=10a and d_f=2.5

(c) non-fractal clustering with the cluster dielectric
constant calculated from the effective medium
approximation. [After ref. (12)].

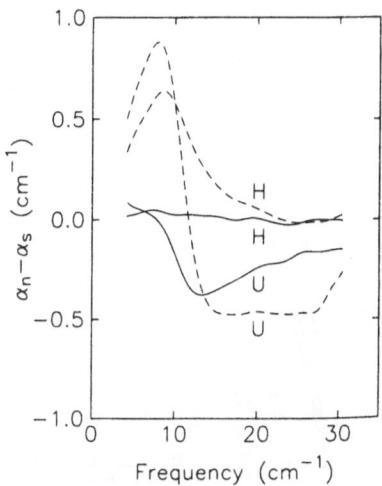

Fig. 8.

Difference absorption $\alpha_n - \alpha_s$ for oxide-coated (solid
lines) and oxide-free (dashed lines) samples before
(U) and after (H) heat treatment. [After ref. (8)].

We may tentatively conclude that only a few of the metal particles
are formed into relatively large fractal clusters, which are
responsible for most of the far-infrared absorption, while most of the
others are present as isolated particles, contributing most of the weight
of the surface plasmon line but relatively little of the far-infrared
absorption.

EFFECT OF SUPERCONDUCTIVITY

The far-infrared absorption of small metal particles can be affected
when they undergo a superconducting transition, as was first shown by
Carr, Garland, and Tanner (15). Typical experiments, due to Curtin et al
(8), are shown in Fig. 8, which plots the difference between the absorption coefficient in the normal and superconducting states for several
samples. Below about 15 cm^{-1}, the normal-state absorption is usually
greater, but above this, the absorption in the superconducting state is
sometimes larger than that in the normal state, especially for clumped
samples.

A plausible way of treating absorption in the superconducting state
is to use the theories described above, but replacing the Drude conductivity of the normal state by the Mattis-Bardeen conductivity. The real
part, σ_1^s, of this conductivity, takes σ_1^s the form

$$\sigma_1^s(\omega)/\sigma_N(\omega) = (1+2\Delta/\hbar\omega)E(k) - 4\Delta K(k)/\hbar\omega \qquad (33)$$

where $E(k)$ and $K(k)$ are elliptic integrals of the argument $k=2\Delta/\hbar\omega$ and
$\sigma_N(\omega)$ is the conductivity in the normal state. The imaginary part of

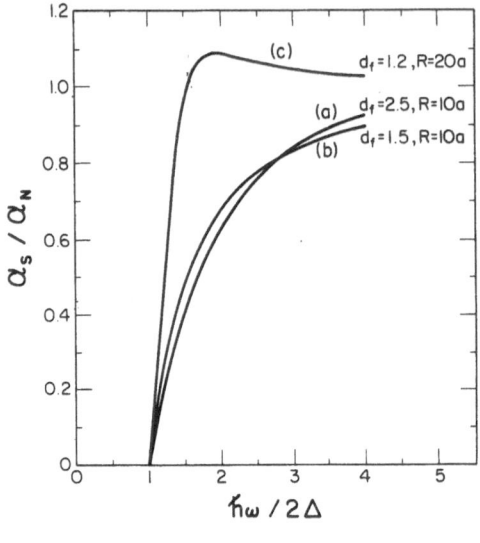

Fig. 9

The ratio $\frac{\alpha_s}{\alpha_n}$ for different values of fractal dimen-

sions and cluster sizes. [After Ref. (12)].

the Mattis-Bardeen conductivity is the Kramers-Kronig transform of this, and the dielectric function is obtained from the complex conductivity by the relation $\varepsilon_s(\omega) = 1+(4\pi i/\omega)\sigma_s(\omega)$.

Because of the gap in the $\sigma_s(\omega)$ for $\hbar\omega<2\Delta$, we expect no absorption in superconducting small particles at frequencies below 2Δ. This accounts for the fact that $\alpha_N-\alpha_s>0$ below these frequencies. In order for $\alpha_N-\alpha_s<0$ above the gap, we must form fractal clusters of very low fractal dimensionality, as illustrated in the model calculations of Hui and Stroud (Fig. 9). These show that the experimental observations of Curtin et al (8) can be explained, in principle, by the effects of clustering in conjunction with the Mattis-Bardeen bulk superconducting conductivity. Similar conclusions were, once again, arrived at by Curtin and Ashcroft (5) using a non-fractal model of clustering. The very low fractal dimension required in this aspect of the model seems somewhat unrealistic to describe clusters in a real composite, though it may be applicable to some aggregates.

QUANTUM SIZE EFFECTS

While it appears, from the foregoing work, that various classical effects are enough to account for most of the "anomalous" far infrared properties of small metal particles, we briefly consider one other possible explanation, particularly germane to this conference – namely, quantum size effects. It is certainly true that, in the limit of

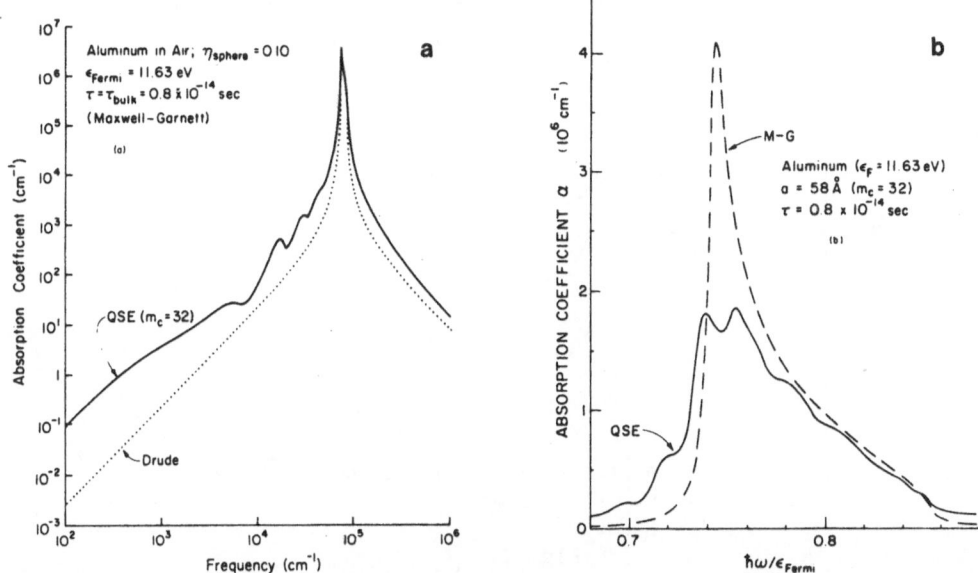

Fig. 10

(a) Quantum Size Effect (QSE) on the absorption
coefficient for a composite of 10% volume fraction η
of Al in air. (b) Absorption coefficient in the
vicinity of the Mie sphere resonance. [After ref.
(6)].

sufficiently small particles, size quantization of electronic energy
levels must cause the properties of individual metal particles to depart
from their bulk. The far-infrared absorption could easily be enhanced,
in principle, because of the breakdown of k-selection rules imposed by
the periodicity of an infinite sample, for example, or because of other
effects.

Several model calculations suggest that such size quantization is not
enough to account for the deviation of experiment from quasistatic predictions. Non-self-consistent calculations of Wood and Ashcroft (6), for
example (Fig. 10), based on size quantization in a cubic particle in
conjunction with a number-conserving relaxation time approximation, show
that the absorption spectrum is slightly broadened near the surface
plasmon line, but that the far-infrared absorption is only enhanced by an
order of magnitude for 100 A particles. This calculation considers only
electric dipole absorption; it appears that no calculations have been
carried out to determine what happens to magnetic dipole absorption when
size quantization is considered.

A more elaborate calculation by Ekardt (16) using a fully self-consistent approach for a spherical particle, based on the Zangwill-Soven
dynamic density-functional method for treating the frequency-dependent
polarizability of closed-shell atoms, also suggests little far-IR enhancement. Ekardt finds hardly any enhancement of the static suscepti-

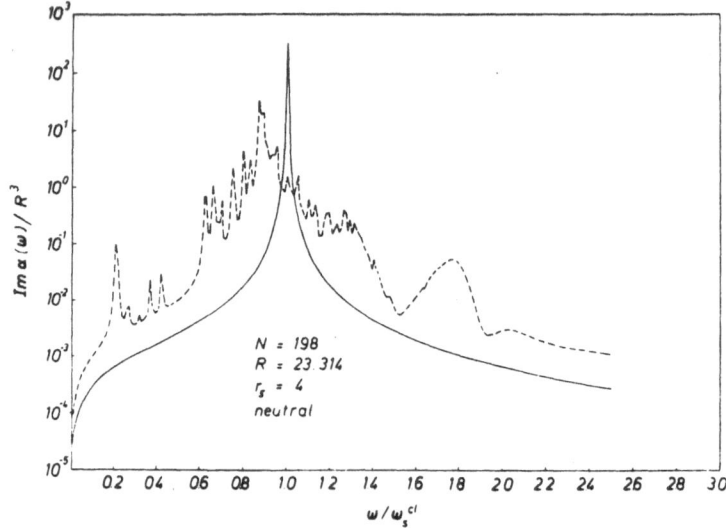

Fig. 11

Imaginary part of the complex polarizability $\alpha(\omega)$ in
units of R^3, calculated within the density functional
theory. The solid line is the local Drude model
result. [After ref. (16)].

bility due to size quantization; analysis of his dynamic susceptibility
as shown in Fig. 11 also suggests less than an order of magnitude enhancement of absorption in the far-IR. Thus his calculations are consistent with those of Wood and Ashcroft.

CONCLUSIONS

    The results presented here, due to the authors and other workers,
cumulatively suggest that the once-anomalous far-infrared absorption is
anomalous no longer: it can be accounted for, both in the normal and the
superconducting state, by basically classical considerations. The most
important of these are clustering, magnetic dipole absorption, and oxide
coatings: together they are probably capable of accounting for most
experimental observations. Quantum size effects, though they certainly
become important for very small particles (20 Å or less), probaby are of
lesser importance for the particles subject to most experimental investigation.

    It is worth remarking, finally, why this problem is important to
understand. The far-infrared "anomaly" may be representative of many
other anomalous enhancements which hve been reported in the small-particle literature: enhanced photoyield, surface-enhanced Raman scattering, and enhanced Van der Waals interactions between small metal
particles. While not all of these have been definitively confirmed
experimentally, it seems likely that some of them result, in part, from

classical effects somewhat related to those discussed here. Thus, understanding one such effect may be a useful advance.

ACKNOWLEDGMENTS

This work was supported by NSF Grant DMR84-14257. The authors are grateful for useful conversations with N.W. Ashcroft, D.B. Tanner, and W.A. Curtin.

REFERENCES

1. D. B. Tanner, A. J. Sievers, and R. A. Buhrman, Phys. Rev. B11:1330 (1975).
2. D. Stroud and F. P. Pan, Phys. Rev. B17: 1602 (1978).
3. E. Simanek, Phys. Rev. Lett. 38: 1161 (1977).
4. P. N. Sen and D. B. Tanner, Phys. Rev. B26: 3582 (1982).
5. W. Curtin and N. W. Ashcroft, Phys. Rev. B31: 3287 (1985).
6. D. M. Wood and N. W. Ashcroft, Phys. Rev. B25: 6255 (1982).
7. R. Ruppin, Phys. Rev. B19: 1318 (1979).
8. W. A. Curtin, R. C. Spitzer, N. W. Ashcroft, and A. J. Sievers, Phys. Rev. Lett. 54: 1071 (1985).
9. S. E. Lee, T. W. Noh, K. Cummings, and J. R. Gaines, Phys. Rev. Lett. 55: 1626 (1985).
10. R. P. Devaty and A. J. Sievers, Phys. Rev. Lett. 52: 1344 (1984).
11. D. R. Bowman and D. Stroud, Phys. Rev. Lett. 52: 299 (1984).
12. P. M. Hui and D. Stroud, Phys. Rev. B33: 2163 (1986).
13. D. A. Weitz and M. Oliveria, Phys. Rev. Lett. 52: 1433 (1984).
14. P. N. Sen, C. Scala, and M. H. Cohen, Geophys. 46: 781 (1981).
15. G. L. Carr, J. C. Garland, and D. B. Tanner, Phys. Rev. Lett. 50: 1607 (1983).
16. W. Ekardt, Phys. Rev. Lett. 52:1925 (1984).

CLUSTER SPECTROSCOPY: VARIATIONS IN IONIZATION POTENTIALS AND SPECTRAL
SHIFTS AS A FUNCTION OF DEGREE OF AGGREGATION AND STUDIES OF CLUSTER
FRAGMENTATION

A. W. Castleman, Jr. and R. G. Keesee

Department of Chemistry
The Pennsylvania State University
University Park, PA 16802

ABSTRACT

     Studies of trends in the variations of spectral features, ionization
potentials, and dissociation processes of clusters as a function of degree
of aggregation are presented. The results bear on such questions as the
changing properties of systems undergoing transitions between the gas and
the condensed phase, as well as the origin of magic numbers. Investigations of the spectral shifts of an electronic transition in a chromophore
such as paraxylene or phenylacetylene show that clusters containing from 3
to 15 argon atoms all undergo a red shift of about 50 cm^{-1} as a limiting
value in the S_1 state. Evidence for spectroscopic changes between the
gaseous and the condensed state is also apparent from the broadening of
linewidths.

     A major advance in the study of unimolecular dissociation has become
available through the use of multiphoton ionization coupled with a
reflectron introduced into the drift region of a time-of-flight mass
spectrometer. Using single and two-color tunable pulsed lasers, the
excess energy introduced into a cluster can be well controlled. The
power of this method is demonstrated by the results of recent investigations of hydrogen bonded clusters such as ammonia and methyl alcohol and
also clusters of rare gas atoms which, following ionization, lead to an
internal ion-molecule reaction and subsequent cluster fragmentation. The
role of dissociation and the influence of the thermochemical stability of
cluster ions in effecting the appearance of magic numbers in certain
cluster distributions is discussed. The application of this method in
determining ionization potentials of probe molecules following successive
clustering with a solvent species is also presented. The results of
studies of dielectrics are contrasted with trends found for alkali metal
systems.

     A final related topic is that of internal ion-molecule reactions
following multiphoton excitation of clusters. A finding of some
importance is an internal Penning ionization process taking place in
certain clusters leading to electron transfer between the chromophore and
the solvent molecules. Findings of a delayed electron transfer reaction,
having implications to the bulk condensed state, are also presented.

Cluster research is a rapidly growing field which offers the exciting prospect of bridging the gap between the gaseous and condensed phase by probing changing properties at the molecular level. Studies of spectroscopic shifts upon successive clustering are especially interesting with regard to the onset of liquid or sold-like features in the spectra (1-6). Investigations of the processes of cluster ionization and dissociation are of particular interest since they contribute to a further understanding of the evolution of changes in ionization potentials as a system approaches the bulk work function. The details of intramolecular energy flow and energy disposal following ionization can also be revealed.

Studies of the dynamics of formation and dissociation, and the changing properties of clusters at successively higher degrees of aggregation, enable an investigation of the basic mechanisms of nucleation to be probed at the molecular level. Work on systems of increasingly higher degrees of aggregation also has a direct bearing on the development of surfaces, and ultimately solvation phenomena and the formation of the condensed state. The progressive clustering of a molecule involves energy transfer and redistribution within the molecular system, with attendant processes of unimolecular dissociation taking place between growth steps. Related processes of energy transfer and dissociation are operative during the reorientation of molecules about ions following the primary ionization event employed in detecting clusters via mass spectrometry.

Investigations of the molecular properties, reactions, and behavior of clusters generally require ionization in one of the steps as either a probe and/or a method of detecting clusters through mass spectrometry. Although ionization can be accomplished through electron impact as well as single photon techniques, resonance enhanced multiphoton ionization often enables selective ionization of clusters in particular states.

Recent advances in the field of molecular beam research, coupled with lasers and time-of-flight mass spectrometry, enable the details of these various processes to be investigated. A major advance in the study of these processes has become available through the use of a reflectron technique introduced into the drift region of a time-of-flight mass spectrometer. Using single and two-color tunable pulsed lasers, the excess energy introduced into a cluster can often be well controlled. The power of this technique is demonstrated by a recent investigation of hydrogen-bonded clusters $(NH_3)_n$ and $(CH_3OH)_n$ which, following ionization, lead to an internal ion-molecule reaction followed by proton transfer and reorientation of the molecular dipoles. Related work on Xe_n^+ is also presented.

More detailed and specific information can be obtained through resonance-enhanced ionization spectroscopy, and is the preferred method when such processes can be readily accomplished. Examples are discussed herein which draw from several different investigations made in our laboratory. In the first example, a detailed investigation of the spectroscopic shifts of two probe molecules, phenylacetylene and p-xylene, clustered by a series of rare gases, CO_2, H_2O, N_2, O_2, NH_3, and CCl_4 shows that clusters of specific composition can be selectively ionized. In the most part, the spectral shifts are toward the red of the main S_1 state of the unclustered parent molecule, although in a few cases, most notably with H_2O, the shifts are to the blue.

In a second example involving p-xylene clustered with argon, the

change in ionizaton potential as a function of the degree of aggregation
is considered. The results provide insight into the extent of
interaction of the ion of the chromophore with the "solvating matrix".
The results are contrasted with trends observed for clusters of sodium
atoms. A problem of long-standing interest is the influence of the
degree of aggregation of the metallic system on the ionization potential.
In this regard, comparison of predictions from theories such as the
Jellium model, classical electrostatics, or quantum calculations with
experimental results are.important. In the case of bulk metals, it is
well known that the adsorption of molecules leads to an alteration
(usually a reduction) in the work function. A problem of some interest
is the degree to which cluster systems which have either been reacted
with molecules or contain adsorbed species reflect such trends.

In another example, cluster ionization is accomplished through high
Rydberg states of the p-xylene chromophore whereby ionization is
accomplished through an intracluster process having analogy to gas-phase
Penning ionization processes. Contrasting results between systems
comprised of p-xylene bound to trimethylamine and to ammonia detail the
molecular processes involved. Evidence for a slow ionization process is
presented.

EXPERIMENTAL

Clusters of the molecules to be studied are formed via adiabatic
cooling from a pulsed nozzle system, with detection of the products of
MPI being made in a time-of-flight mass spectrometer located beyond the
ionization region (7). The combination of ionization and mass selection
has the advantage of direct mass determination of the probed van der
Waals molecules. Because of the weak bonding in complexes, dissociative
ionization may possibly complicate the spectroscopic assignments. In the
present work both single and two-color multiphoton ionization are
utilized. Both methods are utilized in the spectroscopic studies while
two-color techniques are essential for the studies pertaining to
variations in ionization potential. Two-color MPI is accomplished using
a Q-switched Nd:YAG laser to simultaneously pump two tunable dye lasers
whose output is frequency doubled in KDP crystals. The arrangement
allows continuous tunability from the visible down to 216.5 nm, with a
maximum output of 1 to 29 millijoules per pulse (6 ns duration) in the
UV. The results show that fragmentation can be greatly suppressed
through the use of two-color resonance enhanced photoionization whereby
the ionzation process is accomplished with the use of little excess
energy in contrast to one-color experiments.

In some experiments it is necessary to ascertain whether dissociation processes influence the results. In others, an investigation of
dissociation rates is the purpose of the experiment. In a typical
time-of-flight mass spectrometer, either a two element or alternately a
single element accelerating field may be used in the region of ionization. This is followed by a field-free drift region at the end of which
the arrival of the ions is detected. Using the conventional TOF method,
dissociation which occurs with rates in the neighborhood of 10^6 to
10^8 sec^{-1} can be investigated by either of two methods. One involves
analyzing the peak shape (arrival spectrum) of the ions under dual field
accelerating conditions, in which cases a knee is observed since the ions
spend far more time in the first low-field region where ionization is
initiated than in the second high-field region where the bulk of the
acceleration occurs (8). An alternate method is to operate under single

field conditions and deduce rates from the shape of the late arriving tail of the peak (9).

A third method of studying dissociation employs a reflecting field (reflectron). Although originally designed to enhance the resolution of the TOF method (10), a reflectron can also be employed to investigate dissociation in the field-free drift region, so that slower processes may be observed. In the dissociation experiments, the typical photon fluxes are about 10^{16} photons per pulse of approximately 6 ns duration. The ions are accelerated in the accelerating field to about 2 KeV whereafter they enter a field-free region and then are electrically reflected and detected as shown in Figure 1. With appropriate potentials applied to the the reflectron grids, non-dissociating parent ions can be separated from those that dissociate while in the field-free region. A unique identification of these daughter ions can be accomplished by the time separation or by an energy analysis with the reflectron. The separation of the parent and daughter ions occurs as a result of the loss in kinetic energy with essentially no change in velocity of the ion upon dissociation. Hence species with greater kinetic energy (parents) have a longer path to the detector than do the daughter (dissociation) products.

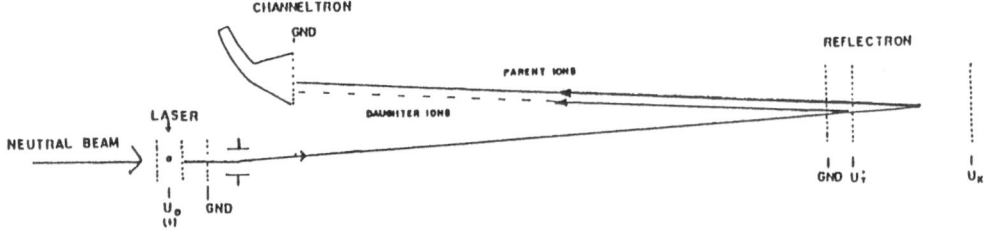

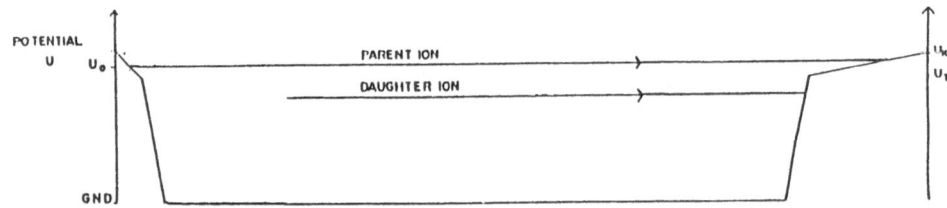

Figure 1. Schematic of time-of-flight mass spectrometer with reflectron where daughter ions are reflected before parent ions. Lower figure depicts the electrostatic potentials.

SPECTRAL SHIFTS

Resonance enhanced multiphoton ionization through the specific excitation of an electronic state of a chromophore contained within a cluster is a powerful method of ascertaining the properties of clusters in relating these to their counterparts in the condensed and isolated gas phases. Generally, the clustering of atoms or molecules onto a chromophore result in a perturbation of the electronic states of that chromophore. The spectral shift of a given electronic transition from that of the isolated chromophore is a measure of the relative differences between the lower and upper states of the energetic perturbation induced by clustering. This is analogous to the spectral shift of electronic transitions observed for molecules in solutions or matrices from their gas-phase transitions. A red shift implies that complex formation has reduced the energy difference between the two states, whereas a blue shift indicates an increase in the difference. The magnitude and direction of the shift are due to a combination of effects, including dispersive and repulsive interactions, hydrogen bonding, and electrostatic forces involving such processes as dipole-induced-dipole or dipole-dipole interactions.

Often revealing are the spectroscopic shifts of an electronic transition in an organic chromophore that result from the perturbations due to clustering (11). We have employed both one- and two-color resonance enhanced multiphoton ionization to investigate the $S_1 \leftarrow S_0$ π-electron transition in phenylacetylene and p-xylene as clusters with various solvents. Single color multiphoton ionization studies of the perturbed $L_b(^1B_2)$ states of phenylacetylene (PA) bound with Ne, Ar, Kr, and Xe were all found to induce a lowering of the S_1 resonance with respect to the ground state. These observed red shifts have been attributed to dispersive interactions with solvent molecules (12,13). In general, a spectral shift is governed by three factors; (1) short-range electronic repulsive interactions which result in a blue shift, (2) electronic dispersive interactions which result in a red shift, and (3) differences of zero point energies between the excited and ground states.

A linear dependence of the spectral shift on the electrostatic polarizability of the rare gas atom has also been found. The sensitivity to atom polarizability is indicative of the important role of dispersive forces in the perturbation of the electronic states in the aromatic molecule. The spectral shift due to solute-solvent interactions, in the Onsager model, contains a term due to dispersion (14). Our results support the finding (12) that in aromatic molecule-rare gas atom systems the spectral shifts are dictated by atom polarizability, i.e., the important role of dispersive forces in the perturbation of the S_1 excited state. Figure 2 displays the results of our study which shows a direct linear dependence of the spectral shift on the electrostatic polarizability of the rare gas atom. The results conform to the Onsager model, but on a microscopic level.

Spectral shifts of phenylacetylene due to aggregation by rare gases are also revealing. Investigations with large-ringed systems have generally shown an approximate additivity of spectral shifts based on the number of rare gas atoms clustered on to the aromatic (12,15). In our own work, we find that this additivity is apparently additive only up to the clustering of two atoms per aromatic ring as seen from the data given in Reference 11. Since the additivity is nearly exact for the two-atom case, the spectral shifts shown in Figure 2 are identical on a spectral shift per atom basis for the phenylacetylene system in the case of the two-atom containing rare gas complex.

Particularly interesting are the trends seen for larger clusters. Figure 3 shows a selected set of data for the spectral shifts relative to the S_1 electronic origin of phenylacetylene for clusters containing four to ten argon atoms. First it is interesting to note that the major feature asymptotically approaches a shift of approximately 50 cm^{-1}. Clearly the additivity rule does not apply. Secondly, the van der Waals modes to the right side of the main resonance begin to fill in for large cluster sizes. The spectra are broadening in analogy to those seen in the condensed phase and the features to the right resemble photon modes for a system of infinite lattice.

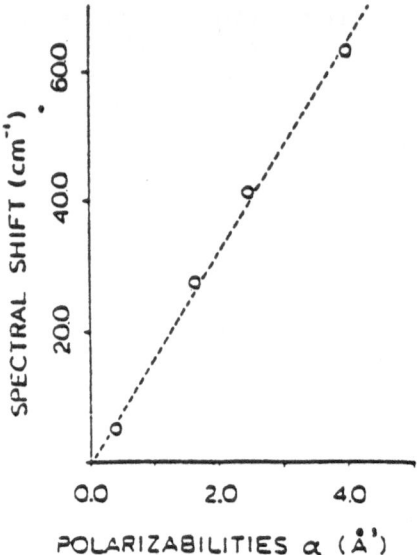

Figure 2. Spectral shift of PA·R (relative to the nascent PA) versus the polarizability of the rare-gas atom. α = 0.40 (Ne), 1.63 (Ar), 2.48 (Kr) and 4.01 Å^3 (Xe).

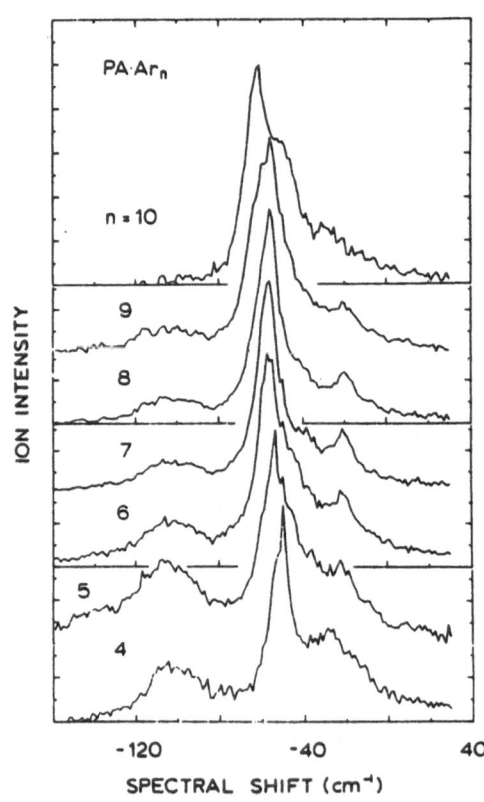

Figure 3. R2PI current versus one-photon energy. The ion currents are recorded at the m/e ratios corresponding to PA·Ar_n (4<n<10). The energy scale is relative to the S_1 electronic origin of PA. The ion current scale is relative and different for each spectrum and p_o = 300 Torr.

Other interesting spectral shifts have been observed for the cluster-
ing of N_2, O_2, N_2O, NH_3, H_2O, CCl_4, and CH_4 to phenylacetylene. In most
cases the main resonance is also red-shifted, although in a few a substantial blue shift is observed, most notably for H_2O. The striking
difference between the isoelectronic molecules H_2O and NH_3 can be
rationalized in terms of the excitation of the π system leading to a
repulsive interaction with the two long-pair electrons of the H_2O molecule
(16).

SHIFTS IN IONIZATION POTENTIAL WITH DEGREE OF AGGREGATION

Organic Chromophores Solvated with Rare Gas Atoms

It is interesting to compare the behavior of the ionization threshold
of neutral clusters (or the appearance potential for the ionized products)
as a function of the degree of aggregation. From a microscopic point of
view, due to the repulsive interactions between the induced dipoles on the
Ar atoms, it is expected that the shift in appearance potential depends
less than linearly on the coordination number n.

The ionization potentials of p-xylene bound with argon ($PX \cdot Ar_n$) were
determined through studies in which the energy of one photon was fixed at
the L_b resonance and the wavelength of the second laser was scanned.
Resonance-enhanced ionization with a single-color laser results in
significant fragmentation due to the fact that the absorbed energy is
substantially above the ionization threshold since the S_1 state lies more
than halfway to the ionization continuum. Cluster fragmentation was found
to be suppressed to a negligible amount in the two-color experiments,
enabling a detailed investigation of the variation in ionization potential
with degree of aggregation to be definitively established.

It is well known that the Stark effect leads to a shift in ionization
potential when measured in an electric field and correction is necessary to
account for shifts on the order of 50 cm^{-1}. The ionization potentials are
found to vary with the square root of the electric field present in the
region of ionization in accordance with expectations and findings of others
(17). Extrapolation to zero field is readily accomplished in view of the
linear dependence and the fact that various cluster systems display lines
of identical slopes in these weakly perturbed rare-gas aggregates.

The shifts in ionization potential of p-xylene in the rare gas
aggregates is shown in Figure 4 for clusters with one to six argon atoms.
The shift in relative ionization potential is observed to display a broadly
linear dependence on the number of argon atoms. The largest deviations
from this trend are observed for the dimer and pentamer. The observed
total shift of about 750 cm^{-1} for the hexamer is to be contrasted with the
matrix isolated value which is about 6000 cm^{-1} for a similar molecule
(benzene) in an argon matrix (18). Evidently, the "local environment" with
which a molecule interacts is relatively large in such a matrix and the
observed shift is far from the expected bulk value. This is in interesting
contrast to the metal systems discussed in the next subsection. When
corrected for ion image potential effects the ionization potential of
aggregates comprised of only a few alkali metal atoms display nearly the
bulk ionization potential of the polycrystalline metallic system (19).

Comparison with Metallic Systems

The problem of nonmetallic-metallic transitions and the factors
responsible for the onset of metallic conductivity are particularly

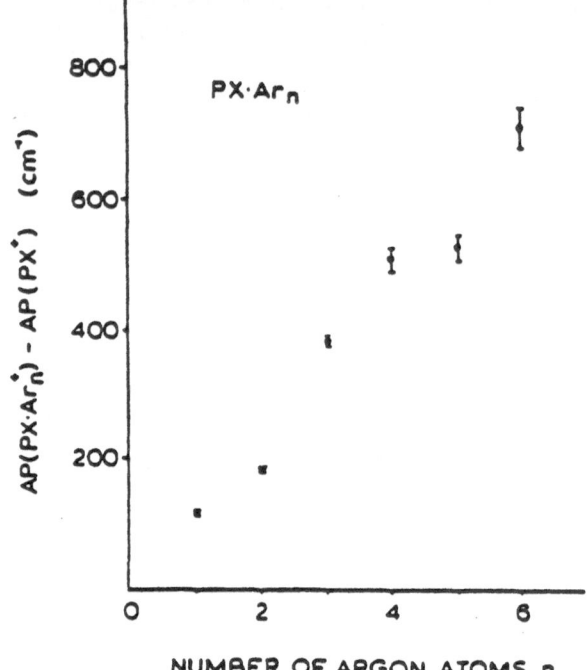

Figure 4. Relative appearance potentials. Field ionization of
$\overline{PX \cdot Ar_n}$ (n=0-6) in a 150V/cm dc field. $AP(PX \cdot Ar_n^+)$ increases with
the coordination number n.

relevant to the field of microelectronics. Germane to the subject matter
is the problem of determining the dependence of metallic conductivity on
size.

Due to their simple one-electron nature, and the ease with which they
can be compared with theoretical calculations, systems comprised of alkali
metals are particularly interesting for study. Photoionization spectra for
Na_x, with x ranging from 1 to 8, have been obtained in our laboratory using
a molecular beam coupled with a UV light source and quadrupole mass
spectrometer detection system (19). Sodium clusters are produced by
vaporizing the metal and expanding it with argon gas at pressures ranging
between 70 and 100 torr through a 300 μm diameter nozzle. Once formed, the
clusters pass through a skimmer into a detection chamber where they are
photoionized and mass analyzed. Photoionization is accomplished by use of
a 500 watt xenon arc lamp. The desired wavelength is selected with a
monochromator. The light intensity on the monochromator is continually
monitored using a thermopile detector and the collected spectra are
corrected for variations in power and the monochromator slit function.

Data from our laboratory (19) agree extremely well with measurements
reported by Schumacher and coworkers (20) for all clusters with the
exception of x = 5, 6, and 8, where our findings point to slightly lower
values. Recent measurements for Na_8 reported by Buttet (21) are in good
agreement with our findings. At the present time, there is ambiguity in
the proper assignment since several possible complicating factors may
exist. In addition to possible influences due to Franck-Condon factors or
hot bands, neither of which are expected to relate to the problem at hand
(19), questions exist concerning the possible existence of isomers or the
possibility that the results can be influenced by fragmentation of larger
clusters (22).

Martins et al. (23) suggest that Na$_6$ should have two isomers separated in energy by only 0.06 eV. The observed difference in ionization potentials is about 0.15 eV if the difference for the two steps in the ionization efficiency curve is attributed to the onset of two different isomers.

Perhaps of most interest are the trends observed in the ionization potentials with cluster size. The experimental data correlate resonably well with the classical electrostatic expression relating the work function of a system of radius r to the bulk work function, W$_\infty$ (see Figure 5). A classical equation relates the influence of particle size on the ion image potential contributions to the work function of a system with spherical symmetry. The equation is derived from classical equations of electrostatics.

$$W(R) = W_\infty + \frac{3}{8}\frac{e^2}{R},$$

where W represents the work function, R the radius of the equivalent sphere supporting elementary charge e, and W$_\infty$ the bulk work function of the polycrystalline metal. The photoionization thresholds of sodium clusters are in good agreement at all but the smallest sizes. A number of other

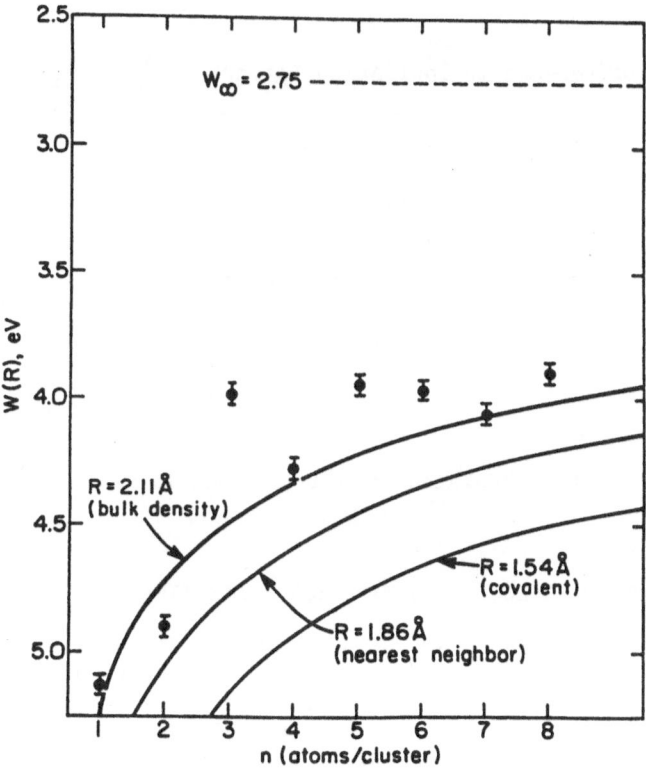

Figure 5. The ionization potential or work function of sodium clusters as a function of size. The curve is based on the classical relationship W(R)=W$_\infty$ + 3e^2/8R where radius R is related to the number of atoms by assuming the bulk density for the clusters.

metallic systems also correlate well with this classical theory. These results indicate that the intrinsic part of the ionization potential is independent of size (24).

Recently, limited data have become available for the photoionization of sub-oxides of alkali metals for further comparison with theory. The values for the metal suboxides are seen to fall dramatically to an essentially constant value for the trimer and tetramer species. The values do not differ greatly from the analogous metallic species, but in the case of sodium they do fall below the bare metal cluster for the tetramer. It has been reported that the work function of sodium is appreciably lowered by the presence of impurities and the discrepancy between early and recent measurements has been traced to surface contamination.

Table I lists the difference in ionization potential between Na_xO and Na_x, where measurements are available from our studies for x up to 4. The ionization potential for NaO is much greater than that for Na. The difference between the oxide and pure species becomes diminishingly small for x=3 and the trend reverses for x=4. The ionization potential for the larger cluster moves into accord with the observation that the presence of adsorbed impurities on polycrystalline sodium surfaces lowers the bulk work function. The potassium system (25), while not displaying this reversal in ionization potential between oxide and pure metal, does show a diminishing trend with cluster size. It is tempting to speculate that there would be a reversal for a larger cluster (see results in Table I).

Table I. Difference in Ionization Potential[a] for M_x and M_xO

	$\Delta \equiv$ IP (M_x) $-$ IP (M_xO)	
x	Na Δ, (eV)	K Δ, (eV)
1	$-$ 1.36	----
2	$-$ 0.20	$-$ 0.65
3	0.08	$-$ 0.35
4	0.33	$-$ 0.02

[a]results calculated from data given in ref. 25.

CLUSTER DISSOCIATION FOLLOWING MULTIPHOTON IONIZATION

A major advance in the study of unimolecular dissociation and the spectroscopy of clusters has become available through the use of multiphoton ionization coupled with a reflectron introduced into the drift region of a time-of-flight mass spectrometer. Using single and two-color tunable pulsed lasers, the excess energy introduced into a cluster can be well controlled. The power of this method is demonstrated by the results of recent investigations of hydrogen bonded clusters which, following ionization, lead to an internal ion-molecule reaction, and cluster fragmentation. Consideration is also given to clusters of xenon atoms. The role of dissociation and the influence of the thermochemical stability of cluster ions in effecting the appearance of magic numbers in certain

cluster distributions is discussed. Results of excitation through the
excited state of a chromophore which leads to internal Penning ionization
is also presented.

INTERNAL ION-MOLECULE REACTIONS AND DISSOCIATION

Ammonia Clusters

Extensive investigations of the dissociation of ammonia clusters
following multiphoton ionization at 266 nm were conducted in our
laboratory (7). Interaction with the laser beam leads to ionization of
one of the molecules in the hydrogen bonded neutral ammonia cluster and
ejection of an electron from the cluster. Due to the large proton
affinity of NH_3 (8.85 eV) and analogous to the well-known gas phase
ion-molecule reaction between NH_3^+ and NH_3, an internal (intracluster)
ion-molecule reaction rapidly forms NH_4^+ with the ejection of NH_2.
Thereafter the ammonia molecules reorient to accommodate the proton.
Unimolecular (unicluster) dissociation may occur as a result of the excess
vibrational energy created during the conversion from a neutral to ionized
system. Collision-induced dissociation is also generally observed as a
result of the interaction of the cluster ions with residual molecules in
the vacuum chamber.

$$(NH_3)_n + h\nu \quad \rightarrow \quad NH_3^+(NH_3)_{n-1} + e \quad\quad\quad (1)$$

$$NH_3^+(NH_3)_{n-1} \quad \rightarrow \quad [NH_4^+(NH_3)_{n-2}]^* + NH_2 \quad\quad\quad (2)$$

$$[NH_4^+(NH_3)_{n-2}]^* \quad \rightarrow \quad NH_4^+(NH_3)_{n-2-m} + mNH_3 \quad\quad\quad (3)$$

The extent of fragmentation is determined through use of the reflectron.
Figure 6a shows a typical ammonia cluster ion experimental TOF without the
reflectron while 6b shows the superimposed daughter distribution which
arises through use of the reflectron. If the reflectron barrier is
lowered sufficiently, only the lower kinetic energy products are
reflected. The non-dissociating ions are thereby eliminated from the
spectrum as shown in the lower part of Figure 6c. Further reduction of
the reflecting potential U_T (with $U_K < U_T$) improves the ability to discern
small contributions from more extensive dissociation.

The contribution of unimolecular compared to collision induced dissociation is determined through experiments in which the pressure in the
field-free region is varied. An extrapolation of the data to a zero of
pressure can be made in a linear fashion under the thin collision
approximation, where, within the uncertainty of the true zero of pressure
the ordinate gives a direct measure of the fraction of dissociation by
unimolecular (evaporative) loss. The component from unimolecular decay
has been found to increase steadily with cluster size from n=4 to 25 all
being of order 10^5 sec^{-1}. Though at first glance the trends are
surprising, they have been found to be consistent with the simple RRK
considerations (7). The implication of our results is that each larger
cluster progressively has more excess internal energy due to factors such
as dielectric relaxation after ionization, latent heat from the formation
of the neutral cluster, and the lower ionization threshold of larger
clusters. Confirmation of this picture follows from the observation that
the rate for the loss of a second ammonia molecule decreases with
increasing cluster size in accordance with expectations for clusters
having little internal energy as a result of the loss of the first ammonia
molecule.

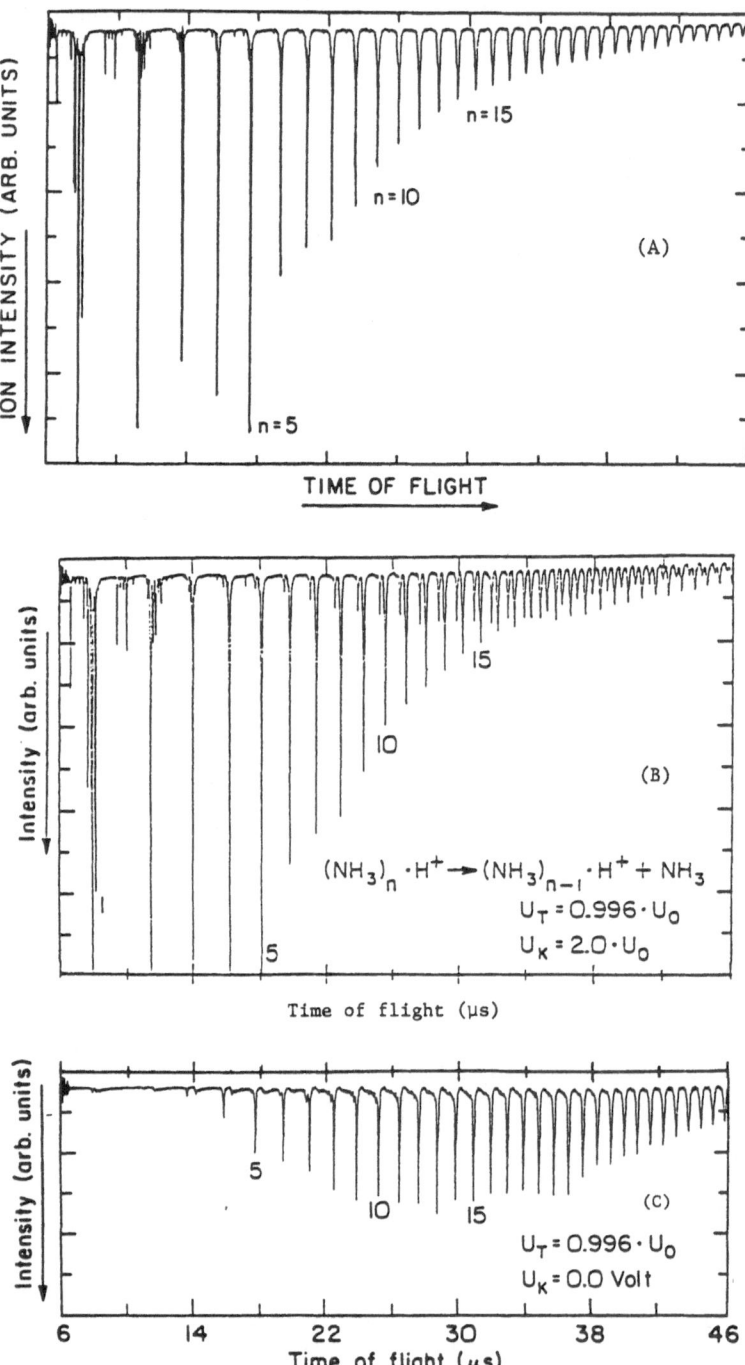

Figure 6. (A) A conventional TOF mass spectrum of $H^+(NH_3)_n$ clusters from the multiphoton ionization of ammonia clusters at 266 nm. (B) Time-of-flight mass spectrum taken using the reflectron. The potentials U_T and U_K are chosen such that all ions are reflected but with the daughter ions arriving at the particle detector earlier than their corresponding parent ions. (C) Time-of-flight mass spectrum taken using the reflectron. U_K is lowered to eliminate the parent ions. Both spectra in Figures 6B and 6C are accumulated ion signals over 2460 laser shots.

The ammonia system is of further interest since $H^+(NH_3)_5$ is distinctive in its relative intensity, displaying one of the so-called "magic numbers"; see Figure 6a. This ion can be considered to be an NH_4^+ ion solvated by one ammonia molecule at each hydrogen. The role of laser fluence for multiphoton ionization at 266 nm has been determined by comparing data taken at much different laser fluences but under nearly identical expansion conditions. The prominence of the protonated pentamer decreases substantially at higher laser fluences providing further evidence for the importance of fragmentation and dissociation in effecting magic numbers in hydrogen bonded clusters (26).

Methanol Clusters

Similar studies to the foregoing were undertaken using 266 nm light to ionize $(CH_3OH)_n$ clusters (27,28). Several processes were observed, the first of which has a direct analogy to the ammonia studies. In particular, following ionization, methanol clusters undergo proton transfer with a concomitant loss of CH_3O. The exothermicity of the proton transfer reaction and subsequent relaxation of the alcohols around the protonated moiety also leads to excess internal energy and dissociation processes. The dissociation processes lead to unimolecular evaporation rates in a manner similar to the ammonia system and the rates are also in the order of 10^5 to 10^6 sec^{-1}. There is no distinct trend with cluster size.

In contrast to the ammonia system, a nonevaporative loss process was also observed, namely the elimination of H_2O from the protonated dimer. This process, however, was not detected for the larger cluster ions. This observation is in direct correspondence with observations (29) of gas-phase ion-molecule reactions

$$CH_3OH_2^+ + CH_3OH \longrightarrow (CH_3)_2OH^+ + H_2O \qquad (4)$$

$$\xrightarrow{M} CH_3OH_2^+ \cdot CH_3OH \qquad (5)$$

and

$$CH_3OH_2^+ \cdot CH_3OH + CH_3OH \xrightarrow{\quad\times\quad} (CH_3)_2OH^+ \cdot CH_3OH + H_2O \qquad (6)$$

$$\xrightarrow{M} CH_3OH_2^+(CH_3OH)_2 \qquad (7)$$

The rate for the H_2O loss from $CH_3OH_2^+ \cdot CH_3OH$ has been measured to be 5.5×10^5 s^{-1}, in good agreement with the range of 10^5 to 10^6 s^{-1} estimated for the gas-phase ion-molecule process.

Based on measurements (30) and estimates for the thermodynamics of various dissociation channels in $H^+(CH_3OH)_3$ one would expect the loss of H_2O from this and larger methanol clusters to be at least as energetically favorable as that for the CH_3OH. Nibbering and coworkers (31) have suggested that an intermediate is important in the rearrangement process which leads to the loss of H_2O. Bowers and his colleagues (29) have suggested that the major product ion in the ion-molecule reaction is the formation of the symmetrical, proton-bound dimer. It may be that rearrangement from one conformation to the other is necessary for the water loss process and that this process can occur readily only in the protonated dimer, perhaps being impeded in higher order clusters due to structural hindrance. Rapid energy dissipation in larger clusters may also reduce the possibility of rearrangement. Hence, these data reflect the importance of structural rearrangements necessary for certain unimolecular processes to be operative, and thereby give further insight into differences which may be seen in the condensed phase compared to the gas for ion-molecule reactions.

Xenon Clusters

Recently, we conducted a multiphoton ionization study of xenon clusters (32). Figure 7 shows the cluster abundances for the system Xe_n^+. Mass spectra with high-energy electron impact ionization are virtually identical (33). This observation indicates the importance of the energy liberated following ionization of the cluster in effecting the magic numbers and size distribution of weakly bound clusters as compared to the energy deposited to initiate ionization. From a study of the peak shapes of the Xe^+, Xe_2^+, and Xe_3^+ in the time-of-flight spectra under single field acceleration, the dissociation rates could be estimated. A tail was found toward longer times which is the result of rapid dissociation of larger clusters. The dissociation rates for Xe_5^+, Xe_4^+, and Xe_3^+ are found to lie in the neighborhood of 5×10^7 sec^{-1}. These are evidently the only species which are metastable in the time domain accessible by this method. The dissociation rates for larger clusters (ten or more atoms) have been reported to be much slower (34).

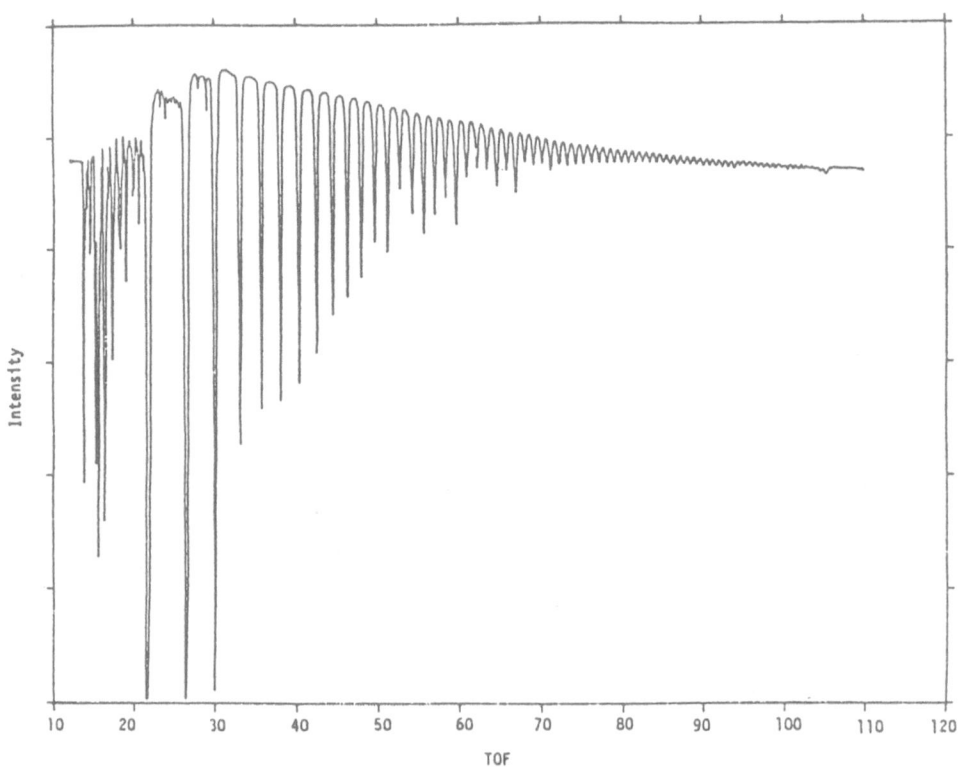

Figure 7. TOF mass spectra of Xe_n^+ obtained by multiphoton ionization of xenon clusters.

INTRACLUSTER "PENNING IONIZATION" AND EVIDENCE FOR
SLOW IONIZATION PROCESSES IN CLUSTERS

Clusters provide interesting systems for comparing ionization and concomitant electron transfer processes for bimolecular processes in the gas phase, including Penning ionization, with analogous ones in the condensed phase. Toward this goal, ionization of clusters comprised of p-xylene (PX) bound to NH_3 and $N(CH_3)_3$ were studied (35) following the absorption of photons through the perturbed S_1 state of p-xylene. An interesting comparison is provided by results of studies involving adducts of p-xylene bound to NH_3 and trimethylamine since the ionization of p-xylene is less than that of ammonia but greater than that of trimethyl-amine. In the case of $PX \cdot NH_3$, ionization by adsorption of a second photon which is adsorbed by the perturbed S_1 state of p-xylene begins near the ionization threshold of p-xylene and leads to the expected cluster ion $PX \cdot NH_3^+$. Two other channels are possible at higher photon energies, namely the formation of NH_4^+ at 0.1 eV above the ionizaton potential of p-xylene and NH_3^+ at 1.8 eV above; NH_4^+ is observed in the two-color experiments at high fluence of the ionizing laser where two photons are absorbed by the S_1 state.

By contrast, absorption into high Rydberg states of p-xylene below its ionization potential in $PX \cdot N(CH_3)_3$ leads to the production of predominantly $N(CH_3)_3^+$ with $H^+N(CH_3)_3$ as a minor product. No $PX \cdot N(CH_3)_3^+$ ion is detectable. One conclusion is that photoexcitation of p-xylene leads to an intercluster ionization process bearing analogy to Penning ionization where the perturbed high Rydberg states of p-xylene interact with the partner molecule $N(CH_3)_3$. A second, and more startling observaton, was the finding of a slow ionization process as evident in the time-of-flight peak shapes shown in Figure 8. Since the laser interacts

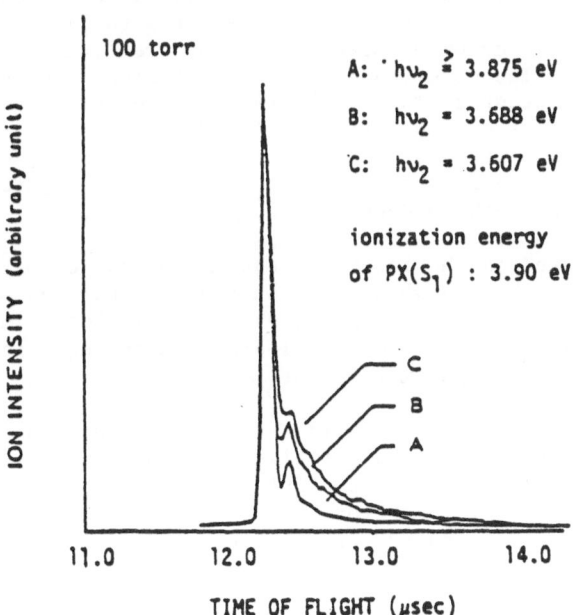

Figure 8. Ion mass peaks at different two-photon energies. Broadenings of TMA^+ ion peaks as a function of the ionization energy. The broadenings in B and C correspond to time constants of 160 ± 20 and 200 ± 20 ns, respectively. The peaks corresponding to $TMA \cdot H^+$ are also observable.

with the molecules in the first of a two-field acceleration region, a long tail is only possible when the ionization process is slow. Fragmentation leads to a knee in the peak shape and not a long tail as observed in the figure. Interestingly, the process is substantially slower with a decrease in the energy of the ionizing photon. Questions arise whether the slow step is associated with the proton transfer channel (i.e., the $(CH_3)_3NH^+$ product) or an electron transfer process (i.e., the $(CH_3)_3N^+$ product). Careful measurements with deuterated species reveal that the tail is largely associated with the electron transfer process. Interestingly, Hatano (36) has found that orientational effects in the liquid phase, where motion is restricted, can lead to a significant reduction in the rate of Penning ionization. Likewise, Harris (37) has evidence for long delays in ionization of NH_3 on silver electrode surfaces. Whether there is some analogy to the foregoing observations is currently unknown. A plausible explanation for the present findings is that a large geometry change is involved in the formation of the trimethylamine ion during the Penning-like ionization process.

SUMMARY AND FUTURE PROSPECTS

The results presented in this paper show that resonance enhanced ionization of clusters provides a detailed way of investigating the molecular aspects of condensation phenomena and the molecular properties of condensed matter at the microscopic level. Information on the spectroscopic shifts of phenylacetylene and paraxylene clustered by a number of solvent molecules are presented. While the spectral features resemble those in the condensed or matrix state, the ionization potentials are found to differ considerably from those of bulk condensed matter.

From a microscopic point of view, due to the repulsive interactions between the induced dipoles on the argon atoms, it is expected that the shift in the parents' potential would depend less than linearly on the coordination number n. Results for paraxylene·Ar_n (n=1 to 6) show a roughly linear dependence on n, with deviations from this trend for the dimer and pentamer. The observed total shift of about 750 cm^{-1} for the hexamer is to be contrasted with the matrix isolated value which is about 6000 cm^{-1} for a similar molecule (benzene) in an argon matrix. An interesting comparison is made with metallic systems where, except for shifts due to the ion image potential, the alkali metal systems display trends that mimic the bulk condensed phase rather closely.

Data is presented on the internal Penning ionization of a cluster following resonance enhanced absorption in a chromophore as a result of electron transfer within the aggregate; surprisingly long time constants ranging up to 200 ns were observed for ionization through excitations to low Rydberg states of the paraxylene chromophore.

Under certain conditions following the ionization of a cluster, reactions proceed between various moieties with the cluster ion that, in terms of products, parallel those resulting from bimolecular gas-phase ion-molecule reactions. In addition, the exothermicity of these reactions and the dielectric relaxation that occurs in newly ionized clusters lead to metastability and dissociation of the clusters. Studies with time-of-flight techniques that have been made on clusters of ammonia, methanol, xenon, and p-xylene·trimethylamine exemplify some of these processes.

Finally, it is evident that continued work on higher order metallic suboxides of varying systems promises to provide further interesting data from which to gain insight into such problems as surface oxidation, the influence of oxygen on the metallic conductivity of thin films, and the chemistry and physics of surfaces in general.

The field of cluster research is rapidly growing and the conceptual gap between the gaseous and condensed phases is closing. Hopefully, an appreciation for the importance of cluster research, the advances which are being made, and the exciting opportunities for fundamental studies pertaining to the dynamics and properties of what is sometimes termed a "fifth state of matter" has been gained.

ACKNOWLEDGMENTS

Support by the Department of Energy, Grant No. DE-AC02-82-ER60044, and the Army Research Office, Grant No. DAAG29-85-K-0215, is gratefully acknowledged.

REFERENCES

1. A. W. Castleman, Jr., in: Electronic and atomic collisions (J. Eichler, I.V. Hertel and N. Stolterfoht, Eds), Elsevier Science Publishers, Amsterdam, pp. 579-590 (1984).
2. A. W. Castleman, Jr. and R. G. Keesee, Chem. Rev. 86, 589 (1986).
3. A. W. Castleman, Jr. and R. G. Keesee, "Clusters: Properties and Formation," Ann. Rev. of Phys. Chem., in press.
4. A. W. Castleman, Jr. and R. G. Keesee, "Clusters: Bridging the Gas and Condensed Phases," Accts. Chem. Res., in press.
5. A. W. Castleman, Jr. and T. D. Mark, in: Gaseous Ion Chemistry/Mass Spectrometry (J. H. Futrell, Ed.) John Wiley and Sons, pp. 259-303(1986).
6. M. F. Vernon, D. J. Krajnovich, H. S. Kwok, J. M. Lisy, Y. R. Shen, and Y. T. Lee, J. Chem. Phys. 77, 47 (1982); R. E. Miller, R. D. Watts and A. Ding, Chem. Phys. 83, 155 (1984); P. M. Dehmer and S. T. Pratt, J. Chem. Phys. 76, 843 (1982).
7. O. Echt, P. D. Dao, S. Morgan, and A. W. Castleman, Jr. J. Chem. Phys. 82, 4076 (1985). See also O. Echt, S. Morgan, P. D. Dao, R. J. Stanley, and A. W. Castleman, Jr. Ber. Bunsenges. Phys. Chem. 88, 217 (1984).
8. J. L. Durant, D. M. Rider, S. L. Anderson, F. D. Proch, and R. N. Zare, J. Chem. Phys. 80, 1817 (1984).
9. H. Kuhlewind, U. Boesl, R. Weinkauf, H. J. Neusser, and E. W. Schlag, Laser Chem. 3, 3 (1983).
10. V. I. Karataev, B. A. Mamyrin, and D. V. Shmikk, Sov. Phys. Tech. Phys. 16, 1177 (1972); V. A. Mamyrin, V. I. Karataev, D. V. Shmikk, and V. A. Zauglin, Sov. Phys. JETP 37, 45 (1973).
11. P. D. Dao, S. Morgan, and A. W. Castleman, Jr., Chem. Phys. Lett. 111, 38 (1984).
12. P. D. Dao, S. Morgan, and A. W. Castleman, Jr., Chem. Phys. Lett. 113, 219 (1985).
13. S. Leutwyler, U. Even and J. Jortner, J. Chem. Phys. 79, 5769 (1983).
14. S. Basu, Advan. Quantum Chem. 1, 145 (1964).
15. D. H. Levy, Ann. Rev. Phys. Chem. 31, 197 (1980); A. Amirav, U. Even and J. Jortner, J. Chem. Phys. 71, 2319 (1979); 75, 3770 (1981); A. M. Griffiths and P. A. Freedman, Chem. Phys. 63, 469 (1981); A. Amirav and J. Jortner, Chem. Phys. 85, 19 (1984); and K. Rademann, B. Brutschy and H. Baumgartel, Chem. Phys. 80, 129 (1983).
16. A.-M. Sapse (personal communication)
17. K. H. Fung, H. L. Selzle and E. W. Schlag, Z. Naturforsch 36a, 1257 (1981).

18. J. Jortner, in: Vacuum Ultraviolet Radiation Physics (E. E Koch, R. Haensel, and C. Kunz, Eds.) Pergamon Press, Oxford, p. 291 (1974).

19. K. I. Peterson, P. D. Dao, R. W. Farley, and A. W. Castleman, Jr. J. Chem. Phys. 80, 1780 (1984).

20. A. Herrmann, S. Leutwyler, E. Schumacher, and L. Woste, Helv. Chem. Acta 61, 453 (1978).

21. J. Buttet, Proc. Int. Symp. on Metal Clusters, Heidelberg, Apr. 7-11, 1986, p. 12.

22. A. W. Castleman, Jr. and R. G. Keesee, "Metallic Ions and Clusters: Formation, Energetics, and Reaction," Zeitschrift fur Physik, in press.

23. J. L. Martins, J. Buttet, and R. Car, Phys. Rev. B 31, 1804 (1985).

24. M. M. Kappes, M. Schar, P. Radi, and E. Schumacher, J. Chem. Phys. 84 1863 (1986).

25. P. D. Dao, K. I. Peterson, and A. W. Castleman, Jr., J. Chem. Phys. 80 563 (1984).

26. A. W. Castleman, Jr., S. Morgan, O. Echt, and P. D. Dao, "Considerations of the Origin of Magic Numbers in Hydrogen Bonded Clusters," to be submitted.

27. S. Morgan, R. G. Keesee and A. W. Castleman, Jr., "Studies of Clusters Using Laser Techniques: Dissociation Processes of Methanol Clusters Following Multiphoton Ionization," Proc. 1986 CRDEC Scientific Conference on Obscuration and Aerosol Research, Aberdeen Proving Ground, MD, June 23-27, 1986, in press.

28. S. Morgan and A. W. Castleman, Jr., "Evidence of Delayed 'Internal' Ion Molecule Reactions Following the Multiphoton Ionization of Clusters: Variation in Reaction Channels in Methanol with Degree of Solvation," submitted to J. Am. Chem. Soc.

29. L. M. Bass, R. D. Cates, M. F. Jarrold, N. J. Kirchner, and M. T. Bowers, J. Am. Chem. Soc. 105, 7024 (1983).

30. R. G. Keesee and A. W. Castleman, Jr., J. Phys. Chem. Ref. Data, 15, 1011 (1986).

31. J. C. Kleingeld and N. M. M. Nibbering, Org. Mass Spectrom. 17, 136 (1982).

32. O. Echt, M. Cook and A. W. Castleman, Jr., "Multiphoton Ionization of Rare Gas Clusters: Xe_n," Chem. Phys. Lett., submitted.

33. O. Echt, K. Sattler, and E. Recknagel, Phys. Rev. Lett 47, 1121 (1981).

34. D. Kreisle, O. Echt, M. Knapp, and E. Recknagel, Phys. Rev. A 33, 768 (1986).

35. P. D. Dao and A. W. Castleman, Jr., J. Chem. Phys. 84, 1435 (1986).

36. Y. Hatano (personal communication); see also T. Wada, K. Shinsaka, H. Namba, and Y. Hatano, Can. J. Chem. 55, 2144 (1977).

37. C. Harris (personal communication) Univ. of California, Berkeley.

TRAPPING SITES AND VIBRATIONAL MODES OF Ag_2 IN RARE GAS MATRICES:

A COMPUTER SIMULATION

Paul S. Bechthold* and Herbert R. Schober

Institut für Festkörperforschung der Kernforschungsanlage Jülich,
D-5170 Jülich, FRG

ABSTRACT

Computer simulations of disilver in fcc rare gas host lattices demonstrate, in agreement with experiment, that in xenon, krypton, and argon matrices one, two, and three trapping sites can be occupied, respectively. They can be identified by the observed and calculated vibrational frequencies. The observed external modes are found to be librations of the Ag_2 molecule.

I. INTRODUCTION

The density of small metal molecules and clusters in free jet expansions is too small for many spectroscopic investigations. Therefore, for such studies the particles are often isolated in rare gas matrices, where they are accumulated, stabilized, and suffer the least perturbations as compared to the gas phase. The perturbations, however, are not negligible and have to be understood in detail. On the other hand the perturbations provide fundamental information on guest-host interactions of metal particles trapped in solids. Thus, from two points of view we need to understand the effects of the matrices.

In recent optical absorption[1-4] and resonance Raman[4-7] measurements on silver particles isolated in rare gas matrices it was found that silver dimers get isolated in one, two, and three differing trapping sites in xenon, krypton and argon matrices, respectively. A single Ag_2 absorption band is observed in a xenon matrix at 393 nm. The corresponding absorption band in krypton matrices shows an asymmetric broadening, indicating a second trapping site. Two clearly split bands are seen in argon matrices. In addition a third absorption band in an argon matrix could be unambiguously attributed to Ag_2 by resonance Raman spectroscopy.[4,7] The resonance Raman spectra show characteristic internal and external modes of the dimers. The internal modes suffer small shifts induced by the matrix cages, while the external modes directly probe the interaction of the molecules with the matrix support at the particular trapping site. The vibrational bands of different sites are sharp, reproducible and clearly separated. This proves that the sites are structurally well defined and that the molecules are not trapped at grain boundaries of the polycrystalline matrix. Based on these premises we can study the matrix effects in a computer simulation of Ag_2 in rare gas matrices. It allows us to identify the geometries of the various trapping sites and also to assign the external modes. The results are summarized below. Further details will be described elsewhere.[8] To our knowledge it is the first identification of trapping sites and external modes of metal molecules in rare gas matrices.

II. COMPUTATIONAL MODEL

Computations are done in two steps: First we calculate the relaxed (meta)stable configurations of a silver dimer embedded in a fcc-rare gas lattice. In the second step the vibrational frequencies and eigenvectors for these relaxed configurations are calculated by diagonalization of the dynamical matrix. To find the relaxed configurations we compute the local minima of the potential energy for a crystallite of approximately 1,500 moveable atoms which is surrounded by an infinite, undistorted fcc-host lattice and includes the silver dimer at its center. The situation is schematically depicted in Fig 1. The change of the ground state kinetic energy (zero point motion) between the various configurations is neglected. For the heavier rare gases (Ar to Xe) it would only marginally shift the relative energies of the various configurations. The calculations were done with a modified version of the program DEVIL, originally developed at AERE Harwell.[9]

The potential energy is assumed to be composed of centrosymmetric pair potentials. The pair potential for the Ag_2 ground state is constructed from gas phase spectroscopic data[10-12] using the Rydberg-Klein-Rees (RKR) method.[13] For the rare gas - rare gas interaction we took a Morse potential[14] because it best reproduced the phonon dispersion curves of the host lattice, but use of a Lennard-Jones (6-12) potential[15] led to essentially the same results. The silver-rare gas interaction potential is unknown and, therefore, was approximated by a purely repulsive Born-Mayer form $V(r)=Ae^{-Br}$, where the parameters A and B were obtained from fits to Thomas-Fermi-Dirac interaction energy calculations.[16,17] This crude approach is justified when the defect properties are mainly determined by the repulsive part of the potential. This is mostly the case.

III. RESULTS AND DISCUSSION

The local symmetries at the trapping sites depend on the alignment of the metal molecules with respect to the crystallographic axes of the host lattice. The molecule can occupy either a single or a double vacancy site in the host lattice. We find in agreement with experiment that one, two and three (meta)stable configurations can be occupied in xenon, krypton and argon matrices, respectively. A single vacancy configuration with D_{4h} symmetry and the molecule aligned in the <100> direction is stable in all

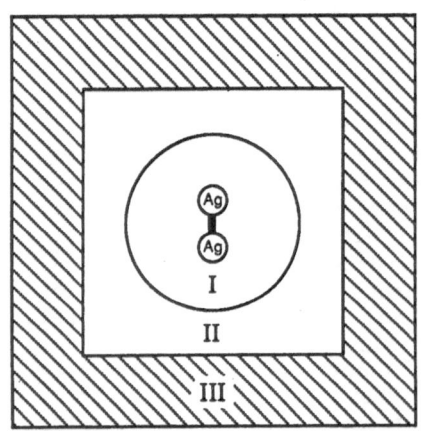

Fig. 1.
Schematic illustration of the computational model. The Ag_2 molecule is initially approximately centered in a crystallite of up to 1,500 atoms (regions I and II). The atomic positions in this crystallite are relaxed in the energy calculations, while the atoms in region III are held at their ideal fcc-positions. Vibrations are calculated for the subcrystallite I of up to 420 atoms with the atoms in II held at their relaxed positions.

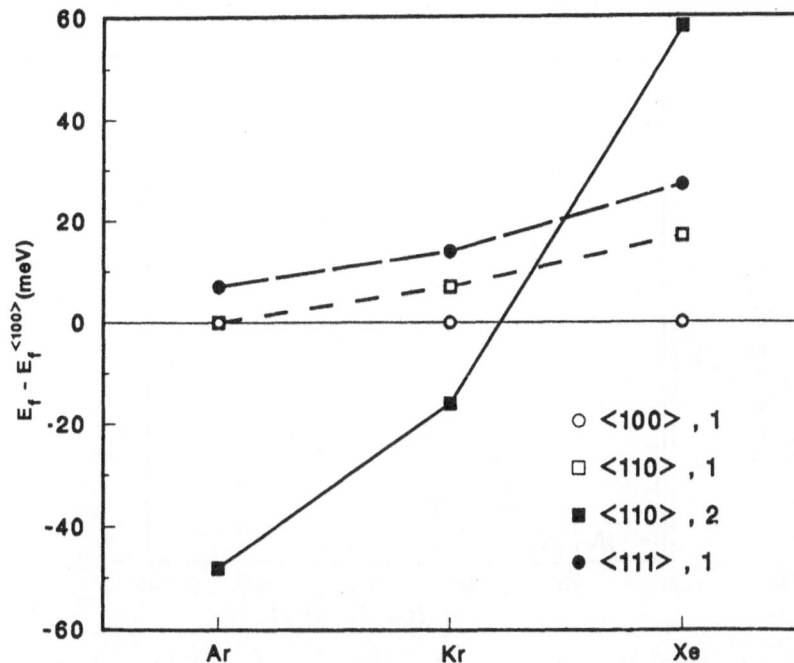

Fig. 2. Minimum relative energies of various configurations of the Ag_2 molecule in Ar, Kr, and Xe matrices, respectively. The legend gives the orientation of the molecule and the number of vacancies occupied. The minimal energies all correspond to centrosymmetric trapping sites of D_{4h}, D_{2h}, and D_{3d} symmetry, respectively. The configurations in the lower part of the figure represent (meta)stable sites. The configurations in the upper part are unstable with respect to small displacements of the molecular orientation, except for the double vacancy site in xenon (see text).

three matrices. The most stable configuration in krypton and argon matrices is the double vacancy-site with D_{2h} symmetry and the molecule aligned along ⟨110⟩. The third stable configuration in argon has D_{2h} symmetry with the molecule aligned along ⟨110⟩ in a single vacancy site. Figure 2 shows for all three host lattices the energies E_f of various configurations relative to the one with D_{4h} symmetry in a single vacancy site E_f⟨100⟩. Plotted are the minimal energies for alignments and vacancy occupations of the molecules as given in the legend of the figure. The most striking feature is the sharp increase of the relative energy of the divacancy configuration for xenon. This is due to the large energy needed to remove the additional host atom in creating the double vacancy and can be understood in terms of the atomic radii. Although metastable this site is energetically so unfavorable that it is not formed. The single vacancy configurations with ⟨111⟩ orientation (D_{3d} symmetry) are always unstable to small distortions and relax to one of the stable configurations. This was found to be true for all configurations of lower symmetry, too, in which the Ag_2 was initially inserted at random positions. The monovacancy D_{2h} and D_{4h} configurations in argon have nearly equal energies.

After finding the (meta)stable configurations we calculated eigenfrequencies and eigenvectors of the dynamical matrix for a crystallite of up to 420 atoms (region I in Fig 1), centered around the Ag_2 and embedded in the relaxed host crystal as obtained in the energy minimization procedure. Here, we consider only those vibrations in which the Ag_2 has a strong

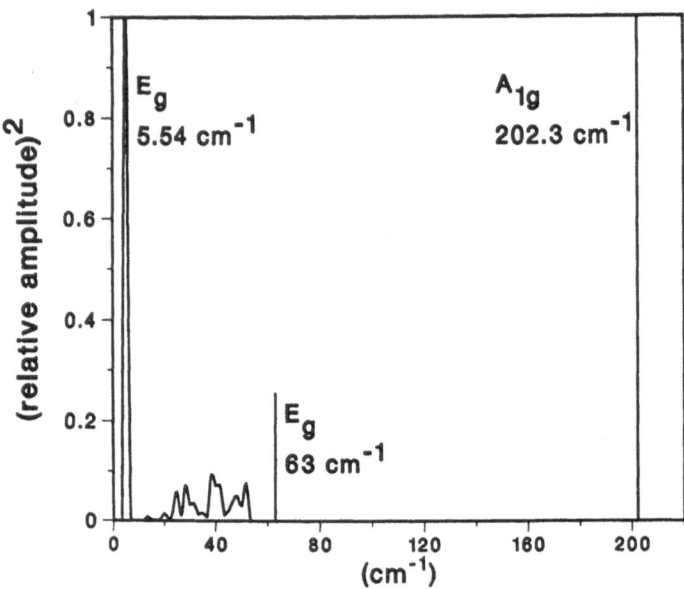

Fig. 3. Spectral distribution of disilver vibrational amplitudes
squared for a molecule trapped in a krypton matrix with
⟨100⟩ orientation in a D_{4h}-monovacancy site.

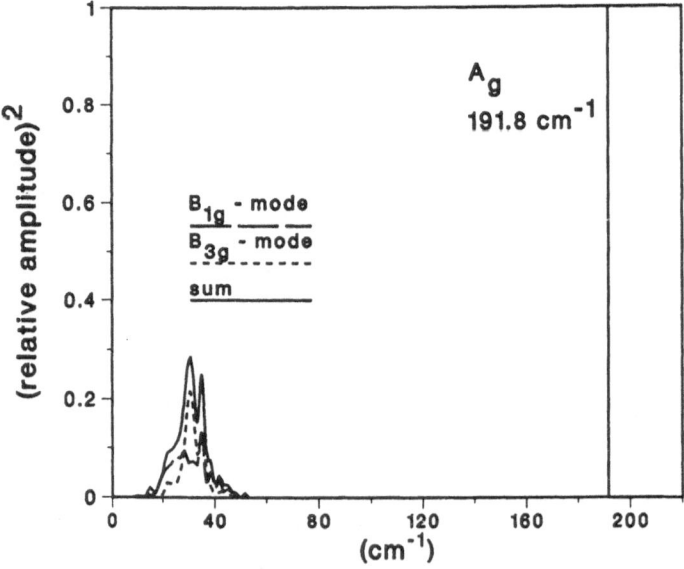

Fig. 4. Spectral distribution as in Fig. 3, but for a molecule
in a divacancy site with D_{2h} local symmetry and ⟨110⟩
orientation.

amplitude and which are in accordance with the Raman selection rules. These are the only ones that may show up in the resonance Raman spectra. They are A_{1g} and E_g modes for sites with D_{4h} symmetry and A_g, B_{1g}, and B_{3g} modes for sites with D_{2h} symmetries.[18,19]

Figures 3 and 4 show the spectral distributions of the squares of the disilver vibrational amplitudes for the two (meta)stable sites in krypton matrices. These are normalized to the square of the lengths of the total respective vibrational eigenvectors. The bond stretching frequency of the molecule in the double vacancy site is close to the gas phase value. In the monovacancy configuration the molecular bond length is slightly reduced because of the pressure exerted by the lattice. Hence the Ag-Ag stretching frequency is slightly increased. Since the stretching frequency is more than three times larger than the maximum phonon frequency of the host lattice, the lattice cannot follow these vibrations and in turn the Ag_2 participates as a rigid molecule in the lattice vibrations. The local compression in the monovacancy configuration causes the occurrence of a localized external mode with a frequency above the band of lattice frequencies and a resonant mode at a very low frequency which is strongly localized around the Ag_2. The lattice compression is much smaller in the divacancy configuration. No localized external mode appears, but there is still a broad distribution of resonant vibrations in the range of the phonon frequencies. Similar results are obtained for the other matrices.

In the elementary classical theory[19] of Raman scattering the contribution of a particular normal mode to the intensity is proportional to the square of its vibrational amplitude. Therefore, in table 1 we compare the calculated frequency distributions with the Raman frequencies observed.

Table 1. Spectroscopic data and mode assignments of Ag_2 molecules isolated in rare gas matrices.

Matrix gas:	Xe	Kr		Ar		
Molecular orientation:	<100>	<100>	<110>	<100>	<110>	<110>
Vacancies occupied:	1	1	2	1	2	1
Site symmetry:	D_{4h}	D_{4h}	D_{2h}	D_{4h}	D_{2h}	D_{2h}
Stretching mode (cm^{-1})						
exp.[a]	198	203.1	194.4	c	193.8	190.5
calc.	A_{1g} 201	A_{1g} 202	A_g 192	A_{1g} 200	A_g 193	A_g 201
External localized mode (cm^{-1})						
exp.[a]	47.5	57	–	c	–	70
calc.	E_g 49.7	E_g 63		E_g 79.8		B_{1g} 70 B_{3g} 70.9
Resonant mode (cm^{-1})						
exp.	–	–	40[b]	–	46[b]	
calc.	E_g 7.3	E_g 5.54	B_{1g} 30[b] B_{3g} 31[b]	E_g 2	B_{1g} 28[b] B_{3g} 28[b]	B_{1g} 3.24 B_{3g} 6.94
Absorption band (nm):	393	389	398	385	410	443

[a]Resolution ± 0.5 cm^{-1}; [b]Center of frequency distribution; [c]Not measured.

As is obvious, we can identify the local symmetry of the Ag_2 molecules with the aid of the calculated spectra. The deviations from the experimental data occurring in the low energy range may be ascribed to the uncertainties of the potentials, particularly the Ag-rare gas potential. Therefore only one major discrepancy has to be explained; it concerns the bond stretching frequency of the dimer trapped in the D_{2h} monovacancy site in argon. In this geometry the molecule is more strongly compressed than in the corresponding divacancy site, where it has the same <110> orientation with respect to the host lattice. Therefore, even with better potentials our model would predict an increase of the vibrational frequency. The experiment, however, shows a decrease, even with respect to the gas phase value of 191 cm^{-1}. Together with the unusual red shift of the optical absorption band with respect to the gas phase value of 435 nm this seems to indicate a change in the electronic structure of the dimer, which is not accounted for by our model.

CONCLUSION

Using a computer simulation and reasonably chosen interaction potentials we have been able to quantitatively interpret the resonance Raman spectra of matrix isolated silver dimers. The technique may be extended to larger particles, once reliable potential energy surfaces are given. Actually, what is needed is the potential surface in the vicinity of the local minima. In connection with matrix spectroscopy the method may then be developed to test ab inito calculated potential energy surfaces.

Acknowledgments: PSB thanks the Chemistry Division of Argonne National Laboratory, particularly S. J. Riley, E. K. Parks, J. Jellinek, K. Liu, L. G. Pobo, W. F. Hoffman, and S. K. Cole for their generous hospitality.

REFERENCES

*Current address: Chemistry Division, Argonne National Laboratory, Argonne, IL 60439, USA

[1] D. M. Gruen and J. K. Bates, Inorg. Chem. **16**, 2450 (1977).
[2] D. Leutloff and D. M. Kolb, Phys. Chem. **83**, 666 (1979).
[3] S. A. Mitchell, G. A. Kenney-Wallace, and G. A. Ozin, J. Am. Chem. Soc. **103**, 6030 (1981).
[4] P. S. Bechthold, U. Kettler, and W. Krasser, Surf. Sci. **156**, 875 (1985).
[5] U. L. Kettler, Ph.D. Thesis, Universität zu Köln (1984); (Berichte der Kernforschungsanlage Jülich No. 1980 (1985)).
[6] P. S. Bechthold, U. Kettler, and W. Krasser, Solid State Comm. **52**, 347 (1984).
[7] P. S. Bechthold, U. Kettler, H. R. Schober, and W. Krasser, Z. Phys. **D3**, 263 (1986); P. S. Bechthold, this conference.
[8] P. S. Bechthold and H. R. Schober, to be published.
[9] H. R. Schober, J. Phys. **F7**, 1127 (1977).
[10] C. M. Brown and M. L. Ginter, J. Mol. Spectr. **69**, 25 (1978).
[11] V. I. Sradanov and D. S. Pešić, J. Mol. Spectr. **90**, 27 (1981).
[12] K. A. Gingerich, J. Crystal Growth **9**, 31 (1971); K. Hilpert and K. A. Gingerich, Ber. Bunsenges. Phys. Chem. **84**, 739 (1980).
[13] H. Telle and U. Telle, Comp. Phys. Comm. **28**, 1 (1982).
[14] H. R. Glyde, J. Phys. **C3**, 810 (1970).
[15] J. H. Jaffe, A. Rosenberg, M. A. Hirshfeld, and N. M. Gailar, J. Chem. Phys. **43**, 1525 (1965).
[16] A. A. Abrahamson, Phys. Rev. **178**, 76 (1969).
[17] I. M. Torrens, Interatomic Potentials, Academic Press, New York (1972).
[18] W. Ludwig, Ergebnisse der exakten Naturwissenschaften **35**, 1 (1964).
[19] E. B. Wilson, J. C. Decius, and P. C. Cross, Molecular Vibrations, McGraw-Hill, London (1955).

LASER-INDUCED FLUORESCENCE AND RAMAN INVESTIGATIONS OF Ag_n ($n \geq 4$)

BY MEANS OF MATRIX ISOLATION SPECTROSCOPY

Ulrich L. Kettler[†], Paul S. Bechthold[‡], and Wolfgang Krasser

Institut für Festkörperforschung, Kernforschungsanlage Jülich

Postfach 19 13, D–5170 Jülich, W. Germany

ABSTRACT

Results of laser-induced fluorescence and resonance Raman investigations of silver microclusters isolated in cryogenic rare-gas matrices are reported. The corresponding emission and Raman bands are correlated with two absorption bands in the green spectral range which have been attributed to silver tetramers or hexamers ($Ag_{4;6}$).[1] The same emission and Raman bands show up at violet excitation, too, giving evidence of a transition to a higher electronic state. The emission bands of one species in Kr and Ar matrices show an indication of a vibrational fine structure. A computer fit to the experimental data gives a vibrational spacing of $\approx 125\,cm^{-1}$ in good agreement with the Raman data. The other species shows modes in the $68-77\,cm^{-1}$ and $150-163\,cm^{-1}$ range depending on matrix gas and the trapping site occupied.

INTRODUCTION

Small silver particles have become one of the most intensively studied cluster systems[2] because they are of fundamental scientific and technological importance. Silver is used as a catalyst for the controlled oxidation of olefins.[3] It is also the most often used substrate for Surface Enhanced Raman Scattering which is induced by aggregates on the bulk metal surface.[4] Recently silver tetramers have attracted particular attention. Roy and Furtak[5] suggested that Ag_4 or Ag_4^{n+} aggregates ($n=1;3$) on the surface of SERS-active silver electrodes are partially responsible for the Raman enhancement. Furthermore, Ag_4 or Ag_4^+ clusters seem to be the most active catalytic centres in silver halide photography.[6] Additional spectroscopic data will be necessary for a more detailed understanding of the processes involved.

Most of our spectroscopic knowledge about silver aggregates arises from studies of such particles when isolated in cryogenic matrices, particularly rare-gas matrices.[1,7] Absorption bands of xenon-isolated silver particles at about 535/557 nm have been tentatively attributed to Ag_4 or Ag_6 from concentration dependence measurements.[1] The assignment to Ag_4 is in good agreement with results of SCF-Xα-SW calculations for a tetrahedral silver tetramer, giving an A ← X transition at 450–550 nm [8,9] and an B ← X transition at 400 nm.[8] — Corresponding absorption bands have been reported at $\approx 510/524$ nm with Kr matrices, at $\approx 494/503$ nm with Ar matrices, and at ≈ 495 nm with Ne matrices.[1,10] Here we correlate laser-induced fluorescence and Raman bands with these absorption bands. — The experimental arrangement has been described elsewhere.[10,11] For excitation we use the emission lines of argon- and krypton-ion lasers and a *cw* dye laser tunable in the 416–510 nm range with two different dyes. The silver concentration is about 1–2%.

[†] Present address

Department of Physics
University of California
Santa Barbara, CA 93106

[‡] Current address

Chemistry Division
Argonne National Laboratory
Argonne, IL 60439

RESULTS

Laser-Induced Fluorescence

Fig. 1 shows emission spectra obtained with Kr matrices. The fluorescence intensity of the bands at 545/625 nm increases when the laser wavelength is changed from 496.51 nm to 514.52 nm (Fig. 1a–c), and a hump is built up peaked at 558 nm. Using the 530.87 nm laser line the fluorescence bands become distinctly weaker (Fig. 1d). The band at 625 nm almost disappears, the hump becomes a weak shoulder, and the luminescence breaks down below 543 nm. As the fluorescence band at 625 nm can hardly be excited by the 530 nm laser line, it is more probably correlated with the 510 nm absorption band than with the one at 524 nm. The same fluorescence bands are also observed with violet (406–430 nm) (e. g., Fig. 2a) and near UV ($\approx$ 350 nm) laser excitation. We also see the corresponding emission bands from Ar and Ne matrices as given in Table 1. The similarity of the fluorescence bands for the three matrices indicates that the same species are involved. — If Kr matrices are produced at substrate temperatures below 12 K one observes a side band of the 625 nm emission at 638 nm (Fig. 2a) which disappears on annealing of the matrix. The new band is probably due to a thermally unstable secondary trapping site of the same cluster that causes the 625 nm emission. Both bands show an indication of a similar vibrational fine structure. Similar observations are made for Ar matrices although the existence of a second trapping site is not so obvious in this case (Fig. 2b). Fig. 2 shows in addition the results of computer fits (thick lines) to the experimental data (dots) of the bands at 625/638 nm and 609 nm for Kr and Ar matrices, respectively. Each of the two superimposed thin, dotted curves — one for each trapping site — is a superposition of several Lorentzians, representing a vibrational progression. Also for the Ar matrix a fit to the data with a single progression is not satisfying. The spacing of the Lorentzians is (125 ± 5) cm^{-1} for both trapping sites in both matrices at a FWHM value of (118 ± 8) cm^{-1}.

With Xe matrices a fluorescence band is observed at about 640 nm when using Ar$^+$ or Kr$^+$ laser excitation between 496.51 nm and 530.87 nm. It cannot be excited with the 568.19 nm laser line. This indicates a correlation with the 535 nm absorption band. The emission is weak with respect to the disilver[10,12] and most of the trisilver[13] fluorescence bands. An emission correlated to the 557 nm absorption band has not yet been identified. If the absorption bands are very strong they are accompanied by a shoulder at about 410 nm on the low-energy slope of the disilver absorption band. The appearance of this shoulder is associated with a weak fluorescence at 460 nm which can be observed at excitation in the 420–430 nm range. It it accompanied by the much stronger luminescence attributed to trisilver and hidden by the disilver emission at 480 nm if excitation below λ=420 nm is used.

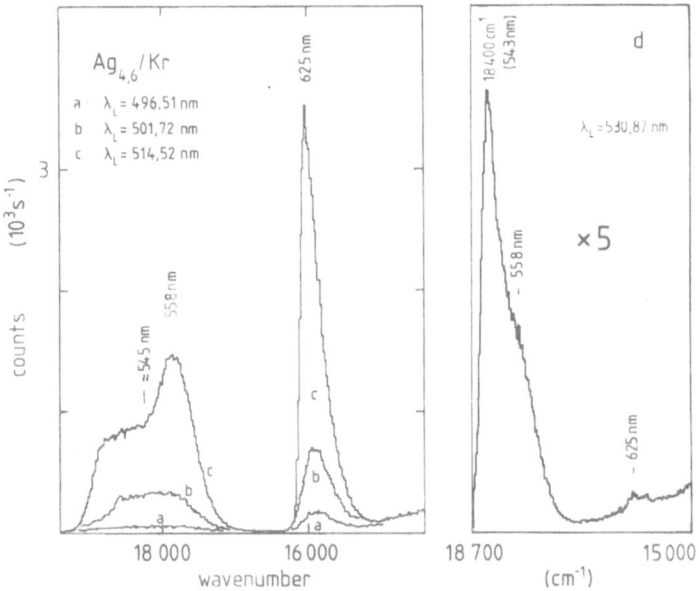

Fig. 1   Fluorescence spectra of Ag$_{4;6}$/Kr for four excitation wavelengths within the A←X absorption bands. As the excitation wavelength approaches the absorption peaks the fluorescence becomes stronger (a–c). The band at 625 nm is almost absent at 530 nm excitation (d), making a correlation to the 510 nm absorption band more likely. Note the fivefold expansion.

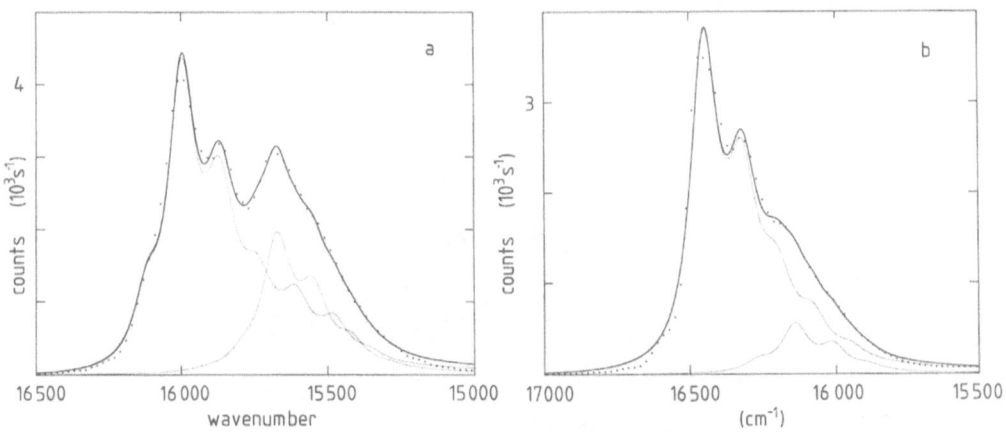

Fig. 2 Computer fits to the experimental data (dots) of the emission bands at 625/638 nm with Kr
matrices (λ_L=415.45 nm)(a) and at 609 nm with Ar matrices (λ_L=406.74 nm)(b). The thin lines
represent vibrational progressions of two different trapping sites, the thick lines their superpositions.
The cluster on the second trapping site in Ar matrices emits a fluorescence peaked at 620 nm.

Raman Scattering

Raman scattering experiments are performed with Xe, Kr, and Ar matrices. Fig. 3 shows the
development of Raman modes from Xe matrices at 123 cm^{-1} and 158 cm^{-1}, respectively, using dye-laser
radiation in the 480.91–493.87 nm range. These two lines are weak or absent at irradiation in the 430–
460 nm range,[13] thus are preresonantly enhanced in Fig. 3. The modes also become strong at violet
irradiation (Fig. 3 left). Closer to resonance, the lines — first the one at 123 cm^{-1}, then (λ_L > 500 nm)
the other one — get weaker because of reabsorption or a decrease in the number of scattering centres
due to the reported phototransformation.[10,14] An additional line at 111 cm^{-1} is the preresonant mode
reported for Ag$_3$ (absorption band at 442 nm).[13] — With Kr matrices and 514.52 nm laser radiation, we
observe a single Raman mode at 121 cm^{-1} on the high-energy slope of a fluorescence band. At lower
temperature we observe a broader structure centred at 123 cm^{-1} with violet laser emission. It is probably
due to the superposition of two modes resulting from the two trapping sites observed in fluorescence.

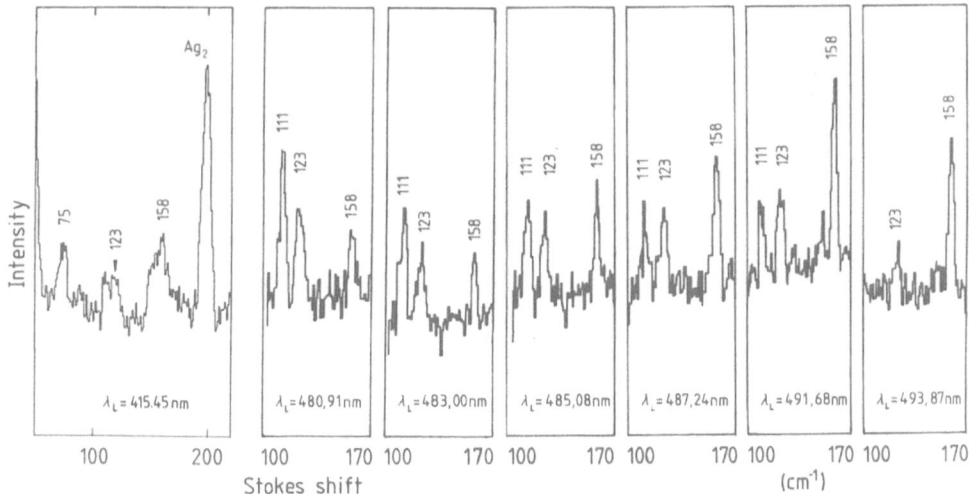

Fig. 3 Development of Raman bands at 123/158 cm^{-1} of a Xe matrix within the preresonance region
of Ag$_{4;6}$. The lines are weaker at 493 nm radiation because of the photosensitivity of the carrier
particles or beginning reabsorption. The single spectrum on the left results from the X-B resonance.

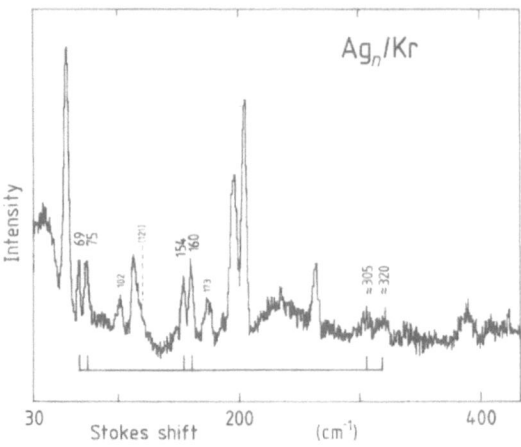

Fig. 4 Resonance Raman spectrum of Ag_n/Kr at Ag_2 and $Ag_{4;6}$ resonance; $\lambda_L = 406.74$ nm. The strongest bands are Ag_2 modes. We see high-energy modes of two trapping sites of $Ag_{4;6}$ at $154/160$ cm^{-1}, their overtones, and their corresponding low-energy modes at $69/75$ cm^{-1}. The mode at 121 cm^{-1} is superimposed by the overtone of a disilver libration and probably by an Ag_3 mode.[11,13,14]

Fig. 4 shows a wide-range resonance Raman spectrum of an annealed Kr matrix using 406.74 nm laser radiation. This wavelength is at the low-energy slope of the disilver absorption bands. Therefore the strongest Raman bands are the internal and external modes of Ag_2 in two different trapping sites, their overtones, and their combination bands.[11,14] In addition we observe lines at $154/160$ cm^{-1} and their weak overtones at $\approx 305/320$ cm^{-1} (resonance condition). These lines are accompanied by two low-energy modes at 69 cm^{-1} and 75 cm^{-1}. Two other modes at $102/173$ cm^{-1} cannot be correlated with the former group of modes and may be due to other aggregates. Under similar conditions with Xe matrices two low-energy modes appear at $68/75$ cm^{-1} together with the lines at $150/158$ cm^{-1}. With Ar matrices and X-A resonance (494 nm absorption band) we observe a line at 126 cm^{-1} and its weak overtone (Fig. 5a). Near X-B resonance modes are observed at $77/163$ cm^{-1} and 126 cm^{-1} (Fig. 5b) which by analogy with the spectra for Kr and Xe matrices may be assigned to the $Ag_{4;6}$ species (The line at 126 cm^{-1} may be partially superimposed by an insufficiently suppressed plasma line of the krypton-ion laser at 408.83 nm).

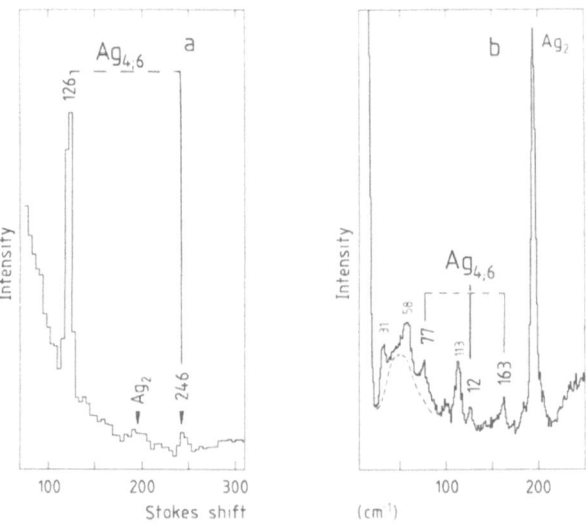

Fig. 5 Resonance Raman spectra of Ag_n/Ar. (a): $\lambda_L = 487.98$ nm. We see a mode at 126 cm^{-1} and its weak overtone. (b): $\lambda_L = 406.74$ nm. The modes at $77/163$ cm^{-1} and 126 cm^{-1} are attributed to $Ag_{4;6}$ by comparison with the spectra of Kr and Xe matrices. The strong line at 194 cm^{-1} is the Ag_2 stretching vibration, the dashed band marks the distribution of disilver librations.[11,14] The lines at $31/58/113$ cm^{-1} belong probably to Ag_3.[15]

Table 1   Absorption, laser-induced fluorescence, and Raman data of Ag_n $(n = 4; 6)$.

matrix gas	Xe		Kr		Ar		Ne
absorption λ/nm							
A ← X	535	557	510	524	494	503	495
B ← X	410 (±5)		(406±5)[a]		(400±10)[a]		(400±10)[a]
fluorescence λ/nm							
A' → X	640		625[b]	≈545	609[b]	≈523	
B ⤳ A' → X			625[b]/638[bc]	558[d]	609[b]/620[bc]	528[d]	602 515
B' → X		460					
Raman[e] $\|\Delta\sigma\|$/cm^{-1}							
X-A resonance	123	158	121		126[f] /246[fg]		
		68/75		69/75		77	
X-B resonance	123	150/158	123[h]	154/160	126	163	
				305/320[i]			

[a] estimated peak position
[b] vibrational fine structure with spacing of (125 ± 5) cm^{-1}
[c] thermally unstable second trapping site
[d] shoulder on the low-energy slope of the Ag_2 emission
[e] ± 2 cm^{-1}
[f] ± 4 cm^{-1}
[g] overtone of the line at 126 cm^{-1}
[h] probably unresolved band from two trapping sites
[i] overtones of the lines at 154/160 cm^{-1}

DISCUSSION

With regard to the systematic blue shifts the fluorescence bands at 640/625/609/602 nm with Xe, Kr, Ar, and Ne matrices, respectively, belong to corresponding trapping sites of the same species (An asymmetric fluorescence band at 637 nm has been reported earlier for Cu_4/Xe.[8]). They are directly correlated with the A ← X absorption bands at 535/510/494/495 nm, respectively, but show up also at violet excitation for Kr, Ar, and Ne matrices*. Therefore, a higher excited electronic state (B) should exist at about 400 nm. This has been predicted from model calculations for a tetrahedral Ag_4.[8] With Kr and Ar matrices the fluorescence investigations prove the existence of secondary, thermally unstable trapping sites of the corresponding species. This matrix splitting has not been observed in absorption spectra. A computer fit to the experimental data of these emission bands gives a vibrational spacing of (125 ± 5) cm^{-1} for both matrices.

The fluorescence bands at 545 nm with Kr and at 523 nm with Ar matrices are correlated with the absorption bands at 524/503 nm, respectively. For Ne matrices the concentration of the species with the 515 nm emission band was too low to be detected in the absorption measurements.

With Xe matrices strong absorption bands in the green spectral range are accompanied by a shoulder at about 410 nm on the low-energy slope of the disilver absorption transition. This shoulder probably represents the B ← X transitions of $Ag_{4;6}$/Xe. The emission band at 460 nm is therefore tentatively assigned to the B' → X emission of the relaxed particle / matrix-cage unit. With other matrices the B ← X bands are more strongly superimposed by the disilver absorption bands, but we estimate their positions as given in Table 1. This results from Raman investigations which give resonant or preresonant conditions in this as well as in the green spectral range, but not in between. All observed lines seem to be (pre)resonantly enhanced, which might be expected at the small concentrations of the clusters involved.

Raman lines at 121–126 cm^{-1} under X-A or X-B (pre)resonance conditions correspond to vibrational fine structures observed in the fluorescence spectra and, therefore, are assigned to the absorption bands as shown in Table 1. The higher vibrational wavenumber for Ar matrices can be explained by stronger repulsive cage forces in the Ar rare-gas lattice. The Raman line with Kr matrices probably is the same one which was tentatively assigned to Ag_3 by Schulze et al.[17] In fact, in their spectrum this line should be preresonantly enhanced when the 488 nm laser line is used.

Raman experiments under X-A preresonance conditions with Xe matrices and under X-B (pre)resonance conditions with Xe, Kr, and Ar matrices show additional modes at 150/158 cm^{-1} for Xe, at

---

* With Xe matrices and violet excitation the 640 nm emission band is strongly superimposed by the trisilver emission bands at 555 nm and 624/659 nm.[13]
The emission band reported at about 2.05 eV by Schrittenlacher et al.[16] using 4.68 eV synchrotron radiation for the excitation of higher Ag_2/Ne states is probably the line we are reporting at 602 nm.

PHOTOELECTRON SPECTROSCOPY OF MOLECULAR CLUSTERS:

N_2 AND N_2O

Frank Carnovale, J. Barrie Peel and Richard G. Rothwell

Department of Chemistry, and Research Centre for Electron
Spectroscopy, La Trobe University, Bundoora, Victoria 3083
Australia

HeI photoelectron spectra of gas phase clusters of
small molecules are presented for the first time. Using a
pulsed jet preparation system the spectra of molecular
beams of N_2 and N_2O respectively are analysed to give data
on the percentage condensation. An extension of the
dielectric polarization relaxation model which considers
spherical clusters shows that shifts in vertical
ionization energies are approximately linear with log N
for clusters of N molecules. The model should be useful
in the $N = 10^2 - 10^4$ size range.

INTRODUCTION

While the study of the electronic structure of atomic cluster species
has seen both experimental and theoretical developments[1,2], comparable
studies of molecular cluster systems have only been undertaken at the
theoretical level for relatively small clusters. Spectroscopic studies
of atomic and molecular dimer species[3] have been fairly numerous in
recent years, but the extension to higher clusters, trimer, tetramer, etc.
has been rare with large clusters considered even less . Theoretical
studies of atomic clusters, particularly of metals, have been extensive[2]
but related experimental observations have been complicated by the physical
reality that clusters are produced as mixtures with a range of cluster
sizes. Mass spectrometric measurements of ionized clusters have provided
much of the experimental data, but vibrational spectroscopy has also
contributed[3]. However it is recognized that ion cluster distributions can
differ from the neutral cluster distributions from which they derive.
As well many measurements only provide indirect evidence of the nature of
the electronic structural effects which influence neutral cluster formation.

Photoelectron spectroscopy using UV radiation has recently been
utilized in the study of rare gas dimers[4] and intermolecular complexes
including van der Waals dimers[5] . Only one example of a trimolecular
system has been reported[6] . While the recent development of resonant
multiphoton ionization photoelectron spectroscopy offers cluster-size
selectivity as does the photo ion-photoelectron coincidence technique ,
the traditional single-photon ionization photoelectron experiment is able
to provide data which is representative of the properties of the neutral
clusters and provides a direct link to the related ionic clusters. As well,

the resulting spectra can be considered as an important addition to the various measurements available for the gaseous species (monomer) through to the liquid[7] and solid (condensed gas)[8] systems. Since gas-phase clusters are known to demonstrate both liquid-like and crystal-like structures[9], direct comparisons of cluster photoelectron spectra with both gaseous monomer and condensed gas spectra are important.

EXPERIMENT

The HeI photoelectron spectra of molecular clusters were measured on a double-chamber spectrometer incorporating a pulsed jet source operated at room temperature . Clusters of N_2 and N_2O respectively were obtained using 10-25% mixtures of the gases in He at stagnation pressures of 1-10 atm. The pulsed valve (Lasertechnics) was operated at pulsewidths of 0.5-2 ms and repetition rates of 15-100 Hz representing duty cycles of approximately 20% at low stagnation pressure ($P_O \sim 1$ atm) and approximately 1% at high stagnation pressure ($P_O \sim 10$ atm). The average working chamber pressure was < 10^{-3} torr. Two conical nozzles of 0.1mm throat diameter were used; nozzle A of 2mm length and conical angle 30°, and nozzle B of 4mm length and conical angle 20°.

The gas density in the region of photoelectron travel was minimized by using a skimmer to select the central portion of the molecular beam into the ionization cell. In addition a liquid nitrogen cryopump, acting as a beam collector, was used in the case of N_2O. While the high gas density in the beam results in a high ion density which retards the photoelectrons, this is measured as less than 0.1 eV. However it causes a diminution of operational resolution from a low pressure value of 30 meV to around 50 meV for the higher stagnation pressures. Though the gas density in the molecular beam is high, the average density in the region of photoelectron travel is small, so that scattering losses of low energy photoelectrons are minimal.

RESULTS

Sequences of HeI PE spectra measured for N_2 and N_2O in He at increasing stagnation pressures are shown in figures 1 and 2. The sharp peaks arising from the monomer species contrast vividly with the broad bands arising from the molecular clusters. The cluster bands are observed to be associated with each monomer band but displaced to lower ionization energy (IE) in each case. This correlation suggests that ionization involves monomer and cluster orbitals of the same origin. By making the reasonable assumption that the photoionization cross-section for each N-molecule cluster is N times that of the monomer, specifically for the first PE band in each case, a comparison of band integrals allows calculation of the fraction of clusters, in mass-weighted terms, present in each mixture. The variation of the percentage of condensation with stagnation pressure is illustrated for the N_2/He and N_2O/He mixtures respectively in figure 3. The contrasting curves are representative of those obtained by calculations based on nucleation theory.

DISCUSSION

1. N_2 Dimer

When the percentage of clustered molecules is around 1% most of the clustered species are dimers with relative amounts of higher clusters being small [4] . While the ionization onset is affected by the presence of small amounts of these higher clusters, the general appearance of the PE bands not associated with monomer species are representative of the dimer. When the N_2

monomer spectrum (P_O = 3.5 atm.) is substracted from the P_O = 4.7 atm.
spectrum of figure 1, the resulting spectrum is essentially that of the
dimer, $(N_2)_2$ (figure 4).

The first band of the HeI PE spectrum of $(N_2)_2$ is broad and unstruc-
tured with a vertical IE (I_v) of 15.2 ± 0.1 eV which is 0.4 eV lower than the
monomer value of 15.58 eV. The onset of ionization measured from the first
band of $(N_2)_2$ is 14.6 ± 0.1 eV in agreement with the appearance energy of
14.69 ± 0.05 eV for $(N_2)_2^+$ obtained using PIMS[10]. This is 0.6 eV lower than
I_v, but may not represent the true adiabatic ionization energy I_A[11].
Experimental data on the geometries of these dimer species are not available,
but recent theoretical calculations are in disagreement concerning the geometry of the neutral dimer molecule. An X configuration is favoured in some
cases[12], but a T configuration in another[13]. By contrast the dimer ion
geometry is proposed to be linear[14]. These calculations indicate however
that $(N_2)_2^+$ has only one strongly bound electronic state. Calculations of

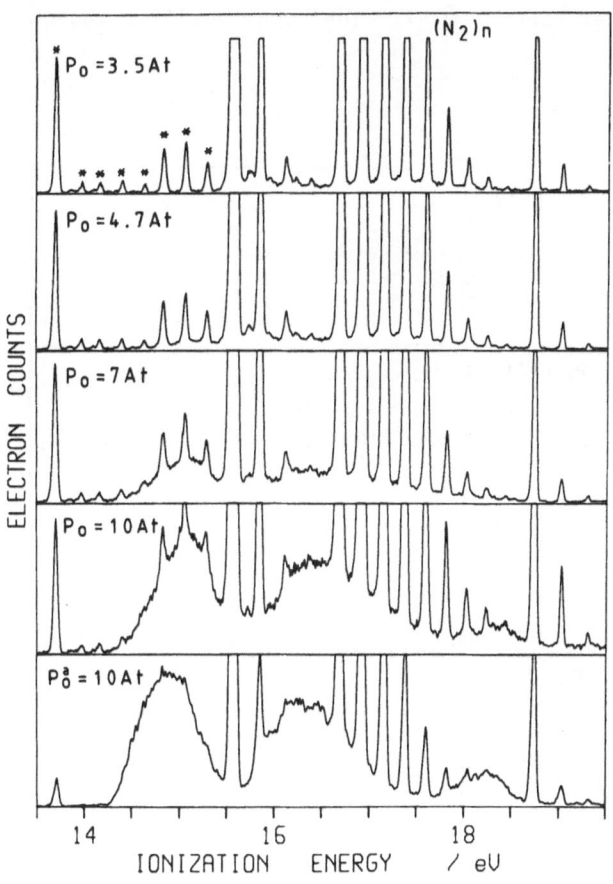

Fig. 1.   HeI photoelectron spectra of 25% N_2 in He mixture at
          stagnation pressures of 3.5 to 7 atm., and 10% N_2 in He
          at 10 atm, using nozzle A. The bottom spectrum is for
          10% N_2 in He at 10 atm using nozzle B. The asterisked
          peaks arise from impurity lines in the He discharge
          source. The spectra are truncated at 2% of the height
          of the first monomer peak, expect for the bottom
          spectrum which is truncated at 10%.

the electronic relaxation energy for $(N_2)_2^+$ representing the $I_V - I_A$ energy difference, based on the above geometries, give a value of 1.0 eV at the MP3/3-21G level , suggesting that the ion appearance energy is higher than I_A.

2. N_2 Clusters

Subtraction of monomer N_2 bands from the high pressure spectrum of figure 1 (P_o^a = 10 atm) gives the spectrum of mixed clusters, $(N_2)_n$, shown in figure 4. This can be compared with the monomer and dimer spectra as well as with a simulated spectrum of condensed N_2, also shown in figure 4. The simulated spectrum is based on the vertical IEs and bandwidths obtained in photoemission measurements at different ionizing energies[15].   Apart from the monomer the widths of corresponding bands are nearly equal, but the mixed clusters spectrum shows IEs higher by 0.8 to 1.0 eV than the condensed N_2 spectrum. The latter is characterised by a gas-to-solid shift of 1.5 eV explained as the effect of electronic relaxation via dielectric polarization of the condensed medium. While the mean cluster size of the mixed clusters measured in figure 4 is not directly known, a theoretical model described later suggests that a shift which is 0.44 of the condensed value correlates with a mean cluster size of $\bar{N}$ ~7. The model verifies the observation that

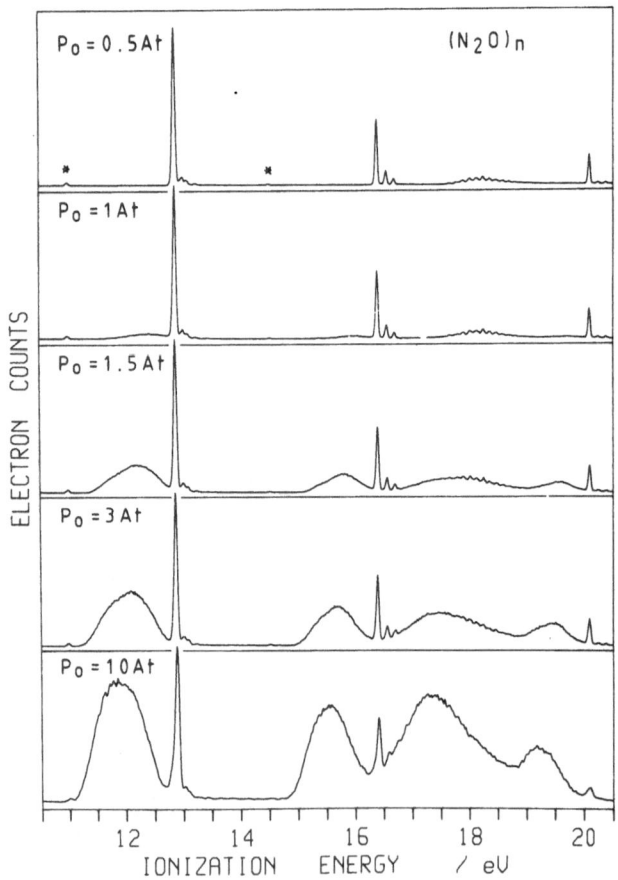

Fig. 2.   He I photoelectron spectra of 10% N_2O in He mixtures at stagnation pressures of 0.5 to 10 atm. using nozzle B.

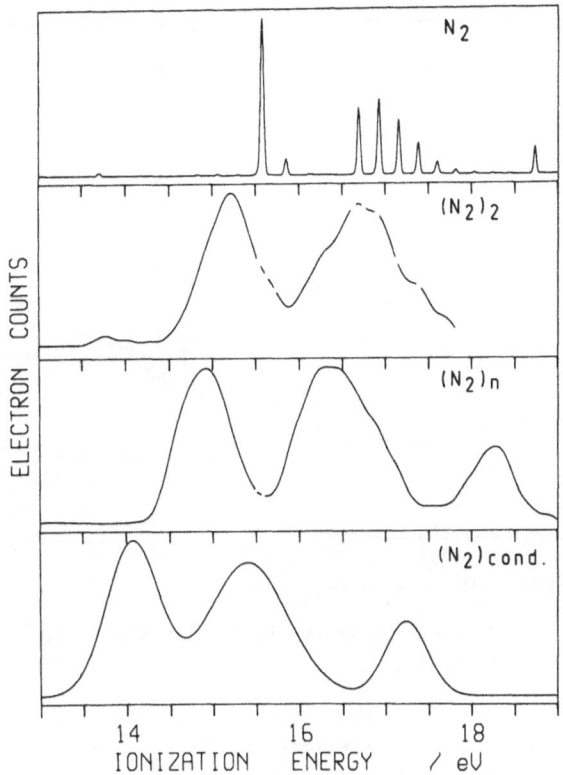

Fig. 3.   Comparison of the HeI photoelectron spectra of the
monomer, dimer, clusters and solid N_2.

the convergence of the dielectric polarization relaxation effect with
increasing cluster size is rather slow. By contrast, the near-equivalence of
the band shapes for dimer, clusters and solid suggests that the $(N_2)_2^+$ species
is the photoionization product in the cluster and solid states. It is
suggested that the dielectric polarization effect occurs around the $(N_2)_2^+$
site and is strongly dependent on cluster size.

3. N_2O Clusters

The HeI spectra of N_2O show increased clustering at high pressures with
up to 86% condensation. Each band group shifts steadily to lower IE with
increased clustering. The 10 atm. spectrum has a first band I_v of 11.9 eV
representing a shift of 1.0 eV below that of the monomer. This is 2/3rds of
the dielectric polarization shift of 1.5 eV for condensed NO. The lowest
value of the ionization onset, 11.0 eV, has clearly not converged to the
limit represented by the solid. Based on the theoretical model described in
the next section, an estimate of the mean cluster size at 10 atm. is $\bar{N}$ ~96.

4. Theory of Relaxation Shifts

The electronic relaxation observed in photoionization studies of a
condensed gas relative to its free molecule has been explained as due to
dielectric polarization around the localized positive hole[15]. A modification
which considers the effect of finite cluster size gives the electronic relaxation experienced around a spherical hole located in different layers of a
spherical cluster. The fractional shift experienced by a surface molecule of

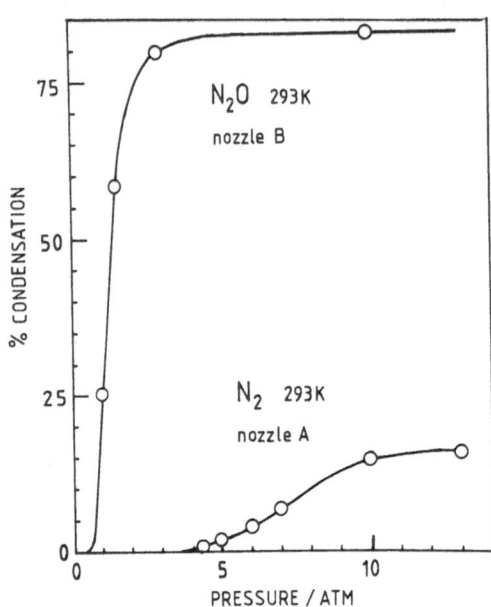

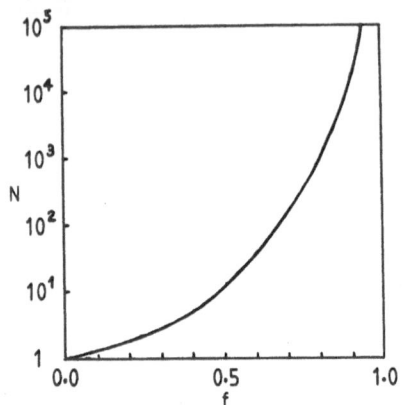

Fig. 5. Variation of relative relaxation shift f in ionization energy with cluster size N.

Fig. 4. Variation of percentage condensation with stagnation pressure for N_2 and N_2O.

a solid is 0.75[16], but for a surface molecule in the 147-molecule icosahedral cluster it is 0.62. The average fractional shift for molecules in this cluster is 0.70. A plot of cluster size N against f, the fractional IE shift calculated by this model is shown in figure 5. The model indicates that f approaches the asymptotic value for the infinite solid for cluster sizes above 10^5.

REFERENCES

1. J. C. Phillips, Chem. Rev., 86, 619 (1986).
2. J. Kontecky and P. Fantucci, Chem. Rev., 86, 539 (1986).
3. F. G. Celii and K. C. Janda, Chem. Rev., 86, 507 (1986).
4. P. M. Dehmer and J. L. Dehmer, J. Chem. Phys. 69, 125 (1978).
5. S. Tomoda, Y. Achiba and K. Kimura, Chem. Phys. Lett., 87, 197 (1982).
6. F. Carnovale, M. K. Livett and J. B. Peel, J. Am. Chem. Soc., 105, 6788 (1983).
7. H. Siegbahn, S. Svensson and M. Lundholm, J. Electron Spectrosc., 24, 205 (1981).
8. M. J. Campbell, J. Liesegang, J. D. Riley and J. G. Jenkin, J. Phys. C., 15, 2549 (1982).
9. L. S. Bartell, Chem. Rev., 86, 491 (1986).
10. S. H. Linn, Y. Ono and C. Y. Ng, J. Chem. Phys., 74, 3342 (1981).
11. H. H. Teng and D. C. Conway, J. Chem. Phys., 59, 2316 (1973).
12. A van der Avoird, P.E.S Wormer and A.P.J. Jansen, J. Chem. Phys., 84, 1629 (1986).
13. H. J. Boehm and R. Ahlrichs, Mol. Phys., 55, 1159 (1985).
14. S. C. de Castro, H.F. Schaefer and R. M. Pitzer, J. Chem. Phys., 74, 550 (1981).
15. F.-J. Himpsel, N. Schwentner and E. E. Koch, Phys. Stat. Solid. B, 71, 615 (1975).
16. W. Keller, H. Morgner and W. A. Muller, Mol. Phys., 57, 623 (1986).

MULTIPOLAR EXCITATIONS IN SMALL METALLIC SPHERES

F. Claro

Oak Ridge National Laboratory
Oak Ridge, TN 37831

R. Fuchs

Ames Laboratory and Iowa State University
Ames, Iowa 50011

ABSTRACT

The polarizability $\alpha_\ell(\omega)$ appropriate to a small metallic sphere is obtained within the semiclassical infinite barrier model, where ℓ is the multipole order. An excitation diagram in the ℓ,ω plane based on the structure of this function is proposed. It represents the spherical analog of the excitation structure of an infinite medium in the k,ω plane.

INTRODUCTION

It was many years ago that Fröhlich predicted that a small dielectric sphere would couple resonantly to an electromagnetic field by means of the excitation of a surface mode.[1] Such mode has been observed in particles of different materials in optical as well as electron scattering experiments.[2,3] A theory that uses a local dielectric function $\varepsilon(\omega)$ predicts an electric resonance at a frequency given by the solution of $\varepsilon(\omega) = -(\ell+1)/\ell$, where ℓ is the pole order of the applied external potential. Thus, in a uniform external field the dipole moment ($\ell=1$) is excited resonantly for $\varepsilon(\omega) = -2$. The width of the resonance is in this model determined by scattering of the surface excitations that sustain it. It is known on the other hand that nonlocal theories introduce a shift in the location of the resonance and modify its wings due to additional excitations including bulk plasmons and electron-hole pairs.[4,5] In the quantum limit of very small spheres (radius $a \lesssim 10$ Å) the spectrum is quite complex and contains many resonances due to a size effect that discretizes the energy levels in the available volume.[6] For not so small spheres ($a \sim 30$ Å or bigger) weak resonances associated with the excitation of bulk plasmons have been found assuming a bulk dielectric function may be used to characterize the response of the metal.[7] In this latter case resonances associated with quantum size effects are naturally absent.

In this paper we introduce the notion of a dispersion law in the frequency ω – angular momentum ℓ, plane. This is a useful adaptation to spherical excitations of the common ω-k representation. Its motivation is that spherical variables are more natural in the description of physical effects induced by geometry in the case of spheres. As we shall see explicitly for the case of not so small spheres a conceptually simple picture in terms of multipolar excitations emerges, emphasizing the main physical effects that take place when the particle is in the presence of an external probe that induces electromagnetic interactions.

THE DIELECTRIC RESPONSE FUNCTION $E(\ell,\omega)$

We assume that the bulk dielectric response of the metal is described by the nonlocal dielectric function $\varepsilon(\vec{k},\omega)$. The solution of the electromagnetic boundary problem for the interaction of the sphere with an external field requires in this case the specification of an additional boundary condition derived from the microscopic properties of the metal at the surface. We here adopt the one appropriate to the semiclassical infinite barrier model.[5] In this model the response of the sphere may be characterized by the sequence of multipolar polarizabilities[7]

$$\alpha_\ell(\omega) = \frac{E(\ell,\omega) - \varepsilon_0}{E(\ell,\omega) + \frac{\ell+1}{\ell}\varepsilon_0} . \tag{1}$$

Here ε_0 is the dielectric constant of the medium the particle is placed in. This formula is identical in form to the corresponding local expression only that the dielectric function of the metal appears in the modified form

$$E(\ell,\omega) = \left[\frac{2}{\pi}(2\ell+1)a \int_0^\infty \frac{j_\ell^2(ka)}{\varepsilon(k,\omega)} dk\right]^{-1} , \tag{2}$$

where $j_\ell(x)$ is the spherical Bessel function of order ℓ. Notice that $E(\ell,\omega)$ equals the usual dielectric function $\varepsilon(\omega)$ when the latter is k-independent (local). We shall adopt it as a characterization of the response of the sphere to an external potential of pole order ℓ and frequency ω.

EXCITATION DIAGRAM IN THE ℓ,ω PLANE

Figure 1 represents the structure in $\alpha_\ell(\omega)$ for a tin sphere of radius 30 Å.[8] Three features are included in this graph. First there is the Fröhlich resonance labeled FR and extending through large values of pole order ℓ. For the case studied it represents the most prominent resonance. It is the only feature present if a local dielectric function is assumed, a case also included in the figure and labeled D since the Drude model was used to obtain it. Notice that nonlocal effects shift the resonance far into the high energy region of the graph. A second feature is the sequence of resonances above ω_p. These correspond to the weak excitation of plasmon standing waves in the presence of the sphere boundary. These resonances are sketched in Fig. 1 by following the evolution of the peaks in $\alpha_\ell(\omega)$. Finally, there is the region where electron-hole pairs are created, delimited in our figure by the rising dotted lines. These are the only excitations possible at low frequencies. Unlike the other cases in which lines in the figure correspond to resonances, here we encounter a broad region where multipolar coupling is possible. The edges were arbitrarily set by the condition that the quantity $Im(E_\ell-1)^{-1}$ reached its maximum value divided by 80 when a relaxation time appropriate for the bulk metal is used.

In the ω vs k representation one gets an approximate expression for the electron-hole edges by requiring that energy and momentum be conserved when the incoming photon takes an electron above the Fermi surface and a hole is left behind. We can here use this same condition to sketch our edges if we keep in mind that the wavelength of an excitation at the surface is approximately the sphere perimeter over ℓ, or $ka = \ell$. Using these relations we get

$$\omega_{\pm} = \frac{\hbar}{ma^2} (\ell \pm 2ak_F)\ell \qquad (3)$$

where k_F is the Fermi wave vector, m the electron mass and ± refers to the left (right) edge. We remark that, as Eqs. (2) and (3) show explicitly the details of the ω vs ℓ graph depend on the radius of the sphere as well as the metal the particle is made of. The curves given by (3) are the rising dashed lines in the figure. The small discrepancy at the right edge is a manifestation of the arbitrariness of the criterion used in drawing the dotted lines, as explained above. The slope of these lines is correctly given by (3), however, providing evidence that the physical picture conveyed by our diagram is essentially correct.

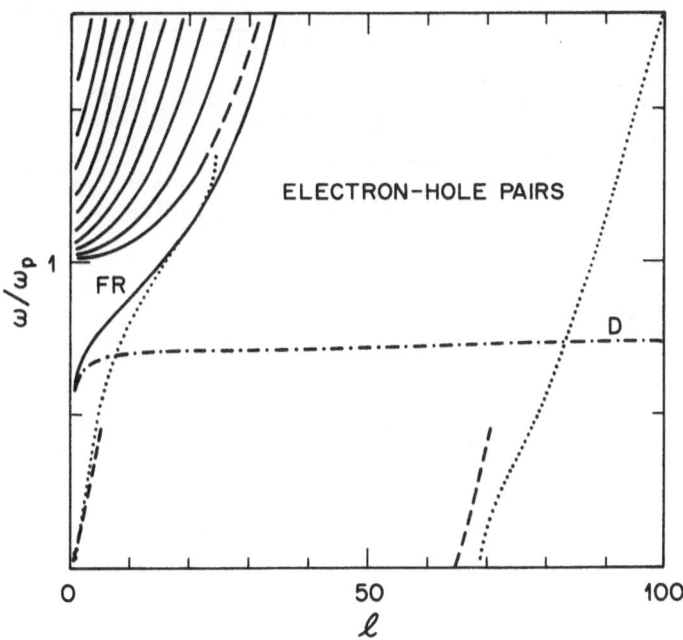

Fig. 1. Dispersion curves of the multipolar resonances for a 30 Å radius tin sphere in the multipole order ℓ - frequency ω, plane. Full lines follow resonances for the Lindhard-Mermin model while the dash-dotted line is the Fröhlich resonance (FR) in the Drude (D) model. The dashed lines rising from the bottom follow Eq. (3).

In optical absorption the electromagnetic field couples to the $\ell=1$ modes if the wavelength of light $\lambda \gg a$, and to higher order modes if this condition is not fulfilled. Similarily, the nonuniform field in the neighborhood of an electron passing near a sphere excites various multipoles with relative amplitudes that depend on the sphere radius. Our diagram may be used to locate the various resonances in each of these cases. For instance, at a given pole order one obtains the features as a function of frequency by simply following a vertical path in the figure. Taking the dipole ($\ell=1$) mode as an example, at very low frequencies one expects electron-hole pairs to be created while a surface plasmon becomes prominent as one approaches ω_p, and finally a sequence of plasmon standing waves are excited above this frequency. A similar pattern describes the quadrupole and higher order excitations, only that the location of the electron-hole edge and of the higher frequency resonances increases in frequency with the value of ℓ. As is usual with this kind of diagrams, Fig. 1 gives no information about the relative strength of the excitations. It is contained, however, in expressions (1) and (2) which should be used in a detail calculation. For small metal spheres, however, the Fröhlich resonance is usually orders of magnitude stronger than the other features represented in the graph.

This research was supported by the USDOE under Contracts W-7405-Eng-82 and DE-AC05-840R21400 and the University of Tennessee—ORNL Distinguished Scientist Program.

REFERENCES

1. H. Fröhlich, "Theory of Dielectrics," Oxford Univ. Press, (1948).
2. For a review see U. Kreibig and L. Genzel, Surf. Sci. 156:678 (1985).
3. P. E. Batson, Phys. Rev. Lett. 49:936 (1982).
4. R. Ruppin, Phys. Rev. B 11:2871 (1975).
5. B. B. Dasgupta and R. Fuchs, Phys. Rev. B 24:554 (1981).
6. W. Eckardt, Phys. Rev. Lett. 52:1925 (1984).
7. R. Fuchs and F. Claro, to be published.
8. The parameters used for tin are $\omega_p = 1.17\times10^{16}$ sec^{-1}, $v_F = 1.24\times10^8$ cm/sec and as a mean free path, the sphere radius.

THEORY OF DAMPING OF EXCITED STATES OF A MOLECULE NEAR A METALLIC CLUSTER

Walter Ekardt

Fritz-Haber-Institut der Max-Planck-Gesellschaft
Faradayweg 4-6, D-1000 Berlin 33

INTRODUCTION

In many problems related to Surface Enhanced Raman Scattering (SERS) one needs to know the lifetime of excited molecular levels (vibronic or electronic) near rough metal surfaces. As the exact dynamical response functions of rough metal surfaces are still unknown at a strictly microscopical level, the problem is usually approached by the application of macroscopic electrodynamics. For instance, one of the model problems which has been studied very often in the past[1-5] is an oscillating point dipole in (electromagnetic) contact with a metallic or dielectric sphere or spheroid, characterized by the macroscopic, local bulk dielectric constant of the material to be described.

However, for very small clusters (modelled by cubes, spheres or spheroids) one might question the use of classical, local bulk dielectric constants for the description of the dynamical response. The reason is that, owing to the finite surface-to-volume ratio, nonlocal surface response properties are expected to play a major role. Among these, the excitation of electron-hole pairs as a dissipative channel in various dynamical processes at surfaces (e.g. SERS) becomes increasingly important. For this reason, the problem of non-radiative damping of excited molecular states near a small metal particle is reconsidered, this time at a truly microscopic level. The particle to which the molecule couples is treated within the selfconsistent spherical jellium background model[6] whose various response functions were studied before[7-10].

Without doubt, the study of a *real* ionic skeleton (and not a positively charged homogeneous background) would be highly preferable. However, the best one can do at this level is the calculation of the *ground* state for a few atoms within the cluster[11] and the study of the dynamical properties for very small molecules[12]. Therefore we believe that the selfconsistent jellium model combined with the successful Time-Dependent Local-Density-Approximation (TDLDA)[13] is a first important step towards a realistic description, at least for alkali metal clusters with a high symmetry and a number of atoms bigger than 8.

Within the nonretarded theory an oscillating point dipole with dipole moment $\vec{p}$ at $\vec{r}_1$ sets up a "bare" electrostatic potential as follows

$$\phi(\vec{r},\vec{r}_1) = \vec{p}\cdot\vec{\nabla}_{r_1} \frac{1}{|\vec{r}-\vec{r}_1|} \tag{1}$$

where $\vec{p}$, in this context, has the meaning of a dynamical dipole transition matrix element from the ground level of the molecule to an excited one (vibronic or electronic). The electronic density $n(\vec{r})$ of the metal particle couples to ϕ via a perturbation Hamiltonian H' in the usual way. Within linear response theory, application of Fermi's Golden Rule leads then to the following transition rate Γ (in Rydberg atomic units)

$$\Gamma = 8 \; \vec{p}\cdot\vec{\nabla}_{r_1} \; \vec{p}\cdot\vec{\nabla}_{r_2} /_{r_2=r_1}$$

$$\sum_{\ell,\tilde{\ell}=0}^{\infty} \quad \sum_{m,\tilde{m}=-\ell,\tilde{\ell}}^{\ell,\tilde{\ell}} \quad \int d\vec{r} \int d\vec{r}'$$

$$B_\ell(r,r_1) \; Y_{\ell,m}(\Omega_1) Y_{\ell,m}^*(\Omega)(-)\operatorname{Im}\chi(\vec{r},\vec{r}';\omega)$$

$$B_{\tilde{\ell}}(r',r_2) \; Y_{\tilde{\ell},\tilde{m}}(\Omega') \; Y_{\tilde{\ell},\tilde{m}}^*(\Omega_2) \tag{2}$$

In this equation, $\operatorname{Im}\chi$ means the imaginary part of the retarded density-denstiy response function of the metal particle's electrons, the $Y_{\ell,m}$ are the spherical harmonics, and the B_ℓ's originate from the multipole expansion of the Coulomb interaction.

If the dipole is not embedded in the electronic tail of the metal particle but located a few atomic units (Bohr) away from the jellium background edge, formula (3) simplifies considerably to the following expression:

$$\Gamma = 4 \sum_{\ell=1}^{\infty} \quad \operatorname{Im}\alpha_\ell(\omega)/R^{2\ell+1} \quad p^2/r_1^3$$

$$(R/r_1)^{2\ell+1} \; \{ \; (\ell+1)^2 \cos^2\theta + P_\ell'(x)/_{x=1} \; \sin^2\theta \; \} \; . \tag{3}$$

Here $\alpha_\ell(\omega)$ is the dynamical ℓ-pole polarizability (studied especially in Refs. 8 and 9), θ is the angle of the dipolar axis measured with respect to the connection line of the center of the sphere to the dipole, and P_ℓ' is the derivative of the ℓth Legendre polynominal.

If the classical expression for $\alpha_\ell(\omega)$ is used in equ. (4),

$$\alpha_\ell(\omega)/R^{2\ell+1} = (\varepsilon(\omega)-1)/(\varepsilon(\omega) + \frac{\ell+1}{\ell}) \tag{4}$$

and if $\varepsilon(\omega)$ is approximated by a Drude form, we recover results already published by Ohtaka and Inoue[5]. However, with our $\alpha_\ell(\omega)$ we are in a position to give a more reliable estimate of the lifetime, simply because

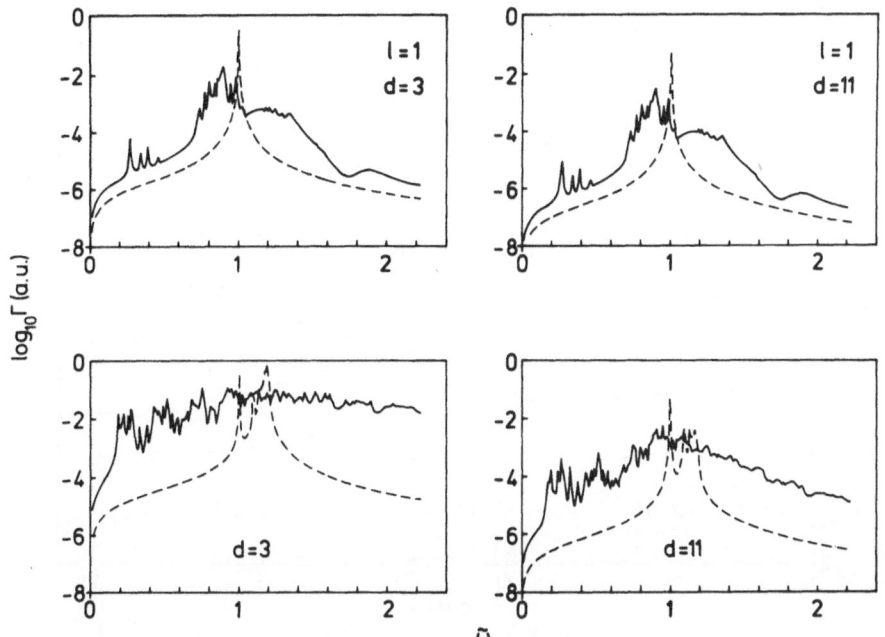

Fig. 1: Frequency dependence of the transition rate Γ, eq. (3), of a point dipole coupled to a jellium sphere corresponding to 92 valence electrons of Na. The dipole moment of the molecule is 1 ea_0. Results are shown for two different distances cl (in a.u.) to the spherical jellium edge and the $\ell=1$ part of the total Γ is shown separately. Γ is in Ry a.u. ($= 2.068 \times 10^{16}s^{-1}$) and the reduced frequency $\tilde{\omega} = \omega/\dfrac{\omega_p^{Na}}{\sqrt{3}}$ with ω_p^{Na} the plasma frequency of Na.

our $\alpha_\ell(\omega)$ are the *microscopic* response functions obtained within the frame of a basically exact theory. The only ingredient which is missing here is the ionic skeleton. But as mentioned above, this should be of minor importance in the case of the alkali metals. For metals like Cu, Ag, or Au the behaviour of the loosely bound s-electrons is very often studied within the jellium model with the proper r_s-value of the s-electronic density. Because Cu, Ag, and Au show the same spherical shell-closing magic numbers[14] as the alkaline metal clusters do[15], it seems to be natural to apply the selfconsistent spherical jellium model to these metals too. Hence, in the example given below we study the lifetime of the CO stretch mode adsorbed on Cu, simply because a number of experimental data exists for this case.

RESULTS AND DISCUSSION

On the experimental side, the following two problems are of utmost interest:
1st, how does Γ depend on the frequency ω, with $\vec{r}_1$ held fixed?
2nd, how does Γ depend on the distance $\vec{r}_1$-R, with the frequency ω held constant?
Theoretically, there is one more question:
3rd, how does the nature of the coupling change from the far-distance region to the near-surface region?
The answers are given in the figures.

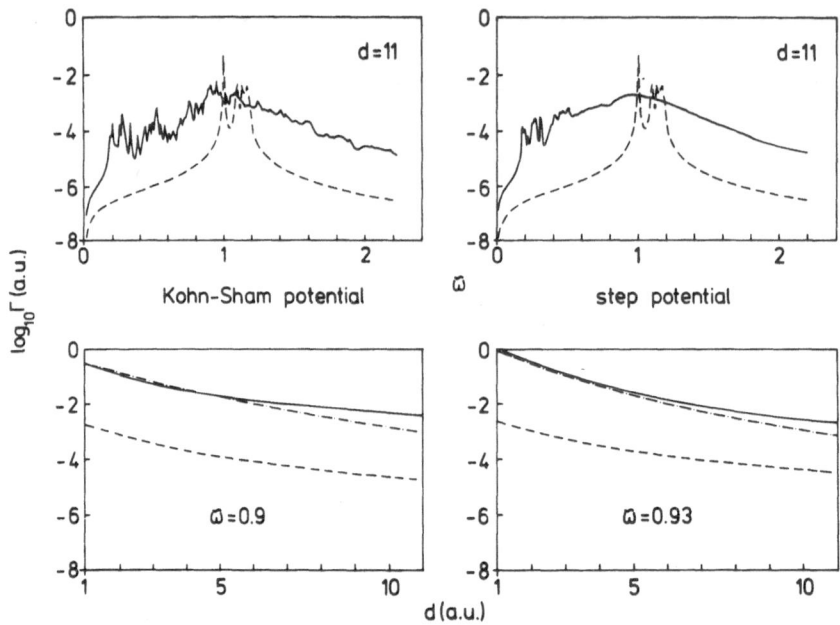

Fig. 2: Frequency dependence of Γ at a given distance d, upper panel, and distance dependence of Γ at a given frequency $\tilde{\omega}$, lower panel. On the left, results are shown for a selfconsistent jellium background potential, on the right the total-energy-minimizing step potential was used. Other parameters are the same as in Fig. 1.

Fig. 1 shows the frequency dependence of the transition rate Γ in atomic units (1 a.u. = $2.068 \cdot 10^{16}$ s^{-1}) for two different molecular distances d (in a.u.) from the spherical jellium edge. The dipole strength assumed is p = 1 ea$_o$, and the dipole orientation angle θ (see eq. (3)) is 0°. The parameters describing the spherical particle are R = 18.057 a$_o$ and r_s = 4 a$_o$, corresponding to 92 valence electrons of Na. The frequency ω is scaled with the classical, dipolar surface plasmon frequency of a "sodium sphere". Hence, $\tilde{\omega}$ = 1 corresponds to ω = 0.2497 Ry $\simeq$ 3.398 eV. The upper panel shows the dipolar contribution and the lower panel shows the total Γ. In each case, the continuous line gives the results of the actual calculation, whereas the dashed line gives Γ following from the classical α_ℓ, eq. (4), using the Drude form for $\varepsilon(\omega)$ with a "numerical" damping γ of 10 meV. For a detailed discussion of this point, the interested reader is referred to Ref. 7.

On comparing the classical "macroscopic" curves with the microscopic ones, the importance of a microscopic description can clearly be seen. It is only in the very-far-distance region that a classical description is approximately sufficient. The reason is that in this case the dominant damping mechanism is the *dipolar* coupling (i.e. the $\ell=1$ part of H') between the molecule and the metal particle. For chemically relevant distances, the most efficient decay channel is the excitation of size-quantized intraband electron-hole pairs which are completely missing in the classical frame.

Fig. 2 gives in the upper panel a comparison of the frequency dependence of Γ, with the distance d held fixed at 11 a.u., for the Kohn-Sham potential and a total energy-minimizing step potential[16]. The lower panel shows the distance dependence of Γ at the dipolar surface plasmon frequency of the Kohn-Sham sphere (left hand side) and of the step-potential sphere (right hand side), in three different approximations;

continuous line: equ. (3); dashed line: $\alpha_\ell(\omega)$ from equ. (4); dot-dashed line: independent electron result. Incidentally, the latter result agrees (for this specific example) more or less with the full TDLDA. On comparing the r.h.s. of the figure with the l.h.s. we see that the much less time-consuming non-selfconsistent total-energy-minimizing step potential is not too bad in describing Γ.

Finally, we want to apply the general result to the special case of the CO stretch vibration. For $\omega = 250$ meV ($\tilde\omega = 0.07$), $p \simeq 0.1$ ea_0, and $d \simeq 3$ a_0, we obtain $\Gamma = 6 \cdot 10^{-7}$ a.u. $= 1.2 \cdot 10^{10}$ s^{-1} which is one order of magnitude less than for the planar case. The reason is simply that the frequency $\tilde\omega = 0.07$ is off-resonant to all the electron-hole pair excitations of the specific example under discussion. On the other hand, for $\tilde\omega = 0.27$ which is typical for a *resonant* coupling to an electron-hole pair, we obtain $\Gamma = 2.3 \cdot 10^{-4}$ a.u. $= 4.6 \cdot 10^{12}$ s^{-1}. This is enhanced with respect to the planar case by a factor of 45. Hence we see that in the case of a very small metallic *particle* Γ shows up a very pronounced resonant character.

ACKNOWLEDGEMENTS

Thanks are due to Prof. Dr. Elmar Zeitler for his continuing interest and support. I am grateful to B. Pettinger for an informative discussion on SERS.

REFERENCES

1. M. Kerker, D.S. Wang and H. Chew, Appl. Optics 24:4159 (1980)
2. J. Gersten and A. Nitzan, J. Chem. Phys. 75:1139 (1981)
3. R. Ruppin, J. Chem. Phys. 76:1681 (1982)
4. C.G. Blatchford, J.R. Campbell and J.A. Creighton, Surf. Sci. 120:435 (1982)
5. K. Ohtaka and M. Inoue, J. Phys. C15, 6463 (1982)
6. W. Ekardt, Phys. Rev. B29:1558 (1984)
7. W. Ekardt, Phys. Rev. Lett. 52:1925 (1984)
8. W. Ekardt, Phys. Rev. B31:6360 (1985)
9. W. Ekardt, Phys. Rev. B32:1961 (1985)
10. W. Ekardt, Phys. Rev. B33:8803 (1986)
11. J. Koutecky and P. Fantucci, Chem.Rev. 86:539 (1986)
12. Z. Levine and P. Soven, Phys. Rev. Lett.50:2074 (1983)
13. A. Zangwill and P. Soven, Phys. Rev.A21:1561 (1980)
14. I. Katakuse, T. Ichihara, Y. Fujito, T. Matsuo, T. Sakurai and H. Matsuda, Int. J. Mas. Sepctr. and Ion Proc. 67:229 (1985)
15. W.D. Knight, K. Clemenger, A.W. de Heer, A.W. Saunders, M.Y. Chou and M.L. Cohen, Phys. Rev. Lett. 52:2141 (1984)
16. W. Ekardt, Z. Penzar and M. Šunjić, Phys. Rev.B33:3702 (1986)

PHOTOINDUCED AGGREGATION AND DISSOCIATION OF Ag AND Au CLUSTERS IN RARE

GAS MATRICES

Rudolf Markus and Nikolaus Schwentner

Institut für Atom- und Festkörperphysik
Freie Universität Berlin, Arnimallee 14
D-1000 Berlin 33, FRG

Absorption bands, emission bands and lifetimes of Ag, Ag_2, Ag_3, Au, Au_2, Au_3 and Au_n clusters are reported. No photosensitive processes of Ag centers[3] in annealed rare gas crystals have been found. Au atoms in Ar crystals are bleached by the formation of Au_2 and Au_3 clusters. In Xe matrix a threshold and a quadratic dependence for the bleaching efficiency of Au atoms are observed.

INTRODUCTION

Many studies of the optical properties of matrix-isolated metal atoms and clusters addressed the problem of matrix-induced changes of absorption spectra. More recently investigations of the energy dissipation of matrix isolated species after optical excitation have been started by fluorescence detection including time-resolved experiments[1]. Important nonradiative processes are the decomposition of larger clusters and perhaps even more interesting the aggregation of atoms or clusters to larger units. Some examples have been reported in the literature[2] and especially for silver clusters a reversible isomerization induced by different photon energies has been observed[3,4]. The reactions require displacements of the products in the lattice of the matrix. A microscopic interpretation of these complex processes is still missing and further experimental data are necessary to characterize these processes in more detail. In this paper results concerning the dependence of the photosensitivity of Ag clusters on the preparation conditions are discussed. Some additional data on the absorption bands, emission bands and lifetimes of Au, Au_2, Au_3 and larger Au clusters are presented. Furthermore new aggregation processes involving Au atoms and dissociation of Au_2 clusters are reported which depend strongly on the matrix.

EXPERIMENT

Absorption spectra have been taken by a 0.25 m monochromator with a XP 2020 Q photomultiplier and a 450 W high pressure Xe lamp. For fluorescence experiments the samples have been excited either by ArF (193 nm) and KrF (248 nm) excimer laser lines (Lambda Physik EMG 201) or by the tunable and, when necessary frequency doubled output of a Lambda Physik FL 2002 dye laser. The fluorescence light has been dispersed by a 0.25 m monochromator and complete spectra are recorded by spatially resolving reticon and vidicon detectors. The vidicon detector is gateable with time windows of 10^{-7}s up to 10^{-4}s. In this way fluorescence spectra, lifetimes and the dependence of the intensity in specific emission bands on the exciting wavelength (excitation spectra) have been taken.

Two different kinds of samples have been used. Thin doped films have been frozen in the standard way onto a transparent LiF window. In addition doped optically clear and free standing rare gas crystals of about 1 cm^3 volume have been grown[5]. The metals have been evaporated for both kinds of samples by the focused excimer laser beam and have been codeposited with the rare gas matrix. A closed cycle refrigerator has been used for cooling of the substrates down to 15-20 K.

RESULTS ON Ag DOPED MATRICES

The fluorescence spectra of Ag doped Xe and Kr crystals showed emission bands and lifetimes of Ag atoms and Ag_2, Ag_3 and Ag_4 clusters which are in agreement with the literature data on wellannealed films[3,6,7]. The reported photosensitive processes[2,7] and the isomerization of Ag_3 clusters[3,4] could not be reproduced in our crystals. This is not in contradiction because the isomerization of Ag_3 centers has been studied on two emission bands which are present in fresh samples before annealing[3,4]. These bands disappear by annealing[3] and the remaining third Ag_3 emission band is also observed in our data on crystals. Evidently these photosensitive processes are restricted to unannealed samples containing special sites whereas the doped crystals correspond to annealed films due to the growing conditions.

ABSORPTION, EMISSION AND LIFETIMES OF Au DOPED MATRICES

Au atoms

The level scheme of Au atoms is included in the right hand part of Fig. 1. The observed $^2P_{1/2}$ and $^2P_{3/2}$ absorption bands are listed in Table 1. They agree well with the literature values[1,8]. Excitation in the vicinity of the resonance transition with the KrF line (248 nm) leads to the $^2D_{3/2} \to ^2S_{1/2}$ and $^2D_{3/2} \to ^2D_{5/2}$ emission bands. The transition energies and lifetimes (Table 1) are in accordance with those of Schrittenlacher and Kolb[1]. The smaller error bars of our lifetime measurements indicate now a monotonic increase from Xe matrices (100 μs) to Ar matrices (180 μs).

Au_2 molecules

At higher Au concentrations additional absorption bands are observed. Two groups of bands are attributed to Au_2 absorptions (Fig. 2 a,c,e and Table 1). One centered around 330 nm (322 Ar, 326 Kr 330 nm Xe) is named B and one around 400 nm (464 Ar, 395 Kr, 391 nm Xe) is named A. The 330 nm band in Xe shows a structure in excitation spectra (320, 332, 337 nm) which is smeared out in the absorption spectra. The identification with Au_2 centers is based on the concentration dependence and on the strong enhancement of these bands in excitation spectra of Au_2 emission bands. In each matrix a typical Au_2 emission band around 550 nm (540 Ar, 540 Kr, 580 Xe) with a lifetime of 35 μs (Table 1) shows up for excitation of any of the Au_2 absorption bands. At shorter wavelength (436 nm) an additional emission band is observed in Xe crystals with a lifetime of 40 μs but only for excitation at the higher energetic B absorption bands. Two transitions at 508 nm and 389 nm of Au_2 molecules in the gas phase are known[9], which can be correlated with the transitions from the 0_g^+ ground state to two 0_u^+ excited states. They are indicated by the arrows A and B in the calculated potential energy diagram[10] shown in the left hand part of Fig. 1. We assign the low energy absorption band which has been named A to the transition A in Fig. 1 because it should correspond to the lowest allowed transition. This leads to a remarkable large and systematically increasing matrix blue shift of 0.22 eV for Ar up to 0.73 eV for Xe. Consequently the bands called B are attributed to the next allowed transition B in Fig. 1 yielding a matrix shift for this transition of about 0.66 eV only weakly depended on the matrix. For the Ar matrix an alternative assignment of transition B to an absorption band at 365 nm reported by Klotzbücher and Ozin[11] would also be possible. The emission band around 550 nm should correspond to the lowest excited state 1_u in Fig. 1 and the transition is indicated by the arrow C. The Stokes shift originates partly from an intramolecular relaxation from 0_u^+ to 1_u and partly from a structural relaxation of the matrix cage. the lifetime in the μs regime reflects a weak 1_u-0_g^+ transition probability. The second emission band in Xe has to start from a state in between the two 0_u^+ states and the most likely transition from an 1_u state is marked by arrow D.

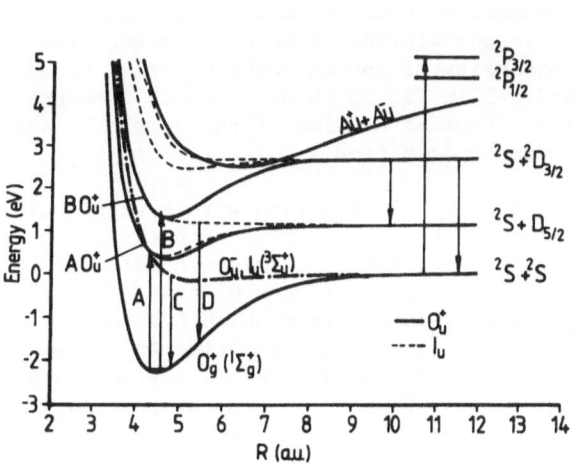

Fig. 1. Potential energy diagram of Au_2 from ref. /10/ with the atomic Au states on the right hand side. The observed transitions are indicated by arrows (see text).

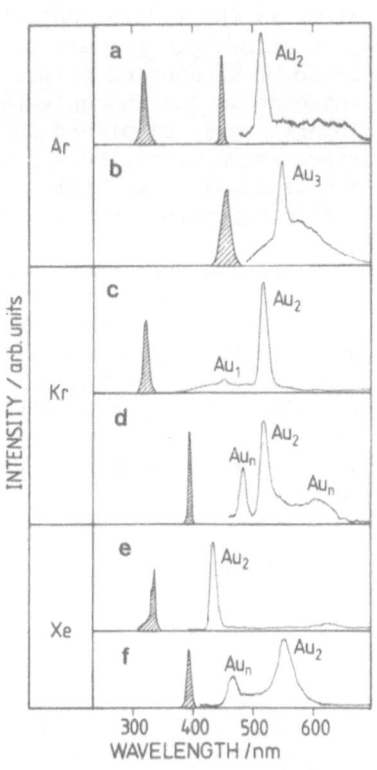

Fig. 2. Absorption (excitation) and emission spectra for Au in Ar (a,b), Kr (c,d) and Xe matrices (e,f) for excitation of Au_2 molecules (a,c,e), and larger aggregates (b,d,f). The absorption and excitation spectra are at the high energy side of a line and are shaded. The emission spectra resulting from excitation into these absorption bands are on the low energy side of the same line.

Au_3 and higher clusters

The Ar_3 absorption band in Ar identified in the literature[11] has also been observed in our excitation spectra. The resulting fluorescence yields an Au_3 emission band at 580 nm with a lifetime of 45 ns. Several additional but unidentified absorption and emission bands in Ar, Kr and Xe matrices are listed in table 1 by Au_n. The emission bands have lifetimes all in the ns regime and they correspond to Au_3 or larger aggregates.

PHOTOSENSITIVE PROCESSES IN Au DOPED MATRICES

In Ar and Xe matrices two rather distinct aggregation processes of Au atoms have been found whereas in Kr matrices no photosensitive Au centers have been observed.

Au in Ar

Illumination with the KrF laser line (248 nm) near the atomic 6s-6p resonance transition leads to a continous bleaching of the corresponding atomic emission band at 454 nm. This decrease in intensity is illustrated by the n=1 curve in Fig. 3. It indicates a real decrease in the concentration of Au atoms in the sample and not only a conversion to other sites with different spectroscopic properties. This effect shows up in crystals and in annealed or unannealed films. The reduction in Au atoms is accompanied by an increase in the concentration of larger clusters like Au_2 and Au_3. This agglomeration is displayed in the intensity of the emission bands of these clusters shown by the curves n=2 and n=3 in Fig. 3. A microscopic interpretation is missing. We exclude a simple thermal heating of the sample because the effect appears down to very low power densities.

Finally the agglomeration can be reversed by illumination with a wavelength of 471 nm as is shown in the final part of Fig. 3. This wavelength is longer than the absorption band A in Fig. 1. The excitation terminates below the O_u^+ state in the repulsive part of the dissociative 1_u and O_u^- states. Thus a dissociation of the Au_2 centers obviously is obtained which has to involve also sufficient energy for a separation of the resulting Au atoms by at least one lattice constant to prevent recombination. This wavelength coincides also with an Au_3 band and it leads to similar processes in Au_3 centers.

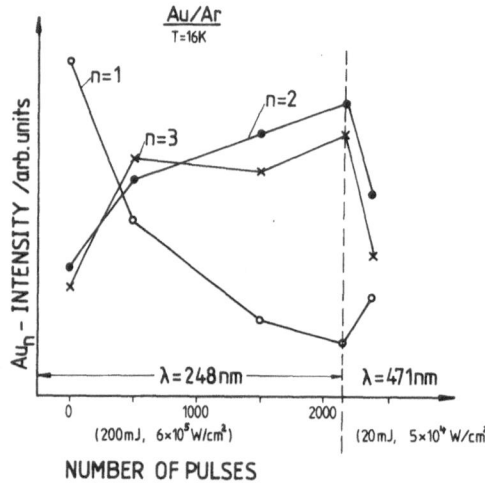

Fig. 3. Decrease of the Au monomere intensity (n=1) in the 454 nm band and increase in the dimer n=2 (540 nm) and trimer n=3 (580 nm) bands versus the number of excimer laser pulses of 248 nm up to the dashed vertical line. Beyond: dissociation of Au_2 and Au_3 with 471 nm and formation of Au.

Au in Xe

Illumination of Au atoms in Xe matrix at the atomic resonance transition initiates a qualitatively different agglomeration process. Fig. 4a shows the resulting fluorescence of a fresh doped Xe crystal with a strong atomic $D_{3/2} \rightarrow S_{1/2}$ transition and a very weak emission band at 372 nm due to an until now not identified larger Au_n cluster. A longer wavelength band of the same Au_n cluster appears at 721 nm which is not shown in Fig. 4. The lifetime in these bands is 200 ns. The fluorescence spectrum 4b taken after illumination with 200 laser pulses demonstrates a decrease of the intensity in the Au emission band and an increase of both Au_n emission bands. Successive repetitions of this cycle 4b-4f cause a bleaching of the Au monomers to a fraction of 10^{-3} of the original amount and the number of Au_n centers increases by a factor of 100. The dependence of the bleaching efficiency on the illumination power density is rather peculiar. Fig. 5 displays the change in the monomer intensity i.e. the amount of bleached Au atoms for each 200 laser pulses versus the power density. The curve shows a threshold behaviour with no effect up to a power density of $3 \times 10^4 W/cm^2$ and an increase with a slope corresponding to the third power of the power density. This slope has to be corrected by one because we detect the emission intensity in a certain volume and the emission intensity has a trivial propertional increase with the power density. Thus the bleaching efficiency itself goes with the second power of the excitation density. This indicates either a basic two photon process or a cooperative process of two excited Au atoms.

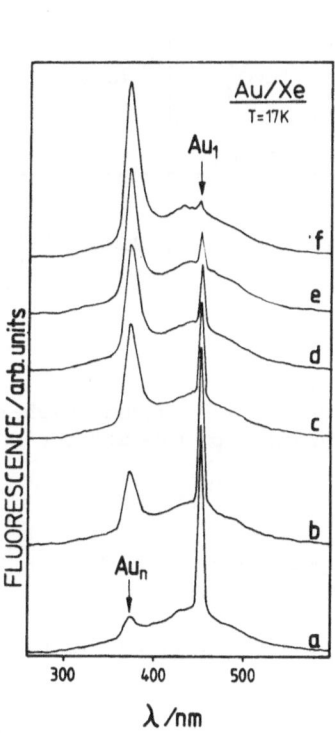

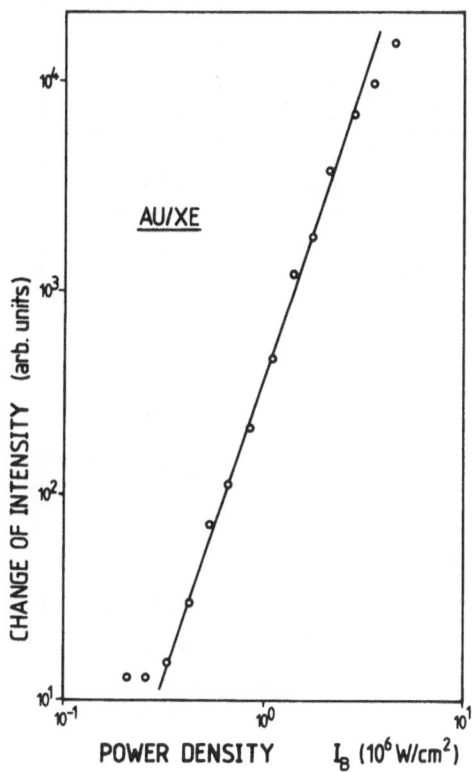

Fig. 4. Emission spectra for Au in Xe for excitation with 248 nm showing an Au monomer band and an unknown Au_n band. a fresh sample; b to f successive bleaching by 200 laser pulses with increasing power density between each spectrum.

Fig. 5. Change in monomer intensity from bleaching cycle to bleaching cycle versus the bleaching power density (see text).

Table 1. Absorption bands, emission bands and lifetimes for Au_n clusters

		Absorption nm	Assignment	Emission nm	Assignment	Lifetime μs
Ar	Au_1	226.0 229.1 255.5	$^2S_{1/2} \to {}^2P_{3/2}$ $^2S_{1/2} \to {}^2P_{1/2}$	454 815	$^2D_{3/2} \to {}^2S_{1/2}$ $^2D_{3/2} \to {}^2D_{5/2}$	180 ± 5
	Au_2	322 465	$X\,0_g^+ \to B\,0_u^+$ $X\,0_g^+ \to A\,0_u^+$	540 540	$1u \to 0_g^+$	35 ± 5
	Au_3	471		580		$(45\pm15)\cdot10^{-3}$
Kr	Au_1			452 818	$^2D_{3/2} \to {}^2S_{1/2}$ $^2D_{3/2} \to {}^2D_{5/2}$	140 ± 5
	Au_2	326 395	$X\,0_g^+ \to B\,0_u^+$ $X\,0_g^+ \to A\,0_u^+$	541 541	$1u \to 0_g^+$	35 ± 5
	Au_n	395		500		
Xe	Au_1	244.3 247.2 271.5	$^2S_{1/2} \to {}^2P_{3/2}$ $^2S_{1/2} \to {}^2P_{1/2}$	454 818	$^2D_{3/2} \to {}^2S_{1/2}$ $^2D_{3/2} \to {}^2D_{5/2}$	100 ± 5
	Au_2	320 332 337 391	$X\,0_g^+ \to B\,0_u^+$ $X\,0_g^+ \to A\,0_u^+$	436 580	$1u \to 0_g^+$ $1u \to 0_g^+$	40 ± 5 $30+8$
	Au_n	391		471 372 721		200×10^{-3}

ACKNOWLEDGEMENT

This work has been supported by the Deutsche Forschungsgemeinschaft via the Sonderforschungsbereich Struktur und Dynamik von Grenzflächen.

REFERENCES

1. W. Schrittenlacher, D.M. Kolb: Ber. Bunsenges. Phys. Chem. 88,492 (1984)
2. S.A. Mitchell, G.A. Ozin: J. Phys. Chem. 88, 1425 (1984)
3. U. Kettler: Diss. Uni Köln 1985
4. U. Kettler, P.S. Bechthold, W. Krasser: Surf. Sci. 156, 875 (1985)
5. O. Dössel, H. Nahme, R. Haensel and N. Schwentner: J. Chem. Phys. 79, 665 (1983)
6. D. Leutloff, D.M. Kolb: Ber. Bunsenges. Phys. Chem. 83, 666 (1979)
7. G.A. Ozin, H. Huber: Inorg. Chem. 17, 155 (1978)
8. D.M. Gruen, S.L. Gaudioso, R.L. McBeth, J.L. Lerner: J. Chem. Phys. 60, 89 (1974)
9. B. Kleman, S. Lindkvist, L.E. Selin: Ark. Fys. 8, 333 (1954)
10. W.C. Ermler, Y.S. Lee, K.S. Pitzer: J. Chem. Phys. 70, 293 (1979)
11. W.E. Klotzbücher, G.A. Ozin: Inorg. Chem. 19, 3767 (1980)

MAGNETO-OPTICAL INVESTIGATION OF Mn AND Cu-Mn SMALL CLUSTERS EMBEDDED IN RARE GAS MATRICES

J.C. Rivoal[a], C. Grisolia[a] and M. Vala[b]

(a) Lab. Optique-Physique, ER 5-CNRS
    10, rue Vauquelin, 75231 Paris Cedex 05 – France

(b) Dept. of Chem., University of Florida
    Gainesville FL 32611 – USA

INTRODUCTION

Among the techniques which are used to probe the structure and the electronic properties of small metal particles, the measurement of the differential absorption of polarized light under an applied magnetic field is one of the most sensitive. Used in conjunction with absorption the magnetic circular or linear dichroism (MCD or MLD) can provide a crucial check of the ground or excited states of clusters[1]. Since the density of small metal particles in a free jet is usually too small to allow absorption measurements in the gas phase, the particles can be trapped in rare gas matrices where they experience the smallest perturbation relative to the gas phase. However the matrix is not inert and it can interact with the atoms[2] or the clusters[3] depending upon the vibrational coupling with rare gas "cage"[2] or the different trapping sites[3]. It is also known that thermal sources produce mostly a beam of atoms and different set-ups have been used to increase the amount of clusters in the emitted beam[4]. Obviously the best way to obtain clusters of selected size is to mass select the particles after they have been produced with sufficient intensity using a sputtering device[5]. Since such a technique was not available to us we used a thermal source and allowed the metal particles to interact in a cold tube before trapping them in argon and krypton matrices. We probed the species formed with an optical set-up that we will describe in the next section. In a second section we will describe and discuss our results on manganese and copper-manganese atoms, molecules or clusters.

EXPERIMENTAL

A schematic of the cryostat and furnace assembly is given in fig. 1.

Two Knudsen cells placed at right angles in the furnace are used to produce by effusion the metal beams. The two beams are allowed to interact together and with the carrier gas between the furnace and the cooled optical window. Before reaching the window they pass through a tube held at liquid nitrogen temperature. The body of the cryostat and the magnet assembly has been purchased from SMC[6] and the CEA[7]. The bottom part and the rotatable sample support were built in our laboratory. The lowest temperature reached by the sample is 4.5 K. It can be raised through a

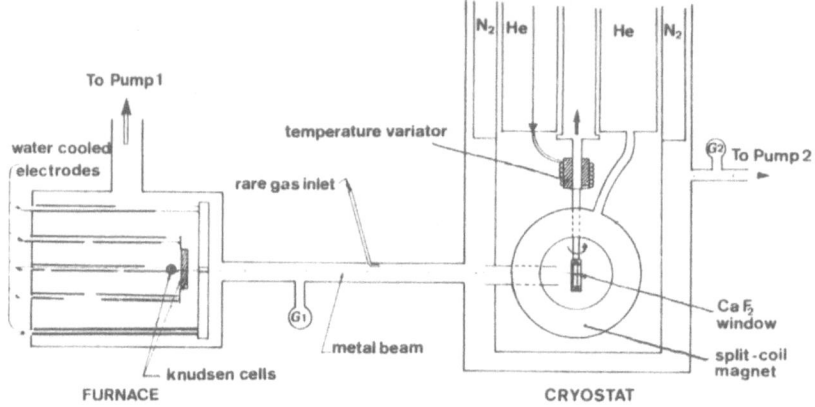

Fig. 1. Schematic of the experimental apparatus used for the codeposition.

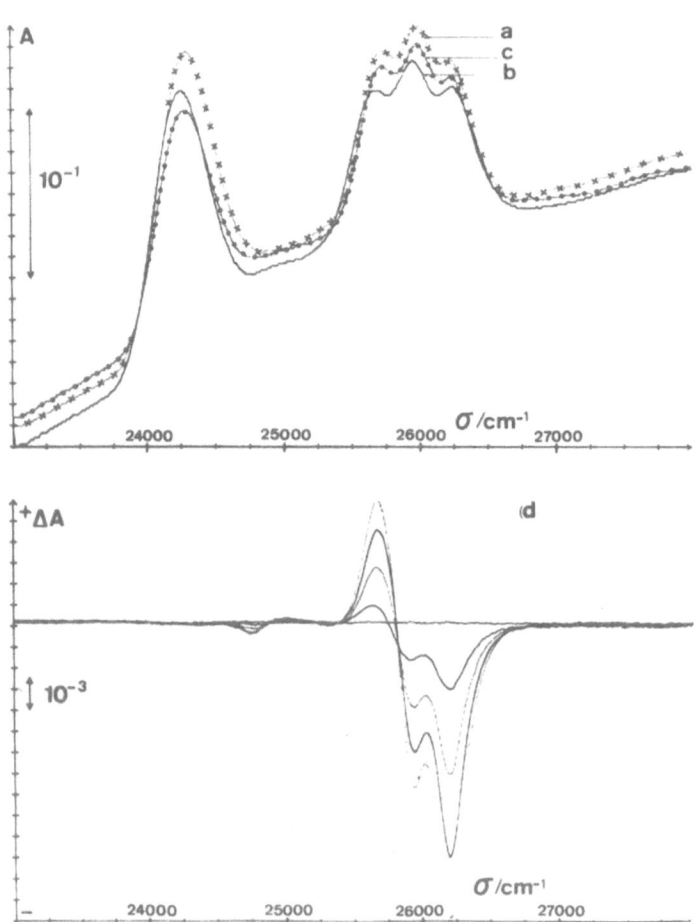

Fig. 2. Selected portion of the spectra of manganese in krypton matrix
after several annealing.
a - (✗✗✗) Absorption spectrum at 5.9 K.
b - (———) Absorption spectrum at 21.5 K.
c - (●●●●) Absorption spectrum at 5.2 K, thirteen hours later.
d - MCD spectra at various temperatures between 6 K and 21.4 K at B = 1.5 T.

temperature variator. The vacuum in the whole assembly is maintained by a diffusion pump which can be connected either on the furnace (pump 1) or onto the cryostat (pump 2) then forcing the metal beams and the rare gas to pass through the nitrogen cooled tube before being trapped on the CaF_2 window.

After the deposition has been made the sample can be rotated by ninety degrees to allow optical studies. The absorption and circular or linear dichroism can be recorded under computer control (PDP-11) using a home-built devide derived from an earlier apparatus[8] equipped with a new photoelastic modulator[9] which can be purchased from Microcontrole[10].

RESULTS AND DISCUSSION

a) Mn Clusters

A selected part of the spectra between 23 000 and 28 000 cm^{-1} of manganese in krypton deposited with the pump in position 1 (fig. 1) is shown in fig. 2. We will not discuss this spectrum in detail since our purpose here is the investigation of clusters. However some remarks can be made :

- The absorption spectrum clearly shows two bands one single centered at 24 250 cm^{-1} and a triplet centered at 26 000 cm^{-1}. Annealing at 21.5 K changes slightly the absorption spectrum narrowing the first band and leaving the second one almost identical (fig. 2b). The spectrum taken 13 hours later (fig. 2c) shows a decrease in the first band and a slight increase in the second band.

- The spectra as a function of the temperature is depicted in fig. 2d. The striking feature is that the first band shows a null signal but the second one shows a linear T^{-1} dependence. Consequently they must correspond to two different species. It can be shown that the second band corresponds mainly to the atom . The first band already present in a freshly prepared matrix corresponds to the Mn_2 molecule or an Mn_x cluster.

- When manganese is deposited in a krypton matrix with the pump in position 2 (fig. 1) the absorption spectrum is almost identical to the one shown in fig. 2, but the MCD signal is changed. The signal of the second band is very similar but the first band experiences a natural circular dichroism. The corresponding signal in zero field is larger than the magnetic signal. Since this phenomenon is amplified when the deposition is made in an argon matrix, we will only discuss this case.

In fig. 3 is shown the absorption and MCD spectra, in the range 21 000 to 41 000 cm^{-1} and at 6.5 K of manganese deposited in an argon matrix with the pump in position 2 (fig. 1). The atom bands around 26 000 and 36 000 cm^{-1} can be identified and correspond to two different sites. The band corresponding to Mn_2 at 29 000 and 30 200 cm^{-1} [11] are also seen by raising the temperature. Three bands at 25 000, 32 000, and 39 500 cm^{-1} correspond to another species and show a distinct circular dichroism (CD) in zero magnetic field (fig. 3b). When the sample is rotated 180°, the CD signal of these 3 bands changes its sign showing that it is not due to optical activity. We conclude that these bands are due to clusters of manganese which are oriented in the matrix. We have also recorded the linear dichroism (LD) of the sample and found that distinct nonzero signal, strongly dependent on the matrix orientation is observed in these three bands. Since the bands corresponding to Mn atoms and Mn_2 molecules are seen in the matrix and since no CD or LD signals can be detected in the corresponding bands we attribute the three

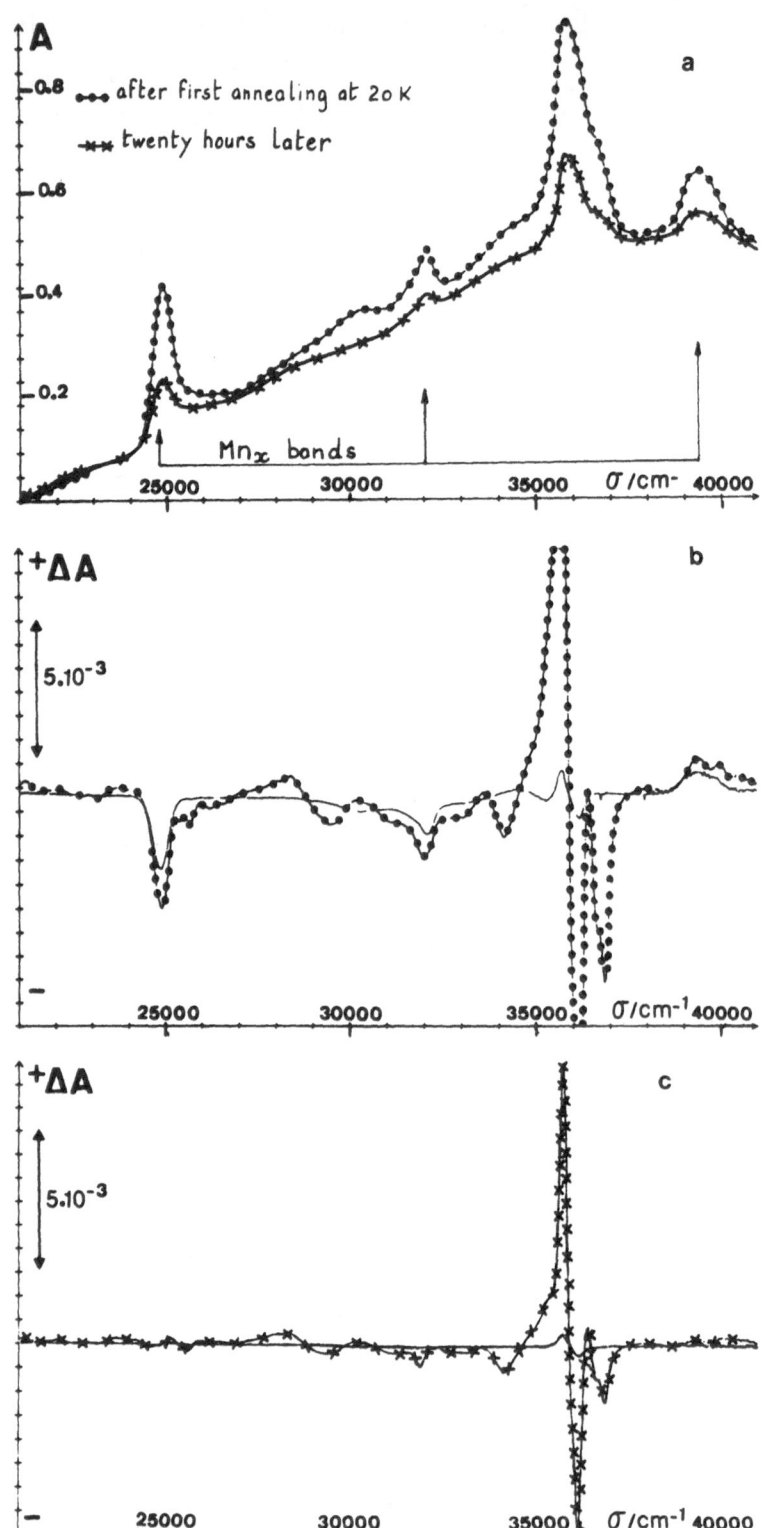

Fig. 3. a – Absorption spectra of manganese in argon matrix at 6.5 K.
b – MCD spectrum at 6.6 K after first annealing (•–•–•) : applied
magnetic field B = 1 T. (——) : B = 0 T.
c – MCD spectrum at 6.8 K twenty hours later. (✹✹✹) : applied
magnetic field B = 1 T. (——) : B = 0 T.

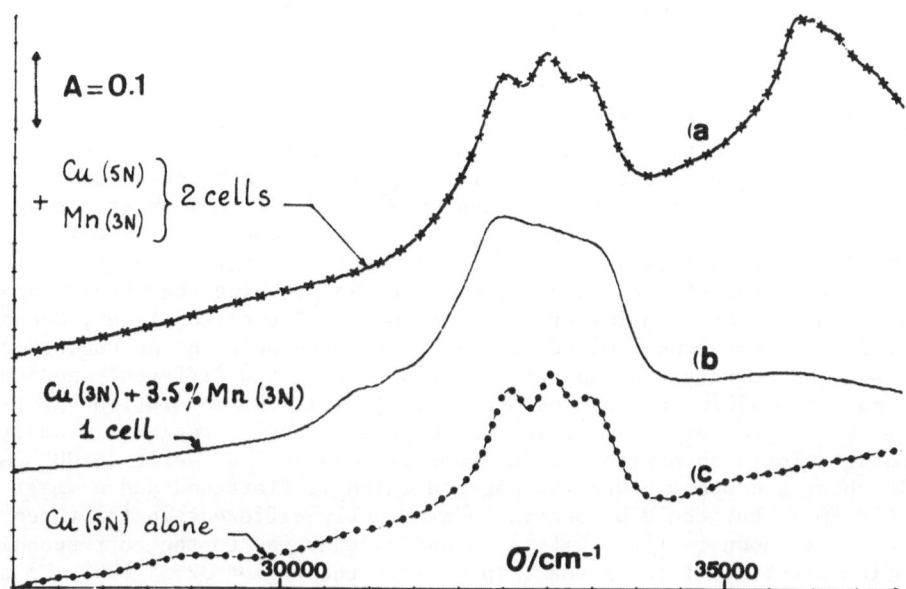

Fig. 4. Absorption spectra of copper-manganese in argon matrix at 6 K.
  a) Copper (5N) alone.
  b) Copper (3N) + 3.5 % manganese (3N) in one cell.
  c) Copper (5N) + manganese 3N) deposited with two cells.

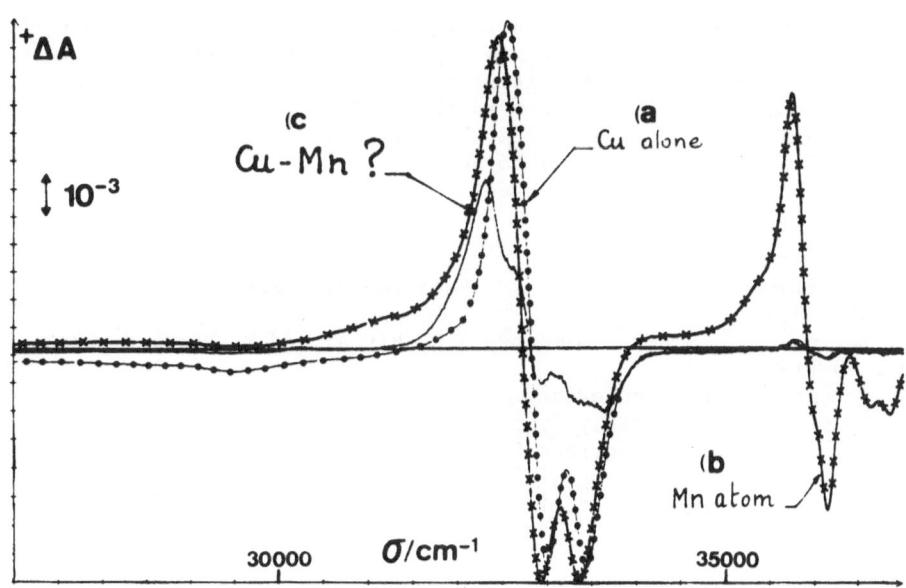

Fig. 5. MCD spectra of copper-manganese in argon patrix at $\simeq$ 5 K.
  Applied magnetic field B = 1 T.
  a)   copper (5N) alone.
  b)   copper (3N) + 3.5 % manganese (3N) in one cell.
  c)   copper (5N) + manganese (3N) deposited with two cells.

bands at 25 000, 32 000, and 39 000 cm^{-1} to an Mn$_x$ cluster with x $\geqslant$ 3. Because Mn$_5$ is easily formed in argon matrices[12] this cluster may correspond to x=5. More work is needed to confirm this point particularly using a source of selected size clusters .

b) Cu-Mn molecule

A large number of heteronuclear molecules can be formed with gas aggregation techniques. Little is however known about their electronic properties[13]. Since we have already investigated copper and manganese species in argon matrices[2,12] we tried to form the Cu-Mn molecule to look at its optical properties. To avoid the formation of clusters we pumped the assembly with the pump in position 1 (fig. 1). We recorded three different spectra (fig. 4). First, the spectrum of the copper alone (fig. 4a) showing the three peaked structure due to the Jahn-Teller coupling in the ^{2}P excited state[2]. Second, the spectrum of the copper mixed with 3.5 % of manganese in one knudsen cell (fig. 4b). And thirdly, the spectrum obtained using two different knudsen cells each containing one of the metals (fig. 4c). The absorption spectra in fig. 4a and 4c are very similar in the copper absorption region. In addition the characteristic absorption of Mn atoms is seen in fig. 4c at 36 000 cm^{-1}. Fig. 4b shows a copper absorption spectra which is flattened and a small band at 36 000 cm^{-1}. But the MCD spectra is distinctly different in the three cases : it is shown in fig 5 with a, b and c referring to the corresponding absorption spectra. If the a and c spectra in the region 32-34 000 cm^{-1} show one positive peak and two negative, we remark that when manganese is present the barycenter of the curve moves to the low energy side. This is enhanced in in fig. 5b where 2 positive peaks are observed instead of one and moreover the peak at low energy has the greatest intensity. We believe that this peak can be due to an electronic transition of the Cu-Mn molecule. The temperature variation of the MCD shows that the ground state of this species is degenerate and because no signal has been observed in a similar ESR experiment[13] we conclude that the ground state of this molecule is $^5\Delta$ or $^5\pi$.

In conclusion we have shown that magneto-optical experiments are more highly sensitive than absorption experiments in investigating the species formed when transition metals are matrix-isolated in rare gases. Small clusters can be formed either in the cell or in the beam and are shown to remain stable in the matrix. The optical and magneto-optical properties of these clusters need to be investigated in more detail to give greater insight into their structure and magnetic properties.

REFERENCES

1. J.Shakhs Emampour, R. Pyzalski, M. Vala and J.C. Rivoal, J. Phys. 45 (1984) 953.
2. K.J. Zeringue, J. Shakhs Emampour, J.C. Rivoal and M. Vala, J. Chem. Phys. 78 (1983) 2231.
3. P.S. Berchtoldt this conference.
4. W. Schulze, H. Abe, Faraday Symp. 14 (1980) 87.
5. P. Fayet and L. Wöste, Surface Science, 156 (1985) 134.
6. Société des Matériels Cryogéniques, 94398 Orly, France.
7. Commissariat à l'Energie Atomique, D. Ph. P.E., 91191 Gif/Yvette, France.
8. M. Billardon, J.C. Rivoal and J. Badoz, Rev. Phys. Appl. 4 (1969) 353.
9. J.C. Canit and J. Badoz, J. Appl. Opt. 22 (1983) 692.
10. Microcontrole, B.P. 144, 91005 Evry Cedex.
11. J.C. Rivoal, J. Shakhs Emampour, K.J. Zeringue and M. Vala, Chem. Phys. Lett. 92 (1983) 313.
12. C.A. Baumann, R.J. Van Zee, S.V. Bhat and W. Weltner Jr., J. Chem. Phys. 78 (1983) 190.
13. W. Weltner Jr., and R.J. Van Zee, Ann. Rev. Phys. Chem. 35 (1984) 291.

MULTIPLY CHARGED CLUSTERS

Olof Echt

Fakultät für Physik
Universität Konstanz
D-7750 Konstanz, West-Germany

ABSTRACT

Weakly bound, multiply charged clusters are not detectable in mass spectra unless their size exceeds a well-defined critical size $n_c(z)$. Values for n_c, with charge states up to $z = 4$, have been reported for van der Waals clusters and, in a few cases, for metal clusters, metal halide clusters and compound clusters. We review the experimental observation of multiply charged clusters and their critical sizes, including the apparently contradicting findings for metal clusters. Theoretical approaches towards a description of the "Coulomb explosion" are summarized. Recent progress in the size analysis of fragments from metastable decay of multiply charged clusters is described in detail.

INTRODUCTION

In consequence of electrical repulsion, a charged spherical mass of liquid, unacted upon by other forces, is in a condition of unstable equilibrium... When (the charge) is great, the spherical form is unstable... Under these circumstances the liquid is thrown out in fine jets, whose fineness, however, has a limit". These words, written more than a century ago by Lord Rayleigh in this Treatise "On the Equilibrium of Liquid Conducting Masses charged with Electricity" /1/ provide an excellent description of what is currently termed "Coulomb explosion" of microclusters[*]. It is intuitively clear that a small, doubly positively charged cluster will disintegrate into singly charged fragments if the cohesive forces

---

[*] The term "Coulomb explosion" has been applied originally to experiments in which very energetic (singly charged) molecules are charge-stripped upon passage through a thin foil and subsequently dissociate into atomic fragments. Symmetries and internuclear separation of the molecules can be determined from the angular distribution of their fragments, see ref. 2.

cannot compensate the Coulomb repulsion between the two holes. In a grossly simplified picture, the maximum separation between the like charges
equals the diameter of the spherical cluster. Upon fission, the charges
will be separated to infinity, thus reducing the Coulomb energy by an
amount of $E_c = e^2/4\pi\varepsilon_o d$ or, if d is measured in Å, by $E_c = 14.40$ eV/d (e
is the elementary charge, ε_o is the vacuum permittivity, and we have
assumed that ε, the dielectric constant of the cluster medium, is 1.0).
Thus, one may expect that doubly or, more generally, z-fold charged clusters will undergo rapid fission if their diameter is less than a critical
size n (z). This is borne out in the experiment, indeed: Multiply
charged small cluster ions A_n^{z+} are conspicuously absent in mass spectra
of van der Waals clusters, but they are observed beyond a certain size
limit. For the rare gases Ar, Kr and Xe, the size limit $n_c(z=2)$ corresponds to a diameter of ~20 Å (calculated from n_c and the bulk density in
a continuum approximation). Critical sizes have been determined for charge
states up to z = 4 /3 - 11/. In these cases, the experimental conditions
(cluster size distribution, energy of the ionizing electrons) had no
discernible influence on the critical sizes.

For metal clusters, however, the experimental results are less clear.
Critical sizes have been reported in a few cases, but multiply charged
clusters as small as the trimer or even the dimer have been observed, too.
It appears that the technique being used to grow and to ionize the clusters has a profound effect on the observability of these species.

This contribution reviews the current literature on multiply charged
clusters. Experimental results for van der Waals clusters, metal halide
clusters, compound clusters and metallic or semiconducting clusters are
presented in chapter II. Chapter III is devoted to a discussion of recent
progress /11 - 15/ in the determination of the size of <u>fragments</u> from
multiply charged clusters. These experiments provide a more detailed
insight into the quantities which determine the stability or lifetime of
multiply charged clusters. Theoretical approaches to this field are summarized in chapter IV. Several additional references to early observations
of multiply charged clusters and macromolecules have been compiled in /16/.

OBSERVATION OF MULTIPLY CHARGED CLUSTERS

<u>Introductory Remarks</u>

Multiply charged atomic clusters, A_n^{z+}, are easily distinguished from
singly charged clusters in mass-resolved spectra, if their size-to-charge
ratio, n/z, is non-integer. Henkes was apparently the first to note the
<u>absence</u> of small multiply charged cluster ions, $z \geq 2$, in mass spectra of
hydrogen clusters /17/. The spectra were resolved for $n/z \leq 80$. The clusters were grown in a jet and ionized with electrons of up to 110 eV
energy; i.e. single electrons could have caused the emission of several
secondary electrons from a cluster. Henkes estimated that doubly charged
hydrogen clusters with $n \leq 1000$ would be unstable. Experimental evidence
for the existence of <u>large</u> multiply charged clusters of hydrogen and
nitrogen was reported by members of the Karlsruhe group about a decade
later /18, 19/. At low electron energy, the unresolved size distributions
of cluster ions showed a single broad maximum at $\bar{n} \sim 10^3 - 10^4$, depending
on source conditions. Several broad, overlapping maxima were obtained at
high electron energy; the multimodal structure could be attributed to
clusters of the same mean size $\bar{n}$, but having different charge states.

Henkes and Isenberg calculated, in a liquid-drop approximation, the size limit of N_n^{2+} from the requirement that the exothermicity of the most favorable dissociation channel be zero /18/. They obtained $n_c(2) = 310$, clusters of that size would most likely disintegrate into $N_{14}^+ + N_{296}^+$.

An experimental identification of doubly charged clusters in mass-resolved spectra was first reported by Sattler et al for Pb, Xe and NaI clusters /3/ (also see: ref. 20). The data showed that the lower size limit, $n_c(2)$, of A_n^{2+} is remarkably well defined. A spectrum of Xe clusters, displayed in Fig. 1, illustrates this point. We find $n_c(2) = 52$, if we define the critical size such that larger doubly charged clusters are observed, whereas smaller ones are not. Clearly, the observability of Xe_{53}^{2+} or, possibly, Xe_{51}^{2+}, will to some extent depend on the experimental conditions (size distribution of neutral clusters, mass resolution, signal-to-noise ratio, electron energy). This dependence, however, is remarkably small, at least for doubly charged van der Waals clusters. Compare, e.g., the nearly identical results for doubly charged water clusters, obtained in four different laboratories /6,8,10/.

One also observes an abrupt increase of the intensity of cluster ions with <u>integer</u> values of n/z beyond the point where odd-numbered doubly charged clusters are detected. Fig.3 (below) displays this more clearly than the spectrum in Fig.1, which is somewhat perturbed by the occurrence of "magic numbers" /21/. This effect demonstrates that <u>even-numbered</u> doubly charged clusters also occur beyond $n_c(2)$.

An unambiguous identification of even-numbered clusters has been accomplished in case of mixed dimers (hetero-clusters) such as $NeKr^{2+}$, $NeXe^{2+}$ and $ArXe^{2+}$ /22/ or isotopically mixed homo-dimers of He /23/, B and C /24/ and Mo /25/. The observation of these small "clusters" (and many other covalently bonded dimers like N_2^{2+}, O_2^{2+}, Cl_2^{2+}) has little if any bearing on our discussion of multiply charged clusters: He_2^{2+}, e.g., is isoelectronic with the strongly bound H_2 molecule, and Mo_2^{2+} is quintuply bonded /26/. Therefore, these ions are metastable or, in some cases, even stable in spite of the strong Coulomb repulsion.

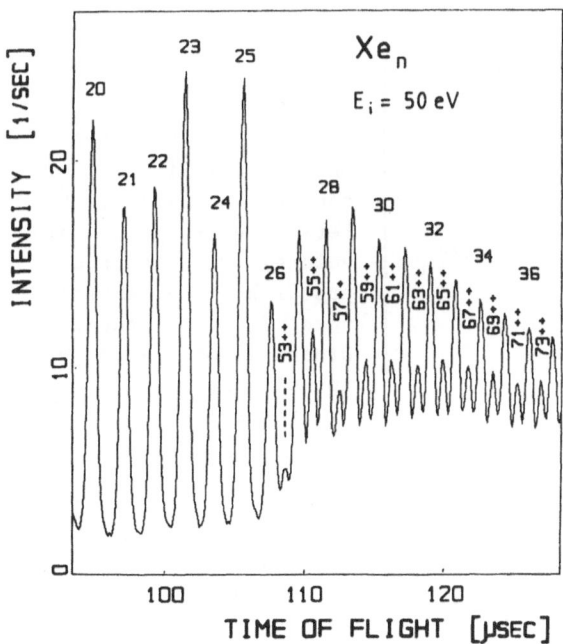

Fig. 1.
Time-of-flight mass spectrum of xenon clusters Xe_n, ionized by electron impact at 50 eV. The peaks due to doubly charged clusters are labelled "n++".

Most of the experimental studies of Coulomb explosion deal with van der Waals clusters. This is mainly due to the ease with which intense beams of large clusters can be generated by adiabatic expansion of a gas (either neat or seeded) /27/. A compilation of experimentally determined critical sizes of doubly charged van der Waals clusters, including hydrogen bonded materials, is presented in Fig. 2. The critical size $n_c(2)$ (denoted n_2 in the figure) has been multiplied by the cube root of the molecular volume (in $Å^3$), calculated in a continuum approximation from the density of the corresponding liquid at its boiling temperature T_b (at standard pressure), and plotted vs. $1/T_b$. In a very simple model /7/, these data points should fall on a straight line which passes through the origin. The values for $n_c(2)$ range from 28 (for $C_2H_3F_2Cl$) to 90 (Ar); the diameters of the "critical" clusters, assumed to be spheres, range from 12.6 Å for H_2O to 22.4 Å for C_5H_{12}. A compilation of the raw data will be published in a forthcoming paper /8,15/.

Clusters in higher charge states can be observed if the electron energy exceeds ~70 eV (depending on the material) and if the source conditions are such that sufficiently large clusters are formed in the jet. Fig. 3 (top) displays a time-of-flight mass spectrum of CO_2 clusters, ionized at 150 eV, with a logarithmic intensity scale /28/. The abrupt increase of the intensity beyond n/z = 22 is due to doubly charged clusters, that beyond 36 is due to triply charged clusters. The peaks of mass-resolved $(CO_2)_n^{2+}$ and $(CO_2)_n^{3+}$ have been connected in the enlarged spectrum (Fig. 3, bottom) by dashed and dotted lines, respectively. Other series of peaks arise from fragment ions such as $(CO_2)_n \cdot O^+$, $(CO_2)_n \cdot O^{2+}$ /4,13,28/. These ions obscure the steep rise in the intensity beyond n_c.

The spectrum in Fig.3 represents one of few cases where clusters in charge states higher than 2 have been resolved. Even at low resolution, however, critical sizes can be determined from the sudden increase of the unresolved background. Earlier results for CO_2 clusters essentially agree with the n_c values obtained from the high-resolution spectrum in Fig. 3.

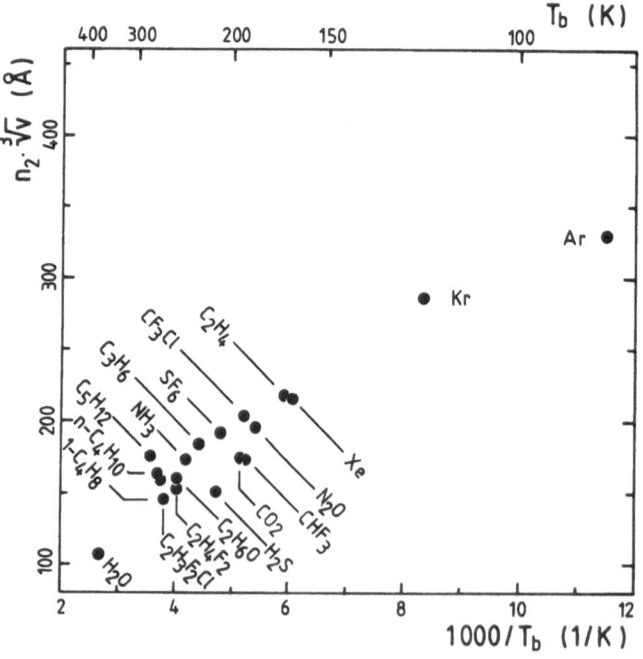

Fig. 2.
Compilation of $n_2 \cdot v^{1/3}$ (n_2 = critical size of doubly charged clusters, v = molecular volume) vs. the reciprocal boiling temperature T_b (at standard pressure) of the corresponding material. Data are taken from /6,8,10/ for H_2O, from /6,11/ for NH_3, from /4,13/ for CO_2, from /3/ for Xe, from /5/ for Kr, from /9/ for Ar, and from /8/ in all other cases.

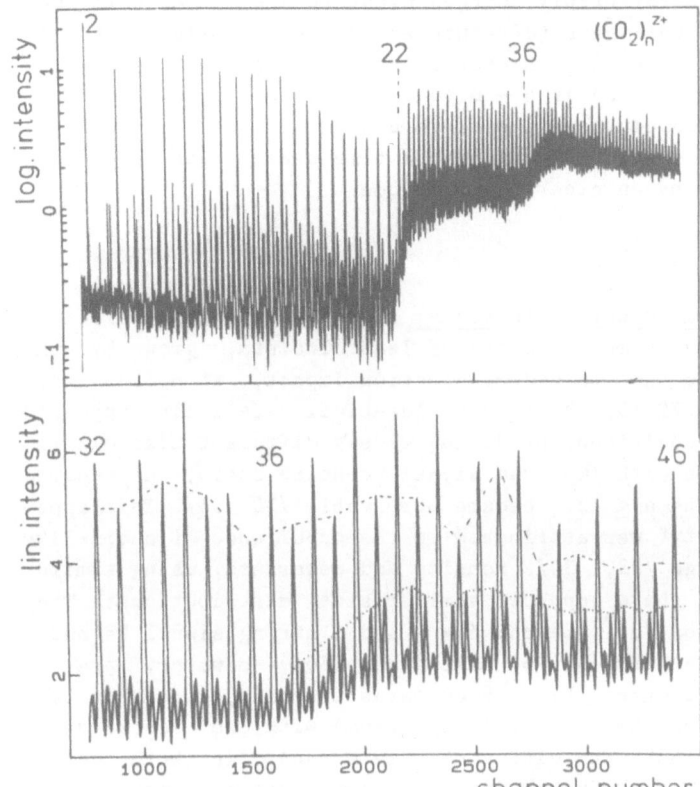

Fig. 3.
Time-of-flight mass spectrum of CO_2 clusters, recorded with a reflectron for improved resolution /28/. Cluster peaks are labelled according to their size n/z. The critical sizes are $n_2(2)=22.2$ and $n_c(3)=36.3$ for doubly and triply charged clusters, respectively. The envelopes of mass resolved peaks from $(CO_2)_n^{2+}$ and $(CO_2)_n^{3+}$ are marked by dashed and dotted lines, respectively.

One has to ascertain, though, that the steps in the size distributions feature a reasonable appearance potential /4, 13/.

The ratios of critical sizes, $n_c(3):n_c(2)$ and $n_c(4):n_c(2)$, show relatively little scatter among the van der Waals clusters (Fig. 4). The average ratios, taken from Fig. 4 (excluding the Pb-value) are 2.2 and 4.1, respectively. This is to be compared with a predicted value of 2.3 and 3.7, calculated from a continuum model which assumes "symmetric" rupture of spherical van der Waals clusters into equally sized, singly charged fragments /29/ (also see chapter 4). The agreement is remarkably good, at least for $n_c(3):n_c(2)$; but the experimentally determined distributions of fragment ions from triply charged clusters (cf. chapter 3) are at odds with the assumption of symmetric fission.

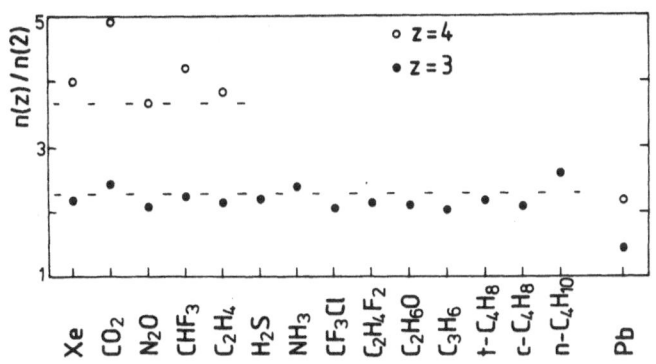

Fig. 4.
Ratios of critical sizes $n_c(z):n_c(2)$, z=3 (full dots) and z=4 (open dots). Data are from /4/ (CO_2), /11/ (NH_3), /30,31/ (Pb), and /8/. Dashed lines indicate predictions /29/.

The critical sizes for triply charged clusters are not as well defined as $n_c(2)$. Recent experiments (cf. chapter 3) have revealed that A_n^{3+}
($A = CO_2$, NH_3 or C_2H_4) have a high probability for fission (i.e. for
emission of charged fragments) if $n \sim n_c(3)$. The lower size limit of
detectable triply charged clusters therefore depends on the time scale of
the mass analysis; it increases by ~10% if the time scale (with respect to
ionization) is increased by an order of magnitude.

Metal and Semiconductor Clusters

Ionization of neutral clusters in beams: Rather well-defined critical
sizes have been observed in mass spectra of lead clusters, grown by the
gas aggregation technique and ionized by electron impact, at $n_c(2) = 30$,
$n_c(3) = 45$, and $n_c(4) = 72$ /3, 30 - 32/. Identical size limits appeared
under different source conditions, producing widely different size distributions. In mass spectra with improved signal-to-noise ratio, however, a
weak signal of Pb_n^{2+}, $7 \le n \le 13$, became observable /30, 33/. This apparent "island of stability" was attributed to the occurrence of chain-like
isomers in this size range /33, 34/. Schulze and coworkers, using similar
experimental conditions, have reported similar size distributions: They
observed a weak signal of Pb_n^{2+} beyond n = 9 and a strong signal beyond n
~ 30 /35/. More recent experiments by this group /36/, however, are at
odds with the former spectra: Fig. 5 compares a mass spectrum of lead
clusters obtained by Schulze et al (top spectrum) with one recorded by
Sattler et al (bottom) (note the different size range). In both cases,
clusters were grown from the vapor in a cold inert gas and ionized by
electrons at 70 eV. The top spectrum shows doubly charged clusters as
small as Pb_3, their intensity becomes very high beyond n ~ 20; whereas the
bottom spectrum is void of Pb_n^{2+} below n ~ 30.

A similarly striking discrepancy has been found for silver clusters.
Hoareau et al observed Ag_n^{2+} beyond $n_c = 18$, the intensity sets in abruptly /37/. Schulze and coworkers reported the observation of these clusters
beyond n = 16 and, additionally, that of Ag_n^{3+} beyond $n_c \sim 33$ /35/. More

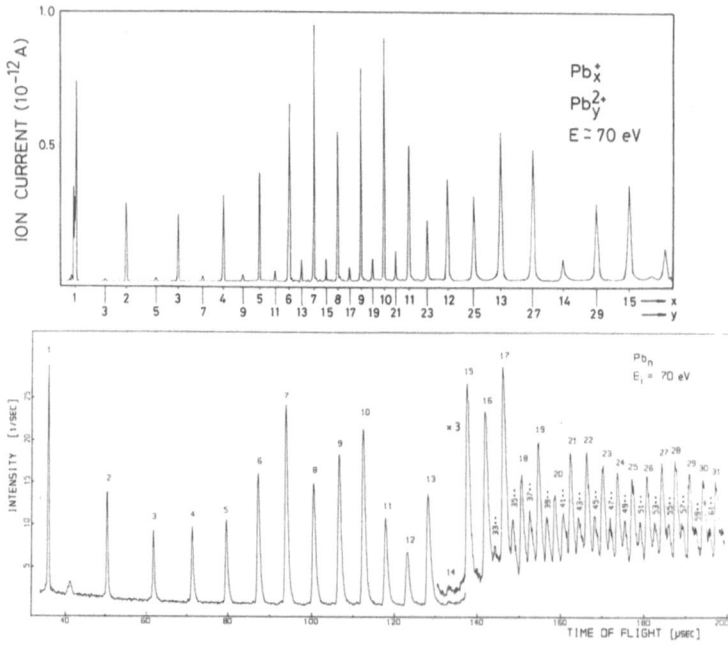

Fig. 5.
Mass spectra of
lead clusters,
ionized by elec-
tron impact at 70
eV, reported by
Schulze et al
/36/ (top spec-
trum) and by Satt-
ler et al /3/
(bottom spectrum).

recently, however, they observed doubly charged clusters as small as Ag_9^{2+}, whereas the lower size limit of triply charged clusters was only slightly shifted to n = 31 /38/.

These shifts in the size limits of doubly charged Pb and Ag clusters are not simply due to improved signal-to-noise ratio in the mass spectra. Schulze et al claim that the size limits mainly depend on the source conditions (mean cluster size in the beam) /38/. This contradicts the findings of Recknagel and coworkers in case of Pb_n^{2+} (see above). We speculate that some of the differences arise from the different electron emission currents in the ionizer. Schulze and coworkers use currents which are 2 orders of magnitude higher than the peak current in the experiments by Recknagel and coworkers. Successive ionization of clusters by two electrons will increasingly contribute to the yield of doubly charged clusters at high emission. In this way, cluster ions below n_c may be possibly formed, because ionization of a structurally relaxed singly charged cluster will result in doubly charged clusters with lower excitation energy, as compared to single-step ionization from neutrals.

Sb_n^{2+} has been reported by Pfau /30/. The ions were observed in a beam-experiment beyond $n_c \sim 28$; the rise in the intensity beyond n_c was not as steep as in the Pb-spectra. Bi_n^{2+} was also detected beyond $n_c \sim 26$ /30/. In this case, the intensity rises very gradually; thus it appears less surprising that Schulze and coworkers were able to observe Bi_9^{2+} and larger ions /36/. They also detected Bi_n^{3+} for $n \geq 38$. Martin has observed doubly charged cesium clusters beyond $n_c = 18$; the intensity of Cs_{19}^{2+} was at least 1.5 orders of magnitude higher than that of the (undetected) ion Cs_{17}^{2+} /39/. A homologous series $(Cs_n \cdot S)^{2+}$ has been detected, too, n = 18 being the smallest observable ion /40/. Doubly charged clusters $Pb_n Eu_5$, $n \geq 6$, have been detected by the same author /41/.

Finally, Hg_n^{2+} has been observed for $n \geq 5$ employing electron impact ionization of Hg clusters in a supersonic beam /20, 42/. The relatively high intensity of these small cluster ions is remarkable in view of the fact that lead, which has about the same surface tension in the liquid, metallic phase, shows a stepwise onset (or increase) of intensity at n = 30 (see above). The spectra of Hg_n^{2+}, though, do not extend to large cluster sizes, therefore there may be another, larger "critical size" $n_c(2)$. The appearance potential of Hg_5^{2+} has been also determined; it is significantly reduced with respect to Hg^{2+}, as to be expected, but well above the first ionization potential of Hg /42/. This fact rules out the importance of successive ionization by two primary electrons.

The first observation of doubly charged clusters produced by one-photon ionization from neutral clusters has been recently reported for Hg_n /43/. The results are remarkably similar to those obtained with electron ionization for similar energies (50 eV). The authors conclude that in both experiments the cluster is ionized in a two-step process, the second ionzation being accomplished by the secondary electron (or photo-electron) within the same cluster /43/.

Cluster ions from field emission: Field emission and related techniques (pulsed-laser stimulated field evaporation, liquid metal ions sources) easily produce atoms in charge states as high as z = 4, if the field strength is sufficiently high. The charge states are probably formed by electron tunneling from the evaporated (singly charged) ion back into the tip /44/. This post-ionization, as opposed to single-step ionization of neutral particles by low-intensity electron impact, appears to be the main reason for the frequent occurrence of small multiply charged clusters. For example, doubly charged trimers have been reported for Ni, W /45/, Au /45, 46/; Sn /47/ (Sn_n^{2+} for $1 \leq n \leq 11$, n odd), Bi /48/, Sb /49/, As /50/. Tsong and coworkers have detected clusters such as Si_n^{2+} (n=3,4,5,6,..) and Si_{13}^{4+} /51/, $He \cdot Me^{2+}$ (Me = Rh, Ir, Pt) and $He \cdot Me^{3+}$ (Me = Mo, W, Re) /52/ and Mo_2^{2+} /25/.

Mass spectra of ions emitted from a liquid metal ion source reveal the existence of the following (and larger) multiply charged clusters: Sn_3^{2+}, Sn_{19}^{4+} /53,54/, Ge_3^{2+}, Ge_4^{3+}, Ge_9^{4+} /55/, Cu_3^{2+}, Cu_{20}^{3+} /53,56,67/. Fig. 6 displays a spectrum of Ge_n^{z+} /55/. Due to the broad mass peaks (caused by the naturally occuring isotopes of Ge), and due to the periodicities which occur in the abundance of Ge_n^{z+} (for fixed z), the extraction of a "critical size", if there is any at all, is nearly impossible.

Other Clusters

Mass spectra of sulphur clusters, ionized by electron impact, show an abrupt onset of intensity for S_n^{2+} beyond n = 20 /39/. This contrasts with the detection of S_2^{2+} and even S_2^{3+} in a spark ion source /57/. Electron impact ionization of $(PbO)_n$, generated by the gas aggregation technique, yields $(Pb_nO_{n-1})^{2+}$ for $n \geq 3$; the intensity is gradually increasing with n /58/. The composition of these ions changes to $(Pb_nO_n)^{2+}$ for $n \geq 13$. Electron ionization of $PbCl_2$ clusters results in $(Pb_nCl_{2n-1})^{2+}$ for $n \geq 11$; again, the intensity rises slowly /58/.

Similar experiments with sodium halides were performed by Pflaum et al /59,60/. The composition of the ions is $(Na_nX_{n-2})^{2+}$, they are observed for n larger than 10 in case of NaI, 8 (NaBr), 14 (NaCl) and 20 (NaF). The intensity rises only gradually with n: this explains the much larger value of $n_c(2)$ being quoted in an earlier report for NaI /3/. The appearance of the doubly charged clusters is ~19eV for NaCl, ~17eV for NaBr and ~14eV for NaI /60/. Within the error bar of ±1 eV, these values are twice the ionization potential of the corresponding monomers. Triply charged clusters have been identified for $n \geq 29$ in case of NaI and $n \geq 37$ in case of NaCl /60/. Quadruply and quintuply charged clusters could also be identified in those parts of the mass spectra where the interference with other mass peaks was small.

Doubly charged carbon clusters have been generated in a spark ion source for all odd values of n, $n \leq 21$, by Cornides and coworkers /61/. The even-numbered clusters, which are much more difficult to identifiy because of the interference with singly charged clusters, were detected for $n \geq 10$. C_n^{2+} ions, formed by laser-induced desorption from graphite into vacuo, have been also reported by Nishina and coworkers /62/. Their mass spectra are strikingly different from those recorded by Cornides et al /61/. The latter ones feature local maxima in the abundance for C_{4n+3}^{2+}, $1 \leq n \leq 4$, whereas the former ones, recorded at low resolution, show

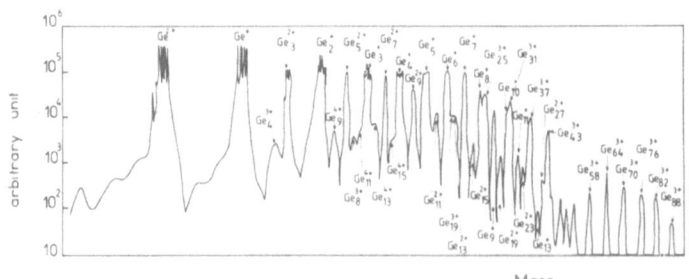

Fig. 6.
Mass spectrum of Ge_n^{z+}, $z \leq 4$, recorded with a liquid metal ion source /55/.

no ions other than C_{3n}^{2+}, $1 \leq n \leq 7$. Clearly, the cluster ions are generated by completely different techniques, but, for <u>singly</u> charged carbon clusters, different techniques have always resulted in qualitatively similar distributions /63/. These remarks express some concern about the experimental evidence for doubly charged carbon cluster <u>anions</u>, also reported by Nishina and coworkers /62/. Resolved, albeit rather broad peaks are attributed to C_{3n}^{2-} ($2 \leq n \leq 9$), no other anions were detected.

The observation of doubly charged cluster anions has also been reported by Märk and coworkers /64/. The anions were formed by low-energy electron attachment to a beam of neutral $(O_2)_n$. They found $(O_2)_n^{2-}$, n=3,5,7,9 /62/. Several tests were made to rule out that these ions were simply due to $(O_2)_n \cdot O^-$, which are produced at elevated electron energies. Experiments with isotopically mixed clusters would be highly desirable to further confirm these unexpected results, but the exceedingly low intensity of these ions may be prohibitive.

Finally, the formation of short-lived, large cluster cations in very high charge states has been reported by Nishina et al (Si_n^{z+}, n >> 10, 1.2 $\lesssim$ n/z $\lesssim$ 1.5) /65/ and, independently, by Joyes et al (Ge_n^{z+} and Si_n^{z+}, z ~ 50, n ~ 10^3) /66,67/. These ions would contain a huge Coulomb energy (a rough estimate for a spherical cluster Si_{100}^{100+} yields 5 keV relative to $100 \cdot Si^+$). They explode before mass analysis can be accomplished; their existence and their properties are derived from large fluctuations which appear in the kinetic energy of very small cluster ions; these ions are believed to be fragments from the "Coulomb explosion" /65-67/.

DISSOCIATION CHANNELS FOR MULTIPLY CHARGED CLUSTERS

A size analysis of product ions from multiply charged clusters has been accomplished only recently. These experiments have been performed in Konstanz, using a time-of-flight (TOF) mass spectrometer, and in Innsbruck, using a double focussing reverse geometry sector field spectrometer /11-15, also see 68,69/. The basic idea of these experiments is as follows: The existence of a critical size $n_c(z)$ in the intensity of A_n^{z+} implies that the lifetime of small clusters ($n < n_c$) is much shorter than the time t_{ms} required for full mass analysis, whereas that of large clusters (n well above n_c) is much longer than t_{ms}. t_{ms} in our TOF spectrometer corresponds to the time required for full acceleration of the ion, t_{ms} ~ $0.2 \sqrt{m/z}$ μs (m = mass of the ion in amu). This is because an ion will contribute to a well resolved mass peak if and only if it does not decompose during acceleration. (Very early dissociations, however, occurring before the ion has moved appreciably, are harmless; the ion will contribute to the mass peak corresponding to the <u>daughter</u> mass. In fact, most of the ions which contribute to resolved mass peaks are believed to be product ions from fast dissociations; this explains the pile-up of intensity at small n usually observed for high electron energies /17,27/).

The above statements about cluster ion lifetime suggest that A_n^{z+}, n being just above n_c, has a lifetime on the order of t_{ms}. This is true indeed for triply (but not doubly) charged van der Waals clusters of CO_2, NH_3 and C_2H_4: Fig. 7 displays three TOF mass spectra of CO_2 clusters /14/. The top one has been recorded in the standard mode; all ions which are accelerated into the drift tube are detected at its end. Any decompositions occurring in the drift tube remain undected, because the possible reaction energy (~1 eV for a "Coulomb explosion", much less for a thermal evaporation of a monomer) is small in comparison to the kinetic energy of the precursor ion (the latter being $2 \cdot z$ keV in our apparatus). In the lower two spectra in Fig.7, however, a repelling potential has been applied to a grid in front of the particle detector. Therefore, only ions with a sufficiently high kinetic energy will be detected. Strictly speaking, the requirement for detection is $(n_f/z_f)/(n_i/z_i) > U_b/U_o$, where U_o is the potential drop in the initial acceleration gap, U_b is the repelling potential, n_i and n_f are the size of the initial ion (at the entrance to the drift tube) and the final ion (at the time when it passes the repelling grid), respectively; and z_i, z_f are the corresponding charge states. Products from singly charged precursor ions cannot pass the grid if $U_b/U_o > 1$ (only their neutral products may pass, giving rise to a background which is independent of U_b).

The large hump in the middle spectrum in Fig.7 therefore arises from metastable clusters with $n_i/z_i \geq 108/3$, $z_i > 1$. The coincidence of this value with the critical size for triply charged clusters ($n_c(3)/3 = 108/3$, see top spectrum) and the dependence of the intensity of the hump on the electron energy (not shown) lead to the assignment $z_i = 3$ /14/. Thus, in agreement with our expectation expressed above, $(CO_2)_n^{3+}$ has a large probability to decompose in the drift tube if n is just slightly larger than $n_c(3)$. The shape of the hump is explained as follows: Smaller triply charged clusters do not reach the drift tube, whereas much larger ones do not decompose in the drift tube.

The charge state of the product ions in question has been analyzed with the double focussing mass spectrometer /13,14/. z_f turns out to be 2. Thus, the ions $(CO_2)_n^{3+}$, $108 \leq n \leq 120$, have a large tendency to dissociate into doubly charged fragments, the size distribution of which can be obtained from the change in the intensity of the hump (Fig.7) upon changing U_b/U_o. The final result is shown in Fig. 8. The distribution shows a maximum at $n_f/n_i \sim 0.92$: The Coulomb explosion of $(CO_2)_n^{3+}$ produces large doubly charged fragments. In other words, small singly charged fragments of size $n_{f2}/n_i \leq 0.08$ are ejected (the "<" sign holds if fission is accompanied by ejection of neutral particles). This explains our failure to

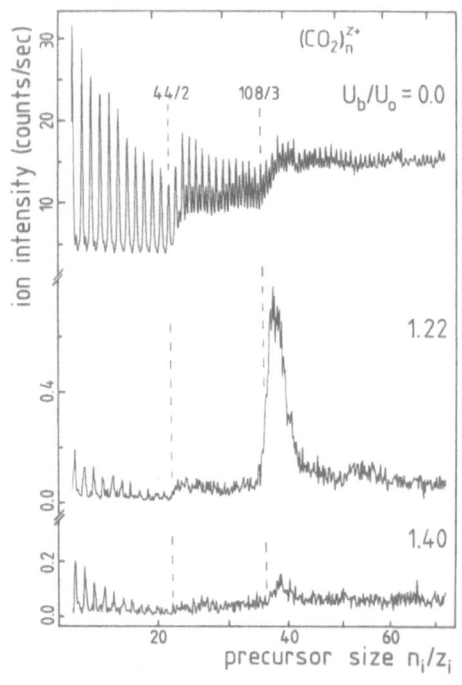

Fig. 7.
TOF mass spectra of $(CO_2)_n^{z+}$ /14/. The top spectrum, recorded in a conventional way, displays the critical sizes $n_c(2) = 44$ and $n_c(3) = 108$. The lower spectra have been recorded with a repelling potential applied to a grid in front of the detector, making possible the detection of energetic product ions. The "hump" in the middle spectrum is caused by products from metastable, triply charged clusters.

detect these latter fragments: They carry only $\leq$ 8% of the initial kinetic energy (which is $z_i \cdot 2$ keV = 6 keV). The recoil energy, ~ 1.6 eV in this case, is mostly imparted to the small fragment. It will add in a random direction; thus the divergence of the beam of small fragments will significantly increase, and most of these ions (~ 90%) will miss the detector.

Dissociation channels of $(NH_3)_n^{3+}$ have been analyzed with the double focussing instrument; the results are very similar to CO_2 /11,14/. Preliminary experiments with C_2H_4 indicate a less asymmetric size distribution of the fragments from $(C_2H_4)_n^{3+}$, $n \sim 110$ /14/.

Doubly charged clusters CO_2 and NH_3 do <u>not</u> undergo fission (into singly charged fragments) in the field free region of the mass spectrometer /11-14/. Instead, these clusters show a high probability to evaporate one or even two neutral monomers. From the absence of doubly charged clusters below $n_c(2)$, however, we know that these undergo fission very rapidly. These seemingly contradicting results can be understood as follows: Those ions which do not explode immediately after their formation will "never" explode, because the initial excitation energy (which results from the ionization and subsequent relaxation of the cluster) is lowered by successive evaporations of monomers. These evaporations will, due to the reduction in size, make the cluster even less stable (i.e. will lower the energy barrier against fission); but the net effect of these two competing processes is obviously a decrease rather than an increase in probability for delayed fission.

THEORETICAL APPROACHES

The most simple-minded model assumes that only those multiply charged ions are observed which are "intrinsically" stable, i.e. those for which fission would be endothermic for all possible reaction channels /3/. Unfortunately, intrinsic stability is neither a necessary nor a sufficient condition for the observability of A_n^{z+}: Within the simple model, and with some further simplifying assumptions, one can show /7/ that all values for $n_c(2)$ of van der Waals clusters should fall on a straight line, if plotted in a fashion similar to Fig.2. The experimental data do indeed follow this trend, but their absolute values are two orders of magnitude smaller than predicted. The shortcoming of this model is obvious: The Coulomb repulsion between like charges is the dominating force for large distances. Thus, formation of a multiply charged cluster in an unfavorable configuration

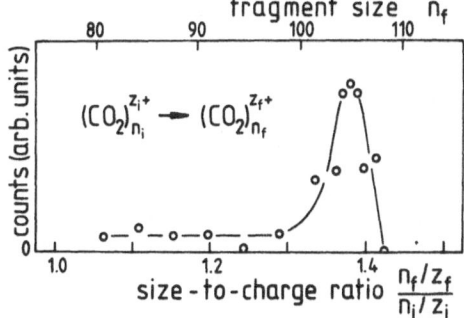

Fig. 8.
Size distribution of doubly charged fragments ($z_f = 2$, size n_f) from metastable triply charged CO_2 clusters ($z_i = 3$, initial size $108 \leq n_i \leq 120$) /12/.

will always lead to prompt explosion. It is the interplay of bonding forces and repulsive forces at short distances which determines whether or not fission of the cluster is impeded by an energy barrier. This implies that the excitation energy in the cluster, directly after ionization, will also affect the observability of multiply charged clusters. This energy will, in general, strongly depend on the method being used to generate A_n^{z+}: vertical ionization from A_n (relaxed) by electron impact or photon ionization, vertical ionization from $A_n^{(z-1)+}$, field detachment, etc. Recent ab-initio calculations for Mg_n^{z+} ($z=0,1,2$, $n\leq 7$) illustrate this point very nicely /70/: Mg_3^{2+} is shown to have a local minimum which can be reached by vertical ionization from Mg_3^+, but not from Mg_3. Similarly, the existence of a metastable configuration, having an energy barrier of ~0.2 eV, has been shown for Na_3Cl^{2+} by Martin /71/. Much larger clusters, however, would be needed to attain intrinsic stability.

Recent calculations of the intrinsic stability of metal clusters have been performed for Na_n^{z+}, Mg_n^{z+} ($z\leq 2$, jellium-type calculations /72/), for Li_n^{z+} ($z\leq 3$, jellium and ab-initio calculations /73/) and for Li_n^{z+} ($z\leq 2$, ab-initio calculations /74/). According to the jellium calculations, intrinsic stability will be achieved only for rather large clusters; a "magic-number" effect (either in the precursor or in the fragment) causes oscillations in the stability /72,73/. Koutecky, however, found that Li_6^{2+} and Li_8^{2+} are stable with respect to all dissociation channels /74/.

A simple theory for the electronic structure of clusters has been presented by Bennemann and coworkers /29/ (also see: /34/). It predicts ratios for critical sizes: $n_c(4):n_c(3):n_c(2) = 3.7 : 2.3 : 1$ for van der Waals clusters, 49:12:1 for clusters with ionic bonding, and 10:4:1 for metal clusters (lead, in particular). These ratios approximately agree with experiment in case of van der Waals clusters (cf. chapter II), but they disagree with the results for lead /30,31/. A different approach has been suggested by Delley /75/. He calculates the _lifetime_ of multiply charged metal clusters with respect to evaporation of monomers as a function of n and z. Taking the time of mass analysis as the minimum lifetime being required for observability, he obtains critical sizes in rough agreement with experiment. Soler et al suggest that the critical sizes for Pb_n^{z+} are controlled by tunneling of electrons between evaporated monomers or dimers, and the remaining large cluster ion /32/. Their values, however, disagree with experiment. Finally, we would like to mention calculations for the stability of charged metallic particles by Iakubov et al /76/ and by Joyes et al /54,77/.

The large number of detailed calculations for multiply charged metal clusters contrasts with our inability to understand, at least in a qualitative way, the factors that control the observability of, say, Pb_n^{2+}. What causes the jump in its intensity beyond n = 30, which mechanism favors the generation of long-lived ions as small as the trimer?

It appears that our understanding of multiply charged van der Waals clusters is developed much better. In this case, various approaches lead to similar values for $n_c(z)$, the agreement with experiment is good, and, last not least, the consistency of the experimental data is much better. - One way to predict critical sizes is to calculate, in a liquid-drop approximation (continuous distribution of matter _and_ charge!), the maximum charge beyond which a small deformation of a spherical droplet of given size will cause an instability. Lord Rayleigh /1/ derives the relation $r_c = (q^2/16\pi\sigma)^{1/3}$ for the critical radius (q = charge in esu, σ = surface tension in cgs units); this is equivalent to $r_c = 16.6 \text{ Å} (z^2/\sigma)^{1/3}$ (z = charge state, σ in dyn/cm). Critical sizes $n_c(2)$, calculated from this relation (taking data for σ and for the density from the bulk liquid) are about a factor of 1.2 to 1.5 larger than the experimental values for Ar, Kr, Xe and H_2O. For metals like Li and Na one obtains $n_c \sim 10$, for Ag, Bi, Hg, Mg and Pb $n_c \sim 5$, and for Mo $n_c \sim 2$. It is surprising (or fortuitous?) that application of Rayleigh's equation to clusters in charge states as small as z=2 yields reasonable values for n_c. One should also realize, however, that the concept of Coulomb explosion becomes akward if the critical size is of the order of 2 or less, because the bonding forces in the dimer may strongly depend on its charge state. Thus, for metals, one should try to determine critical sizes of highly charged clusters.

Some refinements of Lord Rayleigh's treatment are given in /78,79/: Grigor'ev discusses the shape-instability of ellipsoidal droplets, whereas Williams and Salomaa emphasize the difference between the radius of the droplet and the radius of the charge distribution in the droplet.

An especially illuminating study of doubly charged clusters has been reported by Gay and Berne /80/. They view the time evolution of Xe_n^{2+}, n=51 and 55, after vertical ionization of relaxed Xe_n, in a molecular dynamics simulation. The lifetime of Xe_{51}^{2+} with respect to fission is only ~100 ps, whereas Xe_{55}^{2+} does not fracture within the time scale of the simulation. These results, however, strongly depend on the total excitation energy of the cluster ion which, in turn, mainly depends on the initial position of the charges. The above-quoted agreement with experiment ($n_c(2)=52$ /3/) is obtained only if the two holes are formed initially on opposite sites in the cluster. Actually, the initial excitation energy is probably severely underestimated in this study /80/, because the formation of strongly bound dimer ions Xe_2^+ within the cluster /81/ was not taken into account. Thus, the agreement with experiment seems to be fortuitous, unless the initial excitation energy can be lowered rapidly, before fission takes place, by some other mechanism. Evaporation of neutral monomers does provide such a mechanism; it has been shown to occur with a high probability for doubly charged van der Waals clusters (see chapter 3). It is not clear to us whether these events did not occur in the computer simulation, or whether they escaped detection. - Gay and Berne also point out that the probability for a doubly charged cluster, being approximately spherical at time zero, to overcome the fission barrier is high in the first attempt, but decreases thereafter due to energy randomization. This explains our inability to observe delayed fission of doubly charged clusters (chapter 3).

A static model has been developed by Soler and coworkers which accounts for the critical sizes and dissociation channels of van der Waals clusters /12,15/. The total energy of a spherical, charged cluster is calculated within a liquid drop model (continuous distribution of matter and charge); the energy is the sum of a volume term, a surface term (both taken for a neutral droplet) and an electrostatic term. The fission barrier is calculated for a given precursor size and dissociation channel by comparing the total energy before fission with the energy of the (spherical) fragments plus the recoil energy being imparted to them after fission. The distance between the charges at the moment of rupture is taken to be the sum of the fragment radii. (The variation of the shape of the cluster during the fission process yields a more realistic picture; Soler et al have reported such a study for a droplet containing two point charges /82/). The critical size $n_c(z)$ is identified with that cluster which has a zero-eV barrier for the most easily accessible fission channel. The calculated critical sizes agree nicely with experiment; the corresponding most favorable fission channels are close to those being observed for $(NH_3)_n^{3+}$ and $(CO_2)_n^{3+}$ (see chapter 3).

The experimental observation of delayed fission does, of course, show that the randomized excitation energy in the cluster ion helps to overcome non-zero energy barriers. The corresponding shift of n_c to larger values is small, however, if the barrier height rises rapidly with increasing cluster size n. For van der Waals clusters this is the case, indeed /83/. Furthermore, evaporation of monomers can rapidly remove most of the initial excess energy, because the characteristic time of this process is much shorter than that of fission into heavy fragments /80/. For metal clusters, the situation will be different: Most likely they reduce their Coulomb energy by ejection of charged monomers or, perhaps, dimers /32,75/; the characteristic time for this process is the same as that for evaporation. Thus, fission may occur before the initial excitation energy can be lowered, and the observability of multiply charged metal clusters will crucially depend on the way how the cluster ions are formed.

ACKNOWLEDGEMENT

The experimental and theoretical analysis of decomposition channels for multiply charged van der Waals clusters is the result of a close collaboration with K.Leiter and T.D.Märk (Innsbruck), R.Casero, J.J.Sáenz and J.M. Soler (Madrid), and D. Kreisle, T.Leisner and E.Recknagel (Konstanz). This work has been financially supported by the Deutsche Forschungsgemeinschaft and by the Stiftung Volkswagenwerk.

REFERENCES

1. Lord Rayleigh, Phil. Mag. 14: 185 (1882)
2. D.S. Gemmell, Nucl.Instr.Meth. 191: 425 (1981);
   D.S. Gemmell and E.P. Kanter, Physics Today, January 1984, p. S-27;
   I. Plesser, Z. Vager and R. Naaman, Phys. Rev. Lett. 56: 1559 (1986)
3. K. Sattler, J. Mühlbach, O. Echt, P. Pfau and E. Recknagel,
   Phys. Rev. Lett. 42: 160 (1981)
4. O. Echt, K. Sattler and E. Recknagel, Phys. Lett. 90A: 185 (1982)
5. A. Ding and J. Hesslich, Chem. Phys. Lett. 94: 54 (1983)
6. A.K. Shukla, C. Moore and A.J. Stace, Chem.Phys.Lett. 109: 324 (1984)
7. O. Echt, in "Proc. 5th Symp. on Atomic and Surface Physics", F.Howorka, W. Lindinger and T.D. Märk, eds., Obertraun/Austria, 1986, p. 272
8. O. Echt, P. Höfer, M. Knapp, D. Kreisle, A. Reyes Flotte,
   K. Sattler and E. Recknagel, to be submitted to Z. Phys. D
9. P. Scheier and T.D. Märk, J.Chem.Phys., in print
10. D. Dreyfuss and H.Y. Wachman, J.Chem.Phys. 76: 2031 (1982);
    S.S. Lin, Rev. Sci.Instr. 44: 516 (1973)
11. D. Kreisle, K. Leiter, O. Echt and T.D. Märk, Z.Phys. D3: 319 (1986)
12. D. Kreisle, O. Echt, M. Knapp, E. Recknagel, K. Leiter, T.D. Märk,
    J.J. Sáenz and J.M. Soler, Phys.Rev.Lett. 56: 1551 (1986)
13. K. Leiter, D. Kreisle, O. Echt and T.D. Märk, J. Phys.Chem., in print
14. D. Kreisle, Ph.D. Thesis, University of Konstanz, 1986 (unpublished)
15. D. Kreisle, O. Echt, M. Knapp, E. Recknagel, R. Casero, J.J. Sáenz
    and J.M. Soler, in preparation
16. T.D.Märk, in "Book of Invited Papers", 6th Symp. on Elementary
    Processes and Chemical Reactions in Low Temperature Plasmas",
    P. Lukac, ed., MFF, UK, Bratislava, 1986;

T.D. Märk and A.W. Castleman, Jr., Adv.At.Mol.Phys. 20: 65 (1985)

17. W. Henkes, Z.Naturforschg. 17a: 786 (1962)

18. W. Henkes and G. Isenberg, Int.J.Mass.Spectrom.Ion Phys.5: 249 (1970)

19. J. Gspann and K. Körting, J.Chem.Phys. 59: 4726 (1973)

20. An early observation of Hg_n^{2+}, $n \geq 5$, by E. Schumacher and coworkers remained unpublished. See, e.g., M. Hofmann, Ph.D.Thesis, University of Bern, 1980

21. D. Kreisle, O. Echt, M. Knapp and E. Recknagel, Phys.Rev. A33: 768 (1986), and references therein

22. K. Stephan, T.D. Märk and H.Helm, Phys.Rev. A26: 2981 (1982); H.Helm, K. Stephan, T.D. Märk and D.L. Huestis, J.Chem.Phys. 74: 3844 (1981)

23. M. Guilhaus, A.G. Brenton, J.H. Beynon, M. Rabrenović and P. von RaguéSchleyer, J.Phys. B17: L605 (1984)

24. A.Galindo-Uribarri, H.W.Lee and K.H.Chang, J.Chem.Phys.83:3685 (1985)

25. T. T. Tsong, J.Chem.Phys. 85: 639 (1986)

26. J.B. Hopkins, P.R.R. Langridge-Smith, M.D. Morse and R.E. Smalley, J.Chem.Phys. 78: 1627 (1983)

27. O. Echt, M. Knapp, K. Sattler and E. Recknagel, Z.Phys.B53: 71 (1983)

28. T.Leisner, Diploma Thesis, University of Konstanz, 1986 (unpublished)

29. D. Tomának, S. Mukherjee and K.H.Bennemann, Phys.Rev. B28: 665 (1983)

30. P. Pfau, Ph.D. Thesis, University of Konstanz, 1984 (unpublished)

31. K. Sattler, Surf.Sci. 156: 292 (1985)

32. P. Pfau, K. Sattler, R. Pflaum, E. Recknagel, J.M. Soler, G. Diaz and N. Garcia, submitted for publication

33. P. Pfau, K. Sattler, R. Pflaum and E. Recknagel, Phys.Lett. 104A: 262 (1984)

34. S. Mukherjee, D. Tomának and K.H. Bennemann, Chem.Phys.Lett. 119: 241 (1985); also see: K.H. Bennemann, Z.Phys. B60: 161 (1985)

35. I. Goldenfeld, F. Frank, W. Schulze and B. Winter, Int.J.Mass Spectrom. Ion Proc. 71: 103 (1986)

36. W. Schulze and B. Winter, unpublished results

37. A. Hoareau, P. Melinon and B. Cabaud, J.Phys. D18:1731 (1985)

38. W.Schulze, B.Winter, J.Urban and I.Goldenfeld, submitted to Z.Phys.D

39. T.P. Martin, J.Chem.Phys. 81: 4426 (1984)

40. T.P. Martin, Angew.Chem. 98: 197 (1986)

41. T.P. Martin, J.Chem.Phys. 83: 78 (1985)

42. C. Brechignac, M. Broyer, Ph. Cahuzac, G. Delacretaz, P. Labastie and L. Wöste, Chem.Phys.Lett. 118: 174 (1985)

43. C. Brechignac, M. Broyer, Ph. Cahuzac, G. Delacretaz, P. Labastie and L. Wöste, Chem.Phys.Lett., in print

44. D.R. Kingham, Surf.Sci. 116: 273 (1982); G.L. Kellog, Surf.Sci. 120: 319 (1982)

45. W. Drachsel, Th. Jentsch, K.A. Gingerich and J.H. Block, Surf.Sci. 156: 173 (1985)

46. P.Sudraud, C.Colliex and J.Van de Walle, J.de Phys.Lett.40:L207 (1979)

47. A. Dixon, C. Colliex, R. Ohana, P. Sudraud and J. Van de Walle, Phys.Rev.Lett. 46: 865 (1981)

48. L.W.Swanson, A.E.Bell, G.A.Schwind and D.Larson, Proc. 27th Int. Field Emission Symp. (Y. Yashiro and N. Ugata, eds.) Tokyo 1980, p. 418

49. K.Gamo, Y.Ochiai, Y.Inomoto and S.Namba, Proc. 28th Int.Field Emission Symp. (L.W. Swanson and A.E. Bell, eds.) Portland 1981, p. 83

50. T. Sakurai, T. Hashizume, A. Jimbo and T. Sakata, J. de Phys. 45 Coll.C9: 453 (1984)

51. T.T. Tsong, J.Vac.Sci.Technol. B3: 1425 (1985)

52. T.T. Tsong and Y. Liou, Phys.Rev.Lett. 55: 2180 (1985)

53. J. Van de Walle and P. Joyes, J.de Phys. 46: 1223 (1985)

54. P. Joyes and J. Van de Walle, J.Phys. B18: 3805 (1985)

55. J. Van de Walle and P. Joyes, Phys.Rev. B32: 8381 (1985)

56. P. Joyes and P. Sudraud, Surf.Sci. 156: 451 (1985)

57. L.Morvay and I.Cornides, Int.J.Mass Spectrom. Ion Proc. 62: 263 (1984)

58. H.J. Novinsky, R. Pflaum, P. Pfau, K. Sattler and E. Recknagel, Surf.Sci. 156: 126 (1985)

59. R.Pflaum, P.Pfau, K.Sattler and E.Recknagel, Surf.Sci. 156: 165 (1985)

60. R. Pflaum, Ph.D. Thesis, University of Konstanz, 1986 (unpublished)

61. I. Cornides and L. Morvay, Mass Spectroscopy 31: 81 (1983)

62. A. Kasuya and Y. Nishina, Phys.Rev. B28: 6571 (1983)

63. I.Cornides, Int.J.Mass Spectrom. Ion Phys.45:219 (1982); E.A.Rohlfing, D.M. Cox and A. Kaldor, J.Chem.Phys.81: 3322 (1984), and refs. therein

64. K.Leiter, W.Ritter, A.Stamatovic and T.D.Märk, Int.J.Mass Spectrom. Ion Proc. 68: 341 (1986); Int.J.Mass Spectrom. Ion Proc., in print

65. A. Kasuya and Y. Nishina, Phys.Rev.Lett. 57: 755 (1986)

66. P. Joyes and J. Van de Walle, J.de Phys. 47: 821 (1986)

67. P. Joyes, J. Van de Walle and C. Colliex, Ultramicroscopy, in print

68. Evidence for fission of multiply charged water clusters during acceleration has been obtained by energy analysis: M.L.Aleksandrov, L.N.Gall', N.V.Krasnov, Yu.Kusner and V.I.Nikolaev, JETP Lett. 41: 246 (1985)

69. Stace and coworkers /6/ have searched for metastable decay of doubly charged clusters. No fragments were observed because their mass-to-charge ratio is too close to that of the parent ions.

70. G. Durand, J.-P. Daudey and J.-P. Malrieu, J.de Phys. 47:1335 (1986); also see: contribution to this conference

71. T.P. Martin, J.Chem.Phys. 76: 5467 (1982)

72. M.P.Iniguez, J.A.Alonso, M.A.Aller and L.C.Balbás, Phys.Rev.B34: 2152 (1986); M.P.Iniguez, L.C.Balbás and J.A.Alonso, contr. to this conf.

73. B.K. Rao, P. Jena, M. Manninen and R.M. Nieminen, submitted for publication; also see: contribution to this conference

74. J.Koutecký, private communication

75. B. Delley, J.Phys. C17: L551 (1984); J.de Chimie Physique, in print; also see: O.Echt, J.Phys. C18: L663 (1985), and Reply by B. Delley

76. I.T. Iakubov, A.G.Khrapak, L.I. Podlubny and V.V. Pogosov, Sol.State Comm. 53: 427 (1985)

77. P. Joyes and J. Van de Walle, J.de Phys. 47: 789 (1986)

78. A.I. Grigor'ev, Sov.Phys.Tech.Phys. 30: 736 (1985)

79. G.A. Williams and M.M. Salomaa, J.Low Temp.Phys. 57: 539 (1984)

80. J.G. Gay and B.J. Berne, Phys.Rev.Lett. 49: 194 (1982)

81. H. Haberland, Surf.Sci. 156: 305 (1985), and references therein

82. R. Casero, J.M. Soler and J.J. Sáenz, in "Proc. 5th Symp. on Atomic and Surface Physics", F.Howorka, W. Lindinger and T.D. Märk, eds., Obertraun/Austria, 1986, p. 284

83. R. Casero and J.M. Soler, unpublished results

COLLISIONAL ELECTRON TRANSFER TO NEUTRAL CLUSTERS BY HIGH-RYDBERG RARE GAS ATOMS

Tamotsu Kondow

Department of Chemistry, Faculty of Science
The University of Tokyo, Bunkyo-ku, Tokyo 113
Japan

ABSTRACT

High-Rydberg rare gas atoms are allowed to collide with various van der Waals clusters (CO_2, CH_3CN, SF_6, CCl_4 etc), to which the outermost electrons (Rydberg electron) having kinetic energies of 10 meV are found to be collisionally transferred. As a result, many cluster ions which can scarcely be generated by conventional techniques are produced with high efficiency. Systematic analyses of these mass spectroscopic data provide information on the electron affinities of the clusters and relaxation involved in the ionization. The experimental techniques and several typical examples of the analyses will be presented, together with related topics and future prospects.

INTRODUCTION

Electron attachment to a van der Waals cluster has attracted much attention,[1] since negative cluster ions thus produced provide a unique opportunity to investigate electron affinity states of the neutral clusters, and processes of relaxation induced by the electron attachment are by themselves interesting problems of intracluster dynamics. For example, introduction of an electron to a neutral cluster results in an increment in its basicity in analogy with acid-base reactions, and certain relaxation processes can be favored. Indeed, a hydrogen transfer reaction is greatly accelerated by adding an electron to azaaromatic compounds in liquid NH_3.[2,3]

One of the essential problems in these studies is to produce the negative ions efficiently and gently. In order to achieve this goal, a source for very slow electrons of meV-kinetic energies should be developed, since the cross sections for the electron attachment are sizable only when kinetic energies of the attaching electrons are very low, and consequently dissociation associated with the ionization can be minimized.

The present study of the collisional electron transfer by use of high-Rydberg rare gas atoms, Rg^{**}, has been undertaken with this prospect. The high-Rydberg atoms having principal quantum numbers of 20-40 have been used as a source of electrons with near-zero kinetic energy of about 10 meV. In such a high-Rydberg atom, Rg^{**}, its outermost electron (Rydberg electron) is very weakly bound to its core ion by Coulomb interaction since the Rydberg electron moves in a remote orbit with a radius of 10^2-10^3 nm.

Accordingly, the ionization potential and the kinetic energy of the Rydberg electron can be made less than 10 meV. In the collisional processes involving Rg^{**} and a target molecule, M, having a positive vertical electron affinity, the Rydberg electron of Rg^{**} is attached gently to M with a large cross section of 10^{-12}-10^{-13} cm^2. It has been proved that the collision occurs almost exclusively with the Rydberg electron and the core ion simply behaves like a spectator.[4,5] Therefore, the Rydberg electron can be regarded as a free electron which has the same momentum and energy as those of the Rydberg electron.

The present method of ionization has been applied to van der Waals clusters: The clusters bound by van der Waals interaction are chosen, since the weakly-bound van der Waals bonds may be dissociated readily by even a small perturbation associated with the ionization, and can be used for testing how gently these clusters can be ionized by this method.

EXPERIMENTAL

A schematic diagram of the apparatus is shown in Fig. 1. The apparatus consists of (1) a cluster beam source, (2) a triple-grid ion source, (3) a quadruple mass spectrometer, and (4) a data-acquisition system (CAMAC system) based on an LSI-11/23 microcomputer. In an ion source by use of Rg^{**} impact, a cluster beam generated by supersonic expansion is allowed to collide with Rg^{**} produced by electron impact on Rg. Negative cluster ions from the source have been observed by mass spectroscopy.

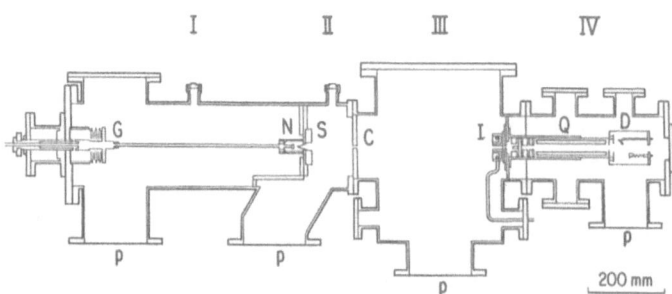

Fig. 1. A schematic diagram of the apparatus. G: inlet of source gas, N: sonic nozzle, S: skimmer, C: collimator, I: ion source, Q: quadrupole mass spectrometer, D: ion detector equipped with a charge conversion dynode, P: oil diffusion pumps.

Cluster Formation

van der Waals clusters are produced by a supersonic nozzle beam technique. The beam source consists of a sonic nozzle having an aperture diameter of 50 μm and a channel length of 0.2 mm (see Fig. 1). A sample gas is seeded in He, Ar or H_2 gas at room temperature with a stagnation pressure of 100 ∿ 3000 torr, and is expanded through the nozzle into a beam expansion chamber (Chamber I). This chamber is maintained at a pressure of 10^{-4} torr by use of a 6" oil diffusion pump. The central portion of the expansion, which contains larger clusters, is sampled by a Beam Dynamics conical skimmer with an entrance hole of 0.31 mm diameter and is injected into a collimation chamber (Chamber II) which is evacuated down to 10^{-5} torr by a 4" diffusion pump. The cluster beam is further collimated by an aperture of 5 mm diameter into an ionization chamber (Chamber III), where a concentric triple-grid ion-source is installed.

Ionization

The clusters are ionized by use of Rg^{**} impact in the ion source. As shown in Fig. 2, the ion source had a housing of 20 mm length and 60 mm diameter, on which three concentric cylindrical grids and filaments are mounted. The grids, G_A, G_B and G_C, are made of stainless-steel mesh with a transparency of 80 %. The four pieces of helical filaments made of thoriated tungsten wire of 0.15 mm diameter form a rectangular square. The cluster beam passing through the collision region is ionized by impact of Rg^{**} atoms or electrons.

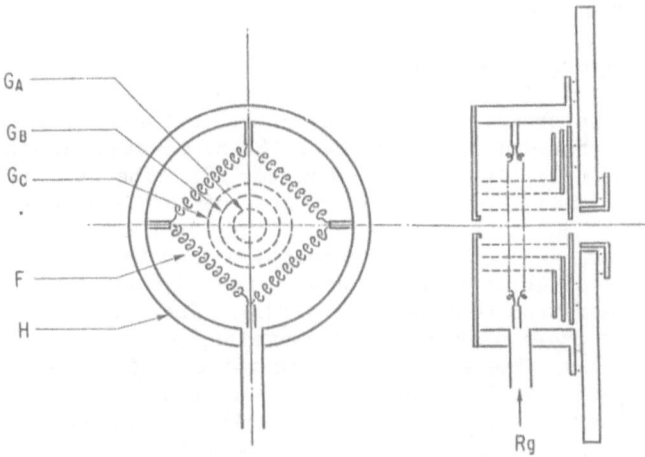

Fig. 2. A schematic diagram of a triple-grid ion-source. Three concentric grids, G_A, G_B, and G_C installed are used for preventing charged species from penetrating into the central collision region. F and H denote filaments and a housing, respectively.

Argon or Krypton gas with more than 99.95% purity are excited in the exterior of G_B by 50 eV-electrons. Ionic species and electrons are retarded by application of appropriate potentials to the three grids, and only neutral species including Rg^{**} are allowed to enter the central collision region. The pressure of the rare gases in the ion source is estimated to be on the order of 10^{-4} torr. In this pressure region, a single collision condition is found to be fulfilled, because mass-selected ion signals are proportional to the rare gas pressures in the ion source. The reaction chamber is evacuated by a 6" diffusion pump with a liquid nitrogen trap. The pressure of the chamber with a background pressure of 10^{-6} torr is increased to about 10^{-5} torr in an operating condition. The principal quantum numbers, n_p, of Rg^{**} are estimated to be in the range of 20-40, since Rg^{**} having $n_p \geq 40$ are ionized by an electric field of about 430 V/cm between G_A and G_B and those having $n_p \leq 20$ are radiationally decayed before arriving at the collision region.[6] The following observations have confirmed that the observed negative ions originate from transfer of the Rydberg electrons to the neutral clusters in collision with Rg^{**}: (1) the intensities of the cluster ions are proportional to the rare gas pressures, (2) the intensities start to rapidly rise in the vicinity of the ionization potential of the rare gases, as increasing electron impact energy for excitation of the rare gases, and (3) the signals tend to be reduced by application of an increasing voltage to G_B; this observation is interpreted as the field ionization of Rg^{**}.

Electron impact ionization of the clusters can also be carried out in the same ion source, where the potentials applied to the grids and the filaments are adjusted so as to allow free electrons to enter the collision region. The average electron energy, $\bar{\varepsilon}$, is obtained from the potential difference between grid G_A and the center pole of the filaments. The energy spread (fwhm) of the electrons is estimated to be about 2 eV at $\bar{\varepsilon} = 4$ eV by comparison of the measured cross section curve for the O^- production from CO_2 and the reported one.[7]

Detection of Negative Cluster Ions

The negative ions produced in the ion source are mass-analyzed by a quadruple mass spectrometer (Extranuclear, 162-8) mounted coaxially with the cluster beam in the detection chamber (Chamber IV). The maximum mass-to-charge ratio of the mass-selected ions is m/z 1650 and a typical mass-resolution is about 300 at m/z 1460. The ions after the mass spectrometer are focused by a lens and converted to positive ions by an ion conversion dynode at a voltage of about 5 keV. Positive ions ejected from the dynode are detected by a Ceratron (Murata, EMS-1081B). The ion conversion dynode made of a stainless steel disk of 22 mm diameter, greatly surpresses noise due to stray electrons. The signal-to-noise ratio is improved by more than three orders of magnitude. The transmission and detection efficiencies of the mass spectrometer and the detector are calibrated by use of the known fragmentation patterns[8] of the positive and negative ions produced by electron impact on perfluorokerosene (PFK). The mass-to-charge ratios are calibrated by the fragment ions of PFK[8] and the cluster ions of CO_2.[9] The detection chamber (Chamber IV) evacuated by a 4" oil diffusion pump exhibits a typical pressure of about 5×10^{-7} torr in an operating condition.

The signals from the detector are amplified by a preamplifier (ORTEC 9301) and registered in a multichannel analyzer (Canberra 3100). The measurement system is controlled by a CAMAC-crate-mounted LSI-11/23 microcomputer.

ELECTRON ATTACHMENT TO A VAN DER WAALS CLUSTER

A Rydberg electron is resonantly transferred to a van der Waals cluster,

$(M)_m$, where m represents the cluster size. The electron is accommodated in an affinity state of $(M)_m$, forming a transient negative cluster ion, $(M)_m^{-*}$. The nuclear configuration of the cluster remains unchanged in the electron transfer process because the electron motion is much faster than the rate of deformation of nuclear geometry in the cluster associated with the electron capture. In order that the electron is captured with high efficiency, the energies of the electron affinity states should be almost equal to the energy of the vacuum level in this resonant process, that is, its vertical electron affinity must not be negative. The kinetic energy of the Rydberg electron is so small that the contribution of the kinetic energy can be ruled out from the present consideration. The cluster system tends to relax toward its ground state by localization of the captured electron and associated nuclear rearrangement. The excess energy generated by the electron attachment is transmitted to the internal degrees of freedom of the cluster, i.e., the cluster is 'heated'. In some cases, evaporation of the component molecules and chemical reactions may occur, by which the excess energy is released out of the cluster. The argument mentioned above and the electron-transfer collisions between electron-attracting molecules and Rg^{**} indicate that these negative cluster ions are produced via the following processes:

$$(M)_m + Rg^{**} \rightarrow (M)_m^{-*} + Rg^+ \qquad (1)$$

$$(M)_m^{-*} \rightarrow (M)_n^- + (m-n)M \qquad (2)$$

$$\rightarrow \text{intracluster reactions} \qquad (3)$$

When $(M)_m^{-*}$ is stabilized, the excess energy generated is transmitted to the internal degrees of freedom of the cluster, and the effective temperature of the cluster increases. If the temperature does not exceed the sublimation temperature of the cluster, no substantial evaporation is expected to occur (non-evaporative). Otherwise, the stabilization energy is released by evaporation of the component molecules (evaporative) or by rupture of the chemical bonds of the component molecules (dissociative).

NEGATIVE CLUSTER-ION FORMATION
TYPICAL EXAMPLES

The collisional ionization by use of Rg^{**} impact has been applied to various van der Waals clusters whose monomer molecules have either positive or negative electron affinities.[9-13] The results obtained are listed in Table 1. As shown in Table 1, the product ions are $(M)_n^-$ except for the clusters of CCl_4 and N_2O. The mass spectra of the cluster ions observed exhibit the following common features: (1) The spectra obtained by use of different rare gases are almost identical, (2) the spectra have a threshold size, n_L, and a broad distribution having a maximum at about n_L=10, (3) several outstanding peaks are discernible at n_M (magic number). In the case of the CO_2 cluster, (4) the intensities of $(CO_2)_n^-$ ions with $11 \leq n \leq 13$ are much weaker than those of the neighboring ions. On the other hand, the clusters of N_2O and CCl_4 are dissociatively ionized and several complex ions such as $(N_2O)_n O^-$, $(N_2O)_n NO^-$ and $(CCl_4)_n Cl^-$, are detected. These experimental features can be explained in terms of the ionization processes: non-evaporative, evaporative and dissociative/intracluster reaction.

CD$_3$CN Clusters - Non-evaporative electron attachment

Ionization of CD_3CN clusters[10], $(CD_3CN)_m$, in collision with Kr^{**} gives negative ions, $(CD_3CN)_n^-$ ($n \geq 11$), as shown in Fig. 3. The threshold size, n=11, is independent of the stagnation pressure, while the size distribution for $(CD_3CN)_n^-$ varies with the stagnation pressure probably because of the change in the size distribution of the neutral clusters by the stagnation pressure.

No ion signal is detected by electron impact on the cluster. This finding indicates that the cross section for formation of a negative cluster ion decreases rapidly with the kinetic energy of the electron to be captured by the neutral cluster.

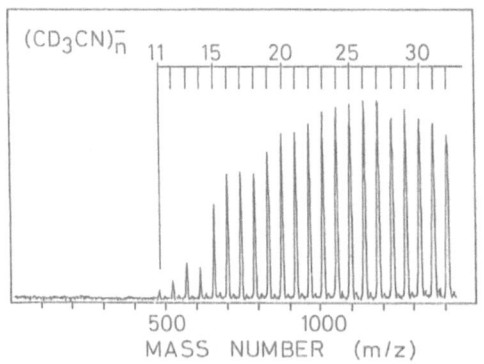

Fig. 3. Mass spectrum of $(CD_3CN)_n^-$ produced by impact of Kr^{**} atoms on the neutral clusters of CD_3CN.

As described in the foregoing, the cluster ions, $(CD_3CN)_n^-$, are formed via the following processes:[9-13]

$$(CD_3CN)_m + Kr^{**} \rightarrow (CD_3CN)_m^{-*} + Kr^+ \qquad (4)$$

$$(CD_3CN)_m^{-*} \qquad \rightarrow (CD_3CN)_n^- \qquad (m=n\geq 11), \qquad (5)$$

where no significant evaporation from $(CD_3CN)_m^{-*}$ is expected for the following reasons. The attached electron is trapped in the cluster probably as a dimeric anion, according to an ESR study on a negatively charged crystal of CH_3CN [14]. In this anion, the two molecules are likely to be oriented antiparallel with each other. The CD_3CN molecules surrounding the trapped electron are polarized and reoriented. The excess energy which is nearly equal to the stabilization energy, is transmitted to at least 6m-6 inter-molecular vibrational and molecular rotational modes, and consequently the effective temperature of the cluster is increased. Statistical distribution of this excess energy among the 6m-6 vibrational modes gives rise to the increment in the effective vibrational temperature of about 240 K for $m\geq 11$. Since the effective vibrational temperature of the original neutral cluster is very low, the effective temperature of the cluster ion is much lower than the boiling temperature of liquid CD_3CN (355 K); the effective boiling temperature of the cluster ion, at which substantial evaporation from $(CD_3CN)_m^{-*}$ is expected, is estimated to be even higher. Therefore, the cluster size is likely to remain essentially unchanged in the process of relaxation.

The sharp rise in the mass spectrum of the negative cluster ions can be ascribed to a sudden increase in the cross section for the electron attachment. The presence of this clear threshold can be explained by the assumption that the vertical electron affinity of $(CD_3CN)_m$, which is known to be negative for the monomer (m=1), turns to be positive at m=11. In this estimation the kinetic energy of the Rydberg electron can be disregarded.

<u>CO_2 Clusters</u> - Evaporative electron attachment

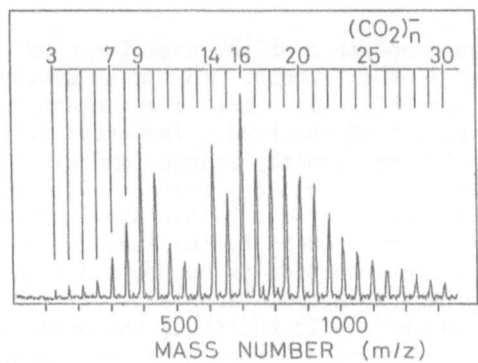

Fig. 4. Mass spectrum of $(CO_2)_n^-$ produced from $(CO_2)_m$ in collision with Kr^{**}.

Figure 4 shows the spectrum of $(CO_2)_n^-$ produced by Kr^{**} impact from $(CO_2)_m$. The spectrum has (1) the threshold size at n=3, (2) a region where the peak intensities are weak ($11 \le n \le 13$), and (3) magic numbers of n=14,16 which are associated with the cluster ions in comparison with those produced by electron impact. The ion, $(CO_2)_{14}^-$, seems to have such an icosahedron structure that a dimeric CO_2 ion is solvated by 12 CO_2 molecules.

Finding 1 indicates that the cross section for the electron attachment increases significantly at $m=m_L$. This sudden increase in the cross section may arise from the following reasons: (a) the vertical electron affinity of $(CO_2)_m$ increases with m and becomes positive beyond m_L, (b) the state density of $(CO_2)_m^{-*}$ which takes part in the electron attachment increases with m, and (c) the lifetime for autodetachment from $(CO_2)_m^{-*}$ also increases with m.

Finding 2 can be explained by assuming that at least 4 CO_2 molecules are evaporated from $(CO_2)_m^{-*}$ in the process of relaxation. Indeed, the effective temperature of the CO_2 cluster ions after the relaxation reaches about 900 K, at which temperature appreciable evaporation is expected. Here, the CO_2 cluster ions are assumed to be stabilized so that the transferred electron is localized in one of the CO_2 component molecules, possibly with a bent structure having the stabilization energy of 3.8 eV.[15,16] This assumption of stabilization is confirmed by theoretical calculations[17,18] on the structures of the cluster ions.

The experimental results are explained, accordingly, in terms of the following evaporative electron attachment:

$$(CO_2)_m + Kr^{**} \rightarrow (CO_2)_m^{-*} + Kr^+ \qquad (6)$$

$$(CO_2)_m^{-*} \rightarrow (CO_2)_{m-4}^- + 4\ CO_2 \quad (m \geq 7) \quad (7)$$

Here, m_L is estimated to be 7, since the number of molecules to be evaporated are at least 4 and $n_L = 3$.

Figure 5 shows a mass spectrum of the negative cluster ions produced from $(CO_2)_m$ by impact of 1 eV-electrons. This mass spectrum is contrasted to that by Kr^{**} impact (see Fig. 4), where larger clusters of $(CO_2)_n^-$ are intense and no fragmented ion is observed. The cross sections for formation of the parent ions, $(CO_2)_n^-$, are greatly reduced and cluster ions with small sizes become more abundant. In addition, fragmented ions, $(CO_2)_n O^-$, are clearly observable. These findings can be explained by introduction of a large amount of kinetic energy brought by the incoming electrons, into the clusters; the attachment cross section is suppressed significantly, and evaporation of the component molecules becomes appreciable with increasing kinetic energy of the electrons. In addition, the depletion region, $11 \leq n \leq 13$, observed in Fig. 4, disappears almost completely in the spectrum of the electron impact ionization, probably because of the extensive evaporation.

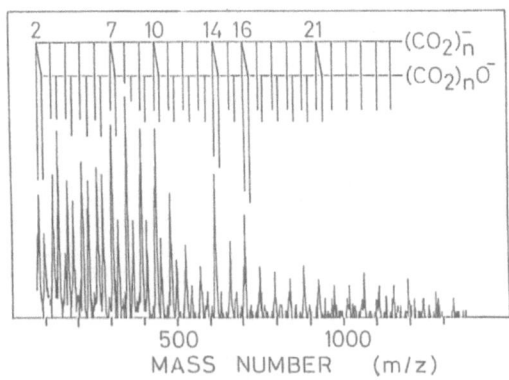

Fig. 5. Mass spectrum of negative cluster ions produced from $(CO_2)_m$ by impact of 1 eV-electrons. Two series of negative ions, $(CO_2)_n^-$ and $(CO_2)_n O^-$, are observed.

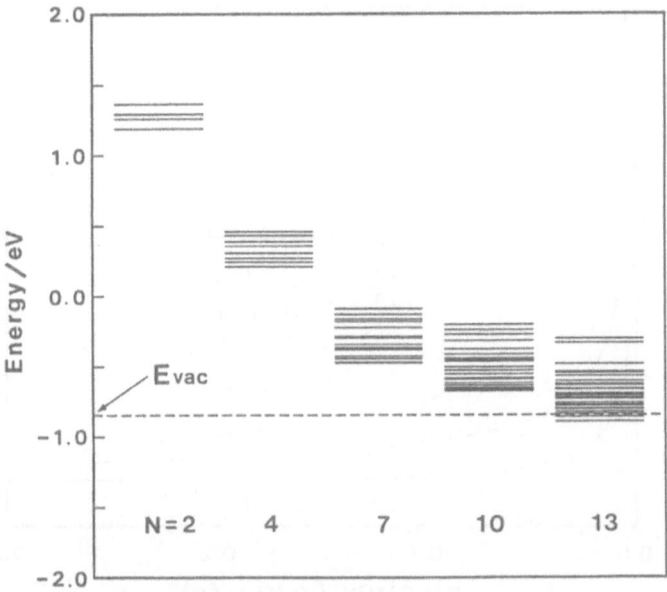

Fig. 6. Affinity levels of $(CO_2)_N$ and $(CO_2)_N^{-*}$ calculated by the
DV-Xα-cluster method, for N=2, 4, 7, 10 and 13.

The affinity levels of $(CO_2)_m^{-*}$ (m≤13) has been calculated by the
DV-Xα-cluster method[47], as shown in Fig. 6, where the cluster structures has
been calculated by the potential-energy minimization method (see Fig.7).[20]
These affinity levels decrease rapidly with increase in its size up to m=7
and tend to level off for m>7. By taking advantage of these electron affinity
levels, the theoretical cross sections for the electron attachment have
been calculated as a function of electron impact energy for the clusters
with different sizes, as shown in Fig.8.[20] The calculated cross section
increases sharply at m=7. This theoretical prediction is consistent with
the experimental finding that m_L is 7.

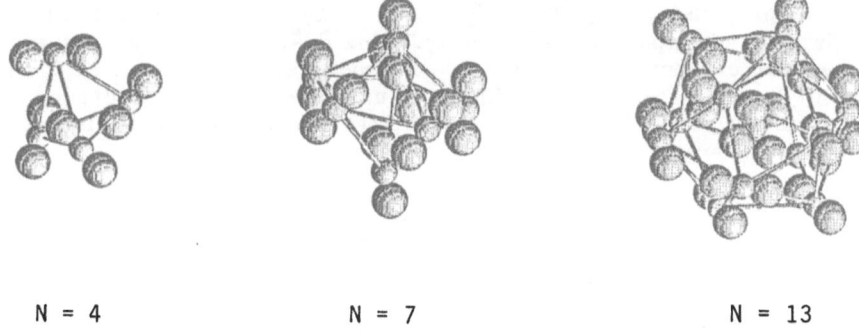

N = 4   N = 7   N = 13

Fig. 7. Structures of the CO_2 clusters calculated by the potential
energy minimization method.

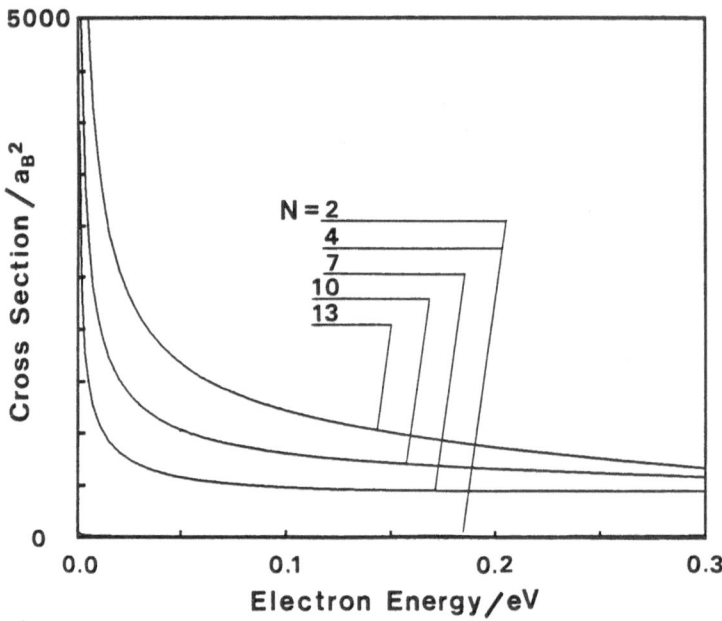

Fig. 8. Cross sections for the electron attachment to $(CO_2)_N$ (N=2, 4, 7, 10) as a function of the kinetic energy of electrons.

NEGATIVE CLUSTER IONS PRODUCED BY OTHER METHODS

Negative cluster ions can also be produced by (1) electron impact, (2) attachment of electrons emitted from metal surfaces thermoionically or by photon impact, (3) charge transfer collisions with translationally 'hot' alkali atoms, etc. In this section, formation of negative cluster ions from gaseous molecular clusters by use of these methods are described in relation to the present method of negative-cluster-ion formation. In addition to these methods, bombardment of solid surfaces by charged particles or intense laser beam have also been investigated extensively.[21,22]

Attachment of 'Almost Zero-Energy' Electrons

Recently, several groups have used 'almost zero energy' electrons to form negative cluster ions from van der Waals clusters.[23-27] In particular, the parent water clusters, $(H_2O)_n^-$ ($n \geq 11$) and $(D_2O)_n^-$ ($n \geq 12$), have been detected. Novel ionic species, such as N_2O^- and CO_2^-, are also generated from the neutral clusters of N_2O and CO_2 respectively[25]. It has also been found that $(O_2)_n^-$ are produced from $(O_2)_m$ by use of very slow electrons, and the cross sections for electron attachment increase sharply with decreasing electron impact energy[26].

Negative Ions Formed in a Multiple-Collision Condition

A method has been developed to employ low-energy electrons emitted from an ordinary heated filament, or from a metal surface thermoionically or radiatively by injection into a condensation zone of a supersonic expansion. Negative ions, $(H_2O)_n^-$, $(D_2O)_n^-$ and $(NH_3)_n^-$, are generated under conditions that the corresponding neutral clusters are grown.[28-30] This method is adequate for growing adiabatically negative cluster ions which are hard to prepare otherwise. Since electrons are attached under multiple-collision conditions, the cluster ions thus produced are 'cold' and carry information on the ground state. Therefore, this ionization is particularly useful for studies of thermodynamic properties of negative cluster ions.

Electron Transfer from Alkali Atoms

The method of endoergic charge transfer from alkali atoms has been applied for production of negative cluster ions from the neutral molecular clusters of SO_2 and Cl_2.[31] Electron affinities of the clusters have been measured as a function of the cluster size. In these processes, electron transfer occurs in the strong transient Coulombic field of the departing alkali cation and, hence, is proved to be adiabatic at least for the dimer.

Table I Negative cluster ions observed

M	Type	Ion(RAI)	n_L	Ion(EI)	n_L
CO_2	E	$(CO_2)_n^-$	3	$(CO_2)_n^-$	2,1[a]
OCS	–	$(OCS)_n^-$	2	no data	–
CS_2	E	$(CS_2)_n^-$	1	$(CS_2)_n^-$	1
				$(CS_2)_n S^-$	0
H_2O	–	$(H_2O)_n^-$	∿13	$(H_2O)_n^-$	11[b]
				$(H_2O)_n OH^-$	0[b]
D_2O	–	no data	–	$(D_2O)_n^-$	12[b]
				$(D_2O)OD^-$	0[b]
CD_3CN	N	$(CD_3CN)_n^-$	11	undetectable	–
CH_3CN+H_2O	E	$(CH_3CN)_n H_2O^-$	∿7	undetectable	–
C_5H_5N	N	$(C_5H_5N)_n^-$	4	$(C_5H_5N)_n^-$	∿3
$C_5H_5N+H_2O$	E	$(C_5H_5N)_n(H_2O)_{n'}^-$	n+n'=4	$(C_5H_5N)_n(H_2O)_n^-$	n+n'=4
SF_6	N	$(SF_6)_n^-$	1	$(SF_6)_n^-$	1
	E			SF_5^-	
N_2O	R	$(N_2O)_n^-$	6	$(N_2O)_n^-$	1[c]
		$(N_2O)_n O^-$	1	$(N_2O)_n O^-$	0, 0[c]
				$(N_2O)_n NO^-$	3,0[c]
N_2O+H_2O	R	$(H_2O)(N_2O)_n O^-$	3	$(H_2O)(N_2O)_n O^-$	1,0[c]
CCl_4	R	$(CCl_4)_n^-$	1	$(CCl_4)_n Cl^-$	0
		$(CCl_4)_n Cl^-$	0		

RAI and EI stand for Rydberg atom and electron impact, respectively.
N: non-evaporative electron attachment, E: evaporative electron attachment, R: reactive electron attachment.

a) A. Stamatovic, K. Leiter, W. Ritter, K. Stephen, and T. D. Märk, *J. Chem. Phys.* 83:2942 (1985).
b) M. Knapp, O. Echt, D. Kreisle, and E. Recknagel, *J. Chem. Phys.* 85:636 (1986).
c) M. Knapp, O. Echt, D. Kreisle, T. D. Märk, and E. Recknagel, *Chem. Phys. Lett.* 126:225 (1986).

FUTURE PROSPECTS

Studies of the electron transfer processes involving gaseous clusters should deal with two fundamental problems: (1) collisional electron transfer to a 'cluster particle' and (2) dynamics of the electron captured in the cluster. The former problem can be treated by use of conventional approaches developed in collision science, but the latter problem seems to be more difficult. In the present stage, photodetachment spectroscopy is a powerful tool to investigate the negative ion states of gaseous clusters and, hence, provides information on the intracluster dynamics.

In the present study, high-Rydberg atoms are estimated to have principal quantum numbers of 25-35, or kinetic energies of the order of 10 meV. The observed cross sections for electron attachment increases with the principal quantum number. Therefore, it is preferable to use high-Rydberg atoms with higher principal quantum numbers. In this respect, it would be profitable to use a laser for production of high-Rydberg atoms, by which a principal quantum number as high as about 80 may be attained.

ACKNOWLEDGEMENTS

The author is grateful to Professors K. Kuchitsu and M. Tsukada, with whom the present study has been undertaken, and to Dr. K. Mitsuke, Messrs. F. Misaizu, H. Tada, and S. Yamamoto for their collaboration. The present work has been supported by a Grant-in-Aid for Scientific Research by the Ministry of Education, Science and Culture of Japan and the Joint Studies Program of the Institute for Molecular Science (1985-1986).

REFERENCES

(1) Y. Hatano, in "Electronic and Atomic Collisions - Invited Papers", D. C. Lorents, W. E. Mayerhof, and J. R. Peterson, eds., Elsevier Science Publishers B. V., Amsterdam (1986) p 153.
(2) W. Lubitz and T. Nyrönen, J. Mag. Res. 41:17 (1980).
(3) R. W. Fessenden and P. Neta, Chem. Phys. Lett. 18:14 (1973).
(4) M. Matsuzawa, in "Rydberg States of Atoms and Molecules", R. F. Stebbings and F. B. Dunning, eds., Cambridge University Press, Cambridge, U.K., (1983) p 267.
(5) B. G. Zollars, C. Higgs, F. Lu, C. W. Walter, L. G. Gray, K. A. Smith, F. B. Dunning, and R. F. Stebbings, Phys. Rev. A. 32:3330 (1985).
(6) T. F. Gallagher, in "Rydberg States of Atoms and Molecules", R. F. Stebbings and F. B. Dunning, eds., Cambridge University Press, Cambridge, U.K., (1983) p 165.
(7) P. J. Chantry, J. Chem. Phys. 57:3180 (1972).
(8) R. S. Gohlke and L. H. Thompson, Anal. Chem. 40:1004 (1968).
(9) T. Kondow and K. Mitsuke, J. Chem Phys. 83:2612 (1985).
(10) K. Mituske, T. Kondow, and K. Kuchitsu, J. Phys. Chem. 90:1505 (1985).
(11) K. Mitsuke, T. Kondow, and K. Kuchiutsu, J. Phys. Chem. 90:1552 (1986).
(12) T. Kondow, in "Electronic and Atomic Collisions - Invited Papers", D. C. Lorents, W. E. Meyerhof, and J. R. Peterson, eds., Elsevier Science Publishers B. V., (1986) p 517.
(13) T. Kondow, J. Phys. Chem. (in press)
(14) F. Williams, and E. D. Sprague, Acc. Chem. Res. 15:408 (1982).
(15) R. N. Compton, P. W. Reinhardt, and C. D. Cooper, J. Chem. Phys. 63:3821 (1975).
(16) J. Pacansky, V. Wehlgren, and P. S. Bayus, J. Chem. Phys. 62:2740 (1975).
(17) Y. Yoshioka and K. D. Jordan, J. Am. Chem. Soc. 102:2621 (1980).

(18) M. Tsukada (unpublished)

(19) H. Adachi, M. Tsukada, and C. Satoko, J. Phys. Soc. Jpn. 45:875 (1978).

(20) M. Tsukada, N. Shima, S. Tsuneyuki, and H. Kageshima, Proceedins on the NEC Symposium on Fundamental Approach to New Material Phases -Microclusters, Tokyo (1986).

(21) J. L. Yntema and P. J. Billquist, Rev. Sci. Instr. 57:748 (1986).

(22) L. A. Bloomfield, M. E. Geusic, R. P. Freemam, and W. L. Brown, Chem. Phys. Lett. 121:33 (1985).

(23) M. Knapp, O. Echt, D. Kreisle, and E. Recknagel, J. Chem. Phys. 85:636 (1986).

(24) A. Stamatovic, K. Leiter, W. Ritter, K. Stephan, and T. D. Märk, J. Chem. Phys. 83:2942 (1985).

(25) M. Knapp, O. Echt, D. Kreisle, T. D. Märk, and E. Recknagel, Chem. Phys. Lett. 126:225 (1986).

(26) T. D. Märk, K. Leiter, W. Ritter, and A. Stamatovic, Phys. Rev. Lett. 55:2559 (1985).

(27) M. Knapp, D. Kreisle, O. Echt, K. Sattler, and E. Recknagel, Surface Sci. 156:313 (1985).

(28) H. Haberland, C. Ludewigt, H. -G. Schindler, and D. R. Worsnop, J. Chem. Phys. 81:3742 (1984).

(29) H. Haberland, H. Langosch, H.-G. Schindler, and D. R. Worsnop, J. Phys. Chem. 88:3903 (1984).

(30) H. Haberland, C. Ludewigt, H. -G. Schindler, and D. R. Worsnop, Surface Sci. 156:157 (1985).

(31) K. H. Bowen, G. W. Liesegang, R. A. Sanders, and D. R. Herschbach, J. Phys. Chem. 87:557 (1983).

SOLVATION OF ALKALI HALIDES IN ALCOHOL CLUSTERS

T.P. Martin and T. Bergmann

Max-Planck-Institut fuer Festkoerperforschung

7000 Stuttgart 80, FRG

The strong electrostatic interaction between Li^+ and I^- ions is screened in the presence of polar solvent molecules, eventually resulting in the dissociation of Li^+-I^- ion pairs. In order to better understand this-solvation process, mass spectrometry experiments and total energy calculations have been carried out on alcohol clusters containing varying amounts of LiI. Clusters of a given size are found to possess a large number of stable configurations all of which are present, simultaneously, in a cluster beam. The form of the most abundant cluster changes with changing temperature in order to simultaneously minimize energy and maximize entropy. Despite this complexity, several general conclusions can be made. Solvation in clusters occurs in two stages with increasing alcohol content. First, ion pairs are isolated from one another. The ion pairs then dissociate into individual ions. At low temperatures the most stable configurations correspond to evenly distributed, symmetric arrangements of alcohol molecules about Li ions. When the number of alcohol molecules per Li ion reaches the value 4, the Li-I distance increases abruptly and the alkali halide fragment "dissolves" in the cluster.

INTRODUCTION

In rapidly developing fields of science, where similar investigations are being made simultaneously by many research groups, a cacophony of terminology often arises, each group drawing words from their experience to describe the same new phenomena. While not an ideal situation, rarely does it lead to serious confusion. Everyone seems to understand, for example, that a metallic cluster does not necessarily conduct electricity. I hope the reader is willing to adopt this lenient attitude concerning terminology while reading this paper on "solvation" in clusters.

It is not possible to describe the structure of disordered condensed matter (liquids, glasses, alloys and mixed crystals) exactly. In response to this challenge enormous effort has been put into each of these fields. Therefore, it is not surprising that we now have clusters with the same names [1]. Investigations on mixed clusters [2-5], alloy clusters [6-10], glass clusters [11-13], and liquid clusters [14-16] will certainly help reveal how condensed matter nucleates and grows. Cluster research may even give some information concerning local order in otherwise disordered solids. The same promise holds with regard to solutions [17] and in particular to electrons [18, 19] and ions [20-35] solvated in polar solvents.

The traditional picture of a salt dissolving in a polar solution is very simple [36]. The salt dissociates into individual ions each of which is surrounded by a tightly bound shell of solvent molecules. However, it is usually not possible to make a direct determination of the number of molecules in the solvation shell. Such a determination is possible by means of mass spectrometry if the ion-shell unit can be produced as a cluster in the gas phase.

Pure solvent clusters have been the subject of a number of investigations [37-39]. Stein and Armstrong [40] examined the structure of water clusters using the technique of electron diffraction in 1972. In this way they were able to determine directly the structure of solvent clusters produced in an expanding beam.

The topic of solvation in polar solvents is not new in cluster science [41-46]. In fact, this topic is one of the first to which cluster techniques have been able to make a substantial contribution. Twenty years ago Kebarle and co-workers [41], using the technique of high-pressure mass spectrometry, were able to obtain values for the free energy changes on successive hydration of alkali ions. In these studies the positive metal ions, e.g. K^+, were obtained by thermionic emission from a filament painted with a suitable salt. It would be useful to extend these studies to solvent clusters that contain not just one metal ion, but a controlled number of alkali halide molecules. In this way it might be possible to observe the solvation of an alkali halide in a polar solvent on a microscopic scale.

In order to apply the methods of cluster physics to the problem of solvation, it is necessary to develop a technique for inserting a controlled amount of a high temperature vapor into a solvent cluster, for example, two NaCl molecules into $(H_2O)_{10}$. Clusters of liquid solvents are normally produced by nozzle beam expansion. Attempts in our laboratory to mix a high temperature vapor into an expanding gas have not been successful. The high pressure gas just blew the vapor away. It seemed to be advantageous to produce the solvent clusters at low pressures under quasistatic conditions. This was done in the following way.

EXPERIMENT

Pure alcohol vapor at a pressure of 200 mbar entered, through a large (10 mm diam.) orifice, a cluster condensation cell cooled with liquid nitrogen. The cell was filled with 1 mbar He gas which served to transport heat from the warm vapor to the cooled chamber walls. Most of the alcohol vapor condensed onto the walls of the chamber. In fact, after completion of an experiment, it was necessary to turn off the vacuum pumps before warming up the condensation chamber in order to avoid large amounts of

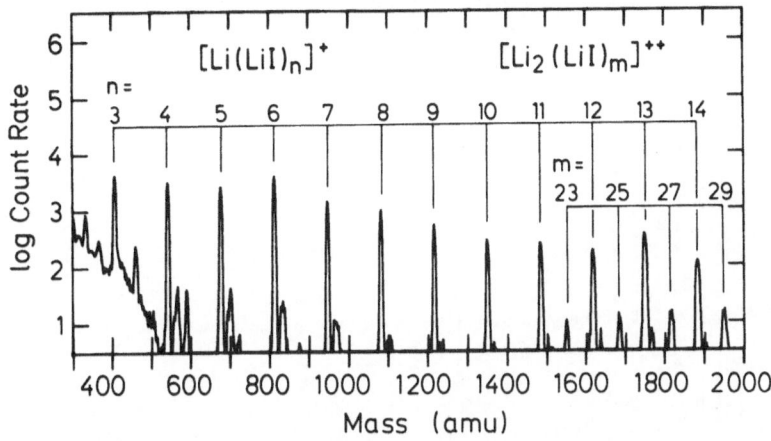

Fig.1 Mass spectrum of clusters formed by quenching LiI vapor in He gas. Each cluster contains one extra Li ion.

alcohol vapor from being pumped through the system. However, enough of the entering alcohol vapor condensed into clusters to provide a large signal in the mass spectrometer. The consensation chamber also contained an oven used to generate a high temperature vapor, e.g. LiI vapor. By varying the temperature of the oven, a controlled amount of LiI could be inserted into the alcohol clusters. The clusters passed through a 3mm diameter hole into an intermediate chamber where excess He gas was pumped off with a 500 l/s turbopump. The cluster beam then passed through a 1mm aperture into a high vacuum chamber containing a quadrupole mass spectrometer pumped with a 360 l/s turbopump. The beam intersected the axis of the spectrometer where the clusters were ionized with 80 eV electrons. Typically, for each experiment two types of mass spectra were recorded, one in the mass range 1-500 amu with a resolution of 1 amu and another one in the mass range 2-2000 amu with a resolution of 4 amu.

EXPERIMENTAL RESULTS

In order to study the solvation of an ionic material in a polar solvent, we have chosen the system LiI-methanol. LiI has several advantages over other alkali halides for such investigations. It is almost isotopically pure, so that each mass peak in the spectrum corresponds to a distinct cluster. More importantly due to its small size the Li^+ ion has an extraordinarily well-defined hydration shell. Previous investigations[47] have shown that pure LiI clusters have the composition $[Li(LiI)_n]^+$, Fig. 1. The clusters contain one extra Li ion, thus providing the net positive charge necessary for mass selection[48]. Clearly, such clusters are produced by fragmentation following the ionization process. For this reason we believe that the mass spectra reflect primarily the properties of charged fragments.

If the LiI oven is turned off and methanol vapor is added to the He cooling gas entering the cluster condensation cell, pure methanol clusters are observed in the mass spectrometer, Fig. 2. The intensities of the mass peaks $(CH_3OH)_n$ for n=1-30 decrease monotonically with increasing cluster size. High resolution spectra indicate that each methanol cluster contains an additional proton which provides the positive charge necessary for mass selection. In addition to the pure alcohol clusters, it is possible to detect clusters containing one water molecule, but only if n is greater than 7.

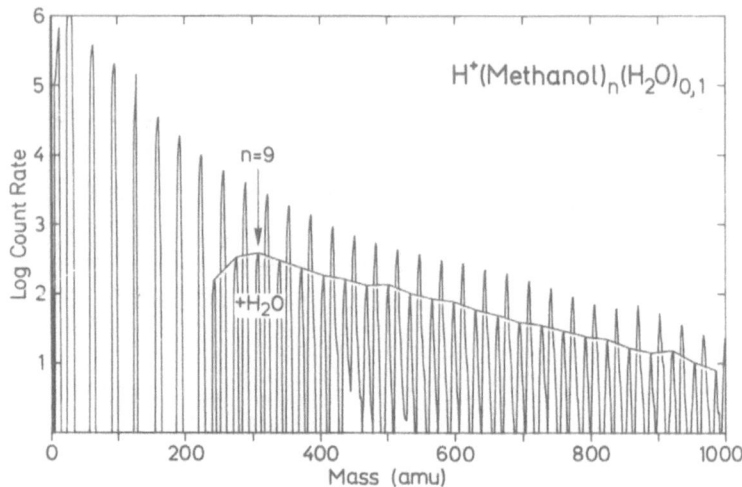

Fig.2 Mass spectrum of methanol clusters. Each cluster
contains one extra proton. The weaker series of
peaks, starting with n=7, corresponds to methanol
clusters containing one water molecule.

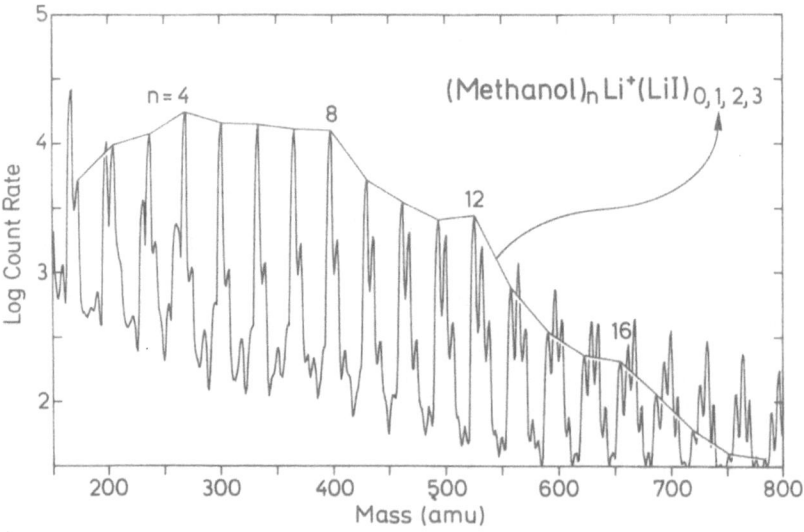

Fig.3 Mass spectrum of methanol clusters containing an
alkali halide fragment with the composition
$Li(LiI)_n^+$. Mass peaks for clusters containing two
Li ions, one I ion and n methanol molecules have
been connected. Notice that the peaks are particularly strong for n equal to 4,8, and 12.

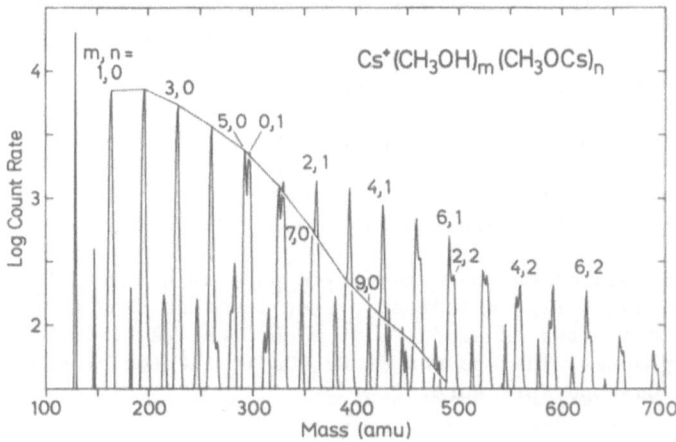

Fig.4   Mass spectrum of alcohol clusters containing a small
amount of cesium.

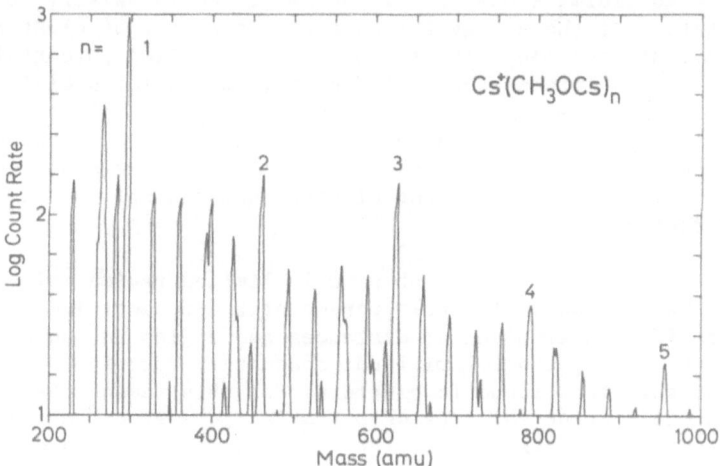

Fig.5   Mass spectrum of alcohol clusters after saturation
with cesium.

If methanol vapor and LiI vapor are mixed in the cluster condensation
cell, mass spectra of the type shown in Fig. 3 are obtained. Methanol
clusters containing from 2 to 20 molecules condense out of the vapor. Each
of these clusters contain either Li^+ or $[Li(LiI)]^+$ or $[Li(LiI)_2]^+$ or
$[Li(LiI)_3]^+$. We have chosen the LiI and methanol partial pressures so that
the strongest peaks correspond to methanol clusters containing $[Li(LiI)]^+$.
Notice that there is some enchanced intensity for clusters containing
4,8,12 and perhaps 16 methanol molecules. Apparently these cluster ions
are particularly stable. What do they look like? Here we should be able to
learn something about the solvation process.

Several possible explanations for the high stability of $[Li(LiI)$
$(CH_3OH)4,8,12]^+$ immediately present themselves. Both calculations[49] and
experiments[50] indicate that the the cyclic tetramer of methanol might have
a high stability. If this stability is not disturbed by the alkali halide,
then the sequence 4,8,12 might indicate the presence of one, two and
three, 4-atom alcohol building blocks. However, the interaction of alcohol
with LiI can be expected to be stronger than the interaction between alco-

hol molecules themselves. Therefore, it is unlikely that this explanation is appropriate.

Another possibility is that the numbers 4, 8 and 12 are connected with highly symmetric arrangements of the alcohol molecules around a central charged core. Since three of the platonic solids (the tetrahedron, the cube and the icosohedron) contain 4, 8 and 12 vertices, this direction of thought is particularly appealing. However, such speculations can serve only as a stimulus to look into the problem further.

Similar techniques can be used to add pure alkali metal to the alcohol clusters. However, Fig. 4 indicates that the resulting clusters are more accurately described as products of a chemical reaction rather than of a solvation process. The content of CH_3OCs in the the cluster increases with increasing Cs vapor pressure, Fig. 5.

MODEL

The clusters under consideration are too small to observe directly. We are forced to investigate their structure by means of total energy calculaions. Mathematically, the problem can be stated very simply: Find all minima on a multidimensional total energy surface. However, this entails computation of the energy for an enormous number of cluster configurations, a feat possible only if a simple interaction potential is used. But here we run into a second difficulty: the varioius minima often differ only slightly in energy. If the interaction potential is too simple, the calculatd absolute minimum will not correspond to the observed clusters. Having these limitations always in mind, we have minimized the total energy of alkali halide-alcohol clusters using simple but tested interaction potentials.

Experiments offer information about only limited regions of configuration space. That is, an empirically parameterized potential may be very good for the spcific intermolecular distances and angles encountered in crystals, but the same potential will not adequately describe the large variety of configurations found in clusters. For this reason there is considerable justification for fitting potential parameters to quantum mechanical interaction energies calculated over a large region of configuration space. Such an approach has been used, for example, in a Monte Carlo study of fixed Li^+ and F^- ions surrouned by 200 water molecules-51-53.

The form of the potential we have used to describe alcohol was originally proposed by Snir et al.[54]. The molecule is composed of 13 centers of interaction, Fig. 6. Six of the centers are positively charged and are located at the known positions of the six atoms in the molecule. The remaining 7 centers are negatively charged and can be conveniently thought of as bond charge and lone pair charge. The bond charges are located on the axes between nuclei. All centers are allowed to interact with one another electrostatically, $q_i q_j / r_{ij}$. The negatively charged centers have an additional interaction with the form:

$$U_{ij} = q_i q_j (A_{ij} \exp(-B_{ij} r_{ij}) - C_{ij}/r_{ij}{}^6) \tag{1}$$

where q_i and q_j are the charges and r_{ij} is the distance between the centers. Without any further assumptions this would lead to 43 fitting parameters, clearly an unmanageable number. Therefore, charges are assigned to each of the centers (Fig. 6) and use is made of the combination rules:

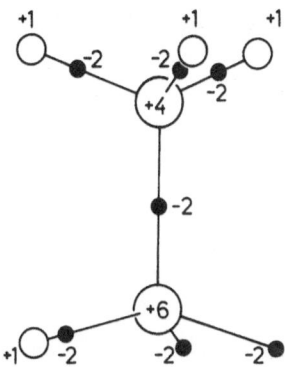

Fig. 6. The alcohol model used consists of 13 interaction centers (ref. 54 and 55). Six centers are located at atomic sites, 5 centers at bond charge sites and 2 centers at lone pair sites.

$$A_{ij} = (A_{ii} A_{jj})^{1/2}, \quad B_{ij} = (B_{ii} + B_{jj})/2, \quad C_{ij} = (C_{ii} C_{jj})^{1/2} \tag{2}$$

thus reducing the number of fitting parameters to 17. We have used the parameters of Marchese et al[55]. The interaction potential between pairs containing either an alkali or a choride ion is assumed to contain an electrostatic Coulomb contribution and a repulsive Born-Mayer contribution[13],

$$U(\text{ions})_{ij} = q_i q_j / r_{ij} + A_{ij} \exp(-B_{ij} r_{ij}). \tag{3}$$

The position and orientation of each rigid methanol molecule have been represented by the three cartesian coordinates defining the center of mass and by three Euler angles. To initiate a minimization run, the coordinates and angles were chosen at random. A fast but rough minimization was first carried out using the steepest descents method. This was followed by a refinement with the Davidson-Flecher- Powell method[56]. After establishing an energy minimum the vibrational frequencies were determined. Starting from a variety of initial configurations, the value of a given energy minimum was reproducible to six digits and the vibrational frequencies to four.

CHARGED CLUSTERS

The stability of a LiI-methanol cluster is determined essentially by three important bonds. The strongest is the Li-I bond. This bond is so strong that the LiI fragment remains essentially undistorted until a large number of methanol molecules have been added to the cluster.

The Li-O bond provides the most important contribution to the solute-solvent interaction. This bond weakens as additional alcohol molecules are forced to share the same Li ion.

The solvent-solvent O-H hydrogen bond is relatively weak but becomes important with increasing methanol content. Simply because there is no more room around the Li ions to allow the formation of an additional Li-O bond. The excess methanol molecules have no alternative but to form hydrogen bonds with existing solvent molecules.

In this paper, results will be presented for calculations on only one type of charged cluster, $(Li_2I)^+(CH_3OH)_n$. Even with this limitation the results must be summarized because of the large number of stable configurations exisiting for each cluster size; 5 configurations for n=2, 11 configurations for n=3, 30 configurations for n=4, etc. After analyzing all

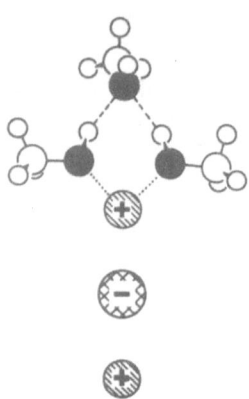

Fig.7 Configuration for a
       local minimum.

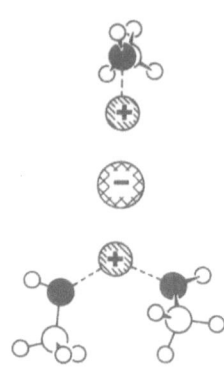

Fig.8 Configuration for a
       global minimum.

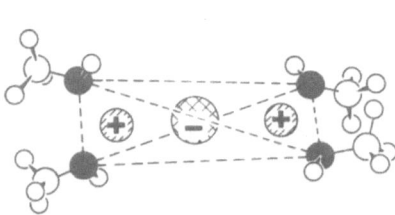

Fig.9 Most stable configu-
       ration $(Li_2I)^+(CH_3OH)_4$

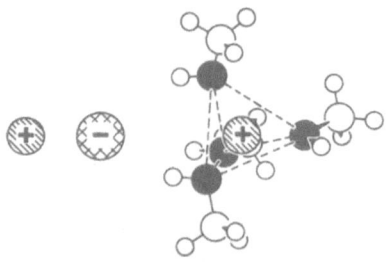

Fig.10 Configuration for a
       local minimum.

this information it has been possible to formulate three "rules" for constructing a cluster of high stability. It should be emphasized that these
rules were not used in our calculations but are rather a qualitative
summary of the results.

1) The $(Li_2I)^+$ fragment remains linear and undistorted for $n < 8$.

2) The solvent molecules should be distributed as evenly as possible
   between the two Li ions.

3) Hydrogen bonding between methanol molecules should be avoided for $n < 8$.

It will be instructive to apply these rules to a simple but nontrivial
example, $(Li_2I)^+(CH_3OH)_3$. One local minimum on the energy surface of this
cluster corresponds to the configuration shown in Fig. 7. This configuration is not energetically favorable because it does not obey rules 2 and
3. The molecules are bunched on one side of the LiI fragment and they form
hydrogen bonds among themselves. The most favorable configuration is shown
in Fig. 8.

Using these rules we can immediately identify which of the more than
30 stable configurations containing 4 methanol molecules is most stable.
It is shown in Fig. 9. This cluster is particularly resistent to dissociation through loss of a solvent molecule, a property that may explain the
strong mass peak of n=4.

A further examination of the local minima for $(Li_2I)^+(CH_3OH)_4$ reveals another structure which, although much less stable, is of considerable interest, Fig. 10. All the methanol molecules have grouped around one end of the alkali halide fragment and one of the Li-I bonds has almost doubled in length. It would appear that a minimum of four methanol molecules are necessary for the solvation of a Li^+ ion. Such solvated ions are found in the most stable configurations only for n 8. The steep decline in the intensity of mass peaks for clusters containing more than eight molecules may very well reflect the onset of solvation.

NEUTRAL CLUSTERS

It should be emphasized that LiI is not a typical alkali halide. Its unusual properties derive from the large difference in the ion radii. If one tries to pack Li^+ and I^- ions together to form a rock-salt crystal structure or to form the usual cluster structures, the problem of I-I overlap immediately becomes apparent. The problem disappears if the coordination of each ion is lowered. Therefore, textbooks often state that LiI "should" crystallize into a four-fold coordinated zinc blende structure rather than the usual six-fold coordinated rock-salt structure. The most stable forms of LiI clusters also tend to be less closely packed than for the other alkali halides. To give an example, consider the structure of the dimer $(LiI)_2$. In general, the dimer has two stable forms, a rhombus and a linear chain. Although the rhombus is the energetically favored form for all other alkali halides, for LiI, the rhombus does not even form a local minimum on the energy surface. The linear chain is the only stable form of $(LiI)_2$. Starting then with this linear form of the bare dimer, what happens as methanol molecules are added to the cluster one at at time?

One might ask the specific question, how many methanol molecules must be added to break a Li-I bond? However, it must be remembered that point-charge interactions are long range and that the ions are bound not merely to nearest neighbors. The danger of using the concept of a local Li-I bond is best illustrated with an example. The linear $(LiI)_2$ dimer can be broken apart at two inequivalent points, either an end ion can be removed, or an ion pair can be removed. The energy required to remove an end ion is 3.6 eV while only 0.62 eV are necessary to symmetrically fragment the cluster into two ion pairs. Such large variations in a "local Li-I bond energy" indicate that such a bonding concept is not useful. The total energy surface of each $(LiI)_2(CH_3OH)_n$ cluster contains many local minima. In the following brief discussion the structure only at apparent global minima will be indicated for each value of n.

Two methanol molecules are sufficient to split the LiI dimer into ion pairs. The most stable configuration is a ring, Fig. 11 composed of alternating LiI and alcohol units. The 0.6 eV lost by fragmentation into ion pairs is more than compensated by the formation of two O-Li and two H-I bonds. In some sense the isolation of ion pairs represents the first stage of solvation. We will see that the next stage, the isolation of individual ions, requires a much larger number of alcohol molecules.

With the addition of 3,4 and 5 alcohol molecules to $(LiI)_2$ there is no qualitative change in the degree of solvations. Even six molecules are insufficient to dissociate the strongly bound ion pairs. Since the Li-O bond is the strongest solvent- solute interaction, it is not without meaning to describe the clusters in terms of the coordination of the Li ions only. In $(LiI)_2(CH_3OH)_6$ the six alcohol molecules distribute themselves evenly between the two Li ions, three molecules to each ion. Following this rule we would expect that for $(LiI)_2(CH_3OH)_7$ the alcohol molecules would be distributed 3 to one Li ion and 4 to the other. That this is

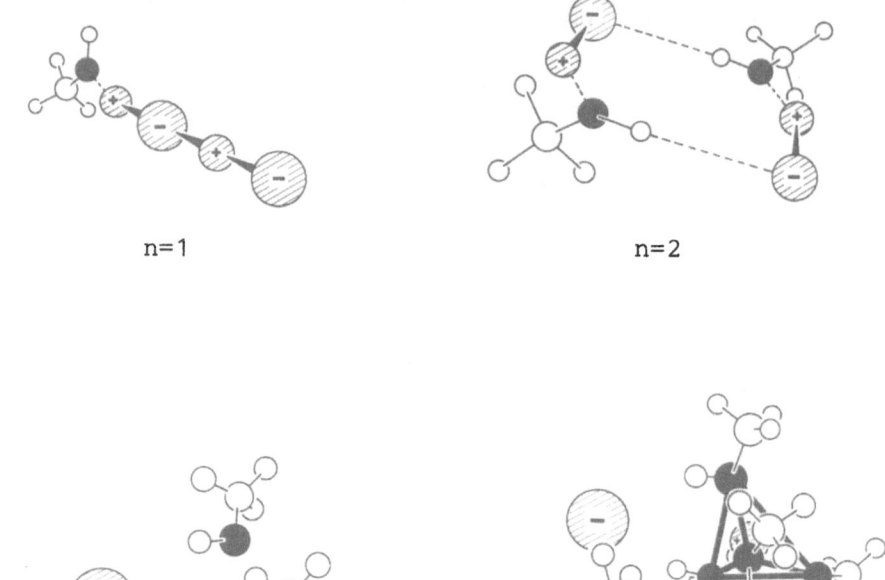

n=1

n=2

n=7

n=8

Fig. 11 Configurations for global minima of $(LiI)_2(CH_3OH)_n$. Ion pair formation at n=2. Complete solvation at n=8.

indeed the case, can be seen in Fig. 11. However, now the second stage of solvation sets in. The ion with coordination 4 dissociates from its counter-ion, resulting in a cluster containing one ion pair and two fully solvated ions. Stated more generally, <u>for clusters with composition $(LiI)_2$ $(CH_3OH)_m$, the second state of solvation starts at n/m greater than 3 and is complete for n/m greater than 4.</u> Just as this rule predicts, $(LiI)_2$ $(CH_3OH)_8$ contains completely solvated ions, Fig. 11.

CATCHMENT AREA

Another important characteristic of a total energy surface is often called the "catchment" region. Imagine rain falling onto the surface. Every raindrop, no matter where it hits the surface, will eventually find its way to a minimum. That area of the surface which collects rain drops for a specific minimum is said to define a catchment area. All of configuration space can be uniquely divided up into a set of such areas. In this way each point in configuration space can be said to belong to a given minimum (except, of course, those points defining the boundry line between catchment areas). In the above discussion the concept of catchment area proved useful in a qualitative description of a total energy surface. We will now attempt to use the concept in a more quantitative way.

Catchment area has a particularly simple meaning for the case of a microcanonical ensemble. In such an ensemble each point in configuration space is equally probable, therefore, the probability of the cluster having a given configuration is simply proportional to the length of a constant energy contour in the corresponding catchment area. This interpretation must be only slightly modified when considering the canonical ensemble. In this case the probability of reaching a given point in phase space must be weighted by $\exp(-H/KT)$. Specifically, the probability that a cluster occupies catchment region "a" relative to the probability that it occupies region "b" is just,

$$P_a/P_b = \int_a \exp(-H/KT)dq^3dp^3 / \int_b \exp(-H/KT)dq^3dp^3 \qquad (4)$$

For the purposes of this discussion we will assume that the integration over momentum coordinates is independent of the configuration since both configurations have the same mass. However, the configurations will not, in general, have the same moment of inertia.

The integration over configuration coordinates is easily carried out if it is further assumed that each catchment region has a paraboli form. This "harmonic" assumption is certainly valid at low temperatures.

$$V_a = E_a + \sum_i \omega_a^2(i)q_a^2(i)/2m \qquad (5)$$

where $\omega_a(i)$ is the ith vibrational frequency in the parabolic catchment region belonging to minimum a. The corresponding displacement amplitude has been denoted with $q_a(i)$. Using this expression the integral in Eq. 4 is easily evaluated to give:

$$P_a/P_b = \exp[(E_b-E_a)/KT] \prod_i \omega_b(i)/\omega_a(i) \qquad (6)$$

Here we have used the classical expression for the energy of a harmonic oscillator in order to retain a simple geometrical description of the probabilities. The quantum mechanical result is obtained if each ω in the product in Eq. 6 is replaced with a corresponding $1-\exp(-\hbar\omega/KT)$. Notice that the relative probability of a cluster being in a given minimum is dependent on the depth of the minimum (through the Boltzmann factor) and on the extend of the catchment region (through the product of inverse frequencies). Just one very low frequency vibration can, at non-zero temperature, stabilize an energetically unfavorable configuration. Now it is possible to state and understand a rather unusual observation. Cluster confirgurations corresponding to shallow minima tend to become more favorable at high temperatures. Cluster configurations corresponding to deep minima are less likely to be found at high temperatures. The reason for this behavior is the following. Energetically unfavorable clusters sitting in shallow minima usually have open structures which possess incipient instabilities. Such clusters resist only weakly certain types of deformations, i.e. they have low frequency vibrations. The entropy generating, low frequency vibrations tend to make the configuration more favorable as the temperature increases. In principal, that configuration with the lowest frequencies will always become the most favorable if the temperature is made high enough. We will now present a specific example to illustrate this point.

Consider once again $(LiI)_2(CH_3OH)_2$. The global minimum on the energy surface of this cluster corresponds to a ring structure. However, this ring was the end result only 5 times in 95 minimizations from random starting configurations. It has a small catchment area. The local minimum corresponds to a chain structure, Fig. 12. This local minimum has a larger catchment area, smaller curvatures, lower frequencies. Therefore, the

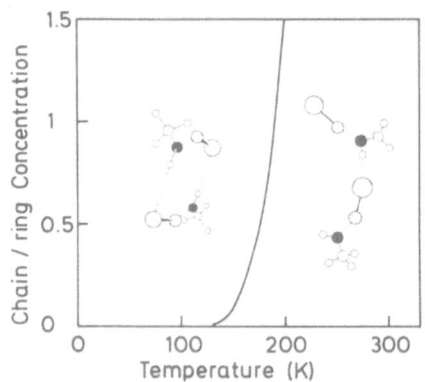

Fig.12 Temperature dependence of the relative concentrations of two different $(LiI)_2(CH_3OH)_2$ cluster configurations. The low frequency chain vibrations
generate entropy at high temperatures.

probability of finding it in the vapor will increase with temperature,
Fig. 12.

CONCLUDING REMARKS

The solvation of LiI in alcohol clusters seems to proceed in two
stages with increasing dilution; a) the isolation of ion pairs, b) followed by the complete solvation of ions. Complete solvation is achieved for
dilutions allowing four alcohol molecules to surround each Li ion. Direct
experimental verification of this conclusion would be difficult. Diffraction techniques are best suited to crystalline samples. If it were posible
to grow and to determine the structure of LiI-alcohol crystals, we would
have unambiguous experimental data for comparison. Recently, such experiments have actually been performed[57] on $Li(MeOH)_4I$. This material turns
out to have the highest ionic conductivity of any known Li compound, a
fact which could make it of great interest for future applications. However, it is the structure of this material that interests us here. Each Li
ion is tetrahedrally surrounded by four alcohol molecules.

It is, without a doubt, convenient to think of a cluster of a given
size as having a specific geometry. Unfortunately, this simplified way of
thinking does not lead to an accurate description of a collection of clusters at finite temperature. The total energy surface of a cluster contains
many local minima, most of which play an important role in determining
cluster geometry. The calculational effort required to characterize all of
these minima is enormous because of the multidimensionality of configuration space. Numerical procedures which are simple in 3 dimensions become
unthinkable in 100 dimensional space. However, some progress can be made
through the use of two-body interaction potentials applied to clusters of
small size.

REFERENCES

1. See conference proceedings published in: Surf. Sci. 106 (1981); 156
    (1985); Z. Phys. D (1986).

2. A. Ding, J. H. Futrell, R. A. Cassidy, L. Cordis and J. Hesslich, Surf. Sci. 156:282 (1985).
3. A.J. Stace, Chem. Phys. Lett. 113:355 (1985).
4. H.P. Birkhofer, H. Haberland, M. Winterer and Dr.R. Worsnop, Ber. Bunsenges, Phys. Chem. 88:207 (1984).
5. J. Diefenbach and T.P. Martin, J. Chem. Phys. 83: 2238 (1985).
6. E.A. Rohlfing, D.M. Cox, R. Petkovic-Loton, and A. Kaldor, J. Phys. Chem. 88: 6227 (1984).
7. M.M. Kappes, P. Radi, M. Schar and E. Schumacher, Chem. Phys. Lett. 119: 11 (1985).
8. T.P. Martin, J. Chem. Phys. 83: 78 (1985).
9. C. Brechignac, Ph. Cahuzac and J. Ph. Roux, Chem. Phys. Letters 127: 445 (1986).
10. D.Schild, R. Pflaum, K. Sattler, E. Recknagel, submitted for publication.
11. T.P. Martin, J. Chem. Phys. 80:170 (1984).
12. J.R. Banavar and J.C. Phillips, Phys. Rev. B 28:4716 (1983).
13. M.R. Hoare, Ann. New York Acad. Sci. 279:186 (1976).
14. J. Gspann, Physics of Electronic and Atomic Collisions, S. Datz (ed.), North-Holland, 1982, p. 79.
15. H. Buchenau, R. Götting, J.R. Minuth, A. Scheidemann, and J.P. Toennies in: Unsteady Fluid Motions, ed. F. Obermaier, G. E. A. Meier, Springer Lecture Notes in Physics (1985).
16. J. Farges, B. Raoult and G. Torchet, J. Chem. Phys. 59: 3454 (1973).
17. U. Even and J. Jortner, J. Chem. Phys. 78:3445 (1983).
18. J. Jortner, Ber. Bunsenges, Phys. Chem. 88:188 (1984)
19. H. Haberland, H.G. Schindler and D.R. Worsnop, Ber. Bunsenges. Phys. Chem. 88: 270 (1984).
20. B.M. Smirov, Sov. Phys. Usp. 20:119 (1977).
21. R.M. Lawrence and R.F. Kruh, J. Chem. Phys. 47:4758 (1967).
22. J. From, E. Clementi and R.O. Watts, J. Chem. Phys. 62:1388 (1975).
23. J.W. Kress, E. Clementi, J.J. Kozak and M.E. Schwartz, J. Chem. Phys. 63:3907 (1975).
24. C.L. Briant and J.J. Burton, J. Chem. Phys. 64:2888 (1976)
25. I.N. Tang, M.S. Lian and A.W. Castleman, jr., J. Chem. Phys. 65:4022 (1976).
26. Gy. I. Szasz, K. Heinzinger and G. Palinkas, Chem. Phys. Letters 78:194 (1981).
27. C.E. Klots, J. Phys. Chem. 85: 3585 (1981).
28. Gy. I. Szasz, K. Heinzinger and W.O. Riede, Z. Naturforsch. 36a:1067 (1981).
29. P.M. Holland and A.W. Castleman, jr., J. Chem. Phys. 76:4195 (1982).
30. J. Chandrasekhar, D.C. Spellmeyer and W.L. Jorgensen, J. Am. Chem. Soc. 106:903 (1984).
31. M.A. Wilson, A. Pohorille and L.R. Pratt, J. Chem. Phys. 83:5832 (1985).
32. A.A. Rashin and B. Honig, J. Phys. Chem. 89:5588 (1985).
33. M. Magnini, M. de Moraes, G. Licheri and G. Piccaluga, J. Chem. Phys. 83:5797 (1985).
34. H. Shinohara, U. Nagashima, H. Tanaka and N. Nishi, J. Chem. Phys. 83:4183 (1985).
35. C.A. Deakyne, M. Meot-Ner, C.L. Cambell, M.G. Hughes and S.P. Murphy, J. Chem. Phys. 84:4958 (1986).
36. J.N. Murrell and E.A. Boucher, "Properties of Liquids and Solutions"; John Wiley and Sons Ltd.: New York, 1982.
37. J.Q. Searcy and J.B. Fenn, J. Chem. Phys. 61:5292 (1974).
38. J.A. Odutola, R. Viswanathan and T.R. Dyke, J. Am. Chem. Soc. 101, 4788 (1979).
39. B.D. Kay and A.J.W. Castleman, jr., J. Phys. Chem. 89, 4867, (1985).
40. G.D. Stein and J.A. Armstrong, J. Chem. Phys. 58: 1999, (1973).

41. P. Kebarle and A.M. Hogg, J. Chem. Phys. 42, 668 (1965).
42. For a recent review see: T.D. Märk and A.W. Castleman, jr., Adv. At. Mol. Phys. 20, 65 (1985).
43. I. Dzichic and P. Kebarle, J. Phys. Chem. 74, 1466 (1970).
44. A.W. Castleman, jr., P.M. Holland, D.M. Lindsay and K.I. Peterson, J. Am. Chem. Soc. 100, 6039 (1978).
45. P. Kebarle, Ann. Rev. Phys. Chem. 28, 445 (1977).
46. C.M. Banic and J.V. Iribarne, J. Chem. Phys. 83, 6432 (1985).
47. J. Diefenbach and T.P. Martin, J. Chem. Phys. 83, 4585 (1985).
48. T.P. Martin, Phys. Rep. 95, 167 (1983).
49. L.A. Curtiss, J. Chem. Phys. 67, 1144 (1977).
50. T.A. Renner, G.H. Kucera and M. Blander, J. Chem. Phys. 66, 177 (1977).
51. H. Kistenmacher, H. Popkie, and E. Clementi, J. Chem. Phys. 58: 5627 (1973).
52. R.O. Watts, E. Clementi and J. Fromm, J. Chem. Phys. 61, 2550 (1974).
53. J. Fromm, E. Clementi, R.O. Watts, J. Chem. Phys. 62, 1388, (1975).
54. J. Snir, R.A. Nemenoff and H.A. Scheraga, J. Phys. Chem. 82, 2497 (1978).
55. F.T. Marchese, P.K. Mehrotra and D.L. Beveridge, J. Chem. Phys. 86, 2592 (1982).
56. J. Stoer: Einfuehrung in die Numerische Matehmatik I; 3rd ed. (Springer, Berlin, Heidelberg, 1979).
57. W. Weppner, W. Welzel, R. Kniep and A. Rabenau, to be published in Angew. Chem.

ON THE ELECTRONIC STRUCTURE OF A SINGLY IONIZED CLUSTER

COMPOSED OF CLOSED SHELL ATOMS OR MOLECULES

Hellmut Haberland[a]

Fachbereich Physik
Universität Kaiserslautern
D-6750 Kaiserslautern
Germany

ABSTRACT

The electronic structure of M_n^+ clusters is discussed, where M is some closed shell atom or molecule which forms an insulator in the bulk. Evidence is presented, that the charge is localized. The electronic structure can be described as $M_j^+ M_{n-j}$, where j is a small number, and very often j=2. The charge localization is similar to that discussed earlier for rare gas clusters.

INTRODUCTION

Numerous experimental and theoretical studies have been performed with closed shell atoms and molecules. To detect them in a mass spectrometer they have to be ionized. The field of cluster ions has recently been reviewed by Märk and Castleman.[1] In this paper we address the problem if the charge is delocalized, i.e. distributed over the entire cluster, or if it is localized over a few atoms and molecules. For rare gas clusters this problem has been discussed erarlier.[2]

The highest molecular orbitals for an M_2 molecule are generally antibonding, due to the closed electronic structure of the M constituents. The bonding orbitals are lower in energy. The lowest electronic state of M_2^+ results from a removal of one of the antibonding orbitals. It is therefore not surprising that the binding energy of M_2^+ is generally much larger than that of M_2. Similarily, among the electronically excited states of M_2^* there are always some of the excimer type, i.e. bound in M_2^* and repulsive in M_2. This leads to the continua in optical emission, which are used in excimer lasers. The bonding is strengthened by the exchange interaction between M and M^+ or M^*. The change in binding energies can be two orders of magnitude. A large reduction of the internuclear equilibrium distance accompanies the increase in binding energy.

These two changes are the driving force for charge localization in a dimer ion for rare gas clusters, recently discussed by the author.[2] For electron impact ionization of an Ar_n cluster the sequence of processes happening in rare gas cluster after ionization can be divided into several steps.[2]

1) $e^- + Ar_n \longrightarrow Ar_n^+ + 2e^-$
   The hole in Ar_n^+ is delocalized at first.

2) $\quad Ar_n^+ \longrightarrow Ar_2^+A_{n-2}$
   The charge localizes in a dimer ion after about 1 ps.

3) $\quad Ar_2^+Ar_{n-2} \longrightarrow Ar_2^+Ar_m + (n-m-2)Ar.$
   Vibrational energy transfer from Ar_2^+ to the rest of the cluster, leading to fragmentation.

The equilibrium internuclear distance in the ionized dimer is about 30% smaller than the neutral one. Once the dimer ion is vibrationally relaxed, the charge is localized, as the transfer integral (the Franck-Condon overlap integral) to neighbouring geometries is exponentially small. The process 3 and part of process 2 have been tested in two recent calculations by Saenz et al.[3], and by Polymeropoulos et al.[4]. A related model was recently applied by Scharf et al. to exciton formation in Ar_{13} and Ar_{55}. It will be pointed out below, that charge localization on a few molecules can be expected for clusters composed of closed shell molecules. Note that these ideas do not apply to closed shell atoms, which form conductors in the bulk. In this case (eg. $Be(2s^2)$ or $Hg(6s^2)$) the s-band overlaps with the p-band in the bulk, and consequently also in a large enough cluster.

Ionization of a dimer composed of two closed shell molecules

It will be pointed out in this part that the change in electronic and geometric structure upon ionization of M_2, M=closed shell molecule, is very similar to that discussed for rare gas dimers discussed above. The two important features are again:

1) The interaction potential is much stronger for M_2^+ compared to M_2.

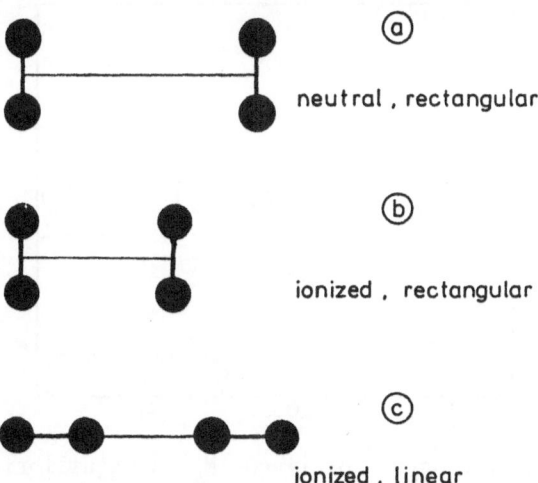

(a) neutral, rectangular

(b) ionized, rectangular

(c) ionized, linear

Fig.1: Equilibrium shapes for neutral (a) and ionized (b) $(N_2)_2$ constrained to a rectangular geometry. The lowest energy has been calculated for the linear geometry of N_4^+ (c). The distance between the two neutral molecules is large, compared to the case, when one of them is ionized.

2) The equilibrium distance between the two moieties is shorter.

a) <u>The electronic structure of N_4^+</u>

The nitrogen molecule dimer will be discussed as one example. The rectangular geometry of Fig.1a has nearly the lowest energy for $(N_2)_2$. The distance between the two centers of mass is computed to be R=3.6Å with a well depth of D=0.015 eV.[6] If $(N_2)_2^+$ is constrained to remain in the same symmetry (Fig.1b) these values change to R=2.36Å and D=0.9 eV. The most favourable geometry of N_4^+ is the linear one, given in Fig.1c. The charge is evenly distributed over the four N atoms. The well depth deepens to D=1.3 eV and the equilibrium distance diminishes to R=1.93Å. The same two effects, well known for rare gas dimers, are also observed for $(N_2)_2$ ionization. Fig.2 shows the potential calculated for linear $(N_2)_2^+$ if the N_2--N_2 distance is varied. States dissociating to a low lying excited state of N_2^+ have been omitted and the R-axis has been shifted by 0.5Å with respect to Fig.6 of ref.7.[8] It can be seen from the data given in ref.7, which have been used to construct Fig.1, that the two molecules still have nearly their asymptotic internuclear distance in N_4^+. Let ψ_1 be the wavefunction for N_2 and ψ_2 that for N_2^+. If the molecules are denoted by a and b, the zero order wavefunction can be written as $\psi_1(a) \psi_2(b)$. Another wavefunction with the same energy is

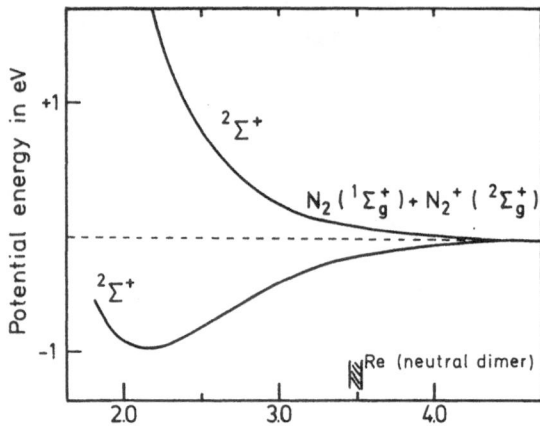

Fig.2: Potential energy diagram[8] of linear N_4. The equilibrium distance of the neutral dimer is indicated. Once the charge is localized on a vibrationally relaxed N_4^+, the overlap integral between the two geometries is very small.

$\psi_1(b)\ \psi_2(a)$. After symmetrization this gives:

$$\emptyset_g = 2^{-1/2} \{\ \psi_1(a)\ \psi_2(b) + \psi_1(b)\ \psi_2(a)\ \}$$

$$\emptyset_u = 2^{-1/2} \{\ \psi_1(a)\ \psi_2(b) - \psi_1(b)\ \psi_2(a)\ \}$$

The potential energy is given by the Coulomb interaction averaged over the electronic wavefunction. As there are two of different symmetry, the potential is split. Obviously this argument is not valid for the interaction of different molecules, e.g. O_2 with N_2^+. Inspecting the N_4^+ potentials calculated by de Castro, Schaefer III, and Pitzer[7] for other geometries one observes that there is always:

    1) one strongly bound state with a well depth of about 1 eV, having

    2) a reduced equilibrium distance.

b) Electronic structure of $(H_2O)_2^+$

Very similar results are observed for the water dimer. Water, being
isoelectronic to neon, shows upon ionization again the two features
discussed above for nitrogen and the rare gases. The third atom
complicates the potential energy surface but does not alter the conclusions. The potential was recently discussed in relation with the
observation of $(H_2O)_n^+$, n>3 clusters, and will not be repeated here.[9,10]

Comparison with experimental data

Castleman and Keesee[11] have recently reviewed the energetics of clustering
of atoms and molecules around ions. The data in table 1 are taken from
their review. The binding energies of M^+M_n with respect to that of M^+M_{n-1}
are given in units of Kcal/mol. This unit is chosen as all the tables
in ref.11 are given in these units.

The first four lines pertain to rare gas clusters. Note the strong
drop in ΔE in going from n=1 to n=2. This shows that M_2^+ is indeed strongly
bound. The value for n=2 is only that expected for charge induced
polarization forces. This indicates strongly, along with the evidence
cited in ref.2, that the charge is indeed localized on the dimer ion.
This has recently been questioned in two theoretical studies.[12,13] The
comparison of lines 3 and 4 reveals that only in the resonant case
(Ne^+ + Ne) the energy increase is large, while in the nonresonant case
(Ar^+ + Ne) it is small. Only in the resonant case the potential is split
due to the g-u symmetry.

In line 5 to 10 there is a drop in the interaction energy of a
factor of 2 to 8 in going from n=1 to n=2. The larger the difference the
more the charge is localized on the dimer. Note that line 10 corresponds
to a negative ion.

In lines 11 to 15 the drop in going from n=1 to 2 is less than a
factor of two. From line 11 and 12 one can conclude that large ionized
O_2 and NO clusters should probably be written as $O_6^+(O_2)_n$ and
$(N_3O_3)^+(NO)_n$. A comparison of line 13 and 14 shows that the
symmetry effect leading to the g-u splitting in linear molecules
still has an influence on the potential for these nonlinear molecules.

Table 1: The binding energies of M^+M_n with respect to that of M^+M_{n-1} are given in units of Kcal/mol. From these data it can be concluded, that the charge is localized on a few M molecules in a large M_n^+ cluster. See text for detailed discussion.

Line	Ion	added neutral	Energy gained by adding n identical neutrals				
			n=1	n=2	n=3	n=4*	n=5
1	He^+	He	54	4			
2	Ne^+	Ne	31				
3	Ar^+	Ne	1.8				
4	Xe^+	Xe	22	7			
5	H_3^+	H_2	10	4	4	2	
6	CO_2^+	CO_2	11.8	3.3	2.8		
7	OCS^+	OCS	17	2			
8	H_2S^+	H_2S	17	3	1.2	1.4	2.6
9	NH_3^+	NH_3	23.1	9.2			
10	SO_2^-	SO_2	13.9	3.5			
11	O_2^+	O_2	10.8	7	3	2	2
12	NO^+	NO	13.8	7.4	4	4	2
13	H_3O^+	H_2O	32	20	18		
14	NH_4^+	H_2O	17	15	13		
15	NH_4^+	NH_3	21.5	16.2	13.5	11.7	7

Conclusions

Charge localization in ionized clusters of closed shell atoms and molecules occurs mainly on the dimer ion, and less often on a larger cluster. It would be very interesting to test these predictions on a more rigorous basis.

Note added after the conference:

The prediction concerning charge localization in $(N_2)_n^+$ clusters was confirmed experimentally by Peel and coworkers (see F. Carnovale, J.B. Peel and R.G. Rothwell, this book).

References

a) permanent address: Fakultät für Physik, Universität Freiburg,
   D-7800 Freiburg

1) T. D. Märk and A. W. Castleman, Adv. Atom. and Molec. Physics 20, 65 (1985)

2) H. Haberland, Surface Science 156, 305 (1985)

3) J. J. Saenz, J. M. Soler and N. Garcia, Chem. Phys. Lett. 114,15 (1985)

4) E. E. Polymeropoulos, S. Löffler and J. Brickmann,
   Z. Naturforschung 40a, 516 (1985)

5) D. Scharf, J. Jortner and U. Landman, Chem. Phys. Lett. 126, 495 (1986)
   and priv. communication

6) R. M. Berns and A. van der Avoird, J. Chem. Phys. 72, 6107 (1980)

7) S. C. de Castro, H. F. Schaefer III, and R. M. Pitzer
   J. Chem. Phys. 74, 550 (1981)

8) The abscissa in Fig.2 has been shifted by 0.5Å with respect to
   Fig.6 of ref.7. Only after this shift Fig.2 is compatibel with
   Fig.5 and table IV of ref.7. There is also a problem with the
   ordinate in Fig.3 and the abscissa of Fig.7 of ref.7.

9) H. Shinohara, N. Nishi, N. Washida, J. Chem. Phys. 84, 5561 (1986)

10) H. Haberland and H. Langosch, Z. Phys. D - Atoms, Molecules and
    Clusters 2, 243 (1986)

11) R. G. Keesee and A. W. Castleman,
    J. Phys. Chem. Ref. Data. 15, 1011 (1986)

12) J. Heßlich and P. J. Kuntz, Z. Phys. D - Atoms, Molecules and
    Clusters 2, 251 (1986)

13) U. Böhmer and S. Peyerimhoff, Z. Phys. D - Atoms, Molecules and
    Clusters, submitted

UNIMOLECULAR DECAY OF SMALL NH$_3$ CLUSTER IONS

W.Kamke, B.Kamke, Z.Wang,
R.Herrmann and I.V.Hertel*

Institut fuer Molekuelphysik
Freie Universitaet Berlin
Arnimallee 14, 1000 Berlin, Germany, F.R.

INTRODUCTION

During the past years a growing interest in the study of molecular cluster ions gave rise to an increasing number of experimental and theoretical work in this field (For a review see e.g. references[1,2]). Beyond the purely mass spectrometric information acquired in electron impact and photoionization studies attention has more and more focussed on questions about the dynamics, i.e. the energy flow within the complex and the dissociation mechanisms. Mass spectra usually do not resemble the primary distribution of the neutral clusters in a molecular beam but are the result of extensive fragmentation of larger clusters[3,4]. Analogously, a photoionization efficiency spectrum (PIES) measured for a given cluster normally is not the true PIES of that cluster ion but is obscured by fragments of larger clusters as has been demonstrated for Ar$_x$ by Dehmer and Pratt[5] and for (N$_2$O)$_x$ by Kamke et al.[6]. Here we report first experimental results on ammonia clusters, focussing on unimolecular reactions.

EXPERIMENTAL

The experimental setup has been described in detail elsewhere[6,7]. Briefly, pure ammonia is expanded through a 70 μm nozzle to produce a molecular beam with a significant fraction of small (NH$_3$)$_x$-clusters, x ranging typically up to 10. Stagnation pressures between 1 and 3 bar were used. The molecules and clusters are ionized by a monochromatized and focussed beam of synchrotron radiation from Berlin's electron storage ring BESSY. The ions are extracted perpendicular to the molecular beam and the light beam, and are sent through a quadrupole mass analyzer. A typical mass spectrum, taken with "white" light ($\lambda \gtrsim 50$ nm), is shown in fig.1. Besides the strong monomer peak and its major fragment NH$_2{}^+$ it clearly shows that the intensity of the homogeneous clusters (NH$_3$)$_x{}^+$, $x \geq 2$, is relatively low, whereas their corresponding fragments (NH$_3$)$_{x-2}$NH$_4{}^+$ make up the major peaks. The prominence of these protonated ions has been discussed previously[8-11] as being due to the intramolecular ion-molecule reaction

---

* permanent address: Fakultaet fuer Physik, Universitaet Freiburg,
Hermann-Herder-Strasse 3, 7800 Freiburg, Germany, F.R.

$$(NH_3)_{x-2} \cdot (NH_3^+ NH_3) \quad ---> \quad (NH_3)_{x-2} \cdot (NH_4^+ NH_2)$$

which, for the dimer (x=2) is exothermic by at least 0.74 eV[9]. Similar behaviour is also observed in other hydrogen-bonded compounds, e.g. in water[12]. The protonated ions are usually considered the main fragments of the unprotonated ions while the latter may be both, either more or less stable parent ions or fragments of larger clusters.

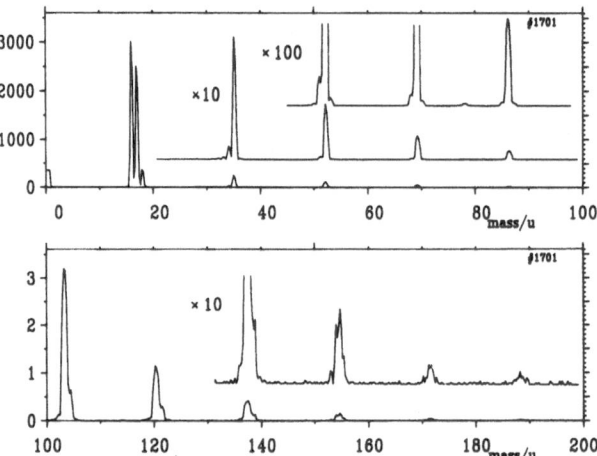

Fig.1. Mass spectrum of an NH_3 molecular beam, photoionized by synchrotron radiation with $\lambda \gtrsim 50$ nm (Monochromator set to zeroth order) Stagnation pressure is 2 bar.

In order to obtain more information about the fragmentation process our ion detection system includes an energy analyzer behind the quadrupole. As described previously[6], only stable ions or such fragment ions which arise from sufficiently fast (typically < 0.2 µs) decay processes are transmitted under normal operating conditions. Thus we call these ions "prompt". Fragment ions born later will have a lower kinetic energy, and will therefore not be transmitted, because their parents have lost a certain fraction ΔE of their kinetic energy due to the mass loss associated with the fragmentation process $m^+ --> m_0^+ + \Delta m$. The energy loss which depends solely on $\Delta m/m^+$ and the parent ion's energy at the moment of decomposition can be compensated for by raising the source potential. All ions fragmenting within the field-free region which follows the accelerating region experience the same energy loss which is specific for a particular decay channel. These "metastable" ions can be selected in the energy analyzer, as well as the prompt ions. Thus we can measure PIES individually for these two groups and, for metastable ions, can also specify the fragmentation channel. The time window which is associated with the metastable processes can be shifted, typically in the range 0.5...5 µs by changing the extraction field strength. If we record the fraction of metastable ions for each flight time we obtain decay curves for the unimolecular channels selected.

PHOTOIONIZATION EFFICIENCY SPECTRA

Photoionization efficiency spectra were taken for various clusters $(NH_3)_x$ and $(NH_3)_{x-2}NH_4^+$ by scanning the monochromator from 50 nm to above threshold. The results for prompt ions are summarized in fig.2. The spectra, except that of the monomer which shows the sharp break at the appearance of the NH_2^+ fragment, show very strong resemblance, indicating similar ionization mechanisms in all clusters. The appearance potentials (AP) are shifted

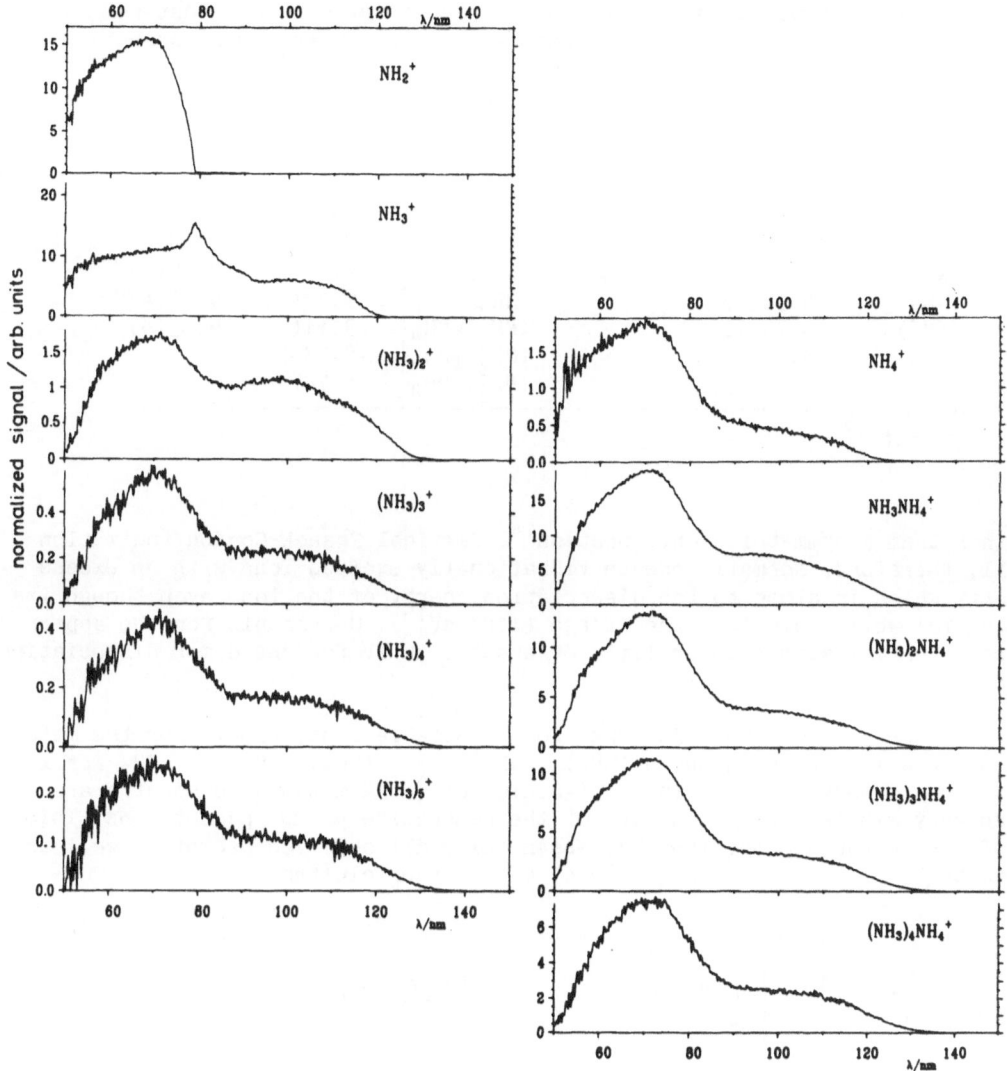

Fig.2. PIES of various unprotonated $(NH_3)_x^+$ and protonated $(NH_3)_{x-2}NH_4^+$ clusters. The spectra are normalized at 80 nm. Stagnation pressure was 1.5 bar. The NH_2^+ PIES is included for comparison.

towards longer wavelengths[7,9] as the cluster size is increased, with the strongest shift between the monomer and the dimer.

The spectra of the protonated cluster species also show a strong resemblance to their respective parents, favoring slightly the higher photon energies ($\lambda \gtrsim 85$ nm). Their AP's lie at slightly higher energies than those of their parents, except for the case of the pentamer where parent and fragment appear at essentially the same energy within the experimental uncertainties.

The AP's are summarized in table 1 in which results of earlier measurements are included for comparison. Our results for the unprotonated clusters are, in general, lower than those of the previous study[9], for the protonated species the values do not differ within the error limits.

The true adiabatic IP of the dimer is especially interesting because the potential well minimum corresponds to a proton-transferred state $NH_2NH_4^+$

Table 1. Appearance potentials (in eV) of various NH_3 clusters. Earlier results of photoionization experiments[9] are included for reference. The maximum deviation in the last digit is given in parentheses.

	this work	other authors		this work	other authors
NH_3^+	10.16	10.16^a			
$(NH_3)_2^+$	9.19(7)	$9.54(5)^b$	$NH_3NH_4^+$	9.54(4)	$9.59(2)^b$
$(NH_3)_3^+$	8.95(15)		$(NH_3)_2NH_4^+$	9.13(7)	$9.15(4)^b$
$(NH_3)_4^+$	8.95(7)		$(NH_3)_3NH_4^+$	9.11(7)	$9.03(4)^b$
$(NH_3)_5^+$	8.85(7)		$(NH_3)_4NH_4^+$	8.78(8)	
			$(NH_3)_5NH_4^+$	8.78(10)	

a ref.[13]
b ref.[9]

rather than a symmetric configuration[14]. Vertical Franck-Condon ionization will, therefore, normally create vibrationally excited ions with an excess energy which is close to the dissociation energy of the ion, even though its potential well depth is rather large (1.07 eV[15]). Our result for the appearance potential sets a lower limit of about 0.35 eV for the dimer dissociation energy.

Figure 3 shows two PIES taken for metastable ions, namely for the metastable fragmentation channels $(NH_3)_{x-2}NH_4^+$ ---> $(NH_3)_{x-3}NH_4^+$ + NH_3 for x=4 and 5. As observed before in for N_2O clusters[6] the metastable spectra are also very similar to the spectra of the respective prompt parent ions. This leads us to the supposition that essentially all observed prompt as well as metastable clusters are fragments of much larger clusters.

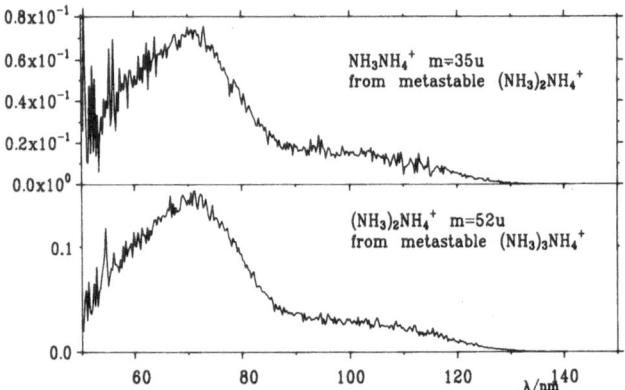

Fig.3. PIES for $NH_3NH_4^+$ and $(NH_3)_2NH_4^+$ after metastable fragmentation of $(NH_3)_2NH_4^+$ and $(NH_3)_3NH_4^+$, respectively The spectra are normalized at 80 nm. Stagnation pressure was 1.5 bar.

ION ENERGY LOSS SCANS

At fixed wavelengths we scanned the ion source potential sweeping over the appropriate settings for the detection of prompt and metastable ions of specific clusters. For $NH_3NH_4^+$ such a scan is shown in fig.4. It clearly demonstrates the appearance of the strong prompt peak at zero energy loss and

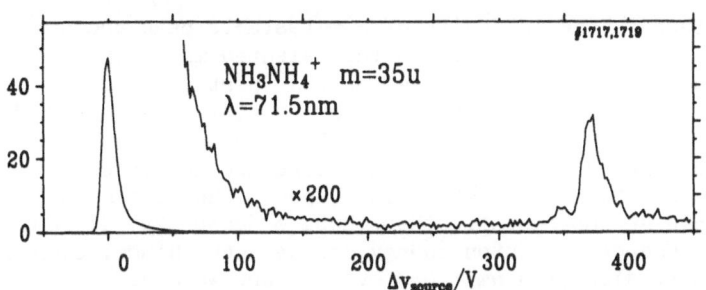

Fig.4. Ion energy loss scan for $NH_3NH_4^+$
at λ=71.5nm. The metastable peaks
at 371 eV and at 349 eV energy
loss correspond to mass losses
of Δm=17u and 16u, respectively.

a weak metastable peak. The latter is, in this case, the result of a unimolecular process with $(NH_3)_2NH_4^+$ as the parent ion. There is a second, even
smaller metastable peak slightly to the left of the larger one. This peak
corresponds to a mass loss of Δm=16u, the loss of an NH_2 molecule from
$(NH_3)_3^+$. It must be concluded from the low intensity of this peak that the
initial emission of an NH_2 molecule takes place very shortly after ionization
in full consistancy with the internal proton-transferred structure of the
stable cluster. Thus this fragmentation channel is not observed very efficiently at typical ion flight times of the order of .5 µs. But, obviously,
the remaining ion is still hot enough to emit, in general probably several,
NH_3 molecules in succesion. This will be a slower process, and it is evident
that it will be detected more efficiently on a metastable time scale.

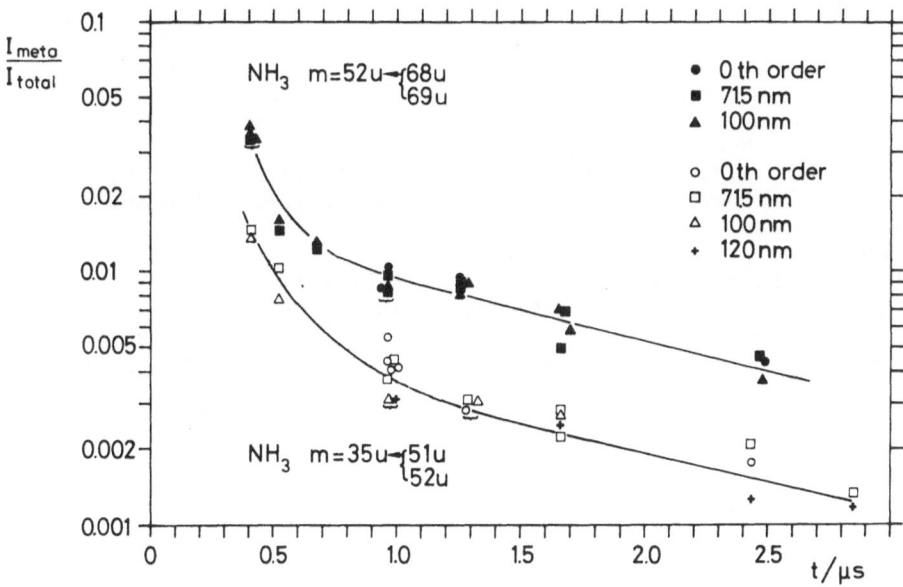

Fig.5. Decay curves for the (unresolved) fragmentation channels
$(NH_3)_3NH_4^+$ and $(NH_3)_4^+$ ---> $(NH_3)_2NH_4^+$, m=69,68u --> m=52u, and
$(NH_3)_2NH_4^+$ and $(NH_3)_3^+$ ---> $NH_3NH_4^+$, m=52,51u --> m=35u.
The data shows no systematic dependence on wavelength.

If we integrate the intensity of a metastable peak and normalize it to the total number of ions detected at the selected mass for a set of ion flight times we obtain a decay curve for that particular decay channel. A preliminary result for the the two NH_3-loss reactions giving $NH_3NH_4^+$ and $(NH_3)_2NH_4^+$ is depicted in fig.5. The NH_2-loss channels have not been separated here. For the smaller cluster the peaks are sufficiently resolved and the result is reported in a forthcoming paper[7]. The data shows that cluster ions of mass 69u are more stable than those of mass 52u. In addition, no systematic dependence on photon energy can be seen. Since, even at the longest wavelength studied (120 nm, 10.3 eV), we are well above threshold for any of the fragmentation processes studied here there is no variation to be expected. However, investigations closer to threshold are planned.

CONCLUSION

Photoionzation efficiency spectra are reported for various ammonia cluster ions $(NH_3)_x^+$, x=2...5, and $(NH_3)_{x-2}NH_4^+$, x=2...6, showing similar behaviour for the different cluster sizes. From the onsets of these spectra appearance potentials are taken to update and expand earlier results. A study of metastable decay cahnnels reveals that, besides a rather fast NH_2 emission from $(NH_3)_x^+$ delayed fragmentation of the type $(NH_3)_{x-2}NH_4^+$ ---> $(NH_3)_{x-3}NH_4^+$ with NH_3-loss is a significant process. This indicates that, after initial emission of an NH_2 molecule, probably several NH_3 molecules follow before the ion is reasonably cold. PIES for the major metastable fragmentation channels giving the fragments $(NH_3)_2NH_4^+$ and $NH_3NH_4^+$ are also reported. The spectra are very similar, supporting the assumption that both classes of ions are mainly fragments of much larger clusters.

ACKNOWLEDGEMENT

This work was supported by the Bundesministerium fur Forschung und Technologie which is gratefully acknowledged.

REFERENCES

1. C.Y.Ng, Molecular Beam Photoionization Studies of Molecules and Clusters, in: "Advances in Chemical Physics", Vol.52, I.Prigogine and S.A.Rice, ed., Wiley, New York (1983)
2. T.D.Märk and A.W.Castleman,Jr., Experimental Studies on Cluster Ions, in: "Advances in Atomic and Molecular Physics", Vol.20, D.R.Bates and R.Bederson, ed., Academic, New York (1985)
3. H.Haberland, Surface Science 156:305 (1985)
4. U.Buck and H.Meyer, J.Chem.Phys. 84:4854 (1986)
5. P.M.Dehmer and S.T.Pratt, J.Chem.Phys. 76:843 (1982)
6. W.Kamke, B.Kamke, H.-U.Kiefl and I.V.Hertel, J.Chem.Phys. 84:1325 (1986)
7. W.Kamke, B.Kamke, Z.Wang, H.-U.Kiefl, R.Herrmann and I.V.Hertel, Z.Phys.D, to be published
8. K.D.Cook and J.W.Taylor, Int.J.Mass Spectrom.Ion Physics 30:345 (1979)
9. S.T.Ceyer, P.W.Tiedemann, B.H.Mahan and Y.T.Lee, J.Chem.Phys. 70:14 (1979)
10. H.Shinohara, N.Nishi and N.Washida, J.Chem.Phys. 83:1939 (1985)
11. O.Echt, P.D.Dao, S.Morgan and A.W.Castleman,Jr., J.Chem.Phys. 82:4076 (1985)
12. S.Tomoda and K.Kimura, Chem.Phys. 82:215 (1983)
13. V.H.Dibeler, J.A.Walker and H.M.Rosenstock, J.Res.NBS A 70A:459 (1966)
14. H.Shinohara, N.Nishi, N.Washida, J.Chem.Phys. 83:1939 (1984)
15. H.Z.Cao, E.M.Evleth, and E.Kassab, J.Chem.Phys. 81:1512 (1984)

ELECTRONIC SPECTRA AND ELECTRONIC RELAXARION IN MERCURY

VAN DER WAALS COMPLEXES STUDIED IN A SUPERSONIC JET

Koji Kaya and Kiyokazu Fuke

Department of Chemistry, Keio University

Yokohama 223, Japan

ABSTRACT

The van der Waals complexes of mercury (Hg) with rare gases (RG = Ne, Ar, Kr, and Xe) and simple molecules (CH_4 and N_2) were prepared in a supersonic jet. By the use of laser induced fluorescence excitation technique, potential energy curves and spectroscopic constants of HG-RG complexes were determined for the $\tilde{X}$, $\tilde{A}$, and $\tilde{B}$ electronic states. Potential depths of the $\tilde{X}$ and $\tilde{B}$ states were found to be linearly proportional to the polarizabilities of individual RG atoms. In the case of HgAr and HgKr, metastable $Hg(^3P_0)$-RG which is optically forbidden from the ground state complex was synthesized and the laser double resonance on the $\tilde{a}$-$\tilde{C}$, $\tilde{a}$-$\tilde{D}$ transitions of the complex has revealed the potential curve of the $\tilde{a}$ state which has double minimum in the case of HgAr. In the Hg-molecule(M) complexes, photo-dissociation reaction accompanied by the electronic relaxation of Hg ($^3P_1$ to $^3P_0$) was found to occur under a collision free condition. van der Waals bending motion as well as the rotational motion characteristic to the Hg-M complex seemed to play a crucial role in the reaction.

INTRODUCTION

Line broadening of mercury resonance line at 253.7 nm ($^3P_1$-$^1S_0$ transition) mixed with high pressure rare gas atoms like Ar or Kr has been the subjet of intensive study since early 1930's in the field of physics as a suitable system to study van der Waals interaction and there have been numerous reports on Hg-RG systems under the static gas condition[1]. Also, in the field of chemistry, mercury photosensitized reaction has been the subject of active investigations because of the importance of the sensitized reaction in the practical organic synthesis as well as fundamental photochemistry.[2] In the photosensitized reaction, an excited Hg atom collides with the target molecule and the energy migration occurs efficiently from the Hg to M through the formation of a collision complex. Especially in the case of $^3P_1$ excited Hg atom, intramultiplet transition from $^3P_1$ to $^3P_0$ is induced by the collisions and subsequent energy transfer to the target molecule becomes very efficient because of the long lifetime of the $^3P_0$ Hg atom.

In order to understand these physical and chemical processes of Hg atom interacting with RG atoms and molecules from microscopic point of view, one wishes to obtain the information on the potential surfaces of Hg-RG and Hg-M collision complexes. However, under static gas condition, the formation of stable collision complexes cannot be expected. In recent years, supersonic expansion technique has been utilized for the production of various weakly bound van der Waals complexes and spectroscopic measurements of supercooled small clusters has been successfully performed by the use of various lasers.

In the present paper, Hg-RG (RG = Ne, Ar, Kr and Xe) were synthesized in a supersonic free jet and potential curves and spectroscopic constants of ground ($\widetilde{X}$) and electronically excited($\widetilde{A}$ and $\widetilde{B}$) states which are asymptotically correlated with $Hg(^3P_1) + RG(^1S_0)$ were determined by the use of laser induced fluorescence(LIF) method. Moreover, by the use of laser double resonance method, the lowest excited state of the HgAr which are correlated with $Hg(^3P_0)$ was characterized. We also synthesized Hg-M (M = CH_4 and N_2) complexes and predissociation reaction ($Hg(^3P_1)$-M = $Hg(^3P_0)$ + M) was studied in detail through this collision complex. The dissociation rate was found to depend strongly on the rotation and bending motion of thecomplex.

EXPERIMENTAL

The details of the experimental set-up was described elsewhere[3]. Briefly, the second harmonic of YAG or/and nitrogen laser excited dye laser was used as an excitation light source. The laser light was focused into an expansion chamber where the gas mixture of heated Hg vapor, RG atoms or M and 3 atm He as a carrier gas was expanded through a pinhole into a vacuum. In the case of LIF experiment using single laser, the intensity of the exciting dye laser was kept constnat of the wavelength and the total fluorescence intensity was monitored as a function of the exciting wavelength, which gives an excitation spectrum and $\widetilde{A}$-$\widetilde{X}$ and $\widetilde{B}$-$\widetilde{X}$ transitions of the Hg-RG and Hg-M complexes were detected by this method.

In the case of HgAr, we have succeeded in the measurement of the $\widetilde{a}$-$\widetilde{C}$ and $\widetilde{a}$-$\widetilde{D}$ transitions where the $\widetilde{a}$ state is optically prohibited from the ground state.[4] $Hg(^3P_1)$ atoms were at first produced by focusing the 253.7 nm dye laser line into the mixture of Hg vapor, 100 Torr nitrogen and 3 atm Ar at the nozzle exit. The nitrogen molecules in the mixture collide with the $^3P_1$ Hg atoms inducing the relaxation of the Hg atoms from $^3P_1$ to $^3P_0$. Then, the $Hg(^3P_0)$ atoms form vdW complexes with Ar in the subsequent expansion. The electronic spectrum of $Hg(^3P_0)$Ar complex was obtained at the 1.5 cm downstream position of the nozzle exit by scanning the second dye laser in the wavelength region of 404.7 nm ($^3S_1$-$^3P_0$) atomic line of Hg atom, which can be assigned to the $\widetilde{C}$ (or $\widetilde{D}$)-$\widetilde{a}$ electronic transition of the complex.

In the case of Hg-M complexes, we have measured at first the LIF spectra of the $\widetilde{A}$-$\widetilde{X}$ and $\widetilde{B}$-$\widetilde{X}$ transitions similarly to Hg-RG complexes by the use of single UV laser in the vicinity of the 253.7 nm resonance line. Because Hg-M complexes are expected to predissociate into Hg. $(^3P_0)$ + M from the $Hg(^3P_1)$-M excited state even under collision free condition, the action spectra of this reaction were measured by the use of "pump-to-probe method" described first by Breckenridge et al. .[5] In this method, the ground state Hg-M complexes were pumped by the excitation laser to the $\widetilde{A}$ or $\widetilde{B}$ state which is asymptotically correlated to the $Hg(^3P_1)$ + M. The Hg atoms produced via the intramultiplet relaxation process from the $hg(^3P_1)$ complex were detected by monitoring the $7 ^3S_1$-$6 ^3P_2$ emission at 546.1 nm of Hg atom following the excitation of generated Hg atoms in the $^3P_0$ state to the $7 ^3S_1$ level by the second probe laser of 404.7 nm. The delay time between two lasers was set at 300 ns. Thus the action spectra were obtained by monitoring the intensity of the 546.1 nm line as a function of the wavelength of the first laser.

RESULTS AND DISCUSSION

HG-RG Complexes

Line broadening of Hg 253. 7 nm line interacted with rare gas atoms has been studied for many years in order to understand vdW interaction. However, because most of the experiments were conducted under high temperature condition, the obtained spectra (either absorption or emission) exhibit congested feature due to the overlap of the bound-bound, bound-free and free-free transitions. In contrast the bulb experiments, recent advance in the supersonic free expansion method enables us to generate supercooled weak vdW complexes where the lowest vibrational level is predominantly populated.

As a result, the LIF excitation spectrum of supercooled HgAr complex in Fig. 1 exhibits extremely simple feature. As seen in the figure, two groups of bands were observed in the longer and shorter wavelength side of the Hg atomic line (253.7 nm) which correspond to $\tilde{A}\,^3 0^+ - \tilde{X}\,^1 0^+$ and $\tilde{B}\,^3 1 - \tilde{X}\,^1 0^+$ transition of HgAr complex, respectively. The 0-0 band of the $\tilde{A}-\tilde{X}$ transition is located at 39194 cm^{-1} and has the 0-1 vibrational spacing of 37 cm^{-1} which decreases to 29 cm^{-1} as the vibrational progression proceeds to $v' = 6$. In the case of the $\tilde{B}-\tilde{X}$ transition, the origin band is assigned to 39482 cm^{-1} and the 0-1 vibrational spacing of 11 cm^{-1} decreases gradually to converge into dissociation continuum as the progression proceeds. From the Birge-Sponer plot of the vibrational spacing of the $\tilde{B}-\tilde{X}$ transition, the energy of the dissociation limit was evaluated to be 39543 cm^{-1}. Then, the dissociation energy D_0 of the $\tilde{B}$ state was determined to be 61 cm^{-1} (= 39543 - 39482 cm^{-1}). Adding the zero vibrational energy of 6 cm^{-1} to the D_0, one obtains the potential depth D_e of the $\tilde{B}$ state to be 67 cm^{-1}. By the similar procedure, D_e value was decided to be 369 cm^{-1} for the A state. The potential depth of the ground state was esimated from the superposition of zero vibrational energy (11 cm^{-1}) and the D_0 value (131 cm^{-1}) which was obtained from the spectroscopic data including LIF and dispersed fluorescence spectra.

Using these spectroscopic constants, one can obtain the information on the potential curves and equilibrium distances r_e of the complex in the $\tilde{X}$, $\tilde{A}$, and $\tilde{B}$ states. However, by the lack of the knowledge on the rotational constants, complete Rydberg-Klein-Rees (RKR) calculation cannot be done and the potential width was exactly calculated by RKR method as a function

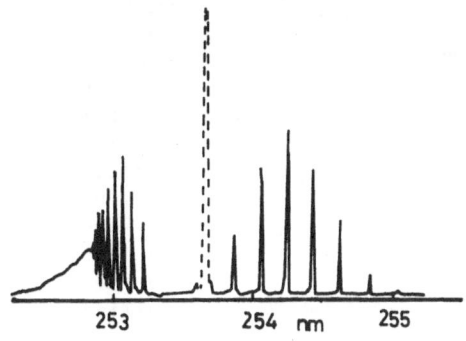

Fig. 1. Laser induced fluorescence excitation spectrum of HgAr vdW complex in a supersonic jet. Dotted line is the $^3P_1 - \,^1S_0$ transition of free Hg atom. The longer and shorter wavelength side of the Hg resonance line correspond to $\tilde{A} - \tilde{X}$ and $\tilde{B} - \tilde{X}$ transition of HgAr, respectively.

of the potential depth. The result indicates that the potential curves
of inididual states can be fit well by the Morse potentials. The r_e
value of the ground state was evaluated following the Kong's rule and the
internuclear distances of the A and B states were determined by simulat-
ing the observed LIF excitation and dispersed fluorescence spectra with
the calculated Franck-Condon factor (FCF). The potential curves thus
determined for HgAr are shown in Fig. 2. As is expected from the analogy
of the other vdW complex like NaAr, the A and B states have π and σ character with respect to the orientation of 6p excited electron of Hg atom
to the Ar atom. Thus, it seems to be natural that the $\tilde{A}$ state is mostly
attractive and has a shorter internuclear distance as compared to the B
state. As is seen in the figure (also see Table 1), the difference in the
internuclear distance between attractive $\tilde{A}$ and repulsive $\tilde{B}$ state is 1.23 A
which seems to manifest itself the nature of weak vdW force.

In addition to these single laser experiment, we have measured the
electronic transition of HgAr from the metastable state denoted as $\tilde{a}$
by the use of double resonance method described in the experimental section.
The $\tilde{a}$ state is optically inaccessible from the ground state and is correlated asymptotically with $Hg(^3P_0) + Ar(^1S_0)$. By the double resonance spec-
troscopy of $\tilde{D}$-$\tilde{a}$ and $\tilde{C}$-$\tilde{a}$ electronic transitions, the a state of HgAr was
found to have double minimum potential whose D_0 values are 97 and 16 cm^{-1},
respectively. This cannot be explained within the framework of simple
van der Waals interaction and the perturbation from upper electronic states
seems to play the important role.

We also have performed LIF experiments on HgNe, HgKr and HgXe vdW
molecules and determined the spectroscopic constants and potential curves
for $\tilde{X}$, $\tilde{A}$, and $\tilde{B}$ states and the results are summarized in Table 1.
As seen in the table, the potential depth D_e of the ground state increases
in going from Ne to Xe complex. This can be interpreted in terms of the
polarizability of the rare gases. This clearly indicates that the interaction force which binds the complexes is the dispersion force. However,
in HgXe, the potential depth in the $\tilde{A}$ state is enormously deeper as compared to the D_e values of other complexes in that state.
This peculiar nature of the HgXe $\tilde{A}$ state seems to be interpreted as a
result of the mixing of the $\tilde{A}$ state with higher excited states like ion
pair state which is expected to be located at relatively lower energy
region as compared other complexes.

Hg-M (M= CH_4 and N_2) complex

Similarly to HgRG complexes,
mercury atom forms vdW complex
with various polyatomic molecules. In this work, we studied
spectroscopic and dynamic behav-
or of Hg-methane and Hg-nitrogen
molecular complexes by the use
of LIF and double resonance
experiments as described in the
previous section. CH_4 and N_2
are known to be effecient
quenchers of Hg ($^3P_1$) atoms
to the lower $^3P_0$ state by
collisions. It is also known
that collisions of RG atoms
with $Hg(^3P_1)$ atom do not induce
this intramultiplet relaxation.
The objective of this experiment is to understand the difference between RG atoms and
polyatomic moleculae in the
role of the quenching reaction

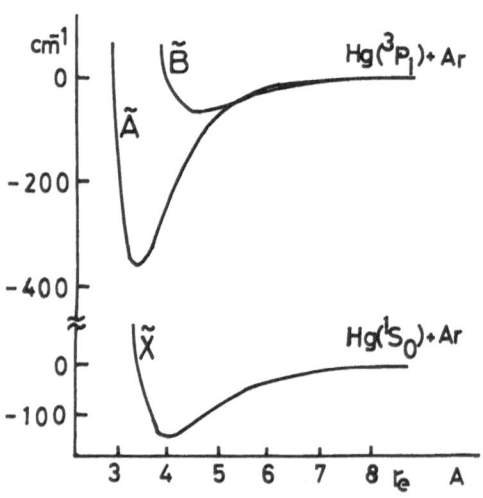

Fig. 2. Potential curves of HgAr
for X, A and B states.

Table 1. Potential depths and equiliblium distances
of Hg-RG complexes in $\tilde{X}$, $\tilde{A}$, and $\tilde{B}$ states.

State		HgNe	HgAr	HgKr	HgXe
X $^1 0^+$	D_e (cm^{-1})	51	142	178	254
	r_e (A)	3.87	4.01	4.07	4.25
A $^3 0^+$	D_e	120	369	517	1457
	r_e	3.12	3.38	3.52	3.25
B $^3 1$	D_e	17	67	96	172
	r_e	4.57	4.66	4.57	4.47

from the comparison of the spectroscopic and dynamical behavior of HgRG
and HgM vdW complexes. These complexes can be considered as intermediate
state of collision induced reaction and the subsequent dynamical behavior
can be observed microscopically by the laser spectroscopic means.

Figure 3 represents LIF excitation spectrum of HgCH$_4$ in the vicinity
of Hg 253.7 nm line. The spectra in $\tilde{A}$-$\tilde{X}$ and $\tilde{B}$-$\tilde{X}$ transition consist of the
progression of vdW stretching mode whose 0-1 spacing are 72 and 21 cm^{-1} for
the former and latter transitions, respectively. In addition to the
stretching, one can observe weak progression of vdW bending mode. The 0-1
spacing of the bending vibration in the $\tilde{A}$-$\tilde{X}$ transition is 23 cm^{-1} and the
spacing decreases rapidly as the progression proceeds. By the use of the
pump-to-probe technique using two dye lasers(see experimental section.),
the the dissociation reaction (Hg($^3P_1$)CH$_4$ -- Hg($^3P_0$) + CH$_4$) was found to
occur even under a collision free condition. The reaction takes place in
the whole region of the $\tilde{A}$-$\tilde{X}$ and $\tilde{B}$-$\tilde{X}$ transition except for the region above
the convergence limit of the $\tilde{B}$-$\tilde{X}$ transition where the complex dissociate into Hg($^3P_1$) + CH$_4$. Figure 4 displays the detailed comparison of the LIF spectrum(Fig.4a) and the action spectrum of the intramultiplet relaxation reaction (Fig. 4b) in the region of v'= 2 (stretching mode) of the $\tilde{A}$-$\tilde{X}$ transition. The figure exhibits the combination bands of v'=2 stretching mode with the vdW

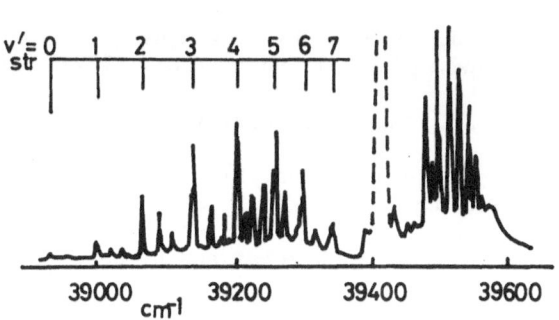

Fig. 3. LIF excitation spectrum of HgCH$_4$.
Group of bands in the lower and higher
energy side of the Hg resonance line (dotted
line) are $\tilde{A}$ - $\tilde{X}$ and $\tilde{B}$ - $\tilde{X}$ transitions, respectively.

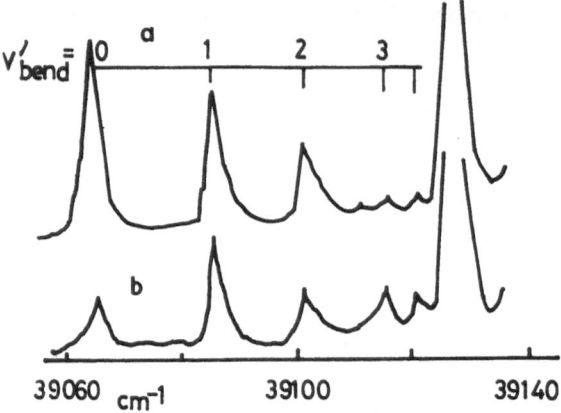

Fig. 4. LIF excitation (a) and $Hg(^3P_0)$ action (b) spectra of $HgCH_4$ in the $v'=2$ region of vdW stretching vibration.

bending mode of around 20 cm^{-1}. From the figures, it can be clearly seen that the excitation of the odd quantum number of bending vibration stimulates appreciably the dissociation reaction. This tendency can be seen in the whole range of the $\widetilde{A}-\widetilde{X}$ transition. Moreover, the LIF and action spectra was found to have different rotational envelope. This indicates that the reaction rate is dependent upon the specific vdW vibration and rotational motion. When one assumes C_{2v} (or C_{3v}) symmetry of the complex, the $\widetilde{A}$ and $\widetilde{a}$ states which are the initial and final state of the reaction belong to A_1 and A_2 symmetry, respectively. Then one needs the perturbation of A_2 type in order to induce the reaction between these two states. Actually, the bending mode (B_1 symmetry) coupled with the rotation perpendicular to the principal axis (B_2) yields A_2 symmetry perturbation. Another possible candidate is the rotational motion around the principal axis (A_2). These two motions are the missing ones in the Hg-RG atom-atom complexes where there occurs no reaction. Similar results were also reach ed in HgN_2 complex.

As a conclusion, we believe that spectroscopic and dynamical behavior observed in vdW molecules provides us with essentially important microscopic informations on the weak vdW interaction and reaction mechanism.

REFERENCES

1. J. R. Fuhr, W. L. Wiese and L. J. Roszman, Bibliography on Atomic Line Shapes and Shifts, Natl. Bur. Stand. (U. S.), Spec. Publ. (1972), Suppl. 1 (1974), and Suppl. 2 (1975).
2. J. C. Calvert and J. N. Pitt, Jr., Photochemistry (Wiley, New York, 1966).
3. K. Fuke, T. Saito, and K. Kaya, J. Chem. Phys. 81, 2591 (1984), K. Yamanouchi, J. Fukuyama, S. Tsuchiya, K. Fuke, T. Saito and K. Kaya, J. Chem. Phys. 85, 1806 (1986).
4. W. H. Breckenridge, O. Kim Malmin, W. L. Nikolai and D. Oba, Chem. Phys. Lett. 59, 38 (1978).

THE THERMODYNAMICS OF MIXED SOLVENT CLUSTER IONS: $K^+(CH_3OH)_n(H_2O)_m$

R. G. Keesee, D. H. Evans, and A. W. Castleman, Jr.

Department of Chemistry
The Pennsylvania State University
University Park, PA 16802

INTRODUCTION

One of the motivations for research on cluster ions is the parallel that the growth of cluster ions has with phenomena such as nucleation[1,2] and solvation.[3,4] For instance, the thermodynamics involved in the clustering of molecules about ions have been related to single ion heats of solvation.[5,6] Most studies of clustering have been concerned with the sequential attachment of one particular species to an ion, as evident in a recent compilation[7] of thermodynamic data on clusters ions. Few studies of competitive clustering by two different solvent species to yield mixed solvent cluster ions have been performed. In such systems, clustering may be enhanced or hindered by solvent–solvent interactions. Furthermore, the environment, as for example the natural atmosphere,[8] is such that mixed cluster ions are prevalent and data for single component systems is inadequate. The purpose of this study is to expand the data base for binary solvent systems and to explore relationships with condensed phase binary solutions. Specifically, we have chosen to investigate the competitive clustering of water and methanol onto the potassium ion K^+.

EXPERIMENT AND RESULTS

The results were obtained by using a high pressure ion source mass spectrometer. The instrument and the experimental procedure for determining the enthalpy, entropy, and free energy changes of clustering reactions have been described previously.[9]

The measured thermodynamic quantities for the $K^+/CH_3OH/H_2O$ system are listed in Table 1. For the association reactions of H_2O onto $K^+(CH_3OH)$ and $K^+(CH_3OH)(H_2O)$, the water partial pressure in the ion source was determined by monitoring the $K^+(H_2O)_n$ intensities and using the well-known equilibrium constants for K^+ hydration[10] to calculate the water partial pressure. This procedure greatly reduced the scatter in the data for these reactions which was apparently caused by absorption of water on the walls of the instrument.

Consistency of the data can be verified whenever the thermochemical properties of a cluster ion can be traced via two independent routes. For example, the formation of $K^+(CH_3OH)_2(H_2O)$ from $K^+(CH_3OH)$, CH_3OH, and

Table. 1. The thermodynamic quantities for the association reaction
I + B → I(B) in the $K^+/CH_3OH/H_2O$ system.

I	B	$-\Delta H°$ (kcal/mol)	$-\Delta S°$ (cal/K mol)	$-\Delta G^o_{298}$ (kcal/mol)	Ref
K^+	H_2O	17.9	21.6	11.5	[10]
$K^+(CH_3OH)$	CH_3OH	18.0	35	7.6	*
$K^+(CH_3OH)$	H_2O	15.6	22.5	8.9	*
$K^+(H_2O)$	CH_3OH	19.6	32.5	9.9	*
$K^+(H_2O)$	H_2O	16.1	24.2	8.9	[10]
$K^+(CH_3OH)_2$	CH_3OH	14.5	28	6.2	*
$K^+(CH_3OH)_2$	H_2O	11.3	18.0	5.9	*
$K^+(CH_3OH)(H_2O)$	CH_3OH	13.5	27.3	5.4	*
$K^+(CH_3OH)(H_2O)$	H_2O	13.1	22.6	6.4	*
$K^+(H_2O)_2$	H_2O	13.2	23.0	6.3	[10]
$K^+(CH_3OH)_3$	CH_3OH	12.5	29	3.9	*
$K^+(CH_3OH)_2(H_2O)$	CH_3OH	12.5	25.6	4.9	*
$K^+(H_2O)_3$	H_2O	11.8	24.7	4.4	[10]

*Present Work

H_2O can be obtained through the summation of data for the pair of equilibria

$$K^+(CH_3OH) + CH_3OH \rightleftharpoons K^+(CH_3OH)_2$$

$$K^+(CH_3OH)_2 + H_2O \rightleftharpoons K^+(CH_3OH)_2(H_2O)$$

to yield −29.3 kcal/mol and −53.0 cal/K·mol, for $\Delta H°$ and $\Delta S°$, respectively. These values compare well with the respective values of −29.1 kcal/mol and −49.8 cal/K·mol obtained from the pair of equilibria

$$K^+(CH_3OH) + H_2O \rightleftharpoons K^+(CH_3OH)(H_2O)$$

$$K^+(CH_3OH)(H_2O) + CH_3OH \rightleftharpoons K^+(CH_3OH)_2(H_2O) \ .$$

In addition to the studied reactions listed in Table 1, the thermodynamic values for other clustering reactions can be determined through the closure of thermodynamic cycles. Figure 1 depicts the clustering pathways with the respective enthalpy changes. The calculated values are indicated with an asterisk.

DISCUSSION

Some general trends in the enthalpy and free energy changes are evident. First, the successive addition reactions exhibit the usual decreasing trend in $-\Delta H°$ and $-\Delta G^o_{298}$ with increasing cluster size. Second, the magnitude of the enthalpy change for association of methanol

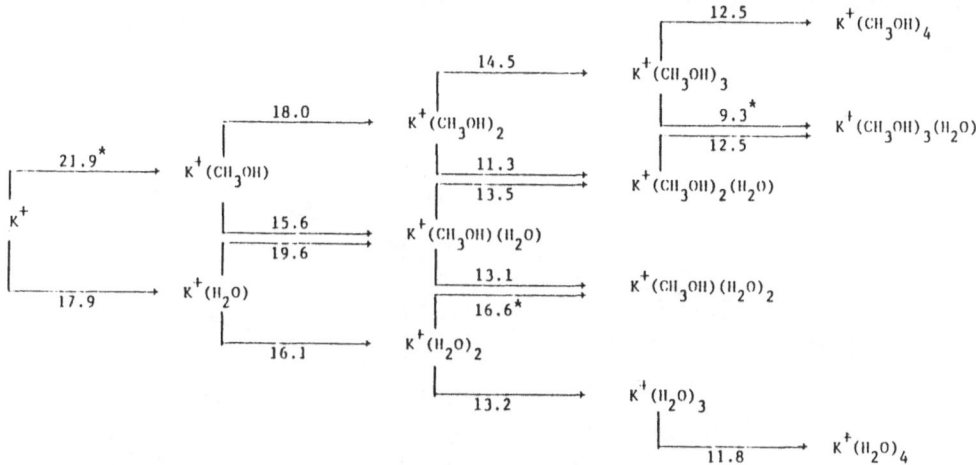

Fig. 1. Enthalpy changes $-\Delta H°$ (in kcal/mol) for clustering reactions of methanol and water onto K^+.

onto a specific ion is consistently greater than that of water. This result cannot be explained by consideration of the ion-dipole interaction alone, since the dipole moment of water is slightly larger than that of methanol (1.85 versus 1.70 debye). The average polarizability of methanol, however, is greater than that of water (3.3 versus 1.45 $Å^3$). The larger polarizability of methanol evidently allows the ion to induce a dipole on methanol which more than compensates for the smaller permanent dipole moment.

In regard to binary solvents, a perspective is gained by considering the total heat of clustering, $-\Delta H^o_{tC}$, for the reaction $I + mA + nB \rightarrow I(A)_m(B)_m$ as a function of the mole fraction of one of the solvents and cluster size. This representation is shown in Figure 2 for data on four different systems. The solid lines indicate the compositionally weighted average of the pure systems for a given cluster size. The two cases for which a strong interaction between the two solvent species is known, i.e., SO_2 in H_2O and NH_3 in H_2O, show a significant deviation of the heat of clustering above the weighted average. In the case of NH_4^+, with NH_3/H_2O, the heat of clustering for $NH_4^+(NH_3)_3(H_2O)$ is greater than that for either of the pure clusters. The upward deviation also appears to increase with increasing cluster size as might be expected since solvent-solvent interactions should play an increasingly important role as the number of solvent molecules increase. For the methanol/water system, in which the heat of mixing in the condensed phase is relatively small, no consistent trends in deviations are noticeable. For benzene/water, for which mutual solubility is limited, a slight downward deviation from the weighted average is apparent. Hence, comparison of the total heat of clustering onto a gas-phase ion with the compositionally weighted average value follows trends that might be expected based on the properties of the condensed binary solvents.

ACKNOWLEDGMENTS

Financial support by the National Science Foundation, Grant No. ATM-82-04010, and the Army Research Office, Grant No. DAAG29-84-K-0087, is gratefully acknowledged.

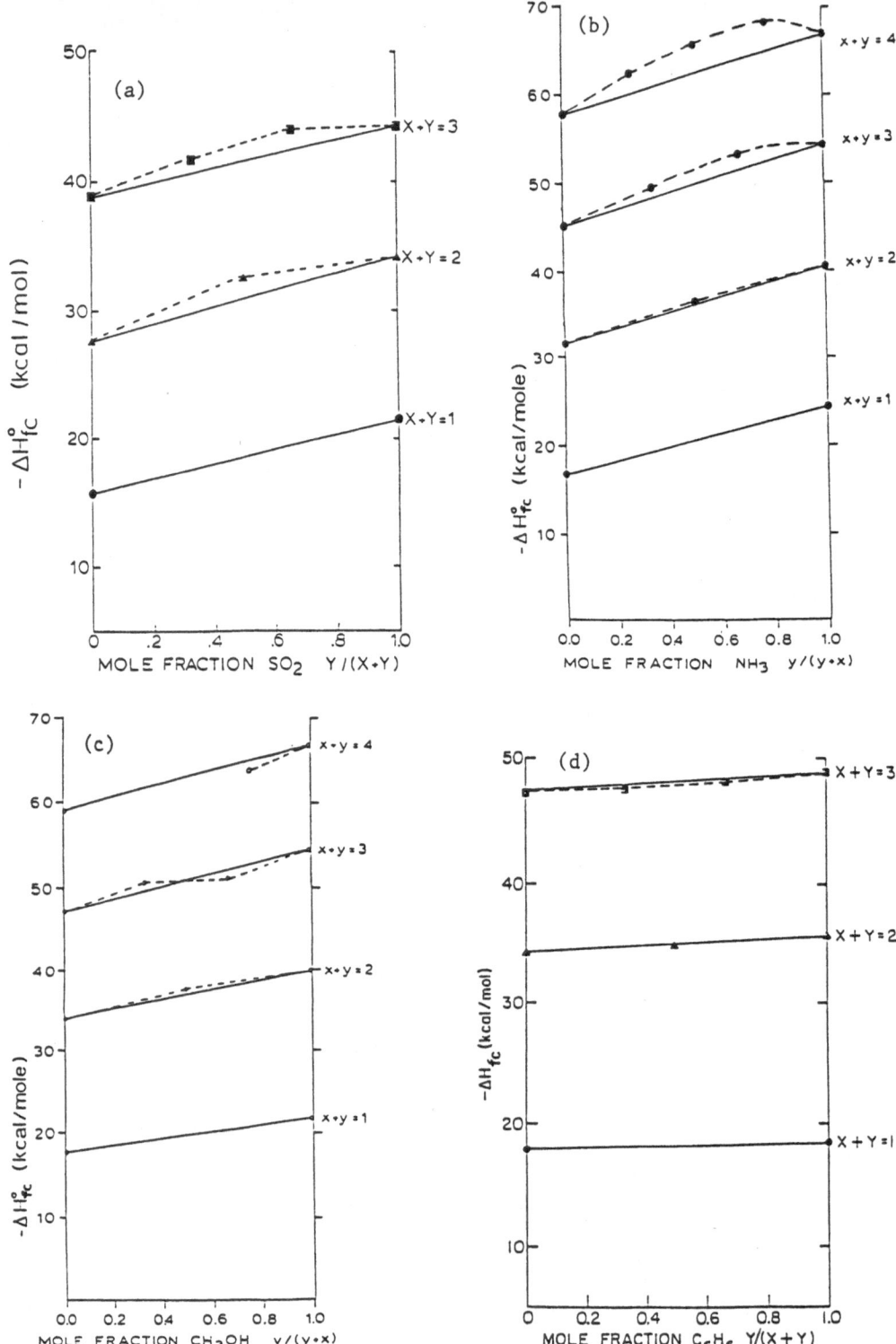

Fig 2. The total heat of clustering, $-\Delta H^{o}_{fc}$, as a function of solvent mole fraction. (a) $Cl^- + xH_2O + ySO_2 \rightarrow Cl^-(H_2O)_x(SO_2)_y$. From ref. 11. (b) $NH_4^+ + xH_2O + yNH_3 \rightarrow NH_4^+(H_2O)_x(NH_3)_y$. Data from ref. 12. (c) $K^+ + xH_2O + yCH_3OH \rightarrow K^+(H_2O)_x(CH_3OH)_y$. (d) $K^+ + xH_2O + yC_6H_6 \rightarrow K^+(H_2O)_x(C_6H_6)_y$. Data from results of switching reactions in ref. 13.

REFERENCES

1. A. W. Castleman, Jr., Adv. Coll. Interface Sci. 10:73 (1979).
2. A. W. Castleman, Jr., P. M. Holland, and R. G. Keesee, J. Chem. Phys. 68:1760 (1978).
3. P. Kebarle, Mod. Aspects Electrochem. 9:1 (1974).
4. A. W. Castleman, Jr. and R. G. Keesee, Acc. Chem. Res., in press.
5. M. Arshadi, R. Yamadagni, and P. Kebarle, J. Phys. Chem. 74:1475 (1970).
6. N. Lee, R. G. Keesee, and A. W. Castleman, Jr., J. Coll. Interface Sci. 75:555 (1980).
7. R. G. Keesee and A. W. Castleman, Jr., J. Phys. Chem. Ref. Data 15:1011 (1986).
8. R. G. Keesee and A. W. Castleman, Jr., J. Geophys. Res. 90:5885 (1985).
9. A. W. Castleman, Jr., P. M. Holland, D. M. Lindsay, and K. I. Peterson, J. Am. Chem. Soc. 100:6039 (1978).
10. S. K. Searles and P. Kebarle, Can. J. Chem. 47:2619 (1969).
11. B. L. Upschulte, F. J. Schelling, R. G. Keesee, and A. W. Castleman, Jr., Chem. Phys. Lett. 111:389 (1984).
12. J. D. Payzant, A. J. Cunningham, and P. Kebarle, Can. J. Chem. 51:3242 (1973).
13. J. Sunner, N. Nishizawa, and P. Kebarle, J. Phys. Chem. 85:1814 (1981).

ELECTRON ATTACHMENT TO VAN DER WAALS CLUSTERS

M. Knapp, O. Echt, D. Kreisle, T.D. Märk[*], and E. Recknagel

Fakultät für Physik
Universität Konstanz
D-7750 Konstanz, West-Germany

ABSTRACT

Electrons of variable energy are attached to preformed clusters of CO_2, N_2O, H_2O and SF_6 in a molecular beam. The shape resonances of CO_2 and N_2O are found to be increasingly redshifted with increasing cluster size. Clusters of CO_2, SF_6 and H_2O feature a resonance for near-thermal electrons; attachment at these energies does not cause intramolecular fragmentation. An additional resonance in the $(N_2O)_n^-$ yield is found at about 0.5 eV electron energy. Delayed autodetachment is suppressed in case of small CO_2 and SF_6 cluster anions, but not for the smallest observable H_2O cluster anions (size $n \geq 12$).

INTRODUCTION

Mass spectrometry of clusters is commonly based on electron impact ionization; i.e. positively charged ions are formed by bombardment with energetic electrons. Formation of cluster anions by electron attachment faces some disadvantages: The electrons should have a low energy (0 to about 10 eV), this seriously limits the electron emission current. On the other hand, electron capture is a resonant process: maxima in the ion yield vs. electron energy often reflect vertical electron affinities or threshold energies for dissociation. The study of these quantities as a function of cluster size can reveal the convergence of cluster properties towards bulk values. One of the particularly intriguing results is the observation of an additional resonance for low-energy electrons in relatively small clusters of O_2 /1/, CO_2 /2/ and H_2O or D_2O /3/. In this contribution we present recent results on electron attachment to clusters of CO_2, N_2O, H_2O, and SF_6.

---

[*] Permanent address: Institut für Experimentalphysik, Leopold-Franzens Universität, A-6020 Innsbruck, Austria

EXPERIMENTAL

Neutral clusters are formed by expansion of an undersaturated gas through a nozzle (diameter typically 0.1 mm) into vacuum. Large clusters can be formed in neat expansions of CO_2, N_2O and H_2O, whereas SF_6 has to be seeded in Xe. The collimated cluster beam is intersected by a pulsed electron beam; the resulting ions are analyzed in a time-of-flight mass spectrometer. The width of the energy distribution can be determined from the apparent width of well-known resonances in the ion yield of SF_6^-, SF_5^-, and O^- (from CO_2). The energy spread (FWHM) is less than 1.0 eV for energies above 1 eV; and slightly larger for very low energies. More details of the experimental set-up are given elsewhere /3,5,6/.

RESULTS AND DISCUSSION

Carbon Dioxide

CO_2 clusters were among the first to be studied by electron attachment with variable energy /7/. The first shape resonance of CO_2, which is observed in the O^- signal at 4.4 eV, finds its analog in the yield of the ions $(CO_2)_{n-1}O^-$, n = 2 and 3. Its position, however, is redshifted by a few tenths of an eV /7-9/. The resonance of larger oxygenated cluster anions appears slightly below 4.0 eV, it is considerably broader than the O^- resonance /6/. Product ions $(CO_2)_n^-$ from "non-reactive" attachment (i.e. no intramolecular fragmentation) are also detected; the monomer ion in this series has been observed only recently /9/. It cannot be formed by attachment to free CO_2 molecules, due to the poor Franck-Condon overlap between the ground states of the neutral molecule and its anion. Therefore, the neutral precursor of CO_2^- is a cluster; the dissociation of CO_2^- into $CO + O^-$, which would proceed within $\sim 10^{-14}$ s in the free molecule, can be quenched by rapid transfer of the excess energy into intermolecular modes. The initial excess energy can be estimated from the difference between the vertical and the adiabatic electron affinity of CO_2, which are -3.8 eV and -0.6(2) eV /10/ respectively. The long-lived ions CO_2^-, on the other hand, cannot contain more than ~ 0.2 eV excitation energy /11/. It should be noted, though, that the formation of long lived $(CO_2)_n^-$ by attachment to $(CO_2)_x$ ($x \geq n$) is rather inefficient for small n.

The size dependence of the resonance in the $(CO_2)_n^-$ yield is displayed in fig. 1. The CO_2^- resonance occurs at ~ 3.4 eV, 1.0 eV below the related O^- resonance. With increasing cluster size, this resonance shifts further to below 3 eV, and it features increasing tailing towards lower energies. The most intriguing finding in fig. 1, however, is the appearance of an additional resonance for electron energies close to zero. The smallest cluster ion that features this low energy resonance is the dimer; a previous study identified the trimer as the smallest observable ion at ~ 0 eV /2/. Our result implies a positive electron affinity for $(CO_2)_2$, irrespective of the neutral precursor of the observed anion. Given the strongly negative adiabatic EA of CO_2, which is -0.6(2) eV /10/, these findings confirm the calculated shift in the AEA between CO_2 and its dimer /12/ and the recent results of Bowen and coworkers who employed photoelectron spectroscopy /13/. The ~ 0 eV resonance becomes increasingly important for larger clusters; clusters larger than n $\sim$ 20 have a big cross section for electron attachment if and only if the energy is close to zero. In view of similar low-energy resonances in other van der Waals clusters we believe that the interaction between the incoming electron and the polarized <u>cluster</u> causes the ~0 eV resonance in large CO_2 clusters. For large enough clusters, this mechanism always provides an attractive interaction /14/ with bound quantum states, even in case of He /15/.

Recent attempts to produce $(CO_2)_n^-$ by electron transfer from high-Rydberg rare gas atoms did not produce these anions below n = 7 /16/, possibly due to limited sensitivity. Mass spectra of $(CO_2)_n^-$, grown by condensation of CO_2 at electrons in a jet /17/, are also strikingly similar to our distributions, as far as the local intensity anomalies are concerned. This suggests that electron attachment to clusters, even at ~0 eV, cannot avoid the problem of (intermolecular) fragmentation for cluster sizes n ≤ 20.

Nitrous Oxide

N_2O is isoelectronic with CO_2; the neutral molecules are linear in their ground state, whereas the anions are bent by ~135° /18/. The barrier of N_2O^- against dissociation into O^- is only ~0.4 eV, an order of magnitude lower than in case of CO_2^-. This low barrier is probably the main reason why the oxygenated cluster anions, $(N_2O)_{n-1}O^-$, form the dominating homologous series of ions in all mass spectra, independent of electron energy. The energy dependence of the ion yield is displayed in fig. 2. A gradual redshift from 2.25 eV for O^- to ~1.5 eV for $(N_2O)_{20}O^-$ is observed.

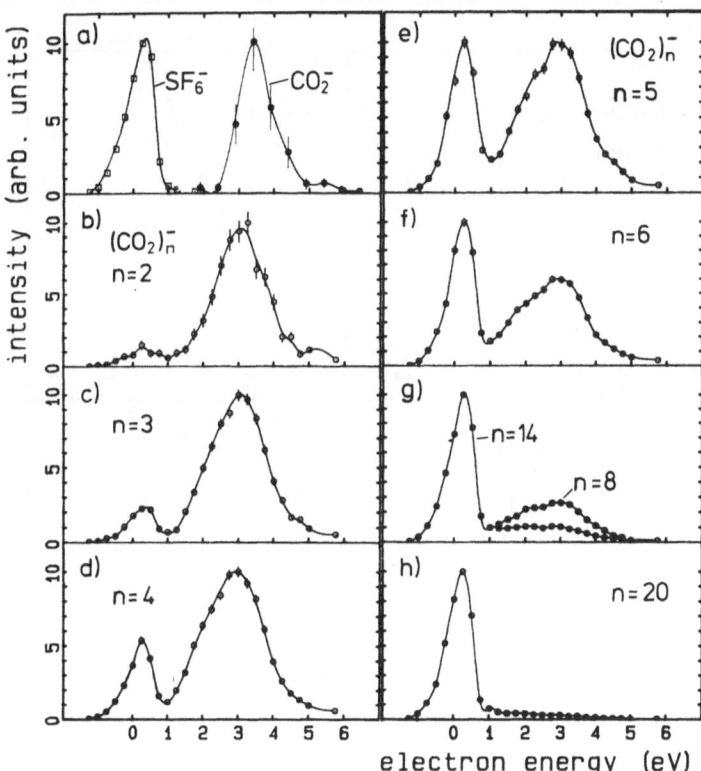

Fig. 1.   Energy dependence of $(CO_2)_n^-$ yield, obtained by electron attachment to CO_2 clusters. Each curve is normalized to the same height at its maximum. The SF_6^- yield (curve a) marks the position of a "zero eV resonance".

The resonance for larger clusters, containing a few hundred N_2O molecules, is shifted even further to ~1.0 eV /6/. The ion series $(N_2O)_n^-$, $n \geq 1$, shows a very similar resonance (the expansion conditions used for recording the data of fig. 2 were unfavorable for the production of N_2O^-; this ion features a well-defined resonance at ~2.0 eV as shown previously /9/). A recent photoelectron spectroscopic study also revealed a remarkably small solvation shift in the vertical detachment energy of N_2O^- and $(N_2O)_2^-$ /19/. An additional resonance is observed at low energies; it is relatively strong for the dimer and larger clusters and seems to occur even in the monomer signal. Its position, however, is at ~0.5 eV, definitely above 0 eV, and it does not depend on the cluster size. The origin of this resonance remains to be explained.

Water

Attempts to attach electrons to water clusters have a long history (see ref. 3 and references therein). Haberland and coworkers succeeded in forming $(H_2O)_n^-$, $n \geq 11$, by spraying low energy electrons into a supersonic jet of water vapor /20/. We have found /3/ that these ions are produced very efficiently by electron attachment to <u>free</u> water clusters if and only if the electron energy is very close to zero. We tentatively assigned this resonance to the interaction between the incoming electron and the polarized cluster, similar as in case of CO_2. In their ground state, excess electrons are self-trapped in liquid water by the induced lattice deformation and local water dipole reorientation. Calculations indicated, however, that neutral, relaxed water clusters would strongly

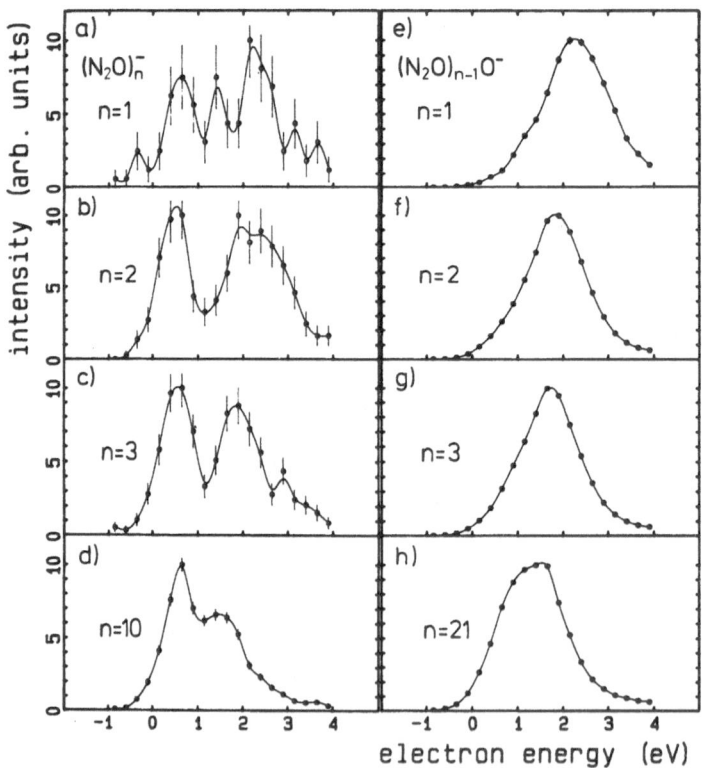

Fig. 2. Energy dependence of the $(N_2O)_n^-$ (left) and $(N_2O)_{n-1}O^-$ (right) yield. The first resonance in the cross section of $(N_2O)_n^-$ definitely occurs <u>above</u> the ~0 eV resonance of SF_6^- (cf. fig. 1).

repel electrons /21/, while the probability of finding local disorder, being favorable for electron localization, would be exceedingly small /22/. But a recent computer simulation demonstrated that the concentration of structural fluctuations with vertical electron affinities as high as +1 eV occur very frequently at room temperature /23/. Thus, direct localization of low energy electrons in free, relaxed water clusters of size n $\geq$ 11 cannot be ruled out.

If a sufficiently large, repelling voltage is applied to a fine nickel mesh in front of the detector, we observe a relatively high intensity of neutrals in the size range $12 \leq n \leq 20$ (precursor size). Comparison with the "direct" mass spectrum reveals that the smallest cluster anions, n ~ 12, have a strong tendency to undergo autodetachment in the drift tube. The probability for this process is $\geq$ 25% up to the 14-mer, it decreases below 1% for the 20-mer and larger clusters. While this trend may be qualitatively explained by the rapid increase of the adiabatic electron affinity of water clusters, the magnitude of this effect is remarkable. Similar experiments on CO_2 and SF_6 clusters show that the probability for autodetachment from a cluster as small as the dimer is only ~2% /6/. The large number of internal degrees of freedom in the water clusters should therefore suffice to suppress autodetachment unless their electron affinity is not much larger than the thermal energy, kT, before attachment. Thus, we estimate that the AEA of the smallest water cluster ions observed in our experiment, is roughly between 0.2 eV (estimated value for kT) and 0.4 eV (estimated heat of sublimation).

Sulphur Hexafluoride

SF_6 is the only molecule in the present study with a strongly positive adiabatic electron affinity (approximately 1.0 eV /24/) and a large cross section for attachment of thermal electrons. Due to its many degrees of freedom, electron attachment results in sufficiently long-lived SF_6 anions, but stable SF_5^- ions are also produced with a high intensity /4/. Attachment to SF_6 clusters proceeds resonantly for energies close to zero eV. The composition of the ions is $(SF_6)_n^-$; the intermolecular fragmentation into $(SF_6)_{n-1}SF_5^-$ is quenched very efficiently.

The size distributions of SF_6 cluster anions and cations, produced by electron attachment and electron impact ionization, respectively, are remarkably similar. The monomer ions, SF_6^- and SF_5^+, respectively, are much more intense than the cluster anions, and local intensity maxima (at n = 13, 59, 78, 91...) are observed in both spectra. Kondow and coworkers have argued that formation of SF_6 cluster anions by collisional electron transfer from high-Rydberg Krypton atoms proceeds without intermolecular fragmentation for cluster sizes larger than n ~ 5 /25/. The same would hold in our experiments, because both, attachment of thermal electrons as well as electron transfer from high-Rydberg atoms, essentially release the same excitational energy (namely, the adiabatic electron affinity). The similarity of the size distributions would then suggest that SF_6 cations are also produced without intermolecular fragmentation, a surprising result in view of the mounting evidence for the important role of fragmentation in electron impact ionization of van der Waals clusters /5, 26/. We assume that the neglect of thermal energy in the cluster before ionization invalidates the conclusions drawn by Kondow and coworkers.

ACKNOWLEDGEMENT

This work was supported by the Deutsche Forschungsgemeinschaft.

REFERENCES

1. T.D. Märk, K. Leiter, W. Ritter and A. Stamatovic, Phys. Rev. Lett.
   55:2559 (1985); T.D. Märk, K. Leiter, W. Ritter and A. Stamatovic,
   Int. J. Mass Spectrom. & Ion Proc., in print
2. A. Stamatovic, K. Leiter, W. Ritter, K. Stephan and T.D. Märk,
   J. Chem. Phys. 83:2942 (1985)
3. M.Knapp, O.Echt, D.Kreisle and E.Recknagel, J.Chem.Phys.85:636 (1986);
   M. Knapp, O. Echt, D. Kreisle and E. Recknagel, J.Phys.Chem., in print
4. L.G. Christophorou, D.L. McCorkle and A.A. Christodoulides,
   ch.6 in "Electron-Molecule Interaction and their Applications",
   vol.1, L.G. Christophorou, ed., Academic Press, Orlando (1984)
5. D. Kreisle, O. Echt, M. Knapp and E. Recknagel, Phys. Rev. A33:768
   (1986); O. Echt, D. Kreisle, M. Knapp and E. Recknagel,
   Chem. Phys. Lett. 108:401 (1984)
6. M. Knapp, Ph.D. thesis, Universität Konstanz, 1986 (unpublished)
7. C.E. Klots and R.N. Compton, J. Chem. Phys. 69:1636 (1978)
8. A. Stamatovic, K. Stephan and T.D. Märk, Int. J. Mass Spectrom. &
   Ion Proc. 63:37 (1985)
9. M. Knapp, O. Echt, D. Kreisle, T.D. Märk and E. Recknagel,
   Chem. Phys. Lett. 126:225 (1986)
10. R.N.Compton, P.W.Reinhardt and C.D.Cooper, J.Chem.Phys. 63:3821 (1975)
11. W.B. England, Chem. Phys. Lett. 78:607 (1981)
12. Y. Yoshioka and K.D. Jordan, J. Am. Chem. Soc. 102:2621 (1980)
13. J.V. Coe, J.T. Snodgrass, K.M. McHugh, C.B. Freidhoff and K.H. Bowen,
   submitted J. Phys. Chem.
14. P.R.Antoniewicz, G.T.Bennett and J.C.Thompson, J.Chem.Phys.77:4573 (1982)
15. M.W. Cole, Rev. Mod. Phys. 46:451 (1974)
16. T. Kondow and K. Mitsuke, J. Chem. Phys. 83:2612 (1985);
   also see: T. Kondow et al., contribution to this conference
17. M.L. Alexander, M.A. Johnson, N.E. Levinger and W.C. Lineberger,
   Phys. Rev. Lett. 57:976 (1986)
18. D.G. Hopper, A.C. Wahl, R.L.C. Wu and T.O. Tiernan,
   J. Chem. Phys. 65:5474 (1976)
19. J.V. Coe, J.T. Snodgrass, C.B. Freidhoff, K.M. McHugh and K.H. Bowen,
   Chem. Phys. Lett. 124:274 (1986)
20. H. Haberland, H.G. Schindler and D.R. Worsnop,
   Ber. Bunsenges. Phys. Chem. 88:270 (1984)
21. N.R. Kestner and J. Jortner, J. Phys. Chem. 88:3818 (1984)
22. M. Tachiya and A. Mozumder, J. Chem. Phys. 61:3890 (1974)
23. J. Schnitker, P.J. Rossky and G.A. Kenney-Wallace,
   J. Chem. Phys. 85:2986 (1986)
24. G.E. Streit, J. Chem. Phys. 77:826 (1982)
25. K. Mitsuke, T. Kondow and K. Kuchitsu, J. Phys. Chem. 90:1552 (1986)
26. A.W. Castleman, Jr., and R.G. Keesee, Chem. Rev. 86:589 (1986)

AN EFFECTIVE HAMILTONIAN STUDY OF MOLECULAR CLUSTERS

S.H. Suck Salk, C.K. Lutrus, and D.E. Hagen

Department of Physics and Graduate Center for Cloud
Physics Research
University of Missouri-Rolla
Rolla, MO

INTRODUCTION

Earlier we reported an improved[1,2] semi-empirical effective
Hamiltonian approach for studying molecular clusters. This method was
shown to be highly successful in removing the problems encountered[3,4] in
treating hydrogen-bonded water clusters by conventional semi-empirical
Hartree-Fock molecular orbital methods. The objective of the present
paper is to present 1) a concise description of the semi-empirical
effective Hamiltonian method in a generalized manner, and 2) results for
the intermolecular binding energies or stabilization energies of large
water clusters which have not been examined thus far.

CONCISE DESCRIPTION OF A SEMI-EMPIRICAL EFFECTIVE HAMILTONIAN

We present a concise description of the semi-empirical effective
Hamiltonian method that was developed earlier[1,2]. Formally we write the
Hamiltonian for the system of a molecular cluster,

$$H = H_o + H' , \tag{1}$$

where H_o is the Hartree-Fock one-electron Hamiltonian, and H', the
residual interaction. They are respectively,

$$H_o = \sum_p \sum_{\alpha_p}^{n_e} [h_{\alpha_p} + \bar{U}_{\alpha_p}] , \tag{2}$$

and,

$$H' = \sum_{p,q} \sum_{\alpha_p < \beta_q}^{n_e} V_{\alpha_p \beta_q} - \sum_p \sum_{\alpha_p} \bar{U}_{\alpha_p} , \tag{3}$$

with

$$h_{\alpha_p} = -\frac{1}{2} \nabla^2_{\alpha_p} - \sum_q \sum_{a_q}^{n_a} Z_{a_q} \; |\vec{r}_{\alpha_p} - \vec{R}_{a_q}|^{-1} \; , \tag{4}$$

and

$$V_{\alpha_p \beta_q} = |\vec{r}_{\alpha_p} - \vec{r}_{\beta_q}|^{-1} \; , \tag{5}$$

where p and q in (2) through (5) are indices which locate molecules in a molecular cluster; α and β, the electron indices; n_e, the total number of electrons in the molecule p, $\vec{r}_{\alpha_p}$, the position vector of the electron α in the molecule p; Z_a, the nuclear charge number of the atom a; $\vec{R}_{\alpha_q}$, the position vector of the atom a in the molecule q and n_a, the total number of atoms in molecule p. $\alpha < \beta$ denotes that interaction between any two electrons is counted only once. $\bar{U}_\alpha$ is the Hartree-Fock potential representing the averaged interaction felt by electron α in the field of the remaining electrons. We rewrite (3) in the form of the pair-wise residual interaction, $U_{\alpha\beta}$,

$$H' = \sum_{p,q} \sum_{\alpha_p < \beta_q} U_{\alpha_p \beta_q} \; , \tag{6}$$

where

$$U_{\alpha_p \beta_q} = V_{\alpha_p \beta_q} - [\bar{U}_{\alpha_p} + \bar{U}_{\beta_q}]/(M-1) \; , \tag{7}$$

with M, the total number of electrons in the molecular cluster.

The i-th 'molecular cluster orbital' is written in terms of the LCAO (linear combination of atomic orbitals) expansion,

$$\psi_i(\vec{r}) = \sum_p \sum_{\mu_p} C_{\mu_p i} \; \phi_{\mu_p} (\vec{r} - \vec{R}_{a_p}) \; , \tag{8}$$

where ϕ_{μ_p} is the μ-th atomic orbital and $C_{\mu_p i}$, the expansion coefficient.

A set of infinite orthonormal electronic configuration states can be constructed using the cluster orbitals defined above. Considering this orthonormal set, we write the projection operator,

$$P = |0\rangle\langle 0| \; , \tag{9}$$

which projects onto the vacuum state $|0\rangle$, and

$$Q = \sum |k\rangle\langle k| \; , \tag{10}$$

onto the remaining excited configuration states $|k\rangle$. They satisfy

$$P + Q = 1 \; . \tag{11}$$

Thus the total state is formally

$$|\psi\rangle = (P + Q)\,|\psi\rangle = \langle 0|\psi\rangle\,|0\rangle + \Sigma\,\langle k|\psi\rangle\,|k\rangle \, , \tag{12}$$

which satisfies the Schrodinger equation,

$$H\,|\psi\rangle = E\,|\psi\rangle \, . \tag{13}$$

The introduction of (9) through (12) into (13) leads to

$$\mathcal{H}_{pp}\,|0\rangle = E\,|0\rangle \, , \tag{14}$$

where

$$\mathcal{H}_{pp} = P\,(H + H\,Q\,(E - QHQ)^{-1}\,Q\,H)\,P \, . \tag{15}$$

Noting that $PHQ = PH'Q$ and $QHP = QH'P$, we rewrite (15) above,

$$\mathcal{H}_{pp} = P\,\mathcal{H}\,P \, , \tag{16}$$

where

$$\mathcal{H} = H + \mathcal{H}' \, , \tag{17}$$

with

$$\mathcal{H}' = H'\,Q(E - QHQ)^{-1}\,Q\,H' \, , \tag{18}$$

or

$$\mathcal{H}' = H'\,Q\,(E - QH_oQ)^{-1}\,Q\,H' + H'\,Q\,(E - QH_oQ)^{-1}\,H'\,x$$
$$x\,(E - QH_oQ)^{-1}\,Q\,H' + \ldots \tag{19}$$

Using (16) and (17) for

$$E = \langle 0|\,\mathcal{H}_{pp}\,|0\rangle \, , \tag{20}$$

we now readily note that

$$E = E_o + \langle 0|\mathcal{H}'|0\rangle \, , \tag{21}$$

where the first term is identified as the H–F (Hartree–Fock) energy, $E_o = \langle 0|H|0\rangle$, and the second term, the correlation energy.

Now introducing (6) and (19), we write

$$\mathcal{H}' = \underset{p,q}{\Sigma}\ \underset{\alpha_p < \beta_q}{\Sigma}\ W_{\alpha_p \beta_q} \, , \tag{22}$$

where

$$W_{\alpha_p \beta_q} = U_{\alpha_p \beta_q} (E-QH_oQ)^{-1} \sum_{r,s} \sum_{\gamma_r \delta_s} U_{\gamma_r \delta_s} +$$

$$+ U_{\alpha_p \beta_q} (E-QH_oQ)^{-1} \sum_{t,u} \sum_{\lambda_t \mu_u} U_{\lambda_t \mu_u} (E-QH_oQ)^{-1} \sum_{r,s} \sum_{\gamma_r \delta_s} U_{\gamma_r \delta_s} +$$

$$+ \ldots \tag{23}$$

Here r, s, t, and u denote molecule indices. Obviously the direct evaluation of (23) is not readily feasible. Integrals involving this term will be subject to semi-empirical evaluation.

Using (1) and (22) above, we rewrite (17),

$$\mathcal{H} = \sum_p \sum_{\alpha_p} h_{\alpha_p} + \sum_{p,q} \sum_{\alpha_p < \beta_q} (V_{\alpha_p \beta_q} + W_{\alpha_p \beta_q}) \ . \tag{24}$$

We apply the variational procedure,

$$\delta E = \delta \langle 0| \mathcal{H}_{pp} | 0 \rangle$$

$$= \delta \langle 0| \sum_p \sum_{\alpha_p} h_{\alpha_p} + \sum_{p,q} \sum_{\alpha_p < \beta_q} V'_{\alpha_p \beta_q} |0\rangle = 0 \ , \tag{25}$$

where

$$V'_{\alpha_p \beta_q} = V_{\alpha_p \beta_q} + W_{\alpha_p \beta_q} \ . \tag{26}$$

This leads to

$$\mathcal{F} \psi_i = \varepsilon_i \psi_i \ , \tag{27}$$

where $\mathcal{F}$ is the effective one-electron Hamiltonian,

$$\mathcal{F} = h + \sum_i [2 J_i - K_i] \ , \tag{28}$$

with

$$J_i = \int \psi_i(\vec{r}_\beta) V'_{\alpha\beta} \psi_i(\vec{r}_\beta) d\vec{r}_\beta \ , \tag{29}$$

and

$$K_i \psi_j(\vec{r}_\alpha) = \int [\psi_i(\vec{r}_\beta) V'_{\alpha\beta} \psi_j(\vec{r}_\beta) d\vec{r}_\beta] \psi_i(\vec{r}_\alpha) \ . \tag{30}$$

ε_i in (27) is the orbital energy corresponding to the molecular cluster orbital ψ_i. For simplicity, the molecule indices p and q are deleted.

Introducing the LCAO representation (8) of ψ_i, we now write the matrix elements of $\mathcal{F}$, (28) between the atomic orbitals, μ_a belonging to the atom a and ν_b belonging to the atom b,

$$\mathcal{F}_{\mu_a \nu_b} = \langle \mu_a |h| \nu_b \rangle + \sum_{c,d} \sum_{\lambda_c \sigma_d} D_{\lambda_c \sigma_d} (\Gamma_{\mu_a \lambda_c}^{\nu_b \sigma_d} - \frac{1}{2}\Gamma_{\mu_a \lambda_c}^{\sigma_b \nu_d}) \ , \tag{31}$$

where

$$D_{\lambda_c \sigma_d} = 2 \sum_j C^*_{\lambda_c j} C_{\sigma_d j} \, , \tag{32}$$

$$\langle \mu_a | h | \nu_b \rangle = \int \phi^*_{\mu_a}(\vec{r}_\alpha) \, h(\vec{r}_\alpha) \, \phi_{\nu_b}(\vec{r}_\alpha) \, d\vec{r}_\alpha \, , \tag{33}$$

and

$$\Gamma^{\nu_b \sigma_d}_{\mu_a \lambda_c} = \int \phi^*_{\mu_a}(\vec{r}_\alpha) \, \phi^*_{\lambda_c}(\vec{r}_\beta) \, V'_{\alpha\beta} \, \phi_{\nu_b}(\vec{r}_\alpha) \, \phi_{\sigma_d}(\vec{r}_\beta) \, d\vec{r}_\alpha \, d\vec{r}_\beta \, . \tag{34}$$

If the multi-center integrals Γ (34) involving $W_{\alpha\beta}$ and thus $V'_{\alpha\beta}$ are determined by a parameterization procedure, the expression (31) simply represents a semi-empirical effective Hamiltonian matrix which incorporates the effects of electron correlation.

PREDICTED INTERMOLECULAR BINDING ENERGIES AND SUMMARY

In order to make a semi-empirical evaluation of the effective Hamiltonian matrix (31), we choose the NDDO (neglect of diatomic-differential overlap) approximation

$$\Gamma^{\nu_b \sigma_d}_{\mu_a \lambda_c} = \delta_{ab} \, \delta_{cd} \, \Gamma^{\nu_a \sigma_c}_{\mu_a \lambda_c} \, . \tag{35}$$

In our approach, the two center-integrals Γ above are then divided into two sets of parametric treatments; one for the intramolecular interaction and the other for the intermolecular interaction[1].

The selected geometric configurations of water clusters are taken from parts of the hexagonal ice (ice I) lattice structure[6]. Geometry optimizations for these geometric structures are made by using the semi-empirical effective Hamiltonian method of I-MNDO[1]. In Fig. 1, the predicted binding energies per bond, B_n, are displayed as a function of cluster size n where n is the number of water molecules in the cluster. It is seen from Fig. 1 that the intermolecular binding energies do not show marked differences among the different geometric configurations of the same cluster size n. The difference is generally within ~ 1 Kcal/mol. It is of note that a similar trend was realized by Clementi[7] who examined various stuctures of water clusters using molecular dynamics calculations based on an analytic potential fit to his ab initio potential surface of the water dimer. The solid curve in Fig. 1 represents an analytic fit to the predicted values using the relation[2], $B_n = B_\infty (1 - 1/n^\alpha)$, where $B_\infty = -12.7$ and $\alpha = .531$. It is interesting to note that $B_\infty = -12.7$ Kcal/mol is remarkably close to the bulk value (-13.4 Kcal/mol) of ice[8].

In summary we have found that 1) the intermolecular binding energies

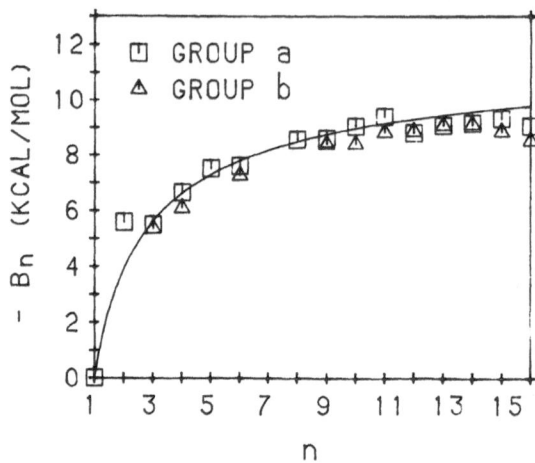

Fig. 1   Binding energy (Kcal/mol) per
bond vs. cluster size n.

of water clusters are not too sensitive to the variation of geometric
structures (see refs. 6 and 9 for the details of geometric structures)
for a given cluster size n, showing the greater the cluster size, the
less the difference, and 2) the binding energies tend to correctly
converge to the bulk limit at a relatively slow rate.

Our present study of binding energies at T = 0°K was to explore only
the intrinsic stability of water clusters, since entropy effects at
finite temperatures are not included.

ACKNOWLEDGEMENT

This material is partially supported by NSF grant ATM82-12328.

REFERENCES

1.   S.H. Suck Salk, T.S. Chen, D.E. Hagen, and C.K. Lutrus, Theor. Chim.
     Acta 70, 3 (1986).
2.   S.H. Suck Salk, and C.K. Lutrus, Phys. Rev. A, submitted.
3.   W. Thiel, Theor. Chim. Acta 48, 357 (1978).
4.   T.J. Zielenski, D.L. Breen, annd R. Rein, JACS 100, 6266 (1978).
5.   J.A. Pople, and D.L. Berveridge, 'Approximate Molecular Orbital
     Theory' (McGraw Hill, New York, 1970).
6.   S. Tomoda, and K. Kimura, Chem. Phys. Lett. 102, 560 (1983).
7.   E. Clementi, 'Lecture Notes in Chemistry' (Springer, New York, 1976).
8.   D. Eisenberg, and W. Kauzmann, 'The Structure and Properties of
     Water' (Oxford Univ. Press, New York, 1969).
9.   D.E. Hagen, L. Jin, M.S. Choe, C.K. Lutrus, T. Oshiro, and S.H. Suck
     Salk, in this issue.

KINETICS OF CLUSTER FORMATION DURING RAPID QUENCHING[a]

S.-N. Yang and T.-M. Lu

Center for Integrated Electronics and Physics Department
Rensselaer Polytechnic Institute, Troy, N.Y. 12181

Abstract

A numerical scheme has been developed to evaluate quantitatively the non-steady state kinetics of cluster generation during the rapid quench of a vapor system. Our method is particularly suitable to deal with the supersonic jet expansion of vapors. Based on this calculation scheme, we have carried out a systematic study on metal cluster formation in nozzle jet beams. We have also introduced the icosahedral structure into our calculation for the cluster size distribution of noble gas elements. Most of the "magic numbers" found experimentally appeared in our calculated spectrum.

I. Introduction

The thermodynamics dealing with the equilibrium properties of phase separation has been rather well developed. The transient process, i.e. the initialization of phase transitions is normally more complicated but of great interest. The subject is intrinsically related to the cluster generation and the small particle physics. The classical theory dealing with the kinetics of this transient process is the liquid-drop nucleation theory [1,2]. Despite its success [3], the theory is still essentially a qualitative treatment and has the following main limitations: 1.The theory is not able to predict cluster size distributions. 2.The theory can not be applied to the cases of very rapid quench (such as jet expansion [4]) where the required quasi-steady state is not established. 3.The theory cannot describe a system with constantly changing temperature and pressure.

In this paper, we would like to report our study on the kinetics of cluster formation and cluster size distributions in the initial stage of condensation based on a numerical scheme we developed.

II. Mathematical Modeling

The mathematical model of the step-wise clustering process involves N coupled differential equations which are described by [1,5]

$$\frac{\Delta f_i(t)}{\Delta t} = C_{i-1}f_{i-1}(t) - E_i f_i(t) - C_i f_i(t) + E_{i+1}f_{i+1}(t), \qquad i = 2,3,\ldots\ldots N \tag{1}$$

where $f_i(t)$ is the instantaneous concentration of size i clusters (i denotes the number of molecules constituting the cluster). The $\Delta f_i(t)$ is the change of $f_i(t)$ during a time interval Δt. The total number of differential equations N is determined by the cluster size range under consideration. The terms on the right-hand side of Eq.(1) represent the following reactions respectively:

1st term: the clusters of size (i-1) catch a monomer and join the group of size i clusters.

2nd term: the size i clusters evaporate a monomer and leave the size i group (the reverse process of the first one).

3rd term: the clusters of size i catch a monomer and become (i+1) clusters.

4th term: the reverse process of the third term.

The first and fourth terms give positive contributions to $f_i(t)$, whereas the second and third terms give negative contributions. C_i and E_i are the condensation rate of monomers on a size i cluster and the monomer re-evaporation rate from a size i cluster respectively. It can be shown, from the gas kinetic theory and the principle of detailed balance [1,5], that

$$C_i = \frac{P \bullet S_i}{(2\pi m k T)^{\frac{1}{2}}}, \tag{2}$$

$$E_i = C_{i-1} \bullet \exp[\frac{\Delta G_{i-1} - \Delta G_i}{kT}], \tag{3}$$

where P is the local pressure, m is the mass of the molecule, k is the Boltzmann constant, T is the temperature, S_i is the effective surface area of the cluster and is approximated by assuming a spherical structure [6], and ΔG_i is the amount of reversible work needed in forming a size i cluster. In the classical treatment [1], the capillarity approximation is commonly adopted and ΔG_i is taken to be:

$$\Delta G_i = -i \bullet k T \bullet \ln[P/P_v(T)] + S_i \bullet \sigma, \tag{4}$$

where $P_v(T)$ is the saturation vapor pressure at temperature T and σ is the bulk surface tension. The first term on the right-hand-side of Eq.(4) can be regarded as the reduction of chemical potential caused by the phase condensation. The second term is the contribution of the surface energy created during the cluster formation.

Based on the above equations, a dynamical simulation of cluster growth may be carried out by numerically integrating Eq.(1) with respect to time t to determine $f_i(t)$ step by step, starting from the initial state, provided that the quench process in known, i.e. T and P of the vapor system are known functions of t. The initial cluster size distribution can be assumed to have either 100% monomers or a Boltzmann distribution [1]. (The effect of these different assumptions on the final cluster size distribution has been found to be insignificant as long as the monomer population dominates the initial (t=0) distribution.)

For the case of isentropic nozzle jet expansion [7], the local temperature and pressure of the vapor during the jet flow can be expressed as functions of the Mach number M (which is defined to be the ratio of the flow velocity V to the sound velocity V_s) as follow:

$$T/T_0 = (1 + \frac{r-1}{2} M^1)^{-1}, \tag{5}$$

$$P/P_0 = (1 + \frac{r-1}{2} M^1)^{-r/(r-1)}, \tag{6}$$

where r is the specific heat ratio, and T_0 and P_0 are the stagnation temperature and pressure respectively.

It is clear that, as the vapor ejects through the nozzle, the flow velocity V increases, and the random thermal energy of the vapor is transformed into the kinetic energy carried by the beam. As a result, the ejected vapor cools itself down. The sound velocity V_s will drop since it is proportional to the square root of T. The increase of V and decrease of V_s cause the Mach number M to drop monotonically as the vapor expands downstream. The Mach number M(x) as a function of the distance x downstream from the nozzle throat can be determined through its relationship with the beam cross sectional area [7], if the boundary layer effect [8] is neglected.

It is noticed from Eqs.(5) and (6) that, not only T, but also P will decrease during the nozzle jet expansion. However, the saturation vapor pressure P_v drops exponentially with temperature T, which is much faster than the decrease of P [9]. Thus, supersaturation condition is eventually reached when P_v becomes smaller than P, and cluster aggregation will be initiated [4,8].

With Eqs.(5),(6) and the form of M(x), T and P at any position along the flow stream can be determined, and so is the supersaturation ratio P/P_v (since P_v is a function of T only). These are all the information concerning the quench process needed for a numerical simulation.

The main principle of our numerical simulation is to break up the time into many short intervals Δt and to approximate $f_i(t+\Delta t)$ by $f_i(t)+[df_i(t)/dt]\bullet\Delta t$, where $df_i(t)/dt$ is evaluated using Eqs.(1) to (4). We assumed a Boltzmann distribution for $f_i(0)$ as the initial condition at $t=0$. Starting from $t=0$, the computer is programmed to execute the following iteration cycle:

1) Choose an appropriate Δt for the time interval suggested by the collision frequency between monomers and clusters.
2) Determine the increment rate of the cluster concentration from the equations and update the cluster size distribution.
3) Plot out the $f_i(t)$ function for every 50 iterations. (The number can be reset according to each individual case).
4) Reduce the vapor density according to the adiabatic expansion and the monomer consumption (for forming clusters) during Δt and then change the temperature, taking into account both the adiabatic cooling and the released latent heat. (Latent heat is assumed to distribute uniformly in the vapor.)
5) If the vapor has expanded into a molecular flow, the final cluster size distribution is printed; otherwise, go back to step 1.

III. Metal Cluster Size Distribution In Nozzle Beams

We first applied our numerical scheme to investigate the metal cluster generation kinetics in nozzle jet beams [10]. In Fig.1, the obtained local P-T variation history (dashed

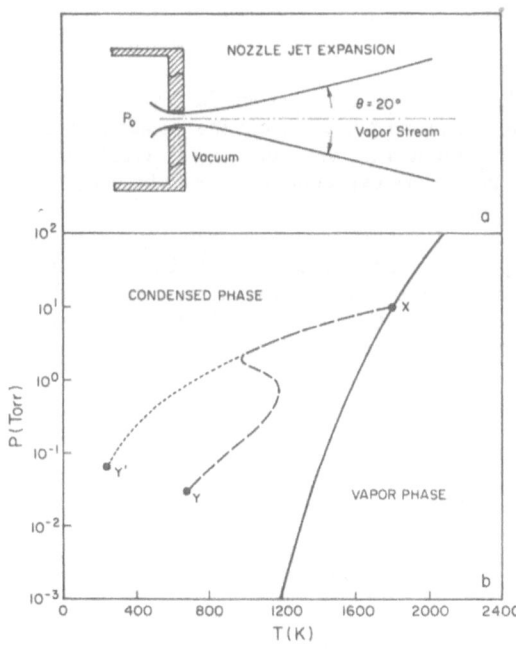

Figure 1: a) A schematic drawing of the nozzle expansion geometry. b) The calculated P vs. T curve (dashed line) for Ag vapor as the expansion proceeds from the crucible nozzle (D=0.2 cm). Should there be no condensation, the expansion would follow the dotted curve (pure adiabatic expansion). The solid curve is the vapor-liquid phase boundary.

curve) for the jet expansion of Ag vapor (through a nozzle with diameter D=0.2cm) has been plotted, together with the Ag P-T phase diagram. The expansion starts at the phase boundary with P_0 taken to be 10 Torr $[=P_v(T=1810K)]$, and the vapor system immediately enters the condensed phase region and becomes meta-stable. After the onset of nucleation, the release of latent heat and the reduction of monomer concentration due to cluster generation cause the P-T curve to deviate from the pure adiabatic expansion route (indicated by the dotted line), and the vapor system is driven back towards the phase boundary. Eventually, the system leaves the deep supersaturation region and the already formed nuclei will grow without further nucleation. This feature is exactly what has been qualitatively predicted many years ago [11]. We found from this study that it is necessary to have this feature appearing in the P-T curve in order to generate a cluster peak (as those shown in Fig.2) in the cluster size distribution spectrum. The cluster growth stops at point Y. At this point, the beam has expanded into molecular flow.

The final mass concentration vs. cluster size i curves are shown in Fig.2 as a function of the stagnation pressure P_0. (Mass concentration is defined to be the cluster concentration multiplied by the cluster size i). It is seen that the mean cluster size increases with the crucible pressure. As the clusters grow larger, the distribution peak broadens while the peak height diminishes. This phenonenon has been confirmed by a recent jet expansion experiment with CO_1 [12]. For sufficiently high P_0, say $P_0 > 10$ Torr, the area under the distribution curve which is less affected by P_0, corresponds to about 25% of the total mass flux. The variation in the expansion angle θ (as defined in Fig.1) has also been found to have a very effective control on the mean cluster size [13].

It is expected that the mean cluster size i_m should be correlated with the terminal Mach number M_t [14]. Because the higher the terminal Mach number, the more the number of collisions in the beam before it expands into the molecular flow region, and, therefore, the larger the mean cluster size. With a minor modification from an earlier theory [14], it can be shown that the terminal Mach number M_t is given by

$$M_t = C \bullet [2 \bullet tg(\theta/2) \bullet K_n]^{-(r-1)/r}, \tag{7}$$

where r is the ratio of the specific heats c_p/c_v, K_n is the Knudsen number defined to be the ratio λ/D with λ ($\propto T_0/P_0$) being the initial mean free path in the vapor and D the nozzle diameter, θ is the expansion angle, and $C \simeq 1.4$ for monotonic gases. In Fig.3, the mean cluster size i_m is plotted as a function of M_t for the jet expansion of saturated Ag vapor. It is seen that i_m increases exponentially with M_t in the range of practical interest. Similar results were obtained from the calculations with Al and Mg vapors.

Recently, conflicting experimental results have been reported on the formation of metal clusters with the mean size greater than 100 atoms/cluster using the jet expansion technique [15-17]. In our own laboratory, we found it was very difficult to create an expansion condition with $P_0 > 10$ Torr and $\theta > 20°$. More experiments have to be done in order to resolve this controversy.

It should be noted that the cluster temperature has been assumed to be the same as the surrounding vapor temperature in our model. However, for the jet expansion with pure vapor (i.e. without carrier gas), the cluster temperature should be higher than the vapor temperature during the growth because of the released binding energy. The energy conservation requirement can be approximately stated as

$$(C_{i-1} \bullet E_i) \bullet u_i = (3/2)k \bullet (T_i - T) \bullet E_i, \tag{8}$$

where C_i and E_i are as previously defined, u_i is the binding energy of an atom in the cluster (which should be a function of cluster size i), and T_i is the average temperature of a size i. Since u_i is usually much greater than $k \bullet (T_i - T)$, T_i can be approximated by setting $C_{i-1}/E_i \simeq 1$, and substituting C_{i-1} and E_i using Eqs.2,3 and 4. This treatment is similar to the one developed by P.G. Hill and co-workers [5]. After replacing T by T_i in the expression for E_i we re-evaluated the metal cluster size distributions. To our great interest, we found that the distribution became log-normal as shown in Fig.4 instead of

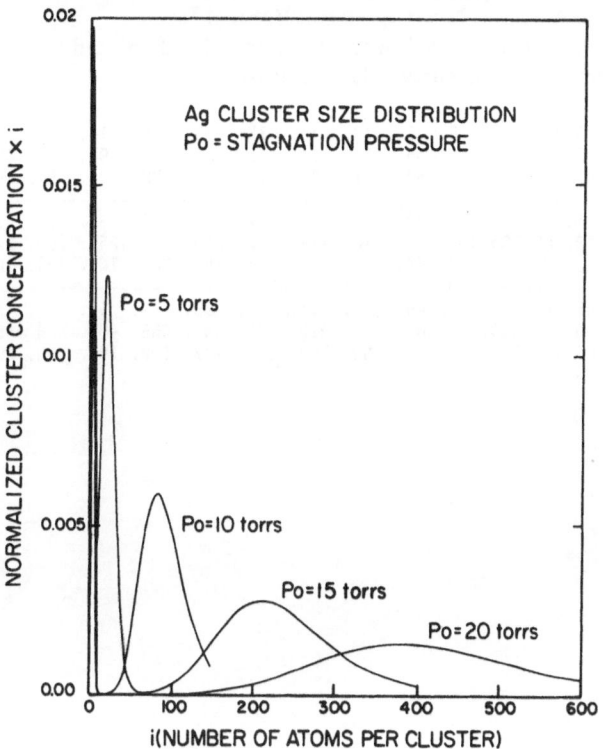

Figure 2:
The calculated Ag cluster size distributions in the nozzle beam (D=0.2 cm) as a function of the stagnation pressure P_0, assuming 20° expansion angle.

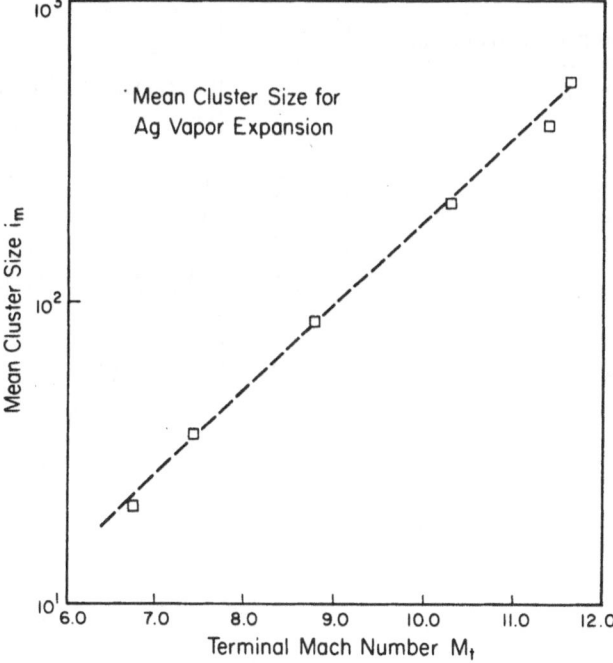

Figure 3:
The calculated average cluster size as a function of the terminal Mach number M_t which contains all the important expansion parameters.

Table 1

The total number of coupled nearest neighbor bonds n_i and the number of broken bonds b_i on the surface for small Ar clusters.

i	1.	2.	3.	4.	5.	6.	7.	8.	9.	10.	11.	12.	13.	14.	15.	16.	17.	18.	19.	20.
ni	0.	2.	6.	12.	18.	24.	32.	38.	46.	54.	62.	72.	84.	90.	96.	102.	108.	114.	126.	132.
bi	12.	22.	30.	36.	42.	48.	52.	58.	62.	66.	70.	72.	72.	78.	84.	90.	96.	102.	102.	108.

i	21.	22.	23.	24.	25.	26.	27.	28.	29.	30.	31.	32.	33.	34.	35.	36.	37.	38.	39.	40.
ni	138.	144.	156.	162.	168.	180.	186.	192.	204.	210.	216.	228.	234.	246.	256.	266.	277.	288.	299.	310.
bi	114.	120.	120.	126.	132.	132.	138.	144.	144.	150.	156.	156.	162.	162.	164.	166.	167.	168.	169.	170.

i	41.	42.	43.	44.	45.	46.	47.	48.	49.	60.	51.	52.	53.	54.	55.	56.	57.	58.	59.	60.
ni	320.	332.	342.	350.	360.	372.	380.	390.	402.	410.	420.	432.	444.	456.	468.	474.	480.	486.	492.	498.
bi	172.	172.	174.	178.	180.	180.	184.	186.	186.	190.	192.	192.	192.	192.	192.	198.	204.	210.	216.	222.

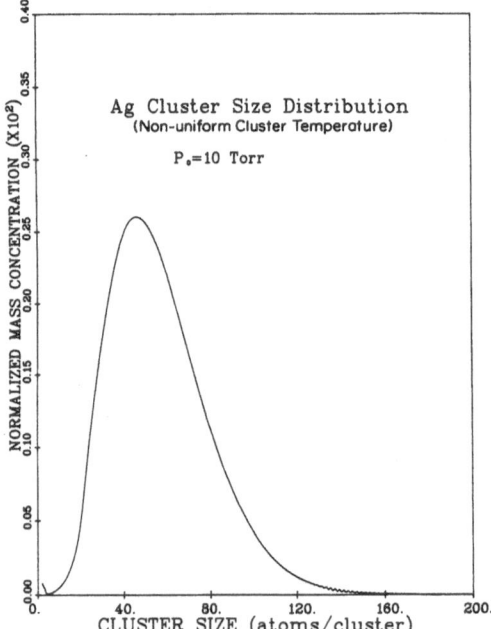

Figure 4: The calculated Ag cluster size distribution after taking into account the temperature difference between the cluster and the vapor. (Laval nozzle, $D = 0.2$ cm, $M_f = 2.5$)

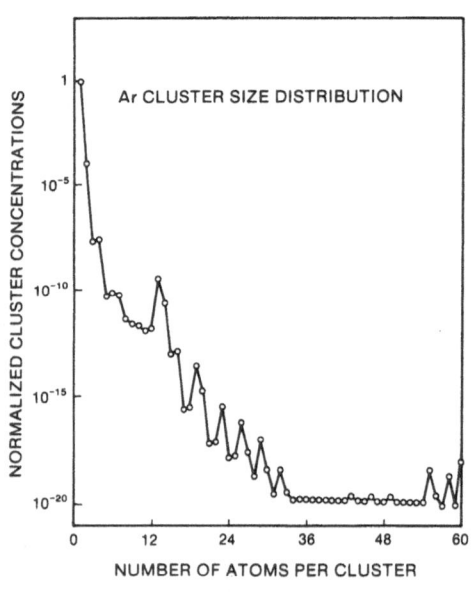

Figure 5: The calculated Ar cluster size distribution generated in a low supersaturated vapor (for the laval nozzle with $M_f = 1.5$).

Gaussian (as in the equal-temperature cases). The log-normal distribution shape is commonly observed in experiments [18,19].

IV. Noble Gas Elements: the "Magic Numbers"

With a proper modification by taking into account the finite size effect for the surface tension of small clusters [20,21], it may be reasonable to a certain extend if one uses the classical ΔG_i expression (i.e. Eq.4) for metal elements. This is because metal clusters produced in nozzle beams are most probably in the liquid state [22]. For noble gas elements, on the other hand, it is found that small clusters are of solid structure [23]. The atomic arrangement in those clusters follows the icosahedral packing [23,24]. Obviously, ΔG_i will no longer be a smooth function of i as Eq.4. It has been shown that the bond-breaking model can be used to describe solid Van de Waals clusters as a first order treatment [25]. Assuming each atom has 12 Van de Waals bonds (a simple geometrical consideration), we have calculated the number of coupled bonds n_i in a size i cluster and the number of broken bonds b_i on the cluster surface with i ranging from 1 to 60 according to the icosahedral packing. The results are listed in table 1. In order to introduce these cluster structure information into the ΔG_i function, we have made a first order correction to the classical theory by modifying Eq.(4) to:

$$\Delta G_i = -i \bullet kT \bullet \ln[P/P_v(T)] + e \bullet b_i, \qquad (9)$$

where e denotes the binding energy of a single Van de Waals bond. The capillary surface energy has been replaced by $e \bullet b_i$. Fig.5 shows our calculated result for the Ar cluster size distribution produced by the jet expansion with a laval type nozzle and under the initial condition equivalent to the one used in Harris' experiment [25]. As should be expected, the distribution is no longer a continuous curve [26]. The local maxima appeared in our calculated spectrum are found to correspond to the "magic numbers" observed in the experiment [25].

V. Summary

A computer dynamical simulation scheme has been developed for investigating the non-steady state nucleation and growth process during the rapid quench of a vapor system. The application of this scheme to the case of metal cluster generation in nozzle beams showed the following characteristics: 1.If both nucleation and growth are operating during the entire cluster formation process, the produced cluster size distribution will be a monotonically decreasing function with increasing size. 2.If further nucleation is prohibited in the later stage of the cluster growth due to the reduction of the supersaturation ratio, a second peak (i.e. beside the one at the monomer concentration) will appear in the cluster size distribution. The size corresponding to the center position of the cluster peak increases exponentially with the terminal Mach number of a jet expansion. 3.The shape of the cluster peak is close to Gaussian if the cluster temperature is assumed to be equal to the surrounding vapor temperature during the growth (resembling the case of cluster growth in a carrier gas medium). For pure vapor expansion where cluster temperature is higher than the temperature of the vapor phase, a log-normal cluster peak shape is observed.

We also introduced the icosahedral structure information into our calculation for Ar cluster size distribution. The result predicted the right "magic numbers" observed in experiments.

[a]This work is supported in part by the Semiconductor Research Corporation.

References

1. See, for example: J. Frenkel, *Kinetic Theory of Liquids*, Clarendon Press, Oxford (1946).

2. S.-N. Yang and T.-M. Lu, J. Appl. Phys. **58**, 541 (1985).

3. M. Volmer and H. Flood, Z. Physik Chemie. **A170**, 273 (1934).

4. For a review, see P.P. Wegerer and B.J.C. Wu, in *Nucleation Phenomena*, Adv. in *Colloid and Interface Sci.*, Vol. **7**, ed. A.C. Zettlemoyer, Elsevier, Amsterdam (1977), p.325.

5. P.G. Hill, H. Witting, and E.P. Demetri, J. Heat Transfer **85**, 303 (1963).

6. R.P. Andres, in *Nucleation Phenomena*, A.C.S. Publication, Washington, D.C. (1966), p.16.

7. E.M. Leonard, *Fundamentals of Thermodynamics*, Prentice Hall Inc., Englewood Cliffs (1957-1958), p.252.

8. S.B. Ryali and J.B. Fenn, Ber. Bunsenges. Phys. Chem. **88**, 245 (1984).

9. S.-N. Yang and T.-M. Lu, Chem. Phys. Lett. **127**, 512 (1986).

10. S.-N. Yang and T.-M. Lu, Appl. Phys. Lett. **48**, 1122 (1986).

11. J. Farges, J. Crystal Growth **31**, 79 (1975).

12. J.M. Soler, N. Garcia, O. Echt, K. Sattler, and E. Recknagel, Phys. Rew. Lett. **49**, 1857 (1982).

13. S.-N. Yang and T.-M. Lu, J. Vac. Sci. & Tech. **B**, Feb. 1987, (to be published).

14. J.B. Anderson and J.B. Fenn, Phys. of Fluids **8** 780 (1965).

15. I. Yamada and T. Takagi, Thin Solid Films **80**, 105 (1981).

16. A. Kuiper, G. Thomas, and W. Schouten, J. Cryst. Growth **51**, 17 (1981).

17. J.B. Theeten, R. Madar, A. Mircea-Roussel, A. Rocher, and G. Laurence, J. Cryst. Growth **37**, 317 (1977).

18. J Gspann, in *Physics of Electronic and Atomic Collisions*, ed. S. Datz, North-Holland (1982), p.79.

19. C.G. Granqvist and R.A. Buhrman, J. Appl. Phys. **47**, 2200 (1976).

20. J.W. Gibbs, *The Collected Works of J. Willard Gibbs, Vol. 1, Thermodynamics*, Longmans and Green, London (1932), p.232.

21. R.C. Tolman, J. Chem. Phys. **17**, 333 (1949).

22. J. Gspann, in *Proc. Int'l Workshop on ICBT (Ionized Cluster Beam Technology)*, Tokyo-Kyoto (1986). p.109.

23. O. Echt, K. Sattler, and E. Rechnagel, Phys. Rew. Lett. **47**, 1121 (1981).

24. A.L. Mackay, Acta Crystallogr. **15**, 916 (1962).

25. I.A. Harris, R.S. Kidwell, and J.A. Northby, Phys. Rew. Lett. **53**, 2390 (1984).

26. For details, see: S.-N. Yang and T.-M. Lu, submitted to Phys. Rev. B.

CLUSTER IMAGING

Klaus Sattler

Department of Physics
University of California, Berkeley
Berkeley, California 94720, USA

ABSTRACT

Cluster beam analysis by time of flight mass spectrometry and simultaneous deposition on a support with subsequent imaging by a transmission electron microscope is described in the first part of the paper. 10-15 Å CsCl clusters, deposited onto glass substrates at room temperature, grow to small particles with thousands of angstroms in size. These results leaded to the use of an instrument with imaging capabilities on a much smaller size scale. The scanning tunneling microscope as a new tool for cluster research could be established. Using this instrument, gold and silver clusters on graphite have been imaged with atomic resolution. Diffusion and growth of individual clusters has been observed. Possible influences of the microscope on the deposited clusters are discussed.

INTRODUCTION

Numerous studies of <u>small particles</u> (20 to 1000 Angstroms in size), supported on solid substrates, using electron microscopy have been reported [1]. Such studies have been done both for fundamental research [2,3] and for applications [4].

In recent years <u>small clusters</u> (containing very few atoms) have been intensively studied in beams and promising results concerning basic science and technology have been gained [5].

Future applications of these investigations require the deposition of the clusters onto solid substrates after their physical and chemical properties have been investigated in beams. These properties however may be considerably changed after deposition because the original sizes and shapes may not be preserved due to diffusion and growth and because geometric and electronic proximity effects due to strong couplings to the substrate may occur.

First we have to answer the question if the sizes of clusters after deposition are different from the sizes in the beam. The beam size distributions can be analysed by mass spectrometry, those on the substrate by electron microscopy. The first part of this paper deals with the deposition of CsCl clusters [6]. The experiments show that clusters with sizes of 10-15 Å in diameter have grown on the substrate to small particles with several thousands of Angstroms in diameter.

In the second part of the paper we take a further step searching for an instrument which can image on a much smaller size scale for direct investigations of nucleation and growth processes of small clusters. The scanning tunneling microscope (STM) for small particle and cluster imaging is described. For Ag and Au coverage, large islands and small clusters are imaged on graphite, with the constant current and the variable current modes, respectively, being applied [7]. Clusters containing a few atoms are imaged with atomic resolution, and diffusion and growth is observed on the atomic scale. Possible influences of the instrument during the imaging process will be discussed.

TEM IMAGING

In order to generate small clusters with narrow size distributions in the beam, an inert gas condensation source (described earlier [8]) has been used. With this method, the vapour effusing from a Knudsen cell is supersaturated in cold He-gas. Suitable growth conditions are choosen to get a cluster beam in the very small size range. The beam is transferred into a vacuum chamber and subsequently ionized by a pulsed electron gun. The charged clusters are extracted by an electrostatic lense system and their sizes are analysed by time of flight mass spectrometry. We show the results of experiments where beam analysis and deposition were performed simultaneously.

The covered glass substrate have been transferred in N_2 atmosphere to the electron microscope (Siemens ELMISCOP 1A). There, a standard technique of preparation has been used: a thin carbon film is deposited onto the cluster specimen, and subsequently the ionic material is solved in an acid or in water. Additionally, small angle shadowing with platinum is applied. These procedures do not change the original cluster sizes.

Figure 1 shows a mass spectrum of CsCl clusters (a) and the corresponding TEM picture (b). The temperature of the He gas in the condensation source was 280 K. Between the mass peaks of singly charged clusters the peaks from doubly charged clusters are displayed. The beam size distribution has its maximum at n≈14, the beam therefore mainly consists of clusters being $d \sim 10$ Å in diameter. The scale of the TEM picture is 10 000 Å, given at the bottom of the picture. Particles with sizes between 2000 Å and 8000 Å are shown.

Figure 2 shows a mass spectrum of CsCl clusters (a) and the corresponding TEM picture (b) with lower He temperature (70K) in the condensation source being applied. The maximum of the TOF spectrum is shifted to higher masses (n=45, $d \sim 15$ Å), but the sizes after deposition are again in the 1000 Å to 10 000 Å range.

The observations show that the original cluster sizes are not preserved after deposition. The small clusters of the beam have changed to particles, thousands of Angstroms in diameter. In addition it is found that different temperatures of the He gas in the source yield different shapes of the particles on the substrate. At 70 K the substrate probably is cooled by the He-gas/cluster mixture, which leads to spherical particles.

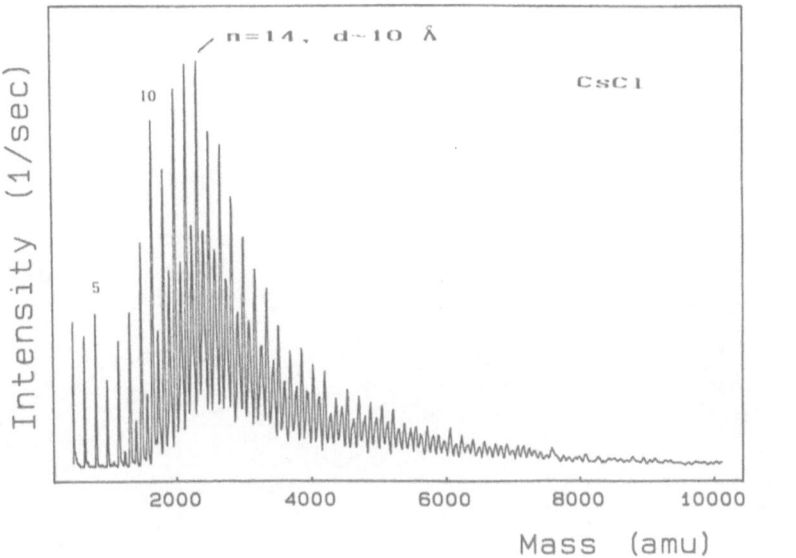

a

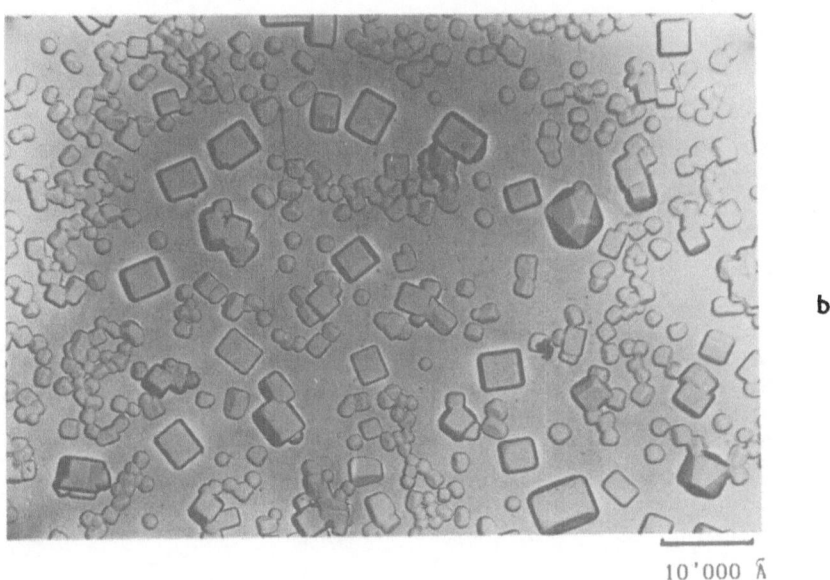

b

Fig. 1.   Deposition of CsCl-clusters onto a glass substrate with simultaneous analysis of the size distribution in the beam: (a) time of flight mass spectrum,   (b) TEM image; He temperature in the condensation cell T_{He}=280 K [6]

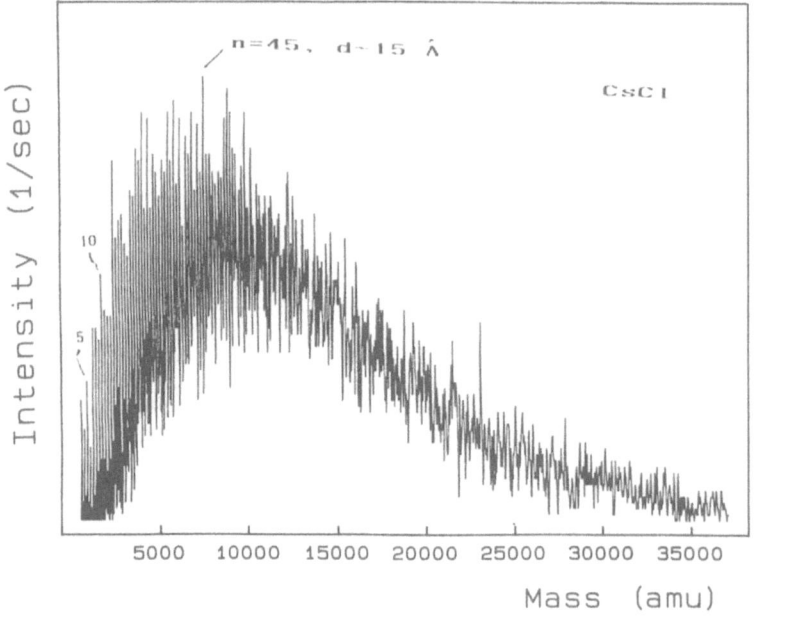

a

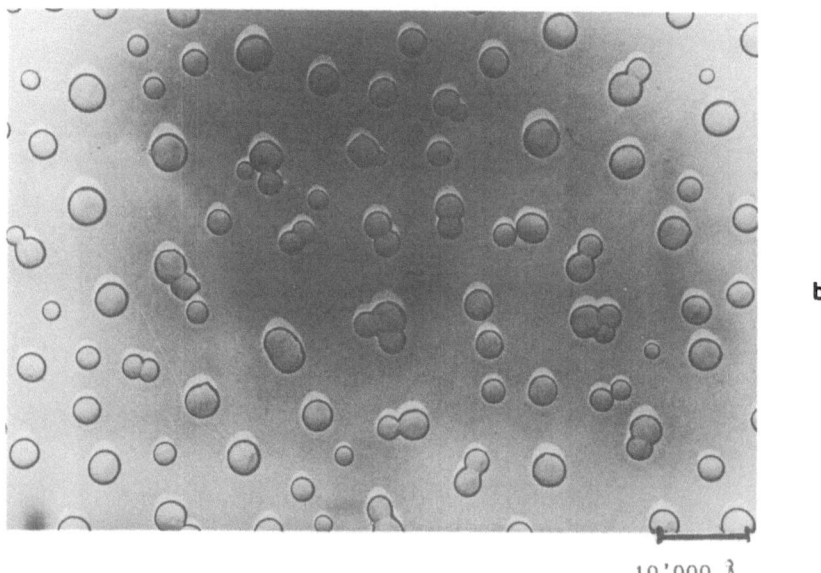

b

10'000 Å

Fig. 2. Deposition of CcCl-clusters onto a glass substrate with simultaneous analysis of the size distribution in the beam: (a) time of flight mass spectrum, (b) TEM image; He-temperature in the condensation cell T_{He}=70 K [6].

In this part of the paper the scanning tunneling microscope (STM) as a new tool for cluster research is discussed. As model systems, Au and Ag clusters are studied on graphite substrates using the STM in air.

In recent years it has been demonstrated that the STM [9] is a powerful new technique for investigations of surface structures. It provides images in real space, in three dimensions, with atomic resolution. It can be applied both for periodic and for non-periodic structures.

The STM has been used to measure the topographic and electronic properties of semiconductors [10] and evaporated metal films [11], and it has been applied to the study of surface roughness [12], superconductivity [13], and charge density waves [14]. The functioning of the STM is not restricted to vacuum conditions. It has been demonstrated that the instrument also operated in air, liquid nitrogen, water or paraffin oil [15].

The STM works by the quantum mechanical phenomenon of tunneling. A metal wire with a sharp tip is mechanically rastered across the surface of the sample. If the tip is close enough, a few Angstroms, the wave functions of the electronic quantum states of an atom at the end of the tip and of those in or ontop of the substrate slightly overlap. This overlap allows electrons to jump across the vacuum gap between the tip and the surface when a voltage is applied between them. Thereby a tunneling current is produced which increases exponentially as the gap distance decreases. This reflects the exponentially damped nature of the electron wave functions in the forbidden gap.

The tunneling current is, according to STM theory [16], directly proportional to the surface electronic charge density at the center of curvature of the tip. The charge density is normally concentrated where the atoms are. However, structures in the STM images may be associated with electronic features alone (charge density waves, or tunneling into bonds).

For gap distances of about 5 Å, an applied bias voltage of several tens of millivolts induces nanoampere current flows. The current typically declines by one order of magnitude for every Angstrom of increase in gap spacing. Therefore electrons tunnel into the outermost part of the spacial charge density distribution of the object to be imaged.

The STM works by two modes of operation. In the topographic mode (slow mode) the tip is maintained at a fixed distance above the surface during scanning by keeping a constant current with an electronic feedback circuit. The trajectory of the tip follows the profile of the surface and therefore three-dimensional (3D) clusters can be imaged. A picture with this mode is taken in a few minutes and therefore it can be applied only for the imaging of clusters with very small diffusion times.

In the variable current mode (fast mode) an adjusted mean distance between tip and surface is held constant and the modulation of the tunneling current is measured while scanning. This mode likely is restricted to two-dimensional (2D) clusters, as 3D-clusters can be destroyed by the tip passing by. In this case the tip can pick up atoms from the clusters. The favorable local atomic structure at the end of the tip therefore can be changed during the experiment and in consequence the contaminated tip has to be replaced by a clean one. The advantage of the variable current mode is that a picture is taken in a few seconds. Fast imaging is required for clusters which move on the substrate. Their jump frequencies have to be small compared to the detection time of the STM in order to get reproducible pictures.

In principle, every metal or semiconductor could be used as a substrate for clusters to be imaged. However, in order to see intrinsic cluster properties the substrate should be chemically inert (i.e. small cluster-substrate interaction) and perfectly flat. If defects are present within the substrate surface, adatoms and adclusters will be trapped at the defect sites and in consequence decorated defects will be imaged instead of the deposited clusters.

Graphite seems to be a suitable substrate material. Both, its electronic and atomic structures have intensively been studied. Localized σ and π states, as well as fairly delocalized interlayer states [17] and empty surface states [18,19] coexist within the range of a few eV around the Fermi energy. The atoms on the cleaved graphite surface (0001 plane) are located at the vertices of the hexagons of a two-dimensional honeycomb pattern with diameter of 2.4 Å. The six carbon atoms making up a ring are not equivalent. The difference stems from the fact that three of the carbon atoms have neighboring atoms in the layer immediately below, whereas the other three do not.

STM studies on bare graphite showed unexpected features ("giant" corrugation, lack of trigonal symmetry) [20,21]. Their interpretation is still a matter of controversy [22-25].

We now describe STM studies of Au and Ag clusters on the surface of highly oriented pyrolitic graphite [7]. The Ag clusters have been generated in Ar-gas before deposition. Conditions have been choosen where 30-100 Å particles are generated. The Au clusters have been formed on the substrate after vapour deposition of single atoms. STM images of hundreds of different areas have been taken which showed a wide variety of

Fig. 3. STM image of a 350 Å Ag particle on graphite, taken in the constant-current mode. The image is 500×600 Å, and the grey scale corresponds to a total height variation of 40 Å dark to light. Size, shape and substructure were stable during the ten minute period of observation in which three images have been taken [7]

of features, ranging in size from a few atoms to many thousands of atoms.

Figure 3 shows an image of a large Ag particle obtained in the topographic mode at a voltage of 16 mV with a tunneling current of several nA. The particle is roughly cylindrical with a diameter of 350 Å and a height of about 30 Å. It is composed of several smaller particles each 30 to 100 Å in diameter. Obviously the deposited particles moved at the substrate and the observed island is the result of coalescence.

Figure 4 shows a Au cluster of about 7 atoms on the clean graphite lattice. The Au atoms arrange independently on the substrate lattice. This and the following pictures have been taken in the variable-current mode. The shadow at the left hand side of the cluster is produced by the imaging process. The feedback circuit is not fast enough to respond. This cluster has been generated on the substrate from single atoms.

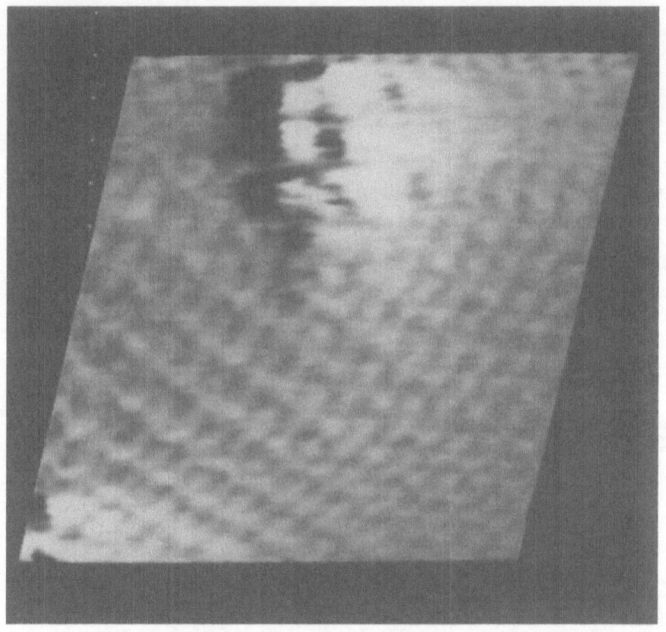

Fig. 4. STM image (24×24 Å, variable-current mode, top view) of a gold cluster containing about 7 atoms [7]

Ag clusters show a tendency to be commensurate with the substrate. In Figs. 5 and 6 the images of bare graphite and of a Ag cluster feature are displayed, respectively. The feature is roughly 15 Å long and 5 Å wide. The substructure can be explained by an array of dimers. The electrons tunnel into covalent bonds and consequently the individual atoms are not imaged separately. The individual dimers are resolved which may be explained by Van der Waals coupling between them.

This area has been observed during 40 minutes while hundreds of scans could be taken. Figs. 7a-c show three pictures taken (from the same region) during these observations: the images b and c where taken 18 min and 23 min, respectively, after a. The images are 24 Å × 24 Å in size. In Figs. 5 and 6 top views and in Figs. 7 a-c, projected views are given.

Fig. 5. STM image (24×24 Å, variable-current mode, top view) of the bare (0001) cleavage plane of highly oriented pyrolitic graphite [7]

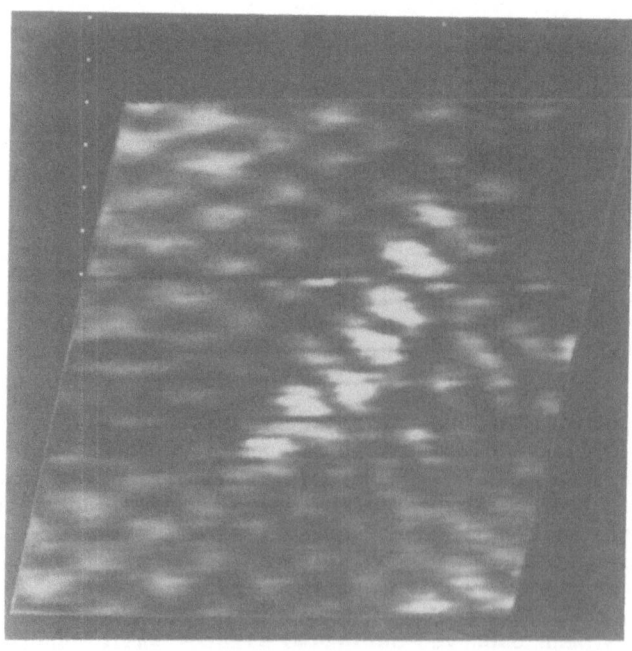

Fig. 6. STM image (24×24 Å, variable-current mode, top view) of a silver cluster on graphite, roughly 15 Å long and 5 Å wide [7]

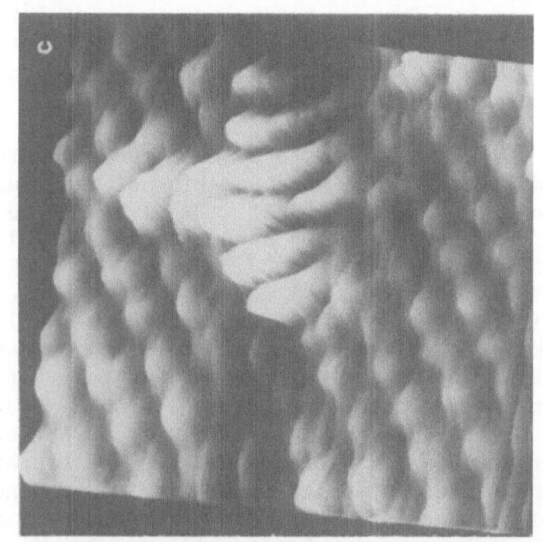
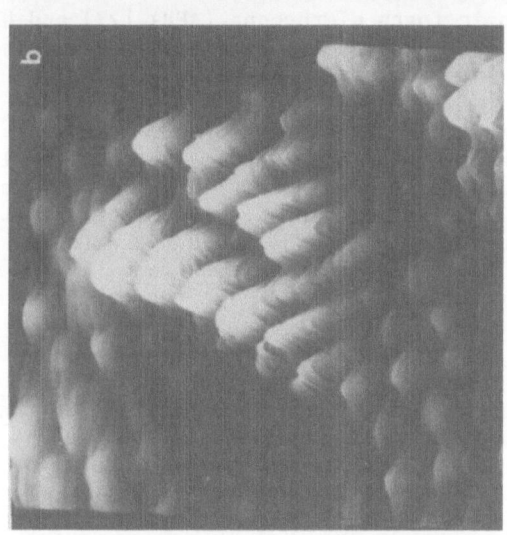
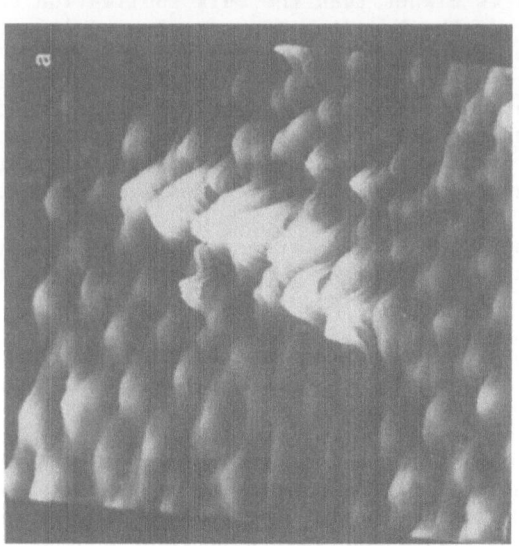

Fig.7. Three STM images (24×24 Å, variable-current mode, projected view) showing the time evolution of the Ag cluster of Fig. 6. The displayed region has been observed during 40 minutes where hundreds of scans could be taken. Single adatoms moved into that region building up a second row of dimers ((b), taken 18 minutes after (a)). Then a sudden change into a Y-shape occured ((c), taken 23 minutes after (a)) which remained stable for 15 minutes [7]

During the time of observation single adatoms moved into that region building up a second row of dimers. Then, a sudden growth event into a Y-shaped structure was observed, which was stable during further 15 minutes. We conclude that the STM yields detailed local information about diffusion and growth of cluster adsorbates.

The different tendency of Ag and Au clusters to be commensurate with the substrate may be explained by different bond energies of the clusters, respectively (Ag_2: 159 ± 6 kJmol^{-1}, Au_2: 221 ± 1 kJmol^{-1}, Ag_3: 253 ± 13 kJmol^{-1}, Au_3: 367 ± 13 kJmol^{-1} [26]). For lower cluster bond energies the cluster-substrate interactions become more important.

Now we consider possible influences of the instrument during the imaging process. For both materials, Ag and Au, clusters in a broad size range and islands with all kinds of shapes were imaged. In the fast mode, many features however did not show atomic resolution or atomic structures were smeared out. This may be explained by proximity effects between the tip and the clusters: For an adjusted distance between the tip and the graphite surface, where individual graphite atoms are imaged, the tip-cluster gap can get small enough that interaction forces on an atomic scale become relevant. The forces acting between tip and sample during scanning are used in the atomic force microscope (AFM) [27] and have been measured recently between a tungsten tip and polycrystalline silver surfaces [28].

Calculations of intermetallic adhesion forces, based on a jellium model of the interaction between two metals, predict attractive tip-sample forces of the order of nN and force gradients of the order of Nm^{-1} under typical tunnel conditions [29]. The measurements [28] show that the force gradient is positive (values between +15 and +50 Nm^{-1} are found) and that it increases as the tip gets closer to the sample surface. In the variable current mode the gap distance changes drastically when the tip moves from an area of the bare substrate to an adatom or cluster. If a 3-4 Å distance to the graphite substrate is adjusted, a silver adatom is imaged by a 1-2 Å gap. At even smaller distances between the two jellium edges (each of which extends about 1 Angstrom beyond the core position of the surface atoms), short range forces like exchange-correlation forces or core repulsion forces become accessible. Then the binding energy of the Ag/W interface becomes relevant. The binding energy of silver atoms on the tungsten surface has been measured [30] by thermal desorption spectroscopy to be 3.15 eV, a value which is higher than the bulk sublimation energy of silver of 2.90 eV. Therefore the tungsten tip can pick up Ag atoms from the clusters when it comes close enough. Furthermore the tip can pull or push the adatoms or clusters during data acquisition. In this case instabilities in the images occur. These often have been observed during the experiment. The pictures shown in Figs. 4 to 7 were selected for presentation because the structure of the displayed clusters was stable over many scans and underwent relatively slow, progressive changes.

In the fast mode movement and growth processes of single adatoms and small clusters have been observed by taking subsequent STM-images (Fig.7). These observations so far are of qualitative nature. However the studies show that the instrument has the capability for detailed quantitative investigations of diffusion and growth processes on the individual level. This opens new possibilities concerning the dynamics of clusters on supports.

For adatoms, dynamic properties can be described by relatively simple statistical mechanical methods [31]. For chusters however the description of the random walk on a corrugated lattice network is complicated since each link in the unit mesh may correspond to a different jump probability, reflecting different activation energies for the various configuration changes of the cluster. Such configuration changes during cluster diffusion have been found with the field ion microscope (FIM) [32]. This method however is only applicable to metal adsorbates of fairly high ionization potentials being adsorbed on highly refractory metal substrates. The coverage must withstand the high fields required for forming the image. Furthermore, electronegative adsorbates are not visible and very electropositive adsorbates are too easily field desorbed. For FIM-studies adatoms have to be strongly adsorbed and this can appreciably distort the substrate lattice in its immediate vicinity. Thus it is quite probable that diffusion of an adatom or a small adcluster involves carrying along this distortion, or "polaron", leading to very low values of the diffusion constant. With increasing cluster size the distortion will diminish because the forces exerted by the additional cluster atoms will tend to restore the symmetry of the original lattice. With the STM, compared to the FIM, systems with much smaller cluster-substrate interactions can be investigated.

Direct observations of mean square displacements of adatoms and adclusters yielding tracer diffusion coefficients, elementary jump distances and jump frequencies would be extremely helpful to a better understanding of nucleation and growth of ultrathin films. For mean displacements of a few Angstroms per second STM studies in the fast mode should give quantitative results concerning the quantities mentioned above. Studies of this kind are in progress.

It is interesting to note that small Ag and Au clusters can be imaged in air. These materials are fairly inert concerning oxidation. Electron microscopy images of Ag-particles on carbon did not change after air exposure [33] and small Ag clusters in beams do not adsorb oxygen [34]. As molecules from other gases however could be adsorbed the data taken in air should not be overinterpreted concerning the properties of the bare clusters.

SUMMARY

The first part of the paper deals with the question of what happens with small clusters after being deposited onto a solid support. The size distribution of CsCl-clusters has been determined by time of flight mass spectrometry in the beam and by transmission electron microscopy on the substrate. It has been found that the 10-15 Å clusters grow to 1000-8000 Å particles on glass substates at room temperature after deposition. These results leaded to the search for an instrument with imaging capabilities on a much smaller size scale. So far, only the field ion microscope (FIM) yielded images of adatoms and adclusters. The FIM however is very limited concerning the substrate and the cluster material, and only strongly adsorbed clusters can be imaged.

The scanning tunneling microscope as a new tool for cluster research is discussed. Au and Ag clusters on highly oriented pyrolitic graphite are imaged. Graphite seems to be a suitable support material because it is atomically flat with a perfect lattice network over thousands of Angstroms and the STM image shows atomic resolution. Thus, the locations of the adclusters can be mapped relatively to the coordinate system defined by the surface crystal lattice.

The two operation modes, being used with the STM, have different applications. In the constant current mode, which takes a few minutes for a picture, 3D-particles can be imaged, but their diffusion constants have to be extremely small. In the variable current mode, a picture takes a few seconds and the clusters can be more mobile to be imaged. However, this mode is restricted to 2D-clusters, especially if we require the display of the substrate lattice as well. The tunneling gap has to be choosen large enough in order to avoid that the tip approaches the cluster to a distance where atomic forces influence the imaging process.

In the constant current mode, Ag and Au particles ranging in size from 10 Å to greater than 800 Å with various shapes have been displayed. In the variable current mode, single adatoms as well as clusters containing a few atoms were imaged with atomic resolution, together with the graphite substrate lattice. Ag clusters show the tendency to be commensurate with the substrate. This shows that the interaction of the Ag coverage with the graphite support is not negligible. The imaged Au clusters were incommensurate with the substrate. The difference between Au and Ag cluster-substrate interaction may be explained by the different bond energies of Ag and Au clusters. For lower cluster bond energies the cluster-substrate interaction gets more important concerning the cluster geometry.

By taking subsequent images from areas with relatively stable clusters, diffusion and growth was observed on the atomic scale. A region, showing a row of (probably) dimers, was observed during 40 minutes and hundreds of fast scan images could be taken. Single atoms moved into that region building up a second row of dimers. Finally, a sudden growth event into a Y-shaped structure was observed. This shows that the STM can be used for detailed studies of the dynamics of supported clusters.

We find with two different methods (TEM and STM) for various cluster materials (ionic clusters, metal clusters) on different substrate materials (glass, graphite) after different preparation techniques (single atom deposition, 10-15 Å cluster deposition, 30-100 Å small particle deposition; imaging in vacuum, imaging in air) that the deposited particles at room temperature diffuse at the substrate growing to "new" adparticles. The diffusion times depend on the individual areas being imaged which could be due to different degrees of perfection of the substrate.

The STM seems also to yield results concerning the character of cluster chemical bonds. For Ag-coverage features with the sizes of dimers have been imaged the single atoms within the dimers not being resolved. We suggest that tunneling into the high charge density region of the covalent bond occurs. This shows the additional capability of the STM for studies of electronic cluster properties.

The author acknowledges a Heisenberg fellowship from the Deutsche Forschungsgemeinschaft.

REFERENCES

1. Proceedings of the "Workshop on Atomic Structure and Properties of of Small Particles", Wickenburg, Arizona, 1986, special issue of "Ultramicroscopy", to appear

2. J.A. Venables, G.D.T. Spiller, and M. Hanbücken, Rep. Prog. Phys. <u>47</u>, 339 (1984)
3. S. Iijima, J. Electr. Micros. <u>34</u>, 249 (1985);
   S. Iijima and T. Ichihashi, Phys. Rev. Lett. <u>56</u>, 616 (1986)
4. "Metal Clusters in Catalysis", H. Knozinger, B.C. Gates, and L. Guczi, ed., (Elsevier, New York, 1985)
5. For reviews of free cluster research see:
   Surf. Sci. <u>156</u>, Parts 1-2 (1985);
   Z. Phys. D - Atoms, Molecules and Clusters - Vol.3 (1986);
   Proc. of the "1st Int. Workshop on Physics of Small Systems", Isle of Wangerooge (FRG) 1986, publ. in "Lecture Notes in Physics", publ. by Springer (1986);
   Proc. of the "1st NEC Symp. on Fundamental Approach to New Materials Phases - Microclusters - , Tokyo/Hakone, Japan, (Springer), 1986
6. R. Pflaum, D. Schild, K. Sattler, E. Recknagel, G. Hegenbart and F. Granzer, to be published
7. D.W. Abraham, K. Sattler, E. Ganz, H.J. Mamin, R.E. Thomson and J. Clarke, Appl. Phys. Lett. <u>49</u>, 853 (1986)
8. K. Sattler, J. Mühlbach and E. Recknagel, Phys. Rev. Lett. <u>45</u>, 821 (1980)
9. G. Binnig and H. Rohrer, Surf. Sci. <u>126</u>, 236 (1983)
10. R.S. Becker, J.A. Golovchenko, E.G. McRae and B.S. Schwartzentruber, Phys. Rev. Lett. <u>55</u>, 2028 (1985);
    R.J. Hamers, R.M. Tromp and J.E. Demuth, Phys. Rev. Lett. <u>56</u>, 1972 (1986)
    J.A. Stroscio, R.M. Feenstra and A.P. Fein, Phys. Rev. Lett. <u>57</u>, 2579 (1986)
11. G. Binnig, H. Rohrer, Ch. Gerber and E. Weibel, Phys. Rev. Lett. <u>49</u>, 57 (1982)
12. R. Miranda, N. Garcia, A.M. Baró, R. Garcia, J.L. Peña, H. Rohrer, Appl. Phys. Lett. <u>47</u>, 367 (1985)
13. S.A. Elrod, A.L. de Lozanne and C.R. Quate, Appl. Phys. Lett. <u>45</u>, 1240 (1984)
14. R.V. Coleman, B. Drake, P.K. Hansma and G. Slough, Phys. Rev. Lett. <u>55</u>, 394 (1985)
15. R. Sonnenfeld and P.K. Hansma, Science <u>232</u>, 211 (1986)
16. J. Tersoff and D.R. Hamann, Phys. Rev. Lett. <u>50</u>, 1998 (1983), and Phys. Rev. B<u>31</u>, 805 (1985)
17. M. Posternak, A. Baldereschi, A.J. Freeman, E. Wimmer and M. Weinert, Phys. Rev. Lett. <u>50</u>, 761 (1983)
18. Th. Fauster, F.J. Himpsel, J.E. Fischer and E.W. Plummer, Phys. Rev. Lett. <u>51</u>, 430 (1983)
19. M. Posternak, A. Baldereschi, A.J. Freeman and E. Wimmer, Phys. Rev. Lett. <u>52</u>, 863 (1982)
20. G. Binnig, H. Fuchs, Ch. Gerber, H. Rohrer, E. Stoll and E. Tosatti, Europhys. Lett. <u>1</u>, 31 (1986)
21. Sang-Il Park and C.F. Quate, Appl. Phys. Lett. <u>48</u>, 112 (1986)
22. J.M. Soler, A.M. Baro, N. Garcia and H. Rohrer, Phys. Rev. Lett. <u>57</u>, 444 (1986)
23. H.J. Mamin, E. Ganz, D.W. Abraham, R.E. Thomson and J. Clarke, subm. to Phys. Rev. B
24. I.P. Batra, N. Garcia, H. Rohrer, H. Salemink, E. Stoll and S. Ciraci, Surf. Sci., in print
25. D. Tomanek, S.G. Louie, H.J. Mamin, D.W. Abraham, R.E. Thomson, E. Ganz and J. Clarke, submitted for publication
26. K.A. Gingerich, I. Shim, S.K. Gupta and J.E. Kingcade Jr., Surf. Sci. <u>156</u>, 495 (1985)

27. G. Binnig, C.F. Quate and Ch. Gerber, Phys. Rev. Lett. 56, 930 (1986)
28. U. Dürig, J.K. Gimzewski, D.W. Pohl and R. Schlitter, submitted for publication
29. J. Ferrante and J.R. Smith, Phys. Rev. B31, 3427 (1985)
30. E. Bauer, H. Poppa, G. Todd and P.R. Davis. J. Appl. Phys. 18, 3773 (1977)
31. A.N. Berker, S. Ostlund and F.A. Putnan, Phys. Rev. B17, 3650 (1978)
32. D.W. Basset, in "Surface Mobilities on Solid Materials" (ed. by Vu Thien Binh), NATO ASI Series, B, Vol. 86, Plenum Press, New York and London, 1983
33. E. Anno and R. Hoshino, Surf. Sci. 144, 567 (1984)
34. D. Cox, talk given at this conference

GAS-PHASE CHEMICAL REACTIONS OF TRANSITION

METAL CLUSTERS WITH SIMPLE MOLECULES

Stephen J. Riley and Eric K. Parks

Chemistry Division
Argonne National Laboratory
Argonne, IL 60439 USA

ABSTRACT

Chemical reactions of isolated transition metal clusters are studied in a laser-vaporization cluster source coupled to a continuous-flow reactor. Detection of reaction products is via laser ionization and time-of-flight mass spectrometry. Experimental probes that have been developed include: 1) kinetics measurements, in which the disappearance of bare cluster signal with increasing reagent gas flow is used to determine absolute reaction rate constants for the addition of the first adsorbate molecule; 2) product composition measurements, in which inferences as to cluster structure and the nature of surface binding sites are derived by determining the total number of adsorbates the clusters can accommodate; 3) laser-induced desorption experiments, from which adsorbate binding energies can be derived; and 4) the observation of actual chemical reactions on cluster surfaces, such as hydrogen/deuterium exchange or adsorbate photochemistry. In addition, a new experimental procedure has been developed that, in a single series of measurements, provides measures of the first three parameters listed above. A review is given of earlier studies of the reaction of iron clusters with hydrogen. More recent results on the reaction of iron clusters with ammonia, and the reaction of ammoniated iron clusters with hydrogen, are also presented.

I. INTRODUCTION

Among the properties of small metal clusters that have recently become amenable to experimental studies are their chemical interactions with simple gas phase molecules. Clusters of transition metal atoms offer obvious models for heterogeneous catalysis, and such studies may some day permit us to design more efficient and selective catalyst preparations. Since small clusters are virtually all surface, their reactions with gas phase molecules mimic, on a molecular scale, the surface chemistry of bulk material. By studying clusters isolated in the gas phase it is possible to probe them on this molecular scale using sensitive, mass-spectrometry based detection techniques. Interactions not localized to the surface, such as cluster oxidation, provide information on metal corrosion and, in some cases, the solid state physics of metal systems.

In the relatively few years of this new science, several experimental procedures have been developed to study cluster chemical properties. In one, the disappearance of bare cluster signal in the mass spectrum is monitored as the quantity of reagent gas interacting with the clusters is increased.[1-3] By appropriate modeling, pseudo first-order reaction rate constants can be extracted that describe the kinetics of the addition of the first adsorbate molecule to the bare cluster. In several instances, it has been found that this measure of bare cluster reactivity shows a dramatic (several orders of magnitude) dependence on cluster size. As chemical reactivity is ultimately determined by electronic structure, attempts have been made to correlate these properties, thus providing more detail about cluster-adsorbate interactions.[1,2] Another experimental technique is to saturate the clusters with reagent molecules.[4] From the composition of the final products (i.e., the number of adsorbate molecules each cluster can accommodate) one can hope to learn something about cluster structure and the nature of the adsorbate binding sites. In the realm of thermodynamic measurements, a technique of laser-induced-desorption has been developed.[5] In this, multiphoton absorption by saturated cluster products leads to adsorbate desorption. From the number and energy of the photons absorbed and the number of species desorbed, adsorbate binding energies can be determined using appropriate statistical (RRKM) modeling. Results of experiments like these will be reviewed.

Recently we have developed a new experimental procedure that, in favorable cases, allows essentially simultaneous determination of all three properties outlined above: kinetics, composition, and thermodynamics. The procedure consists of determining the average number of reagent molecules taken up by the clusters as a function of reagent pressure in the reaction region. These measurements are carried out over a very wide range (many orders of magnitude) in reagent pressures. As will be presented below, for the reaction of iron clusters with ammonia we find several regions of distinctly different behavior in the dependence of the average number of adsorbed ammonias on ammonia pressure. By appropriate modeling of this behavior it is possible to determine cluster reactivity as a function of coverage, adsorbate binding energy both as a function of coverage and independent of any sort of statistical model, and saturated cluster composition.

Another development is the modification of our cluster chemical reactor to permit two-reagent studies. As an example of such studies the reaction of ammoniated iron clusters with hydrogen will be discussed. We will begin with a brief description of the experimental apparatus and a review of recent results on the reactions of iron clusters with hydrogen.

II. EXPERIMENTAL

The heart of the apparatus is a continuous flow-tube-reactor (FTR) coupled to a laser-vaporization cluster source. A schematic is shown in

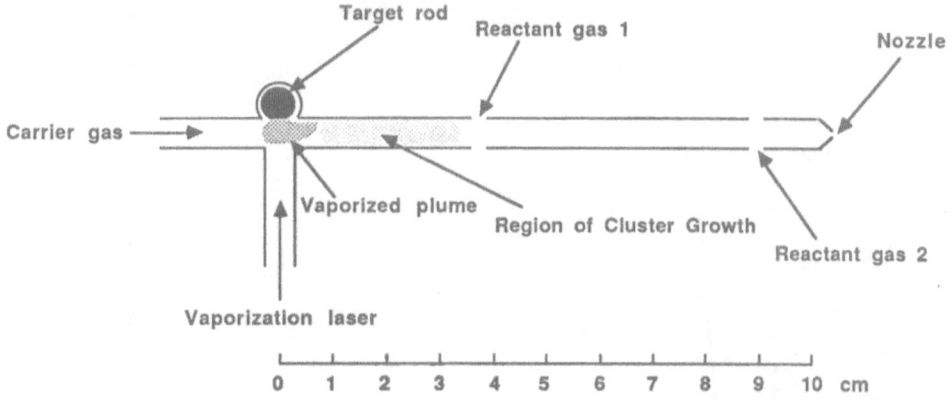

Fig. 1. A schematic diagram of the laser-vaporization cluster source
and flow-tube reactor.

Fig. 1. Pulses from an excimer laser are focused onto a metal rod
located within a 3 mm dia. channel down which He carrier gas is continuously flowed. The vaporized metal is rapidly cooled by the gas, and
clusters grow as the metal moves down the channel. We have performed
extensive studies to characterize this source, and conclude that at low
He pressure (~20 torr) the principal mechanism for cluster growth is by
monomer addition. The He pressure also controls the rate at which monomers diffuse to the wall of the channel, and if this rate is sufficiently high, cluster growth will terminate within a few centimeters of
the metal rod. Under appropriate conditions, we can assure that clusters have both stopped growing and, in addition, have cooled to room
temperature, before the reactant gas is injected into the channel at a
point 3.7 cm downstream of the metal rod. This use of continuous flows
for both the He carrier gas and the reagent, instead of pulsed gas
injection, is crucial to these experiments, because it allows us to
establish controlled and reproducible reaction conditions such as reagent temperatures, pressures and interaction times. The current version of the source, as shown in Fig. 1, permits the introduction of a
second reactant gas at a point further downstream.

Clusters and their reaction products exit the FTR through a 1.0 mm
dia. nozzle into vacuum, and are formed into a molecular beam. The species are detected in a laser-ionization time-of-flight mass spectrometer
located 42 cm downstream of the nozzle. Various excimer lasers are used
for ionization. In general, their fluences are kept below 1 mJ/cm^2 to
minimize multiphoton processes. The mass spectrometer has a resolution
of ~500. To take advantage of this, and to allow study of clusters containing up to 300 atoms, isotopically pure metal rods are used.

III. A REVIEW OF RECENT RESULTS

A. Kinetics

Figure 2 shows a typical rate constant plot for a cluster-reagent
system in which there is a strong dependence of reactivity on cluster

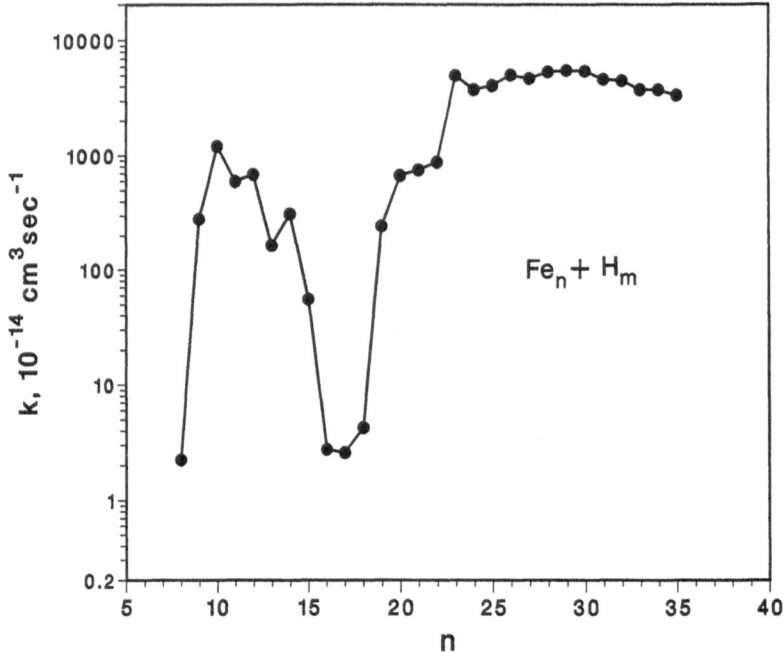

Fig. 2. Cluster size dependence of the rate constants
for the reaction of iron clusters with hydrogen.

size.[6] For the reaction of iron clusters with H_2, the absolute rate
constant for the addition of the first H_2 is seen to vary by over three
orders of magnitude between Fe_8 and Fe_{23}. Other experimental measurements suggest that the large dips in reactivity shown in Fig. 2 for Fe_8
and near Fe_{17} arise from an activation barrier for the breaking of the
H_2 bond, a prerequisite for actual H atom chemisorption. The final jump
in rate constant between Fe_{22} and Fe_{23} is to a value close to that predicted by the bulk iron sticking probability. Thus we can say that in
terms of their reactivity towards hydrogen, iron clusters start behaving
like bulk metal abruptly at Fe_{23}.

Similar cluster-size dependences of chemical reactivity have been
reported for other systems.[2,3] In some cases, an anticorrelation
between cluster rate constants and ionization potentials has been
noted.[2] This sort of information provides us with valuable insight into
the detailed electronic interactions that govern cluster reactivity.

B. Product Composition

When clusters are allowed to react with an excess of reagent,
saturated products are formed whose composition (the number of adsorbates on a given cluster) gives a measure of the number of binding sites
on a cluster. Such information may be able to give us some insight into
cluster structure, or at least the structure of the underlying metal
framework. Figure 3 shows typical results for hydrogenated iron clusters.[4] As has been discussed in Ref. 4, a detailed examination of these
results indicates that smaller iron clusters appear to be rather rigid
in structure, tend to be roughly spherical in shape, and have their surfaces essentially covered with hydrogens (i.e., one hydrogen atom for

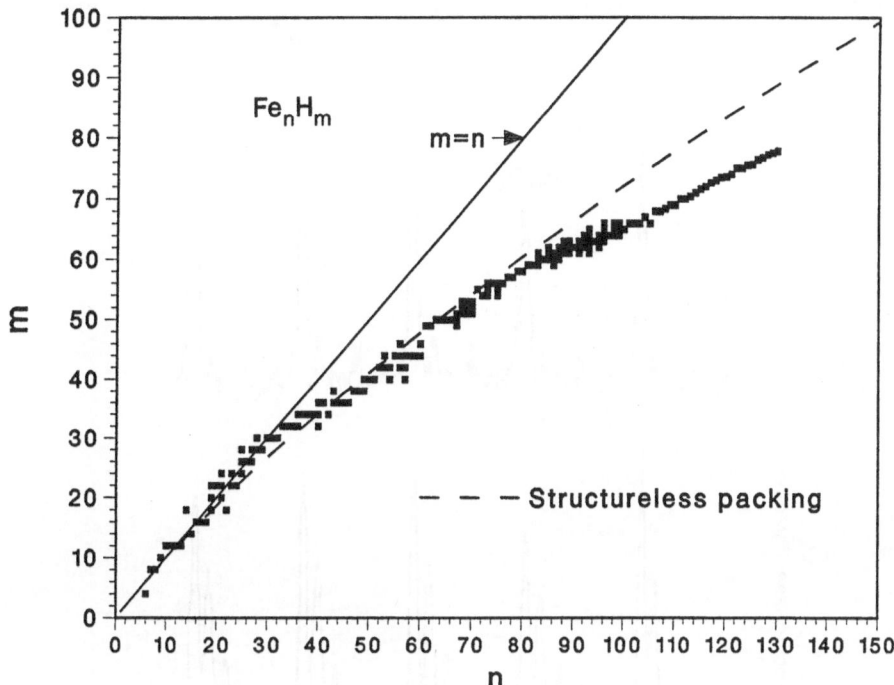

Fig. 3. The composition of fully hydrogenated iron clusters. For $n \leqslant 60$ the plotted symbols show actual m values of species having considerable intensity in the mass spectrum. Above 60, the centers of the symbols give the centers of the observed but unresolved mass peaks. Between 60 and 100 the height of a symbol represents the width of the mass peak, while above 100 only the peak centers are indicated. The structureless packing model assumes that the ratio of surface to total metal atoms in a cluster of n atoms is $1-[1-(4\pi/3n)^{1/3}]^3$. (Adapted from Ref. 4.)

every surface iron atom). More recent work, however, has shown that larger iron clusters, up to Fe_{264}, are not fully covered with hydrogens, while nickel clusters are. (The drop below full coverage for iron is seen in Fig. 3 to begin near Fe_{80}.) Such differences may help to explain differences found in the catalytic behavior of these metals.

C. Laser-induced Desorption

One essential assumption made in the experiments on product composition is that the laser ionization used in the mass spectrometer detector does not produce an ion distribution that is significantly distorted from the distribution of neutral species exiting the FTR. One type of distortion is due to differences in ionization probability for different neutral species. This usually doesn't seriously affect experiments where simply the existence or non-existence of a given species is in question, or the kinetics experiments where one looks at the disappearance of a given species, but it is important when a comparison of the intensities of different product species is required. To recover the correct product distribution in such cases, the dependence of ionization

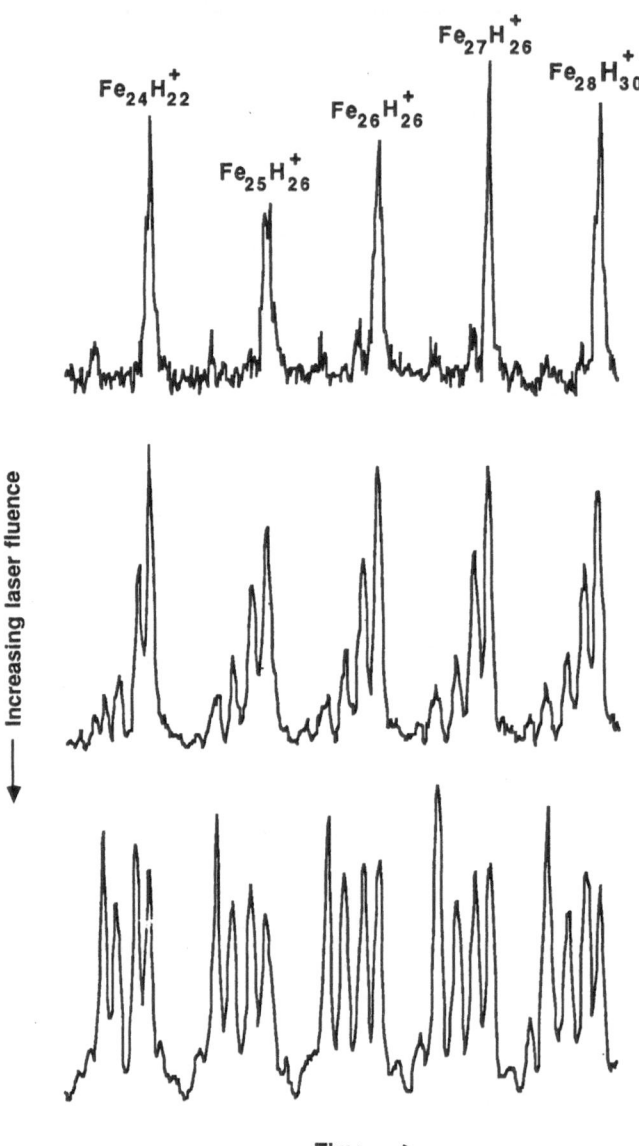

Fig. 4. A portion of the time-of-flight mass spectrum of hydrogenated iron clusters showing the effects of increasing ionizing laser fluence. Each new feature at higher fluence represents the loss by desorption of discrete numbers of hydrogen molecules.

probability on the number of adsorbed species must be determined. Another distortion of great concern is fragmentation in the ionization process, especially by multiphoton absorption.[7] As mentioned above, fragmentation can usually be controlled by keeping the laser fluence sufficiently low. However, under suitable conditions, multiphoton processes can actually be used as a probe of the thermodynamics of cluster chemical reactions. For example, the absorption of extra photons above the one needed to ionize the cluster will deposit extra energy into the cluster. If the cluster contains adsorbed species, this extra energy may lead to desorption of some of the species. From the number and energy of the extra photons absorbed and the number of species desorbed an RRKM analysis can be used to estimate the adsorbate binding energy.

Figure 4 shows this effect for hydrogenated iron clusters. Using an ArF excimer laser (6.42 eV photon energy) for ionization, it is seen that with increasing laser fluence additional peaks appear in the mass spectrum corresponding to absorption of additional photons and loss of certain numbers of hydrogen molecules. Analysis of results like these provide binding energies as a function of cluster size and coverage.[5] Similar experiments have been done for hydrogenated nickel clusters and ammoniated iron clusters.

D. H/D Exchange

If isolated transition metal clusters are to serve as model systems for heterogeneous catalysts, an obvious experiment is to demonstrate actual catalytic activity on cluster surfaces. This we have done for the exchange of hydrogen and deuterium on iron clusters. When deuterated iron clusters are mixed with hydrogen in the FTR at elevated temperatures, we see new peaks in the mass spectrum indicative of the exchange of deuteriums on the cluster by hydrogen, and the desorption of HD molecules. (Exchange may proceed at room temperature, but apparently too slowly to occur within the time the reagents are in the FTR.) We model the exchange by using the well-known Langmuir-Hinshelwood mechanism, in which exchange occurs between chemisorbed atoms on the surface, not via a gas phase-surface exchange. The essence of our assumption is that there is complete statistical randomization of the chemisorbed atoms prior to any desorption of a new molecule. Thus we say that for a general cluster $Fe_n D_{m-x} H_x$ having m adsorbate atoms we have the reaction scheme

$$Fe_n D_{m-x} H_x + H_2 \xrightarrow{k_{H_2}} Fe_n D_{m-x} H^{\ddagger}_{x+2} \longrightarrow \begin{cases} Fe_n D_{m-x} H_x + H_2 \\ Fe_n D_{m-x-1} H_{x+1} + HD \\ Fe_n D_{m-x-2} H_{x+2} + D_2 \end{cases}$$

with a similar reaction with D_2 having addition rate constant k_{D_2}. The metastable intermediate $Fe_n D_{m-x} H^{\ddagger}_{x+2}$ has three decay pathways.[2] We say that the relative probabilities p of these three pathways are determined purely statistically, i.e.,

$$P_{H_2} = \frac{j_H(j_H-1)}{j(j-1)} \qquad P_{HD} = \frac{2j_H j_D}{j(j-1)} \qquad P_{D_2} = \frac{j_D(j_D-1)}{j(j-1)}$$

where j_H and j_D are the number of H and D atoms on the metastable intermediate, and $j = j_H + j_D = m + 2$. The only other assumption is that k_{H_2} and k_{D_2} are independent of cluster composition, i.e., the value of x. Figure 5 shows the comparison of this model with the experimental exchange spectra for Fe_{13}. Clearly the model, with the only adjustable

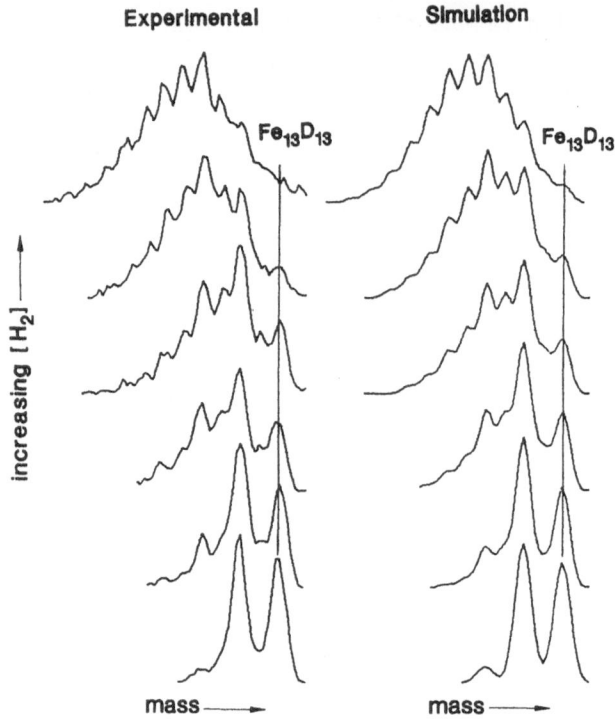

Fig. 5. A comparison of the measured mass spectra for H/D exchange on Fe_{13} with the Langmuir-Hinshelwood simulation described in the text. The simulated spectra have been convoluted over experimental width functions to match the resolution of the experiment. The two large peaks in the bottom (no H_2) spectra are $Fe_{13}D_{12}$ and $Fe_{13}D_{13}$.

parameters the addition rate constants k, provides good fits to the observed spectra. In fact, the statistical aspects of this simulation are so important that the fit is relatively independent of the k values used. An attempt to simulate the data with the other principal exchange mechanism, the Eley–Rideal process in which a gas–surface exchange reaction occurs, was unsuccessful. No choice of adjustable parameters could come close to fitting the observed spectra. This analysis is a good example of how mass spectrometry provides new insight into surface phenomena. While there is nothing new about the Langmuir-Hinshelwood mechanism, the ability to actually monitor the composition of the metal-containing species as the exchange reaction proceeds, and thus prove the exchange mechanism, is an exciting development.

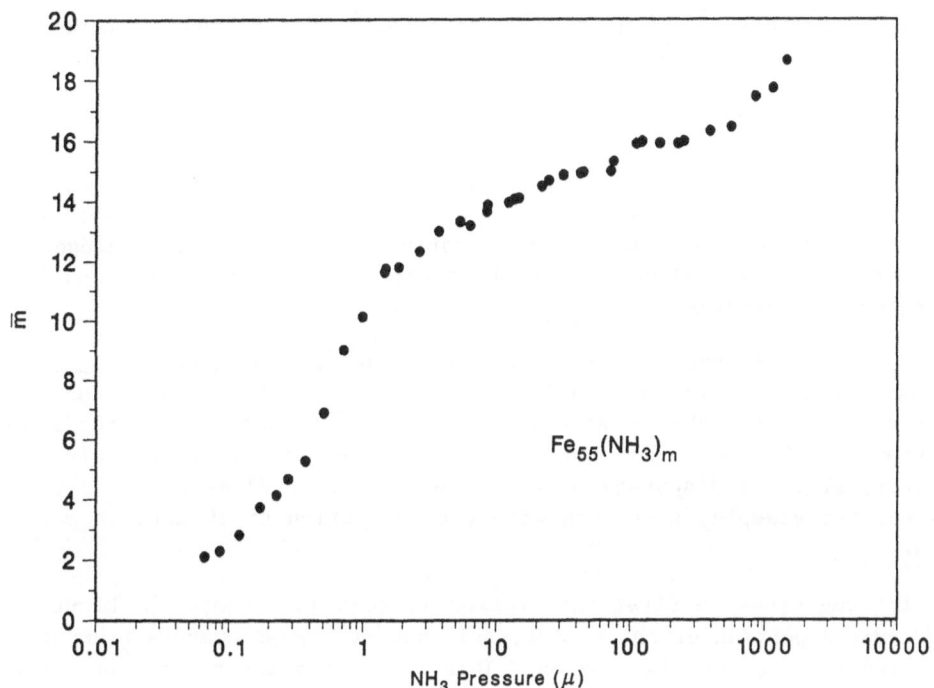

Fig. 6. The $\bar{m}$ plot for Fe_{55}. Note the logarithmic pressure scale.

IV. THE REACTION OF IRON CLUSTERS WITH AMMONIA

A. The $\bar{m}$ Plot: A New Experimental Procedure

Our studies of the reactions of iron clusters with ammonia have prompted us to develop a new way of probing cluster chemical properties. In essence, the technique consists of measuring the uptake of adsorbate by the clusters over a very wide range in reagent pressure in the FTR (five orders of magnitude for the ammonia-iron system). The uptake is approximated from the average number of molecules bound to a given cluster for a given pressure. For the ammonia-iron case, we denote the general product as $Fe_n(NH_3)_m$, and calculate the average m value from

$$\bar{m} = \sum_m m I_m / \sum_m I_m .$$

Here I_m is the intensity of the $Fe_n(NH_3)_m^+$ peak in the TOF mass spectrum. An example of the ammonia pressure dependence of $\bar{m}$ for Fe_{55} is shown in Fig. 6. There are clearly regions of different behavior in this plot. At the lowest ammonia pressures, $\bar{m}$ increases rapidly with pressure. Our studies indicate that the reaction is essentially kinetically limited in this regime. Collisions between ammonia and clusters are relatively rare, but the adsorption cross sections are very large. Once adsorbed, an ammonia does not desorb within the FTR. The behavior in this pressure region can be modeled to determine reaction cross sections, and the dependence of these cross sections on cluster coverage.

At a pressure of 1.5 μ, the $\bar{m}$ plot takes a sudden break to a region of weaker dependence on ammonia pressure. We believe that in this region there is essentially an equilibrium between ammonia and ammoniated clusters in the FTR, that is, ammonia desorption and readsorption occur. This portion of the curve is essentially an equilibrium adsorption isotherm, and from the dependence of $\bar{m}$ (essentially a measure of cluster coverage) on pressure, the free energy change for the adsorption process can be determined. From an estimate of the entropy change in the reaction one can then extract the binding energy of ammonia and its dependence on coverage.

At an ammonia pressure of about 100 μ, the $\bar{m}$ curve shows a flattening out (or an inflection point) characteristic of the formation of the saturated product. The $\bar{m}$ value in this region gives the saturated composition, and from this one can go on to consider binding sites, cluster geometry, etc., as discussed above for hydrogenated clusters. The Fe_{55} cluster, for example, saturates with the adsorption of 16 ammonia molecules.

In some cases, a first flat region is seen just above the kinetically limited portion of the $\bar{m}$ plot, followed at higher ammonia pressures by a rise to a second flat region. This represents the filling of first strongly bound sites, then more weakly bound sites. For example, we find that Fe_{13} has four strongly bound and two weakly bound sites for ammonia. Such observations will provide us with insight into cluster structure and the nature of binding sites on cluster surfaces.

At still higher pressures, $\bar{m}$ again increases rapidly with pressure (see Fig. 6), indicative of the formation of a second, physisorbed layer of ammonias due to an increasing degree of cooling in the expansion out of the nozzle. Data such as shown in Fig. 6 has been collected and reduced for iron clusters from Fe_2 to Fe_{85}. Other experiments have demonstrated that ammonia binds to iron clusters nondissociatively, i.e., ammonia decomposition on the cluster surface does not occur within the FTR.

B. The Saturated Products: Evidence For a Phase Change

From the saturated regions of plots such as shown in Fig. 6 the compositions of saturated ammoniated iron clusters can be determined. The results are shown in Fig. 7 for Fe_2 through Fe_{100}. For comparison, a curve showing one-third the number of surface iron atoms is also shown. Clearly, ammonia binds at a ratio greater than one molecule for every three surface metal atoms for small clusters, decreasing to less than one for every three for larger clusters. For bulk iron, the ratio is more like one for every six.[8] We interpret this behavior as indicative of the curvature of the cluster surface. Steric hindrance between adjacent ammonias increases as the cluster size increases and its surface becomes more planar. As a result, fewer ammonias can be accommodated.

Another feature evident in Fig. 7 is a distinct dip in the $\bar{m}$ plot at Fe_{15} and Fe_{61}. The dip at 61 is caused by the formation of an extremely stable species, $Fe_{61}(NH_3)_{16}$. In fact, a detailed inspection of the TOF mass spectra in this region shows a surprising result. At an

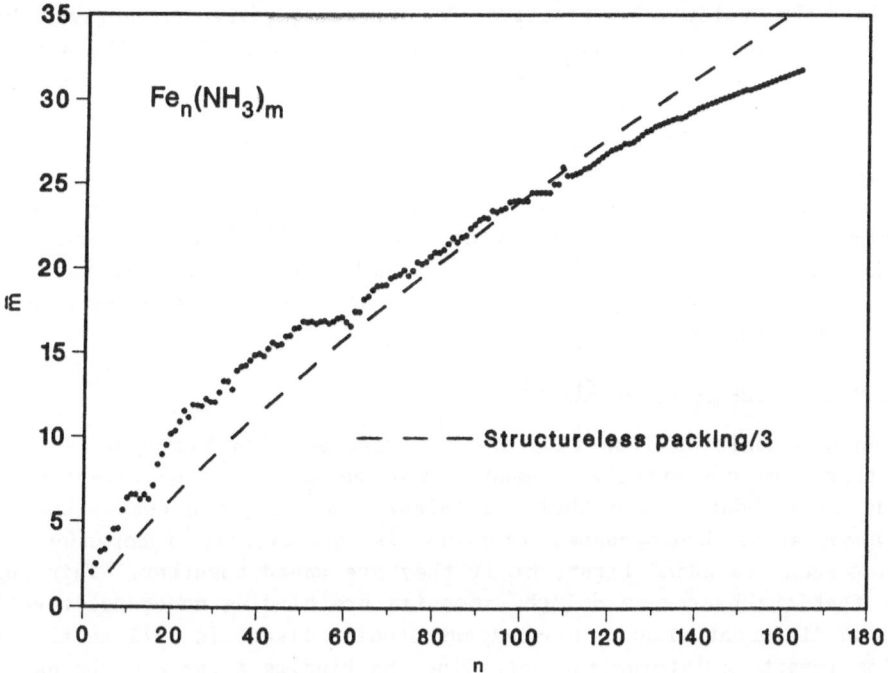

Fig. 7. The dependence of $\bar{m}$ on n for $Fe_n(NH_3)_m$ clusters. The
dashed curve is one third of the analogous curve in
Fig. 3.

ammonia pressure of 224 μ, Fe_{61} has a major peak in the spectrum corresponding to 16 ammonias, and a peak with one third the intensity having
18 ammonias. If this behavior implies an equilibrium between the two
species, then at a higher ammonia pressure one would expect conversion
to the more ammoniated product. Instead, as the ammonia pressure is
increased the more ammoniated species decreases in intensity, leaving
only the $Fe_{16}(NH_3)_{16}^+$ peak. Our best explanation is that the iron
framework is undergoing some sort of structural change with cluster size
in this region. The Fe_{61} structure that is becoming more stable with
increasing cluster size prefers to bind 16 ammonias, while the less
stable form wants 18. There is presumably some barrier to the conversion from the less to the more stable form. As the ammonia pressure is
increased, reaction within the FTR is more rapid, and the cluster will
be heated to higher temperatures during the adsorption process. This
heating will facilitate the conversion to the more stable form. The
evidence is that such a structural transformation (perhaps a phase
transition) begins at Fe_{55} and is completed by Fe_{65}.

V. THE REACTION OF AMMONIATED IRON CLUSTERS WITH HYDROGEN

A. Kinetics

Like bare iron clusters, fully ammoniated iron clusters show a
strong cluster size dependence of their reactivity with hydrogen, with
several orders of magnitude change in rate constant within a narrow
cluster range. However, the dips in reactivity seen for bare clusters
at Fe_8 and Fe_{17} (see Fig. 2) are shifted to smaller cluster sizes. For

example, the relatively reactive Fe_{13} becomes very unreactive when ammoniated. In general, the absolute rate constants for fully ammoniated clusters tend to be an order of magnitude smaller than for bare clusters. The adsorbed ammonia obviously inhibits collisions of the hydrogen with the iron framework, reducing the clusters' reactivity. Kinetics studies of partially ammoniated iron clusters show that the shift of unreactivity from Fe_{17} to Fe_{13} is essentially a continuous function of the extent of ammoniation. Clearly, the binding of ammonias to the clusters dramatically changes their electronic structure and hence their chemical properties. Further experiments are being done to investigate these effects.

B. Where Are the Binding Sites?

When ammoniated iron clusters are reacted with hydrogen, we find that they bind essentially as many hydrogens as bare iron clusters do, and do so without losing their ammonias. In fact, the composition of the ammoniated, hydrogenated clusters is practically independent of which reagent is added first, or if they are added together. This suggests that the hydrogens and the ammonias are binding noncompetitively, i.e., on different sites. Two-reagent studies like this will be able to provide important information regarding the binding sites and the nature of the chemical interactions between small metal clusters and adsorbates.

VI. CONCLUSION

As the present review hopefully demonstrates, transition metal clusters provide a rich, fascinating, and often confusing arena for studying fundamental chemical processes. In addition to offering ideal testing grounds for surface science, solid state physics and heterogeneous catalysis, clusters turn out to be very interesting chemical reagents in their own right. By reacting in ways never seen before they challenge our chemical intuition and encourage us to broaden our appreciation of the intricacies of chemistry. We expect that the study of cluster chemistry will provide many years of fruitful and interesting work, and we look forward to it with great anticipation.

ACKNOWLEDGMENTS

We wish to thank Drs. K. Liu, S. C. Richtsmeier, G. C. Nieman, W. F. Hoffman, and P. S. Bechthold for help in gathering and analyzing the data presented here, and Drs. Liu, Nieman, A. F. Wagner, and J. Jellinek for useful discussions. This work is supported by the U. S. Department of Energy under contract W-31-109-Eng-38.

REFERENCES

[1] S. C. Richtsmeier, E. K. Parks, K. Liu, L. G. Pobo, and S. J. Riley, J. Chem. Phys. 82, 3659 (1985).
[2] a) R. L. Whetten, D. M. Cox, D. J. Trevor, and A. Kaldor, Phys. Rev. Lett. 54, 1494 (1985); b) R. L. Whetten, M. R. Zakin, D. M. Cox, D. J. Trevor, and A. Kaldor, J. Chem. Phys. 85, 1697 (1986).

[3] a) M. D. Morse, M. E. Geusic, J. R. Heath, and R. E. Smalley, J. Chem. Phys. **83**, 2293 (1985); b) P. J. Brucat, C. L. Pettiette, S. Yang, L.-S. Zheng, M. J. Craycraft, and R. E. Smalley, J. Chem. Phys. **85**, 4747 (1986).

[4] E. K. Parks, K. Liu, S. C. Richtsmeier, L. G. Pobo, and S. J. Riley, J. Chem. Phys. **82**, 5421 (1985).

[5] K. Liu, E. K. Parks, S. C. Richtsmeier, L. G. Pobo, and S. J. Riley, J. Chem. Phys. **83**, 2882 (1985); erratum J. Chem. Phys. **83**, 5353 (1985).

[6] The rate constants shown in Fig. 2 are somewhat different from those initially reported in Ref. 1. The principal change is in the earlier assumption that the cluster flow velocity in the FTR is equal to the average He flow velocity. In fact, as further experiments have shown, the clusters tend to stay in the middle of the FTR channel, where their velocities are roughly twice the average He velocity.

[7] P. J. Brucat, L.-S. Zheng, C. L. Pettiette, S. Yang, and R. E. Smalley, J. Chem. Phys. **84**, 3078 (1986).

[8] M. Weiss, G. Ertl, and F. Nitschke, Appl. Surf. Sci. **2**, 614 (1979).

METAL CLUSTERS: SIZE DEPENDENT CHEMICAL AND ELECTRONIC PROPERTIES

D. M. Cox, M. R. Zakin, and A. Kaldor

Corporate Research
Exxon Research and Engineering Co.
Annandale, New Jersey 08801

I. Introduction

Compared to atomic or bulk materials, small isolated gas phase metal clusters have recently been shown to have individually unique properties which not only vary with cluster size, i.e. number of atoms in the cluster, but also vary dramatically from one metal to another [1-9]. Thus the study of the "exotic" chemical and physical properties of such species has attracted much interest from both experimentalists and theoreticians.

In a sense clusters containing a well-defined number of metal atoms can be considered as unique molecules and as such they may be expected to possess individually unique chemical, electronic or magnetic properties. In this paper we will be concerned with clusters which are "naked" in the sense that initially they contain no ligands, i.e. they are highly coordinatively unsaturated. This suggests that they might be highly reactive. What is observed is that - yes - certain size clusters do react readily, but other size clusters are relatively unreactive toward the same reagent. On the other hand with different reagents a different reactivity pattern typically emerges. Such size-sensitive behavior suggests that clusters have the potential for high specificity in reactions.

Next we note that for clusters containing less than 100 atoms most atoms of the cluster are "surface atoms". For instance if one simply assumes bulk densities and spherical clusters, a 100 atom cluster would have about 72 surface atoms. A 20 atom cluster on the other hand would have about 18-19 surface atoms. Obviously if the cluster structure is planar or chain-like, all atoms should be considered as surface atoms. Thus probably with reasonable justification it has been suggested that the study of clusters should be viewed as the study of "molecular surfaces"[10].

In this paper we shall limit the discussion primarily to results from experiments performed in our laboratory since other papers in this volume review work from other laboratories. For hydrogen activation and chemisorption, we shall present experimental results demonstrating how the chemisorption depends sensitively on cluster size and metal type for different transition metals. We will also show how hydrogen chemisorption does (or does not) correlate with a particular electronic property, the threshold ionization energy of the bare cluster. We shall outline a model of hydrogen activation based upon simple frontier orbital considerations. Such a model is useful in that it can qualitatively explain the general trends observed in such bond activation reactions. Next we will discuss the chemisorption of carbon monoxide on a large number of transition metal clusters. From these measurements we are able to deduce, within the context of a kinetic model, a relative ordering of the metal-CO bond strengths. Finally we will end with a discussion of experiments in which infrared multiple photon dissociation of cluster-adduct species is used to probe the chemisorbed state of an adduct on gas phase metal clusters. Such experiments now allow us to address the question of the structure of the metal cluster-adsorbate complex which becomes critical to understanding how cluster chemistry relates to "traditional" chemistries (e.g. surface science, organometallic reactions, isolated molecule reactions).

II. Experimental Technique

Only a general overview of the experimental techniques will be given here. For the interested reader further details are readily available from the references[1-3,7-11]. Figure 1 shows schematically the chemical reactor addition to the cluster source used in the gas phase reaction studies. In operation a high pressure pulse of helium gas is present at the metal rod when a high power vaporizing laser is fired. The metal vapor is entrained in the helium carrier and confined to flow through a narrow channel where cluster growth occurs. The clusters then expand into the reactor tube where a second pulsed valve injects either helium or a mixture of helium and reactant. The reaction is quenched when the cluster/helium/reactant mixture expands into vacuum at the exit of the reaction tube. The resulting beam is collimated and subsequently interrogated downstream by photoionization time of flight mass spectrometry.

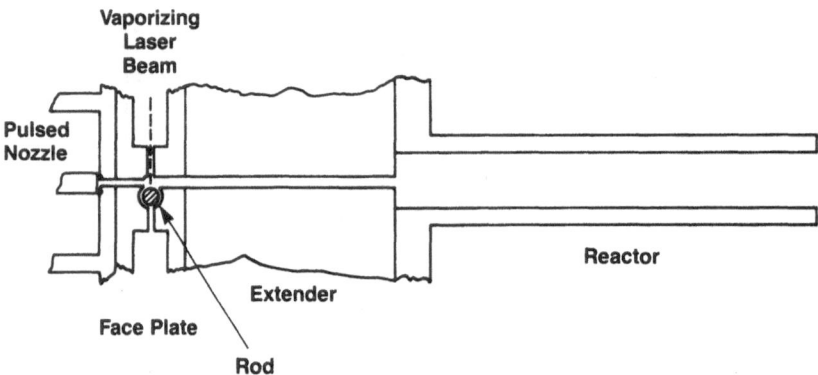

Figure 1.   Scale-drawn schematic of the essential features of the cluster source and chemical reactor system. The channel for the nozzle to the rod is 0.1 cm diameter, opening to 0.2 cm diameter from the rod to end of the 1.27 cm wide faceplate. The extender and reactor tubes as shown are 0.2 cm diameter by 3.81 cm long and 1 cm diameter by 7.6 cm long, respectively. Different length and diameter extender and reactor tubes are easily installed.

Reaction is presumed to occur if the bare cluster ion signal is depleted upon addition of reactant and/or if new product peaks are detected in the mass spectrum. Inherent in such measurements are the following major assumptions: The reaction must be relatively fast, since the residence time of clusters in the reactor is estimated to be only on the order of 100 usec. During this time typically 10 to 1000 cluster-reactant collisions and about 10,000 helium collisions might occur. The products must be collisionally stabilized on a time scale rapid compared to unimolecular decay in order to survive and escape the reactor. Similarly products must survive the transit time to the detection zone (~500 usec in our apparatus) after exiting the reactor, and in addition must survive the photoionization process (i.e. product ions observed are assumed to be parent ions). The reaction kinetics for chemisorption of a reagent molecule R onto a cluster M_n is described by equ. 1 and 2,

$$M_n + R \xrightarrow[\xleftarrow{\;\;k_{n'}\;\;}]{\;\;k_n\;\;} M_nR^* \tag{1}$$

$$M_nR^* + He \xrightarrow{\;\;k_s\;\;} M_nR + He , \tag{2}$$

where k_n is the rate constant for the addition reaction, $k_{n'}$ is the unimolecular decay rate of the M_nR^* species and k_s is the collisional stabilization rate. Assuming the steady state approximation can be applied to M_nR^*, recognizing that the reactant and helium concentrations are both in excess and that $[M_n]_f/[M_n]_0 = I_f/I_0$, it is easily shown that

$$- \ln(I_f/I_0) = k_n [R] t (k_s[He]/(k_s[He] + k_{n'})) \tag{3}$$

where t is the residence time in the reactor, the quantities in square brackets are the concentrations of the respective species in the reactor and I_f and I_0 are the bare cluster ion signals after and before reaction, respectively. The last expression in parentheses on the right hand side of equ. 3 represents the competition between unimolecular decay and collisional stabilization of M_nR^*. If the collisional stabilization rate is dominant then equ. 3 simply becomes

$$-\ln(I_f/I_0) = k_n [R] t \tag{4}$$

and we find that an experimentally determined quantity, $-\ln(I_f/I_0)$, is directly proportional to the pseudo first order rate constant, k_n. Inherent in the derivation of equ. 4 are the assumptions that the reactant concentration [R] is sufficiently high that it is little changed in the reaction, and that all clusters have the same residence time, i.e. the proportionality constant [R]t in equ. 4 is the same for all n, so the relative rate constant for different cluster sizes can be obtained from measurement of the survival fraction for the n-atom cluster, I_f/I_0. A further assumption is that although the effective temperature for all bare clusters is not well known, it is the same for each cluster prior to reaction. For our present pulsed reactor experimental setup, absolute rate constants are relatively uncertain since reactant concentration and cluster residence time can only be crudely estimated. Absolute rate constants have been reported by one research group [6] through use of a continuous flow system for the reactant and helium buffer but still using pulsed laser vaporization of metal. A comparison of H_2 chemisorption on iron clusters measured independently by three different research groups shows good qualitative agreement[5-7].

Keep in mind that rate constant measurements reported here reflect only the initial reaction step i.e. addition of the first reactant molecule to the cluster $M_n + R \longrightarrow M_nR$. Although observation of product ions is taken as additional confirmation that reaction has occurred, we do not use these ion signals to estimate the reactivity. This is because the ionization cross section for the bare cluster and the same cluster with a reactant molecule chemisorbed onto it may differ and thus give rise to different detection efficiencies. We have shown that the ionization threshold of a cluster with chemisorbed hydrogen increases substantially over that of the bare cluster, and that the direction of this change (increase for hydrogen [12] and decrease for ammonia [13]) can be predicted from simple frontier orbital considerations[12]. Similarly the rates for chemisorption of a second, third, etc. molecule onto an already reacted cluster are not yet known. Further experiments designed to probe the ionization thresholds, cross section variation, and subsequent reactivity for clusters containing adsorbates are in progress.

III. Chemical Reactivity

Many classes of cluster reactions are just now beginning to be studied. For instance we have performed initial survey studies of oxidation reactions on clusters of Al, Fe, Pt, Pd and Ag; H-H, N-N, C-O and C-H bond activation reactions on many different metal clusters, desulfurization reactions on Al and Fe clusters, and dehydrogenation reactions on Nb, Pt and Rh clusters.

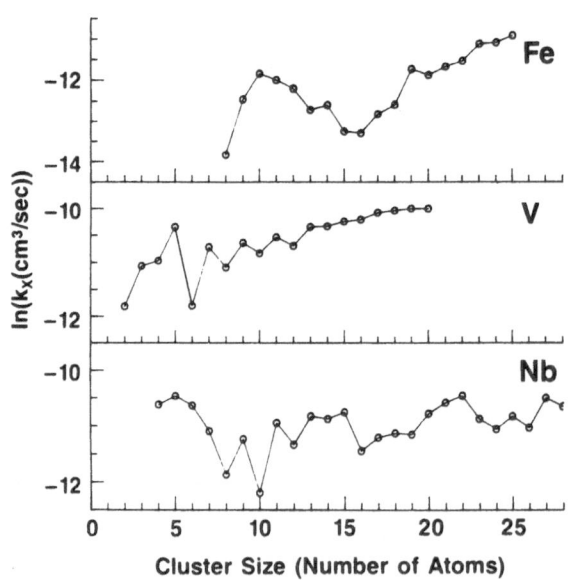

Figure 2.   The logarithm of the rate constant, $\ln(k_n)$, for reaction toward deuterium is plotted as a function of cluster size for iron, vanadium and niobium clusters. The reactivity data (note the log scale for the abscissa) is presented in this fashion in order to better observe the correlation with the ionization threshold energies shown in fig. 4 and predicted from equ. 5. The absolute rate constants for our data are determined by normalizing our value at Fe_{10} with that measured by Richtsmeier et al.[6].

In this section we will describe results from di-hydrogen bond activation reactions of clusters of different transition metals. In figure 2 we compare the iron [7], vanadium [8] and niobium [8] cluster reactivity toward molecular hydrogen. In all cases the reactivity is observed to vary dramatically and non-monotonically with cluster size, and to exhibit a unique size-dependent reactivity for each metal. For instance, note the oscillatory behavior observed for iron clusters which exhibits a local maximum in reactivity near Fe_{10} followed by a local minimum near Fe_{16-17}. For V and Nb we note that as for iron, certain clusters are considerably less reactive toward hydrogen than others. For instance most vanadium clusters containing and even number of atoms are less reactive than their odd atom neighbors, i.e. V_6, V_8, V_{10}, and V_{12} exhibit considerably reduced reactivity compared to V_5, V_7, V_9, and V_{11} and V_{13}. For niobium Nb_8, Nb_{10}, Nb_{16}, Nb_{24}, and Nb_{26} exhibit considerably reduced reactivity toward hydrogen than neighboring clusters. As a contrast note that the small niobium clusters exhibit an enhanced reactivity.

In addition, several orders of magnitude variation in reactivity occurs between the least and most reactive clusters, with the most dramatic variation in reactivity occurring for clusters containing fewer than about 30 atoms. For instance about three orders of magnitude variation in reactivity is found between Fe_8 and Fe_{25} and about 1.5 orders of magnitude variation between Fe_{11} and Fe_{15}. In contrast the reactivity is high and changes little for iron clusters larger than about 23-25 atoms [6]. It has been suggested that this is the cluster size for which the onset of bulk-like behavior occurs, since hydrogen chemisorption on bulk iron is facile (and dissociative). Applying a similar interpretation to our results on vanadium and niobium would suggest that bulk-like behavior is achieved about V_{17} or V_{18}, while for niobium it must occur for clusters containing greater than 28 atoms.

Above we have shown that the chemisorption of the first hydrogen molecule on metal clusters exhibits a pronounced dependence on cluster size. In this section we briefly address the question of how much hydrogen can be chemisorbed upon a given size cluster, i.e. what is the saturation coverage which can be obtained and how does it vary with cluster size. The saturation coverage of iron clusters has been investigated in detail by Parks et al [14]. Their results suggest that for iron clusters dissociative chemisorption terminates when each metal SURFACE atom has chemisorbed one hydrogen atom. For clusters containing less than 30 atoms, essentially all metal atoms are surface-like with at most one or two atoms possibly in the "interior" or "bulk-like". Thus for Fe_{30} and smaller clusters, essentially a saturation coverage of one hydrogen atom per metal atom is observed. For clusters ranging in size from 30 to 80 atoms, Parks observes that the 1:1 ratio is achieved only if one counts surface atoms. We have measured the hydrogen uptake for small vanadium clusters n<30, and find that the small vanadium clusters take up about 1.4 hydrogen atoms per metal surface atom, somewhat larger than that measured for the small iron clusters. Hydrogen uptake experiments on the larger vanadium clusters is presently being investigated.

To this point we have presented results for hydrogen chemisorption on Fe, V, and Nb clusters and have shown that there is a very pronounced variation with cluster size and metal type. Such a pronounced size-selectivity in itself is quite exciting. However another property, the ionization threshold energy, is also found to exhibit a strong non-monotonic variation with cluster size. Interestingly, it was found [7] that for Fe clusters larger than eight atoms, clusters with low (high) ionization threshold energies are those which exhibit high (low) reactivity toward di-hydrogen. The explanation put forth to account for this anticorrelation was that hydrogen chemisorption required charge transfer from the metal cluster for the activation of the H-H bond, with charge transfer being more facile the lower the ionization threshold. A simple empirical model was proposed in which the IP and chemisorption rate constant k_n of the n-atom cluster are related according to

$$k_n = A \exp [-\epsilon(IP_n - E_0)/kT] \qquad (5)$$

in close analogy to an Arrhenius expression for an activated process. The term $\epsilon (IP_n - E_0)$ may then be interpreted as something akin to an activation energy E(act). The relative rate between two different size clusters n and n* is then expressed as

$$k_n/k_n{}^* = \exp [-\epsilon(IP_n - IP_n{}^*)] \qquad (6)$$

which allows scaling of ratios of rate constants to differences in IPs. The IP-ln(rate) empirical fit of the experimental IP and relative reaction rate constants is found to be quite good for iron clusters containing more than 8 atoms[7]. Ionization threshold measurements have now been extended to V [8] and Nb [8,9] clusters and these results together with those for iron are compared with their corresponding rate constants in figure 3. We note a striking correspondence between high ionization threshold and low reactivity for clusters larger than about eight atoms in size for all three metals.

However, we now observe that certain smaller (n<8) clusters which have higher ionization thresholds also exhibit considerable reactivity, exactly opposite to the behavior observed in the larger clusters. This is particularly dramatic for the small niobium clusters n<7, but Fe_4 (see ref. 5) and V_5 also have considerably higher reactivity than would be predicted by the simple empirical model of metal to H_2 charge transfer originally proposed to explain the IP-ln(rate) correlation observed for the larger clusters. In the next section we outline the general features of a model which can account for the activation of hydrogen on both large and small metal clusters.

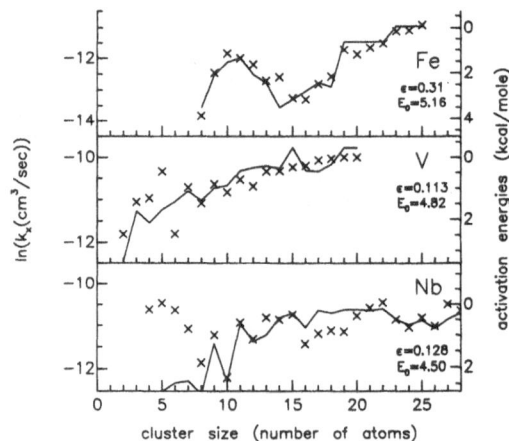

Figure 3. Comparison of rate constants (crosses) and bare cluster ionization thresholds (solid line) for iron, vanadium and niobium. The ionization potential is given by $IP=E_0+E(act)/\epsilon$ (see equ. 5). The constant A is estimated by assuming the large clusters have no barrier toward hydrogen chemisorption and setting it equal to the rate constant measured for large iron clusters[6]. E_0 and ϵ are determined by least squares fitting the ionization threshold energies to the rate constants assuming the constant A is independent of cluster size. The values of E_0 and ϵ for each metal are given on the plot. The same value of the constant A is assumed for iron, vanadium and niobium.

IV. Hydrogen Activation by Metal Clusters

In this section we will discuss the various factors believed to be most important in the H-H bond activation by metal clusters. We will present a simple frontier orbital model whose primary features are: (1) Attractive metal cluster donor interactions between the high lying unoccupied H_2 σ^* antibonding molecular orbital and the highest lying occupied metal orbital M'(HOMO) at the Fermi energy. (2) Attractive metal acceptor interactions between the occupied H_2 σ bonding orbital and the lowest unoccupied metal orbital M(LUMO). (3) Repulsive (Pauli repulsion) interactions between the filled σ H_2 orbital and filled cluster valence orbitals.

Interactions (1) and (2) are attractive and lead to H-H bond weakening as well as increased M-H_2 bonding. This is because in interaction (1) charge density is transferred into the antibonding H_2 σ^* orbital, whereas in (2) charge density is transferred out of the bonding H_2 σ orbital. The repulsive interaction is directly responsible for the barrier to hydrogen chemisorption and dominates the energetics in the approach to the transition state which leads to dissociative chemisorption. It develops at large interaction distances. Elongation of the H-H bond occurs only at shorter distances when it is possible to transfer charge either into the H_2 σ^* or from the H_2 σ. Without (3) the combination of (1) and (2) would lead to facile (non-activated) dissociative chemisorption with the formation of two new low-lying (relative to the Fermi energy) metal-hydride bonds. The energetic cost in overcoming the activation barrier (3) and in enhancing metal to hydrogen charge transfer (1) is expected to be smallest (and thus the dissociative adsorption rate largest) when the cluster IP is smallest. This is easily rationalized by recognizing that the metal valence orbitals primarily will fall within an band of energies significantly higher in energy than that of H2 σ. Lastly, but importantly, we note that the M(LUMO) and M'(HOMO) must be of the appropriate symmetry to interact with the H_2 σ and the H_2 σ^* orbitals, respectively. If this symmetry condition is not satisfied then the next highest or lowest lying orbital of the proper symmetry must be considered.

Considering just these three frontier orbital interactions, we are now able to qualitatively explain the major features of the experimental results. First let us consider the IP-ln(rate) anticorrelation for larger clusters. This can be explained by recognizing that as the IP decreases the energy difference between H_2 σ and the metal cluster HOMO M', $E(\sigma^* - E(M')$, becomes progressively smaller and the energy difference between the metal cluster LUMO M and H_2 σ, $E(M)-E(\sigma)$, becomes progressively larger. Applying first order perturbation theory to this process, Shustorovich [15] has derived a simple expression for the stabilization of an occupied metal orbital M' due to mixing with the σ^* which can be written as

$$Q_M' \sim (\beta^*)^2 / [E(\sigma^*) - E(M')] \tag{7}$$

Similarly, stabilization of the H_2 σ due to mixing with an unoccupied metal orbital M is written as

$$Q_H \sim (\beta)^2 / [E(M) - E(\sigma)] \tag{8}$$

β and β^* are the interaction matrix elements between a metal orbital (M or M') and the H_2 σ and σ^* respectively. In general it is assumed that E(M) and E(M') are near the Fermi level with E(M)>E(M'). Donation to σ^* will dominate when $Q_M'/Q_H > 1$. This can occur even though the energy denominator $E(\sigma^*)-E(M')$ may be greater than $E(M)-E(\sigma)$ if β^* is sufficiently greater than β. This appears to be true for metal surfaces [16], but such conditions will become progressively harder to satisfy for small clusters with high IPs. Equ. 7 shows that the stabilization of the occupied metal

747

orbital M' depends inversely on the energy separation between $E(\sigma^*)$ and $E(M')$. Recognizing that $E(M')$ will vary with cluster size and is related to the ionization threshold energy, we predict a direct correlation between the ability of the cluster to donate charge to the H_2 σ^* (hydrogen activation) and the energy of M' (the cluster IP), thus an anticorrelation with ionization threshold energy.

For the smallest clusters the ionization threshold energy has become sufficiently large that donation from H_2 σ to M will dominate. The stabilization due to this interaction will increase as the IPs increase (see Equ. 8) and in the simplest picture the chemisorption rate should increase with increase in IP. However, symmetry considerations are expected to play an increasingly important role as the cluster size decreases. This is simply due to the fact that the density of states near the Fermi energy becomes increasingly sparse as the cluster size decreases. Thus finding a metal orbital of the appropriate symmetry near the Fermi energy becomes increasingly less probable. Thus it may not be surprising that small clusters (<8-10 atoms) may be perverse in their reactivity pattern as a function of cluster size and metal type. For example, the pentamer of vanadium exhibits pronounced reactivity whereas the iron pentamer is highly unreactive. Just the opposite behavior is observed for their respective tetramers. For niobium on the other hand the small 4<n<8 clusters are nearly as reactive as the larger clusters. In each instance the IPs are sufficiently high that little reactivity would be predicted based only on the metal to hydrogen charge transfer picture used to explain the behavior of the larger clusters. Small clusters in particular should be ideally suited for in depth theoretical studies as has recently been carried out for aluminum clusters [17].

Finally let us consider the observation that above a certain cluster size the chemisorption rate reaches a maximum value and does not vary further with cluster size. This is precisely what would be expected for a non-activated process where the barrier is non-existent or extremely small and results in facile chemisorption. Harris and Anderson [18] have examined the conditions whereby the activation barrier may be alleviated. They point out that when H_2 closely approaches a metal surface, destabilized metal orbitals may relax by populating entirely new metal orbitals which interact less repulsively with H_2. In particular they presented a case for 4s orbital occupancy shifting into 4d orbitals, and found a resultant decrease in the repulsive barrier.

We note that a number of theoretical investigations of the activation of H_2 by metal clusters and surfaces have been reported in recent years. These works are briefly reviewed in a companion article [19] in which electronic factors important in the dissociative chemisorption of H_2 on aluminum clusters and Al_6 in particular have been studied in detail. In addition the charge state of the cluster (anion, neutral or cation) is found to have only a small effect on the activation barrier, which suggests that similar size-selectivity might be expected for charged and neutral clusters, particularly if the orbital being ionized is not one which is participating in the reaction as is found for the Al_6-H_2 reaction. Thus the observation in a recent FT-ICR experiment [20] that Nb_7^+, Nb_8^+, and Nb_9^+ exhibit a similar size selectivity for H_2 dissociative chemisorption as observed for neutral niobium clusters should not be surprising. Similarly FT-ICR experiments do not necessarily represent firm evidence that the IP-ln(rate) correlation noted for Fe, Nb and V is fortuitous[20]. The rate enhancements [21] observed for ionic clusters relative to neutral clusters may simply reflect enhanced reaction cross sections common in ion-molecule reactions and similarly cannot be taken as firm evidence against an IP-ln(rate) correlation.

V. Chemisorption of Carbon Monoxide

The chemisorption of CO has been measured [22] in our laboratory for first row transition metals V, Fe, Co, Ni and Cu; second row metals Nb, Mo, Ru, and Pd; and third row metals W, Ir, and Pt as well as for Al. Our results show that CO is readily chemisorbed on most transition metal clusters containing three or more atoms and that the reactivity of the larger (n > 4) clusters is comparable, nearly independent of cluster size or metal type. This is in direct contrast to the very pronounced size selectivity discussed above for hydrogen chemisorption and suggests that CO chemisorption is a non-activated reaction.

The CO chemisorption on niobium and cobalt clusters has been reported previously [5]. The two sets of measurements are in reasonable agreement for clusters containing 3 or more atoms. However under our particular experimental conditions, the cobalt atom and dimer also exhibit depletion whereas no depletion was observed on these two species in the earlier measurements.

In these studies we have purposely held all detection, reaction, and vaporization conditions as constant as possible and changed only the metal rod being vaporized. Dependent on the metal, the onset of reactivity may begin with the atom, the dimer, the trimer or even a larger size cluster. Why should some metals show reaction for all size metal clusters (including the atom) whereas others only show reactivity above a certain minimum cluster size? By appealing to the the kinetic model described in section III, we explain such results as a competition between collisional stabilization versus unimolecular decay in the reactor, and from this to infer an ordering of metal-CO bond strengths as a function of the metal type.

To see this let us examine equ. 3 in more detail. The term in brackets on the right hand side of equ. 3 expresses the relationship between the collisional stabilization rate, k_s[He], and the unimolecular decay rate $k_{n'}$. For a fixed set of pressure conditions in the reactor we may reasonably expect the collisional stabilization rate is very nearly constant independent of cluster size (ignoring the slow increase associated with growth in the physical size of the clusters). However the unimolecular decay rate depends very strongly on cluster size, decreasing rapidly with increasing number of atoms. This is most easily seen by considering the simple RRK prediction for $k_{n'}$,

$$k_{n'} = A[(E-E_0)/E]^{s-1} \qquad E > E_0 \qquad (9)$$

where A is the frequency factor (typically around 10^{14} sec^{-1}), s is the number of modes (3n for the M_nCO^* species), E is the overage vibrational energy in the M_nCO^* species and E_0 is the critical energy needed for decomposition, which we shall assume for this simple argument is the cluster-CO bond strength assumed equal to the heat of adsorption. Assuming A does not change dramatically with cluster size, we see the term in brackets rapidly becomes smaller as n increases, quickly reducing the unimolecular decomposition rate for larger clusters. For E_0 = 0.5 eV, T=400K and n=3, the unimolecular decomposition rate as predicted from equ. 9 is about 2×10^8 sec^{-1} or on the order of the collision frequency. Thus trimers with metal-CO bond strengths $\leq$ to 0.5 eV should decompose in the reactor and no reaction would be observed. On the other hand if E_0 is 1.0 eV, the trimer decomposition rate would be only about 10^6 sec^{-1} and products should be stabilized i.e. reaction will be observed. Thus for a given bond strength, Eo, a minimum cluster size exists below which the cluster-reactant complex will unimolecularly decompose and above which it will be stabilized.

In applying this model to CO chemisorption studies on different transition metals we have assumed that the metal-CO bond strength varies little with cluster size and the CO chemisorption is non-activated. For one set of experimental conditions we find that reactivity is first observed for atoms of Pt and Co; dimers of Pd, trimers of V, Ni, Nb, Ru, and Ir; tetramers of W; pentamers of Cu and Mo, hexamers of Fe and Al. Thus we propose that the metal-CO bond strengths be ordered according to: Co, Pt, Pd > V, Ni, Nb, Ru, Ir, W > Mo, Fe, Cu, Al Interestingly this ordering is consistent with heats of desorption of CO from metal crystals and surfaces. For instance heats of desorption greater than 30 kcal/mole have been reported for Co [23a], Pt[23b], Pd[23c], W[23d,23e], Ru[23f], Nb[23e], Ir[23g], and Ni[23h]; whereas heats of desorption of less than 25 kcal/mole have been reported for Fe[23i], Mo[23j], Cu[23k], and Al[23l]. By carefully controlling the reactor buffer gas pressure, i.e. collisional stabilization rate, a much less coarse ordering of bond strengths should be possible.

VI. IR Photodissociation Spectroscopy of Metal Clusters

We have applied the techniques of infrared multiple photon dissociation spectroscopy to metal clusters in order to demonstrate that vibrational spectra of molecules chemisorbed upon metal clusters can be obtained as a function of cluster size and the number of chemisorbed molecules [24]. We have chosen to study the reaction of methanol on metal clusters, because high resolution electron energy loss (EEL) spectra of methanol on single crystal metal surfaces[25] show a strong IR absorption between 1000 and 1050 cm^{-1}, a frequency range accessible with a CO_2 laser. Somewhat arbitrarily we have chosen iron as the metal. The EEL spectrum of methanol on iron single crystal surface exhibits a strong absorption near 1020 cm^{-1} [26]. The surface experiments have been taken as evidence that at 300 K the methanol has decomposed into atomic hydrogen and methoxy with the 1020 cm^{-1} feature assigned to the C-O stretch of the methoxy group. From time of flight mass spectra alone, we cannot directly demonstrate that the analogous dissociative chemisorption is taking place on the iron clusters. However the combination of time of flight mass spectrometry and infrared photodissociation does allow measurement of the strong IR adsorption features of the cluster-adsorbate species.

Iron clusters react readily with methanol as shown in the photoionization time of flight mass spectrum in Fig 4. Note that mass peaks assigned to Fe_n and $Fe_n(CH_3OH)_y$ adducts are easily detected. This spectra shows that in sequential reactions more than one methanol can react with a cluster. As shown in the lower trace of Fig 4, when an appropriately timed high power pulsed CO_2 laser operated at 984.5 cm^{-1} irradiates the molecular beam, the $Fe_n(CH_3OH)_y$ adducts decompose as evidenced by loss of nearly all the ion signal on these mass peaks. As reported earlier [24], bare iron clusters are transparent at all CO_2 laser frequencies we have examined, whereas the $Fe_n(CH_3OH)_y$ exhibit strong depletion near 984 and 1075 cm^{-1}, but are transparent near 940 and 1040 cm^{-1}. Table I summarizes results from experiments with isotopically substituted methanol. These results are consistent with the assignment of the 984 cm-1 feature as the methoxy stretch. Initially from experiments with CH_3OH and CH_3OD, we proposed [24] that the 1075 cm^{-1} feature might be assigned to a bridge-bonded hydrogen stretch[27]. However the non-observation of IR decomposition near 1075 cm-1 with the CD_3OH species now makes this assignment doubtful.

Table I. IR frequencies for which decomposition of methanol isotopes
chemisorbed on iron clusters is observed.

Isotope	Frequency (cm-1)	
CH_3OH	984	1075
CH_3OD	984	----
CD_3OH	955	----

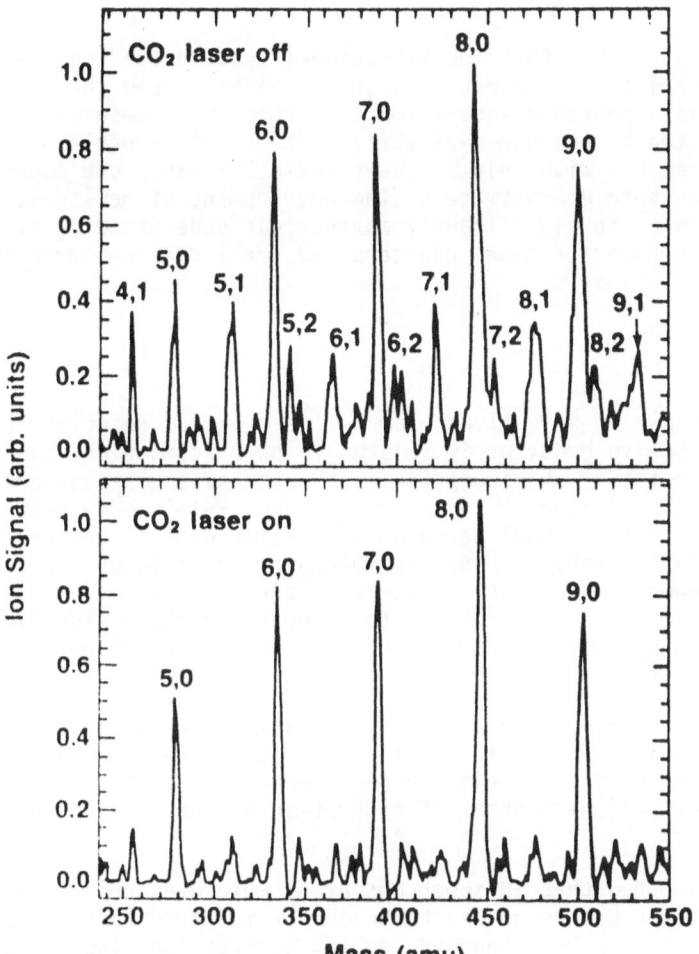

Figure 4. Time of flight mass spectra for the iron cluster-methanol
complexes. The upper trace shows the reference spectrum obtained from
the reaction of methanol with iron clusters. The peaks are labeled
according to (n,y) corresponding to masses equivalent to the chemical
composition $Fe_n(CH_3OH)_y$. The upper trace is taken with the CO_2 laser
blocked. The lower trace is taken with the CO_2 laser unblocked. The CO_2
laser frequency is 984 cm^{-1}.

The IR laser-induced decomposition of iron cluster-methanol complexes exhibits a threshold behavior on laser fluence consistent with a multiple photon absorption process. This is similar to results obtained on large molecules which have a large number of low-lying vibrational levels[28]. As IR laser fluence is increased the dissociation spectra exhibit red-shifted broadening expected in multiple photon dissociation processes. The IR dissociation spectra of iron clusters containing one or two chemisorbed methanols are found to be nearly identical suggesting that the first two chemisorbed methanols are bonded in a similar fashion to the cluster. In addition the IR multiple photon dissociation spectrum is independent of n, the number of Fe atoms in the cluster, in the range $5 <= n <= 12$. This suggests that the nature of the CH_3OH chemisorbed state is independent of cluster size.

These results show that the vibrational spectra and thus the structure of molecules chemisorbed on metal clusters can be probed through use of infrared multiple photon dissociation spectroscopy. These early experiments have been limited by the non-availability of broadly tunable high power infrared lasers, but with only a line tunable CO_2 laser the potental of such an approach has been demonstrated. The development of new tools such as this will allow us to significantly advance our understanding of the structure and bonding of metal clusters and metal clusters containing adsorbates.

VII. SUMMARY

In this paper we have given examples of chemical reaction studies in which size-sensitive behavior is related to bond activation reactions involving dissociative chemisorption. In such cases a strong correlation is found between the cluster IP and ln(rate). We explain such behavior in terms of a qualitative model based upon frontier orbital interactions between the cluster orbitals and the molecular orbitals of the reactants, and develop general guidelines in terms of charge transfer, Pauli repulsion, and orbital symmetry considerations to account for size-selective behavior versus non size-selective behavior and to predict activation barriers as a function of cluster size.

On the other hand we show that for a facile non-activated reaction such as CO chemisorption, a kinetic model can be developed from which relative CO-metal cluster bond strengths can be inferred and that such bond strength ordering agrees well with heats of adsorption on the corresponding metal surface.

Finally we show that IR spectroscopy of molecules chemisorbed onto clusters can be probed using the technique of infrared multiple photon dissociation. Such a technique by probing cluster and cluster-adsorbate vibrational frequencies. has general applicability for helping to advance our understanding of metal-metal bonding in metal clusters themselves as well as bonding of adsorbates to clusters. It is limited at present by the fact that only in limited frequency ranges are sufficiently powerful IR lasers readily available.

Such experiments only give a limited perspective to this rapidly growing field of gas phase cluster chemistry and physics. The broad character of this field is clearly and dramatically illustrated by the diversity and complexity of problems addressed by the various papers published in this volume. However there does appear to be a need for stronger interaction between theory and experiment particularly on problems which both can effectively attack. It is through such interactions that even more rapid progress should be possible.

REFERENCES

1. E. A. Rohlfing, D. M. Cox, and A. Kaldor, Chem. Phys. Lett. 99, 161 (1983).
2. E. A. Rohlfing, D. M. Cox, and A. Kaldor, J. Phys. Chem. 88, 4497 (1984)
3. E. A. Rohlfing, D. M. Cox, A. Kaldor and K. H. Johnson, J. Chem. Phys. 81, 3846 (1984)
4. M. E. Geusic, M. D. Morse, and R. E. Smalley, J. Chem. Phys. 82, 590 (1985).
5. M. D. Morse, M. E. Geusic, J. R. Heath, and R. E. Smalley, J. Chem. Phys. 83, 2293 (1985).
6. S. C. Richtsmeier, E. K. Parks, K. Liu, L. G. Pobo, and S. J. Riley, J. Chem. Phys. 82, 3659 (1985).
7. R. L. Whetten, D. M. Cox, D. J. Trevor, and A. Kaldor, Phys. Rev. Lett. 54, 1494 (1985).
8. D. M. Cox, R. L. Whetten, M. R. Zakin, D. J. Trevor, K. C. Reichmann, and A. Kaldor, AIP Conference Proceedings no. 146, Adv. in Laser Science I, Eds. W. C. Stwalley and M. Lapp, American Institute of Physics, (1986)
9. R. L. Whetten, M. R. Zakin, D. M. Cox, D. J. Trevor, and A. Kaldor, J. Chem. Phys. 85, 1697 (1986)
10. (a) A. Kaldor, E. Rohlfing, and D. M. Cox, Laser Chem. 2, 185 (1983). (b) D. M. Cox, E. A. Rohlfing, D. J. Trevor, and A. Kaldor, J. Vac. Sci. Technol. A2, 812 (1984). (c) A. Kaldor, D. M. Cox, D. J. Trevor and R. L. Whetten, ACS Symposium Series No. 288, "Catalyst Characterization Science", Eds. M. L. Deviney and J. L. Gland (1985).
11. (a) E. A. Rohlfing, D. M. Cox, and A. Kaldor, J. Chem. Phys. 81 3322 (1984). (b) E. A. Rohlfing, D. M. Cox, R. Petkovic-Luton, and A. Kaldor, J. Phys. Chem. 88, 6227 (1984) (c) R. L. Whetten, D. M. Cox, D. J. Trevor, and A. Kaldor, J. Phys. Chem. 89, 566 (1985)
12. M. R. Zakin, D. M. Cox, R. L. Whetten, D. J. Trevor, and A. Kaldor, Chem. Phys. Lett. submitted
13. S. J. Riley, E. K. Parks, K. Liu, and L. G. Pobo, Am. Chem. Soc. 191st annual meeting, paper 160, Div. of Coll. and Surf. Chem., April, 1986.
14. E. K. Parks, K. Liu, S. C. Richtsmeier, L. G. Pobo, and S. J. Riley, J. Chem. Phys. 82, 5470 (1985).
15. (a) E. Shustorovich and R. C. Baetzold, Science 227, 876 (1985). (b) E. Shustorovich, J. Phys. Chem. 87, 14 (1983). (c) E. Shustorovich, R. C. Baetzold, and E. L. Muetterties, J. Phys. Chem. 87, 1100 (1983).
16. J. Y. Saillard and R. Hoffmann, J. Am. Chem. Soc. 106, 2006 (1984).
17. (a) T. H. Upton, Phys. Rev. Lett. 56, 2168 (1986). (b) T. H. Upton, J. Chem. Phys. submitted
18. J. Harris and S. Andersson, Phys. Rev. Lett. 55, 1583 (1985)
19. T. H. Upton, D. M. Cox, and A. Kaldor (These proceedings).
20. J. M. Alford, F. D. Weiss, R. T. Laaksonen and R. E. Smalley, J. Phys. Chem. 90, 4480 (1986).
21. P. J. Brucat, C. L. Pettiette, S. Yang, L.-S. Zheng, M. J. Craycraft, and R. E. Smalley, J. Chem. Phys. 85, 4747 (1986).
22. D. M. Cox, K. C. Reichmann, D. J. Trevor and A. Kaldor, in preparation
23. (a) H. Papp, Ber. Bunsenges. Phys. Chem. 86, 555 (1982). (b) P. Imnbihl, M. P. Cox, G. Ertl, H. Muller, and W. Brenig, J. Chem. Phys. 83, 1578 (1985). (c) P. W. Davies and R. M. Lambert, Surf. Sci. Lett. 111, L671 (1981). (d) D. Brennen and F. H. Hayes, Phil. Trans. Roy. Soc. A258, 347(1965). (e) G. A. Samorjai, Chemistry in Two Dimensions: Surfaces, Cornell University Press, Ithaca, NY (1981). (f) H. Pfnur, P. Feulner, H. A. Engelhardt and D. Menzel, Chem. Phys.

Lett. <u>59</u>, 481 (1978); E. D. Williams and W. H. Weinberg, Surf. Sci. <u>82</u>, 93 (1979). (g) J. L. Taylor, D. E. Ibbotson, and W. H. Weinberg, J. Chem. Phys. <u>69</u>, 4298 (1978); B. E. Nieuwenhuys and G. A. Samorjai, Surf. Sci. <u>72</u>, 8 (1978); M. E. Thomas, H. Poppa, and G. M. Pound, Thin Solid Film, <u>58</u>, 273 (1979). (h) G. Wedler, H. Papp and G. Schroll, Surf. Sci. <u>44</u>, 463 (1974); C. R. Helms and R. J. Madix, Surf. Sci. <u>52</u>, 677 (1975). (i) T. J. Vink, O. L. J. Gijzeman and J. W. Geus, Surf. Sci. <u>150</u>, 14 (1985). (j) J. W. Erickson and P. J. Estrup, Surf. Sci. <u>167</u> 519 (1986). (k) J. C. Tracy, J. Chem. Phys. <u>56</u>, 2748 (1972); K. Horn, M. Hussain and J. Pritchard, Surf. Sci. <u>63</u>, 244 (1977). (1) C. B. Bargeron and B. H. Nall, Surf. Sci. <u>119</u>, L319 (1982).

24. M. R. Zakin, R. O. Brickman, D. M. Cox, K. C. Reichmann, D. J. Trevor, and A. Kaldor, J. Chem. Phys. <u>85</u>, 1198 (1986).

25. see for example, J. Hrbek, R. A. dePaola, and F. M. Hoffmann, J. Chem. Phys., <u>81</u>, 2818 (1985) and references therein.

26. P. H. McBreen, W. Erley, and H. Ibach, Surf. Sci. <u>133</u>, L469, (1983).

27. A. M. Baro and W. Erley, Surf. Sci. <u>112</u>, 1759 (1981).

28. T. G. Dietz, M. A. Duncan, R. E. Smalley, D. M. Cox, J. A. Horsley, and A. Kaldor, J. Chem. Phys. <u>77</u>, 4417 (1982).

ACTIVATION AND CHEMISORPTION OF HYDROGEN ON ALUMINUM CLUSTERS

T. H. Upton, D. M. Cox, and A. Kaldor

Corporate Research Science Laboratories
Exxon Research and Engineering Company
Annandale, New Jersey   08801

Abstract: We present results from theoretical and experimental investigations of the chemical (H_2 activation) and electronic properties (ionization potentials) of aluminum clusters. The chemisorption of H_2 on aluminum clusters exhibits a remarkable sensitivity to the number of metal atoms in the cluster. Al_6 is the smallest cluster for which chemisorption of H_2 is observed experimentally and for which a stable dissociately chemisorbed state for H_2 is predicted. For clusters containing more than 6 atoms, the reactivity decreases rapidly with increasing cluster size. For the bare aluminum clusters, theoretical predictions and experimental measurements of ionization thresholds are in good agreement. Using the reactive Al_6 cluster as model, we discuss how electronic factors influence H_2 dissociative chemisorption on metals. We find that while charge transfer from the cluster to the H_2 antibonding orbital is important, the activation barrier is dominated by repulsive interactions between the H_2 and the cluster. The charge state of the cluster (anion, neutral or cation) has only a small effect on the activation barrier, which suggests that similar size selectivity might be expected for charged and neutral clusters.

I. Introduction

Recent studies have shown that the chemisorption of hydrogen on metal clusters exhibits a remarkable sensitivity to both the number, n, of metal atoms in the cluster as well as the metal type [1]. For example, certain size niobium clusters, Nb_8, Nb_{10}, and Nb_{16} in particular, are significantly less reactive toward molecular hydrogen than other niobium clusters [1c-e],whereas for cobalt, Co_6 and Co_7 are the least reactive clusters [1c,1e]. For iron clusters, reactivity toward hydrogen is found to be an oscillatory function of the number of iron atoms in the cluster with Fe_6, Fe_7, and Fe_8 being the least reactive of the small clusters, while Fe_{15}, Fe_{16}, and Fe_{17} are the least reactive of the mid-sized clusters [1a-c]. Oscillations in the reactivity are found to correlate well with metal cluster ionization thresholds for clusters containing eight or more atoms, and a simplistic model based upon charge transfer for hydrogen activation has been proposed to explain this correlation [1a]. A similar correlation of ionization threshold energies with reactivity has now been observed for niobium and vanadium clusters [1d,2]. Thus the electronic character of the cluster appears to be important in controlling hydrogen chemisorption.

In this paper we extend the scope of these observations to include aluminum. We consider both the detailed reactivity of aluminum clusters towards hydrogen and, in some depth, the electronic features that control activation of the H_2 molecule during dissociation. In the next sections we summarize results from theoretical and experimental investigations showing how hydrogen chemisorption and ionization threshold energy vary with cluster size. Experimentally, we show that the hydrogen chemisorption is greatest for Al_6. No reaction is observed for clusters containing less than 6 atoms. For clusters containing more than 6 atoms the reactivity decreases rapidly with increasing cluster size becoming unmeasurable for clusters containing more than about 20 atoms [3]. Consistent with the experimental observations, we find in theoretical calculations that aluminum clusters containing less than 6 atoms are thermodynamically unstable upon adsorption of 2 hydrogen atoms. A six atom cluster containing two hydrogen atoms is found to be stable. For the bare aluminum clusters the theoretical predictions [4] and the experimental measurements of the ionization thresholds are in good agreement.

Following the discussion of size-dependent data, we present a brief review of theoretical models for H_2 dissociation on metals and an extended discussion of the electronic factors controlling this process on the reactive Al_6 cluster. We have used this cluster as a model, intending to focus on the most general features of the dissociation process and present concepts of potential relevance beyond this specific cluster system. In particular, we show that while charge transfer from the cluster to the H_2 antibonding orbital is important, the activation barrier is dominated by repulsive interactions between the H_2 molecule and the cluster. We also find that large variations in the cluster ionization potential (induced by forming positive and negative cluster ions) have only a small effect on the activation barrier, but that the effect is consistent with the experimental observations for iron, niobium, and vanadium cited above.

II. Aluminum Cluster Ionization Thresholds: Experimental

The experimental techniques used to create and detect metal clusters have been described in several previous publications [5] and will be summarized only briefly here. Metal clusters are formed by the sudden condensation of atomic vapor produced by pulsed laser evaporation of metal substrates inside the throat of a high pressure pulsed nozzle. After exiting the source region the clusters are collimated into a molecular beam and detected by photoionization time of flight mass spectrometry. The ionization thresholds are bracketed by measuring the dependence of the cluster ion signals on ionizing photon energy and intensity.

Figure 1 shows the aluminum cluster photoionization time of flight mass spectra for two different photon energies, 6.42 and 7.87 eV, at low photon flux. The details of the experimental conditions are given in the figure caption. Note that all aluminum clusters are detected with 7.87 eV. In addition all cluster ion signals vary linearly with the 7.87 eV photon flux. This is taken as evidence that all Al_n clusters have ionization threshold energies below 7.87 eV. A much different mass spectrum is obtained when the ionizing laser photon energy is reduced to 6.42 eV, as seen in the lower panel of Figure 1. At low 6.42 eV photon flux no signal is observed on Al_3, Al_4, and Al_5. This is evidence that these clusters have ionization threshold energies above 6.42 eV. In addition, only a very weak ion signal can be detected on Al_8 suggesting that the ionization threshold energy for Al_8 is quite close to 6.42 eV; while the ion signals on Al_9 and Al_{13} are fairly weak suggesting that their ionization threshold energies are close to but below 6.42 eV. Note that Al_7 and Al_{14} appear as locally intense peaks with the 6.42 eV ionization. We suggest that Al_7 and Al_{14} may be considered somewhat "magic" in the sense that they probably have somewhat higher ionization cross sections at 6.42 eV than their nearby neighbors. It is uncertain as to whether they are more (or less) stable (neutals or ions) than other nearby clusters. Note that with 7.87 eV ionizing laser photon energy little evidence of their "magic" character remains. In addition to ionizing with 6.42 and 7.87 eV photons, 6.5, 6.0, and 5.9 eV were also used in order to better bracket the ionization threshold energies. From these measurements we are able to bracket the ionization threshold energies for aluminum clusters as follows: Al_4 has the highest ionization threshold falling between 6.5 and 7.87 eV. Al_3 and Al_5 have ionization thresholds between 6.42 and 6.5 eV. Al_8 is quite near but probably slightly above 6.42 eV, whereas Al_9 and Al_{13} are also near but slightly below 6.42 eV. Al_2, Al_6, and Al_7 have threshold energies between 6.0 and 6.42 eV where as Al_{10}, Al_{11}, Al_{12} have ionization threshold energies between 5.9 and 6.42 eV. Larger aluminum clusters all have ionization threshold energies below 6 eV. Thus small aluminum clusters n=2-9 all have ionization threshold energies above that of the atom, 5.98 eV, [6] in contrast to transition metal clusters where the ionization threshold energies typically fall below that of the corresponding transition metal atom [1d,2,7]. This behavior has been attributed to the high density of 3p bonding orbitals in the smaller Al clusters [4].

III. Reactivity of Aluminum Clusters with Deuterium: Experimental

The experimental arrangement used to study chemisorption of reagent molecules on neutral metal clusters have been described previously [1,8]. Briefly a reactor tube is attached to the end of the narrow tube where cluster condensation occurs. A second gas pulse containing either pure helium or a helium/reactant mixture is injected into the reactor at a time synchronized to coincide with the passage of the metal clusters through the reactor. The bare clusters and clusters containing chemisorbed molecules exit the reactor tube, are collimated into a molecular beam and are detected by photoionization time of flight mass spectrometry. The ionizing laser in these reaction studies is an excimer laser operated on either the F_2 (7.9 eV) or the ArF (6.42 eV) laser transition.

The experimental measure of reactivity is the survival fraction of the bare aluminum cluster ion signal. Reactivity is measured as a function of cluster size and reactant molecule. Inherent in these measurements are the following major assumptions:

i) The bare cluster ion signal is proportional to the concentration of that cluster in the reactor and thus the ratio of bare cluster ion signals with and without reactant injected into the reactor is equal to the ratio of the cluster concentration before and after reaction.
ii) The collisional stabilization rate in the reactor is much faster than the unimolecular decomposition rate and thus pseudo first order reaction kinetics are valid.
iii) The photoionization process itself is non-destructive and produces predominately parent ions.

If such assumptions are valid, it has been shown [1c,9] that the rate constant, k_n, for the initial chemisorption step where the first molecule chemisorbs onto the bare cluster can be calculated from

$$k_n = -\ln\{I_f / I_o\} / [D_2] \, t$$

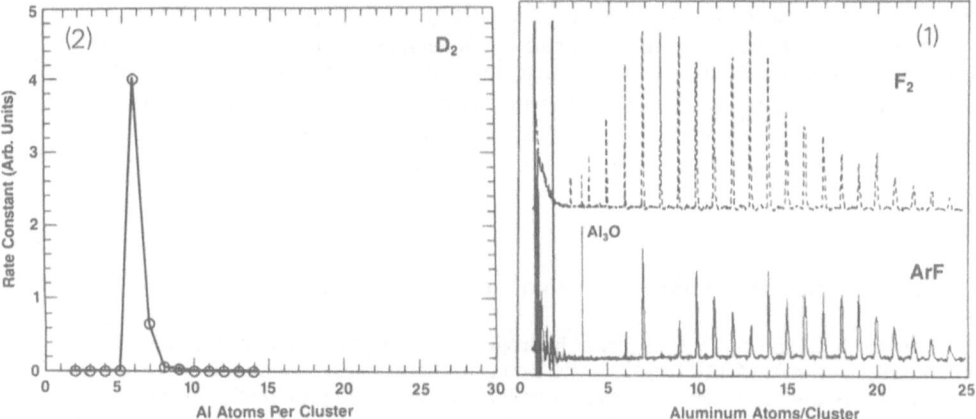

Figure 1. The photoionization time of flight mass spectrum for aluminum clusters at two different ionizing photon energies. The dashed upper trace is obtained for 7.87 eV photons (F_2 excimer laser line, 0.12 mJ/cm²) and for 6.42 eV photons (ArF excimer laser line, 0.33 mJ/cm²). The fall off in cluster ion signal toward the low and high mass side reflects the transmission function through two vertical deflection plates used to compensate for the component of translational energy of the clusters perpendicular to the direction of ion extraction. The two spectra are normalized to the ionizing laser fluence, but the 6.42 eV spectrum has been scaled by an additional factor of 6.9 in order better compare shapes.

Figure 2. The relative rate constant of aluminum clusters toward deuterium is plotted as a function of cluster size. The vertical scale is the value of $-\ln[I_f / I_o]$ divided by the ratio of deuterium to helium in the mix fed into the reactor.

where I_f and I_0 are the final and initial ion signals, respectively, $[D_2]$ is the concentration of the reactant, molecular deuterium in this case, and t is the residence time of the clusters in the reactor. In our pulsed reactor the absolute reactant concentration and interaction time can only be roughly estimated. However for a fixed set of operating conditions, all clusters experience the same conditions and thus the relative rate constant, $\ln\{I_f / I_0\}$, between different size clusters can be measured.

The chemisorption of deuterium on aluminum clusters is very cluster size specific as shown in Figure 2 where the relative rate constant is plotted as function of cluster size. We see that Al_6 is the most reactive aluminum cluster, observe no evidence of reaction for clusters containing less than 6 atoms, and find that the reactivity drops dramatically for clusters with more than 6 atoms. For those clusters for which reaction is measured, product ions of the form $Al_nD_2^+$ are observed.Similar results are obtained for hydrogen chemisorption except that the relative rate constants are about a factor of 1.7 ± 0.4 higher than for deuterium.

In an attempt to determine if larger aluminum clusters form stable Al_nD_y products, D_2 was added directly to the helium carrier gas [10]. In this instance product peaks, Al_nD_y, n=6-7, y=1-2; n=8-12, y=1-5; n=13-21, y=1-3; are detected (7.87 eV ionizing laser energy). Even though product ion peaks can now be observed out to n=21 under these conditions, such product ion peaks become progressively less intense as cluster size increases with those for n>15 being extremely weak compared to the corresponding bare cluster ion peaks. This is consistent with the larger aluminum clusters becoming substantially less reactive with increasing cluster size as observed above for molecular hydrogen chemisorption. Finally detection of y>2 product peaks on all reactive clusters except Al_6, suggests that Al_6D_y y>2 either are not formed or, if formed, are thermodynamically unstable and decompose, are not detected (ionization threshold higher than 7.87 eV), or are unstable upon ionization.

We note that the cluster size dependent hydrogen chemisorption observed for aluminum clusters differs greatly from that observed on transition metals. For the transition metals [1] a strong size dependent chemisorption typically is observed for the smaller clusters with less than 30 atoms. However the larger transition metal clusters are found to be the most reactive in contrast to the non-reactive behavior of the larger aluminum clusters.The chemisorption of aluminum clusters for different molecules (D_2O, O_2, CH_3OH, CH_4, D_2, CO and H_2S) in each case exhibits a different and unique dependence on both the rate and the cluster size. It is interesting that Al_6 is also the most reactive cluster towards CO. However, in contrast to the results with H_2, both smaller and larger clusters are found to react readily with CO. With water, Al_{10} is found to be the most reactive cluster, while smaller clusters are relatively inert and larger ones are reactive. These results as well as those for the other molecules listed above are discussed more fully elsewhere [3].

IV. Theoretical Aspects of H_2 Adsorption on Al_3-Al_6

The structural and energetic effect of hydrogen chemisorption on metal clusters has been recently discussed for Al [4,11], Li [12a], and Be [12b]. Numerous other studies of hydrogen chemisorption employing clusters as surface models have also been presented [13]. The computational methods used in this study to obtain the isolated Al [14] and Al_nH_2 [4] structures have been discussed elsewhere [14] and are only summarized here. Self-consistent wavefunctions for Al_n and Al_nH_2 were obtained in which electron-correlation effects were included self-consistently in all (3s and 3p) cluster valence orbitals (a generalized valence bond or GVB wavefunction [14]). Extensive electron correlation effects were subsequently included via POL-CI calculations in which a high order of excitation was allowed among the GVB natural orbitals. Such a procedure should be adequate to discriminate successfully among structures and provide accurate thermodynamic estimates for adsorption energies. Limitations arise from the need to optimize structures: we employed a gradient optimization technique [14] which yields accurate minimum energy structures, but is sensitive to the optimization starting point. We cannot exclude the possibility that more favorable structures were missed by this procedure.

Cluster	Geometry	H Atom Sites	H_2 Binding Energy (eV)	Bond Length Change (Å)[a]	Shortest R(Al-H)
Al_3	triangle	bridge	-1.09	0.16	1.82
Al_4	"puckered" rhombus	threefold	-0.85	0.25	1.75
Al_5	square pyramid	threefold	-0.73		2.06
	"	bridge	-0.64	0.16	1.83
Al_6	octahedron	opposite threefold	-1.53	0.30	1.78
	"	apex-sharing threefold	0.10	0.20	2.03
	"	bridge-sharing threefold	0.59		1.90

[a] largest change in Al-Al bond lengths due to hydrogen adsorption (unless change < 0.1 Å)

The geometric structures and thermodynamics of H_2 chemisorption for the Al_nH_2 clusters are summarized in Table I [4]. We found that the heat of adsorption increases with cluster size, but only found a single Al_6 structure among those considered for which chemisorption is exothermic. While an exhaustive search over possible structures was impractical, these findings are consistent with the experimental observation above that Al_6 is the smallest cluster that will adsorb H_2. The results in the table imply that the failure of smaller clusters to adsorb H_2 is a thermodynamic limitation. The large variations in ΔH values are mostly associated with electronic changes required of the cluster in order to accomodate the formation of Al-H bonds. The cluster orbitals are diffuse, extending over the entire cluster [4] and assuming shapes that are consistent with the 1s,1p,1d... characterization of the electron droplet or shell model [15]. To form bonds with approaching H atoms, which are more electronegative than the Al clusters, these delocalized bonding orbitals must mix with one another in order to increase orbital amplitude in the region of the H binding sites. If the orbitals that mix are degenerate and equally occupied, there is no cost in this mixing. For most of these clusters this was not the case however. In some cases, the Al cluster could only bond to both H atoms after first promoting the cluster to an electronic excited state. The excitation energy required translates directly into loss of adsorption energy. In other cases, ground state orbitals were adequate for chemisorption, but still required changes in the occupation of the orbitals (e.g. mixing among singly and doubly occupied orbitals) that had an associated energetic cost. These orbital changes also altered the character of the cluster bonding orbitals themselves in such a way that the geometry of the cluster framework was itself altered. This implies a competition between the strength of the Al-Al bonds and the Al-H bonds. In even the worst case (Al_3) the H atom binding energies are comparable to the cluster cohesive energy per atom [4,14], providing a thermodynamic driving force to alter the cluster framework as needed to accomodate the H atoms. Only the most stable Al_6H_2 structure was essentially immune from these problems. As discussed below, the highest occupied orbitals of the Al_6 octahedron are two nearly degenerate 1d orbitals (t_{2g} symmetry), with lobes extending from threefold faces. These orbitals mix to form the Al-H bonds with almost no energy cost and only very small structural changes in the cluster framework. It is dangerous to extrapolate beyond this short series, nevertheless the Al_6 octahedron does appear to be special in the Al_n series. Its electronic structure is ideally suited to the binding of two H atoms, as well as the activation of the H_2 molecule (see below). By symmetry, there are twelve pairs of edge-sharing threefold sites that are nearly degenerate and dissociation should be possible at each edge. Thus the probability of reaction should be unusually high.

Even though Al_6 appears 'special', why the adsorption of H_2 terminates abruptly beyond Al_6 remains an open question. Less extensive calculations on hydrogen chemisorption for larger Al clusters have been reported [11]. Binding energies of 0.92 eV per H atom for $Al_{13}H_2$ and 0.85 eV for $Al_{13}H_6$ were calculated for an HCP geometry of the Al_{13} cluster, in which the H atoms were found to penetrate the cluster 'lattice'. These energetics appear consistent with experimental observations, but the authors did not discuss the origins of the weak Al-H bonds.

V. A Model for the Activation of the H-H bond by Aluminum Clusters

A. A Brief Review of Previous Models

The activation of H_2 by metal clusters and surfaces has been the subject of a number of theoretical investigations in recent years. Earlier efforts were largely the result of applying molecular concepts to surface problems (with appropriate modification) such as symmetry relations [16a] and simple valence bond (LEPS) hamiltonians [16b-c]. More recent efforts have built on these with discussion of the role of Pauli repulsion [17], initial and final state mixing [18], 'correlations' [19], charge transfer [20,21], and dynamical effects [20, 22]. By way of introduction, we will briefly discuss a few representative models that capture the essential physical effects that collectively are responsible for dissociative adsorption of H_2.

The simplest view of the factors responsible for dissociative adsorption results from an application of first-order perturbation theory, as discussed most recently by Shustorovich [21]. A central issue addressable via this approach is the role of charge transfer from the metal to the H_2 σ^* in facilitating dissociation. The author points out that the elevated energetic position of the σ^* does not preclude its participation in dissociation, and derives simple expressions for the stabilization of an occupied metal orbital M' due to mixing with the σ^*, which we write as,

$$Q_{M'} \approx \beta^{*2} / \delta E_{M'}$$

and conversely, stabilization of the H_2 σ due to mixing with an unoccupied metal orbital M:

$$Q_H \approx \beta^2 / \delta E_M$$

In these expressions, β and β^* are matrix elements between a metal orbital (M or M') and the H_2 σ and σ^*, respectively, $\delta E_{M'} = E(\sigma^*) - E(M')$, and $\delta E_M = E(M) - E(\sigma)$. In general, we assume that $E(M)$ and $E(M')$ are near the fermi level, and $E(M) > E(M')$. Thus when $\beta^*/\beta > 1$, donation to σ^* will dominate (i.e. $Q_{M'}/Q_H > 1$) as $E(M')$ approaches $E(\sigma^*)$, which is usually less true for metal clusters than for metal surfaces. Shustorovich points out however that when the H-H axis is perpendicular to the surface, $\beta^*/\beta < 1$ except for small molecule-surface separations [23].

An alternative many-electron view is provided by Upton [18a], who has shown that when the transition state is viewed as a superposition of initial (H-H and M-M bonds) and final (two H-M bonds) states, the Pauli principle introduces specific requirements on the total wavefunction that govern the orbital shapes. Specifically,

$$\psi(TS) = \psi(initial) + \psi(final)$$
$$= A\{(\phi_H\phi_M\phi_{H'}\phi_{M'}) (2(\alpha\alpha\beta\beta+\beta\beta\alpha\alpha)-\alpha\beta\alpha\beta-\beta\alpha\beta\alpha-\alpha\beta\beta\alpha-\beta\alpha\alpha\beta)\}$$

where ϕ_H, and $\phi_{H'}$ are orbitals localized on the individual H atoms, and ϕ_M, and $\phi_{M'}$ are the metal orbitals to which the H atoms are bonded at the transition state. The spin function shown requires that ϕ_H and ϕ_M be triplet coupled to one another and thus orthogonal. The same is true of $\phi_{H'}$ and $\phi_{M'}$. In calculations employing this representation, it was found that the H_2 σ^* is forced to become populated (e.g. $\phi_{H'}$ evolves to $\phi_{H'-H}$) in order to achieve this orthogonality. Thus from both the many electron and perturbation theory views, the σ^* plays an important role.

The two approaches differ however, in that the many electron picture finds the energetics in the <u>approach</u> to the transition state to be dominated by Pauli repulsion. In the approach, the total wavefunction is well-described by $\psi(initial)$ and only very close to the transition state does the mixing with $\psi(final)$ become important. Since $\psi(initial)$ contains separate H-H and M-M bonds, (Pauli) repulsion between these bonds (they must remain orthogonal) drives the energy higher. A similar picture has been explored recently by Harris and Andersson [17], who point out that Pauli repulsion is given approximately by,

$$E_P = \sum_i^{occ} \varepsilon_i' - \sum_i^{occ} \varepsilon_i$$

where the ε_i are orbital energies of the free metal (cluster) and the ε_i' are metal orbital energies that result when an adsorbate is brought up next to the surface and the metal orbitals are allowed to relax and become orthogonal to the (frozen) adsorbate orbitals. They point out that for many metal systems this expression must be altered to account for the fact that as the orbitals relax, it may become energetically favorable to populate entirely new metal orbitals that interact less repulsively with the adsorbate. An example of such shifts in occupation was presented in which dissociation of H_2 over Ni produce 4s-> 3d occupation shifts and a resultant small barrier.

Similar effects have been noted for H_2 interacting with Sc^+ [24], and apparently in another study of H_2 on Ni [19] though in that study the presence of a small barrier was not described in these terms.

Finally, we note that various attempts have been made to relate the size of an activation barrier to the heat of adsorption of the dissociated atoms. Shustorovich [25], using a Morse potential representation for H-H and M-H bonds (similar to that employed in LEPS studies [16b,c]) has derived an expression relating the activation energy ΔE^* to the atomic heats of adsorption Q_i and the molecular dissociation energy D_{ij} as

$$\Delta E^* = D_{AB} - (Q_A + Q_B) + Q_A Q_B/(Q_A + Q_B).$$

This expression results from constraining the system at the transition state to conserve the original bond order of the adsorbing molecule (i.e. bond order=1). The bond order is distributed over the vanishing adsorbate bond and the forming metal-adsorbate bonds, but it makes no assumption about the bond order of the affected metal-metal bonds. While this might be a valid approximation for some metal surfaces (supporting data suggests that it is), it is questionable whether this approach may be extended to small metal clusters where the effect of the adsorbate on metal-metal bonding is likely non-negligible.

B. Activation of H_2 by Al_6

In this section we will explore a particular reaction pathway leading to the stable Al_6H_2 cluster from isolated H_2 and the reactive Al_6 cluster. The purpose of these calculations is not to fully characterize the dissociation process, since to do so would require consideration of a prohibitively large number of possible reaction pathways. Rather, we wish to determine some general features about the nature of the interaction between H_2 and Al clusters, and identify some of the electronic requirements for a favorable interaction path.

The interaction geometry used in the calculations is shown schematically in figure 3. In this pathway, the H_2 molecular axis is aligned above and perpendicular to an Al-Al nearest neighbor bridge separating adjacent threefold faces of the Al_6 cluster. The H_2 molecule was allowed to approach the cluster along a line connecting the H-H and Al-Al bond midpoints. As the H_2 molecule approaches the cluster, the H-H separation increases greatly (at some cost in energy), allowing the H atoms to take up positions near the centers of the adjacent threefold sites. The ground singlet and triplet states of Al_6 are distorted slightly (a Jahn-Teller distortion from a t_{2g}^4 occupation in the perfect octahedron) and upon H_2 chemisorption this distortion increases further. To separate the effects of cluster distortion from bond activation, we have frozen the Al_6 cluster in the geometry that it assumes in Al_6H_2. This distortion raises the energy of Al_6 by only about 0.2 eV from the isolated Al_6 geometry. The character of the Al_6 orbitals remains unchanged.

The highest occupied orbitals of the isolated (and distorted) Al_6 cluster are also shown in figure 3a-b. In calculations for isolated Al_6 the lowest state was found to be a triplet state, with one electron in each of these orbitals. Almost degenerate with this was a singlet state in which both electrons occupied the the a_1 orbital [26]. As no low-lying triplet Al_6H_2 states were found, we have considered here only the interaction of H_2 with singlet Al_6 to produce the ground state of Al_6H_2. By symmetry, as the reaction proceeds the H_2 σ orbital will interact with the Al_6 a_1 orbital shown, and the H_2 $\sigma*$ orbital will interact with the Al_6 b_2 orbital. As shown below, the essential elements of the reaction may be understood by consideration of only these four orbitals.

In figure 4, a potential energy surface for the dissociative adsorption reaction is shown. The surface is characterized by a rapid variation in energy with respect to H-H distance at large

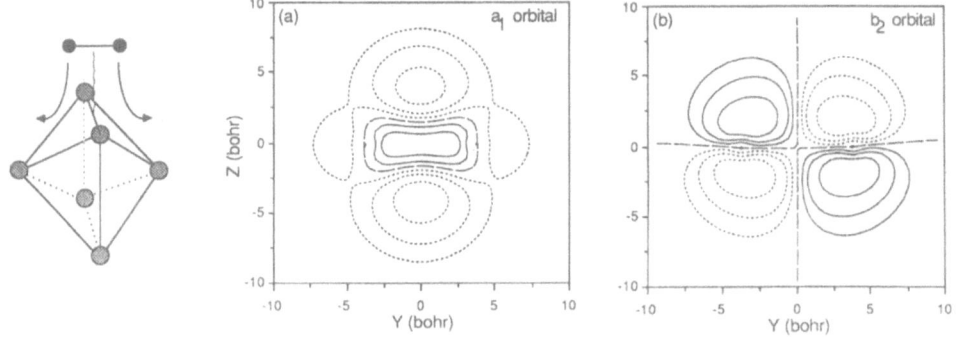

Figure 3. The interaction geometry and plots of the two highest occupied orbitals in Al_6 (see text). The plotting plane is coincident with the reaction plane. Contours (dotted, dashed, and solid for negative, zero, and positive, respectively) are plotted for amplitudes of .01, .02, .04, etc.

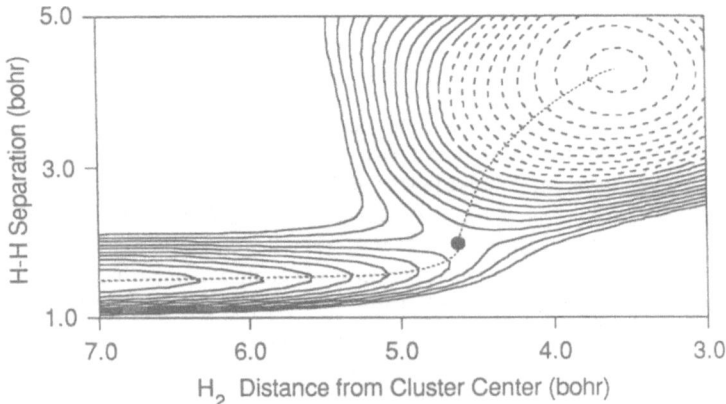

Figure 4. A potential energy surface for the reaction between H_2 and Al_6, showing an approximate reaction coordinate (dotted line). Contours are spaced by 0.1 eV. The transition state is marked by the filled black circle. Dashed contours are energies below the separate molecule limit, solid contours above.

H_2 - Al_6 separation (the unperturbed H_2 bond), a saddle point (the transition state), and a broad minimum at large H-H separation (the final H atom positions in Al_6H_2). A 'reaction coordinate' [27] connecting these points is marked in the figure. Evidently, the H_2 is able to approach to within a short distance of the cluster before the H-H bond begins to elongate ($r(H_2$ - bridge) $\approx$ 2.15 a_0). Even though the H_2 separation is unaffected over this distance, the total energy of the H_2 -Al_6 system rises. This energetic barrier has grown to more than half of its total value of 0.78 eV before the H-H bond abruptly begins to elongate. The principle source of the energetic barrier is Pauli or exchange repulsion. The filled H_2 σ orbital is not orthogonal to the a_1 cluster orbitals, but is required to become so by the Pauli principle. The cluster orbitals, being more polarizable than the H_2 σ orbital, are able to adjust to achieve orthogonality. As discussed in the last section, this adjustment nevertheless costs energy, and the energetic cost increases as the H_2 molecule approaches.

As noted above, when the H_2 molecule is far from the cluster, the H_2 σ and cluster a_1 (fig. 3a) orbitals are filled while the H_2 σ^* and cluster b_2 (fig. 3b) orbitals are empty. At the transition state, the situation is altered as depicted in figure 5. As the H_2 molecule approached the cluster, the a_1 orbital was destabilized by Pauli repulsion. Ultimately, it became more favorable for electrons in the highest cluster orbital to shift from the a_1 to the b_2 orbital in order to reduce Pauli repulsion. Thus at the transition state, the occupied orbitals are the H_2 σ and the cluster b_2 orbitals in figure 5. This shift in occupation is of the same origin as that described by Harris and Andersson [17] (see above). There are also more subtle changes from the separated molecule limit that are revealed in the detailed shape of these orbitals. The cluster b_2 orbital has delocalized onto the H_2 molecule. To do so it has partially occupied the H_2 σ^* orbital. In a similar fashion, the H_2 σ orbital has delocalized onto the cluster and in this case partially occupied the a_1 orbital. This may be seen qualitatively by comparing the cluster component of the orbital in fig. 5a with the a_1 orbital (fig. 3a), and may also be verified numerically. On balance, delocalization from the metal into the σ^* dominates leading to a net negative charge of $\approx$ 0.15 electrons/H atom (by Mulliken populations) at the transition state. The result is consistent with the perturbation theory view presented in the last section with a_1=M and b_2=M'.

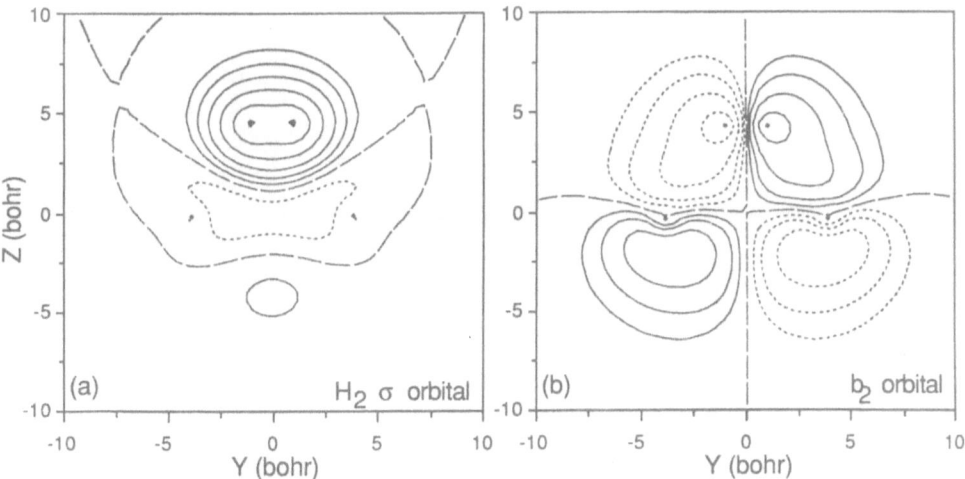

Figure 5. The two reacting orbitals of the Al_6-H_2 complex at the transition state, plotted as in figure 3. The H atom positions are indicated by dots at z $\approx$ 5.0 a_0. The bridge Al atoms are above and below the plotting plane at y=0, z=2.46 a_0.

D. Effect of Cluster Charge on Dissociative Adsorption

The calculations characterizing the interaction between H_2 and neutral Al_6 illustrate that both long range Pauli repulsion and short range H-H bond elongation induced by metal -> σ^* donation play a role in dissociation of H_2. We may further clarify the relative importance of these two factors by altering the character of the cluster orbitals that participate in the dissociation process (e.g. those in fig. 3). A simple way in which to achieve this modification is by introducing a charge into the system. By removing an electron from Al_6, the remaining electrons are more tightly bound. This would be expected to both reduce the polarizability of the cluster and make it less favorable for the cluster to donate charge to the H_2. Conversely, adding an electron to Al_6 would be expected to increase the polarizability and the tendency towards charge transfer. There will be other electrostatic effects associated with the presence of a charge in the system; we will address these in a later section. It is important to note that the introduction of charge must be done in such a way as not to alter the number of cluster electrons actually interacting directly with the adsorbate. This is easily achieved: the reaction occurs entirely in a plane passing through the center of the Al_6 cluster, so adding or removing an electron from orbitals that possess a node in the reaction plane satisfies this requirement.

Potential energy surfaces resulting from carrying out the reaction on cationic, and anionic Al_6 are compared with the neutral surface in figure 6. The same interaction geometry was used in each case. Clearly, the effect of including the charge has only a modest effect on the reaction energetics. Indeed, to within the accuracy of the calculations, we find the geometric coordinates of the transition state to be the same in each case. While this consistency may well be fortuitous, it is surprising that the transition states are at all similar given the range of cluster orbital binding energies represented by these systems (the orbital energy of the highest occupied cluster orbital is 2.6, 6.2, and 10.4 eV for the anion, neutral, and cation, respectively). Further, the position of the onset of H-H bond elongation is essentially the same in each case, as is position of the minimum corresponding to adsorbed H atoms. The activation barriers grow in a comparable way as the H_2 approaches each cluster, but possess final heights that differ by a small but (kinetically) significant amount ($\Delta E^* = 0.72, 0.78$, and 0.86 eV for anion, neutral, and cation, respectively). Barrier heights correlate only crudely with the H atom adsorption energies ($0.78, 0.91, 0.92$ eV for cation, neutral, and anion, respectively), as some studies have suggested for metal surfaces [25] (see above).

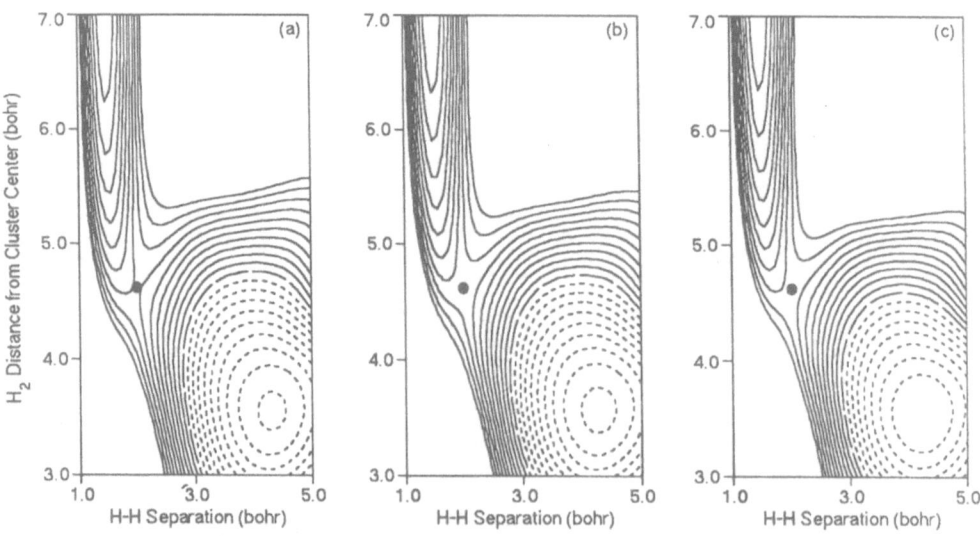

Figure 6. A comparison of potential energy surfaces for the charged Al_6-H_2 systems: (a) anion; (b) neutral; (c) cation. Plotting parameters are as in figure 4.

A different view of the competing factors determining the dissociation process is provided by figure 7, where three different measures of the reaction are shown as a function of the reaction coordinate (a path for each system analogous to that marked in fig. 4 for the neutral). In fig. 7a a trace of the change in total energy for the $H_2 + Al_6^x$ (x=-1,0,1) system is shown from the separate molecule limit to the transition state via the reaction coordinate. From this view it is evident that the discrimination between activation barriers begins while the H_2 is still far from the surface, and the only significant operative interaction at this distance is Pauli repulsion. The energetic ordering reflects the difference in polarizabilities of the clusters. The anion orbitals, since they are the most polarizable of the three Al_6 systems, are able to adjust to become orthogonal to the incoming H_2 at lesser energetic cost than either the neutral or the cation.

The effect of the other major factor in H_2 bond activation, charge transfer from the metal to H_2, is illustrated by the remaining panels in the figure. Figures 7b shows the net charge transfer per H atom from the cluster as a function of distance along the reaction coordinate and clearly indicates that this charge transfer is a far shorter range phenomenon than the buildup of repulsive interactions. The short range of such charge transfer processes is consistent with the perturbation theory view above [21] and has been noted elsewhere [17]. Significant charge transfer occurs only after the barriers are partially formed and the ordering of barrier heights well-established. Surprisingly, the charge transfer buildup at short range is almost <u>independent</u> of cluster charge both in rate of buildup and final magnitude at the transition state. Moreover, the onset of this charge transfer is not signalled by any important change in the way that the activation barrier forms in figure 7a. However, it does correlate very strongly with the rate at which the H-H bond distance changes, as plotted in figure 7c. Only a single curve is indicated in figure 7c; the rate of bond elongation was virtually identical (within our ability to extract a reaction coordinate from the potential energy surfaces) for all three cluster systems. A clear causal relationship between charge transfer and bond elongation is suggested by the data, but neither of these seems to be obviously related to the actual barrier heights.

The computed activation energies correlate well with cluster ionization potential, and demonstrate the same type of relationship between electron binding energy and reaction rate (i.e. binding energy is proportional to log(rate)) that has been shown to be present experimentally over a wide range of cluster sizes for transition metal systems [1a,1d,2]. It is surprising

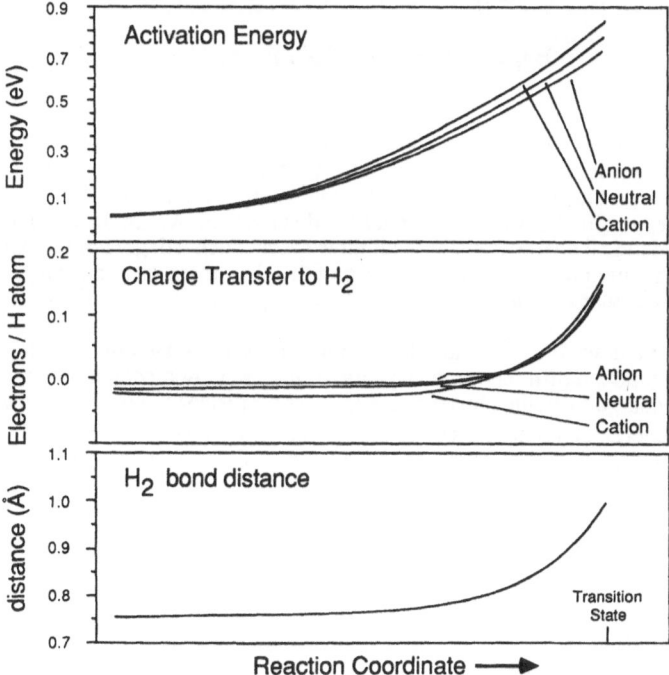

Figure 7. Three views of the Al_6-H_2 systems as a function of the reaction coordinate (see figure 4): (a) activation energy; (b) charge transfer from Al_6 to H_2; (c) H_2 bond distance change.

however, that the enormous range in ionization potentials present in the charged Al_6 clusters ($\approx$ 8 eV) has a relatively small effect on the barriers ($\approx$ .15 eV). It is also surprising that the range in ionization potentials has little effect on the degree of charge transfer to H_2 near the transition state. Part of the explanation no doubt lies in the fact that these are <u>charged</u> systems, and additional electrostatic effects are present in this theoretical study that were not present in the neutral transition metal experiments. Two principal effects deserve mention. First, the presence of a charge on the Al_6 cluster should induce a dipole moment on the incoming H_2 species. In an actual experiment, this dipole might well lead to reorientation of the adsorbate, a different interaction geometry, and an enhancement of the reaction cross section as is well-known in ion-molecule reaction studies. In our idealized study, the dipole must be induced by polarization of the H_2 electron density normal to the molecular axis and should be small. Evidence for this polarization in the calculations is present in fig. 7b at large interaction distances. In each case the H_2 shows a small positive charge at large distance due to orbital mixings induced by Pauli principle effects and basis set superposition errors [28]. This charge is largest for the Al_6^+-H_2 interaction since the H_2 density is polarized <u>towards</u> the cluster and experiences larger Pauli repulsion at each distance than the other systems. For Al_6^--H_2, the H_2 density polarizes <u>away</u> from Al_6 and the positive charge (and Pauli repulsion) is smallest.

The second, and probably more important, electrostatic interaction arises at close range when charge transfer to the H_2 becomes favorable. At a given interaction distance, the relative degree of charge transfer is principally determined by the difference between cluster ionization potentials: the cation will delocalize less onto H_2 because the IP of the delocalizing cluster orbital is largest. For neutral clusters then, we would not expect to see a result similar to fig. 7b; charge transfer should be largest and occur at greater distance for the cluster with the smallest orbital energy for the delocalizing orbital (figs. 3b,5b). An additional factor must be considered however when the IP difference is induced by the presence of a cluster charge. A dipole interaction results from the charge transfer and in a simple point dipole approximation will take the form:

$$E_{dipole}(anion) \approx -\delta q^2 / R + \delta q / R$$

$$E_{dipole}(neutral) \approx -\delta q^2 / R$$

$$E_{dipole}(cation) \approx -\delta q^2 / R - \delta q / R$$

where δq is the charge transferred to the H_2 and R is the effective charge separation. There is an additional energetic term for the charged systems that is attractive for the cation (thus enhancing the probability of charge transfer) and repulsive for the anion (diminishing charge transfer). These terms counter the effect of differing orbital IP's discussed above. In these calculations the net effect is nearly identical charge transfer in all three cases. Such a degree of similarity may well be fortuitous; the essential point is that the presence of the charge serves to diminish the effect that such large IP differences might have been expected to produce on the activation barriers.

In summary then, we find that the activation barrier is dominated by Pauli repulsion that develops at large interaction differences. Elongation of the H-H bond occurs only when it is possible to transfer charge from the cluster to the H_2 at short distance. We believe these conclusions to be general, and indeed they are consistent with previous work on different metal systems [17, 19, 24]. For neutral clusters, we should expect the energetic cost for both effects to be smallest (and thus the dissociative adsorption rate to be largest) when the cluster IP is smallest. When the difference in orbital IPs is magnified by the presence of positive or negative charges on the cluster, the rate difference will not necessarily be correspondingly magnified since dipole interaction terms will counter the effect of the magnified IP differences at short range. It is important to note that such general statements are clearly subject to exceptions: here we have considered a case where electronic 'conditions' are ideal for bond activation. Should the orbital occupations or symmetries differ, or the cluster state densities be sparse such that orbitals of appropriate symmetries are inaccessible, the resulting rates may be drastically affected. We would expect such problems to be most pronounced for smaller clusters. In this regard, we note that while our conclusion that an IP-log(rate) correlation has generality is based on study of Al_6, the small Al clusters as a group do not obey this relation. We take this as further evidence that for these clusters, hydrogen chemisorption is thermodynamically, rather than kinetically limited. In a similar vein, a recent FT-ICR experiment [29a] has shown that the relative reaction rates for H_2 dissociation for Nb_7-Nb_9 are largely independent of the presence of a positive charge on the cluster. Such results are not necessarily surprising in light of the calculations presented here, especially in the event that the orbital being ionized in the experiment is not one of those participating in the reaction, as was true here. The experiments do not represent firm evidence that the IP-log(rate) correlation noted for Fe clusters [1a] (and confirmed here) is fortuitous as was suggested [29]. Observed rate enhancements upon cluster ionization for N_2 adsorption [29b] likewise may well result from enhanced reaction cross sections that are common in ion-molecule reactions (due to induced dipoles as noted above), and similarly do not represent evidence against an IP-log(rate) relation.

References

1. For H_2 chemisorption on transition metal clusters see (a) R. L. Whetten, D. M. Cox, D. J. Trevor, and A. Kaldor, Phys. Rev. Lett. 54, 1494 (1985). (b) S. C. Richtsmeier, E. K. Parks, K. Liu, G. Pobo, and S. J. Riley, J. Chem. Phys. 82, 3659 (1985). (c) M. D. Morse, M. E. Geusic, J. R. Heath, and R. E. Smalley, J. Chem. Phys. 83, 2293 (1985). (d) D. M. Cox, R. L. Whetten, M. R. Zakin, D. J. Trevor, K. C. Reichmann, and A. Kaldor, AIP Conference Proceedings No. 146, Adv. in Laser Science-I, Nov. 1985, Eds. W. C. Stwalley and M. Lapp, AIP, New York (1986). (e) M. E. Geusic, M. D. Morse, and R. E. Smalley, J. Chem. Phys. 82, 590 (1985).
2. R. L. Whetten, M. R. Zakin, D. M. Cox, D. J. Trevor and A. Kaldor, J. Chem. Phys. 85, 1697 (1986).
3. D. M. Cox, R. L. Whetten, D. J. Trevor, and A. Kaldor, to be published.
4. T. H. Upton, Phys. Rev. Lett. 56, 2168 (1986).
5. For example see (a) T. G. Dietz, M. A. Duncan, D. E. Powers, and R. E. Smalley, J. Chem. Phys. 74, 6511 (1981). (b) D. L. Michalopoulos, M. E. Guesic, S. G. Hansen, D. E. Powers, and R. E. Smalley, J. Phys. Chem. 86, 2556 (1982).(c) V. E. Bondybey, J. Phys. Chem. 86, 3396 (1982). (d) E. A. Rohlfing, D. M. Cox and A. Kaldor, Chem. Phys. Lett. 99, 161 (1983).(e) A. Rohlfing, D. M. Cox, and A. Kaldor, J. Phys. Chem. 88, 4497. (1984).
6. C. E. Moore, Nat. Stand. Ref. Data Ser., Nat. Bur. Stand. (U.S.), NSRDS-NBS 35 (1971).
7. E. A. Rohlfing, D. M. Cox, A. Kaldor and K. H. Johnson, J. Chem. Phys. 81, 3846 (1984).
8. (a) R. L. Whetten, D. M. Cox, D. J. Trevor, and A. Kaldor, J. Phys. Chem. 89, 566 (1985). (b) M. E. Geusic, M. D. Morse, S. C. O'Brien, and R. E. Smalley, Rev. Sci. Instr. 56, 2123 (1985).
9. D. M. Cox, K. C. Reichmann, D. J. Trevor, and A. Kaldor, submitted for publication.
10. Clusters containing an odd (as well as an even) number of D atoms are produced when deuterium is added to the carrier gas because deuterium is also decomposed in the vaporization region producing the highly reactive atomic species. Such effects have been observed on iron clusters (E. K. Parks, K. Liu, S. C. Richtsmeier, L. G. Pobo, and S. J. Riley, J. Chem. Phys. 82, 5470 (1985).
11. H. Partridge and C. W. Bauschlicher, J. Chem. Phys. 84, 6507 (1986).
12. (a) B. K. Rao, P. Jena, and M. Manninen, Phys. Rev. Lett. 53, 2300 (1984). (b) C. W. Bauschlicher, Chem. Phys. Lett. 117, 33 (1985).
13. see for example (a) T. H. Upton and W. A. Goddard III, Phys. Rev. Lett. 42, 472 (1979). (b) S. G. Louie, ibid. 42, 476 (1979).(c) C. Umrigar and J. W. Wilkins, ibid. 54, 1551 (1985). (d) P. Nordlander, S. Holloway, and J. K. Norskov, Surf. Sci. 136, 59 (1984). (e) H. Nakatsuji and M. Hada, J. Amer. Chem. Soc. 107, 8264 (1985). (f) J. Garcia-Prieto, M. E. Ruiz, and O. Novaro, J. Amer. Chem. Soc. 107, 5635 (1985). (g) H. O. Beckmann and J. Koutecky, Surf. Sci. 120, 127 (1982). (h) P. Cremaschi and J. L. Whitten, Phys. Rev. Lett. 46, 1242 (1981).
14. T. H. Upton, J. Chem. Phys., submitted, and references within.
15. W. Knight, K. Clemenger, W de Heer, W. Saunders, M.-Y. Chou, and M. Cohen, Phys. Rev. Lett. 52, 2141 (1984).
16 (a) C. F. Melius, Chem. Phys. Lett. 39, 287 (1976). (b) J. H. McCreery and G. Wolken, Jr., J. Chem. Phys. 64, 2845 (1976). (c) V. I. Avdeev, T. H. Upton, W. H. Weinberg, and W. A. Goddard III, Surf. Sci. 95, 391 (1980). (d) A. Gelb and M. J. Cardillo, Surf. Sci. 64, 197 (1977).

17. J. Harris and S. Andersson, Phys. Rev. Lett. $\underline{55}$, 1583 (1985).
18. (a) T. H. Upton, J. Amer. Chem. Soc. $\underline{106}$, 1561 (1984). (b) A. K. Rappé and T. H. Upton, J. Amer. Chem. Soc. $\underline{107}$, 1206 (1985).
19. P. E. M. Siegbahn, M. R. A. Blomberg, and C. W. Bauschlicher, J. Chem. Phys. $\underline{81}$, 2103 (1984).
20. (a) S. Holloway and J. W. Gadzuk, J. Chem. Phys. $\underline{82}$, 5203 (1985). (b) J. W. Gadzuk and S. Holloway, Chem. Phys. Lett. $\underline{114}$, 314 (1985). (c) D. K. Bhattacharyya, J.-T. Lin, and T. F. George, Surf. Sci. $\underline{116}$, 423 (1982).
21. (a) E. Shustorovich, J. Phys. Chem. $\underline{87}$, 14 (1983). (b) E. Shustorovich and R. C. Baetzold, Science $\underline{227}$, 876 (1985). (c) E. Shustorovich, R. C. Baetzold, and E. L. Muetterties, J. Phys. Chem. $\underline{87}$,1100 (1983).
22. C.-Y. Lee and A. E. DePristo, J. Chem. Phys. $\underline{84}$, 485 (1986); $\underline{85}$, 4161 (1985).
23. this conclusion is the result of simple geometric arguments that are specific to nd orbitals. It is possible that less severe criteria are appropriate for other orbital symmetries or types.
24. A. K. Rappé and T. H. Upton, J. Chem. Phys. $\underline{85}$, 4400 (1986).
25. E. Shustorovich, Surf. Sci. $\underline{150}$, L115 (1985).
26. in fact the occupation is best described as $(\lambda_1 a_1^2 - \lambda_2 b_2^2)(\alpha\beta - \beta\alpha)$ where $\lambda_1 > \lambda_2$.
27. obtained by following the 'steepest descent' path from the transition state the initial and final states.
28. because of the large number of calculations required for this study, the basis set employed was more limited than in the study of Al_n and Al_nH_2 (see refs. 4,12). The Al basis was reduced to (3s,2p) and the H basis to (2s), with all other procedures remaining the same. This basis will have a limited ability to describe the polarization effects present in this system.
29 (a) J. M. Alford, F. D. Weiss, R. T. Laaksonen, and R. E. Smalley, J. Phys. Chem. $\underline{90}$, 4480 (1986). (b) P. J. Brucat, C. L. Pettiette, S. Yang, L.-S. Zheng, M. J. Craycraft, and R. E. Smalley, J. Chem. Phys. $\underline{85}$, 4747 (1986).

UNENHANCED RAMEN SPECTROSCOPY

OF CO ADSORBED ON NICKEL PARTICLES

E. B. Bradley and H. A. Marzouk

Department of Electrical Engineering
University of Kentucky
Lexington, Kentucky

Introduction

Our understanding of chemical bonding at surfaces has increased by using a variety of experimental techniques to study well-defined surfaces in high vacuum where surface impurities can be controlled. Studies of adsorption on these surfaces provide model systems that can be studied systematically to determine the evolution of molecular surface structure and bonding as environmental parameters are changed. A technique recently used to study chemical bonding on well-defined surfaces is normal unenhanced Raman spectroscopy (NURS). This technique was developed at the University of Kentucky, and it was used to obtain Raman spectra of molecules of low polarizability adsorbed on the non-enhancing metal surfaces Ni(111) and Ni(100). Raman signals from these adsorbates could be detected from one reflection off the metal surface. The technique is non-destructive of the surface species and low laser power (100-300 mW at the sample surface) is used.

The development of this capability is important because now surfaces need not be tailored to cause surface enhancement of Raman signals. Such tailoring may cause changes in the surface chemistry and reactivity.

We also have the opportunity with NURS to expand our knowledge of the adsorption by metals of reactant gases at higher pressures where commercial catalysis often occurs. One may follow with Raman the evolution of molecular surface structure and bonding on single crystals beginning at low pressure (UHV) and continuing on to higher gas pressures where other surface analysis techniques fail. This advantage of Raman spectroscopy is important because at pressures up to several atmospheres, kinetic measurement of catalyzed reactions of synthesis gases on well-defined metal surfaces show that these data are revalent to catalytic processes occurring over high surface area supported catalysts.[1.] Thus an important connection has been established between the behavior of commercial catalytic processes and the microscopic structure of the surface of a catalyst.

Fischer-Tropsch catalysis has been chosen for study because of its clear technological importance. The methanation catalyst $NiO/Cr_2O_3/MgSiO_3$ has been studied by Raman spectroscopy as prepared and after methanation. The rationale for this study is comparison and contrast of CO adsorption on suspended nickel particles (in catalyst) with CO adsorption on clean, oriented single crystals of nickel. The Raman spectra were recorded at more realistic CO exposures (10^6L) than is often the case in single crystals studies, although unenhanced Raman spectroscopy can detect vibrational modes of surface adsorbed CO at residual gas pressures of 10^{-9} Torr.

Attempts to correlate results for catalyst and crystal are often complicated. At a resolution of 1 cm^{-1} one records on crystals many CO modes due to linear and bridged CO, whereas for the catalyst these modes are unresolved or are absent.

The most important application of Raman spectroscopy to surface studies is in vibrational analysis of the adsorbate. The analysis is available only in a limited number of cases. Surface bonding and reactivity can be studied by Raman, and because of its intrinsically high resolution (1 cm^{-1}) Raman spectroscopy provides information on multiplicity of adsorption states. Isotope-induced shifts of CO vibration frequencies were also recorded and used to verify assignments.

Bonding to transition metal surfaces not only varies across the periodic table, but also the fundamental vibration frequency is altered by bonding to different crystallographic planes of a metal surface (ref) but nevertheless, studies of gas adsorption on single crystal surfaces provide "model" systems that can be controlled and studied systematically. The assignment of the vibrational modes is important for characterizing the surface structure of adsorbed molecular species and must be accounted for in any explanation of the reactivity of the metal surface.

The problem of correlation of single crystal results with actual catalysts is further complicated by the following: When a molecule is adsorbed on a surface the rotational and translational degrees of freedom become vibrational degrees of freedom. Thus there are 6 (or 5) so-called "frustrated" translations and rotations, termed external modes, that can mix with the former internal modes of the same symmetry. In principle, this mixing can modify the frequency of the internal modes, even if there is no change in the force constants of the molecular system. The number of bands appearing in the vibrational spectrum will depend on the symmetry that is dictated by the molecular structure and the surface site, i.e., the molecular point group of the adsorbate complex. The number of Raman bands actually observed will depend on the appropriate selection rules that account for site symmetry.

The "parallel" vibrations are Raman active because the selection rules depend upon change in the polarizability of adsorbed species. Complete vibrational and structural assignment cannot be made without knowledge of perpendicular and parallel modes. Thus Raman, unlike infrared spectroscopy, is not bound by the metal surface selection rule and is complementary to infrared spectroscopy. Infrared spectroscopy, though not a pressure limited technique, is bound by the metal surface selection rule and "parallel" modes are not recorded.

Experimental

Catalyst - Catalysts were transferred to the Raman spectrometer under nitrogen atmosphere, then placed in a rotating sample cell to avoid excessive surface heating by the beam from an Ar^+ laser, 100-300 mW. All Raman spectra for catalysts and single crystals were recorded at 5145 $\overset{o}{A}$ and 4880 $\overset{o}{A}$. Spectra were recorded on a Jarrell-Ash model 25-300 Raman spectrometer controlled by a PDP/11 computer. The dark count is 1 cps or less and the computer was programmed for 1-2 cm^{-1}/step with count times of 10-20 seconds per step.

Single Crystals - Ni(111) and N(100). The crystals were cut with a diamond saw, polished mechanically in a water-alumina slurry, and orientation checked by X-ray diffraction. Planes are to within $\pm 1.5^o$. SEM analysis of surface showed only strong nickel peaks. The crystal was placed in a UHV chamber and Ar^+ bombarded at 2KV for 20 minutes with the sample at 200 C. At a sample temperature of 300 C, a number of reduction cycles were done for 15 minutes, each at an H_2 pressure of 4 X 10^{-5} Torr.

A residual gas analyzed showed traces of H_2O, CO, CO_2, O_2 and H_2 in the sample chamber: CO (research grade) was admitted to the cell through a variable leak valve and a UHV gas transfer system at controlled exposures of up to 10^6 L. ($1L = 10^{-6}$ Torr -sec.). The laser beam is incident on the crystal at $83°$ with respect to the surface normal and scattered light is collected $20°$ off normal to the crystal.

Data and Discussion

At 1 cm^{-1} resolution and 10^6 L exposure of CO the following Raman bands were recorded from the prepared catalyst and the single crystals, each sample at 298 K. The other catalyst (after methanation) was not exposed. All Raman bands are listed in cm^{-1}.

Catalyst

As-prepared (exposed to CO)	330, 540, 585, 595, 820, 1590, weak, broad band to 1950
After methanation	490, 530, 540, 595, 1370, 1590

Single Crystals

Ni-C Stretch

	on-top	two-fold bridge	four-fold bridge
Ni(111)	460	420	--
Ni(100)	670	512, 534, 542, 554	442, 472, 499

CO stretch

	on-top	two-fold bridge	four-fold bridge
Ni(111)	2092, 2143	1980, 1991, 2009	--
Ni(100)	2036, 2071, 2098, 2111	1909, 1957, 1971, 1968	1745, 1775, 1830

The Raman band at 820 cm^{-1} in the as-prepared catalyst is due to CrO_4^-. The band at 595 cm^{-1} is present in both catalysts at roughly equal intensity and is the Ni-C stretching mode. The 330 cm^{-1} band is the Ni-CO bond stretching vibration, while after methanation (with residual surface carbon) the bands due to the Ni-C stretch appear at 490, 530, and 540 cm^{-1}. It is thought that these bands are due to a mix of surface Ni-C and Ni-CO (re-absorbed CO) on the catalyst. The band at 1590 cm^{-1} is due to a graphite binder in the catalyst. The 540 and 585 cm^{-1} bands in the as-prepared sample are due to terminal OH^- and to hydrogen-bonded H_2O, respectively. The 1370 cm^{-1} mode is likely due to an H species $\nu(Ni_4-H)$ coadsorbed with CO or oxygen. The weak, broad bands we observe from 1600-1950 cm^{-1} on the exposed, as-prepared catalyst are unresolved CO stretches of two-fold and four-fold bridge configurations of CO adsorbed on Ni crystallites.

Reference

1. D. W. Goodman, R. D. Kelly, T. E. Madey and J. T. Yates, Jr., Chap. 1 in Hydrocarbon Synthesis from Carbon Monoxide and Hydrogen, ed. E. L. Kugler and F. W. Steffgen, Advances in Chemistry Series 178, ACS, Washington, DC, 1979.

THE SURFACE CHEMISTRY OF OSMIUM CARBONYL CLUSTERS SUPPORTED ON SILICA :

AN XPS INVESTIGATION

C. Furlani, R. Zanoni, [1]C. Dossi and [2]R. Psaro

Dipartimento di Chimica, Università di Roma "La Sapienza"
Roma, Italy
[1]Dipartimento di Chimica Inorganica e Metallorganica and
[2]Centro C.N.R., Milano, Italy

ABSTRACT

An XPS investigation on silica-supported Osmium carbonyl clusters is reported. Different sequences of surface reactions have been followed, evi-
dentiating structural modifications of the species involved. The assignment
of Os surface compound obtained via oxidation or reduction treatments has
been proposed on the basis of a previous identification of particular Os
4f binding energy regions for pure molecular compounds. The possibility
of a controlled reduction of supported Os species is proposed on the basis
of a range of b.e. values obtained for supported bare metal clusters. Results of an extension of present data to a catalytic test of 1-butene isomerization are discussed.

INTRODUCTION

Supported metal catalysts are difficult to characterize because they
are nonuniform, consisting of metal crystallites of various sizes and shapes
on high surface area inorganic oxides[1]. In the last 10 years researchers
have investigated supported molecular clusters as model catalysts and as
catalyst precursors[2]. Metal carbonyl clusters have been used for the preparation of supported metal species as potential Fischer-Tropsch catalysts[3].
Supported triosmium carbonyl clusters have been deeply investigated mainly
because they may be grafted on various substrates without breaking of the
metal frame. Infrared, UV-visible and Raman spectroscopies[4], TEM[4] and
EXAFS[5] studies have been reported on these systems. Knözinger et al.[6] reported only a few data using XPS. In a preliminary XPS work[7], we investigated selected pure Osmium carbonyls, demonstrating the existence of different Os $4f_{7/2}$ binding energy (b.e.) regions related to Osmium oxidation
states, cluster nuclearity and electronic effects of the ligands. In the
present work, we have critically evaluated the applicability of XPS for

studying the interaction of Osmium carbonyl clusters with silica. The main goals were to follow in situ surface reactions and the formation of supported metal particles by controlled thermal decomposition of Osmium carbonyls.

EXPERIMENTAL

The SiO_2 used as support was non-porous Aerosil 200 from Degussa with a nominal surface area of $\sim$ 200 m^2/g. The support was impregnated with solutions of $Os_3(CO)_{12}$ in dichloromethane; the mixture was stirred for 2 hr at 298 K under Ar. Subsequent evaporation to dryness was carried out in vacuum (0.133 Pa) for 16 hr at 298 K; infrared spectra were found characte-ristic for physisorbed cluster[8]. The chemisorbed cluster $|HOs_3(CO)_{10}(OSi)|$ was prepared by refluxing in dry n-octane a solution of $Os_3(CO)_{12}$ with SiO_2. Weights of $Os_3(CO)_{12}$ and of support were such as to give 2.0% by weight of Os in all samples. XPS measurements were carried out on a VG ESCA 3 photoelectron spectrometer using AlKα exciting line (hν = 1486.6 eV). Thermal treatments in different pure gases (100 kPa) or in vacuum (10^{-5} Pa) were conducted directly in the preparation chamber of the XP spectrometer. Results are referred to a Si 2p b.e. taken at 103.5 eV and are considered accurate to $\pm$0.2 eV.

RESULTS AND DISCUSSION

Characterization of physisorbed and chemisorbed species

In a preliminary study on selected pure osmium carbonyls[7], we demonstrated the existence of four different spectral regions for Os $4f_{7/2}$ ionization peaks of reported compounds, respectively related to: 1) bulk Os (50.7 eV); 2) Os(0) clusters: $Os_3(CO)_{12}$ and $Os_6(CO)_{18}$ (51.6, 51.5 eV); 3) Os carbonyl clusters with a formal charge per Os atom equal to +2/3 a.u. : $|HOs_3(CO)_{10}(OPh)|$ and $H_2Os_3(CO)_{10}$ (52.1, 51.9 eV); 4) mono and binuclear Os(II) carbonyls: $Os(CO)_4Cl_2$, $|Os(CO)_3Cl_2|_2$ (53.4, 53.1 eV). These same spectral regions are still represented after interaction with the SiO_2 support, by compounds reported in table. Physisorption results only in a weak perturbation ($\pm$ 0.3 eV) of Os 4f b.e.s. This result is coherent with the substantial constancy in νCO values from IR spectra of these physisorbed species with respect to pure compounds, as already found in a preceding study[8]. The chemisorption process between $Os_3(CO)_{12}$ and silica surface at 373-423 K gives rise to the grafted cluster $|(\mu-H)Os_3(CO)_{10}$ $(\mu-OSi\leqslant)|$ resulting from the oxidative addition of a silanol group to the Os-Os bond. The XP data are in agreement with this surface structure already proposed by different research groups[5,6,8]. After chemisorption, a clear stuctural modification of the $Os_3(CO)_{12}$ molecule is evidentiated by the Os 4f b.e. value of 53.2 eV. This value lies between typical Os(0) and Os(II) supported carbonyls b.e.s. (51.2, 53.4 eV) and results quite close to the value of 52.1 eV for the molecular analogue compound $|HOs_3(CO)_{10}$ (OPh)|. Assuming the surface structure of a bicapped cluster, each Os atom should bring a formal charge of 2/3 = 0.67 a.u.; this value is coherent with our XPS findings, since a slope of $\sim$1 eV per unit charge has already been reported[9]. Subsequent pyrolysis generates new osmium carbonyl surface species, the structure of which is object of a controverse interpretation in the literature[8, 10]. We suggest the formation of oxidised osmium carbonyl species, most probably $Os(II)$ units, from a comparison of their Os 4f b.e. and reported values for supported mono and dinuclear $Os(II)$ carbonyl compounds (see table). These results agree with the previously reported infrared characterization[8]. We will refer to these oxidised species as $|Os(CO)_x(OSi\leqslant)_2|_n$. The obtainment of SiO_2 supported Os (bare metal clusters) will be discussed in the next section, in a comparison with all reported reduced samples.

Table. Osmium $4f_{7/2}$ binding energies for pure and supported Os carbonyl compounds.

SAMPLE	Os $4f_{7/2}$	Ref.
$Os_3(CO)_{12}$	51.6	7
Os powder	50.7	This work
$|HOs_3(CO)_{10}(\mu\text{-}OPh)|$	52.1	7
$Os(CO)_4Cl_2$	53.4	7
$|Os(CO)_3Cl_2|_2$	53.1	7
$Os_3(CO)_{12} + SiO_2$ (a)	51.6	This work
$Os_3(CO)_{12}$ on SiO_2 (b)	51.2	" "
$|HOs_3(CO)_{10}(OSi\leqslant)|$ (c)	52.3	" "
$Os(CO)_4Cl_2$ on SiO_2	53.4	" "
$|Os(CO)_3Cl_2|_2$ on SiO_2	53.1	" "
$|Os(CO)_x(OSi\leqslant)_2|_n$ (d)	53.0	" "
Metallic osmium on SiO_2 (e)	51.2	" "

a) mechanical mixture

b) $Os_3(CO)_{12}$ physisorbed on SiO_2 at 298 K

c) $Os_3(CO)_{12}$ chemisorbed (see experimental section)

d) $|HOs_3(CO)_{10}(OSi\leqslant)|$ after heating under vacuum (10^{-3} Pa) to 473 K for 72 hrs

e) $|Os(CO)_x(OSi\leqslant)_2|_n$ after reduction in H_2 (100 kPa) at 573 K for 8 hr

Surface reactions conducted in situ

The knowledge at a molecular level of the reactions which occur between metal carbonyl clusters and the chemical groups which are present at the surface of a support personifies a model of surface reactivity and surface structure in heterogeneous calatysis.

Os 4f spectra reported in fig. 1 show the evolution of physisorbed species through a sequence of surface reactions conducted in situ. An easy and complete conversion from the physisorbed (fig. 1a) to the grafted cluster (fig. 1b) is evidentiated by the reported b.e. value of 52.3 eV obtained after a slow heating in Ar at 373 K. A structural modification of the cluster metal cage happens at 473 K (fig. 1c);Os 4f b.e. suggests an increase in the mean oxidation state of Os atoms, probably subsequent to the establishment of new Os-O bonds. The final rupture of the triatomic units into Os(II) carbonyl species is indicated by the value of 53.1 eV in fig. 1d, obtained after exposure to O_2. In indipendent experiments we obtained the same value for samples of physisorbed and chemisorbed $Os_3(CO)_{12}$ directly exposed to O_2. The cycle of surface reactions was terminated with a reduction under H_2 (fig. 1e) of freshly formed $|Os(CO)_x (OSi\lessapprox)_2|_n$ which brought to metal particles on SiO_2.

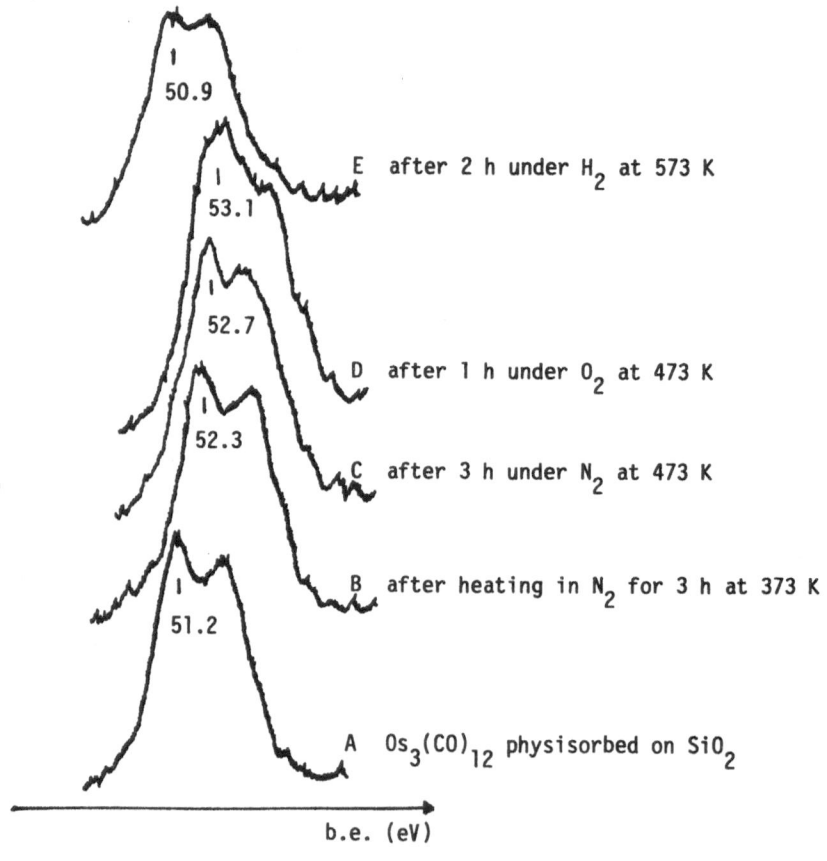

50.9

E after 2 h under H_2 at 573 K

53.1

52.7

D after 1 h under O_2 at 473 K

52.3

C after 3 h under N_2 at 473 K

B after heating in N_2 for 3 h at 373 K

51.2

A $Os_3(CO)_{12}$ physisorbed on SiO_2

b.e. (eV)

Figure 1.   Variation of Os 4f b.e. of a sample of $Os_3(CO)_{12}$ physisorbed upon various treatments conducted in situ.

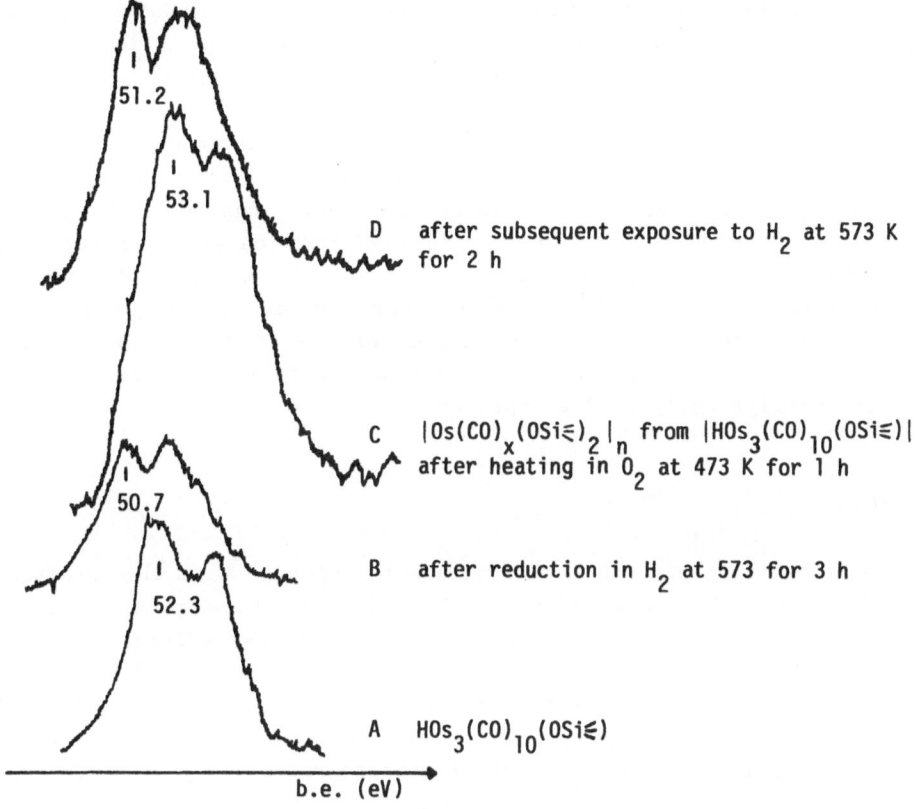

51.2

53.1

D   after subsequent exposure to H$_2$ at 573 K
     for 2 h

C $|Os(CO)_x(OSi\lessgtr)_2|_n$ from $|HOs_3(CO)_{10}(OSi\lessgtr)|$
     after heating in O$_2$ at 473 K for 1 h

50.7

B   after reduction in H$_2$ at 573 for 3 h

52.3

A $HOs_3(CO)_{10}(OSi\lessgtr)$

b.e. (eV)

Figure 2.   Variation of the Os 4f b.e. of the chemisorbed species $|HOs_3$
$(CO)_{10}(OSi\lessgtr)|$ upon various thermal treatments.

In fig. 2 are shown the results of the reduction using different precursors, the molecular grafted cluster (fig. 2b) and the oxidised units
(fig. 2d). The lowest reported Os 4f b.e. for reduced species falls at
50.7 eV (fig. 2b), a value coincident with that obtained for a sample of
pure Os powder. This result allows us to make the hypothesis that the
overall effect on Os b.e. due to the interaction of Os metal particles
with SiO$_2$ is weak; this assumption was proven valid for physisorbed species collected in table. Thus, we correlate the value of 50.7 eV with
large metal clusters or crystallites on SiO$_2$. Within this assumption, all
the positive shifts from 50.7 eV in this series of reduced samples can be
related to the physical state of Os aggregates and, in particular, to a
decrease in the dimensions of bare Os clusters. The existence of a b.e.
range (50.7-51.2 eV) for our reduced species demonstrates the possibility
of a controlled reduction. An additional effect was put in evidence by
exposing a freshly in situ reduced sample of $|HOs_3(CO)_{10}(OSi\lessgtr)|$ to CO at
298 K. The resulting Os 4f b.e. (51.4 eV) and the increase (from 1.2 to
1.4%) in Os/Si atomic ratio (as determined from XPS peak areas) concord
with the formation of a more dispersed phase on the SiO$_2$ surface, with the

breaking of large clusters into low-nuclearity species. Successive H_2 treatment restored a lower value (50.9 eV), paralleled by a decrease in the Os/Si atomic ratio to 1.2%, both indicating that reaction with CO is easily reversed in our experimental conditions.

We have recently investigated the isomerization of 1-butene with silica supported osmium catalysts[11], which have been characterized by infrared and XP spectroscopies. The active species, after the catalytic tests, showed the same value of 52.7 eV, using different catalyst precursors. We observe exactly the same b.e. value after in situ thermal treatment of the physisorbed $Os_3(CO)_{12}$ (fig. 1c). Further work is now in progress to perform the catalytic tests in the chamber of the XP spectrometer. The main goals are to follow the surface reactions and to characterize the nature of the catalytically active species.

CONCLUSIONS

A multitecnique approach to the study of supported molecular clusters seems, at present, the only way of reaching a deep knowledge of these model systems. In our work we explored the possibilities offered in this area by XPS, in conjuction mainly with IR spectroscopy, when the tecnique is applied in a sufficiently wide and systematic way. Even if a number of results are not conclusive, XPS has offered a continuous monitoring on more or less subtle structural effects in the course of surface reactions.

REFERENCES

1. J.R. Anderson, in "Structure of Metallic Catalysts" (Academic Press, New York 1975).
2. D.C. Bailey, S.H. Langer, Chem. Rev. 81, 109 (1981).
3. J. Zwart, R. Snel, J. Mol. Catal. 30, 305 (1985).
4. J. Schwank, L.F. Allard, M. Deeba, B.C. Gates, J. Catal. 84, 27 (1983).
5. S.L. Cook, J. Evans, G.S. Mc Nulty, G.N. Greaves, J. Chem. Soc., Dalton Trans. 7 (1986).
6. H. Knözinger, Y. Zhao, B. Tesche, R. Barth, R. Epstein, B.C. Gates, J.P. Scott, Faraday Discuss. Chem. Soc. 72, 54 (1982).
7. R. Zanoni, V. Carinci, H. Abu-Samn, R. Psaro, C. Dossi, J. Mol. Struct. 131, 363 (1985).
8. R. Psaro, R. Ugo, G.M. Zanderighi, B. Besson, A.K. Smith, J.M. Basset, J. Organometal. Chem. 213, 215 (1981).
9. I.V. Lin'ko, B.E. Zaitsev, A.K. Molodkin, T.M. Ivanova, R.V. Linko, Russ. J. Inorg. Chem. 28, 857 (1983).
10. G. Collier, D.J. Hunt, S.D. Jackson, R.B. Moyes, I.A. Pickering, P.B. Wells, A.F. Simson, R. Whyman, J. Catal. 80, 154 (1983).
11. R. Psaro, C. Dossi, A. Fusi, R. Ugo, G.M. Zanderighi, P. Doldi, V. Ragaini, R. Zanoni, in "Proc. Vth Int. Symp. on Relations between Homogeneous and Heterogeneous Catalysis", edited by Yu. Yermakov and V. Likholobov (VNU Science Press, Utrecht 1986)

CHEMISTRY OF SMALL METAL CLUSTER IONS

Luke Hanley, Stephen Ruatta, and Scott Anderson

Department of Chemistry
State University of New York at Stony Brook
Stony Brook, N. Y. 11794-3400

INTRODUCTION

In this paper we report absolute cross sections for reactions of small aluminum cluster ions (Al^+_{1-5}) with several neutral reagents. Dependence of reactivity on cluster size and collision energy has been measured in a tandem mass spectrometer equipped with radio frequency (rf) ion guides and a unique source of thermalized metal cluster ions.

This work extends our original study of the same reactions[1] on an earlier, single mass spectrometer instrument. In that study, we found that the aluminum dimer ion was more reactive than either the atomic or trimer ions in the formation of addition products. We also observed extensive cluster ion fragmentation at higher collision energies, which also was promoted by increasing internal energy in the cluster ion reagents. The experiments discussed here are largely in agreement with our previous results; however, mass-selection of our primary ion beam reveals a rich variety of additional reactions which depend strongly on collision energy. In particular, we observe strong interplay between 'chemical' product channels and collision induced dissociation (CID). In the CID experiments, very specific fragmentation channels dominate for different size clusters.

EXPERIMENTAL

Metal clusters are produced by sputtering a high purity target with 10 keV Ar ions from a homemade saddle field ion gun. Sputtered cluster ions are injected into a radio frequency cooling trap constructed in the form of a labyrinthine channel of total length 67 cm, filled with buffer gas to ca. 2mTorr. The cluster ions diffuse through this rf-trap, and in the process are thermalized. While we have no means of measuring the clusters' internal energy distribution, we believe that essentially complete thermalization is achieved. In our previous work[2] on cooling of cluster ions in a <u>linear</u> rf-ion trap, we found that although most of the clusters were translationally thermalized, a significant fraction of 'hot' ions were transmitted. The labyrinthine design eliminates this problem.

The cooled cluster ions are mass-selected by a Wien filter and injected into a set of two octapole ion guides, where the desired collision energy is set. Reactions occur in a scattering cell wrapped around the second ion guide which is filled to 5.0 x 10⁻⁴ Torr with reagent gas. Ionic products and unreacted reagent clusters are collected by the ion guide, then mass analysed and counted by a quadrupole mass filter and Daly detector. Cross sections for reactions of Al_{1-5}^+ with O_2 and D_2O were measured at collision energies of 0.3, 1.0, 3.0, and 5.0 eV. Collision induced dissociation of the clusters with argon was studied at energies of 0.3, 1.0, 1.5, 2.0, 2.5, 3.0, and 5.0 eV. Due to uncertainty in determining the effective length of our scattering cell, the absolute scale for the cross sections could be in error by as much as 50%. In our current geometry there is insufficient differential pumping between the scattering cell and the rest of the machine, which leads to the appearance of false products in a few cases. In addition, while use of ion guides insures efficient collection of product ions, transmission through the quadrupole mass filter for ions of different mass and scattering angles may not be equal. Modifications to improve differential pumping and increase Wien filter resolution, and experiments to quantify these effects are in progress.

RESULTS AND DISCUSSION

The cross sections for all reaction channels greater than 0.05 $Å^2$ are plotted in Figs. 1-3.

Collision Induced Dissociation

The top row of plots in Fig. 1 shows the collision energy dependence of the major CID channels for Al_{2-5}^+. While all possible fragment clusters are observed for each reagent cluster, strong propensity is observed for production of Al^+ and Al_{n-1}^+ in each case. This suggests that the primary CID process at low collision energies may be loss of a single atom. For CID of the trimer, the dominant ion product at all collision energies is the dimer, suggesting that the dimer may have a lower IP than the atomic ion, at variance with calculations of Upton[3]. For the tetramer and pentamer, the atomic ion product dominates at low collision energies, while at 5.0 eV, the intensities of Al^+ and Al_{n-1}^+ are equal. The fact that Al_4^+ dominates CID of the pentamer to such an extent, suggests that Al_4^+ may be a particularly stable species. Begemann et al.[4] have reported CID studies of internally hot sputtered aluminum cluster ions with argon and oxygen performed at 1.8 keV translational energy. In contrast with our results, they see a relatively smooth distribution of all possible fragment cluster ions, peaking at the atomic ion in most cases.

Reactions with Oxygen

Fig. 2 shows the full distribution of products observed for reactions of Al_{1-5}^+ with O_2; the collision energy dependence of the major channels is shown in the second row of Fig. 1. For reaction of the atomic ion, AlO^+ formation is observed with an apparent threshold around 3.0 eV. This is in agreement with previous more accurate results[5], except that our apparent threshold is too low, probably due to reactions outside the scattering cell. The dimer ion also adds an oxygen atom to form Al_2O^+, however the collision energy dependence of the cross section suggests that this is an exoergic process in contrast to the atomic ion reaction. For larger clusters, Al_2O^+ is the only oxygen addition product observed. No Al_nO^+ products are observed, in agreement with data of Jarrold and Bower[6].

For the dimer through pentamer ions, production of bare aluminum cluster fragment ions is a major channel. Comparison with the results for CID by Ar suggests that two mechanisms are important in these processes. At low collision energies, bare clusters must be produced through chemical reactions, while at high energies a simple CID mechanism becomes important.

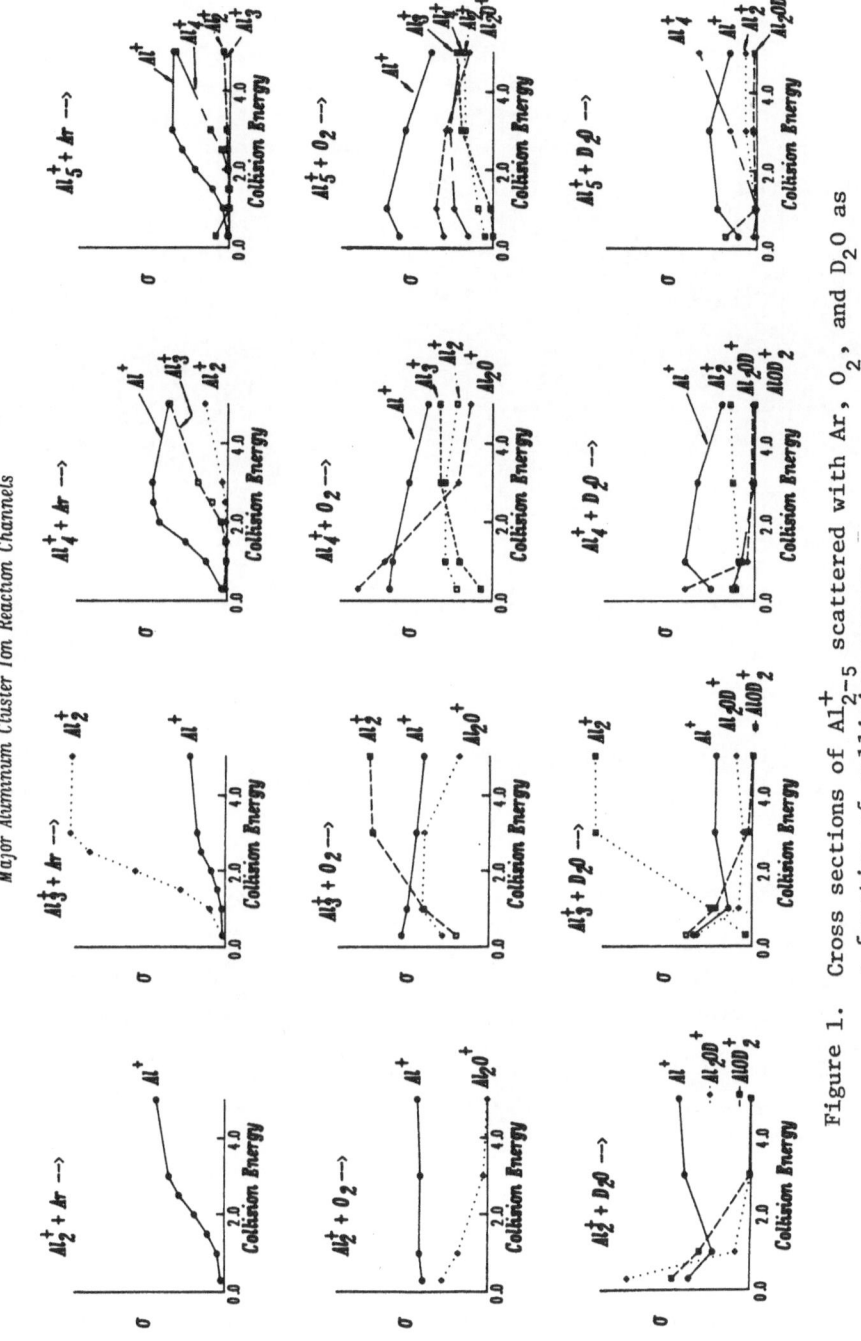

Major Aluminum Cluster Ion Reaction Channels

Figure 1. Cross sections of Al_{2-5}^+ scattered with Ar, O_2, and D_2O as a function of collision energy.

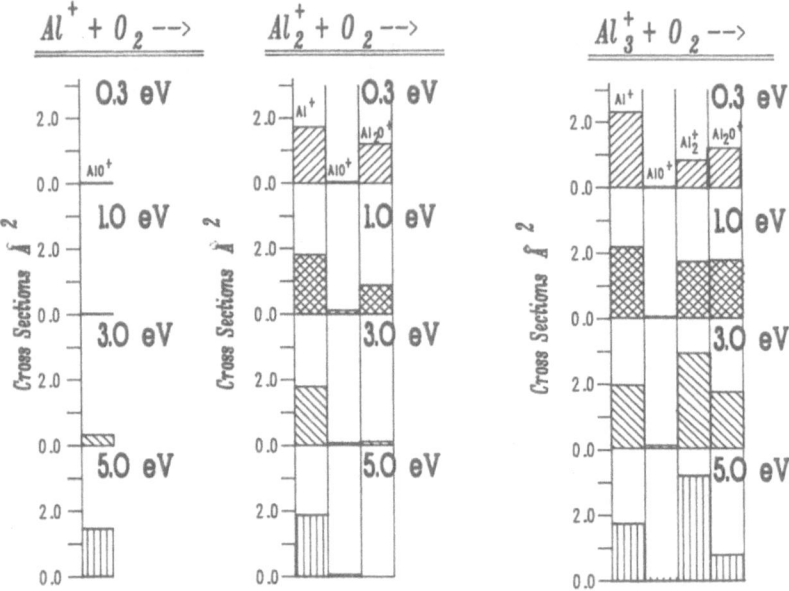

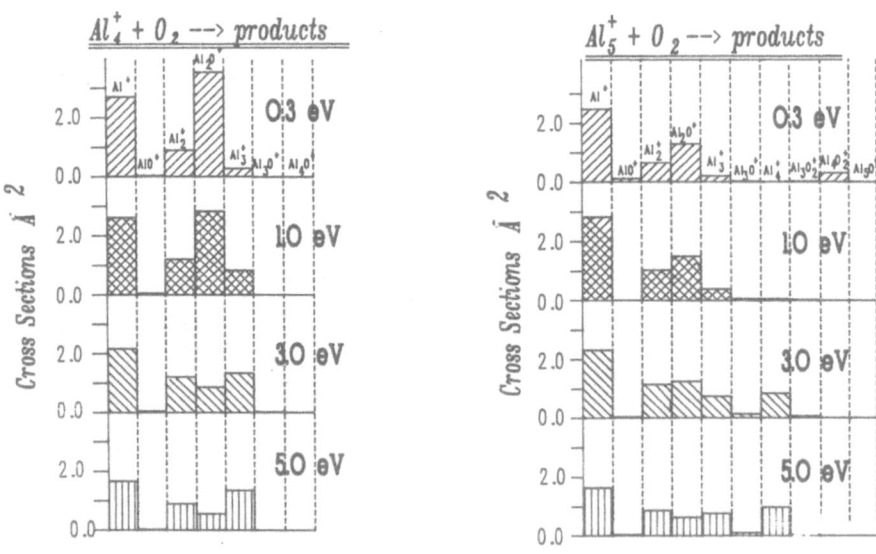

Figure 2. Exact cross sections of all product channels for reactions with O_2.

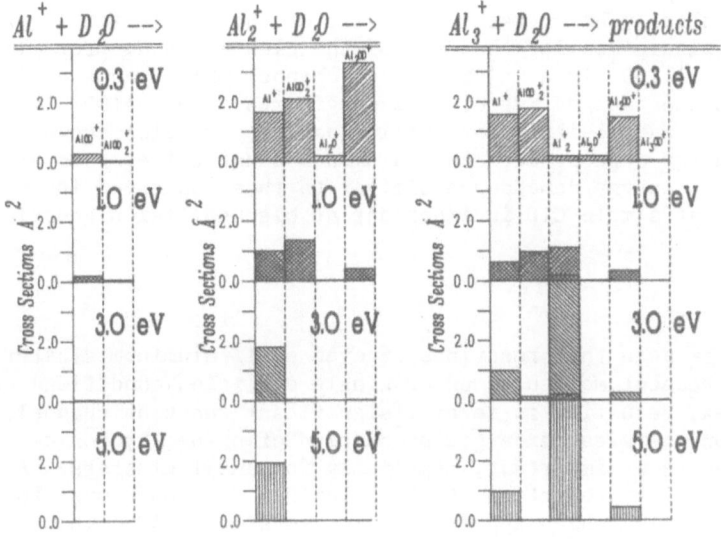

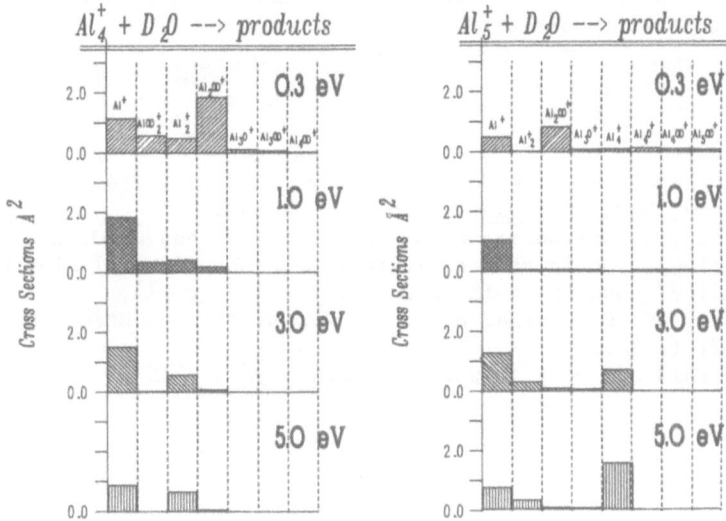

Figure 3. Exact cross sections of all product channels for reactions with D_2O.

Reactions with Water

Fig. 3 shows the full distribution of products observed for reactions of A_{1-5}^+ with D_2O; the collision energy dependence of the major channels is shown in the third row of Fig. 1.

While the pattern of reactions is quite complicated, there are several striking results. The overall reactivity of the atomic ion is at least an order of magnitude smaller than any of the cluster ions. For all reagent cluster ions, Al_2OD^+ is a major product channel which shows collision energy dependence suggestive of an exoergic reaction. $AlOD_2^+$ is also a major channel, particularly for small clusters, and also shows a 'exoergic' collision energy dependence. Bare aluminum cluster fragments are produced with collision energy dependence similar to those observed in argon CID, suggesting that simple CID is important at high collision energies.

CONCLUSIONS

It can be seen that reactions of even small aluminum cluster ions with simple reagent molecules under single collision conditions can be quite complex, resulting in several significant reaction channels. At low collision energies, exoergic production of oxide, hydroxide, and hydrate products is important, especially for small clusters. At high collision energies production of bare aluminum fragment ions proceeds through a mechanism which appears similar to simple CID. These results emphasize the importance of being able to measure accurate cross sections over a range of collision energies in studies of reaction mechanisms.

ACKNOWLEDGEMENT

This work was supported by the US Office of Naval Research under contract number N00014-85-K-0678.

REFERENCES

1. L. Hanley and S. L. Anderson, Chem. Phys. Lett., 129:429 (1986).
2. L. Hanley and S. L. Anderson, Chem. Phys. Lett., 122:410 (1985).
3. T. H. Upton, Phys. Rev. Lett., 56:2168 (1986).
4. W. Begemann, S. Dreihofer, K. H. Meiwes-Broer, and H. O. Lutz, "Experiments On Sputtered Clusters As Probe of Metal Cluster Ion Stability", published in this volume.
5. M. E. Weber, J. L. Elkind, and P. B. Armentrout, J. Chem. Phys., 84:1521 (1986).
6. M. F. Jarrold and J. E. Bower, J. Chem. Phys., 85:5373 (1986).

REACTIONS OF DIATOMIC AND TRIATOMIC METAL CLUSTERS OF IRON AND COPPER WITH DIHYDROGEN IN INERT MATRICES

R.H. Hauge, Z.H. Kafafi,[†] and J.L. Magrave

Rice Quantum Institute & Department of Chemistry
Rice University
Houston, TX 77251

ABSTRACT

Diatomic copper was found to react successively with up to three dihydrogen molecules to form Cu_2H_2, Cu_2H_4 and Cu_2H_6, respectively, on an inert matrix surface at 12-15K. Diatomic iron was found to be totally unreactive. Triatomic copper was observed to react spontaneously with molecular hydrogen. Dihydrogen complexes were observed for Cu_2H_4, Cu_2H_6 and Cu_3H_2 species. A reaction of molecular hydrogen with Fe_x where x is 3 or 4 was also observed.

INTRODUCTION

Reactions of molecular hydrogen with small metal clusters is emerging as a particularly interesting area for critical tests of theory with respect to reaction barriers and types of possible metal-hydrogen bonding which includes weak interaction of molecular hydrogen, terminal hydride-like interaction and one, two and three-fold bridging sites as well as various interstitial sites for the larger clusters. Indications of complexation of molecular hydrogen in organometallic compounds has been observed,[1-8] and very recently a molecular hydrogen complex with a Pd atom has also been matrix isolated.[7] Such compounds can be considered similar to physisorbed molecular hydrogen on metal surfaces. Hydrogen atoms bound in terminal and bridging positions have been studied for many organometallic species.[10-14] These studies provide much useful structural and spectroscopic information with respect to atomic hydrogen interaction with atomic and multiatomic metal cluster sites. Recent work with laser vaporization and supersonic jet expansion has demonstrated the highly selective reaction behavior of molecular hydrogen with free metal clusters of different sizes ranging from 2-200 atoms.[15-17] This work has also suggested that the observed reactions are due to chemisorption of molecular hydrogen since the reaction occurs at room temperature where physisorbed molecular hydrogen is not expected to be stable. Since the present study deals with the reactions of diatomic and triatomic metals

†Present Addresss: Naval Research Laboratory, Code 6551, Washington, DC 20375-5000

with molecular hydrogen, it is appropriate to review the observations of recent supersonic beam studies with respect to these species. The results of the supersonic beam experiments can be easily summarized since no diatomics were observed to react and all triatomics other than copper did react with molecular hydrogen.

In contrast FTIR matrix isolation studies which we report here in part have shown that a number of diatomics do, in fact, react readily at 12-15K as do the triatomics, including copper. This report discusses our initial studies of copper and iron diatomic and triatomic reactions with molecular hydrogen.

Iron and copper were vaporized from a resistively heated tantalum foil furnace. Both were contained in an alumina cell. Deposition rates were measured with a quartz crystal microbalance. Ratios of metal to argon varied from 0.1 to 60 parts per thousand. Deposition was carried out for thirty minutes at 12-15K and at a rate of approximately 150 A/sec. The addition of H_2 and D_2 was monitored by the rise in background pressure. In general, it was found that the system had sufficient background hydrogen such that the addition of one molecule of hydrogen occurred without additional hydrogen. Information regarding the multisurface matrix apparatus can be found in a previous publication.[18]

Discussion of Results

I.   Iron Reactions with Molecular Hydrogen

Cocondensation of iron atoms with dihydrogen at the concentrations indicated in Figure 1 shows no reaction until the 2-6% metal concentration region is reached. A complex set of features is observed to grow over the region of 1000 to 1400 cm^{-1} with the two most intense features located at 1355 and 1270 cm^{-1}, respectively. They shift to 960 and 915 cm^{-1}, respectively, with dideuterium. Figure 1 also illustrates the growth of the known $Fe_2(H_2O)$ complex.[19] It is evident that its growth does not parallel the growth of the 1270 and 1355 cm^{-1} features. It has also been shown that the addition of more hydrogen does not affect the formation of $Fe_2(H_2O)$. This is taken to indicate that diiron does not react with dihydrogen since it does not compete with water for the available diiron. A log plot of the ratio of the absorbance of the 1270 cm^{-1} peak to the absorbance of $F_e(H_2O)$ gives a slope of 2.8 when plotted relative to the log absorbance $(Fe(H_2O))$. A similar plot for $Fe_2(H_2O)$ gives a slope of 1.2 for the same set of depositions. This result suggests that the 1270 and 1355 features should be assigned to a triiron or tetrairon but clearly not to a diiron species. Variation of the dihydrogen flux at the matrix surface over a wide range does not appreciably affect the extent of reaction. This suggests that the Fe_3 or Fe_4 cluster is reacting with a single dihydrogen since addition of more than one dihydrogen should be favored by the higher hydrogen flux. It will be shown that this is indeed the case for copper cluster reactions. We thus suggest that the reaction product is either Fe_3H_2 or Fe_4H_2 where the hydrogen atoms are in bridging positions.

II.   Copper Reactions with Molecular Hydrogen

Figure 2 illustrates a metal concentration study with a constant dihydrogen concentration. Peaks labeled I, II, and III vary in the same manner with copper atom concentration. Peak IV is enhanced at the highest copper concentration relative to the others. A plot of the log of the absorbance of the 1350 cm^{-1} (I) peak relative to the log of the copper

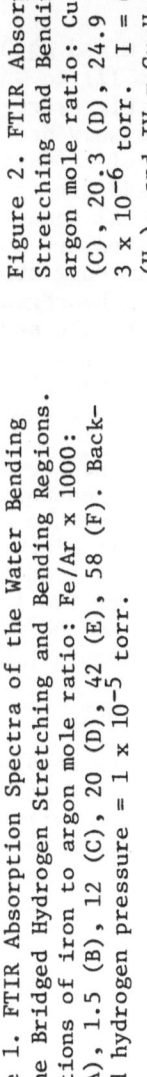

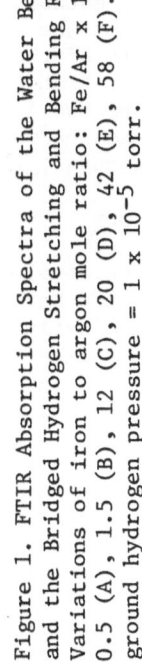

Figure 2. FTIR Absorption Spectra of the Bridged Hydrogen Stretching and Bending Regions. Variation of copper to argon mole ratio: Cu/Ar x 1000: 4.9 (A), 10.4 (B), 15.2 (C), 20.3 (D), 24.9 (E). Background hydrogen pressure = 3 x 10⁻⁶ torr. I = Cu₂H₂, II = Cu₂H₂(H₂), III = Cu₃H₂(H₂)₂ and IV = Cu₃H₂.

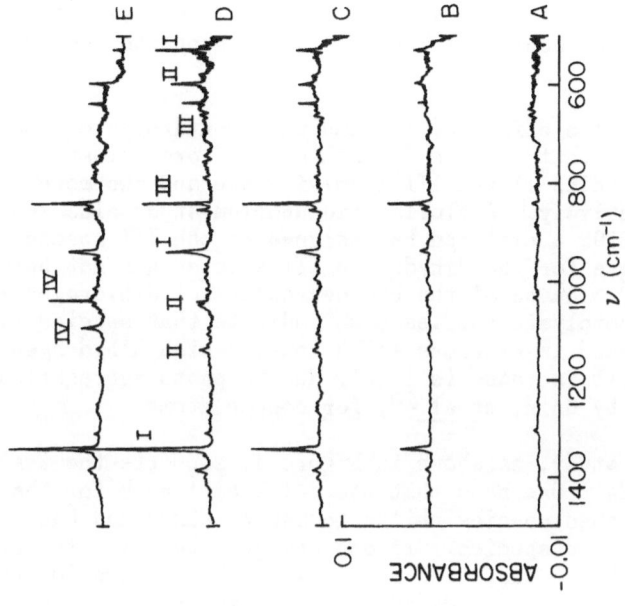

Figure 1. FTIR Absorption Spectra of the Water Bending and the Bridged Hydrogen Stretching and Bending Regions. Variations of iron to argon mole ratio: Fe/Ar x 1000: 0.5 (A), 1.5 (B), 12 (C), 20 (D), 42 (E), 58 (F). Background hydrogen pressure = 1 x 10⁻⁵ torr.

atom concentration gives a slope of 2.3. This result suggests that the sets of peak labeled I, II and III are dicopper species and that peak IV may be assigned to a tricopper species. A set of broad features are also observed at 3352 and 3232 cm^{-1}, respectively. We have tentatively assigned these features to a dihydrogen tricopper complex. A future paper will discuss their photochemical behavior which supports the tricopper assignment.

Figure 3 shows the effect of increasing the hydrogen concentration. It clearly indicates that species I contains the lowest number of dihydrogens with species II and III involving one and two more dihydrogens, respectively. A similar dideuterium study also indicates that a feature at 2192.4 cm^{-1} can be assigned to the III species. A corresponding feature for the dihydrogen III species has not been clearly identified possibly because of the broader nature of dihydrogen features in this region. Photolysis studies also indicate that species III can be dramatically increased in matrices which contain high dihydrogen concentrations. This increase is likely due to photo-aggregation processes as noted by Ozin, et al.[20], for copper atoms.

A mixed H_2/D_2 study, as shown in Figure 4, supports the assignment of species III to Cu_2H_6. One sees that the 841.3 cm^{-1} peak for the completely hydrogenated species shifts to 849.7, 862.7 and 648.9 for the increasingly deuterated species. If one assigns the 841.3 feature to a motion of bridging hydrogens, then successive substitution for the two complexed dihydrogen groups by dideuterium is expected to produce the two additional features as shown near the 841.3 feature. Thus, the detection of the feature at 2192.4 cm^{-1} combined with the above observations strongly suggests that species III is indeed a $Cu_2H_2(H_2)_2$ with one set of bridging hydrogens and two sets of equivalent dihydrogen groups, a diborane-like structure where the two extra pair of hydrogens are side-on molecularly complexed rather than terminally bound. This also implies that species II and I contain one set of bridging hydrogens and one and no dihydrogen groups, respectively.

The number of hydrogen atoms associated with the tricopper species is less well established, however, experiments indicate that species IV persists at the lowest hydrogen concentrations and thus is most probably the result of the addition of a single dihydrogen. Photolytic studies have shown that the species responsible for the 3232 cm^{-1} feature converts in part to the tricopper IV species which suggests that it may also be considered the result of the interaction of one dihydrogen with tricopper where, however, the dihydrogen is complexed with retention of an H-H bond.

Conclusions

These studies have shown that dihydrogen spontaneously reacts with Cu_2, Cu_3 and Fe_3 or Fe_4 at 12-15K on an inert matrix surface. Diiron (Fe_2) is clearly unreactive toward molecular hydrogen. Table I presents the frequencies which have been assigned to the various species resulting from the respective reactions. It is clear that the first dihydrogen, when it dissociatively reacts, prefers multimetal bonding (bridge bonding) to single metal atom bonding (terminal bonding).

One also notes an apparent difference in observed reactivity of the copper clusters from that for jet-cooled beam work where no reactivity was observed for copper clusters ranging from Cu_2 to Cu_{13}. A similar lack of reactivity has been noted for diatomic nickel and cobalt species in beam studies.[15-16] It seems likely that the higher cooling rates available on an inert surface with respect to a gas phase system may be responsible for observations of the apparent differences in diatomic reactivity. A second

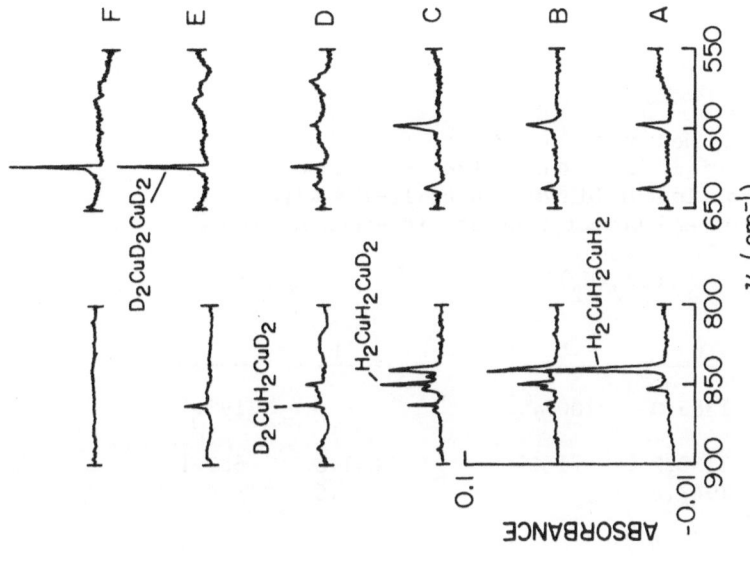

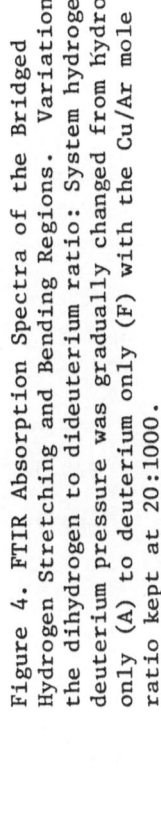

Figure 4. FTIR Absorption Spectra of the Bridged Hydrogen Stretching and Bending Regions. Variation of the dihydrogen to dideuterium ratio: System hydrogen/deuterium pressure was gradually changed from hydrogen only (A) to deuterium only (F) with the Cu/Ar mole ratio kept at 20:1000.

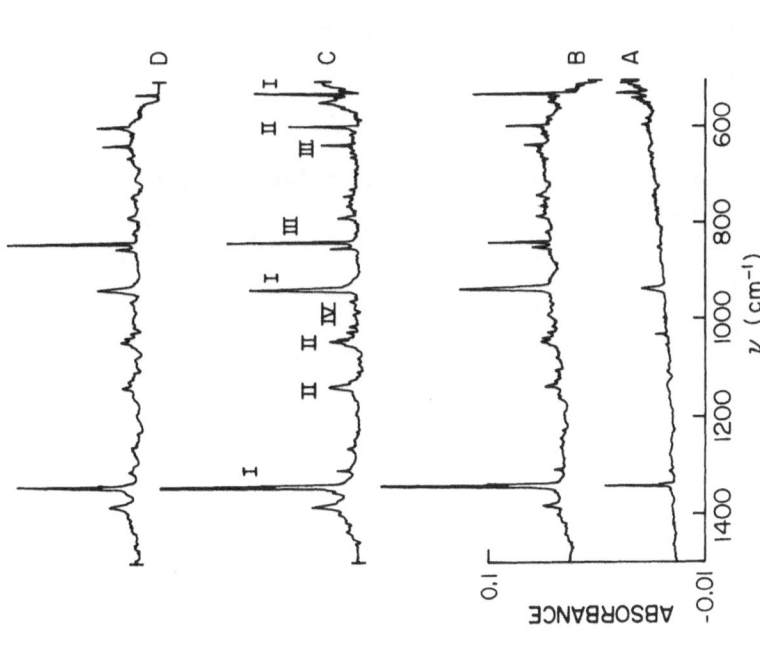

Figure 3. FTIR Absorption Spectra of the Bridged Hydrogen Stretching and Bending Regions. Variation of the hydrogen to argon concentration: System hydrogen pressure (torr): 2×10^{-7} (A), 6×10^{-7} (B), 1×10^{-6} (C), 9×10^{-6} (D). Copper/argon mole ratio is 23:1000. I = Cu_2H_2, II = $Cu_2H_2(H_2)$, III = $Cu_2H_2(H_2)_2$ and IV = Cu_3H_2.

Table 1. FTIR Measured Frequencies (cm^{-1}) of
Species Which Result from Spontaneous
Reaction of Dihydrogen at 12-15K with
Iron and Copper Clusters in Argon Matrices

Cu_2H_2		$Cu_2H_2(H_2)$*		$Cu_2H_2(H_2)_2$*	
H	D	H	D	H	D
1341.7	975.3	1385.0	1007.9	–	2192.4
938.7	687.7	1138.2	713.3	841.3	623.1
528.0	–	1047.8		637.6	–
		597.3			

$Cu_3(H_2)$*		Cu_3H_2		Fe_3H_2 or Fe_4H_2	
H	D	H	D	H	D
3351	2448	1082.1	781	1355	960
3232	2355	1039.7	755.7	1270	915
		537.1			

*Parentheses indicate that the hydrogens are molecularly complexed
to Cu_2 and Cu_3 clusters.

792

difference between the matrix and gas phase system reaction condition is the temperature at which the reaction is occurring, i.e., 12-15K versus 300K for the gas phase reaction. As pointed out by Smalley, et al.[15], reaction at room temperature precludes the observation of species with stabilities less than approximately 13 kcal. This may also provide an explanation for the lack of observed reaction of copper clusters at room temperature since the heat of formation of chemisorbed dihydrogen may be less than 13 kcal for all copper clusters. In any case it appears important for beam experiments to investigate more fully the apparent lack of reactivity of diatomic metals with hydrogen since one can expect theory to initially focus on their reactivity as a test of theoretical methods.

Acknowledgements

The authors would like to acknowledge financial support by the National Science Foundation and the Robert A. Welch Foundation for this work.

References

1. G.J. Kubas, R.R. Ryan, B.I. Swanson, P.J. Vergamini, and H.J. Wasserman, J. Am. Chem. Soc. 106, 451 (1984).

2. R.L. Sweany, J. Am. Chem. Soc. 107, 2374 (1985); R.L. Sweany, Organometallics 5, 387 (1986).

3. R.K. Upmacis, G.E. Gadd, M. Poliakoff, M.B. Simpson, J.J. Turner, R. Whyman, and A.F. Simpson, J. Chem. Soc., Chem. Commun., 27 (1985).

4. S.P. Church, F.W. Grevels, H. Hermann, K. Schaffner, J. Chem. Soc., Chem. Commun. 30 (1985).

5. R.H. Morris, J.F. Sawyer, M. Shirolian, J.D. Zubkowski, J. Am. Chem. Soc. 107 5581 (1985).

6. G.J. Kubas, C.J. Unkefer, B.I. Swanson, and E. Fukushima, J. Am. Chem. Soc., 108, 7000 (1986).

7. G.A. Ozin and J. Garcia-Prieto, J. Am. Chem. Soc. 108, 3099 (1986).

8. R.K. Upmacis, M. Poliakoff and J.J. Turner, J. Am. Chem. Soc. 108, 3645 (1986).

9. R.H. Crabtree, M. Lavin, L.J. Bonneviot, J. Am. Chem. Soc. 108, 4032 (1986).

10. H.D. Kaesz and R.B. Saillant, Chem Reviews, Vol. 72, 231 (1972).

11. D.S. Moore and S.D. Robinson, Chem. Soc. Rev. 12, 415 (1983).

12. G.A. Ozin and C. Gracie, J. Phys. Chem. 88, 643, (1984); G.A. Ozin and J.G. McCaffrey, ibid 88, 645 (1984).

13. G.A. Ozin and J.C. McCaffrey, J. Am. Chem. Soc. <u>106</u>, 807 (1984).

14. R.L. Rubinovitz and E.R. Nixon, J. Phys. Chem. <u>90</u>, 1940 (1986).

15. M.D. Morse, M.E. Geusic, J.R. Heath, and R.E. Smalley, J. Chem. Phys. <u>83</u>, 2293 (1985).

16. S.C. Richtsmeier, K.E. Parkas, K. Kiu, L.G. Pobo and S.J. Riley, J. Chem. Phys. <u>82</u>, 3659 (1985).

17. R.L. Whetten, D.M.Cox, D.J. Trevor and A Kaldor, Phys. Rev. Letters <u>54</u>, 1494 (1985).

18. R. H. Hauge, L. Fredin, Z. H Kafafi and J.L. Margrave, Appl. Spectrosc. <u>40</u>, 588 (1986).

19. J.W. Kauffman, R.H. Hauge and J.L. Margrave, J. Phys. Chem. <u>89</u>, 3541 (1985).

20. G.A. Ozin, H. Huber, D. McIntosh, S. Mitchell, J.G. Norman, Jr., and L. Noodleman, J. Am. Chem. Soc. <u>101</u>, 3504 (1979).

MOLECULAR BEAM RELAXATION SPECTROSCOPY APPLIED TO ADSORPTION KINETICS ON SMALL SUPPORTED CATALYST PARTICLES

Claude R. Henry, Claude Chapon and Christian Duriez

CRMC2-CNRS, Campus de Luminy - case 913

13288 Marseille Cedex 9 - France

A new apparatus performing adsorption-desorption kinetics experiments on small supported particles, by the MBRS technique, is presented. The clusters are prepared by in-situ condensation of a collimated Pd beam on MgO(100). A nozzle molecular beam source gives a directed beam of CO (or He) which is chopped at frequencies from $\simeq$ 0 up to 400 Hz. The molecules re-emitted from the sample are synchronously detected by a mass-spectrometer. The signal is analysed in time or in frequency with in line FFT. Angular distribution of the scattered beam is obtained by a 2nd, rotatable, mass-spectrometer. An example of Pd deposit and typical angular and temporal measurements are presented.

I. INTRODUCTION

The understanding of surface reactions, at the atomic level, is a necessary step in fundamental catalysis studies. They are generally done in using surface science techniques. For the special case of reaction-kinetics the Molecular Beam Relaxation Spectroscopy technique (MBRS) has proven its ability in resolving complicated mechanisms[1]. The main interest of this technique is in measuring rapid surface reaction rates (e.g. 10^5 s^{-1}) and thus evidencing elementary steps : adsorption, desorption, surface diffusion...[2]. Up to now MBRS has been only applied on extended single-crystal surfaces. The first attempt to use this technique on small clusters has been recently done in studying adsorption-desorption kinetics of cadmium on gold clusters deposited on NaCl[3,4]. The atomic beam impinging on the substrate, kept at a given temperature, is modulated and the Cd atoms reemitted from the sample, in a given direction, are synchronously detected by a quadrupole mass-spectrometer. Thus we obtain a typical signal (see Fig. 4) which exhibit a steep increase/decrease followed by a smooth increase/decrease when the beam is on/off. The slope of the steep variations is nearly equal to that of the impinging pulse rise. These parts of the signal correspond to (immediately) reflected atoms and to desorbed ones at a rate much higher than the modulation frequency. The smooth variations are due to atoms desorbing at a rate of the order of the modulation frequency. For a first order desorption and at small coverage these smooth variations are exponential and the time constant gives the mean life-time of the adsorbed atoms. The ratio between the maximum desorbed flux (at time $\simeq$ 0) and the signal amplitude gives the associated adsorption probability. For an heterogeneous surface we expect several adsorption sites corresponding to several life-times. For a given one, τ, if the modulation frequency, ω, is such as $\omega\tau \lesssim 1$ the desorption rate from this channel, and then τ, will be measurable. On the contrary if $\omega\tau \ll 1$ the desorption component of the signal will be confused with the reflection component. At last if $\omega\tau \gg 1$ the desorbed component is completely demodulated and gives only a dc contribution to the signal. These effects have been used in the preceding studies[3,4] in order to separate the desorption from the clusters and from the substrate. In addition the measurement of the proportion of the impinging atoms desorbing from the clusters, as a function of their mean size, has allowed to measure the mean diffusion length of Cd atoms adsorbed on the NaCl substrate[3].

This analysis of the signals in the time domain is long (many experiments at different modulation frequencies) and requires a good signal/noise ratio (S/N) which is attained by signal averaging[2]. Moreover it is limited to simple first order mechanisms. For complicated ones a frequency domain analysis is more adapted. In this last case one calculates the Fourier transform of the signal. After deconvolution of the experimental contributions one obtains the transfer function which is characteristic of the reaction mechanism[1]. However Fourier transform has been little used in comparison with lock-in detection although it gives only one point of the transfer function. In this paper we will describe a new apparatus specially designed for MBRS measurements on small deposited particles and using the recent improvements of this technique. We will also present some first results obtained with this apparatus for CO adsorption on Pd clusters deposited on MgO(100).

II. EXPERIMENTAL

2.1. General Design

The basic ideas in designing this apparatus were : in-situ deposition of the clusters on a clean single crystal face, ultra-clean molecular beam source, large modulation frequency range, angular and temporal measurements, temporal and frequency analysis of the signal. Fig. 1 shows a cross section of the assembly. It is composed of three uhv-chambers : preparation (2), reaction (4) and molecular beam source (11) chambers.

2.2. Preparation Chamber

A pressure of 1×10^{-9} Torr is obtained by ion and Ti pumping. Single crystal substrates are cleaved in vacuum or in air (and then outgased) before to be transferred in the main chamber. A carbon evaporator is available to prepar carbon replica for TEM observation of the clusters.

2.3. Reaction Chamber

After bakeout, a pressure of 5×10^{-10} Torr is obtained in the main chamber by a 300 1/s ionic pump and a liquid nitrogen (LN) cooled titanium pump. When a CO molecular beam is used the pressure increases depending of the beam diameter, for example with a 3 mm beam (on the sample) pressure remains smaller than 1.2×10^{-9} Torr. With an He beam ion pumping cannot be used so it is replaced by a 1000 1/s LN-baffled diffusion pump. With the same beam diameter pressure goes up to 3×10^{-9} Torr. The sample (6x6mm) (5) is fixed on a precision manipulator and can be heated up to 600°C. The Pd source is composed by a Knudsen cell (6) and an LN-cooled collimator (7) giving a spot of 5 mm on the sample. Its intensity, typically 1×10^{13} cm^{-2} s^{-1}, is in-situ controlled by a quartz monitor. During the Pd evaporation pressure goes up to 2×10^{-9} Torr. A gold Knudsen cell is also available for the characterization of the substrate surface by decoration[5]. Two quadrupole mass-spectrometers serve as detectors. One of them (21), movable in the

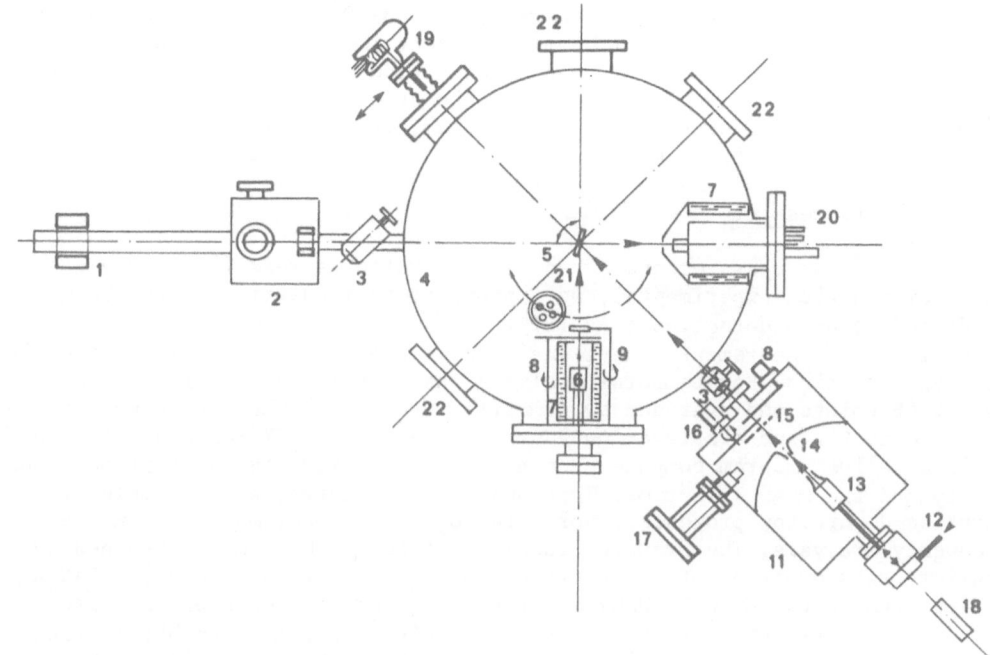

Fig. 1. Experimental assembly cross section. (1) transfer device, (2) preparation chamber, (3) straight-through valve, (4) reaction chamber, (5) sample, (6) Pd source, (7) LN-cooled collimator, (8) shutter, (9) quartz monitor, (11) nozzle source, (12) gas inlet, (13) nozzle, (14) skimmer, (15) collimators, (16) chopper, (17) gaz analyzer, (18) laser, (19) stagnation gauge, (20) detector, (21) rotatable detector, (22) view ports.

incidence plane over 180°, gives the angular distribution of the scattered beam. Its crossed beam ionization chamber has a collimating slit of 2 mm width distant of 55 mm from the sample. The other detector (20), immobile but more sensitive, is used for temporal measurements. Its 3 mm entrance hole is 40 mm far-away from the sample.

2.4. Molecular Beam Source

The molecular beam is obtained by a supersonic expansion through a 0.2 mm nozzle (13). A 0.5 mm skimmer is situated in front of the nozzle at a distance variable up to 5 cm. The beam source, bakeable to 200°C, is pumped by two LN-baffled oil diffusion pumps of 750 l/s. Five different beam diameters (1-5mm on the sample) can be selected during experiments by moving TEM diaphragms (15) through the beam. Smallest sizes are set for angular measurements and largest for temporal ones. A 3 mm diaphragm is placed between the source and the main chamber in order to hold a large difference of pressure between the 2 chambers. A laser beam (18), centered on the molecular-beam axis, makes easy the alignment of the diaphragms and the sample, relatively to the beam. The beam can be chopped at low frequency (≤ 10 Hz) by an electromagnetic shutter or at higher frequency (8-400 Hz) by a slotted disc driven by a bakeable synchronous motor (16). A 2 slots (50 % duty cycle) disc is used for MBRS measurements and a 1 slot (0.55 % duty cycle) disc for measurements of the velocity distribution in the incident beam. An electro-optical device gives a reference signal in phase with the beam.

The profile and the intensity of the beam are measured by a capillary-stagnation detector, located in the main chamber and mounted on an precision XY carriage (19). For the CO beam the typical intensity on the sample (623 mm far away from the nozzle) is 1.3×10^{14} cm^{-2} s^{-1} and the Mach number is 3.5. Thus for a 3 mm beam the direct impinging beam on the sample is 1000 times more intense than those yielded by the isotropic CO pressure. For suitable stagnation pressures the stability of the beam intensity is better than 0.6 % over 30 mn. The He beam, used in diffraction experiments, is about two times more intense.

2.5 Signal Processing

In angular measurements, when the S/N ratio is weak (it is the case in He diffraction experiments), we improve it by modulation of the beam at 40 Hz and lock-in detection (PAR 121).

In MBRS measurements the signal can be processed either in time or in frequency domains. For temporal analysis the signal waveform is sampled over 2048 points and averaged by summation (PAR 4202) then it is entered in a microcomputer and displayed on a graphic terminal (TEKTRONIX 4012). Then an interactive program computes the adsorbed quantity, the adsorption probability and the mean life-time. This analysis is restricted to simple adsorption-desorption processes. For more complex mechanisms we can use a frequency analysis. The transfer function of the mechanism is obtained by Fourier transformation of the averaged waveform[1]. For this we use a PASCAL FFT program on the COOLEY-TUKEY algorithm[6]. Extensive tests of this program on synthetic signals show[7], for 1024 points, a precision better than one % up to the 30th harmonic. Practically the main problem is to obtain a good S/N ratio[8]. Thanks to the great stability of our modulation it is possible to achieve good S/N ratios after long averaging times. Thus for the quantitative determination of the parameters of the reaction mechanism (reaction probability, reaction rate...) we prefer to use a two phase lock-in detector (EGG 5206) which gives the phase and the amplitude variations of the fundamental of the signal as a function of 1 parameter (generally the substrate temperature). This can be done more rapidly than by FFT.

III. RESULTS

3.1 Characterization of the Pd Clusters

We have shown in a previous study[9], of the nucleation and growth of Pd clusters on a (100) epitaxially grown MgO film, that at substrate temperatures between 300 and 400°C the particles are in good epitaxy and that it is possible to vary their sizes, their number being constant. Thus we can change the mean size of the particles from 0.8 to 5 nm and independently their density from 1.0 to 3.2×10^{12} cm^{-2}. These conditions being not fulfilled in previous studies of CO adsorption on Pd clusters[10,11], it was not possible to conclude definitely upon the origine of the increase of adsorption probability as the size decreases. Fig. 2 shows an example of Pd deposit on a (100) MgO single crystal face. The MgO crystal being air-cleaved, we use He diffraction as a check of the substrate cleanness. Indeed the first order diffraction peaks are well developed only after heating of the substrate to 500°C, in vacuum during at least 5 hours. This treatment was also necessary to observe epitaxy of the clusters. Electron diffraction (see Fig. 2b) indicates that most of the particles are (100) oriented and some of them are in (110) orientation. These facts are supported by direct visualization of atomic plans in HRTEM[12]. The density and the mean size are nearly the same but the size dispersion is slightly greater than on an epitaxial MgO film. Another difference is that some alignments of particles (probably due to the presence of steps) are visible on the cleaved surface leading to local variations of the particle density.

Another problem in working with small particles, is to be sure that the clusters do not suffer any change during the experiments. Then we have

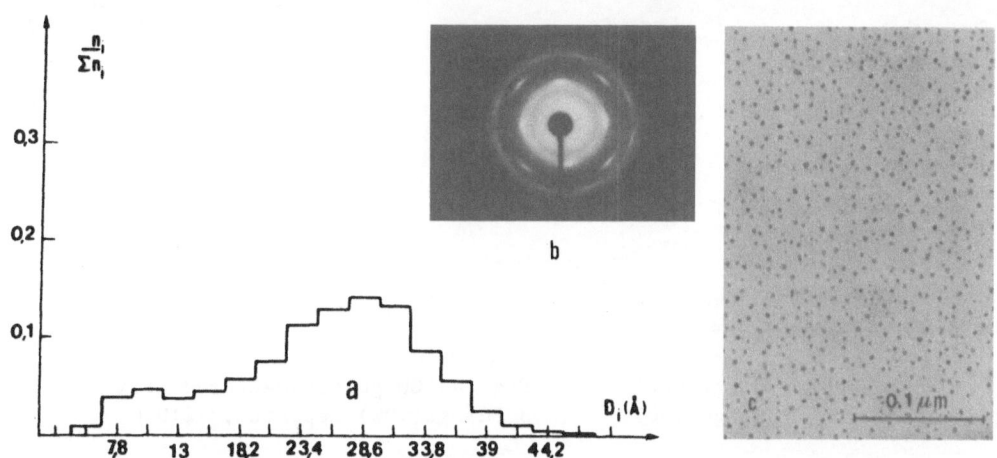

Fig. 2. 400°C Pd deposit on MgO(100). (c) micrography ; (b) diffraction
pattern ; (a) size histogram (mean size : 2.53 nm, size dispersion :
0.82 nm, density : 1.11×10^{12} cm^{-2}, covered area : 6.2 %).

submitted samples with various particle mean sizes to annealing at 400°C
during 1 hour in vacuum and in CO atmosphere (Pco = 1×10^{-5} Torr). In order
to detect small variations in the size distribution we prepare at each experiment two identical deposits, one them being annealed. TEM observation
does not show variations of the size histograms or of the orientation of
the clusters. Thus we conclude that Pd clusters are stable during CO adsorption experiments. This result is also supported by in-situ TEM observations[13], but this would not be true in the presence of oxygen where
coalescence and faceting have been observed[13,14].

3.2 Angular Distribution

Fig. 3 shows the angular distribution of CO on MgO(100). It is nearly cosine excepted a small increase near the specular direction. It is difficult to ascertain that CO molecules are desorbed or diffusively reflected.
The lobular contribution (near the specular direction) proves that some molecules are reflected. Time domain analysis, in pulsed regime, does not
give any measurable life-time at 20°C in agreement with the value of 9 kcal/
mole calculated for the adsorption energy of CO/MgO[15]. In the presence of
Pd clusters the angular distribution of CO is only slightly modified.

3.3. Temporal Measurement

Fig. 4 shows an example of single pulse (without averaging) of CO
reemitted from a particulate Pd deposit on MgO(100). The steep variations
of the signal when the beam is on or off show that, in this case, 64 %
$(1-\alpha)$ of the incident beam is reflected or desorbed from the MgO substrate.
The smooth increase gives a life-time of 1.5 s for CO molecules adsorbed
on the Pd particles. The smooth decrease indicates a smaller value because,
at the end of the pulse, the CO coverage on the Pd is about 0.2 and then
the desorption energy is smaller than at about zero coverage [16].

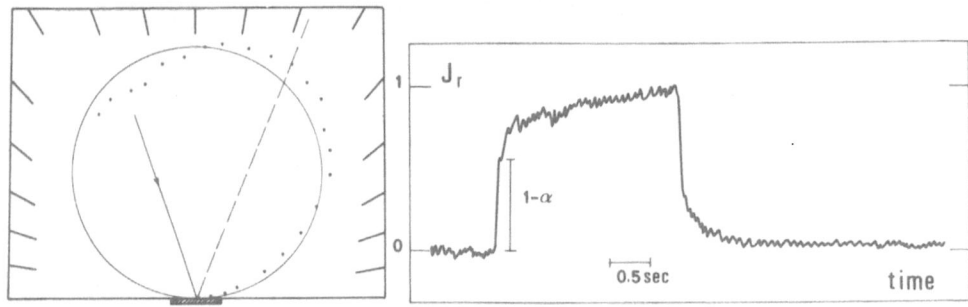

Fig. 3. RT angular distribution of CO scattered from MgO(100).

Fig. 4. CO pulse scattered from a Pd/MgO(100) deposit at 210°C.

IV. CONCLUSION

In this paper we have described a new apparatus which allows adsorption-desorption kinetics measurements on supported clusters by the MBRS technique. Angular and temporal measurements are available. Time and frequency domain analysis of the signals allow the determination of the reaction mechanism. Clean molecular beam source, in-situ preparation of particles, and high direct/isotropic fluxes ratio are the required conditions to obtain significant measurements. Particle size effects can be studied only with reproducible collections of clusters, mean size and number of which being independently changed. Furthermore the stability of the clusters during the adsorption experiments must be checked. First results on CO adsorption on Pd particles, deposited on MgO(100), show that the adsorption energies of CO molecules on the clusters and on the substrate are sufficiently different in order to separate these two processes.

REFERENCES

1. M.P. D'Evelyn and R.J. Madix, Surf. Sci. Rep. 3 (1983) 413.

2. C.R. Henry, C. Chapon and B. Mutaftschiev, Surf. Sci. 163 (1985) 409.

3. C.R. Henry and C. Chapon, Surf. Sci. 156 (1985) 952.

4. C.R. Henry, C. Chapon and C. Claeys, Bull. Soc. Chimique 3 (1985) 325.

5. C.R. Henry and C. Chapon, J. Physique 46 (1985) 1217.

6. J.W. Cooley and J.W. Tukey, Math. Computation 19 (1965) 297.

7. C. Duriez, Diplôme d'Etude Approfondie Marseille 1986.

8. H.H. Sawin and R.P. Merrill, J. Vac. Sci. Technol. 19 (1981) 40.

9. C. Chapon, C.R. Henry and A. Chemam, Surf. Sci. 162 (1985) 747.

10. S. Ladas, H. Poppa and M. Boudart, Surf. Sci. 102 (1981) 151.

11. E. Gillet, S. Channakhone and V. Matolin, J. Catal. 97 (1986) 437.

12. J.M. Penisson, C.R. Henry, C. Chapon and G. Nihoul, to be published.

13. K. Heinemann, T. Osaka and H. Poppa, Ultramicroscopy 12 (1983) 9.

14. M.F. Gillet and S. Channakhone, J. Catal. 97 (1986) 427.

15. E.A. Colbourn and W.C. Mackrodt, Surf. Sci. (1984) 391.

16. T. Engel, J. Chem. Phys. 69 (1978) 373.

STUDIES OF THE REACTIVITY AND COLLISION INDUCED DISSOCIATION OF

METAL CLUSTERS: ALUMINUM CLUSTER IONS, Al_n^+ (n=3-26), WITH OXYGEN

Martin F. Jarrold and J. Eric Bower

AT&T Bell Laboratories
600 Mountain Avenue
Murray Hill, NJ 07974

ABSTRACT

The reactivity and collision induced dissociation of size selected metal cluster ions with up to twenty-six atoms have been investigated using a low energy ion beam apparatus. Recent results on the collision induced dissociation of bare aluminum cluster ions; on the reactions between aluminum cluster ions and oxygen; and on the collision induced dissociation of aluminum clusters with one and two chemisorbed oxygen atoms are summarized and discussed.

INTRODUCTION

An important aspect of the current research on metal clusters is investigating their chemistry. Besides the obvious commercial applications in areas such as catalysis there is considerable fundamental interest in this topic as well. There is particular interest in investigating the change in chemical properties with cluster size and studying the emergence of bulk surface chemistry for the larger clusters. Clusters have been used for some time by theoreticians as models for surface processes, ultimately they may prove useful as experimental models as well.

For our studies of the chemistry of metal clusters we chose to work with ions rather than neutral species because of the ease of selecting the cluster size before the reaction. We report here a summary of the work we have performed in the last few months investigating the chemistry of aluminum cluster ions with up to twenty-six atoms. Specifically, we will discuss the collision induced dissociation of bare aluminum cluster ions, the reactions of aluminum cluster ions with oxygen, and the collision induced dissociation of aluminum cluster ions with one or two chemisorbed oxygen atoms. Since all the work discussed here is very recent some of the results presented are still preliminary.

Aluminum clusters have been investigated on several occasions in the past. Recent work includes studies of the unimolecular dissociation of aluminum cluster ions generated by sputtering [1]; an investigation of the reactions of small aluminum cluster ions, Al_n^+ (n=1-3), with a number of simple molecules [2]; a study of the magnetic properties of neutral aluminum clusters [3]; and there have been several recent theoretical studies of aluminum clusters [4-6], including calculations on the chemisorption of oxygen onto an Al_{13} cluster [6].

EXPERIMENTAL METHODS

The reactions were studied using a low energy ion beam apparatus. The clusters were generated by pulsed laser vaporization [7] of an aluminum rod in a continuous flow [8] of helium buffer gas. The clusters were ionized, in the buffer gas, by a 2-3 kV electron beam. After ionization the cluster ions and buffer gas expand into the vacuum chamber. Since ionization occurs before the expansion, cold metal cluster ions are generated. We use a high repetition rate vaporization laser (200 Hz) which results in a nearly continuous beam of metal cluster ions (40% duty factor). After the expansion, a specific cluster size is selected by a quadrupole mass filter, focussed into a low energy ion beam, and passed through a gas cell where the collision gas or neutral reactant is introduced. After passing through the gas cell the products and unreacted ions are focussed into a second quadrupole mass spectrometer where they are analyzed. At the end of the second quadrupole the ions are detected by an off-axis collision dynode and dual microchannel plates.

RESULTS AND DISCUSSIONS

A) Collision Induced Dissociation of Bare Aluminum Clusters

We have investigated the collision induced dissociation of aluminum cluster ions, Al_n^+ (n=3-26), by argon. Four products were observed: Al_{n-1}^+, Al_{n-2}^+, Al_{n-3}^+, and Al^+. The product branching ratios, measured with a center of mass collision energy of 5.25eV, are shown in Fig. 1. The branching ratios for Al_{n-3}^+ are not shown because it is a minor product. The main product is Al_{n-1}^+ for the larger clusters and Al^+ for the smaller ones. The transition between production of Al^+ and Al_{n-1}^+ as the main product is quite sharp and occurs around Al_{15}^+.

The excitation in low energy collision induced dissociation is deposited mainly as vibrational energy. Assuming that the subsequent dissociation of the excited cluster is

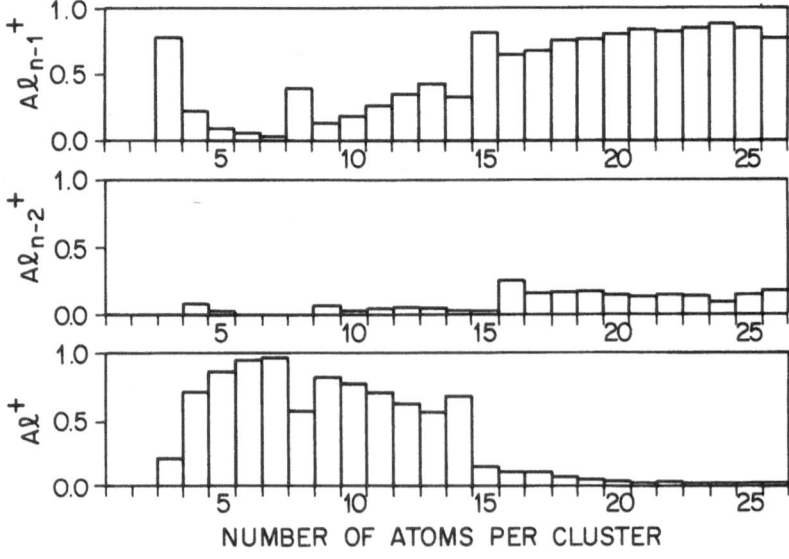

Fig. 1   Product branching ratios for the collision induced dissociation of bare aluminum cluster ions by argon.

statistical the product branching ratios should be dominated by the most stable products. Recent calculations for aluminum clusters [4] with up to six atoms indicate that their ionization potentials are all slightly larger than that of an aluminum atom. Since the work function of bulk aluminum is considerably less than the ionization potential of the atom we know that with increasing cluster size the ionization potential must drop below that of the atom as it approaches the bulk work function. We suspect that the relatively sharp transition observed in the product branching ratios at Al_{15}^+ occurs at the point where the ionization potential of the clusters falls below that of the atom. Thus for clusters, Al_n^+, with $n \geq 15$ the lowest energy products are $Al_{n-1}^+ + Al$, and for clusters with $n < 15$ the lowest energy products are $Al^+ + Al_{n-1}$.

The cross sections for collision induced dissociation rise to a maximum at between Al_6^+ and Al_9^+ and then decrease with increasing cluster size. Minima in the cross sections occur at Al_7^+, and Al_{13}^+ and Al_{14}^+. Studies of the pressure dependence of the cross sections indicate that for the smaller clusters, a single collision is sufficient to cause dissociation. However, for the larger clusters multiple collisions are required. These results indicate that more energy is required to dissociate the larger clusters. This at least partly reflects the decrease in the dissociation rate (for a given internal energy) as the cluster size increases, which is due to distributing the energy amongst more internal modes. However, it could also be due to an increase in cluster binding energy. Calculations for clusters up to Al_6 indicate that they are bound by only 0.9-1.6eV [4], which is considerably less than the bulk value (3.358eV). So an increase in binding energy with cluster size is expected.

Al_7^+ and Al_{14}^+ are "magic numbers" in the mass spectra of aluminum cluster ions recorded by us and other workers [3,9]. There are indications in the product branching ratios that Al_7^+ and Al_{14}^+ are slightly favored products. For example, production of Al_{n-1}^+ from Al_8^+ and Al_{15}^+, and production of Al_{n-2}^+ from Al_9^+ and Al_{16}^+ are slightly favored processes. Also, as we discussed above, minima occur at Al_7^+, and Al_{13}^+ and Al_{14}^+ in the collision induced dissociation cross sections. These results suggest that Al_7^+ and Al_{14}^+ are slightly more stable than their neighbors.

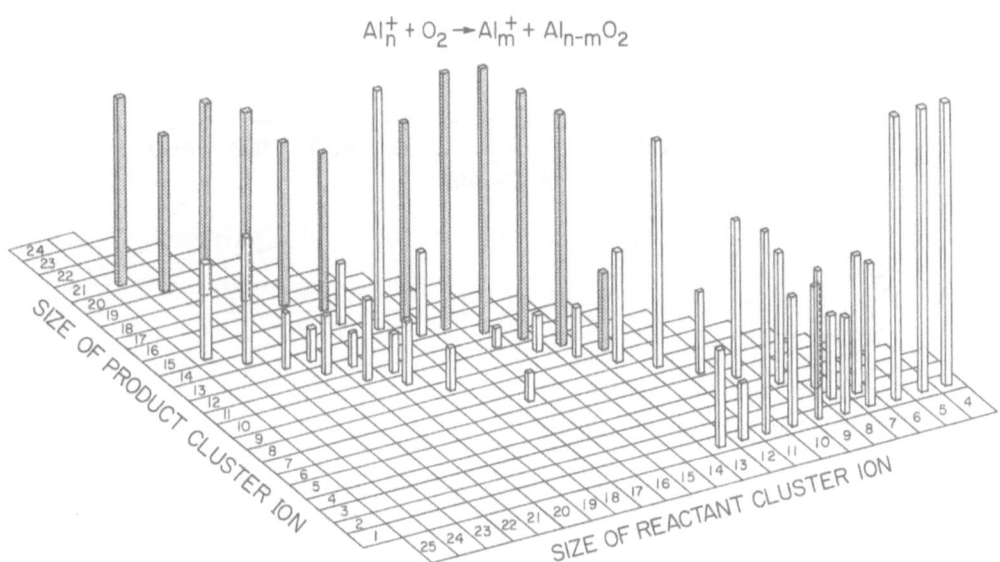

$$Al_n^+ + O_2 \rightarrow Al_m^+ + Al_{n-m}O_2$$

Fig. 2   Histogram showing the main products from the reactions between aluminum cluster ions and oxygen.

B) Reactions of Aluminum Cluster Ions with Oxygen

A histogram of the products arising from the reactions between aluminum cluster ions and oxygen, at a center of mass collision energy of approximately 1eV, is shown in Fig. 2. For clarity only the main products are shown. Almost all of the products arise from aluminum cluster fragmentation and oxygen containing product ions generally constitute only a few percent of the total product intensity. For the smaller clusters formation of Al^+ is an important product. Starting with Al_7^+ loss of five aluminum atoms, to yield Al_{n-5}^+, becomes a major product channel. At Al_{13}^+ a transition occurs and loss of four aluminum atoms becomes the main product channel. This continues up to Al_{25}^+ with one exception. Al_{19}^+ looses five aluminum atoms (to produce the "magic number" Al_{14}^+) rather than four. For clusters with seventeen and more atoms loss of ten aluminum atoms becomes important. Studies of the pressure dependences of the product branching ratios indicate that Al_{n-10}^+ products arise from sequential reactions. For example, for Al_{20}^+, the Al_{10}^+ product arises from the reaction between the primary Al_{16}^+ product and oxygen. Thus the reaction sequence is

$$Al_{20}^+ \xrightarrow{-4Al} Al_{16}^+ \xrightarrow{-6Al} Al_{10}^+$$

Different numbers of aluminum atoms are lost in the primary and secondary processes; although Al_{16}^+, like Al_{20}^+, loses four aluminum atoms in the primary reaction when it is a reactant ion. Similar results were observed for all the larger clusters. The differences between the product distributions for primary and secondary reactions probably occur

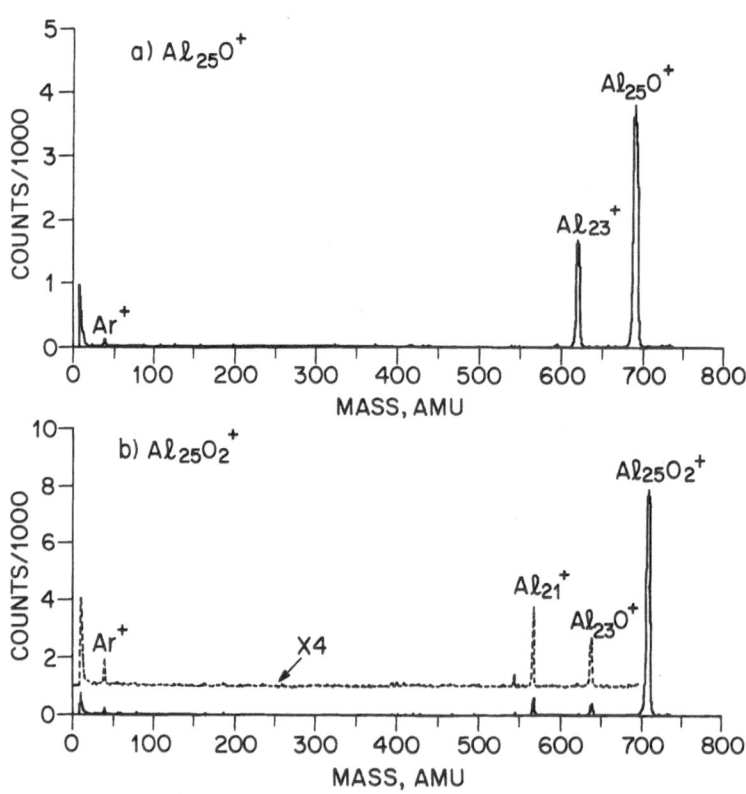

Fig. 3   Mass spectra measured for the collision induced dissociation of a) $Al_{25}O^+$ and b) $Al_{25}O_2^+$. The Ar^+ peaks and the peaks at close to zero mass are artifacts.

because the product ions are internally excited by the initial reaction with oxygen. To test this idea the reactions with oxygen were studied at higher collision energies. As expected the product distributions changed. For example, Al_{20}^+ which preferentially loses four aluminum atoms with a center of mass collision energy of 1eV, loses five aluminum atoms when the collision energy is raised to 4eV. This is an interesting example of the influence of energy on a chemical reaction.

The cross sections for the reactions between aluminum cluster ions and oxygen increase with increasing cluster size. The cross sections are a few $\overset{\circ}{A}{}^2$ for the smaller clusters and rise to over $50\overset{\circ}{A}{}^2$ for the larger ones. The rate of increase of the reaction cross section with cluster size approximately follows the geometric size of the clusters.

Up to this point we have discussed the fate of the aluminum cluster ions but we have not considered the neutral products. There appears to be nothing particularly unique about Al_4O_2, Al_5O_2 and Al_6O_2 as possible neutral products. In the next section we describe some experiments designed to determine the nature of the neutral products and elucidate the reaction mechanism.

C) Collision Induced Dissociation of Aluminum Clusters with Chemisorbed Oxygen

Aluminum clusters with chemisorbed oxygen were generated by adding a trace (10-100ppm) of oxygen to the buffer gas. The results of collision induced dissociation of $Al_{25}O^+$ and $Al_{25}O_2^+$ by argon at a center of mass collision energy of 5.25eV are shown in Fig. 3. For $Al_{25}O^+$ the main product, Al_{23}^+, arises from loss of an Al_2O molecule. For $Al_{25}O_2^+$ products due to the loss of one Al_2O molecule ($Al_{23}O^+$) and two Al_2O molecules (Al_{21}^+) can be seen. Similar results were obtained for the other clusters studies. As discussed above low energy collision induced dissociation produces predominantly the most stable products. The products from $Al_{25}O^+$ and $Al_{25}O_2^+$ arise from loss of Al_2O molecules. The driving force for these reactions must be the stability of the Al_2O species. Al_2O has been observed in high temperature studies of aluminum oxide and is very strongly bound [10].

Collisional activation of $Al_nO_2^+$ and the reaction between Al_n^+ and O_2 are closely related processes. If similar products result from these processes, then they proceed through common intermediates. Similar products are observed, although they are not identical probably because of energy differences. Thus, these results suggest that the $Al_n^+ + O_2$ reaction proceeds through a long-lived intermediate complex where the oxygen is chemisorbed on the cluster and the resulting $Al_nO_2^+$ undergoes statistical dissociation.

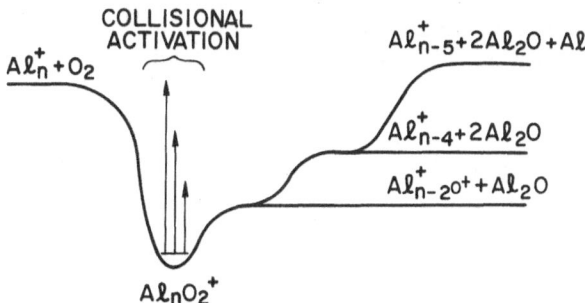

Fig. 4   Schematic potential energy surface for the $Al_nO_2^+$ system (for $n > 13$).

A schematic potential energy surface for the $Al_nO_2^+$ is shown in Fig. 4. Chemisorption of O_2 onto the aluminum cluster must be fairly exothermic because it supplies sufficient energy to "evaporate off" two Al_2O molecules. This highlights an important difference between surface reactions and cluster reactions, for surface reactions the heat of chemisorption is distributed throughout the bulk, for isolated clusters it remains localized in the cluster and results in dissociation. After the loss of two Al_2O molecules the reaction stops for the larger clusters, resulting in an Al_{n-4}^+ product. However, for the smaller clusters, which are probably less strongly bound, there is sufficient energy to fragment even further by loss of an aluminum atom to yield Al_{n-5}^+.

CONCLUSIONS

We have briefly summarized our recent work. Our main results and conclusions are:

1. Collision induced dissociation of Al_n^+ results in mainly Al^+ for the smaller clusters and mainly Al_{n-1}^+ for the larger ones. The change in product distribution occurs at Al_{15}^+ which is probably where the ionization potential of the clusters drops below that of the atom.
2. There is evidence from the collision induced dissociation studies that Al_7^+ and Al_{14}^+ (the "magic numbers") have enhanced stability.
3. The reactions between Al_n^+ and O_2 results in Al_{n-4}^+, Al_{n-5}^+ and Al^+ as the major products. Al_{n-4}^+ is the main product for the larger clusters with the exception of Al_{19}^+ which loses five atoms to produce the "magic number" Al_{14}^+.
4. Collision induced dissociation of aluminum clusters with chemisorbed oxygen results in loss of Al_2O molecule(s). These are probably the same neutral products that are formed in the reaction between Al_n^+ and O_2.
5. The reaction between Al_n^+ and O_2 proceeds through a long-lived intermediate complex in which the oxygen is chemisorbed on the aluminum cluster and the complex then undergoes statistical dissociation.
6. Collision induced dissociation has proved to be a useful technique to obtain qualitative thermochemical information about metal cluster ions.

ACKNOWLEDGEMENTS

We are grateful to J. S. Kraus for designing and installing our computer system; and to T. Insano, R. Tronio, F. D'Amato and the other members of the Murray Hill Development Shops for constructing portions of the experimental apparatus.

REFERENCES

1. W. Begemann, K. H. Meiwes-Broer and H. O. Lutz, Phys. Rev. Letts., 56, 2248 (1986).
2. L. Hanley and S. L. Anderson, (in press).
3. D. M. Cox, D. J. Trevor, R. L. Whetten, E. A. Rohlfing and A. Kaldor, J. Chem. Phys., 84, 4651 (1986).
4. T. H. Upton, Phys. Rev. Letts., 56, 2168 (1986).
5. C. W. Bauschlicher and L. G. M. Pettersson, J. Chem. Phys., 84, 2226 (1986).
6. H. Partridge and C. W. Bauschlicher, J. Chem. Phys., 84, 6507 (1986).
7. T. G. Dietz, M. A. Duncan, D. E. Powers and R. E. Smalley, J. Chem. Phys., 74, 6511 (1981).
8. S. C. Richtsmeier, E. K. Parks, K. Liu, L. G. Pobo, and S. J. Riley, J. Chem. Phys., 82, 3659 (1985).
9. M. E. Geusic and R. R. Freeman, (unpublished results).
10. J. Drowart, G. DeMaria, R. P. Burns and M. G. Inghram, J. Chem. Phys., 32, 1366 (1960).

PHYSICAL PROPERTIES OF HIGH-NUCLEARITY METAL CLUSTER COMPOUNDS

L.J. de Jongh*, H.B. Brom*, G. Longoni[+], P.R. Nugteren*, B.J. Pronk*, G. Schmid°, H.H.A. Smit*, M.P.J. van Staveren* and R.C. Thiel*

* Kamerlingh Onnes Laboratorium, Rijksuniversiteit Leiden Nederland
+ Chimica Inorganica e Metallorganica, Università degli studi Milano, Italia
° Anorganische Chemie, Universität Essen, W-Deutschland

Polynuclear metal cluster compounds provide an interesting class of model systems for small metal particles with chemisorbed molecules. They are composed of large macromolecules consisting of a metal core of a certain number (n) of metal atoms, to which core a "shell" of ligands is coordinated. In fig.1 the carbonyl cluster $[Ni_{38}Pt_6(CO)_{48}H]^{5-}$ is shown as an example[1]. In going from compound to compound the type of metal atom can be varied, as well as the number n of atoms in the metal core. For Au, Pt, Ru and Rh, the maximum value of n reached at this moment is already n = 55, namely in $Au_{55}(PPh_3)_{12}Cl_6$ (fig.2) and related Pt, Ru and Rh cluster compounds[2]. For Pd a metal core as large as n = 570 has recently

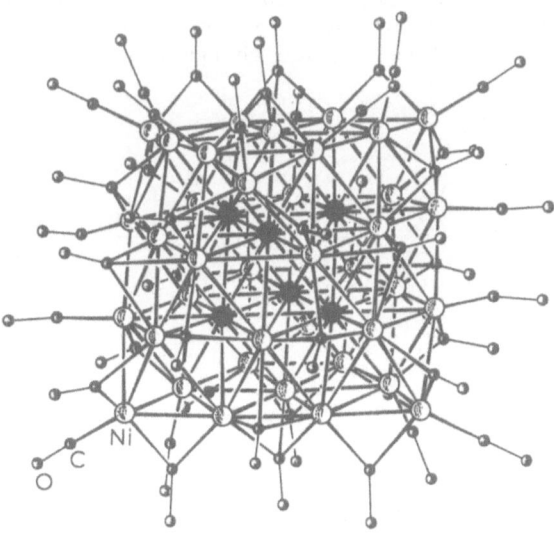

Fig. 1 Structure of the $(Ni_{38}Pt_6(CO)_{48}H]^{5-}$ anion occurring in $[Ni_{38}Pt_6(CO)_{48}H](NEt_4)_5$. The six Pt atoms (shaded), together with the 38 Ni atoms form the metal core of the cluster molecule.

been reported[3]. The advantages of these materials for fundamental research in small metal particle physics are clearly: (i) complete homogeneity of the metal cluster size in each sample; (ii) possibility to study the properties as a function of cluster size by comparing compounds with different n; (iii) availability of macroscopically large (0.01 - 10 g) samples for a variety of physical experiments.

Problems of fundamental interest in the physics of small metal particles are[4]: the transition from nonmetallic to metallic behavior as the cluster size is increased, the relation with surface physics and -chemistry, the magnetic and the (super)conducting properties, the phonon spectra, etc. In our group at Leiden we have recently started a systematic study of the physical properties of these polynuclear metal cluster compounds. We combine the measurements of macroscopic thermodynamic variables like the magnetization, susceptibility, specific heat and electrical conductivity, with microscopic probes like ESR, NMR and Mössbauer-Effect studies on the electronic and nuclear spins of the metal atoms.

The electrical properties of some of these materials are very interesting. In $Au_{55}(PPh_3)_{12}Cl_6$ the electrical resistivity is relatively small at room temperature[5]; it is of a similar order of magnitude as for a typical semiconductor. Fig.3 shows the temperature dependence of the resistivity (R), which is seen to be very well described by the law: $R \propto \exp(T_0/T)^{\frac{1}{2}}$, with the parameter $T_0 = 5.7 \times 10^4$ K. Interestingly, the same $\exp(T)^{-\frac{1}{2}}$ law is observed in a variety of other (random) discontinuous metal systems, for example in cermets (finely dispersed metal particles in a dielectric), in quasi 1-d conductors, and very recently[6] in a 3-d metallic system claimed to exhibit a metal-insulator transition of the Anderson type. It is not clear yet why such a diversity of materials should show the same type of temperature dependence for the resistivity. In all cases, however, the conductivity at low temperatures is expected to be dominated by thermally activated hopping of the charge carriers, from site to site or from molecule to molecule. It seem logical therefore, to assume that the conductivity in $Au_{55}(PPh_3)_{12}Cl_6$ is also attributable to intermolecular hopping or tunneling processes. The nature of these processes certainly deserves further systematic investigation. In particular the electron-phonon coupling may play an important role here, since the average dwelling time of the extra charge on the molecule will be long enough for it to distort. In this respect there will be strong analogies with the well-known small-polaron problems[7] from solid state physics and the intermediate valence charge-transfer problems considered in solid state chemistry[8].

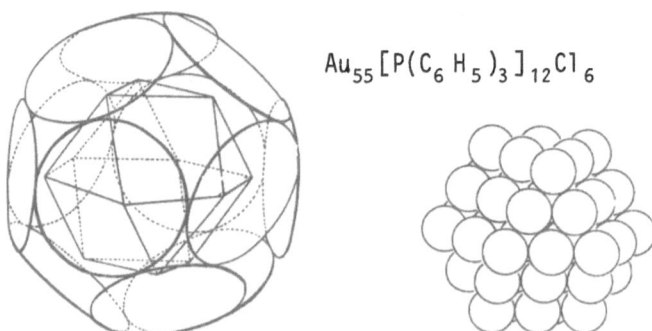

$$Au_{55}[P(C_6 H_5)_3]_{12}Cl_6$$

Fig. 2 Schematic figure of the cuboctahedral metal core of 55 Au atoms (right) in $Au_{55}(PPh_3)_{12}Cl_6$, and the way in which the core is surrounded by the large phosphane ligands (left).

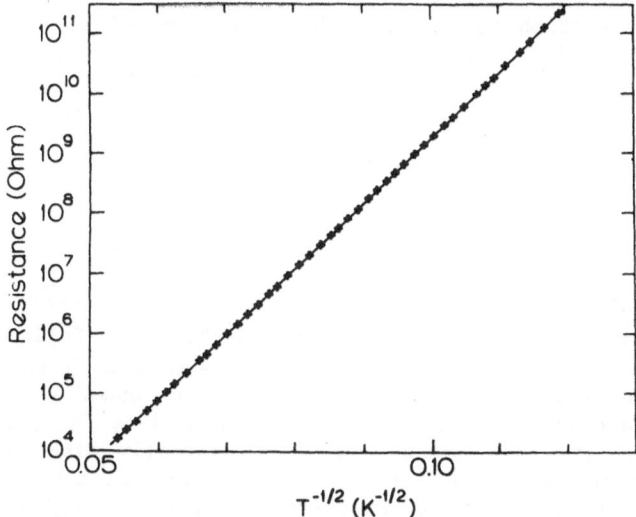

Fig. 3   Resistance of a pressed $Au_{55}(PPh_3)_{12}Cl_6$ powder pellet plotted on a logarithmic scale against $T^{-\frac{1}{2}}$.

The ^{195}Au Mössbauer-spectrum of $Au_{55}(PPh_3)_{12}Cl_6$ at T = 4.2 K has been studied earlier at Nijmegen[9]. The Mössbauer effect provides a microscopic probe by which one may study the chemical environment of the metal atoms via their nuclear spins. Indeed, the spectrum for the Au_{55} cluster could be decomposed into four different contributions related to the 13 Au atoms at the center of the cuboctaedron, to the 24 uncoordinated peripheral atoms, to the 12 peripheral atoms coordinated to the PPh groups and to the 6 peripheral atoms coordinated to the Cl atoms. We have recently studied the spectra as a function of temperature in the range 0.5 K - 30 K, thereby obtaining additional information on the Debije-Waller factors associated with these four different Au sites. Thus the Mössbauer effect provides a microscopic method to study the phonon excitations of the metal cluster itself, i.e. apart from those of its environment. In addition, in case of magnetic clusters the Mössbauer effect can also be used to study the magnetic properties.

The information on the phonon spectra is important to understand the bulk specific heat measurements that we have so far performed on the Au_{55} cluster and on the $[Ni_{38}Pt_6(CO)_{48}H](NEt_4)_5$ compound. In both cases the low temperature (T < 10 K) specific heats show strong deviations from the usual T^3 Debije behavior. Although over a certain temperature range the data appear to follow a T^2-law, one cannot expect this to be related to 2-dimensional surface (Rayleya) waves, in view of the fact that the phonon wavelengths at these low temperatures exceed by far the perimeter of these clusters. Although we have not yet a definitive explanation, we remark that the results appear to be describable by a collection of Einstein oscillators, that can be related to the measured Debije-Waller factors in case of the Au_{55} cluster.

In NMR one also uses the nuclear spin of the metal atoms e.g. Pt to probe their physical and chemical properties. The NMR resonance line in insulating compounds is determined by the chemical shifts, whereas in bulk Pt metal the conduction electrons produce a much larger additional shift, the well-known Knight-shift. In principle this provides one of the most direct ways to decide whether a particle is metallic or not. A second criterion is provided by the temperature dependence of the nuclear-spin lattice relaxation time (T_1), which in a metal should follow the Korringa relation ($T_1 T$ = constant) and which in a dielectric is very much longer. Preliminary measurements on $[Pt_{38}(CO)_{44}H_2](PPN)_2$ and smaller Pt carbonyl clusters show resonance positions in the range of known chemical-shifts, and also T_1's which do not follow the Korringa law. It is obvious that an explanation of these phenomena has to deal with the fact that even for Pt_{38} all but six Pt atoms are coordinated by the CO ligands.

The study of the magnetic behavior of metal clusters in general is quite important since the electronic structure is directly reflected in the magnetic properties. From our first magnetic susceptibility and high-field magnetization measurements[10] we may mention the following salient features. In case of clusters of nonmagnetic metals such as Pt, Ru, Rh, the magnetic susceptibility (χ) is quite small but shows an upturn in the lowest temperature range (T ≲ 50 K). Fig.4 shows data for $[Pt_{38}(CO)_{44}H_2]$ $(PPN)_2$ to illustrate this point. The χ is basically diamagnetic except for the Curie-type tail observed for $T \to 0$. The same behavior has been reported by Teo et al.[11] for six Pt carbonyl clusters ranging from Pt_6 to Pt_{38}. From the Curie constants the magnetic moment per Pt cluster can be estimated (assuming $S = \frac{1}{2}$) to be about $0.2 \mu_B - 0.3 \mu_B$ per cluster, fairly independent of cluster size. For $Rh_{55}(PPh_3)_{12}Cl_6$ we see essentially the same behavior. For $Ru_{55}(PBu_3)_{12}Cl_{20}$, on the other hand the χ is already positive at all temperatures and the Curie-term is much more pronounced. The χ curve for Ru_{55} is in fact quite similar to that for $[Ni_{34}(CO)_{38}C_4H](NEt_4)_5$ shown in fig.5. For the larger Ni clusters the moment per cluster is of the order of a few μ_B, whereas it decreases to much smaller values for low-nuclearity clusters such as $[Ni_{12}(CO)_{21}H_2]$ $(PPh_3CH_2Ph)_2$ and $[Ni_9(CO)_{17}C](NBu_4)_2$. We note that bulk Ni metal has a moment of $0.6 \mu_B$ per atom, or about $20 \mu_B$ for a cluster of Ni_{34}. Thus there is a very strong reduction of the Ni moment in these cluster compounds, which is attributed to the coordination by the ligands. The low moment is confirmed by the low-temperature magnetization curves measured in very high magnetic fields. The example of $[Ni_{38}Pt_6(CO)_{48}H_2](NEt_4)_4$ is given in fig.6. Of importance here is that the magnetization does not saturate in fields of a few tesla, as would be the case for paramagnetic impurities or ferromagnetic fine Ni metal particles. There is a roughly linear increase observed up to the highest attained fields. Quite similar behavior has been observed for Co_{55}, where again the magnetic moment is

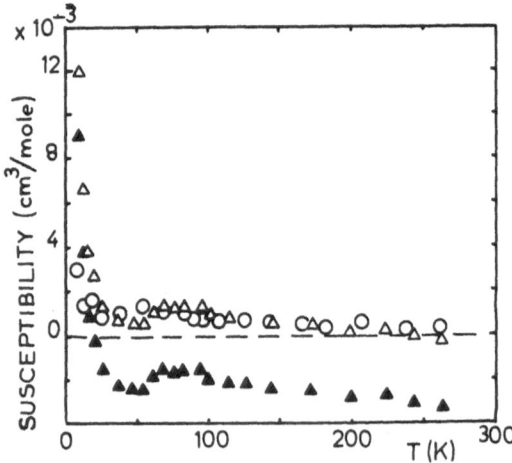

Fig. 4 Magnetic susceptibility of $[Pt_{38}(CO)_{44}H_2](PPN)_2$ versus temperature. Shown are the raw data (▲) and those corrected for the estimated diamagnetism of the compound (△). Also shown are corrected data for $[Pt_{26}(CO)_{32}](PPh_4)_2$ (O).

only a few μ_B per cluster, instead of the 94 μ_B expected for a bare Co_{55} particle on basis of the bulk Co metal moment of 1.7 μ_B/atom.

The strong reduction of the magnetic moment can be attributed to the presence of the ligands. For "naked" small particles of magnetic metal (Fe,Co,Ni) values for the magnetic moment per atom comparable to the bulk value have been found. Also, different types of molecular orbital calculations agree on that even for small clusters of the magnetic metals there is a sizable number of unpaired spins, resulting in a magnetic moment of the same order as in the bulk. The effect of the bonding of the ligands upon the energy level structure of the bare metal cluster is to destabilize a certain number of orbitals[12,13]. These loose their electrons which pair-off the unpaired spins, thereby reducing the magnetic moment with respect to the bare metal cluster. The process can be viewed as a molecular analogue of the high-spin/low-spin transition in e.g. ionic Fe^{2+} compounds. In the smallest clusters (n $\leq$ 8) the reduction of the moment can be complete and the cluster molecules are often diamagnetic. This agrees with the general rule that in chemical bonding the participating components try to reach closed-shell configurations. As the metal core becomes larger, however, complete spin-pairing is no longer possible, and a magnetic moment appears that grows with increasing n. Finally, as n becomes very large the quenching of the moment by the attached ligands reduces to a small surface effect.

With these few examples we hope to have shown that the polynuclear cluster compounds provide very interesting materials for solid state research, and that besides the chemistry, there are a sizable number of fundamental physical problems to study as well. As a first conclusion from our measurements we would like to state that as a function of particle size a sharp metal-nonmetal transition in these very small particles probably does not occur. Instead, there appears to be a wide intermediate range of cluster sizes in which the properties are neither metallic, nor atomic, but are instead described by (macro)molecular physics. Novel and interesting phenomena are met in this field, for which we have recently proposed the name **"meta-metallic"** phase[5,14].

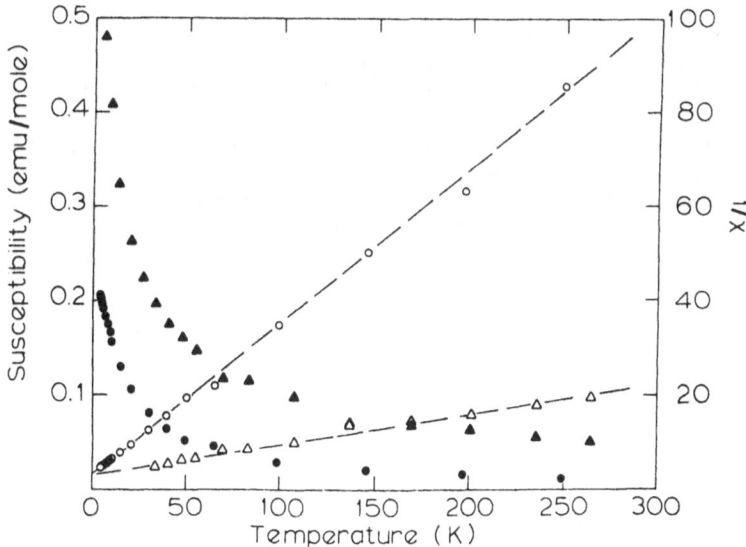

Fig. 5 The (inverse of) the molar magnetic susceptibility of $[Ni_{34}(CO)_{38}C_4H](NEt_4)_5$ ($\triangle$,$\blacktriangle$) and of $[Ni_{38}Pt_6(CO)_{48}H_2](NEt_4)_4$ ($\circ$,$\bullet$) versus the temperature (corrected for diamagnetism).

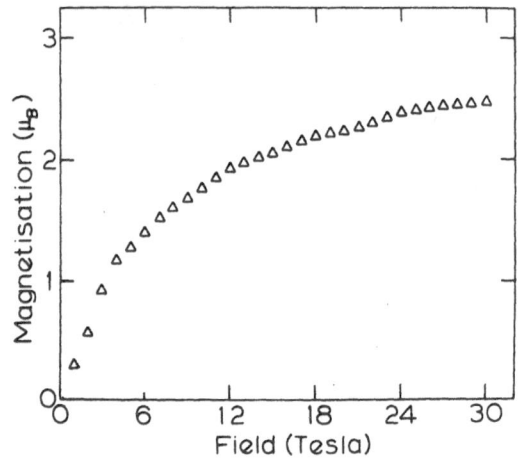

Fig. 6 High-field
magnetization curve for
$[Ni_{38}Pt_6(CO)_{48}H_2](NEt_4)_4$

Acknowledgements - The work in Leiden is part of the research program of the "Stichting voor Fundamenteel Onderzoek der Materie" (Foundation for Fundamental Research on Matter) and was made possible by financial support from the "Nederlandse Organisatie voor Zuiver-Wetenschappelijk Onderzoek" (Netherlands Organization for the Advancement of Pure Research). The investigations are sponsored by the Leiden Materials Science Group ("Werkgroep Fundamenteel Materialen Onderzoek"). The research program is also sponsored by the Commission of the European Communities.

REFERENCES

1. A. Ceriotti, F. Demartin, G. Longoni, M. Manassero, M. Marchionna, G. Piro and M. Sansoni, Angew. Chem. 24:697 (1985).
2. See e.g. G. Schmid in "Structure and Bonding" 62: 52, Springer Verlag, Berlin Heidelberg (1985).
3. M.N. Vargaftik, V.P. Zagorodnikov, I.P. Stolyarov, I.I. Moiseev, V.A. Likholobov, D.I. Kochubey, A.L. Chuvilin, V.I. Zaikovsky, K.I. Zamaraev, and G.I. Timofeeva, J. Chem. Soc., Chem. Commun., 937 (1985).
4. For recent reviews see e.g. W. Halperin, Rev. Mod. Phys. 58:533-606 (1986) or J.A.A.J. Perenboom, P. Wijder, and F. Meier, Phys. Rep. 78: 173 (1981).
5. M.P.J. van Staveren, H.B. Brom, L.J. de Jongh and G. Schmid, Solid State Commun. 60: 319 (1986).
6. P. Nédallec, A. Traverse, L. Dumoulin, H. Bernas, L. Amaral and G. Deutscher, Europhys. Lett. 2:465 (1986).
7. T. Holstein, Ann. Phys. 8:325 and 343 (1959); I.G. Austin and N.F. Mott, Adv. Phys. 18:41 (1969).
8. K.Y. Wong and P.N. Schatz, Progr. Inorg. Chem. 28:369 (1981).
9. G. Schmid, R. Pfeil, R. Boese, F. Bandermann, S. Meyer, G.H.M. Calis and J.W.A. van der Velden, Chem. Ber. 114: 3634 (1981).
10. B.J. Pronk, H.B. Brom, L.J. de Jongh, G. Longoni and A. Ceriotti, Solid State Commun. 59:349 (1986).
11. B.K. Teo, F.J. Disalvo, J.W. Waszczak, G. Longoni and A. Ceriotti, Inorg. Chem. 25:2262 (1986).
12. D.R. Salahub and F. Raatz, Int. J. Quantum Chem. S18:173 (1984); Surface Science 146:L609 (1984).
13. G.F. Holland, D.E. Ellis, and W.C. Trogler, J. Chem. Phys. 83:3507 (1985).
14. L.J. de Jongh, Int. Conf. on Molecules, Clusters and Networks in the Solid State, Birmingham, July 1986.

INFRARED MATRIX-INVESTIGATIONS OF THE INTERACTION OF

OXYGEN WITH IRON ATOMS AND AGGREGATES

Ulrich L. Kettler[†], Paul H. Barrett[†], and Ralph G. Pearson[*‡]

Department of Physics[†] and Department of Chemistry[‡]

University of California, Santa Barbara, CA 93106, USA

ABSTRACT

We report on infrared absorption investigations of the reactivity of Fe_n ($n=1,2,...$) with molecular oxygen in low temperature argon matrices. Besides the known ν_{O-O} IR line of iron peroxide $Fe(O_2)$[1] at $956\,cm^{-1}$ we observe that the band at $\approx 873\,cm^{-1}$ which has been reported for FeO[2,3] becomes strong at *higher* iron concentrations. So far the occurrence of FeO has only been reported when atomic oxygen is formed in the device or at the matrix surface, e. g. by the reaction of O_2 with excited rare-gas atoms from a hollow-cathode discharge. In this case also O_3 lines have been reported. As we see no O_3 lines, it is believed that Fe_2 is capable of O-O bond rupture resulting in a reaction product of $2\,FeO$. — A new band grows at moderate iron concentration at the expense of $Fe(O_2)$. It is assigned to the ν_{O-O} mode of a FeO_2Fe molecule. A band at $1\,098\,cm^{-1}$ is attributed to $Fe(O_2)_2$.

INTRODUCTION

The interaction of iron atoms and aggregates with oxygen is of importance from different points of view. The reaction of oxygen with Fe atoms coordinated in biological systems has been the subject of several studies.[4-7] For possible catalytical application the interaction of O_2 with iron aggregates is more likely of interest than with atomic iron. Futhermore the different mechanisms of forming FeO and other iron-oxygen complexes under low temperature conditions are potentially of great astrophysical importance because of the high cosmic abundances of both Fe and O.

A vibrational mode at $(872\pm1)\,cm^{-1}$ has been attributed to the ground state of FeO from both gas phase[8] and matrix investigations.[1-3] Other bands at $969/956/946\,cm^{-1}$ have been reported with matrices and assigned to FeO_2 molecules of different structure.[1] The wavenumbers of these bands lie between the two regions of the metal hyperoxides ($Me(O_2)_2$, $\approx 1\,100\,cm^{-1}$) and the superoxides ($Me-O-O$, $\approx 850\,cm^{-1}$). Hyperoxides have been reported for Cu,[9] Pt, Pd, Rh, and Ni.[10] Bondings of the iron superoxide type have been reported for biological complexes.[7] — IR studies of the reactivity of molecular oxygen with iron atoms *and* aggregates are the subject of this report.

EXPERIMENTAL

Detailed descriptions of the IR vacuum system have been published elsewhere.[11] The samples are prepared by codeposition of Fe and an Ar/O_2 mixture onto a NaCl window which is kept at $\approx 15\,K$ by an Air Products Displex closed-cycle refrigerator. Iron is evaporated from iron powder in an alumina crucible held in a resistively heated tantalum furnace. Temperatures are measured with an optical pyrometer and are typically $1300-1450\,°C$. Infrared spectra are recorded on an Perkin-Elmer 683 spectrometer. — Background experiments without oxygen or iron do not show any of the reported IR bands.

RESULTS AND DISCUSSION

Fig. 1 shows the IR spectra in the $1000-850\,cm^{-1}$ region for increasing iron concentration from top to bottom. The O_2 concentration is kept constant at 0.5%. Fig. 2 shows the spectra in the same wavenumber region for decreasing oxygen concentration from top to bottom. The crucible temperature is $1400\,^\circ$C. Under these conditions there is a good chance for the formation of iron dimers, at least at low oxygen concentration. Decreasing oxygen concentration makes essentially the same effect as increasing iron concentration.

At higher iron/lower oxygen concentration new bands develop, especially below $910\,cm^{-1}$. The changes in band intensities can be summarized as follows.

The intensity of the band at $955\,cm^{-1}$ $(\pm2\,cm^{-1})$ (C in Fig. 1,2) is proportional to the relative iron concentration (at low Fe concentration). It appears at low oxygen *and* low iron concentrations and is very sharp with a half width of $\leq1\,cm^{-1}$. Furthermore this band grows (at low iron concentration) upon annealing of the matrix to 30 K. The attribution to a (side-bonded) iron peroxide molecule,[1] $Fe(O_2)$, is therefore reasonable.

Modes at $946\,cm^{-1}$ and $969\,cm^{-1}$ have tentatively been assigned to a bent and a linear FeO_2 molecule of the structure O-Fe-O, respectively,[1] formed by the reaction of one iron atom with two oxygen atoms. The presence of atomic oxygen is usually recognized by the presence of O_3 bands in the IR spectra. As we do not see any O_3 bands[12] we do not expect to have significant amounts of atomic oxygen in our matrices. Therefore these bands should be absent in our spectra.

The spectroscopic results agree quite well concerning the $946\,cm^{-1}$ band, but not with that at

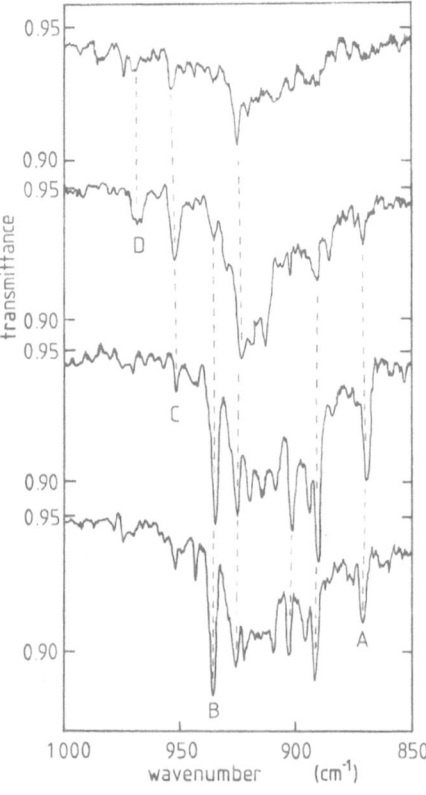

Fig. 1   Infrared spectra of a Fe/Ar matrix with 0.5% O_2 for increasing iron concentration from top to bottom. At higher Fe concentration the FeO band at $873\,cm^{-1}$ appears (A). The band B is attributed to $\nu_{O-O}(FeO_2Fe)$, whereas C is the ν_{O-O} mode of $Fe(O_2)$.

about 970 cm^{-1} (D in Fig. 2). At an oxygen concentration $\geq 5\%$ we see a broad, structured band with at least four peaks spaced by 3–4 cm^{-1}. The positions are $\approx 969/972/976/980$ cm^{-1}. As we did no isotopical substitution for O_2 these lines cannot be interpreted as an isotopical splitting. Upon annealing of the matrix the bands change their relative intensities. Therefore the presence of different trapping sites cannot be excluded. Taking our results into account we cannot agree with the previously reported structure for this species. It should be a molecule with at least one O–O bond. As the bands do not appear at low oxygen concentration[1] the molecule possibly contains more than one dioxygen group.

The band at 938 cm^{-1} (B) is reported for the first time. It becomes strong at increasing iron/lower oxygen concentration whereas that of $Fe(O_2)$ at 955 cm^{-1} (C) decreases (Fig. 1, 2). Therefore it should contain two iron atoms. It is even more stable upon warming the matrix than the 955 cm^{-1} band, and it can be enhanced thereby at the expense of $Fe(O_2)$ if both species are present. We suggest a formation by the secondary reaction

$$Fe(O_2) + Fe \rightarrow FeO_2Fe .$$

Such a structure has been reported for other metals. In these cases it has been noted that a Me_2-O_2 molecule shows a lower energy mode than MeO (Me=Si, Li),[13,14] whereas MeO_2Me and especially MeO_2 molecules show a higher energy mode than MeO (Me=Li, Na).[14,15] The formation of LiO_2Li

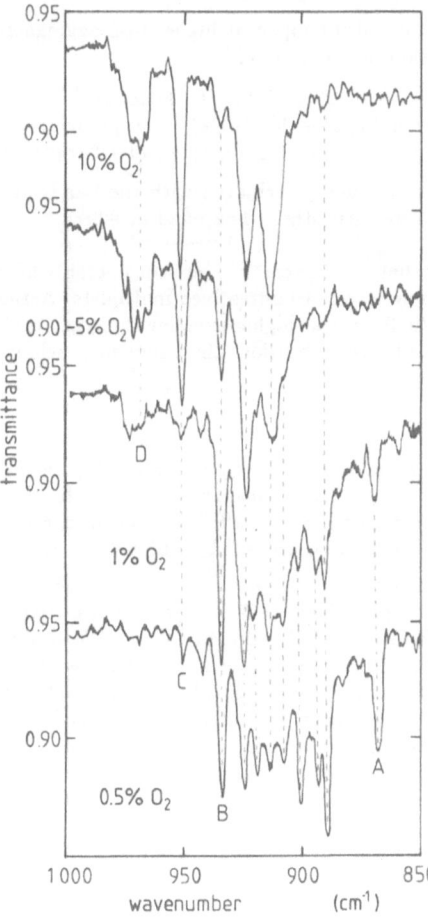

Fig. 2   Infrared spectra of a Fe/Ar matrix for different O_2 concentrations. The iron concentration is approximately 1%. At decreasing oxygen concentration the band at 955 cm^{-1} (C) weakens in intensity in favour of a new band at 938 cm^{-1} (B). This band is attributed to a FeO_2Fe molecule. At low oxygen concentration the same bands below 910 cm^{-1} appear as in Fig. 1.

and NaO_2Na has been observed in oxygen-doped argon matrices, but not at all in pure oxygen matrices. In our case the $938\,cm^{-1}$ band (B) lies between the $\nu_{O-O}(Fe(O_2))$ (C) and the FeO modes (A). This and the development of the line during warming the matrix favours the interpretation as a $\nu_{O-O}(FeO_2Fe)$ mode.

The band at $873\,cm^{-1}$ (A) becomes strong at low oxygen/high iron concentration. Its intensity is nearly proportional to the second order of the relative iron concentration. The same Fe concentration dependence is found for the bands at $894/899/905\,cm^{-1}$. Therefore Fe_2 should be involved in the formation of these species.

The $873\,cm^{-1}$ band has been attributed to FeO from both matrix and gas phase investigations[1-3,8]. We have no reason to disagree with this assignment, so this band is assigned to the diatomic iron oxide FeO on the basis of agreement, although we do not expect to have atomic oxygen in our matrices.

Looking at the bond energies of Fe_2 ($96\,kJ/mol$)[16], O_2 ($494\,kJ/mol$), and FeO ($403\,kJ/mol$)[17] we get

$$E_b(O_2) > E_b(FeO) \quad \text{but}$$
$$E_b(Fe_2) + E_b(O_2) < E_b(2\,FeO)\ ,$$

so that

$$Fe + O_2 \ \not\rightarrow\ FeO + O$$

shall not react exothermically, but

$$Fe_2 + O_2\ \rightarrow\ 2\,FeO$$

may. This O-O bond rupture may also happen at higher iron aggregates. This interpretation is in full agreement with the conclusion of Chang et al.[1]

At dioxygen concentrations of 10–20% a band at $1098\,cm^{-1}$ becomes strong. This is exactly the region expected for the iron hyperoxide $Fe(O_2)_2$. Hyperoxides have been reported for Cu ($1106/1090\,cm^{-1}$),[9] Pd ($1111\,cm^{-1}$), Ni ($1062\,cm^{-1}$), Pt ($1050\,cm^{-1}$), and Rh ($1045\,cm^{-1}$).[10]

No concentration dependence could be achieved with the bands at $909/916/927\,cm^{-1}$ because they are strongly superimposed by each other. The band at $909\,cm^{-1}$ appears at higher iron/lower oxygen concentration. The bands at $927\,cm^{-1}$ and $916\,cm^{-1}$ are strong at higher oxygen concentration. These bands — as well as that at $1098\,cm^{-1}$ — are considerably broader than those for FeO_2 or FeO_2Fe, suggesting that they may be due to unresolved multiplets. Actually we observe at least two lines at $924/929\,cm^{-1}$ and $914/919\,cm^{-1}$ at higher iron/lower oxygen concentration or after warming the matrix, but because of the different conditions the bands may belong to other species.

CONCLUSION

The observed spectra may be accounted for by assuming that the initial species arriving at the matrix surface are individual Fe atoms, Ar atoms, and O_2 molecules. Fe aggregates, most probably Fe_2, may be formed at the matrix surface at sufficient high iron concentration. In contrast Fe_2 formation at the matrix surface would be decreased by a high O_2 concentration reacting rapidly with Fe atoms.

The appearance of a band reported for FeO at *higher* iron concentration implies that an iron aggregate, most probably Fe_2, is capable of O-O bond cleavage. We observe several other bands at the same iron concentration so that several oxides are formed with at least two iron atoms involved in their forming. The formation of FeO is expected to take place at the matrix surface during evaporation, because warming the matrix never increases the intensity of the FeO band. That means neither Fe_2 nor O_2 can be mobilized by warming the matrix. During evaporation however, the mobility of Fe_2 and O_2 at the matrix surface is high enough to lead to a reaction. Once the Fe-O bonds are formed, the intermediate product may be split by electrostatic forces, i.e. two FeO molecules are formed and no $(FeO)_2$, which should show a different vibrational wavenumber. The O-O bond cleavage by Fe_2 may be analogous to the C-C cleavage in alkanes by small nickel clusters.[10]

For one species which appears at moderate Fe concentration, a FeO_2Fe structure is suggested on the basis of agreement with Li and Na.

The band at $1098\,cm^{-1}$ corresponds to the $1062\,cm^{-1}$ band of $Ni(O_2)_2$.

A complete analysis of our spectra including isotopic substitution for O_2 and Mössbauer studies will be published later.

ACKNOWLEDGMENT

This work was supported by a grant from the U.S. Department of Energy.
(Contract DE-AS03-76SF00034)

REFERENCES

1. Chang, S., Blyholder, G., Fernandez, J., *Inorg. Chem.* **20**, 2813 (1981)
2. Abramowitz, S., Acquista, N., Levin, I. W., *Chem. Phys. Lett.* **50**, 423 (1977)
3. Green, D. W., Reedy, G. T., *J. Mol. Spectrosc.* **78**, 257 (1979)
4. Valentine, J. S., *Chem. Rev.* **73**, 235 (1973)
5. Dunn, J. B. R., Shriver, D. F., Klotz, I. M., *Biochemistry* **14**, 2689 (1975)
6. Collman, J. P., Gange, R. R., Reed, C. A., Halbert, T. R., Lang, G., Robinson, W. T.,
   J. Am. Chem. Soc. **97**, 1427 (1975)
7. Kurtz, Jr., D. M., Shriver, D. F., Klotz, I. M., *J. Am. Chem. Soc.* **98**, 5033 (1976)
8. Barrow, R. F., Senior, M., *Nature* (London) **223**, 1359 (1969);
   West, J. B., Broida, H. P., *J. Chem. Phys.* **62**, 2566 (1975)
9. Ozin, G. A., Mitchell, S. A., García-Prieto, J., *J. Am. Chem. Soc.* **105**, 6399 (1983)
10. Klabunde, K. J., *Chemistry of Free Atoms and Particles*,
    Academic Press, New York (1980) (and references therein)
11. Peden, C. H. F., Parker, S. F., Barrett, P. H., Pearson, R. G., *J. Phys. Chem.* **87**, 2329 (1983)
12. Andrews, L., Spiker, Jr., R. C., *J. Phys. Chem.* **76**, 3208 (1972)
13. White, D., Seshadri, K. S., Dever, D. F., Mann, D. E., Linevsky, M. J.,
    J. Chem. Phys. **39**, 2463 (1963);
    Seshadri, K. S., White, D., Mann, D. E., *J. Chem. Phys.* **45**, 4697 (1966);
    Andrews, L., *J. Chem. Phys.* **50**, 4288 (1969)
14. Anderson, J. S., Ogden, J. S., Ricks, M. J., *Chem. Commun.*, 1585 (1968);
    Anderson, J. S., Ogden, J. S., *J. Chem. Phys.* **51**, 4189 (1969)
15. Andrews, L., *J. Phys. Chem.* **73**, 3922 (1969)
16. Gingerich, K. A., *Symp. Faraday Soc.* **14**, 109 (1980)
17. Smoes, S., Drowart, J., *High Temp. Sci.* **17**, 31 (1984)

ON THE DISPERSION AND CHARACTERISTICS OF PLATINUM-TIN BIMETALLIC CLUSTERS

Jitendra Kumar[*] and M. Ghosh[+]

Indian Institute of Technology, Kanpur-208016 (India)
*at the Materials Science Programme
+at the Department of Chemical Engineering

ABSTRACT

Metallic clusters of platinum in combination with tin dispersed over alumina exhibit interesting catalytic properties for reforming reactions by providing selectivity advantages and lower rates of deactivation. To understand these aspects, model samples of platinum-tin supported on alumina have been studied in detail using transmission electron microscopy and diffraction techniques. Clusters are dispersed on alumina by vacuum evaporation method while maintaining the substrate at 25° and $350^{O}C$ during deposition of tin and platinum respectively. Pt-Sn system is found to be quite complex in as deposited state but after reduction in hydrogen shows the formation of bimetallic clusters of PtSn intermetallic alongwith a small quantity of SnO. These phases remain intact during sulfurization (i.e. heating in thiophene-hydrogen mixture). However, reduction in hydrogen for extended period leads to disappearance of SnO without affecting PtSn. The accelerated deactivation intentionally caused by passing methylcyclohexane (MCH) decomposes PtSn into platinum clustres with simultaneous formation of SnO_2. Microstructural observations reveal that sulfur addition helps in maintaining high dispersion of bimetallic clusters for prolonged period by suppressing the coarsening process. The electronic interaction, dilution effects and modifications of support, resulting due to addition of tin, taken together seem to be responsible for improved selectivity towards aromatic hydrocarbons and better stability of $PtSn/Al_2O_3$ system in reforming processes.

INTRODUCTION

Metallic clusters of platinum alone or in combination with a second element such as rhenium, iridium or tin dispersed over alumina support exhibit excellent catalytic properties for reforming reactions [1-3]. Addition of second component offers selectivity advantages and better stability over Pt/Al_2O_3 system [4-7]. Also, reforming catalysts display better selectivity for longer period when sulfurized [8,9]. However, the mechanism of such a behaviour is not well understood as yet. This is probably due to the complex nature of heterogeneous catalysis in general and reforming process in particular. In order to obtain information regarding origin and nature of curious and useful characteristics of bimetallic system, the present investigation was undertaken, in which, platinum in combination with tin supported on alumina films has been studied by transmission electron microscopy as such and after reduction, sulfurization and deactivation treatments. While an important industrial reaction viz. dehydrogenation of methylcyclohexane is chosen for deactivation, thiophene is used as sulfur bearing compound for sulfurization treatment.

EXPERIMENTAL DETAILS

Alumina support films (thickness~15nm) were prepared by anodization method [10] using 3% tartaric acid solution of pH 5.5 as electrolyte and transferred onto gold specimen grids. Suitable amounts of tin and platinum were deposited over alumina by vacuum evaporation technique at a pressure ~10^{-5} torr by maintaining the substrate at 25° and $350^{o}C$ respectively. Since platinum-tin system shows [11] best performance in respect of catalytic activity and selectivity when made by introduction of tin before that of platinum, evaporation was carried out mainly in this order. Specimens thus prepared were placed in a ceramic boat for inserting into the reactor tube of an electric furnace for various treatments. Hydrogen was passed over the specimen at the flow rate of 50 ml/min for 7 hours at $450^{o}C$ for reduction. During sulfurization, apart from hydrogen, calculated amount of thiophene was injected at interval of 5 minutes to maintain its concentration at 10 ppm. This operation was carried out for 7 hours at $450^{o}C$. To reduce the possibility of condensation of thiophene on quartz tube, hydrogen gas was passed through a preheater before allowing it to enter the reactor. Further, to determine the role of sulfur or thiophene, another batch of samples were reduced in hydrogen for 14 hours. For deactivation process, temperature was set at $450^{o}C$ and nitrogen to methylcyclohexane ratio was kept at 1:5. Nitrogen was introduced only to reduce the partial pressure of hydrogen produced during the reaction. This accelerated the reaction and caused an appreciable deactivation of the catalyst within reasonable time (i.e. 7 hours only). All model specimens were examined in a Philips EM301 transmission electron microscope (TEM) at 100 kV to obtain information regarding the morphology, size, dispersion and structural crystallography of dispersed metallic clusters on alumina support before and after subjecting to treatments as outlined above.

RESULTS AND DISCUSSION

Formation of bimetallic clusters

For preparing platinum-tin alumina specimens, tin is evaporated first while platinum afterwards by raising the substrate temperature and maintaining it at $350^{o}C$. Under these conditions, tin clusters ought to be in molten state and therefore could be quite susceptible to alloy formation with finely divided platinum. The question then arises as to whether the resulting samples display the formation of intermetallic compound(s) or still correspond to metallic clusters of individual metals namely, platinum and tin. Careful analysis of the complex electron diffraction pattern of these samples indeed reveal the emergence of PtSn intermetallic alongwith some unknown phase(s). On reduction, the diffraction patterns became less complex and showed the coexistence of PtSn intermetallic and SnO. Needless to say that when deposition is carried out in reverse order (i.e. first platinum and then tin), instead of intermetallic, clusters of individual components are formed. Fig. 1(a, b) shows the microstructures of Pt-Sn/Al_2O_3 samples as such and after reduction in hydrogen at $450^{o}C$ for 7 hours. The corresponding diffraction pattern is given in Fig.1c. Table 1 lists the interplanar spacings alongwith the miller indices of diffraction lines.

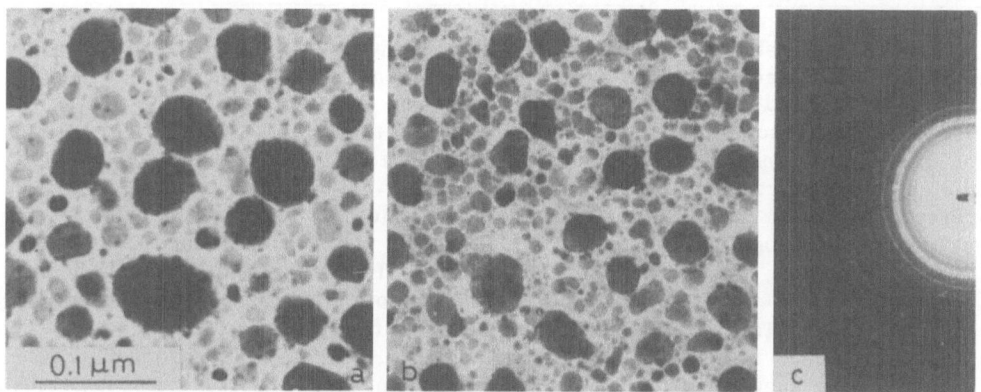

Fig. 1   Microstructures and electron diffraction pattern of platinum-tin/ alumina catalyst; a) as prepared, b) and c) after reduction in hydrogen for 7 hours at 450°C.

Table 1.   Interplanar spacings and miller indices of diffraction rings recorded in platinum-tin system after reduction

Ring no.	at 450°C for 7 hours			at 450°C for 14 hours		
	d(nm)	PtSn hk.l	SnO hk.l	d(nm) observed	d(nm) calculated	PtSn hk.l
1.	0.353	10.0	–	0.356	0.356	10.0
2.	0.298	10.1	101	0.295	0.298	10.1
3.	0.269	–	110	0.216	0.216	10.2
4.	0.245	–	002	0.205	0.206	11.0
5.	0.217	10.2	–	0.178	0.178	20.0
6.	0.204	11.0	102	0.169	1.169	20.1
7.	0.181	–	112	0.161	0.162	10.3
8.	0.169	20.1	–	0.149	0.149	20.2
9.	0.161	10.3	211			
10.	0.149	20.2	202,103			
11.	0.135	00.4	–			
12.	0.130	21.1	–			

PtSn: hexagonal (NiAs type)
a=b~0.411 nm, c~0.544 nm
SnO : tetragonal structure
a=b~0.380 nm, c~0.484 nm with
Z=2 and space group P4/mmm

These results indicate that reduction treatment of the original Pt-Sn/Al$_2$O$_3$ samples leads to stabilization of two phases namely, PtSn and SnO. Also, they correspond to hexagonal structure (NiAs type) with lattice parameters a = b~ 0.411 nm, c ~0.544 nm, and tetragonal structure having a = b ~ 0.380 nm, c ~0.484 nm, Z = 2 respectively. At this stage, it will be worthwhile to compare the impregnation method for preparing PtSn/Al$_2$O$_3$ with the vacuum evaporation technqiue employed here.   In the impregnation method[5], stanic chloride is added to alumina and calcined before introduction of chloropla- tinic acid.   After addition of platinum containing acid, the product is calcined again and then reduced at 450°C for 6-9 hours.   In contrast, in the present work, both tin and platinum have been deposited in succession at substrate temperature 25° and 350°C respectively and reduced at 450°C for 7 hours.   The method is unique in the sense that it leads to the formation of bimetallic clusters in PtSn/Al$_2$O$_3$ system.   In fact, the phases present include PtSn and SnO.   The emergence of SnO reflects that tin is in bivalent state, in confirmity with Mossbauer study of Pt-Sn/Al$_2$O$_3$ catalyst prepared by impregnation method[5].

Effects of sulfurization, prolonged reduction and deactivation

Fig. 2 (a,b) shows the microstructures resulting on sulfurization with thiophene and deactivation by methylcyclohexane respectively. For second batch, the initial samples were first reduced in hydrogen for 14 hours at 450°C and then allowed to undergo deactivation treatment, mainly, to determine the sulfurization effect. Fig.2 (c,d) depicts electron micrographs of reduced and deactivated samples respectively. Two diffraction patterns of interest are shown in Fig.2 (e,f). Curiously enough, sulfurization treatment does not change the nature of diffraction pattern and hence the phases as well. However, the reduction at 450°C for longer period (i.e. 14 hours) at a stretch leads to disappearance of SnO without affecting PtSn intermetallic (Fig. 2e and Table 1). The intertness of PtSn towards thiophene during sulfurization and its ability to remain intact even after prolonged reduction treatment suggest that addition of tin imparts stability to Pt/Al_2O_3 reforming catalyst through formation of some intermetallic (as e.g. PtSn, in this case). The deactivation process causes precipitation of platinum with simultaneous conversion of SnO to SnO_2 (Fig. 2f and Table 2). The rings are invariably broad indicating the presence of small crystallites and/or strained particles. The particle size distributions obtained from Figs. 2b and d show a substantial increase in the number density of clusters in size range 1.5-3.0 nm as a result of emergence of platinum clusters during deactivation.

Table 2. Interplanar spacings and miller indices of diffraction rings observed in reduced, sulfurized/without sulfurized $Pt-Sn/Al_2O_3$ system after deactivation

Line no.	Interplanar spacings d(nm)	hkl	
		Pt	SnO_2
1.	0.332	–	110
2.	0.263	–	101
3.	0.227	111	–
4.	0.196	200	–
5.	0.176	–	211
6.	0.139	220	–
7.	0.118	311	–

Pt : face-centred cubic strucutre a = 0.392 nm, Z = 4 and space group Fm3m

SnO_2 : tetragonal strucutre a = b ~0.474 nm, c~0.319 nm with Z= 2 and space group P4/mmm

Implication to catalysis

The reforming reactions require catalysts (e.g. Pt/Al_2O_3) that are bifunctional; while the hydrogenation and dehydrogenation activities are performed by metal component, acidic sites present on the carrier surface are responsible for catalyzing hydrocarbon rearrangements of a type commonly observed in acid catalysis[3]. Moreover, the support is generally covered by the metal to an extent of about 1% or less only. Under prevailing conditions, therefore, carbonaceous residues get invariably deposited on the catalyst surface due to dehydrogenation of chemisorbed hyrocarbons to highly saturated species and subsequent condensation or polymerization[3]. The deactivation of catalyst is usually caused by three processes viz. coke deposition, poisoning and coarsening of clusters. The addition of tin decreases coke formation and enhances the selectivity for aromatization of the

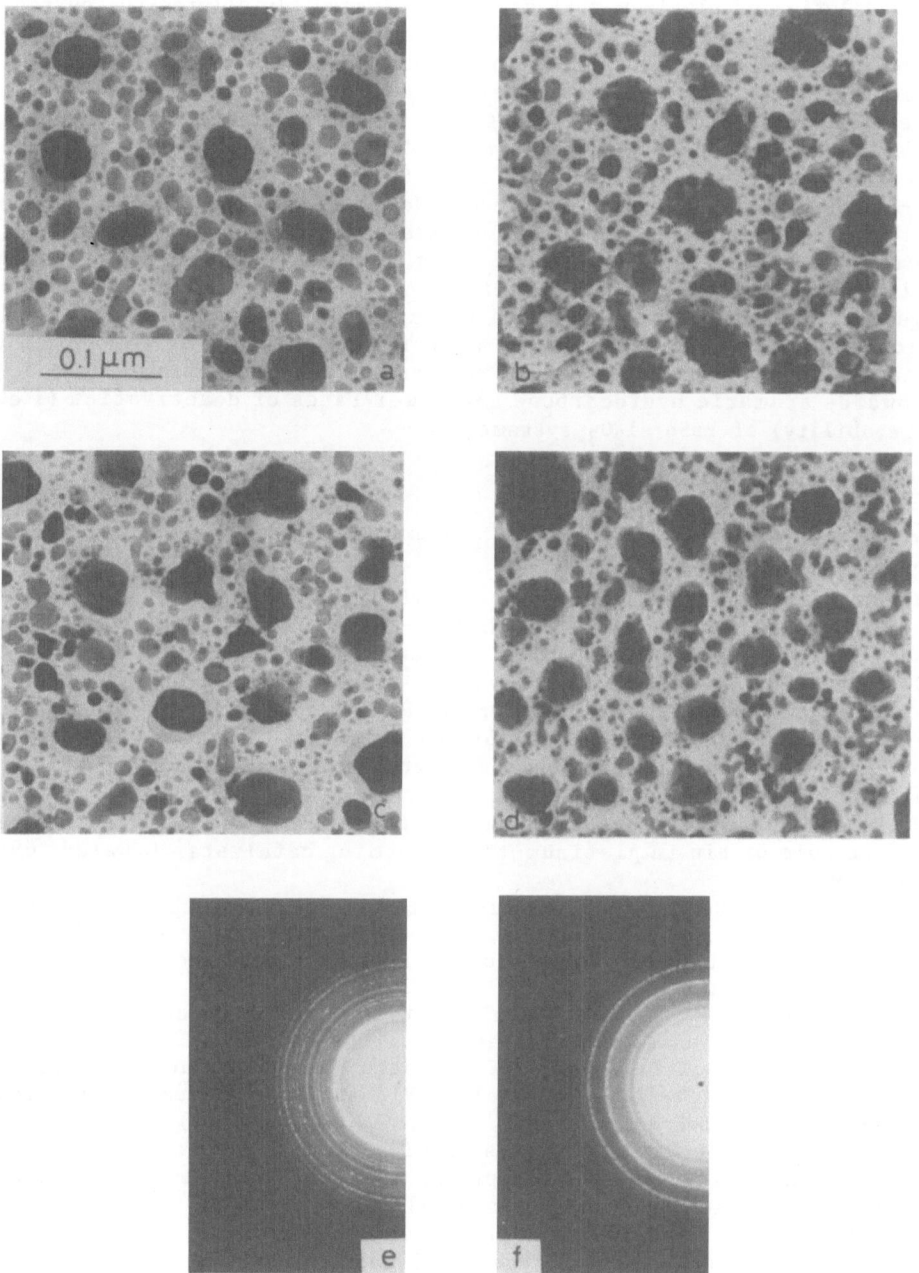

Fig. 2. Electron micrographs and diffraction patterns of platinum-tin/alumina system after treatment(s) at 450°C;
   (a) reduction and sulfurization, each for 7 hours,
   (b) and (f) reduction, sulfurization and deactivation, each for 7 hours,
   (c) and (e) reduction for 14 hours, and
   (d) reduction for 14 hours and deactivation for 7 hours in succession.

Pt/Al_2O_3 system. The origin of these beneficial effects is attributed to electron transfer from tin to platinum[12], weakening of Pt-C bond due to electron enrichment of platinum[6], poisoning of acidic sites of alumina by tin[5] and dilution of platinum surface[4]. Accordingly, tin dilutes the surface in such a way that number of sites containing three or more adjacent platinum atoms required[13] for the formation of carbonaceous residues decreases. The dilution effect is quite obvious from the present investigation as PtSn intermetallic is formed which has altogether different distribution of metal species in comparison to pure platinum. The emergence of SnO supports Burch's observation[14] that part of tin in Pt-Sn catalyst exists in oxidation state +2 after normal reduction treatment. Moreover, tin may act as a promotor rather than poison because of being present in a non-metallic state[14]. It, possibly, also modifies the acidic sites of the support somehow. Therefore, it may be concluded that electronic interaction, dilution effects (i.e. geometrical consideration) due to formation of PtSn intermetallic, and modifications of acidic sites of the support and presence of some tin in oxidation state +2 taken together are responsible for the improved selectivity towards aromatic hydrocarbons and lower rates of deactivation (i.e. better stability) of $PtSn/Al_2O_3$ system.

REFERENCES

1.   V. Haensel, G.R. Donaldson, Platforming of pure hydrocarbons, Ind. Eng. Chem. 43: 2102 (1951)
2.   J.K.A. Clarke, I. Manninger and T. Baird, Preparation of supported platinum-gold, platinum-tin and rhodium-copper catalysts and some tests with n-hexane/hydrogen reactants, J. Catal. 54: 230 (1978)
3.   J.H. Sinfelt, "Bimetallic catalysts : Discoveries, Concepts, and Applications," John Wiley, New York (1983)
4.   F.M. Dautzenberg, J.N. Hella, P. Biloen and W.M.H. Sachtler, Conversion of n-hexane over monofunctional supported and unsupported Pt-Sn catalysts, J. Catal. 63 : 119 (1980)
5.   R. Bacaud, P.Bussiere and F. Figueras, Mossbauer spectra investigation of the role of tin in platinum-tin reforming catalysts, J. Catal. 69: 399 (1981)
6.   R. Burch and L.C. Garla, Platinum-tin reforming catalysts II. activity and selectivity in hydrocarbon reactions, J. Catal. 71 : 360 (1981)
7.   B. Coq and F. Figueras, Conversion of methylcyclopentane on platinum-tin reforming catalysts, J.Catal. 85 : 197 (1984)
8.   G. Leclercq and M. Boudart, Catalytic hydrogenation of cyclohexene IV. effect of sulfur on supported platinum, J. Catal. 71 : 127 (1981).
9.   C.R. Apesteguia, C.E. Brema, T.F. Garetto, A. Brogna and M. Parera, Sulfurization of Pt/Al_2O_3-Cl catalyst VI. sulfur platinum interaction studied by infrared spectroscopy, J. Catal. 89: 52 (1984)
10. L. Young, " Anodic Oxide Films," Academic Press, New York (1961)
11. B.A. Sexton, A.E. Hughes and K. Forger, An x-ray photoelectron spectroscopy and reaction study of Pt-Sn catalyst, J. Catal. 88 : 466 (1984)
12. B.H. Davis, G.A. Westfall, J. Watkins and J. Pezzanite, Jr., Paraffin dehydrocyclization VI. the influence of metal and geseous promoters on aromatic selectivity, J. Catal. 42 : 247 (1976)
13. J.K.A. Clarke, Selectivity in catalysis by alloys, Chemical Reviews 75 : 291 (1975)
14. R. Burch, Platinum-tin reforming catalysts I. the oxidation state of tin and the interaction between platinum and tin, J. Catal. 71 : 348 (1981)

BEHAVIOUR OF PLATINUM AND PALLADIUM MICROCLUSTERS ON ALUMINA SUPPORT FILMS

Jitendra Kumar and Rakesh Saxena

Indian Institute of Technology Kanpur,
Materials Science Programme, Kanpur-208016, India

ABSTRACT

Microclusters (typical diameter~4-12 nm) of platinum and palladium have been dispersed over alumina support films by vacuum deposition of metal at substrate temperature $350^{\circ}C$ under vacuum (~10^{-5} Torr) and characterized in detail by transmission electron microscopy and diffraction techniques. An interesting observation encountered is that clusters of platinum and palladium are highly reactive and lead to hydride formation during heat treatment in hydrogen. The phases formed are PdH_x (x<1) possessing f.c.c. structure having a~0.403 nm, γ-Pd_3H_4 simple cubic with a ~0.299 nm and a new hydride of platinum based on a body centred cubic unit cell having a~0.310 nm. In view of the catalytic applications of platinum and palladium-alumina systems, coarsening and redispersion of microclusters have been examined and discussed with specific reference to deactivation and rejuvenation. It is illustrated that deactivation is caused not just by coarsening of clusters but by interaction between the catalyst and gaseous components as well (i.e. through hydride or oxide formation in reducing or oxidizing conditions, described as reactive coarsening or reactive redispersion).

INTRODUCTION

Metallic clusters are becoming increasingly important due to their unusual physical and chemical properties and use as supported metal catalysts in a number of chemical reactions [1-4]. The origin of their remarkable behaviour lies in the increase in surface-to-volume ratio with decrease in cluster size. In fact, a surface-to-volume ratio has the 1/r dependence for spherical clusters of diameter 2r. Consequently, surface effects become important and play a dominant role in determining the characteristics of clusters below a certain size (say, 10nm). Studies have, therefore, been undertaken in the recent past to gather information on various systems with an objective of providing a better understanding of cluster properties. The development of modern sophosticated equipments for both preparation of samples under well specified conditions and characterization down to atomic dimensions have contributed a great deal in handling such systems [5]. Nevertheless, any particular technique chosen for the study invariably puts limitation on the nature of the samples. Hence, investigations have been largely pursued on model specimens [5]. In the present work also, an attempt has been made in this direction, in which, model specimens of two catalytic materials namely, platinum and palladium, supported on alumina have been prepared by an alternative method and studied as such and after heat treatment in hydrogen by transmission electron microscopy and diffraction techniques. The preparation method undertaken essentially consists of vacuum deposition of ultrathin films (thickness determined by loading requirement) of metal on substrates held at elevated temperatures under vacuum ($\sim 10^{-5}$ Torr or better). The technique is unique[6] in the sense that it directly yields a fine dispersion of clusters and eliminates altogether the need for the existing practice of heating the deposited film in hydrogen, and thus avoids possible contamination resulting due to gaseous uptake during the heating step. Also, the problem related to catalytic application, i.e. coarsening and redispersion of microclusters in platinum-and palladium-alumina systems has been examined and discussed with specific reference to deactivation and rejuvenation.

EXPERIMENTAL DETAILS

Alumina support films of thickness ~ 15 nm were prepared by an anodization method using 3% tartaric acid solution of pH 5.5 as electrolyte, and transferred onto gold specimen grids [6-8]. Suitable amounts of platinum and palladium (mean thickness 1.5 and 0.7 nm respectively) were then deposited on alumina under vacuum ($\sim 10^{-5}$ Torr) by raising the substrate temperature to 350°C to obtain reasonable dispersion of clusters in terms of both the size and the number density [6,9]. A Philips EM 301 transmission electron microscope (TEM) was used at 100 kV to examine the morphology, dispersion and structural crystallography of the resulting clusters on alumina support films. Samples were subsequently heat treated in hydrogen at 600°C for 8 hours. For this, specimens were placed in a ceramic boat and introduced into the reactor tube of an electric furnace. The hydrogen gas flow was maintained at 45 ml/ minute. The samples were examined again in the TEM for changes and possible reactions that may have taken place in the heat treatment step.

RESULTS AND DISCUSSION

Formation of clusters

Platinum and palladium samples prepared by vacuum deposition at a pressure of $\sim 10^{-5}$ Torr while maintaining the alumina substrate at 350°C show emergence of uniformly distributed clusters possessing island type microstructures (Fig. 1). The corresponding electron diffraction patterns reveal that clusters belong to face-centred cubic phases with lattice parameters a=0.392 nm for platinum and 0.389 nm for palladium in respective system. These results clearly demonstrate that dispersion of platinum and palladium clusters can indeed be realized by an alternative method in which the substrate is held at elevated temperatures during vacuum deposition of the respective metal. Hence heat treatment in hydrogen usually carried out to produce clusters after deposition of metals can be eliminated altogether. It should be noted that the nature of dispersion, in fact, depends on several parameters viz. the metal itself, the amount of material deposited (or loading requirement), evaporation rate, substrate details and temperature, metal-substrate interactions, operating pressure, the residual gases present etc. As reported elsewhere [10], vacuum deposition at a substrate temperature of 350°C gives rise to not only metallic clusters but also to the appeearence of two oxide phases (Cu_2O and NaCl types) in case of palladium. The microstructure (Fig.1b) contains some dark regions corresponding to thin, well spread, single crystals of PdO (NaCl-type). Obviously, palladium metal is quite susceptible to oxidation at $\sim 10^{-5}$ Torr. This behaviour indicates that palladium is more active in this finely dispersed state. It is believed that emergence of oxide phases can be avoided by carrying out evaporation under ultra high vacuum conditions.

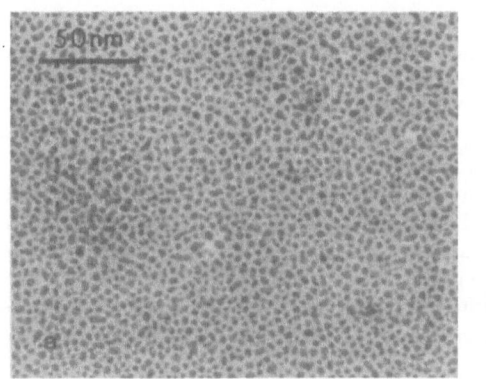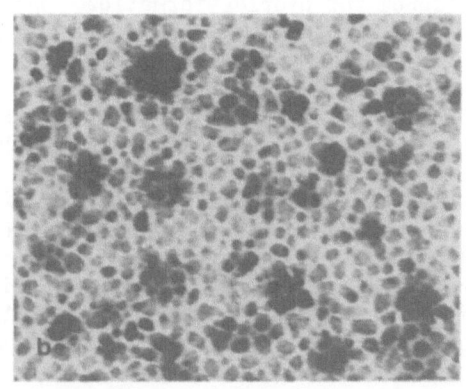

Fig. 1. Transmission electron micrographs showing clusters as dispersed over alumina at substrate temperature 350°C in platinum (a) and palladium (b) systems.

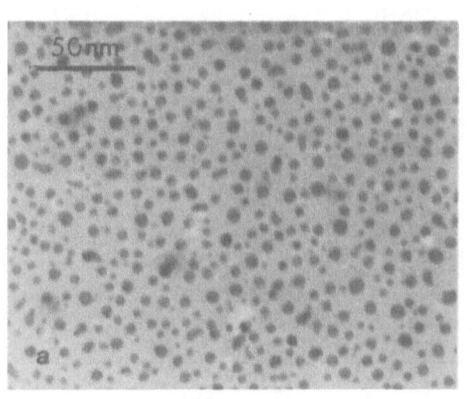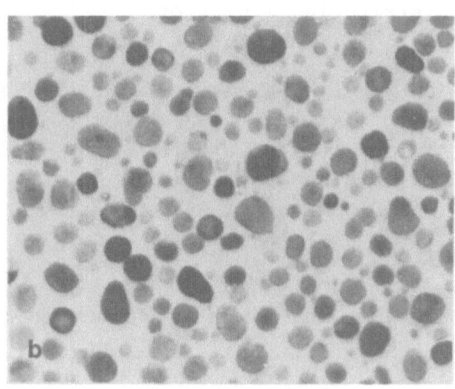

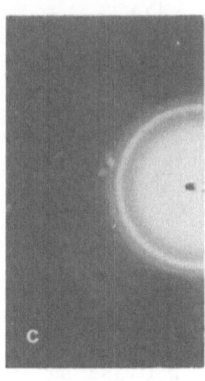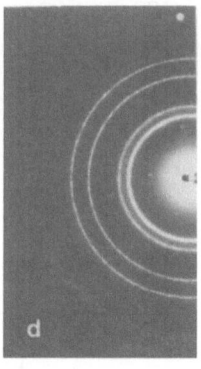

Fig. 2. Transmission electron micrographs and diffraction patterns of clusters after heat treatment in hydrogen at 600°C for 8 hours in platinum (a,c) and palladium (b,d,e) systems

Evidence for hydride formation

Fig.2 (a,b) shows representative electron micrographs of platinum and palladium systems observed after heat treatment in hydrogen (flow rate ~45 ml/minute) at 600°C for 8 hours. Table 1 provides information as deduced from micrographs and corresponding histograms in respect of size, distribution and surface area of clusters for virgin as well as treated samples. Notice that heat treatment in hydrogen causes coarsening of clusters in both systems; while the average particle diameter increases to 1.4 times in platinum, it is almost doubled in palladium. Also, it is found that the treatment reduces the oxides found in virgin samples and leads to the formation of respective hydrides. It is, therefore, appropriate to name the phenomenon of cluster enlargement as "reactive coarsening" since it occurs through hydride formation.

Table 1. Dispersion parameters of clusters in platinum- and palladium-alumina systems before (a) and after (b) heat treatment in hydrogen at 600°C for 8 hours

		Pt/Al_2O_3		Pd/Al_2O_3	
		(a)	(b)	(a)	(b)
Number of clusters counted		1190	625	995	635
Distribution peak	clusters %	46.2	36.0	22.1	20.7
	size range (nm)	4–5	6–7	10–12	15–17
Average cluster size (nm)		5.2	7.2	10.7	20.4
Average surface area per unit mass (m^2/g)		53.8	38.8	46.6	24.5

Table 2. Electron diffraction data for palladium and platinum systems after heat treatment in hydrogen at 600°C for 8 hours

Ring no.	d(nm)	Palladium		d(nm)	Platinum	
		hkl			hkl	
		PdH_x	$\gamma-Pd_3H_4$		Pt	Pt-hydride
1.	0.300	–	100	0.226	111	–
2.	0.286	110	–	0.219	–	110
3.	0.233	111	–	0.196	200	–
4.	0.214	–	110	0.155	–	200
5.	0.201	200	–	0.139	220	–
6.	0.150	–	200	0.125	–	211
7.	0.143	220	–	0.118	311	–
8.	0.122	311	211	0.109	–	220
9.	0.116	222	–	0.098	400	310
10	0.107	–	220	0.090	–	222
11.	0.100	400	–	0.083	–	321
12.	0.095	–	310	0.078	–	400
13.	0.092	331	–			
14.	0.090	420	–	Pt:f.c.c., a=0.392 nm		
15.	0.086	–	222	Pt-hydride:cubic, a~0.310nm		
16.	0.082	422	–	PdH_x:cubic, defect NaCl-type		
				a= 0.403 nm		
				$\gamma-Pd_3H_4$: cubic, a = 0.299 nm		

In the case of palladium, analysis of diffraction patterns revealed that the resulting clusters correspond to PdH$_x$ (x<1), possessing a defect NaCl (Fig. 2d) type structure [11,12] having a lattice parameter a = 0.403± 0.005 nm, alone, or to PdH$_x$ (x<1) and a high temperature γ-Pd$_3$H$_4$ simple cubic phase [13] having a = 0.299±0.005 nm with no trace of pure metal (Fig. 2e, table 2). In PdH$_x$ (x<1), hydrogen occupies octahedral voids present in face-centred cubic configuration of palladium atoms. Since the H/Pd ratio is less than unity, not all octahedral sites are filled. This allows hydrogen to get distributed amongst various available positions in a random manner. As a consequence, weak lines appear in the diffraction pattern which correspond to mixed miller indices and originate from hydrogen atoms alone. The diffraction rings of γ-Pd$_3$H$_4$ are strong for h+k+l = even, suggesting that the phase is based on a body-centred cubic unit cell. In the platinum system (Fig. 2c and table 2), a survey of the interplanar spacings reveals the presence of the metal and another cubic phase with lattice parameter a= 0.310±0.005 nm. The reflections follow the characteristic h+k+l = even. Since palladium exhibits the formation of hydrides under similar conditions, it is believed that the new phase observed in the case of platinum also corresponds to platinum hydride. Further, the composition Pt$_3$H$_4$ may be assigned to this new phase in view of the results discussed above and information available about palladium[13]. Needless to say that this behaviour is rather unusual as hydrogen is known[14] to have very little solubility in bulk platinum. Also, there is no information available so far concerning the crystal data of any hydride of platinum. The present observation, therefore, provides first evidence for the formation of platinum hydride and a new datum on its crystal structure. The hydride formation seems to be supported by the finely dispersed state (i.e. microclusters) of platinum. Whether such a phase exists in bulk platinum is yet to be established.

Coarsening and redispersion

The effects of heat treatment in air at 400°C have been dealt with in detail elsewhere[6,9] and can be summarized for the present discussion as follows: There occurs pronounced coarsening of clusters in the platinum/ alumina system in air. Also, under identical treatment conditions, coarsening is more effective in air than in hydrogen. The resulting product clusters correspond to a number of oxide phases which include PtO (NaCl-type), PtO$_2$ and Pt$_3$O$_4$. On the other hand, clusters get redispersed (i.e.become smaller in size with simultaneous increase in number density) in case of palladium. The phases present are PdO (tetragonal with a= b ~ 0.303 nm and c ~0.532nm) and metallic palladium. Also, redispersed clusters group into a narrow size range around 3 nm in contrast to the initial gaussian distribution with the peak at 12 nm.

It is now obvious that for platinum, coarsening of clusters occurs in both air and hydrogen. On the other hand, in palladium, while coarsening takes place in hydrogen, redispersion is observed in air. This unusual behaviour of palladium i.e. redispersion in an oxidizing and coarsening in a reducing atmosphere is believed to be due to the very nature and properties of its oxides and hydrides. Our observations clearly demonstrate that platinum and palladium clusters no longer remain metallic when heat treated in air or hydrogen. This supports our objection to the usual method (involving vacuum deposition of metal on a substrate at room temperature and subsequent heat treatment in hydrogen at elevated temperatures) of obtaining metallic clusters for model studies. Obviously, the resulting product will be contaminated and may even correspond to altogether new compound(s)/ phase(s), as revealed in the present case on heat treatment in hydrogen. These results are of significance in catalysis where metallic clusters of platinum and palladium find applications in reactions involving oxidizing and reducing atmospheres. In such situations, the deterioration of catalytic properties on prolonged usage will be determined not just by coarsening of clusters (so as to reduce the total surface area and hence the active sites) but by possible reactions that occur between the catalyst and gaseous components present. For example, deactivation may take place through reactive coarsening as found in the case of platinum in both air and hydrogen and palladium in reducing atmosphere. It may also occur on redispersion through oxide formation as observed in case of palladium. If the heat treatment in hydrogen is a necessary step in the preparation of industrial catalysts, our observations suggest that metal and support in conjunction with hydrogen are responsible for the catalytic action.

The deactivation is not a desirable process as it influences the catalyst life and, therefore, be kept at a bare minimum somehow. This can be achieved by suitable additives or by using more stable systems as catalysts. Alternatively, one may think to develop rejuvenation processes for the aged (i.e. deactivated) catalysts. An ideal process will be one which ensures recovery of all critical characteristics; that means restoration of not only the metallic clusters but also their size and number density. However, causes of deactivation need to be ascertained before proceeding in the development of a rejuvenation process. The results presented here have relevance in the sense that redispersion is found to occur in oxidizing atmospheres in palladium. Therefore, if controlled reduction of redispersed oxide clusters can be done without causing coarsening, the palladium system may be rejuvenated.

REFERENCES

1. J.R. Anderson, "Structure of Metallic Catalysts," Academic Press, London (1975)
2. J.F. Hamilton and R.C. Baetzold, Catalysis by small metal clusters, Science 205: 1213 (1979)
3. J.T.Waber,"Characterization and Behaviour of Materials with Submicron Dimensions," World Scietific, Singapore (1985)
4. "Metal Clusters," M. Moskovits, Ed., John Wiley, New York (1986)
5. H. Poppa, Model studies in catalysis with UHV-deposited metal particles and clusters, Vacuum 34: 1081 (1984)
6. R. Saxena, Electron microscope studies of platinum and palladium catalysts on carbon and alumina support films, M.Tech. Thesis, Indian Institute of Technology Kanpur (1983)
7. G.Hass, On the Preparation of hard oxide films with precisely controlled thickness on evaporated aluminium mirrors, J. Opt. Soc. America 30: 532 (1949)
8. L. Young, "Anodic Oxide Films," Academic Press, New York (1961)
9. Jitendra Kumar and Rakesh Saxena, Studies of platinum and palladium model catalysts on carbon and alumina support films I. Dispersion of fine particles, II. reactive coarsening and redispersion of particles, Applied Catalysis, to be published
10. Jitendra Kumar and Rakesh Saxena, Formation of NaCl and Cu_2O type oxides of platinum and palladium on carbon and alumina support films, Less-Common Metals, to be published
11. W.B. Pearson," A Handbook of Lattice Spacings and Structure of Metals and Alloys," Vol.2, Pergamon Press, Oxford (1967)
12. Y.P. Khodyrev, R.V. Baranova, R.M. Imamov and S.A. Semiletov, Electron diffraction study of β-fcc palladium hydride, Kristallografiya, 23:1046 (1978)
13. R.V. Baranova, Y.P. Khodyrev, R.M. Imamov and S.A. Semiletov, Crystal structure of palladium hydride with a primitive cubic lattice (a=0.2995 nm), Kristallografiya 25:1290 (1980)
14. W.M. Mueller, J.P. Blackedge and G.G. Libowitz, " Metal Hydrides," Academic Press, New York (1968)

STRUCTURE OF METAL CLUSTERS - A HIGH RESOLUTION ELECTRON MICROSCOPY STUDY OF OSMIUM CLUSTERS ON MAGNESIUM OXIDE

N.J. Long*, B.C. Gates+, M.J. Kelley+# and H.H. Lamb+

*Center for Solid State Science, Arizona State University, Tempe
AZ 85287
+Center for Catalytic Science & Technology, Department of
Chemical Engineering, University of Delaware, Newark, DE 19716
#Engineering Technology Laboratory, Experimental Station, E.I.
du Pont de Nemours & Co., Inc., Wilmington, DE 19898

ABSTRACT

High resolution electron microscpy has been used to obtain direct structure images of small aggregates of osmium dispersed on MgO. Some molecular clusters of $Os_{10}C(CO)_{24}^{2-}$ were deposited directly onto the MgO support and others were synthesised on the support from chloroosmic acid. The electron micrographs show a distribution of aggregate sizes in the latter sample, the larger aggregates having the osmium hexagonal-close-packed structure. Also evident are very small particles ranging in size from one or two atoms up to several atoms, with structures which cannot readily be interpreted as being hcp. The Os_{10} clusters deposited on the support do not have the normal osmium hexagonal-close-packed structure, nor can the images be interpreted as being face-centred-cubic. The images can be interpreted on the basis of the assumption that the clusters have an hexagonal-close-packed structure with a larger c/a ratio. There is a strong interaction between the MgO and the Os aggregates, which are aligned parallel to the MgO {002} planes. The strong interaction could be responsible for the distortion of the cluster away from the normal bulk osmium hexagonal-close-packed structure.

INTRODUCTION

Ultra-high resolution electron microscopy (HREM) is an essential technique for characterising ultradispersed metals on supports because it can give direct structure images, in projection, of the heavy metal atoms within individual groupings of atoms. From electron micrographs, estimates can be made of the size distribution, the structure and the particle-support interaction.

We have used HREM to examine the structures obtained when the osmium carbonyl cluster $Os_{10}C(CO)_{24}^{2-}$ is supported on MgO. The HREM results, when interpreted in the context of EXAFS and infrared data to be reported elsewhere,[1] show that the decaosmium clusters can be deposited intact on the oxide surface and, further, that they can be synthesised directly on the surface from a chloroosmic acid precursor, although with this method a variety of related structures are ultimately formed on the surface under the analysis conditions. The osmium entities all show a preferential alignment along a well-defined crystallographic direction of MgO.

EXPERIMENTAL

Sample preparation and handling

Osmium clusters on MgO, which were the precursors of the supported particles observed by HREM, were prepared by two methods. Sample A was synthesised by the reductive carbonylation of Os(IV) on MgO [derived from aqueous H_2OsCl_6 and MgO (MCB chemicals)] with an equimolar CO + H_2 mixture at 275°C and atmospheric pressure. This treatment was inferred to produce predominantly $[Os_{10}C(CO)_{24}]^{2-}$ on MgO; the decaosmium cluster could be isolated from the solid by extraction with an aqueous acetone solution of $[(PPh_3)_2N\ Cl]$ (PPN^+Cl^-).[1,2] Sample B was prepared by the adsorption of $[Et_4N]_2[Os_{10}C(CO)_{24}]$ from dry tetrahydrofuran solution onto MgO (derived from $Mg(OH)_2$) which had been pretreated at 800°C in vacuo. Both materials were found to decompose over a period of weeks when stored in contact with air. Thus some degradation may have occurred during the time which elapsed before HREM could be performed, despite shipping and storage under dry nitrogen. Therefore the structures observed on the MgO surfaces may not be those that were present immediately after sample preparation; thus it is not advisable to compare the sample preparation methods on the basis of the images reported here.

Electron microscopy

The powders were ground in an agate pestle and mortar and dispersed on holey carbon films supported on copper grids using a dry preparation technique, i.e. without a solvent dispersion medium. The characterisation was done with a 400-kV JEM 4000EX ultra-high resolution electron microscope (point-to-point resolution better than .17nm) fitted with a fibre-optically coupled image intensifier, a TV camera, and a YAG single crystal scintillator as the transmission screen. This combination permitted the easier use of low beam currents during the location of areas of interest, thereby minimising the total electron dose. Lattice fringe images from the MgO ({200} = 0.2105nm) were used, whenever possible, as a calibration for magnification and for optical diffractograms obtained from the larger metal aggregates. Comparisons were made against the normal bulk osmium structure, which is hexagonal-close-packed (hcp) (a = 0.2733nm, c = 0.43191nm, c/a = 1.58).

The microscope was also operated at voltages lower than the maximum of 400kV to determine the effect of electron energy on the stability of the clusters. The micrograph shown in fig. 1, for example, was taken at 300kV. There was little, if, any difference between any of the accelerating voltages in terms of sample stability, although there was an expected decrease in the resolution of the microscope at lower voltages; for example, at 200kV the small clusters could be seen, but it was difficult to discern any internal structure within them. Sample charging under the electron beam was a serious problem for all the preparations examined and was the principle limitation in obtaining high-resolution images.

RESULTS

Sample A

High resolution electron micrographs of sample A revealed that there were aggregates with a range of sizes present, all < 3nm. The features observed for this sample preparation are the following:
1. very small (one or two atom) osmium entities with very low contrast, which are still visible because of the low contrast from the MgO support (fig. 1);
2.a) small aggregates, less than 1nm in diameter, some of which appear to be located preferentially along surface steps and ledges;
  b) some of these smaller aggregates exhibit unusually strong image

contrast relative to that arising from the larger ones, or from the mono- or diatomic clusters, as shown in fig. 2;

3.a) relatively large osmium aggregates with a hexagonal-close-packed structure viewed along the [$1\bar{2}13$] zone axis, giving rise to atom rows comprising the {$0\bar{1}1\bar{1}$}, {$1\bar{1}0\bar{1}$} and {$10\bar{1}0$} type planes (illustrated schematically in fig. 3).

b) alignment of these aggregates with the MgO <200> directions (as seen with reference to the {200} fringes and surface relief along the cube faces).

Other micrographs showed these same features to a greater or lesser extent; evidently there are at least three different types of osmium entity: (1) mono- or diatomic residual osmium, (2) small clusters, presumably the Os_{10} metal core of $[Os_{10}C(CO)_{24}]^{2-}$, and (3) larger particles of hcp osmium metal which have grown in a pronounced surface epitaxial relationship with the MgO support.

Figure 4 represents a different area of the specimen, at first glance including only fairly large structures. Closer examination shows that some of these aggregates appear to consist of a core and an outer region. The origin of these outer layers is not easily determined, since they could be an oxide layer or be related to the original ligand groups (the ligands are usually believed to disappear very rapidly on exposure to the electron beam and vacuum system). Also seen in fig. 4 is a larger particle (1.6 x 2.9 nm) (marked A) with a strong contrast giving a lattice image that is inconsistent with hcp osmium. The image at first looks like an [$0\bar{1}11$] projection, but the predominant atom rows make angles of 75° instead of the expected 81° for {$1\bar{1}0\bar{1}$} and {$10\bar{1}1$}.

Sample B

An example of the images obtained for this sample is shown in fig. 5. Again there are a number of pertinent features:

1. there is only a small variation in aggregate size;
2. there are no large aggregates of hcp osmium;
3.a) the angles between the rows of atoms are incompatible with normal hcp osmium projections (or any cubic projection).

b) there is a strong preferential alignment of the particles along the MgO <200> planes.

DISCUSSION

The results show clearly that extremely small osmium aggregates can be dispersed with a high degree of uniformity onto MgO, and structure images of them can be obtained by high resolution electron microscopy. The structure of the smaller entities is not immediately apparent, but there is evidence suggesting the presence of hexagonally-close-packed osmium with the normal osmium lattice parameters together with a second structure for the larger aggregates. A simple calculation shows that the abnormal angles observed for this second structure could be explained by an 18% increase in the c/a ratio of hexagonal-close-packed Os. This would give the observed values for the angles between {$10\bar{1}1$}-type planes in an <$0\bar{1}11$> projection and only a small change in the angles for the <$1\bar{2}13$> projection, viz:

Zone axis	hcp osmium c/a=1.58	c/a=1.86	observed
[$0\bar{1}11$]	[$1\bar{1}0\bar{1}$]^[$10\bar{1}\bar{1}$]=81°	76°	~75°
[$1\bar{2}13$]	[$0\bar{1}1\bar{1}$]^[$1\bar{1}0\bar{1}$]=52°	54°	~54°

It is easier to measure the angles between rows of atoms for such small particles than to measure the atomic spacing, therefore it is not possible to measure the lattice spacings accurately enough to detect the 3% change in the {$10\bar{1}1$} spacing that would occur if there was an 18% increase in the c/a ratio with no change in the a parameter. We stress that this interpretation

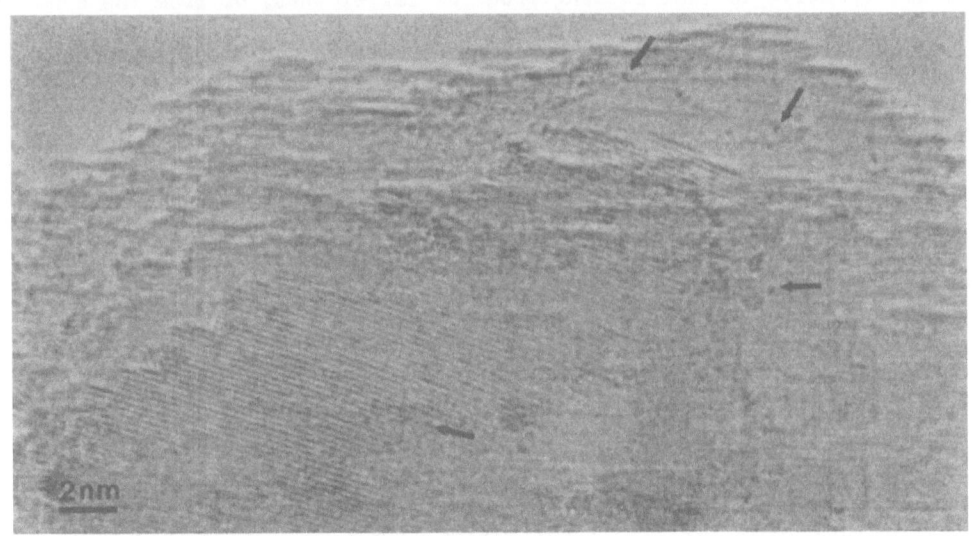

Fig. 1 HREM image of Sample A, showing mono- and diatomic Os clusters on MgO.

Fig. 2 Os aggregates in sample A less than 1nm in diameter. Note preferential alignment of aggregates on steps and ledges in the MgO, and the strong contrast exhibited by some of these aggregates (see text).

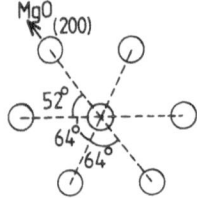

Fig. 3 Schematic representation of hcp Os viewed along $\langle 1\bar{2}13 \rangle$

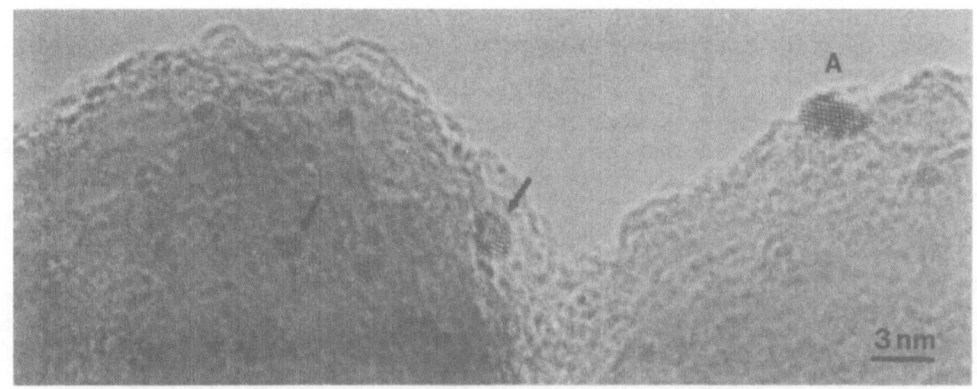

Fig. 4   HREM image of area in sample A showing large aggregates with
pronounced inner core and outer layers (arrowed).   The large particle
marked A shows a lattice image inconsistent with hcp or fcc Os.

Fig. 5   HREM image of Sample B showing narrow size distribution of aggregate
sizes and aggregate alignment parallel to the MgO (200) type planes.

should be treated with caution since there are other possibilities not considered here in detail. For example, the observed strong interaction between osmium and MgO may induce distortion in the osmium aggregates sufficient to change their structures; the aggregates may have a completely different structure to either hexagonal-close-packed or face-centred-cubic which cannot be readily identified from these results.

Petford-Long, Long, Smith, Wallenberg and Bovin[3] observed that small particles of ruthenium can exist with both hexagonal-close-packed and face-centred-cubic structures, and that the two structures can be seen in the same particle. Thus it is possible for small particles to exist in highly non-equilibrium structures when they are both small and subject to strong external influences (in the case of ruthenium quoted above the samples were subjected to very high beam doses whilst supported on amorphous carbon films).

The smaller aggregates pose a difficult problem because there are so few atoms present that the images are not readily interpretable by direct measurement. Suitable image processing may improve this situation. However, in view of the errors in measurement, it would appear that the small clusters observed in Sample A may also be described as an hcp structure with an expanded c/a ratio, although the hypothesis that they represent genuine "$Os_{10}C$" metal cores cannot be ruled out. The enhanced image contrast observed from the smaller aggregates relative to the larger ones suggests that these are three-dimensional, in contrast to the larger structures, which are more raft-like.

If one assumes that all the aggregates have a hexagonal-close-packed structure, then the preferred alignment that is observed can be described as $\langle 10\bar{1}1 \rangle Os // \langle 200 \rangle MgO$, which is also indicated by the similarity in the lattice spacings, since $\{10\bar{1}1\}Os = 0.215$nm and $\{200\}MgO = 0.2105$nm. The strong interaction between osmium and MgO support could easily distort the osmium atoms into a better fit when the clusters are small.

CONCLUSIONS

HREM has been used to image very small osmium aggregates produced by two different routes. The sizes range from one or two atoms up to 3nm, but there is a similarity between them all in that they have a strong tendency to align themselves along the MgO $\langle 200 \rangle$ directions. The structure of the aggregates appears to be a form of hexagonal-close-packed osmium with an increased c/a ratio relative to that found in bulk osmium.

ACKNOWLEDGEMENTS

The electron microscopy was performed at the Arizona State University National Facility for High Resolution Electron Microscopy, supported by grant DMR-8306501 from the National Science Foundation.

REFERENCES

1. H.H. Lamb et al., to be published.
2. H.H. Lamb, T.R. Krause and B.C. Gates, J. Chem. Soc. Chem. Commun.: 821 (1986).
3. A.K. Petford-Long, N.J. Long, D.J. Smith, L.R. Wallenberg and J.-O. Bovin, Proc. of this conference: (1986).

REACTIVITY OF SIZE-SELECTED SILICON CLUSTERS AS STUDIED BY
FOURIER TRANSFORM MASS SPECTROMETRY

M. L. Mandich, W. D. Reents, Jr., and V. E. Bondybey

AT&T Bell Laboratories
Murray Hill, NJ 07974

ABSTRACT

The chemistry of positively and negatively charged bare silicon cluster ions is investigated using laser evaporation coupled with Fourier transform ion cyclotron resonance mass spectrometry. Systematic trends in the rates and products with cluster size indicate that small ionic silicon clusters have two distinct types of reactivity. These two types are related to different dangling bonds of the clusters involving either a divalent or a trivalent (or charged) silicon center. This interpretation agrees well with ab initio electronic structure calculations as well as known reactivity of divalent and radical organosilicon centers. "Magic numbers" and reaction thermodynamics are of little heuristic value in explaining these trends in the reaction rates and product distributions.

INTRODUCTION

Small gas phase clusters comprise a unique class of highly reactive compounds. Often composed entirely of surface atoms, clusters have numerous dangling bonds resulting from coordinative unsaturation. Thus, clusters are likely to have a rich chemistry which changes with cluster size and composition. One major challenge in studying cluster reactivity is to be able to unambiguously monitor thermal bimolecular reaction rates and products for each cluster size. We have recently shown that this can be accomplished by using laser evaporation to form a variety of ionic clusters in-situ at the trapping cell of a Fourier transform ion cyclotron resonance mass spectrometer (FTMS).[1-2] This technique is ideal for measuring cluster reactivity since the clusters can be easily mass selected and then stored for seconds under single collision, essentially thermal conditions. This experimental approach has been used to study the exothermic reactions of positively and negatively charged silicon cluster ions with a variety of neutral reagents.[3] These include organic molecules such as CH_3SiH_3,[1] $CH_3C{\equiv}CH$, $CH_2{=}C{=}CH_2$, and $CH_3C(O)CH_3$, as well as inorganic species: NO_2,[4] WF_6,[5] XeF_2[6] and $C{\ell}F_3$. The kinetics and product distributions of these reactions provide an arsenal of information for comprehending the fundamental nature of silicon cluster chemistry. We find that that there are at least two different classes of silicon cluster reactivity. These distinct types are related to reaction at either divalent or radical silicon centers on the clusters.

EXPERIMENTAL METHOD

Silicon cluster ions are created by laser evaporation of a stationary bulk silicon target located just outside the ion cell of a modified Nicolet FT/MS-1000 Fourier transform mass spectrometer. The resulting trapped negative or positive ion populations consist of clusters containing up to seven or nine atoms, respectively.[1] The ion cell and method of laser vaporization are depicted in Figure 1. In this configuration, the cell consists of two sides which are separately pumped. The enormous advantage of the dual cell is that it prevents contamination of the target surface by the reagent gas. When the target and reactant gas are left in contact, the initial cluster population is altered: the larger clusters decrease in intensity and product-like functionalized clusters appear directly from laser vaporization.

The exothermic ion-molecule reactions of the clusters are studied by first isolating a single cluster size from the total ion population using standard double resonance techniques. The bimolecular reaction rates and products of these selected clusters are then monitored by recording the mass spectra of the ions in the cell at a sequence of reaction times. The time dependences of the normalized reactant and product ion intensities afford a means for determining reaction rates and for differentiating between primary versus secondary and higher order products (Figure 2). All reactant ion intensities are single exponential and scale linearly with the pressures ($\sim 10^{-6} - 10^{-7}$ Torr) in these experiments. The reaction times are sufficiently long to allow for collisional and radiative relaxation of any excess translational and internal energy that may be present initially in the cluster ions.

RESULTS AND DISCUSSION

1.) Trends in the Reaction Probabilities As a Function of Cluster Size

The reaction rates of the ionic silicon clusters vary greatly with cluster size and charge as well as with the neutral reagent. In order to understand the magnitudes of the trends, it is more useful to compare reaction probabilities per collision instead of the reaction rates. These probabilities are calculated from the measured reaction rates divided by the ion-molecule collision rates[7] and are plotted for several silicon

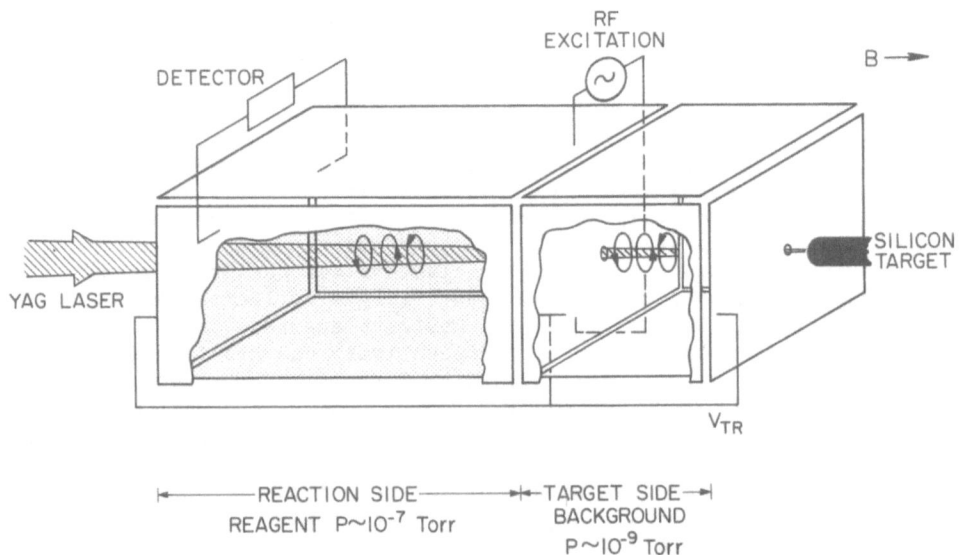

Figure 1. FTMS dual cell depicting the stationary sample target for Nd:YAG laser evaporation (532 nm, $\sim 3 \times 10^8$ W/cm^2). The FTMS is typically operated with a solenoid magnetic field strength of 2.96T, trapping potentials of +/- 0.6 V, and with a M/ΔM >10,000.

cation reactions in Figure 3. The size dependences of the reaction probabilities show several pronounced trends. First, the probability of reaction for positive silicon cluster ions decreases with increasing cluster
size. Second, amongst the reaction probabilities for the various reagents
in Figure 3, CH_3SiH_3 is unique. It only reacts with silicon cluster
cations containing five or less atoms. Third, a similar comparison for
the anions shows a much less pronounced decrease in rate with increasing
cluster size.

2.) Trends in the Reaction Products As a Function of Cluster Size

The product distributions show conclusively that ionic silicon clusters
do not indiscriminately bind neutral reagents, shred them and then extrude
random products. Typically, one or two products dominate the distribution
for a given reagent, particularly for clusters containing four or more
atoms. For example, in the reactions of positively and negatively charged
clusters with NO_2, the overwhelming fate of the silicon clusters is to be
etched, atom by atom, down to the atomic ion:

$$Si_n^{\pm} + NO_2 \rightarrow Si_{n-1}^{\pm} + SiO + NO \qquad (1)$$

Excluding electron transfer, this single reaction channel is responsible
for more than 85% of all of the reaction products for most of the clusters. Similarly, fluorine abstraction dominates the silicon cluster
cation reactions with WF_6 and

$$Si_n^{+} + WF_6 \rightarrow Si_nF^{+} + WF_5 \qquad (2)$$

is the only reaction for silicon clusters containing four or more atoms.

CH_3SiH_3 shows a different behavior from the reagents where reactivity is
seen for all cluster sizes.[1] Si_{1-4}^{+} react to form H_2;

$$Si_n^{+} + CH_3SiH_3 \rightarrow Si_{n+1}CH_4^{+} + H_2 \qquad (3)$$
$$\rightarrow Si_{n+1}CH_2^{+} + 2H_2 .$$

At Si_5^{+}, however, an abrupt switch occurs to loss of radical neutral products:

$$Si_5^{+} + CH_3SiH_3 \rightarrow Si_6CH_5^{+} + H \qquad (4)$$
$$\rightarrow Si_6H_3^{+} + CH_3 .$$

Another abrupt change happens after Si_5^{+} and Si_{6-7}^{+} are not observed to
react at all with CH_3SiH_3.

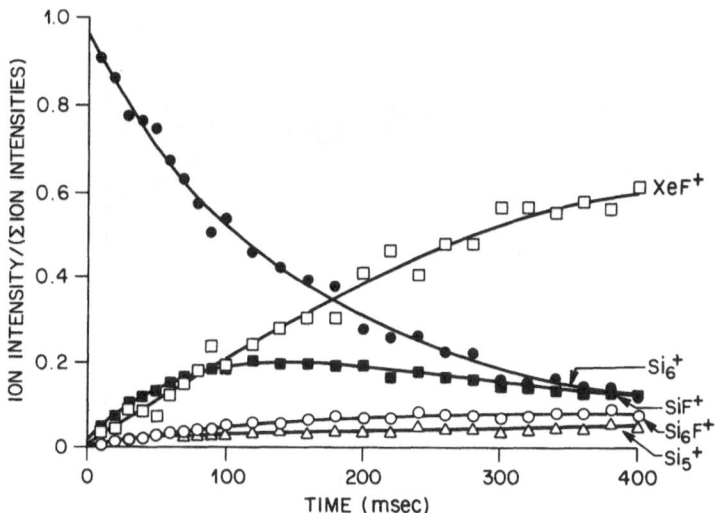

Figure 2. Time evolution of product formation in the reaction of Si_6^{+} with
XeF_2. Note that the curvatures indicate that SiF^{+}, Si_6F^{+}, and Si_5^{+} are
primary products whereas XeF^{+} is a secondary product.

3.) Correlation of Reactivity with Various Thermodynamic Quantities

Energetic differences between possible exothermic products for various cluster sizes are poor indicators of either reaction rates or product distributions. Although only exothermic reactions are observed, the major product is often less favored thermodynamically. This is illustrated in the product branching ratio for the reaction of Si_2^+ with XeF_2:[6]

$$Si_2^+ + XeF_2 \rightarrow Si^+ + SiF_2 + Xe \qquad 10\% \qquad \Delta H = -134 \text{ kcal/mole} \qquad (5)$$
$$\rightarrow SiF^+ + SiF + Xe \qquad 73\% \qquad \Delta H = -127 \text{ kcal/mole}$$
$$\rightarrow Si_2F^+ + XeF \qquad 17\% \qquad \Delta H = \ ?$$

The SiF^+/Si^+ product ratio of seven is counter to a Boltzmann prediction which would favor formation of Si^+ by a factor of 10^5.

Si_4^+ and Si_6^+ have been identified as "magic numbers" on the basis of their prominence in silicon cluster distributions[8] and in fragmentation patterns following photoinduced[9] or metastable unimolecular decomposition.[10] The reasons for the appearance of these magic numbers is currently a matter of controversy, however, recent theory shows that Si_4^+ and Si_6^+ represent local thermodynamic maxima with respect to the total binding energy per atom in the cluster.[11] The presumed stability of Si_4^+ and Si_6^+ is not reflected in either their relative reaction rates or product distributions. Table 1 shows the calculated exothermicities, rates and product fractions for the silicon cluster cation "etching" reaction in equation 1. No changes in products or minima in the rates are seen for these "magic clusters". In fact, the rate increases slightly for Si_6^+ and then decreases for Si_7^+ despite a large exothermicity change from -0.3 kcal/mole to -26 kcal/mole for the respective reactions.

4.) Proposed Mechanism Involving Divalent and Radical Silicon Centers

Trends in the reaction probabilities and product distributions indicate that silicon cluster ions display two distinct types of reactivity. One class is observed for the positively charged silicon clusters in reactions with organosilane and organic reagents, e.g., CH_3SiH_3 and $CH_3C \equiv CH$. These

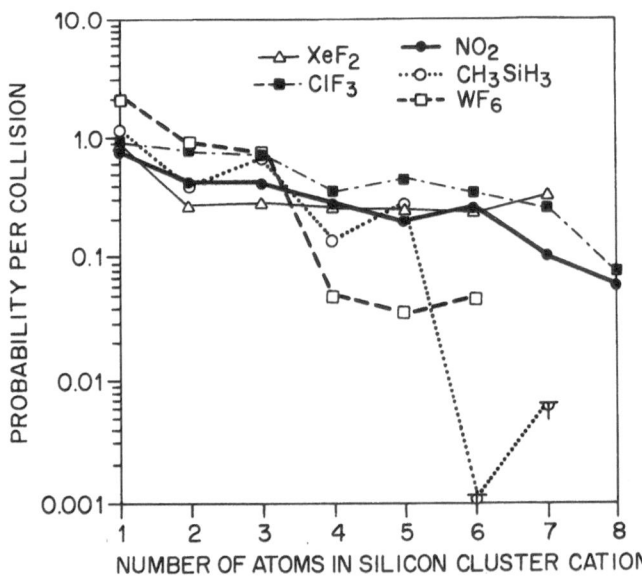

Figure 3. Probability of reaction per collision derived from the measured reaction rate divided by the ion-molecule collision rate. The errors are typically $\pm$ 10-20%. The rates shown for the reaction of Si_{5-7}^+ with CH_3SiH_3 are upper bounds; no reaction products were detected.

Table 1. Silicon cluster size dependence of product fraction, rate and energetics for the reaction:[4]

$$Si_n^+ + NO_2 \rightarrow Si_{n-1}^+ + SiO + NO$$

Cluster Ion	Product Fraction (%)	Rate Constant (× 10¹⁰ cm³ molecule⁻¹ sec⁻¹)	Estimated ΔH_{rxn}^a (kcal/mole)	Ionization Potential Si_n (eV) [b]
Si_2^+	61	3.4 ± 0.9	-35	7 4 ± 0.3
Si_3^+	58	3.1 ± 0.3	-28	7.9
Si_4^+	97	2.0 ± 0.1	+6.6	7.6
Si_5^+	95	1.4 ± 0.2	-42	7.8
Si_6^+	100	1.8 ± 0.4	-0.3	7.5
Si_7^+	76	0.7 ± 0.3	-26	(c)
Si_8^+	100	0.4 ± 0.2	(c)	(c)

a. Reaction thermochemistry calculated using theoretically derived silicon cluster cation fragmentation energies taken from Reference 11.
b. Ionization potentials taken from Reference 11 except for Si_2 which references an experimentally derived number in Reference 12.
c. Theoretically derived numbers for this value not available.

reactions occur only with clusters containing five or fewer atoms and usually involve addition of a moiety such as $-SiCH_2$ or $-CH_2$ to the cluster accompanied by elimination of a stable neutral molecule such as H_2 or C_2H_2. The second class pertains to almost <u>all</u> positive and negative cluster ions in reactions with halogen or oxygen containing inorganic molecules, e.g. NO_2 and $C\ell F_3$. The reaction products typically result from atom transfer of $-F$, $-C\ell$, or $-O$ to the ionic cluster and production of neutral radicals such as WF_5 or NO. In this reactivity class, the rate decreases with increasing size but reactions occur for all cluster sizes.

An explanation for two types of reactivity is that the silicon clusters have two different silicon centers as possible reaction sites. The first is a divalent (silylene) silicon center which has a localized lone pair of non-bonding electrons. The second center contains a localized radical electron at either a charged or trivalent silicon atom in the cluster. The chemistry of divalent and radical silicon centers has been exhaustively studied both theoretically and experimentally. <u>Direct insertion</u> of divalent silicon centers into Si-H, C=C, C=O and C≡C bonds is well-established.[13] The reaction products of the Si_n^+ clusters with CH_3SiH_3 and the other organic reagents indicate that these reactions are similarly initiated by an insertion of a divalent silicon center on the cluster. For example, the first reaction step with CH_3SiH_3 is Si-H insertion to form a $[Si_{n-1}Si(-H)(-SiH_2CH_3)]^+$ intermediate; this is probably followed by a second Si-H or Si-CH₃ insertion via another lone pair of electrons on a neighboring silicon atom.[1] The typical chemistry of radical silicon centers involves <u>atom abstraction.</u> Oxygen and halogen abstractions by the $Si_n^{+/-}$ cluster from NO_2, WF_6, etc., are representative of a large class of silicon radical reactions which are well known in the chemistry of reactive silicon compounds.[14] For example, the first step in the $Si_n^{+/-}$ reactions with NO_2 is probably coupling of a radical electron on the cluster with an electron on one of the oxygen atoms of NO_2, forming a terminal silicon-oxygen covalent bond. NO is lost followed by extrusion of the highly labile SiO.[4]

The effects of cluster size on radical and silylene reactivity reflect the electronic structures of the silicon clusters as a function of size. The silicon centers in diatomic silicon are monovalent and are entirely divalent in three atom and four atom silicon clusters. Note that the divalency of the smallest clusters does not rule out radical reactivity for the ionic silicon clusters which inherently contain an unpaired electron. With five and more atoms, the silicon atoms in the clusters become increasingly more bonded to each other; divalent centers are replaced by trivalent and tetravalent centers.[11,15-16] Even very large clusters do not become totally tetravalent since, by analogy to silicon surfaces, they are expected to have trivalent silicon centers among the surface atoms. Thus, increasing the cluster size has a detrimental effect on silylene reactivity which requires a divalent silicon center, but not radical reactivity which is not as restrictive. This explains why Si_{6-7}^+ do not react with CH_3SiH_3, etc., and why radical reactivity with NO_2, WF_6, etc., persists for all cluster sizes.

CONCLUSIONS

1. Positively and negatively charged silicon cluster ions react exothermically with a variety of neutral reagents.
2. The reaction rates and product distributions show pronounced trends as a function of cluster size. One class of reactions only occurs for Si_{1-5}^+ and is described by an insertion process which usually leads to elimination of stable neutral products. All clusters participate in the other class of reactions which is typified by atom abstraction and loss of neutral radicals.
3. The two classes of reactivity correspond to two distinct reactive centers on the silicon clusters. One contains a localized pair of electrons, the other has a radical electron. The reactivity of these two distinct sites is consistent with well known silylene and radical reactions in organosilicon chemistry as well as theoretical descriptions of the cluster structures and bonding.
4. Relative stabilities of "magic number" clusters have only minor effects on reaction rates and product distributions.
5. FTMS is a powerful technique for studying the chemistry of naked silicon clusters. This approach is quite versatile and is easily extended to studies of other metal and nonmetal bare clusters.

REFERENCES

1. M. L. Mandich, W. D. Reents, Jr. and V. E. Bondybey, J. Phys. Chem., 90:2315 (1986).
2. V. E. Bondybey, W. D. Reents, Jr., and M. L. Mandich, submitted to Chem. Phys. Lett.
3. Another study of silicon cluster cation reactions has recently been reported by W. R. Creasy, S. W. McElvany, and A. O'Keefe, Proc. of the 34Th Ann. Conf. on Mass Spectrom. and Allied Topics, June 9-13, 1986.
4. M. L. Mandich, V. E. Bondybey, and W. D. Reents, Jr., submitted to J. Chem. Phys.
5. W. D. Reents, Jr., M. L. Mandich, and V. E. Bondybey, to be published in Chem. Phys. Lett.
6. W. D. Reents, Jr. A. M. Mujsce, V. E. Bondybey and M. L. Mandich, submitted to J. Chem. Phys.
7. T. Su and M. T. Bowers, in Gas Phase Ion Chemistry, Vol. 1, M. T. Bowers, ed., Academic Press, New York, 1979.
8. See, for example, L. A. Bloomfield, M. E. Geusic, R. R. Freeman, and W. L. Brown, Chem. Phys. Lett., 121:33 (1985).
9. L. A. Bloomfield, R. R. Freeman, and W. L. Brown, Phys. Rev. Lett., 54:2246 (1985).
10. W. Begemann, K. H. Meiwes-Broer, and H. O. Lutz, Phys. Rev. Lett., 56:2248 (1986).
11. K. Raghavachari and V. Logovinsky, Phys. Rev. Lett., 55:2853 (1985).
12. R. D. Levin and S. B. Lias, National Stand. Ref. Data Ser., Nat. Bur. Stand., Washington D. C., 1982.
13. Y. Tang, in Reactive Intermediates, Vol. 2, R. A. Abramovitch, ed., Plenum Press, New York,1982.
14. J. Wilt, in Reactive Intermediates, Vol. 3, R. A. Abramovitch, ed., Plenum Press, New York, 1983.
15. D. Tomanek and M. A. Schluter, Phys. Rev. Lett. 56:1055 (1985).
16. G. Pacchioni and J. Koutecky, J. Chem. Phys. 84:3301 (1986).

THEORETICAL INVESTIGATION OF SURFACE RELAXATION EFFECTS ON CHEMISORPTION OF ATOMIC OXYGEN ON ALUMINUM CLUSTERS

Gianfranco Pacchioni and Piercarlo Fantucci

Dipartimento di Chimica Inorganica e Metallorganica
Centro CNR, Università di Milano, I-20133 Milano, Italy

ABSTRACT

The electronic structures and properties of Al clusters containing up to 49 atoms were studied by means of a modified version of the MO-LCAO INDO method. The Al clusters have been employed to model the oxygen chemisorption on the tetrahedral cavities of the Al (111) surface. The effect of surface relaxation on the barrier to penetration of the oxygen atom below the surface has been investigated.

I. INTRODUCTION

The interaction of molecular oxygen with Al surfaces is a very intensively studied phenomenon.[1] Experimentally, there is general agreement about the fact that under oxygen exposure an ordered overlayer of O atoms is formed above the Al (111) surface.[1] Nevertheless, many aspects of the problem are still unsolved. For instance, it has been suggested that the overlayer of O atoms can coexist with an underlayer located in the tetrahedral or octahedral cavities of the Al (111) surface. However, there is no information on the height of the energy barrier to oxygen penetration as well as on the influence of the relaxation of the surface on this barrier.

In this paper we report preliminary results of semiempirical MO-LCAO calculations on the interaction of Al clusters sections of the Al (111) surface with atomic oxygen. In the first part of the paper we discuss the electronic structures and properties of bare Al_n clusters (n = 2-49) while in the second part are reported the results of the calculations on the interaction of atomic oxygen with Al clusters.

II. THEORETICAL METHOD

The calculations of Al_n and Al_n-O systems were performed according to a modified version[2] of the approximate INDO method.[3] According to the present version, the one-center integrals are obtained by non-empirical Effective Core Potential (ECP) calculations.[4] This assures that the ECP atomic energies are exactly reproduced in our INDO scheme.

Table 1 - Electronic properties of Al clusters

Cluster[a]	Ground state	E_T (au)	D_e/n (eV)	IP[b] (eV)	ANNN[c]
Al	^2P	-1.8822	...	5.68	...
Al$_2$	$^3\Sigma_g^-$ ($D_{\infty h}$)	-3.8136	0.67	6.37	...
Al$_3$ (3,0)	$^2A_1'$ (D_{3h})	-5.7718	1.14	6.13	2.0
Al$_4$ (3,1)	$^3A_2$ (C_{3v})	-7.7268	1.34	6.44	3.0
Al$_7$ (3,3,1)	$^2A_1$ (C_{3v})	-13.6606	1.89	6.33	4.3
Al$_{10}$ (7,3)	$^3A_1$ (C_{3v})	-19.5375	1.95	5.22	4.8
Al$_{13}$ (3,7,3)	^2E (D_{3h})	-25.5290	2.22	5.31	5.5
Al$_{19}$ (12,6,1)	$^2A_1$ (C_{3v})	-37.3615	2.29	5.34	5.5
Al$_{22}$ (12,7,3)	^3E (C_{3v})	-43.2963	2.33	5.30	6.1
Al$_{31}$ (12,7,12)	$^4A_2$ (D_{3d})	-61.0189	2.34	4.75	6.6
Al$_{37}$ (6,12,7,12)	$^4A_2$ (C_{3v})	-73.0590	2.51	5.06	7.0
Al$_{49}$ (27,19,3)	$^2A_2$ (C_{3v})	-96.6870	2.48	...	7.1
Al$_{49}$ (3,12,19,12,3)	$^2A_2$ (D_{3d})	-96.8081	2.54	4.40	7.6
Al$_{49}$ (6,12,19,12)	$^4A_2$ (C_{3v})	-96.8047	2.54	...	7.7
Al bulk		...	3.41	4.24	12.0

[a] The Al-Al distance was fixed at 5.4 au. [b] Computed as $E(Al_n^+) - E(Al_n)$
[c] ANNN = average number of nearest neighbors.

The two-center Coulomb integrals involving p and d orbitals are computed over s-type functions whose exponents and contraction coefficients are determined in such a way to reproduce the radial distribution of the corresponding atomic orbitals. The one-electron two-center terms are approximated as

$$(\mu_A | h(1) | \nu_B) = \beta_{AB} (\mu_A | \nu_B)$$

where the β quantity, specific to a given two-center interaction, is obtained by fitting theoretical "ab initio" or experimental data for diatomic molecules (e.g. Al_2, AlO).

The molecular wavefunction and energy are computed according to a symmetry restricted and spin unrestricted iterative procedure. This new parametrized INDO method has reproduced with high accuracy several properties (e.g. ground state, dissociation energy, equilibrium geometry, excitation energy) of Al_n and Al_n-O obtained with accurate "ab initio" calculations.[5-9] Further details concerning the theoretical method are reported elsewhere.[2]

III. ELECTRONIC STRUCTURE OF Al CLUSTERS

The transition from the Al_2 molecule to medium-sized Al clusters is accompanied by important changes in bonding nature as well as in electronic properties. The MO eigenspectrum tends to a continuum as the cluster size increases. This explains the existence of many close-lying electronic states and justifies the preference for high-spin configurations observed in some cases (Table 1).

The trend of the binding energy per atom (D_e/n) versus cluster size shows a rapid increase of the cluster stability for aggregates containing less than 10 atoms (Table 1). For larger systems the D_e/n grows in a rather monotonic way. The cluster stability is not only a function of the size but also of the compactness. In fact, among the three Al_{49} clusters considered, the less compact $Al_{49}(27,19,3)$ cluster (in parenthesis is the number of Al atoms in each layer) is 3.16 eV less stable than the other two structures considered (Table 1).

The cluster stability converges to the cohesive energy of the bulk metal in a regular way. Nevertheless, the binding energy per atom of the largest cluster considered is still 1 eV smaller than the experimental value for the bulk.

The plot of D_e/n versus the parameter $n^{-1/3}$ allows to report in one picture all values from the isolated atom to the bulk (Fig. 1). The $n^{-1/3}$ parameter is proportional to the surface-to-volume ratio. The theoretical curve is therefore the straight line joining the reference atomic to the bulk cohesive energy. Quite surprisingly the dependence of the cluster binding energy per atom on $n^{-1/3}$ is linear and the calculated points lie very close to the theoretical curve (Fig. 1). The fitting curve extrapolated to the bulk situation gives a cohesive energy, 3.59 eV, which is only 5% larger than the experimental data (3.41 eV). It is therefore possible to estimate the cluster size required to reproduce a given portion of the metal cohesive energy. In order to obtain the 90% of the bulk stability a cluster of 976 atoms is required. Assuming for this cluster a spherical shape and an fcc packing of atoms, this corresponds approximately to a small particle of 30 Å of diameter. The

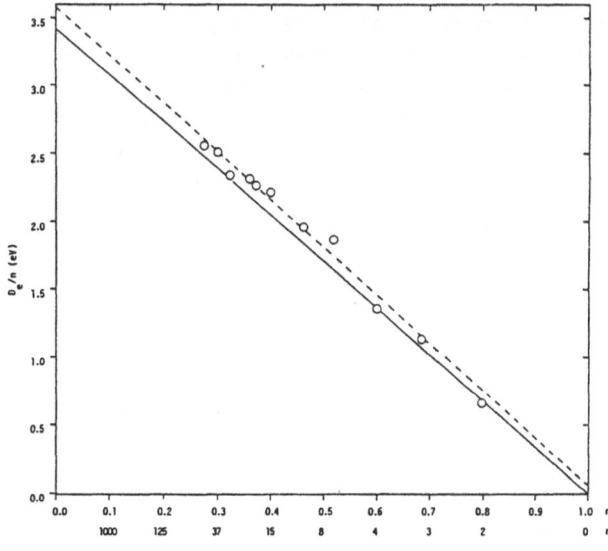

Fig.1 - Dissociation energy per atom (D_e/n) of Al clusters versus cluster size. The solid line represents the theoretical curve joining the limit values of the single atom (0 eV) and bulk metal (3.41 eV). The dotted-dashed line represents the least square fitting curve of the computed D_e/n values (o).

number of atoms necessary to further approach the bulk value increases very rapidly: with about 8000 atoms one has the 95% of the cohesive energy, while only a cluster of 100.000 atoms will possess the 99% of the metal stability.

The ionization potential (IP) of metal clusters is often taken as a measure of the convergence of cluster to bulk properties. In fact, an important criticism to the cluster-surface analogy is related to the position of the Fermi level (the HOMO of the cluster) which, in clusters, can be different than in surfaces. As predicted by chemical intuition, there is a decrease of the IP with increasing cluster size. The IP of Al_2 (6.37 eV) is 2 eV higher than in Al_{49} (4.40 eV). This latter value is only 0.16 eV larger than the experimental work function of the Al (111) surface (Table 1). It follows that the convergence of the cluster IP to the metal work function is even more rapid than the convergence of the cluster stability to the bulk cohesive energy.

A better understanding of the charge distribution on the cluster surface is of fundamental importance for the employ of clusters to study chemisorption phenomena. In order to rationalize the charge fluctuations occuring in Al clusters we have identified two groups of "surface" and "bulk" atoms according to the coordination, with the result that a flow of charge occurs from the surface to the bulk of the cluster. The charges on the surface are in general small (0.03-0.11) while an average charge -0.15 is found on the bulk atoms.

According to these and to other results not reported here for brevity the use of clusters to model chemisorption processes on Al surfaces seems practicable provided that some caution is paid for in the interpretation of the results. Any conclusion concerning the cluster-surface analogy must be drawn with some care since even the properties of clusters containing 50 atoms are far to be converged and still depend on the cluster shape.

IV. OXYGEN CHEMISORPTION ON Al CLUSTERS

We have studied the chemisorption of atomic oxygen on Al_4(3,1), Al_{10}(3,7), Al_{22}(12,7,3), Al_{31}(12,7,12), and Al_{49}(27,19,3) clusters modeling the tetrahedral cavities of the Al (111) surface. The potential energy curve for the interaction of oxygen with the Al_{49} cluster is shown in Fig. 2. It is possible to see that the aluminum-oxygen interaction is very strong, being the corresponding binding energy of the order of 10 eV (Fig. 2). This result is not surprising considering that the dissociation of the diatomic Al-O molecule requires 5.27 eV and that when oxygen occupies the open positions on the Al (111) surface three Al-O bonds are formed.

The maximum bond strength is found when oxygen is 1.6-1.8 au above the surface (Fig. 2), in reasonable agreement with the commonly accepted height of oxygen from Al (111) surface.[1] Also the theoretical vibrational frequencies, about 500 cm^{-1}, are close to the experimental value (588 cm^{-1}).[1]

By changing the dimensions of the bare metal cluster the chemisorption properties do not change too dramatically; for instance, on going from Al_{22}-O to Al_{49}-O only minor changes in the cluster-adsorbate properties are found. In other words, the chemisorption properties seem to be strongly dependent on the structure and geometry of the few Al atoms

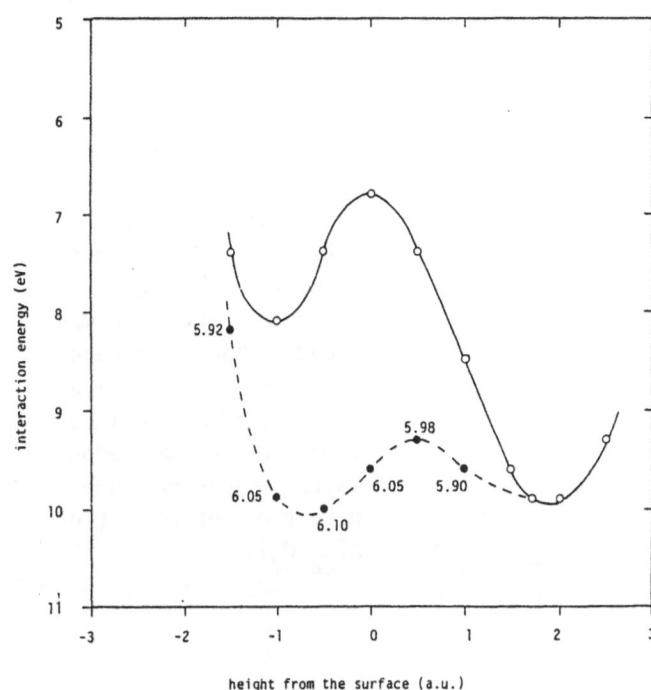

Fig. 2 - Potential energy curves for oxygen interaction with Al_{49} (27,19,3) cluster. o——o rigid lattice (R(Al-Al) = 5.4 au); ●- - -● surface relaxation (the optimized Al-Al distances of the three-hollow site are reported).

defining the chemisorption site, whereas the effect of the first and second shell of neighboring atoms, albeit quantitatively not negligible, does not change dramatically the shape of the interaction energy curve.

When the oxygen atom moves towards the cluster an energy barrier is encountered (Fig. 2). This barrier, of about 3 eV in all clusters considered, prevents an easy penetration of oxygen into the bulk metal. Once the barrier is passed over, the oxygen is attracted into the tetrahedral cavity below the surface where is strongly bound to the Al atoms. Therefore, an ordered underlayer of oxygen atoms can exist at about 1 au below the top layer.

Of course there are several different factors influencing the height of the energy barrier, but a direct estimate of each contribution represents a formidable task for the experimentalist. Here, we have investigated three phenomena which can in principle determine the amount of energy required to penetrate into the bulk. The effects considered are the relaxation of the surface, the first-second layer expansion, and the cooperative effect generated by several O atoms adsorbed on the surface (this is the so called "coverage effect" and was studied by means of the $Al_{22}-O_7$ cluster where six fixed O atoms represent the ordered overlayer of chemisorbed oxygen above the surface and the seventh O atom is free to penetrate below the Al top layer).

Among these three effects, the surface relaxation is clearly the most relevant. When the Al atoms on the surface are free to move from their rigid lattice positions the three-hollow site increases its area (Fig. 2) favouring the diffusion of oxygen into the metal. This process alone reduces the energy barrier from 3 eV to 0.6-0.8 eV (Fig. 2).

The other two effects considered, namely the firs-second layer expansion and the coverage dependence, contribute to reduce the energy barrier but to a smaller extent (by about 1/3, Fig. 3). The overall effect of these contributions will be to considerably lower the barrier to oxygen penetration, and a crude estimate of the

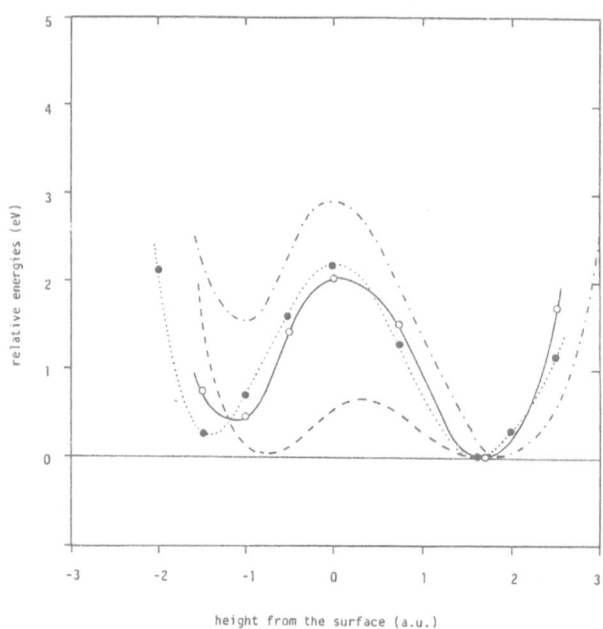

Fig. 3 - Effects influencing the energy barrier to oxygen penetration below the Al surface. The zero reference energy is that of the minimum above the surface. - · - · - rigid lattice; - - - surface relaxation; ●········● second layer expansion; o——o coverage effect (for $Al_{22}-O_7$).

height of the barrier could be approximately 0.5 eV. An energy of this order of magnitude is not in disagreement with the only reported experimental attempt to evaluate this quantity (about 1 eV).[1]

In conclusion, the present work confirms the idea that the interaction of oxygen with the Al (111) surface proceeds through three steps, the formation of a chemisorbed overlayer, the incorporation on below-surface cavities and finally the oxide formation. During the transition from the first to the second step the rearrangement of the metal surface plays an essential role.

ACKNOWLEDGEMENT - This work was supported by the Italian CNR.

V. REFERENCES

1. I.P. Batra and L. Kleinman, J. Electr. Spectr. Rel. Phen., 33, 175 (1984).
2. G. Pacchioni, P. Fantucci, to be published.
3. J.A. Pople, D.L. Beveridge, and P.A. Dobosch, J. Chem. Phys., 47, 2026 (1967).
4. J.P. Barthelat, Ph. Durand, and A. Serafini, Mol. Phys., 33, 159 (1977).
5. G. Pacchioni and J. Koutecký, Ber. Bunsenges. Phys. Chem., 88, 242 (1985).
6. M.E. Schwartz and C.M. Quinn, Surf. Sci., 106, 258 (1981).
7. P.S. Bagus, I.P. Batra, C.W. Bauschlicher Jr., and R. Broer, J. Electr. Spectr. Rel. Phen., 29, 225 (1983).
8. B.N. Cox and C.W. Bauschlicher Jr., Surf. Sci., 115, 15 (1982).
9. H. Partridge and C.W. Bauschlicher Jr., J. Chem. Phys., 84, 6507 (1986).

KEYWORDS: Aluminum/ Electronic structure/ Stability/ Ionization poten-
Chemisorption/ Surface relaxation

THE STUDY OF INTERACTIONS BETWEEN IRON PARTICLES ENCAPSULATED IN ZEOLITES OF THE FAUJASIT-TYPE AND THE MATRIX

Ahadul I. Quazi* and F. Schmidt

Institut für Physikalische Chemie
Universität Hamburg
Bundesstr. 45, D-2000 Hamburg 13, FRG

ABSTRACT

$Fe(CO)_5$·Zeolite-adducts were thermally decomposed under vacuum and under argon gas atmosphere. While small as well as relatively large iron clusters were formed under vacuum decomposition, Fe(II) and $[Fe_3(CO)_{11}]^{2-}$ species were formed under argon gas atmosphere. In order to identify the formed particles, a variety of measurements was carried out, such as thermal desorption spectroscopy (TDS), Mössbauer Effect spectroscopy (MES), IR-spectroscopy and magnetic measurements.

INTRODUCTION

Zeolites, due to their immense surface area, their defined cage structure and crystalline structure, are suitable supports for metals as well as for model catalysts [1,2]. Various attempts were made to encapsulate iron clusters in the pores of the zeolites of the Faujasit-type. Reduction of Fe(II) exchanged zeolites was found to be impossible with molecular H_2 [3,4], whereas reduction with sodium vapors may lead to modified metal phases. Therefore, the thermal decomposition of $Fe(CO)_5$·zeolite-adduct was chosen as an alternative route.

EXPERIMENTAL

I. Materials

The zeolites used had the following chemical compositions:

NaX: $Na_{88}Al_{88}Si_{104}O_{384}$·240 H_2O

NaY: $Na_{55.5}Al_{55.5}Si_{136.5}O_{384}$·240 H_2O

The zeolites were treated with 0.1 M NaCl solution to remove cation

* Present address: Department of Chemistry, University of Michigan, Ann Arbor, Michigan 48109, U.S.A.

deficiencies, washed, air-dried and stored over saturated NH$_4$Cl solution.
Before leading with Fe(CO)$_5$, the zeolites were degassed in situ at 673 K
for about 48 hours, at 10^{-6} Torr, at a heading rate of 2°/min. Ironpenta-
carbonyl from Ventron (99.5%) was distilled in the dark and stored over
molecular sieve Fa.

II. Methods

1) TDS. The NaX·Fe(CO$_5$) was decomposed thermally under argon gas
atmosphere at a constant heating rate of 0.2°/min by means of an EUROTHERM
regulator. The pressure-temperature curve was recorded by means of multiple pen x,t-recorder.

2) Mössbauer-Effect-Spectroscopy (MES). The ordinary Co-Cu sources
were used for recording the Mössbauer spectra. In order to fit the spectra, the widely spread LSQ-method was applied.

3) IR-Spectroscopy. The IR-spectra were recorded by means of an IR
model 225 from Perkin-Elmer.

4) Magnetic measurements. Magnetic measurements of the samples were
carried out at 293 K as well as at 4.2 K by means of a FONER vibration
magnetometer.

RESULTS AND DISCUSSION

Fig. 1 shows the Mössbauer spectra of samples 1 and 2, prepared by
the thermal decomposition of NaX·Fe(CO)$_5$ and NaY·Fe(CO)$_5$ under vacuum
(10^{-6} Torr), respectively. The spectra are similar and both of them
consist of a sextet as well as a singlet. While the sextets with IS = 0
(Table 1) represent relatively large iron clusters, the singlets with
IS = 0.19 mms^{-1} (Table 1) represent relatively small clusters in the pores
of the zeolites. The areas of the singlets to the total area are directly
proportional to the Sanderson electronegativity (EN$_s$) of the corresponding
zeolite type (Table 2).

Completely different particles were formed when NaX·Fe(CO)$_5$-adduct
was thermally decomposed under argon gas atmosphere (sample 3). The
reaction took place in a single step, as shown in Fig. 2. The activation
energy of the reaction was 110 KJ/mol, which is in good agreement with
the results reported by different authors [5]. The results of TDS show
that only 2.5 CO/Fe were set free during the reaction. The Mössbauer
spectra of sample 3 were recorded at 293 K as well as at 4.2 K (Fig. 3).
The Mössbauer parameter of sample 3 are shown in Table 3. The spectra
are very similar and both of them consist of three doublets. Subsystem
II (Table 3, Fig. 3) with an isomer shift of 0.82 mms^{-1} and a quadrupole
splitting of 0.56 mms^{-1} is in good agreement with those of trigonal coordinated Fe(II) ions in A-type zeolite [6]. We assign subsystem III
(Fig. 3, Table 3) with IS = 1.12 mms^{-1} and a quadrupole splitting of
1.83 mms^{-1} Fe(II) ions in X-type zeolite. The Mössbauer parameters of
sub-system I (Fig. 3) with an isomer shift of -0.10 mms^{-1} and a quadrupole
splitting of 0.58 mms^{-1} are in good agreement with those of the
$[Fe^{2+}][Fe_3(CO)_{11}]^{2-}$ system [7].

Fig. 4 shows the IR-spectrum of sample 3. The stretching vibrations
of CO appear at 1650 cm^{-1}, 1900 cm^{-1} and at 1980 cm^{-1}.

The magnetic measurements of sample 3 (Fig. 5) carried out at 293 K
as well as at 4.2 K, show that the sample is of superparamagnetic nature,

i.e., the 4s electrons of Fe were not free, which must be free in order to show a ferro-magnetic behavior.

The results of TDS show that only 2.5 CO/Fe were set free during the reaction. It can be seen from the area ratios of the Mössbauer parameter (Table 3) that 66% of the total area were taken by subsystem I (Fig. 3), the particles containing CO. This corresponds to 3.78 CO/Fe (i.e., 2.5 CO/ .66 Fe) in the sample. The isomer shift of this subsystem is in good agreement with that of $[Fe^{2+}][Fe(CO)_3]^{2-}$ (IS = -0.10 mms^{-1}). The IR spectrum of the same sample show bridge as well as terminal CO's (Fig. 4). Since the magnetic measurements show superparamagnetic behaviour (i.e., no free 4s electrons in Fe atoms), we conclude that $[Fe^{2+}][Fe(CO)_{11}]^{2-}$ species were formed during the thermal decomposition of NaX·Fe(CO)$_5$-adduct under argon gas atmosphere. It has already been mentioned that the Mössbauer parameter of subsystem II as well as III (Table 3, Fig. 3) represent Fe(II) ions in the pores of zeolites of Faujasit-type.

The following reaction scheme (scheme 1) is proposed for the formation of Fe(II) and $[Fe_3(CO)_{11}]^{2-}$ species in sample 3:

$$4 \text{ NaX·Fe(CO)}_5 \xrightarrow{\text{Pathway 1}} 4 \text{ Fe} + 20 \text{ CO} \longrightarrow Fe_3(CO)_{12} + Fe(0) + 8 \text{ CO}$$

Pathway 2

$$[Fe_3(CO)_{11}]^{2-} + Fe^{2+} + 9CO \qquad\qquad Fe^{2+} + [Fe_3(CO)_{11}]^{2-} + 9CO$$

Scheme 1

The above reaction is strongly believed to take place by pathway 1 because of the following reasons:

a) At about 315 K a rapid increase in pressure was observed (Fig. 2a), from which it can be concluded that at first Fe(CO)$_5$ molecules were decomposed to form Fe(0) and CO.

b) After this reaction phase a mild increase in pressure was observed. This step can be explained by the formation of $[Fe_3(CO)_{11}]^{2-}$ and CO as a result of the decomposition of assumed Fe$_3$(CO)$_{12}$.

c) The results of TDS show that E_A of the reaction was 110 KJ/mol. The energy of dissocation of the Fe-CO bond in Fe(CO)$_5$ molecule was reported to be 117 KJ/mol [8], which is within the limits of experimental error, in good agreement with that determined by us. The determined energy of activation is a good evidence to believe that Fe(CO)$_5$ molecule was at first decomposed to form Fe(0) + 5 CO.

d) Finely distributed Fe(0) atoms are easily oxidized.

CONCLUSION

Only iron clusters were formed when NaX·Fe(CO)$_5$ and NaY·Fe(CO)$_5$ adducts were thermally decomposed under <u>vacuum</u>. In both cases relatively small as well as relatively large iron clusters were formed. The amounts

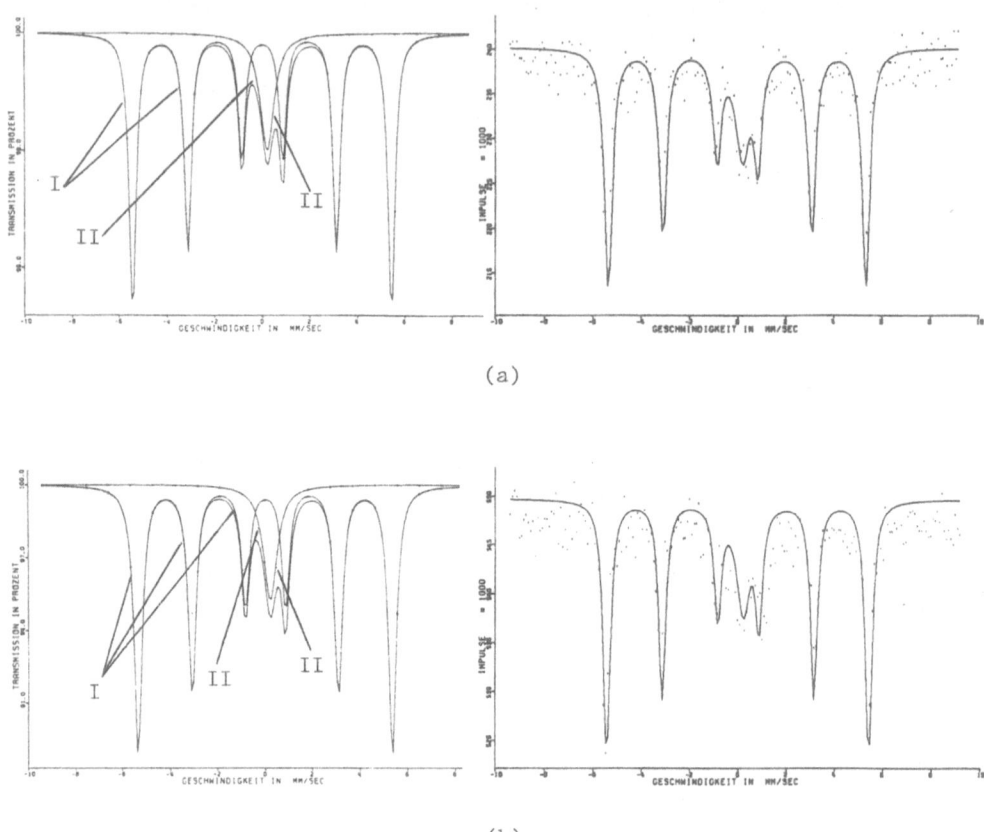

Fig. 1. Mössbauer spectra of samples 1 and 2, prepared by the thermal decomposition of NaX·Fe(CO)₅ and NaY·Fe(CO)₅ adducts, recroded at 293 K. (a) Sample.2 ≅ NaY; (b) Sample 1 ≅ NaX.

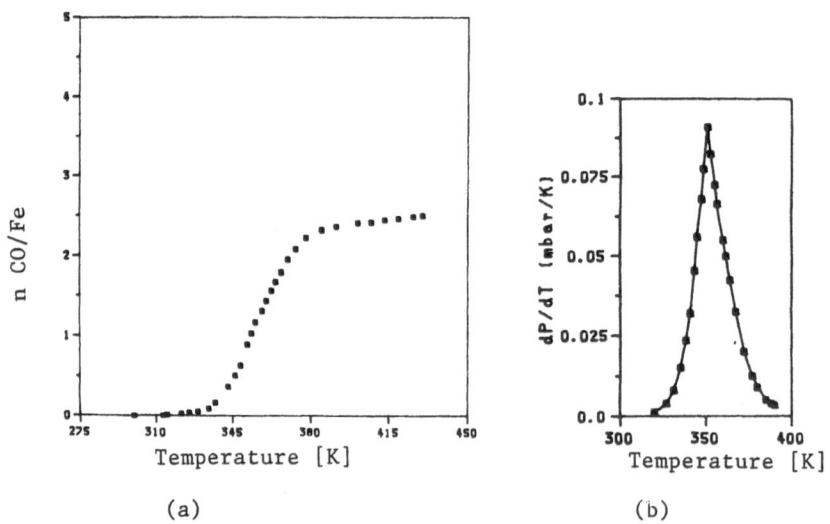

(a)   (b)

Fig. 2. TDS of sample 3, prepared by the thermal decomposition of NaX· Fe(CO)₅ adduct under argon gas atmosphere. (a) P vs. T; (b) dP vs. dT.

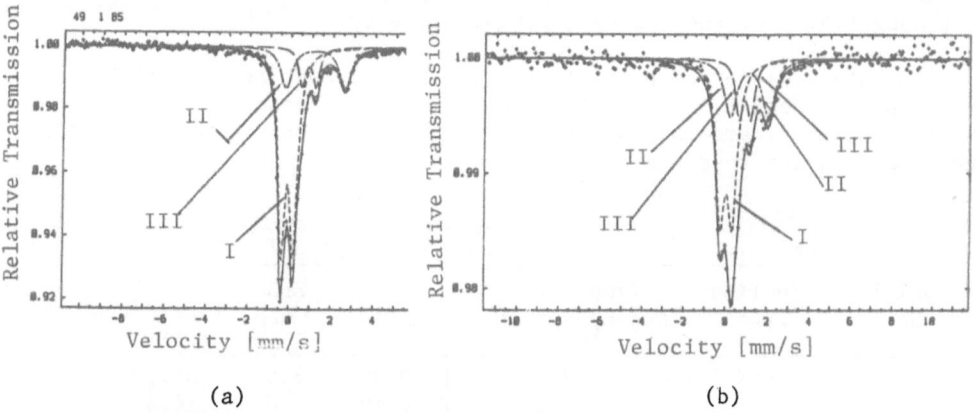

(a)   (b)

Fig. 3.   Mössbauer spectra of sample 3, recorded (a) at 4.2 K, (b) at
293 K.

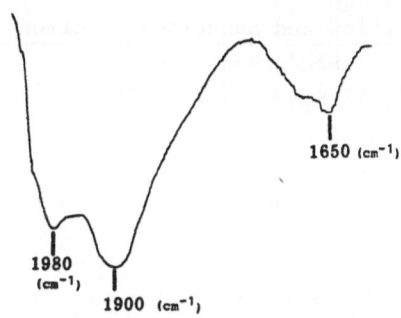

Fig. 4.   Infrared spectrum of sample 3.

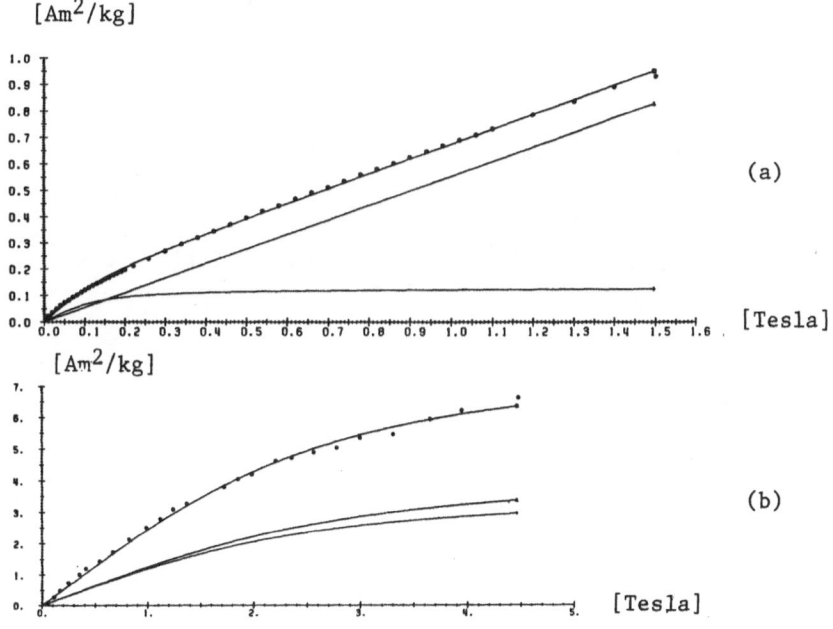

Fig. 5.   Magnetic measurements of sample 3, carried out (a) at 298 K,
(b) at 4.2 K.

of formed small iron clusters are directly proportional to the Sanderson electronegativity of the corresponding zeolite type.

Completely different particles, i.e., Fe(II) and $[Fe_3(CO_{11})]^{2-}$, were formed when an adduct of $NaX \cdot Fe(CO)_5$ was thermally decomposed under argon gas atmosphere. There is good evidence to believe that Fe(O) and $Fe_3(CO)_{12}$ were formed as intermediate products.

Table 1. Mössbauer Parameter of Samples 1 and 2

Sample No.	Zeolite Type	Temp. of Decomposition [K]	IS [mms^{-1}]	HFS [KG]	Area/ Total area [%]
1	NaX	373	I 0.00	355	85.60
			II 0.19	0	14.40
2	NaY	373	I 0.00	354	84.97
			II 0.19	0	15.03

Table 2. Area Ratios and Sanderson Electronegativity

Sample No.	Sample Type	Sanderson Electronegativity [EN$_S$]	Area Ratios (AR) [%]	EN$_S$/AR [%]
1	NaX	3.24	14.40	22.50
2	NaY	3.56	15.03	23.68

Table 3. Mössbauer Parameter of Sample 3

	293 K				4.2 K			
	IS [mms^{-1}]	QS [mms^{-1}]	Γ [mms^{-1}]	Area [%]	IS [mms^{-1}]	QS [mms^{-1}]	Γ [mms^{-1}]	Area [%]
Sub-system I	0.04	0.59	0.53	54	−0.10	0.58	0.43	66
Sub-system II	0.82	0.56	0.44	17	0.87	0.67	0.39	12
Sub-system III	1.12	1.83	0.72	29	1.18	2.82	0.67	22

REFERENCES

1. Th. Bein, P. A. Jacobs and F. Schmidt, in: "Metal Microclusters in Zeolites," P. A. Jacobs et al., ed., Elsevier Scientific Publ. Co., Amsterdam (1982).
2. Th. Bein, F. Schmidt and W. Gunsser, Surface Sci. 156:57 (1985).
3. Y. Y. Huang and J. R. Anderson, J. Catal. 40:143 (1975).
4. R. L. Garten, W. N. Delgass and M. J. Boudart, J. Catal. 18:90 (1970).
5. P. Engelking and W. J. Lineberger, J. Am. Chem. Soc. 101:5569 (1970).
6. Zi Gao and L. V. C. Rees, in: "Zeolites" Vol. 2 (April) (1982).
7. A. Vertes, L. Corecz and K. Burges, in: "Mössbauer Spectroscopy," Elsevier Publishing Co., Oxford (1979).

ROLE OF ENSEMBLES IN CATALYSIS BY METALS

Guy Antonin Martin

Institut de Recherches sur la Catalyse
2, avenue Albert Einstein
69626 Villeurbanne, Cédex - France

ABSTRACT

Among the recent ideas in catalysis by metals, the concept of ensemble appears to be one of the most promising. According to this view the active site is considered as a 2 x D cluster, whose size varies with the reaction considered, in a metallic environment. This model originates from Group VIII-IB alloys experiments, where an active element is diluted in an inactive IB matrix, from kinetics data and from adsorption experiments where the magnetic properties of metals are measured. This is illustrated by Ni and Ni-Cu catalysts and reactions including hydrogenation of benzene, D_2-hydrocarbon exchange reaction and hydrogenolysis. The ensemble model allows a description of rates which is quantitative over a wide range of experimental conditions using only one adjustable parameter, which can be identified to the nuclearity (number of atoms) of the ensemble, associated with an energetic term. This model yields a quantitative description, with the very same remarkable few number of adjustable parameters, of apparently different aspects of catalysis such as size-sensitivity and poisoning. Finally, the field of application, the limit and the future of the ensemble model are tentatively outlined.

INTRODUCTION

Finely divided metals, particularly those of groups VIII and IB are used extensively as catalysts either pure or as alloys, associated with supports (alumina, zeolite, monolith cordierite, silica, activated carbon, silica-alumina) and promoters (alkalis, chlorine ...). They are mainly used in petrochemistry and for automobile exhaust pipe, and to a less extent in chemistry and refining. The sale of metallic catalysts in western countries in 1981 is valued at $ 0.3 billions, a figure which is the fourth of the total market of catalysts, which is forecast to grow at an annual rate of 2 %[1]. The economic importance of catalysts (including metals, oxides, sulfides or zeolites) can also be illustrated by the fact that more than nine-tenths of the chemicals are manufactured through a catalytic process. The understanding of catalysis, however, still remains in the infancy.

According to Boudart[2], in catalysis by metals, the main challenge to scientists is the catalytic _specificity_: for one given reaction, the activity can vary by more than 10 powers of ten when going from a metal to another one. Comparisons of activity between different metals are now available and the understanding of these patterns remains the ultimate goal of catalytic models.

Another point has focused attention: the catalytic activity may also depend on the size of the metal particle and on the nature of crystallographic planes which are exposed. This is the _size_ and _structure-sensitivity_ problem, which, at present time, has not received a fully satisfying answer. Researches on metal alloy catalysts[3,4] has led to the following interesting discovery: the activity for reaction A may be much more affected than that for reaction B by alloying two metals, leading to change in the catalytic _selectivity_. In this respect, VIII-IB alloys experiments are of special interest since an active element, the group VIII metal is diluted in an inactive matrix, the IB metal, providing a direct insight into the nature of the active site and its nuclearity.

From the chemical engineering point of view, phenomena such as ageing, mainly due to sintering and _poisoning_ are of particular importance. As a matter of fact they have a direct bearing on the cost of a process. By now, they are receiving a particular attention from a macroscopic point of view, aiming at modelling poisoning process. The analysis on a microscopic scale is also developing with some attempts in bridging the gap between both approaches.

As can be seen, scientists are faced to a large number of data paralleling a variety of concepts and theories, more or less swinging from electronic to geometric "theories". Various recent concepts, such as catalysis over the second adsorbed layer[6], extractive chemisorption, direct interaction between reactants (CO) and promoters[7] are breaking through. However, among these relatively new ideas, the concept of ensemble, appears to be one of the most promising and unifying guideline in catalysis by metals. In this paper, this model which states that the active site in catalysis by metals is not an isolated atom but rather an ensemble composed of a certain number of adjacent free metal atoms, is dealt with into more details: its bases are first described, and as a second step recent developments including size-sensitivity and poisoning are considered. Finally the field of application, the limit and the future of the model are tentatively outlined.

EXPERIMENTAL BASES OF THE ENSEMBLE MODEL

This model is probably to be traced to the "multiplet" theory of Balandin[8], who proposed in 1929 that a reacting molecule can be simultaneously adsorbed on several metal atoms ("katalysierende Molekül gleichzeitig an mehreren Atomen von mehreren Punkten der katalysatorfläche angezogen wird"). The other aspect of this theory -the group of adsorbing atoms (multiplet) should have the correct geometry and lattice spacing to accomodate reacting molecules and to accelerate their dissociation- has attracted most of the attention of scientists owing to its relevance to the structure-sensitivity of catalytic reactions, and has shadowed the first aspect of this theory which according to us still remains of present interest.

Early support for an ensemble effect derived from the works on

VIII-IB alloys, where UPS and chemisorption data indicate that the identity of metal atoms in the alloy are hardly altered by changing the nature of their neighbors[9], in full agreement with current theories of metal alloys such as the coherent potential approximation. The term of ensemble was first proposed by Dowden[10] to represent a group of adjacent atoms A at the surface of an alloy AB. The influence of neighboring atoms –the ligand effect-, however, was not overlooked[11], but it was assumed to be of less importance. Another experimental approach of the ensemble model emerged from the work of Frennet and its associates[12]. It is based on a rigorous kinetic analysis of some H_2-hydrocarbon reactions, where, surprisingly for the first time in catalysis, experimental surface coverages were introduced in rate equations, resulting in the following evidence: hydrocarbon adsorption in presence of hydrogen requires "landing sites" composed of up to 8 metal atoms (for instance for the case of CH_4 over Rh).

Another independent and simultaneous approach stemmed from a systematic study of chemisorption over Ni and from the determination of the "number of bonds" between the surface and adsorbed species with the help of magnetic methods[13]. This was the starting point of further works performed in our group in the seventies, which, associated with the alloy and the kinetic approach, led to a better understanding of various facets of catalysis by metals in the light of the ensemble model.

The Magnetic Approach

The principle of the method[13] and the most important results have been described elsewhere[14]. Let us briefly recall that it has been postulated that each metal atom of a ferromagnetic substrate, e.g. Ni, which is involved in the bonding with an adsorbate, ceases to participate in the collective ferromagnetism[13]. The decrease of saturation magnetization per adsorbed molecule, α, divided by the magnetic moment of one metal atom μ gives what was termed by Selwood,[13] the bond number n

$$n = \alpha/\mu$$

Experimental evidences of the validity of the bond number calculation were provided in the course of a study of CO chemisorption over Ni-Cu/SiO_2 alloys by infrared spectrometry and magnetic methods,[11]: the bond numbers obtained by the latter technique corresponding to linear (2050 cm^{-1}) and bridged (1940 cm^{-1}) species was 1 and 2, respectively, in good agreement with ir data. The bond number approach was also substantiated by UPS and theoretical considerations. As a matter of fact, adsorption of various gases over Ni results in an increase of the density of state at the foot of the d-band and in a decrease of the density of state at the Fermi level $n(E_F)$[15]. When the local density of state $n(E_F)$ of Ni atoms decreases below a certain critical value given by the Stoner criterion of ferromagnetism, $U.n(E_F) > 1$ (U, coulombic interaction), their magnetic moment would cease to participate in the collective ferromagnetism. It turns out that the bond number as defined by Selwood is the number of metal atoms strongly perturbed from the electronic viewpoint.

Chemisorption of ethane and benzene, among other hydrocarbons, was investigated by magnetic methods[14]. Figure 1 shows the variation of the bond number involved in ethane chemisorption over Ni/SiO_2 catalysts with the temperature of treatment.

The bond numbers obtained over pure Ni are large, 6 and 12, showing that ethane adsorption requires large ensembles. They probably correspond to a partially dehydrogenated species ($n = 6$) and to a completely dehydrogenated and cracked state ($n = 12$) into elemental carbon bonded to 3 nickel atoms and atomic hydrogen bonded to one nickel atom. Addition of Cu (alloying) or H (chemisorption) to Ni inhibits the chemisorption of ethane and increases the temperature of complete cracking, presumably for geometric reasons (dilution of the active phase).

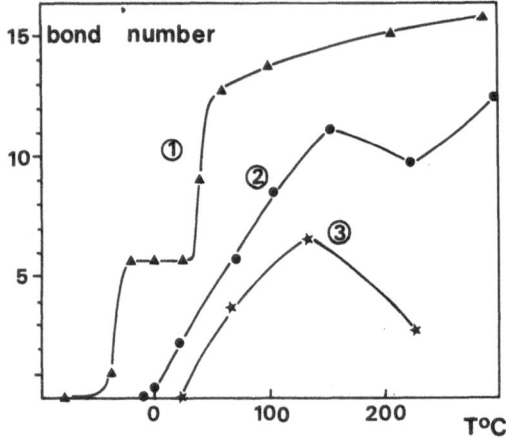

Fig. 1. Bond number (as deduced from magnetism) of ethane adsorbed over Ni/SiO$_2$, Ni–Cu/SiO$_2$ and Ni/SiO$_2$ precovered with hydrogen (curves 1, 2 and 3, respectively) as a function of the treatment temperature.

It is striking to notice that preadsorbed hydrogen over Ni seems to play the same role towards ethane adsorption as copper in Ni–Cu. This was confirmed in the same paper[14] by the study of benzene chemisorption: on pure nickel, benzene is adsorbed with a corresponding bond number of 8; presence of preadsorbed hydrogen or addition of copper results in the very same bond number for benzene adsorption, $n = 3$. This idea has also received a support from a calorimetric study of hydrogen adsorption over nickel surface[17]: the data indicate that the surface of a Ni–Cu alloy has the same reactivity with respect to hydrogen as a pure nickel surface containing a fractional coverage of preadsorbed hydrogen.

It was tempting to examine the relevance of these results to catalysis, particularly to the following catalytic reactions involving these hydrocarbons, benzene hydrogenation (reaction 1) and ethane hydrogenolysis (reaction 2).

$$C_6H_6 + 3 H_2 \rightarrow C_6H_{12} \tag{1}$$

$$C_2H_6 + H_2 \rightarrow 2 CH_4 \tag{2}$$

Both reactions are indeed representative of two families of reactions. The first group which includes mainly hydrogenation reactions, is characterized by a small structure-sensitivity, a moderate specificity, a small sensitivity to alloying, and particular kinetic features (order in hydrogen near unity, order in hydrocarbon near zero, small apparent activation energy). In the second group of reactions (hydrogenolysis, isomerization, D_2-hydrocarbon exchange) the activity varies strongly with the nature of the metal, the surface structure or as a result of alloy formation. The kinetic parameters are also quite different from those of group I: the activation energies are large, the orders with respect to hydrocarbons generally positive and the order in hydrogen negative.

On the basis of magnetic data, we have proposed the following assumptions for catalytic reactions (1) and (2):
(i) the intermediate of benzene hydrogenation is the species corresponding to n = 3, since it is the observed one in presence of hydrogen;
(ii) the intermediate of ethane hydrogenolysis[14] corresponds to n = 12 (this species is the only one which gives methane after a subsequent hydrogenation);
(iii) adsorbed hydrogen on nickel (coverage, θ_H) plays the same role as copper in Ni-Cu alloys ([Cu], atomic fraction).
If we are right in these conjectures, and if adsorbed H or Cu in Ni-Cu alloys are randomly distributed, we can anticipate that the reaction rates of both reactions varies as:

$$r \sim (1-[Cu])^N \tag{3}$$

$$r \sim (1-\theta_H)^X \tag{4}$$

with N = X = n, equalling 3 and 12 for benzene hydrogenation and ethane hydrogenolysis, respectively. This was the starting point of a new experimental research aiming at substantiating these predictions which is briefly described in the following sections.

The Alloying Approach

The first step of this study consisted in preparing silica-supported Ni-Cu alloys, 6 nm in diameter, with a surface composition nearly equal to the bulk one[18]. As a second step, the kinetic parameters (apparent activation energies, orders) were measured. They were found independent of the copper atomic fraction, indicating that the nature of the active sites remains unchanged by alloying: the rate variations shown in Figure 2 are reflecting a change of the number of active sites rather than a modification of their chemical and energetic properties. Data of Figure 2 show that the rate of both reactions varies as $(1-[Cu])^N$ with N = 3 and N = 12 for benzene hydrogenation and ethane hydrogenolysis, respectively. This is exactly what was expected from previous speculations on the basis of magnetic data. We examine in the next section how far adsorbed hydrogen plays the same role as Cu in alloys, and to what extent the rate varies as $(1-\theta_H)^X$.

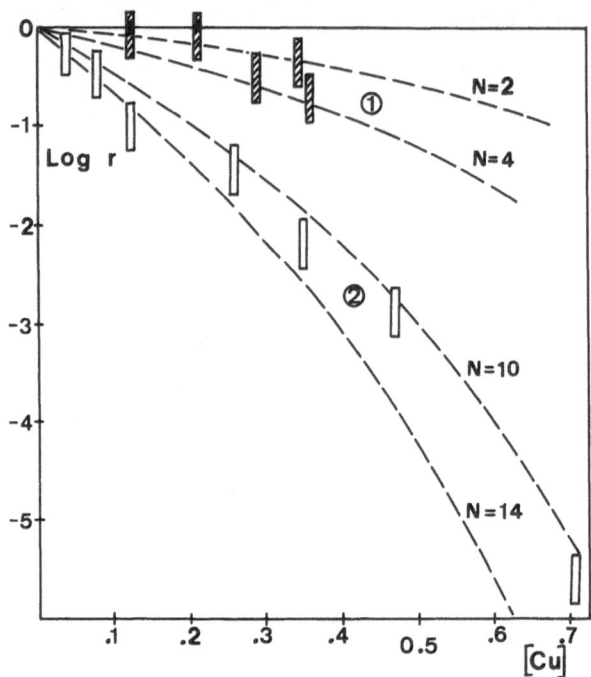

Fig. 2. Variations of Log rate of benzene hydrogenation and
hydrogenolysis of ethane (curves 1 and 2, respectively) as a
function of $[Cu]$, the copper concentration of the Ni-Cu
alloys. The dashed lines are calculated assuming that rates
vary as $(1-[Cu])^N$.

The Kinetic Approach

Our aim was to express the reaction rate as a function of the
surface concentration of reactants. The situation was deliberately
simplified by working in conditions where the hydrocarbon coverage was
negligible. This was achieved at low hydrocarbon partial pressure for
ethane hydrogenolysis and at temperature less than 400 K for benzene
hydrogenation.

As a second step, the hydrogen coverage was measured as a function
of pressure P_H and temperature T: $\theta_H(T, P_H)$.

The third step consisted in measuring the variation of the rate, r,
as a function of T and the partial pressure of reactants P_H, P_{ETH} or P_B

over a large range of experimental conditions. This kinetic study led to apparent activation energies and orders vayring with both P and T, indicating that the power law, universally utilized in catalysis, is physically meaningless.

Finally, the substitution of $\theta_H(T,P_H)$ in $r(T,P_H)$ was possible and the following equations were derived[19,20]:

$$r = k\ e^{-E_0/RT}\ (1-\theta_H)^X P_H P_B{}^\varepsilon \qquad (5)$$

$$r = k\ e^{-E_0/RT}(1-\theta_H)^X P_{ETH} \qquad (6)$$

with $X = 4$ and 15, and $E_0 = 8$ and 14 kcal/mole (1 cal $= 4.18$ Joule) for benzene hydrogenation and ethane hydrogenolysis, respectively. It should be noted that these kinetic parameters are T and P independent. These experimental equations, which model correctly the complex variation of r over a large range of experimental conditions with the help of only two adjustable parameters, are in full agreement with our predictions.

We summarize the physical significance of this type of equations very briefly since it has been discussed elsewhere[19,20]: for the case of ethane hydrogenolysis, kP_{ETH} was found nearly equal to the number of ethane molecules colliding with the nickel surface. Thus, we have proposed to describe ethane hydrogenolysis as a two-bodies activated process between a gaseous ethane molecule and the active site consisting of an ensemble composed of X adjacent nickel atoms free from adsorbed hydrogen, whose concentration is $(1-\theta_H)^X$.

A series of papers of this laboratory[19-24] extended this approach to various reactions which were described by the ensemble model, using only two adjustable parameters, the number of atoms in the ensemble, and an energetic parameter, the true activation energy, E_0. They are listed in Table 1.

Table 1 shows that within 30 %, X, N and n have the same value, pointing out the coherence of the proposed hypotheses. The size of the ensemble is low for reactions of group 1 (benzene hydrogenation) and large for group 2. This explains satisfactorily most of the features of both groups. As a matter of fact, in group 2, the larger N, the larger the variation of Log $(1-\theta_H)^N$ with $1/T$, E_θ, and the larger the apparent activation energy E_a which is related to E_θ and E_0 by the Temkin's relation[25], $E_a = E_0 + E_\theta$. Similarly, large N values lead to large variations of $(1-\theta_H)^N$ with P_H, yielding highly negative orders in hydrogen. One can also expect for large N-values, an important sensitivity to alloying. The reverse is obviously true for reactions of group 1. The ensemble model seemed to us sufficiently coherent and versatile to examine how far other catalytic features, such as structure or size-sensitivity and poisoning can be accounted for within its framework. This is the purpose of the next section.

Reactions	X ± 1	N ± 1	n ± 1	$E_0 ± 2$ (kcal/mole)
Hydrogenation				
1 $C_6H_6, H_2 = C_6H_{12}$	4	3	3	8
D_2 exchange				
2 $CH_4, D_2 = CD_4$	7.4	5	7	15
3 " $= CH_3D$	3	3	–	18
4 $C_2H_6, D_2 = C_2H_4D_2$	3.6	–	–	14
5 " $= C_2H_2D_4$	5.4	–	6	13
6 " $= C_2D_6$	7.5	–	–	11
Hydrogenolysis				
7 $C_2H_6, H_2 = 2CH_4$	15	12	12	14
8 $C_3H_8, H_2 = CH_4 + C_2H_6$	17	12	–	14
9 " $= 3CH_4$	24	17	17	9

RECENT DEVELOPMENT OF THE ENSEMBLE MODEL

Size-Sensitivity

It is striking that decreasing the size of nickel particles below 5 nm is equivalent from the viewpoint of activity and selectivity to add copper to nickel. This is illustrated in Table 2 where the sign of the variations of catalytic activity A (i) with particle size D, $\alpha = dA/dD$, (ii) with the copper atomic fraction in the Ni-Cu alloy $\beta = dA/d[Cu]$ can be compared.

Table 2. Signs of variation of activity with Ni particles size (α) and with the copper in the Ni-Cu alloy (β)

Reactions (see Table 1)	Particle size sensitivity $\alpha = dA/dD$	Alloying sensitivity $\beta = dA/d[Cu]$
react. 1 at T < 400 K	+	–
react. 1 at T > 400 K	–	+
react. 2-9	+	–
$CO + 3H_2 = CH_4 + H_2O$	+	–
$CO + 4H_2 = C_2H_6 + H_2O$	+	–

It can be clearly seen that α has a sign opposite to that of ß. Moreover, data of Figure 3 shows that CO hydrogenation over Ni into methane, is less size-sensitive than CO hydrogenation into ethane which, according to alloying data, requires larger ensemble (12 and 20 adjacent metal atoms, respectively). This is a general trend: the larger the ensemble site the larger the activity decrease on smaller particle.

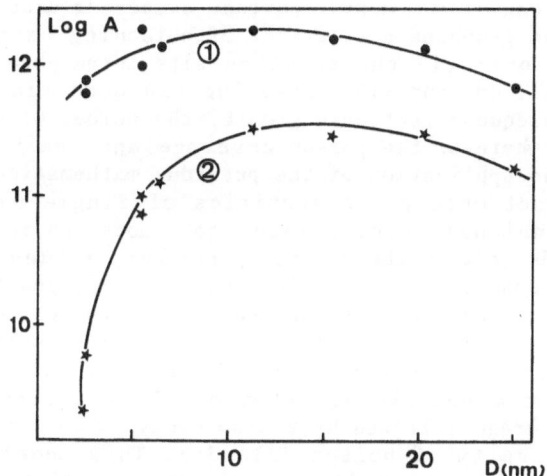

Fig. 3. Activity of Ni/SiO$_2$ catalysts for CO hydrogenation into CH$_4$ (curve 1) and into ethane (curve 2) as a function of particle size.

This observation, already partially arrived at by Boudart[2], has prompted us to calculate the concentration of ensembles composed of X adjacent free atoms at coverage Θ as a function of particle size D[26]. It was additionally assumed that each particle has the same coverage, a condition which is probably fulfilled when the heat of adsorption decreases with coverage. The calculations were done using elementary statistics[27]. The equations thus obtained predict a decrease of activity when D decreases below 5 nm. The larger the ensemble, the larger the size-sensitivity, in quantitative agreement with experimental data. Another interesting feature of this model is that the rate variation depends also on Θ: the larger Θ, the greater α, suggesting that the size-sensitivity of a given reaction can be increased when operating at high pressure and(or) low temperature.

A scrutiny of literature data, however, reveals that in some catalytic systems, the activity increases when D decreases, in contrast with the prediction of the ensemble model. For these cases, it can be postulated that some other phenomena such as electron transfer between the metal and the support, or enhanced activity of atoms in a particular topological situation (edges, corners ...) are operating, adding their effect to the ensemble size-sensitivity. It is possible to discriminate between these various types of sensitivity since the electronic or topoligical contribution is expected to be T and P independent, in contrast with that due to the ensemble effect. Works in this field are still rare and appears to be desirable.

Metal Poisoning

In catalysis by metals, it is known that reactions belonging to group 2 (hydrogenolysis) are much more sensitive to poisoning than those of group 1 (hydro-deshydrogenation ...). It seemed to us of interest to examine to what extent the ensemble model is also able to deal with poisoning which is one of the most important causes of deactivation. As a first stage, we have proposed a treatment of poisoning based on the two following assumptions: (i) the reaction site is an ensemble, (ii) the poison atoms form islands and the poisoning proceeds via a nucleation mechanism and a subsequent isotropic growth, the number of poison islands remaining constant whatever the poison coverage and their distribution being random[28]. The application of the previous mathematical treatment[27] describing the exact occupation statitics of single particle on an infinite two-dimensional lattice leads to a model where the catalytic activity (A) depends only on the reduced parameter, γ, depending itself on the size of the ensemble N, the poison coverage θ_p, and the density of poison islands. The model predicts that the larger the ensemble size, the more marked the poison effect. It also predicts at increasing θ_p either a linear decrase of A when γ is small (small ensemble, small density of poison islands) or a non-linear behavior characterized by a steep decrease at low coverage followed by a slow decay when γ is large (large ensemble, large density of poison islands). This model accounts for experimental data already available, confirming the versatile and fruitful character of the ensemble model.

DISCUSSION

In the ensemble model, the possible large number of atoms in the active site has focused the attention. The group of Frennet[12] which has shown that CH_4 adsorption requires an ensemble of 8 adjacent Rhodium atoms, has introduced the notion of "landing site": the high number of atoms composing the ensemble (landing site) does not mean necessarily that all the atoms are involved in the bonding. We feel that Frennet is right, at least for the case of CO adsorption and hydrogenation. As a matter of fact, it has been observed that the relative concentration of the bridged adsorbed CO molecule bonded to two adjacent Ni atoms decreases much more rapidly than $(1-[Cu])^2$, the probability for finding two adjacent Ni atoms. Mathieu and Primet[29] have proposed that the active site could be an ensemble of two adjacent Ni atoms more or less surrounded with other nickel atoms. In the same vein, we have discussed the following apparent paradoxical situation[30]: the active site of the methanation reaction is an ensemble composed of 14 Ni atoms according to Ni-Cu experiments while 4 Ni atoms suffice to dissociate CO into C + O. We have suggested that CO dissociation requires 4 Ni atoms, electronically unperturbed by the presence of adjacent Cu atoms; such an ensemble composed of 4 Ni atoms surrounded with Ni atoms leads to an ensemble composed of ca. 15 adjacent Ni atoms, in good agreement with our observations.

Burch[4] has extended our hypothesis to the case of hydrocarbon conversion. We feel that if this would be true, we would observe, like for CO hydrogenation, a bond number value, n, clearly smaller than N and

X, the number of metal atoms in the ensemble. Table 1 shows clearly that this is not correct: the number of Ni atoms involved in the active site as deduced from kinetic experiments and alloying data is similar to that deduced from magnetic experiments.

We would rather suggest another explanation, still open to discussion: when a gas molecule (ethane for example) collides with an ensemble composed of a certain number of free adjacent atoms, it is successively dehydrogenated step by step. If the size of this ensemble is too small, the reaction does not proceed up to its ultimate stage, the C-C bond breaking and the adsorbed ethane molecule can desorb after having exchanged some of its hydrogen atoms with those of the adsorbed pool. This situation can arise from the fact that the time of sojourn of H atoms is larger than the time required for an elementary step (C-H or C-C bond rupture). It can also be said that a small ensemble is rapidly saturated with fragments of ethane molecule; the continuation of further elementary steps would require either a hydrogen desorption or an atom migration. The corresponding activation energies of these processes would make these steps less probable. If now the size ensemble is large enough (N > 12) the reaction proceeds up to the C-C bond cleavage and the formation of CH_4 is then possible. As can be seen the ensemble model does not contradict the principle of microreversibility. It also implies that all the metal atoms composing the ensemble are occupied by fragments of the reacting hydrocarbon molecule.

Another point deserves further comments: what is the interplay between the specificity of metals toward a reaction and the size of the ensemble? This question arises from the fact that the specificity of metals is larger for reactions of group 2 which are associated with large ensemble. Frennet has tentatively speculated that small variations of coverage when going from one metal to another yield large rate variations owing to the large exponant of $(1-\theta)^X$. This interesting approach of the specificity, however, requires further experimental work, particularly by comparing the coverage of reactants over various metals. In this respect the comparison of Ni and Pt which have very different activities toward ethane hydrogenolysis seems promising, keeping in mind that for the case of Pt, the possibility of catalytic reactions over the second layer[6] can confuse the issue.

From another standpoint the ensemble model probably requires further elaboration. It does not suffice for the ensemble to have a certain minimum size; it probably should have definite or demanding requirements in terms of the type and arrangements of the atoms. This question which has already been tackled in this paper will probably attract more attention in the future.

CONCLUSION

It can be concluded that a quantitative solution to the problems of catalytic activity, of size, alloying and poisoning sensitivity in catalysis by metals is slowly emerging. These various aspects can be rationalized by the ensemble model, according to which each catalytic reaction can be properly described with only two parameters: the true activation energy and the nuclearity of the active site, considered as an surface cluster composed of adjacent free metal atoms.

REFERENCES

1. Chem. Eng. News, Feb. 17 (1986).

2. M. Boudart, "Proc. Sixth Int. Cong. Catalysis", G. C. Bond et al., ed., The Chemical Society, London, (1976), p. 1.

3. J. H. Sinfelt, J. L. Carter, and D. C. Yates, J. Catalysis 24:283 (1972).

4. V. Ponec and W. M. H. Sachtler, J. Catalysis 24:250 (1972).

5. R. Burch, Acc. Chem. Res. 15:24 (1982).

6. G. A. Somorjai, "Proc. 8th Int. Cong. Catalysis", Verlag Chemie (1984), p. I-113.

7. W. M. H. Sachtler, "Proc. 8th Int. Cong. Catalysis", Verlag Chemie (1984), p. I-151.

8. A. A. Balandin, Z. Phys. Chem. 132:289 (1929).

9. Y. Soma-Noto and W. M. H. Sachtler, J. Catalysis 34:162 (1974).

10. D. A. Dowden, "Proc. 5th Int. Cong. Catalysis", Amsterdam (1972), p. 621.

11. J. A. Dalmon, M. Primet, and G. A. Martin, Surf. Sci. 50:95 (1975).

12. A. Frennet, G. Liennard, A. Grucq, and L. Degols, J. Catalysis 53:150 (1978); Surf. Sci. 80:419 (1979).

13. P. W. Selwood, "Chemisorption and Magnetization", Academic Press, N.Y. (1975).

14. J. A. Dalmon, J. P. Candy, and G. A. Martin, "Proc. Sixth Int. Cong. Catalysis", G. C. Bond et al., ed., The Chemical Society, London (1976), p. 903.

15. See e.g. H. Conrad, G. Ertl, J. Küppers, and E. E. Latta, Surf. Sci. 58:578 (1976).

16. M. C. Desjonquères and F. Cyrot-Lackmann, Surf. Sci. 80:208 (1979).

17. J. J. Prinsloo and P. C. Gravelle, J.C.S. Faraday I 76:2221 (1980).

18. J. A. Dalmon, G. A. Martin, and B. Imelik, Japanese J. Appl. Phys. 2:261 (1974).

19. J. A. Dalmon and G. A. Martin, J. Catalysis 66:214 (1980).

20. G. A. Martin, J. Catalysis 60:345 (1979).

21. C. Mirodatos, J. A. Dalmon, and G. A. Martin, J. Catalysis, in press (1986).

22. H. F. Leach, C. Mirodatos, and D. C. Whan, J. Catalysis 63:138 (1980).

23. M. F. Guilleux, J. A. Dalmon, and G. A. Martin, J. Catalysis 63:138 (1980).

24. M. Temkin, Acta Physicochimica Soviet Union 3:313 (1935).

25. G. A. Martin, J. A. Dalmon, and C. Mirodatos, "Proc. 8th Int. Cong. Catalysis", Verlag Chemie, Berlin (1984), p. IV-371.

26. E. Miyazaki and I. Yasumori, J. Math. Phys. 18:215 (1977).

27. G. A. Martin, Surf. Sci. 162:316 (1985).

28. M. V. Mathieu and M. Primet, Surf. Sci. 58:511 (1976).

29. J. A. Dalmon and G. A. Martin, "Proc. 7th Int. Cong. Catalysis", Kodansha ed., Tokyo (1981), p. 402.

CLUSTER CALCULATIONS FOR DIFFUSION ON

AND IN TRANSITION METALS

Jan Andzelm and Dennis Salahub

Département de chimie
Université de Montréal
Montréal, Québec, Canada H3C 3J7

ABSTRACT

Local Density calculations have been performed, using a Gaussian representation of the wave function and a model potential for the core electrons, for clusters representing a hydrogen atom diffusing through palladium and rhodium and for a CO molecule diffusing over Pd(100).

For the Pd(111) + H system, represented by a ten-atom cluster, the relative energies of the various H-atom sites are correctly given and the calculated diffusion barier (0.32eV) is in encouraging agreement with its experimental counterpart (0.23eV). For rhodium the calculated barrier is much greater (0.91eV). Calculations for rhodium at the palladium lattice spacing (about 2% greater) yield 0.71eV so that the increase relative to palladium is roughly one-third structural and two-thirds electronic.

Several cluster models, containing up to 17 atoms, have been chosen to model the atop (A), bridge (B) and 4-fold centered (C) sites for CO adsorption on Pd(100). The B site is correctly found to be the most stable, followed closely (within ~ 0.15eV) by C, A being significantly higher (~1.0eV). The molecule stands perpendicular to the surface at all three sites; however, along a diffusion path between B and C it tilts by as much as ~ 20°, showing a tendency for the carbon to point back towards the bridge site. The barrier for diffusion is calculated to be about 0.35eV, a reasonable value.

These "dynamic" results complement previous results on the equilibrium properties (geometries, ionization potentials, vibrational frequencies, nature of the bonding) and further help to delimit the domain of the method.

INTRODUCTION

Calculating and understanding the dynamics for a molecule impinging on a surface, scattering from it, sticking to it, diffusing over it, reacting with other molecule on it, and/or desorbing from it, represents a most formidable challenge. At the heart of any dynamics calculation lies an interaction potential. While intermolecular potentials suitable for elastic and inelastic scattering calculations have been given much study[1] and are reasonably easy to formulate, the situation for reactive processes is in a much more primitive state. This is particularly true for the important case of transition metal surfaces, where the nature and energetics of the metal-adsorbate bonding are only beginning to be understood. The few dynamics calculations which have been attempted for such systems[2] involve empirical potentials adjusted to agree with available experimental information or with energies for a few points calculated from quantum chemistry. The ultimate goal, of course, would be to calculate both the potential surface and the dynamics from first principles. Although this goal will not likely be realized for some years to come, progress is being made on several fronts. In our view, one of the most encouraging has been the application of local density functional theory to cluster models of chemisorption systems[3-5]. Equilibrium properties (geometries, ionization potentials, vibrational frequencies) are calculated with sufficient accuracy to support the interpretations derived from the calculations. Although the method shows a tendency to overestimate absolute binding energies[6,7] which will ultimately require more work on non-local corrections[6,8], there is growing evidence that the relative energies may be reasonably given for geometries which do not differ too widely. The points along a diffusion path fall into this category since, although old bonds are broken and new ones formed, the diffusing species is kept in contact with the system through which it is diffusing. In this paper we will describe our first attempts to calculate the energies that govern diffusion on and through transition metals.

Surface diffusion is an important step in many surface processes[9]. Until recently most of the relevant experiments were done using the field emission technique[10]. The surface diffusion rates were measured for several atoms and molecules adsorbed, primarily on tungsten surfaces. More recently, laser-induced thermal desorption emerged as a powerful technique of measuring surface diffusion coefficients[11,12]. Microscopic interpretations of the activation energy for surface diffusion have been proposed for several systems including CO on Cu(100)[11], H on Ni(100)[12] and H on Ru(001)[13]. These experiments, along with other modern surface techniques such as ESDIAD[14] or HREELS[15] allow the determination of the binding geometries and binding energy differences, results which may be compared with those from quantum chemistry.

On the quantum chemical side, attempts to calculate the relative energies of different adsorption sites on cluster models using first principles ("ab initio" or density functional) approaches have been relatively few. Refs. 16 to 19 cover the principal methods and should provide an entry into the literature. The ab initio work of Upton and Goddard for H atoms diffusing between four-fold sites on Ni(100) by way of twofold bridges predicted an activation energy of 9 kcal/mol for a rigid surface. The authors state that if the effects of thermal motion of the surface atoms were included this would probably decrease by as much as one half. The measured value is 4 kcal/mol[12] - highly encouraging for the use of quantum chemistry in this area. The Amsterdam group[17] has applied the Discrete Variational Local Density method for CO adsorbing at several sites on several different Cu clusters (up to 7 atoms). They found good agreement with a number of experimental properties for top-site adsorption; however, the adsorption energy (calculated using a transition - state procedure) erroneously favors the hollow site. The extent to which this reflects errors in the cluster model, the computational approach, or the Local Density approximation itself is, at the moment, not clear. Cao et al[18], also using the DVM approach, studied CO chemisorption on Ru and Ru-Cu clusters. Among other things they predict that, on the trigonal Ru(0001) surface adsorption over the 3-fold hollow is 0.8eV more stable than on top of a Ru atom. Persson and Müller[19] used the Linear Muffin Tin Orbital variant of local density theory to study the energetics and properties of CO on Al(111), represented by a ten-atom cluster. They considered atop, two-fold and threefold sites and the diffusion path joining them. Interestingly (see further below) the CO stands perpendicular to the surface at these sites but tilts at the lower-symmetry points along the diffusion path.

In what follows we will concentrate on our own recent efforts to extend the realm of applications of the LCGTO-MP-LSD method beyond the immediate neighborhood of potential energy minima. These efforts have focused on two diffusion systems, H atoms diffusing into and through a Pd cluster, with calculations on Rh for comparison, and a CO molecule diffusing over cluster models of Pd(100).

REMARKS ON CLUSTER MODELS AND THE LCGTO-MP-LSD METHOD

In recent years Local (Spin) Density (LSD) calculations of potential energy surfaces have proven to provide reliable results, at the same time being less demanding then "ab initio" (HF + CI) type calculations[3-7], [17-21]. In the present paper we are using the LCGTO-MP-LSD method (see references 3 and 7 for a description of this approach and comparisons with other methods). An advantage of the LCGTO-MP-LSD scheme lies in utilizing highly accurate model potentials[21] to replace core electrons. The relatively smooth valence density and the corresponding exchange-correlation potential can then be efficiently fitted using Gaussian type functions. Once the fitting is done the total energy is calculated analytically which provides accurate energies at a cost of calculating $\sim N^3$ integrals (N = size of the basis set). The full strength of the LSD method emerges if one investigates clusters of transition metal atoms. In order to obtain meaningful results one has to consider d or even f electrons, which are highly correlated and may possess magnetic properties. Moreover, if one wishes to model surfaces and study adsorption, then the question arises of convergence of the cluster model toward the infinite limit. It is in principle simple to show convergence; one adds atoms until no significant further changes in the properties of interest are observed. In practice, however, this implies the ability to treat large clusters at an appropriate level of theory. The rate of convergence will depend on the system, on the properties studied and, unfortunately, may also depend on computational and methodological approximations (Ref. 22 will provide an entry to the literature for the cluster convergence question in ab initio calculations; and Ref.23 to the related question of cluster embedding).

We will give elsewhere[24] a detailed discussion of the cluster size and shape-dependency of our own calculations along with the dependence of various properties on the quality of the basis set and model potentials used. Here we will only mention a few generalities of relevance to the present cases, which have been gleaned from an extensive series of test calculations:

  i)   Adsorption geometries and vibrational frequencies are rather insensitive to cluster size (and in good agreement with their experimental counterparts).
  ii) Binding energies converge more slowly (and appear to be somewhat too large).
  iii) To obtain reasonable convergence in the binding energy for CO on the bridge site of Pd(100), the two atoms of the bridge were described with a high quality (16-valence electron, pds) model potential (MP) and an extensive (4s, 4p, 3d) basis set including four s, four p and three d contractions.
  iv) Some, but not necessarily all[24] of the nearest neighbors of the bridge atoms have to be present in the cluster and the d

electrons for these atoms have to be treated explicitly, albeit at a somewhat lower level (10-electron, ds, MP and d single-zeta basis). Using an s- only model potential for these atoms leads to very large changes in the adsorption energy; implying that a reasonable description of d-d bonding is necessary and that the chemisorption disturbs the d-d bonds.

v) When comparing energies of different sites, or points along a diffusion path, they should - as far as possible - be treated equivalently and, preferably, all sites should be included in the same cluster in order that edge effects may cancel to the maximum possible extent. For such complex situations, involving several surface atoms in contact with the adsorbates, the "convergence problem" is aggravated - one inevitably must treat several transition metal atoms very accurately and this is where the LSD methods can be used to advantage. The cluster models and the computational approximations discussed in the next section were arrived at with the above considerations in mind. In the end they must be judged by the variety of properties which they explain or predict. We turn now to a discussion of these predictions and explanations.

DIFFUSION OF H ATOMS THROUGH PD AND RH CLUSTERS

The diffusion of hydrogen atoms through palladium and rhodium clusters has been studied using a 10-atom model (Fig.1) of the fcc(111) surface and underlying interstices. This system makes a good starting point for the study of diffusion because we have well-defined, high-symmetry adsorption and absorption sites. We can treat all atoms at a high level of basis set (d-triple-zeta) and model potential (16-valence electron, pds). Since we are using the same cluster to study all adsorption sites the possible errors arising from "physical" edge effects, from basis set deficiencies, from differences in the quality of the fits or from BSSE (basis set superposition error) will tend to cancel.

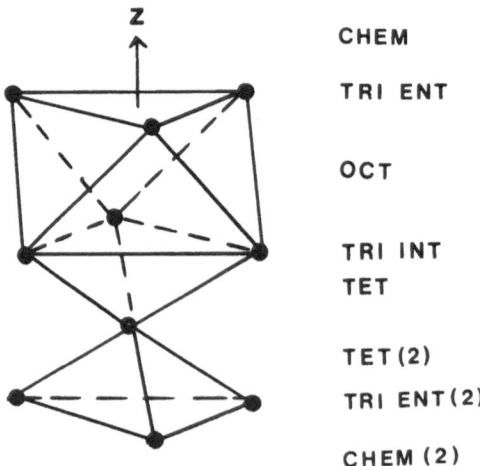

Fig. 1 A 10-atom cluster representing the fcc(111) surface and various sites on a diffusion path.

The energy variations as hydrogen approaches Pd(111) or Rh(111) and diffuses through the solid are presented in Fig.2. Agreement, for the Pd cluster, with known experimental energy differences is very good[5] if we compare with data from chemisorption and bulk experiments. The octahedral site is found to be the most stable interior site and it is above the chemisorption site by 0.25eV (exp. 0.26eV). The calculated diffusion barrier, 0.32eV, is in acceptably close agreement with its experimental counterpart (0.23eV).

We note in passing that, recently, electron-stimulated desorption of H from Pd(111)[25] revealed a strong temperature dependence of the proton yield at a coverage of two-thirds of a monolayer. A thermodynamic interpretation led to the conclusion that the sub-surface occupation of H is energetically favored over surface-site occupation. A theoretical study using a semi-empirical embedded-atom method[26] concluded that

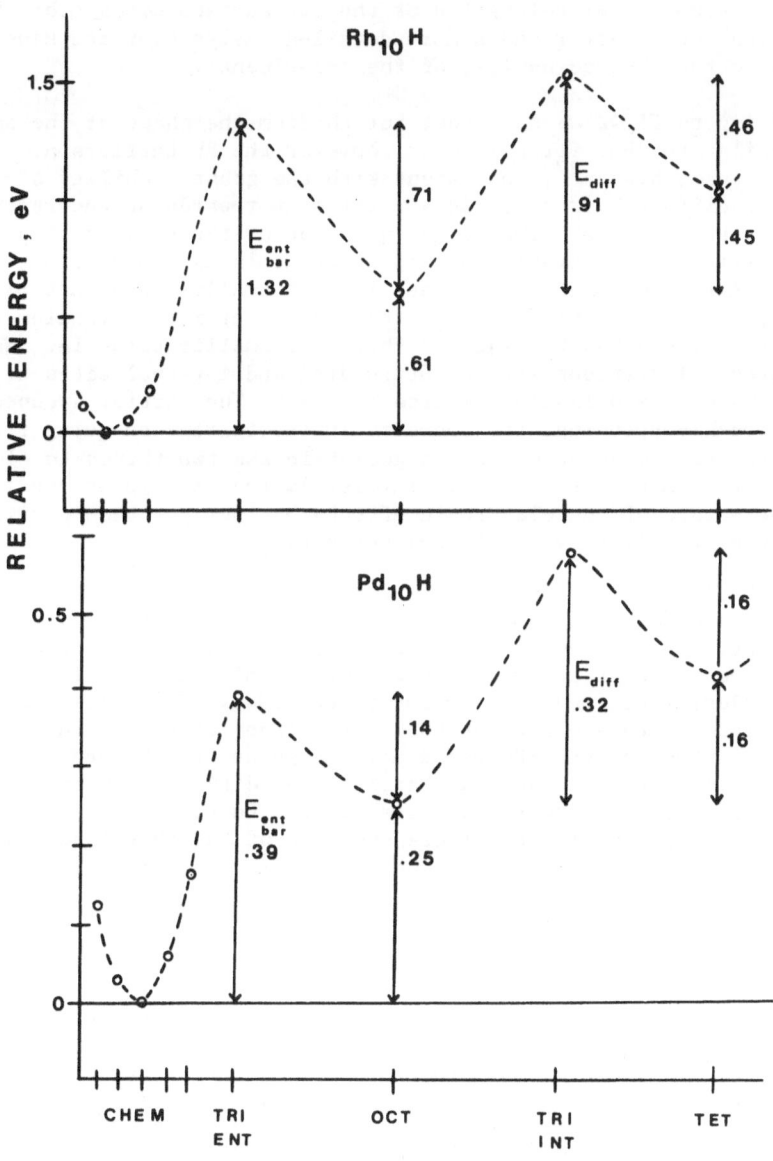

Fig. 2 Calculated energy variations for a hydrogen atom at various sites in Rh_{10} and Pd_{10}.

Table 1. Relative energies of H in $Rh_{10}H$ cluster.

	ΔE	$\Delta E(-BSSE)$
Chemisorption	0.00	0.00
tri ent	1.48	1.32
oct	0.56	0.61
tri int	1.67	1.52
tet	0.95	1.06

the surface and subsurface sites lie within 0.006eV of each other. This would imply that sub-surface sites, occupied at high coverage, are quite different from either bulk octahedral (or tetrahedral) sites, and the unrelaxed (bulk geometry) sub-surface sites present in our cluster. To carry the question further would require an examination of H-H inter-actions, the geometrical relaxation of the sub-surface sites, the effect of "embedding" our cluster and a more detailed analysis of the kinetics, as opposed to the thermodynamics, of the experiments.

Returning to Fig.2 we note that for rhodium the shape of the path is very similar to that for palladium; however the Rh barriers are about a factor of three higher[27], consistent with the greater ability of palladium to diffuse hydrogen. As a first step towards an understanding of this property we have examined the question of the relative importance of geometric and electronic factors in determining the barrier heights. The rhodium lattice constant is ~ 2% smaller than that of palladium so a facile explanation of the higher barrier in rhodium would simply be that the H has to "sqeeze" through a smaller triangle. We have repeated calculations for the octahedral and internal sites of $Rh_{10}H$, but using the palladium lattice constant. The barrier dropped to 0.7eV so, roughly speaking, the increase in the barrier on going from Pd to Rh may be attributed one-third to geometric and two-thirds to elec-tronic effects. Analysis of the energy levels and wave functions to de-termine the nature of the electronic effects has not yet been carried out. We hope it will prove to be enlightening.

We will close this section on a technical note. The BSSE is always present in basis set dependent calculations. However, if a basis set has a relatively extensive core and the fitting functions are of good quality, it can be kept under control. For example in our calculation on a Pd_8CO cluster we found this error to be 0.04eV. The relative binding energies with and without BSSE correction for $Rh_{10}H$ are reported in Table 1. Although the BSSE would not change any of the conclusions of our study it does influence the barriers for diffusion by as much as 10%. This error appears to be much smaller for Pd clusters[5] and is likely to be connected with the increased size of the Rh valence orbitals which can now more easily penetrate into the core region of the neighbours.

DIFFUSION OF CO ON PD CLUSTERS

Recently[4] we performed LCAO-MP-LSD model potential calculations for CO interacting with Pd clusters chosen to model the bridge site of the (100) surface. The character of the chemisorption bond which emerged from these calculations was discussed. It was postulated that the 4σ and 5σ levels contribute significantly to the bond. The contributions from the $2\pi^*$ orbitals were found in the d band and the main peak was localized ~ 3eV above the Fermi level. The σ-donation, π- back-bonding mechanism was clarified. The initial σ-repulsion affect destabilizes one of the σ-type d levels and this antibonding level appears above the Fermi level. Then the electrons of this level are transferred to new lower-lying π-type levels, thus enhancing back donation to 2π*. In addition the 5σ orbital polarizes strongly towards the oxygen atom i.e. mixes in 4σ character. In order to quantify these effects, in the present paper we report calculations with forced-occupancy of the σ-type antibonding level. This is clearly the reverse of the above mechanism and as a consequence the 2π* back-donation should be weaker. We obtained a binding energy of 1.0eV which should be compared with the value of 2.6eV for the fully relaxed, ground state configuration[4]. Therefore most (1.6eV) of the binding energy is gained through this change of configuration.

The adsorption energies (ΔE) reported in Ref.4 show a rather strong dependence on the cluster size, dropping from 4.8eV for Pd_2CO, to 3.8eV for Pd_4CO to 2.6eV for Pd_8CO. The experimental initial heat of chemisorption is 1.6eV. We have further investigated the convergence of ΔE for the bridge, B, position by performing calculations for $Pd_{14}CO$ (top row of Fig. 3) which contains all first neighbors of the bridging atoms. The ΔE obtained is 2.7eV, nearly the same as for Pd_8CO. Preliminary indications[24] are that adding some second neighbors, to form a $Pd_{20}CO$ cluster, and explicitly including only the sp "conduction" electrons, reduces ΔE to 2.1eV. This longer-ranged "conduction" electron contribution to ΔE should be quite similar for various adsorption sites so we believe the clusters shown in Fig. 3 are adequate for the purpose of mapping out the energetics of diffusion. A more complete discussion of the convergence of various properties with cluster size will be given elsewhere[24].

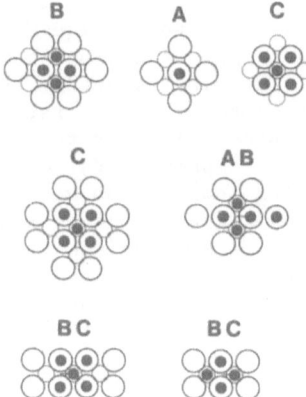

Fig.3 . Some cluster models used for the bridge (B), atop (A) and 4-fold centered (C) sites on Pd(100) and diffusion paths between them. The Pd atoms designated by a heavy dot are treated in the "best" manner (16 valence electron, pds, model potential; d-triple-zeta basis). The others are treated at a somewhat lower level (10 valence-electron, ds ,model potential; d-single-zeta basis).

Table 2. Equilibrum properties for CO adsorption at B,C and A
positions on Pd clusters.

		d_{C-Pd}	d_{C-S}	d_{C-O}	ω_{CO}	ω_{CO-S}	ΔE
		(Å)	(Å)	(Å)	(cm^{-1})	(cm^{-1})	(eV)
B	$Pd_{14}CO$	1.91	1.32	1.18	1826	501	2.7
C	Pd_9CO	2.73	0.88	1.21	1621	290	2.6
A	Pd_9CO	1.87	1.87	1.15	2064	402	1.7

The first question to be answered in a study of the diffusion path
is the relative energy of various sites, A(atop), B(bridge), C(4-fold
centered). Calculations have been performed using various clusters with
and without the adsorbate. In the latter case the ghost CO basis set
was retained, so the possible BSSE error was always corrected. The
clusters used are displayed in Fig.3 and some results are presented in
Table 2. As one can see, adsorption at the C position brings elongation
of the CO bond relative to the bridge site value and diminishes the CO
vibrational frequency. The effect of atop (A) adsorption is the
reverse; the CO bond shrinks and the vibrational frequency rises. These
features correspond well to CO properties at hollow, atop and bridge
positions known experimentally[28] and correlate with π^* populations
increasing in the order A,B,C.

As usual it is somewhat more difficult to evaluate the quality of
the calculated binding energies, even in a relative sense. It is very
similar for B and C positions and much smaller for A. We have carried
out a number of tests to establish the sensitivity of ΔE to the model
and computational details. Since the $Pd_9CO(C)$ cluster is missing many
neighbors of the five atoms that "touch" CO we performed calculations on
$Pd_{17}CO$ with C_{4v} symmetry at the equilibrum geometry of the Pd_9CO
cluster (Fig.3). We obtained a binding energy, 2.5eV, slightly lower
but still near 2.6eV. A more detailed comparison of B and C sites will
be presented below where the same cluster and geometry will be used to
study both.

We repeated calculations for A using a Pd_9 cluster but with the
best pds model potential for all the atoms. The CO basis set was augmented with additional diffuse p and s functions and the auxiliary set
was correspondingly extended. The binding energy changed by merely
0.1eV, showing that the effect of basis set is rather small. Finally we
set up a Pd_{10} cluster (Fig.3) which allows us to study adsorption at A
and B positions in parallel. We compared total energies of the CO
structures at their corresponding equilibrium geometries from independent cluster calculations. The A adsorption is destabilized by about
0.9eV with respect to B. Therefore we conclude that the LSD calculations provide a very unstable A position. Although there is no direct
experimental evidence against this last conclusion it is perhaps
somewhat surprising in view of the smaller anisotropy of CO adsorption
on other surfaces[28,29]. A comparison with nickel is interesting – the
diffusion activation energy for CO/Ni(100)[30] was found to be 0.2eV and
the diffusion path from bridge to top position was suggested. In
support of the relative instability of the A site for CO on Pd(100) one
can mention the semiempirical calculations by Doyen and Ertl[31] which are
in general agreement with our findings. The B and C positions are close
in energy and A is destabilized by ~ 0.4eV. Recent SCFMO pseudopotential calculations[32] resulted in a much more uniform potential
surface (within 0.1eV); however, the inclusion of correlation effects

and the saturation of the results with respect to cluster size would be required before definite conclusions could be reached. LSD calculations on the Ni—CO system would be very interesting and we hope to perform them in the future. On the Pd(100) surface it is generally accepted that the B adsorption dominates[28,33-35] at low coverage. There is a possibility of coexisting B and A species on Pd surfaces but this was observed at high coverage on supported Pd surfaces[36]. It is possible that in this case the adsorption energy at A-top may be comparable to B adsorption and we plan to investigate this effect.

We assume now that the possible diffusion on the Pd(100) surface takes place along a line connecting the B and C positions. The $Pd_{11}CO$ cluster used to study the diffusion path of CO is shown in Fig.3. We follow the path by placing the CO at the B site then between B and C sites at $M(\frac{1}{4})$, $M(\frac{1}{2})$, $M(\frac{3}{4})$ and finally at the C site. In each of these positions the distances of C from the surface (h) and the C—O bond length (d) are optimized assuming the molecule is adsorbed perpendicular to the surface. The possibility of tilted structures of CO was studied by optimizing the angle θ at fixed optimal h and d. We expect our search for the optimum geometry is accurate to within 0.05eV. In Fig. 4

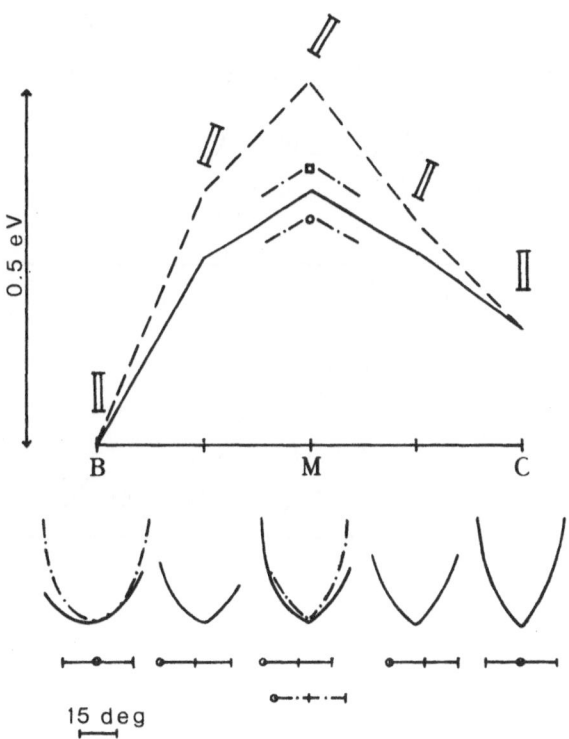

Fig. 4 Energy variation along a diffusion path between B and C sites for CO on Pd(100). (———) Pd_{11}+ CO, CO perpendicular to surface (———) Pd_{11}+ CO, CO at optimum tilt, (—·—□—·—) Pd_8 +CO, CO perpendicular to surface, (—··—○—··—) Pd_8+ CO, CO at optimum tilt. The lower part of the figure shows energy variation with tilt angle, the circle indicating zero tilt; (———)Pd_{11}+CO, (—·—)Pd_8CO.

the potential energy is presented. For movement in the perpendicular configuration a diffusion barrier of 0.5eV exists between the absolute minimum at the B site and the local minimum at C which is 0.16eV higher. The bending of CO is favorable energetically and it diminishes the barrier by as much as 0.14eV. We repeated the calculations using the Pd_8CO cluster which is now centered at the B site. We again found the diffusion barrier, at a somewhat lower height of 0.4eV for the perpendicular configuration; 0.3eV if tilting is allowed. So we have to attribute ~.1eV to cluster size and shape effects. In both cases however we found an unambiguous lowering of the barrier upon tilting of CO. Let us now summarize the diffusion process of CO along the B—C line.

CO stands perpendicularly at the B position. When the system is heated CO can move towards the C site where it is again perpendicular. On its way the carbon end of CO points towards the global minimum at the B position and gradually straightens up while approaching the C site. The tilted form of CO was also very recently found for CO diffusing on an Al surface[19]. We can rationalize why it happens by comparing the θ angle calculated so that the carbon end of the molecule points back to the bridge center, with the one obtained from the cluster calculations.

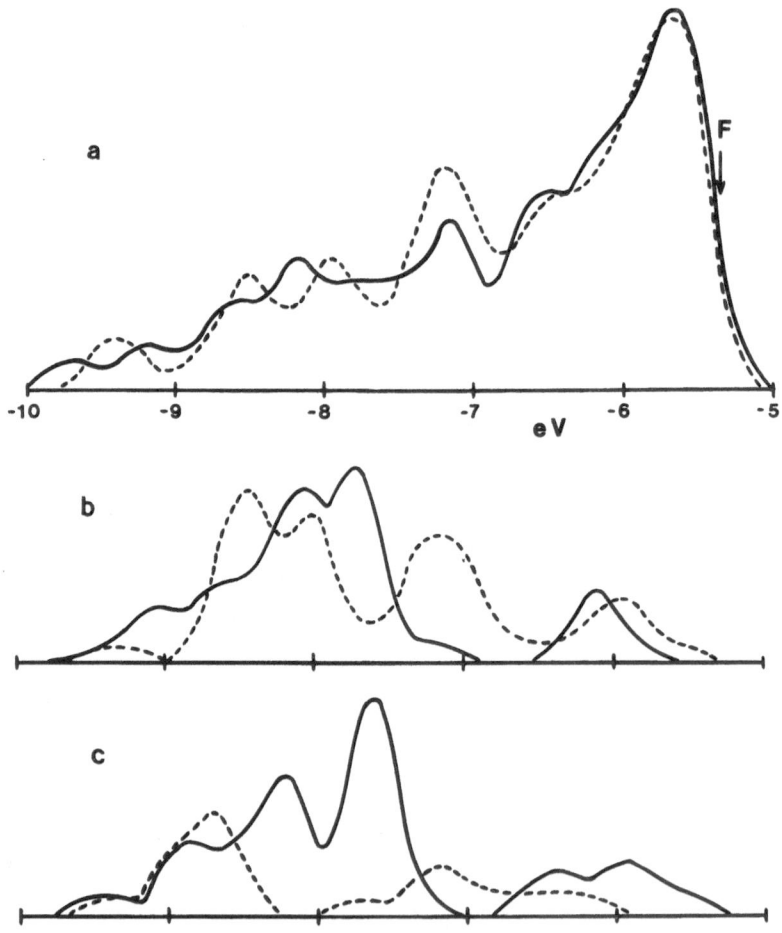

Fig. 5 a) Total density of states: (_____) $Pd^{11}CO$ (B) and (---) $Pd^{11}CO$ (C). b) Projected density of states on CO: (_____) $Pd_{11}CO$ (B) $Pd_{11}CO$ (C). c) Projected density of states on CO (_____) Pd_8CO (B) and (---) Pd_8CO (A).

The "geometrical" angle is 15°,28° and 43° for M($\frac{1}{4}$), M($\frac{1}{2}$) and M($\frac{3}{4}$) positions and this should be compared with the cluster results of 18°, 20° and 10°, respectively. The calculated angle depends somewhat on the cluster used (20° for $Pd_{11}CO$ and 14° for Pd_8CO). Although we can not give accurate angles, as this would require more accurate geometry optimization and likely larger clusters, we think that the qualitative picture will remain valid. The CO molecule tends to follow the B orientation presumably because it maximizes overlap of both 5σ and 2π* orbitals with the two Pd atoms at the B site. Apparently the influence of the B site dominates at M($\frac{1}{4}$) and then vanishes when CO approaches the C site.

As a final topic we will attempt to rationalize the differences in the binding at A,B and C sites by discussing the electronic structure by means of density of states (DOS) plots.The total DOS (Fig.5a) is plotted for B and C adsorption on the Pd_{11} cluster. Although one can see a slight shift of the total DOS towards lower energies for the B case and there are some changes in the shape of the plot which are caused by levels with CO character, the significance of these changes is not obvious from the total DOS per se. In fact it is presented primarily to situated the palladium d band. Projected densitites of states yield local information and are more informative. Those shown in Figs. 5b and 5c are obtained by weighting the DOS with the CO contributions, measured as a sum of Mulliken charges on C and O. These essentially display the "backbonding" component of the Pd-CO interaction - examination of the individual orbitals shows that the major peaks correspond almost exclusively to π* contributions, with some π contributions in the lower part of the band and smaller σ* terms scattered throughout. We first note, to help establish the appropriateness of the analysis, that the PDOS for the two B site calculations are essentially similar to each other and that there are distinct differences amongst the PDOS for different sites. We also note that the net Mulliken charges on CO are +0.04 for A, -0.26 for B and -0.51 for C, which provides a comparative measure of the net donation versus backdonation. The PDOS for the A site (Fig.5c) is much less intense than that for B or C, indicating a lower participation of π*, consistent with the low net CO population. The bonding, inner CO-like orbitals 1π and 5σ are somewhat (~ 0.06eV) more stable for the A site, compared with B. Although we are well aware of the possible dangers of basing arguments about total energies on one-electron eigenvalues, this may be an indication that the "donation" component of the bonding is stronger for the A site. In the end; however, the back-bonding dominates in the comparison and B is much more stable (by 0.9eV). The comparison between B and C (Fig. 5b) shows the importance of considering both donation and back-bonding components. In this case the bonding 1π and 5σ levels are stabilized more for the B adsorption, suggesting an increased strength of σ-donation. On the other hand, the PDOS show a somewhat greater 2π* contribution for C, consistent with the net Mulliken charges (-0.51(C), -0.26(B)). In rough terms then it would appear that the σ bonding is stronger for B, the π backbonding is stronger for C, the net result being similar binding energies for the two sites.

Perhaps further insight into the relative energies could be obtained by examining the PDOS for the palladium atoms involved in the bond, before and after chemisorption, an analysis which we hope to perform in the near future.

CONCLUDING REMARKS

The subjects discussed in this paper represent our first attempts to address questions related to surface dynamics using cluster models and the LCGTO-MP variant of local density functional theory. We realize that we have touched on only a small corner of the field and that even for the systems considered only small sections of the potential surfaces have been examined. Much remains to be done; however, at the current level of theory and methodology the results are already sufficiently accurate to make meaningful direct comparison with experimental data for extended systems and to provide encouragement for future work. Such work should certainly include relaxation and vibration of the surface atoms and the coupling of these with the various adsorbate motions.

There is much current interest and activity aimed at improving the theory (non-local corrections), and the methods used to implement it (gradient calculations, relativistic programs, perhaps the new "combined" molecular dynamics - LSD approach[37]). We are optimistic that this effort will bear fruit in terms of even more accuracy and ease of computation. Taken together with the steady advance in computer technology the "pie-in-the-sky" dream of computing and understanding surface dynamics from first principles may not be so unrealistic after all.

ACKNOWLEDGEMENTS

N.A. Baykara and S.Z. Baykara have contributed substantially to the hydrogen diffusion project. We have had many interesting discussions with Dr. Francis Raatz on the Pd-CO project. We are grateful to the Natural Sciences and Engineering Research Council of Canada, to the Fonds FCAR of the government of Québec, and to the Institut Français du Pétrole for support of this work, and to the Centre de Calcul de l'Université de Montréal for supplying computing resources.

REFERENCES

1. J.C. Tully, Ann. Rev. Phys. Chem., 31, 319 (1980); D.C. Clary and A.E. DePristo, J. Chem. Phys., 81, 5164 (1984).
2. J.-H. Lim and B.J. Garrison, J. Chem. Phys., 80, 2904 (1984); C.-Y. Lee and A.E. DePristo; J. Chem. Phys., 84, 485 (1986); J.G. Lauderdale and D.G. Truhlar, J. Chem. Phys., 84, 1843 (1986).
3. D.R. Salahub "Applied Quantum Chemistry" V.H. Smith, K. Morokuma and H.F. Schaefer III eds., Reidel, Dordrecht (1986) p.185 and references therein.
4. J. Andzelm and D.R. Salahub, Intern. J. Quantum Chem., 29, 1091 (1986).
5. N.A. Baykara, J. Andzelm, D.R. Salahub and S.Z. Baykara, Intern. J. Quantum Chem., 29, 1025 (1986).
6. A.D. Becke, J. Chem. Phys., 84, 4524 (1986).
7. D.R. Salahub, Adv. Chem. Phys., 69, (1987), in press, and references therein.
8. D.C. Langreth and M.J. Mehl, Phys. Rev., B28, 1809 (1983).
9. A.G. Noumouets and Y.S. Vedula, Surface Sci. Rept. 4, 365 (1985).
10. R. Lewis and R. Gomer, Surface Sci. 17, 333 (1969). M. Tringides and R. Gomer, Surface Sci. 155, 254 (1985). J.R. Chen and R. Gomer, Surface Sci. 81, 589 (1979).
11. R. Viswanathan, D.R. Burgess Jr., P.C. Stair and E. Weitz, J. Vac. Sci. Technol 20, 605 (1982).
12. S.M. George, A.M. DeSantolo and R.B. Hall, Surface Sci. 159, L425 (1985).
13. C.H. Mak, J.L. Brand, A.A. Deckert and S.M. George, J. Chem. Phys. 85, 1676 (1986).

14. T.E. Madey, J. Vac. Sci. Technol. $\underline{A4}$, 257 (1986).

15. S.L. Tang, M.B. Lee, Q. Y. Yang, J.D. Beckerle and S.T. Ceyer, J. Chem. Phys. $\underline{84}$, 1876 (1986).

16. T.H. Upton and W.A. Goddard III, Phys. Rev. Lett., $\underline{42}$, 472 (1979); and in "Chemistry and Physics of Solid Surfaces", Vol III, R. Vanselow and W. England, eds. (CRC Press, Boca Raton, 1982).

17. D. Post and E.J. Baerends, J. Chem. Phys., $\underline{78}$, 5663 (1983).

18. P.-L. Cao, D.E. Ellis, A.J. Freeman, Q.-Q. Zheng and S.D. Bader, Phys. Rev., $\underline{B30}$, 4146 (1984).

19. B.N.J. Persson and J. E. Müller, Surface Sci., $\underline{171}$, 219 (1986).

20. J.P. Dahl and J. Avery, eds, "Local Approximations in Quantum Chemistry and Solid State Physics", Plenum, New York, 1984.

21. J. Andzelm, E. Radzio and D.R. Salahub, J. Chem. Phys., $\underline{83}$, 4573 (1985).

22. C.W. Bauschlicher Jr., J. Chem. Phys., $\underline{84}$, 250 (1986); Chem. Phys. Lett., $\underline{129}$, 586 (1986).

23. W. Ravenek and F.M.M. Geurts, J.Chem. Phys., $\underline{84}$, 1613 (1986).

24. J. Andzelm and D.R. Salahub to be published.

25. G.D. Kubiak and R.H. Stulen, J. Vac.Sci Technol. $\underline{A4}$, 1427 (1986).

26. S.M. Foiles and M.S. Daw, J. Vac. Sci. Technol. $\underline{A3}$, 1565 (1985).

27. J. Andzelm, N.A. Baykara, S.Z. Baykara and D.R. Salahub, unpublished

28. S. Ishi, Y. Ohno and B. Viswanathan, Surface Sci. $\underline{161}$, 349 (1985).

29. E. Shustorovich, Surface Sci. Rept. $\underline{6}$, 1 (1986).

30. D.A. Mullins, B.Roop and J.M. White, Chem. Phys. Lett., $\underline{129}$, 511 (1986).

31. G. Doyen and G. Ertl, Surface Sci. $\underline{69}$, 157 (1977).

32. A. Gavezzotti, G.F. Tantardini and M. Simonetta, Chem. Phys. $\underline{105}$, 333 (1980).

33. R.J. Behm, K. Christmann, G. Ertl and M.A. VanHove, J. Chem. Phys., $\underline{73}$, 2989 (1986).

34. A. Ortega, F.M. Hoffmann, A.M. Bradshaw, Surface Sci., $\underline{119}$, 79 (1982).

35. A. Brown, J.C. Vickerman, Surface Sci., $\underline{151}$, 319 (1985).

36. P. Gelin, J.T. Yates Jr., Surface Sci., $\underline{136}$, L1 (1984).

37. R. Car and M. Parrinello, Phys. Rev. Lett., $\underline{55}$, 2471 (1985).

LOCAL DENSITY MODELS FOR BARE AND LIGATED

TRANSITION METAL CLUSTERS

D.E. Ellis[1], H.P. Cheng[1], and G.F. Holland[2]*

[1]Department of Physics and Astronomy
and Materials Research Center
[2]Department of Chemistry
Northwestern University
Evanston, IL 60201

INTRODUCTION

Transition metals exhibit interesting and technically useful chemical interactions both among themselves and with other atoms. Their high strength and corrosion resistance makes them essential structural materials, and their chemical reactivity and selectivity give them a central role as catalytic promoters of reactions. These special properties are popularly associated with the presence of a semilocalized, partially occupied nd electronic configuration. However, the nature of the metal-metal bond, and the precise character of electronic interactions associated with cohesion and bonding in the transition metals (TM) has not yet been fully elucidated. In particular, the properties of TM surfaces and small particles (which are essentially all surface) are at present very poorly understood. There are, of course, large scale experimental and theoretical efforts underway to understand and control electronic properties of these materials.

In this brief report, we discuss some recent results in theoretical modeling of TM clusters and ligated species, in the context of the self-consistent-field local density (LD) theory. The work reported here was carried out using the discrete variational (DV-Xα) method with numerical atomic/ionic basis functions, which is fully described in the literature[1-5]. The energy levels, charge and spin distribution, and cohesive energies of bare metal particles are discussed in Sec. II. In Sec. III, the effects of coordination to ligands such as CO and C_6H_6 are examined and the results of both experimental and theoretical studies are briefly presented. Finally, in Sec. IV, we consider relativistic effects on the platinum system, and give some preliminary results in the Dirac-Slater scheme.

*Present address:
Department of Chemistry
University of California
Berkeley, CA 94700

Dimers of the first transition metal (TM) series provide a natural starting point for discussing properties of bare particles, for which some experimental data are also available. The incomplete 3d shell and small number of atoms leads to a dense set of multiplet levels lying close to the ground state. This level density and open shell structure has made it difficult for the traditional methods (configuration interaction, valence bond theory) to obtain reliable results[6-9]. On the other hand, the rather high level density is coarsely averaged by the effective exchange and correlation potentials of local density (LD) theory and surprisingly accurate results for cohesive energy and bond length have been calculated.

In the celebrated case of Cr_2 the self-consistent-field LD results provide an "antiferromagnetic" broken symmetry solution. Here, one-electron orbitals of opposite spin are partially localized on different centers for bond lengths near equilibrium[10]. In the case of Cu_2, with the d^{10} configuration, exchange energy plays no special role and a simple diamagnetic solution is found[5]. In Fig. 1, we show a binding energy curve for the nickel dimer, in both spin-restricted and unrestricted models. Starting with the nominal d^8s^2 atomic configuration, and using an extended 3d, 4spdf numerical basis, we find a net d population of 9.35, with minority spin contributions in the diffuse 4sp components near equilibrium. With an atomic moment of 1.07 μ_B, this represents a "low spin" system where spin polarization leads to a small but significant contribution to cohesion.

Let us turn now to trimers and tetramers, using the $5d^{10}6s^0$ platinum atom as an example. The results presented here were all obtained in the nonrelativistic (NR) LD model, which is certainly inadequate for a quantitative description of cohesion. However, we think it is useful to have reference calculations in order to clearly separate relativistic factors from other features (e.g. expansion of 5d orbitals relative to 4d) specific to the atomic number. In Table 1 are given the energy levels in the vicinity of the Fermi energy, the Mulliken populations, binding energy per pair of atoms, and equilibrium interatomic distances for dimer, equilateral triangle, and regular tetrahedron. The binding energies reported are with respect to the spin-restricted $d^{10}s^0$ Pt atomic configuration. The binding energy curves $E_b(R)$ were fitted to a sum of pairwise interactions of the form

$$E_b(R) = \sum_{i<j} E_2(R_{ij}) \tag{1}$$

where

$$E_2(R) = E_0 - A\,e^{-\beta(R-R_0)} + \frac{C}{R^\alpha} \tag{2}$$

These pairwise potentials were found to be remarkably similar for the three clusters, leading to the hope that, with slightly larger number of atoms, one may extrapolate to alloy and surface environments.

It is well known that a sum of pairwise interactions is inadequate for accurate simulation of bulk metal vibrational and thermodynamic properties. The additional "many-body" interactions can be modelled in many ways, and no general consensus seems to exist on optimal procedures. In our present approach, the "many-body" interactions appear as corrections to the summed two-body interactions optimized for the dimer. Thus, in Fig. 2, the deviation between the binding energy curves for different shapes of the four-particle system provides data for the determination of corrections to the E_2 sum of Eq. (1). We have found that the "embedding atom" approach derived from considerations of an isolated atom sitting in a uniform electron gas[12] provides a useful starting point for construction of a reasonably accurate correction scheme. Provisionally, we take a density-dependent parametrized form $F = F(\rho, Z)$ and obtain a representation of the function from data such as shown in Fig. 2. This function is then added to the pairwise interactions of Eq. (1). A full report on this subject will be presented elsewhere.[13]

LIGATED CLUSTERS

In this section, we wish to discuss an aspect of TM particles familiar to inorganic chemists – the structure and properties of cluster compounds. The most famous examples are the $M_n(CO)_m$ carbonyls and their derivatives for which stable single crystals can be produced for X-ray and spectrographic studies. Two primary questions about the electronic structure arise immediately: what is the nature of metal-metal bonding in the metallic "core", and how is the core modified by interaction with ligands? These questions are amenable to both experimental and theoretical attack, and we have presented data for $Ru_3(CO)_9$ and the analogous Os compound as well as the cobalt dodecacarbonyl $Co_4(CO)_{12}$ system[14]. When treated in isolation, the metal fragment bears some resemblance to the bulk metal – s to d conversion and presence of an extended nearly uniform "conduction electron" density in the interior of the particle. Generally speaking, the metal atoms bear a magnetic moment not very different from that of the isolated atom, since with few bonding partners it is energetically favorable to produce localized spin-polarized orbitals. With the addition of ligands, the magnetization is suppressed, essentially in a local atom by atom response, and the saturated system is typically diamagnetic.[15]

Modifications of the core density occur upon substitution by more or less electrophilic ligands, such as halides, hydrocarbons and phosphines. In Fig. 3 are shown the benzene-substituted "pi-capped" compound $Co_4(CO)_9(C_6H_6)$ and the corresponding "tripod" compound $Co_4(CO)_9(C_6H_6)[(PH_2)_3CH]$. The tripod compound is experimentally attractive, in that the phosphine "legs", often taken as triphenylphosphine, stabilize the heavier TM compounds without significantly distorting their geometry or spectroscopic properties. From another point of view, we can consider this species as the beginning of a detailed model for the attachment of a metal particle to a surface. Changes in the metal core density of states and the carbonyl- derived level structure are surprisingly small when comparing substituted to nonsubstituted species. As seen in Fig. 4, the additional hydrocarbon and phosphine-derived states essentially interleave the pre-existing metal and CO band structures with a minimum of modification. This result is consistent with the known minimal changes in spectroscopic properties of the substituted compounds.

Electrochemical measurements on these and related compounds show a strong correlation between the electrochemical redox potentials and the net charge (electron withdrawal) on the metal core[16]. This suggests that it may be possible to use LD methods to aid interpretation of electrochemical properties of ligated metal particles in solution. Unfortunately, very little work has been done in this area up to the present because of the perceived difficulties in modelling the solvent environment.

RELATIVISTIC STUDIES

The self-consistent-field Dirac-Slater (DS) model provides a fully relativistic analog to nonrelativistic (NR) local density approaches. In our implementation of the DS model, four-component solutions of free atoms/ions embedded in a potential well are used as basis functions in a variational solution of the Dirac equation. The use of numerical solutions of a central field Dirac problem bypasses all the well-known numerical and convergence difficulties associated with the use of analytic bases.[17] Since the wavefunctions are complex with mixed spin, the computation time in DS calculations is somewhat greater than for the NR case, but it is entirely feasible to treat systems with, say N < 30, heavy atoms on present day machines.

In Table 2, are given the one-electron energy levels of both NR and DS solutions for the platinum atom. These data illustrate the major consequences of relativistic effects, increase of binding of core levels, and rearrangement of valence levels due to inner-level shielding changes. In particular, the relative motion of 5d, 6s, and 6p levels in Pt can be

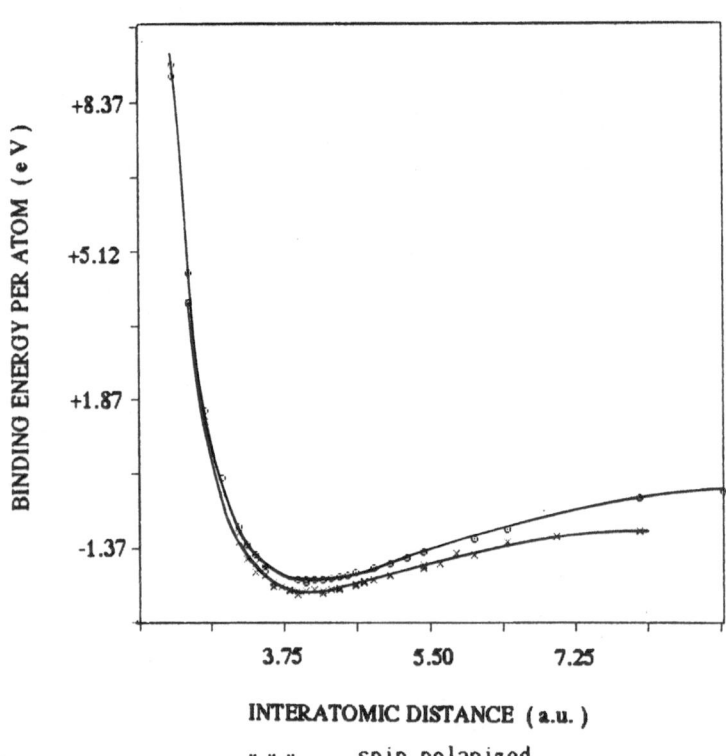

1.   Binding energy versus interatomic distance (a.u.) for Ni_2;
   o = non—spin polarized,
   x = spin polarized.

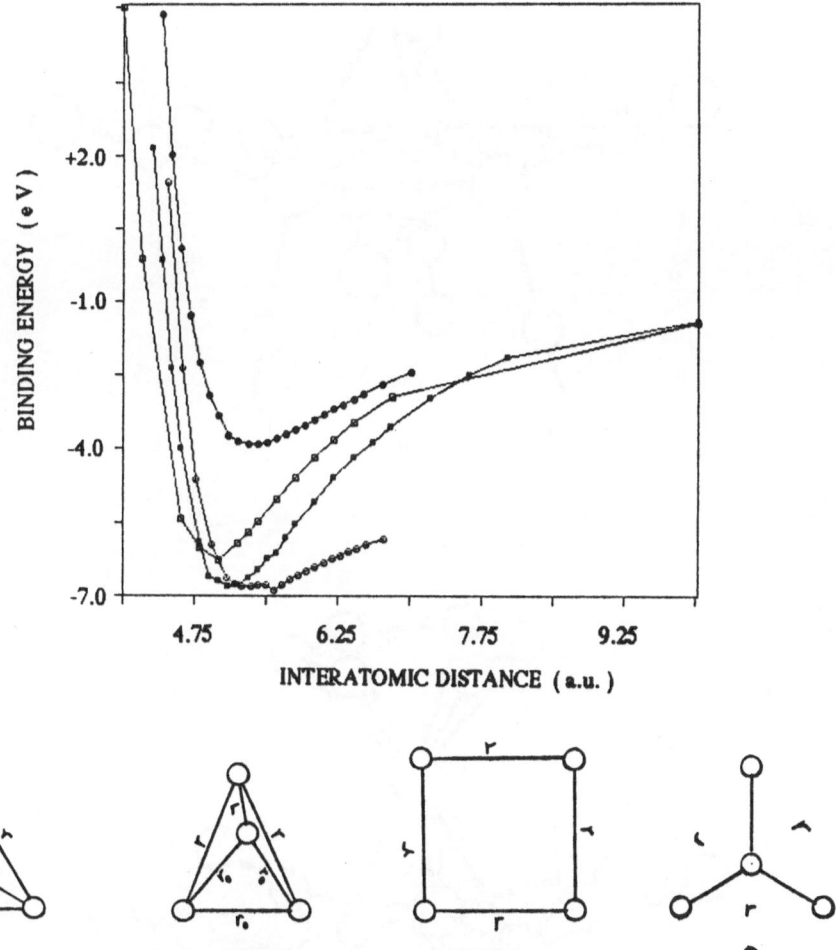

2. Variation in binding energy of the Pt_4 particle for different
    geometrical configurations. Binding energies are here calculated
    with respect to zero energy at $R \rightarrow \infty$. Note that the distorted
    tetrahedron does <u>not</u> dissociate to free atoms, since three basal
    plane Pt are held fixed.

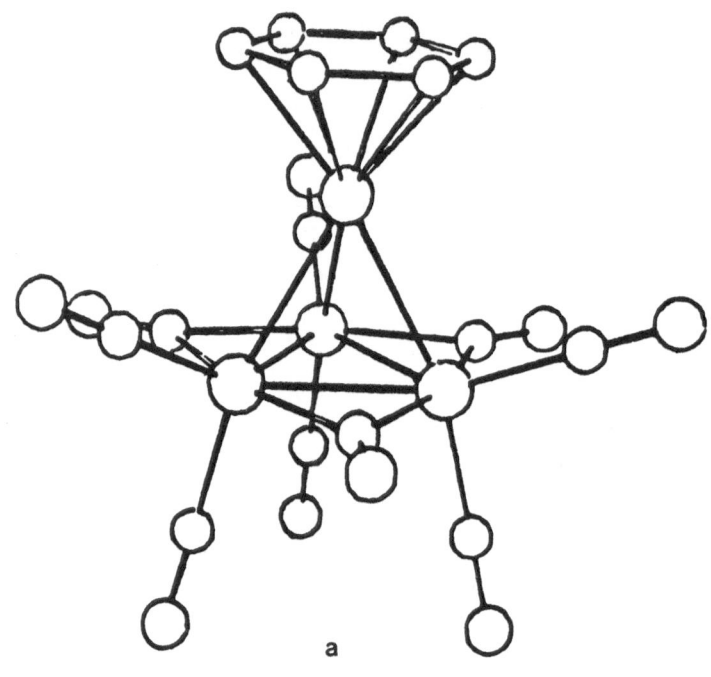

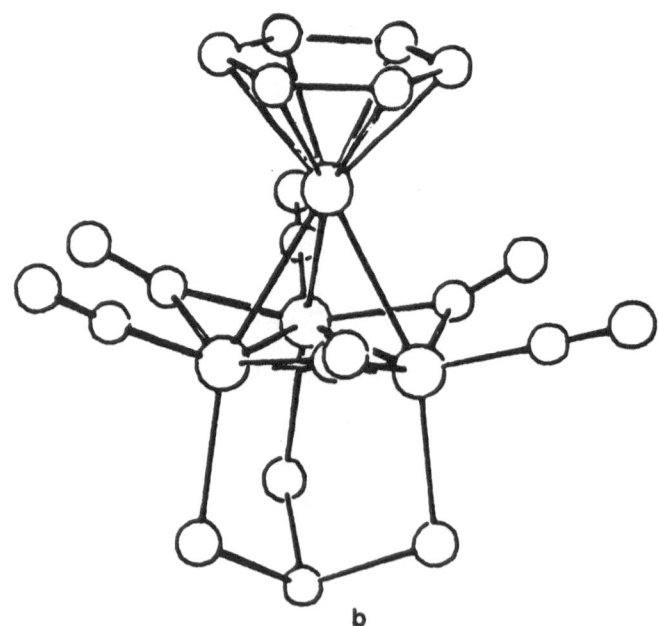

3. Nuclear arrangement for :
   (a) $Co_4(CO)_9(C_6H_6)$ and
   (b) $Co_4(CO)_9(C_{6s}H_6)[(PH_2)_3CH]$.

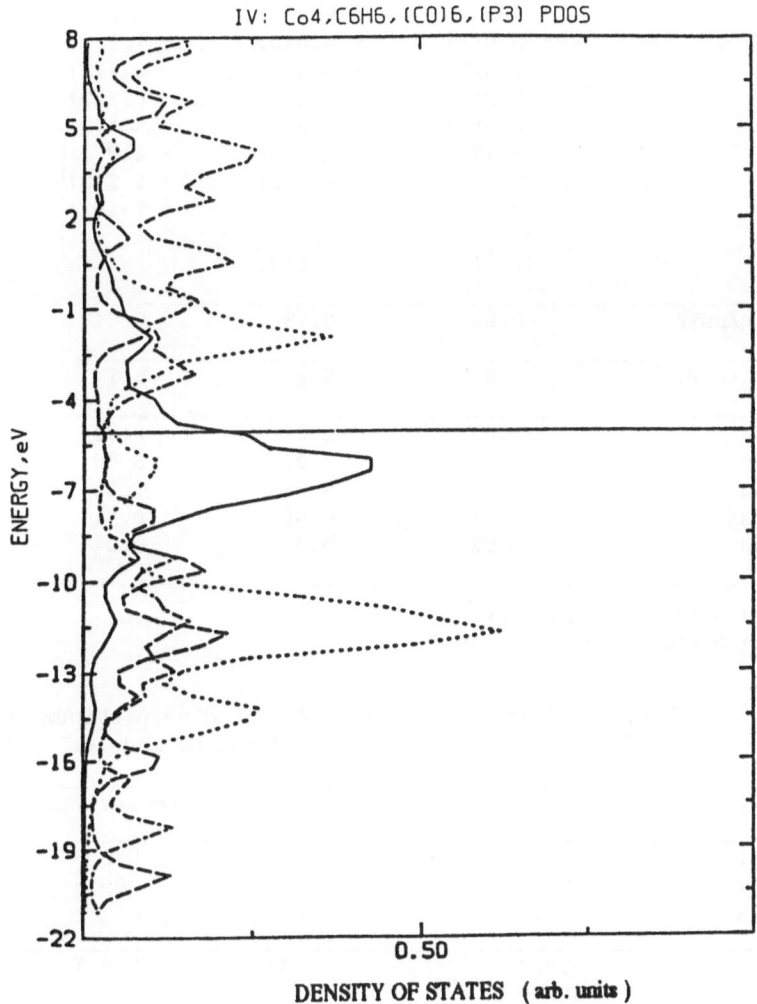

4. Partial densities of states for the cobalt tripod compound (Ref. 15).
   Co_4 - solid curve
   $(CO)_6$ - dotted
   $\eta-C_6H_6$ - dashed
   $(PH_2)_3CH$ - dot-dashed

TABLE 1 – Valence energy levels, binding energy (eV), Mulliken orbital populations, and equilibrium interatomic distances for Pt_2, Pt_3, and Pt_4. A nonrelativistic spin restricted model with $\alpha=0.7$ was used.

	Pt_2	Pt_3	Pt_4
ϵ_i	−75.4(2)[a]	−77.8(3)[a]	−78.7(4)[a]
	−46.5(6)	−48.7(9)	−49.6(12)
	− 4.3	− 6.7	− 8.2
	− 3.8(2)	− 6.0(3)	− 7.1(3)
	− 3.4(2)	− 5.4	− 6.7(2)
	− 3.2(2)	− 5.3(2)	− 6.2(3)
	− 2.8(2)	− 4.6(2)	− 5.0(8)
	− 2.7[b]	− 4.4(3)	− 4.9
	− 2.2	− 4.3(2)	− 4.2(3)[b]
	− 0.1	− 3.8[b]	− 1.1(2)
		− 3.7	
		− 0.8(2)	
E_b/pair	0.80	0.79	0.91
R_e(a.u)	5.4	5.2	5.1
5s	2.00	2.00	2.00
5p	6.00	6.00	6 00
5d	9.87	9.53	9.43
6s	0.11	0.44	0.54
6p	0.02	0.03	0.04

a) orbital degeneracy
b) highest occupied level

TABLE 2 – Orbital and total energies (eV) for d^8s^2 platinum atom ($\alpha = 0.7$, spin restricted) in NR and DS models.

level	relativistic	level	nonrelativistic
4s 1/2	695.6	4s	578.5
4p 1/2	587.4	4p	485.6
3/2	496.4		
4d 3/2	318.8	4d	314.7
5/2	302.0		
4f 5/2	74.6	4f	86.3
7/2	71.1		
5s 1/2	103.8	5s	83.1
5p 1/2	69.0	5p	53.9
3/2	53.8		
5d 3/2	8.2	5d	9.4
5/2	6.8		
6s 1/2	5.9	6s	4.6
6p 1/2	1.1	6p	0.9
3/2	0.5		
E_{tot} (x10^5)	5.0192786		4.7169316

888

TABLE 3 – Valence energy levels, binding energy (eV), and Mulliken orbital populations for Pt_2 in the relativistic DS scheme. Bond length was chosen as R=4.5 a.u. A spin restricted model with $\alpha=0.7$ was used. The deep-lying levels are 5p in character.

ε_i	ε_i	ε_i	ε_i
-66.7	-10.0	-5.7	-1.7
-66.6	- 7.9	-5.4	-1.3
-51.9	- 7.4	-5.1	-0.2
-51.6	- 6.9	-5.1[a]	
-51.5	- 6.9	-4.5	
-51.5	- 6.0	-3.7	

$E_b = -2.2$ eV

Populations	
$5p_{1/2}$	2.00
3/2	4.00
$5d_{3/2}$	3.85
5/2	4.76
$6s_{1/2}$	1.13
$6p_{1/2}$	0.09
3/2	0.18

a) last occupied level

TABLE 4 – Valence energy levels (eV) and Mulliken populations for Pt_3H in equilateral geometry, R(Pt-Pt)=R(Pt-H)=4.95 a.u. A nonrelativistic spin restricted model with $\alpha=0.7$ was used.

ε_i	ε_i	ε_i
-76.3(3)[a]	-8.3	-4.2(2)
-48.0	-7.0	-4.0
-47.7(2)	-6.3(2)	-4.0(2)
-47.5	-5.8	-3.4[b]
-47.5(2)	-5.6	-3.3
-47.5	-5.3(2)	-2.4(2)
-47.4	-4.3(2)	-0.5

Populations

Pt	5s	2.00	H 1s	1.12
	5p	6.00		
	5d	9.06		
	6s	0.61		
	6p	0.29		

a) Orbital degeneracy
b) Last (singly) occupied level

889

expected to lead to significant differences in bonding schemes found in molecular calculations. The 5d-6s "gap" is reduced from the NR value of 4.8 eV to 0.8 eV for the DS model, strongly hinting at increased s-d hybridization in relativistic schemes.

Valence energy levels and populations for Pt_2 in the DS model are given in Table 3 for a somewhat arbitrarily chosen interatomic distance. Aside from the separation of the (nl) Mulliken populations into the (nlj) relativistic sublevels, we can clearly see the realignment of d- versus s-occupancy in the two models. We observe a significant increase in binding, to ~ 2.2 eV, which is however still less than the experimental value of 3.6 eV[11]. Further calculations are in progress to check consequences of basis set optimization, potential expansion, and other computational parameters.

Finally, in Table 4, we present some preliminary data for the Pt_3H molecule which is part of an ongoing study of reactivity of small particles. In addition to charge transfer of ~ 0.1 e onto the hydrogen atom, we note a transfer of 5d → 6s character among the Pt atoms. Thus it appears that the d-d and d-s interactions among the metals may be strongly modified by the presence of bonding ligands.

ACKNOWLEDGEMENTS

This work was supported in part by the NSF through the Northwestern University Materials Research Center, grant No. DMR82-16972.
G.F. Holland acknowledges support by the NSF through grant No. DMR82-14966. DEE thanks the Allied-Signal Engineered Materials Research Center for use of their facilities.

REFERENCES

1. E.J. Baerends, D.E. Ellis and P. Ros, Chem. Phys. 2:41 (1973).
2. A. Rosen, D.E. Ellis, H. Adachi and F.W. Averill, J. Chem. Phys. 65: 3629 (1976).
3. B. Delley and D.E. Ellis, J. Chem. Phys. 76:1949 (1982).
4. D.E. Ellis and B. Delley, "Local Density Approximations" in "Quantum Chemistry and Solid State Physics", ed. by J.P. Dahl and J. Avery Plenum, NY (1984).
5. B. Delley, D.E. Ellis, A.J. Freeman, E.J. Baerends, and D. Post, Phys. Rev. B27:2132 (1983).
6. M.M. Goodgame and W.A. Goddard, III, Phys. Rev. Lett. 48:135 (1982).
7. J. Harris and R.O. Jones, J. Chem. Phys. 70:830 (1979).
8. P. Joyes and M. Leleyter, J. Phys. B6:150 (1973).
9. M.F. Guest, I.H. Hillier, and C.P. Garner, Chem. Phys. Lett. 48:587 (1977).
10. B. Delley, A.J. Freeman, and D.E. Ellis, Phys. Rev. Lett. 50:488 (1983).
11. S.K. Gupta, B.M. Nappi, and K.A. Gingerich, Ame. Chem. Soc. Inorg. Chem. 22:996 (1981).
12. M.S. Daw and M.I. Baskes, Phys. Rev. B29:6443 (1984).
13. D.E. Ellis and H.P. Cheng, manuscript in preparation.
14. B. Delley, M.C. Manning, D.E. Ellis, J. Berkowitz, and W.C. Trogler, Inorg. Chem. 21:2247 (1982).
15. G.F. Holland, D.E. Ellis, and W.C. Trogler, J. Chem. Phys. 83:3504 (1985).
16. G.F. Holland, D.E. Ellis, and W.C. Trogler, J. Am. Chem. Soc. 108:1884 (1986).
17. A. Rosén and D.E. Ellis, J. Chem. Phys. 62:3039 (1975); D.E. Ellis, J. Phys. B10:1 (1977); D.E. Ellis, in Actinides in Perspective, ed. by N.M. Edelstein, Pergamon, NY (1982); D.E. Ellis and G.L. Goodman, Intl. J. Quantum Chem. 25:185 (1984); D.E. Ellis in Handbook on the Physics and Chemistry of the Actinides, ed. by A.J. Freeman and G.H. Lander, North Holland, Amsterdam (1985).

ACTIVATION OF C-H AND C-C BONDS IN ALKANES BY HETERODINUCLEAR METAL CLUSTER IONS IN THE GAS PHASE

Y. Huang, S.W. Buckner and B.S. Freiser[*]

Department of Chemistry, Purdue University

W. Lafayette, IN 47907

We report on the special reactivity of $RhFe^+$, $RhCo^+$, and $LaFe^+$ with alkanes and oxygen containing compounds. These three clusters are the first heterodinuclear transition metal cluster ions observed to activate C-H and C-C bonds in alkanes. All three cluster ions react with alkanes larger than methane mainly via dehydrogenation. In this respect, the reactivity of $RhFe^+$, $RhCo^+$, and $LaFe^+$ is more like that of Rh^+ and La^+ than that of Fe^+ and Co^+. $RhFe^+$ and $RhCo^+$ show similar patterns of reactivity with multiple dehydrogenation of alkanes larger than ethane, while $LaFe^+$ shows less extensive dehydrogenation. n-Pentane and cyclopentane react with $RhFe^+$ and $RhCo^+$ to split the cluster and form $RhC_5H_6^+$ and $RhC_5H_5^+$, presumably the cyclopentadiene- and cyclopentadienyl-rhodium cations. These reactions imply $D°(Rh^+\text{-cyclopentadienyl}) > 189$ kcal/mol. Other thermochemical limits are reported including $D°(RhCo^+\text{-}CH_2) > 93$ kcal/mol, $D°(RhFe^+\text{-}CH_2) > 78$ kcal/mol, and $D°(RhFe^+\text{-benzene}) > 60$ kcal/mol. The special reactivity is discussed in terms of reduced overlap in cluster ions containing a second or third row metal bound to a first row metal relative to clusters containing two first row metals. Finally, oxide abstraction is observed for reaction of the clusters with ethylene oxide suggesting $D°(MM'^+\text{-}O) > 89$ kcal/mol.

INTRODUCTION

The study of bare atomic metal ions and small metal ion clusters has recently become the focus of intense investigation due to their importance in such areas as organometallic chemistry and catalysis [1-3]. In general, gas phase atomic transition metal ions have been found to be very reactive with alkanes, exhibiting both C-H and C-C bond activation [3]. In contrast the dimeric ions $FeCo^+$ [4], Co_2^+, Mn_2^+ [5], VFe^+ [6], $CuFe^+$, $ScFe^+$, $TaFe^+$, and $NbFe^+$ [7] are unreactive with alkanes. This behavior is exactly reversed for that observed for neutrals where, for example, the homonuclear dimers of iron and nickel are reactive towards alkanes while the atomic species are not [8-10]. Interestingly, Co_2CO^+ [11] and Co_2Fe^+ [12] react with alkanes predominantly by attacking C-H bonds. The reactivity of $Co_2(CO)^+$ was attributed to polarizing effects of the CO ligand, while for the trimer species electronic factors probably play a key role. As part of our continuing effort to understand these variations in reactivity, we wish to report here on three bare dimeric cluster ions, $RhFe^+$, $RhCo^+$, and $LaFe^+$, which are observed to react with alkanes.

EXPERIMENTAL

All experiments were performedon a prototype Nicolet FTMS-1000 Fourier transform mass spectrometer equipped with laser ionization for production of metal ions [13]. The clusters were prepared and isolated using a method previously described [14]. Briefly, laser generated metal ions undergo clustering reactions (1)-(5) with $Fe(CO)_5$ and $Co(CO)_3NO$ in the gas phase eliminating one or more carbonyl groups. Subsequent collision-induced dissociation

$$La^+ + Fe(CO)_5 \xrightarrow{100\%} LaFe(CO)_3^+ + 2CO \qquad (1)$$

$$Rh^+ + Fe(CO)_5 \xrightarrow{6\%} RhFe(CO)_3^+ + 2CO \qquad (2)$$

$$\xrightarrow{94\%} RhFe(CO)_2^+ + 3CO \qquad (3)$$

$$Rh^+ + Co(CO)_3NO \xrightarrow{39\%} RhCo(CO)_2NO^+ + CO \qquad (4)$$

$$\xrightarrow{61\%} RhCo(CO)NO^+ + 2CO \qquad (5)$$

(CID) of the product ions strips off the remaining CO and NO ligands to generate the bare metal cluster ions, which are then isolated by swept double resonance ejection pulses [15,16].

$Fe(CO)_5$ and $Co(CO)_3NO$ were introduced into the cell via a pulsed valve [17] in order to prevent subsequent reaction of the isolated cluster ions with any background metal carbonyl neutral gas. Organic and oxide reagent gases were present in the cell at a static pressure of $\sim 2 \times 10^{-7}$ torr. A static pressure of argon was introduced to bring the total pressure to $\sim 1 \times 10^{-5}$ torr. Argon served both as a collision gas for CID and as a buffer gas to collisionally cool any excited cluster ions. Under the present conditions, the bare cluster ions undergo ~ 150 thermalizing collisions with background argon and reagent before isolation. This is expected to cool most of the ions [18], but a small population of nonthermal ions cannot be completely ruled out.

RESULTS AND DISCUSSION

Primary product distribution for the reactions of $RhFe^+$, $RhCo^+$, and $LaFe^+$ with linear, cyclic, and branched alkanes are shown in Table I. All three cluster ions react with alkanes larger than methane, mainly via dehydrogenation. In this respect the reactivity of the cluster ions more closely resembles that of Rh^+ [19] and La^+ [20] than Fe^+ [21] and Co^+ [21]. Multiple dehydrogenation is more prevalent for the rhodium cluster ions than for $LaFe^+$. C-C cleavage products are observed in those limited cases where the C-C bond is expected to be particularly susceptible to attack.

Interestingly, pentane and cyclopentane react with $RhFe^+$ and $RhCo^+$ to split and cluster and form, presumably, $Rh(cyclopentadiene)^+$ and $Rh(cyclopentadienyl)^+$. From $D^o(Rh^+-Fe) = 69 \pm 5$ kcal/mol [22], and $D^o(Fe-H) = 29 \pm 3$ kcal/mol [24], values of $D^o(RhM^+-cyclopentadiene) > 67$ kcal/mol and $D^o(Rh^+-cyclopentadienyl) > 189$ kcal/mol are implied from the reaction of $RhFe^+$ with pentane [25], in accord with $D^o(Rh^+-cyclopentadienyl) > 111$ kcal/mol reported earlier [26].

The reactivity of $RhFe^+$ and $RhCo^+$ with ethylene oxide and O_2 is interesting in as much as Rh^+ [27], Fe^+ [28], and Co^+ [29] do not abstract oxygen from either compound. This behavior, however, has been noted earlier for other dimeric and trimeric metal cluster ions [30]. Oxygen abstraction from ethylene oxide occurs for both $RhFe^+$ and $RhCo^+$ indicating $D^o(RhM^+-O)$ (M=Fe,Co) > 89 kcal/mol [25]. Direct oxygen abstraction by the rhodium cluster ions from O_2 is not observed, possibly suggesting $D^o(MM'^+-O) < 119$ kcal/mol. However, the presence of competing reactions makes this upper limit uncertain. These values as well as other thermochemical values derived from the reactions of these cluster ions are listed in Table I.

Table I. Primary Product Distributions and Pertinent Thermochemical Limits for Reactions of MM'^+ with Various Hydrocarbons.

Reactant	Product	$LaFe^+$	$RhFe^+$	$RhCo^+$	Implications (if observed)[a,b,c]
CH_4		NR	NR	NR	
C_2H_6	$MM'C_2H_4^+ + H_2$	100	100	100	$D°(MM'^+-ethene) > 33$ kcal/mol
C_3H_8	$MM'C_3H_6^+ + H_2$	100	92	83	$D°(MM'^+-propene) > 30$ kcal/mol
	$MM'C_3H_4^+ + 2H_2$		8	17	$D°(RhM^+-allene) > 69$ kcal/mol
$n-C_4H_{10}$	$MM'C_4H_8^+ + H_2$	20			$D°(LaFe^+-butene) > 30$ kcal/mol
	$MM'C_4H_6^+ + 2H_2$	80	100	100	$D°(MM'^+-butadiene) > 56$ kcal/mol
$n-C_5H_{12}$	$MM'C_5H_8^+ + 2H_2$	100	69	10	$D°(RhM^+-cyclopentadiene) > 67$ kcal/mol
	$MM'C_5H_6^+ + 3H_2$		6	20	
	$MC_5H_6^+ + H_2 + (M'+H_2)$[d]			24	
	$MC_5H_6^+ + 2H_2 + (M' + H_2)$[d]		10	37	
	$MC_5H_5^+ + 3H_2 + M'H$		15	9	$D°(Rh^+-cyclopentadienyl) > 189$ kcal/mol
$n-C_6H_{14}$	$MM'C_6H_{10}^+ + 2H_2$	100	40	27	
	$MM'C_6H_8^+ + 3H_2$		44	73	
	$MM'C_6H_6^+ + 4H_2$		16		$D°(RhFe^+-benzene) > 60$ kcal/mol
$c-C_3H_6$	$MM'C_3H_4^+ + H_2$	62	63	59	$D°(MM'^+-allene) > 33$ kcal/mol
	$MM'C_2H_2^+ + CH_4$	38	37		$D°(MM'^+-acetylene) > 24$ kcal/mol
	$MM'CH_2^{+\cdot} + C_2H_4$			41	$D°(RhCo^+-CH_2) > 93$ kcal/mol

(Cont'd)

Reactant	Product	LaFe+	RhFe+	RhCo+	Implications (if observed)[a,b,c]
c-C$_4$H$_8$	MM'C$_4$H$_6$+ + H$_2$	75	90	100	
	MM'C$_4$H$_4$+ + 2H$_2$	25	10		
	MM'C$_2$H$_4$+ + C$_2$H$_4$				D°(RhFe+ –ethene) > 45 kcal/mol
c-C$_5$H$_{10}$	MM'C$_5$H$_6$+ + 2H$_2$	100	14	21	D°(MM'+ –cyclopentadiene) > 50 kcal/mol
	MC$_5$H$_6$+ + H$_2$ + (M' + H$_2$)^d		15	65	
	MC$_5$H$_5$+ + 2H$_2$ + M'H		72	14	D°(Rh+ –cyclopentadienyl) > 162 kcal/mol
c-C$_6$H$_{12}$	MM'C$_6$H$_8$+ + 2H$_2$	100	35	100	
	MM'C$_6$H$_6$+ + 3H$_2$		65		D°(MM'+ –benzene) > 49 kcal/mol
i-C$_4$H$_{10}$	MM'C$_4$H$_8$+ + H$_2$	100	29	100	
	MM'C$_4$H$_6$+ + 2H$_2$		71		
neo-C$_5$H$_{12}$	MM'C$_5$H$_8$+ + 2H$_2$	57			
	MM'C$_5$H$_6$+ + 3H$_2$	43			
	MM'C$_4$H$_6$+ + CH$_4$ + H$_2$		100	100	
ethylene oxide	MM'O+ + C$_2$H$_4$	e	38	55	D°(MM'+ – O) > 85 kcal/mol
	MM'CH$_2$+ + CH$_2$O		62	45	D°(MM'+ –CH$_2$) > 78 kcal/mol
O$_2$	M+ + M'O$_2$				
	MO+ + M'O	100	100	100	

a – Structures are assigned as reasonable as opposed to proven. b – Ref. 34. c – Observed reactions assumed exothermic or near thermoneutral. d – Assumed product, FeH$_2$. e – LaFe+ not isolable due to reaction of La+ with ethylene oxide.

In general, secondary reactions of these cluster ions are slower relative to the primary reactions. This is in contrast to the behavior of the previously reported cluster ions which were unreactive with alkanes, but which in their reactions with alkenes show greater reactivity in their secondary reactions relative to their primary reactions [4,6,7].

The enhanced reactivity of these cluster ions with simple alkanes clearly is related to the fact that they involve a second or third row transition metal bound to a first row metal. Although $NbFe^+$ and $TaFe^+$ are unreactive with alkanes, they also follow this trend in so much as their reactivity with alkenes is considerably greater than that of VFe^+. The combination of a first row and a second or third row metal may involve decreased orbital overlap due to different constituent orbital size and energy and thus, atoms in the cluster would behave more like the individual atomic metal ions. A similar situation might exist when a ligand is bound to a metal cluster ion. Bonding by a ligand to a metal in the cluster could draw electron density from that metal out of the metal-metal bonding orbitals and cause the other metal to behave more like the constituent atomic metal ion, with increased reactivity.

Acknowledgement is made to the Division of Chemical Sciences in the Office of Basic Energy Sciences in the United States Department of Energy (DE-AC02-80ER10689) for supporting the transition metal ion research and to the National Science Foundation (CHE-8310039) for continued support of the Fourier transform mass spectrometry instrumentation. S.W.B. gratefully acknowledges the Purdue Research Foundation for providing fellowship support.

REFERENCES

1) Cotton, F.A.; Walton, R.A.; "Multiple Bonds Between Metal Atoms", Wiley, 1982.
2) Smalley, R.E.; "Supersonic Cluster Beams: An Alternative Approach To Surface Science" ,Comparison of Ab-Initio Quantum Chemistry With Experiment,R.J. Bartlett ed.,(D. Reidel Publ. Co.).
3) For a recent review of gas phase transition metal ion chemistry see: Allison, J.; Prog. Inorg. Chem.; 34 (1986).
4) Jacobson, D.B.; Freiser, B.S.; J. Am. Chem. Soc. , 107, 1581 (1985).
5) Freas, R.B.; Ridge, D.P.; J. Am. Chem. Soc., 102, 7129 (1980).
6) Hettich, R.L.; Freiser, B.S.; J. Am. Chem. Soc., 107, 6222 (1985).
7) Tews, E.C.; Lech, L.M.; Buckner, S.W.; Freiser, B.S.; in preperation.
8) Barrett, P.H.; Pasternak, M.; Pearson, R.G.; J. Am. Chem. Soc., 101, 222 (1979).
9) Davis, S.C.; Klabunde, K.J.; J. Am. Chem. Soc., 100, 5974 (1978).
10) Davis, S.C.; Severson, S.J.; Klabunde, K.J.; J. Am. Chem. Soc., 103, 3024 (1981).
11) Freas, R.B.; Ridge, D.P.; J. Am. Chem. Soc., 106, 825 (1984).
12) Jacobson, D.B.; Freiser, B.S.; J. Am. Chem. Soc., 106, 5351 (1984).
13) Cody, R.B.; Burnier, R.C.; Reents, W.D., Jr.; Carlin, T.J.; McCrery, D.A.; Lengel, R.K.; Freiser, B.S.; Int. J. Mass Spectrom. Ion Phys., 33, 37 (1980).
14) Jacobson, D.B.; Freiser, B.S.; J. Am. Chem. Soc., 106, 4623 (1984).
15) Comisarow, M.B.; Marshall, A.G.; J. Chem. Phys., 64, 110 (1976).
16) Cody, R.B.; Burnier, R.C.; Freiser, B.S.; Anal. Chem., 54, 96 (1982).
17) Carlin, T.J.; Freiser, B.S.; Anal. Chem., 55, 571 (1982).
18) Ahmed, M.S.; Dunbar, R.C.; "Collisional Quenching of Photo-Excited Bromobenzene Ions"; presented at the 34th Annual Conference on Mass Spectrometry and Allied Topics, Cincinnati, OH, June 8-13, 1986.
19) Byrd, G.D.; Freiser, B.S.; J. Am. Chem. Soc., 104, 5944 (1982).
20) Huang, Y.; Freiser, B.S.; Organometallics, submitted.

21) Byrd, G.D.; Burnier, R.C.; Freiser, B.S.; J. Am. Chem.Soc., 104, 3565 (1982).

22) Jacobson, D.B.; Freiser, B.S.; in preparation.

23) Halle, L.F.; Klein, F.S.; Beauchamp, J.L.; J. Am. Chem. Soc., 106, 2543 (1984).

24) Sallans, L.; Lane, K.R.; Squires, R.R.; Freiser, B.S.; J. Am. Chem. Soc., 107, 4379 (1985).

25) Supplementary thermochemical information taken from: Rosenstock, H.M.; Draxl, D.; Steiner, B.W.; Herron, J.T.; J. Phys. Chem. Ref. Data, Suppl. I, 6 (1977).

26) Jacobson, D.B.; Byrd, G.D.; Freiser, B.S.; Inorg. Chem., 23, 553 (1984)

27) Byrd, G.D.; Ph.D. Thesis,1982, Purdue University.

28) Jackson, T.C.; Jacobson, D.B.; Freiser, B.S.; J. Am. Chem. Soc., 106, 1252 (1984).

29) Armentrout, P.B.; Halle, L.F.; Beauchamp, J.L.; J. Chem. Phys., 76, 2449 (1982).

30) Jacobson, D.B.; Freiser, B.S.; J. Am. Chem. Soc., 108, 27 (1986).

DISSOCIATIONS OF MANGANESE CARBONYL CLUSTER IONS

R. B. Freas, J. H. Callahan, M. M. Ross and A. P. Baronavski

Chemistry Division
Naval Research Laboratory
Washington, D. C.

INTRODUCTION

The structures and reactivities of cluster ions have been studied by a variety of mass spectrometric techniques. Among the methods which have been employed are ion cyclotron resonance (ICR) [1], multiphoton ionization/dissociation time of flight (MPD/TOF) mass spectrometry [2], and tandem mass spectrometry using magnetic sector instruments [3,4]. These techniques often provide complementary information. ICR has proven useful in studies of the formation and reactivities of clusters. This is due to the ability to trap ions for significant lengths of time, as well as to eject all ions from the trap other than the species under study [4]. Tandem mass spectrometric techniques are useful because particular clusters can be mass selected and then fragmented with a collision gas [4]. This can provide valuable structural information. Recent work in MPD/TOF has given considerable insight into the photochemistry and structures of several cluster systems [2]. Information about the energetics of cluster formation and about both ground and excited states can often be obtained.

Due to their potential catalytic properties, there has been considerable interest in the chemistry of metal carbonyl clusters. Since they are readily introduced into the gas phase as ions, they are amenable to mass spectrometric analysis. Here we report the results of studies of the reactivities and structures of manganese carbonyl clusters. ICR was used to study the formation of clusters, while tandem mass spectrometry combined with collision-induced dissociation (CID) was used to characterize structures. These studies showed that clusters with small electron deficiencies had low reaction rate constants and tended undergo collision-induced dissociation with the loss of large stable groups, such as $Mn(CO)_5$. By contrast, clusters with large electron deficiencies were more reactive and tended to lose only carbonyls upon collision. Additionally, the results of preliminary photodissociation studies of manganese carbonyl ions are also reported. It was found that fragmentation by photodissociation was dependent of several factors, including the absorptivity of the parent ion, the wavelength of excitation, and the initial internal energy of the ion.

EXPERIMENTAL

ICR mass spectrometric studies reported here were performed at the University of Delaware, as described elsewhere [1,4]. Double resonance ICR techniques were used to identify the product ions of various reactant ions. Ions were formed by 80-eV electron ionization of $Mn_2(CO)_{10}$, and typical experimental pressures were 2×10^{-6} Torr. Tandem mass spectrometric experiments were carried out using a VG Analytical ZAB-2F reverse geometry double focussing mass spectrometer. In the CID studies, ions were mass selected by the magnetic sector and underwent collisions with helium in the field-free region between the magnet and the electrostatic analyzer (ESA). The fragment ions were identified by ion kinetic energy scans. The collision gas pressure was maintained so as to attenuate the primary beam intensity by 70-80%. In the photodissociation (PD) studies, an argon ion laser (Coherent Innova 100-20) or the output of a tunable dye laser (Coherent CR-699) pumped by an argon ion laser was aligned to overlap coaxially with the ion beam in the second field-free region of the spectrometer. Ions were mass selected by the magnet, and PD products were analyzed by scanning the ESA. In order to separate fragmentation resulting from photodissociation from that occurring due to unimolecular fragmentation (which arises as a result of the deposition of excess internal energy during ionization), the laser beam was chopped (at 2000 Hz) and the detector output was monitored with a lock-in amplifier. The manganese carbonyl ions for the CID and PD studies were generated by 70 to 150 eV electron ionization of $Mn_2(CO)_{10}$.

RESULTS

Clustering reactions in the manganese carbonyl system were studied using ICR. The most abundant ions from the initial ionization process are $[Mn_x(CO)_y]^{.+}$, where x=1-2 and y=0-5. Subsequently, these ions reacted with neutral $Mn_2(CO)_{10}$, forming cluster ions. These reactions occurred with the concurrent loss of carbonyls (or $Mn(CO)_5$); these losses are necessary (in the absence of collisional stabilization by a third body) to carry off excess energy and prevent cluster decomposition. For example, Mn_2^+ reacted with $Mn_2(CO)_{10}$ to form $[Mn_4(CO)_8]^{.+}$ and $[Mn_4(CO)_9]^{.+}$. Both of these species subsequently underwent unimolecular decomposition with the loss of one carbonyl. This suggests that the product ion results from the stepwise loss of CO from an $[Mn_2-Mn_2(CO)_{10}]^{.+}$ metastable complex. Other types of clusters are formed by the reaction of $[Mn(CO)_y]^+$ (y=1-5) and $[Mn_2(CO)_z]^+$ (z=1-5) to form $[Mn_3(CO)_x]^+$ and $[Mn_4(CO)_x]^+$ (x=8-12,15). Further reactions included the formation of clusters with up to 8 Mn atoms (although their precursors could not be unambiguously identified). Additionally, $[Mn(CO)_5]^+$ and $[Mn_2(CO)_5]^+$ were observed to react with $Mn_2(CO)_{10}$ to form $[Mn_3(CO)_{15}]^+$ and $[Mn_4(CO)_{15}]^+$. In the absence of collisional stabilization by a third body (the 10^{-7} Torr pressures at which these studies were carried out is too low to allow collisions in the lifetime of the complex), radiative emission is the most likely mechanism for stabilization of these latter complexes.

The structures of clusters formed in ion-molecule reactions were subsequently studied using collision-induced dissociation (CID) combined with tandem mass spectrometry. Mass-selected cluster ions were fragmented and analyzed by kinetic energy scans. CID studies of mass-selected clusters indicated that there were two major types of decomposition processes. For example, the primary CID process observed with $[Mn_2(CO)_{10}]^{.+}$ and $[Mn_3(CO)_{15}]^{.+}$ was the loss of $Mn(CO)_5^.$. This suggested that both clusters had chain-like structures (as opposed to a possible triangular structure with 3 Mn-Mn bonds in $[Mn_3(CO)_{15}]^{.+}$). The results also suggested that the $Mn(CO)_5$ subunit is stable (although its dissociation after cleavage cannot be ruled out). In contrast to these results, the CID spectra of many other clusters were characterized primarily by the loss of a single carbonyl or by loss of successive carbonyls. The CID of $[Mn(CO)_5]^+$ produces Mn^+ as the primary product ion, while CID of $[Mn_2(CO)_5]^+$ results in the loss of a single carbonyl as the most abundant fragment ion. Other coordinatively unsaturated clusters, such as $[Mn_3(CO)_9]^{.+}$, $[Mn_3(CO)_{11}]^{.+}$ and $[Mn_3(CO)_{12}]^{.+}$ also dissociate primarily with the loss of one carbonyl.

The results of the cluster formation and CID studies suggest a general trend. Based on previously determined reaction rate constants [1], electron deficiencies (using the modified 18-electron rule [4]) and the CID data, the results indicate that clusters with small electron deficiencies ($[Mn_2(CO)_{10}]^{.+}$, $[Mn_3(CO)_{15}]^{.+}$ and $[Mn_4(CO)_{15}]^{.+}$) do not readily react to form higher clusters, and undergo fragmentation primarily by the loss of coordinatively saturated, stable, neutral fragments ($Mn(CO)_5^.$ and $Mn_2(CO)_{10}^.$). By contrast, clusters that have large electron deficiencies readily undergo reactions to form larger clusters and tend to lose one carbonyl by the CID process. These results are summarized in Table 1 (from references 1 and 4).

Table 1

Cluster Ion	Rate Constant	Electron Deficiency	M-M Bonds	%M-CO Dissociation	Major CID Fragments
$[Mn_2(CO)_{10}]^{.+}$	<0.2	1/2	1	42	$Mn(CO)_5.$
$[Mn_3(CO)_9]^+$	-	3-1/3	3	94	CO
$[Mn_3(CO)_{10}]^+$	1.4	2	4	35	$Mn(CO)_5.$
$[Mn_3(CO)_{11}]^{+.}$	-	2	3	70	$Mn(CO)_5.,CO$
$[Mn_3(CO)_{12}]^+$	-	1-1/3	3	95	CO
$[Mn_3(CO)_{15}]^+$	<0.2	0	2	20	$Mn_2(CO)_{10}$
$[Mn_4(CO)_8]^{.+}$	6.9	4-1/4	6	63	$Mn_2(CO)_2,CO$
$[Mn_4(CO)_9]^{.+}$	6.1	3-3/4	6	60	$Mn(CO)_5.,CO$
$[Mn_4(CO)_{10}]^{.+}$	10.0	3-1/4	6	75	CO
$[Mn_4(CO)_{11}]^{.+}$	9.9	2-3/4	6	98	CO
$[Mn_4(CO)_{12}]^{.+}$	7.3	2-1/4	6	96	2CO
$[Mn_4(CO)_{15}]^{.+}$	0.3	1-1/4	5	57	$Mn_2(CO)_{10}$

Photodissociation studies of mass-selected cluster ions of manganese carbonyls were also conducted, in order to obtain additional structural and chemical information, as well as to compare the PD and CID processes. Because of low signal intensities, it was not possible to observe photodissociation in all of the ions for which CID information has been obtained. However, results are reported for some representative manganese carbonyl ions. For example, the unimolecular, CID and PD spectra were compared for the $[Mn_2(CO)_{10}]^{.+}$ ion. Unimolecular fragmentation occurs either by loss of a carbonyl or by cleavage of a Mn-Mn bond (with the loss of $Mn(CO)_5.$). The former process requires roughly 1 eV of energy [5], while the latter requires roughly 0.7 eV of energy [5], suggesting that approximately 1 eV of excess internal energy is deposited into the ion during during electron ionization. The CID spectrum of $[Mn_2(CO)_{10}]^{.+}$ shows considerably more fragmentation (Figure 1). The major loss is due to the cleavage of an Mn-Mn bond

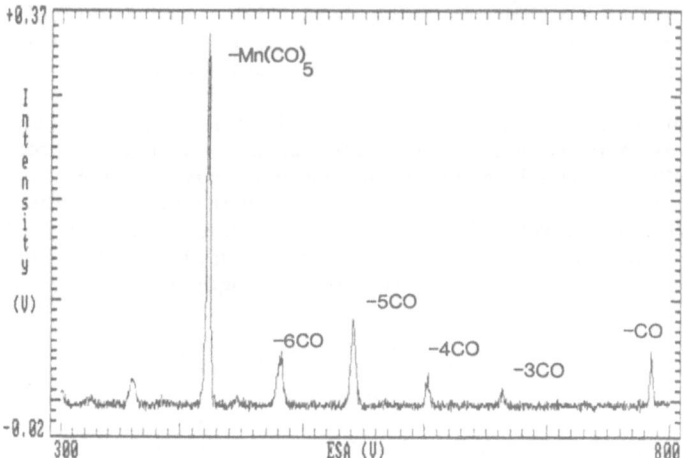

Figure 1. CID Spectrum of $[Mn_2(CO)_{10}]^{.+}$. (Ion kinetic energies are proportional to the ESA voltages on the x-axis).

(with the loss of $Mn(CO)_5.$), but there is also a series of sequential carbonyl losses. This is consistent with the deposition of higher quantities of energy through collision (than is deposited during ionization). The photodissociation spectrum at 514 nm (1W laser power) is different from both the unimolecular and CID spectra (Figure 2). The

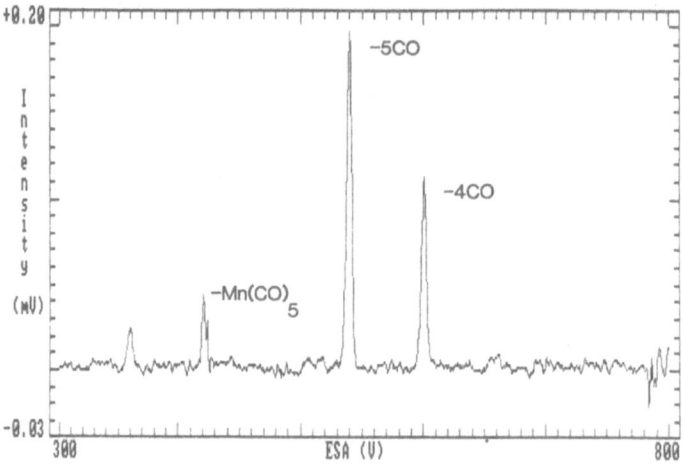

Figure 2. Photodissociation Spectrum of $[Mn_2(CO)_{10}]^{.+}$ (514 nm, 1 Watt)

major fragmentations result in the loss of 4 CO and 5 CO, as well as the loss of $Mn(CO)_5^.$. Additionally, there is evidence for photodissociation of the $[Mn_2(CO)_9]^{.+}$ ion (a product of the unimolecular fragmentation of $[Mn_2(CO)_{10}]^{.+}$). By mass selecting this ion, it was found that three and four carbonyls were lost from the cluster. The resultant products overlap with photodissociation products from $[Mn_2(CO)_{10}]^{.+}$.

The photodissociation process is dependent on several factors. The photodissociation intensity is affected by the absorptivity of the parent ion. This was assessed by monitoring the decrease in the parent ion signal with the laser on, as compared to the signal level when the beam is blocked. For the $Mn_2(CO)_{10}$ ion, the spectrum showed that absorption reached a maximum at 514 nm (all spectra were normalized to both laser power and parent ion intensity) and decreased at longer and shorter wavelengths (these studies were limited to the visible wavelength region). This represents a considerable red-shift in the absorption maximum for the parent ion as compared to that for neutral $Mn_2(CO)_{10}$, which has an absorption maximum near 330 nm.

The types of fragmentations observed are also dependent on the excitation wavelength. For example, as photon energy decreases, the fragmentation process favors a lower energy pathway. This is mainly manifested by a decrease in the ratio of $[Mn_2(CO)_5]^{.+}$ (-5 CO) to $[Mn_2(CO)_6]^{.+}$ (-4 CO), and an increase in the relative abundance of $[Mn_2(CO)_7]^{.+}$ at longer wavelengths. This behavior is shown in Figure 3, where the relative abundances of photodissociation products are shown plotted against photon energy. The trends consistently show that shorter wavelengths result in loss of more carbonyl ligands.

The internal energy of the parent also has a significant effect on its photodissociation behavior. Under normal conditions (after running for several hours), the electron ionization source heats up to approximately 200°C. Under these conditions, the loss of 5 CO is considerably favored over the loss of 4 CO. If the same spectrum is obtained just after the source is turned on (the source temperature is approximately 70°C), the ratio of the loss of 5 CO to the loss of 4 CO is roughly equal. If the internal energy of the ions is lowered still further (by the use of a buffer gas at 0.15 Torr to collisionally cool the ions), the loss of 4 CO is considerably favored. Additionally, if the ratio of the loss of 5 CO to the loss of 4 CO is monitored with time (as the source heats up), it is observed that the higher energy fragmentation pathway becomes favored over the lower energy process. Thus, although it cannot be measured exactly in this case, internal energy has a significant effect on the PD spectrum.

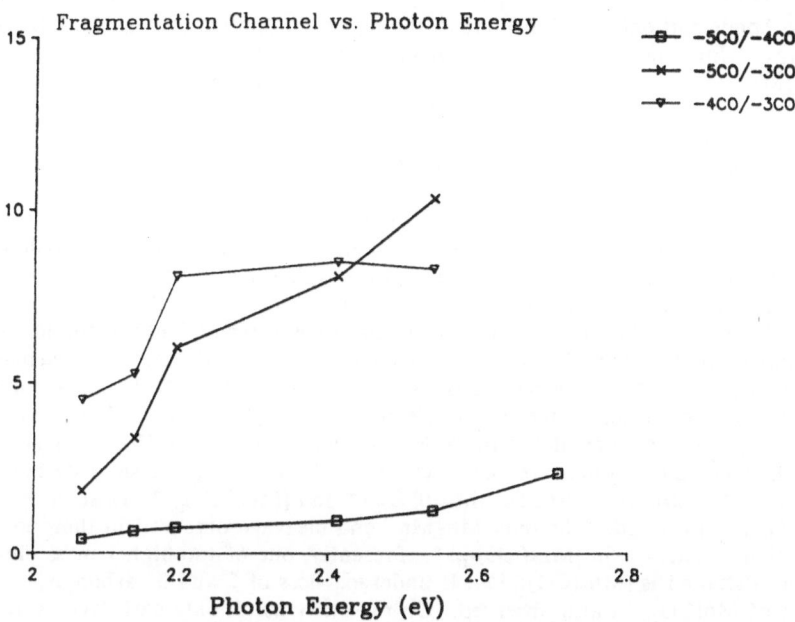

Figure 3. Effect of Photon Energy on the Fragmentation Pathway (symbols represent the ratios of the indicated products).

These results give some insight into the photodissociation behavior of manganese carbonyl ions. As in other systems, ions undergo absorption from the ground state to an excited electronic state. The ion may undergo radiationless decay to a vibrationally excited ground state, resulting in fragmentation. Energy in excess of that needed to induce dissociation will be partitioned into the fragments, which may undergo additional fragmentation. Alternatively, initial excitation may occur directly to a dissociative excited state, which would result in the cleavage of a bond. Subsequently, excess energy could induce unimolecular fragmentations in the products. It is not possible to directly confirm either mechanism from the results cited here. However, the results are relatively consistent with the expected energetics of photodissociation. If the initial cleavage is at a Mn-CO bond, roughly 1 eV is required [5]. Since the ion absorbs photons with 2.0-2.8 eV of energy, and already possesses internal energy (1 eV) from the ionization process, there is sufficient excess energy in the products to result in additional fragmentations. This is consistent with the loss of several carbonyls. However, if each bond cleavage requires 1 eV, it is difficult to reconcile the loss of as many as 5 CO. This suggests that the cleavage of Mn-CO bonds may require successively less energy as more are lost, or that the bond energies cited in the literature are higher than is actually the case (in fact, one report cites an average Mn-CO bond energy of 0.7 eV in the electron impact studies of the loss of the first five carbonyls [6], which would also be consistent with a varying bond strength). The variation of the extent of fragmentation with wavelength and temperature is also reasonable. Excitation with higher photon energies would be expected to result in the deposition of larger quantities of energy into the ion (as long as it absorbs at the wavelength in question), ultimately favoring the loss of more carbonyls. Similarly, increasing the internal energy of the ion results in the absorption of photons from higher ground state vibrational levels, which again would result in the deposition of larger quantities of energy into the parent ion and subsequent fragments.

Photodissociation of $[Mn_2(CO)_{10}]^{.+}$ also induces the cleavage of an Mn-Mn bond to yield an $[Mn(CO)_5]^+$ fragment. This should require roughly 0.7 eV of energy [5]. The remainder of the energy from the photon is most likely carried off as internal energy in the fragments (the neutral fragment may undergo carbonyl loss). There is also the possibility that the appearance of $[Mn(CO)_5]^+$ arises from the successive loss of 5 CO and the cleavage of an Mn-Mn bond. The 2.0-2.8 eV photons may be sufficient to cleave five carbonyl bonds and one Mn-Mn bond if the parent ion possesses considerable initial internal energy. Unfortunately, it is difficult to accurately characterize the relative intensity of $[Mn(CO)_5]^+$ (which would lend insight into this possibility, especially with temperature and wavelength variations) because it is formed as unimolecular fragmentation product and is itself photodissociated (resulting in the addition of an unknown negative signal to the PD signal for $[Mn(CO)_5]^+$).

Due to the difficulty in obtaining sufficient signal intensities from many cluster ions or fragment ions, only limited attempts have so far been made to correlate PD behavior with cluster structure. Most of the structural studies to date have been conducted at a single wavelength (514 nm) at high laser powers. For cluster ions of the type $[Mn_2(CO)_y]^{\cdot+}$ (where y=1-5), absorption of a 514 nm photon results only in the loss of carbonyls. For example, PD of $[Mn_2(CO)_2]^{\cdot+}$ is characterized by the loss of 2 carbonyls. Despite the fact that Mn_2^+ can be photodissociated at 514 nm, Mn-Mn bonds are not cleaved in unsaturated clusters where carbonyls are lost (i.e. $[Mn_2(CO)_2]^{\cdot+}$ is not cleaved to yield $[Mn(CO)_2]^+$). For ions of the form $[Mn_2(CO)_y]^{\cdot+}$ (where y=6-10), primary fragmentations are again through the loss of carbonyls. $[Mn_2(CO)_6]^{\cdot+}$ loses 1 carbonyl to form $[Mn_2(CO)_5]^{\cdot+}$, while the losses for $[Mn_2(CO)_7]^{\cdot+}$ and $[Mn_2(CO)_9]^{\cdot+}$ are such that only $[Mn_2(CO)_5]^{\cdot+}$ is formed. The only Mn-Mn bond cleavage observed in these lower molecular weight clusters is in $[Mn_2(CO)_{10}]^{\cdot+}$. Presently, one of the higher mass clusters which has been studied is $[Mn_3(CO)_{10}]^{\cdot+}$. It undergoes loss of 2 and 3 carbonyls, although loss of $Mn(CO)_5^{\cdot}$ is also observed. On the other hand, only carbonyls were lost from $[Mn_4(CO)_y]^{\cdot+}$ (where y= 8, 9 and 10).

CONCLUSIONS

The results presented here have given some insight into the structures and reactivities of carbonyl clusters. Clearly, electron deficient clusters are more reactive toward higher cluster formation and tend to lose single carbonyls more readily than less electron deficient clusters. The latter tend to be unreactive and lose stable moieties (i.e. $Mn(CO)_5$) through the cleavage of Mn-Mn bonds.

The results of the photodissociation studies are less conclusive, but suggest that photodissociation results in processes similar to those observed in CID. The range of fragments observed in PD does not appear to be as great, however, because a well defined amount of energy is deposited into the ion. Additionally, the extent of fragmentation observed in PD is not as great as in CID, primarily because less energy is deposited (in these studies) when photons are absorbed. The limited extent of these studies does not allow identification of the specific nature of the photodissociation process (whether it involves absorption to a dissociative state or radiationless decay to an excited ground state). Nevertheless, the effects of internal energy and wavelength are consistent with a process in which excess energy (beyond that which goes into the initial cleavage) results in further dissociation. The PD results also indicate that it may be possible to more readily control energy deposition in fragmentation studies of structure using photons. Whether photodissociation studies such as these will be useful in structural characterization or in the study of the photochemistry of cluster ions will require further study.

REFERENCES

1. Meckstroth, W. K.; Freas, R.B.; Reents, W. D.; Ridge, D. P. Inorg. Chem. 1985, 24, 3139-3146.
2. Leopold, D. G.; Vaida, V. J. Am. Chem. Soc. 1984, 106, 3720-3722.
3. Freas, R. B.; Campana, J. E. J. Am. Chem. Soc. 1986, 108, 4659-4661.
4. Freas, R. B.; Campana, J. E.; Ridge, D.P.; submitted for publication.
5. Connor, J. A. Topics Current Chem. 1977, 71, 22.
6. Svec, H. J.; Junk, G. A. J. Am. Chem. Soc. 1967, 89, 2836-2840.

d BAND OF NICKEL CLUSTERS ON SiO$_2$ AMORPHOUS SUBSTRATE : ELECTRON SPECTROSCOPIES AND REACTIVITY

A. Masson, B. Bellamy, Y. Hadj Romdhane, V. de Gouveia
and M. Che*

Laboratoire de Physico-chimie des Surfaces, UA 425, CNRS, ENSCP
Université P. et M. Curie, 75005 Paris, France

*Laboratoire de Réactivité de Surface et Structure, UA 1106, CNRS
Université P. et M. Curie, 75252 – Paris Cedex 05, France

INTRODUCTION

The intrinsic size effect of particles on chemical reactivity is both and important problem in itself and in heterogeneous catalysis by metal.

If one try to explicit the concept of size, this has to be replace by : a minimum number of atoms necessary to start a reactivity (1) change in electronic properties (2) surface relaxation (2) and even variation of crystalline structure (4).

In order to answer some of these questions we have developped by atomic beam techniques condensed on a surface a model catalyst (5).

The nucleation and growth process of metallic phase and the in situ characterization is followed by electron spectroscopies. We reported results on nickel and we conclude for intrinsic size effect concerning the hydrogenolysis of n-butane.

EXPERIMENTAL

Starting with UHV conditions, a thin film of amorphous silica (200 Å) is formed by thermal oxydation on a (100) surface of single crystal of silicon. A calibrated kundsen cell delivering a flux of $5 \ 10^{13}$ atom.cm^{-2} sec^{-1} is used to obtain coverage of nickel between 10^{14} until $5 \ 10^{15}$ atoms cm^{-2}. Electron spectroscopies AES, UPS follows the change in electronic properties.

Finally the sample is magnetically translate under vacum in a chemical reaction in which hydrogenolysis is performed. The three connected chambers of this machine have been described in detail elsewhere (6).

RESULTS AND DISCUSSION

Nucleation and growth of the metallic phase :
When nickel vapour is condensed onto insulating substrate thermodynamic arguments indicate that the growth mechanism is of Volmer-Weben type
i.e. characterized by a distribution of three dimensional nuclei (7).

Using Auger electron spectroscopy in integral mode it is possible to
follow this mechanism as it is shown on fig.(1).

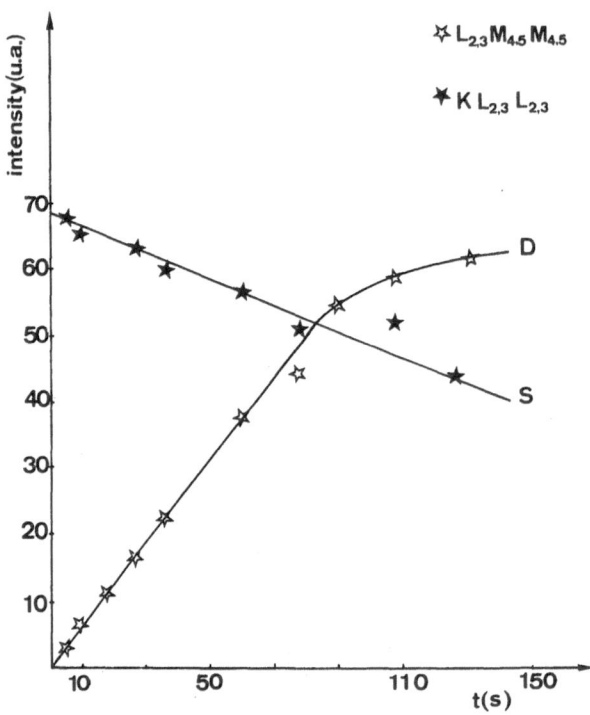

Fig. 1. Variation of the Auger intensity peaks of nickel and oxygen
as a function of time of evaporation : flux of 5 10^{13} atom.
cm^{-2} sec^{-1}.

Here, the peak intensity of oxygen transition (K. L_{23} L_{23}) of the SiO_2
substrate on one hand and the nickel transition (L_{23} M_{45} M_{45}) on the
other hand versus time of evaporation have been choosed, the total coverage
of nickel being calibrated by Rutherford backscattering measurements.

ELECTRON SPECTROSCOPIES

We have reported in fig.(2) and fig.(3) the Auger core valence valence
transition (L_{23} M_{45} M_{45}) and the valence band in UPS for different nickel
coverage, that is for decreasing size between spectra F to A (see legend
in the figures). One must immediately concluded that in both ensemble of
spectra changing of size is changing in the electronic distribution of the
valence band of the nickel particles. Moreover concerning the core valence
valence transition are very similar to the one of BENNETT concerning Ni
alloys (8) the interpretation could be related to change in d electron

density (9). As for the UPS-spectra the change of the density of states at the Fermi level is similar to experiments of Wertheim and Al (10) on Pd and typical of redistribution in the d electron density as an intrinsic effect.

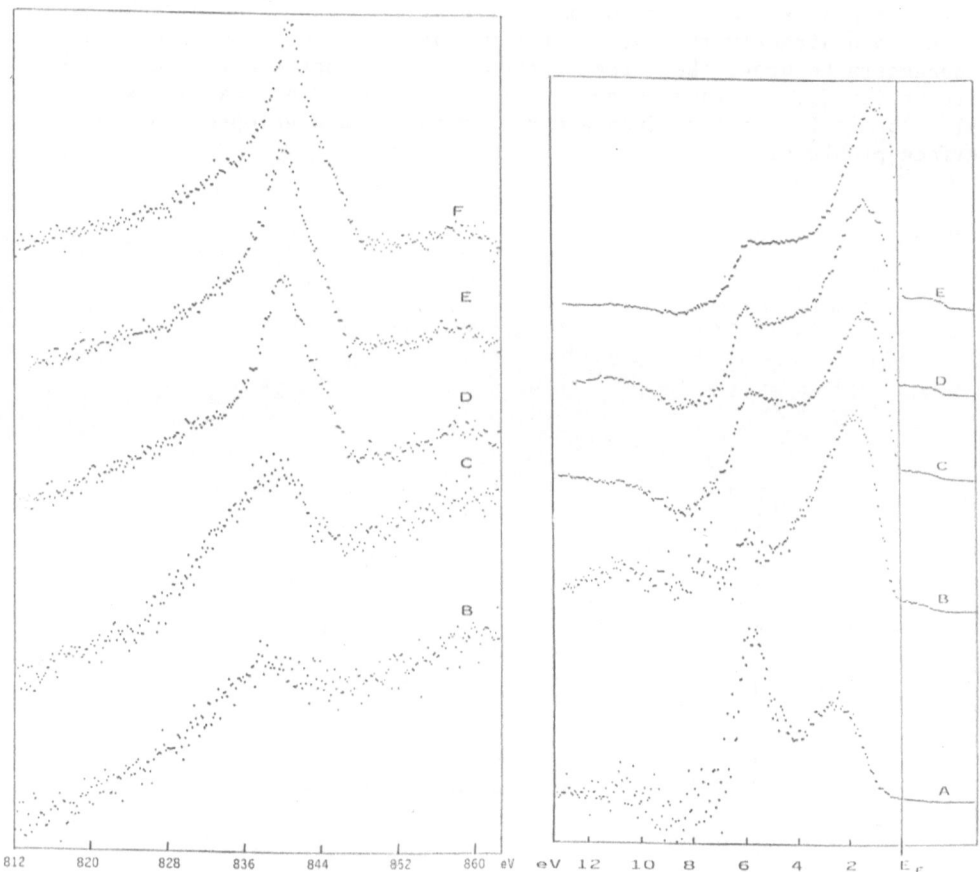

Fig. 2. L_3VV nickel Auger transition as a function of nickel coverage
*A. 2,6 10^{14} atom.cm^{-2}, (see fig.3)
B. 8 10^{14}
C. 1,5 10^{15}
D. 4,7 10^{15}
E. 9 10^{15}
F. 1,16 10^{16}

Fig. 3. Valence band of nickel with UPS as a function of nickel coverage (see fig.2)

CATALYSIS

The hydrogenolysis on n-butane has been choosed as a chemical probe with nickel, as it is well known that this reactivity is size sensitive (11). Two main results are shown in fig.(4).

First a minimum number of nickel atoms is necessary to brake the carbon carbon bond, a very simple calculation with and equilibrium form for the nickel particle give 30 atom at the surface.

Secondly a maximum in the curve is obtained for cluster without bulk electronic properties. Comparison with bigger particles and single crystal face are reported elsewhere (12).

CONCLUSION

By using nickel clusters of monitored size by atomic beam techniques we have shown directly for the first time that a minimum number of atoms are necessary to brake the carbon carbon bond and that electronic effect has to be invoque in order to explain the maximum in the reactivity of a complex chemical reaction. Such a model approach is developped now for interface problems.

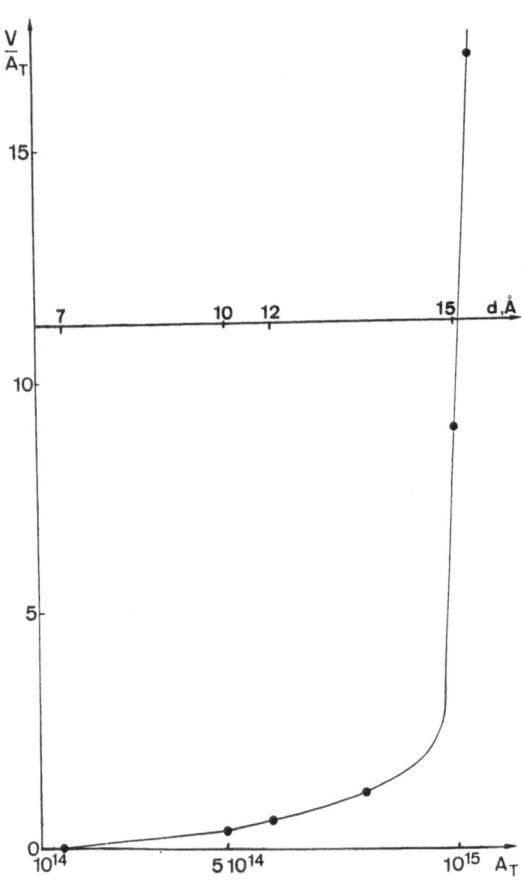

Fig. 4. Variation of the experimental parameter $\dfrac{V}{A_T}$ as a function of coverage and particle diameter. V = rate of conversion of $C_n H_n$ molecules.

REFERENCES

1. G.A. Martin, J.A. Dalmon and C. Mirodatos, Bull. Soc. Chem. Belg. <u>88</u>, 559, 1979.
2. F. Cyrot-Lackmann, S.N. Rhana, J.B. Buchet and J. Buttet, Surf. Science 127, 165, 1983.
3. L.D. Marks, W. Heine, J. Catalysis 94, 570, 1985.
4. M. De Crescenzi, P. Diociauiti, P. Picozzi and S. Santucci, Phys. Ret. B. 34, n°6, 4334, 1986.
5. A. Masson, B. Bellamy, G. Colomer, M. M'Bedi, P. Rabette and M. Che, Proc. 8th Inter. Congr. on Catalysis 4, 333, 1984.
6. A. Masson, B. Bellamy, Y. Hadj Romdhane, M. Che, H. ROULET and G. Dubour, Surf. Science 173, 479, 1986.
7. A. Masson, NATO A57 SERIE E, 104, 295, 1986.
8. P.A. Bennett, J.C. Fuggle, F.U. Hillebrecht, A. Lenselink and G.A. Sawatzky, Phys. Rev B. 27, 2194, 1983.
9. G. Treglia, M.C. Desjonqueres, F. Ducastel and D. Spanjaard, J. Phys.C solid state phys. 14, 4355, 1981.
10. C.K. Wertheim, S.B. Dicenzo and D.N.E. Buchanan, Phys. Rev.B. 33, n°6, 5384, 1986.
11. M. Boudart, Int. Congr. of Catalysis Londres 1, 1977.
12. A. Masson, G. Bellamy, Y. Hadj Romdhane, V. De Gouveia and M. Che, Surf. Science (in the press).

IN SITU EXAFS STUDIES OF SMALL Fe PARTICLES USED AS CATALYSTS FOR NH$_3$ SYNTHESIS

W. Niemann, B.S. Clausen, and H. Topsøe

Haldor Topsøe Research Laboratories, DK-2800 Lyngby, Denmark

ABSTRACT

Supported Fe particles used as catalysts for ammonia synthesis are studied by X-ray absorption measurements in a newly developed in situ cell under real reaction conditions. Starting with the precursor, various reduction stages and the final reaction conditions at high pressure and high temperature (80 bar, 400°C) have been investigated. The reduction of iron oxide and the growth of Fe particles can be directly observed in the x-ray absorption fine structure. Simultaneously, the activity of the catalyst is measured. A comparison with the absorption fine structure of an iron foil measured at the same temperature gives indications for the size of the small Fe particles in the catalyst.

INTRODUCTION

The geometrical and electronic structure of small metal clusters and crystallites has been an active area of research for both the theoreticians and the experimentalists /1/. This activity has been motivated primarily by the large technological importance of metal clusters, particularly in heterogeneous catalysis. Effects of particle size on various properties, e.g. binding energy per atom /2/ and catalytic activity /3/ have been clearly demonstrated. A variety of surface sensitive techniques has been applied to catalytic systems. However, the great majority of experimental investigations in the past was done on precursors, model compounds or catalysts taken out of the reactor volume. Techniques and spectroscopies commonly used to characterize catalysts, e.g. volumetric adsorption techniques, transmission electron microscopy (TEM), infrared spectroscopy (IR) and X-ray photoelectron spectrocopy (XPS) fail when matching the real reaction conditions. Thus, there is presently a poor knowledge of the geometrical and electronic properties of catalysts at reaction conditions which normally involve high pressure and high temperature. These conditions require a special sample cell equipped with windows (e.g. Be) which allow the transmission of spectroscopic information. Consequently only techniques like x-ray diffraction (XRD), Mössbauer spectroscopy (MS) or x-ray absorption spectroscopy (XAS) can be applied for direct in situ studies of catalysts. When doing these measurements it is crucial to be able to simultaneously record the catalytic activity.

In continuation of our earlier XAS work on catalysts (see, e.g., refs. /4/ and /5/) we report the first XAS in situ measurements on a NH_3 catalyst performed with a new in situ cell /6/ under real reaction conditions. Starting with the precursor, various stages of the reduction procedure and the final reaction conditions are investigated. X-ray absorption near edge structure (XANES) as well as extended x-ray absorption fine structure (EXAFS) have been measured at the different temperature stages (30°C → 400°C). Simultaneously, the activity of the catalyst has been measured. As EXAFS is a technique which probes the local structure around the absorbing Fe atoms, these investigations directly demonstrate the reduction of the precursor and the growth of small metallic Fe particles which are the active phase of the catalyst.

EXPERIMENTAL

XAS measurements with the in situ cell have been performed at the EXAFS II beamline of the Hamburger Synchrotronstrahlungslabor HASYLAB at Deutsches Elektronen-Synchrotron DESY (Hamburg, FRG). At this beamline a high-flux x-ray beam is achieved by focussing the synchrotron radiation emitted from the DORIS storage ring in a 1:1 image onto the sample /7/. A Si 111 double-crystal monochromator (D = 3.135 Å) was used. More details can be found in refs. 7 and 8. Absorption measurements were performed in a standard transmission configuration by monitoring the intensities I_1 and I_2 of the monochromatic x-ray beam before/behind the sample with ionisation chambers. For the experiments at the Fe K-edge (7.11 keV), the ionisation chambers were filled with 150 hPa of nitrogen. The EXAFS in situ cell was mounted separately to avoid thermal loading of the ionisation chambers. The energy resolution was $\Delta E = 1.5$ eV.

In the past several in situ cells for XAS measurements have been described in the literature /9,10,11/. We have developed a small catalytic reactor which allows simultaneous measurements of catalytic activity and XAS /6/. It is also well suited for other techniques like XRD and MS. In this cell, real reaction conditions (T ≤ 500°C, p ≤ 100 bar) can be obtained. This type of catalyst cell is not restricted to one specific type of reaction. The cell consists of stainless steel walls with two Be windows for the transmission of the x-ray beam. Two tubes are connected to the cell to allow input of the synthesis gas and output of the reaction products. Great care has been taken to ensure real plug flow conditions within the reactor volume which is filled with a few milligram of catalyst powder. If necessary, dilution material can be added to the sample. The Be windows are protected by thin windows of carbon. A versatile control unit operated by a microcomputer supervises and changes temperature, gas flow, gas composition and pressure in the reactor cell according to the requirements of different catalytic systems.

The catalyst used for these first in situ studies is an ammonia catalyst. The precursor has been prepared by coprecipitation of Fe nitrate and Al nitrate and calcination afterwards. Potassium was added by impregnation with KNO_3. A few milligram of the catalyst powder together with 50 % of dilution material (graphite) have been installed in the reactor volume to give an optimum absorption of $\mu \cdot d = 1$ as required for the EXAFS measurements /12/ (μ = absorption coefficient, d = thickness). A constant flow gas mixture of $3H_2$ and N_2 has been exposed to the sample. The reduction of the oxidic precursor has been performed by slowly increasing the temperature from 30°C to the maximum temperature of 400°C. At various temperature stages, the temperature increase has been stopped for a while to record XANES and EXAFS spectra at each of these stages. Finally the gas pressure has been increased from initially 1 bar to a maximum pressure of 80 bar.

RESULTS

Activity

The activity of the catalyst has been determined by measuring the percentage of ammonia in the outlet by use of a commercial infrared gas analyser. The minimum reduction temperature for the production of detectable amounts of NH_3 was found to be 330°C. More detailed activity studies will be reported in a later publication.

XANES

In figure 1 XANES measurements at the Fe K-edge for various reduction stages are displayed. The precursor is characterized by a small pre-edge shoulder, a strong white line and a pronounced dip at 35 eV above the threshold energy. By increasing the temperature, the intensity of the white line and the depth of the 35 eV dip decrease. The intensity of the pre-edge shoulder obviously increases. For $T \geq 350$°C the XANES structures closely resemble the corresponding structures of the iron foil. This indicates that the reduction process is terminated at about 350°C.

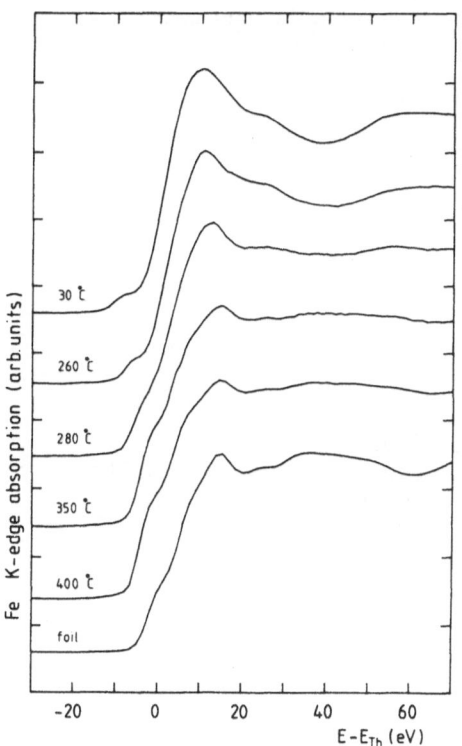

Fig. 1.: X-ray absorption near edge structures (Fe K-edge) of the NH_3 catalyst at various reduction stages and of a Fe foil.

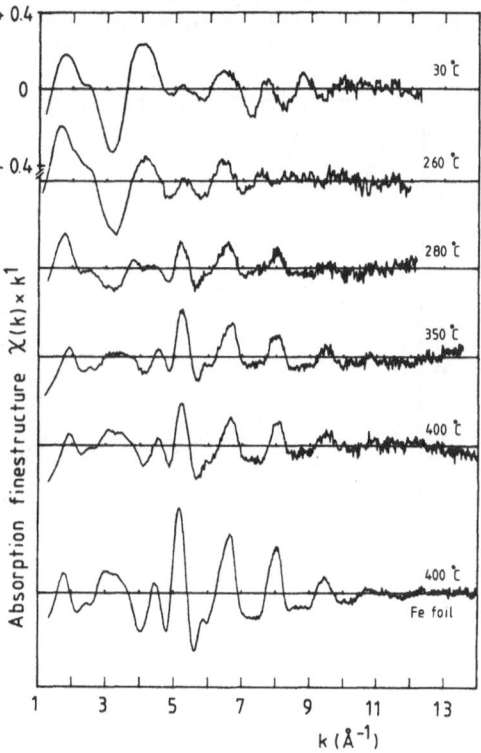

Fig. 2.: Weighted fine structures $X(k) \cdot k^1$ of the NH_3-catalyst and of a Fe foil.

The EXAFS were analyzed according to standard procedures as described in detail in the literature /12,13/. Figure 2 shows the k^{1}-weighted fine structures of the catalyst at various reduction stages and of a Fe foil kept at 400°C. All spectra have been plotted on the same scale. Up to 260°C the catalyst shows the same fine structures as the precursor. Reduction in amplitude may be due to an increased Debye- Waller factor. At 280°C, very different structures appear which belong to the first small Fe clusters that already have grown. For T $\geq$ 350°C the EXAFS of the catalyst closely resembles the EXAFS of the Fe foil except for a large reduction in amplitude. This shows that the reduction process is terminated at around 350°C. This is in accordance to the XANES results.

Figure 3 shows the Fourier transforms (with gaussian windows) of the spectra displayed in fig. 2. In the initial state, peaks at 1.4 Å and 2.5 Å appear. They are due to Fe-O and Fe-Fe distances in the precursor. With increasing temperature, the Fe-O peak gradually decreases and new Fe-Fe peaks at 2.1 Å and 4.5 Å grow up. From the inspection of the Fourier transforms it is suggested that the reduced catalyst has the same bcc structure as the metal foil has it. However, the peak areas are very different as can be seen in figure 4 where the Fourier transforms of $\chi \cdot k^{2}$ of the 400°C catalyst and of the 400°C foil are displayed (square windows). The EXAFS oscillations of the catalyst and the peak areas of the Fourier transform are reduced by a factor of 2 compared to the Fe foil.

For a bcc structure, the first peak in the Fourier transform contains information on two close-lying shells. The r-separation is $\Delta r = 0.38$ Å for Fe which is not resolved by the Fourier transform. So the backtransform has to be fitted with a two shell model. According to recent studies

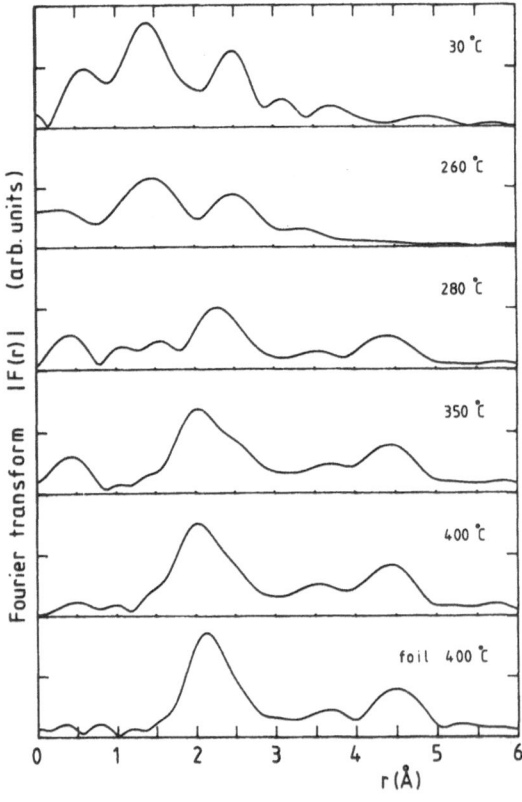

Fig. 3.:
Fourier transforms of the spectra shown in fig. 2 (arb. units).

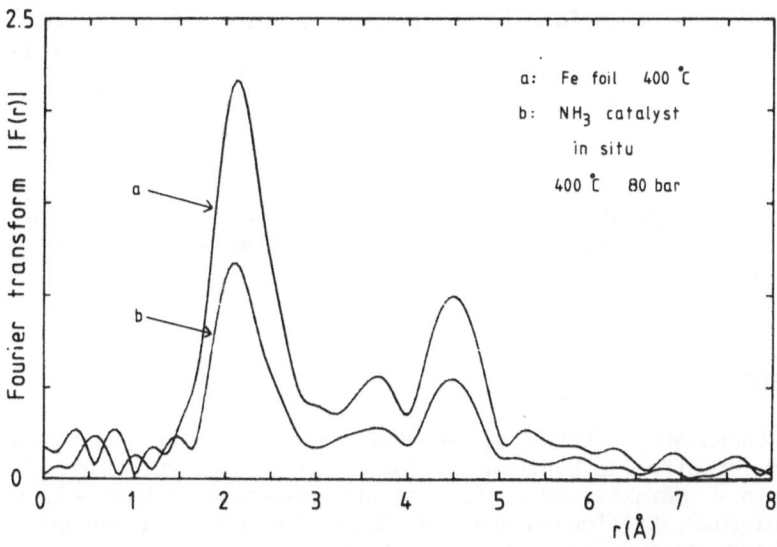

Fig. 4.: Fourier transform of the catalyst at reaction conditions and of a Fe foil at 400°C.

/14/ experimental phases and amplitudes from neighbouring elements in the periodic system with $Z \leq 2$ can be taken without introducing remarkable errors. So experimental phases and amplitudes from Co have been used for fitting all the Fe-Fe EXAFS. Results for the fits of the Fe foil at 400°C and the catalyst sample (at 400°C and 80 bar) are given in Table 1.

Table 1. Results of the EXAFS-fits

Fit parameter	Fe metal 400°C		NH$_3$ catalyst 400°C	
	1.shell	2.shell	1.shell	2.shell
N	8.0*	6.0*	3.8	2.4
R (Å)	2.46$^+$	2.84$^+$	2.46	2.85
$\Delta\sigma^2$(Å)	0.9×10^{-2}	1.0×10^{-2}	0.8×10^{-2}	0.8×10^{-2}

*: Parameter was kept fixed; $^+$: XRD data: 2.48 Å/2.86 Å

DISCUSSION

It is a very remarkable fact that low coordination numbers are found from the in situ EXAFS data of the reduced catalyst the precursor of which was prepared by coprecipitation. Using the formulae given in /15/ for the determination of size-dependent coordination numbers a particle size of 20 to 30 atoms follows. By earlier non-in situ XRD measurements on reduced NH$_3$ catalysts (with magnetite precursor) stored in pentane particle sizes of approx. 360 Å have been found /16/. Non-in situ XRD measurements on this coprecipitated NH$_3$ catalyst resulted in a particle size of approx. 160 Å /17/. On the other hand, no change in amplitude by a factor 2 between the Fe bulk EXAFS and the EXAFS of 160 Å particles should be observed. In situ XRD measurements on this "coprecipitated" catalyst are in progress to study this problem.

In the Fourier transform of the catalyst, the 4.5 Å peak containing the 4th and 5th Fe shell is also very pronounced. This shows that there is some "medium"-range order in the sample and that the small cluster model may be too simple. Possibly, percolation structures have been created during the reduction process. In the percolation case the EXAFS amplitudes are significantly reduced because of the enhanced transmission of the sample /18/. So additionally to the small cluster model a percolation model could also explain the low coordination numbers and the occurrence of higher shells. Intercalation effects with carbon can be ruled out because no significant change in Fe-Fe bond lengths is observed. Furthermore no Fe-C EXAFS could be detected.

ACKNOWLEDGEMENT

The authors are grateful to the staff of HASYLAB for experimental support and for offering beamtime at the EXAFS II station. Valuable discussions with J. Nørskov and P. Stoltze are gratefully acknowledged. We are grateful to G. Steffensen and J.W. Ørnbo for technical assistance and to E. Pedersen for catalyst preparation.

REFERENCES

1. M.G. Mason, Phys. Rev. B27:748 (1983).
2. R.C. Baetzold, Inorgan. Chem. 20:118 (1981).
3. P. Stoltze and J.K. Nørskov, Phys. Rev. Lett. 55:2502 (1985).
4. B.S. Clausen, H. Topsøe, R. Candia, J. Villadsen, B. Lengeler, J. Als-Nielsen and F. Christensen, J. Phys. Chem. 85:3868 (1981).
5. B.S. Clausen, B. Lengeler, B.S. Rasmussen, W. Niemann and H. Topsøe, Journ. Physique, in press
6. B.S. Clausen, W. Niemann and H. Topsøe, to be published
7. W. Malzfeldt, W. Niemann, R. Haensel and P. Rabe, Nucl. Instr. Meth. 208:359 (1983).
8. W. Niemann, Ph.D. thesis, University of Kiel, FRG, (1985)
9. F.W. Lytle, G.H. Via and J.H. Sinfelt, in: "Synchrotron Radiation Research", ed. by H. Winick and S. Doniach (Plenum Press, New York, 1980), p. 401.
10. D.C. Koningsberger and J.W. Cook, Jr., in "EXAFS and Near Edge Structure", ed. by A. Bianconi, L. Incoccia, S. Stipcich (Springer-Verlag, Berlin 1983), p. 412.
11. R.A. Dalla Betta, M. Boudart, K. Foger, D.G. Löffler and J. Sanchez-Arrieta, Rev. Sci. Instr. 55:1910 (1984).
12. P.A. Lee, P.H. Citrin, P. Eisenberger, B.M. Kincaid, Rev. Mod. Phys. 53:769 (1981).
13. B. Lengeler, P. Eisenberger, Phys. Rev. B21:4507 (1980).
14. B. Lengeler, Journ. Physique, in press
15. R.B. Greegor, F.W. Lytle, Journ. Catalysis 63:476 (1980).
16. A. Nielsen and H. Bohlbro, J. Am. Chem. Soc. 74:963 (1952)
    A. Nielsen, "An investigation on promoted iron catalysts for the synthesis of ammonia", (J. Gjellerups Forlag, Kopenhagen 1968).
17. P. Stoltze, private communication (1986).
18. W. Niemann, to be published.

CURRENT VIEWS ON ASTROCHEMISTRY

B. E. Turner

National Radio Astronomy Observatory
Edgemont Road
Charlottesville, VA 22903

INTRODUCTION

Astrochemistry is the study of the formation and destruction of interstellar and circumstellar molecules. Such molecules have been a major field of astrophysical investigation for about 17 years. As of 1987 January, 66 species are identified, and the discovery rate is about three per year. The known species are listed in Table 1 along with the type of region in which they are found. Important areas of study are the distribution and abundances of molecules in external galaxies, the morphology and evolution of galactic molecular clouds and their relation to star formation, interstellar chemistry, and the distribution of isotopes throughout the galaxy and their implications for cosmology and the nucleosynthesis history of the galaxy. Here, we shall discuss the current status of interstellar (and circumstellar) chemistry. For this purpose, a brief summary of the physical conditions characterizing the several morphological regimes of the interstellar medium (ISM) is useful.

Morphology of the ISM*

Current observations suggest that the ISM is comprised of several different regimes. (1) <u>"Coronal" gas</u> is a very hot (10^5–10^6 K), tenuous (~10^{-2} cm^{-3}) component revealed by absorption lines of highly ionized atoms observed in the UV; it possibly accounts for 20 to 50% of the ISM by volume. (2) <u>Intercloud gas</u> is slightly cooler (10^3–10^4 K) and denser (0.1–1 cm^{-3}) and may be in pressure equilibrium with the coronal gas, in which case it would occupy roughly 30% by volume of the ISM. (3) <u>Diffuse interstellar clouds and "cirrus" clouds</u> range from densities of ~0.02 cm^{-3} where hydrogen (and other elements) are still atomic in form and the extinction is $\lesssim$ 0"1, to densities of ~10^2 cm^{-3}, where

---

*Astrophysical nomenclature used here is as follows: Densities are given as particles per cm^3; when the gas is molecular, a particle is an H_2 molecule since hydrogen is 10^4 times more abundant than any other molecule. Distances are in parsecs (pc) or astronomical units (AU); 1 pc = 3.26 light years; 1 AU = 1.5 x 10^{13} cm. Opacities (extinctions) are measured in magnitudes (m); the attenuation factor is defined as $10^{0.4m}$, or 5 magnitudes is a factor 100. One solar mass = 2 x 10^{33} gm.

hydrogen is molecular, and extinctions reach ~2^m. Temperatures are of order 100 K, and cloud masses of a few hundred solar masses are typical. Although molecules (H_2, CO, OH) make their first appearance in these clouds, they contain more atomic than molecular gas. (4) Molecular clouds represent the dense, cool component of the ISM and, although filling only ~2% of it by volume, they account overall for at least 50% of its mass and play a dominant role in the dynamics and kinematics of the ISM. In these clouds, hydrogen is essentially all molecular in form, the onset occurring quite sharply at an extinction of ~2^m. Molecular clouds fall roughly into two types: cool dark clouds, and warm clouds, the latter usually being associated with regions of ionized atomic hydrogen (HII regions) surrounding, or adjacent to, hot young stars whose UV radiation ionizes the gas. Cool dark clouds are typically a few hundred solar masses, ~1 pc in size, and have overall densities ~10^3 cm^{-3}, with "cores" up to 10^4 cm^{-3}. They are not associated with formation of massive early-type stars and thus have no significant internal sources of heat. Being opaque to UV radiation, they are apparently heated by cosmic rays and cooled by rotational excitation of CO, a balance which endows them with a narrow range of temperatures, ~5-15 K. Warm molecular clouds associated with HII regions are by comparison much more massive (10^4-10^7 solar masses), and much larger (~3-100 pc in extent). While their overall densities are ~10^2-10^3 cm^{-3}, they contain numerous and much denser core regions (10^4-10^{10} cm^{-3}). The warm cores of these clouds are usually quite localized near regions of massive star formation; either embedded protostellar objects or expanding HII regions created by newly formed massive stars within or at the edges of these clouds heat the surrounding dust grains and gas to temperatures as high as 70 K over quite large regions, and much higher over smaller regions. The regions far from the heated cores of these warm clouds have properties more like the cold dark clouds. (5) Circumstellar envelopes (CSEs) of gas and dust surround highly evolved stars which are losing mass in the form of stellar winds to the ISM. Temperatures and densities within CSEs range from as high as 2000 K and ~10^{15} cm^{-3} close to the star, to ~20 K and < 100 cm^{-3} at the outermost extremes, where the interstellar UV field dissociates the molecules. CSEs are essentially entirely molecular except at the outermost extremes, and divide into "carbon-rich" and "oxygen rich" categories (C/O > cosmic, C/O < cosmic respectively) as do the central evolved stars. The chemistry of CSEs differs between these categories, and in general from that of the ISM.

Interstellar Molecular Chemistry: General Guidelines

Interstellar chemistry appears to be very complex; the formation and destruction of molecules can involve atoms, ions, free radicals, molecular ions, neutral molecules, solid surfaces, cosmic rays, energy from shocks, and radiation fields. In principle, molecules may be formed in situ within the dark, opaque clouds where they are presently seen, either in the gas phase, or on surfaces of grains, or by disruption of grains that produce the extinction. Or, they may form in the atmosphere of cool stars or presolar nebulae embedded within these clouds, and be injected into the clouds. Finally, they could form in and be ejected from the CSEs of evolved cool stars which lose mass to the ISM.

Fortunately, a few observed properties of interstellar molecules allow us to narrow these choices. (i) The lifetimes of typical interstellar molecules (CO, OCS, H_2CO, H_2O) are only about 100 years against photodissociation by the UV field in unshielded regions of space. Thus molecules cannot travel across open regions of space, and therefore must form within the dark clouds, or CSEs where they are now seen. (ii) While isotope ratios in interstellar molecules are not usually precisely terrestrial, they contrast sharply with the highly non-terrestrial ratios that characterize evolved stars known to lose mass. Again, therefore, we conclude that molecules seen in dense molecular clouds cannot have originated in the atmospheres of evolved stars. Molecules observed in CSEs around evolved stars do reflect the stellar isotope ratios. (iii) The considerable number and abundance of molecular ions shows that ion-molecular gas-phase reactions must play an important, possibly dominant part in interstellar chemistry. (iv) Laboratory and theoretical work indicate that interstellar grains must be important as sinks of molecules and of the refractory elements, since the grains readily adsorb many species under interstellar conditions, at rates at least comparable with ion-molecule formation rates. Experimental evidence that molecules can be formed on grain surfaces and released into the gas phase is much less conclusive. Only H_2 seems necessarily to be formed on grain surfaces. (v) Noticeably different molecular abundances are observed toward high-temperature (few hundred K) protostellar cores, suggesting that either high-temperature gas-phase chemistry ("shock" chemistry) or disruption of grains is important.

The chemical characteristics of the observed molecules have so far offered less conclusive clues. Organic compounds dominate inorganic ones in number and complexity. All of the (13) inorganic compounds are small molecules, with NH_3 the largest at 4 atoms. Among the organic species, 22 are stable, and include most well-known terrestrial classes (alcohols, aldehydes, ketones, acids, amides, esters, ethers, parafin derivatives, and acetylane derivatives). Twenty-nine of the organic compounds are unstable (radicals, ions, or long-chain species) and are dominated by linear, long-carbon-chain species. No branched-chain species are yet known, and only two ring compounds (both three-membered). The elements are represented roughly according to their cosmic abundances, at least for the seven currently observed in interstellar molecules (H, O, N, C, S, Si, Cl), but the absence of Mg, P, Fe, in molecules is noteworthy, since $[Mg] \approx [Fe] \approx [Si] \approx [S]$ cosmically. The lack of N-O bands is also striking.

The failure to observe some of the more complex molecules in the cold dark clouds (Table 1) does not necessarily indicate a "simpler" chemistry there than in warm clouds; both the lower excitation and the lower column depth in the cold clouds mean significantly lower signal levels for a given fractional abundance. On the other hand, most of the long-carbon-chain species are seen in very cold cores of cold dark clouds but not in warm cloud cores, indicating a definite difference in the chemistry. In addition, CSEs have their own chemistry, in which molecular ions are conspicuously absent.

Currently suggested mechanisms of interstellar molecule formation include (i) gas-phase, two-body reactions at low temperatures (< 100 K); (ii) gas-phase, two-body reactions at high temperatures (100-3000 K, as occurs in shocks); (iii) surface catalysis on dust grains; (iv) disruption of grains by shocks or by UV radiation; (v) formation under conditions of thermo-dynamic equilibrium (dense, inner regions of CSEs).

ION-MOLECULE CHEMISTRY

At the low temperatures and densities of interstellar space, gas phase reactions are restricted to those that involve only bi-molecular collisions, and negligible activation energies. They must also be exothermic, except in shocked regions. Those with reasonable rates (i.e., the Langevin rate) include ion-molecule reactions (including electron-recombination reactions) and a few neutral-neutral reactions that involve abundant species. Only these gas phase reactions can be significant on the time scale of the lifetime of a molecular cloud. Ion-molecule chemistries, driven by cosmic ray ionization of H_2 in cold, weakly ionized H_2 clouds (rate $\simeq 10^{-17}$ molecule^{-1} sec^{-1}) were first proposed by Herbst and Klemperer (1973) and Watson (1974). Early models predicted sizeable abundances of HCO^+ and N_2H^+ before these species were identified, and also large abundances for many of the other simple molecules in Table 1. Subsequent increasingly complex models by many have been quantitatively successful in explaining all of the observed molecular ions, and many of the other observed molecules. There appears to be little doubt that ion-molecule chemistry is the dominant chemistry of the cold dark clouds and also of the bulk of the warm clouds with the exception of their energetic, star-forming cores. Ion-molecule chemistry appears not be relevant in CSEs, where no molecular ions are observed.

The basic scheme of interstellar ion-molecule chemistry starts with the cosmic ray ionization of H_2. Although H_2 is overwhelmingly more abundant than all other species, no two- or three-body gas-phase reaction is sufficiently rapid under typical interstellar conditions. Thus H_2 forms on grain surfaces, from which its unique volatility allows it to escape to the gas phase at low temperatures. The ionization of H_2 produces H_2^+ and H^+. These and He^+ (also highly abundant) react quickly with H_2 to form H_3^+, the "central" ion in the ion-molecule scheme. H_3^+ reacts with CO, N_2, CS, ... to produce HCO^+, N_2H^+, HCS^+. Other highly important ions produced directly from H_3^+ such as H_3O^+, NH_3^+, CH_3^+, go on to produce the hydroxyl, amine, and hydrocarbon families in dense clouds. The electron fraction, in which all negative charge resides, is typically $10^{-8}-10^{-7}$ in dense clouds, and balances the positive charge of the molecular ions and of singly ionized metals. The metal abundances in dense clouds are currently unobservable but possibly very low, as they may be locked up on the dust grains.

Not all simple molecules are produced by ion-molecule reactions. Species such as CO, N_2, O_2, CN are produced largely by many neutral-neutral reactions which, although slower, result in significant abundances because destruction rates for these species are also slow. Most neutral molecules are, however, produced from ions by the fast dissociative recombination-reactions with electron, and by charge transfer reactions (especially with H_2).

Radiative association reactions are also important; since their reaction rates increase with increasing size of the fragments involved, they may be particularly important for the synthesis of larger molecules such as CH_3OH, H_2CCO, CH_3CHO, $(CH_3)_2O$, etc. The limitation of these reactions are presently unknown, but they are speculated to supercede the non-radiative association reactions in producing molecules of $\gtrsim 4$ or 5 atoms in the cold dark clouds. It is even possible that they may eventually build "mushy" grain-sized agglomeration ($\sim 10^{-5}$ cm size) of water, alcohols, aldehydes, etc.

Despite the considerable success of ion-molecule interstellar chemistry, it entails many uncertainties: (i) The majority of rate constants used in the models have not been measured, especially at low (< 77 K) temperatures, and are usually assumed to have the Langevin rate. (ii) Models are sensitive to input atomic abundances. Solar abundances, and various "depleted" abundance ratios measured for diffuse interstellar clouds from UV absorption lines, are used. Nothing is known of the atomic abundances in dense clouds, where differential adsorption onto grain surfaces is surely important. (iii) The time-dependence of the interstellar chemistry is not tractable without tying the chemistry to the detailed physical evolution of the cloud, about which little is known.

Given such uncertainties, it is surprising how successful ion-molecule chemistry has proven to be, at least for the simpler (< 4-atom) species. There is general agreement between predicted and observed abundances for most simple molecules containing first-row elements, in diffuse and in cold dark clouds. Not only the abundance ratios of the molecular ions and of species such as OH, HNC, HCN, H_2O, CN, but also their predicted dependence upon overall particle density, are borne out by observations.

Perhaps the greatest success of ion-molecule chemistry has been the explanation of the large concentrations of deuterium observed in several species on cold cloud cores. The observed ratios, $DCO^+/HCO^+ \simeq$ 0.01, $N_2D^+/N_2H^+ \approx NH_2D/NH_3 \approx DCN/HCN \sim 0.001 - 0.01$, even though $D/H = 2 \times 10^{-5}$ cosmically, are predicted naturally from the few ion-molecule reactions that govern these species, provided that the fractional ionization is very low ($< 10^{-8}$), which in turn occurs only for the lowest temperatures ($\lesssim$ 15 K) at which metals (which carry much of the positive charge) freeze onto grains. The large ratios observed for DC_3N/HC_3N, DC_5N/HC_5N, $HDCO/H_2CO$ strongly suggest that these species are also governed by ion-molecule reactions regardless of which ion pathway dominates their production.

A reasonable summary of the current status of ion-molecule models for simple species is found in the models of Graedel et al. (1982). These studies reveal a few problems for ion-molecule chemistry as well as the many successes. For example, several N-bearing molecules (e.g., N_2H^+, NH_3) are not predicted to be as abundant as observed, and it is possible that ion-molecule theory incorrectly treats chemistry beginning with atomic N. This may also be indicated by its failure to predict the rarity of N-O bonds among interstellar species, or the observed rather low abundance of NO itself. Moreover, one or two simple species such as CH, CH^+ have poorly predicted abundances unless the endothermic reaction $C^+ + H_2 \rightarrow CH^+ + H$ is invoked, thus requiring high temperatures and shocks (peculiar velocities often observed for CH^+ gas actually do suggest independently the possibility of shocks). Perhaps the most striking class of problems are illustrated by the sulfur-bearing molecules such as H_2S and SO, which seem to require high temperature reactions to explain their observed abundances. These observations indicate that ion-molecule chemistry has limitations, which can best be delineated by further observations of the chemistry of cold dark clouds, where high-temperature processes can be ruled out.

SHOCK CHEMISTRY

Supersonic velocities are common in massive star forming regions within warm molecular clouds, and in several diffuse clouds upon which supernova explosions have impinged. These velocities, which give rise to shocks, are observed directly in the spectra of many atomic and molecular species. Models of such regions, as well as the observed ionization and excitation equilibrium of atomic and molecular species indicate temperatures as high as 2000 K and densities at least as high as 10^7-10^{11} cm^{-3} (in star forming regions). The cooling time for such gas (to ~ 100 K) is a few hundred years, much longer than reaction times at these temperatures and densities. Higher temperatures in molecular gas (100 to 500 K) may also result from heating of the gas by imbedded protostellar objects. In either case, endothermic reactions, and ones with activation barriers may now occur, so that "ion-molecule" abundances become seriously altered, the more so the higher the temperature. In the case of passing shocks, the subsequent rapid cooling of the gas may cause the "shock chemistry" products to become "frozen" for some time in the post-shock material.

Numerous shock chemistry models have been made (e.g., Mitchell 1984). They are similar to ion-molecule models in consisting of complicated networks of chemical reactions built into extensive computer codes, but in addition to allowing many more reactions, they must couple the shock kinetic equations in with the chemical network. Models so far have dealt only with strictly dissociative shocks, rather than magnetohydrodynamic shocks which are even more complicated, including among other things the effects of "magnetic precursors" which cause the shock acceleration and compression of material to occur in a more continuous manner, with the result that molecules can elude dissociation at higher shock velocities.

Because of these complexities, shock chemistry models are at best rudimentary so far, but have nevertheless explained a few observational results, which eluded ion-molecule predictions. These are the large abundance of CH^+ in diffuse clouds, of CH toward certain very high density warm molecular cloud cores, and of the highly refracting species SiO seen only in regions where there is direct observational evidence of shocks. Observed abundances of sulfur species also seem to require the high temperatures. Curiously, there is no evidence as yet that any of the simpler species listed in Table 1 are produced by shock chemistry, but not by ion-molecule chemistry. Shocks appear only to modify their abundances in many, but not all, cases. The role of shock chemistry in producing the more complex species listed in Table 1 is completely unknown. Many of these complex species are in fact observed in only two regions (Orion, and near the center of the Galaxy) which contain not only strong shocks, but also higher excitation conditions and larger columns of gas than any other sources, and perhaps even unusual atomic abundances. It is now known that the molecular chemistry varies sharply on small size scales within these regions, but the reason is unclear. Even if shocks can be established as responsible, it remains to determine whether the resulting high temperatures, or the sputtering of molecules from grains (a process that does not presently allow meaningful predictions) are the fundamental processes. The overall role of shocks in astrochemistry has yet to be evaluated.

DUST GRAINS AND ASTROCHEMISTRY

Grains affect interstellar chemistry in a variety of ways. They extinguish starlight and thus protect molecules in cloud interiors from photodestruction. They lock up significant fractions of reactive atoms, especially the refractive elements, which are therefore depleted in the gas phase and are less readily available for gas phase reactions. Grains in dense clouds are known to accrete molecular mantles, which may possibly undergo further processing, and which eventually are returned to the gas, either near hot stars, or when the cloud is dissipated. In some processes, the precise nature of the grain surface is unimportant; it merely provides an inert surface upon which atoms and molecules can stick and undergo further processing. In other cases, specific catalytic reactions may occur at particular types of sites on grains of specific composition. Finally, grains may be disrupted by shocks or by UV radiation, returning large numbers of molecular species to the gas phase.

From the 1930's onward, increasing knowledge of the depletions of gas phase elements, as well as of the optical properties of various solids, has led to increasing understanding of grain composition. The most recent depletion determinations suggest that silicates and oxides make up 70% of grains by mass and take up 75% of the depleted oxygen. Carbonaceous material must also be important. These are the "basic" grain compositions, on top of which mantles of accreted molecules may or may not occur, depending on the temperature.

Catalytic Reactions on Grain Surfaces

With the exception of H_2, it is not clear whether molecules form and are returned to the gas phase from the surfaces of grains. Certainly, the four important processes in any model of surface catalysis are: (i) adsorption of a radical or atom from the gas; (ii) migration over the surface via quantum-mechanical tunnelling; (iii) reaction with a second absorbed atom or molecule; (iv) ejection or evaporation of the product molecules back to the gas phase. The details of each of these processes depend critically on the temperature, composition, and physical properties of the surface. Independent of the composition, extensive laboratory work suggests that in cold clouds (T $\lesssim$ 20 K) the binding energy to grains of saturated molecules, and probably also of atoms, will be weak, corresponding to physical adsorption. The reason is that even if grains do not initially contain ice mantles or low-binding monolayers of H_2, they will soon accumulate them. On initially more strongly binding surfaces, evaporation is not possible, so condensation of H_2 or of ices of H_2O, NH_3, CH_4 will occur, on top of which the binding is weak. In warm clouds in which temperatures exceed ~80 K, ice mantles or H_2 layers evaporate, surfaces are chemically active, and atoms or radicals will bind tightly, although saturated molecules may still be only weakly bound.

Do the four basic processes occur at rates sufficient to compete with gas phase reactions as a source of interstellar molecules? Certainly process (i) does, as may be calculated from the known number densities and sizes of grains, the gas temperatures, and the fact that laboratory-measured sticking probabilities are at least 0.2 and probably close to unity under interstellar conditions. Measured mobilities are also large enough but definitely depend on species. Activation barriers are probably sufficiently small for reactions between atoms or radicals, but not necessarily between a radical and a saturated species, and not between saturated species. The biggest problem is the ejection of the products. The heat of formation is likely adequate for physically adsorbed saturated species, is uncertain for physically adsorbed atoms and radicals and for semi-chemically adsorbed saturated species on active surfaces, and is insufficient for chemically adsorbed atoms and radicals. Alternative ejection mechanisms, such as photodesorption, and thermal pulses caused by photons or cosmic rays may at best suffice under special conditions.

Because of all of these uncertainties, modelling of catalytic processes (e.g., Allen and Robinson 1977) has yielded highly varied predictions, and none of the models matches the observational picture very well, nor appears capable of the degree of specificity needed to describe, for instance, the different relative molecular abundances found in the cold vs. warmer molecular clouds.

For cold clouds, we can cite a number of problems for grain catalysis that are independent of model details. (i) The relative abundance of several species (HCN, HNC, CO, H_2CO), is higher in cold clouds than in warm clouds, while surface formation rates should decrease with decreasing temperature. (ii) Molecular ions certainly are not expected products of surface reactions. (iii) Enhanced deuteration, as observed in the coldest cloud cores, is unexpected for surface reactions because surface deuterium-exchange reactions have measured activation energies of ~1000 K. (iv) The enhanced abundance of the long-chain cyanopolyyne species in cold cloud cores is unexpected for surface reactions because these species are expected to bind much more tightly to grains than most species. (v) under conditions of physical adsorption, where surface mobilities are unrestrained, diatomic hydrides should form predominantly over other species because of the large H abundance, so that $OH/NH/CH \approx O/N/C$ is expected. This is far from observed. For these and other reasons grain catalysis in cold clouds does not seem indicated for the majority of observed species.

For warm clouds (T $\gtrsim$ 80 K, where ice mantles sublime) the basic composition of the grains becomes relevant. Relatively little pertinent laboratory data exist. In <u>silicate</u> surfaces Fisher-Tropsch type reactions proceed efficiently at T $\gtrsim$ 500 K, but not at the lower temperatures typical of warm clouds. Further, many species produced (e.g., complex rings) as well as relative abundances as a function of C-number, bear little resemblance to the interstellar situation. On laboratory <u>graphite</u> surfaces at 78 K, CO, CO_2 and hydrocarbons are synthesized from H, N, O, S, but a carbon atom is removed from the graphite for each molecule formed, a process that would soon erode away interstellar grains, which actually appear instead to be accreting with time in cool clouds. <u>Metallic oxide</u> grains (MgO/SiO/FeO/CaO) are well known catalysts which at low temperatures (77 K) appear able to form species such as H_2O, HCO, NH_3, H_2CO with high efficiency (e.g. Duley and Millar 1978). In diffuse interstellar clouds such species could be photodissociated to yield CO, OH, NH, etc. In dense clouds, however, models suggest the active surface sites of oxide grains appear to become poisoned and may only be able to form molecules with several heavy atoms, which could not be released to the gas. Thus it appears that oxide grains are more likely to account for the depletion of the elements in the ISM (models of which agree well with observed depletions) than for production of gas phase molecules.

In summary, the catalytic formation of interstellar molecules on grain surfaces does not appear likely (except for H_2) when T $\lesssim$ 80 K, which comprises the cold dark clouds and most of the warm ones that are heated by embedded protostars rather than by shocks. Recent observational evidence is accumulating that as much as 50% of volatile molecules such as CO, H_2O, NH_3 are locked up in the mantles of grains in the vast majority of molecular clouds. This seems to support the arguments based on specific abundances that molecules do not come off grains under normal conditions. It is much harder to decide whether catalysis is important for grains in higher temperature regions, but one notes that temperatures between 80 K (ice mantle sublimation) and $\gtrsim$ 500 K (indicating the presence of shocks, which would have volatized the grains and made previous catalysis irrelevant) represent a very small range of possible physical conditions, which would not be expected to characterize many interstellar regions. In addition, the few difficulties encountered at present by gas-phase models of the chemistry are not alleviated by invoking grain surface reactions.

The Disruption of Interstellar Grains

<u>Ultraviolet Photoprocessing.</u> Observational evidence has accumulated within the past few years that there is a sufficient flux of high energy UV photons in molecular clouds to modify grain mantles, which at low temperatures consist mainly of CO and ices of H_2O, NH_3, CH_4, species

922

which have accreted from the gas phase. A longstanding question is why, in dense cold clouds lacking other forms of energy, all of the gas is not accreted onto grains on time scales of $\lesssim 10^6$ years, which is comparable to or shorter than the dynamic lifetime of the cloud against gravitational collapse ($\sim 10^7$ yr) or the lifetime against cloud-cloud collisions or disruption by the formation of a nearby hot star (10^7-10^8 yr). One answer to this dilemma has been provided by the extensive laboratory work of Greenberg and collaborators (cf. Greenberg 1986) on the effect of irradiating simulated grain mantles at T = 10 K with UV radiation. Such radiation modifies the chemistry of the mantles, in particular creating many radicals which are shown at these low temperatures not to diffuse readily within the mantle and continuously react with each other, but rather to remain largely trapped in local sites, thus storing up increasing amounts of chemical energy. Now, if the temperature of the grain is suddenly increased, as in a sufficiently energetic grain-grain collision which occurs every $\gtrsim 10^5$ yr, the sudden increase in the rate of radical reactions generates heat faster than it can be conducted away, thus leading to further release of trapped radicals and a runaway explosive chain reaction. This mechanism for depletion of grain mantles appears to be adequate to maintain the gas phase species against depletion onto the grains. IR spectra of laboratory photoprocessed mantles agree much better with interstellar IR spectra than do spectra of unprocessed mantles, and indicate that complex organic refractory residues are formed in the mantles along with the volatile ice mantles. The composition of the latter is modelled to vary over the cloud lifetime, initially being dominated by H_2O, NH_3, CH_4 ices, later evolving to mostly H_2O, CO, and CO_2. Gas phase and mantle relative abundances differ greatly on this picture, for example the H_2O mantle abundance greatly exceeding the gas phase abundance, so that H_2O is largely formed on the grains. Despite the laboratory basis for these processes, the question whether interstellar grain mantles can actually store up the required density of radicals to produce explosive disruption remains unanswered. In any case, it has not yet proved possible to characterize the products of the explosively disrupted mantles. There is a possibility, however, that some of the products are more complex than those so far identified in cold interstellar clouds. Further laboratory and observational work are needed to decide whether the "continuous" desorption or ejection of molecules by photoprocessing of grain mantles can solve the depletion problem for gas phase molecules, or whether this is accomplished by major dynamical events such as star formation or cloud-cloud collisions.

Evaporation and Sputtering of Grains

The rich chemistry of the energetic core of the Orion star forming cloud, its proximity to Earth, and the success of interferometric techniques at millimeter wavelengths in recent years, have taught us much about the intricate and sharp variations of interstellar chemistry with variations in physical conditions over small distances. In the Orion core, for example, it is possible to distinguish the chemistry of regions at ~ 200 K, apparently heated by embedded protostars from those at ~ 2000 K, which are heated by powerful shocks.

Grains whose temperatures range from ~ 80 K to a few hundred K will suffer evaporation of their volatile mantles and perhaps some of their organic residue, but the underlying grain (silicate, metal oxide) will remain intact. In the very dense star forming cores, grain concentrations are large, and evaporation of their mantles may well dominate the chemistry of the core. In addition to many complex molecular species so far observed only in the Orion core and in the "similar" Sgr B2 molecular cloud near the galactic center, the recent observation of copious quantities of H_2O and NH_3, and of their deuterated analogues in the hot Orion core provides the first strong indication that grain mantle evaporation is occurring. Ion molecule reactions fail to predict large H_2O, NH_3 abundances, and apparently higher temperature gas phase reactions also have difficulty. Quite certainly, gas phase reactions cannot produce the large NH_2D/NH_3 and HDO/H_2O ratios observed at the higher Orion core temperatures, because deuteration relies on the zero-point energy differences between deuterated and undeuterated species, the effect of which diminishes with increasing temperature. Conversely, grain mantles are expected to contain an overabundance of D relative to H, because H_2 can be released from grains more easily than HD. Photoprocessing then converts the excess D into an excess of deuterated molecules in the mantles. Excessive deuteration of interstellar molecules is thus a firm prediction of both ion molecule chemistry at $T \lesssim 20$ K and of grain mantle evaporation at $T \gtrsim 80$ K, and is not expected to occur at intermediate temperatures. Observations confirm these predictions. The more complex species seen only in the hot core of Orion may therefore be surmised to be products of grain mantle evaporation, and include EtCN, CH_3CN, $(CH_3)_2O$, CH_3OHCO, VyCN, etc. If grain mantle evaporation indeed produces the large H_2O, NH_3 abundances seen in hot cores, then the smaller relative abundances seen in cold clouds implies that explosive disruption of photoprocessed mantles is unimportant.

Grains subjected to strong shocks, and attendant high ($T \gtrsim 2000$ K) temperatures will lose not only their volatile mantles and organic residues, but also the underlying silicate, carbonaceous, or metal oxide structure will be atomized. Virtually nothing is known of this process in dense molecular clouds, because SiO and SiS are the only observed molecules that are possibly relevant to this process. Aside from CSEs, these highly refractory species are seen only in very hot energetic star forming regions, and it is assumed that they result from grain disruption. One of the few additional clues to the chemistry of shock-disrupted grains is that MgO/SiO << Mg/Si is observed in these regions, and has been argued as consistent with (but not indicative of) the hypothesis of metal oxides as basic grain constituents. The major obstacles to progress in this area are (i) almost nothing is known about the gas phase chemistry of molecules containing Si, Mg, Fe, Ca, etc., (ii) microwave spectroscopy exists for only very few hydrides or oxides of the refractory elements, these being the most likely compounds of these elements to survive grain disruption or to reform quickly in the post shock gas; (iii) the refractory atoms themselves are not currently observable in these dense, hot, opaque regions, their fine-structure transitions (if any) lying in regions of the far-IR that are blocked by the earth's atmosphere. Some additional information about shock-induced grain destruction comes from optical observations over the past 50 years of absorption lines of atomic Na, Ca, Ti, etc., in diffuse clouds, seen toward hot stars. These observations have shown that Ca, in particular, is normally severely depleted in "quiescent" diffuse clouds, but is relatively undepleted in certain diffuse clouds that exhibit peculiar velocities apparently produced by shocks. If the shock velocity exceeds ~ 20 km/s, the grain cores seem to be completely atomized, although the shock disruption might possibly produce small molecules which are then photodissociated.

Perhaps the basic question in all of astrochemistry is: which came first, grains or molecules? Although interstellar molecules cannot have formed in stellar atmospheres, interstellar grains can. Are all grains originally formed in stars? If grains are then needed to form interstellar molecular clouds, and such clouds form stars, then how did the stars originally form? The basic nature of grain cores, and their rate of destruction and formation, is central to answering these questions. It has been suggested (Greenberg 1986) that grain evolution is cyclic, with grains finding themselves alternately in the diffuse ISM and in dense molecular clouds. Grains in the diffuse ISM get captured by dense clouds, in which they accrete mantles. Dense clouds form stars which disrupt the clouds, returning the surviving grains to the diffuse ISM where at least their volatile mantles, and some of their organic refractory material are eroded by photoprocesses. Pure silicate grains, unprotected by residual organic material, would be destroyed at a rate 100 times faster than they could possibly be regenerated by the known mass loss rate of stars to the ISM. This implies that only a small fraction of diffuse clouds can undergo atomization of their silicate grains by shocks (as seems to be observed), or else that silicate grains can somehow condense out of the diffuse ISM. What is completely unknown, however, is whether silicate grains are commonly disrupted in dense star forming regions. This is where the astrochemistry of the refractory second-row elements, so far unstudied, must provide answers.

THE CHEMISTRY OF CIRCUMSTELLAR ENVELOPES

Although CSEs form as a result of mass loss of evolved stars to the ISM, the molecules formed in these shells are not returned to the ISM, but are photodissociated at the outer regions of the envelopes. CSEs exhibit a wide range of physical conditions, from those of the inner envelope (T ~ 1000-3000 K, density $\sim 10^{10}$-10^{15} cm^{-3}) to those of the shell edge (T ~ 10 K, density $\sim 10^2$ cm^{-3}) which are similar to those in interstellar clouds. In the inner envelope, the high temperature and density mean that molecules form under thermo-chemical equilibrium conditions. Ion-molecule reactions are unimportant. The expansion of the inner shell gas to the outer edges is so fast that the "equilibrium" molecular abundances of the inner envelope are thought to be "frozen in" throughout the envelope (e.g., McCabe et al. 1979). Freeze-out is thought to occur at densities less than $\sim 10^{10}$ cm^{-3} at which point grains are believed to start forming as well. Molecular species throughout the shell should therefore be stable ones, not radicals or molecular ions which make their appearance only at the outer regions as a result of photoprocesses.

Of course, the details of the chemistry depend on whether the shell is O-rich or C-rich. Thermo-chemical equilibrium models for C-rich envelopes are largely tailored to conditions in the nearby object IRC 10216, by for the most extensively observed. The abundances of the simplest species containing C, N, O, H are well predicted, including CO, C_2H_2, HCN, CH_4, HC_3N. Almost all oxygen is tied up in CO. The ratio HNC/HCN is correctly predicted as $\lesssim 0.01$, very different from the value of ~1 observed (but not well explained) in interstellar clouds. Several other species, including NH_3, HC_5N, and most radicals, have predicted abundances considerably smaller than observed. By contrast, SiS and SiO are calculated to be 10^3 times more abundant than observed. The discrepancy concerning Si species is probably explained by most of the silicon being locked up in grains, consistent with the 11 µm spectral feature observed in IRC 10216; or the molecules may have condensed onto grain surfaces. The observed high radical abundance is explainable by photodissociation at the shell edges, which is not treated in the models. However, the discrepant predictions of other species probably indicates a failure of the "freeze-out" models for at least some species in C-rich shells.

Freeze out models are also utilized for O-rich shells, where reactions involve mostly H, O, N, Si, and S. All availble carbon is locked up in CO. The chemistry differs from C-rich envelopes also because inner shell temperatures are much higher (~3000 K vs. $\lesssim$ 1000 K). Many fewer molecules have so far been observed in O-rich CSEs, mainly CO and NH_3 in the inner envelope (via IR techniques), and CO, H_2O, OH, SiO throughout the envelope. OH, H_2O, and SiO all exhibit maser amplification, making accurate abundance determinations difficult, but indicating larger abundances than occur in C-rich envelopes, as predicted by the models. Nearly all Si should be in the form SiO, and all oxygen in CO, OH, H_2O, as observed. Other than these simple conclusions, little can be said to compare observations and theory, so that the freeze-out model has not yet posed any definite problems.

In summary, "freeze-out" chemistry probably dominates molecule formation throughout a good portion of both C-rich and O-rich shells. However, the effects of photodissociation at the shell edge, as well as differential condensation into grains, are not currently well understood although they may well control the chemistry in the outer layers of the shells. The role of shocks will also have to be evaluated, since observational evidence for them has recently been found. Finally, although the molecular gas phase chemistry of CSEs is isolated from the ISM, that of grains is not. These grains may well pass into the general ISM via the mass loss of these stars. Their nature is unfortunately poorly understood. Based on probable identifications of a very few IR spectral features, grains are thought to be mainly silicates (e.g., Mg_2SiO_4, Mg_2SiO_3) in O-rich envelopes, and graphite as well as SiC in C-rich envelopes. Various speculations involving carbyne, amorphous carbon ("soot") and even organic polymers have been made, but supporting evidence from observed IR spectra is at best inconclusive.

CONCLUSIONS

In this cursory review, many important subjects in astrochemistry have been omitted: isotope ratios in molecules, and their bearing on the nucleosynthesis history of the Galaxy; cometary chemistry and the possible connection between interstellar chemistry and that of the protosolar nebula, and perhaps even the origin of life; the 50-year-old mystery of the "diffuse interstellar bands"; stellar atmosphere chemistry; meteorite chemistry; early Universe chemistry; planetary atmosphere chemistry. To better understand any one of these areas requires advances in the laboratory, in the areas of microwave spectroscopy, the synthesis of terrestrially unstable compounds, rate constants for reactions of ions, radicals, and other types of molecules at all temperatures, but especially low ones, photoprocesses (rates, oscillator strengths), collision processes (selection rules, rates), surface chemistry, and possibly even the properties of large "cluster" molecules (e.g., polycyclic aromatic hydrocarbons such as C_{60}) which have recently been suggested as possible forms for small (~10 Å) interstellar grains which seem to occur in the tenuous "cirrus" component of the ISM. The many chemistries relevant to the ISM are clearly interrelated. To disentangle them will also require greatly refined observational data, at all wavelengths (many of which are accessible only from space) and at all spatial size scales (requiring major advances in interferometry, a technique only in its infancy at wavelengths shorter than 1 cm). The ultimate unification of early Universe chemistry, interstellar chemistry, circumstellar and stellar chemistry, cometary and meteorite chemistry (together with planetary atmosphere chemistry) is a challenging task which requires research in many different branches of physics, chemistry, and astronomy. Its successful completion will teach us much about the Universe and its evolution.

Table 1. Known Interstellar Molecules (January 1987)

No. of Atoms	Molecule	Opt/UV/IR	Radio Spectrum	Type of Objects[a]			
2	H₂ hydrogen	UV IR	-	1	2	3?	4
	OH hydroxyl radical	UV	Λ-doubling, rotational	1	2	3	4
	CH⁺ methylidyne ion	Opt	-	1			
	CH methylidyne	Opt	Λ-doubling	1	2		4
	CO carbon monoxide	UV IR	rotational	1	2	3	4
	CN cyanogen radical	Opt	rotational	1	2	3	
	CS carbon monosulfide	IR	rotational	1	2	3	
	NO nitric oxide		rotational		2	3	4
	C₂	IR	-	1	2	3	
	SO sulfur monoxide		rotational	1	2	3	
	NS nitrogen sulfide		rotational		2		
	SiO silicon monoxide	IR	rotational		2	3	
	SiS silicon monosulfide		rotational		2	3	
	HCℓ hydrogen chloride		rotational		2		
3	HCO⁺ formyl ion		rotational	1	2		4
	N₂H⁺ diazenylium		rotational	1	2		
	H₂O water		rotational		2	3	4
	HCN hydrogen cyanide	IR	rotational	1	2	3	4
	HNC hydrogen isocyanide		rotational	1	2	3	
	CCH ethynyl radical	IR	rotational	1	2	3	
	HCO formyl radical		rotational		2		
	OCS carbonyl sulfide		rotational		2		
	H₂S hydrogen sulfide		rotational	1	2	3	
	SO₂ sulfur dioxide		rotational	1	2	3	
	HCS⁺ thioformyl ion		rotational	1	2		
	SiC₂ silicon dicarbide radical		rotational	1	2	3	
	H₂D⁺		rotational	1			

(Cont'd)

Table 1. — Continued

No. of Atoms	Molecule	Opt/UV/IR	Radio Spectrum	Type of Objects[a]
4 NH_3	ammonia	IR	inversion	1 2 3 4
H_2CO	formaldehyde		rotational, K-doubling	1 2 3 4
HCCH	acetylene	IR	–	1 3
C_3N	cyanoethynyl radical		rotational	1 3
HNCO	isocyanic acid		rotational	1 2
H_2CS	thioformaldehyde		rotational, K-doubling	1 2
HCNS	isothiocyanic acid		rotational	2
$HOCO^+$	protonated carbon dioxide		rotational	2
C_3H	propynylidyne		rotational	1 3
C_3O	tricarbon monoxide		rotational	1
$HCNH^+$	protonated hydrogen cyanide		rotational	2
H_3O^+?	protonated water?		rotational	2?
5 CH_4	methane	IR	rotational?	1 2? 3
C_4H	butadiynyl radical		rotational	1 2 3
HC_3N	cyanoacetylene		rotational	1 2 3
H_2CCO	ketene		rotational	2
NH_2CN	cyanamide		rotational	2
CH_2NH	methanimine		K-doubling	2
HCOOH	formic acid		rotational, K-doubling	2
SiH_4	silane	IR	–	
C_3H_2	cyclopropynylidene		rotational	1 2 3
6 CH_3CN	methyl cyanide		rotational	1 2 3
NH_2CHO	formamide		rotational, K-doubling	2
CH_3OH	methanol		rotational, K-doubling	1 2
CH_3SH	methyl mercaptan		rotational	1 2
C_2H_4	ethylene	IR	–	3
C_5H	pentynylidyne radical		rotational	3

No. of Atoms	Molecule	Opt/UV/IR	Radio Spectrum	Type of Objects[a]
7 CH_3CCH	methyl acetylene		rotational	1 2
CH_3CHO	acetaldehyde		rotational, K-doubling	1 2
NH_2CH_3	methylamine		rotational	2
CH_2CHCN	vinyl cyanide		K-doubling	1 2 3
HC_5N	cyanobutadiyne		rotational	1 2 3
8 $HCOOCH_3$	methyl formate		rotational	2
CH_3C_3N	methylcyanoacetylene		rotational	1
9 CH_3CH_2OH	ethanol		rotational	2
CH_3CH_2CN	ethyl cyanide		rotational	2
CH_3OCH_3	dimethyl ether		rotational	2
HC_7N	cyanohexatriyne		rotational	1
CH_3C_4H	methyldiacetylene		rotational	1 3
10 None				
11 HC_9N	cyano-octatetrayne		rotational	1 3
12 None				
13 $HC_{11}N$	cyanotetracetylene		rotational	1 3

[a]1: cold dark clouds and/or diffuse clouds; 2: warm molecular cores in Giant Molecular Clouds; 3: circumstellar envelopes; 4: external galaxies.

REFERENCES

Allen, M., and Robinson, G. W. 1977, The Molecular Composition in Dense
    Interstellar Clouds, Ap. J. 212:396.
Duley, W. W., Millar, T. J., and Williams, D. A. 1978, Molecule
    Production on Interstellar Oxide Grains, Mon. Not. Roy. Astr. Soc.
    185:915.
Herbst, E., and Klemperer, W. 1973, The Formation and Depletion of
    Molecules in Dense Interstellar Clouds, Ap. J. 185:505.
Gammon, R. H. 1978, Chemistry of Interstellar Space*, Chemical and
    Engineering News 56:21.
Graedel, T. E., Langer, W. D., and Frerking, M. A. 1982, The Kinetic
    Chemistry of Interstellar Clouds, Ap. J. (Suppl.) 48:321.
Greenberg, M. J. 1986, Basic Laboratory Studies of Grains, in IAU
    Symposium No. 120, Astrochemistry, ed. M. S. Vardya and
    S. P. Tarafar (D. Reidel).
McCabe, E. M., Connon-Smith, R., and Clegg, R.E.S. 1979, Molecular
    Abundances in IRC 10216, Nature 281:263.
Mitchell, G. F. 1984, Effects of Shocks on Molecular Composition of a
    Dense Interstellar Cloud, Ap. J. (Suppl.) 54:81.
Thaddeus, P. 1981, Radio Observations of Molecules in the Interstellar
    Gas*, Phil. Trans. Roy. Soc. Lon. A303:469.
Turner, B. E. 1980, Interstellar Molecules*, J. Mol. Evol. 15:79.
Turner, B. E., and Ziurys, L. M. 1987, Interstellar Molecules and
    Astrochemistry*, in Galactic and Extragalactic Radio Astronomy,
    eds. K. I. Kellermann and G. L. Verschuur, Springer-Verlag.
Watson, W. D. 1973, The Rate of Formation of Interstellar Molecules by
    Ion-Molecule Reactions, Ap. J. (Letters) 183:L17.

*General Review

PARTICIPANTS

John Allen
Division of Chemistry
University of Maryland
College Park, MD

Young K. Bae
Su-Machi
Tsuchiura-Shi, Ibaraki-ken 300
Japan

Francois Amar
Department of Chemistry
University of Maine
Orono, Maine 04469

John T. Bahns

Aldo Amore
ITSE A.d.R.
Via Colaria Km 29.500 C.P. 10
Monterontondo Stazione 00016
Italy

Luis C. Balbas
Department Optica
University Valladolio
Valladolio, Spain

Scott L. Anderson
Chemistry Department
SUNY
Stony Brook, NY 11794

A. P. Baronavski
Chemistry Division
Naval Research Lab.
Washington, D. C. 20375-5000

R. P. Andres
School of Chemical Eng.
Purdue University
Chemical and Metallurgical Eng.
   Bldg.
West Lafayette, IN 47907

Yael Barshad
Isotope Department
The Weizmann Institute
Rehovot 76100
Israel

T. Araya
Mec. Eng. Res. Lab
Hitachi Ltd.
502 Kandat
Tsuchiura-shi Ibaraki-ken 300
Japan

L. S. Bartell
Department of Chemistry
University of Michigan
Ann Arbor, Michigan 48109-1055

Paul S. Bechthold
Chemistry Division
Argonne National Lab.
Argonne, Il 60439

Scott Berry
AT&T Bell Labs
Murray Hill, N.J. 07974

Donald E. Beck
Department of Physics and Lab.
   for Surface Studies
University of Wisconsin-Milwaukee
Milwaukee, Wisconsin 53201

Estela Blaisten-Barojas
Instituto de Fisica
UNAM, Ap. Postal 20-364
Mexico D.F. 01000
Mexico

Thomas Beck
Department of Chemistry
University of Chicago
Chicago, Il 60637

L. A. Bloomfield
Department of Physics
University of Virginia
Charlottesville, VA 22901

Benjamin Bederson
Department of Physics
New York University
4 Washington Place
New York, NY 1003

Kit Bowen
Department of Chemistry
Johns Hopkins University
Baltimore, MD 21218

R. E. Benfield
University Chemical Lab.
Univ. of Kent at Canterbury
Canterbury, Kent CT2 7NH
U. K.

E. B. Bradley
453 Anderson Hall
University of Kentucky
Lexington, KY 40506-0046

Jerry Bernholc
N. C. State Univ.
Box 8202
Raleigh, N. C. 27695

Catherine Brechignac
Lab. Aime Cotton
Campus Orsay
CNRS II Bat. 505
91405
France

E. R. Bernstein
Chemistry Dept.
Colorado State Univ.
Ft. Collins, Colo. 80523

M. Broyer
Universite Claude Bernard
Lyon 1
Laboratorie de Spectrometrie
Batiment 205
43, Blvd. du 11 Novembre 1918
69622 Villeurbanne Cedex
France

R. S. Berry
Department of Chemistry
University of Chicago
Chicago, IL 60637

P. J. Brucat
Dept. of Chemistry
University of Florida
Gainesville, FL 32611

934

Steve Buckner
Department of Chemistry
Box 338
Purdue University
West Lafayette, Indiana 47907

Eugene Choi
Chemical and Metallurgical Bldg.
Purdue University
West Lafayette, IN 47906

Philippe Cahuzac
Lab. Aime-Cotton NRS 11 Bat 505
91405 Orsay Cedex
France

Mikael Ciftan
U. S. Army Research Office
P. O. Box 12211
Research Triangle Park
N. C. 27709

J. H. Callahan
Chemistry Division
Naval Research Lab
Washington, D. C. 20375-5000

Francisco Claro
Facultad de Fisica
Universidad Catolica de Chile
Casilla 6177
Santiago 22, Chile

A. W. Castleman, Jr.
Dept. of Chemistry
152 Davey Laboratory
Pennsylvania State University
University Park, PA 18803

Franco Cocchini
Scuola Normale Superiore
Pisa 56100
Italy

R. L. Champion
Physics Department
College of William and Mary
Williamsburg, VA 23185

Dr. S. Keith Cole
Argonne National Lab.
Chemistry Division, D-200
Argonne, Il 60534

Andrew Chatelain
Inst. Phys. Exp.
Fed. Inst. of Tech.
1015 Lausanne
Switzerland

Charles Conover
Department of Physics
McCormick Road
Charlottesville, VA 22901

Sandeep Chaube
Physics Department
Virginia Commonwealth Univ.
Richmond, VA 23284-0001

Brian Constance
Physics Department
Virginia Commonwealth University
Richmond, VA 23284-0001

Donald Cox
Exxon Research
Rt. 22, East
Annandale, N.J. 08801

J. Dorantes-Davila
Av. Instituto Politenico
Nacional No. 2508
Mexico D. F. 07000
Mexico

F. Cyrot-Lackmann
Lab. D'etudes des Proprietes
    Elec. des Solides
25, Avenue des Martyrs
B. P. 166
38042 Grenoble Cedex
France

Michael A. Duncan
Department of Chemistry
University of Georgia
Athens, GA 30602

Joe Darby
Department of Energy, Materials
    Science Division
Washington, D. C. 20585

R. W. Dwyer
Philip Morris Research
Box 26583
Richmond, VA 23261

J. A. Darsey
Department of Physical Sciences
Tarleton State University
Stephensville, TX 76402

Robert A. Eades
Signal Research Center Inc.
50 E. Algonquin Road
Box 5016
Des Plaines, IL 60017-5016

Guru Das
Physical Chemistry Division
National Chemical Laboratory
Pune - 8
India

Olof E. Echt
Universitat Konstanz, Fak. Physik
D-7750 Konstanz
West Germany

L. J. de Jongh
Kamerlingh Onnes Lab.
Der Rijksuniversiteit Leiden
Nieuwsteeg 18
Leiden, Netherlands

Alan S. Edelstein
Metal Physics Branch
Code 6632
Dept. of the Navy
Naval Research Laboratory
Washington, D. C. 20375

Stephanie Di Cenzo
41 Evergreen Road
Summit, N. J. 07901

Walter P. R. Ekardt
Fritz-Haber-Institut der M.P.G.
Faradayweg 4-6, D-1000
Berlin 33
West Germany

D. E. Ellis
Department of Physics & Astronomy
Northwestern University
Evanston, IL 60201

John L. Fry
University of Texas at Arlington
Arlington, TX 76019

J. D. Eversole

Pei Yu Gao
Fritz-Haber-Institut der MPG
Abt. Elektronenmikroskopie
Faradayweg 4-6
D-1000 Berlin 33 (Dahlem)
West Germany

David W. Ewing
Department of Chemistry
John Carroll University
Cleveland, Ohio 44118

Michael E. Geusic
Room 4C 330
AT&T Bell Labs
Crawfords Corner Road
Hoimclel, N.J. 07733

J. Farges
Groupe des Agregats Atom. & Mol.
Lab. de Physique des Solides
Universite de Paris Sud, F-91405
Orsay, France

Robert H. Gowdy
Physics Department
Virginia Commonwealth Univ.
Richmond, VA 23284-0001

Liu Feng
Physics Department
Virginia Commonwealth University
Richmond, VA 23284-0001

Jurgen Gspann
Universitat Karlsruhe, Postfach
3640
D-7500 Karlsruhe
West Germany

Bradley F. Feuston
Department of Physics
University of Cincinnati
Cincinnati, Ohio 45221

Bingling Gu
Tsinghua University
Physics Department
Beijing, China

Friedrich W. Froben
Institute fur Molekulphysik
Fachbereich Physik
Freie Universitat Berlin
Arnimallee 14
D-1000 Berlin 33, Germany

T. W. Haas
Graduate Engineering Cooperative
Virginia Commonwealth University
Richmond, VA 23284-0001

H. Haberland
Albert-Ludwigs-Univ.
Fakultat fur Physik
Postanschrift: Hermann-Herder
     Strasse 3
7800 Freiburg, West Germany

Peter Hackett
Laser Chemistry Group
National Research Council
Division of Chemistry
Ottawa, Canada K1A OR6

David Hagan
Physics Department
Virginia Commonwealth University
Richmond, VA 23284-0001

D. E. Hagen
109 Norwood Hall
University of Missouri
Rolla, MO 65401

Timur Halicioglu
Stanford University
Department of Materials Science
Stanford, CA 94305

Andrew Hall
Physics Department
VCU
Richmond, VA 23284

Isaac A. Harris
Department of Physics
College of Arts and Sciences
University of Rhode Island
Kingston, R.I. 02881-0817

Laszlo Harsanyi
University of Michigan
Ann Arbor, Michigan 48109

David Hartman
Humanities and Sciences
Virginia Commonwealth University
Richmond, VA 23284-0001

John D. Head
Department of Chemistry
2545 The Mall
University of Hawaii
Honolulu, Hawaii 96822

Claude R. Henry
CRMC2-CNRS-Campus Luminy
Case 9-13
13009 Marseille-Cedex 9
France

Ajit Hira
Department of Physics
University of Texas at Arlington
Arlington, TX 76019

M. J. Histed
Division of Chemistry
National Research Council
Ottawa Ontario K1J7H3
Canada

James A. Howard
Division of Chemistry
Montreal Road
Ottawa Ontario K1J7H3
Canada

Sumio Iijima
R&D Corporation of Japan
c/o Department of Physics
Meijo University 501, 1-chome
Tempaku-ku, Nagoya 408
Japan

Wolfgang D. Kamke
Inst. fur Molekulphysik
Fachbereich Physik
Arnimallee 14
D-1000 Berlin 33
Germany

Matt Janssens
c/o R. P. Andres
School of Chemical Engineering
Purdue University
West Lafayette, Indiana 47907

M. M. Kappes
Inst. fur Anorganische und
    Physikalische Chemie
Universitat Bern
Ch-3500 Bern 9
Freistrasse 3, Switzerland

Martin F. Jarrold
AT&T Bell Laboratories
600 Mountain Avenue
Murray Hill, N.J. 07974-2070

A. Kasuya
The Research Inst. for Iron,
Steel and other Metals
Tohoku University
Sendai, 980
Japan

J. Jellinek
Chemistry Div.
Argonne National Lab.
9700 So. Cass Avenue
Argonne, IL 60439

Koji Kaya
Department of Chemistry
Keio University
3-14-1 Hiyoshi
Kohoku-ku, Yokohama 223
Japan

Puru Jena
Department of Physics
Virginia Commonwealth University
Richmond, VA 23284-0001

Robert G. Keesee
152 Davey Laboratory
University Park, PA 16802

Jia Luo
School of Physics
Georgia Institute of Technology
Atlanta, GA 30332

Hugh P. Kelly
Department of Physics
University of Virginia
Charlottesville, VA 22901

Z. K. Kafafi
National Bureau of Standards
Bldg. 222
Rm. A. 165
Gaithersburg, MD 20899

Neil R. Kestner
Chemistry Department
LA State University
Baton Rouge, LA 70803

Ulrich L. Kettler
Department of Physics
University of California
Santa Barbara, CA 93106

Tamotsu Kondow
Department of Chemistry
Faculty of Science
University of Tokyo
Hongo, Tokyo 113
Japan

Shiv Khanna
Physics Department
Virginia Commonwealth University
Richmond, VA 23284-0001

J. Koutecky
Fachbereich Chemie
Institut fur Physikalische Chemie
Freie Universitat, Berlin
Takustrasse 3
1000 Berlin 33, West Germany

P. K. Khowash
Department of Physics
Northwestern University
Evanston, IL 60201

Henry Krakauer
College of William and Mary
Department of Physics
Williamsburg, VA 23185

B. R. King
Department of Chemistry
University of Georgia
Athens, GA 30602

Jitendra Kumar
Indian Inst. of Tech, Kanpur
Materials Science Programme
ACMS Bldg.
Kanpur, 208016   India

Andrew Kirkwood
Department of Chemistry
University of Toronto
Toronto, Ontario M5S 1A1
Canada

P. Labastie
9 Avenue Condorcet
Villeurbanne F69100
France

Barry M. Klein
Code 4680
Naval Research Lab.
Washington, D. C. 20375

K. Laihing
Department of Chemistry
University of Georgia
Athens, GA 30602

Cornelius Klots
Oak Ridge National Laboratory
Oak Ridge, Tennessee 37830

Uzi Landman
GA Tech
School of Physics
Atlanta, GA 30332

B. C. LaRoy
Philip Morris Research
P. O. Box 26583
Richmond, VA 23261

Hans O. Lutz
Fakultat fur Physik
Universitat Bielefeld
4800 Bielefeld 1
West Germany

D. Levandier
Department of Chemistry
University of Waterloo
Waterloo, Ontario N2L 3G1
Canada

Emmett Maddry
Physics Department
Randolph Macon College
Ashland, VA 23005

Derek M. Lindsay
Chemistry Department
City College
Convent Avenue at 138th St.
New York, N.Y. 10031

Jean-Paul Malrieu
Lab. Physique Quantique
University P. Sabatier
Toulouse, France 31062

A. Cliff Lilly
Research Fellow
Philip Morris Research Center
Richmond, VA 23261-6583

M. L. Mandich
AT&T Bell Laboratories
Rm. 1D-251
600 Mountain Avenue
Murray Hill, N.J. 07974

N. J. Long
Center for Solid State Science
Arizona State University
Tempe, AZ 85287

R. Markus
Inst. fur Atom und Festkorperphysik
Freie Universitat Berlin
Arnimalee 14, 1000 Berlin 33
West Germany

Zhiwei Lu

G. A. Martin
Inst. de Recherche sur la Catalyse
F-69626 Villeurbanne Cedex
France

Timothy Lucas
Physics Department
VCU
Richmond, VA 23284

T. P. Martin
Max Planck Inst. fur
    Festkorperforschung
Heisenbergstr. 1
7000 Stuttgart 80
West Germany

Jose Martins
University of California, Berkeley
Department of Physics
Berkeley, California 94720

P. K. Misra
Department of Physics
The University of Rhode Island
Kingston, RI 02881-0817

Albert Masson
ENSCP-11, rue Pierre et Marie
    Curie
Paris 75005
France

Pedro A. Montano
Department of Physics
West Virginia University
Morgantown, W.VA 26506

Ruth McDiarmid
1800 G Street, N.W.
Washington, D.C. 20551

Michael D. Morse
Department of Chemistry
Henry Eyring Building
University of Utah
Salt Lake City, Utah 84112

Kevin McHugh
Johns Hopkins University
Department of Chemistry
Baltimore, MD 21218

Martin Moskovits
Department of Chemistry
University of Toronto
Toronto MSS 1A1
Canada

Michael Mehl
Naval Research Lab.
Code 4684
Washington, D.C. 20375

Sugata Mukherjee
Ecole Polytechnique Federale
de Lausanne
PHB-Ecublens
CH-1015 Lausanne
Switzerland

Karl-H. Meiwes-Broer
Universitat Bielefeld
Fakultat fur Physik, DO
Universitatsstrabe
D-4800 Bielefeld 1
West Germany

Dr. Wilhelm Niemann
Haldor Topsoe A/S
DESY/HASYLAB
Notkestrasse 85
D-2000 Hamburg 52
West Germany

Gordon A. Melson
Department of Chemistry
Virginia Commonwealth University
Richmond, VA 23284

Yuichiro Nishina
Research Inst. for Iron,
    Steel & other Metals
Sendai, Japan 980

Joseph Nuth
Division of Chemistry
University of Maryland
College Park, MD

Seung Park
School of Chemical Engineering
Purdue University
Chemical & Metallurgical Eng. Bldg.
W. Lafayette, Ind. 47907

M. Oda
Hayashi-Ultrafine Particle Proj.
Research Development Corp. of Japan
Formation Process Laboratory
523 Yokota, Sanbu-machi, Sanbu-gun,
Chiba 289-12

Eric Parks
Building 200
Argonne National Laboratory
Argonne, IL 60439

Shuhei Ohnishi, NEC
Fundamental Res. Labs.
Miyazaki 4-1-1, Miyamae
Kawasaki, Kanagawa 213
Japan

J. Barrie Peel
Chemistry Department
La Trobe University
Bundoora, Victoria
Australia 3083

Fulvia Orsini
Centro Studi Sostanze Organiche
    Naturali del CNR
Dipartimento de Chimica Organica
    Industriale
Milano, Italy

Gary V. Pfeiffer
Dept. of Chemistry
Ohio University
Athens, Ohio 45701

R. S. Ott
Department of Physics
University of Texas at Arlington
Arlington, TX 76019

Rainer Pflaum
Fak. f. Physik
Bucklestr University. KN
D-7750 Konstanz
West Germany

Gianfranco Pacchioni
Department of Inorganic Chemistry
University of Milan
Via Venezian 21
20133 Milano
Italy

Donald H. Phillips
Mail Stop 234
NASA-Langley Research Center
Hampton, VA 23665

Risto Pankaluoto
Department of Physics
Northeastern University
360 Huntington Avenue
Boston, Mass. 02115

J. C. Phillips
AT&T Bell Laboratories
Research, Physics Division
600 Mountain Avenue
Murray Hill, NJ 07974

943

Rob Phillips
Department of Physics
Washington University
St. Louis, MO 63130

Krishnan Raghavachari
AT&T Bell Labs
600 Mountain Avenue
RM 1A-362
Murray Hill, N.J. 07974

Patricia L. M. Plummer
Department of Physics and Chemistry
University of Missouri
Columbia, Missouri 65211

M. V. Ramana
NASA-Langley Research Center
Hampton, VA 23665

Lynmarie Posey
Yale University
Department of Chemistry
225 Prospect St.
New Haven, CT 06511

Bijan K. Rao
Department of Physics
Virginia Commonwealth University
Richmond, VA 23284

Mehernosh Press
Physics Department
Virginia Commonwealth University
Richmond, VA 23284

A. K. Ray
Department of Physics
University of Texas at Arlington
P. O. Box 19059
Arlington, TX 76019

James G. Pruett
4440 Warrensville Center Road
Cleveland, Ohio 44128

E. Recknagel
Facultat fur Physik
Universitat Konstanz
D-7750 Konstanz
West Germany

James Y. Quagliano
Department of Chemistry
Virginia Commonwealth University
Richmond, VA 23284

Richard Reid
VPI & SU
1212 3rd St.
Radford, VA 24141

A. Quazi
Department of Chemistry
The University of Michigan
Ann Arbor, Michigan 48109

Stephen J. Riley
9700 W. Cass Avenue
Argonne, IL 60439

Jean C. Rivoal
Lab. Optique Physique
10 rue Vauquelin
75231 Paris
France

D. R. Salahub
Department of Chemistry
University of Montreal
C.P. 6210, Succ. A.
Montreal, Quebec
Canada H3C 3V1

Celeste M. Rohlfing
Group T-12, MS J569
Los Alamos National Lab.
Los Alamos, NM 87545

Sung-Ho Salk
109 Norwood Hall
University of Missouri-Rolla
Rolla, MO 65401

Eric A. Rohlfing
Div. 8353
Sandia National Labs
Livermore, CA 94550

Cameron Satterthwaite
Professor Emeritus
Virginia Commonwealth University
Richmond, VA 23284-0001

Arne Rosen
Department of Physics
Chalmers University of Technology
S-41296 Goteburg
Sweden

Klaus Sattler
Department of Physics
University of California
Berkeley, CA 94720

Frederick Rothworf
BDM Corporation
7915 Jones Branch Drive
McLean, VA 22102-3396

Winston Saunders
Institute de Physique Exper.
Ecole Polytechnique Federale
    de Lausanne
CH-1015 Lausanne
Switzerland

Susumu Saito NEC
Fundamental Research Lab.
Miyazaki 411, Miyamae
Kawasaki, Kanagawa 213
Japan

Daniel A. Scarpiello
Catalysis Research
GRI
8600 West Bryn Mawr Avenue
Chicago, IL 60631

Sanshiro Sako
Department of Physics
Faculty of Education
Mie University
Kamihama Tsu Mie 514
Japan

Gunter Schmid
University Essen
Inst. fur Anorganische Chemie
4300 Essen I
West Germany

945

John C. Schug
Department of Chemistry
VPI & State University
Blacksburg, VA 24061

David J. Singh
Department of Physics
College of William & Mary
Williamsburg, VA 23185

Ahmed Y. Serageldin
Department of Physics
Northeastern University
360 Hunington Avenue
Boston, Mass. 02115

R. E. Smalley
Department of Chemistry
Rice University
P. O. Box 1892
Houston, Texas 77251

Natthi Sharma
Physics Department
VCU
Richmond, VA 23284

Peter Stampfli
FUB FB20 WES
Arnimallee 14
D-1 Berlin 33
Germany

G. K. Shenoy
Material Science & Tech. Div.
Argonne National Lab.
Argonne, IL 60439

C. E. Stronach
Department of Physics
Virginia State University
Petersburg, VA 23803

Donald D. Shillady
Chemistry Department
Virginia Commonwealth University
Richmond, VA 23284

D. G. Stroud
Department of Physics
Ohio State University
Columbus, Ohio 43210

Irene Shim
Chemistry Department B
Technical University of Denmark
DTH 301, DK-2800
Lyngby, Denmark

Bruce Taggart
The BDM Corporation
7915 Jones Branch Drive
McLean, VA 22102

S. Shimamura
Department of Applied Science
Faculty of Engineering
Yamaguchi University
Tokiwadai, Ube 755
Japan

James Terner
Chemistry Department
Virginia Commonwealth University
Richmond, VA 23284-0001

946

Gerard J. Torchet
University de Paris Sud
Lab. Physique des Solides
Bat 510
Orsay 91405
France

Thomas H. Upton
Exxon Research & Engineering
Rt. 22, East
Annandale, NJ 08801

Frank V. Trager
Physikalisches Institute
Philosophenweg 12
D-6900 Heidelberg
West Germany

Martin Vala
University of Florida
Department of Chemistry
Gainesville, Fla. 32611

Joe Trivisonno
Div. of Materials Research
National Science Foundation
Washington, D.C. 20550

Edward J. Valente
Department of Chemistry
Box 4065
Mississippi College
Clinton, MS 39058

T. J. Tseng
Department of Physics
Chung Yuan Christian University
Chung-Li
Taiwan, R.O.C. 32023

Lidia M. Vallarino
Chemistry Department
Virginia Commonwealth University
P. O. Box 2006
Richmond, VA 23284-0001

Patrice Turchi
Lawrence Livermore Laboratory
Dept. of Materials Science-L-280
P. O. Box 808
Livermore, CA 94550

Stephen P. Walch
NASA Ames Research Center
RTC 230-3
Moffett Field, CA 94035

Robert J. Turnbull
Department of Electrical and
    Computer Engineering
University of Illinois at Urbanna
    Champaign
1406 W. Green St.
Urbanna, IL 61801-2991

L. R. Wallenberg
Inorganic Chemistry 2
Chemical Center
P. O. Box 124
S-221 00 Lund
Sweden

B. E. Turner
National Radio Astronomy
    Observatory
Edgemont Road
Charlottesville, VA 22903-2475

Russell Walstedt
AT&T Bell Labs., Research Physics
    Division
600 Mountain Avenue
Murray Hill, N.J. 07974

William Weltner, Jr.
Department of Chemistry
University of Florida
Gainesville, FL 32611

Roberto Zanoni
Departimento di Chimica
Universita La Sapienza
p. le Aldo Moro 5
00185 Rome
Italy

Gunter K. Wertheim
AT&T Bell Laboratories
RM 1C-425
Murray Hill, NJ 07974

Eugene Zaremba
Department of Physics
Queen's University
Kingston, Ontario
Canada K7L 3N6

R. G. Wheeler
University of Georgia
Department of Chemistry
Athens, GA 30602

Klaus Ziock
Lawrence Livermore National Lab.
Mail Code L401
P. O. Box 808
Livermore, CA 94550

Henry Windischmann
Standard Oil Research & Development
4440 Warrensville Center Road
Cleveland, Ohio 44128

Ellis B. Zuckerman
Department of Chemistry
Box 2006
VCU
Richmond, VA 23284

S. N. Yang
Center for Integrated Electronics
Department of Physics
Rensselaer Polytechnic Institute
Troy, N.Y. 12181

unenhanced Raman spectroscopy,
771

Copper Clusters,5,9-12
band gap,4,10
negative cluster ions,2,4-6
photo electron spectroscopy,3,
6-8
time of flight analysis,6,8

Damping of Excited States,605
surface enhanced Raman scattering,605
time dependent local density
theory,605,606,608

Discrete Variational Method,881
ligated transition metal
clusters,881,883
Pt_2, Pt_3, and Pt_4,885

Electron Impact Fragmentation,341,
343-344
Bi_2,344,345
Bi_4,344
magnetic polarization,345

Electronic Excitation,169
alkali halide clusters,178-180
electron solvation,182
molecular dynamics,169
QUPID,169,175-177
rare gas clusters,172-174
water clusters,180-183

ESR,353
alkali metal clsters,354-356
Be particles,479,481
carbon clusters,361-362
transition metal clusters,
357-360

Evaporation,499,501
binding energy,503
cluster growth,500
Na clusters,499,503
thermal desorption,502,503

Fe_n (also see iron particles),
463,467,813,909
EXAFS,909,910,912
ferromagnetic order,464
geometry,469
Hubbard model,463
in NH_3 synthesis,909,911
infrared spectroscopy,813,814,
815
ionization potential,467,469,
470

magnetic moments,463,465
O_2 bond rupture,816
reactivity with O_2,813
temperature effects,466
vibrational modes,814-816

Gas Phase Reactions,727-729
ammonia,735,737
H-D exchange,733,734
hydrogen,730,732,737
kinetics,729,737
transition metal clusters,727

Giant Atom Model,115
alkali atoms,117
hydrocarbons,118
magic number,115,116
rare gas atoms,116
shell model,116
transition metal atoms,117

Gold Clusters,61
multiple expansion cluster
surface,61-63
size distribution,65
TEM study,64

Gold Crystals,73
diffraction patterns,75-76
high resolution electron micrography,74-76
structure,74-76

Graph Theory,79
gas phase metal clusters,81
post-transition metal clusters,
81
topology,80

H_2, Reactions,787
chemisorption in Al_n,755,758,
759
Cu_2, Cu_3 clusters,787,788,
790,792
Fe_2, Fe_3 clusters,788,790,792
infrared spectroscopy,789,791

Helium Clusters,199
Cs atom scattering from,199
quantum fluid,199

Hg Van der Waals Complex,681,
682,685
electronic spectra,683,685,686

High Resolution Electron Microscopy,127,831,835,836
aggregation of osmium in,833,
835

ERRATUM

Because of a printer's error, page 594 has been left blank.
The correct text of page 594 appears below:

154/160 cm^{-1} for Kr, and at 163 cm^{-1} for Ar matrices, respectively. These "doublets" probably result from different trapping sites, because the modes do not always appear together. Another double structure is seen at low wavenumber for Xe and Kr matrices, and correspondingly a single line for Ar matrices. These modes appear together with the lines at higher wavenumber and therefore are probably due to the same species. They are correlated with the 557/524/503 nm absorption bands for Xe, Kr, and Ar matrices, respectively. — In a recent paper Roy and Furtak[5] assigned Raman modes at 70 cm^{-1} and 160 cm^{-1}, which are observed from electrochemically roughened silver electrodes in connection with SERS studies, to Ag$_4$ particles on the electrode surface. These SERS data may correspond to our resonance Raman data. The mode in the 110 cm^{-1} region reported for SERS-active silver electrodes is possibly the same line we attribute to Ag$_3$.[12]

Although we still cannot prove definitively the actual size or molecular structure of the clusters involved, the correlation of the spectroscopic data leads us closer to the identification and characterization of the carriers.

ACKNOWLEDGEMENT

We acknowledge the support of M. Campagna and T. Springer and the technical assistance of H. Ervens. ULK wishes to thank the Phys. Dept. and the Chem. Dept., UCSB, for supporting the preparation of this contribution, and V. I. Srdanov for helpful discussions.

REFERENCES

1. Schulze, W., Becker, H.-U., Abe, H., *Chem. Phys.* **35**, 177 (1978);
    Ber. Bunsenges. Phys. Chem. **82**, 138 (1978);
    Becker, H.-U., PhD Thesis, Techn. Univ. Berlin (1978);
    Ozin, G. A., Huber, H., *Inorg. Chem.* **17**, 155 (1978);
    Klotzbücher, W. E., Ozin, G. A., *Inorg. Chem.* **19**, 3776 (1980)
    Schulze, W., Abe, H., *Faraday Symp. Chem. Soc.* **14**, 87 (1980);
    Mitchell, S. A., Ozin, G. A., *J. Phys. Chem.* **88**, 1425 (1984)
2. Ref. 3–23 in[14]
3. Kilty, P. A., Sachtler, W. M. A., *Catal. Rev.-Sci. Eng.* **10**, 1 (1974) (and references therein)
4. Otto, A., in: *Light Scattering in Solids IV*, Cardona, M., Güntherodt, G., (eds.),
    Topics in Applied Physics Vol. 54, p. 289, Springer, Berlin (1984)
5. Roy, D., Furtak, T. E., *Chem. Phys. Lett.* **124**, 299 (1986)
6. Fayet, P., Granzer, F., Hegenbart, G., Moisar, E., Pischel, B., Wöste, L.,
    Phys. Rev. Lett. **55**, 3002 (1985); *Z. Phys. D* **3**, 299 (1986)
7. Ozin, G. A., Mitchell, S. A., *Angew. Chem. Int. Ed. Engl.* **22**, 674 (1983);
    Weltner, W., van Zee, R. J., *Annu. Rev. Phys. Chem.* **35**, 291 (1984)
8. Ozin, G. A., Mitchell, S. A., McIntosh, D. F., Mattar, S. M., García-Prieto, J.,
    J. Phys. Chem. **87**, 4651 (1983)
9. Baetzold, R. C., *J. Chem. Phys.* **55**, 4363 (1971);
    Baetzold, R. C., Hamilton, J. F., *Prog. Solid St. Chem.* **15**, 1 (1983)
10. Kettler, U., PhD Thesis, Univ. zu Köln (1984)
    [*Berichte der Kernforschungsanlage Jülich* **1980** (1985)]
11. Bechthold, P. S., Kettler, U., Krasser, W., *Solid State Commun.* **52**, 347 (1984);
    Surf. Sci. **156**, 875 (1985)
12. Leutloff, D., Kolb, D. M., *Phys. Chem.* **83**, 666 (1979);
    Ozin, G. A., *Faraday Symp. Royal Soc. Chem.* **14**, 7 (1980);
    Mitchell, S. A., Farrell, J., Kenney-Wallace, G. A., Ozin, G. A.,
    J. Am. Chem. Soc. **102**, 7702 (1980);
    Mitchell, S. A., Kenney-Wallace, G. A., Ozin, G. A., *J. Am. Chem. Soc.* **103**, 6030 (1981);
    Kolb, D. M., in: *Matrix Isolation Spectroscopy*, NATO Adv. Study Inst. Series C, Vol. 76,
    Reidel, Dordrecht (1981)
13. Kettler, U., Bechthold, P. S., Krasser, W., *Surf. Sci.* **156**, 867 (1985);
    Chem. Phys. Lett. **131**, 213 (1986)
14. Bechthold, P. S., Kettler, U., Schober, H. R., Krasser, W., *Z. Phys. D* **3**, 263 (1986)
15. Kettler, U. L. (unpublished data)
16. Schrittenlacher, W., Rotermund, H. H., Schroeder, W., Kolb, D. M., *Surf. Sci.* **156**, 777 (1985)
17. Schulze, W., Becker, H.-U., Minkwitz, R., Manzel, K., *Chem. Phys. Lett.* **55**, 59 (1978)